RADIOGRAPHY IN THE DIGITAL AGE

SECOND EDITION

RADIOGRAPHY IN THE DIGITAL AGE

Physics—Exposure—Radiation Biology

By

Quinn B. Carroll, M.Ed., R.T.

CHARLES C THOMAS • PUBLISHER, LTD.
Springfield • Illinois • U.S.A.

Published and Distributed Throughout the World by

CHARLES C THOMAS • PUBLISHER, LTD.
2600 South First Street
Springfield, Illinois 62704

ISBN 978-0-398-08096-9 (Hard)
ISBN 978-0-398-08097-6 (Ebook)

First Edition, 2011
Second Edition, 2014

Library of Congress Catalog Card Number: 2014015954

With THOMAS BOOKS careful attention is given to all details of manufacturing and design. It is the Publisher's desire to present books that are satisfactory as to their physical qualities and artistic possibilities and appropriate for their particular use. THOMAS BOOKS will be true to those laws of quality that assure a good name and good will.

Printed in the United States of America
UB-R-3

Library of Congress Cataloging-in-Publication Data

Carroll, Quinn B., author.
Radiography in the digital age : physics, exposure, radiation biology / by Quinn B. Carroll. — Second edition.
p. ; cm.
Includes index.
ISBN 978-0-398-08096-9 (hard) — ISBN 978-0-398-08097-6 (ebook)
I. Title.
[DNLM: 1. Radiography. 2. Physics. 3. Radiobiology. 4. Radiographic Image Enhancement. 5. Radiology—methods. 6. Technology, Radiologic. WN 200]

RC78.7.D53
616.07'572—dc23

201401594

REVIEWERS

C. William Mulkey, PhD, RT(R)

Dean, Dept. of Radiologic Sciences
Midlands Technical College
West Columbia, North Carolina

Philip Heintz, PhD

Director, Biomedical Physics, Dept. of Radiology
University of New Mexico Medical Center
Albuquerque, New Mexico

Donna Endicott, MEd, RT(R)

Director, Radiologic Technology
Xavier University
Cincinnati, Ohio

Dennis Bowman, AS, RT(R)

Clinical Instructor
Community Hospital of the Monterey Peninsula
Marina, California

Miranda Poage, PhD

Associate Professor, Biology
Midland College
Midland, Texas

Dedication

To Jason and Stephanie,
Melissa and Tim,
Chad and Sarah,
Tiffani and Nate,
Brandon, and Tyson,
a most remarkable family,
and to my cherished wife, Margaret,
who made it possible for them all
to come into my life.

PREFACE

New to This Edition

This edition was peer-reviewed by five colleagues who brought many valuable corrections and improvements to the text. Several sections have been deleted, moved, or reorganized to provide smoother transitions and development of the topics, with particular focus on the digital imaging chapters. Material on *rescaling* the digital image has been greatly strengthened, and new graphs have been added that make histogram analysis and errors much easier to grasp.

The large chapter (Chapter 29) on digital image processing was split into two chapters, "Digital Image Preprocessing" and "Digital Image Postprocessing" (Chapters 30 and 31), rewritten for the student to more easily assimilate these concepts even while assuring a thorough overview. Important new practical material has been added on the limitations of digital features such as smoothing and edge enhancement, with direct implications for clinical practice. A new chapter (Chapter 36) is dedicated to PACS (picture archiving and communication systems) from the perspective of what a practicing staff radiographer should know.

The large chapter (Chapter 13) on qualities of the radiographic image was divided into two chapters, "Visibility Qualities of the Image" and "Geometrical Qualities of the Image" (Chapters 13 and 14) and revised for easier reading. The math review chapter (Chapter 3) includes a section on basic graphs. Along with material on the x-ray beam spectrum, a new section titled "Understanding the Digital Histogram" has been added, which includes foundational support exercises directly related to the later chapters on digital image processing.

A glossary of technical radiographic and digital imaging terms has been added for quick reference. In addition, a deliberate effort has been made to include the content areas identified in the Curriculum Guide published by the American Society of Radiologic Technologists, and to address the Standard Definitions published by the American Registry of Radiologic Technologists.

Scope and Philosophical Approach

The advent of digital radiographic imaging has radically changed many paradigms in radiography education. In order to bring the material we present completely up to date, and in the final analysis to fully serve our students, much more is needed than simply adding two or three chapters on digital imaging to our textbooks.

First, the entire emphasis of the *foundational* physics our students learn must be adjusted in order to properly support the specific information on digital imaging that will follow. For example, a better basic understanding of waves, frequency, amplitude and interference is needed so that students can later grasp the concepts of spatial frequency processing to enhance image sharpness. A more thorough coverage of the basic construction and interpretation of graphs prepares the student for histograms and look-up tables. Lasers are also more thoroughly discussed here, since they have not only medical applications, but are such an integral part of computer technology and optical disc storage.

Second, there has been a paradigm shift in our use of image terminology. Perhaps the most disconcerting example is that we can no longer describe the direct effects of kVp upon image contrast. Rather, we can only describe the effects of kVp upon the subject contrast in the remnant beam signal reaching the image detector, a signal whose contrast will then be drastically manipulated by digital processing techniques. Considerable confusion continues to surround the subject of scatter radiation and its effects on the imaging chain. Great care is needed in choosing appropriate terminology, accurate descriptions and lucid illustrations for this material.

The elimination of much obsolete and extraneous material is long overdue. Our students need to know the electrical physics which directly bear upon the production of x-rays in the x-ray tube—they do not need to solve parallel and series circuit problems in their daily practice of radiography, nor do they need to be spending time solving problems on velocity. MRI is briefly overviewed when *radio* waves are discussed under basic physics, sonography is also discussed under the general heading of *waves*, and CT is described along with attenuation coefficients under digital imaging.

It is time to bring our teaching of image display systems up to date by presenting the basics of LCD screens and the basics of quality control for electronic images. These have been addressed in this work, as part of eleven full chapters dealing specifically with digital and electronic imaging concepts. If you agree with this educational philosophy, you will find this textbook of great use.

Organization

The basic layout is as follows: In Part I, "The Physics of Radiography," ten chapters are devoted to laying a firm foundation of math and basic physics skills. The descriptions of atomic structure and bonding go into a little more depth than previous textbooks have done. A focus is maintained on *energy* physics rather than mechanical physics. The nature of electromagnetic waves is more carefully and thoroughly discussed than what most textbooks provide. Chapters on electricity are limited to only those concepts which bear directly upon the production of x-rays in the x-ray tube.

Part 2, "Production of the Radiographic Image," presents a full discussion of the x-ray beam and its interactions within the patient, the production and characteristics of subject contrast within the remnant beam, and the proper use of radiographic technique. This is conventional information, but the terminology and descriptions used have been adapted with great care to the digital environment.

Part 3, "Digital Radiography," includes nine chapters covering the physics of digital image capture, extensive information on digital processing techniques, and the practical application issues of both CR and DR.

Part 4, "Special Imaging Methods," includes chapters on mobile radiography, digital fluoroscopy and an extensive chapter on quality control which includes digital image QC. Finally, Part 5 consists of five chapters on "Radiation Biology and Protection," including an unflinching look at current issues and practical applications.

Feedback

For a textbook to retain enduring value and usefulness, professional feedback is always needed. Colleagues who have adopted the text are invited to provide continuing input so that improvements might be made in the accuracy of the information as well as the presentation of the material. Personal contact information is available in the *Instructor and Laboratory Manual* on disc.

This is intended to be a textbook written "by technologists for technologists," with proper focus and scope for the practice of radiography in this digital age. It is sincerely hoped that it will make a substantial contribution not only to the practice of radiography and to patient care, but to the satisfaction and fulfillment of radiographers in their careers as well.

Instructional Resources

Instructor Resources CD for Radiography in the Digital Age. This disc includes the answer key for all chapter review questions and a bank of over 1450 multiple choice questions with permission for instructors' use. It also includes 35 laboratory exercises with 15 demonstrating the applications of CR equipment. The manual is available only on disc from Charles C Thomas, Publisher.

PowerPoint Slides on Disc. *PowerPoint*™ slides are available for classroom use, covering the entire textbook and as many as four courses in a typical radiography curriculum. The four titles are available from Charles C Thomas, Publisher:

The Physics and Equipment of Radiography
Principles of Radiographic Imaging
Digital Image Acquisition and Display
Radiation Biology and Protection

Student Workbook for Radiography in the Digital Age. This 312-page supplement covers everything in the textbook and as many as four courses in a typical radiography curriculum It is deliberately organized in a concise "fill-in-the-blank" format that provokes students to participate in class without excessive note taking. Questions focus on key words that correlate perfectly with the above slide series. Available from Charles C Thomas, Publisher.

DVD Mini-Lessons. To assist the instructor on particularly difficult digital topics, a series of 20-minute video mini-lessons are available from Digital Imaging Consultants that correlate with and supplement *Radiography in the Digital Age*. Video object-lessons are combined with lucid graphics and clear, progressive explanations to make difficult material "click" for the student. Visit the website at radiographyceusonline.com

ACKNOWLEDGMENTS

Many thanks to the reviewers for the second edition, C. William Mulkey, Donna Endicott, Philip Heintz, Dennis Bowman, and Miranda Poage, who provided many improvements for content, organization and readability.

Special thanks to Georg Kornweibel and Doctor Ralph Koenker at Philips Healthcare, and to Gregg Cretellen at FujiMed for their sustained assistance. Thanks also to Lori Barski at Carestream Health (previously Kodak). All were extremely helpful in obtaining images and a good deal of information related to digital imaging and processing. Doctor J. Anthony Seibert at the University of California Davis Medical Center was generous with his time and expertise, as well as providing energy-subtraction images. His help was greatly appreciated.

Some material was adopted and adapted from contributing authors to my textbook, *Practical Radiographic Imaging* (previously *Fuchs's Radiographic Exposure, Processing and Quality Control*). They include Robert DeAngelis, BSRT in Rutland, Vermont, Robert Parelli, MA, RT(R) in Cypress, California, and Euclid Seeram, RTR, MSc, in Burnaby, British Columbia, Canada. Their contributions are still greatly valued.

Many photographs and radiographs were made available through the gracious assistance of several local radiographers including Kathy Ives, RT, Steven Hirt, RT, Jason Swopes, RT, Phil Jensen, RT, Sungunuko Funjiro, RT, Trevor Morris, RT, and Brady Widner, RT, all graduates whom I proudly claim. Thanks, in particular, to William S. Heathman, BSRT, my colleague in radiography education for many years.

Photographs were also helpfully provided by Fyte Fire and Safety in Midland, Texas and Apogee Imaging Systems in Roseville, California, and made available in the public domain by the U.S. Army and U.S. Navy.

Trevor Ollech at Charles C Thomas, Publisher rendered many dozens of pieces of artwork for this project. His many hours of patient labor are greatly appreciated, and the quality of his illustrations is self-evident throughout the book.

I appreciate the long-suffering editorial assistance of Lynn Bergesen and the support of Michael Payne Thomas at Charles C Thomas, Publisher.

Without the gracious assistance of all these individuals and companies, the completion of this work would have been impossible.

On a more personal note, I wish to express appreciation for the professional support and loyal friendship of Doctor Eileen Piwetz, which never waivered over 25 years, along with my love and admiration for all my colleagues in health sciences education, who, often against all odds, make miracles happen on the "front line" every day.

CONTENTS

PART II: PRODUCTION OF THE RADIOGRAPHIC IMAGE

RADIOGRAPHY IN THE DIGITAL AGE

Part I

THE PHYSICS OF RADIOGRAPHY

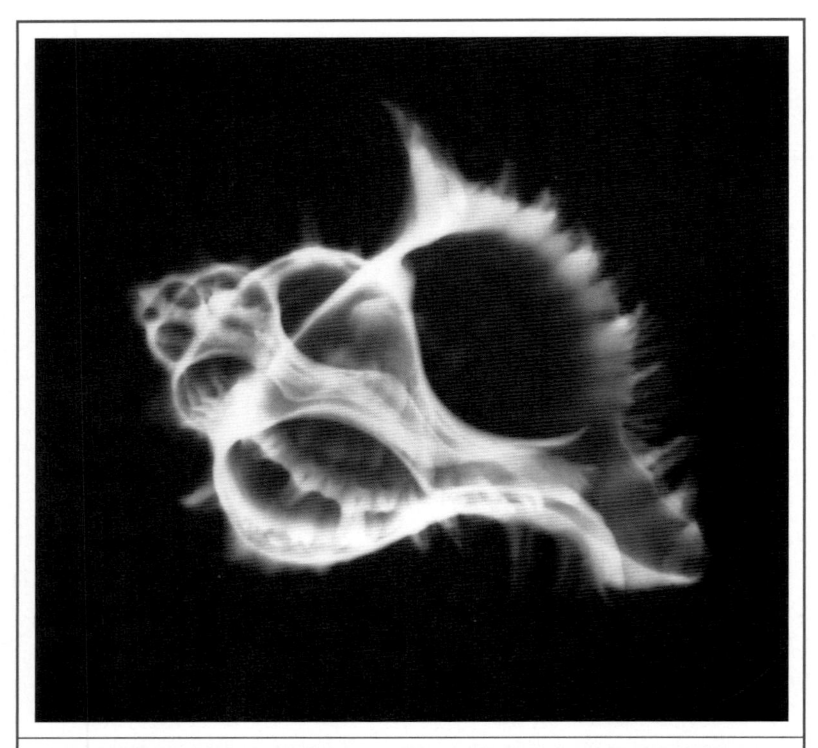

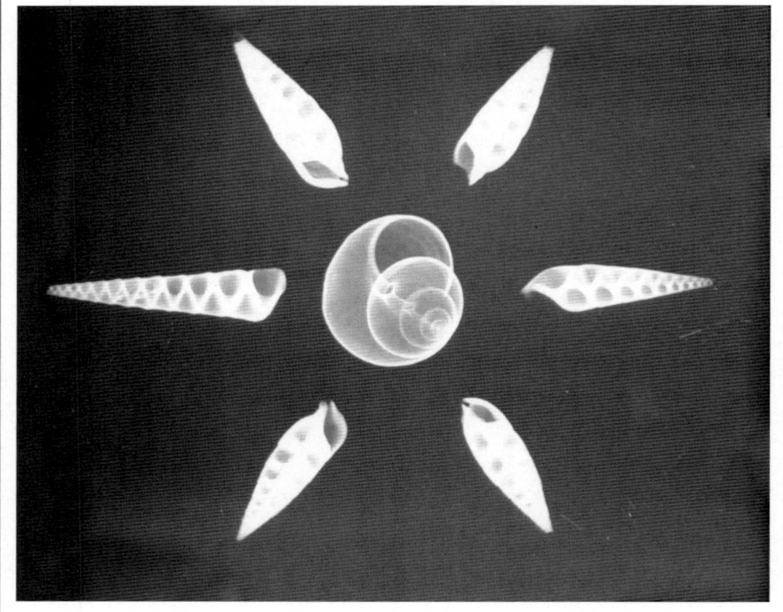

Radiographs of various seashells.

Chapter 1

INTRODUCTION TO RADIOGRAPHIC SCIENCE

Objectives:

Upon completion of this chapter, you should be able to:

1. List the foundational principles of the scientific method and how they relate to the standard of practice for radiographers.
2. Describe landmark events in the development of medical radiography, with particular focus on those that brought about reductions in patient exposure.
3. Overview landmark events in the development of modern digital radiographic imaging.
4. Present a scientifically balanced perspective on the hazards of radiation in our environment and workplace.
5. Understand and appreciate the ALARA philosophy in modern radiographic imaging.

THE SCIENTIFIC APPROACH

Radiography is a branch of the modern *science* of medicine. Science is objective, observable, demonstrable knowledge. Try to imagine your doctor engaging in practices that were not grounded in scientific knowledge! What is it that sets science apart from art, philosophy, religion and other human endeavors? There are actually several foundational principles to scientific method. It is worthwhile to give a brief overview of them. They include:

Parsimony: The attempt to simplify concepts and formulas, to economize explanations; the philosophy that simple explanations are more likely to be true than elaborate, complex ones.

Reproducibility: The requirement that proofs (experiments) can be duplicated by different people at different times and in different locations with precisely the same results.

Falsifiability: The requirement that any theory or hypothesis can logically and logistically be proven *false*. Anything that cannot be proven false is not science, but belongs in another realm of human experience.

Observation: The requirement that experiments and their results can be directly observed with the human senses.

Measurability: The requirement that results can be quantified mathematically and measured.

As a fun practice exercise, consider the following three statements. Which one is scientific?

1. *The moon is made of green cheese.*
2. *Intelligent life likely exists elsewhere in the universe.*
3. *Albert Einstein was the greatest physicist in the twentieth century.*

The most scientific statement is No. 1. Even though it may not be a true statement, it is nonetheless a statement that can be (and has been) proven false with modern travel technology, it is simple, and experiments proving that moon rocks do not consist of green cheese can be reproduced by anyone, anywhere on earth with the same, observable, measurable results. Statement No. 2 may be true or false, but *cannot be proven false*, because to do so would require us to explore every planet in the entire universe, documenting that we have looked in every crevice and under every rock. It may be classified as a philosophical statement, but not as a scientific one. Statement No. 3 is, of course, a simple matter of personal opinion that depends upon how one defines the word "greatest." It is a historical statement that defies standardized measurement or observation.

Perhaps the strongest aspect of the scientific method is that when it is used properly, it is *self-correcting*. That is, when a theory is found to be wrong, that field of science is expected to be capable of transcending all politics, prejudice, tradition and financial gain in order to establish the new truth that will replace it. Sometimes this process is painful to the scientific community, and it has been known to take years to complete. But, at least it presupposes a collective willingness to accept the *possibility* that a previous position may have been wrong, something one rarely sees in nonscientific endeavors.

This principle of *self-correction* is nicely illustrated in the story of Henri Becquerel and the discovery of natural radioactivity, related in the next section. Also demonstrated in both his story and that of Wilhelm Roentgen, the discoverer of x-rays, is the fact that many scientific truths are discovered by accident. Nonetheless, it is *because* scientific method is being followed, not in spite of it, that they have occurred, and *through* scientific method that they come to be fully understood.

How does this scientific approach apply to radiography, specifically? Even though some aspects of radiography, such as positioning, are sometimes thought of as an art, the end result is an image that contains a quantifiable amount of diagnostically useful details, a measurable amount of information. Image qualities such as contrast, brightness, noise, sharpness and distortion can all be mathematically measured. Even the usefulness of different approaches to positioning are subject to measurement through repeat rate analysis. In choosing good radiographic practices, rather than relying on the subjective assertion from a cohort that, "It works for me," important matters can be objectively resolved by simply monitoring the repeats taken by those using the method compared to those using another method. By using good sampling (several radiographers using one method and several using another over a period of weeks), reliable conclusions can be drawn.

The standard of practice for all radiographers is to use good common sense, sound judgment, logical consistency and objective knowledge in providing the best possible care for their patients.

A BRIEF HISTORY OF X-RAYS

It is fascinating to note that manmade radiation was invented *before* natural radioactivity was discovered. If this seems backward, it is partly because x-rays were discovered by accident. In the late 1800s, Wilhelm Conrad Roentgen (Fig. 1-1) was conducting experiments in his laboratory at Wurzburg University in Germany. It had been discovered that a beam of electricity (glowing a beautiful blue in a darkened room) could be caused to stream across a glass tube. With strong enough voltage, the electricity could be caused to "jump" from a negatively-charged *cathode* wire across the gap toward a positively-charged *anode* plate, although most of it actually struck the glass behind. Since they were emitted from the cathode, these streams of electricity were dubbed *cathode rays*.

Several researchers were studying the characteristics of cathode rays. These glass tubes, known as Crookes tubes, came in many configurations. Figure 1-2 shows several that Roentgen actually used in his experiments. If most of the air was vacuumed out of the tube, the cathode rays became invisible. (It was later understood that they were in fact the electrons from the current in the cathode, far too small for the human eye to see, and that the blue glow was the effect from the ionization of the air around them.)

Other researchers had noticed that the glass at the anode end of the tube would fluoresce with a greenish glow when the cathode rays were flowing. They began experimenting with placing fluorescent materials in the path of the beam. They learned how to deflect the beam at right angles with a plate so it could exit the tube through a window of thin aluminum. In this way, cards or plates coated with different materials could simply be placed alongside the tube, in the path of the electron beam, to see how they fluoresced. Researchers learned to surround the tube with black cardboard so as to not confuse any light that might be generated within the tube with the fluorescence of the material outside the tube.

This was the type of experiment Roentgen was engaged with on November 8, 1895, when he noticed that a piece of paper laying on a bench nearby was glowing while the tube was activated in its black cardboard box. This paper was coated with barium platinocyanide, but it was not in the direct path of the cathode rays (electron beam).

Roentgen quickly realized that there must be some other type of radiation being emitted from the tube, other than the electron beam. He dubbed this radiation as "x" indicating the unknown. This radiation seemed to be emitted in all directions from the tube and was able to affect objects such as the plate at some distance. Placing various objects between the tube and the plate, he saw that they cast partial shadows on the glowing screen, while lead cast a solid shadow, stopping the mysterious rays altogether. He deduced that they traveled in straight lines and were able to penetrate less dense materials. During the following days, Roentgen conducted brilliant experiments delineating the characteristics of the x-rays.

Early in his experiments, he was astonished to see the image of the bones in his own hands on the screen, while the flesh was penetrated through by the x-rays. The field of radiography was born when he placed his wife's hand in front of the screen and allowed the screen's fluorescent light to expose a photographic film for about four minutes (Fig. 1-3). Along with three other radiographs, this image was

Figure 1-1

Wilhelm Conrad Roentgen, discoverer of x-rays.

Figure 1-2

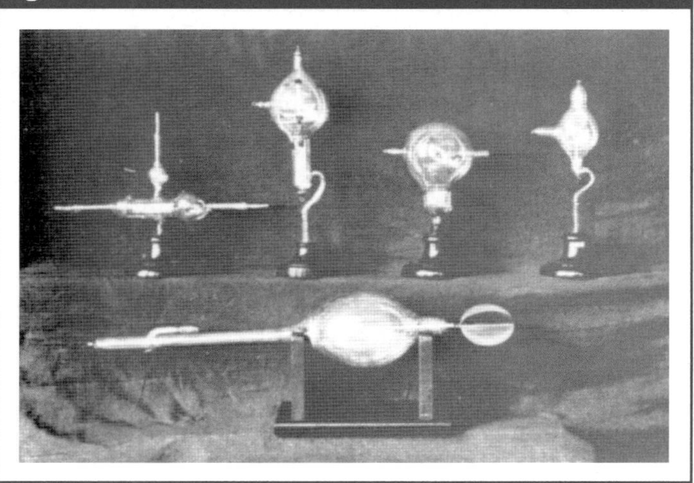

Photograph of Crookes tubes employed by Roentgen in his experiments on cathode rays, which led to the discovery of x-rays. (From Quinn B. Carroll, *Practical Radiographic Imaging*, 8th ed. Springfield, IL: Charles C Thomas, Publisher, Ltd., 2007. Reprinted by permission.)

Figure 1-3

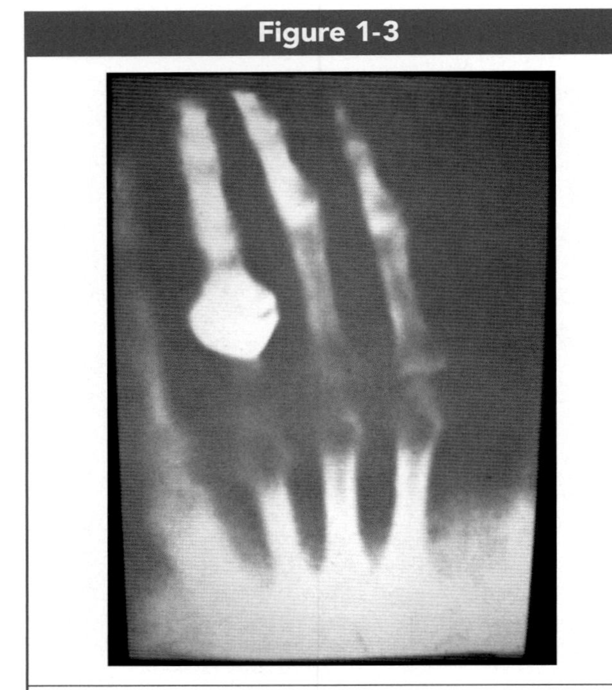

The first radiograph, showing the hand of Marie Roentgen with her wedding band, took over 4 minutes to expose.

published two months later in his paper, "On a New Kind of Rays," introducing the process of radiography to the world. With uncommon modesty, Roentgen refused to patent his radiographic process for commercial gain, showing great character to match his tremendous scientific acumen.

However, the discovery was truly accidental, as many scientific discoveries have been, taking an unexpected turn even while scientific method is rigorously followed. It was accidental because Roentgen was investigating the effects of the *cathode rays* or electron beam upon fluorescent materials, and was not expecting to find an object fluorescing outside of that beam of electrons.

It was in the following year, 1896, that Antoine Henri Becquerel, a French physicist, discovered natural radioactivity. Inspired by Roentgen, he hypothesized that crystals which phosphoresce ("glow in the dark") after absorbing light might also emit x-rays at the same time. He thought he had proven his theory when a phosphorescing crystal exposed a photographic plate wrapped in black paper. He wanted to repeat the experiment with a crystal known to phosphoresce for only 1/100th second, but was frustrated when cloudy weather prevented him from letting the crystal absorb some sunlight to begin. He placed the wrapped-up photographic plate and the crystal in a dark drawer. Later, on a pure whim, he developed the old plate. To his great surprise, it was darkened with exposure. He realized that "x-rays" must have been continuously emitted by the stone while it was in the drawer, rather than being emitted only along with phosphorescent light. Thus, another happy accident led to more accurate knowledge.

As the process of self-correcting scientific investigation continued in the following years, it was found that Becquerel's natural radiation consisted not strictly of x-rays, but of *three* distinct types of radiation. These were named *alpha*, *beta* and *gamma* rays. Using magnets and electrodes to deflect their paths, physicists were able to prove that alpha rays consisted of extremely heavy particles with positive electric charge, and beta rays consisted of very light particles with negative charge (electrons). Gamma rays were, in their nature, essentially the "x-rays" that Becquerel was looking for, but they had far higher energy than those produced by Roentgen's x-ray machines. These high energies gave them different abilities than x-rays, and made them unsuitable for producing radiographs, warranting their own distinct name, *gamma rays*.

Because of their brilliant investigative work, both Roentgen and Becquerel received Nobel Prizes. Our understanding of the atom developed hand-in-hand with our understanding of radiation. Ernest Rutherford, a British physicist, found that the alpha particle was identical to the nucleus of a helium atom. He proved the existence of the proton and predicted the neutron. Einstein discovered the photoelectric effect and much of his work built upon Roentgen, Becquerel, Rutherford and others. Thus, Wilhelm Roentgen "began a revolution in modern physics that was to include the quantum theory, radioactivity, relativity, and the new Bohr atom."[1] Figure 1-4 shows one of the first x-ray machines, installed at Massachusetts General Hospital in 1896.

[1] Encyclopedia Americana, Vol. 24, p. 68, 1970.

Figure 1-4

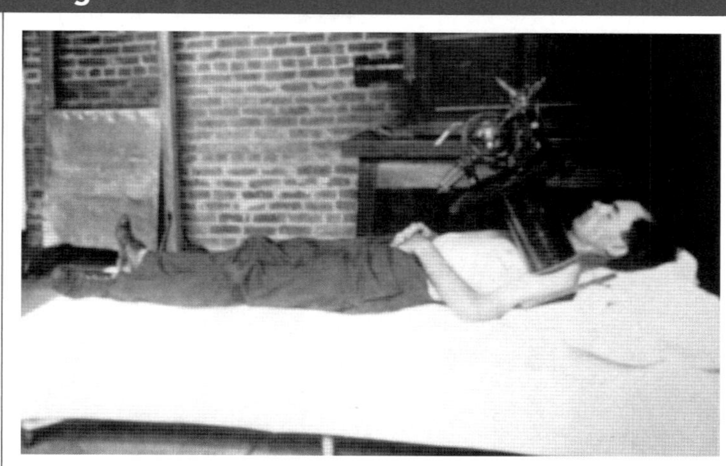

The first x-ray unit installed at Massachusetts General Hospital in 1896. Note that although a lead cone was installed to reduce scatter radiation to the image, there is no lead housing around the x-ray tube to protect personnel from primary radiation emitted in all directions. (From Ronald Eisenberg, *Radiology: An Illustrated History*. Philadelphia, PA: Elsevier Health, Inc., 1992.)

THE DEVELOPMENT OF MODERN IMAGING TECHNOLOGY

Within one year of Roentgen's discovery, in 1896, the great American inventor Thomas Edison developed a device he called a "fluoroscope." A simple fluorescent screen in a light-tight viewing cone made of metal, it allowed a doctor to view the patient's body under x-ray examination in dynamic real-time, that is, in motion and immediately as things happened. This imaging process has since been known as *fluoroscopy*.

For over fifty years, no improvement was made on this basic concept; fluoroscopic screens were simply suspended above the patient while an x-ray tube under the table projected the beam upward through the patient to the screen. The x-ray room had to be darkened for viewing the screen. Unfortunately, very high x-ray techniques were required to make the screen glow bright enough. And, these were multiplied by cumulative exposure times of several *minutes*, as compared to the fractions of a second required by still radiographs. Exposures to the doctors and technologists could be very high indeed, and exposures to the patients were excessive, limiting fluoroscopic procedures to extreme medical need.

Finally, in 1948, John Coltman developed the electronic image intensifier, a modern example of which is shown in Figure 1-9. Described in a later chapter, this device converts incident x-rays into an electron beam, which can then be both focused and sped up by using electrically charged plates. When these accelerated electrons strike the small fluorescent screen at the top of the tube, the brightness of the light emitted can be as much as 5000 times increased. This invention reduced fluoroscopic techniques to much less than one-hundredth of those previously used, perhaps the greatest single improvement in patient exposure in the history of radiography.

A few major historical inventions improving the efficiency and safety of the x-ray tube bear mention: In 1899, just four years after the discovery of x-rays, a dentist named William Rollins developed the concepts of both x-ray filtration and collimation. His filters, aluminum plates placed in the beam, drastically reduced radiation exposure to patients, while his "diaphragms," lead plates with apertures in them used to constrict the area of the x-ray beam, significantly reduced radiation to both workers and patients.

In 1913, William Coolidge used tungsten to produce an x-ray tube filament that could withstand extreme temperatures. This allowed electrons to be "boiled off" of the cathode in a process called *thermionic emission*, prior to exposure. Every time the radiographer "rotors," this process takes place, so that when the exposure switch is engaged, electrons do not have to be "kicked out" of the filament wire, but are already free to move across the tube as the high voltage pushes them. Figure 1-5***A*** shows the first mass-marketed Coolidge tube alongside a modern x-ray tube.

Figure 1-5

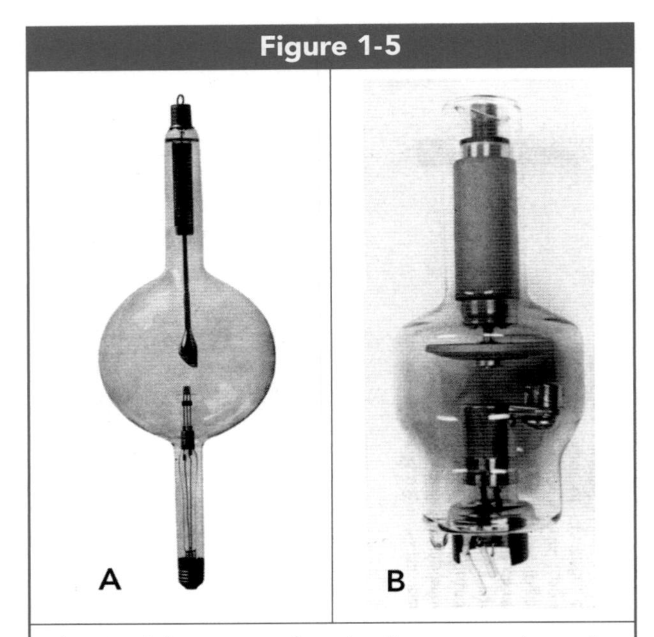

The Coolidge x-ray tube, the first x-ray tube to be mass-marketed, ***A***, and a modern Jackson focus tube with rotating anode, ***B***. Both tubes are shown with the *anode* above and the *cathode* filament below. (From Quinn B. Carroll, *Practical Radiographic Imaging*, 8th ed. Springfield, IL: Charles C Thomas, Publisher, Ltd., 2007. Reprinted by permission.)

In 1929 the first rotating anode x-ray tube was introduced (Fig. 1-5***B***), allowing the tremendous heat from the impact of electrons to be dispersed around the outside of the anode disc, rather than all impacting in one spot which might then warp or melt. This allowed much higher techniques to be used. The "Jackson focus tube" used negatively-charged pits in the cathode to surround the filament with negative charge. This caused the electron stream, as it left the filament, to be compressed such that the electrons converged to a small "focal spot" on the anode. This greatly enhanced image sharpness.

In 1921 the Potter-Bucky grid was introduced. Lead strips in the grid greatly reduced scattered radiation before it reached the film, but also left white "grid lines" as an artifact in the image. The Potter mechanism used a motor to oscillate the grid back and forth during exposure to blur out these grid lines.

In 1942, Russell H. Morgan demonstrated the first "phototimer" or automatic exposure control device, helping to standardize more consistent exposures. The same year, the first automatic film processor was developed. Up to this time, films had to be manually immersed in developer solution, fixer solution, and wash water, then dried for up to 30 minutes!

Most progress in the field has occurred in the form of innovations to the image-receiving plate or device. Less than a year after Roentgen discovered x-rays, Michael Pupin of Columbia University sandwiched radiographic film between two fluorescent screens, inventing the first "screen cassette." Figure 1-6 shows the first radiograph of a hand using Pupin's screen cassette, with his signature at the bottom. The material for the screens, calcium tungstate, had been developed by Thomas Edison after evaluating over 5000 chemicals for their light-producing capability when stimulated by x-rays. Since a light wave has much less energy than an x-ray, these screens would emit hundreds of light rays from a single x-ray hit. In this way, the effect of the exposure upon the film was magnified, and less technique could be used.

The film inside the screen cassette, then, was mostly exposed to light. This great invention reduced the exposures required to produce an image to less than 1/50th those previously required. For still images, this was the greatest reduction in patient dose in the history of radiography, and was the last significant advancement in capturing the image until digital imaging was developed. This means that for almost 80 years, from 1896 to 1974, there was no significant improvement upon Pupin's calcium tungstate screen cassette for capturing the x-ray image!

Finally, in the mid-1970s, two significant changes occurred: Intensifying screens using "rare earth" chemical compounds were developed. These compounds were better both at absorbing x-rays and at converting them into light, and reduced radiation exposure to patients to less than one-half of the previous levels. At the same time, the first computerized processing of the still x-ray image was being developed. It was dubbed *computed radiography* or "CR," and is discussed in the next section.

It has only been since the mid-1960s that other medical imaging modalities began to multiply. Ultrasound became established as a medical imaging tool in 1966. Computed tomography (CT) and magnetic resonance imaging (MRI) both came out in 1973. Spiral CT became available in 1990, and multislice CT in 1998.

THE DEVELOPMENT OF DIGITAL IMAGING

Digital fluoroscopy was first demonstrated in 1979. TV camera tubes, already in use for many years, could convert the light image from the image intensifier into electrical current. All that remained was for analog-to-digital converters to be refined and coupled to the TV camera tube in order to digitize the information, and for computer technology to develop to the point where huge amounts of these electronic signals could be fed into the computer as images.

Three years later, in 1982, the introduction of digital picture archiving and communication systems (PACS) revolutionized the storage, management and access of digital images. Coupled with *teleradiology*, the ability to send electronic images anywhere in the world, these systems have vastly improved health care efficiency, allowing almost instant access and correlation of images and patient information to individual physicians in their offices and homes.

Computed radiography or "CR" became commercially available in the early 1980s, but was at first fraught with technical problems. The process still employed a cassette with fluorescent screens in it, but the new fluorescent materials used were able to reemit their light a *second* time when stimulated with a laser beam. This allowed the reemitted image to be scanned with light-sensitive diodes, converting the information into electrical signals that could be measured and then stored by a computer.

Unfortunately, on average, CR *increased* required radiographic techniques as much as double. Thus, with respect to patient exposure it might be considered a step backwards. Nonetheless, CR helped pave the way for *direct digital radiography* (DR), and has since been greatly refined. In fact, it continues to have some advantages over DR, particularly in regard to mobile or "portable" procedures.

Finally, direct-capture digital radiography (DR) was first demonstrated in 1996. It uses miniature electronic x-ray detectors. This allows the x-ray image to be captured directly by the electronic elements with no intermediate steps (such as converting it into light first). Both CR and DR are currently in use.

While CR was nicknamed "filmless radiography," DR was dubbed "cassetteless radiography," since Pupin's screen-and-film cassette was completely done away with. One might ask why these advancements took so long to come about. The development of CR had to wait until computer power increased to the point where computers could handle large files of high-resolution images. But, DR had to wait for the technology of *miniaturization* to bring along electronic detector elements both small enough and cheap enough that an entire plate (14" × 17", or 35 × 42 cm) could be covered with hardware *detector elements* (dels) smaller than the resolution of the human eye at normal reading distance.

The main advantage of all digital imaging is something called *postprocessing*. Postprocessing means that the contrast, brightness and several other aspects of the image can be manipulated and changed *without repeating the original x-ray exposure*. Not only has this saved millions of dollars by reducing repeated procedures in medical imaging departments, but it

Figure 1-6

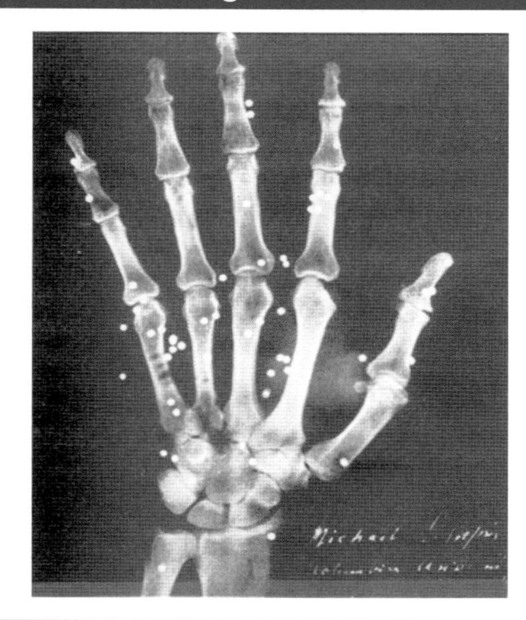

The first radiograph of a hand using Michael Pupin's fluorescent screen cassette to enhance the x-ray exposure, made in 1896, less than one year after Roentgen discovered x-rays. It shows buckshot embedded in the patient's hand from a shotgun. (From Quinn B. Carroll, *Practical Radiographic Imaging,* 8th ed. Springfield, IL: Charles C Thomas, Publisher, Ltd., 2007. Reprinted by permission.)

has dramatically reduced overall radiation exposure to the public as patients. This is the very goal of medical radiography, to *maximize diagnostic information while minimizing radiation exposure to the public*. It is amazing to consider the progress that has been made in just over one century since Wilhelm Roentgen discovered x-rays.

LIVING WITH RADIATION

Radiation is all around us in nature. In its broadest sense, anything which transfers energy through space from one point to another may be called radiation. Usually we think of this energy as *radiating* outward from a central source. It can be carried by *particles*, by mechanical *waves in a medium*, or by *electromagnetic waves*.

Examples of particulate radiation include the *alpha* and *beta* particles detected by Henri Becquerel and identified by Ernest Rutherford. Rutherford established that the alpha particle consisted of two protons and two neutrons, identical to the nucleus of a helium atom, and that the beta particle was identical to an electron (but traveling at very high speed).

Figure 1-7

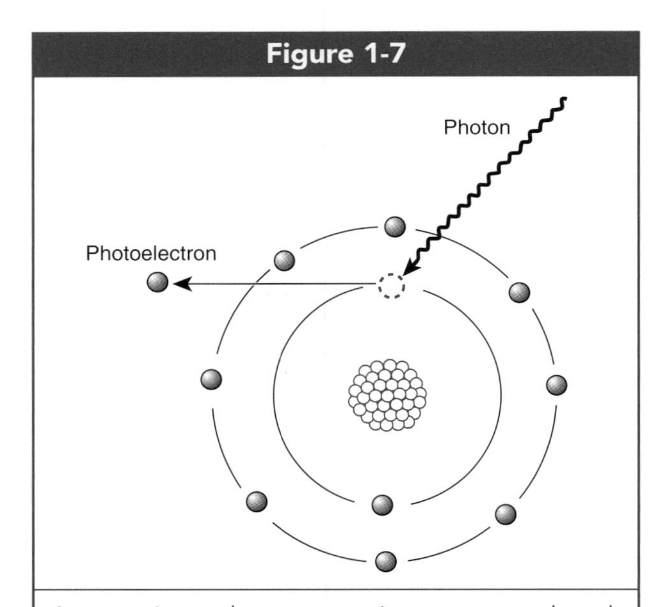

An x-ray is an electromagnetic wave; even though it has no mass, it carries enough energy to knock electrons out of atoms, which can lead to chemical changes.

Both, although smaller than an atom, were physical objects with mass. Since the amount of energy these particles carry is manifested by their speed, or *motion*, it would be categorized as a form of kinetic energy.

An example of mechanical waves in media is sound. For humans, sound normally consists of organized compressions and expansions of *air*, but for whales it consists of orderly compressions and expansions of *water*. There must be a *medium*, or matter, for the sound to travel in. In outer space, if it were possible to survive without a space suit, you could be inches away from another person, yelling at the top of your lungs, and the other person would not hear a thing because there are no molecules between you for the sound to travel in. Since it involves the organized *movement* of molecules, sound might also be considered a form of kinetic energy.

In contrast, electromagnetic waves are best considered as a form of potential energy. Although they are able to transfer energy from one place to another, they do so without the movement of any physical object, particle or molecule. They have no mass. Rather, they consist of the fluctuation of electrical and magnetic *fields* such as the pulling force you feel from a magnet. At different levels of energy, they take on different characteristics of behavior. This gives us the distinctions that we then label as *visible light*, *infrared light*, *ultraviolet light*, *microwaves*, *radio waves*, *x-rays*, *gamma rays*, and *cosmic rays*.

You can see that most types of radiation, such as sound and light, are harmless. A few, including particulate radiations and x-rays, are capable of *ionizing atoms* in any type of material, which means that they can eject electrons from the atoms (Fig. 1-7). This is potentially harmful because it can lead to chemical changes. In turn, chemical changes can cause biological changes, including diseases such as cancer.

The important point to make here is that more than three-fourths of all the radiation we receive comes from nature rather than from technology. It is something we have always lived with, even before the discovery of x-rays. We are *irradiated* by minerals in the ground, by minerals within our own bodies, by the sun and even by stars and novas in space (Fig. 1-8). We are irradiated by the bricks and cement in our homes, made from natural substances. The

Figure 1-8

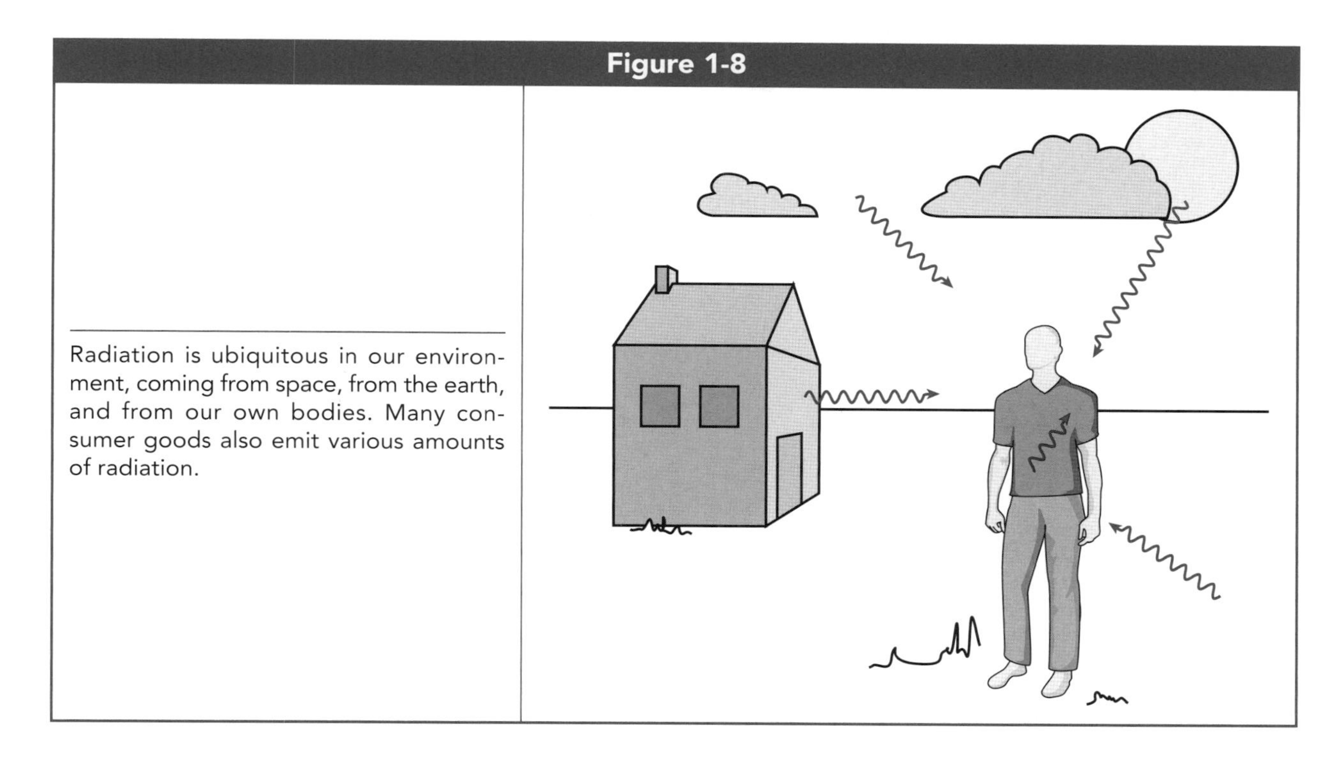

Radiation is ubiquitous in our environment, coming from space, from the earth, and from our own bodies. Many consumer goods also emit various amounts of radiation.

water we drink has trace amounts of radioactive minerals in it. Even bananas have an unusual amount of radioactive potassium in them!

The radioactive gas, *radon*, has become a serious problem in various geographic regions of the world, especially in buildings and houses with basements, where it can accumulate from poor ventilation. Simple tests are available to check the radon levels in your basement. Radon accounts for more than two-thirds of all natural radiation exposure to the human population.

A surprising array of everyday products expose us to radiation. Televisions, smoke detectors and even glossy magazines give us small amounts, while such things as mantles for camping lanterns and cigarette smoke can expose portions of the body to large amounts of radioactivity.

Historically, there have been a few truly catastrophic radiation events such as the meltdown of the Chernobyl nuclear power plant in the Ukraine in 1986 and the dropping of two atom bombs on Hiroshima and Nagasaki, Japan, in 1945. These events resulted in tens of thousands of deaths.

However, very minor events have also been grossly exaggerated by the sensationalist media and activist groups, in order to fuel an irrational and paranoid fear of all things nuclear. For example, the worst nuclear power plant accident in the history of the United States was a partial core meltdown at the Three-Mile Island plant in 1975. A small amount of radioactive steam was emitted into the air. Compared to *50 million* curies of radioactivity released at Chernobyl, the radiation from Three-Mile Island measured *17* curies. Yet, due to media and political hype, it resulted in the cancellation or postponement of dozens of nuclear plants over the next few decades. Only about one-fifth of all electricity generated in the United States is from nuclear power plants; by comparison, France gets over 95 percent of its electricity from nuclear power.

The applications of nuclear science in both industry and medicine have benefitted mankind in many ways. Roentgen's discovery of x-rays continues to play an essential role in medical diagnosis and treatment, with the benefits far outweighing the risks. However, radiation must be managed with a healthy respect and with common sense, so that exposures to both patients and workers are kept *ALARA*, "as low as reasonably achievable."

Accidents and diseases in early radiation workers lead speedily to the recognition that protective aprons, gloves and barriers made of lead were

Figure 1-9

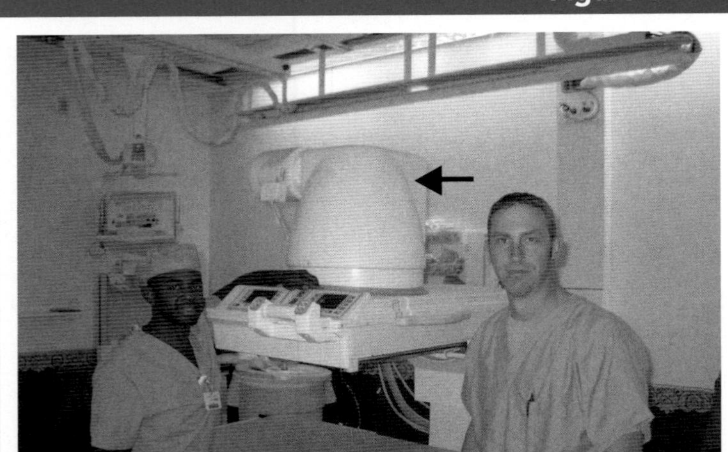

The modern image intensifier, (*arrow*), had the greatest impact on reducing medical patient dose in the history of radiography. Modern radiography is classified as a safe profession. With the use of common sense and the ALARA philosophy, occupational levels of exposure to radiation are very minimal indeed. (Courtesy of Phil Jensen, R.T., and Sungunuko Funhiro, R.T.)

needed. Today, regulations ensure that these protective barriers and devices meet minimum standards. Combined with much more efficient x-ray equipment, these methods of protection have reduced the typical exposure for radiographers from their work to a level roughly equivalent to the amount of naturally-occurring radiation exposure we all accumulate each year. The incidence of various cancers among radiographers is no different than that for the overall population.

Radiography is classified as a "safe" profession, with associated risks closer to those of a secretary or school teacher than to those of heavy industry or chemical workers (Fig. 1-9). By simply following policies and using good common sense, radiographers have nothing to fear from their occupation, but can rather look forward to an engaging and interesting career that greatly benefits their fellow man.

SUMMARY

1. Radiography is founded upon the principles of scientific method. The standard of practice for radiographers is to use good common sense, sound judgment, logical consistency and objective knowledge.
2. X-rays were discovered by Wilhelm Conrad Roentgen on November 8, 1895. Natural radioactivity was discovered by Antoine Henri Becquerel in 1896. Both scientists received the Nobel Prize.
3. Historically, the greatest reduction in patient exposure came with the invention of the image intensifier for fluoroscopy in 1948. The greatest reduction in patient exposure for static ("still-shot") radiographs was effected with the invention of the fluorescent "screen-cassette" by Michael Pupin in 1896.
4. Digital imaging was made possible by the development of PACS systems and CR imaging in the early 1980s, and DR in 1996—all due to advances in computer power and miniaturization technology.
5. Radiation is all around us in our everyday lives, and is largely misunderstood by the public. While the hazards of excessive radiation must be respected, radiography is a safe profession.
6. The professional charge of radiographers is to maximize diagnostic information while minimizing radiation exposure to the patient and personnel.

REVIEW QUESTIONS

1. That simple explanations are more likely to be true than complex ones is the scientific principle of:

2. The strongest aspect of the scientific method is that it is expected to be self-_______________.

3. When Roentgen accidentally discovered x-rays, he was investigating the properties of _______________ rays.

4. What are the three types of radiation discovered by Henri Becquerel?

5. Dynamic, real-time imaging with an image intensifier is known as:

6. Digital radiographic imaging had to wait for what two technological developments to occur?

7. The main advantage of digital radiographic image is post-_______________.

8. What are the three broad categories of radiation:

9. All of the physical, chemical and biological changes that can be caused by x-rays are due to their ability to _______________ atoms of any material.

10. A radiographer's average annual occupational exposure to radiation is about equal to _______________ radiation accumulated each year.

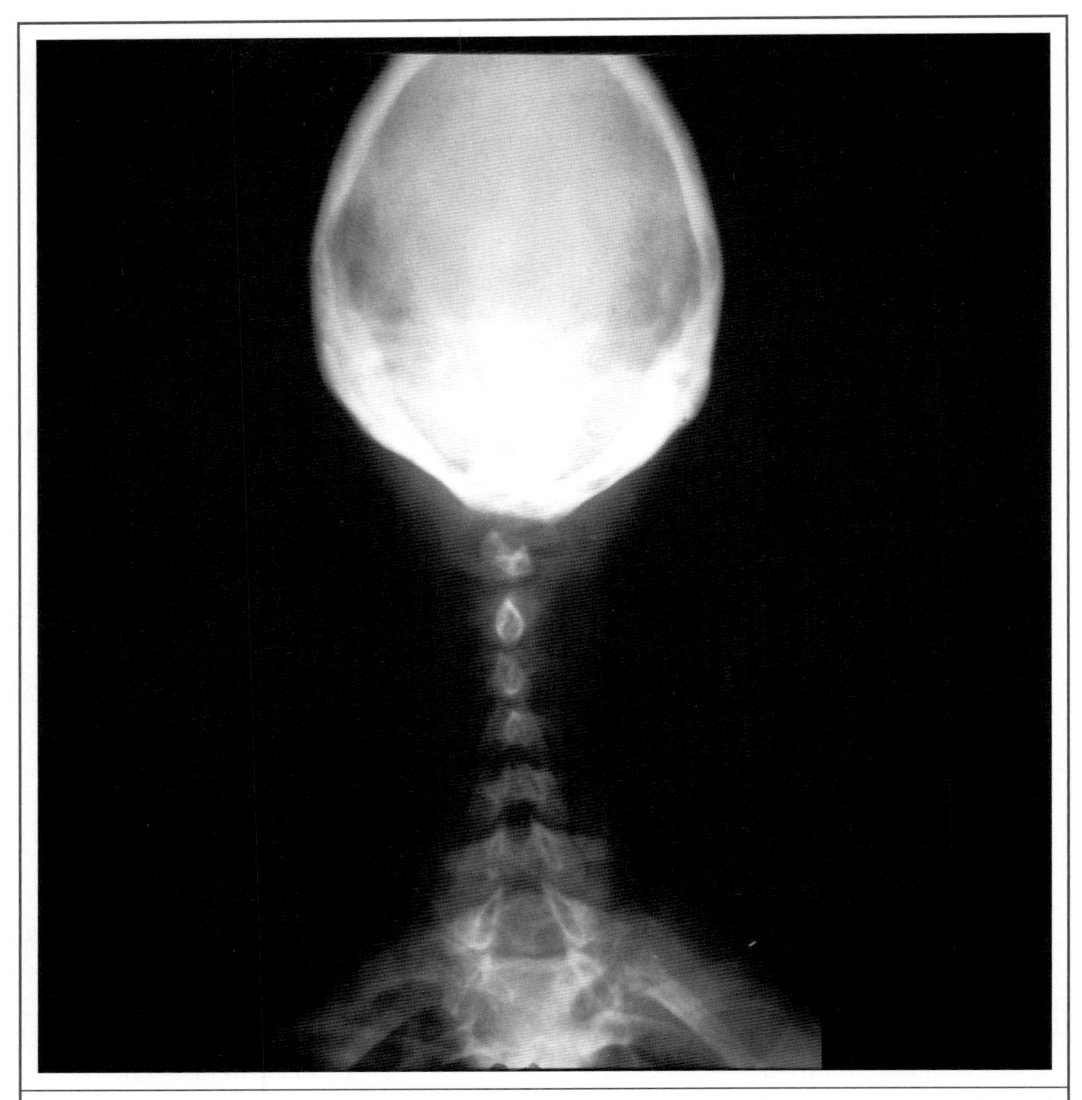

An accidental double exposure superimposing two oblique cervical spine projections "burned out" the facial masses, leaving a double-image of the posterior skull that looks "alien."

Chapter 2

BASIC PHYSICS FOR RADIOGRAPHY

Objectives

Upon completion of this chapter, you should be able to:

1. List and define each of the three base quantities and the four fundamental forces in physics.
2. State the conversions for basic units between the *Systeme International* and the traditional British system, and the value of the most used mathematical prefixes for units.
3. List the types of energy and give medically pertinent examples of *transducers.*
4. Explain the *law of conservation of energy*, forces, and *conserved and non-conserved quantities.*
5. Describe how *kinetic* and *potential energy* relate to electron positions within the atom.
6. Define the basic states of matter, and how heat causes them to transition from one to another.
7. List and define the three methods of heat transfer, and how *each* is applied in the x-ray tube.

THE BASE QUANTITIES AND FORCES

Parsimony, the philosophy of simplifying concepts, explanations and formulas as much as possible, was discussed in Chapter 1 as one of the fundamental tenets of scientific method. William of Ockham, an English philosopher in the 1300s, stated it this way: "Entities must not be multiplied beyond what is necessary." This quote came to be known as *Ockham's razor*, to be applied by scientists whenever an explanation or theory became so tortuous as to beg credibility.

A classic historical example was when Copernicus challenged the teachings of the Greek Ptolemy that the Earth was at the center of the solar system. As the church clung to this idea, in order to explain the apparent motions of Mars and other planets in the sky, its wise men had to invent increasingly complicated overlapping orbits called *epicycles*—circles within circles within circles. The truth was much simpler—just place the Sun at the center of the system, and all of the apparent motions of the planets made perfect sense as the Earth "caught up" and passed them in its own orbit.

We should add to the great success stories in scientific simplification the reduction of all measurements to just *three* standards, and the reduction of all forces in the universe to just *four*. The three standards of measurement are:

1. Time
2. Length
3. Mass

The standard unit for time is the **second**. Originally defined by the motion of the Earth around the Sun, it is now based upon the number of times an atom of cesium vibrates in that time, an incredibly reliable and consistent number. The standard unit for length is the **meter**. Originally defined by lines marked just over three feet apart on a metal bar, it is now defined as the distance light travels in 0.0000033 seconds.

You will note that the meter, at 39.3 inches, is very close to the 40-inch standard distance for the x-ray tube above the x-ray table used in the United States. As a practical matter, this is a level slightly above the average radiographer's head where gauges can be read and buttons easily reached. But it is also one meter above the table, consistent with countries using the metric system.

The *mass* of an object is the amount of matter that it contains. In a hypothetical oversimplification, one could obtain this measurement by counting all of the "particles" that make up the object. The standard unit for mass is the **kilogram**. It was originally based upon the amount of water contained in a vessel of 1000 cubic centimeters volume (about a 4-inch, or 10-cm cube) at a specific temperature. The mass of just one of these cubic centimeters of water is *one gram*.

Although it is used as a unit of weight in most countries, in this strict context the kilogram is the *amount of matter*, not how much it weighs. The difference between mass and weight is *location*. For example, assume that my body has 70 kilograms of matter in it. On the Earth it weighs 154 pounds (I wish!), but if I go to the moon with its weaker gravity, I will weigh only about 27 pounds. Yet, on the moon my body *mass* is still 70 kilograms.

All other measurements are taken in *derived units*. No matter what is being measured, they can always be reduced to these three *fundamental units* in relation to each other. For example, speed is expressed as *length over time* or *length divided by time*, such as miles per hour or kilometers per second. The density of an object is its *mass over volume*, such as grams per cubic inch, but volume is based on cubing a length such as inches or meters, so in effect this is *mass divided by length*.

Ever more complex measurements such as temperature can still be so reduced. Note that the traditional thermometer consists of a glass tube with liquid mercury inside. As temperature increases, the mercury expands so that it has a lower density (mass/volume), and the top of the column of red liquid moves *up* the tube by a certain *length*. By these changes in density, volume and length, we obtain a measurement of how hot it is.

From Einstein we learn that the entire universe may be defined as a continuum of space-time containing only mass and energy. These are all fundamental quantities: Space is measured by lengths, and time and mass are already fundamental units. By $E = mc^2$, we learn that energy (E) and mass (m) are interchangeable. Mass can be converted to energy and vice versa. Every mass has an energy equivalent, and every energy has a mass equivalent measured in . . . you guessed it . . . grams.

All of the mass in the universe interacts as bodies of matter by only four fundamental forces. A force may be thought of as anything that exerts a *push* or a *pull* on something. These four fundamental forces are:

1. Gravity
2. The weak nuclear force
3. Electromagnetism
4. The strong nuclear force

That's it! All of the physical interactions that take place in the entire universe can be reduced to these four forces. Electromagnetism encompasses all electrical and magnetic phenomena. The weak nuclear force is responsible for the radioactivity that Becquerel discovered in some materials. And, the strong nuclear force holds protons and neutrons within the nucleus of an atom.

Each of these forces, in the order listed above and in Figure 2-1, is *magnitudes* stronger than the preceding force. The strength of a force may be measured by how *small* a particle it can affect. Gravity is the weakest force, so weak that large amounts of mass are required in order to "feel it" or see its effects. Your body feels "weighty" only because the mass of the *entire earth* is pulling on it.

Yet, when it comes to pulling your laundry fresh from the dryer, a piece of lint will stick to a shirt, defying the gravity of the whole earth that is trying to pull it down to the floor. This "static cling" is due to just a small amount of electrical charge on the lint. Likewise, the smallest magnets overcome gravity by keeping papers and photos posted on your refrigerator door. Both electrical charges and magnets are billions of times stronger than gravity.

The intensity of the "weak" nuclear force falls between that of gravity and electricity. It operates at distances equivalent to the diameter of one or two nuclear particles in an atom. The *strong* nuclear force is millions of times stronger. It is peculiar to think that, electrically, the nucleus of an atom is made up of *all positive charges*, which we know *repel* each other. Something much stronger than electrical charge must overcome that electrical repulsion in order to pull many dozens of these positive charges into such close proximity to each other and lock them into place. If it were not for the strong nuclear force, matter in the universe could not exist in any organized fashion. Figure 2-1 summarizes these concepts.

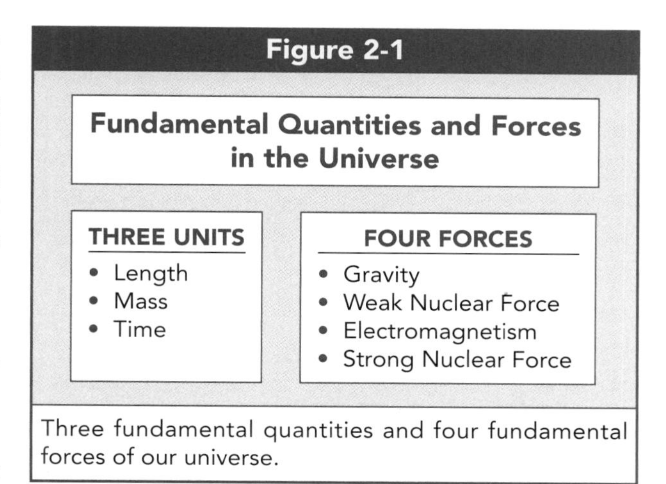

Figure 2-1
Fundamental Quantities and Forces in the Universe
THREE UNITS
• Length
• Mass
• Time
FOUR FORCES
• Gravity
• Weak Nuclear Force
• Electromagnetism
• Strong Nuclear Force

Three fundamental quantities and four fundamental forces of our universe.

UNIT SYSTEMS

The metric system of units is known throughout the world as the SI system (Le Système International). It lends itself to scientific notation, and is less cumbersome than the old British units because *all* of the unit conversions within the SI system are based on multiples of ten, rather than 12 inches to a foot, 3 feet to a yard, 16 ounces to a pound, and so on. Ironically, while Britain itself has long since converted to the metric system, Americans still cling to the old British units. The following examples may help those using the British units to visualize metric lengths:

One meter (m) is about 3 inches longer than one yard

One centimeter (cm) is about the width of your smallest fingernail

One millimeter (mm) is about the width (diameter) of a pinhead

Practice making mathematical conversions between British and SI units is provided in the next chapter. Four simple conversions are sufficient for our purposes. They are as follows:

For length: 1 inch = 2.54 centimeters (or 25.4 millimeters)
1 meter = 39.3 inches
1 mile = 1.61 kilometers

For mass: 1 kilogram = 2.2 pounds

The standard unit for time, seconds, is abbreviated ***s***. The meter, the standard unit for length, is abbreviated ***m***. For mass, the unit *gram* is abbreviated ***g***, and the unit *kilogram* (1000 grams) is abbreviated ***kg***.

Table 2-1 presents Greek prefixes to designate very large and very small numbers, along with their abbreviations. This particular range of units was selected because they are widely used in radiography, computer science and general physics. These prefixes and abbreviations should be committed to memory.

In Table 2-1, note that capitalizing the abbreviation makes a difference; for example, the upper case ***M*** means *mega*, while the lower case small ***m*** means *milli*.

A shorthand form of scientific notation is used in Table 2-1: The magnitude of each unit given simply as the number 10 raised to an exponential power. This is actually shorthand for the expression "1×10^x" where "x" is the power. In other words, all these units start with a 1 or 1.0. To see what the number in the table looks like written out longhand, if the power of 10 is a positive exponent, simply move the decimal that many places to the right from 1.0, adding zeros as needed. Thus, 10^3 designates 1.0 with the decimal moved three places to the right for 1000. If the exponent is a negative number, move the decimal that many places to the left, also adding zeros as needed, and starting from 1.0. For example, 10^{-3} moves the decimal from 1.0 to .001, which is read as "one-thousandth."

The magnitudes listed in Table 2-1 are in the *generic* units: *meters*, *volts*, *hertz*, *bytes*, and *seconds*. For example, the first line in the table is stating that there are 10^3 *meters* in a kilometer, *not* that there are 10^3 kilometers in a meter. The generic unit is meters. The fourth line is stating that there are 10^{12} computer *bytes* in a terabyte. The generic unit is *bytes*.

A negative exponent simply expresses the term as a fraction. When changing from 10^6 to 10^{-6},

Table 2-1

Greek Prefixes and Abbreviations

Prefix	Abbreviation	Example	Magnitude	Read as:
kilo-	k	km = kilometer	10^3	Thousands
mega-	M	MV = megavolts	10^6	Millions
giga-	G	GHz = gigahertz	10^9	Billions
tera-	T	TB = terabytes	10^{12}	Trillions
centi-	c	cm = centimeters	10^{-2}	Hundredths
milli-	m	ms = milliseconds	10^{-3}	Thousandths
micro	μ	μm = micrometers (microns)	10^{-6}	Millionths
nano-	n	nm = nanometers	10^{-9}	Billionths
	Å	Angstrom	10^{-10}	Ten-billionths
pico-	p	pm = picometers	10^{-12}	Trillionths

"millions" becomes "millionths," "billions" become "billionths," and so on, as these numbers are verbally expressed. In Table 2-1, interpret the "Read as" column as follows: The first entry, "kilo" means "thousands," so a kilometer is a thousand meters. The bottom half of the table is a tad trickier. "Centi" means "hundredths," so a centimeter is a hundredth of a meter. This means it takes 100 *centimeters* to make one *meter*. "Milli" means "thousandths," so a millimeter is a thousandth of a meter—it takes 1000 millimeters to make a meter, and so on. A single atom is about one-tenth of a nanometer in diameter. This is one Angstrom.

All kinds of electromagnetic radiations consist of *ripples* similar to water waves. From one ripple to the next, it takes one "crest" or peak and one "trough" or dip to make up a single *wavelength*. One water wavelength can be from a few inches long in your bathtub to many feet long in the ocean. The wavelength of a typical radio wave is about one-half mile long. The wavelength of the microwaves in your microwave oven is about one centimeter, the width of your smallest fingernail.

By comparison, the wavelength of an x-ray is incomprehensibly small. A special unit for measuring such extremely small lengths is the *Angstrom*, abbreviated Å. The angstrom is 10^{-10} meters, read as "one ten-billionth of a meter." Thus, the angstrom falls in Table 2-1 at the bottom between *nanometers* and *picometers*. The wavelengths of x-rays used in medical diagnosis range from 0.1 Å to 0.5 Å, one-tenth to one-half of an angstrom ($\frac{1}{10}$ to $\frac{1}{2}$ the size of an atom). Restated, this is from 10 to 50 *billionths* the diameter of a pinhead, very difficult to visualize!

The next chapter explains simple methods for converting units like these, and how scientific notation can help.

THE PHYSICS OF ENERGY

X-rays and all other types of radiation are forms of energy. Energy may be defined as the ability to cause a change in the motion or state of an object—that is, the ability to *do work*. The generic unit for energy is the *joule*. One joule is roughly enough energy to get a one-pound object moving about 10 miles (16 km) per hour.

Energy belongs to a class of concepts that physicists call *conserved quantities*. This means that, within a closed system, the total amount of energy is always constant—new energy cannot appear "from nothing," nor can any energy be destroyed or disappear within the system. What it *can* do is change from one form to another.

A lightbulb is a closed system. Given a certain amount of electricity passing through it, that energy can be changed from electrical energy into light energy or heat energy. But it must always be accounted for, it must always sum up to the same total. This is referred to as the *Law of Conservation of Energy*.

(You might ask for an example of a *non-conserved quantity*. The concept of a *force* would fit this description; by using a lever, with the pivot located correctly, the force of your arms pushing down on one end with 10 pounds can be multiplied to 100 pounds of force on the other end of the lever. Forces do not follow a law of conservation. Many quantities in physics are conserved quantities, but not all.)

The entire universe is a closed system. The total amount of energy contained within it can never change. The energy can change forms, from sunlight to electricity to heat, for example. But it cannot just disappear out of existence. Einstein found that one of the forms of energy is matter:

$$E = mc^2$$

This famous formula is so simple that it epitomizes the principle of *parsimony* discussed in Chapter 1. Yet it packs all kinds of profound implications, one of which is that energy and matter are interchangeable. "$E = m$" tells us that energy and matter are essentially the same thing. But when they change back and forth, the calculation must be multiplied or divided by a *huge* number, c^2, the speed of light squared. In meters per second, this number comes to 90,000,000,000,000,000 or 90 quadrillion.

When matter is converted into energy, an infinitesimal amount of matter will produce incredible power (90 quadrillion times its "weight"). This is the basis of the nuclear bomb. Going the other direction, from energy to matter, we find that an enormous amount of energy is required to produce even the smallest particle. This is why particle accelerators ("atom smashers") have to be miles long and use gigavolts or teravolts of energy. You could think of matter as "compressed" or "condensed" energy. A good way to visualize the creation of matter is that energy must be "compressed" by a factor of 90 quadrillion in order to become an object that has weight.

The *rest energy* of any particle is the energy that would be released if it were "annihilated," or changed into pure energy, discounting any speed or motion it already had. The annihilation of a single electron could produce 510 thousand volts of electricity, that of a single proton 938 million volts (Fig. 2-2). The annihilation of all the atoms in a human body would produce electrical voltage equal to 4 with 37 zeros behind it!

Some of the different forms that energy can take include *mechanical* energy—the energy of movement and position, *chemical* energy such as that contained within a battery, *electrical* energy such as the electricity released when the battery is connected in a flashlight and it is turned on, *thermal* energy or heat such as that emitted by the flashlight bulb, *electromagnetic* energy such as the light also emitted by the flashlight bulb, and *nuclear* energy created by the weak and the strong nuclear forces within the heart of an atom.

Any device that can change one form of energy into another is called a *transducer*. You may recognize this as the device used by sonography technologists to produce ultrasound images. It is so named because

Figure 2-2

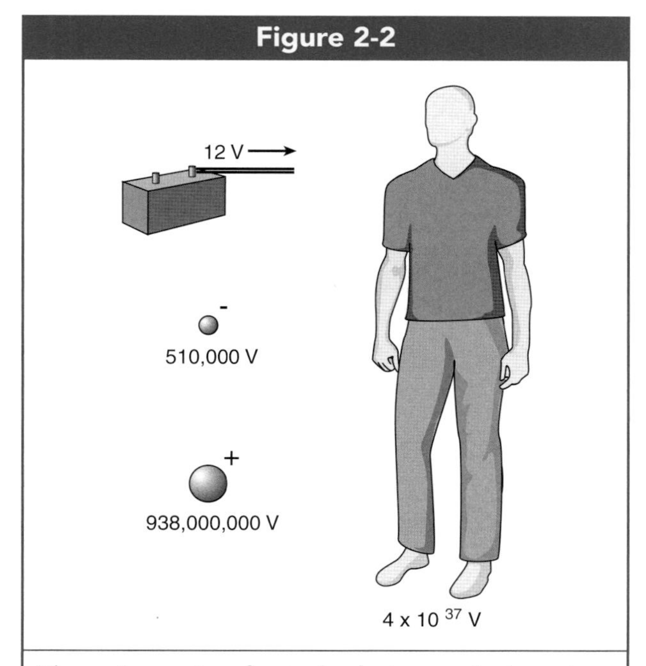

The *rest energies* of a single electron, a single proton, and a single human body, in volts. (For comparison, a car battery generates 12 volts.)

Figure 2-3

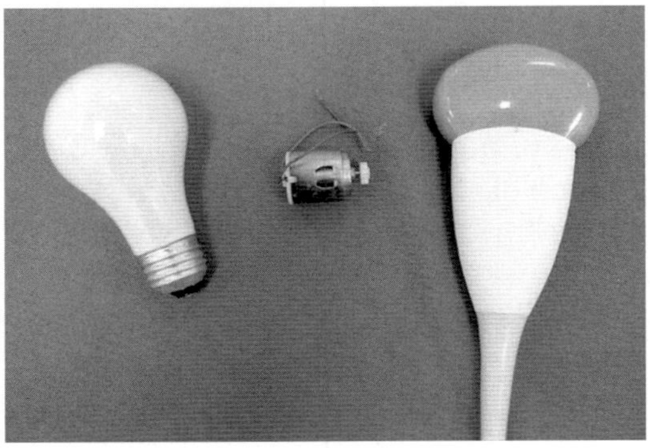

Examples of *transducers:* A light bulb, a small electric motor, and an ultrasound transducer.

a crystal in it converts pulses of electricity into pulses of mechanical sound waves that travel through the body. The term *transducer*, however, has much broader meaning. Gas-powered engines, light bulbs, electrical generators and motors, and x-ray tubes are all transducers (Fig. 2-3).

Mechanical energy can be further divided into two classifications important to understanding the production of x-rays. They are *kinetic* energy (KE) and *potential* energy (PE). While kinetic energy is the energy of *motion*, potential energy is the energy of *position*. According to the law of conservation for mechanical energy, the sum of an object's kinetic energy and its potential energy must always add up to the same amount, but the two *can* trade off.

For example, if you lift your textbook and hold it 3 feet above your desk, it acquires potential energy from the motion energy "spent" by the muscles in your arm. Let's assume that, given the particular weight of the book, its potential energy is 1 joule (1J) relative to your desktop. This means that, if you drop the book, it has the *potential* to speed up to 1J of kinetic energy by the time it hits the desk. The kinetic energy is represented by the actual speed, the potential energy is represented by the potential to speed up (due to the force of gravity).

At any point, as the textbook falls, its kinetic energy and potential energy must sum up to the same total, 1J. When the book is one-third of the way down, with two feet remaining to go, what are its kinetic and potential energies?

Answer: *KE = 0.33 J (one-third of the original PE of 1J)*
PE = 0.67 J (the remaining 2/3 of original PE)
____
Sum = 1.0 J

When the book is two-thirds of the way down, with one foot remaining to go, it has sped up more, and now has KE = 0.67J and PE = 0.33J (Fig. 2-4). At the last possible moment before it strikes the desk, it has a speed equivalent to 1J of kinetic energy, and its PE is zero relative to the desk, since it cannot speed up any more.

Figure 2-4

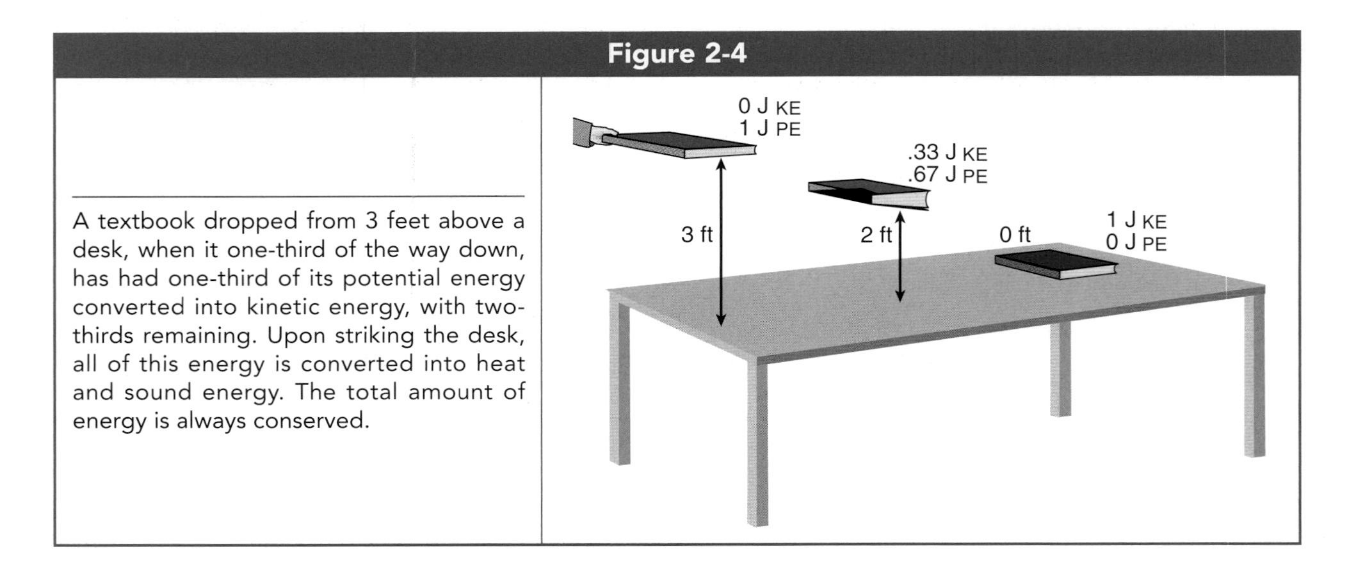

A textbook dropped from 3 feet above a desk, when it one-third of the way down, has had one-third of its potential energy converted into kinetic energy, with two-thirds remaining. Upon striking the desk, all of this energy is converted into heat and sound energy. The total amount of energy is always conserved.

"But wait!" you exclaim, "Where does the kinetic energy go after the book has stopped on the desk?" At this point, it cannot go back into potential energy unless the book is lifted again. The PE is zero, and the KE is zero since it is still. The answer is that this energy has now been converted into yet two *other* types of energy—*heat* energy which increases the temperature of your desk, and *sound* energy in the form of a loud *whack*. Technically, the heat and sound are both still forms of kinetic energy, only at the molecular level, because the temperature of your desk depends upon the *movement* of the molecules in it, and the sound represents the mechanical *movement* of air molecules striking your ear.

Eventually, all organized forms of energy degenerate into heat energy, but the sum is always the same. The law of conservation prevents the energy from ever disappearing.

In radiography, the law of energy conservation is most important in understanding both the production of radiation and the absorption of radiation within atoms. Electrons with negative charge are situated in specific energy states, similar to orbits at set distances from the nucleus. The positive charge of the nucleus tends to pull them down toward it, just as earth's gravity pulls your textbook down to the desk.

The electrons can be lifted to higher orbits by absorbing x-rays or light. This represents changing *electromagnetic potential energy*, the energy of the x-ray or light wave, into the potential energy of the physical position of the electron higher above the nucleus. The total amount of energy is conserved. Also, electrons may drop down to open vacancies in orbits closer to the nucleus. When they do, they lose *potential energy*, the energy of their position relative to the nucleus. The law of conservation states that this energy cannot disappear, but must be converted into some other form of energy. In this case, *positional potential energy* is converted into *electromagnetic potential energy* as a light wave or x-ray is emitted from the atom (Fig. 2-5).

You can see that understanding radiation and understanding the atom are closely interconnected. A single x-ray or light ray is called a *quantum* (*quanta* for plural). *Quantum physics* is the study of all these interactions and relationships between the atom and waves of radiation.

Figure 2-5

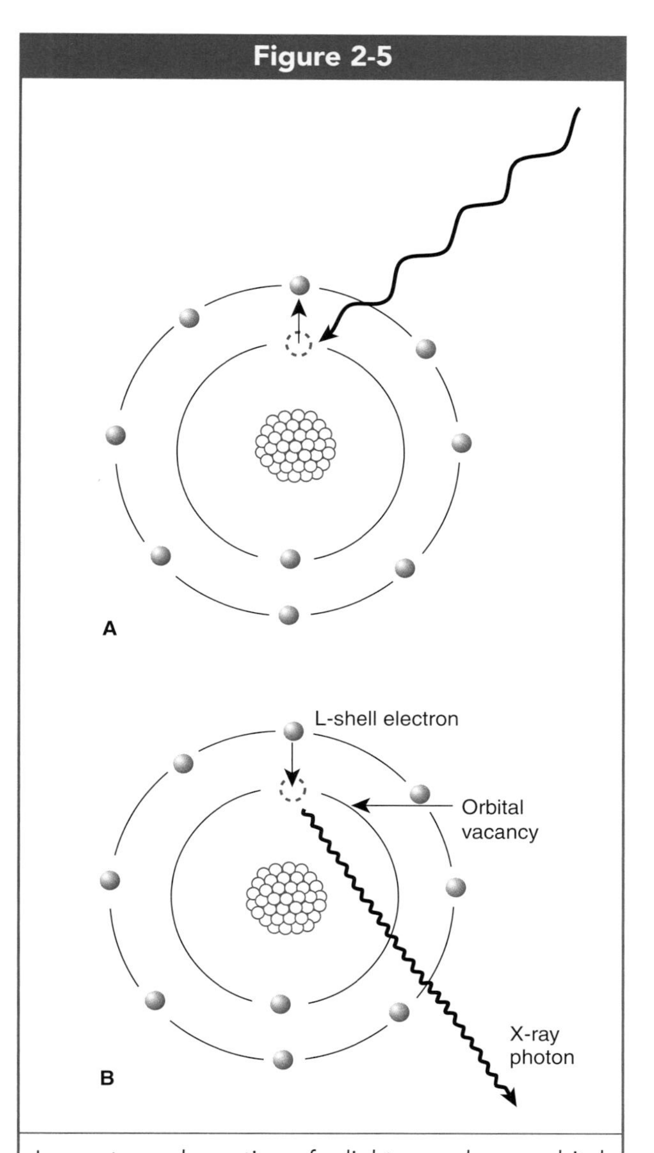

In an atom, absorption of a light wave by an orbital electron can lift it up to a higher "orbit," **A**, increasing its *potential energy* relative to the nucleus. When an electron drops down to a lower "orbit," its potential energy must be reduced by emitting a light wave (or x-ray), ***B***.

HEAT AND STATES OF MATTER

All materials have a normal *state* which we might describe as their condition "at room temperature," some solid, some liquid, and some gas. Nonetheless, *all* of them can be caused to change from their "normal" state into other states by the processes of heating and cooling. These topics are pertinent to radiographers for several reasons, the foremost being an understanding of the x-ray tube, yet others relate to such diverse things as catheters and automatic injectors, spinal injuries and even blood circulation.

The familiar states of matter can be precisely defined by asking only two questions: (1) Does the material have a definite, fixed *shape*; and (2) Does the material have a definite, fixed *volume*? (Volume is the *amount of space* the material takes up or occupies.) With this simple scheme, we can define solids, liquids and gases as follows:

Solids have: –Definite shape
–Definite volume

Liquids have: –Indefinite shape
–Definite volume

Gases have: –Indefinite shape
–Indefinite volume

A solid, such as a marble, maintains both its shape and volume regardless of what *container* it may reside in, a large room, a small box, or a large syringe. A liquid takes on the shape of its container—if it is in a syringe or cup, it will be cylindrical, but it can be placed into a square container and acquire the shape of a cube. Its shape is indefinite. However, its *volume* is definite, meaning that a liquid cannot be substantially *compressed* or *expanded* and remain liquid.

The resistance of liquids to compression explains many things including how an entire car can be lifted by a one-foot diameter cylinder at the gas station, how the brakes on your car work, how the human circulatory system works, why "slipped disks" occur in the spinal canal, why a syringe works, and why your tube of toothpaste will rupture if you squeeze it hard enough with the cap on.

The great French physicist Blaise Pascal discovered that, because liquids will not compress, any pressure applied to a liquid will be *transmitted undiminished* throughout the liquid. Imagine a pipe with a solid cylinder or piston at each end. Only, somewhere in the middle the pipe *expands* from a 1-inch diameter to 20 inches (Fig. 2-6). It is filled with liquid. Now, to simplify the math, imagine the whole pipe system is square rather than round. If you press on the cylinder at the small end of the pipe with just 10 pounds of force, the *pressure* is 10 pounds per square inch. By Pascal's law, the pressure at the other end must also be 10 pounds per square inch. But at this end of the pipe, the square piston is 20 inches across, so its surface area is $20 \times 20 = 400$ square inches. What is the total force of this piston? It is 10 pounds per square

Figure 2-6

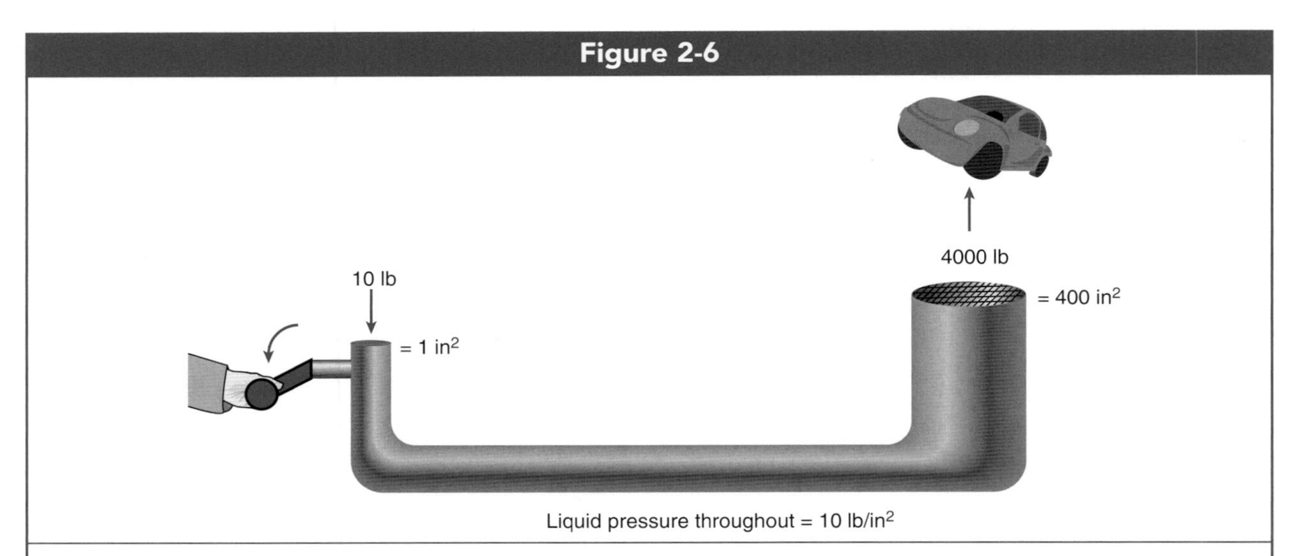

In this liquid-filled system, a force of 10 pounds applied at the small cylinder can be leveraged up to 4000 pounds at the large cylinder, because the *pressure* is conserved throughout the system at 10 pounds per square inch. The pressure is equal throughout because liquids *will not compress*. This is the principle upon which liquid pumps (such as the human heart) work.

inch *times* 400 square inches, or 4000 pounds! We can now lift a car or put on your brakes. Remember that force is *not* a conserved quantity. It can be multiplied by levers or expanding pipe systems.

"Slipped disks" and ruptured toothpaste tubes are the result of pressure being transmitted undiminished *throughout* a liquid. Because the liquid will not compress, it must *go somewhere* when unrelenting pressure is applied. In the human spine, the intervertebral disks are made of a semi-liquid *nucleus pulposis* (the "pulpy center") surrounded by a fibrous ring, the *annulus fibrosis*. When a skydiver lands too hard on the ground, the sudden vertical jolt tries to compress these disks. The semi-liquid center will not compress, so it must go somewhere. Just as your toothpaste will rupture the tube to escape, the nucleus pulposis ruptures through the outer fibrous ring into the spinal canal, pushing against the spinal cord and causing pain.

Gases, like liquids, have indefinite shape and take on the shape of their container. But, they *also* take on the *volume* of their container. Since gases have indefinite *volume*, they *can* be compressed or expanded. The air in your classroom could all be compressed to fit into a small tank. (Since sound requires air to travel in, this would spare you listening to your instructor! But, this might not be an altogether positive result.)

The human heart is a *liquid pump*. When it contracts, the blood must flow out from it because blood is a liquid and refuses to compress. It is possible for a very large bolus of *air* to be accidentally injected into a patient by leaving a running IMED pump below the level of the patient for a long time. When the heart contracts down on this air, it simply compresses and then expands again with the heart. The liquid blood stops flowing because no substantial pressure is being transmitted. The heart muscle itself receives insufficient oxygen, and soon the patient has a heart attack. Serious consequences to ignoring the laws of physics! The same type of thing happens when you get an "air lock" in the cooling system of your car—the water pump can no longer move the liquid coolant through the engine.

Upon heating, when a solid *melts*, like ice, it undergoes a change of state to a liquid. *Freezing* is when a liquid changes into a solid. *Condensation* is when a gas changes into a liquid, such as when it rains, and *evaporation* describes a liquid changing into a gaseous state. Most substances have *thresholds* for changing state, specific temperatures called *boiling points*, *freezing points*, *condensation points* and *evaporation points*. No matter how long you wait, water will not freeze at 33 degrees F (1° C), nor even at 32.1°. Nor will water boil at 211 degrees (99° C).

Some materials, called *resins*, do *not* have thresholds to change state. Catheters are made of resins. As you heat them up, they become softer and softer. They will gradually become a liquid rather than suddenly melting. This allows us to reshape catheters by heating them up, molding them how we want, and then quicky cooling them. Resins gradually become harder at colder temperatures.

Fluids refer to any substance that *flows*. Although we frequently use the term *fluid* when we mean *liquid*, this is imprecise since wind currents of air and other gases also flow. All liquids and gases are fluids.

Heat is defined as the *flow* of internal energy from one object or molecule to another. Internal energy has both kinetic energy and potential energy components. Its kinetic energy is in the form of *temperature*, determined by the average movement or vibration of the molecules. Its potential energy is in the form of *state*, whether solid, liquid or gas. Note that most solids are denser than liquid and will drop to the bottom of a liquid. Likewise, most gases will drop when they condense into a liquid, such as rain condensing from the vapor of a cloud. When a liquid boils into steam, a gas, it *rises* up into the air. Therefore, the state of a substance has a good deal to do with its higher or lower *position*, hence, its potential energy.

When a substance is heated, the in-flowing energy can be used to increase either its temperature or to change its state. At the very moment when water boils, the heat energy is being used to raise its position up into the air *rather* than to increase its temperature. Once it is in the air, however, further heating will result in increasing the temperature of the steam.

A popular quote from the days when radiographic films had to be developed in a darkroom was, "Don't open the door, you'll let all the *dark* out!" This is funny because we intuitively know that *dark* is really just the lack of light.

In a similar vein, for physicists, the concept of *cold* is really just the *lack of heat*. Remember that energy cannot disappear. It can only change form or flow from one place to another. Whenever something cools down, its heat energy *must* be flowing into another object or substance nearby, thus heating that substance up. The atmosphere actually feels a tad warmer when it begins to snow because as the water freezes into snowflakes, its heat energy escapes to warm the air around the snowflakes. Your refrigerator can only cool food by pumping its heat outside the box, thus *heating your kitchen*. This is why it feels hot above and behind the fridge. So in the strictest sense, there is no such thing as cooling, only heating. Ice does not really cool your drink, rather, your drink heats the ice! Proof? Observe that the ice melts.

There are three ways in which the flow of energy can occur, resulting in heating. They are:

1. Conduction
2. Convection
3. Radiation

The *conduction* of heat occurs when a hot object or substance comes in direct *contact* with a cooler one. The strongly vibrating molecules of the hot object frequently "bump into" those of the cooler object, causing them to jostle about more themselves, and thus raising the temperature.

Materials that transmit heat readily are called *thermal conductors*. Ironically, the tile on a bathroom floor feels cooler than a rug because the tile is a *good* thermal conductor; It conducts body heat so quickly away from your feet that you feel *your feet*, not the tile, cooling. If you step onto a rug in the same room, the rug is at the same temperature as the tile. Yet, the rug *feels* warmer than the tile because the rug is a poor thermal conductor—it takes heat away from the bottom of your feet more slowly. Materials that are poor at conducting heat are called *thermal insulators*.

Convection refers simply to stirring hot molecules into cooler ones through the *mixing of fluids*. As kinetic motion is transferred from hotter molecules bumping into the many cooler ones around them, the *temperature* (a measurement of molecular motion) of the originally hot substance drops. Heat energy is dispersed as it spreads out among more molecules.

Radiation here refers particularly to electromagnetic waves passing through space. The nature of these waves will be fully described later, but along with x-rays and ultraviolet light, they include visible light and infrared radiation which we feel warming us from the sun. This provides us another example of the distinction between heat and temperature: Although radiation is a method of *heat transfer* or the flow of energy, the radiation itself has no temperature.

Infrared radiation is a form of "light" waves emitted by the hot sun. Since there are (almost) no molecules or atoms in space, there is nothing there to be heated, yet the electromagnetic waves of infrared light pass through that cold space on their way to Earth. When they strike your skin and are absorbed by your molecules, electromagnetic energy is converted into kinetic energy (Fig. 2-7). Your own molecules begin to vibrate more with the extra energy. This molecular vibration or motion can then be measured as the increased *temperature* of your skin. It is your *skin* that is "hot," not the infrared radiation that brought the energy from the sun through space to you.

An x-ray tube and its surrounding housing use *all three* methods of heat transfer in order to cool down as quickly as possible, so the tube can be used to make the next exposure. Each time billions of electrons bombard the anode to make an x-ray exposure, the

Figure 2-7

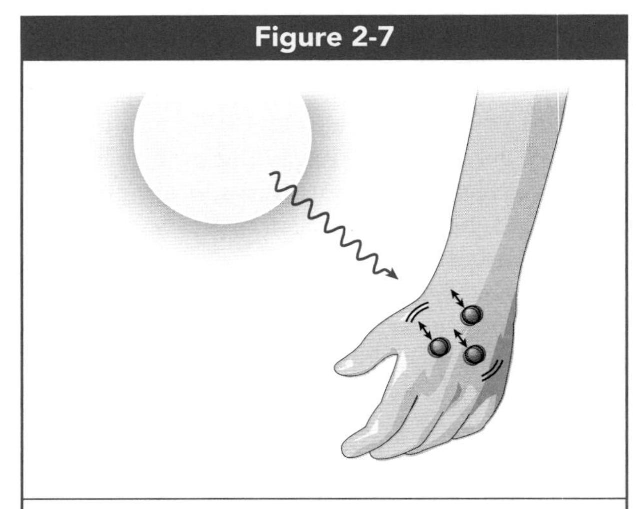

Electromagnetic radiation, such as light from the sun, has no temperature of its own, but it raises the temperature of your skin as the energy deposited there causes the body's molecules to "shake" more.

friction heats the anode disc to white-hot temperatures. First, this heat is *conducted* away from the anode disc by flowing down the metal shank that holds it, and on out into the glass of the tube and brackets of the tube housing. Second, most of the heat energy is emitted as electromagnetic *radiation*. Only a small portion of this is x-rays—most of the radiation consists of light, infrared, and ultraviolet rays. Some of the radiation escapes all the way out of the tube housing or is allowed through the "window" directed toward the patient, but most of it is absorbed by the lead tube housing and glass of the tube itself.

By *conduction*, heat is passed from the outer glass of the x-ray tube to a layer of cooling oil around the tube. In some machines, the oil is cooled by a fan and then circulated around the tube (Fig. 2-8). As heated oil is stirred and mixed with cooled oil, heat dispersion by *convection* occurs. Heat is conducted from oil to the lead outer housing, and thence to the air in the room. It may then be further dispersed by *convection* as the air in the room circulates. *Most of the heat dispersion from an x-ray tube is from various forms of electromagnetic radiation* being absorbed by various objects in all directions from the tube.

On the opposite end of the temperature spectrum from super-heated x-ray tubes are the *super-cooled* coils of an MRI (magnetic resonance imaging) machine. In order to generate the tremendous magnetic fields required for MRI scans, ceramic coils must be cooled to the point where there is almost no resistance to the electricity passing through them. The electrical coils are placed inside a chamber called a *dewar* which contains liquid helium. This, in turn, is surrounded by a second dewar containing liquid nitrogen. Amazing temperatures as low as –452° F (–269° C) are achieved, only 7 degrees F, or 4 degrees C, respectively, above *absolute zero*, the temperature at which all molecular motion would completely stop.

Much of the technology of radiography and related medical imaging requires a basic understanding of these fundamental concepts of energy and heat.

SUMMARY

1. All measurements in physics can be reduced to the three *fundamental quantities* of time, length and mass.
2. All forces in the universe can be reduced to four *fundamental forces*: gravity, the weak nuclear force, electromagnetism, and the strong nuclear force. Each is several magnitudes stronger than the previous. Without the strong nuclear force, matter could not exist in any organized form.

Figure 2-8

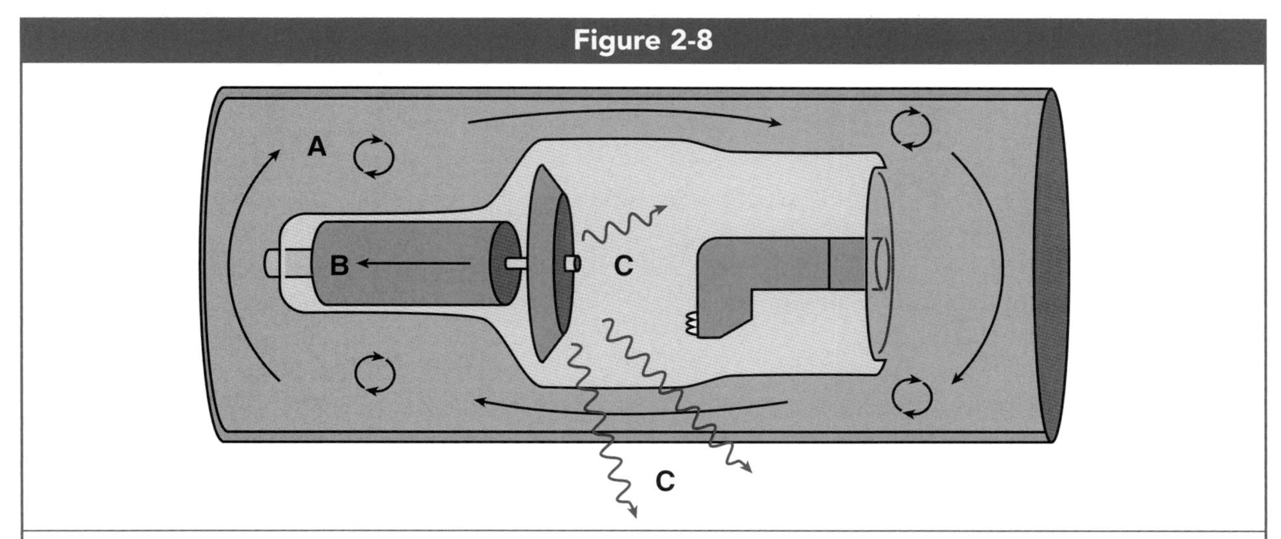

Convection can be used to disperse heat from an x-ray tube by circulating cooled oil around the tube, ***A***. Some heat also escapes by conduction down the shaft of the anode, ***B***, but most is emitted in the form of electromagnetic radiations, ***C***.

3. The metric or SI system of units is more simple and easy to use than conventional British units. Its fundamental units are the *second* for time, the *meter* for length, and the *gram* or *kilogram* for mass.
4. The total *energy* within a closed system is a *conserved quantity* and follows the *law of conservation*. It can be changed (by *transducers*) into many different forms, including *matter*.
5. When electrons within an atom are lifted from their orbital shells by an x-ray, *electromagnetic* potential energy is converted into the potential energy of their physical position "above" the nucleus. When their position drops back down, the change in positional energy must be emitted as electromagnetic energy to obey the law of conservation.
6. States of matter are defined by their conservation of *shape* and *volume*. Liquids conserve only their *volume*, and so they cannot be compressed. Therefore, any pressure applied to a liquid is transmitted undiminished throughout the liquid.
7. Most substances have thresholds for changing their state. *Resins* do not.
8. Heat is the *flow of internal energy*, manifest in *temperature* (kinetic E), and *state* (potential E). It can be transferred through conduction, convection or electromagnetic radiation. An x-ray tube uses all three methods to disperse its heat.

REVIEW QUESTIONS

1. When an astronaut travels from the earth to the moon, his/her *mass* will __________ (increase, decrease, or remain the same).
2. What is defined as an object's *mass* divided by its *volume*?
3. Every energy level has a __________ equivalent measured in grams.
4. Which of the four fundamental forces is responsible for natural radioactivity?
5. Which metric unit is about the size of a pinhead?
6. Which unit prefix means "millionths"?
7. X-ray wavelengths are best measured using what special unit?
8. What is the generic unit of energy?

(Continued)

REVIEW QUESTIONS *(Continued)*

9. As a book is dropping onto a desk, __________ energy is converted into __________ energy.

10. As an electron drops down into a lower atomic shell, its *positional* form of potential energy is emitted as an __________ form of potential energy.

11. The heart is only able to pump blood because as a liquid, the blood will not __________.

12. Heat energy is manifest as a change in *temperature* OR as a change in __________ .

13. List the three methods of heat transfer.

14. What is defined as the temperature at which all molecular motion would theoretically stop?

15. What proper term prefix describes *billionths* of a volt?

16. 80 mm = how many meters?

17. 200 microseconds = how many milliseconds?

18. 500 micrograms = how many kilograms?

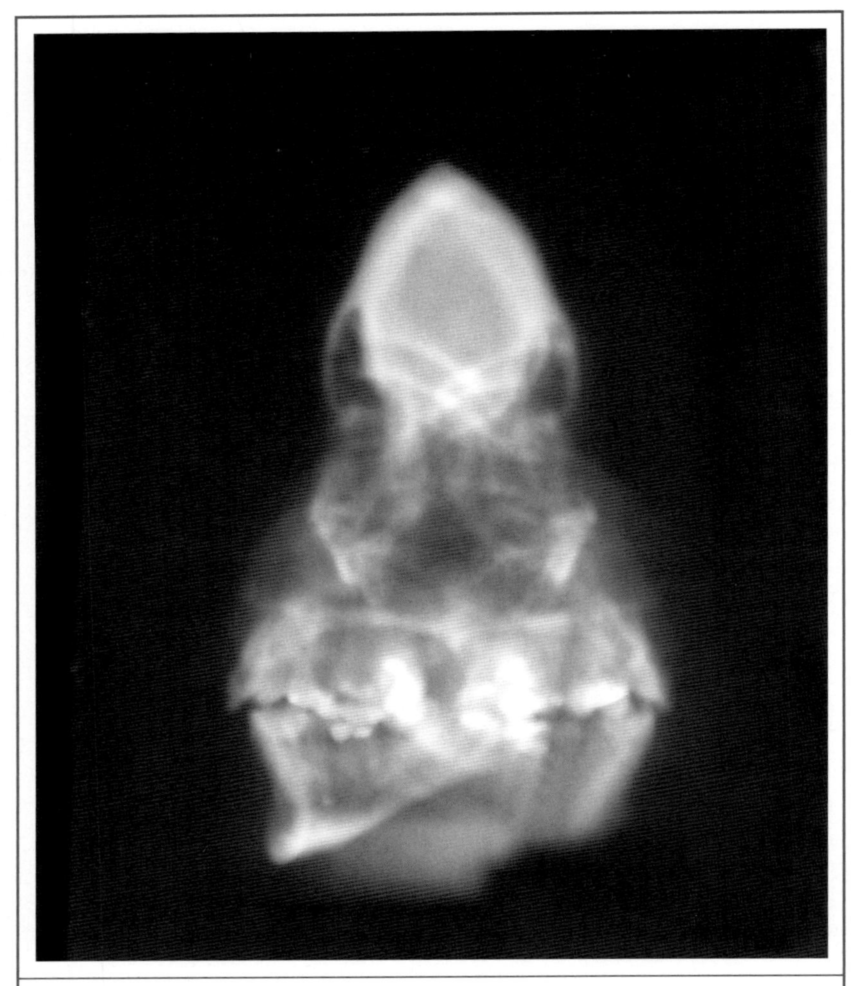

An accidental double exposure superimposing two lateral sinus projections "burned out" the posterior portion of the skull, leaving a double-image of the facial mass.

Chapter 3

UNIT CONVERSIONS AND HELP WITH MATH

Objectives:

Upon completion of this chapter, you should be able to:

1. Describe the basic order of operations in math.
2. Appreciate and *apply* scientific notation.
3. Appreciate and *apply* dimensional analysis to make difficult unit conversions.
4. Understand calculations for *areas* and *volumes*, and how they relate to heating and cooling.
5. Become adept at using the *inverse square law* and its applications for radiography.
6. Properly construct graphs from data, and correctly interpret various types of graphs.
7. Describe the characteristics of various types of graphs including in particular the x-ray beam spectrum and the digital histogram.

This chapter provides a review of basic math concepts pertinent to radiography, how to make unit conversions in the simplest possible ways, and how scientific notation makes the handling of very large and very small numbers *easy*. Graphs have always been important to the understanding of the x-ray beam and its production, electrical currents that flow through the x-ray machine, the heating and cooling of the equipment, and the processing and quality control of radiographic images. The use of histograms in digital imaging has only made the ability to interpret graphs more important than ever. A solid foundation for that skill will be developed here.

Care has been taken to include *only* those mathematical operations that you will need later on in dealing with radiographic machinery, techniques, and radiation protection. Therefore, you should strive for a thorough understanding and memorization of everything presented in this chapter.

If you would like to test yourself to see which of these sections you might need to review, do the Practice Exercise 3.1 at the end of this chapter. An answer key is provided in the Appendix. If you miss any of those problems, be sure to review the related section.

MATHEMATICAL TERMINOLOGY

The distinctions between phrases such as "directly related" and "directly proportional" are often confusing for students. The following provides clarification for several of these terms.

"DIRECTLY PROPORTIONAL" means that an increase in variable "A" causes an exactly identical increase in factor "B." If A doubles, B doubles. The same is true for decreasing the number: If A is cut to $\frac{1}{3}$, B is cut to $\frac{1}{3}$.

Example: mAs is directly proportional to image density: Triple the mAs, and the radiograph turns out exactly 3 times darker. Cut the mAs in half, and the radiograph turns out one-half as dark.

$$A = B$$
$$2A = 2B$$

It does not matter if "B" is being multiplied by some factor.

In the formula, $A = \frac{1}{4}B$, "B" is still directly proportional to "A," because if "A" is doubled, "B" will also double:

Example: For A = 2, 2 = ¼ B, B = 8
For A = 4, 4 = ¼ B, B = 16
*Note that when "A" doubled from 2 to 4, "B" also doubled from 8 to 16. They are directly proportional.

"EXPONENTIALLY PROPORTIONAL" or "EXPONENTIALLY RELATED" means that one or the other variable is squared or cubed in the formula: A small change in that variable will cause a big change in the other factor.

Example: Exposure intensity is exponentially related to kVP as:

$$kVp^3 = I$$

Triple the kVp, and the exposure increases $3^3 = 27$ times.

In the formula, $A = \frac{1}{4} BC^3$, "A" is directly proportional to "B," but "A" is exponentially proportional to "C."

"DIRECTLY RELATED" means only that an increase in factor "A" causes an increase in variable "B," that is, they go up or down together. However, it does not specify how much, whether the relationship is proportional or exponential.

In the formula, $A = \frac{1}{4} BC^3$, "A" is directly related to both "B" and "C"

"INVERSELY PROPORTIONAL" means that the two variables change by exactly the same magnitude, but in opposite directions: When one goes up, the other must go down by the same factor. If "A" doubles, "B" must be cut to ½ . If "A" is cut to ⅓, "B" must triple.

Example: In the formula, A = BC, "B" and "C" are inversely proportional: If "B" were cut to ½, in order for "A" to remain the same, "C" must double so that "B" and "C" cancel.

BASIC OPERATIONS

Converting Fractions to Decimals

Long-hand fractions are rarely seen on radiographic equipment. But we frequently express various ratios verbally as fractions, and may have occasion to convert these into decimal numbers. To do so, simply divide the denominator into the numerator, using a calculator as needed.

Example: Convert ⅖ into its decimal equivalent:
2 divided by 5 = 0.4

Converting Decimals and Percentages

Percentages are used a great deal in radiography. To convert a decimal number into a percentage, simply move the decimal point two places to the right.

Example: Convert 0.756 into a percentage.
0.756 = 75.6%

To convert a percentage into a decimal number, move the decimal point two places to the left.

Example: Convert 32.8% into a decimal number.
32.8% = 0.328

Extent of Rounding

How many decimal places should an answer be rounded to? A good rule is to round to the same number of decimal places as the entry with the least number of decimal places to the right of the decimal point. For most applications, more than two digits to the right ("hundredths") is rarely needed.

Example: Solve and round 243.87 + 96.5017.
The entry 243.87 has only two digits to the right of the decimal point.

233.87
96.5014

Answer: 330.3714

Round to: 330.37

Order of Operations

In solving all algebra problems, it is critical to remember the *order of operations.* If it is not strictly followed, the wrong answer will result. The order of operations is as follows:

A. Any operation within parentheses, such as (2 + 5), must be carried out prior to Part B for the formula as a whole. Within the parentheses, follow the order of operations in Part B.

B. 1. Apply all exponents first
   2. Then carry out all multiplications and divisions.
      –It does not matter which of these two precedes the other.
   3. Last, carry out all additions and subtractions.
      –It does not matter which of these two precedes the other.

For example, in solving the following equation, the order of steps is listed below:

$$X = \frac{3(2 + 8)^3}{5^2} - 60$$

1. Parentheses:
   Sum 2 + 8, because it is in parentheses,
   Answer = 10
2. Exponents:
   Cube the sum in parentheses, (10 × 10 × 10),
   Answer = 1000
   Also, square the 5 underneath, (5 × 5),
   Answer = 25
3. Multiplication and Division:
   Multiply the top number in the ratio by 3,
   Answer = 3000
   Also, divide out the entire ratio, (3000/25),
   Answer = 120
4. Addition and Subtraction for the formula as a whole:
   Subtract 60 from the solved ratio (120 – 60)

Final Answer = 60

Algebraic Operations

The following rules provide a brief review of basic algebraic operations:

1. When the unknown "x" is multiplied by a number, divide both sides of the equation by that number

$$ax = b$$

$$\frac{ax}{a} = \frac{b}{a}$$

$$x = \frac{b}{a}$$

2. When a number is added to the unknown "x," subtract that number from both sides of the equation. When a number is subtracted from the unknown "x," add that number to both sides of the equation.

$$x + a = b$$

$$x + a - a = b - a$$

$$x = b - a$$

3. When both sides of an equation are in the form of proportions (ratios), cross-multiply and then solve for "x."

$$\frac{x}{a} = \frac{b}{c}$$

$$\frac{x}{a} \times \frac{b}{c}$$

$$xc = ab$$

$$x = \frac{ab}{c}$$

4. When two numbers are added or subtracted within parentheses, and the whole set is multiplied by another factor, rewrite the phrase by multiplying *each* number by the factor and dropping the parentheses.

$$x(a + b) = xa + xb$$

Rules for Exponents

An exponent or power by which a number is raised expresses how many times the number is multiplied *by itself*. The binary number 2^4 is 2 × 2 × 2 × 2, in four repetitions. Using exponents makes it easier to multiply and divide large and small numbers. As long as the base number is the same (2 in the above example), the following three rules apply.

1. To multiply, *add* the exponents.

   Example: $10^3 \times 10^5 = 10^{(3+5)} = 10^8$

   Example: $10^3 \times 10^{-6} = 10^{[3+(-6)]} = 10^{(3-6)} = 10^{-3}$

2. To divide, *subtract* the exponents.

   Example: $10^7/10^5 = 10^{(7-5)} = 10^2$

   Beware that subtracting a negative number is equivalent to adding it.

   Example: $10^3 \div 10^{-6} = 10^{[3-(-6)]} = 10^{(3+6)} = 10^9$

3. To raise an number with an exponent by yet another power, multiply the exponents.

   Example: $(10^3)^4 = 10^{(3 \times 4)} = 10^{12}$

When the base numbers are different, multiply or divide the base numbers first, then add or subtract the exponents.

Example: $10^2 \times 8^4 = (10 \times 8)^{(2+4)} = 80^6$

Example: $10^2/8^4 = (10 / 8)^{(2-4)} = (1.25)^{-2}$

CONVERTING TO SCIENTIFIC NOTATION

In keeping with the spirit of *parsimony* discussed in Chapter 1, the whole purpose of scientific notation is to *simplify* the expression and calculation of large, unwieldy numbers and infinitesimally small numbers. Scientific notation is an extremely useful tool for scientists. Both very large and very small numbers are used frequently in radiography. By taking the time to fully understand this section, you can save yourself a lot of grief doing math problems for technique, physics, equipment and radiation biology.

The format for scientific notation is to express a number as some quantity multiplied by a power of 10, that is, a quantity times ten raised to a specific power, such as 5×10^3. As you will see, using this format actually makes calculations of large and small numbers much *easier*.

To convert any decimal number into scientific notation, first move the decimal point, either to the left or to the right, to the position following the first nonzero digit. If the decimal point is moved to the left, the power of 10 will be the number of places the decimal point was moved. If the decimal is moved to the right, the power of 10 will be the number of places the decimal was moved expressed as a *negative* number.

Examples:

1. Express 80.25 in scientific notation:

   Answer: Moving the decimal point one place to the left, $= 8.025 \times 10^1$

2. Express 0.025 in scientific notation:

   Answer: Moving the decimal point two places to the right, $= 2.5 \times 10^{-2}$

If there are no zeros in the number, or if the decimal point is not moved, the number may be expressed by itself times ten to the *zero* power, 10^0.

Any number can be converted into scientific notation by simply adding "$\times 10^0$" behind it.

Example: Express 5.53 in scientific notation:

Answer: 5.53×10^0

However, there is also the question of reducing an unwieldy number of digits in front of the decimal place. After all, this was the original goal, to make the number more manageable. Therefore, as a general rule it is good form not to have more than two digits in front of the decimal point (and never any zeros in front of the decimal point).

Example: Express 9573.8 in good scientific notation format:

Answer: Move the decimal point two places to the left, and raise the power of 10 by two (starting from 10^0)

$= 95.738 \times 10^2$

To convert a scientific notation into a regular decimal number to see what the number looks like written out longhand, just perform the above operations in reverse—if the power of 10 is a positive exponent, move the decimal that many places to the *right*, adding zeros as needed. Thus, 2×10^3 designates 2.0 with the decimal moved three places to the right, or 2000. If the exponent is a negative number, move the decimal that many places to the left, also adding zeros as needed. For example, 6×10^{-3} moves the decimal from 6.0 to .006.

Finally, note that any scientific notation for the number 1 followed only by zeros can be reduced to a kind of shorthand by leaving out the one and simply stating the number 10 raised by the appropriate power.

Examples: 1×10^8 can be expressed as just 10^8

1×10^{-5} can be expressed as just 10^{-5}

This shorthand format was used for all of the SI units described in Table 2.1 in Chapter 2.

CALCULATING WITH SCIENTIFIC NOTATION

To add or subtract large and small numbers, just use their conventional form, carefully aligning the decimal points. Convert any numbers given in scientific notation into decimal numbers.

Scientific notation is designed to make multiplication and division easy for very large and very small numbers in any combination, (two large numbers, two small numbers, or a large and a small number).

To multiply:

1. First, multiply the numbers preceding the "times" signs.
2. *Add* the exponents from the bases of 10 (see Rules for Exponents).
3. The answer is expressed as the product from Step #1 times 10 raised to the sum from Step #2.

Example: Multiply $(3 \times 10^6) \times (2 \times 10^3)$

Step #1: $3 \times 2 = 6$ for first expression
Step #2: Add Exponents: $6 + 3 = 9$

Answer: 6×10^9

To divide:

1. First, divide the numbers preceding the "times" signs.
2. *Subtract* the exponents from the bases of 10 (see Rules for Exponents).
3. The answer is expressed as the ratio from Step #1 times 10 raised to the difference from Step #2.

Example: Divide $\dfrac{3 \times 10^6}{2 \times 10^3}$

Step #1: $3/2 = 1.5$ for the first expression
Step #2: Subtract Exponents: $6 - 3 = 3$

Answer: 1.5×10^3

BEWARE of negative exponents; Remember that subtracting a negative number makes it an added number, as follows:

Example: Dividing with a negative denominator:
$\dfrac{3 \times 10^6}{2 \times 10^{-3}}$

Step #1: $3/2 = 1.5$
Step #2: Subtracting the exponents:
$6 - (-3) = 6 + 3 = 9$

Answer: 1.5×10^9

CONVERTING UNITS WITH DIMENSIONAL ANALYSIS

Dimensional analysis, in spite of its intimidating name, is a wonderfully simplified method to help convert units that are widely disparate, such as centimeters into miles, or yards into millimeters. The beauty of the method is that it lets you break the process down into a series of steps using unit conversions you are comfortable with. In fact, you *choose* the unit conversions you can already easily do. The more basic your chosen unit conversions, the more steps are required, but each step in itself is much easier, and you can always arrive at the answer.

Aside from the British unit conversions you learned in grade school (12 inches to a foot, and so on), the only other prerequisite knowledge you need is a good recollection of the units presented in Chapter 2.

To begin, on a piece of paper, sketch out a grid made of a single horizontal line with a series of vertical slashes, like this:

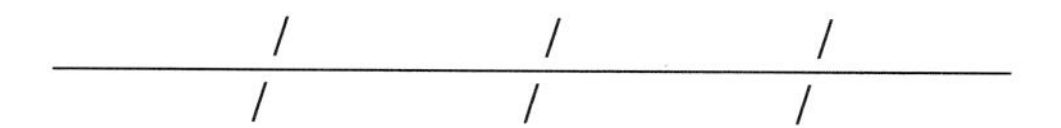

Leave plenty of room between the slashes to enter your numbers and units. Place the starting quantity and its units in the first box above the line. Choose a series of simple unit conversions with which you *are* familiar that will lead from the starting units to the converted units. For example, it may not be immediately obvious to you precisely how many feet there are in 1 meter, but you *are* familiar (from Chapter 2) with how many centimeters are in an *inch*, so begin with this. You also know how many inches are in a foot and how many feet are in a yard, so this is the pattern you will follow. (If you know how many inches are in a yard, you can skip a step, but it doesn't matter how many steps you take, only that you are careful that each step is formatted correctly.)

In setting this up, there is only one *critical* rule to know:

The top unit must always become the *bottom* unit in the next bracket.

Lay out your series of units before inserting any numbers to be sure this rule is carefully followed. We will use the above example of a centimeters to yards conversion to illustrate.

Example: How many meters are there in 30 feet?

Your layout should look like this:

30 feet /	inches /	centimeters /	meters
/	feet /	inches /	centimeters

Note that feet, which we begin with, is at the bottom in the next expression, inches then follows dropping underneath, then centimeters. In each expression, *it does not matter which way you are converting*, the previous top unit must be placed underneath. For example, the second set has inches on top and feet underneath; *it does not matter whether you are asking how many feet there are in an inch, or how many inches there are in a foot, just so "feet" is placed at the bottom.*

Now with the layout completed, fill in the numbers with which you are familiar. In the above example, the second expression can be read as "inches over feet" or "inches per feet." Consider this as if it were asking "How many inches in one foot?"—write "12" on top and "1" on bottom. The third expression is also laid out this way—it already seems to read, "How many centimeters to an inch?" Write "2.54" on top and "1" on bottom.

However, in this regard the *last* expression seems upside down; that is, it seems to be asking, "How many *meters* in one *centimeter*?" The answer is one-hundredth or 0.01, but this is not likely to be the way you learned this conversion. What you *are more familiar with* is that it takes 100 centimeters to make a meter. Just write in what you are comfortable with, but *leave the units where they are written*, with centimeters on the bottom. Write "100" on the bottom and "1" on the top in this case. *It does not matter that the "1" is on top and the larger number on bottom—"centimeters"—must remain underneath because it was on top in the last expression.* Your filled-in grid should look like this:

30 feet / 12 inches / 2.54 centimeters / 1 meters
/ 1 feet / 1 inches / 100 centimeters

Now follow these simple math operations:

1. **Multiply everything across the top.**
2. **Multiply everything across the bottom.**
3. **Divide the top product by the bottom product (unless the bottom is "1").**

For the above example, the answer is as follows:

For the top: $30 \times 12 \times 2.54 \times 1 = 914.4$

For the bottom: $1 \times 1 \times 100 = 100$

For the final ratio: $\frac{914.4}{100} = 9.14$

Finally, allow all units to cancel, leaving just the last unit on the top of the grid. For the above example, "feet" are canceled on top by "feet" underneath in the second set, inches cancel out, and centimeters cancel out. The only unit that doesn't show up both on top and on bottom is the last unit on top, *meters*.

Answer: 9.14 meters

Use the following two exercises to practice:

1. If there are 5,280 feet in one mile, how many inches are there in 2 miles? Use scientific notation to report your answer.

   Solution: 2 miles / 5280 feet / 12 inches
   / 1 miles / 1 feet

   $2 \times 5280 \times 12 = 126{,}720$ inches

   or 12.672×10^4 inches

   (After moving the decimal four places to the left, the zero at the end may simply be dropped.)

2. Convert 9,144 cm into feet.

   Solution: 9144 cm / 1 inches / 1 feet
   / 2.54 cm / 12 inches

   $9144 \times 1 \times 1 = 9144$

   $2.54 \times 12 = 30.48$

   $\frac{9144}{30.48} = 300$

   Answer: 300 feet

Using Table 2-1

If you use the magnitudes directly from Table 2-1, be sure to plug them into the grid *with the* generic unit, *not with the unit being defined in the first column of the table.* The generic units listed as examples in Table 2-1 are *meters, volts, hertz, bytes,* and *seconds.* For example, the table states that the prefix "centi" has a magnitude of 10^{-2}. This means that a centimeter is 10^{-2} *meters,* (not that a meter is 10^{-2} centimeters). The generic unit is meters. To plug this unit into the conversion grid, it should read 10^{-2} *meters* to make one centimeter. An example follows:

1. How many microvolts are there in 3 megavolts?

   Solution: 3 megavolts / 10^6 volts / 1 microvolt
   / 1 megavolt / 10^{-3} volts

   Note that the magnitudes from Table 2-1 are both applied to the generic unit "volts," one on top, and one on bottom of the formula.

   $3 \times 10^6 \times 1 = 3 \times 10^6$ (Just leave it in scientific notation)

$$1 \times 10^{-3} = 1 \times 10^{-3}$$

$$\frac{3 \times 10^{6}}{1 \times 10^{-3}} = 3 \times 10^{9}$$

Answer: 3×10^{9} or 3 trillion microvolts

Combining scientific notation with dimensional analysis, you have powerful tools for dealing with any scientific problems involving large and small numbers.

AREAS AND VOLUMES

In order to understand the concentration of x-rays at different distances from the x-ray tube (the inverse square law), each field size must be considered as an *area*. In order to understand why, after an x-ray exposure is made, it takes a thick anode disc longer than a thin anode to cool down, the *volume* of each anode must be taken into consideration. Radiography students must get comfortable with the basic math of areas and volumes.

Whereas straight-line distances are measured in *linear* units (inches, centimeters), areas must be measured in *square* units (square inches, square centimeters). Volumes are measured in *cubic* units (cubic inches, cubic centimeters). Always remember to report areas and volumes in their correct units. They can be abbreviated by using the related exponent or power alongside the abbreviation for the unit:

5 square inches = 5 in^2

6 cubic centimeters = 6 cm^3 = 6 cc

Note that the special medical unit "cc" stands for "cubic centimeters," a volume unit widely used on syringes. One cc of volume contains precisely 1 milliliter of liquid (one-thousandth of a liter). So, 6 cc of iodine is the same as 6 ml of iodine.

The *liter* is the metric system unit for liquid volume. It is just over one quart, 1.056 quarts. For the purpose of visualizing liters, you can imagine a roughly equal number of quarts.

Always visualize *square inches* as a flat area made up of 1-inch *squares*. Always visualize *cubic* centimeters as a volume made up of stacked rows and columns of 1-cm *cubes*.

The most important thing to remember in working with areas and volumes is that you must square or cube the *calculations*, not just the units.

If we ask how many square inches there are in a square foot, the answer is not 12 (as for a linear foot), but 12^2 or $12 \times 12 = 144$ in^2. If we ask how many cubic centimeters there are in one cubic meter, the answer is not 100 (as for a linear meter), but 100^3 or $100 \times 100 \times 100 = 1{,}000{,}000$ cm^3. Always square or cube the calculations along with the units. A bit of practice is provided at the end of this chapter.

To really grasp the inverse square law or the heating and cooling of x-ray tubes, one must appreciate the interrelationships between linear, area and volume measurements as an object becomes larger or smaller. To do this, we will examine their *change ratios*, as illustrated in Figure 3-1.

1. What is the ratio of change between a 6-inch line and a 12-inch line:

   Answer: $\frac{12}{6} = \frac{2}{1} =$ 2:1 ratio

2. What is the ratio of change between a 6-inch SQUARE and a 12-inch SQUARE:

   Answer: $\frac{12^2}{6^2} = \frac{144}{36} =$ 4:1 ratio

Figure 3-1

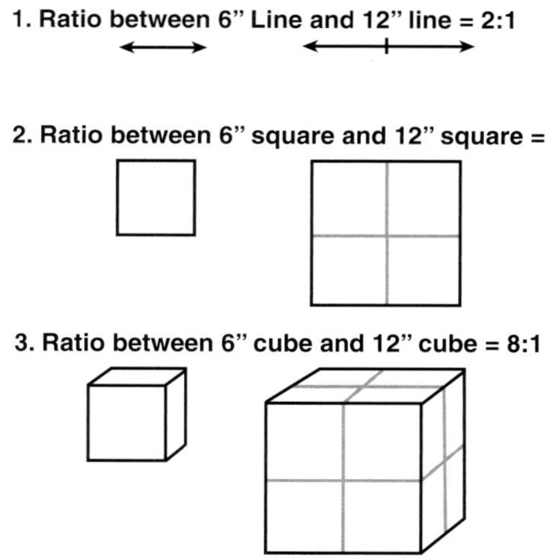

The *change ratio* between a 6" line and a 12" line is 2:1. Between a 6" square and a 12" square it is 4:1. Between a 6" cube and a 12" cube it is 8:1.

3. What is the ratio of change between a 6-inch CUBE and a 12-inch CUBE:

$$\textit{Answer: } \frac{12^3}{6^3} = \frac{1728}{216} = 8{:}1 \text{ ratio}$$

Note the relationship between these ratios: When an object is doubled in length and width, its *surface area* increases by 2^2 or 4 times; when it is doubled in length, width, and height, its *volume* goes up by 2^3 or 8 times (Fig. 3-2). We find the same relationship in *inverse proportion* when the size of an object goes down. When its dimensions are cut in half, the surface area will be ¼, and the volume will be ⅛. Now here is the key point:

Volumes increase and decrease *faster* than areas.

Imagine you get put on *KP* duty ("kitchen patrol") in the Army. You have to peel one of two baskets full of potatoes. One basket has exactly 10 pounds of small potatoes, and the other holds exactly 10 pounds of large potatoes (Fig. 3-3). Would you rather peel the *small* potatoes or the *large* ones?

To save work, you want the *large* ones. Here's why: The amount of *peel* is based on surface area, while the weight in *pounds* is based on volume. Smaller potatoes have less *peel* by a factor of a *square*, but they have less pounds by a *cubed* factor. The amount of peel decreases "*less* quickly" than the pounds. Therefore, the small potatoes have more *peel per pound.*

Figure 3-2

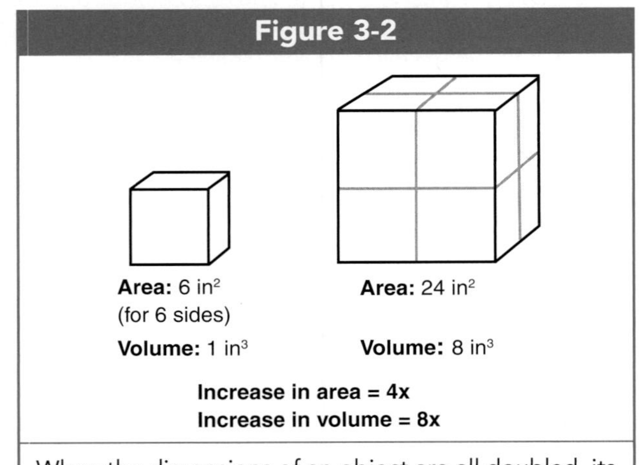

When the dimensions of an object are all doubled, its surface *area* increases by 2 *squared*, but its *volume* increases by 2 *cubed*. Since the surface goes up less than the volume, heat contained within it takes longer to escape through the surface. This applies to the heating and cooling of x-ray tube anodes.

Figure 3-3

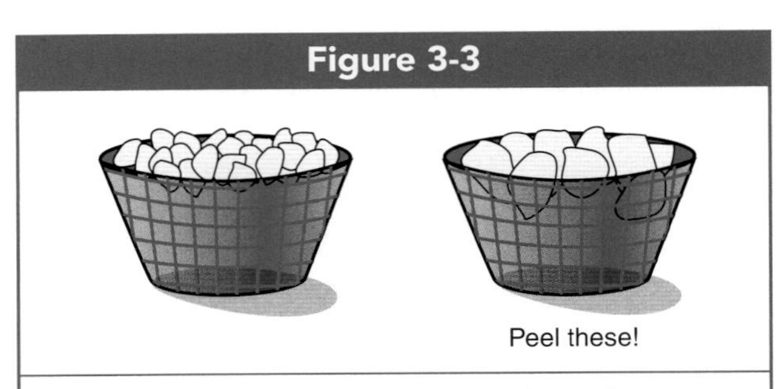

By the same principle, a bushel of *small* potatoes actually has more peel-per-pound than a bushel of *large* potatoes.

This is the principle by which small insects can float on water; compared to a human, their weight is diminished much more than their surface. So, there is less weight "per square inch" (or less *pressure*) where they contact the water, and the surface-tension of the water is able to buoy them up.

The heating and cooling of an object, discussed in Chapter 2, take place only through its *surface areas*. The ability of a hot x-ray tube anode to cool down is a *square* relationship. However, the ability of that object to *store* heat is based on its volume, and therefore is a *cube* relationship. As it gets larger, the ability of an x-ray tube anode to store heat goes up "more quickly" than its ability to *emit* that heat. Therefore, a thick anode takes much longer to radiate the heat out.

Anything that takes longer to cool also takes longer to heat, because no matter which way the heat is flowing, in or out, it must go through the surface area of the object. Hence, larger potatoes not only take longer to cool down, but they also take longer to cook in the first place.

To summarize, large x-ray tube anodes take longer both to heat and to cool, because their surface area has not increased as much as their volume, so their capacity to *emit* heat has not increased as much as their capacity to *store* heat, when compared with smaller anodes.

THE INVERSE SQUARE LAW

The inverse square law is a law of *areas*. As x-rays spread out, we measure the way they are less concentrated across an area measured in square inches or square centimeters. Most forces and ALL types of radiation follow the Inverse Square Law, because they spread out *isotropically* (evenly in all directions across all areas). The inverse square formula tells us how the intensity or concentration of these forces changes at different distances from their source.

For example, in Figure 3-4 we see the change in surface *area* over which a radiation beam is spread out when the distance from the source of radiation is doubled. At a particular distance, ***d***, the radiation has spread out over an area of one square inch. When the distance is doubled to ***2d***, we see that this area does not increase by a factor of 2, but rather by a factor of 2^2 or *4 times*. The opposite is also true—if the distance were cut to one-half, the area of spread would be reduced to $\frac{1}{2}^2$ or ¼.

What we are particularly interested in is how this geometry affects the *concentration* of radiation exposure. To illustrate, observe Figure 3-5 in which we assume the intensity of radiation at distance ***d*** to be 16 x-ray photons, indicated by dots. This amount is concentrated in one square inch of surface, and this ratio is an expression of radiation *exposure*. What will be the radiation exposure if the distance is doubled? At this distance, ***2d***, we see that the 16 x-ray photons have evenly spread out over 4 square inches of area, making the new intensity of radiation *4 x-rays per square inch*. This is the expression of the new exposure, which is ¼

Figure 3-4

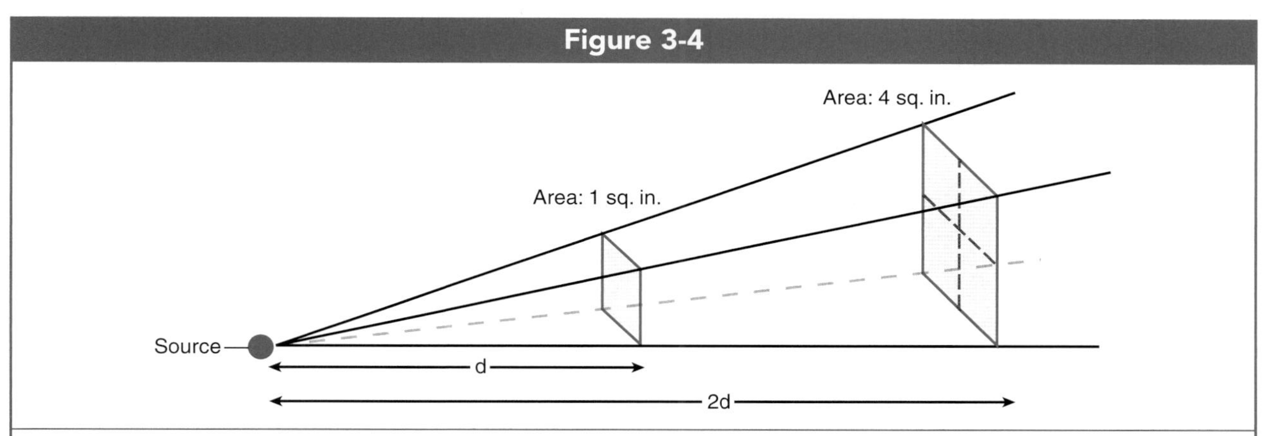

Anything which *radiates* isotropically outward from a central point will spread out according to the *square* of the increasing distance. At twice the distance, it will be spread out over 2^2 or 4 times the area.

Figure 3-5

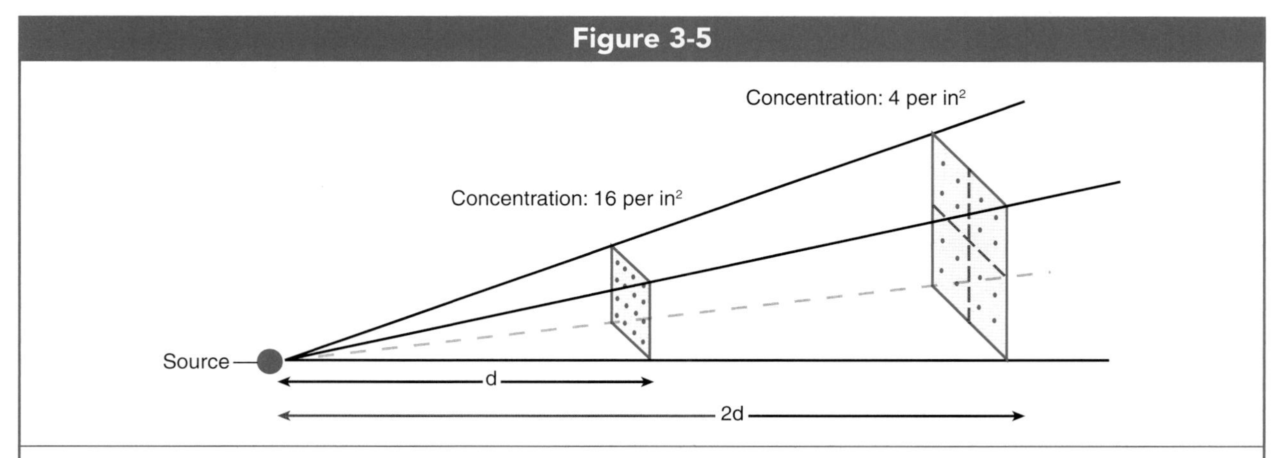

The Inverse Square Law, based on the increasing *area* at greater distances, states that the *concentration* of radiation will be inversely proportional to the square of the distance. At twice the distance, the radiation will be $\frac{1}{2}^2$ or ¼ as intense, (here, 4 dots per square rather than 16).

what it was at distance ***d***. Thus, we see an inverse square relationship between exposure and distance.

One can mentally solve these problems when the distance changes are simple ratios, by just applying *the name of the law* to the *ratio of distance change*. For example, if the distance is increased by 5 times, just take the ratio 5, invert it for ⅕, and square it for 1/25th. The new exposure will be 1/25th the original. If the distance is reduced to ⅓, invert this value for 3/1 or 3, and square it for 9. The radiation exposure will be 9 times the original.

As with all formulas, there are several ways to set the inverse square formula up depending on what one is specifically solving for. Here, we are particularly interested in how changes in distance from an x-ray tube or other source of radiation affect the amount of radiation exposure received by a patient, a radiographer, or the image receptor plate during an x-ray study. For this purpose, the best format for the ISL formula is:

$$\frac{I_O}{I_N} = \frac{(D_N)^2}{(D_O)^2}$$

where *I* is the intensity of x-rays, expressed either as a *rate* of exposure or a total *quantity* of exposure, *D* is the distance from the source of radiation (such as the x-ray tube) to the person or to the image receptor plate, *o* stands for "old" or "original," and *n* for new.

The formula states, then, that the *ratio* of the original radiation intensity to the new intensity is equal to the ratio of the *new* distance squared to the *original* distance squared. Note that the "old" and the "new" values are inverted between the two sides of the equation. The *name of the law*, the "inverse square" law, tells you what must be done with the distances: They must be inverted ("new" over "old"), and they must each be squared. If the original quantity of radiation is known, we can calculate what the new radiation exposure will be at a different distance.

Example: An x-ray detector receives an exposure of 25 mR from an x-ray tube that is 40" away. Using the same technique factors, if the tube is moved to 72", how much exposure will the detector receive?

SET-UP: $$\frac{25\text{mR}}{X} = \frac{72^2}{40^2}$$

$$\frac{25}{X} = \frac{5184}{1600}$$

Cross-multiplying: $(25)1600 = 5184(X)$

$40{,}000 = 5184(X)$

Dividing both sides by 5184 to isolate X:

$X = 40{,}000/5184 = 7.7$ mR exposure at film

GRAPHS

Modern radiographers must be fluent at reading the information from graphs. They must understand graphic representation not only well enough to interpret a graph, but also well enough to plot raw data obtained from measurements in graphical format. Most graphs are based on a horizontal axis labeled "x" and a vertical axis labeled "y." However, for most radiography applications, these designations are substituted with the two specific quantities being described, for example, technique versus exposure, or radiation dose versus leukemia rate.

Let us begin by constructing a simple graph from data. Use the data from Table 3-1 which hypothetically presents the percentage of 1000 mice that come down with leukemia as the entire population is exposed to increasing amounts of radiation.

To construct a graph from this data, first draw the "x" and "y" axes in a large "L" shape. In order to label the two axes, it is usually desirable to have that quantity which is actually being determined or studied on the *vertical* or "y" axis. The central question above is not how much radiation the mice received, but how many of them developed leukemia from the radiation. Plot the percentages of leukemia vertically and the doses of radiation horizontally for this graph (Fig. 3-6).

Take note of the range of both quantities in the table of data: What is the highest number in each set? You must choose a convenient scale that allows the data to fill the graph, but also allows enough space in the graph to plot all of the data points accurately. For the above data set, the highest percentage of leukemia is 100 percent: If you make this the highest vertical point of the graph, and tick off marks at every 20 percent, placing the marks ½-inch apart, the graph will be 2½ inches tall. You would not want a graph any smaller than this because it will be difficult to accurately plot the points on it. You may wish to make it larger.

Table 3-1

Hypothetical Percentage of Mice Population Acquiring Leukemia from Radiation Exposure

Radiation Dose (rad)	Percentage of Mice with Leukemia
20	0%
40	0%
60	5%
80	13%
100	26%
120	39%
140	52%
160	65%
180	78%
200	88%
220	95%
240	98%
260	100%
280	100%

Figure 3-6

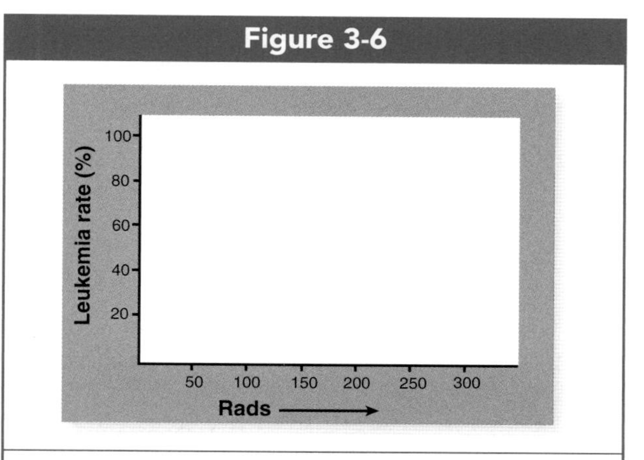

Proper layout for a graph based on the data presented for the incidence of leukemia in mice receiving increasing radiation doses.

Figure 3-7

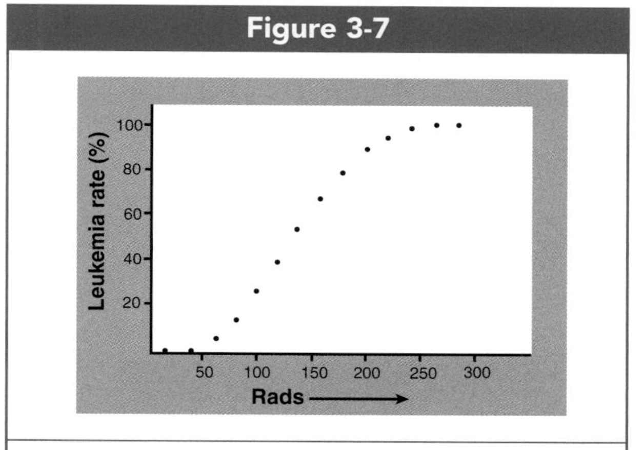

Data points plotted from Table 3-1 for the incidence of leukemia in mice receiving increasing radiation doses.

The highest dose in the above data table is 280 rads. For the horizontal or "x" axis, round the end point up to an even 300 rads. Note that if you tick off marks ½-inch apart for every 20 rads, like you did for the percentages, the graph will end up 7½ inches long. This would be an unwieldy, long rectangle. By changing this scale to 50 rads for every ½-inch tick mark, the graph will be 3 inches long, closer to a square shape and easier to read. Generally, it is desirable to have the two axes of a graph close to the same length. Labeling the vertical tick marks in twenties, and the horizontal ones in fifties, the resulting layout should look like Figure 3-6.

Now plot each data point from the table. Note that percentages which fall between the tick marks in the graph must be carefully estimated as to where they fall, (Fig. 3-7). The 5 percent point (for 60 rads) should fall ¼ of the way from zero to the 20 percent tick mark: For the 52 percent point (at 140 rads), find a point half-way between the 40 percent and the 60 percent tick marks, then bring it *up* slightly.

Finally, a *smooth* "best-fit" line should be drawn through the plotted data points. Figure 3-8 illustrates what is meant by "smooth"—the line must be a *curve*, not a series of straight lines. This is essential to being able to read the graph accurately. Begin with a pencil—you will find that it usually requires several attempts to get the line just right, then you can ink it in.

Note that *a best-fit line does not necessarily touch every data point plotted.* Frequently there will be flukes in some of the measurements, resulting in one

Figure 3-8

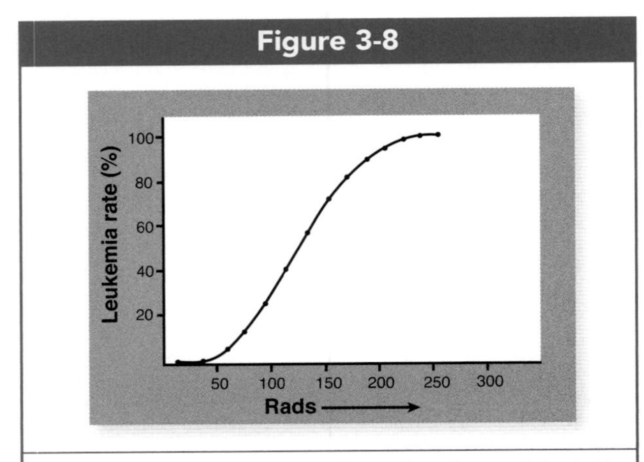

"Best-fit" curve generated from this data (Table 3-1). The curve must be smooth, not a series of straight lines with angled points.

Figure 3-9

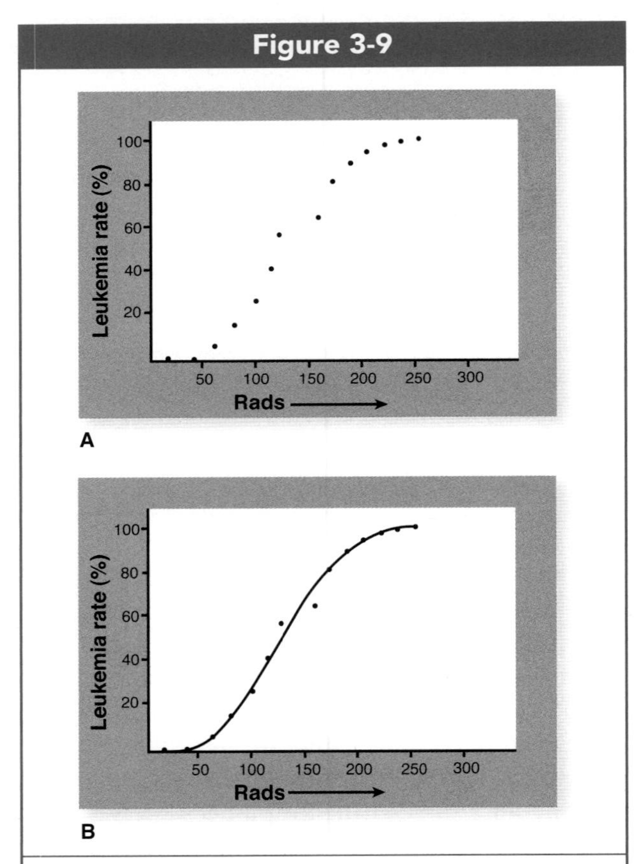

Plotted data points, *A,* and "best-fit" curve, *B,* for the same data with two "fluke" readings, a high one at 58% for 140 rads exposure, and an unusually low one at 63% for 160 rads. This is the most accurate curve for interpretation.

or two plotted points that fall way out of line with the rest of the points. This is common, and is due to the fact that we cannot always control all of the conditions around an experiment that might affect the results. These are flaws in the data, but can be compensated for by effectively "averaging" these points with all of the others. This is what a *best-fit* line does.

For example, in the table for leukemia in mice, let's replace the reading at 140 rads with an unusually high number of 58 percent and the reading at 160 rads with an unusually low number of 63 percent. The resulting data points for the graph are shown in Figure 3-9***A***. In Figure 3-9***B***, the *best-fit* line is drawn in correctly to pass somewhat to the *right* of the high point, and somewhat to the *left* of the low point, because they are not in line with the great majority of the data points.

Reading a Graph

In making conclusions from this study, the question might be asked, "How many rads of radiation was required to cause leukemia in one-half of the population of mice?" To read and interpret the graph, find the 50 percent point on the vertical axis and use a ruler to draw a *perfectly horizontal* line to intersect the curve. From this intersection, use a ruler to draw a *perfectly vertical* line down to the horizontal axis of the graph (Fig. 3-10). Read this point on the horizontal axis to answer the question in units of rads: The answer is 128 rads.

Figure 3-10

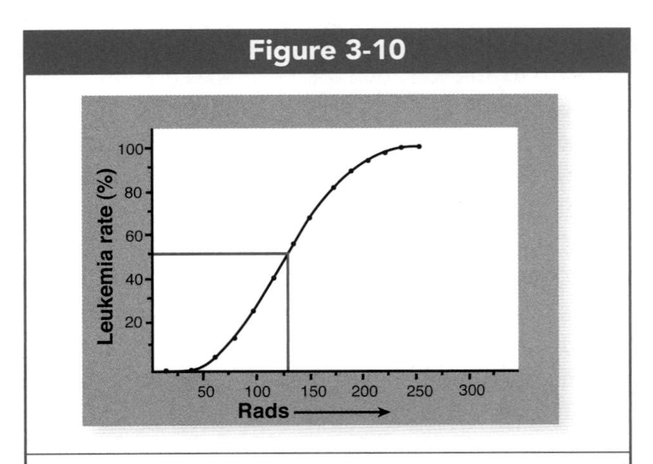

Proper interpretation of the graph for the question, "How many rads were required to cause one-half of the population of mice to manifest leukemia" results in an accurate answer of 128 rads.

With a single glance at a graph like that in Figure 3-8, one can immediately characterize the relationship between the two quantities being studied as a *linear* or a *nonlinear* relationship, and as a *threshold* or *nonthreshold* relationship.

For standard graphs, if the entire constructed curve is a straight line, the relationship is *proportional* as defined at the beginning of this chapter. If the tick-marks on the x and y axes of the graph have identical magnitudes (they go up by the same amount), and the straight line lies at a 45-degree angle, the relationship between the two variables is *directly proportional* as defined at the beginning of this chapter. An *inversely proportional* relationship will be graphed as a straight line at 45 degrees, but slanting *down* from left to right as the graph is read.

Nonlinear relationships are represented by any shape of line that is not straight throughout, including *exponential* relationships. The particular nonlinear shape in Figure 3-8 is common, and is named a *sigmoid* or "s-shaped" curve.

A *nonthreshold* curve has its beginning point precisely at zero on the graph. The meaning of this is that there is no amount of the variable listed on the horizontal axis that is so small as to not have the effect listed on the vertical axis. For example, in our study of radiation and leukemia in mice, it would mean that no matter how small the dose of radiation, *some* leukemia might still be caused within a very large population of mice. Any amount has some effect.

Figure 3-8 is a *threshold* curve. No effect is measured until the radiation doses reaches a specific amount between 40 and 60 rads. The interpretation is that at any amount of radiation *below* that point, the mice are "safe" from the effect, leukemia. Mice may be exposed to small amounts of radiation without concern for leukemia being induced by it.

On a standard graph, an exponentially increasing quantity, such as a population of rabbits, will appear as a curve that climbs more steeply with each data point, Figure 3-11. Natural radioactivity in substances such as uranium decreases in an *inversely exponential* fashion, graphed in Figure 3-12. One of the advantages of graphs in representing data is that one can identify these various types of relationships at a glance, by the shape of the curve.

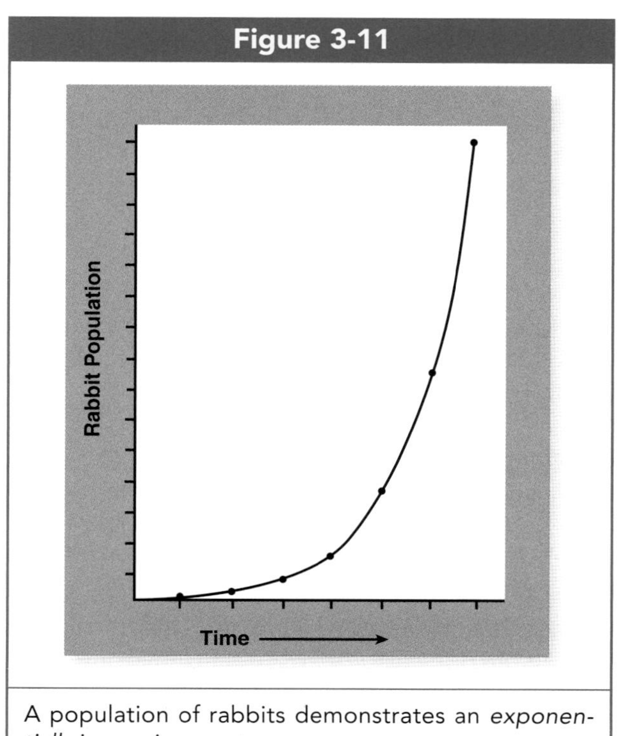

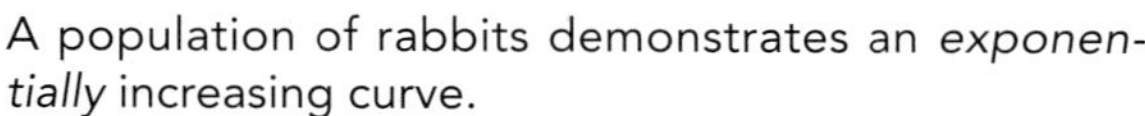
A population of rabbits demonstrates an *exponentially* increasing curve.

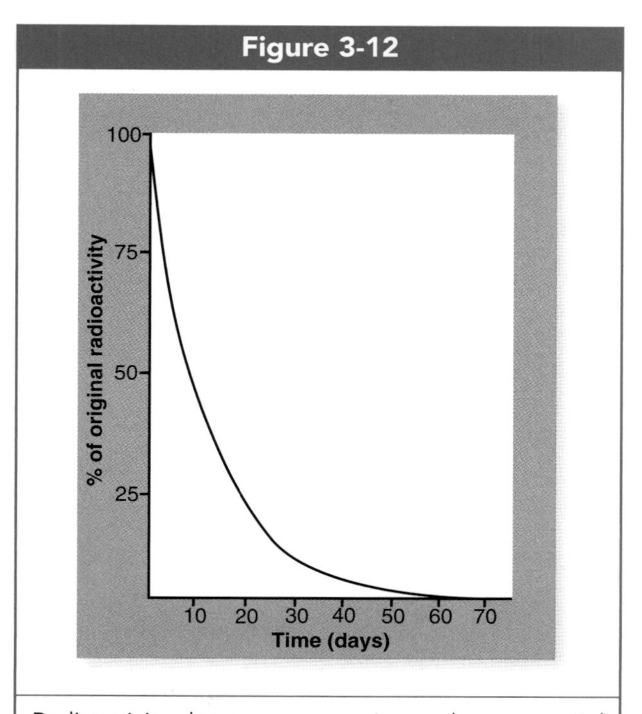

Radioactivity demonstrates an *inversely exponential* decrease over time.

A second type of graph that pertains to radiography and medical science is the *bell-curve* graph, one in which the data points result in a curve shaped like a bell. To illustrate, let us study how tall the students are within a large class of 200 (the course meets in an auditorium!). Taking simple tape-measurements, we produce the data in Table 3-2.

When these data points are plotted and a smooth, *best-fit* curve is drawn in, the graph appears as Figure 3-13. This familiar shape, the *bell-curve*, describes many types of data, especially sociological statistics, in which the samples are distributed around some *average* measurement. Fixed distances from the average, called *standard deviations*, are set by their percentage of distance from the mid-point to the extremes. The important thing to note about a bell curve is that most of the samples (usually 68%) fall within the range of just one standard deviation from the average, and 95 percent fall within the *norm* or "normal" which is typically defined as falling within two standard deviations from the average. This leaves less than 5 percent that could appropriately be described as "abnormal" or outside the norm.

Figure 3-13

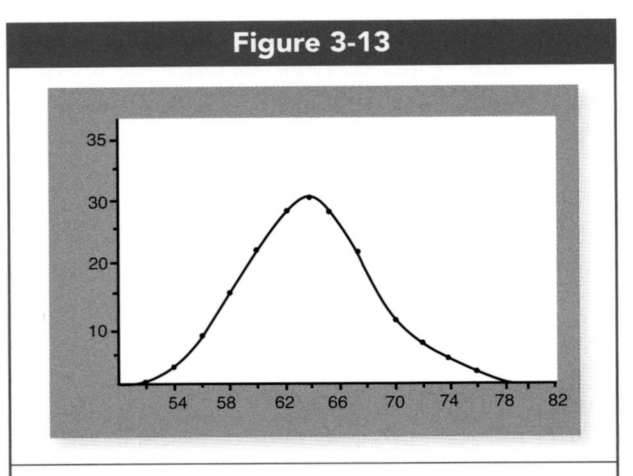

Bell-curve generated for the data presented on the height of students in a class.

Table 3-2

Hypothetical Number of Students in a Class Measured by Height

Height	Number of Students	Height in Inches
4'4"	1	52"
4'6"	4	54"
4'8"	8	56"
4'10"	14	59"
5'	22	60"
5'2"	27	62"
5'4"	30	64"
5'6"	27	66"
5'8"	21	68"
5'10"	12	70"
6'	9	72"
6'2"	4	74"
6'4"	2	76"

For the bell-curve graph in Figure 3-13, the sum of 200 students participating in the study is represented by the *total area under the curve*. The average height of students in this class can be found graphically by a *vertical line which divides that area into two equal halves*.

Imagine that we have only the graph in Figure 3-13 to look at, without the table of data. To read the graph, we should be able to pick any particular height, such as 6 feet, and ask, "How many students are there in the class at that height?" To answer this, find 72 inches along the bottom of the graph, and, using a ruler, extend a *perfectly vertical* line up until it intersects the curve. From this point, use a ruler to extend a *perfectly horizontal* line over to the vertical axis and read that number—there are 9 students that fit this height.

Understanding the X-Ray Beam Spectrum Curve

A graph of the energies within a diagnostic x-ray beam is constructed in just the same way as the above graph of "tallness" within a class of students. Instead of asking how many students there are in the class at each height, we ask the question, "How many x-rays are there in the x-ray beam at each *energy* level. Energies, measured in kilovolts (kV) are listed

Figure 3-14

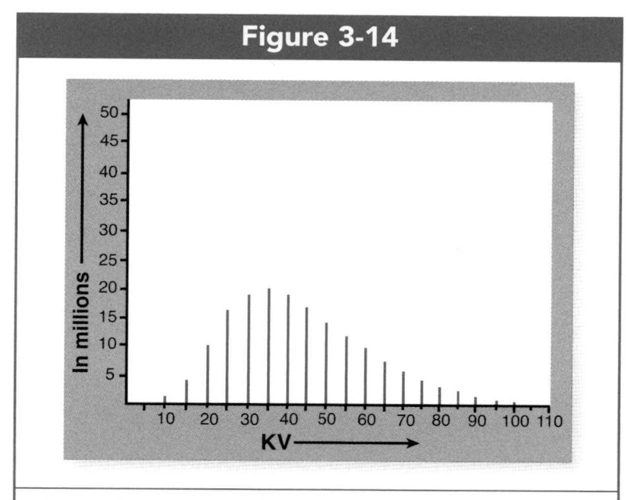

Plot of the quantity of x-rays possessing each kV level within a typical x-ray beam.

Figure 3-15

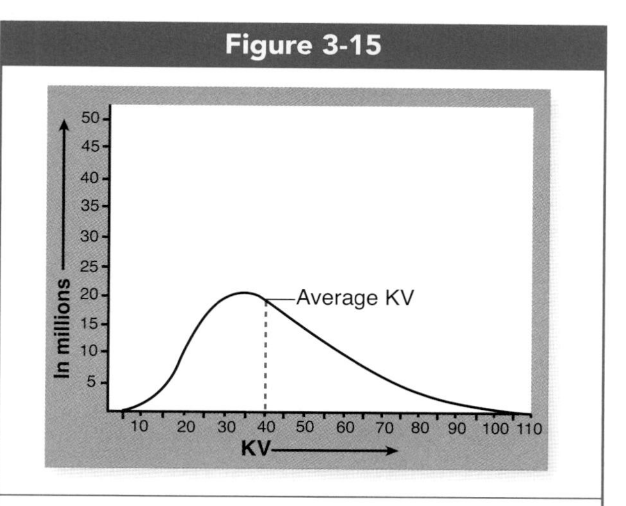

Best-fit bell-type curve for a typical x-ray beam, delineating the *average* kV (dotted line).

horizontally along the x axis. How many individual x-rays there are in the beam *at each energy* are plotted as vertical points (Fig. 3-14). When these points are all connected with a *best-fit* curve, we obtain a graph like Figure 3-15.

Graphs of the energies within diagnostic x-ray beams are somewhat unique: They take on the general appearance of a bell-curve, and the average energy for the x-ray beam can indeed be found by a vertical line dividing the area under the curve into two halves of equal area (Fig. 3-15). But, in the pure sense this is not a bell-curve that precisely follows the rules of standard deviation. Rather, it is an *inversely exponential* graph (Fig. 3-12) which has been truncated or cut off on the left-hand side.

At the moment x-rays are produced within the anode, most of them have extremely low energies, and the higher the energy, the fewer of them there are. The resulting graph slopes gently downward in concave fashion, shown in Figure 3-16. However, in order to reach the patient and imaging plate or cassette, the x-rays must pass through the glass of the x-ray tube, aluminum filters, and other materials as they escape the tube housing. These materials absorb the lowest-energy x-rays. When these low-energy x-rays are removed from the graph, the curve representing the remaining, useful x-ray beam takes on the shape of a lopsided bell (Fig. 3-16).

The vertical line defining the average energy of the entire beam still divides the area under the curve in half, but since the curve is lopsided to the left, the average falls roughly one-third of the way from the minimum energy to the maximum energy. The maximum energy is represented by the rightmost point of the curve, where it returns to zero on the x axis. There are no x-rays in the beam above this energy. This point is controlled by the *kVp* or "*kV-peak*" setting on an x-ray machine console. Given the graph, what general statement can you make about the *average* energy of an x-ray beam

Figure 3-16

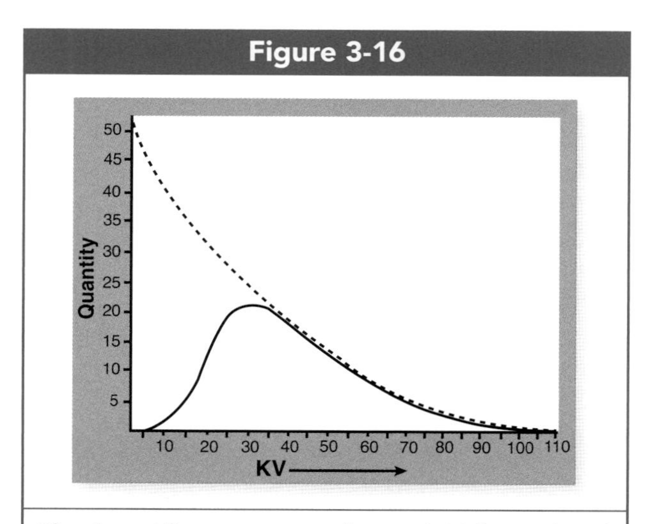

The dotted line represents *all* x-rays initially produced in the x-ray tube anode. The solid line represents the spectrum of x-rays emitted from the x-ray tube *after* the effects of filtration, which removes the lowest energies.

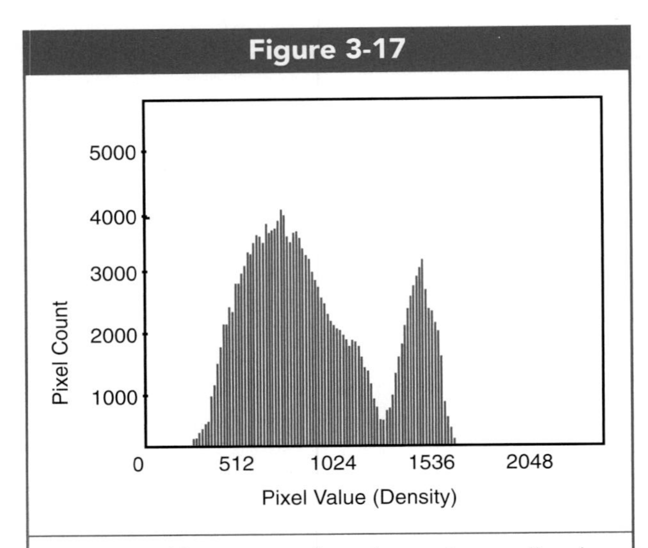

The original histogram of any image is actually a bar graph indicating the pixel count for each gray level or density.

compared to the *kVp* you have set? *Answer: Generally, the average energy of an x-ray beam is roughly one-third of the kVp.*

In interpreting the x-ray beam spectrum graph, we would say that the *right-most* point of the curve is determined by the kVp. The *left-most* point of the curve is determined by the *filtration* placed in the beam, which eliminates x-rays having low energies. The overall *height* of the curve is controlled primarily by the *mAs* setting, which determines how many x-rays are produced overall.

Now, let's practice graphic interpretation by looking more specifically at Figure 3-15. Is this a *threshold* curve? What is the lowest energy, in *kV*, of any x-rays in this beam? What does this mean about those x-rays originally produced with energies below that amount?

Answers: This is a threshold curve. The lowest energy of any x-rays in this beam is 5 kV. There are no x-rays in the beam having less than 5 kV of energy, because they have been taken out of the beam by filtration.

In Figure 3-15, how many x-rays in this beam have 15 kV of energy? Find 15 kV along the bottom of the graph, and use a ruler to extend a *perfectly vertical* line upward until it intersects the curve. From that point, extend a *perfectly horizontal* line over to the vertical axis and read that number. The answer is 4 million.

How many x-rays in this beam have 45 kV of energy? How many x-rays in this beam have 75 kV of energy? The answers are 17 million and 6 million respectively. Again, note that most x-rays in the beam average around ⅓ of the set kVp, and only a few are close to the kVp.

Graphs of the x-ray beam spectrum will be used extensively later on, so it is important that you become adept at plotting and interpreting these types of graphs.

Understanding the Digital Histogram

Digital images are typically displayed within an "image review screen" that includes control buttons for adjusting and annotating the image along with a histogram of the image. As shown in Figure 3-17, the data set for the initial histogram actually forms a *bar* graph. (Note in Figure 3-14 that this was technically true for the x-ray beam spectrum graph as well.)

As was the case for the x-ray beam spectrum, a "best-fit" curve for the histogram is constructed by connecting the uppermost points of the vertical bars of the graph as shown in Figure 3-18. Radiographers must understand the basic implications of the resulting histogram shape, since it affects both digital processing and the calculation of the exposure indicator for each radiograph taken.

On the original bar graph histogram shown in Figure 3-17, the height of each vertical bar represents a simple count of the *number of pixels (picture elements)* in the image containing each pixel value, that is, each *gray level* or *density*. To reinforce this concept, use a straight-edge on the graph to answer the following two questions:

1. The largest pixel count in the "main lobe" of this histogram occurs at a pixel value of about 683. How many pixels are demonstrating this gray level?
2. The histogram dips to a low-point between the "main lobe" and the "tail spike" at a pixel value (density) of about 1170. How many pixels are demonstrating this particular brightness?

The answer to the first question is that there are about 4,125 pixels holding a value of 683. (This peak is approximately one-eighth of the way from 4000 to 5000.) The answer to question #2 is that there are

Figure 3-18

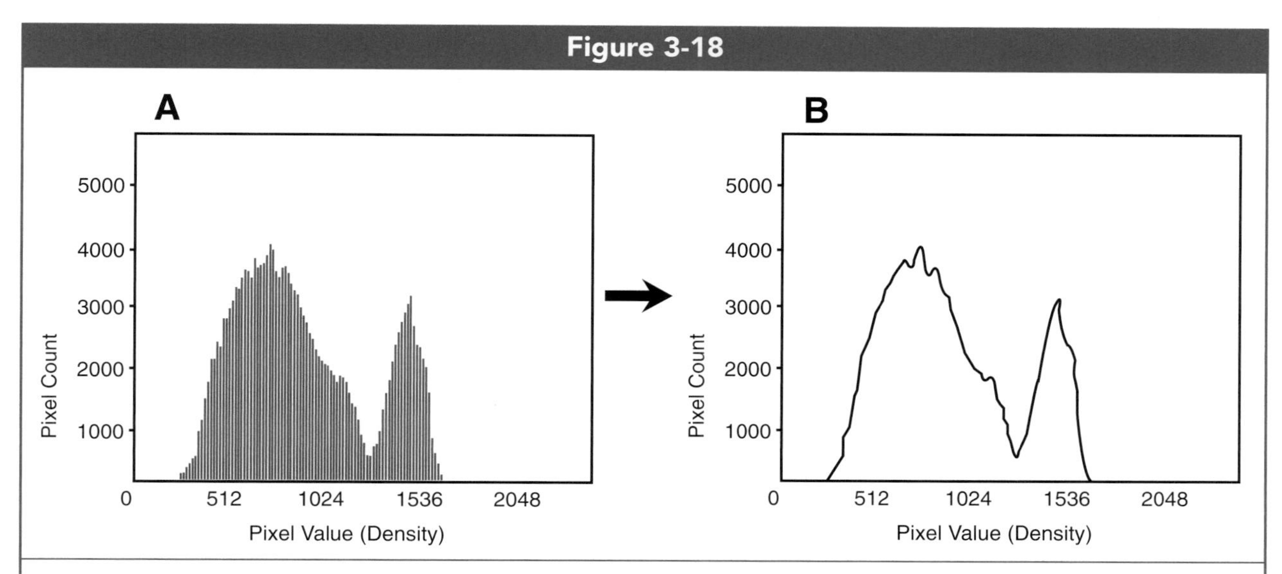

The displayed histogram is constructed by using a "best fit curve" to connect the uppermost points of each bar on the original graph.

about 500 pixels demonstrating this brightness level of 1170.

It has become the standard histogram format to plot image densities progressing from light to dark, left to right. Figure 3-19 helps visualize this relationship by shading the histogram. This histogram is typical for a chest radiograph, so we expect to see the largest number of pixels at medium to dark densities representing the lung fields, represented by the peak labeled ***B***. To the left of this, we expect fewer pixels representing the lighter areas of the mediastinum and heart. Note the distinctive "tail spike" to this curve that represents the raw radiation exposure areas to the sides of the chest and above the shoulders.

Figure 3-20 illustrates a histogram with three distinct "lobes." These are the pixel counts for a radiograph of the abdomen with a large bolus of barium in the stomach or intestine, and a small amount of raw radiation area to the sides of the body part. The number and general shape of the lobes are characteristic of each general type of body part—chest, abdomen, skull, extremities, or mammogram which presents primarily soft tissue.

Histogram analysis is an essential part of digital image processing and is performed on *every* digital image before it is displayed. The computer must identify landmarks in the data set based on expected troughs or peaks in the shape of the histogram. A full discussion of histogram analysis is presented in Chapter 30.

Figure 3-19

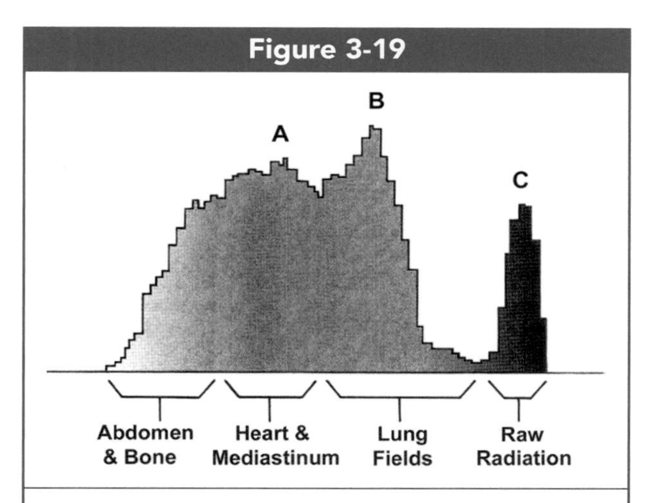

Histogram for a typical chest radiograph showing how pixel brightness is plotted from lighter to darker densities left-to-right. The *main lobe* of the histogram includes lighter densities of the abdomen and spine, and mid-level densities of the heart and mediastinum. It also includes the darker densities of the lung fields, showing a peak at ***B*** because of the large area they cover within the field. The *tail spike*, ***C***, represents raw exposure to the detector above the shoulders and to the sides of the chest. (From *Understanding Digital Radiograph Processing*, Midland, TX: Digital Imaging Consultants, 2013. Reprinted by permission.)

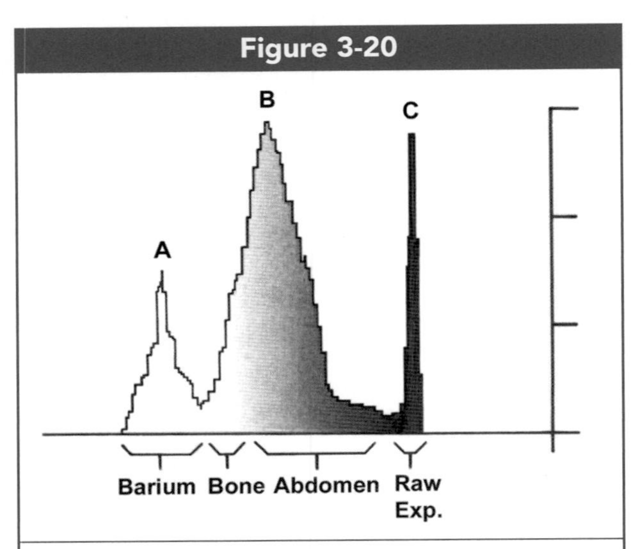

Figure 3-20

The distinctive three-lobe histogram for an abdomen with a large bolus of barium included in the stomach or intestines, and some raw exposure to the sides. (From *Understanding Digital Radiograph Processing*, Midland, TX: Digital Imaging Consultants, 2013. Reprinted by permission.)

SUMMARY

1. The general mathematical order of operations is to apply exponents first, multiply or divide, then add or subtract.
2. Using scientific notation makes calculations easier. Any number can be converted into scientific notation by simply adding "$\times 10^0$" behind it.
3. When multiplying numbers in scientific notation, *add* the exponents, when dividing, *subtract* the exponents.
4. Dimensional analysis simplifies unit conversions. The main rule is to be sure the *top* unit always goes to the *bottom* in the next expression.
5. When calculating for areas, both units and calculations must be *squared*. When calculating for volumes, both units and calculations must be *cubed*.
6. Larger potatoes, and larger x-ray tube anodes, take longer both to heat and to cool, because the *areas* of surface through which radiation passes change *less drastically* (by the square) than the *volumes* that contain the heat (by the cube).
7. The inverse square law is a law of areas, which states that the intensity of radiation will change according to the *inverse square* of the distance from the radiation source.
8. Graphs must use a "best-fit" line in order to accurately represent data. The ability to interpret various graphs is very important for the radiography student.
9. A *beam spectrum* graph for an x-ray beam plots the amount of x-rays emitted from the x-ray tube (vertically) against the various energies contained within the beam (horizontally). The typical graph shape for a diagnostic beam forms a lopsided bell-curve, peaking at about one-third of the peak energy of the beam (kVp).
10. A digital image *histogram* is actually a bar graph that plots pixel counts (vertically) against the range of pixel values present within the image. These graphs can have from one to three "lobes" representing those pixel values or densities that predominate the image.

REVIEW QUESTIONS:

Practice Exercise 3-1

1. Convert 3/7 into its decimal equivalent:

2. Convert 0.34 into a percentage.

3. Solve for X:

$$X = \frac{2(8+5)^3}{32} - 76$$

4. Solve for X:

$$\frac{X}{4} = \frac{81}{9}$$

5. Solve for X: $X + 17 = -23$

6. $6(3 + 8) =$

- Convert the following measurements into full scientific notation. Don't forget to include the units in your answers:

7. 0.3 mg =

8. 0.000008 mm =

9. 640,000 kg =

10. 53,000,000,000 lb =

- SOLVE THE FOLLOWING PROBLEMS USING SCIENTIFIC NOTATION:

11. $(2.1 \times 10^5) \times (5.4 \times 10^{-2}) =$

12. $(8.1 \times 10^5) / (2.4 \times 10^{-2}) =$

13. What is the "longhand" expression for your answer to #11:

14. What is the "longhand" decimal expression for your answer to #12:

(Continued)

Practice Exercise 3-1 *(Continued)*

- First convert the following units into full scientific notation, then insert these into the dimensional analysis grid and solve. All expressions in the grid and answers must be in full scientific notation:

  15. Convert 2300 ms into hours:

  16. Convert 8.5 mV into μV:

  17. Convert 10 inches into angstroms (Å) [Recall 1Å = 10^{-10} m]

  18. How many square inches are there within 3 square feet?

  19. How many cubic *centimeters* are there within 2 cubic meters?

20. When the distance from a radiation source is cut to ¼, precisely how much more or less radiation exposure will there be?

21. If the radiation dose to a patient was 10 mR at a 40-inch distance from the x-ray tube, what would the dose be if the distance were reduced to 30 inches?

22. If the radiation dose to a patient is 10 mR at a distance of 100 cm, what would it be if the x-ray tube is moved back to 150 cm?

23. Referring to the graph in Figure 3-21, at what radiation dose will 80 percent of the population in this sample come down with leukemia?

24. Is the relationship in Figure 3-21 threshold or nonthreshold, and is it linear or non-linear?

Figure 3-21

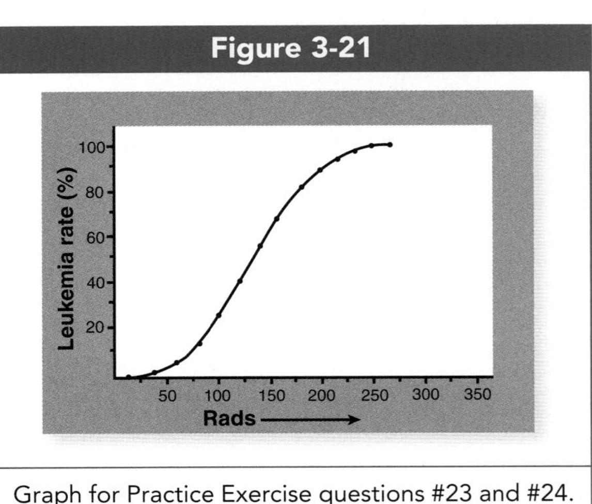

Graph for Practice Exercise questions #23 and #24.

(Continued)

Practice Exercise 3-1 *(Continued)*

25. Referring to the x-ray beam spectrum graph in Figure 3-22, what is the significance of point *A*?

26. Referring to the graph in Figure 3-22, how many x-rays in this beam possess 65 kV of energy?

Figure 3-22

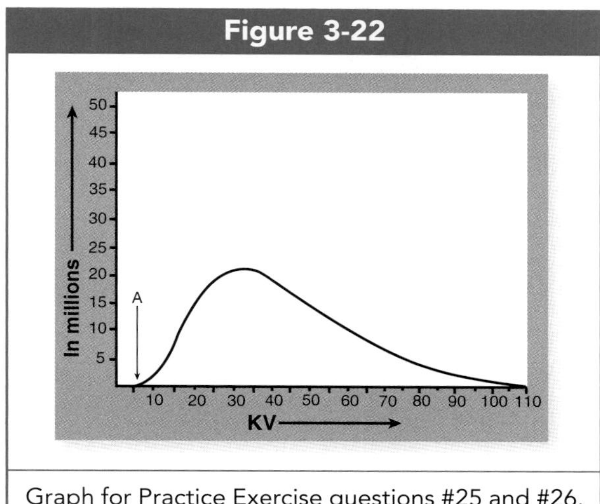

Graph for Practice Exercise questions #25 and #26.

27. Referring to the histogram in Figure 3-23, how many pixels are demonstrating a gray level or pixel value of 1366 (at the peak of the "tail lobe")?

28. Referring to the histogram in Figure 3-23, how many pixels are demonstrating a brightness level or pixel value of 200?

Figure 3-23

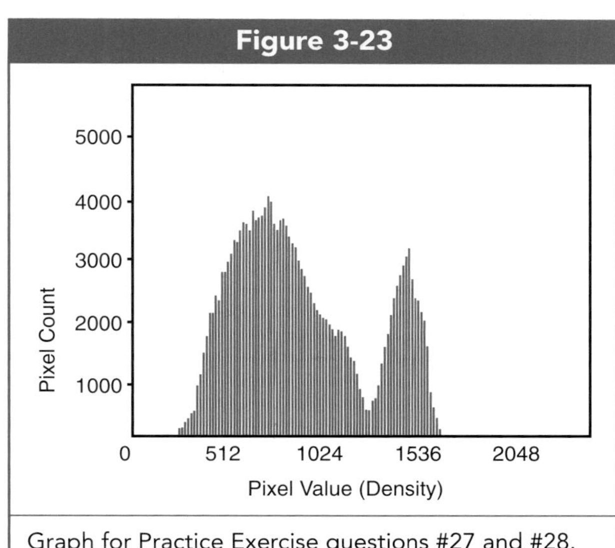

Graph for Practice Exercise questions #27 and #28.

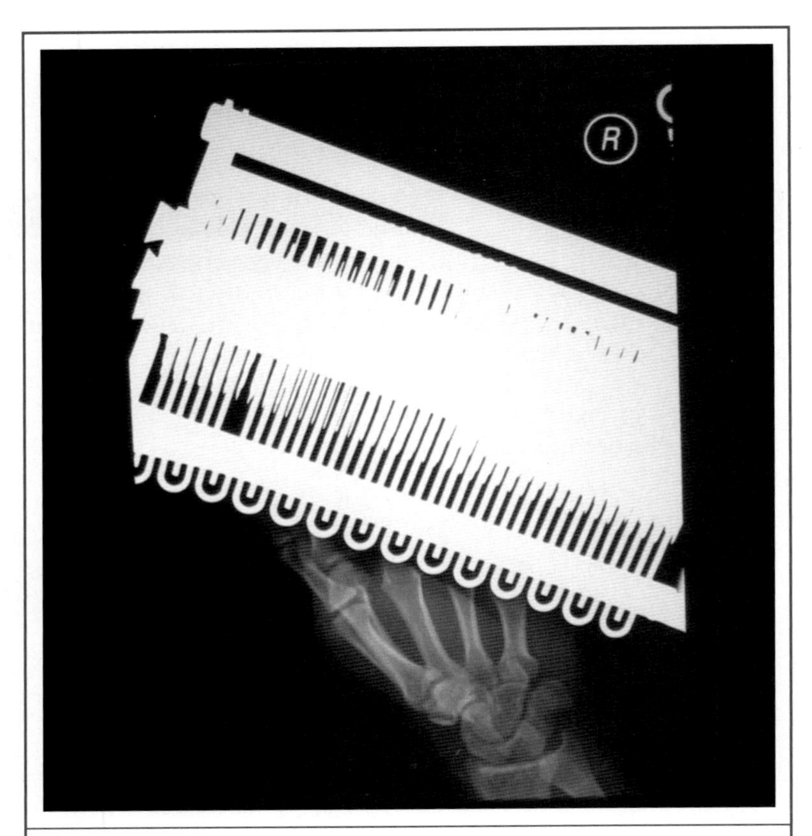

A butcher got his hand caught in an electric meat grinder. The feed mechanism had to be detached from the machine to bring him into the emergency room with the hand as intact as possible.

Chapter 4

THE ATOM

Objectives:

Upon completion of this chapter, you should be able to:

1 Define chemical elements, compounds, and mixtures, atoms and molecules.
2. Describe the structure of the atom, its nucleus, electron "shells" and suborbitals.
3. Describe the *neutron* and how it relates to radioactive decay.
4. Explain the *2N*^{2} and *octet* rules and how they relate to *periodicity* in the chart of elements.
5. Describe the two major forms of chemical bonding between atoms.
6. Define *ionization* and how it affects chemical bonds and disease.
7. Interpret nuclear notation for atomic mass and atomic number.
8. Describe the progression of the neutron number as atomic number increases.
9. Define nuclear *fission* and *fusion*.
10. Define and give examples of isotopes, radioisotopes, and isomers.
11. Explain the emission of alpha, beta and gamma radiation and their effects on the nucleus.

MATTER

Matter is anything that has shape or form and occupies space. It is the substance of which all physical objects are composed. All matter has *mass*, a quantity which, under the influence of gravity, gives the matter *weight*. In Chapter 2 we discussed how, in accordance with Einstein's equation, mass and energy can be interchanged. Perhaps the primary distinction between mass and energy is that a mass, however small, can be *weighed* when it is in a gravity field, whereas forms of pure energy, such as electromagnetic waves, cannot be weighed.

The general term *substance* is used to describe any material with a definite and constant composition. The simplest form of any substance is an *element*. An element is a substance that cannot be broken down into any simpler substance by ordinary means, that is, by mechanically smashing it or by chemical interactions. An element is comprised of a group of identical atoms that share specific chemical behaviors. The *atom* is the smallest single unit of an element that retains those chemical behaviors.

Chemical behavior refers to the making or breaking of actual connections between atoms in which they physically *share their components* (that is, the *electrons* in their outermost shells actually spend time around each atom, exchanging back and forth). One might consider them "stuck together."

A *compound* describes a substance in which atoms of different elements are chemically bound together. A *mixture* describes the combination of two or more substances in such a way that they are *not* chemically bound together. Mixtures are mechanically stirred together, but given time, they tend to separate. You can easily tell whether a particular liquid is a compound or a mixture by waiting to see if it separates: If it is a mixture, the heavier substance will "settle" to the bottom of the container. An example is powdered chocolate mixed into milk. A compound will not separate solely on the basis of passing time.

Any time two or more atoms are chemically bound together, they make up a *molecule*. A molecule is the smallest unit of any chemical compound. However, note that to have a molecule, it does not necessarily require two different *elements* to bind, only two *atoms*. This is because the atoms of some elements can bind together with others of the same element. An example is the oxygen we breath, O_2—this molecule is made of two oxygen atoms connected. Thus, it is possible to have a *molecule* of an element and also a *molecule* of a compound.

Barium sulfate suspension, commonly used in radiography for studies of the gastrointestinal tract, provides a perfect example of several of these concepts: The liquid that the patient drinks for an upper GI study is actually a *mixture* of barium sulfate molecules suspended within water molecules (Fig. 4-1). (The barium sulfate molecules are the larger, more complex ones.) If it is left to sit for a few hours, the barium sulfate, which is actually a powder, will settle to the bottom of the cup, separating from the water molecules because it is not chemically bound to them.

The smallest unit of the barium sulfate powder is a single molecule (Fig. 4-2). This molecule of barium sulfate is also a *compound*, since it is made up of more than one element. Examining the molecule in Figure 4-2, we see that it is specifically composed of one atom of the element barium, one atom of the element sulfur, and *four* atoms of the element oxygen. This formula is indicated by the chemical abbreviation for barium sulfate, $BaSO_4$, where Ba = barium, S = sulfur, O = oxygen. Each of the atoms in the molecule is physically connected to it, *chemically bound* to the other atoms by the sharing of electrons in the outermost shells. If this powder is ground down

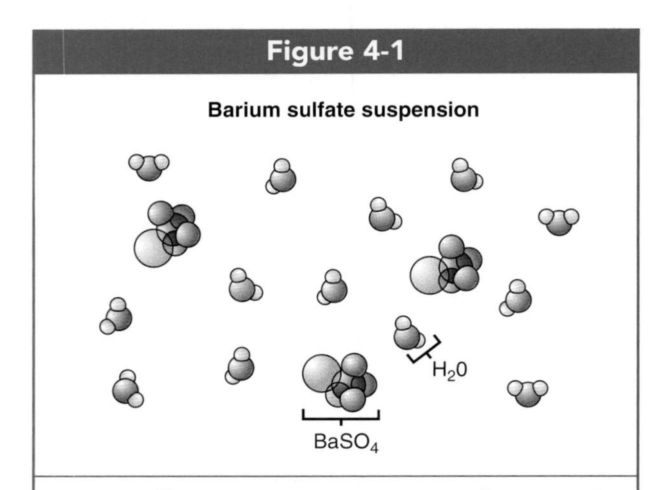

Barium sulfate suspension is a mixture of water molecules and barium sulfate molecules which never chemically bind together. Left to sit, the two substances will separate.

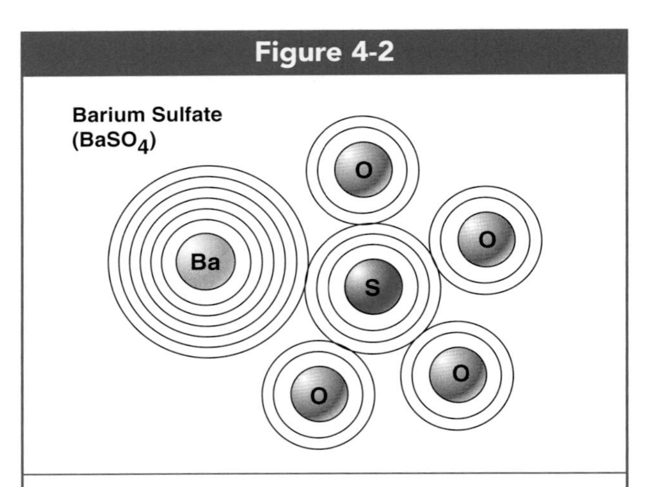

Pure barium sulfate is a molecule comprised of six atoms, but only three elements. The molecule can be broken down into its constituent elements by chemistry.

mechanically by a mortar and pestle, or smashed by a hammer, these atoms will still not separate from the molecule. However, they can be *chemically* separated from each other by interacting with other chemicals.

A single atom of the element barium is separated from the molecule and shown in Figure 4-3. It has an *atomic number*, or *Z number*, of 56 which means that there are 56 positive charges within the central nucleus. It is these 56 positive charges that determine the atom's identity as the element barium. Even other chemicals cannot alter this configuration of 56 positive charges in the nucleus of a barium atom. It cannot be broken down by "ordinary" everyday means.

There are about a dozen elements which radiographers should be familiar with, enough to know their abbreviations and the Z numbers. These are listed in Table 4-1. Note that for some elements the abbreviation is an upper case letter with a lower case letter. The formula for calcium tungstate is given below. (Calcium tungstate was the compound Thomas Edison found in 1896 to be most effective at converting x-rays into light for Pupin's screen cassettes.) From the formula, how many atoms make up one molecule of calcium tungstate? How many *elements* are there in the compound calcium tungstate?

$$CaWO_4$$

Answer: There are *six* atoms in a calcium tungstate molecule; one calcium, one tungsten, and *four* oxygen atoms, but there are only three elements.

Figure 4-3

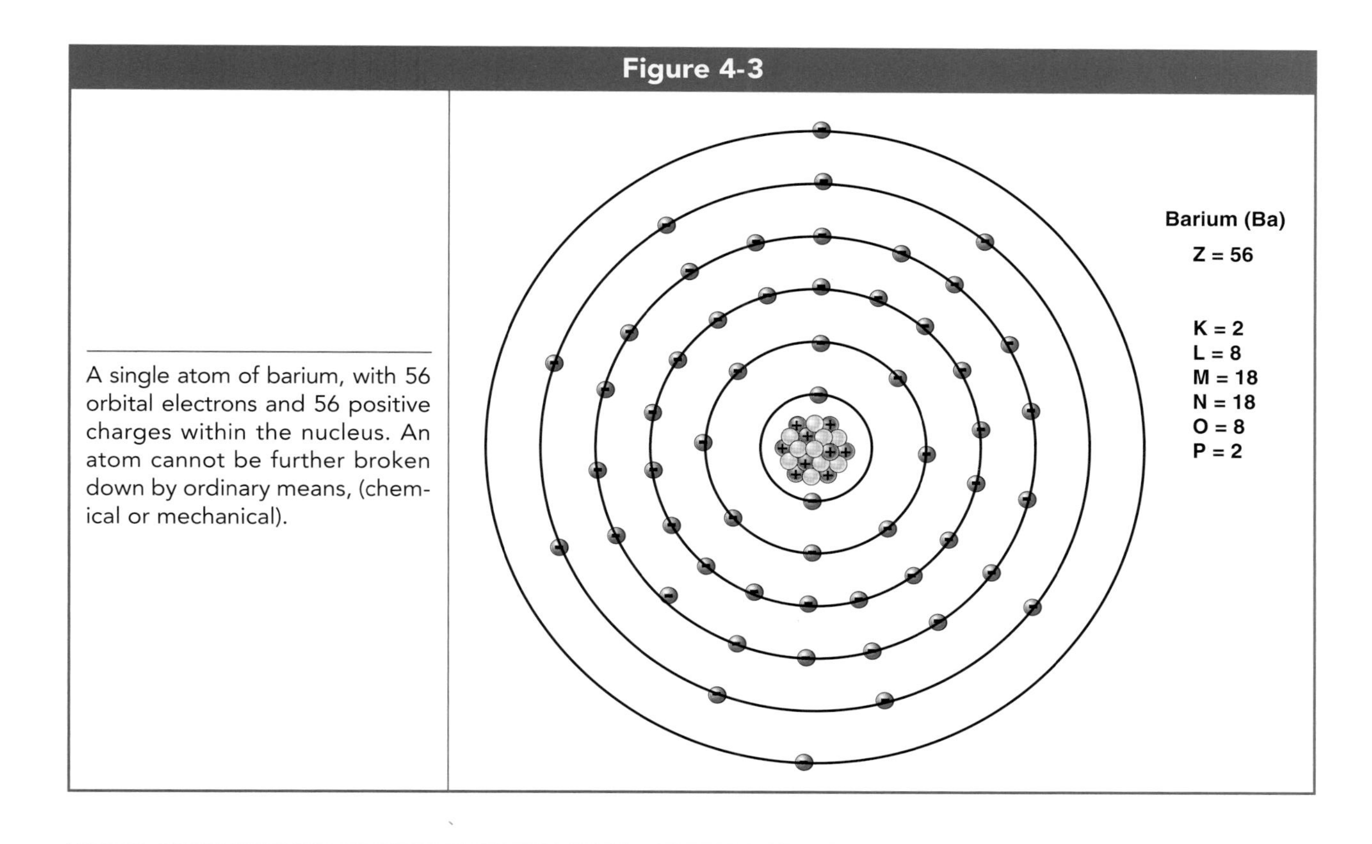

A single atom of barium, with 56 orbital electrons and 56 positive charges within the nucleus. An atom cannot be further broken down by ordinary means, (chemical or mechanical).

PHYSICAL STRUCTURE OF ATOMS

In 1911, the British physicist Ernest Rutherford completed a series of experiments that established the modern concept of the atom—a very dense *nucleus* with a positive electrical charge surrounded by a cloud of negatively-charged electrons. Shortly thereafter the Danish physicist Niels Bohr showed that, for a particular atom, the electrons could only exist at certain, set energy levels, placing them in prescribed orbits at set distances from the nucleus. Today we know from quantum physics that those energy levels are better represented by volumes of space with specific shapes around the nucleus, called *orbitals* (Fig. 4-4). Electrons can actually be located anywhere within these orbitals at any time.

For the purposes of learning radiography, however, it is best to use the *Bohr* model of the atom with its picture of electrons inhabiting circular orbits like planets around the sun (Fig. 4-5). A fascinating and important aspect of atoms is that, like the solar system, they are

Table 4-1

Elements Radiographers Should Know

Application	Element	Abbreviation	Atomic (Z) Number
Basic	Hydrogen Helium	H He	1 2
In the body	Carbon Oxygen	C O	6 8
X-ray filter	Aluminum	Al	13
In the body	Calcium	Ca	20
Contrast agent	Iodine	I	53
Image receptor	Barium	Ba	56
X-ray tube	Tungsten Rhenium	W Re	74 75
Shielding.	Lead	Pb	82
Radioactive	Uranium	U	92

Figure 4-4

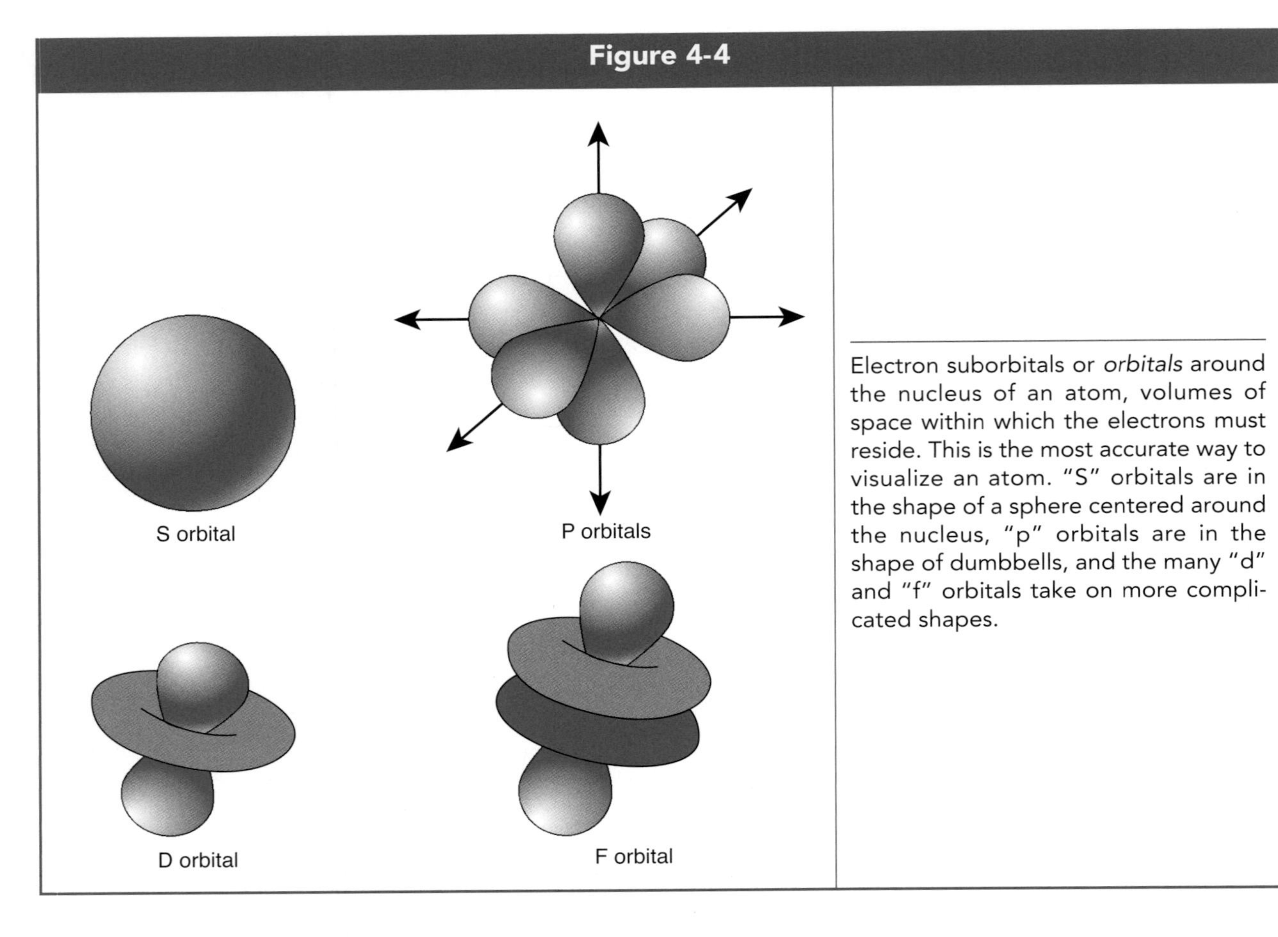

Electron suborbitals or *orbitals* around the nucleus of an atom, volumes of space within which the electrons must reside. This is the most accurate way to visualize an atom. "S" orbitals are in the shape of a sphere centered around the nucleus, "p" orbitals are in the shape of dumbbells, and the many "d" and "f" orbitals take on more complicated shapes.

Figure 4-5

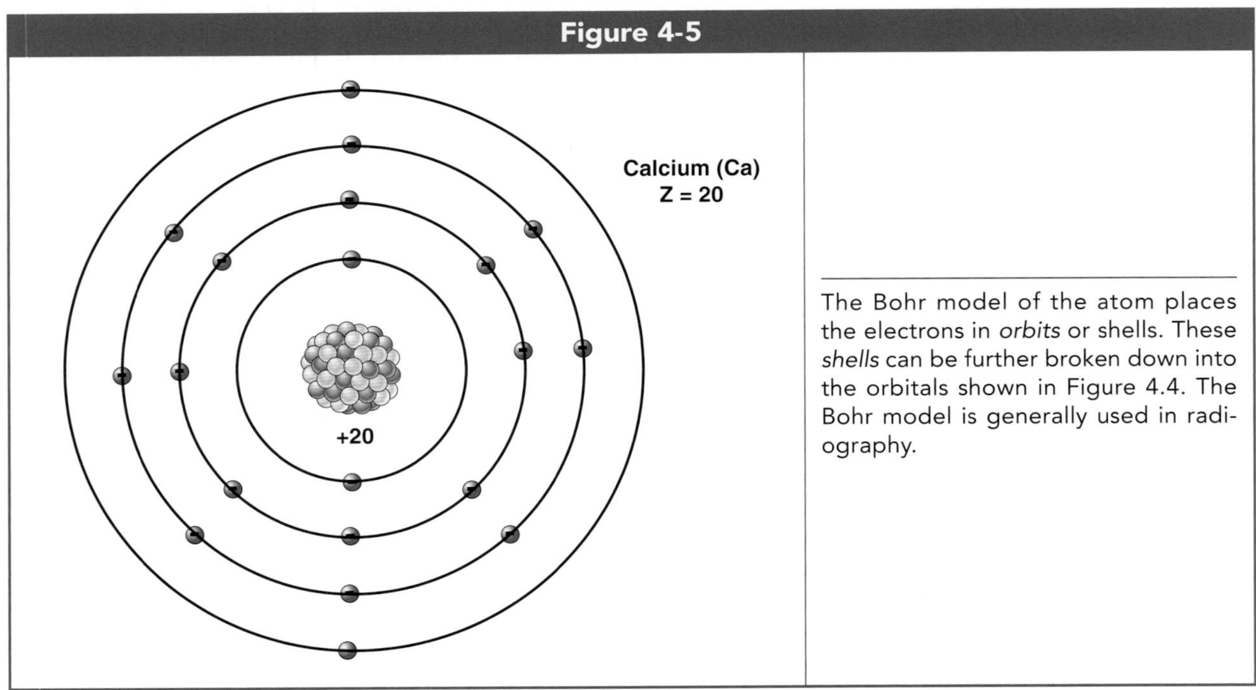

The Bohr model of the atom places the electrons in *orbits* or shells. These *shells* can be further broken down into the orbitals shown in Figure 4.4. The Bohr model is generally used in radiography.

mostly composed of empty *space*. If the nucleus of an atom were the size of a marble and placed at one end of a football field, the nearest electron would be at the other end of the field and would be the size of a grain of sand!

Electrical charge plays a crucial role in the structure and behavior of atoms. For example, given the above statement that atoms are mostly space, one might wonder why we cannot walk through walls, since both our bodies and the walls are made of atoms, and therefore, mostly space. The answer is that, even though an atom may be electrically balanced overall, the positive charges are all in the nucleus and the negative charges out in the orbits or "shells." So when two atoms approach each other, what they each "see" is only the outermost "skin" of negative charge. Negative charges repel each other, and in solid objects the atoms are locked together in their positions, presenting a negatively-charged "wall" or "skin" to the other object. The electrical repulsion between the two objects makes them seem solid.

Rutherford identified the positive charges in the nucleus as large particles called *protons*. Remember, it is the atomic number (Z) which determines the identity of an atom or the element it belongs to. The Z number is the number of *protons* within the nucleus.

Each proton is nearly 2000 times the physical size and weight of a single electron. Yet the magnitude of its positive charge exactly cancels the negative charge on an electron. This is one of the continuing mysteries in science, how two fundamental particles can have the same magnitude of charge (+1 and –1), and one of them has 2000 times the mass of the other.

In comparing the mass or "size" of subatomic particles, physicists normally use the proton as the standard. Its mass is defined as one *atomic mass unit, (amu)*. Rather than state that a proton is 2000 times bigger than an electron, it might be more appropriate, then, to say that an electron has a mass of about 1/2000th amu.

Dmitri Mendeleyev, a Russian chemist, formulated the *law of periodicity* which states that the properties of elements are *periodic* functions of their atomic weight. That is to say, these properties *repeat* themselves at specific periods as the Z number increases. Armed with the conviction that there is order in nature, he organized the *periodic table of the elements* (Table 4-2, p. 58), later refined but still in use today. Each column in the table, called a *group*, represents increasing Z number as one reads the table left-to-right. Each row, called a *period*, repeats the cycle of properties.

Rutherford also predicted that another nuclear particle, with neutral (or no) electrical charge, would eventually be proven to exist. He was right, and the particle was named the *neutron*. The mass of a neutron is slightly heavier than that of a proton. In fact, it is roughly equal to the mass of a proton plus the mass of a single electron. This helps to explain its neutral electrical charge. With a large enough "atom smasher," a neutron could be created by smashing a single electron into a proton with such force that they would "stick together." In doing so, the resulting particle would have a net electrical charge of –1+1 = 0, and weigh slightly more than a proton (Fig. 4-6). With a bit of rounding out, we could say that both protons and neutrons are roughly 2000 times more massive than a single electron.

Neutrons are relatively *unstable*—they can *decay* and frequently do, especially when they are outside of an atomic nucleus, falling apart into their component proton and electron. Since the electron is much lighter, it is ejected at high speed from the larger particle (Fig. 4-7). This emission of high-speed electrons from a nucleus is called *beta radiation*, and is one of the forms of natural radioactivity. Each time a neutron decays, a proton is left behind. The ramifications of this will be discussed further on. Protons can also decay hypothetically, but their stability is about equal to the age of the universe (!), so we need not concern ourselves with proton "decay." Neutron

Figure 4-6

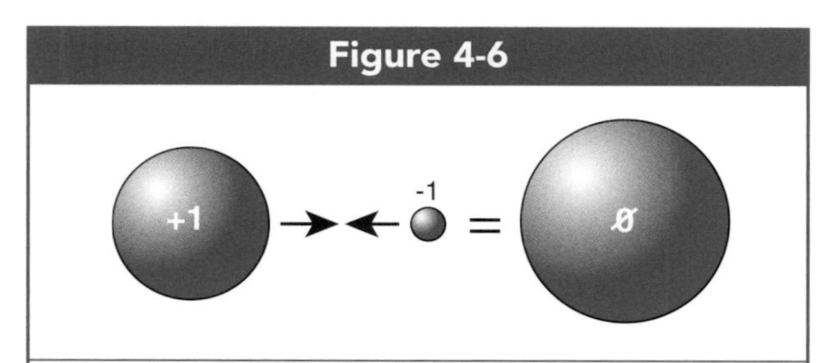

The neutron is electrically neutral because it is essentially comprised of a proton added to an electron, so that their respective electric charges of +1 and –1 cancel each other out. Its total mass is slightly greater than that of a proton.

Table 4-2

PERIODIC TABLE OF THE ELEMENTS

Period	I	II											III	IV	V	VI	VII	VIII
	Alkali Metals	Alkaline-earth Metals	Transitional Elements														Halogens	Noble Gases
1	1 H 1.00797																	2 He 4.0026
2	3 Li 6.939	4 Be 9.0122											5 B 10.811	6 C 12.0111	7 N 14.0067	8 O 15.9994	9 F 18.9984	10 Ne 20.183
3	11 Na 22.9898	12 Mg 24.312											13 Al 26.9815	14 Si 28.086	15 P 30.9738	16 S 32.064	17 Cl 35.453	18 Ar 39.948
4	19 K 39.102	20 Ca 40.08	21 Sc 44.956	22 Ti 47.90	23 V 50.942	24 Cr 51.996	25 Mn 54.9380	26 Fe 55.847	27 Co 58.9332	28 Ni 58.71	29 Cu 63.54	30 Zn 65.37	31 Ga 69.72	32 Ge 72.59	33 As 74.9216	34 Se 78.96	35 Br 79.909	36 Kr 83.80
5	37 Rb 85.47	38 Sr 87.62	39 Y 88.905	40 Zr 91.22	41 Nb 92.906	42 Mo 95.94	43 Tc 99	44 Ru 101.07	45 Rh 102.905	46 Pd 106.4	47 Ag 107.870	48 Cd 112.40	49 In 114.82	50 Sn 118.69	51 Sb 121.75	52 Te 127.60	53 I 126.9044	53 Xe 131.30
6	55 Cs 132.905	56 Ba 137.34	* 57-71	72 Hf 178.49	73 Ta 180.948	74 W 183.85	75 Re 186.2	76 Os 190.2	77 Ir 192.2	78 Pt 195.08	79 Au 196.967	80 Hg 200.59	81 Ti 204.37	82 Pb 207.19	83 Bi 208.980	84 Po 210	85 At 210	86 Rn 222
7	87 Fr 223	88 Ra 226.03	+ 89-103															

* Rare earths (Lanthanide Series)	57 La 138.91	58 Ce 140.12	59 Pr 140.91	60 Nd 144.24	61 Pm 145	62 Sm 150.36	63 Eu 151.96	64 Gd 157.25	65 Tb 158.93	66 Dy 162.50	67 Ho 164.93	68 Er 164.26	69 Tm 168.93	70 Yb 173.04	71 Lu 174.967
+ Actinide Series	89 Ac 227.03	90 Th 232.038	91 Pa 231.036	92 U 238.029	93 Np 237.048	94 Pu 244	95 Am 243	96 Cm 247	97 Bk 247	98 Cf 251	99 Es 254	100 Fm 257	101 Md 258	102 No 259	103 Lr 260

Upper number is atomic number. Lower number is atomic mass averaged by isotopic abundance in the earth's surface, expressed in atomic mass units (amu).

decay, then, is responsible for much of the radioactivity occurring in natural substances.

We complete the detailed picture of the Bohr atom as a dense, massive nucleus, made of both protons and neutrons, and comprising some 99.9 percent of the total mass of the atom, with extremely small, light electrons in orbits that are at a considerable distance (Fig. 4-5).

Electron Configuration

In its normal state, when an atom is not electrically charged, the number of electrons in the "shells" is equal to the number of protons in a nucleus. Up to seven orbits or shells themselves are arranged like the layers of an onion, in concentric circles (according to the Bohr model). The shells are frequently labeled alphabetically, beginning with "K" for the first shell, "L" for the second, and so on to the letter "P" for the seventh shell found in very large atoms (Fig. 4-3).

The electrons must be arranged according to two specific rules. They are:

1. The *maximum* number of electrons which can occupy a shell under *any* conditions is equal to $2N^2$, where "N" is the shell number.
2. The *outermost* shell of the atom can never hold more than 8 electrons. This "octet rule" overrides rule #1.

The shell number is referred to by physicists as the *principle quantum number*. Table 4-3 gives the maximum number of electrons for each shell, using the $2N^2$ rule. You can see that it increases in an exponential fashion for each shell.

The chemical behavior of an atom is determined *only* by the number and configuration of electrons in the *outermost* shell. By rule #2 above, the outermost shell is considered *filled* when there are 8 electrons in it. As these 8 electrons repel each other by their negative charge, they become evenly distributed around the nucleus. This sets up a homogenous or smooth negative "screen" around the nucleus, shielding other electrons outside the atom from the positive pull of the nucleus.

Because of this *screening* of the positive pull of the nucleus, neutral atoms with exactly 8 electrons in the outermost shell have no tendency at all to connect with other atoms. They are said to be chemically *inert*. Interestingly, they all turn out to be gases. In the periodic chart of the elements in Table 4-2, these inert elements all fall into the 8th vertical column, or *group number 8*. Referred to as the Noble gases, they include the familiar helium and neon, then argon and krypton, xenon which is used in nuclear medicine, and radon which we have described as a highly radioactive gas that can accumulate in unventilated basements. Since inert elements do not combine with other elements to form compounds, you will never hear of "helium chloride" or "neon oxide."

On the periodic chart (Table 4-2), if we move to the next "larger" atom to the right of any inert element in group 8, a proton must be added to the nucleus bringing the Z number up by one, and a corresponding electron would be added in the shells. The octet rule (rule #2) forbids 9 electrons in the

Figure 4-7

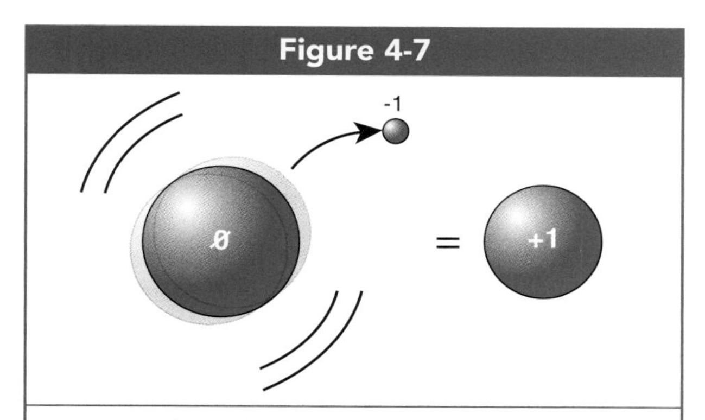

Neutrons decay by emitting a high-speed electron, called a *beta particle*. By removing the negative charge component of the neutron, a positively-charged proton is left behind.

Table 4-3

Maximum Number of Electrons by Atomic "Shell" Based on $2N^2$ Rule

Shell	Principle Quantum Number	$2N^2$ Number
K	1	2
L	2	8
M	3	18
N	4	32
O	5	50
P	6	72

outermost shell, so the only way the new electron can be accommodated is to begin a new shell around the atom. This is reflected on the chart by starting a new row or *period*.

For example, *neon* is a noble gas with Z# = 10, having 10 protons in the nucleus. It is in group 8 because its outermost shell, the "L" shell or shell #2, is filled with 8 electrons. (By the $2N^2$ rule, the first two electrons are in the first shell and the remaining 8 in the second shell.) Sodium, with Z# = 11, is the next "larger" atom in the sequence. But it can only accommodate the next electron by starting a new shell. So an atom of sodium has *three* shells, placing sodium in the third row on the chart, but it falls in *group #1*, the first column, because the third shell, now the *outermost* shell, has one single electron in it. It is the octet rule that accounts for the *periodic* or repeating layout of the chart of the elements.

You will find on the periodic chart that the first period consists only of group 1 and group 8; this is because the first shell "K" is *filled* with only two electrons according to the $2N^2$ rule. The octet rule is not broken because it sets only a *maximum*, not a minimum, that a shell can hold. In the second period and for the second shell, both rules set the limit at 8.

From the fourth period down, we see a block of columns called "transitional" elements inserted in the middle, creating subgroups that total more than 8. This block includes the "lanthanium" and "actinium" series that are broken out below. The chart does this because, once the 4th shell is filled *as an outermost shell* with its 8 electrons (at iron, "Fe", Z# = 26), the next "larger" atom, cobalt (Z# = 27), is able to put what would have been the 9th electron in that shell *back* into the 3rd shell; the 3rd shell is no longer the *outermost* shell, therefore it can now hold its $2N^2$ limit or 18 electrons. It reaches this capacity at Krypton gas (Z# = 36). At this point, no more electrons may be added without starting a new 5th shell. For clarification, let's list the configuration of the electrons in Krypton gas, in Table 4-4.

Rubidium, the next element in the chart at Z# = 37, must begin a 5th shell for the next electron. The 3rd or M shell has reached its $2N^2$ capacity and cannot take any more electrons under any circumstance. The 4th or L shell has reached its capacity *as an outer shell*, and cannot take any more electrons until another shell is added, such that it becomes an *inner shell*.

Table 4-4

Configuration of Electrons for Krypton

Shell	Number of Electrons	Restricted by Rule
K (1)	2	$2N^2$
L (2)	8	$2N^2$
M (3)	18	$2N^2$
N (4)	8	Octet
Total electrons:	36	

Larger atoms become more complicated. The barium atom, shown in Figure 4-3, adds two electrons to a new shell "P" before bringing the "N" shell up to 18 electrons. But, the two rules, the $2N^2$ rule and the octet rule, are never broken.

CHEMICAL BONDING

Covalent Bonding

The *orbitals* illustrated in Figure 4-4 may be considered as subgroups of the electron shells. For this reason, they are more properly called *suborbitals*. Each suborbital can contain two and only two electrons. The reason for this relates to magnetism: Two magnets can be laid side-by-side against each other, without repelling each other, only *if* the north and south poles are on opposite ends (Fig. 4-8). Each electron is like a tiny magnet with north and south poles. All electrons have a property called *spin* which correlates to their magnetic poles. If we attempt to place two electrons with the same spin into the same orbital, they will repel each other. They must have *opposite spins*, effectively one with magnetic north pointing *up*, the other with north pointing *down*, as shown in Figure 4-8. Every shell can be divided into its suborbitals of electron pairs, with larger shells accommodating more and more suborbitals (Fig. 4-9).

Thus, each orbital represents a *pair* of vacancies in which electrons may reside. When there is an even number of electrons in the outermost shell of an atom, all of the orbitals have their pairs of electrons evenly distributed. This sets up a homogenous or smooth "screen" of negative charge around the nucleus, shielding other electrons outside the atom from the positive pull of the nucleus. However, when there is an *odd* number of electrons in the outermost shell, there must be an orbital which contains only one electron within its pair of vacancies.

By definition, an *odd* number of electrons cannot be *evenly distributed* within orbitals. When there is only one electron in an orbital, the portion of that orbital space opposite to the electron is left open. It effectively leaves a *hole* in the screen of negative charge around the atomic nucleus, a defect in the cloud, where the positive pull of the nucleus can be "felt" by any outside electrons or other atoms (Fig. 4-10). The nuclear pull felt through this defect is relatively weak and very localized. Therefore, atoms with it will "stick together" when these areas on their surfaces happen to come in contact with each other, but for this to happen the atoms must essentially "bump into each other" by chance.

Covalent bonding occurs whenever two atoms come together because they both have this "defect" of an odd number of electrons. They literally *share* their unpaired electrons: When the first atom borrows the other's unpaired electron, its suborbitals are all filled and its outer screen of negative charge is completed or evenly distributed. The other atom then takes its turn borrowing the unpaired electron of the first atom. *Both* of the electrons, one from each atom, go back and forth, hence the term *co-valent*. Both electrons spend time around each atom—you might think of them as making "circle-eights" around the two atoms at high speed, creating the appearance that both atoms have filled all of their orbitals.

Covalent bonding is the most common form of chemical bonding between atoms, and is relevant to understanding of the effects of x-rays. There are three other types of bonding: ionic bonding, proton bridges and Van Der Waals forces. The last two types are rare in occurrence and of minor significance to radiography. Ionic bonds, however, are important to the understanding of the biological and photographic effects of x-rays.

Ionic Bonding

An *ion* is any electrically charged particle, be it a speck of charged dust, a single electron, or an atom with a net charge produced by an imbalance between its protons and electrons. It is possible for an atom to acquire, from a chemical interaction with

Figure 4-8

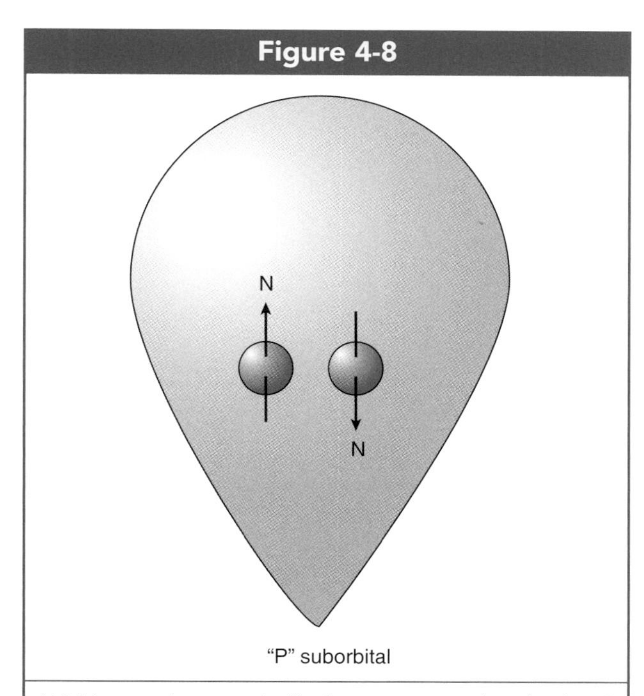

Within an electron shell of an atom, each suborbital can accommodate only two electrons having opposite *spins*, one with its "north pole" pointing up and the other pointed down, so they do not magnetically repel each other.

Figure 4-9

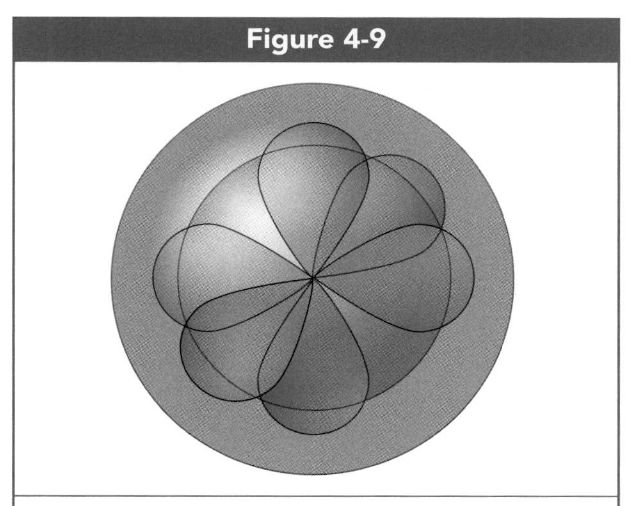

In the atom, larger shells accommodate more suborbitals. Shown here are only two "s" orbitals and a set of "p" orbitals.

Figure 4-10

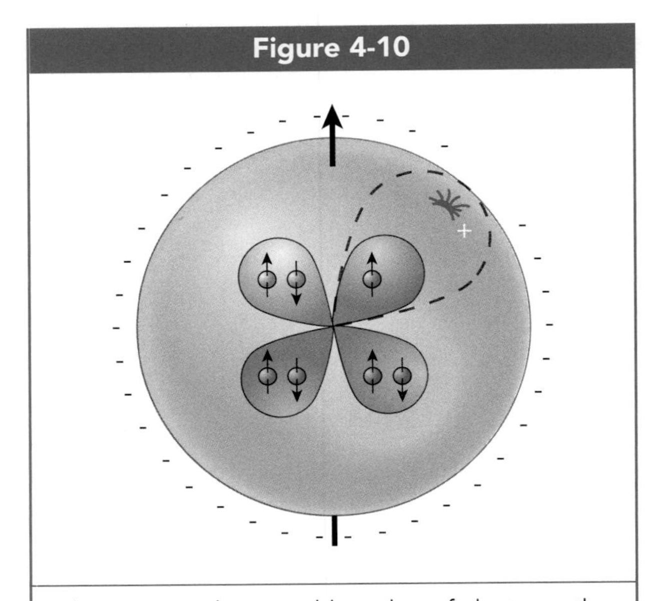

When an atom has an odd number of electrons, they cannot be evenly distributed within the suborbitals. This leaves a defect or *hole* in the screen of negative charge surrounding the nucleus (upper right). The slightly more positive charge at this hole leads to *covalent bonding* with other atoms.

another atom, an *extra* electron in its shells such that the total of electrons outnumbers the sum of protons in the nucleus. Such an atom will have a net charge of minus 1. It is also possible for an atom to *lose* an electron through chemical processes *or* by some types of radiation, including x-rays, that have enough energy to eject electrons out of its shells. Such an atom, with a net electrical charge of plus 1, is considered a *positive ion*.

In a fluid, if they are in the general area of each other, negative and positive ions will migrate toward each other due to the force of electrical attraction. When they come into contact, a strong chemical bond is formed when the positive atom "steals" or takes the extra electron from the negative atom and positions it into one of its own orbitals. This "taking" or "giving" of a single electron in ionic bonding is contrasted with the mutual "sharing" of *two* electrons in covalent bonding.

Covalent bonding occurs because of the flaws in the *regional* distribution of charge up close around each atom. The negative outer "curtain" of charges screening the pull of the nucleus from the outside world has a small flaw in a specific area where the pull of the nucleus might be "felt" by another atom. The same atom *at a distance* appears electrically neutral since it possesses a balanced number of protons and electrons. Ionic bonding, however, involves net electrical charges on the atoms creating pushing or pulling forces which can be "felt" at much longer distances. Atoms with an opposite charge anywhere in the general area will be strongly drawn toward opposite ions.

In ionic bonding, whole atoms feel this electrical *force* pulling them toward each other from a distance, and the bond formed when they come together is very strong. In contrast, for *covalent bonding* to occur, two or more atoms with unfilled suborbitals must effectively "bump into each other" by accident or by human design. Speaking at the molecular level, we might think of covalent bonding as a short-distance phenomenon, and ionic bonding as a long-distance phenomenon. Figure 4-11 illustrates the difference between covalent bonding and ionic bonding.

IONIZATION

The acquisition or loss of an electron by an atom is termed *ionization*. Figure 4-12 illustrates how the ionization of an atom by an x-ray might be visualized. As a form of electromagnetic energy, an x-ray is a wave that has no mass or weight, so it is not quite accurate to think of it like an object that physically "knocks" the electron out of its orbit like two pool balls colliding.

Rather, think of how a water wave can knock you over. It is not strictly the water knocking you over—after all, if the water were still, there would be no problem. Rather it is the *disturbance* in that water, the wave itself, that carries sufficient energy to make you feel pushed back. In a similar way an x-ray carries a lot of energy, but as a disturbance in electrical and magnetic fields rather than in water or other matter; and, even though the x-ray has no mass, it is able to affect the electron because the electron *itself* has electrical and magnetic fields around it. The energy of the x-ray is *deposited* into the electron. The electron then has too much energy to remain in its orbit or shell.

Figure 4-11

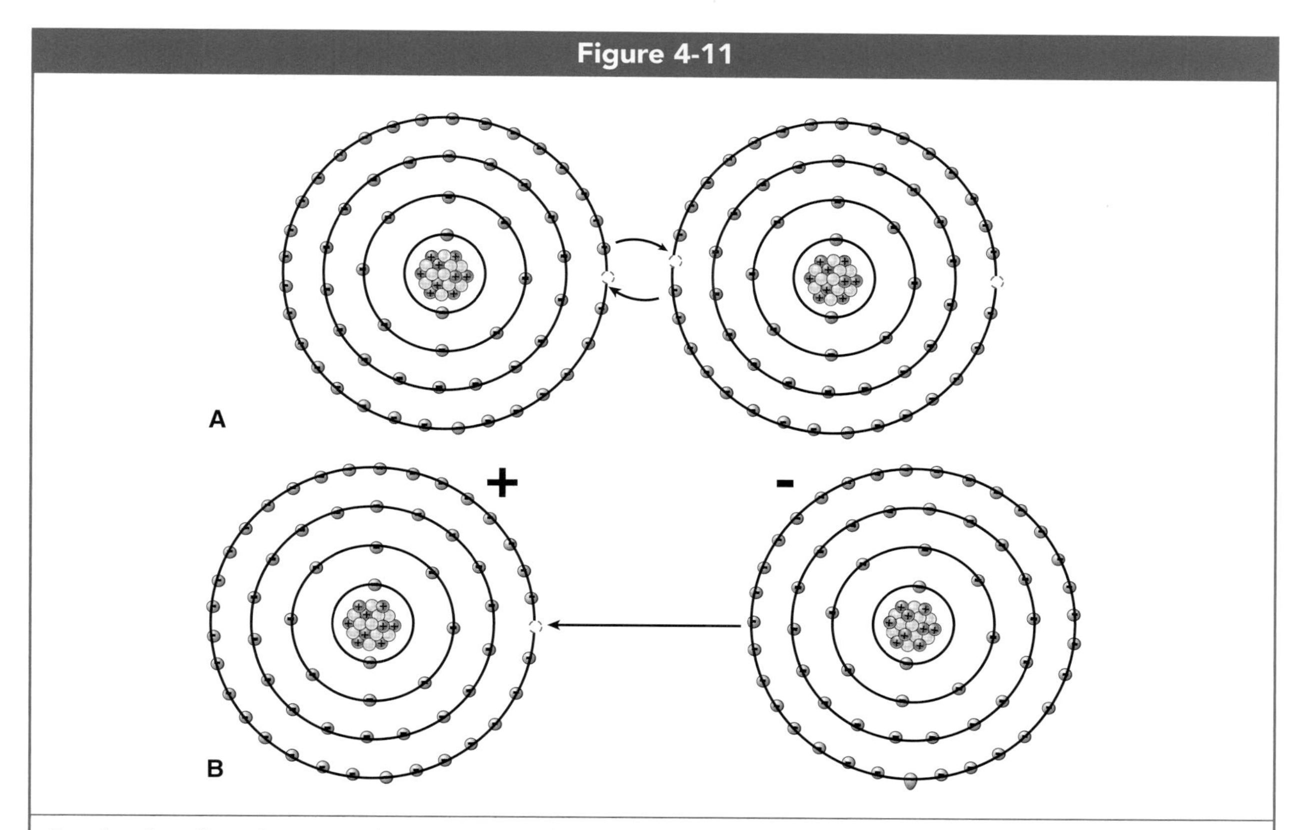

Covalent bonding, ***A***, occurs when two atoms *share* a pair of electrons to fill a suborbital. Ionic bonding, ***B***, occurs when two atoms have opposite net electrical charges due to an imbalance between the protons in the nucleus and the electrons in the shells. Because it involves an electrical charge on the entire atom, the force generated can pull atoms together from a distance, and the bond formed is much stronger than a covalent bond.

When additional energy is imparted to a physical object in orbit around the earth, such as a satellite, it will use the extra energy to speed up in its motion. The increased centrifugal force will cause it to "sling" out of orbit. A specific shell or orbit around the nucleus of an atom is synonymous with a certain energy level. If one of its electrons acquires extra energy, it cannot remain in that shell or orbit, but is raised up out of it just as heated steam raises up out of water. This is the proper way to understand Figure 4-12.

The ionization of an atom by an x-ray results in the creation of an *ion pair*, since both the emitted electron and the atom left behind then possess electrical charges, one negative, one positive. Both ions can lead to chemical changes. The ionized atom may form an *ionic bond* with an atom of opposite charge. The electron may be acquired by another positively-charged atom, neutralizing its charge, and thereby

Figure 4-12

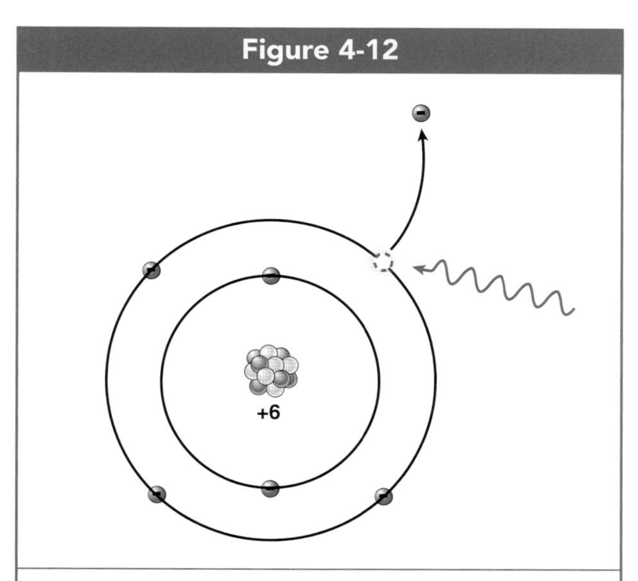

Ionization of an atom by an x-ray, leaving the atom with a net positive charge and making it an *ion*.

breaking its bond with another atom. New chemical compounds form, which can be the basis for diseases such as cancer.

When numerous ionizations occur, freeing up many electrons, those electrons can be attracted to a positively-charged plate and then flow down a wire, creating electrical current. This is the fundamental basis of most imaging and detection devices for x-rays, including modern digital image receptors.

Ionization can also occur from physical collisions between particles. Indeed, this type of ionization is one of the processes that takes place in the x-ray tube anode during the production of the x-ray beam. A high-speed electron from the hot cathode filament is accelerated toward the anode disc. When this electron smashes into the disc, it can physically collide with an orbital electron in a tungsten or rhenium atom (Fig. 4-13) causing it to ricochet out of the atom. This ionizes the atom, momentarily leaving it with a positive charge. In this particular case, the ionizing event is caused by two physical objects colliding just like two pool balls, one at high speed (the cue ball) causing the other to be moved out of its place.

Figure 4-13

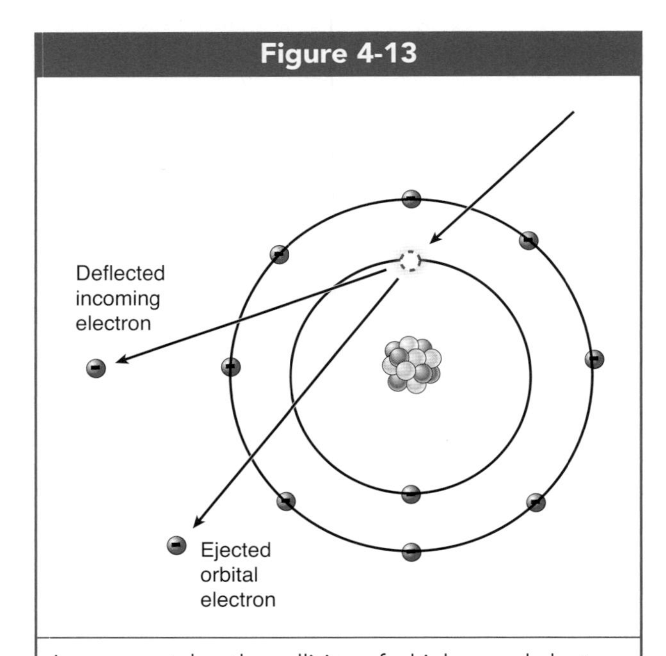

In an x-ray tube, the collision of a high-speed electron from the cathode with an orbital electron in the anode causes it to ricochet out of its atom, thus ionizing the atom.

STRUCTURE OF THE NUCLEUS

All of the large particles comprising the atomic nucleus are collectively referred to as *nucleons*. These include the positively-charged protons and the neutrons with their neutral state of electric charge. As we progress up the periodic chart to "larger and larger" atoms, we find that in the nucleus the normal number of neutrons increases more quickly than the number of protons. The first element, hydrogen, normally consists of a single proton and *no* neutron. Helium normally has an equal number of protons and neutrons, two of each. Balanced numbers continue up the chart until we reach chlorine (Z# = 17), where we begin to see more neutrons than protons in the average nucleus for that element. Nickel has 2 "extra" neutrons, and copper 5. Bromine typically has 35 protons, but 44 neutrons. The ratio of neutrons to protons continues to increase exponentially: Lead averages 125 neutrons to only 82 protons, and the most common form of uranium has 146 neutrons combined with only 92 protons. Why does this progression occur?

Remember that protons, having like positive charges, repel each other electrically. It takes more nuclear force to hold them together because this electrical resistance must be overcome. Neutrons do not repel each other, so less energy is needed to hold them together.

The most stable state for the atomic nucleus is its *lowest energy* state. For a given total number of nucleons, it takes less energy to hold a large nucleus together if there are more neutrons than protons, than it does if their numbers are equal. Therefore, the most stable state for a large nucleus to be in is to have a predominance of neutrons.

The sum total of protons plus neutrons, that is, the number of all *nucleons*, is abbreviated *A*. This is called the *atomic mass*.

(You may note that the periodic chart gives *atomic weights* under the symbol for each element, and that these are not whole numbers, but include decimal points, making them *fractions*. This is because there are additional exchange particles in the nucleus, collectively called *gluons* (!), mediators of the strong nuclear force, which flow between all the protons and neutrons to hold the nucleus together. Gluons add to the actual weight of the nucleus. Sometimes the terms

Figure 4-14

In nuclear *fission*, a high-speed neutron can split an atomic nucleus into two, releasing large amounts of energy.

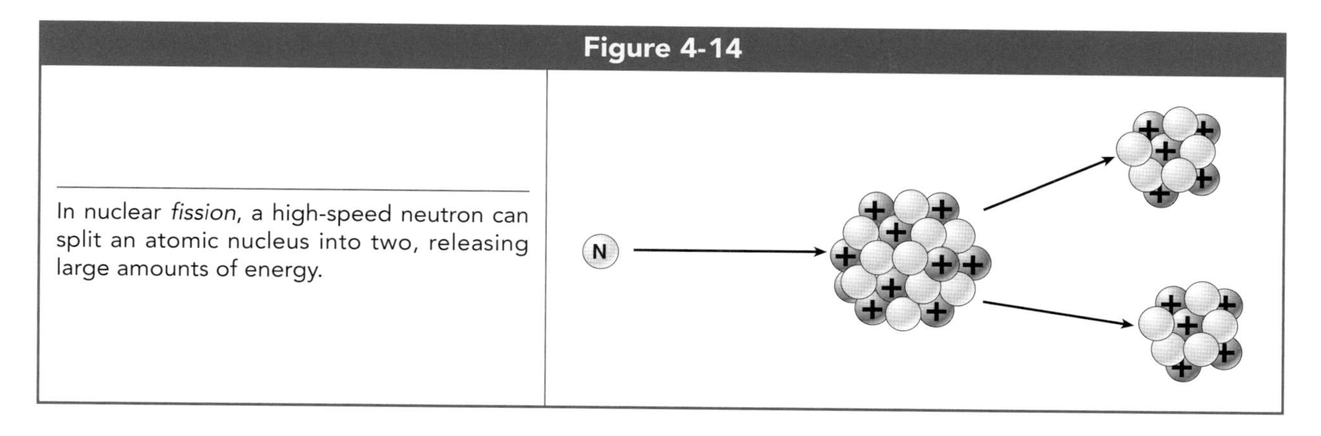

atomic weight and *atomic mass* are used interchangeably, but strictly speaking, they are not the same thing. For our purposes here, we are interested only in the atomic mass, which is always a whole number.)

Each element can be found in different forms based on altering the number of neutrons present. These variations are called the *isotopes* of each element. A kind of shorthand notation has been developed for use in discussing isotopes: The abbreviation for the element (from the periodic chart) is written with the *A* number, the atomic mass, in superscript above it. This *A* number may appear either in front of or behind the element symbol, but must be above in superscript, as follows:

$$X^A \quad \text{or} \quad {}^{A}X$$

Where "X" is the element: For example, C^{14} (or ${}^{14}C$) is the shorthand for the isotope carbon 14. The most common form of carbon has an *A* number of 12 (representing the sum of 6 protons and 6 neutrons). C^{14} is an isotope of carbon with the unusual number of 8 neutrons, or two "extra" neutrons, making the total number of nucleons 14. But, the element is still identified as carbon, as long as the number of *protons* remains at 6.

The number of neutrons within the nucleus is abbreviated *N*. We have learned that the *atomic number* is the number of protons and is abbreviated Z. If the atomic mass and the atomic number are known (from the periodic chart), one can find the neutron number by the formula $A - Z = N$, (total nucleons – protons = neutrons).

In shorthand format, the atomic number *Z* is always written in front of the element in subscript (at the bottom). The format is:

$${}^{A}_{Z}X$$

In this format, showing both the *A* and the *Z* numbers, carbon 14 is written as:

$${}^{14}_{6}C$$

From the notation for this isotope of carbon, how is the number of neutrons determined? The answer is by subtracting $A - Z = 8$. Since both the *A* and the *Z* numbers are written into the symbol, the *N* number is implied by these shorthand notations, and is not necessary to write out.

Nuclear *fission* describes the "splitting" of atomic nuclei into smaller fragments, Figure 4-14. In nuclear reactors, this is accomplished by striking those nuclei with high-speed neutrons. This must be done with extreme energy in order to overcome the nuclear binding energy that holds the nucleus together. Nuclear *fusion* is the forcing together of two smaller nuclei to form a single, larger nucleus (Fig. 4-15). This is what occurs in our sun and in other stars, as

Figure 4-15

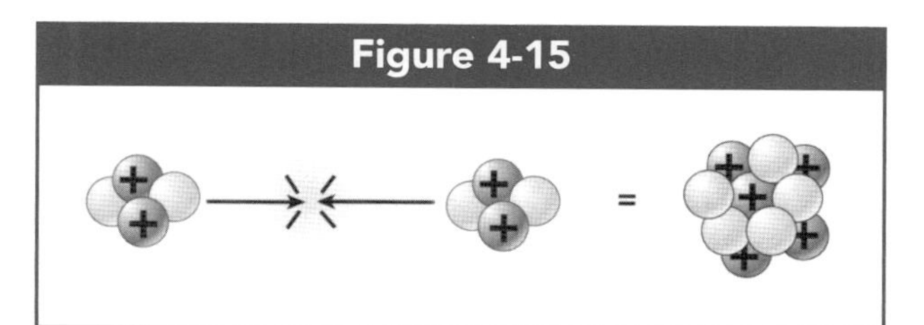

In nuclear *fusion*, two nuclei are brought together with such force that they stick together, forming a larger nucleus. To accomplish this, the electrical force must be overcome so that the strong nuclear force can take over. Large amounts of energy are released, as evidenced by our sun.

hydrogen nuclei are fused together to make helium atoms. Again, extreme energies are required—the mutual electrical repulsion of protons must be overcome to force these nuclei to "stick" together. Both processes, fission and fusion, result in the release of large amounts of energy that can be used for power.

RADIOACTIVITY

We have indicated that for each particular element, or for a particular number of protons, there is an "ideal" number of neutrons to have in the nucleus which results in the most stable configuration, positioning the nucleus as a whole at its lowest possible energy state. This is called its *ground state*, and all nuclei seek out that lowest energy state, just as all objects seek out their lowest point in a gravity field. It is possible for an atom to have too few or too many neutrons in its nucleus. It becomes *unstable*.

When an atomic nucleus becomes unstable, it means that it is spending too much energy trying to keep the nucleons together. It is in a "hyper" state, and its tendency is to get rid of the excess energy. This energy will be emitted in the form of radiation, either particulate or electromagnetic. The nucleus seeks a more stable, lower state of energy. If it emits a particle, it loses mass, so it no longer has so much "weight" to hold together. But, remember that mass and energy are interchangeable, so losing mass is equivalent to losing energy. It may also emit excess energy in the form of an electromagnetic wave, such as a gamma ray. Any *unstable* nucleus is also a *radioactive* nucleus.

We define a *radionuclide* as any atom which is radioactive. We define an *isotope* as any atom with an unusual number of neutrons for that element. A *radioisotope*, then, is any atom which is radioactive *because* it has too few or too many neutrons.

Any number of neutrons *different* from the usual is not necessarily too few or too many. For each element, there is a *range* of neutrons that can be held within the nucleus without making it unstable or radioactive. Hydrogen normally has only a single proton and no neutrons (Fig. 4-16A). It may, however, have a single neutron join its proton without becoming unstable. This form of hydrogen is called *deuterium*

Figure 4-16

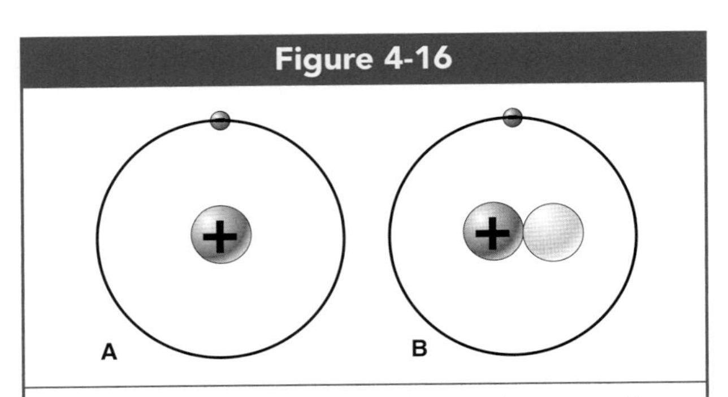

Normal hydrogen consists of a single proton for a nucleus, ***A***. Deuterium, ***B***, has a neutron added into the nucleus, but is a stable *isotope*.

because it has a total of 2 nucleons (Fig. 4-16B). Deuterium is a simple isotope of hydrogen, not a radioisotope, because it is stable.

It is also possible for hydrogen to acquire 2 neutrons for a total of 3 nucleons (Fig. 4-17). This substance, called *tritium*, is a radioisotope rather than a simple isotope. It is *unstable* because 2 neutrons is too many for this nucleus to "hang onto" and remain stable. Due to its radioactivity, it is used in the "hydrogen bomb." Very "large" atoms have a larger range of stability—they may be able to lose one or two neutrons, or gain three or four neutrons, and all of these isotopes of that element may be stable.

The loss of any mass or energy from the nucleus of an atom is referred to as *radioactive decay*. As will be shown shortly, this decaying process can alter the number of *protons* in the nucleus, thus changing the actual identity of the element. The term *transmutation* describes the changing of one element into another by radioactive decay.

Figure 4-17

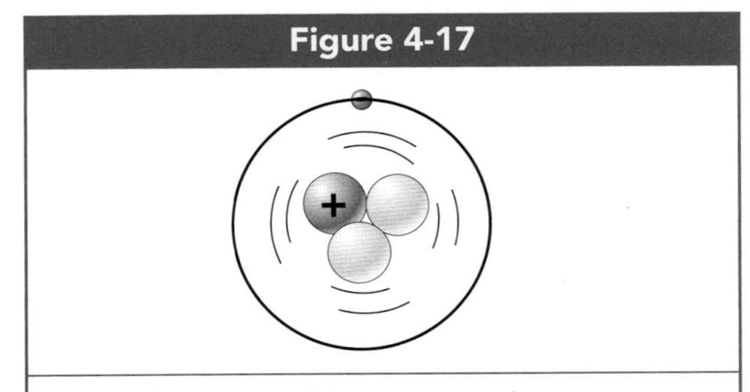

Tritium is a form of hydrogen with two neutrons added into the nucleus. It is unstable and emits radiation as a *radioisotope*.

Radioactive elements seek a more stable nuclear configuration by emitting radiation. Specifically, there are three types of naturally occurring radiation which all allow the nucleus to lose some of its energy or mass; they are *alpha*, *beta*, and *gamma* radiation.

Alpha radiation consists of large *alpha particles* which are emitted by very unstable nuclei. Each alpha particle consists of two protons combined with two neutrons. This is a large "chunk" of the nucleus that breaks off (Fig. 4-18). Since it carries two protons away from the nucleus, the *Z* number of the element drops by two. This transmutation leads to a new element that falls *two columns to the left* on the periodic chart of elements (Fig. 4-19).

As an example, the radioactive element *uranium* can undergo a whole series of transmutations, each of which moves us two columns to the left on the periodic chart, from alpha emission. The uranium (Z = 92) becomes radioactive thorium (Z = 90), which emits another alpha particle to become radium (Z = 88). Radium decays into radon gas (Z = 86) and then radioactive polonium (Z = 84). Finally, the polonium gives off an alpha particle to become lead (Z = 82) which is stable.

Note that the emission of an alpha particle also reduces the *A* number, the atomic mass of an atom, by 4. These changes in the nucleon count are distinctive of alpha emission—that is, we know that an alpha particle has been emitted when the A number drops by 4 and the Z number drops by 2.

When a nucleus is moderately unstable, the most efficient means of reaching its ground state may be to emit *beta radiation*, which consists of small particles identical to electrons but traveling at very high speed. We are used to thinking of electrons as only being in the shells of an atom—how can *electrons* originate from the *nucleus*? Remember that a neutron is slightly heavier than a proton, and may be thought of as consisting of a proton plus an electron whose negative and positive charges cancel each other out. A neutron can decay by separating these charges. Since the negative charge is carried by a much smaller, lighter particle, the electron, it is easier to move than the huge proton. Therefore, the heavy proton remains behind in the nucleus, and the "newly created" little electron flies out of the nucleus at high speed (Fig. 4-20). This is beta radiation.

SIDEBAR 4-1: The *alchemists* of medieval times mixed various chemicals together in an attempt to create gold by *chemical* transmutation. This never worked because all chemical interactions are based upon the configuration of outer shell electrons rather than the nucleus. It *is* hypothetically possible to create gold by nuclear transmutation from radioactive decay, but is not economical nor practically feasible for physicists to do this.

Now let us consider the transmutation effects of beta emission. A neutron has decayed into a proton. In counting the numbers of particles within the nucleus, it is as if a neutron has been *exchanged* for a proton. Although the neutron number will go down by 1, the Z number *which identifies the element* will increase by 1 from the gain of a proton. The element actually climbs *up* the periodic chart to the next column on the right (Fig. 4-21). These changes in the nucleon count are distinctive of beta emission—we know that a beta particle has been emitted when

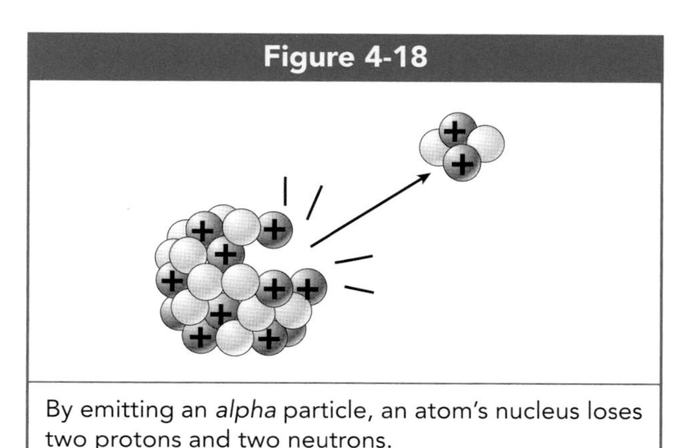

Figure 4-18

By emitting an *alpha* particle, an atom's nucleus loses two protons and two neutrons.

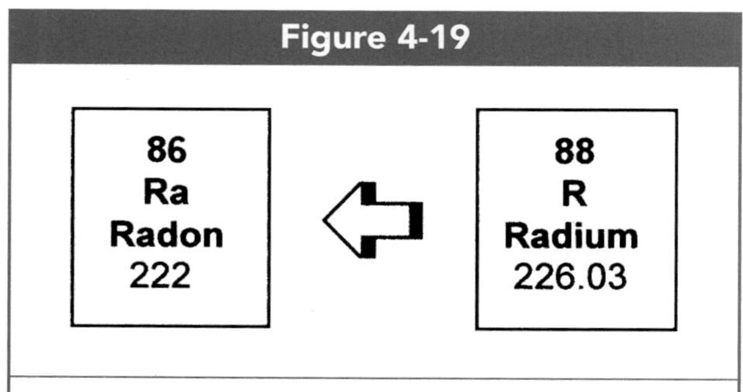

Figure 4-19

With alpha emission, an element is *transmutated* into the element two columns to the left in the periodic chart.

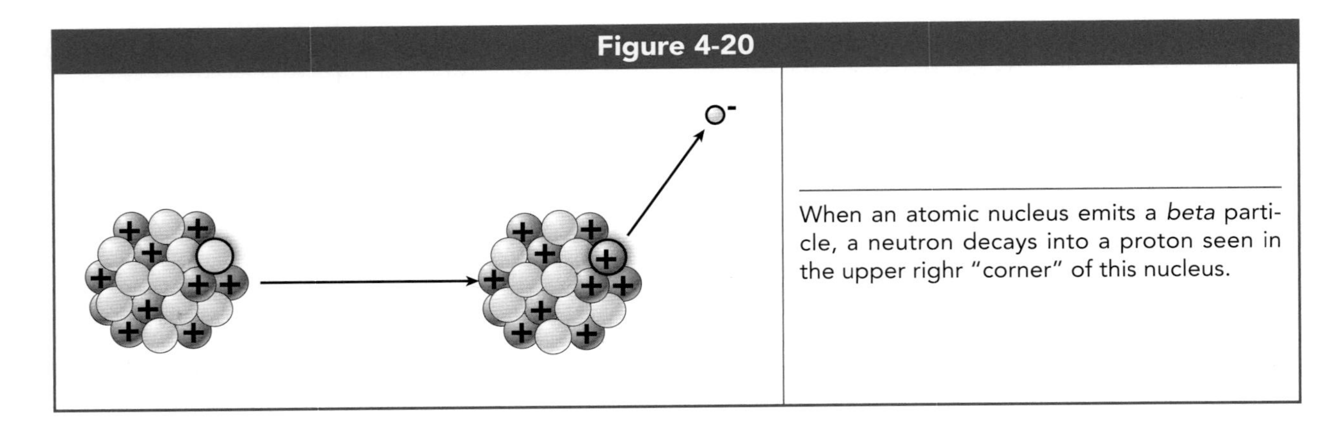

Figure 4-20

When an atomic nucleus emits a *beta* particle, a neutron decays into a proton seen in the upper righr "corner" of this nucleus.

the Z number increases by 1 and the N number drops by 1.

Tritium, the isotope of hydrogen shown in Figure 4-17, provides a simple example of transmutation by beta emission. This isotope is unstable and radioactive, and seeks a lower energy state for its nucleus. It does this by emitting a beta particle, shown in Figure 4-22. Note that in doing so, it becomes a new element, helium, since the proton count has gained one. On the periodic chart, helium is in the next column to the right from hydrogen.

The resulting nucleus in Figure 4-22 looks unusual, since there is only one neutron with two protons. It *is* unusual, because normally, helium has two of each. This is an isotope of helium called helium-3, abbreviated ^{3}He, which is "missing" a neutron when compared to its most common form. Nonetheless, this configuration of the nucleus is at a lower energy state than the previous hydrogen-3. The beta particle has removed some mass and energy from the nucleus, leaving it in a more stable state than before.

Figure 4-21

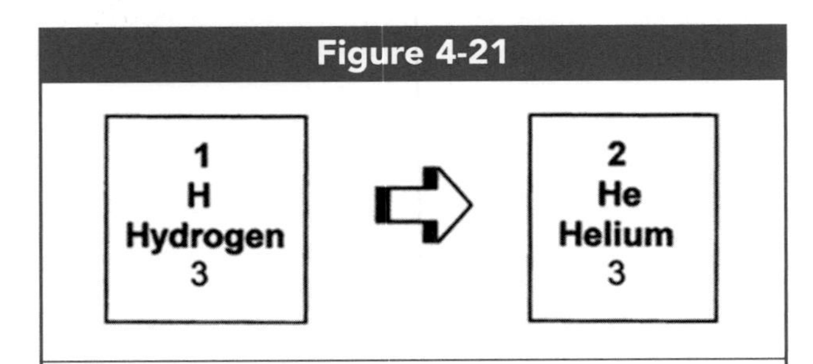

Tritium decay: By beta emission, an element is transmutated into the element one column to the *right* in the periodic chart.

A moderately unstable atomic nucleus may also emit radiation in the form of *gamma rays*. A gamma ray is not a particle, but a form of electromagnetic radiation. It consists of a disturbance or wave within an electrical or magnetic *field*, such as the pull you might feel from a magnet. This wave, of itself, has no mass or weight. It does not carry any particles away from the nucleus, only excess energy. For gamma emission, there is no change in either the atomic mass *A*, nor in the atomic number *Z*, of the atom. Transmutation does not occur. Only energy is lost in the form of a wave (Fig. 4-23).

It is possible for an atom to simply have too much energy contained within its nucleus. This frequently occurs after an alpha particle has been emitted, and there is some "leftover" nuclear energy which is no longer needed to hold the remaining configuration of the atomic nucleus together. Remember that particles are held in the nucleus by the strong nuclear force, a kind of nuclear *glue*. After a very radioactive atom emits an alpha particle (2 protons and 2 neutrons), the amount of this "glue" left behind in the nucleus is slightly more than necessary to hold together the remaining nucleons.

An element in this state is referred to as an *isomer*, designated by a small superscript "m" after its abbreviation. An example is technetium 99m, abbreviated Tc^{99m}, which is commonly used in nuclear medicine. Isomers always become more stable by emitting a *gamma ray* in order to dispose of this excess energy. There is no change in the physical make-up of the nucleus—the number of both protons and neutrons remains the same. The gamma ray is simply an electromagnetic wave, with no mass, which carries away energy just as infrared radiation carries

Figure 4-22

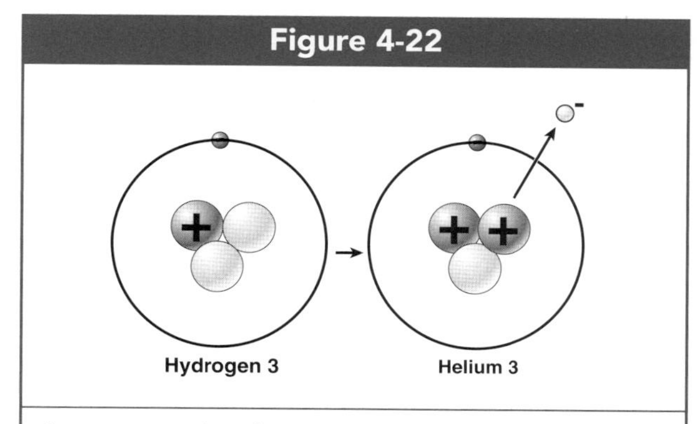

As an example of transmutation by beta emission, tritium, an isotope of hydrogen, becomes helium-3. Even though this is an isotope of helium, it is still more stable than tritium.

Figure 4-23

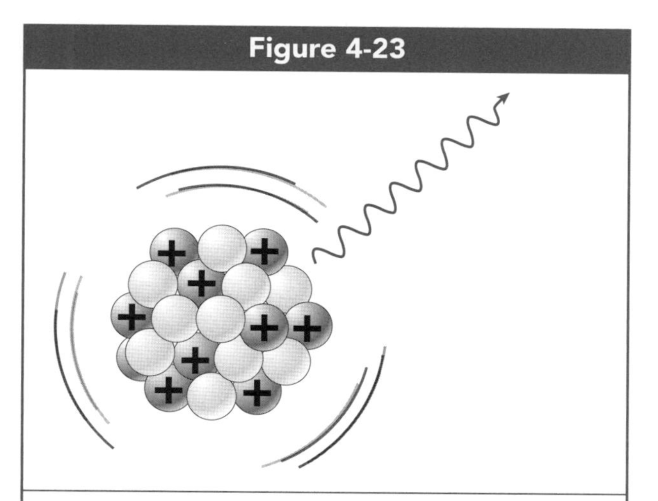

In *gamma* emission, an atomic nucleus rids itself of excess energy in the form of an electromagnetic wave. Transmutation does not occur since there is no change in the particles making up the nucleus.

heat energy away from a hot object (Fig. 4-23). Thus, the energy contained within the nucleus drops to a more stable state.

SUMMARY

1. Chemical compounds are composed of more than one element chemically bound together. Any single unit of two or more atoms chemically bound together is a molecule.
2. Atoms are composed of a nucleus of protons and neutrons, with electrons widely spaced in volumes of space called *suborbitals* or *orbitals*. The number of protons (Z) defines the element.
3. The neutron is effectively composed of a proton plus an electron, and can radioactively decay into these particles.
4. The chemical behavior of atoms is based on the configuration of electrons in the outermost shell. The electron capacity of any shell is determined by the $2N^2$ rule and the octet rule. The $2N^2$ rule also determines the *law of periodicity* by which the periodic chart of the elements is organized.
5. Atoms can be chemically bound in four ways, the most important of which are *covalent bonding* in which a pair of electrons are "shared," and *ionic bonding* in which a single electron is donated to a positively-charged atom.
6. *Ionization* of atoms can occur either from certain electromagnetic waves such as x-rays, or from physical collisions of particles, and results in chemical changes that can cause disease.
7. *Isotopes* are atoms with an unusual neutron (*N*) number, radioisotopes are unstable and radioactive because this number is outside the range that nuclear energy can hold together. The *atomic mass* (*A*) is the total number of nucleons (protons plus neutrons).
8. Unstable atoms emit alpha, beta, or gamma radiation, the first two of which *transmutate* the element, moving it up or down the periodic chart.

REVIEW QUESTIONS

1. How many *atoms* are there in a molecule of $NaHCO_3$?

2. What is the *Z* number of iodine?

3. What is the abbreviation for tungsten?

4. The horizontal rows in the periodic table of the elements are called:

5. In the periodic chart of elements, the group numbers are based on:

6. What is defined as the volume of space within an atom in which a *pair* of electrons must reside?

7. Compared to an electron, how much more massive is a neutron?

8. Within each shell of an atom, which of the two rules for electron configuration always takes precedence?

9. In an atom of calcium, how many electrons are there in the *N* shell?

10. Which type of chemical bonding between atoms occurs because of a "defect" in the distribution of negative charge that forms a smooth screen around the atom?

11. Why is an ionized atom able to affect other atoms from a substantial distance?

12. In the anode disc of an x-ray tube, what "knocks" electrons out of their normal shells in the tungsten atoms?

13. What is defined as an atom which is unstable due an *N* number well outside the normal range?

(Continued)

REVIEW QUESTIONS *(Continued)*

14. What is the atomic *mass* of $^{131}_{53}I$?

15. How many neutrons are there in $^{44}_{20}Ca$?

16. What is defined as the "splitting" of an atomic nucleus?

17. When a beta particle is emitted from a radioactive atom, what happens to its atomic number?

18. When gamma radiation is emitted from a radioactive atom, what happens to its *A* number?

19. As a result of an alpha particle being emitted from an atom, which way and how far does the atom move on the periodic chart?

20. The lowest energy level possible for a particular atomic nucleus is called its:

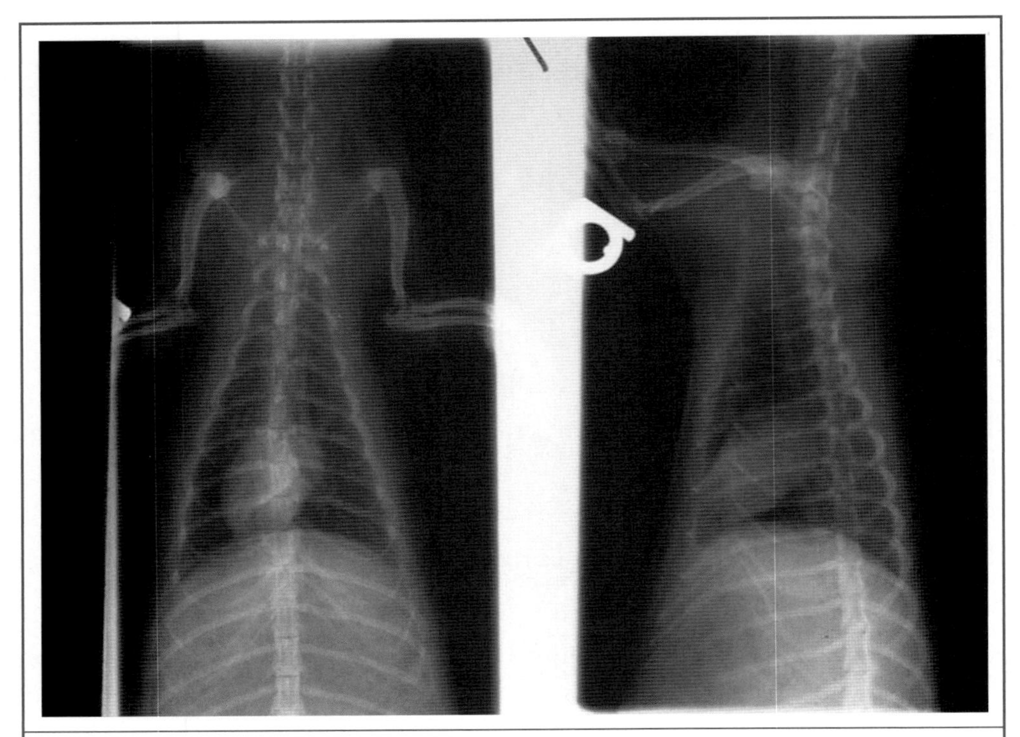

Chest radiographs on a ferret.

Chapter 5

ELECTROMAGNETIC WAVES

Objectives:

Upon completion of this chapter, you should be able to:

1. List and define the four common characteristics of all waves. Define transverse and compressional waves.
2. Solve wave formula problems for frequency, wavelength and speed.
3. Solve electromagnetic wave formula problems for frequency and wavelength.
4. Combining the Planck formula and the wave formula, solve for minimum x-ray wavelength and for kVp.
5. Explain how electromagnetic waves are created, and their magnetic and electric components.
6. List the eight types of radiation comprising the electromagnetic spectrum, landmark energies and wavelengths, and the energy and wavelength ranges for diagnostic x-rays.
7. Overview the use of waves in MRI and sonography, and for lasers.
8. Discriminate between the characteristics and behavior of light waves and x-rays.
9. Describe how the *differential absorption* of x-rays results in radiopaque and radiolucent portions within a radiographic image.
10. Overview the *dual nature* of subatomic particles and waves, and how their *quantum* characteristics can best be visualized in the x-ray beam.

WAVES

A wave is a disturbance in any medium (such as water), which transports energy from one place to another without causing any permanent change in the medium itself. The sound of your voice consists of organized waves traveling in the medium of air. When you finish talking, the disturbance of the air molecules subsides and the air remains as it was before, with no permanent change.

There are two general types of waves: transverse, and longitudinal or compressional. For a *transverse* wave, the displacement of the medium is *perpendicular* to the direction the wave is traveling. This is shown in Figure 5-1. A wave in water is a transverse wave: Although the wave travels horizontally, the water itself is displaced up and down. We call the upward displacement of the water *crests*, and the downward displacement *troughs*. Since the upward and downward displacement of the crests and troughs is vertical, it is perpendicular to the horizontal direction in which the wave travels.

For a *longitudinal* or *compressional* wave, the displacement of the medium is *parallel* to the direction the wave is traveling (Fig. 5-2). Sound is a compressional wave—it can be defined as an organized series of compressions and expansions of air molecules traveling from the speaker to the listener. As the waves travel to the listener, so do the expansions and compressions, in the same direction. Thus, these displacements of the air molecules are parallel to the direction the waves are traveling.

All waves have four measurable qualities about them which the radiographer should understand. These four characteristics are:

Figure 5-1

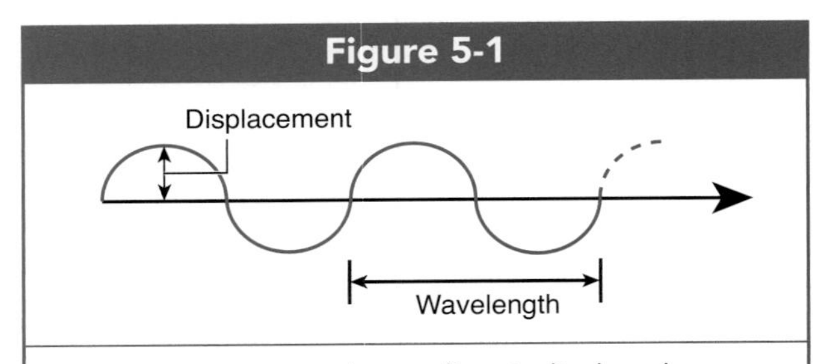

In a transverse wave, the medium is displaced *perpendicular* to the wave's direction of travel. In a water wave, the water is displaced up and down, while the wave moves horizontally.

1. Speed
2. Amplitude
3. Wavelength
4. Frequency

The *amplitude* of any wave is the maximum displacement of the medium from its equilibrium position. For a wave in water, you may think of this as how *high* each wave crest is from where the level water would lie perfectly still and undisturbed. Note that this is not the distance from the bottom of a trough to the top of a crest, but the maximum distance of a single trough or crest from the undisturbed water level.

The amplitude of a compressional wave is more difficult to visualize because it is measured in the same direction the wave is traveling. For a sound wave in air, the amplitude is the maximum displacement of molecules in a compression or in an *expansion from where they would be if the air were undisturbed.*

Amplitude is associated with how "strong" a wave is. A taller water wave carries more water and strikes

Figure 5-2

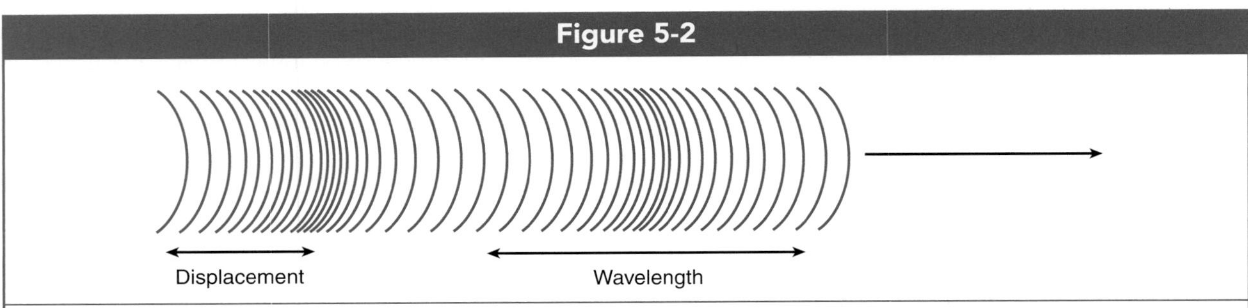

In a *longitudinal* or *compressional* wave, the medium is displaced parallel with the wave's direction of travel. For sound in air, the compressions and expansions of air molecules move along with the wave.

Figure 5-3

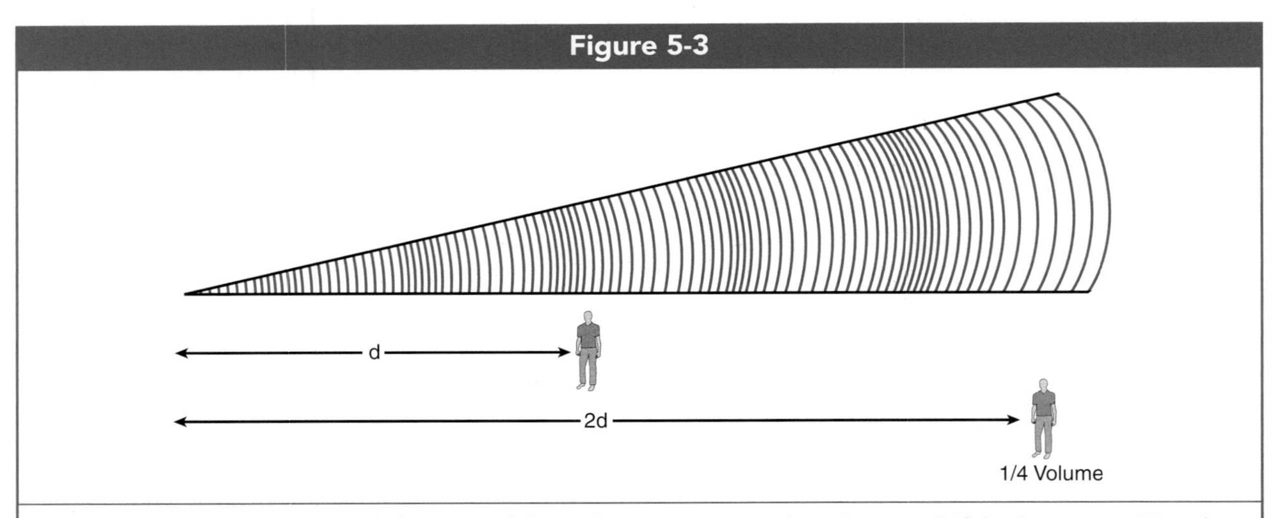

The loss of intensity or *amplitude* for waves follows the inverse square law. For sound, if the listener positions herself twice as far away from the source, the sound will be ¼ as loud.

us with more force. For an x-ray beam, the amplitude relates to the intensity of the beam (controlled by the mA). For the sound waves from your stereo, amplitude translates into *volume* or loudness. As waves spread out at increasing distance from their source, their amplitude drops. Water waves lose height as they spread. Sounds are not as loud at greater distances. This loss of amplitude follows the same *inverse square law* as x-rays do (Fig. 5-3). For each doubling of the distance traveled, the amplitude drops to one-quarter of the original.

(It is interesting to note that whenever waves can be *prevented* from spreading out, their amplitude will be maintained. When your stereo speaker is placed at one end of a hallway with all of the doors shut, the loudness of the music can be transmitted undiminished to a room at the other end, because the narrow hallway prevents the sound waves from spreading out.)

The *wavelength* of any wave is defined as the distance between two like points, such as the distance from the top of one crest to the top of the next crest. For a compressional wave, it can be measured from the mid-point of one compression to the mid-point of the next compression. Any starting point along the wave form can be used, as long as the length is measured to the next identical point (Fig. 5-4).

One completion of the wave form before it repeats itself is called a *cycle*. Each cycle consists of two *pulses*, one positive and one negative. For a transverse wave, the positive pulse would be a "crest," and the negative pulse would be a "trough," as shown in Figure 5-5. For a compressional wave, the positive pulse is a compression, and the negative pulse is an expansion.

The term "cycle" has the same origin as the word "circle." Figure 5-6 shows why: If we start at the beginning of a crest for a transverse wave, draw one complete cycle, and then slide the negative pulse to the left under the positive pulse, trough under crest, we get a diagram of a circle. Each cycle in a series of waves completes one circle or one waveform.

The *frequency* of a series of waves is defined as the number of cycles that pass by a given point *each second*. If you are standing in the water at a beach, the frequency is the number of crests in the water that strike you each second. The physics unit for frequency is the *hertz*, abbreviated *Hz*, and defined as one cycle per second.

$$1 \text{ Hz} = 1 \text{ cycle / second}$$

Now, clearly, the frequency of a series of waves will depend upon both the speed of the waves and their wavelength. They are all interrelated. For example, if the waves at the beach come in at higher speed, more of them will strike you per second. However, this is not the only way to get a higher frequency—the other way would be to make the *wavelengths shorter*. That is, if the waves are traveling at the same speed, but the waves themselves are shorter waves, more of

Figure 5-4

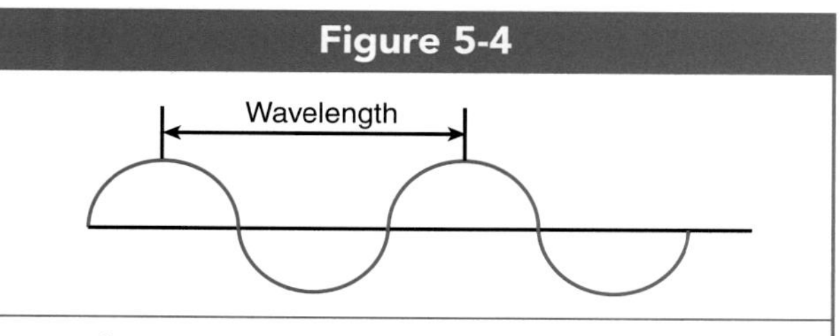

Wavelength is measured between two similar points along the waveform.

Figure 5-5

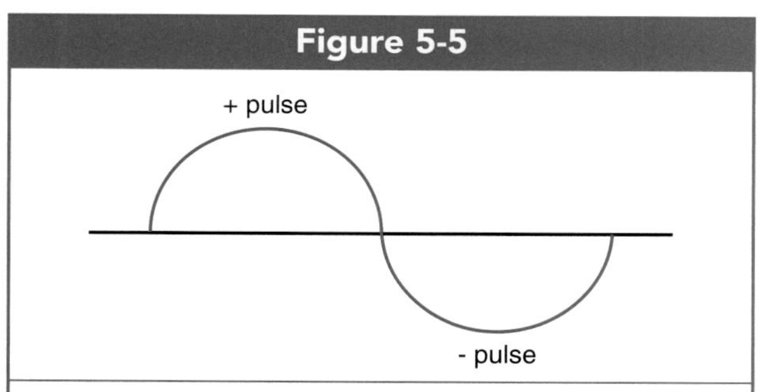

For a transverse wave, each full cycle consists of a positive *pulse*, the "crest," and a negative pulse, the "trough."

Figure 5-6

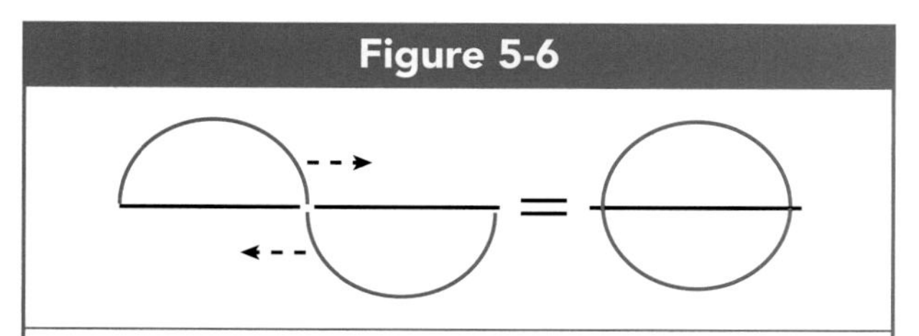

The term *cycle* is derived from the fact that when a positive pulse and a negative pulse are superimposed, they form a complete circle.

Figure 5-7

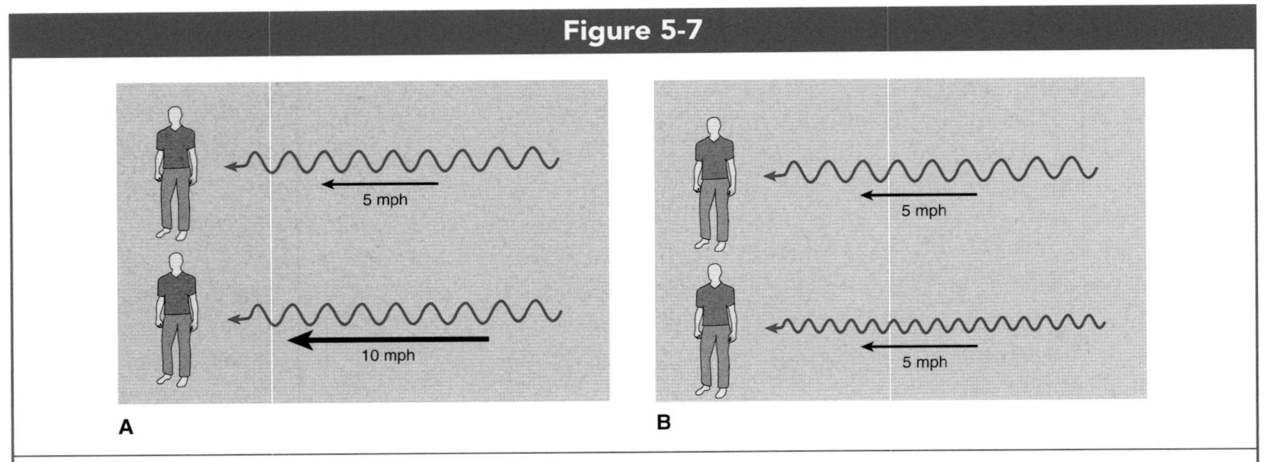

There are two ways to increase the *frequency* of waves: By increasing their speed, ***A***, or by shortening the wavelengths while the speed is maintained, ***B***.

them will strike you per second (Fig. 5-7). These relationships between speed, frequency, and wavelength are all summed up in the *wave formula*:

$$v = f\lambda$$

where v is the velocity or speed, f is the frequency, and λ represents the wavelength. (The Greek letter *lambda* is written λ and is equivalent to the English letter "L," which you may associate with "length" or wavelength.) The formula states that the velocity of the waves will always equal the product of their frequency times their wavelength.

Let's take a minute to translate the implications of this simple formula into English: If one of the variables is *fixed*, that is, it cannot be changed, you can immediately see the relationship between the other two variables by covering up the fixed one with your fingertip.

Figure 5-8

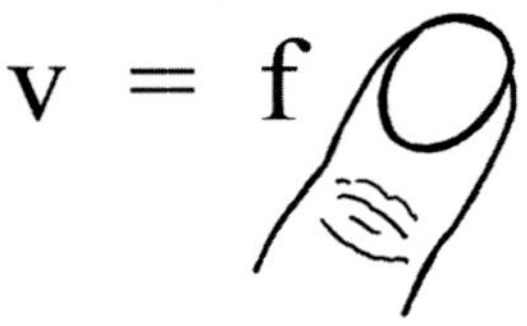

$$v = f$$

To see the relationship between velocity and frequency in the wave formula, cover the wavelength λ with your finger. This can be done with other formulas. Here, it shows that for waves, frequency is *proportional* to velocity.

For example, for a given fixed wavelength, what is the relationship between velocity and frequency? Covering up the λ in the formula with your fingertip (Fig. 5-8), you see that velocity and frequency are *directly proportional*. If the speed is doubled, so must be the frequency. That is, if the waves are traveling twice as fast, then twice as many of them will strike you per second. You can visualize this intuitively. *Speed and frequency are directly proportional to each other.*

For a given fixed frequency, what is the relationship between velocity and wavelength? Covering up the *f* in the formula with your fingertip, you see that velocity and wavelength are also *directly proportional*. If the speed is doubled, in order to have the same number of waves striking you per second they would have to be twice as long in order to cancel out the speed.

For a given fixed speed, what is the relationship between frequency and wavelength? Covering up the *v* in the formula with your fingertip, imagine it to be the number "one." You see that frequency and wavelength, when multiplied, must always equal "1." They are *inversely proportional* to each other: If frequency is doubled, the wavelength must be cut in half. If the wavelength is doubled, then the frequency must be cut in half. That is, if the waves are twice as long, yet traveling the same speed, then half as many of them will strike you per second. You can also visualize this intuitively. *Wavelength and frequency are inversely proportional to each other.*

To find the speed of a series of waves, multiply the frequency and the wavelength:

$$v = f\lambda$$

To find the frequency of a series of waves, divide the velocity by the wavelength:

$$f = v / \lambda$$

To find their wavelength, divide the velocity by the frequency:

$$\lambda = v / f$$

In memorizing this relationship, a helpful pattern to take note of is that the speed v is never divided *into* something, that is, v always stays on top of the equation no matter what you are solving for. When setting these problems up, a helpful visual aid is the *T-triangle* shown below, a triangle with a "T" placed inside it, which is then filled in with "v" in the upper compartment and "f" and "λ" in the two lower compartments.

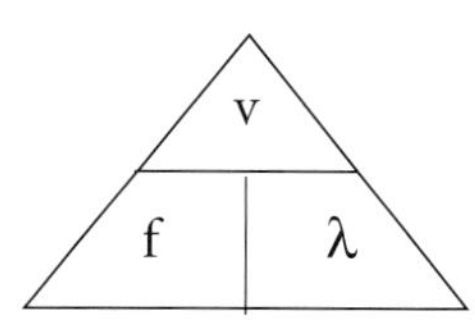

Simply cover up the factor you are solving for with the tip of your finger, and the diagram shows you how to set up the solution. For example, when solving for the frequency, cover up the "f" in the T-triangle and you see that v is to be divided by λ. To solve for the velocity, cover it up and you see that f and λ are side-by-side, indicating that they are to be multiplied. Let's try a couple of practice exercises:

Practice Exercise #1:

Standing in the water at the beach, you note that one wave strikes you every 2 seconds. You estimate that the crests of the waves are about 5 feet apart. What is the *speed* of this series of waves in feet-per-second?

Solution: By covering the "v" in the T-triangle, we see that the solution is to simply multiply the frequency times the wavelength. However, a conversion must be made first: Note that the frequency is given as one wave every 2 seconds. The units here are actually "seconds per wave" rather than "waves per second." Remember that the unit for frequency, the Hertz, is defined as "cycles per second." Frequency must always be stated in the unit Hertz.

We can make this conversion either logically or mathematically: Logically, we would ask, "If one wave hits me every 2 seconds, what fraction of that wave hits me every second?" The answer is one-half of a wave or 0.5.

Mathematically, this would be set up as a simple ratio, stating, "1 wave per 2 seconds equals how many waves per second," as follows:

$$\frac{1 \text{ wave}}{2 \text{ seconds}} = \frac{X \text{ waves}}{1 \text{ second}}$$

$$\frac{1}{2} = \frac{X}{1}$$

Cross-multiplying: $2X = 1$

Dividing by 2: $\frac{2X}{2} = \frac{1}{2}$

$X = \frac{1}{2}$ ***or 0.5 Hz***

Now, following the T-triangle, we can simply multiply 0.5 Hz times the wavelength of 5 feet, to obtain the velocity:

Answer: The speed of the waves is 2.5 feet per second

Practice Exercise #2:

The waves at the beach are traveling at a speed of 2 meters per second. The crests of the waves are 20 meters apart.

A. What is the *frequency* of these waves—how many waves will strike you each second?
B. How many seconds will there be between each wave striking you?

Solution A: Covering the "f" in the T-triangle, we see that the solution is to divide the speed by the wavelength, as follows:

$$\frac{2 \text{ meters per second}}{20 \text{ meters}} = \text{Frequency in Hz}$$

$$\frac{2}{20} = 0.1 \text{ Hz}$$

Answer A in Hertz: 0.1 Hz
That is, 0.1 waves, or one-tenth of a wave, strikes you per second

Solution B: Now, to solve for how many seconds between each wave, we must make a unit conversion from Hertz (waves per second) to seconds per wave as follows:

$$\frac{0.1 \text{ waves}}{1 \text{ second}} = \frac{1 \text{ wave}}{X \text{ seconds}}$$

$$\frac{0.1}{1} = \frac{1}{X}$$

Cross-multiplying: $0.1X = 1$

Dividing by 0.1: $\frac{0.1X}{0.1} = \frac{1}{0.1}$

$$X = \frac{1}{0.1} \text{ or 10 seconds}$$

Answer B in seconds per wave: It takes 10 seconds for each wave to pass by.

THE ELECTROMAGNETIC WAVE FORMULA

Electromagnetic waves include radio and TV waves, radar, microwaves, infrared light, visible light, ultraviolet light, x-rays, gamma rays and cosmic rays. These are all a form of *transverse* waves, and so they behave just like water waves in every way except for one important distinction: Waves in water can travel at different speeds, but electromagnetic waves *always travel the same speed* in a vacuum such as outer space, a velocity we commonly refer to as *the speed of light.*

This speed is 3×10^8 meters per second, or about 186,000 miles per second. For humans this speed is almost unimaginable. To try to visualize it, note that a rocket traveling at the speed of light could circle the earth *seven times* in a single second (over *25 thousand* times in an hour)! Because the speed of light is unique and constant, we abbreviate it with the special designation *c* rather than *v* (for *velocity*). The *T-triangle* for solving wave problems will have this *c* on top, as follows:

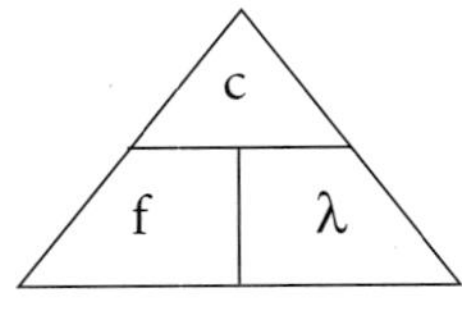

The different iterations of the wave formula for a series of electromagnetic waves, then, are as follows:

$$c = f\lambda$$
$$f = c / \lambda$$
$$\lambda = c / f$$

Because the value of *c* is unchanging, it is actually simpler to set up electromagnetic wave problems when compared to other types of waves, especially if you memorize the value of *c*: To solve for the frequency of any electromagnetic wave, just divide 3×10^8 by the wavelength in meters. To solve for the wavelength, just divide 3×10^8 by the frequency in meters per second.

$$f = \frac{3 \times 10^8}{\lambda}$$

$$\lambda = \frac{3 \times 10^8}{f}$$

Note that the formula to solve for *c* is not given here—you never need to solve mathematically for the speed of electromagnetic radiation—it is always 3×10^8 m/s, which you should commit to memory.

One aspect of the electromagnetic wave formula helps simplify units: Remember that the unit for frequency is the Hertz, defined as one cycle *per second.* Since the speed of light is also usually given *per second* rather than per hour or per minute, you will find that when solving problems, the *seconds* in the speed cancel out the *seconds* in the frequency. This leads to two conclusions: First, the frequencies of electromagnetic waves will always be reported in hertz units; Second, whatever unit of distance is used in presenting the speed (meters, centimeters, or miles), this will always be the unit you report for the wavelength. Unit conversions are often unnecessary. For example, if the speed of light is given in miles per second, 186,000 mi/s, the wavelength would be reported in miles. If the wavelength is given in centimeters, the speed could be reported as 3×10^{10} cm/sec.

Remember, too, that the frequency and the wavelength are always *inversely proportional* to each other. If one doubles, the other is cut in half. If one is cut to one-third, the other is tripled. For rounded-out problems like these, it is often not necessary to even bother dividing into the speed *c*.

Since the speed of electromagnetic waves is a very large number and many of the wavelengths are in excruciatingly small numbers, you should use scientific notation to make your calculations much simpler. For these problems, *always convert all of the numbers into scientific notation* that are not already in that format. (See Chapter 3 for help.) Now, let's try a few practice exercises.

Practice Exercise #3:

A particular shade of green light has a wavelength of 5×10^{-7} meters. How many waves of this green light strike you each second, i.e., what is its frequency in hertz?

Solution: Covering the "f" in the T-triangle, we see that the solution is to divide the speed by the wavelength, as follows:

$$f = \frac{3 \times 10^{8}\ m/s}{5 \times 10^{-7}\ m}$$

$$3/5 = 0.6$$

$$10^{8}/10^{-7} = 10^{15}$$

Answer: 0.6×10^{15} *cycles per second or* 0.6×10^{15} *Hz*

Note that meters in the equation cancel out, leaving only the designation "per second." This may be read as "times per second" or "cycles per second," which is, by definition, already in the unit hertz.

This format should be reduced to eliminate the decimal point, as

$$6 \times 10^{14}\ Hz$$

To express this verbally, we might reduce the exponent to 10^{12} *which can then be expressed as "trillions." It would then read*

$$600 \times 10^{12}$$

That is, 600 trillion waves of this green light strike you each second.

Practice Exercise #4:

The frequency of red light is 460 trillion cycles per second, or 4.6×10^{14} hertz. What is the wavelength of this electromagnetic wave?

Solution: When solving for electromagnetic frequency or wavelength, always simply divide the other one into the speed of light, as follows:

$$f = \frac{3 \times 10^{8}\ m/s}{4.6 \times 10^{14}\ cycles/s}$$

$$3/4.6 = 0.65$$

$$10^{8}/10^{14} = 10^{-6}$$

Answer: 0.65×10^{-6} *meters*

Note the canceling out of units in the equation: The numerator is in meters per second, and the denominator is in hertz or cycles per second. The per second portion of these units cancels out as follows:

$$\frac{meters/s}{cycles/s} = \frac{meters}{cycles}$$

or in other words, meters per cycle. This is the wavelength in meters.

Now, this format should be reduced to eliminate the zero, as

$$6.5 \times 10^{-7}\ Hz$$

This is 650×10^{-9} *meters or 650 nanometers. (See Table 2.1 in Chapter 2.)*

THE PLANK FORMULA

Max Planck was a German physicist who might be considered the "father of quantum theory." He hypothesized that the energy in the shells of atoms is found only in discrete packets called *quanta* (for *how much*?). That is, it is found only in certain predictable amounts. Each of these amounts also had a waveform with a predictable *frequency*. He was way before his time, and his ideas were rejected at first. But, Einstein used them to describe the photoelectric effect, and others later proved their validity and usefulness in describing the universe, for which Planck received a Nobel Prize.

Like Einstein, one of Planck's major contributions to our understanding came down to a very simple formula:

$$E = hf$$

In which ***E*** is the amount of energy in a particular quantum, ***f*** is the frequency of its waveform, and ***h*** is Planck's constant (4.15×10^{-15} volt-seconds). Planck's constant is essentially a unit conversion factor which allows volts to be converted into hertz and vice versa. It is not critical for radiographers to know this value, but it is important to understand the relationship described in Plank's formula: If you cover Planck's constant *h* with your finger, you will see that it simply states that energy is *directly proportional* to frequency. An x-ray is an example of a quantum. If the energy (the voltage) of an x-ray doubles, then its frequency as an electromagnetic wave must also double.

Now, remember from the electromagnetic wave formula that frequency and the wavelength are always *inversely proportional* to each other—if the frequency doubles, the wavelength must be one-half as long. We can derive, then, that Planck's formula could be written to indicate that *the wavelength of an electromagnetic wave, such as an x-ray, is* inversely proportional to *its energy*:

$$E = \frac{h}{\lambda}$$

The one is always divided into Plank's constant to find the other:

$$\lambda = \frac{h}{E}$$

For x-rays, the higher the voltage, (kVp), the shorter the waves. Shorter wavelengths are better able to penetrate through the human body to the detector plate. Therefore, a higher kVp setting results in better penetration.

We have stated that all electromagnetic waves including x-rays travel at the speed of light in a vacuum, *c*, which cannot be changed. In the electromagnetic wave formula, we see that wavelength and frequency can be calculated from each other by dividing into *c*. In Planck's formula, we see that wavelength can also be divided into *h* to obtain the voltage. We may multiply *c* and *h* to obtain a constant that simplifies the conversion between voltages and wavelengths, as follows:

$$h \times c = 12.4$$

when the unit for energy is *kilovolts*, and the unit for wavelengths is *angstroms*, both commonly used in radiography. This makes for a very handy tool: *To obtain the kilovoltage or the wavelength of an x-ray from each other, simply divide into 12.4.*

$$kV = \frac{12.4}{\lambda}$$

$$\lambda = \frac{12.4}{kV}$$

An actual x-ray beam consists not of one energy, but of millions of quanta or photons having a spectrum of different energies from zero up to the set kV*p*, with the *p* meaning the *peak* energy or highest energy within the beam. The *peak* energy would represent the *shortest* of the wavelengths within the beam, or the minimum wavelength.

For a single particular x-ray, we may calculate the wavelength from the kV. But, for a *beam* of x-rays, when kV***p*** is used rather than kV, the result must be expressed as the *minimum* wavelength. Let's try applying the formula:

Practice Exercise #5:

What is the minimum wavelength of an 80 kVp x-ray beam?

Solution: $\frac{12.4}{80} = 0.15$

Answer: 0.15 Angstroms

Practice Exercise #6:

What is the kV of a single x-ray having a wavelength of 0.2 Angstroms?

Solution: $\frac{12.4}{0.2} = 62$

Answer: 62 kV

THE NATURE OF ELECTROMAGNETIC WAVES

Recall that one of the four basic forces in the universe is the electromagnetic force. It was once believed that electricity and magnetism were separate phenomena, but they seemed to behave almost identically—different electrical charges attracted each other just like the north and south poles of two magnets, and like charges repelled each other just as similar poles of magnets pushed each other away. It was later proven that the push and pull of charges and magnets are actually two different manifestations of the same basic force.

Any electrically charged particle which is *moving* generates a magnetic field around it along with its electrical field. These two fields are always oriented perpendicular to each other. If you imagine the magnetic lines of force shaped like elephant ears and lined up with the north-south axis, you can picture the electrical field expanding as rings around the "equator" of the particle, like the rings of Saturn (Fig. 5-9).

Figure 5-9

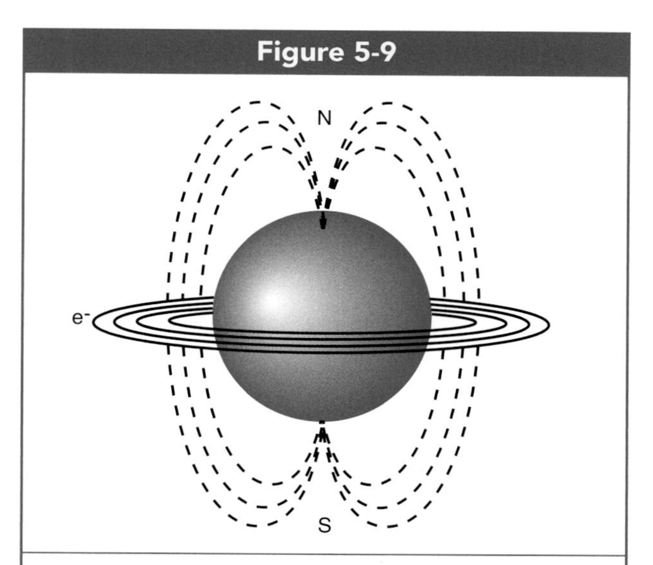

The electrical lines of force surrounding an electron are in the shape of "equatorial" rings, while the magnetic lines of force surrounding an electron are in the shape of elephant ears. Aligned with the magnetic poles of the electron, the magnetic lines are always *perpendicular* to the electrical lines.

In an atom, each electron and proton has a characteristic called *spin*, which, although it is not strictly the same as a spinning top, nonetheless constitutes a kind of *movement* for the particle. Therefore, electrons do not have to be traveling down a wire in the form of electricity to possess a magnetic field—when they seem to us to be stationary, they still have spin. This generates the magnetic field illustrated in Figure 5-9.

Now, imagine grabbing hold of the electron in Figure 5-9 and vigorously shaking it up and down: What is happening to the two fields that surround it? Both the electric field and the magnetic field are also being shaken up and down. An oscillating disturbance has been set up in these fields.

How is this disturbance experienced, or how can we detect it? The answer is: The same way we detect a disturbance in a swimming pool, by the waves it creates. A person with a plunger in the middle of the pool might be slapping the water up and down, but the waves created by this action travel *horizontally* outward to the edges of the pool. This is precisely how electromagnetic waves are created (Fig. 5-10).

An example is your local radio station (Fig. 5-11). At the station, a very large antenna is charged with electricity. Using AC or alternating current, the electrons in the antenna are made to jostle up and down. This movement sets up two huge fields around the antenna, one from the electrical charge and the other from the magnetic pull. But, these fields are not stationary—they are moving up and down with the electricity in the antenna. Waves in these fields travel out horizontally until they strike the antenna of your car radio. The electrons in your car antenna are pulled up and down by these wobbling fields, creat-

Figure 5-10

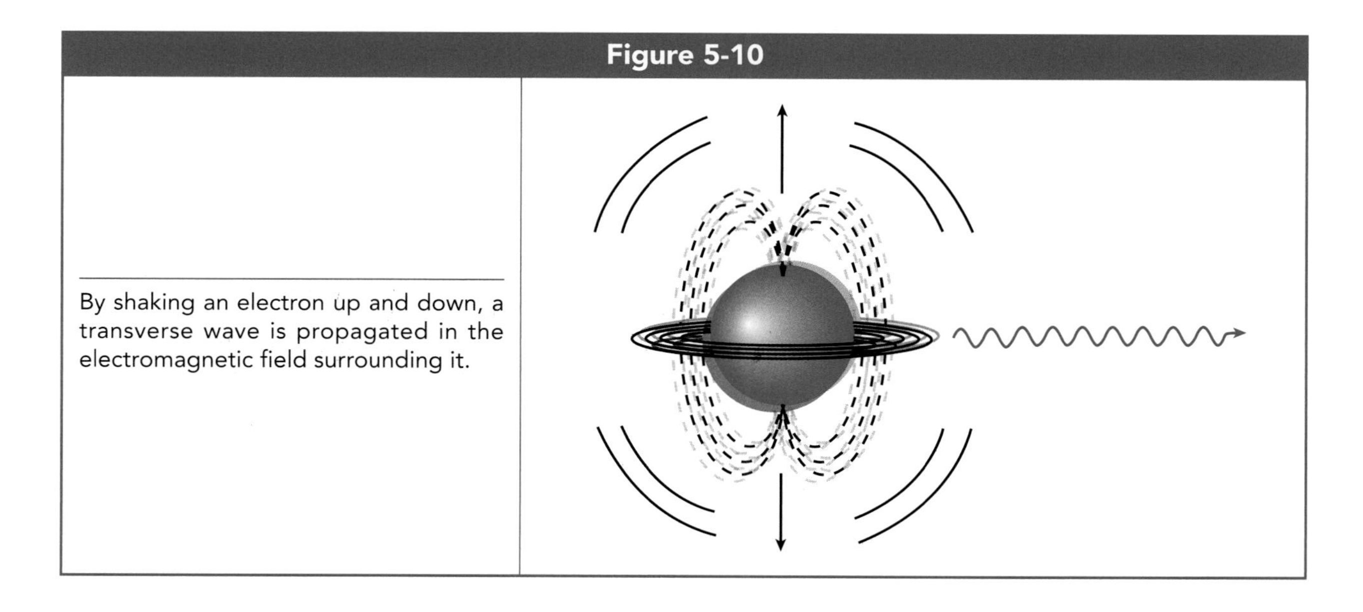

By shaking an electron up and down, a transverse wave is propagated in the electromagnetic field surrounding it.

Figure 5-11

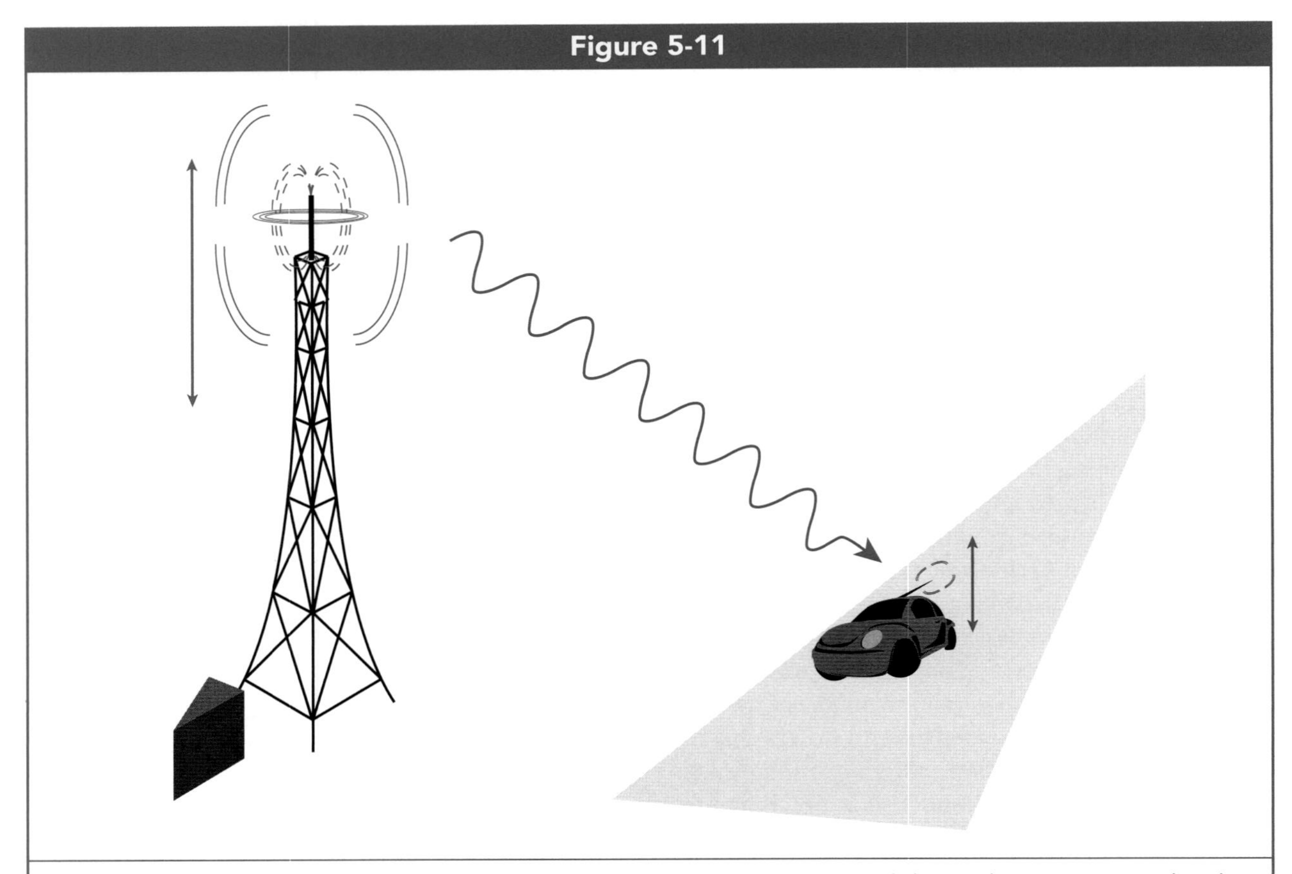

The strong electromagnetic field around a radio station's antenna moves up and down. This wave causes the electrons in your car antenna to move up and down with it, inducing electrical current in your radio.

ing electrical current in your radio which is then used to cause its speakers to vibrate.

As we have previously discussed, radio waves, x-rays and all other kinds of electromagnetic waves are of the *transverse* type, so their amplitude modulates up and down perpendicular to the direction they travel. However, what makes electromagnetic waves different from water waves and other transverse waves is that *electromagnetic waves are double-waves.*

Remember that, along with the electrical field, there is also a magnetic field always associated with moving electrical charges. There must be a wave in both fields. Referring again to Figure 5-9, we see that since the magnetic field and the electrical field are always perpendicular to each other, *the waves created in them must also be perpendicular to each other.* Our final conception of what an electromagnetic wave looks like is presented in Figure 5-12. It consists of two distinct transverse waves which are nonetheless closely associated and perfectly synchronized together as they travel along in a direction perpendicular to *both* of their amplitudes. One represents the disturbance in the magnetic field, the other the disturbance in the electrical field.

THE ELECTROMAGNETIC SPECTRUM

A *spectrum* describes any phenomenon that can be measured in an orderly, continuous progression of minute degrees, or a broad *range* of values. Visible light can be broken down by a prism into a spectrum of many colors separated by very small degrees. So can the entire range of electromagnetic radiations, from those having extremely small amounts of energy to those with very high energies, from unimaginably short wavelengths to wavelengths measured in miles, and from low frequencies to incredibly high frequencies.

We find that, although this spectrum is a continuous progression, certain groupings can be distinguished by the way that these waves *react* with the atoms and particles in their environment. For example, visible light cannot penetrate through the human torso, yet x-rays can. Radio waves will bounce off of the ionosphere, but visible light passes through it. Microwaves are absorbed by food, thereby heating it up, but cosmic rays will go right through it without affecting it. Visible light is absorbed by clouds on an overcast day, but you still get sunburned because *ultraviolet* light, which causes sunburns, penetrates through the clouds.

The operational concept here is the principle of *resonance*. We can use music to illustrate: We know that when a particular tuning fork is struck, the sound waves emanating from it can cause another tuning fork to also vibrate *if that fork is tuned to the same pitch*. A guitar or piano string has the same effect on other strings which are at set intervals of the pitch played, but not on those strings which are tuned too high or too low.

All substances, molecules, atoms and even subatomic particles have some type of resonant frequency to which they are "tuned." This is called their *natural frequency*. Electromagnetic waves will interact with those substances that resonate with the frequency of the waves, but not with those that are "tuned" too high or too low. By "interact with," we mean that the waves will be *absorbed* by the substance. Thus, we find different kinds of radiation able to penetrate certain substances and absorbed by others. From these distinctions, we derive the spectrum in Table 5-1 of electromagnetic waves.

Table 5-1 gives the energies and the wavelengths of each type of electromagnetic radiation. The frequencies have been left out to simplify the table. But, remember that the frequencies are inversely proportional to the distances given in the "wavelength" column, and have just as impressive a range of values.

A few points from Table 5-1 are worth mentioning to develop an appreciation for the ranges involved. For example, we see that the energies range all the way from millionths of a volt for radio waves up to billions of volts for cosmic rays. Note that the entire spectrum of visible light is encompassed in only a one-volt range between 2 and 3 volts. In rough approximations, we can say that red light has about 2.0 volts, orange 2.2 volts, yellow 2.4, green 2.6, blue 2.8, and violet 3.0 volts, thus covering the rainbow. For comparison, note that it takes two standard batteries to generate the *3 volts* needed to run a typical flashlight.

Figure 5-12

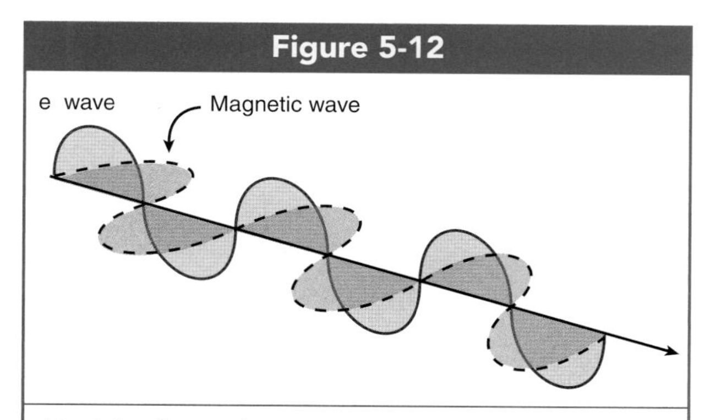

Model of an electromagnetic wave—a "double-wave" whose magnetic component is always perpendicular to its electrical component.

X-rays range in the tens-of-thousands to the hundreds-of-thousands of volts, but those specifically used in medical diagnosis are in the range of 30,000 to 150,000 volts. As far as wavelengths are concerned, it is interesting to note that the "average" radio wave is about one-half mile long, whereas the microwaves in your oven are about one centimeter, the width of your smallest fingernail: Infrared light waves are smaller than a pinhead, and visible light waves about the size of the *point* of a pin. The wavelength of x-rays reaches distances smaller than an atom, yet gamma rays and cosmic rays are smaller still. You should commit to memory the energy range and the wavelengths (one-tenth to one-half an angstrom) for diagnostic x-rays.

Finally, note that electromagnetic waves can be modified in several ways. On your radio dial, the letters *AM* stand for "amplitude modulation," *FM* stands for "frequency modulation." Each station sends out a fixed wave called the *carrier wave*, and the waves that represent specific sounds "ride on" the carrier wave. For an AM station, when the sound wave is added to, or "stacked on" the carrier wave, it results in the complex wave in Figure 5-13 in which the amplitude or height of the waves fluctuates. For an FM station, adding the sound wave to the carrier

Table 5-1

The Electromagnetic Spectrum

Radiation	Energy	Wavelength
Cosmic Rays	Billions of Volts	In quadrillionths of a mm (10^{-15} mm)
Gamma Rays	Hundreds of Thousands to Millions of Volts	Up to a Billionth of a mm (10^{-9} mm)
X-Rays	Tens of Thousands to Hundreds of Thousands of Volts	In Millionths of a mm (Fractions of an Angstrom) (10^{-6} mm)
Diagnostic X-Rays	**30,000–150,000 Volts**	**0.1–0.5 Angstroms**
Ultraviolet Light	Hundreds of Volts	In Tens of Nanometers
Visible Light	A Few Volts: (Violet ≃ 3V, Red ≃ 2V)	In Microns
Infrared Light	Thousandths of a Volt	In Hundreds of Microns
Microwaves	Ten Thousandths to Millionths of a Volt	One Centimeter or Less
Radio Waves (Radar, TV)	Millionths to Billionths of a Volt	In Kilometers (Fractions of a Mile)

wave results in the complex wave in Figure 5-14 in which the frequency and wavelengths fluctuate. You may have noticed that FM stations tend to carry much farther distances than AM.

Your television is essentially a *radio* receiver. AM radio waves carry the picture by controlling the brightness of the color elements in the screen, and FM radio waves are used to carry the sound. Radar is an acronym for *radio detection* and *ranging*, the use of radio waves to determine how far an airplane or other object is by bouncing the waves off of it and timing how long it takes for them to return. Microwave ovens use a range of wavelengths identical to radar, but employ either an electron *resonator*, or a *maser* (from *microwave-amplification by stimulated emission of radiation*) which results in synchronizing the waves (Fig. 5-15). Just as synchronizing two water waves results in a single taller wave,

Figure 5-13

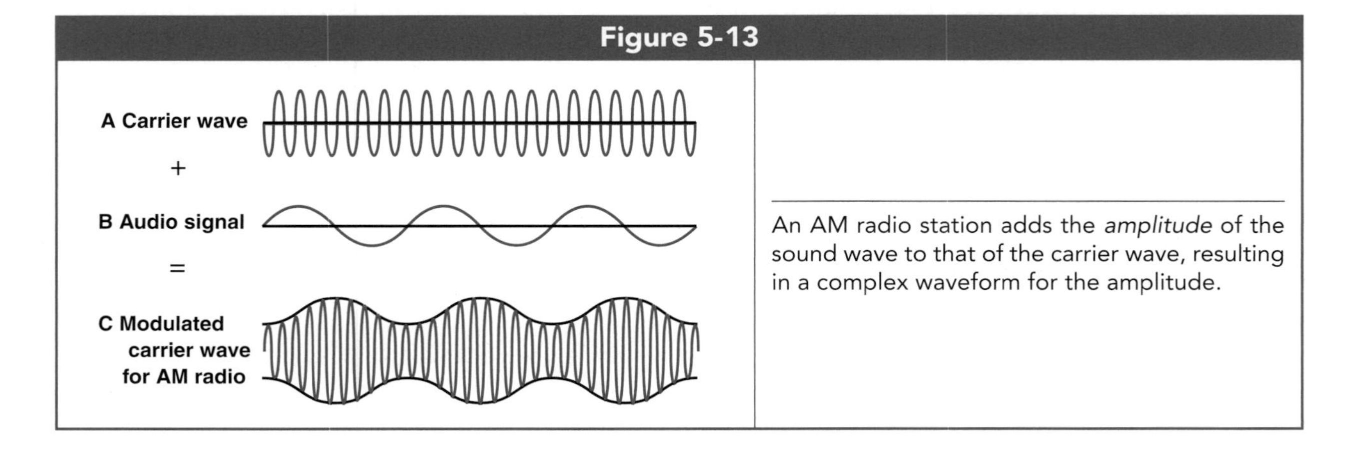

An AM radio station adds the *amplitude* of the sound wave to that of the carrier wave, resulting in a complex waveform for the amplitude.

Figure 5-14

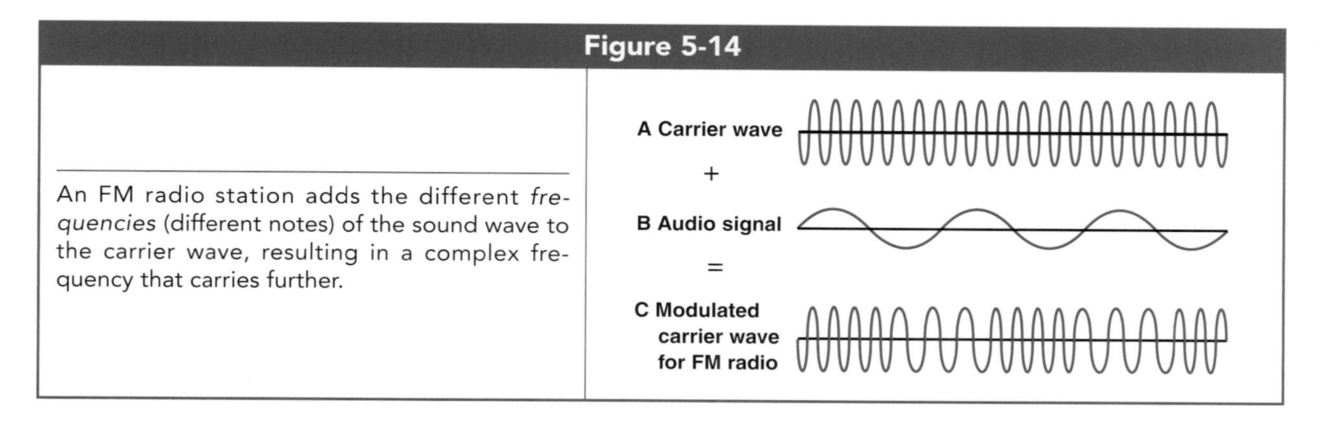

An FM radio station adds the different *frequencies* (different notes) of the sound wave to the carrier wave, resulting in a complex frequency that carries further.

synchronizing microwaves results in a boosted amplitude which causes more effective heating of food.

Infrared light is so named because it's energy is *below* that of visible red light. Infrared lamps give off both red light, which you can see from the glow of the bulb, and infrared light which is invisible. Infrared has more inherent energy than microwaves, so it is powerful enough to warm up food without having to be synchronized.

Ultraviolet light is so named because its energy is *above* that of visible violet light. Due to its high energy, it can cause many minerals to fluoresce "in the dark." It is essential to photosynthesis in plants. When standing in the sunlight, it is the infrared radiation from the sun which you feel heating your skin, but it is the ultraviolet which can penetrate through clouds and cause severe sunburn. Ultraviolet causes tanning and increased freckles from chemical changes in the pigments of the skin. While moderate amounts of ultraviolet radiation stimulate the production of vitamin D in the skin, excessive amounts from frequent use of "tanning booths" contribute to skin cancer and should be avoided.

Figure 5-15

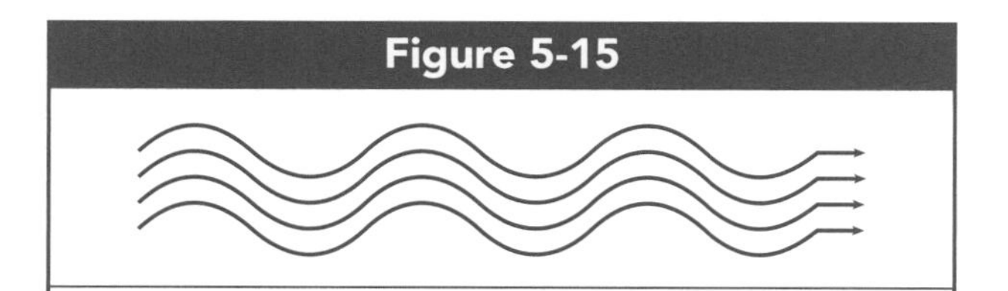

Microwaves are *synchronized* radio waves. By emitting them *in phase* with each other, their amplitudes are added together to carry much more energy (enough to heat food).

MEDICAL APPLICATIONS OF WAVES

Electromagnetic and sound waves find many applications in medical imaging. In addition to the x-rays used in radiography and computerized tomography (CT), and the gamma rays used in nuclear medicine and radiation therapy, waves are also used in magnetic resonance imaging (MRI), ultrasound, and laser surgery, for which brief descriptions will be given here. Radiographers should have enough foundational knowledge to answer basic questions from patients about these modalities.

Magnetic Resonance Imaging (MRI)

In magnetic resonance imaging, the human body is induced to emit *radio* waves, which are then detected and used to create an image. This requires several steps, as follows:

1. The patient is placed inside a superconducting magnet, magnetizing hydrogen atoms in the body such that their protons point toward the north pole of the external magnetic field (Fig. 5-16***A***). For a particular strength of magnetic field, the protons "wobble" or *precess* at a known rate.
2. To initiate an image, the machine emits a burst of radio waves bombarding the body from the side of the magnetic field. The frequency of these radio waves is set to precisely match the rate at which hydrogen protons are known to "wobble" or *precess*. At this frequency, the radio waves "knock over" the magnetized, spinning hydrogen protons. (This occurs because the

Figure 5-16

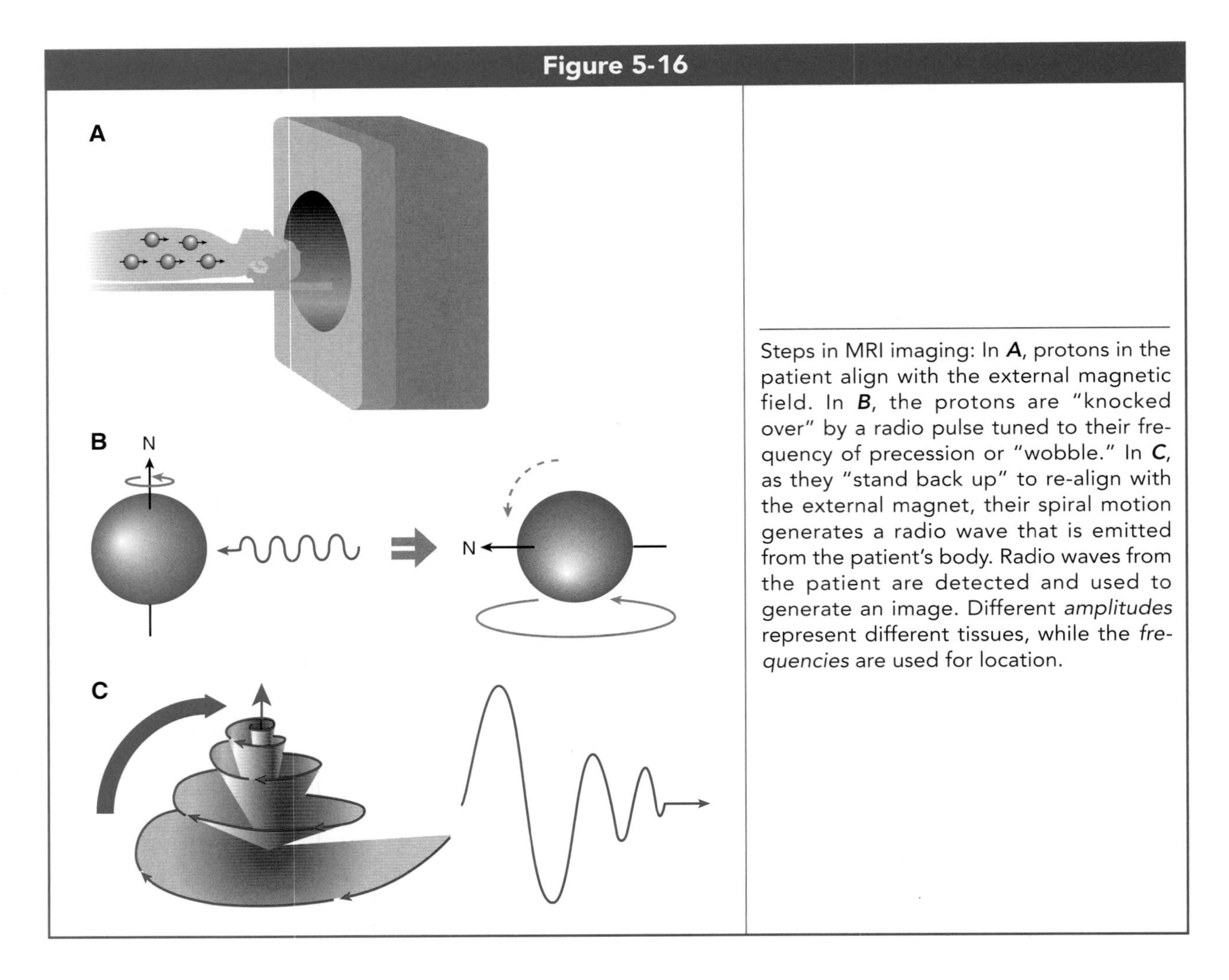

Steps in MRI imaging: In ***A***, protons in the patient align with the external magnetic field. In ***B***, the protons are "knocked over" by a radio pulse tuned to their frequency of precession or "wobble." In ***C***, as they "stand back up" to re-align with the external magnet, their spiral motion generates a radio wave that is emitted from the patient's body. Radio waves from the patient are detected and used to generate an image. Different *amplitudes* represent different tissues, while the *frequencies* are used for location.

protons absorb the radio energy, i.e., the radio waves are set to *resonate* with the hydrogen.) (Fig. 5-16***B***.)

3. The radio burst is then shut off and the antennas of the MRI machine go into "listening" mode.
4. With the radio burst off, the protons "stand back up" in order to realign to magnetic north. But, because of their natural *precession* or "wobble," they stand back up in a *spiral* motion (just the opposite of what a spinning top does as it slows down and spirals down to the floor). Since protons are spinning charged particles, they each have electromagnetic fields around them. The spiral movement sets up an oscillating disturbance in these fields. This is a radio wave (Fig. 5-16***C***), referred to as the *return signal.*
5. The radio frequency (RF) antennas of the MRI machine pick up these radio waves from the patient's body. The *location* in the body from which a particular radio wave is emitted is encoded as follows: The magnetic field around the patient is graduated in strength vertically, horizontally and transversely. The *frequency* of the radio wave emitted depends upon the strength of the magnetic field. Therefore, different frequencies of radio waves originate from different locations within the body. The strength or *amplitude* of each radio wave is characteristic of the type of tissue emitting the wave.
6. The frequency and amplitude data are fed into a computer, which then translates this data into locations and tissue types to generate an image.

Ultrasound

The process of generating ultrasound (sonography) images is much simpler than that for MRI. Sound waves follow the basic laws of *reflection* and *refraction*, discussed in the following section on light.

Sound is composed of *compressional* waves traveling in the molecules of a substance. In air, sound travels at 340 meters per second, or about 1000 feet per second. (Thus, it takes 5 seconds for the sound from a lightening bolt to travel one mile or approximately 5000 feet.) When sound waves encounter any *interface* (boundary) between two different types of material, they will be *reflected* as an echo. By timing the return of the echo, multiplying this time by the known speed of the sound, then dividing it in half (to account for the "one-way" distance rather than the "round-trip" distance), the distance to the interface can be calculated.

Let us illustrate the echo time with a simple example. Standing on one ridge of a canyon, you find that it takes 4 seconds for the echo of your shout from the opposite canyon wall to be heard. Using the speed of sound in air given above, how far is the opposite ridge of the canyon? First, multiply the speed of sound by 4 seconds for the round-trip distance. The answer is 1360 meters or 4000 feet. Now divide this in half for the one-way distance of 680 meters or 2000 feet.

This is just how an ultrasound machine "knows" where to place a dot on the image screen, by timing the return signal or echo from each boundary or interface between different types of tissue. How does it know how *bright* to make each dot so that different types of tissue are represented? This is based on the *amplitude* of the return echo, which depends on the tissue's ability to transmit, attenuate or disperse the sound waves. This measurement of amplitude must be taken after accounting for the expected reduction due to the *inverse square law*, which can be used to predict the loss from normal spreading-out of the sound waves. Having performed this calculation for the distance the sound traveled, any further loss of signal would have to be due to the attenuation or dispersal of the sound by the tissue (Fig. 5-17).

The only difference between the audible sound waves we use to communicate with and ultrasound waves is that the frequencies of ultrasound are very high, above the range of human hearing. Physicists define sound with a frequency exceeding 20,000 hertz as *ultra*-sound. At these frequencies, sound waves will be reflected from interfaces between bone and soft tissue or substantially different soft tissues based on their fluid content, but not by an interface with *air*. Gases such as air scatter the ultrasound waves in random directions, destroying any possibility of acquiring an image. It is for this reason that a liquid gel must be placed between the patient's skin and the ultrasound transducer, so that no air pockets can develop.

Lasers

Lasers are used in the radiology department in computerized radiography processors, in film digitizers, laser film printers and optical disc reading. They are used as a scalpel for surgery in the eye. They are used in dermatology, gastroenterology, otolaryngology, urology and pulmonary medicine. They are used in your home and car to play (and burn) compact discs and DVDs.

By synchronizing waves of light, an intense beam of *concentrated* light energy can be produced that

Figure 5-17

To generate an ultrasound image, *location* (distance ***d***) is determined by timing the return echo. Having adjusted (by inverse square law) for loss of amplitude due to the round-trip distance, remaining differences in signal *intensity* represent the sound-reflectivity of different tissues.

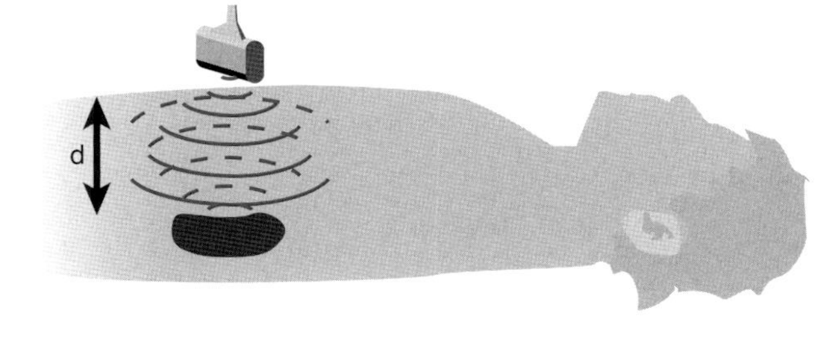

d

follows a nearly parallel, narrow path. The *laser* (*light amplification by stimulated emission of radiation*) was invented in 1960.

When an atom is *excited*, it can release its excess energy by the *spontaneous emission* of incoherent light (light emitted in any direction). In *stimulated emission*, additional energy *pumped* into an atom that is already excited triggers it into releasing its own energy as light. In this case, most of the light produced has the same frequency *and* travels in the same direction as the triggering light.

In radiology, the helium-neon laser is most frequently used, which emits a particular frequency of red light. The gas is contained in a cylindrical *resonant cavity*. The shape and size of this chamber correspond to the type of gas, such that the red light waves produced are not only all of the same frequency, but are also *in phase* with each other, or "synchronized," Figure 5-18. This enhances their *amplitude* in just the same way that synchronizing radar waves results in high-amplitude microwaves. In this way, a coherent beam of intense light is produced.

In order to synchronize the light waves, they must all be of the same wavelength and frequency, giving the laser beam a specific color. This synchronization also allows beams of light to be produced which are highly *directional*, forming a focused beam with sides that stay almost parallel. This makes it possible to create laser beams narrow enough to read microscopic tracks on a CD or DVD disc, or concentrated enough to cut even diamonds.

There are three essential components to every laser system:

1. The *medium* provides the atoms that will be stimulated. Mediums can be solid, gaseous or liquid. Solid mediums include crystals such as the ruby crystal used in the first laser, glass, or semiconductor materials. Gas lasers can use argon, nitrogen, carbon dioxide, helium or neon. The most commonly used laser in radiology is the helium-neon laser which emits red light.
2. The *power source*, or *pumping source*, produces intense flashes of visible or ultraviolet light, or bursts of electric current which stimulate the medium to emit its characteristic light waves.
3. The *resonant cavity* can be in the shape of a cylindrical chamber or prism with reflective surfaces at each end. As the emitted light reflects back and forth, the chamber begins to resonate at the same frequency, resulting in the amplification of all light waves at this specific frequency. An aperture centered at one end allows a narrow beam of this intensified, coherent light to escape in the form of the laser beam.

These three components are generically illustrated in Figure 5-19. A brief description of four applications of laser technology within a medical imaging department follows.

Computed Radiography (CR) Readers

Computed radiography (CR) was initially called *photostimulable phosphor digital radiography* or *PSPDR* because it uses an image receptor plate that can be stimulated to emit light by a laser beam. This plate is coated with a fluorescent phosphor layer not unlike the *intensifying screens* that were formerly used with film-based radiography. When the plate is exposed to x-rays, many electrons are "shaken" out of their atoms and immediately fall back into their shells, emitting fluorescent light. But, some of the freed electrons are *trapped* in specific locations within the phosphor's molecular structure, where they remain until the plate is processed in the CR *reader*. In this

Figure 5-18

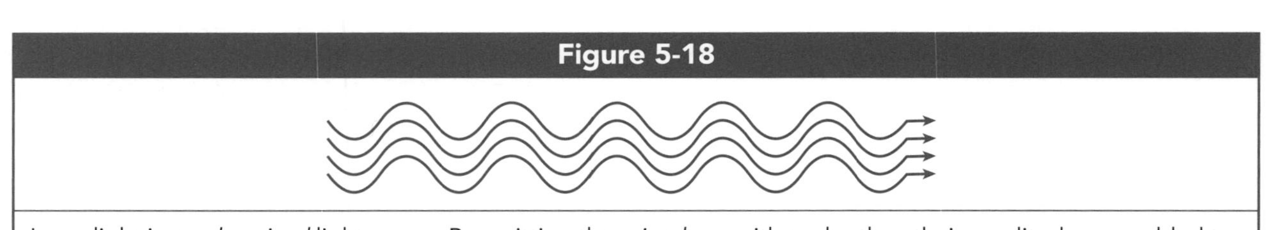

Laser light is *synchronized* light waves. By emitting them *in phase* with each other, their amplitudes are added together to carry much more energy.

Figure 5-19

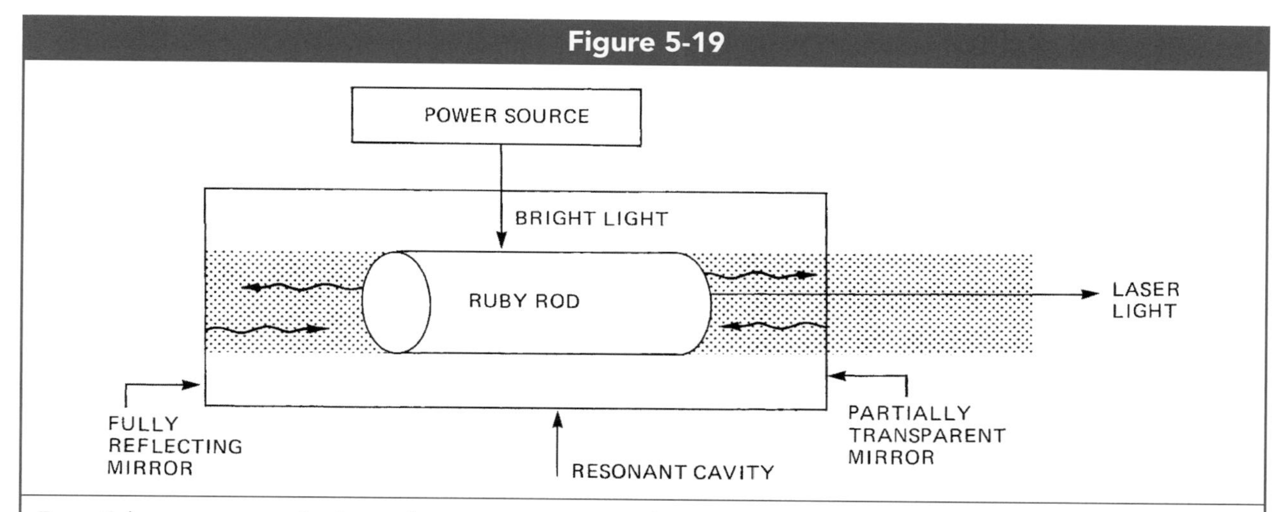

Essential components of a laser, the power source, medium (a ruby rod), and resonant cavity. (From Quinn B. Carroll, *Practical Radiographic Imaging*, 8th ed. Springfield, IL: Charles C Thomas, Publisher, Ltd., 2007. Reprinted by permission.)

device, the plate is scanned by a laser beam. The added energy from the laser beam frees these trapped electrons so that they can fall back into their atoms, causing the plate to glow for the *second time.*

Photomultiplier tubes detect this light and transform it into electric current. The electrical signals are then converted into digital form by an analog-to-digital converter. The resulting data are processed by a computer and stored on magnetic or optical media as digital images. Details of this process are described in the following chapters.

Laser Film Digitizers

A laser film digitizer (Fig. 5-20), converts x-ray films into digital images. The basic layout is illustrated in Figure 5-21, in which the film is scanned line by line by a laser beam. As the laser light passes through the radiograph, the various densities in the image partially attenuate it, resulting in different intensities of light that are detected by a bundle of optical fibers. These pass the light energy to a photomuliplier tube where it is converted into electrical current and amplified. An ADC converts these signals into digital data which are fed into a computer for processing and storage.

Note that film digitizers literallly follow the steps illustrated in Figure 29-6 (Chapter 29) for digitizing analog images: *Scanning*, *sampling*, and *quantizing*

Figure 5-20

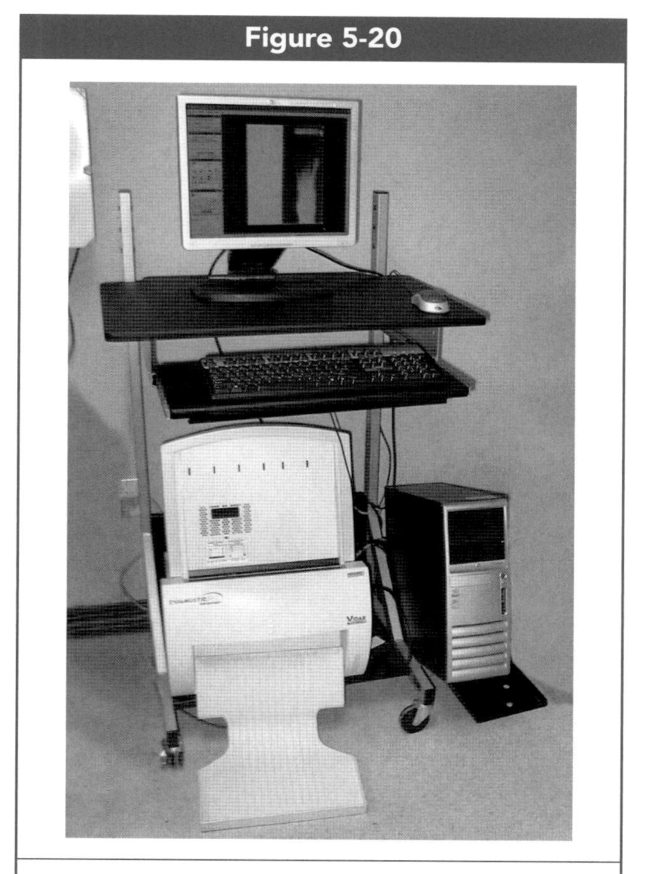

X-ray film digitizers (shown here), laser printers, and optical disk storage all use laser light, an electromagnetic wave. (Courtesy, Kathy Ives, R.T.)

Figure 5-21

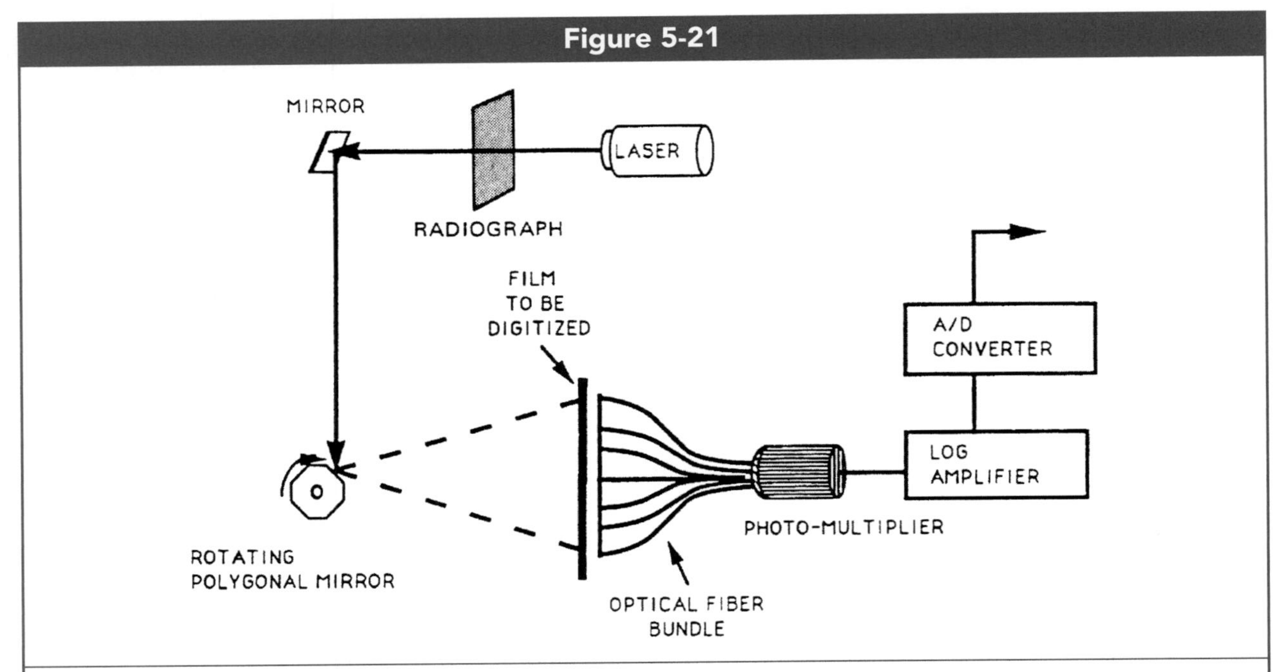

Diagram of a film digitizer using a laser beam. (From R.E. Greene & J.W. Oestman (1992). *Computed Digital Radiography in Clinical Practice.* Thieme Medical Publishers, Inc. Reproduced by permission.) (From Quinn B. Carroll, *Practical Radiographic Imaging,* 8th ed. Springfield, IL: Charles C Thomas, Publisher, Ltd., 2007. Reprinted by permission.)

the film-based image. This process is used not only for archival storage of digitized x-ray films, but also for the *transmission* of images from the radiology department to other departments and locations.

Laser Film Printers

Laser film printers are widely used in medical imaging departments to produce hard copies of radiographs, CT scans and MRI scans. These printers typically use a helium-neon laser or a solid state diode laser to write digital data onto special film. The printer first takes the digital data for a particular image from the computer and converts it into electronic signals that are sent to the laser. As shown in Figure 5-22, the laser then projects the varying intensities of laser light produced by those signals onto the film pixel by pixel, scanning back and forth through the mechanism of a system of rotating mirrors. As this transverse scanning projection progresses, the film itself is indexed along line by line using a series of rollers.

The special transparent film is coated on one side with a carbon-based emulsion. *Heat* from the laser beam causes carbon molecules in the emulsion to turn black, developing an image. Even in a "filmless" department, there are likely to be instances where at least one printer capable of producing high-quality hard copies of images will be needed. Laser printers can be directly networked into the PAC system.

Optical Disc Reading and Writing

Optical discs are widely used for storing digital radiographic images. (Their more familiar forms are the *CD* and the *DVD.*) Optical discs are fully described in Chapter 28 on *Computer Basics.* A very thin, high-intensity laser beam is used to write data onto the disc. Bursts of heat generated from the laser beam etch *pits* into the reflective surface of the disc. The series of pits and flat spaces between them, called *lands*, represent a binary code of *1's* and *0's.*

To read the optical disc as it spins, a low-intensity laser beam is reflected off of the surface to strike a light-sensitive detector. The *pits* in the surface disperse the laser beam so that it effectively *misses* the detector. The series of light bursts and pauses that is detected constitutes a binary code of digital data that can be used to build up and reconstruct an image.

Figure 5-22

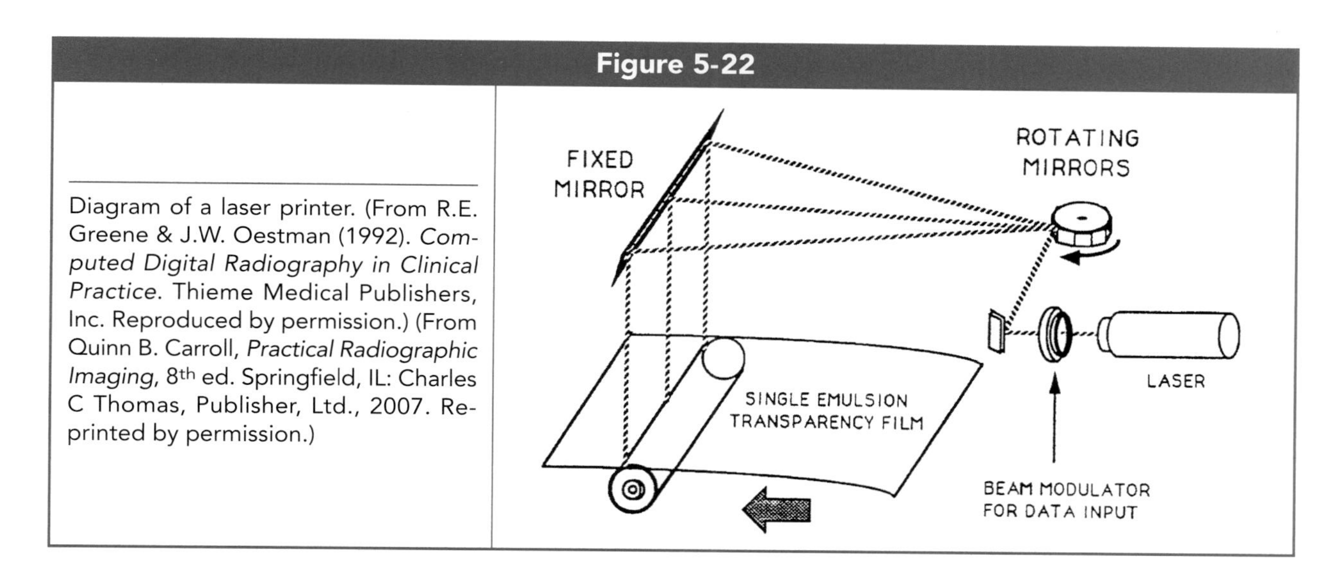

Diagram of a laser printer. (From R.E. Greene & J.W. Oestman (1992). *Computed Digital Radiography in Clinical Practice.* Thieme Medical Publishers, Inc. Reproduced by permission.) (From Quinn B. Carroll, *Practical Radiographic Imaging*, 8th ed. Springfield, IL: Charles C Thomas, Publisher, Ltd., 2007. Reprinted by permission.)

CHARACTERISTICS OF VISIBLE LIGHT VS. X-RAYS

Visible light consists of all the colors of light possessing energies between 2 and 3 volts. Isaac Newton discovered that white light is created by combining all of these colors. Light is a double-transverse wave just as x-rays are. It is an electromagnetic wave with much more energy than radio waves, but less than x-rays. Several distinctions between light and x-rays bear mention, in relation to their penetration, transmission, reflectivity and refraction characteristics.

Light follows the *Law of Reflection*, which states that when it encounters the interface or surface of a mirror or other shiny object, it will reverse direction such that the *angle of reflection is equal to the angle of incidence* (Fig. 5-23). X-rays are not reflected by a normal mirror, but pass right through it. This distinction allows a flat mirror to be placed in the middle of a typical x-ray collimator in order to provide a field centering light—placed at a 45-degree angle, this mirror reflects light from a bulb mounted to the side to be projected downward, while the x-ray beam itself, coming down from directly above the mirror, passes through it with only the slightest filtration effect, but with no change in direction (Fig. 5-24). (If the light bulb were placed in the

Figure 5-23

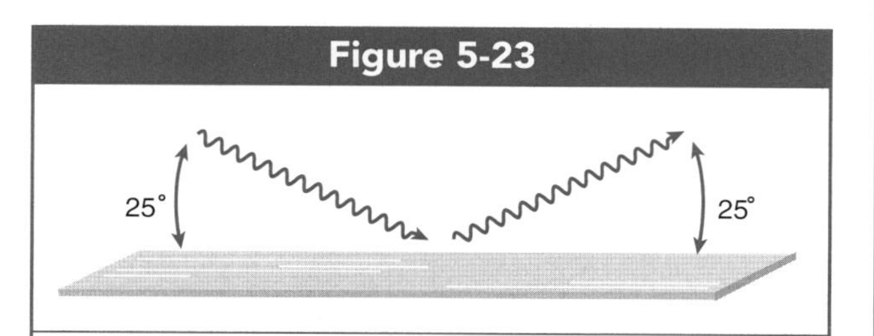

The law of light reflection: The angle of reflection is equal to the angle of incidence.

Figure 5-24

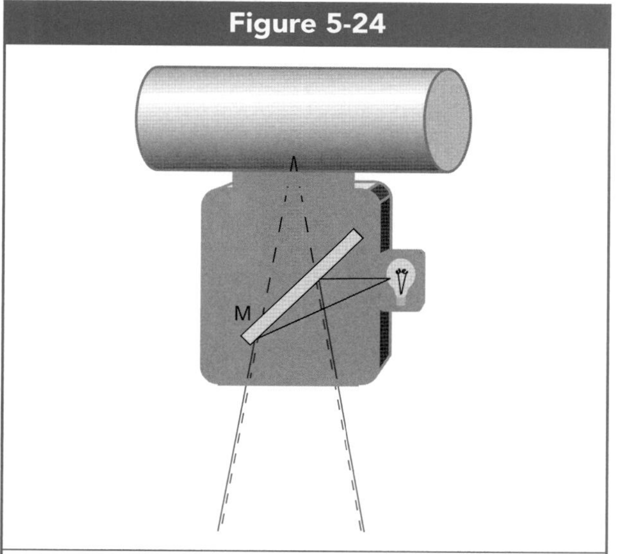

The 45-degree mirror, ***M***, in an x-ray collimator reflects the field light (solid lines) downward from the side of the collimator, but x-rays from the tube (dashed lines) are not reflected by it and pass right through with only slight attenuation.

Figure 5-25

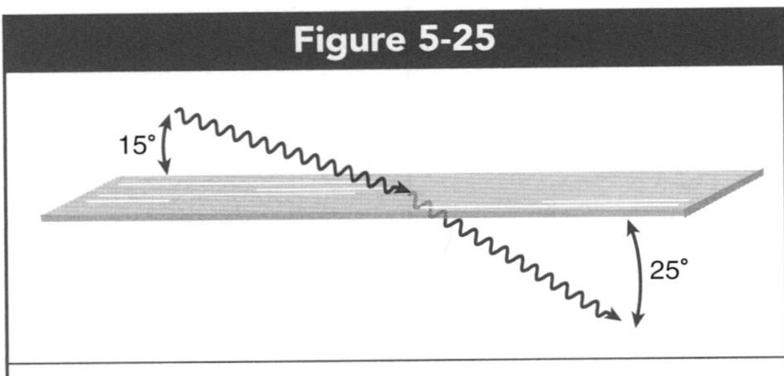

Refraction is the bending of light as it passes through an *interface* between two substances.

middle of the collimator, its metal components would absorb enough x-rays to cause artifacts.)

Light can also be easily *refracted* by glass lenses, whereas x-rays cannot. Refraction is the *bending* of light *as it passes through an interface* between two materials (Fig. 5-25). It is this bending of light as it passes from water into air that makes a person's legs look shorter than normal when he is standing in a pool. Since any change of interface causes refraction, light passing through a typical lens is refracted *twice*, once when it passes from the air into the glass, and again when it passes from the backside of the glass into the air. The result is that a biconvex lens, such as that in a magnifying glass, can be used to *focus* light. X-rays will pass right through the glass of a lens without changing direction, and cannot be thus focused.

The *dispersion* of light occurs when an organized sequence of refractions through a prism causes the colors of light to separate in orderly fashion. X-rays are unaffected in their direction by a prism. *Diffusion* or *scattering* describes the *random* refraction or reflection of waves. The projection of any organized image, such as a shadow, is rendered ineffectual by this randomization of the projected beam. Air has this effect on an ultrasound beam. When the ultrasound waves reach a pocket of air, such as the lungs, rather than reflecting directly back at the interface, the sound waves are diffused and scattered throughout the gas and no longer contribute to any organized image. This information is lost.

Absorption refers to an x-ray or a ray of light being completely stopped. *Transmission* refers to its passing completely through without any loss of energy or intensity. However, it is possible for only a portion of the energy from a single x-ray to be absorbed through a scattering event (Fig. 5-26). Sometimes the term *attenuation* is used to describe this partial absorption.

There are trillions of x-rays in a typical x-ray beam. In producing medical radiographs, it is ideal for the intensity of the x-ray beam to be *attenuated*. It is precisely because only a portion of the x-ray beam is absorbed and part of it is transmitted through the body that an image is obtained. In this case, *attenuation* refers to a portion of the intensity or quantity of the overall x-ray beam being absorbed (Fig. 5-27). *Differential attenuation* refers to the different ratios of absorption and transmission that are characteristic of each different tissue in the body.

In regard to the transmission of light, different materials are classified as *opaque* when very little light passes through, and as *transparent* or *translucent* when light can pass through easily. Similar terms have been adopted to describe the effect of different materials on x-rays: A *radiopaque* material is one that does not allow x-rays to pass through easily. The term *radiolucent* describes a material that x-rays can easily pass through (although this word, by direct translation of its roots, is a misnomer meaning "radiation-light," and might rather have been dubbed "trans-radiant.")

Generally, when a dark or dense image is seen against a background of light, such as the writing on

Figure 5-26

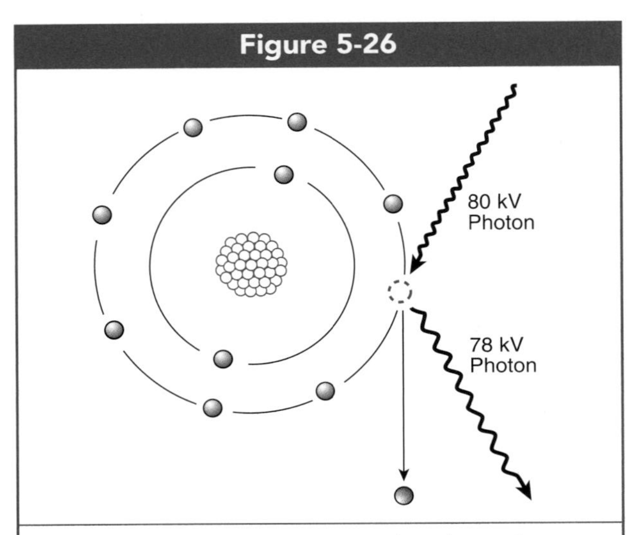

Attenuation of a single x-ray implies that only a portion of its energy was absorbed by an orbital electron in a "scattering event."

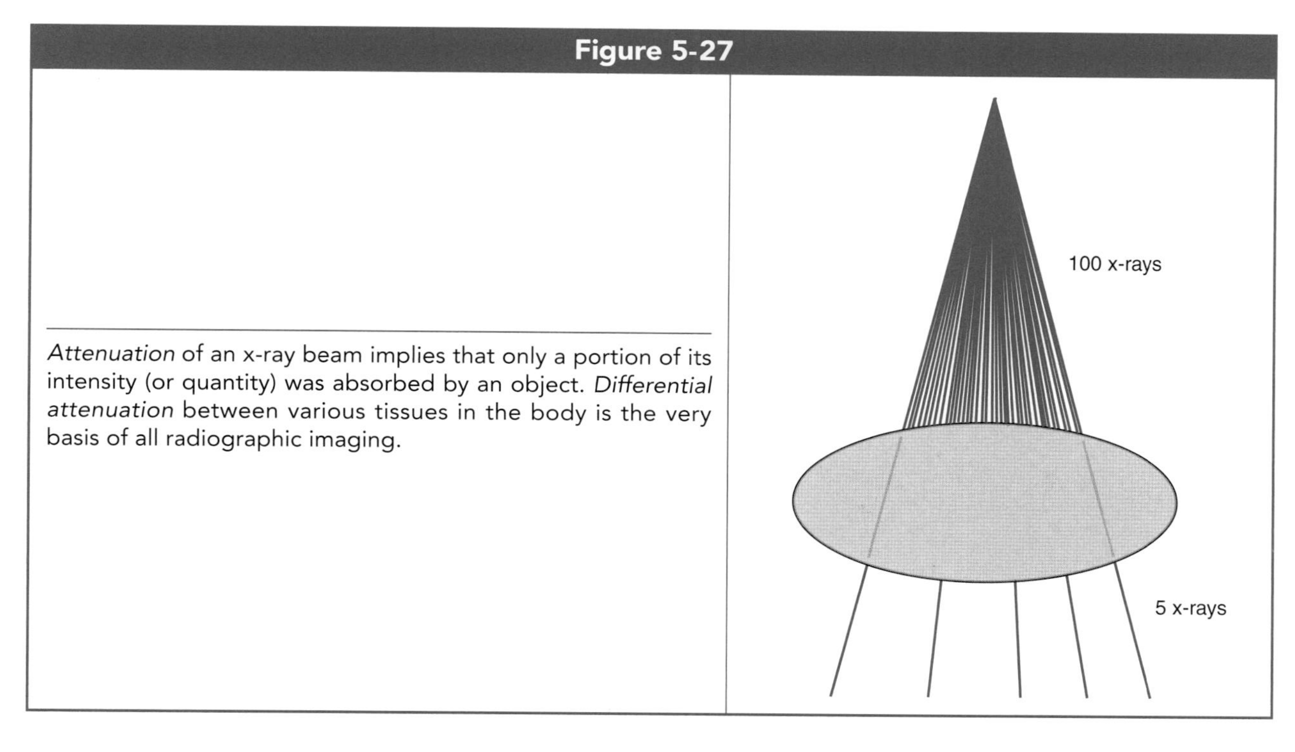

Attenuation of an x-ray beam implies that only a portion of its intensity (or quantity) was absorbed by an object. *Differential attenuation* between various tissues in the body is the very basis of all radiographic imaging.

this page, it is considered a *positive* image. Conventional radiographic images are considered *negative* images, since physically denser materials such as bone do not appear darker or "radiographically denser" in the image, but are reversed, appearing lighter against a dark background. This can be confusing, because radiopaque tissues result in translucent images on a sheet of "hard-copy" print-out film, and radiolucent materials, those that allow x-rays through, result in opaque densities on such a film (Fig. 5-28).

DUAL NATURE OF ALL MATTER AND RADIATION

In 1887, Heinrich Hertz (for whom the unit of frequency is named) discovered that when ultraviolet light shines upon an electrode plate made of any of a number of different types of metals, an electric discharge from the plate can be detected. That is, as long as light is striking the plate, electrons are emitted from its surface. This phenomenon was dubbed *photoelectric emission*. Two related processes were soon established: In the *photoconductive effect*, a plate is better able to conduct electricity as long as light is striking it (this is the basis of the "electric eye"), and in the *photovoltaic effect*, electrical voltage is generated in the region between two different materials as long as light is striking it.

Albert Einstein provided a full explanation of all three processes when he published a paper on the *photoelectric effect* in 1905. He won his only Nobel prize for this paper on the photoelectric effect even though it was one of three papers published that year which also introduced his more famous theory of special relativity.

We have described light and x-rays in this chapter as peculiar types of *waves*. And yet, in the photoelectric effect, they seem to behave more like *particles*. For example, we see that a single electron within an atom can absorb an entire photon (quantum) of light or x-ray. If you picture an expanding wave front such as a water wave, it is hard to visualize this entire wave being pulled into a single particle. If we imagine the photon as a small blob, a contained "bundle" or "packet" of energy, it is easier to imagine the electron "swallowing it whole." But, when we think of a small, contained bundle of energy, it sounds suspiciously like a *particle*.

Furthermore, in photoelectric interactions, electrons are "knocked out" of atoms and emitted from

Figure 5-28

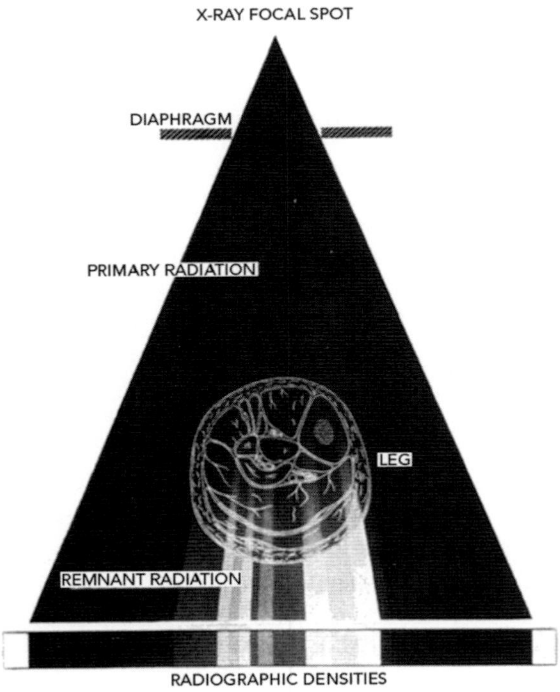

X-RAY FOCAL SPOT
DIAPHRAGM
PRIMARY RADIATION
LEG
REMNANT RADIATION
RADIOGRAPHIC DENSITIES

Radiolucent tissues allow x-rays through, resulting in dark areas in the image. *Radiopaque* tissues, such as bones, stop x-rays from reaching the detector plate, resulting in light areas in the image. From Quinn B. Carroll, *Practical Radiographic Imaging*, 8th Ed. Springfield, IL: Charles C Thomas, Publisher, Ltd., 2007. Reprinted by permission.

the material. This aspect is also easier to understand when the incoming photon is visualized as a particle colliding with the electron like two balls colliding on a pool table (Fig. 5-29). This also helps in understanding the Compton effect, in which only part of the energy of the incoming photon is absorbed by the electron, as a pool ball *glancing off* another such that both "particles" are sent off at angles from the original direction of the cue ball.

On the other hand, experiments showed that by passing a beam of light through two slits in a partition, an *interference pattern* was generated on the wall behind (Fig. 5-30). This only made sense if light was just like a water wave which would be broken into two waves when passing through the slits. These two waves would then interfere with each other whenever the crest of one intersected the trough of the other, thus canceling each other out at those points.

So, were photons light particles? Or, were they waves? There was solid evidence for *both* propositions.

Louis Victor de Broglie, a French physicist, made the brilliant leap that if very small wavelengths sometimes behaved like particles, perhaps very small (subatomic) particles might also sometimes behav- like waves. Remember that Einstein's equation $E = mc^2$ had shown all forms of matter to be a manifestation of energy, and that extremely high energies were required to form just a small amount of mass,

Figure 5-29

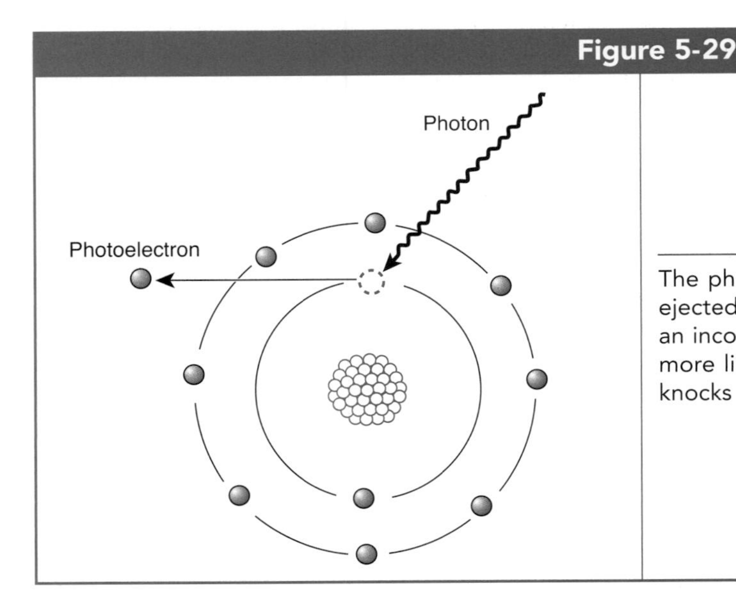

The photoelectric effect, in which an orbital electron is ejected from its atom by absorbing *all* of the energy of an incoming light photon. Here, light is found behaving more like a *particle* than like a wave, as if a "collision" knocks the orbital electron out.

Figure 5-30

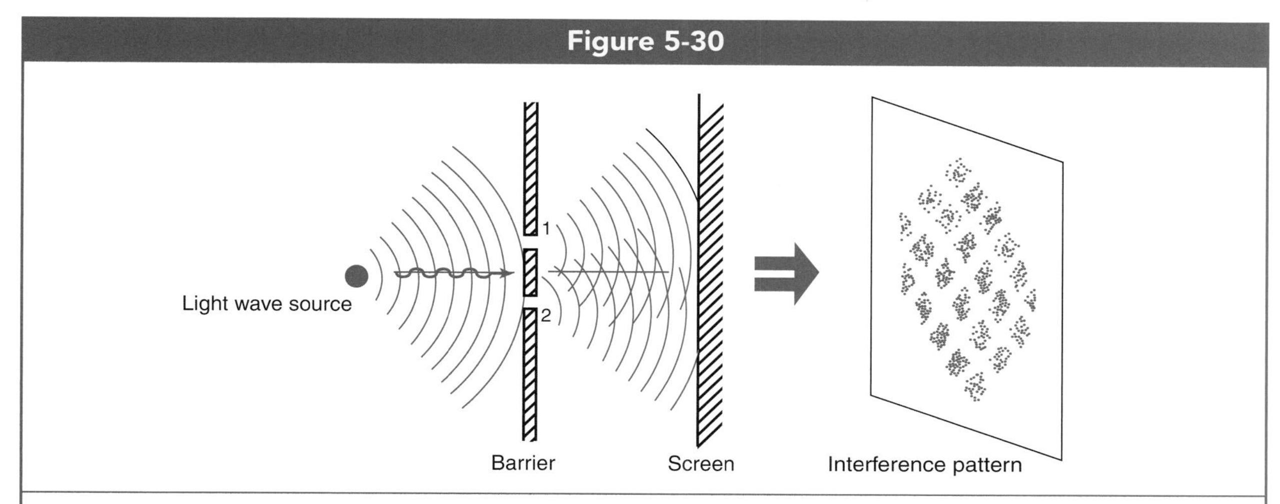

The two-slit experiment shows light behaving like a wave, which can be broken into two waves by the slits and caused to create an interference pattern on the screen behind them.

such as an electron. In turn, Max Plank had demonstrated that high energies are associated with high frequencies, and hence, short wavelengths. Could it be that if $E = mc^2$, then $f = mc^2/h$ where h is Plank's constant? If so, the mass and weight of an electron would also have an associated *wave function* with an assigned frequency and wavelength.

Several experiments subsequently confirmed this to be the case. In one, a beam of electrons was directed toward a luminescent screen that would glow when they struck it, but a metal plate with two holes in it was placed in the way, as shown in Figure 5-31. An interference pattern with many luminescent spots was generated on the screen behind the plate. Somehow, the two beams of electrons were *interfering with each other* after passing through the two holes. If they were strictly particles, there would be no interference and only two luminescent spots would result. Yet *within* each area of the pattern where the electrons were distributed, distinct spots could be

Figure 5-31

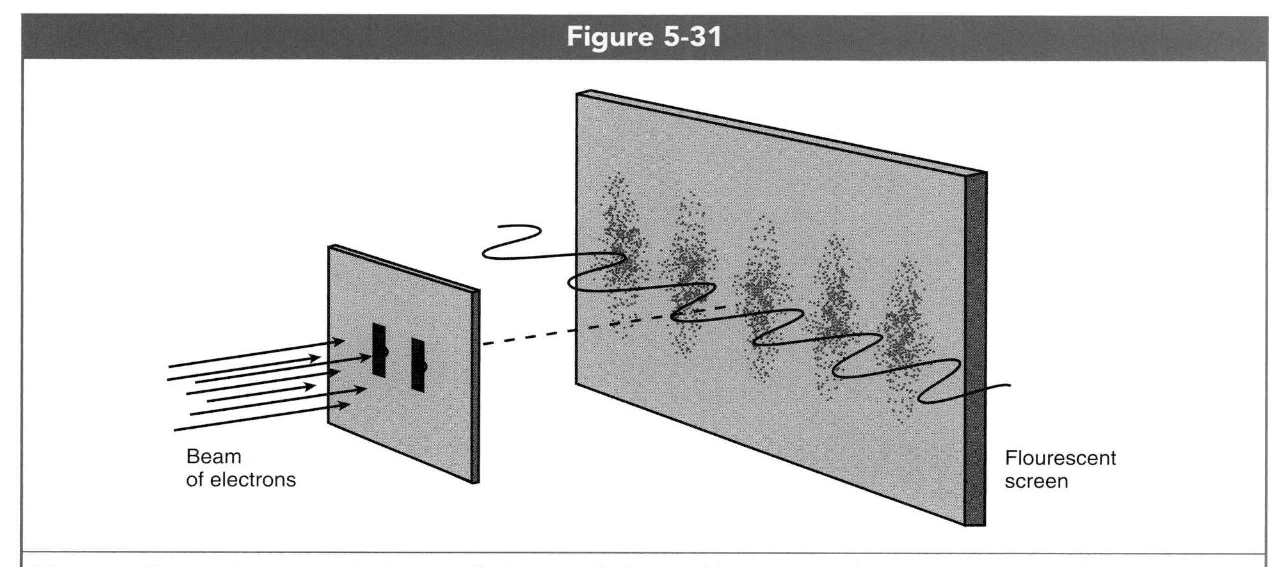

The two-slit experiment caught beams of *electrons* behaving like waves and generating an interference pattern. Yet, within the pattern, spots where individual electrons had struck as particles could be identified.

identified where each individual electron had collided with the screen like a particle (Fig. 5-31).

The mathematical applications of all this new *quantum theory* explained heretofore unsolved problems with our understanding of the Compton effect, the photoelectric effect and many other aspects of the subatomic world, including Plank's restriction of electrons in orbits to only certain discrete, quantized amounts of energy.

If an electron can act either like a particle or like a wave, what would determine *when* it acts like a particle and *when* it acts like a wave? This depends on the specific method being used to *observe* or detect the electron, or, in other words, how a particular experiment is set up. When it comes to the subatomic world, the experiment itself affects the results. For example, let's say you want to "see" an electron to study it; to do this, you must shine light on it. Light is a wave which can interfere with or augment other waves, including the *wave function* of the electron. Indeed, the effect of shining light on an electron is to smooth out its wave function, such that it then behaves as a particle. It is believed that electrons exist in a wave state *until* they are observed!

The entire weird subatomic world consists of "particle-waves" which can be laid out in a spectrum of energies as demonstrated in Table 5-1. Note in Figure 5-32 that those waves with extremely low energies take on the behavior of *fields*, such as a field of magnetic pull or electrical attraction, those with intermediate energies generally behave like *waves*, and those with high energies tend to behave more like *particles*. As we ascend the scale of energies, the associated wavelengths become smaller and smaller, until, visually, they become a *blur* which appears as a solid line. We might think of this as the point where the energy itself takes on measurable mass and we can begin to detect its *weight* as a physical object.

In conclusion, the strange but substantiated truth of the subatomic world is that particles sometimes behave like waves, and electromagnetic waves sometimes behave like particles. When it comes to electromagnetic waves, the higher the energy of a wave, the greater its tendency to behave like a particle. We can state the following key points:

1. Visible light usually behaves like a wave, but sometimes like a particle.
2. X-Rays, with higher energy, behave like a *particle* most of the time, but can behave like a wave.
3. What determines which behavior they manifest is the specific conditions around them, that is, the method used to observe or detect them.

Figure 5-32

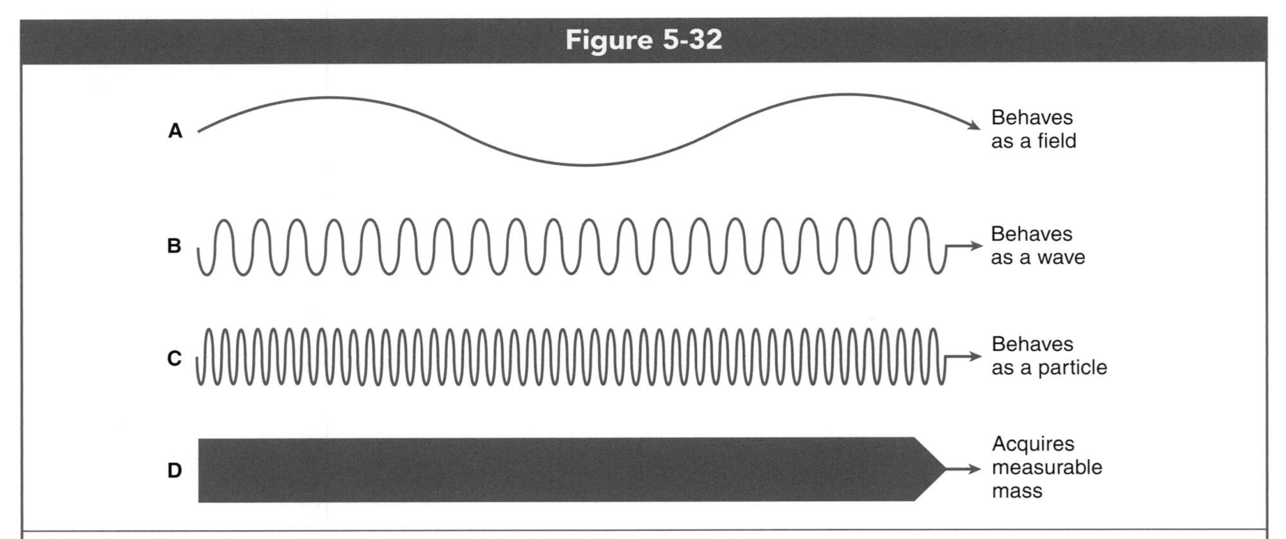

As energies increase from ***A*** to ***D***, very long wavelengths behave like *fields*, intermediate wavelengths behave like *waves*, very short wavelengths begin to behave like *particles*, and objects with extremely short wave functions, with very high energy "compressed into a small space," take on measurable mass and can be weighed.

What would an accurate sketch of an x-ray beam look like? How can the dual nature of the x-ray photons (quanta) best be visualized, with each one appearing as a self-contained "particle" but also possessing a wavelength? Since x-rays behave like particles most of the time, a stream of continuous waves is probably not the best model. The "corpuscular" model shown in Figure 5-33***A*** makes a good illustration, in which each photon appears as a blob or corpuscle containing a specific wavelength within it. Or, as in Figure 5-33***B***, each photon might be depicted as a separated "chunk" of a wave, long enough to show the wavelength, but short enough to indicate that it occupies a specific area in space. These may be about as good an illustration as we can invent to indicate the dual nature of the particle-wave called an *x-ray*.

Figure 5-33

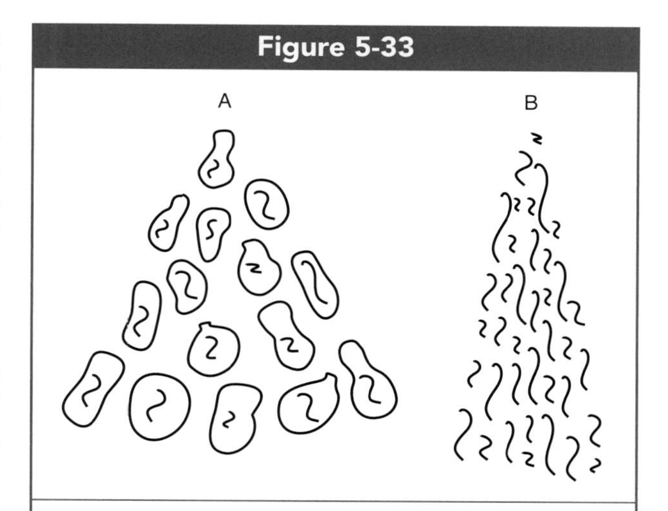

Two more accurate ways of visualizing an x-ray beam: ***A***, the corpuscular model, in which the beam consists of photons or "bundles of energy" which each have an associated wavelength, or ***B***, waves which are not continuous but fragmented into individual segments representing photons.

SUMMARY

1. Waves can be transverse or compressional. All waves have speed, amplitude, wavelength and frequency. Frequency is proportional to speed and inversely proportional to wavelength.
2. Electromagnetic waves are transverse disturbances in electrical and magnetic fields, always perpendicular to each other. Their speed is fixed at ***c***. Their frequency is proportional to their energy, while their wavelength is inversely proportional to their energy.
3. For an x-ray beam, the minimum wavelength is inversely proportional to the kVp set at the console.
4. The electromagnetic spectrum is made up of eight radiations which differ in their penetration and absorption properties for various materials, based on resonance with the natural frequency of the material.
5. In medical practice, MRI uses the resonance of radio waves with precessing protons to create an image, ultrasound uses the reflectivity of sound waves, and lasers use synchronized, amplified light.
6. Lasers are widely employed in medical imaging. They use a power source to "pump" energy into the stimulable atoms of a medium so that they give off light. A resonant cavity amplifies and synchronizes the light waves of a specific frequency, which are then emitted through an aperture.
7. Unlike light, x-rays cannot be reflected, refracted, or dispersed by mirrors, lenses or prisms. As they pass through the body, they are *attenuated* in intensity or energy by interacting with atoms.
8. Radiographs are negative images consisting of lighter radiopaque areas and darker radiolucent areas.
9. Since matter and energy are interchangeable, at the subatomic level particles have wave functions and sometimes behave like waves, while high-energy electromagnetic waves such as x-rays often behave like particles.
10. The term *photon* (or *quantum*) is used to describe the particle-like bundle of energy which constitutes an individual x-ray.

REVIEW QUESTIONS

1. Which characteristic of a wave is associated with its intensity or strength?

2. What are the two broad categories of waves?

3. In music, the note *middle C* has a frequency of 262 hertz. If the speed of sound in air is 340 meters per second, what is the wavelength of *middle C*?

4. If the wavelength of a homogeneous beam of x-rays was reduced to one-half the original wavelength, exactly how would the number of waves striking you per second change?

5. The dial on your AM radio is set at 11, which is 1100 kilohertz, to tune in to a local rock-and-roll station. (Speed of light = 3×10^8 m/s.) What is the wavelength of the carrier wave?

6. The wavelength of red light is approximately 650 nanometers (650×10^{-9} meters). What is the frequency of this electromagnetic wave? (Speed of light = 3×10^8 m/s.)

7. What is the minimum wavelength of a 75-kVp x-ray beam?

8. The source of electromagnetic waves is a vibrating or oscillating __________.

9. Which electromagnetic radiation has an energy of 2 to 3 volts?

10. Which electromagnetic radiation has a wavelength about the width of your smallest fingernail?

11. A body tissue which allows most x-rays to penetrate through it is described in one word as:

12. Sonography (ultrasound) images are made possible because when the sound waves reach an interface, they__________, a characteristic which x-rays do not share.

13. List four different applications for lasers in the medical imaging department:

14. What are the three required components for every laser?

15. Laser printers use the heat from a laser beam to turn molecules of __________ black within the emulsion of the film.

(Continued)

REVIEW QUESTIONS *(Continued)*

16. The electrical wave and the magnetic wave components of an x-ray are always __________ to each other.

17. What two terms describe the *random* refraction and reflection which ultrasound waves undergo when they encounter an air pocket?

18. Experiments confirm that electrons have a wave function because two beams of electrons can __________ with each other.

19. Visible light behaves like __________ in most circumstances, whereas x-rays behave like __________ in most circumstances.

20. In the diagram below (Fig. 5-34), what is the wavelength of this wave?

Figure 5-34

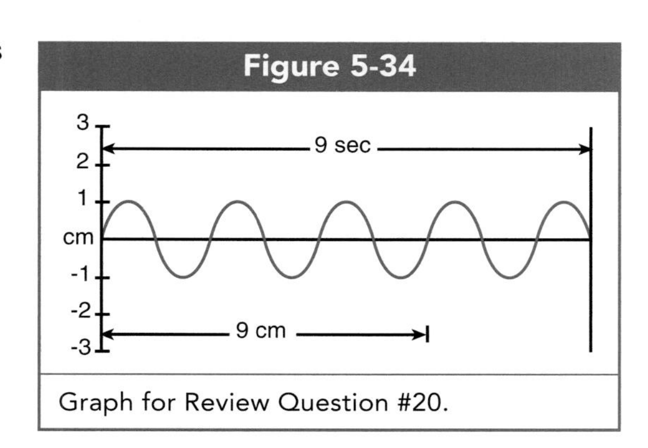

Graph for Review Question #20.

21. In Figure 5-35 below, how fast does this wave travel through its medium?

22. Determine the frequency for the wave shown below in Figure 5-35:

Figure 5-35

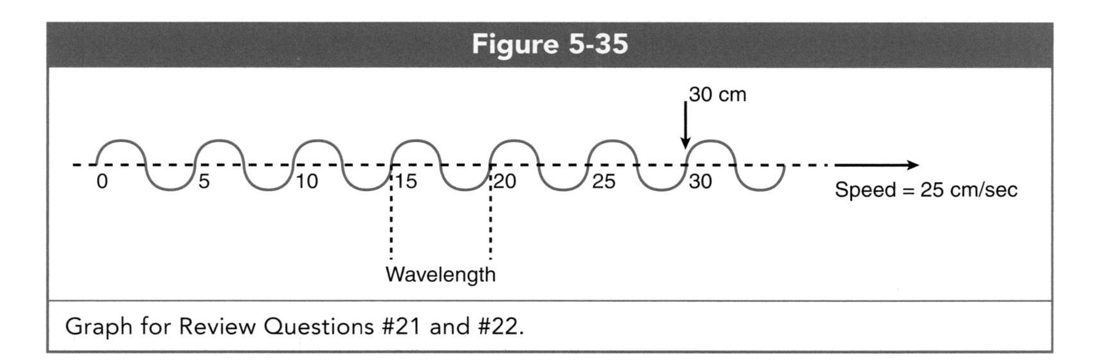

Graph for Review Questions #21 and #22.

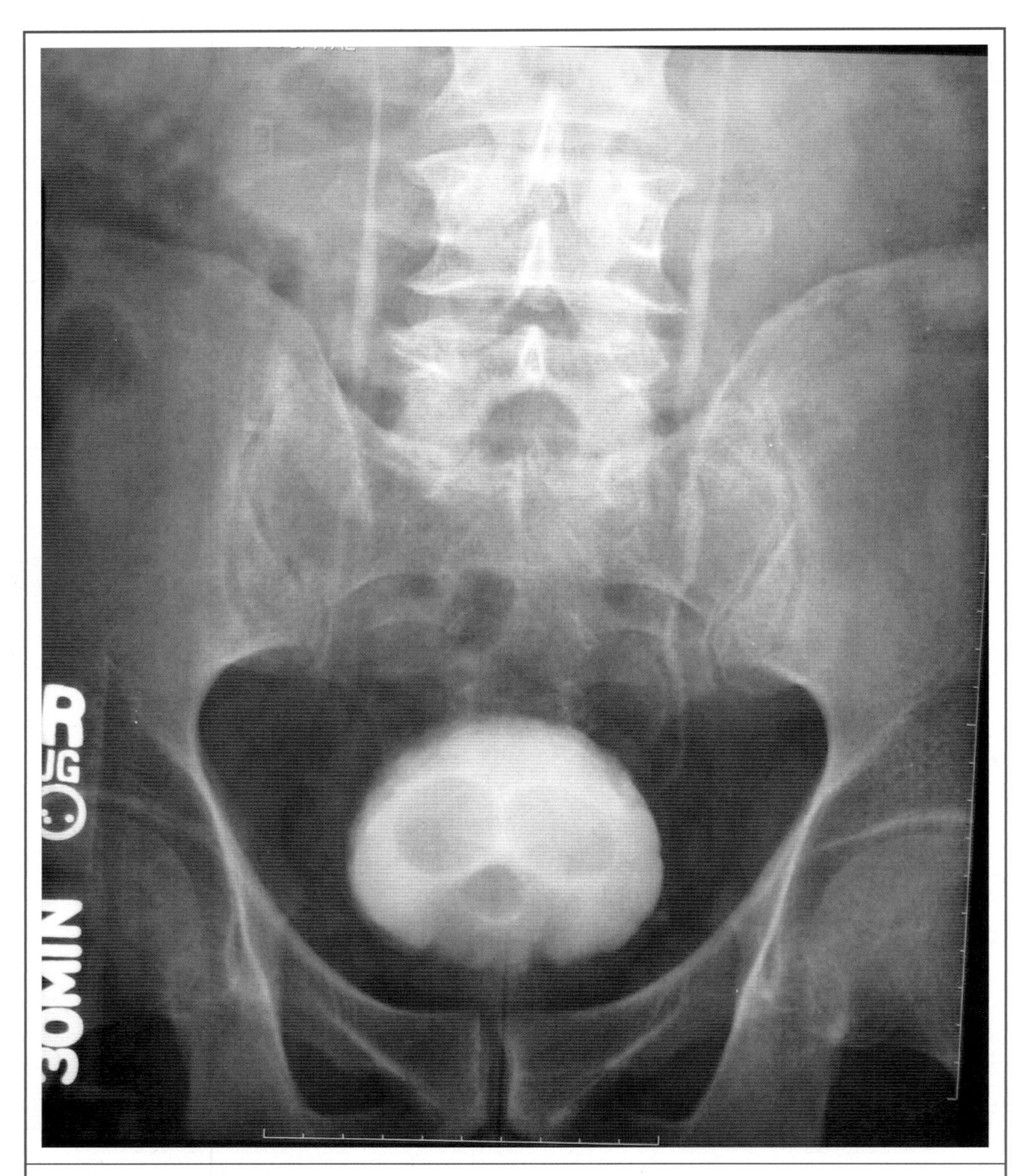

The "skull in the bladder." Rectosigmoid gas bubbles superimpose the bolus of iodine in the bladder, and the rugal folds of the bladder floor create the appearance of upper teeth.

Chapter 6

MAGNETISM AND ELECTROSTATICS

Objectives:

Upon completion of this chapter, you should be able to:

1. Define the *magnetic moment* of particles, the *magnetic dipole* of atoms, and *magnetic domains* in materials.
2. List and describe the four types of materials according to their magnetic properties.
3. Define magnetic *permeability* and *retentivity.*
4. Describe the effects of different materials on magnetic lines of force.
5. Solve simple Gauss' law formula problems for the effects of pole strength and distance.
6. State the *five laws of electrostatics.*
7. Explain why only negative charges move in a solid, and why they distribute themselves evenly only on its surface.
8. Describe the effects of charge and distance on electrical force.
9. Define the three methods of *electrification.*
10. Define *potential difference* and *electromotive force.*
11. Solve for charge distribution when objects of varying charge come into contact.
12. Explain *polarization* and *grounding.*
13. Describe how the *pocket dosimeter* uses the principles of the electroscope to detect and measure radiation.

The earliest known experiment in electromagnetism was conducted in 1819 by Hans Christian Oersted, a Danish physicist who discovered that a magnetic needle such as that used in a compass was deflected by a nearby electrical current (Fig. 6-1). He concluded that, in addition to the electrical field which caused a pushing or pulling force due to their charge, *all moving electric charges develop magnetic fields around them.*

A commonplace example of this phenomenon is the static you may hear on a radio or TV in your car as you pass under large power lines at some intersections, especially if you are listening to an AM radio station. The movement of electrical current in these cables generates a powerful magnetic field which reaches for several feet around them (Fig. 6-2). We have discussed in the last chapter how the electromagnetic waves coming from a radio station cause the electrons in the antenna on your car to oscillate up and down, which in turn creates an electrical signal within the car radio. The large magnetic field around power lines also affects these electrons, disturbing their orderly oscillation, and causes static noise to come out of the speaker. In the same way, an activated cell phone disturbs nearby computer speakers.

The discovery of Oersted would eventually lead to the development of both motors (based on the ability of electrical current to move magnets) and generators (based on the ability of moving magnets to induce electrical current), topics to be discussed in the next chapter.

Also introduced in the last chapter was the concept that particles within an atom have a property called *spin*, which, although it does not precisely match the full rotations of a planet like the earth,

Figure 6-1

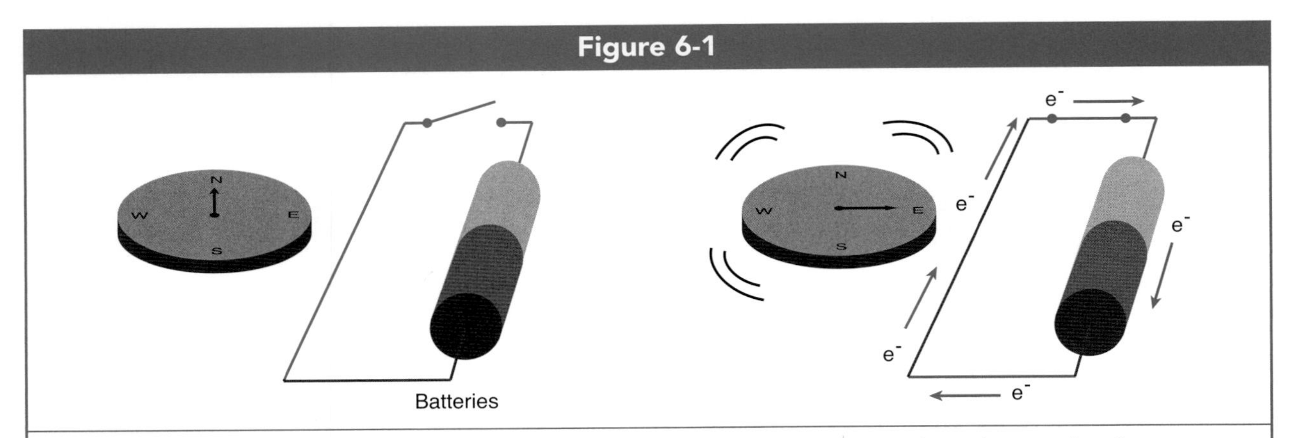

Hans Oersted discovered that a flowing electrical current nearby (right) will deflect the needle of a compass. Therefore, the moving electrical charge must be generating a magnetic field around it.

Figure 6-2

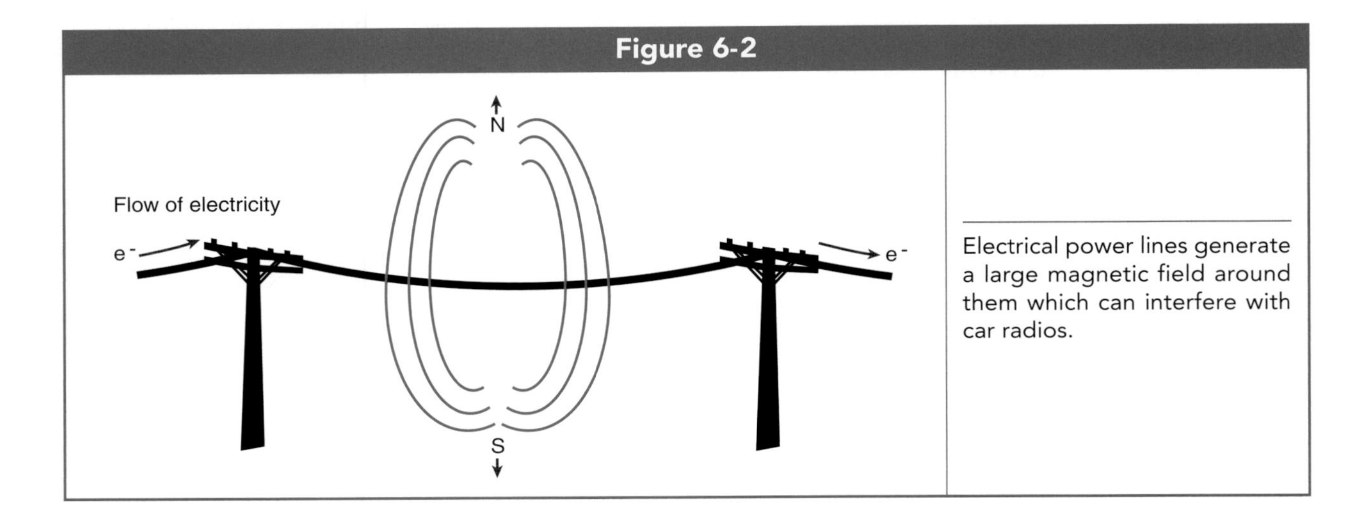

Electrical power lines generate a large magnetic field around them which can interfere with car radios.

Figure 6-3

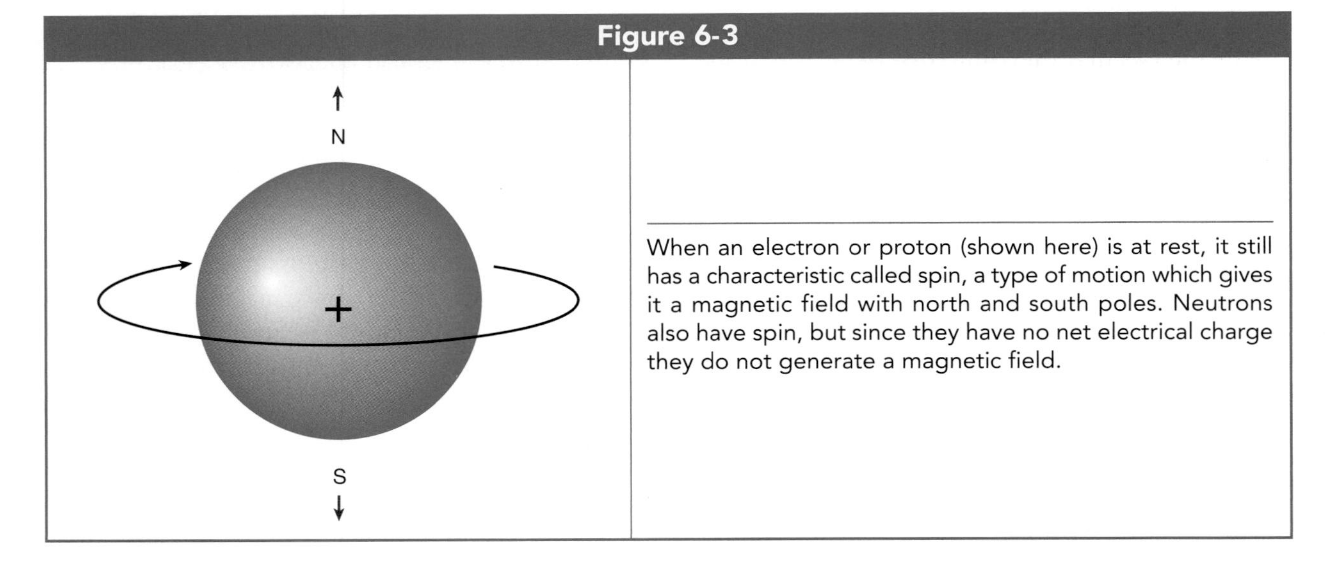

When an electron or proton (shown here) is at rest, it still has a characteristic called spin, a type of motion which gives it a magnetic field with north and south poles. Neutrons also have spin, but since they have no net electrical charge they do not generate a magnetic field.

Figure 6-4

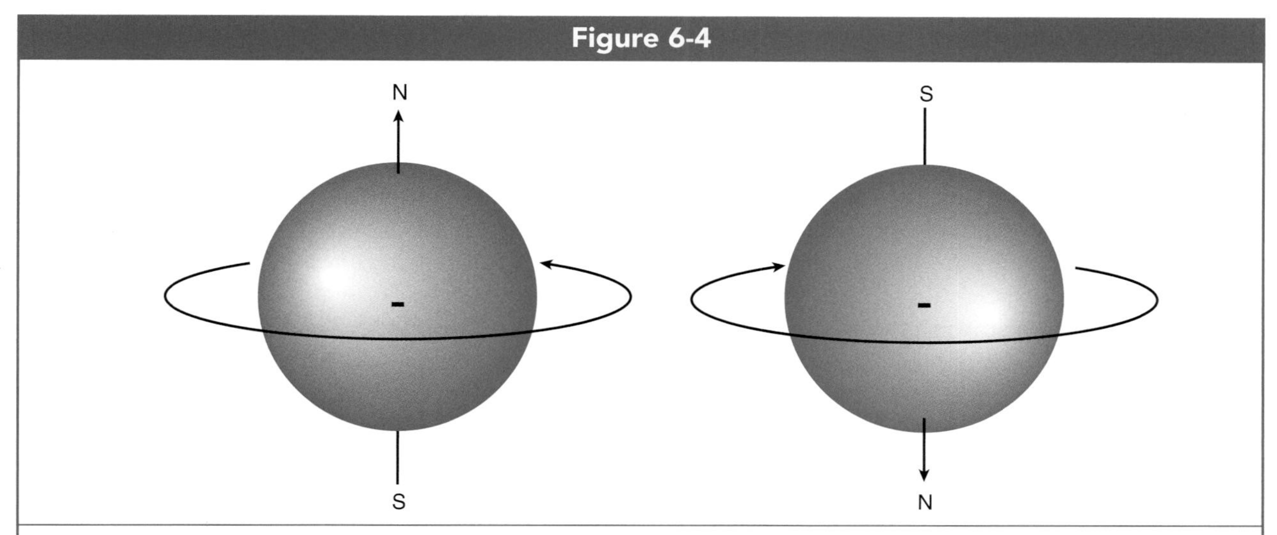

An electron with reversed spin may be thought of as having its magnetic poles flipped "upside down," or its *magnetic moment* inverted. In a sub-orbital filled with its pair of electrons, one is always inverted so that their magnetic moments cancel each other out.

nonetheless generates angular momentum and may be considered a type of *motion*. Therefore, each charged particle within an atom, be it a proton or an electron, develops a magnetic field around it, with a north pole and a south pole (Fig. 6-3). (A neutron also has spin, but because it is electrically neutral, it does *not* develop a magnetic field.)

By using a somewhat simplified view of spin, we can imagine that electrons which spin in a counter-clockwise "rotation" may be defined as having their *north pole* pointing upward, while those with spin in a clockwise "rotation" have their north pole pointing downward (Fig. 6-4). The direction of spin determines the orientation of the magnetic field, called the *magnetic moment*.

An entire atom may develop its own systemic magnetic field. Remember that electrons are arranged as a *pair* in each orbital, an orbital being a subdivision of a "shell." Each specific orbital can accommodate one electron with its magnetic moment pointing "up," and one electron with its magnetic moment pointing "down." In this case, the magnetic moments of the two electrons cancel each other out and no net magnetic field remains. But, if an atom has an *odd number of electrons*, there will be at least one orbital with a single electron whose magnetic moment is *not* cancelled. This generates a small net magnetic field around the entire atom, called a *magnetic dipole* (Fig. 6-5). If the one unpaired electron has counter-clockwise spin with its north pole pointing up, then the

Figure 6-5

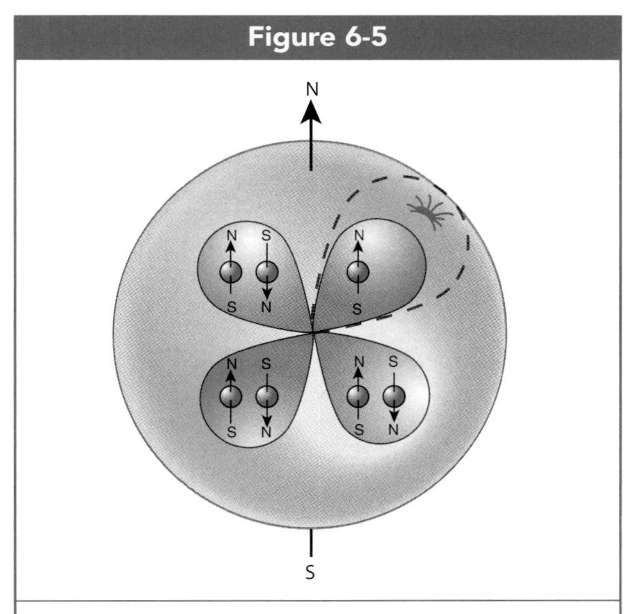

A *magnetic dipole* is an entire atom with a magnetic field. If there is an odd number of electrons, the unpaired electron will not have its magnetic moment cancelled out, and the atom as a whole develops a weak magnetic field with the north pole in the same direction as that electron.

entire atom will have a weak magnetic field with its north pole pointing upward, and a *net* spin that is counter-clockwise.

Now, in easily magnetized materials such as iron, we find that even when an iron bar has *not* been yet magnetized, the atoms within small regions of the bar have a tendency to "line up" with each other such that their net spins are in the same direction, clockwise or counterclockwise, and their north poles all point the same way (Fig. 6-6***A***). These small regions of aligned atoms are called *magnetic domains*. However, the bar as a whole will not act as a magnet because the random alignment of the several magnetic domains tends to cancel one another out.

MAGNETS

Iron, nickel and other materials which are easily magnetized are called *ferromagnetic*, (*ferrum* = "iron"). When these materials are placed within an *external* magnetic field, their magnetic domains all tend to line up with the field, especially when the material is hammered or tapped while in the field (Fig. 6-6***B***). *Lodestones* are ferromagnetic rocks that behave like weak *natural* magnets because many of their magnetic domains have aligned with the earth's magnetic field.

Figure 6-6

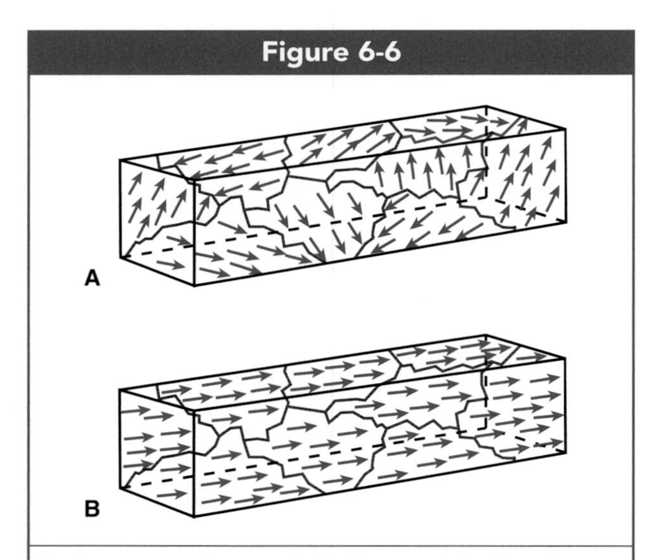

In ferromagnetic materials, ***A***, *magnetic domains* are regions in which the atoms tend to line up their magnetic dipoles. When the material is magnetized, ***B***, all of the regional domains line up uniformly with an external magnetic field.

The *ease* with which a material can be penetrated by an external magnetic field (and thus magnetized) is called its *permeability*. A permeable material may be thought of as *magnetically soft*. The ability of the material to retain or *hold onto* its magnetism over time is called its *retentivity*. Unfortunately, most ferromagnetic materials with high permeability tend to have low retentivity. Once they have been magnetized, if they are hammered or tapped *outside of an external magnetic field*, they quickly lose their magnetism as the magnetic domains are jolted back into random directions.

Most *permanent* or *artificial* magnets are made from mixtures of iron, nickel, cobalt and other substances in order to improve their retention of the field. They are often then more *magnetically hard*, and can only be magnetized by very strong magnetic fields, but most of these will have high retentivity and remain magnetized for long periods of time. Of course, the ideal alloy is a mixture of metals that allows both high permeability and high retentivity. To magnetize a ferromagnetic object, a strong permanent magnet may be repeatedly stroked alongside it in one direction.

Most materials are nonmagnetic and unaffected by magnets. Certain elements, such as oxygen and sodium, have magnetic dipole atoms but do not form magnetic domains. They are called *paramagnetic* materials, and are only slightly attracted to strong magnets. A few substances, such as glass and water, are actually *repelled* away from magnets and are referred to as *diamagnetic*.

A fascinating aspect of magnets is that they always remain *dipolar* or *bipolar*, retaining two poles. If a magnet is broken into two smaller magnets, each will develop its own north pole and south pole. Even a single atom has its two poles. Unlike electrical charges, which we have been able to separate into individual positive and negative particles, no *monopole* having only north or south properties has ever been discovered.

In addition to natural magnets and artificial (or permanent) magnets, a third type of magnet can be manmade by using only electricity. As Oersted

Figure 6-7

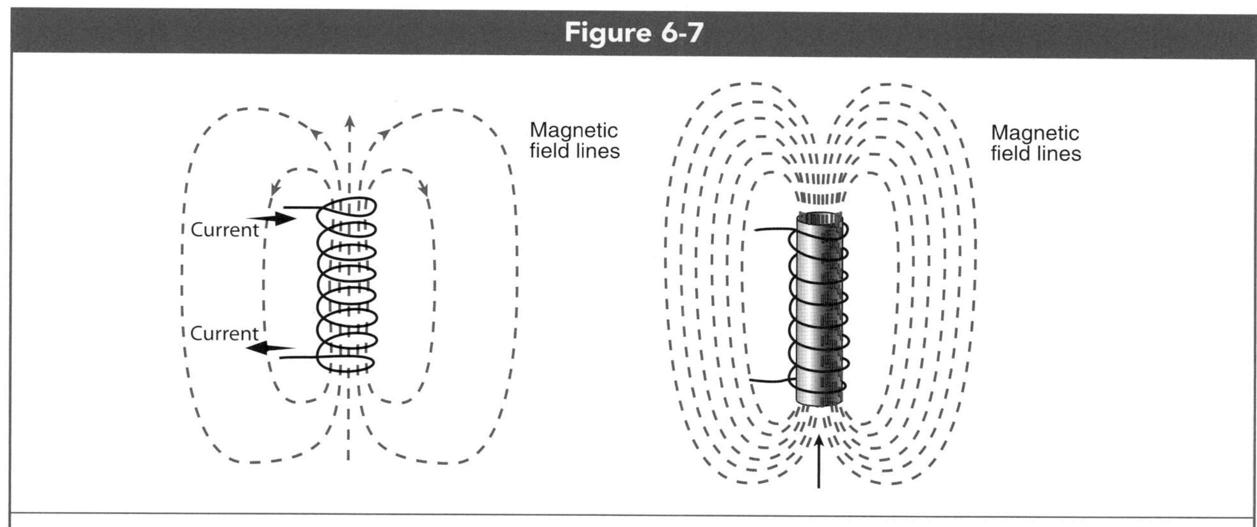

Inserting an iron core into a coil of wire carrying electrical current intensifies the strength of the magnetic field formed around it.

discovered, all electrical currents induce a magnetic field around them. By passing electricity through a *coil* of wire (or *solenoid*), a strong *electromagnet* is created with north and south poles. Its strength can be further multiplied by inserting an iron bar within the coil (Fig. 6-7). When the direction of electrical current is reversed, the north and south poles also reverse their position. We might say that an electromagnet has almost no retentivity, since it immediately loses its magnetic field when electrical current stops flowing through it.

If magnetic fields are always associated with electrical current, it should come as no surprise that electrical currents are also associated with magnetic fields. Indeed, very small electrical currents have been detected flowing within permanent bar magnets.

MAGNETIC FIELDS

Although the field around a magnet is normally invisible to the naked eye, it can be mapped out by shaking a thin layer of very fine slivers of iron onto a piece of paper with a magnet held underneath. A magnificent three-dimensional view of the magnetic field can be obtained by surrounding it with a viscous fluid containing fine slivers of iron as demonstrated in Figure 6-8.

These experiments show that the magnetic field follows distinct *lines of force* which take on the shape of elephant ears, with the magnetic force flowing *outward* from the north pole, then circulating around and flowing *into* the south pole. (In fact, this is how the poles are defined as *north* and *south*.) The

Figure 6-8

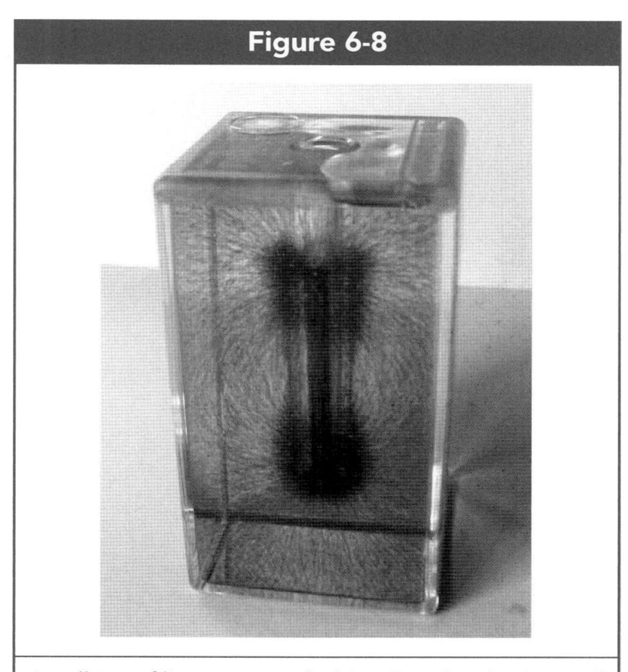

Small iron filings suspended in oil make the lines of magnetic force visible around a bar magnet.

magnetic field is strongest and most concentrated at the poles, indicated by the lines of force merging closer together.

Remember, too, that these lines of force will always be *perpendicular* to the lines of force of associated electrical fields. (Figure 5-9 in the previous chapter.)

Figure 6-9 illustrates how the magnetic field lines of force are affected when the opposite poles of two magnets approach each other, and when different materials are inserted between the magnets. Note that ferromagnetic materials such as iron cause the lines of force to concentrate, indicating an increase in strength of the field. Nonmagnetic materials have no effect on the lines of force. Diamagnetic materials such as glass weaken the field, causing the lines of force to deviate.

Figure 6-9

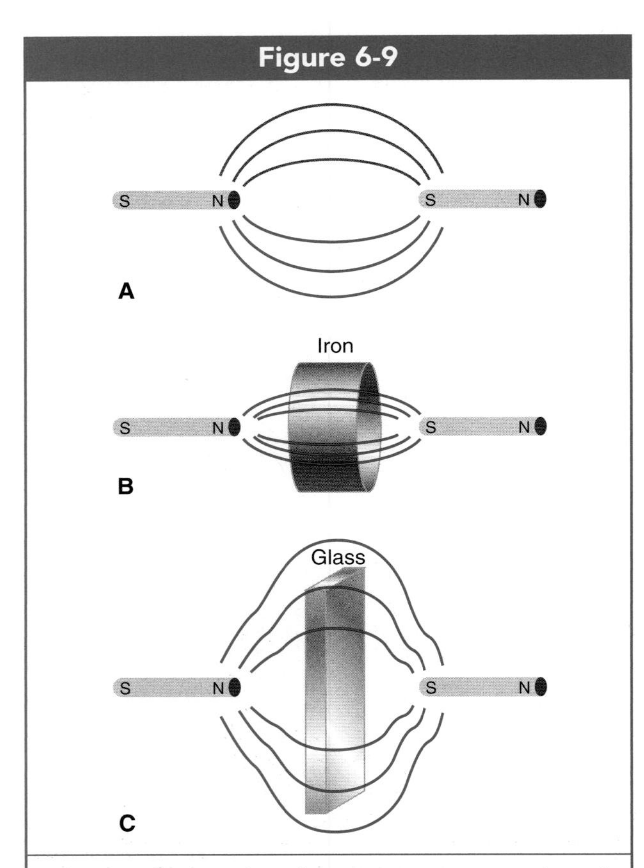

When opposite poles of two magnets approach each other, the lines of magnetic force between them are (***B***) concentrated and strengthened by ferromagnetic materials placed between them, and (***C***) deviated and weakened when diamagnetic materials intervene. Non-magnetic materials have no effect.

Magnetic fields follow two basic laws of behavior. The first is that like poles repel each other, and opposite poles attract. It is interesting to note that the north pole of a compass needle is indeed the magnetic north of the needle itself, and so it should be attracted to and point toward the earth's *south* magnetic pole. In fact, it does. Geologists have found from magnetic strata that the earth's magnetic field reverses position every 200,000 years or so. Currently, it is "upside down." That is, at the present time the earth's north *geographic* pole is actually its *south magnetic pole!* Thus, the north pole of a compass needle does point toward its opposite.

The second law, called *Gauss' Law*, states that the strength or intensity of the magnetic field is proportional to the product of the pole strengths, and inversely proportional to the *square* of the distance between them (which you may recognize as the *inverse square law*). Gauss' formula summing up these relationships is stated:

$$F = k \frac{P_1 P_2}{D^2}$$

Where ***F*** is the amount of force of attraction *or* repulsion, ***P*** is the strength of the pole from each magnet, ***1*** and ***2***, and ***D*** is the distance between the poles. (The constant ***k*** adjusts for units, normally gauss to newtons which is the generic unit for force. If the force between the magnets is measured in gauss, the same as the pole strengths, then the value of ***k*** is 1 and ***k*** can be ignored.)

What are the *units* with which pole strength is measured? The most convenient unit for typical bar magnets is the *Gauss*, abbreviated G. The strength of a magnetic cabinet door latch is on the order of 1000 G. By comparison, the earth's magnetic field is about 1G at the poles and ½ G at the equator. The superconducting magnets used in magnetic resonance imaging (MRI) and other scientific applications such as large particle accelerators require a much larger unit. The magnetic strength of these devices is measured in *Tesla*, abbreviated *T*. One Tesla is equal to 10,000 Gauss. A typical medical MRI unit generates a magnetic field between 1.0 and 1.5 Tesla. Both units are named for physicists who made important discoveries relating to magnetism.

Let's try a couple of exercises to be sure you understand Gauss' formula:

Practice Exercise #1:

Two magnets are positioned 1cm apart with their south poles facing each other. Magnet A has a strength of 50 Gauss measured at 1 cm from either pole. Magnet B has a strength of 20 Gauss. By Gauss' formula, what is the total force of repulsion with which the two magnets are pushing each other away?

Solution: With k =1 for units in gauss, and for a set, unchanged distance, the total force generated is as follows:

$$F = P1 \times P2$$
$$= 50 \times 20$$
$$= 1000$$

Answer: The total force of repulsion is 1000 Gauss.

Practice Exercise #2:

The above two magnets are now repositioned with their south poles 3 cm apart. What is the total force of repulsion between them at this new distance?

Solution: This is an inverse square law problem which should be set up as follows: (See Chapter 3 for help.)

$$\frac{F_o}{F_n} = \frac{(D_n)^2}{(D_o)^2}$$

$$\frac{1000}{X} = \frac{3^2}{1^2} = \frac{9}{1}$$

Cross multiplying: $9X = 1000$

$$X = \frac{1000}{9}$$

$$X = 111.11$$

Answer: The total force of repulsion is 111 Gauss.

ELECTROSTATICS

Electrostatics is the study of electrical charges *at rest*, or *static electricity*, as distinguished from regularly moving electrical current. An object becomes *electrified* any time it develops either an excess of electrons or a deficiency of electrons. (This is directly related to the *ionization of atoms* discussed in Chapter 4.) This build-up will be discharged in the form of a spark when the object comes close to anything that will conduct electricity to or from the ground. Lightening is an identical discharge of static electricity, but, of course, on a much larger scale.

The ground of the earth acts as a kind of *infinite reservoir* for electrical charge—large amounts of negative charge can be discharged when electrons flow down into the ground. Objects with positive charge can also *receive* large numbers of electrons from the ground when they come in contact with it or are connected to it by a conductor such as a wire or metal rod. We refer to this making contact as *grounding* the object. By grounding, any static electricity built up on an object is effectively dissipated.

The unit for electrical charge is the *Coulomb*, named after a pioneer in the study of electrostatics. One Coulomb of negative charge represents 6.3×10^{18} electrons—this is slightly more than *6 billion billion* electrons. Sparks jumping to or from your fingertips may discharge only a microcoulomb of charge, yet this still represents an incredible number of electrons, 6.3×10^{12} or 6 *trillion* electrons.

Remember that electrical charge can be either negative or positive. A *positive* charge of 1 Coulomb would mean that the object was *deficient* of 6.3×10^{18} electrons. This will still result in a spark when another object approaches, but the *flow* of electrons will be from the other object to the positively-charged one.

In fluids, free electrons will drift toward a positive plate, and whole atoms that are positively charged will drift toward a negative plate. However, in *solids* the atoms are locked into place, with the positively-charged protons embedded deep within their nuclei. Only the electrons that are loosely bound in the outermost shells of the atoms are free to move.

The Five Laws of Electrostatics

Five fundamental laws govern the behavior of electrical charges. They are:

1. Like charges repel each other, opposite charges attract.
2. In solid objects, only negative charges (electrons) can move.
3. In solid objects, electrical charges exist *only on the surface*. This is because of the first law. Since they repel each other, they will not remain inside an object but will move as far away from

each other as they can, to the outer surface, where they will then be distributed evenly at the maximum distance from each other that the surface area allows (Fig. 6-10).

4. In solid objects, electrical charge will concentrate at the greatest curvature of the surface. As shown in Figure 6-10, this is entirely due to geometry. Charges will be evenly distributed across the *surface* area, but where the surface curves, we find them more concentrated within the *volume* of space, such as per *cubic* millimeter.
5. Coulomb's law states that the amount of force generated by electrical repulsion or attraction is proportional to the product of the two charges, and inversely proportional to the *square* of the distance between them, (the inverse square law). Coulomb's formula is:

$$F = k \frac{Q_1 Q_2}{D^2}$$

Figure 6-10

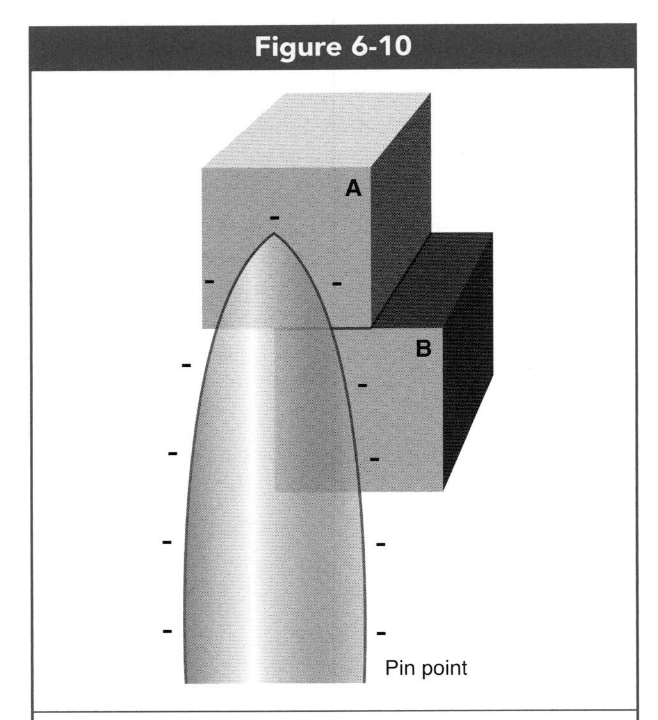

Static electrical charge distributes itself evenly along the *surface* of a solid object. This results in higher concentration of charge *per cubic millimeter* at the greatest curvature of the surface. Note that at the point of this pin, there are 3 negative charges within a cubic millimeter, ***box A***, while along the sides there are only 2 charges per cubic millimeter, ***box B***.

Where F is the force, Q is the amount of charge on each object, 1 and 2, and D is the distance between the objects. (The constant k adjusts for units, Coulombs to volts or Newtons.)

ELECTRIFICATION

Objects can become *electrified* in three different ways: by friction, by contact, and by induction. Some materials are composed of atoms that have very loosely bound outer shell electrons. Electrification by friction occurs when such a material is rubbed against another material. Both materials may have been electrically neutral at first, but loosely-bound electrons are stripped from the first (leaving it with a positive charge), and deposited onto the other material (giving it a negative charge). We hear the crackling sound of discharges of static electricity when removing clothing from a dryer as the clothes rub together. Combing your hair when it is very dry produces a crackling sound as sparks jump back and forth. Negative charges accumulate on the comb, leaving your hair positively charged. The hair stands on end as the positive charges repel each other and push the individual hairs apart.

Discharges of static electricity in or around image processing equipment can cause artifacts in the image. There are three ways to minimize static discharges: First, insulator (or *dielectric*) materials such as rubber, plastic or oil can be placed around electrical wires and devices; second, *grounding* metal structures by providing a conductive path (wire) allows any build-up of charge to flow to the ground before causing a spark; and third, keeping the humidity high (above 40%) allows water molecules from the air to attach to charged surfaces, neutralizing the charge—water molecules are *bipolar*, with one side slightly more positive and the other more negative, so one side or the other of the molecule is always attracted to local charges.

Electrification by contact, or *conduction*, occurs whenever *a potential difference* exists between two objects that touch each other. Electrical charge at rest is a type of *potential energy*, specifically, the potential to cause electrons to move or electricity to flow. Any difference in charge is also a difference in *potential*.

SIDEBAR 6-1: When x-ray images were recorded on film, an ever-present challenge was preventing artifacts, such as those in Figure 6-11, which were caused by static electricity discharges as film was slid across counter tops and into cassettes, or moved by conveyor belts into processors. By keeping the air humidity aove 40%, static was kept at a minimum.

Figure 6-11

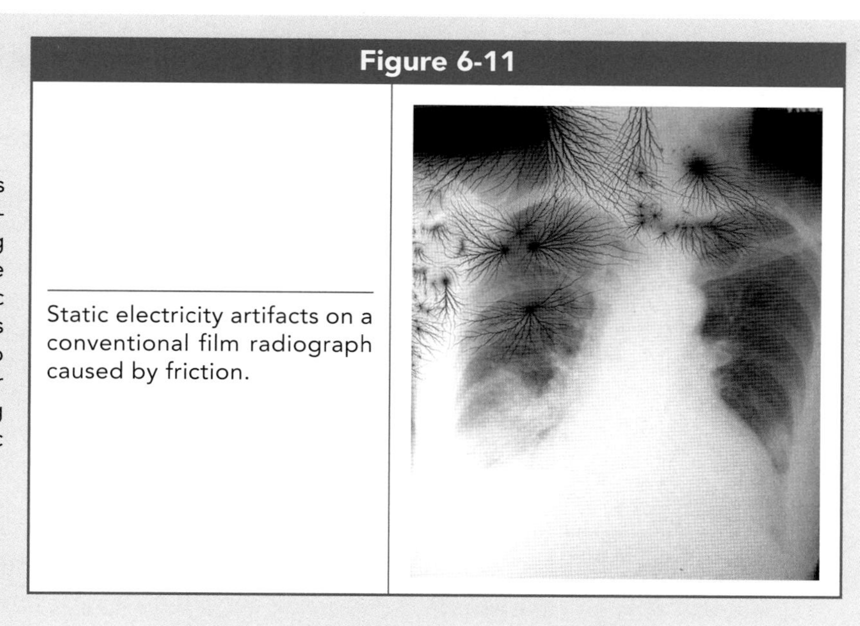

Static electricity artifacts on a conventional film radiograph caused by friction.

A difference in electrical charge between two objects will cause electrons to *move* to new positions when the objects touch—negative electrons will always flow toward those surfaces that are *relatively more positive*. We learned in Chapter 2 that anything which can cause another object to move is a *force*. The force created by any potential difference is referred to as *electromotive force*, or *EMF* (from *electro* for electron, and *motive* for motion). The unit for both potential difference and electromotive force is the *volt*. The higher the voltage, the greater the force tending to push electrons from one point to another.

The movement of electrons from one object to the other will always be such that their positions are spread out as far as possible from each other across the surface *as if the two objects were one object*. The result is that when the objects are then separated, the *charge will always be equalized at the lowest possible value*. For example, if there are 10 excess electrons on one object and 6 excess electrons on the other, when they come in contact the sum of 16 electrons will spread out across both objects, so that when they are separated there will be 8 excess electrons on each object.

Remember that for an exchange of electrons to take place it is not necessary that one object be negatively charged and the other positively charged—only that there is *any difference in potential or charge*. One object could be charged and the other be neutral; or, both could be positively charged, but one more than the other—in this case, electrons will move from the *less positive* object to the *more positive* object. As long as there is any difference at all, electrons will reposition themselves on contact to equalize the charge on the two objects. Figure 6-12 demonstrates several scenarios for this equalizing of charge.

The term *induction* comes from the verb *to induce*. In social terms, we induce a person to do something when we persuade them to rather than using physical force which would involve touching them. Electrification by induction occurs when a charged object *induces* charge in another object nearby without directly touching it.

Imagine that a metal bar with a large negative charge on it is brought near the end of another metal bar which is electrically neutral. Outer shell electrons which are loosely bound in the atoms at the end of the neutral bar will be repelled by the negative charge nearby. They will begin to move away from it and toward the far end of the bar, where they will accumulate. This leaves the near end of the bar with a positive charge, since it is now deficient of electrons (Fig. 6-13). Technically, the overall bar still has a balance of charge, but it has become *polarized*, with negative charge accumulated at one end (or pole) and positive charge at the other end.

Figure 6-12

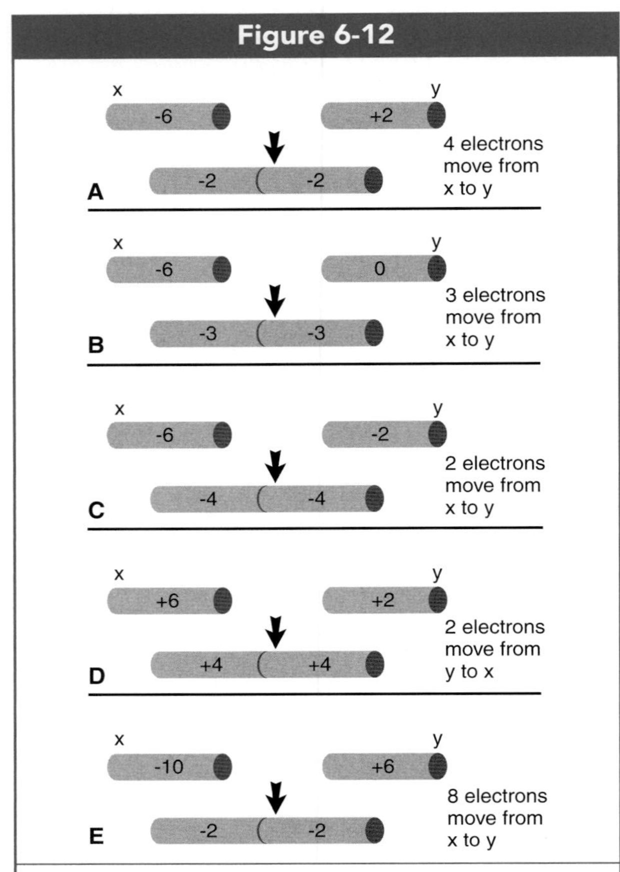

Equalization of electric charge by the contact of two bars with: ***A***, charges of –6 and +2; ***B***, –6 and zero (neutral); ***C***, –6 and –2; ***D***, +6 and +2; ***E***, –10 and +6.

Suppose that the negatively charged end of this bar is now *grounded* by placing it on a metal table or other conductor that allows its electrons to flow to the ground. When it is lifted back up, the entire bar will be left with a net positive charge (Fig. 6-14). We may say that the bar has been electrified by induction, since the original charged object never touched it but was only brought near to it.

Using an Electroscope to Detect Radiation

A simple device to detect electrical charge can be made by placing two strips of metal foil close together with their flat sides facing each other, and connecting them at the top to a common metal bar or plate, as shown in Figure 6-15. The strips must be suspended from a cork or rubber stopper that acts as an insulator, so the strips do not become accidentally grounded. Any device based on this basic layout is called an *electroscope*. Variations of the electroscope are widely used in devices for detecting radiation, and so it is pertinent for radiographers to know how an electroscope works.

If a charged object is brought near the top plate or bar of an electroscope, electrification by *induction* takes place in the leaves of metal foil. For example, if a negative charge is held over the device, electrons within the bar and foil will be repelled toward the bottom of the leaves. The bottoms of the two leaves of foil, having like charges, will be repelled from each other and begin to move away from each other. It can then be observed that the further these foil leaves separate, the stronger must be the charge on the object above the electroscope.

In order to detect radiation, the electroscope is first prepared by placing a positive charge on it. One way this can be done is by touching it to the positive end of a battery. With positive charge distributed throughout the foil leaves, they will separate (Fig. 6-15***B***). Whenever radiation enters the electroscope, it will *ionize* molecules in the air surrounding the foil by knocking electrons out of atoms in the air

Figure 6-13

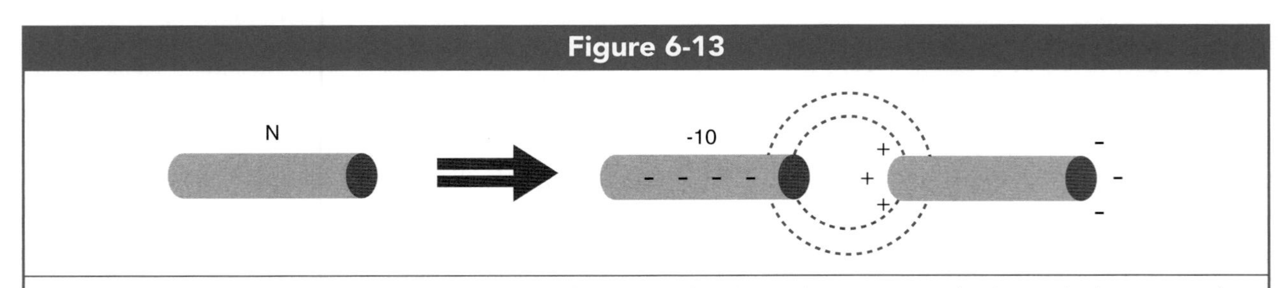

Electrical polarization of a metal bar by induction from a nearby object that is negatively-charged. Electrons within the bar are repelled to the opposite end, leaving both ends of the bar in a charged state.

Figure 6-14

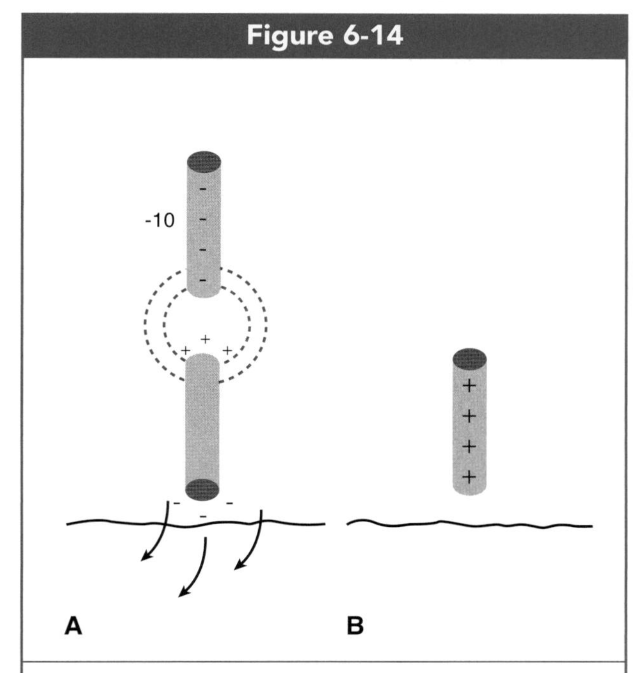

Touching the negative end of the bar to the ground while it is polarized allows the electrons to flow out, leaving the entire bar with a positive charge.

Figure 6-15

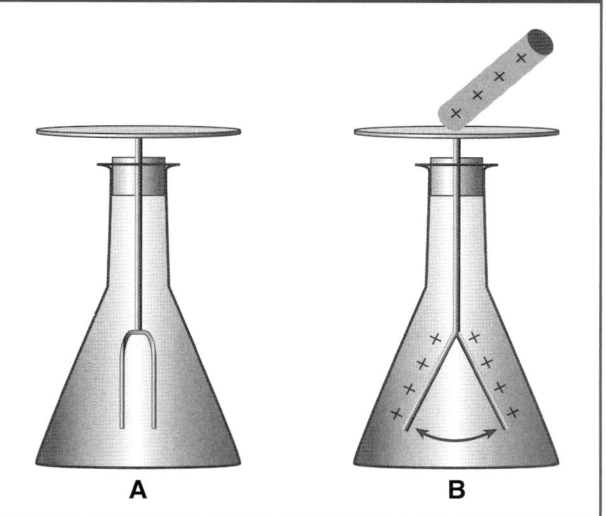

A simple electroscope consisting of two aluminum foil leafs attached to a conductor plate above the glass jar. A rubber or cork stopper is used as an insulator holding the metal parts in place. A positive charge placed on the electroscope causes the foil leaves to separate from mutual repulsion.

(Fig. 6-16*A*). These freed electrons will be attracted to the positively-charged foil and move toward it. As the electrons contact the foil leaves, they cancel out positive charge by filling gaps in those atoms that are missing electrons in the foil. Thus, as more and more electrons are freed from the air, the positive charge on the foil is neutralized, and there is less force of repulsion between the two leaves. The two leaves, by spring tension, then tend to fall back toward each other as shown in Figure 6-16***B***.

Figure 6-16

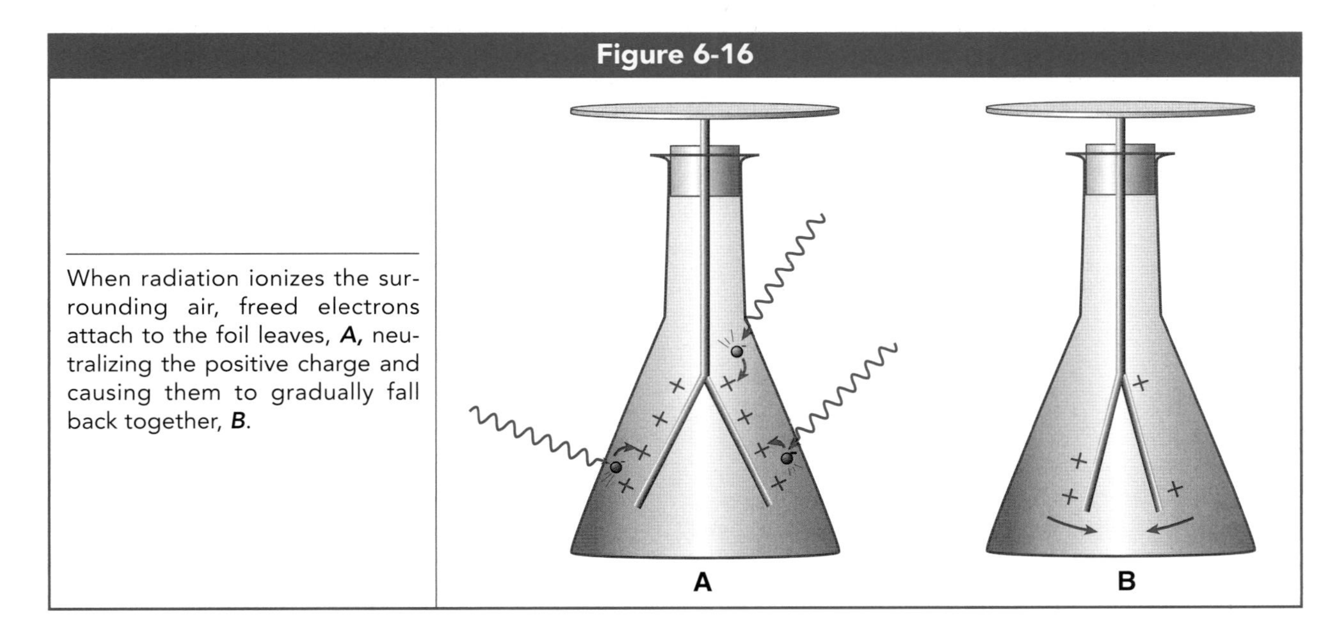

When radiation ionizes the surrounding air, freed electrons attach to the foil leaves, ***A***, neutralizing the positive charge and causing them to gradually fall back together, ***B***.

Figure 6-17

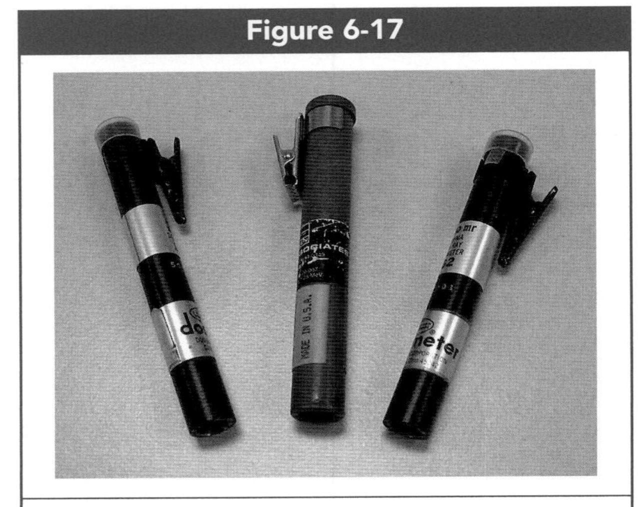

Photograph of pocket dosimeters, which operate on the basis of the electroscope.

The simplest example of a radiation detection device using the electroscope is the personal pocket dosimeter (Fig. 6-17). In this device one "foil leaf" is connected to the wall of the chamber and the other is left free to move. When the dosimeter is given a positive charge by touching it to a battery, the free leaf is repelled away from the wall. The end of this leaf appears as a fiber when one looks into the end of the chamber through a small lens (Fig. 6-18). A full positive charge moves this fiber over to the *zero* point on a scale painted onto the lens. As radiation frees electrons from the air in the chamber and the charge on the fiber is neutralized, it moves back toward the opposite wall and thus across the scale indicating the amount of radiation that has been received.

Figure 6-18

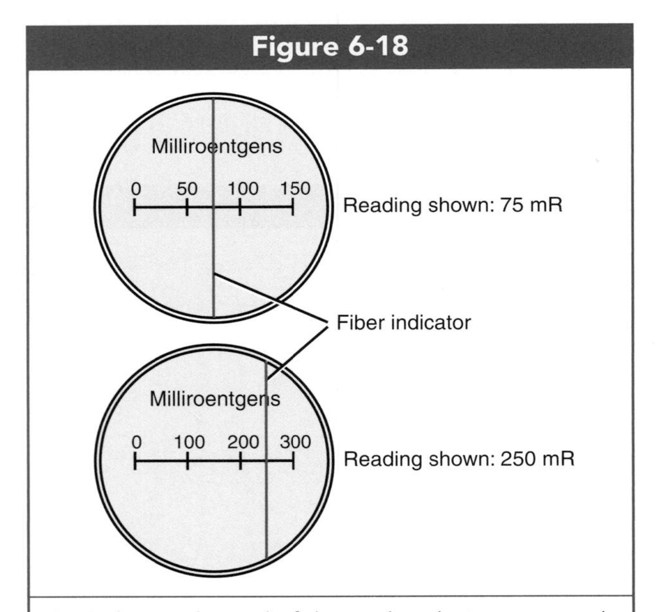

A window at the end of the pocket dosimeter reveals the moving fiber of the electroscope *end-on*. As electric charge is dissipated by radiation exposure, the fiber moves across a scale to indicate the amount of radiation received.

We see that a fundamental knowledge of electrostatics is a necessary part of understanding not only the electronic equipment that produces x-rays, but also the devices that detect them.

SUMMARY

1. All moving electrical charges develop a magnetic field around them. Because electrons and protons have *spin*, they also develop a *magnetic moment* with north and south poles.
2. Atoms with an odd number of electrons develop a net spin, and become *magnetic dipoles* with north and south poles.
3. In ferromagnetic materials, *magnetic domains* align upon magnetization by an external field. Their high permeability makes them easy to magnetize.
4. An electromagnet has almost no retentivity, since it immediately loses its magnetic field when electrical current stops flowing through it.
5. Magnetic fields follow Gauss' law, which states that their strength is proportional to the product of the pole strengths, and inversely proportional to the square of the distance between the poles.
6. In solid objects, only negative electrical charges can move, and they distribute themselves evenly on the surface.
7. Electrification occurs whenever an object acquires either an excess or a deficiency of electrons. The three methods of electrification are friction, contact, and induction.
8. Differences in electrical charge between two objects create a *potential difference*, which in turn generates an *electromotive force (EMF)* that causes electrons to move.

9. Charges will always equalize between two objects that come in contact with each other.
10. The ionizing effect of radiation in air frees electrons, which can be used to neutralize a positive charge placed on metal foil leaves. This is the principle of the pocket dosimeter, which uses an electroscope to measure radiation exposure.

REVIEW QUESTIONS

1. What are the *two* requirements for a subatomic particle to acquire a magnetic moment?
2. What are the *four* classifications of materials, relative to their magnetic permeability?
3. Where would the earth's magnetic field be most concentrated?
4. What are the three general types of magnets?
5. Two magnets separated by 1 cm exert a force of 2 Gauss on each other. Like poles are facing each other. If the magnets are pulled to 4 cm separation in distance, how much force will there be between them, and will it be repulsive or attractive?
6. What unit is defined as 6 billion billion charges collected on an object?
7. The principle reservoir for the deposit of excess electric charge is:
8. What are the three methods of electrifying an object?
9. Two protons are held close together and then released. Will they move toward or away from each other, and will they speed up, slow down, or continue at the same velocity as they move?
10. If an electron and a proton separated by 4 picometers exert a force of 3 millivolts on each other, when they are moved closer to just 1 picometer apart, how much force will there be between them?
11. What are the two reasons why only electrons move in charged, solid objects?
12. Two charged objects, one with a charge of –20, the other with a charge of –2, come into contact. What will the charge on each object be when they are separated again?
13. When a charged object comes close to an uncharged object but does not touch it, what is the term for the separation of charges in the second object which results?
14. As x-rays ionize atoms of air in a pocket dosimeter, what happens to the electrical charge placed on the metal foil leaves of the electroscope?

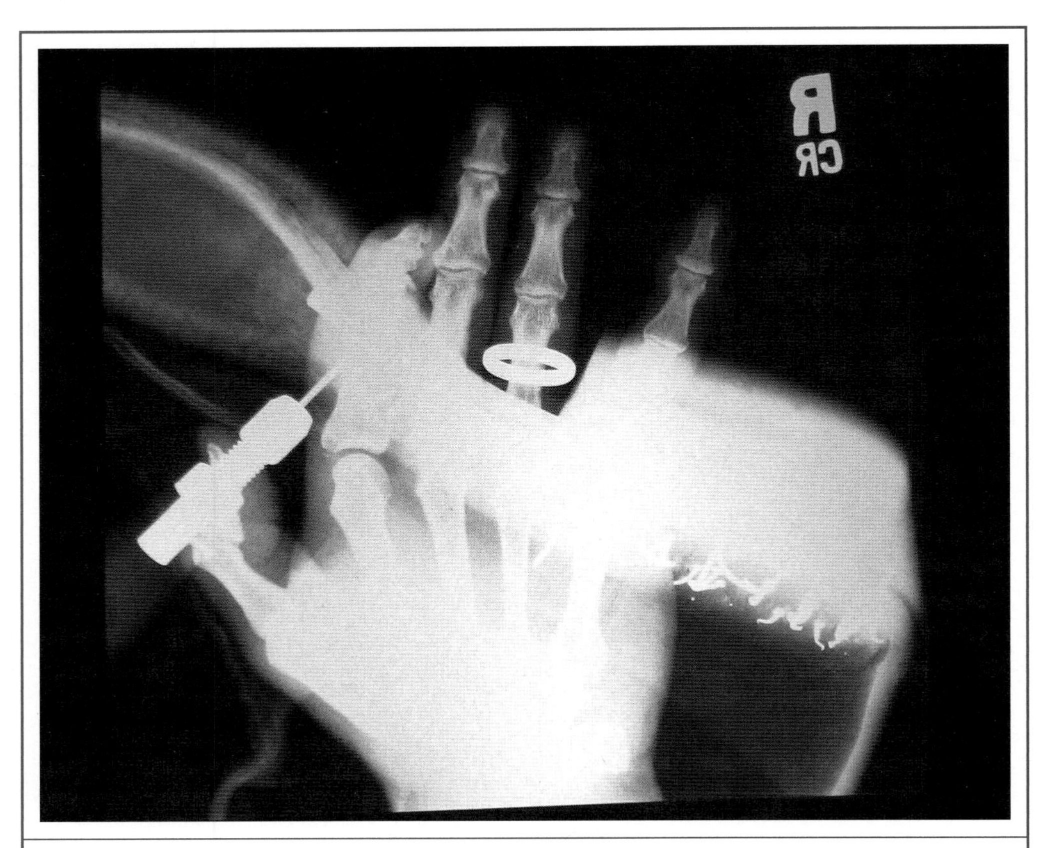

A shoe cobbler inserting a large auger through a leather boot impaled the index finger upon the needle. The patient was brought into the emergency room with the boot still attached.

Chapter 7

ELECTRODYNAMICS

Objectives

Upon completion of this chapter, you should be able to:

1. Define electrical current and its units, and solve for total *mAs* and coulombs of charge spent.
2. Define and give examples of electrical conductors, dielectrics, and semiconductors.
3. List the three components of every electrical circuit.
4. List the three changes in a conductor that affect its electrical resistance.
5. Describe the two basic types of circuits.
6. Define the three characteristics of electricity and their units.
7. Solve simple *Ohm's law* problems for current, voltage and resistance.
8. Define electrical *power* and its conservation law.
9. Solve simple *power law* problems for power, current, voltage and resistance.
10. Draw the wave forms for AC and DC electricity, and explain their shape.
11. Distinguish between the effects of electrical *current* and actual electron movement.
12. Define electrical frequency.
13. Describe *electromagnetic induction*, and how it applies to both generators and motors.
14. Explain how an *induction motor*, such as the one in the x-ray tube, works without magnets.
15. Define step-up, step-down, and auto-transformers.
16. Solve transformer law problems for voltage and amperage.

ELECTRICAL CURRENT

Electricity refers to the continuous flow of electrons along the surface of a conductor. *Electrodynamics* is the study of flowing electrical current. Materials may be classified according to their *conductivity*, or ability to transmit electrical current through them. Copper is used for most electrical wiring because of its extremely high conductivity. Silver, aluminum, brass and several other metals and materials including *water* allow electricity to pass through easily, and are called *conductors*. In these materials the outer-shell electrons of the atoms are very loosely bound, and flow when a potential difference (or *voltage*) is present. Although we usually think of electrical current as a steady flow down the wire like water through a pipe, in reality the electrons leap-frog from one atom to the next in sequential fashion as shown in Figure 7-1.

Materials that resist the flow of electricity are referred to as *dielectrics* or *insulators*. Rubber is an insulator commonly used around electrical wires for protection from electrical shock. High-voltage devices such as large transformers and x-ray tubes are immersed in oil for safety. Glass, plastics and dry wood are insulators. In these materials the electrons are strongly bound to the molecules and are not free to flow.

Semiconductors are a very important component of computers and other high-tech electronic devices. A semiconductor is a material in which the conduction

Figure 7-1

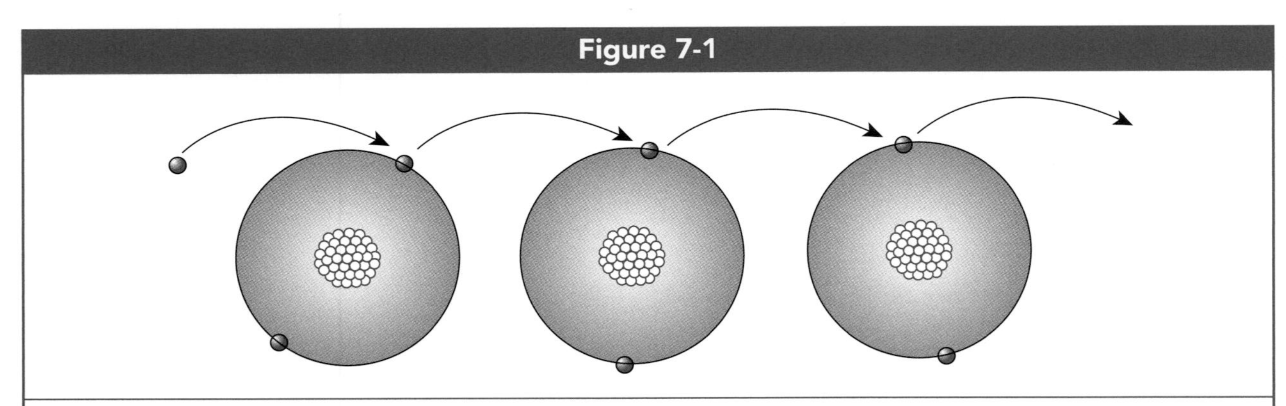

Electrical current in a wire actually consists of the outer orbital electrons of its atoms "leap-frogging" from one atom to the next in sequence.

of electricity depends on specific conditions, such as being hot or cold, or whether they already have an electric charge placed upon them.

The *rate of flow* of electricity is referred to as *current*. As a rate, it is a quantity divided by *time*, such as electrons per second, or Coulombs per hour. The unit for electrical current is the *Ampere*, abbreviated *A*, and often shortened to the word *amp*. It is specifically defined as a current flow of one Coulomb per second. Remember that the Coulomb is a unit of electrical charge that consists of 6.3×10^{18} electrons. If this many electrons are sitting at rest as a static charge built up on an object, it is one Coulomb, but if this many electrons *pass by you* each second flowing down a wire, it is one Ampere.

The milliamp, abbreviated *mA*, is often referred to in radiography. It is only one-thousandth of an amp, but still represents a lot of electrons (6.3×10^{15}). In fact, it only takes a bit more than one-half an mA, 0.6 milliamps, to lock up the muscles in the human body and cause death. So, the amp and the milliamp may be thought of as relatively large units of electricity.

The total amount of x-ray exposure delivered to a detector plate or cassette is controlled by the *mAs* set at the console of the x-ray machine. For convenience, radiographers often pronounce this term like the word "mass," but properly understood it is the mA-s or *milliampere-seconds* of electricity used to produce the exposure, derived from multiplying the mA station used times the length of the exposure time in seconds. Remember that mA is a *rate*, so multiplying the rate of electrical flow times the seconds of duration gives the *total amount of electricity used* during the exposure.

This is precisely the same math you would use when traveling: The *rate* of speed in miles-per-hour times the number of hours you drive will yield the *total miles driven* (60 miles per hour times 3 hours = 180 miles). When we multiply mA X s, we get an indication of the total number of electrons used (or the number of *coulombs* used), for the entire exposure. This, in turn, determines the end result of the total radiation exposure produced.

Practice Exercise #1

Using the definition for one *milliamp* given above, calculate the total number of electrons used for an x-ray exposure taken at 300 mA and 0.2 seconds.

Solution: First, the total mAs must be calculated by multiplying the mA by the exposure time. This number must then be multiplied by the number of electrons per mA given above, as follows:

$$300\ mA \times 0.2\ s = 60\ mAs$$

$$60 \times 6.3 \times 10^{15} = 378 \times 10^{15}$$

Answer: The total number of electrons used was 378×10^{15} *electrons.*

Practice Exercise #2

For the above exposure, calculate the total number of coulombs of electric charge used.

Solution: The resulting number of electrons could be converted into Coulombs, but, realizing that one milliampere represents one milli-coulomb per second, it is easier just to multiply the milli-coulombs per second by the exposure time and report the answer in milli-coulombs:

300 mA = 300 milli-coulombs per second

300 mC/s × 0.2 s = 60 mC

60 mC = .06 Coulombs (moving the decimal 3 places)

Answer: The total amount of electric charge used was 60 milli-coulombs or .06 Coulombs.

ELECTRICAL CIRCUITS

An electrical *circuit* is a *circle*, in which the wires that leave from a battery or generator must pass through the devices being operated and all the way back to the battery or generator. Not only does the negative end of a battery *push* electrons out into the wire, but the positive end of the battery must *pull* electrons in, to replace them, from the other end of the wire. If you cut through a pipe carrying a water current, the water continues to flow and just dumps out of the chopped-off end; but, if you cut a wire carrying an *electrical* current, the flow completely stops. It does not "dump out" of the end of the wire. A completed circle of conductors is necessary for electricity to flow.

A battery does not *create* electrons, it only *pumps* them out one end and pulls them in from the other. The layers of chemicals in the battery are "stacked" in such a way that electrons are pushed toward the negative terminal and pulled from the positive terminal. This pumping action cannot take place unless the circuit is completed all the way around and back to the battery.

An electric *switch* is nothing more than a method of breaking the circuit. When the switch is "opened," it disconnects the wire and electrical current stops flowing. When it is "closed," it simply creates a bridge that reconnects the wire.

Any meaningful electrical circuit must have three components:

1. A conductor, which is the source of electrons that are free to flow as current. It is the *wire*, not the battery, that provides these electrons. A switch is generally placed somewhere in this wire to open and close the circuit.
2. A source of EMF (electromotive force). Remember that for electrons to move, there must be a *potential difference* present, a difference in positive or negative charge. This creates a kind of electrical *pressure*, which pushes the electricity from the more negative charge and pulls it toward the more positive charge. This pressure comes from a battery or a generator.
3. A device to be operated, which uses up some of the energy from the electricity and thus acts as a *resistor* in the circuit.

The conducting wire itself can create significant resistance to the flow of electricity if it is very long, too narrow, or made of poorly conducting material. This resistance is primarily due to *friction*, which makes the wire get hot. Energy is lost from this heating, so it is important to make sure that a wire is well constructed and thick enough in diameter to handle the electrical load that will be placed upon it.

Imagine a typical garden hose with water flowing through it. If the hose is made very long, the water must overcome more friction all along the inside walls (the lumen) of the hose, and resistance is increased. By the time the water reaches the end of the hose, either the flow of the current will be reduced, or the pressure must be turned up back at the faucet to maintain the flow. The same results occur if the water is squeezed into a hose with a narrower diameter—the flow out of the end of the hose will be reduced from higher friction, or the water within the tube must speed up (which represents increased pressure).

This is precisely what happens with electricity. When resistance is increased, either the flow of current will be reduced or the pressure must increase to maintain the flow. To summarize, three types of changes increase electrical resistance within a wire. They are:

1. Longer length
2. Narrower diameter
3. Poor conducting material or poor construction

There are two general types of circuit layouts, *series* and *parallel*. In a series circuit, all of the devices are connected *in a row* within the same line or wire, one after another. Cheap strands of Christmas

Figure 7-2

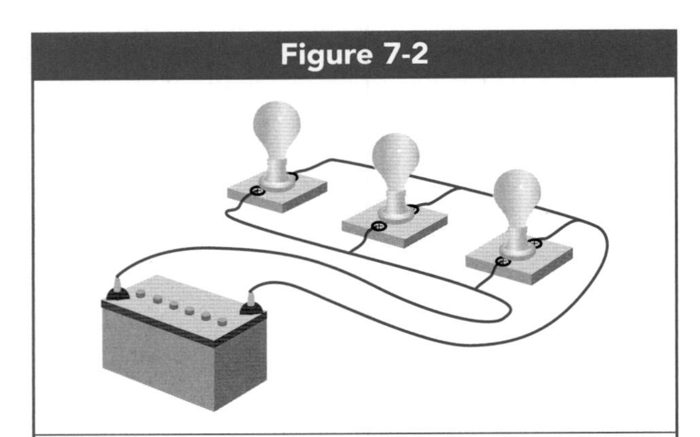

A parallel circuit, in which *branches* are split off of the original wire to each of the devices to be run, such as light bulbs.

lights are sometimes arranged this way. If any one of the light bulbs burns out, it breaks the circuit and the entire strand of lights no longer works. To avoid having one device which burns out affect other devices, the wiring in the walls of your home is connected in *parallel* circuits (Fig. 7-2). The parallel circuit uses *branches* of wire, which split off from the original incoming wire and connect across to the outgoing wire, for each device or plug. If the device in one branch burns out or has a short circuit, the other branches are not affected.

CHARACTERISTICS OF ELECTRICITY

All electricity flowing in circuits has three characteristics. They are current, resistance, and electromotive force (or potential difference). We have described the unit *ampere* for measuring current. In formulas, this unit is often abbreviated with an *I* (for intensity) rather than an *A*. The unit for electrical resistance is the *ohm*. It is abbreviated with the Greek letter *omega*, Ω, or with an *R* (for *resistance*). To get an idea of the magnitude of the *ohm* unit, about ten feet (3 meters) of typical wire offers about one ohm of resistance to the flow of electricity.

The *volt* is the unit used for electrical pressure, electromotive force or potential difference. In formulas it may be abbreviated with a *V* or with an *E* (for *energy*). The dictionary definition for one volt is that it is sufficient electrical force or pressure to push *one ampere* of current through *one ohm* of resistance (or ten feet of wire). Table 7-1 summarizes these three qualities of electricity, their units and abbreviations.

The relationships between electrical resistance, pressure and flow are summed up in a simple formula known as Ohm's law:

$$V = IR$$

where *V* is the voltage, *I* is the amperage and *R* is the resistance present in a circuit. The three relationships represented in this formula can be broken down and stated "in English" by setting at a fixed amount each of the values in turn (mathematically giving it a value of "1"), and reading out the relationship between the other two factors. To do this, just place a finger over the fixed value (Fig. 7-3). Let's work from right to left, starting out by covering up the *R*; we find that, for a given, fixed amount of resistance in a circuit, the voltage and the amperage are proportional (V = I). In other words, as the electrical pressure increases, so does the current. This seems intuitively true—pushing harder results in more flow.

Now cover the *I*, and we find that voltage and resistance are also proportional to each other. In "English," this simply indicates that in order to keep a certain, fixed amount of current flowing, the more resistance there is, the harder we have to push. Since they are directly proportional, we can be more specific, and state that if the resistance is doubled, the pressure must also be doubled to maintain the current.

Covering the *V*, or rather, imagining it to be unity or a "1," we see that amperage and resistance are *inversely* proportional. In other words, if the resistance is doubled, and the pressure cannot be increased to compensate, the amount of current flowing will be reduced to one-half. Likewise, for a certain fixed

Table 7-1

Characteristics of Electricity

Characteristic	Unit	Abbreviation
Current	Amps (Amperes)	I (or A)
PD or EMF	Volts	V (or E)
Resistance	Ohms	Ω (or R)

amount of pressure, if the resistance is cut in half, the current will double because it is easier for it to flow.

The *T-triangle* which was previously used with the wave formula is also very helpful with Ohm's law problems. By covering up the value you are solving for with your fingertip, it shows how to set up the formula.

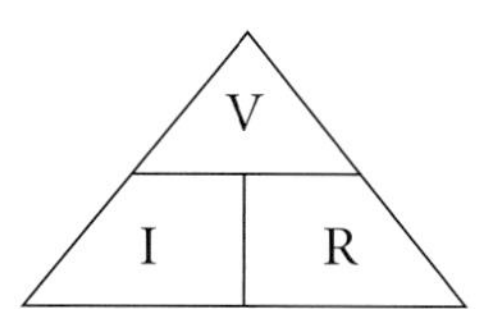

To solve for the voltage (covering up the "V"), multiply I times R, the amperage times the resistance. To solve for the amperage flowing in a circuit, divide V by R. To find the resistance, divide V by I. It is helpful to note that the voltage is always *on top* of the equation, that is, the voltage never gets divided *into* the other factors. Thus, the three variations of the formula, to solve for each factor, are as follows:

$$V = IR$$
$$I = V/R$$
$$R = V/I$$

The following practice problems provide some practice applying Ohm's law and the associated T-triangle.

Practice Exercise #3

A light bulb in your house has 30 ohms of resistance. Four amps of electrical current are flowing through it. What is the *voltage* from the wires in your house?

> *Solution: By covering the* V *in the T-triangle, we see that the solution is to multiply the amperage times the resistance:*
>
> $$I \times R = V$$
> $$4 \times 30 = 120$$
>
> *Answer: The electrical pressure in your house is 120 volts.*

Figure 7-3

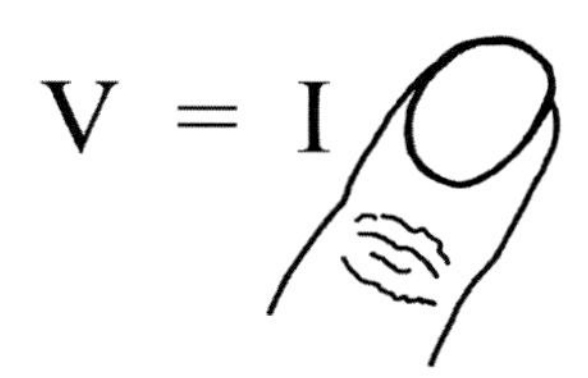

To see the relationship between voltage and current in the Ohm's law formula, cover the resistance ***R*** with your finger. This can be done with other formulas. Here, it shows that at a given set resistance, current is *proportional* to voltage.

Practice Exercise #4

A hair dryer creates 14 ohms of resistance. The voltage from the plug in your house at this time is about 112 volts. How many *amps* of current are flowing through the hair dryer?

> *Solution: By covering the* I *in the T-triangle, we see that the solution is to divide the voltage by the resistance:*
>
> $$V / R = I$$
> $$112 / 14 = 8$$
>
> *Answer: The electrical current flowing in the hair dryer is 8 amps.*

Practice Exercise #5

A toaster has 7.5 amps of *electricity* flowing through it. The voltage from the plug in your house today is about 115 volts. What is the resistance of the toaster?

> *Solution: By covering the R in the T-triangle, we see that the solution is to divide the voltage by the amperage:*
>
> $$V / I = R$$
> $$115 / 7.5 = 15.33$$
>
> *Answer: The electrical resistance of the toaster is 15.33 ohms.*

ELECTRICAL POWER

Power is generally defined as the *rate* at which work is done. Imagine two cars climbing a long hill, an "economy" car with a four-cylinder engine, and a sports car of about the same weight with an 8-cylinder engine. Both cars can do the same amount of work to reach the top of the hill, only the sports car can do it much faster. The sports car has more power (horsepower). Even an athlete on a bicycle can achieve the same amount of work, only he must take much more time than either of the cars.

The amount of work done is essentially synonymous with the amount of *energy* spent. Therefore, we may also think of power as the *rate* at which energy is spent. The bicyclist, the economy car, and the sports car all expend the same amount of energy to reach the top of the long hill, but the more powerful devices expend that amount of energy much more quickly.

$$P = E / t$$

where P is power, E is energy and t is time.

The total power of a system can be obtained by multiplying the intensity or *quantity* of objects or entities being moved by the *force*, the speed, or the quality with which they are being moved. Given a fixed, certain amount of power, you can move several cars very slowly, or a few cars very quickly. Within an x-ray beam, the total power is represented by multiplying the *intensity* of the beam, that is, the number of x-rays flowing per second, times the overall *voltage* or energy which they carry; this is the *quantity* times the *quality* of the x-ray beam.

Electrical power, then, is the *rate* at which electricity can do work, or the rate at which electrical energy is spent. The unit for electrical power is the *Watt*, abbreviated *W*. Each month the electrical power company bills your home for the number of *kilowatt-hours* used. The kilowatt-hours are calculated by multiplying the total current used, based on the number of appliances you had running and how long they were run, times the voltage with which the power company "pushed" that electricity.

Electrical power is related to voltage and amperage by the formula:

$$P = IV$$

As with Ohm's law and the wave formula, the *T-triangle* is also helpful here. In this case, it reminds us that *P* always "stays on top," so it is never divided into either of the other two variables. As with the other formulas, cover the factor you are solving for and it shows you what to multiply or divide.

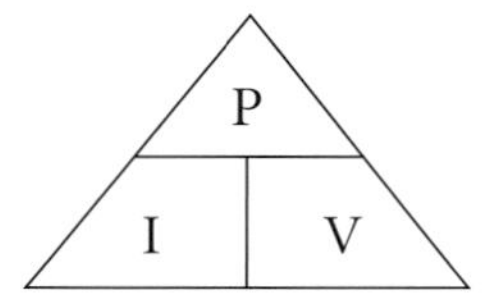

To solve for the amperage divide the power by the voltage:

$$I = P / V$$

To solve for the voltage, divide the power by the amperage:

$$V = P / V$$

Practice Exercise #6

A 40-watt light bulb is plugged into a 120-volt socket. How much current, in amps, is flowing through the light bulb?

Solution: By covering the I *in the T-triangle, we see that the solution is to divide the power by the voltage:*

$$P / V = I$$

$$40 / 120 = 0.33$$

Answer: The current flowing through the light bulb is .33 amps.

If the voltage for a particular circuit is unknown, but the resistance is known, the following variation of the power formula can be used:

$$P = I^2 R$$

The power is the product of the amperage *squared* times the resistance. This iteration of the formula is obtained by simple substitution, replacing the *V* in the original power formula with *IR* from Ohm's law, as follows:

$V = IR$ by Ohm's law

$P = (V)I$ by the power law

Substituting IR for V, $P = (IR)I$

$P = I^2 R$

Practice Exercise #7

A 40-watt light bulb has a current of 0.33 amps flowing through it. What is the resistance of the light bulb?

Solution: $P = I^2 R$

$40 = (0.33)^2 R$

Squaring first, $40 = 0.109 R$

$40 / 0.109 = R$

$367 = R$

Answer: The resistance in the circuit is 367 ohms.

WAVE FORMS OF ELECTRICAL CURRENT

There are two general ways in which electricity can move in a wire or cable. They are called *direct current* (DC) and *alternating current* (AC). In direct current, the electrons flow in a steady stream in one single direction. The batteries in a flashlight generate DC current, always "pushing" the electrons away from the negative terminal and "pulling" them back toward the positive terminal. In alternating current, the individual electrons actually *oscillate* or "vibrate" back and forth, left to right, then right to left. This is the type of current produced by most electrical "generators" such as those at a power plant, and the type that flows in the wires of your home. Both DC and AC electricity are used within an x-ray tube during the production of x-rays, so it is important for radiographers to develop a good understanding of them.

When we observe direct current closely, we see that it takes about an hour for an individual electron to travel down the wire a distance of *one inch*. This seems very slow, but remember that along the conductive surface of one inch of wire are millions of atoms which the electron has traveled through, jumping from the outer shell of one atom to the next in sequence. More importantly, even though the individual electrons may seem to travel very slowly, the *effect* from moving the electrons travels at *the speed of light*, and so seems virtually instantaneous. When you turn on a light switch, the bulb in the ceiling instantly lights up—you do not have to wait for the electrons at the switch to "make it" all the way to the bulb.

To understand how the effect of current can be instantaneous, fill a small tube with ping pong balls or a larger tube with tennis balls as shown in Figure 7-4. When you push a new ball into the tube, the ball at the opposite end instantly pops out. Each ball has only traveled one ball-length down the tube, yet the *effect* of inserting the new ball was observed at the opposite end of the tube instantly. Furthermore, it does not matter how long the tube is—it could be miles long, but as long as it is filled with balls, the effect at the far end will always be instantaneous.

In a similar fashion, when a power plant five miles away from your home "pushes" a single electron into a wire, each electron in the wire instantly moves down the wire to the next atom, and in your home the effect of this movement is experienced immediately (if the switch is on). It is this *effect* that we call electrical current. It is not quite the same thing as the actual movement of the electrons, but it is *caused* by the movement of the electrons.

Figure 7-4

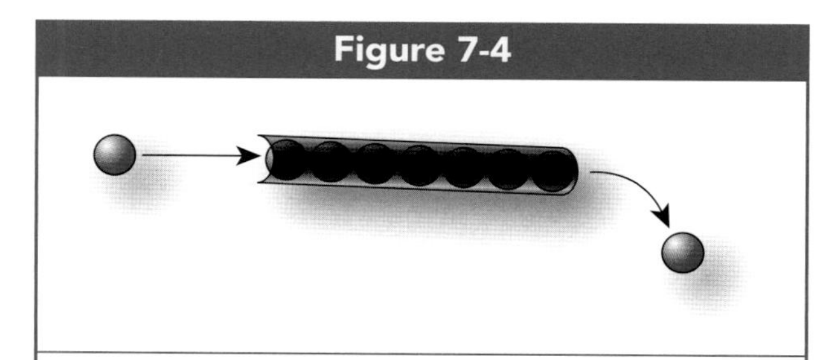

When a tube is filled with balls, pushing another ball into one end *instantly* causes a ball on the opposite end to pop out. In the same way, when a power plant "pushes" a single electron into the end of a wire, even though it may be miles long, the effect of the electrical current is conducted *instantly* to your house.

Indeed, Benjamin Franklin defined electrical current as flowing from the positive charge toward the negative charge, and electricians have thought of it this way ever since. This seems backward to the student, for we know that electrons are repelled away from the negative charge and pulled toward the positive. But, if you remember that electrical "current" refers to the *effect* of the movement rather than the actual movement of electrons, then it does not really matter. For example, we can think of our analogy in Figure 7-4 either as *pushing* one ball into the tube which forces a ball to pop out the other end, or as *pulling* a ball out from the other end which "sucks in" a new ball from our end. The effect still occurs regardless of the specifics of how the balls move.

The flow of electrical current can be graphed by plotting the voltage "pressure" or *speed* of the electrons against time. The graph in Figure 7-5 is representative of the DC current flowing in a typical flashlight. Stacking two 1.5-volt batteries inside the flashlight results in a total electromotive force (EMF) of 3 volts. However, when the flashlight is switched on, the electron flow does not instantly jump to its maximum speed at 3 volts of pressure. Just like a car, the current must first accelerate to reach its maximum speed. We see this on the graph

Figure 7-5

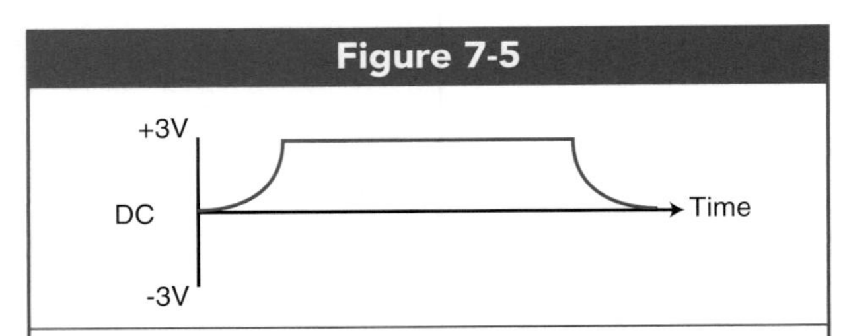

Graph of direct current flow for a typical flashlight, speeding up as the EMF rises to 3 volts when the switch is turned on, then flowing at steady speed and voltage until it is turned off. At steady current flow, the magnetic field generated around the flashlight is also constant and does not move.

as a rising of the voltage from zero to 3 volts, which takes a few milliseconds. Once it reaches 3 volts, which is all the batteries can produce, the current flows at a steady speed, represented by the horizontal portion of the curve on the graph. This is just as if you had set your car on "cruise control." During this time, the application of a steady force keeps the speed and voltage constant, and the bulb in the flashlight has a steady glow. When the switch is turned off, the electrons quickly slow down to zero speed, but again, this takes a few milliseconds for them to decelerate to a complete stop.

Now, note in Figure 7-5 that the graph is laid out to also show a *negative* 3 volts. This is done to accommodate the possibility that current can also flow in the *opposite direction*. If you were to turn the batteries around in the flashlight, and rig the connections at the end of the batteries so the wires made good contact with the batteries, the flashlight will work just as before and the bulb will burn just as bright with the current flowing "backwards." There are still 3 volts of force pushing the electricity along; But, to indicate the reversed flow of the current on the graph, the curve would be drawn *under* the axis line, dipping down to minus 3 volts and holding steady until the switch was turned off. With this understanding, we can turn our attention to graphing AC electricity.

A typical graph for the alternating current (AC) in a typical home is shown in Figure 7-6. First, note the labeling at the left of the graph. Most appliances in the western hemisphere operate on 60-hertz current and a voltage of 110 to 120. Recall from Chapter 5 that the *hertz* is the unit for the frequency of a series of waves (see Figure 5-7). Alternating current oscillates back and forth in the wire. It flows as a series of *pulses*, one "positive" indicating its forward direction, and the next reversed in direction or "negative." The negative pulses are graphed underneath the axis line (Fig. 7-6).

Just as with the transverse waves described in Chapter 5, in AC electricity each *pair* of pulses, one positive and one negative, make up one *cycle* of current. When we state that the frequency of the AC current graphed in Figure 7-6 is 60 hertz, we mean that there are 60 cycles per second moving in the wire. With this in mind, *how long does each* pulse *last*? The answer is 1/120th of a second or 0.00833 second. Remember that each cycle consists of 2 pulses, so there are 120 pulses of electricity per second, half traveling in one direction, and half in the other.

This current wave form is characteristic of electrical generators such as those used in most power plants. All of these generators begin with some method of mechanically *spinning* a rotor device on an "axle," whether it be from water falling over a paddle wheel in a dam, or from burning coal to make steam which thrusts a paddle wheel.

Two physicists, the American Joseph Henry and the great English physicist Michael Faraday, independently proved that what Hans Oersted had discovered about electricity affecting magnets could be reversed: A moving magnetic field could be used to induce electricity to flow in a nearby wire. The

Figure 7-6

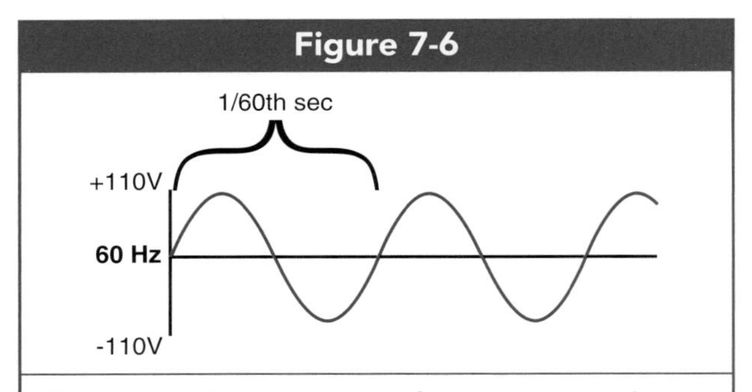

Graph of typical AC current for an American home, flowing at a frequency of 60 hertz. Each second, 60 pulses of electricity flow *away* from the power plant, and 60 pulses flow *back toward* the plant, each reaching 110 volts only to drop back to zero. At all points on the graph, electrons are either speeding up or slowing down, such that the magnetic field generated around the wire is constantly blooming or collapsing.

paddle wheels from dams or steam engines can, in turn, be used to spin the rotor device to move that magnetic field.

We will simplify this picture somewhat by imagining that the rotor is a simple bar magnet that is spinning within a coil of wire. If this is done *perpendicular* to the wire, electrical current will be induced, but in the peculiar way shown in Figure 7-7. Note that with the spinning rotor, for the first half of the spin cycle the *north* pole of the magnet passes from left to right relative to the wire, but that for the second half of the cycle, when the north pole swings down under the axle, it is now passing *right to left* relative to the same wire. This means that the magnetic field around the magnet is constantly moving across the wire, but *reversing its direction* every half-cycle. If the magnet spins 60 revolutions per second, 120 pulses of electricity will be produced with every other one *reversing direction.*

Figure 7-7

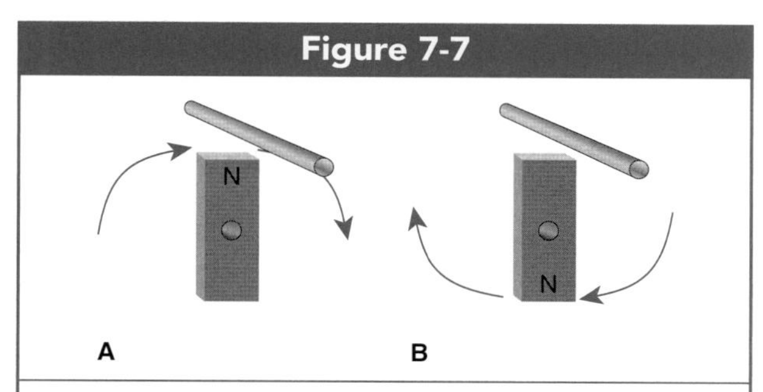

A simplified AC generator using a spinning magnet near a wire. During the first half of a cycle, the *north* pole of the magnet passes left-to-right relative to the wire, but during the second half of the cycle the north pole is passing right-to-left relative to the wire. The wire will experience pulses of induced electricity that flow in reversing directions.

Figure 7-8 aligns a diagram of the above magnet's spin cycle with the AC waveform produced. The magnet first begins moving clockwise from a fixed horizontal position with the north pole to the left (point A). As it moves to a vertical position near the wire (point B), we see from the graph that the voltage rises to +110 volts while the electrons *accelerate.* At point B, the electrons have reached their highest speed. Then, as the magnet's north pole moves on to a horizontal position the effect of its movement upon the wire is lessened. The deceleration of the electrons is represented by the dropping portion of the curve. At point C, the magnet is furthest from the wire and no effect results, so the movement of electrons comes to a stop and the voltage has returned to zero.

From point C to point D, the north pole of the magnet is moving in the opposite direction, and the *south* pole passes by the wire. This induces current to flow the other way. It does so by gradually accelerating again until the magnet is nearest at point D, where the electrons reach their maximum speed and voltage is diagramed at *minus* 110 volts. Finally, the magnet continues a clockwise spin to its original horizontal position, point E, during which time the electrons decelerate to zero motion and voltage returns to zero, ready to begin the next cycle.

Although the electrons must stop at the end of each pulse in order to reverse direction, this happens so quickly that for all practical purposes, we may say that AC electricity "never stops moving." More importantly, as you examine the graph in Figure 7-8, it becomes apparent that AC electricity never stops *changing speed:* It is always either accelerating or decelerating. This is the principle on which the electrical generator works. The principle that constantly changing magnetic fields generate steady electricity, and the formulas associated with it, are known as Faraday's law because he did not hesitate in publishing a thorough treatise on the subject, while the American Henry procrastinated.

The electrical current in a fixed radiographic room is normally AC, and this is what is supplied to the x-ray machine. However, in an x-ray tube, electrons must flow from a thin filament on one end to the thick anode disc on the other end. If electrical current were to flow "backwards" at any time across an x-ray tube, it would vaporize the thin filament. To prevent this, AC electricity must be somehow converted to a DC form before it reaches the x-ray tube.

By using a clever system of electrical *gates*, this can be done. This device, called a *rectifier*, will be described in the next chapter, but the resulting current waveform will look like Figure 7-9. Those pulses of electricity that would have gone the wrong direction have been turned around. The graph now shows all pulses above the axis line, indicating that they are all traveling in the same direction. This AC current has been *rectified* or "made straight." Rectified AC is also

Figure 7-8

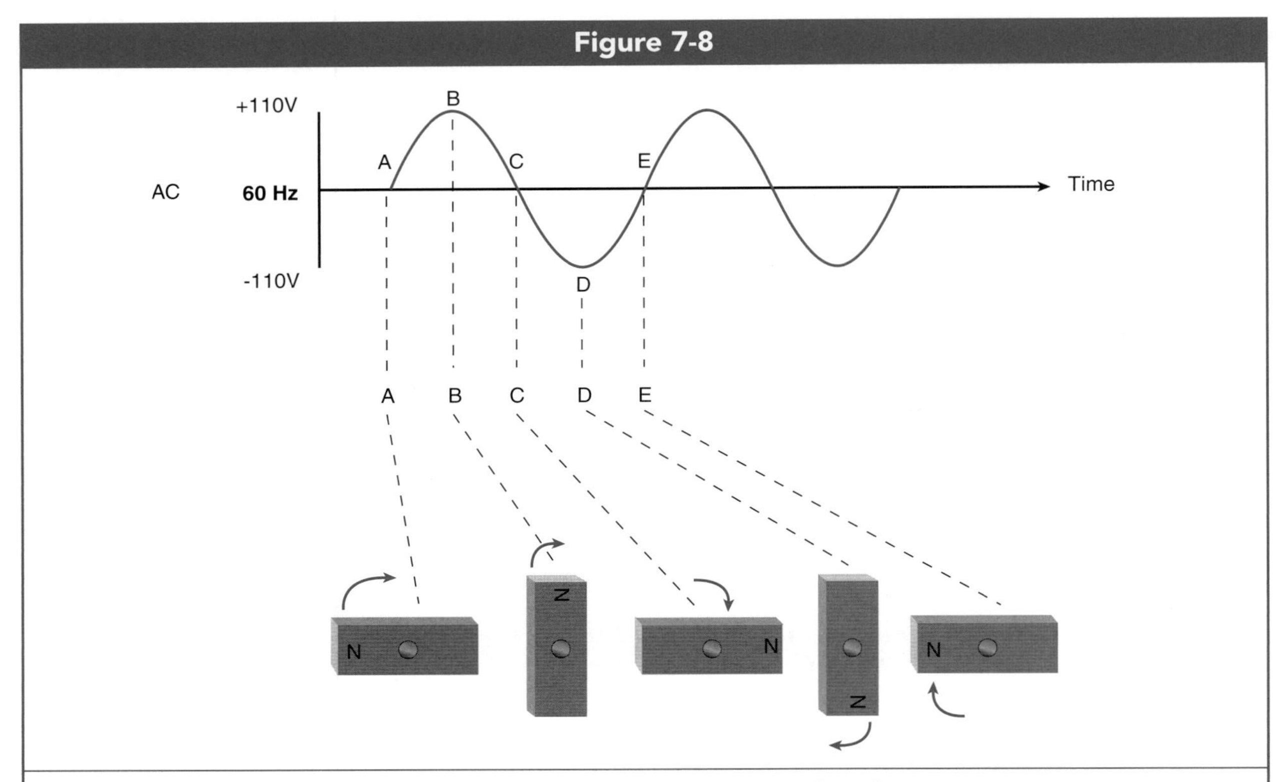

Graph correlating the movement of the magnet with the voltage wave form for an AC generator. Points A to B, acceleration; Point B, maximum speed and voltage; Points B to C, deceleration; Point C, reversal of direction; Points C to D, acceleration in the opposite direction; Points D to E, deceleration to original position.

called *pulsed DC* electricity, which is perhaps a better term when we compare the waveform in Figure 7-8 with that of regular DC in Figure 7-5.

ELECTROMAGNETIC INDUCTION

Whereas electrical *generators* apply Faraday's law to create AC electricity from a moving magnetic field, electrical *motors* apply Oersted's law, using AC electricity to cause a magnet to move (Fig. 7-10). A generator converts mechanical energy (movement) into electrical energy, while an electric motor converts electrical energy into mechanical energy or movement.

Every motor consists of two major components: The *rotor* and the *stator*. The magnetized *rotor*, as the name implies, rotates or spins when the motor is turned on. An axle from the rotor can then be

Figure 7-9

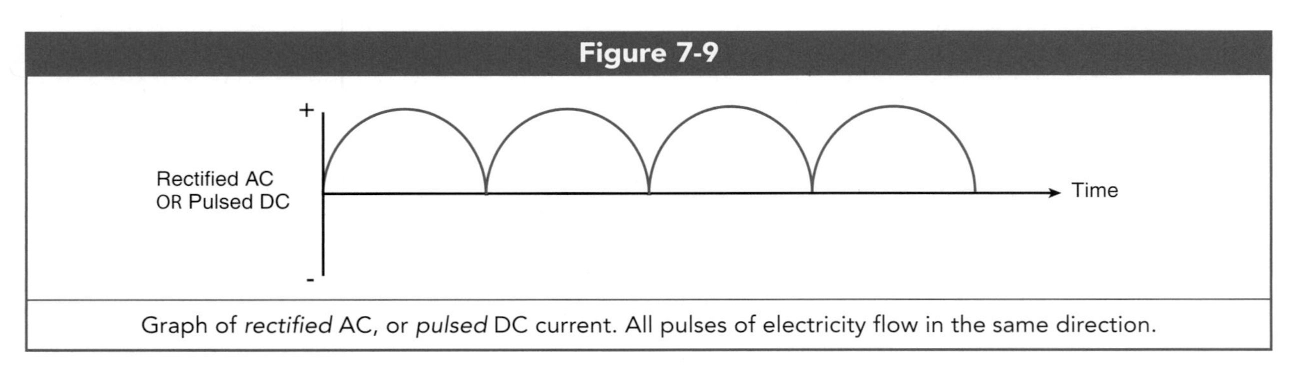

Graph of *rectified* AC, or *pulsed* DC current. All pulses of electricity flow in the same direction.

Figure 7-10

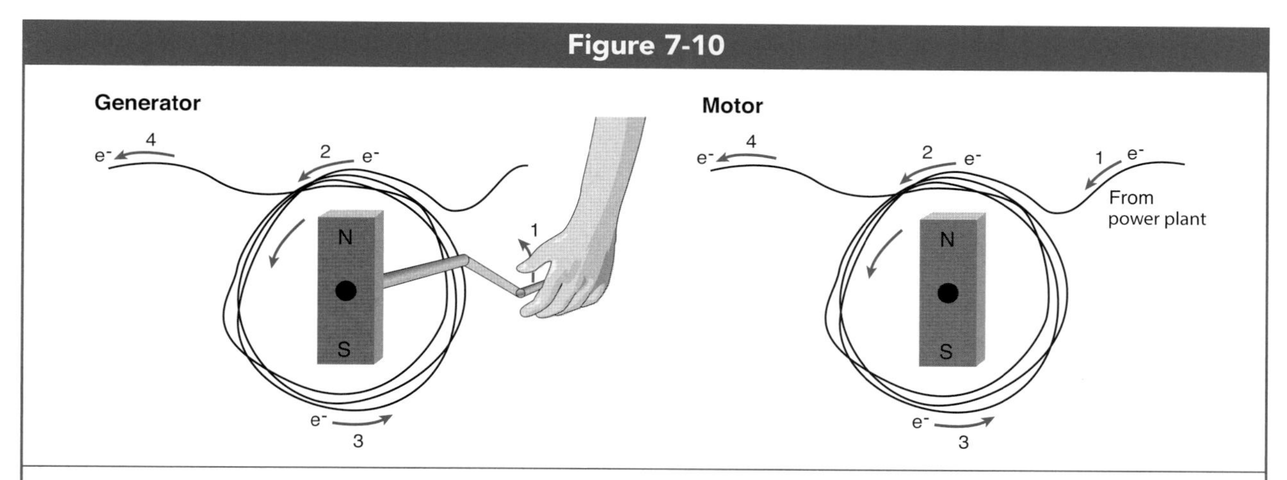

While turning the crank on a generator may use a moving magnet to cause electricity to flow, (left), a *motor* using AC electricity causes a magnet to spin because of the constantly changing magnetic field around the electric wires (right).

attached to anything, such as the wheels of a toy car, that we wish to spin. The *stator* is the stationary portion of the motor. It consists of coils of wire that surround the rotor. Current is passed through the stator to cause the rotor to spin.

In a simplified view, we might think of a motor as a device in which the magnet of the rotor tries to "follow" the electricity circulating around it, and a generator as a device in which the electrons in the wire try to "follow" a spinning magnet (Fig. 7-10). However, it is more accurate if we can visualize what the actual magnetic *field* is doing around the stator and rotor. Furthermore, as we do so, we find that an actual magnet is not needed to make an electric motor, because of the principle of *mutual induction*. A motor can be devised using only coils of wire for both the rotor and the stator. Such a motor is called an *induction motor*.

This is depicted in Figure 7-11 which shows two wires near to each other with AC electricity passing

Figure 7-11

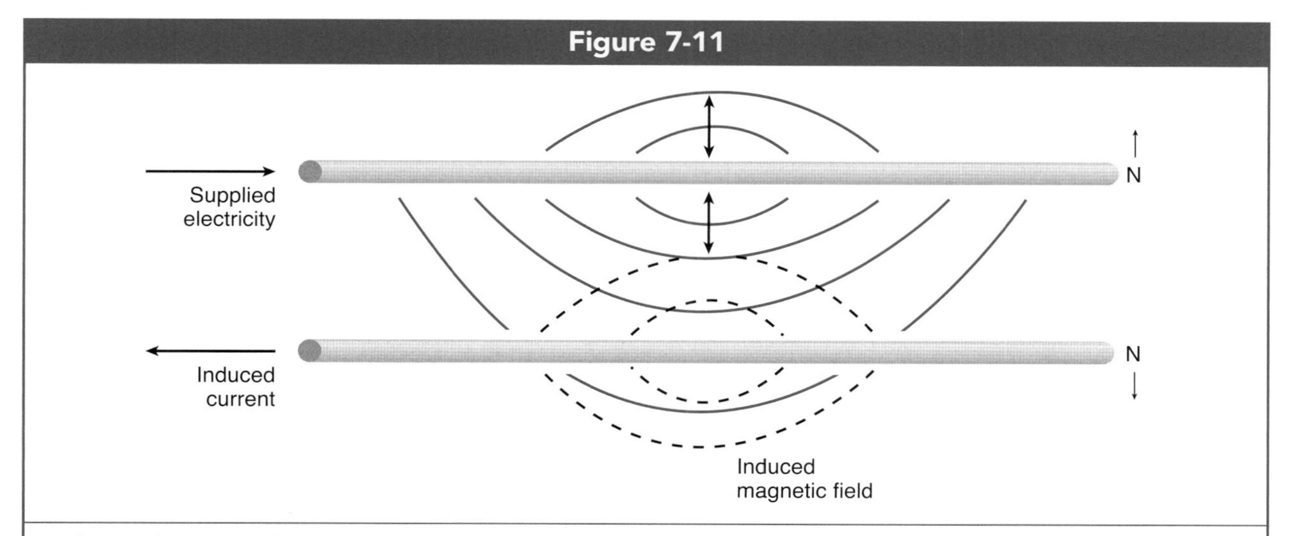

With AC electricity, the constantly blooming and collapsing magnetic field around a wire will *induce* electrical current to flow in a nearby wire. This second wire now develops a magnetic field of its own, effectively *becoming* a magnet. In this way, induction motors need no actual bar magnet to run.

through one of them. As the first pulse of electricity begins to accelerate (Fig. 7-8), a magnetic field begins to *bloom* around the wire, expanding like a balloon being blown up. The faster the electrons flow (and the greater the voltage behind them), the stronger the magnetic field becomes. A stronger magnetic field is *larger*, so the field actually grows as the electricity accelerates. As it expands, we can imagine the magnetic lines of force *crossing* the other wired nearby. This constitutes a moving magnetic field and induces an electrical current to also flow in the second wire.

Now, with current flowing in the second wire, it develops *its own magnetic field* and effectively becomes a magnet. This magnetized coil of wire, the *rotor*, will be "pushed" by the magnetic field of the original electrical current in the stator, and begin to spin. Heinrich Lenz, a German-Russian physicist, discovered that *the induced current* (in the second wire) always flows in the *opposite direction* from the original current (in the first wire). This means that the magnetic field around the rotor will also oppose the magnetic field of the stator.

When you hold two magnets "backwards" to each other so that the like poles face each other, north to north or south to south, they will repel one another. In the same way, in the induction motor, the magnetic field of the stator repels the "backward" magnetic field of the rotor, causing it to spin. Without Lenz's law, the rotor would not spin.

Returning our attention to the blooming magnetic field around the stator, we find that no sooner does it reach its maximum size than it begins to then collapse as the electrons decelerate in the second half of an electrical pulse (Fig. 7-8). At the end of this pulse, the field has completely collapsed, but instantly the electrons begin to accelerate in the opposite direction. A magnetic field with *opposite polarity*, with north facing the opposite way, begins to *bloom*. As it reaches its maximum diameter, it begins to collapse, only to start a new cycle. At all times the magnetic field is either *growing* or *collapsing*. This keeps the magnetic lines of force constantly moving across the second wire, which continues to induce electricity to flow in it.

Imagine the same scenario with two wires, only using *DC* electricity. By Figure 7-5, we see that the electrons only accelerate once, when the switch is turned on. During this acceleration, the magnetic field around the first wire would bloom with its lines of force crossing the second wire. For an instant, current would be induced in the second wire, but then it would *stop*. This is because when the electrons in the first wire are flowing steadily in one direction and the voltage is constant, the magnetic field *stays* expanded at its maximum diameter. Like a balloon that has been blown up and then tied, it is no longer *moving*.

Faraday's discovery was that a *moving* magnetic field induces electricity. Even though a magnetic field may be present around the second wire, no matter how strong it is, if it is holding still no induction will occur in the second wire. Hence, the second wire does not develop its own magnetic field and is not induced to spin. For an induction motor to work, *AC electricity is required.*

Within the stator, a *series* of coils is fired in sequence, which effectively keeps the "north pole" of its magnetic field circulating. The "north pole" of the rotor is repelled by it, causing the rotor to spin.

Nikola Tesla, a Yugoslav immigrant to America, refined and perfected the AC motor. He had been fired by inventor Thomas Edison, who insisted that DC electricity was the best way to power an entire city. Tesla eventually proved that AC generators are much more efficient for this purpose.

One last electronic invention is essential to x-ray equipment—*the transformer*. Like the electric motor, it operates by electromagnetic induction and requires alternating current (AC). Michael Faraday, in his investigations into the electromagnetic induction of generators and motors, accidentally assembled the first transformer. The purpose of the transformer is to change the intensity of voltage or the amperage of the electrical current.

Recall that by placing a bar of iron within an electrical coil a strong electromagnet may be constructed. The magnetic lines of force are concentrated by the ferromagnetic material of the core. If the iron core is extended and curved into a continuous loop as in Figure 7-12, there are no ends to the bar from which magnetic field lines can escape. The field then *follows* the core all the way around the loop, confined to the iron core. If a neutral wire is then wrapped around the other side of the loop of iron, the moving magnetic field of AC electricity passing through the first coil will induce a current to flow in the secondary loop, just as occurs in a motor.

Figure 7-12

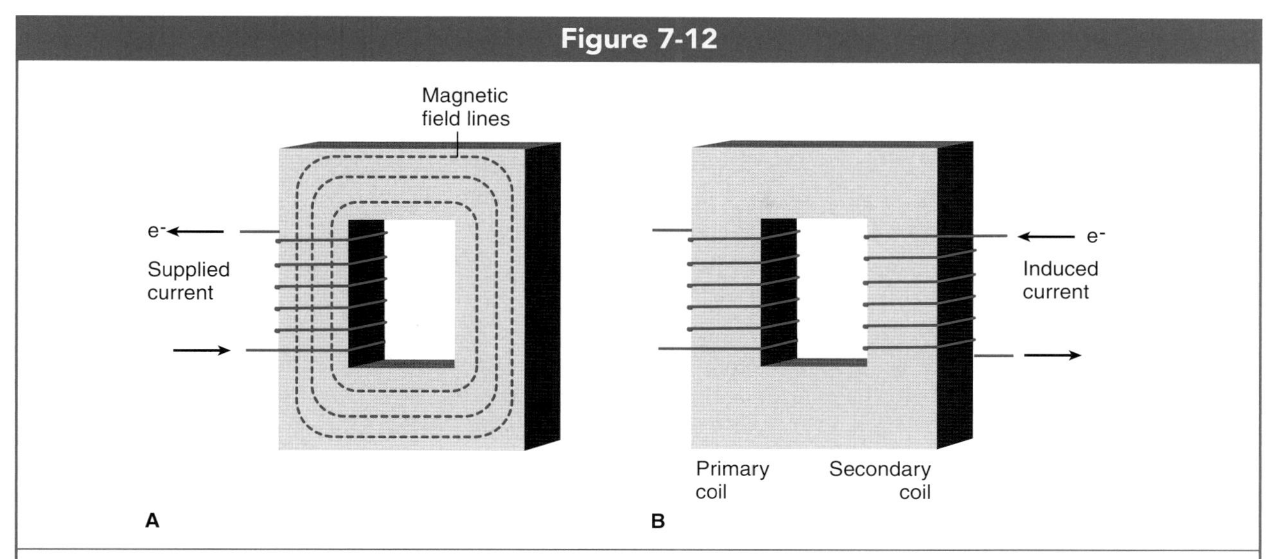

A, In a typical AC transformer, the iron core of an electromagnet is formed into a square loop, making the magnetic field follow all the way around. In *B*, if a coil of neutral wire is wrapped around the opposite side of the core, electricity will be induced in it by the constantly moving magnetic field.

The key principle behind the transformer is that, *if the secondary coil has more turns or windings than the primary coil, voltage is increased proportionately in the secondary coil.* For example, if the secondary coil has twice as many windings as the primary coil, the voltage generated in the secondary coil will be doubled. This directly proportional relationship is expressed in the *transformer law*, which is stated:

$$\frac{N_s}{N_p} = \frac{V_s}{V_p}$$

where N is the number of turns or windings in each coil, V is the voltage, and the subscript *s* and *p* indicate the secondary and the primary coils. The quantity N_s / N_p is known as the *turns ratio*, expressed as $N_s : N_p$. If the secondary coil has 100 windings and the primary has 5, the turns ratio is 20:1 (after reducing 100:5). The primary voltage will be multiplied by this amount. Let's try a simple practice exercise to reinforce the transformer law:

Practice Exercise #8

The secondary coil of a transformer has 20,000 turns; the primary coil has 500 turns. If 120 volts is supplied to the primary coil, what will be the resulting voltage in the secondary coil? Also, what is the turns ratio?

Solution:

$$\frac{N_s}{N_p} = \frac{V_s}{V_p}$$

$$\frac{20{,}000}{500} = \frac{X}{120}$$

Cross-multiplying:

$$20{,}000\,(120) = 500\,X$$

$$2{,}400{,}000 = 500\,X$$

$$\frac{2{,}400{,}000}{500} = X$$

$$4800 = X$$

Answers: The voltage in the secondary coil is 4,800 volts. This is 40 times the original voltage of 120, because the turns ratio is 40:1.

Now you might ask, doesn't this defy the law of conservation of energy? You cannot increase voltage without something else being reduced as a consequence. Higher voltage cannot come "from nowhere." This is correct. Earlier we defined *power* as the rate at which work is done and energy is spent. The total energy of an electrical system is a conserved quantity—it cannot be increased or reduced, but only change form. Some power is lost in the form of heat. What remains is the product of voltage times amperage; this is the power law:

$$P = IV$$

Voltage and amperage are *not* conserved quantities—they can be *leveraged* up or down, as long as the total power remains the same. But, you can see from the power formula that it will be the *amperage* which drops in consequence of increasing the voltage, and vice versa.

These are inversely proportional, so when any transformer doubles the voltage in the secondary coil, the *amperage* must be cut in half. If the voltage increases by tenfold, the amperage must be reduced to one-tenth. *The change in amperage is inversely proportional to the turns ratio of the transformer.*

Practice Exercise #9

For the transformer described in Practice Exercise #8, if 600 mA is fed into the primary coil, what is the resulting secondary amperage, in *mA*?

Solution: This was determined to be a 40:1 ratio transformer, so the resulting amperage will be 1/40th of the original.

$$\frac{600}{40} = 15$$

Answer: The amperage in the secondary coil is 15 mA.

With this inverse relationship in mind, how would we build a transformer for the purpose of *increasing the amperage*? The answer is to have the turns ratio be a fraction of one, by winding *fewer* turns in the secondary coil than in the primary. If the primary coil has 20 windings, but the secondary coil only has 10, the voltage will drop to one-half, but the *amperage* will double in turn.

Transformers are named according to their effect on the voltage; thus, a transformer that increases the voltage is called a *step-up* transformer, and one that decreases the voltage is called a *step-down* transformer. Remember that a *step-down* transformer is needed when we wish to *increase the amperage.*

A special type of transformer called an *autotransformer* is used when only a very small step-up or step-down in voltage is required. Rather than mutual induction, the autotransformer (*auto* from *self*) operates on the principle of *self-induction* by using only one coil of wire. Incoming voltage is supplied to this coil, and a secondary circuit is connected to it by *taps* which can be slid up or down the coil as shown in Figure 7-13. By tapping off electricity from only some of the coils, a fraction of the original voltage is produced within the secondary circuit in accordance with the transformer law formula. If there are 100 windings, and the secondary taps are connected between #1 and #75, three-fourths of the primary voltage will result.

$$\frac{75}{100} = \frac{3}{4}$$

The coil of an autotransformer can also be extended beyond the incoming primary taps, and the secondary taps can be slid a certain amount *above* the primary connections to increase voltage by a limited amount (Fig. 7-14).

On a typical x-ray machine console, the *major* kVp (voltage) can be selected in multiples of 10, and the *minor* kVp in steps of 1 or 2. These kVp selection controls set the taps on the *autotransformer*. Once this amount has been selected (from the original 220 volts supplied by the power company), the electrical current is then passed through a *step-up* transformer to obtain the high-voltage current needed to produce x-rays. If the step-up transformer has a turns ratio of around 1000:1, then when 80 volts is tapped off at the autotransformer the end result after passing this through the step-up transformer is a current having 80,000 volts or 80 kV.

A step-down transformer is used in a separate circuit to increase the *amperage* supplied to the filament

Figure 7-13

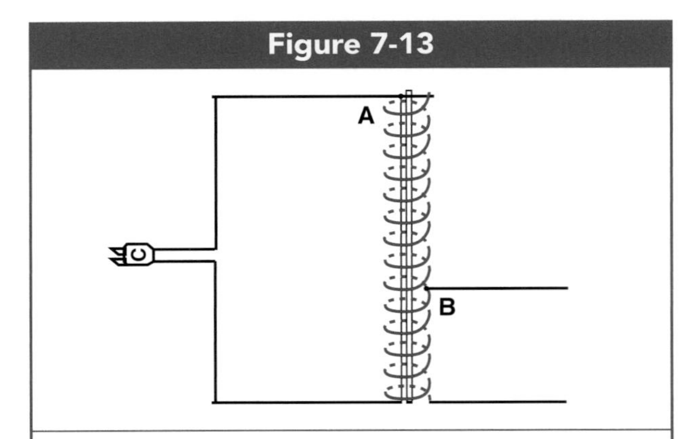

An autotransformer uses taps that can be slid up or down the primary coil of wire. By *self-induction*, the number of coils tapped from the coil will induce a secondary voltage that is a fraction of the original. Here, only the first five of 15 coils are tapped, (***B***) The resulting voltage will be one-third of the original.

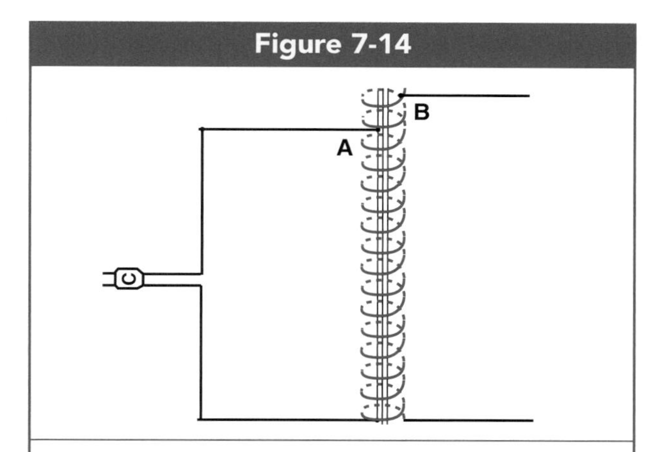

Figure 7-14

By extending the coils above the primary taps, secondary voltage can be induced that is higher than the primary voltage, to a limited amount. The kVp contol on an x-ray machine adjusts an autotransformer.

of the x-ray tube. This high-amperage current generates a great deal of friction from the high number of electrons being crowded into and through the thin filament wire. This is ideal for *heating* the filament to extremely high temperatures which begin to "boil" electrons off of the filament wire, making them available to be thrust across the x-ray tube to the anode when the high-voltage current from the step-up transformer is also applied.

Generators, motors, rectifiers, and all three types of transformers are all essential components of x-ray machines, and batteries are used in mobile (portable) units. A basic layout of how these are all arranged in the x-ray machine circuit and how they work together in the production of x-rays is the topic of the next chapter.

SUMMARY

1. Electrical conductors have loosely-bound outer shell electrons that are free to flow as electrical current; dielectrics or insulators resist electrical current, while semiconductors allow it to flow only under specific conditions.
2. For a radiographic exposure, the product of the electricity rate in mA and the exposure time yields the total *mAs*, an indication of the total amount of electrical charge (in coulombs) used.
3. To complete an electrical circuit, a source of electrons which is the conductor must be connected to a source of EMF and a resistive device in an unbroken circle.
4. Poor conductors, narrower wires, and longer wires all increase electrical resistance, heating up the circuit. As resistance increases, EMF must be increased or the current will be reduced.
5. For a given voltage, resistance and amperage are inversely proportional to each other.
6. Electrical power is the rate at which work is done or energy is spent. It is the product of the voltage times the current.
7. For a given amount of electrical power, voltage and amperage are inversely proportional to each other.
8. DC electricity always moves electrons in the same direction, while AC oscillates them.
9. The *effect* of moving electrons, called current, moves at the speed of light.
10. Electrical current with a frequency of 60 hertz consists of individual pulses 1/120 second long, half of which travel each direction.
11. A rectifier bridge can be used to reverse every other pulse of AC current, resulting in *rectified AC* or *pulsed DC* current.
12. All devices which work by electromagnetic induction require AC current in order to keep the magnetic fields moving.
13. Electrical generators convert mechanical energy into electricity, while electrical motors convert electricity into mechanical movement. Both operate on the principle of electromagnetic induction.
14. Because of *mutual* induction, an induction motor can be constructed without magnets, but with only coils of wire.
15. Transformers also operate by electromagnetic induction, and can be used to step up or step down the voltage in proportion to their *turns ratio*, or to modify current inversely to the turns ratio.
16. The kVp control at the console of an x-ray machine is an *autotransformer* which works on the basis of self-induction.

REVIEW QUESTIONS

1. What unit is defined as the rate of one coulomb per second?

2. How many electrons are used with a radiographic technique of 200 mA at 0.8 seconds?

3. What are the three components of an electrical circuit?

4. What three changes in a conductor increase its electrical resistance?

5. What are the two general types of electrical circuits?

6. If a circuit has 110 volts and 5 ohms, what is the amperage flowing through it?

7. If a circuit has a resistance of 33 ohms and a current of 200 amps, what is the potential difference?

8. If circuit has a potential difference of 65 kilovolts and a current of 200 milliamperes, what is the resistance?

9. An x-ray machine is said to have a 30 kilowatt generator. If the maximum x-ray tube voltage is 150 kV, what is the available tube current?

10. A 100-watt light bulb has 2 amps of current flowing through it. What is its resistance?

11. Faraday's law states that electrical current will flow through a conductor if it is placed in a ___________ magnetic field.

12. AC current is required for any device that operates on the basis of electromagnetic ___________.

(Continued)

REVIEW QUESTIONS *(Continued)*

Figure 7-15

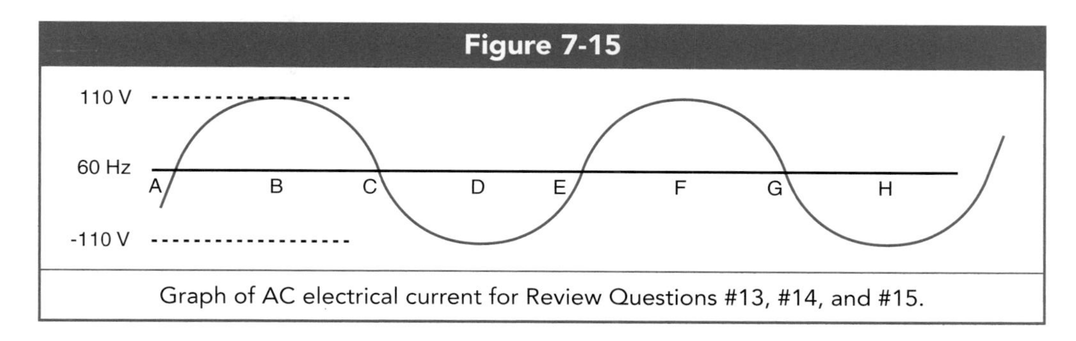

Graph of AC electrical current for Review Questions #13, #14, and #15.

13. In Figure 7-15, how is the movement of electrons changing from points ***D*** to point ***E***, when compared with points ***C*** to ***D***?

14. In Figure 7-15, what is the voltage at point ***D***?

15. In Figure 7-15, how much time has elapsed between point ***A*** and point ***E***?

16. What type of electrical current waveform is transmitted across an x-ray tube from cathode to anode?

17. What are the two components of every motor?

18. What type of motor is used to rotate the x-ray tube anode?

19. What type of transformer is used in the filament circuit of an x-ray machine so the filament can be heated?

20. A step-up transformer has 18,000 turns in the secondary coil, and 600 turns in the primary coil. If 220 volts are supplied to the primary coil, what will the output voltage be?

21. An autotransformer operates on the principle of:

22. An autotransformer has 800 windings. If 240 volts are supplied to it, and the secondary taps are connected between windings #1 and #500, what will the output voltage be?

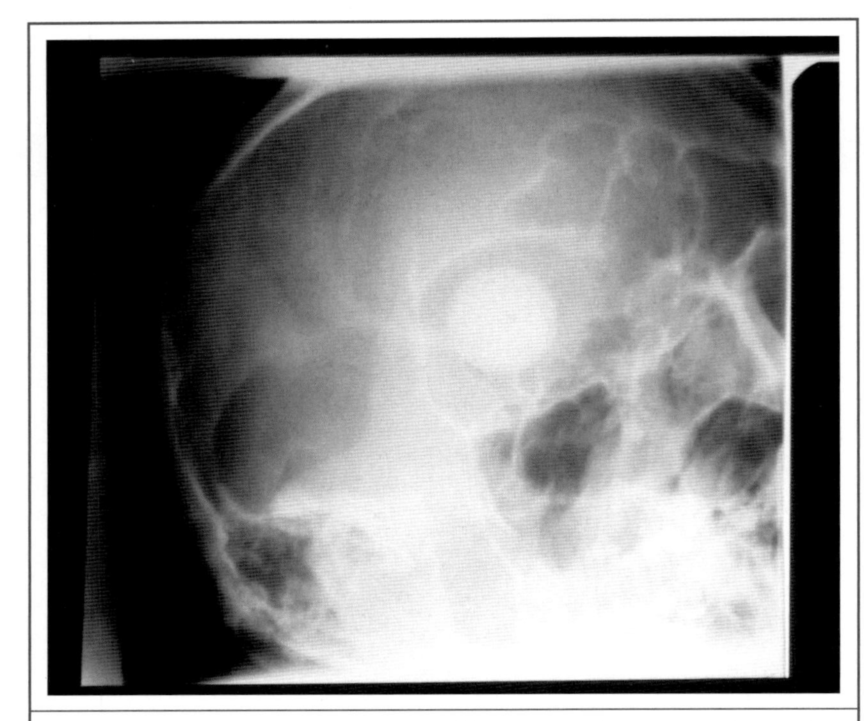

A Stenver's projection for the temporal bone was taken on a patient with a glass eye (light circular area).

Chapter 8

X-RAY MACHINE CIRCUITS AND GENERATORS

Objectives:

Upon completion of this chapter, you should be able to:

1. Sketch the basic layout of an x-ray machine circuit, showing in proper order the three main transformers, rectifier bridge, mA control, kVp meter and mA meter.
2. Describe the waveform and voltage changes between each major component of the circuit.
3. Explain how a modern solid-state diode prevents electricity from flowing "backward" on its way to the x-ray tube.
4. In the filament circuit, explain how high amperage is produced, how it results in *thermionic emission* at the filament of the x-ray tube, and when it occurs.
5. Explain why the mA and kVp meters are connected where they are in the circuit, and why they are connected in different manners.
6. Describe higher-power generators increase both the *quantity* and the *quality* of electrical current.
7. Give the percentage ripple and *average kV* as a percentage of the kVp for each type of generator.
8. Solve for the power rating in kilowatts for single-phase and 3-phase x-ray generators.
9. List the three general types of exposure timers.
10. Explain the *principle* on which an AEC is able to automatically terminate the exposure.
11. State the functional steps for the AEC circuit to work, and what the *thyratron* is.
12. Explain why the AEC *cannot* compensate for any changes in grid or image receptor.

A BASIC X-RAY MACHINE CIRCUIT

Building on the information from the last chapter, Figure 8-1 is a simplified schematic of a complete x-ray machine. The main circuit is composed of three general sections: The control console, the high-voltage section, and the x-ray tube circuit. To the far left you can see the incoming lines from the power plant which normally carry AC electricity at 220–240 volts. The power switch to the whole x-ray machine is shown as a break in the wires.

Not shown is a large circuit breaker for protection, which kicks off if there is an overload or short circuit in any part of the system, thus protecting delicate electronic components from overheating. Also not shown is an important device called a *line-voltage compensator*. Fluctuations in the current provided by the power company combine with the various sections of a hospital using large amounts of electricity to make the incoming voltage unreliable. It is essential that it be constant so that radiographic techniques may be set with confidence. This device measures the incoming voltage from the power line,

Figure 8-1

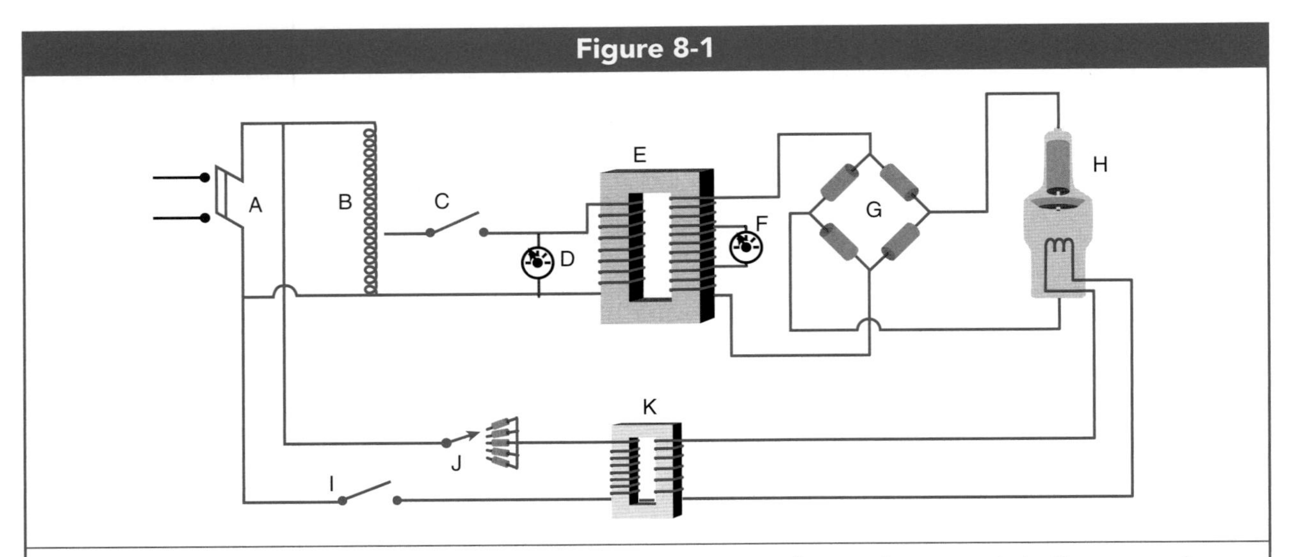

Simplified schematic of a complete x-ray machine circuit: ***A***, incoming lines and power switch; ***B***, autotransformer for kVp selection; ***C***, exposure switch with timer; ***D***, pre-reading kVp meter; ***E***, step-up transformer for high voltage; ***F***, mA meter; ***G***, rectification bridge; ***H***, x-ray tube; ***I***, rotor switch; ***J***, bank of resistors for mA selection; and ***K***, step-down transformer.

and automatically compensates by adjusting the autotransformer slightly up or down.

The *kVp major* and *kVp minor* controls on the x-ray machine console actually set the autotransformer, shown next in the diagram from left-to-right. It is safer to do this *before* stepping up the voltage into the tens of thousands, so the kVp settings on the console really read out what the kilovoltage *will be* after it passes through the step-up transformer, rather than the actual voltage coming from the autotransformer. Typically, if the incoming voltage is 220 V, the actual output of the autotransformer ranges from 100 to 400 volts.

An exposure switch and timing system is normally placed after the autotransformer in the circuit. There are several types of timers, and these will be discussed later. The electrical current next passes through the high voltage transformer.

The high voltage transformer is a step-up transformer with a turns ratio normally between 500:1 and 1000:1. This and another transformer for the filament circuit may be housed in a large box near the wall in the x-ray room, from which the large cables to the x-ray tube extend and are suspended along the ceiling. Figure 8-2 is a cut-away view of one of these larger transformer units, which are filled with oil for both electrical and heat insulation. Figure 8-3 shows the voltage waveforms for the current entering and exiting the high voltage transformer. Note that both are unrectified AC current, since transformers must have AC current to operate on the principle of electromagnetic induction. The only difference in the two waveforms is the *amplitude* or height of the waves, which indicates increased voltage.

Rectification

The final step in the main circuit is to have this high-voltage AC current *rectified* so that the electrons will always enter the x-ray tube in the right direction, at the filament end of the tube. One format for a *rectifier bridge* that could be used in a single-phase x-ray machine is illustrated in Figure 8-4. A minimum of 4 diode rectifiers is required for full efficiency. Each diode acts as a gate that only allows electricity to flow through it in one direction, indicated by the arrows. Note that in this diagram, no matter which way electrons flow *into* the rectifier bridge, they can only exit the right-hand side by flowing toward the x-ray tube filament.

Modern solid-state diodes are made from semiconductor crystals, such as silicon. These semiconductor crystals come in two types: *n-type* and *p-type*. In an n-type crystal, loosely-bound electrons are free

Figure 8-2

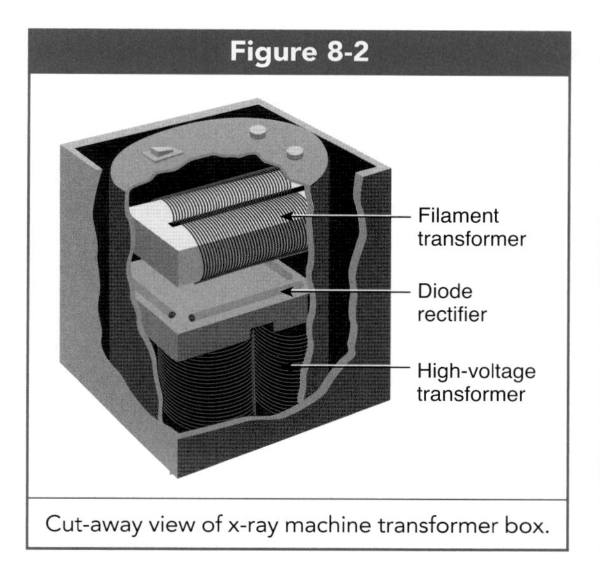

Cut-away view of x-ray machine transformer box.

Figure 8-3

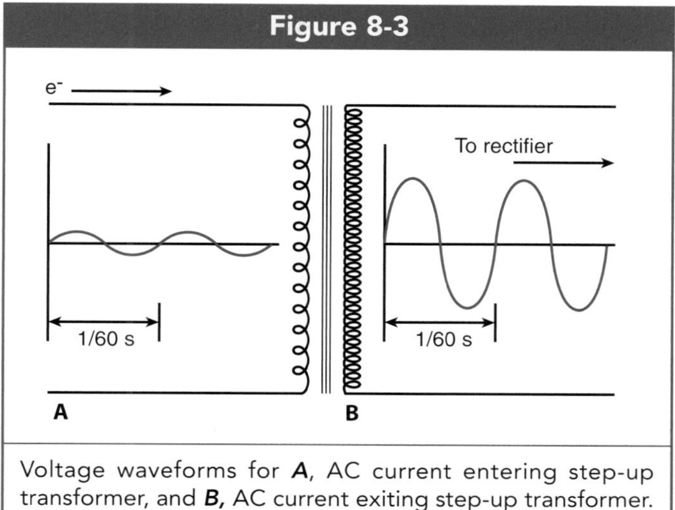

Voltage waveforms for ***A***, AC current entering step-up transformer, and ***B***, AC current exiting step-up transformer.

to flow. A p-type crystal, on the other hand, has spaces or *holes* in the outermost shells of its molecules which can also drift. The diode is constructed by connecting an n-type crystal to a p-type crystal, forming an *n-p junction* (Fig. 8-5).

When electricity approaches the *n* side of the diode, its electrons are repelled from the electricity toward the n-p junction. The holes on the *p* side of the crystal also drift toward the junction, repelled by the positive charge of the wire connected to that side. As long as electrons and holes are huddled together along the n-p junction, Figure 8-5***A***, a *potential bridge* is formed and electricity can flow through the diode. When AC current reverses direction, however, the negative electricity flows up to the *p* side of the diode and pulls the positively-charged holes toward the wire. The opposite wire, being positive, also pulls the electrons on the *n* side toward it. This *polarizes* the diode with the holes at one end and the electrons at the opposite end (Fig. 8-5***B***). As long as the holes and electrons are separated from the n-p junction, electricity cannot flow through the diode.

Three-phase machines require 6 or 12 diodes. Rectification is the last step in preparing high-voltage electricity to enter the x-ray tube.

The Filament Circuit

Early x-ray tubes required very high voltages to get electrons to spark across the tube to the anode. However, in 1913, William David Coolidge, an American physicist, discovered a more efficient way of producing x-rays. He found that if tungsten were used for the x-ray tube filament, it could be heated *before* making an exposure, to "boil" off electrons in a process called *thermionic emission* (from *thermo* = "heat," and *ion* for the negatively-charged electrons). He had used the same process to improve Edison's light bulb. (Edison had used carbonized cotton for filaments; Coolidge introduced the tungsten filament.)

In the x-ray tube, this process of thermionic emission freed electrons from the filament wire *prior to* exposure which greatly improved efficiency. Otherwise, some of the energy from the high-voltage current would be expended knocking electrons free from the filament, leaving less energy to thrust them across

Figure 8-4

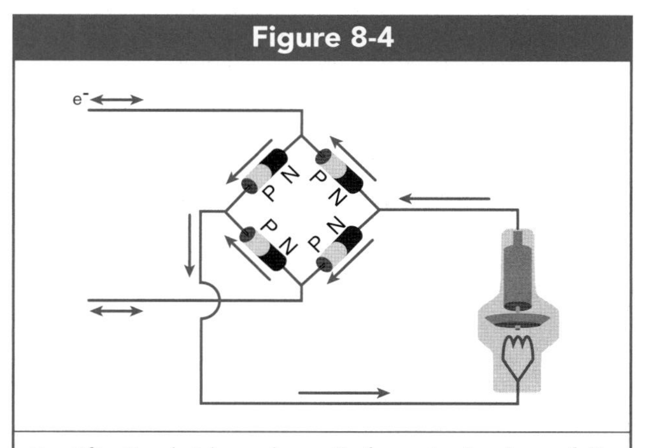

Rectification bridge schematic for a single-phase, fully rectified x-ray machine.

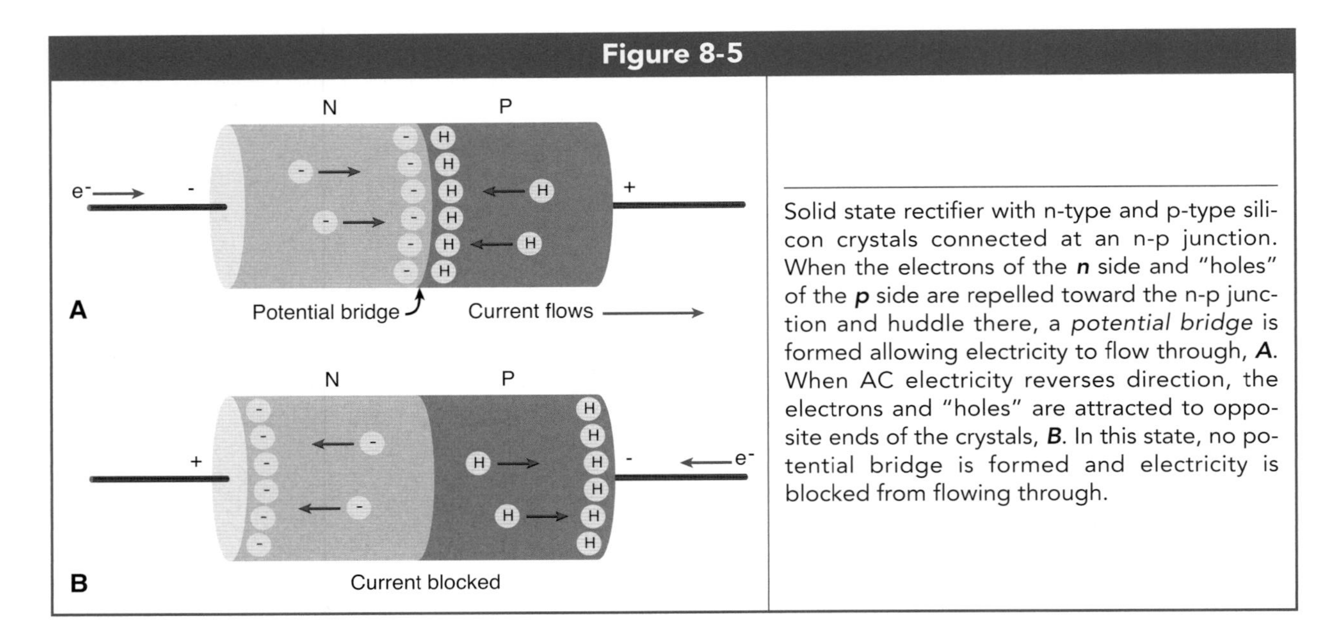

Figure 8-5

Solid state rectifier with n-type and p-type silicon crystals connected at an n-p junction. When the electrons of the ***n*** side and "holes" of the ***p*** side are repelled toward the n-p junction and huddle there, a *potential bridge* is formed allowing electricity to flow through, ***A***. When AC electricity reverses direction, the electrons and "holes" are attracted to opposite ends of the crystals, ***B***. In this state, no potential bridge is formed and electricity is blocked from flowing through.

the tube. For thermionic emission to occur, the filament had to be brought to an extremely high temperature. To heat the wire, rather than use high voltage which would push electrons through the filament *faster*, what is needed is *high amperage*, which results in millions more electrons trying to crowd their way through the filament, thus causing great friction.

Also, the filament wire is much thinner than the regular wire carrying electricity into it, which further increases friction in order to heat the wire. (To visualize this effect, imagine tripling the number of cars on a 5-lane freeway, perhaps during rush hour, and then closing three lanes! As all these cars are forced into just one or two lanes, more collisions *will* occur.) With high-amperage current entering the thin filament, electrons bump into and jostle against each other so much that a tremendous amount of friction is produced, heating the wire.

Hence, in a modern x-ray tube, *two* currents enter the filament, one designed to heat it using high amperage, and a later one with high voltage to produce the actual exposure. The high-amperage current used to heat the filament is applied *whenever the "rotor" button is held down* at the console. Note that once the "rotor" button is pushed, the x-ray machine will not make the exposure until a few seconds have elapsed. At this time a "ready" indicator light will come on and you may hear a click. This delay is necessary because it takes a few seconds for the high-amperage current to heat up the filament. The filament must be brought up to the full temperature corresponding to the mA station that has been selected. The exposure can then be made.

(The "rotor" button also performs two other functions: It may start moving the grid in the Potter-Bucky mechanism prior to exposure, and it spins the x-ray tube anode up to full rotation speed for the exposure, which will be discussed in the next chapter. It gets its name from this last function, since the anode is connected to the rotor of an AC motor.) It is important for radiographers to remember that *every time you "rotor," the filament is being heated.* When the rotor button is held down for long periods of time without making an exposure, it causes unnecessary wear and tear to the x-ray tube filament (and to anode bearings), reducing x-ray tube life.

Now, for this heating of the filament, another entire circuit, called the *filament circuit*, is required. This is shown in Figure 8-1. The filament circuit must be separate from the main circuit because both high voltage and high amperage cannot be produced in the same circuit, given a limited amount of total electrical power available. A *step-down transformer* must be used in the filament circuit. The turns ratio is usually around 1:44, that is, there are 44 windings in the *primary coil* for every winding in the secondary coil. The effect of this is to divide the incoming voltage from 220 down to about 5 volts. However, the *amperage*, which is what we are interested in here, is *multiplied* by a factor of 44. This produces a current of up to 5 amps (at 5 volts), plenty to heat up the filament wire.

This filament-heating current enters one end of the filament and exits the other—it does *not* jump across the x-ray tube (as the high-voltage current does). Therefore, it is unnecessary to rectify the filament current, and no rectifier is present in the filament circuit in Figure 8-1.

On the x-ray machine console, there are typically seven or eight *mA stations* to choose from. When you select an mA station, an electrical tap in the filament circuit makes contact with a particular electronic *resistor*. Resistors employ various types and degrees of poorly-conducting materials. Remember that, by Ohm's law, when the voltage is fixed (at 220 in this case), as the resistance increases the amperage goes down:

$$220 = IR$$

An array of different resistors has been found to be the most efficient way to adjust the mA stations. To summarize, the only devices required in the filament circuit are the rotor switch connection, an mA station selector, and a *step-down* transformer (Fig. 8-1).

Meters

In the schematic diagram of an x-ray machine circuit in Figure 8-1, you will find two meters, an *mA meter* and a *kVp meter*. A device which reads out the *amperage* of the electrical current is called an *ammeter*. If a meter is simply placed *in series* at some point in a single wire, so that electricity must flow through it to continue on, it simply "counts" the electrons flowing through that particular portion of the circuit. This quantity of current flowing is measured as a *rate* of electricity per second, or the *amperage*.

In contrast, a meter connected *in parallel*, that is, branched *across the circuit* between the wire going into a resistive device and the wire coming back from it, as in Figure 8-6***B***, will measure the voltage pressure because it is *comparing* the electrical flow between two points in the circuit, before and after it has passed through a resistor (such as the x-ray tube).

Meters can be placed anywhere in the x-ray circuit to measure the current and voltage before and after it passes through the autotransformer, the step-up transformer, the resistors in the mA selector or the step-down transformer. But, what the radiographer is primarily interested in is the *end result*, which is the amount of current and voltage actually transferred cross the x-ray tube to produce x-rays in the tube anode. Only these two meters are typically found on the x-ray machine console.

It can be hazardous to have the extremely high voltage generated by the step-up transformer passing through a meter at the console which a person can touch. Therefore, a *pre-reading* voltmeter is used for the kVp read-out. The meter actually measures the amount of voltage selected from the autotransformer

Figure 8-6

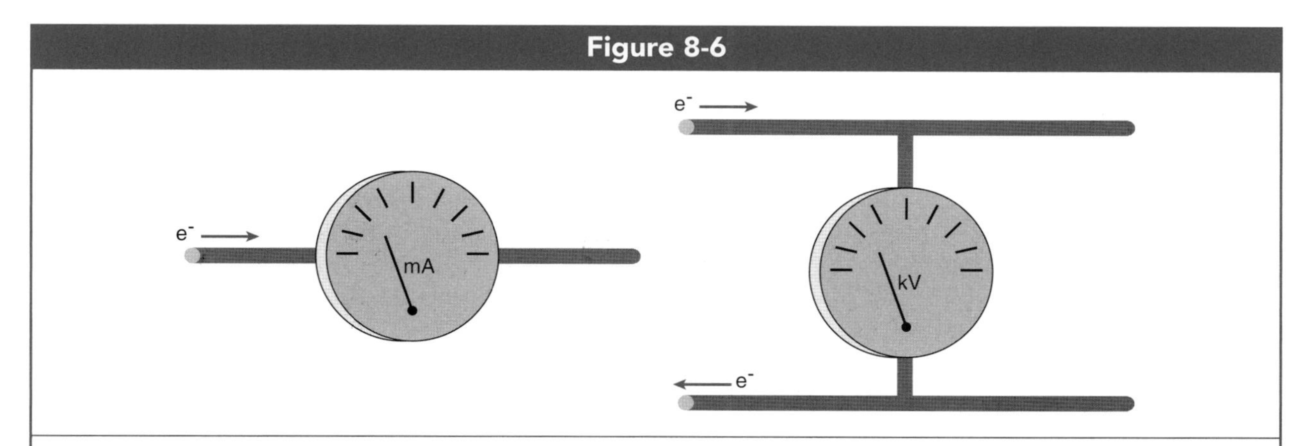

A meter connected in *series* (left) will simply "count" electrons flowing through it and measure milliamperage; but a meter connected in *parallel* (right) compares the difference between the electricity entering the x-ray tube and the electricity exiting the tube—this is the potential difference or voltage.

before it is stepped up, but the scale on the read-out is simply adjusted to account for the turns ratio of the step-up transformer, so that it reads what the kVp *will be* when it reaches the x-ray tube. The kVp meter is seen in Figure 8-1 connected in parallel between the autotransformer and the step-up transformer.

The mA meter on the console measures the milliamperage of the current which actually *flows across the x-ray tube*, not the current that heats up the filament. This can be somewhat confusing, because the filament current *controls* the resulting mA across the tube: The filament current, set by the mA station selected, determines how hot the filament burns. In turn, the temperature of the filament determines the quantity of electrons "boiled off" of it and available to move across the tube to the anode. However, the number of electrons "boiled off" is but a fraction of the current passing through the filament itself. For example, an actual filament current of 5 amps might boil off 500 *milli*-amps of electrons which will be thrust across the x-ray tube when the exposure switch is fully engaged.

It is this milliamperage which we wish to measure. Therefore, the mA meter can be seen in Figure 8-1 connected in series in the main circuit rather than in the filament circuit, placed between the step-up transformer and the x-ray tube.

X-RAY MACHINE GENERATORS

More efficient means of producing a steady, high-voltage current across the x-ray tube have been devised over the years, accomplished primarily by the way in which the high-voltage transformer is wired into the circuit and the complexity of the rectifier bridge. The specifics are beyond the scope of what a radiographer needs to know. But, it is useful to understand the voltage *waveforms* produced by these different types of generators, because they have a direct impact upon the setting of radiographic techniques.

To begin with the simplest scenario, note that it is possible to have an x-ray machine with *no* rectifiers. In this case, the x-ray tube acts as its own rectifier because it is difficult (but not impossible) for electricity to jump from a cold and thick anode backwards across the tube to the filament. Referred to as *self-rectification*, this format is employed in some portable and dental x-ray machines.

Self-rectification puts the expensive x-ray tube at some risk at higher kilovoltages, so it is generally better to place one, two or three rectifying diodes between the high-voltage transformer and the x-ray tube. This type of circuit is referred to as a *half-wave rectified* system. All the rectifiers do in this system is *block* those pulses of electrical current which are going the wrong way from reaching the x-ray tube. The resulting waveform for both self-rectification and half-wave rectification is diagrammed in Figure 8-7.

Half-wave rectification and self-rectification share an efficiency problem in that they only take advantage of one-half of the electricity available. For 60 hertz current, only 60 pulses of electricity reach the x-ray tube per second, rather than 120. To compensate so that a sufficient number of x-rays are produced, radiographic techniques must be doubled, either by using twice the mA, or twice the exposure time which also risks patient movement.

Full-wave rectified systems contain at least four rectifying diodes as shown in Figure 8-4. The resulting waveform for the electricity is shown in Figure 8-8, in which those pulses of current which originally were traveling in the wrong direction now appear above the axis of the graph, and all 120 pulses per second of incoming AC current are utilized in the exposure.

All of the foregoing waveforms are referred to as *single-phase* power. In single-phase equipment, the voltage waveform drops to zero at certain points. The x-rays produced when the voltage is near zero are of little value because of their low penetrability. The *average kilovoltage*, shown in Figure 8-8, is only approximately one-third of the peak kilovoltage or kVp. Furthermore, the maximum amount of current

Figure 8-7

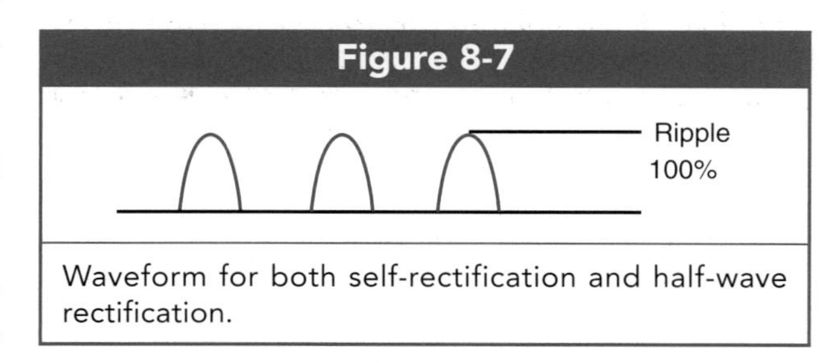

Waveform for both self-rectification and half-wave rectification.

that is utilized in single-phase power is 120 pulses per second. In order to multiply the number of pulses to obtain a more continuous flow of current, and, more importantly, to maintain a higher *average* voltage, *three-phase* generators were developed.

By the use of multiple coils in the step-up transformer, we are able to produce *three* distinct sets of electrical current that are sequenced over time such that the pulses of electricity overlap each other (Fig. 8-9). If the first pulse of electricity is thought of as the first half of a 360-degree cycle, or 180 degrees, then an overlapping pulse begins one-third of the way through it, or at 60 degrees. Yet a third current begins its cycle two-thirds of the way through the first pulse, or at 120 degrees. Each second 360 pulses are produced.

This results in an effect called *ripple*, in which the voltage never drops to zero, but oscillates between 86 percent and 100 percent of the set kVp (Fig. 8-10). We say that this waveform has 14 percent ripple. A single-phase machine has 100 percent ripple (Fig. 8-8), since the voltage varies all the way from zero to the set kVp. The *average voltage* is now one-third of the way from 86 percent to 100 percent, or about 91 percent of the kVp set. This provides higher overall penetration for the x-ray beam and more exposure to the detector plate.

In addition, note that there are now six overlapping pulses of electrical current, instead of two pulses, which peak before the end of the first cycle. The shaded areas in Figure 8-11 illustrate the additional current that has been added where a single-phase waveform would be dropping to zero. We say that for a given set mA station, the *effective mA* has increased. Current is now *constantly* flowing to the x-ray tube, and at higher average voltage, so that literally *more x-rays are produced* each second. Not only does the x-ray beam have higher penetration, but it *also* has more x-rays in it. This further increases exposure to the detection plate or cassette.

For a three-phase unit, both *quantity* and *quality* of the x-ray beam have been improved. Within the *remnant beam of x-rays* behind the patient this results in an overall doubling of the exposure which reaches

Figure 8-8

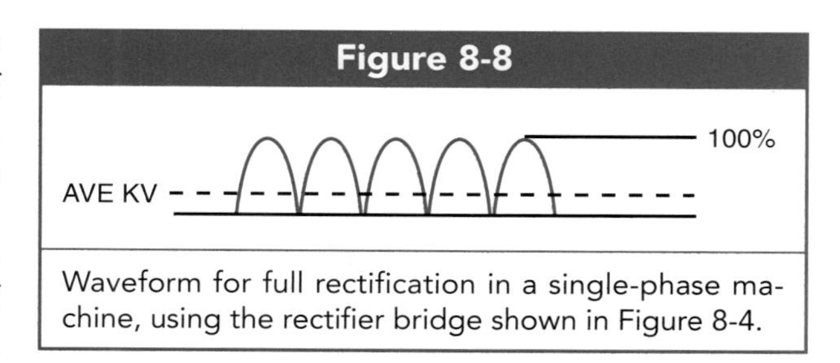

Waveform for full rectification in a single-phase machine, using the rectifier bridge shown in Figure 8-4.

Figure 8-9

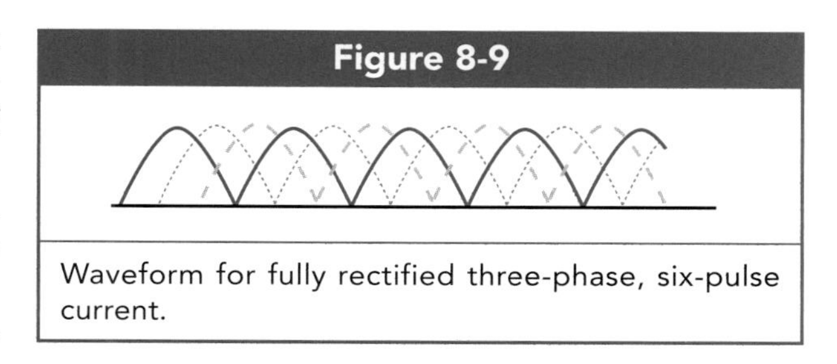

Waveform for fully rectified three-phase, six-pulse current.

Figure 8-10

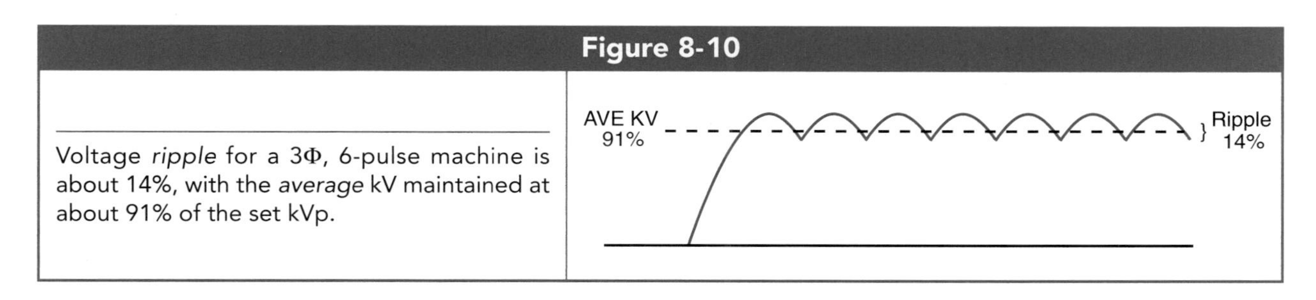

Voltage *ripple* for a 3Φ, 6-pulse machine is about 14%, with the *average* kV maintained at about 91% of the set kVp.

Figure 8-11

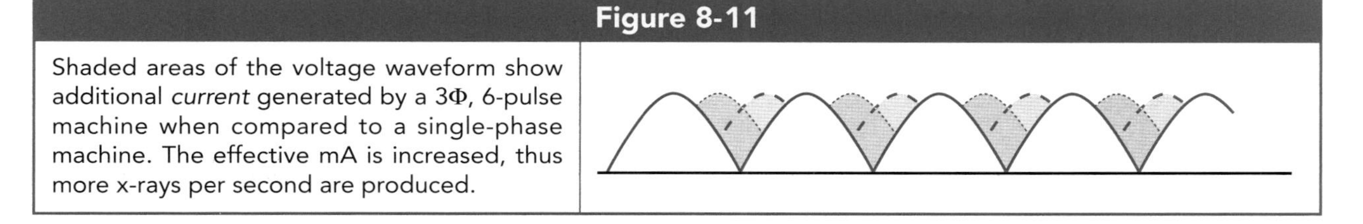

Shaded areas of the voltage waveform show additional *current* generated by a 3Φ, 6-pulse machine when compared to a single-phase machine. The effective mA is increased, thus more x-rays per second are produced.

the plate or cassette. Therefore, when changing from a single-phase machine to a three-phase machine, overall radiographic techniques can be *cut in half* and still maintain the same end exposure to the detector. This also reduces radiation exposure to the patient. The amount of mAs used may be cut in half, or the kVp may be decreased by 10. The reduction in mAs is recommended because it reduces patient exposure more than the equivalent reduction in kVp.

Further modifications of the three-phase generator allow even more efficiency by doubling the number of pulses generated in each cycle from six to twelve, for a total of 720 pulses per second. The waveform for this "three-phase, 12-pulse" machine is shown in Figure 8-12. The voltage ripple is reduced to 4 percent, so the *average* voltage is now about 97 percent of the peak voltage (kVp).

The reduction in radiographic technique from a 3-phase 6-pulse machine to a 3-phase 12-pulse machine is slight. Compared to a single-phase machine, the kVp may now be reduced by 12 (as opposed to 10). For practical purposes, the *mAs* may be approximately cut in half for either three-phase machine when compared to a single-phase machine.

High-frequency generators have increased in popularity and are common in battery-powered mobile units. Before stepping up the voltage, they alter the waveform of the incoming electrical current and convert it from 60 hertz to a much higher frequency ranging from 500 Hz to 25,000 Hz. The resulting waveform is illustrated in Figure 8-13. Voltage ripple is less than 1 percent, so the waveform is nearly constant.

This remarkable efficiency in maintaining high voltage accounts for the fact that radiographic techniques for most mobile x-ray units tend to be considerably lower than those in fixed radiographic rooms. For example, a nongrid AP chest exposure in a fixed room normally uses around 80 kVp, whereas the AP chest technique for a typical mobile unit uses around 70 kVp. This is partly because for the portable machine the *average kV is nearly equal to the kVp set*, whereas the average kV for a fixed room is 86 to 96 percent of the set kVp depending on the phase. For small extremities, the mAs setting on a mobile unit is usually less than that in a fixed room.

With higher-power generators, mA stations as high as 800 to 1200 mA are available. This allows exceedingly short exposure times which are particularly helpful in pediatric and interventional radiography. Three-phase generators are more expensive to install than single-phase machines, but can have lower maintenance costs. High-frequency generators are less costly than three-phase equipment, yet they achieve higher efficiency, and take up much less space. They are becoming the dominant form.

The electrical power rating of a generator or transformer is used as an indication of its overall quality, and is normally expressed in units of kilowatts (kW). By the power law, the watts can be calculated by multiplying the amperage times the voltage, (P = IV). The standard for comparing x-ray machines is to use the maximum mA station available at 100 kVp, with an exposure time of 0.1 second. Note that in multiplying mA and kVp, the *milli-* and the *kilo-* prefixes cancel each other out to where the power is expressed in watts; however, power ratings are usually expressed in *kilo*-watts, so this figure must be divided by 1000.

Figure 8-12

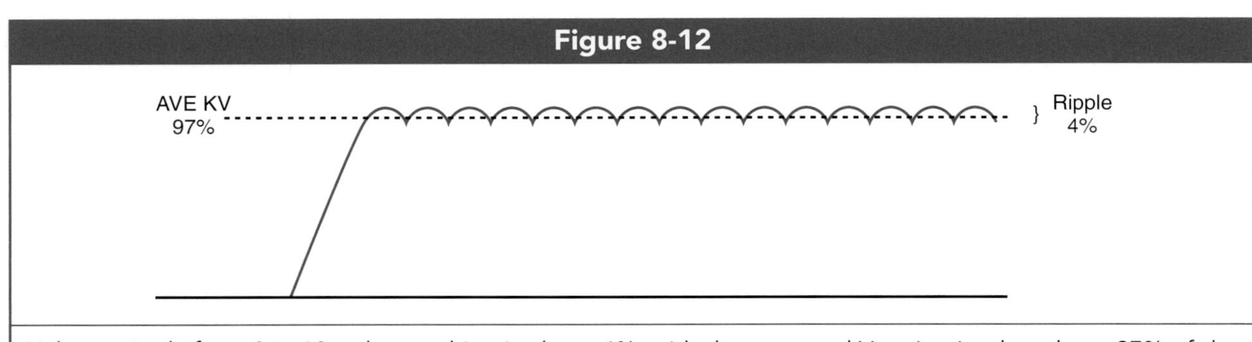

Voltage *ripple* for a 3Φ, 12-pulse machine is about 4%, with the *average* kV maintained at about 97% of the set kVp.

Figure 8-13

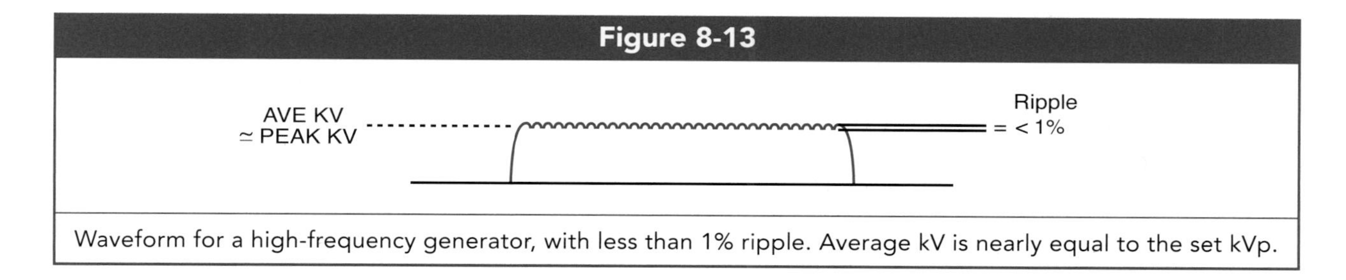

Waveform for a high-frequency generator, with less than 1% ripple. Average kV is nearly equal to the set kVp.

$$\text{Power rating in kW} = \frac{\text{mA} \times \text{kVp}}{1000}$$

Because of the tremendous voltage ripple in single-phase equipment, it is much less efficient in overall power. The power rating formula for a single-phase machine adds a constant, 0.7, to indicate this loss of efficiency:

$$\text{Single-phase power rating in kW} = \frac{0.7 \text{ x mA} \times \text{kVp}}{1000}$$

EXPOSURE TIMERS

Three general types of exposure timers are used in modern x-ray equipment for "manually-set" techniques. The synchronous timer spins the axle of a special motor at precisely 60 revolutions per second. The manually set exposure is shut off after the machine effectively "counts" the correct number of revolutions in the synchronous motor. Such units have a minimum exposure time of 1/60 second, and all available exposure times are multiples of 1/60.

The most common "manual" timers are electronic timers. They are based upon the time it takes to fully charge a special capacitor through different amounts of electrical resistance. When the capacitor's limit is reached, a surge of electricity is allowed to "run off" from it. This pulse of electrons charges an electromagnetic that pulls open the exposure switch, shutting off the exposure. Electronic timers are extremely accurate and can produce exposure times as short as 1 millisecond.

Commonly found on mobile x-ray units is the *mAs timer*. On these units, the highest safe mA is automatically selected for a particular kVp setting. The machine then monitors the actual mAs accumulated during the exposure through a meter located in the high-voltage section of the circuit, and terminates the exposure time when the preset mAs is reached.

Automatic Exposure Controls (AEC)

Automatic exposure controls, or AECs, were developed for the purpose of achieving more consistent exposures, reducing repeated exposures, and ultimately reducing radiation exposure to patients. The first automatic exposure terminating device was the *phototimer*, developed in 1942. Phototimers, which measured light emitted from a fluorescent screen, are no longer in use, but the term *phototiming* is still frequently used as a verb by radiographers to refer to the use of the AEC.

The AEC is based on the simplest of concepts: Radiation reaching the detection plate or cassette behind the patient is detected by ion chambers or solid-state devices, which generate electricity from it. The charges from this electricity are simply *counted* until they reach a preset amount, at which time the AEC system shuts off the exposure. We will detail how this works in basic terms.

In Chapter 6 we introduced the *pocket dosimeter* as a radiation detection device in which electrons, freed from the ionization of air molecules by radiation, attach themselves to positively-charged strips of metal foil. Imagine a device in which these strips of metal foil are replaced by a wire which is connected to a battery (or generator) such that it maintains a *continuous* positive charge. Electrons freed from air molecules by radiation will not only be attracted to the wire and attach themselves to it, but they will then continue to flow down the wire toward the positive terminal of the connected battery (Fig. 8-14). Thus, flowing *electrical current* has been induced by the ionization of air from radiation.

Figure 8-14

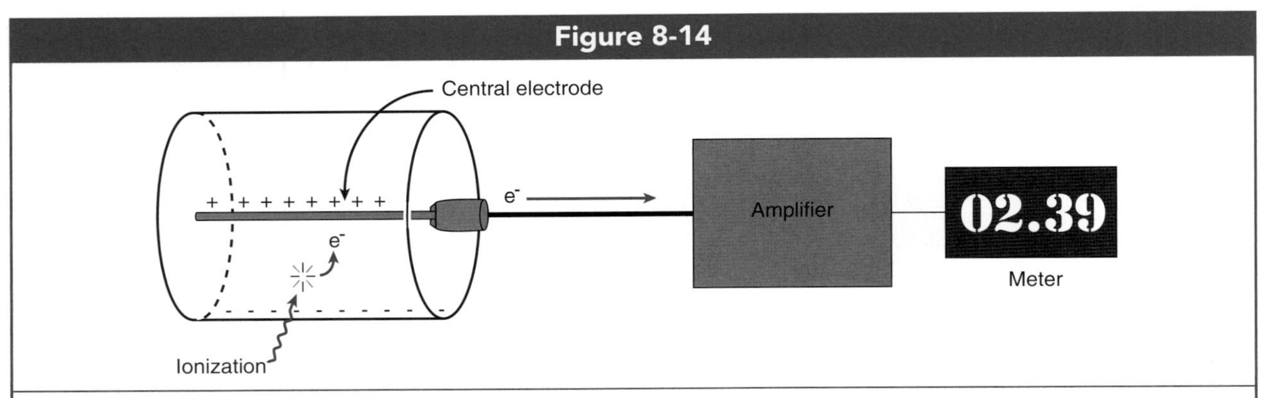

An ion chamber works on the same principle as an electroscope (Figure 6-16 in Chapter 6). Continuous radiation frees electrons from the air, which then flow down a positively-charged anode wire, forming electrical current.

In most AEC systems, the radiation detection device is a gas ion chamber type designed as a thin, flat rectangular box with air trapped between two sheets of aluminum. It can be placed within an x-ray tabletop in front of the bucky tray or detection plate without creating an artifact on images. It is connected to the circuit by extremely thin wires that also barely show up on an image. Normally, there are three such chambers arranged in a triad pattern. Figure 8-15 shows the location of these chambers, as small rectangles, demarcated on a typical "chest board."

An electrical *capacitor* is a device designed to store up electrical charge. In an AEC system, electricity from the detector cells flows to a capacitor and accumulates there as negative charge. A connected device called a *thyratron* presets the amount of charge this capacitor will hold. When the capacitor reaches this amount, the stored-up electricity is allowed to discharge from the capacitor, causing a surge of electricity down the AEC circuit (Fig. 8-16). This electrical surge charges up an electromagnet that is set close to a switch in the main x-ray circuit. The energized electromagnet pulls the switch in the main circuit open, terminating the exposure.

Figure 8-15

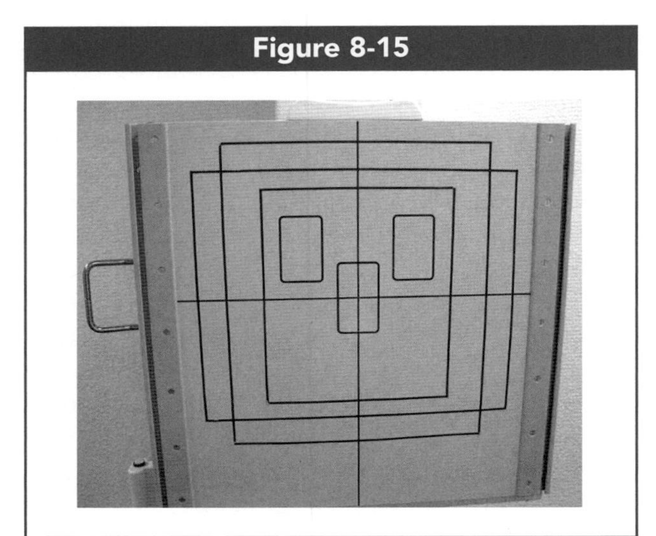

Typical location of three ion chambers in a "chest board." These can be seen on a very low-technique radiograph of the "raw" x-ray beam.

There are two switches in the main circuit, one which the radiographer must hold down to start and continue the exposure, and another which remains closed until the AEC pulls it open. Current can only flow as long as *both* switches are closed (turned "on").

How does the thyratron "know" when to discharge the capacitor? When the x-ray machine is installed, the electrician uses step-wedge penetrometers and phantom models as absorbers in the x-ray beam to simulate anatomy. The ideal exposure is determined by experimentation and extrapolation. He then sets the thyratron to the corresponding amount of electrical charge that resulted in a proper exposure.

Everything else is "automatic" in the following sense: There is no need for the AEC circuit to actually "time" anything—all it knows how to do is *count* electrical charge, and shut off the exposure when the preset amount is reached. Suppose the patient is rolled up onto her side; since the anatomy is much thicker, less of the x-ray beam is able to penetrate through the patient in lateral position. The *rate* of

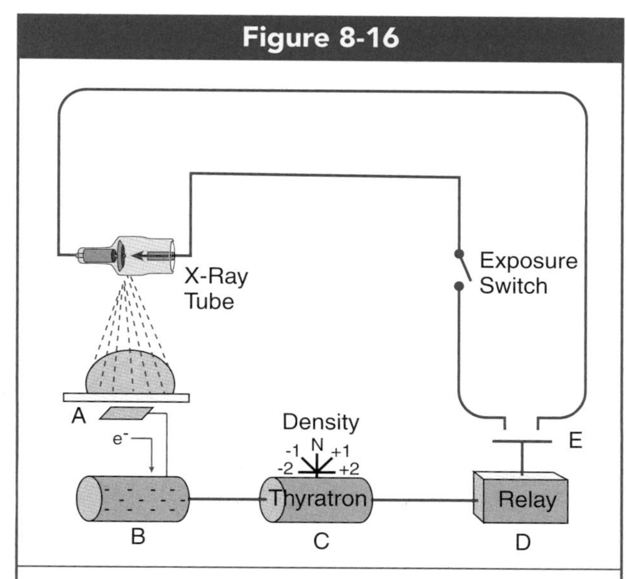

Figure 8-16

Simplified circuit for automatic exposure control (AEC). ***A***, ion chambers, ***B***, capacitor, ***C***, thyratron for density control, **D**, magnetic relay to pull open exposure switch ***E***.

x-rays reaching the detectors behind the patient is reduced. Consequently, the *rate* of electrical charge flowing into the capacitor is also lessened. It simply *takes longer* for the capacitor to reach the preset amount. Therefore, the exposure runs a longer time before it is shut off by the AEC.

When a larger patient is placed on the x-ray table, the same scenario occurs—it takes longer for the preset amount to be reached. If thinner anatomy is radiographed, the exposure *rate* to the detectors increases, the capacitor reaches its limit sooner, and the exposure is shut off earlier.

On the x-ray machine console, there is a "density" control for the AEC system. This readjusts the *thyratron* higher or lower by set amounts. The central point, usually designated with an *N*, is the original setting made by the installing electrician. Most density controls readjust this by steps of 25 percent each, but some vary in format. *The density control is the only logical way to alter the resulting exposure while using the AEC.* That is, as long as the AEC is engaged, changing the mA or even the kVp will not be effective in altering the exposure. Rather, the thyratron must be reset by using the density control. The fine points of using AEC technique will be reserved for a later chapter.

Any time the image receptor system is changed, the AEC system must be reset by a qualified serviceman. The system automatically compensates for any changes that occur *in front of the table*, because these all affect the *exposure rate reaching the detectors*. But, grids, cassettes, and detector plates lie *behind* the AEC detectors, so the system has no way of "knowing" when they are changed.

The AEC must not become a "crutch" for radiographers who wish to avoid setting "manual" techniques. While it does improve consistency for most procedures, there are a number of procedures and circumstances for which it is not appropriate. The radiographer must know when to turn the AEC off, and must still have the mental math skills to estimate and adjust radiographic techniques "manually." Good judgment must be used in all cases.

SUMMARY

1. The main components of the high-voltage circuit of an x-ray machine are an autotransformer, a step-up transformer, and a rectifier bridge which ensures that the high-voltage current crosses the x-ray tube in the right direction from cathode to anode.
2. The filament circuit consists of various resistors and a step-down transformer, designed to provide the high amperage necessary to heat the filament by friction.
3. Each time the "rotor" button is held down, the filament is heated and the anode spins. Excessive rotoring should be avoided because of the wear and tear it causes in the x-ray tube.
4. A bridge of at least four rectifying diodes is required for full-wave rectification.
5. *Average kV* for single-phase (1Φ) machines is about one-third of the kVp, for 3Φ6p machines it is 91 percent of the kVp, for 3Φ12p it is 97 percent, and for high-frequency generators it is 99 percent. Higher-power generators also produce higher effective mA. Radiographic techniques can thus be reduced.
6. The power rating for an x-ray generator in kilowatts is calculated by multiplying the maximum mA times the kVp (X 0.7 for 1Φ), divided by 1000.

7. The most common type of x-ray timers are electronic timers that use capacitors to store charge corresponding to the set time for the exposure.
8. Automatic exposure controls (AECs) typically use gas ion chambers to convert x-rays to electrical charge which is stored on a capacitor. When the limit preset at the thyratron is reached, this charge is released to trip off the exposure switch.
9. Any change which reduces the rate of x-ray exposure *in front* of the AEC, such as rolling the patient up onto his/her side, makes the AEC stay on longer before the limit is reached. Changes *behind* the AEC cells cannot be compensated for and will result in an incorrect amount of exposure.

REVIEW QUESTIONS

1. In the high-voltage x-ray machine circuit, what is the very last thing that must be done with the electrical current before it reaches the x-ray tube?
2. The step-up transformer is between the __________ and the __________.
3. The step-down transformer is between the __________ and the __________.
4. What is the typical range of output voltage from the autotransformer?
5. What is the typical turns ratio for the step-down transformer in the filament circuit?
6. In a solid-state diode, if electrons drift in n-type silicon, what drifts in p-type silicon?
7. Why can't electricity pass backward through the solid-state diode?
8. Why is high amperage, rather than high voltage, needed for heating of the filament?
9. After depressing the exposure switch all the way, why does the x-ray machine prevent any exposure until the "ready" light comes on?

(Continued)

REVIEW QUESTIONS *(Continued)*

10. On most x-ray machines, when the mA station is selected, what particular type of electrical device in the circuit is being selected?

11. When the kVp is selected, what electrical device is being adjusted?

12. Voltage meters must be connected in which way in the circuit?

13. Does the mA meter measure the filament current or the tube current?

14. What is defined as the "boiling off" of electrons from a heated filament wire?

15. Why is self-rectification dangerous to the x-ray tube?

16. Compared with a 1Φ machine, radiographic techniques can generally be changed by how much for 3Φ and high-frequency machines?

17. Higher-power generators are more efficient because they have a *lower* percentage of what effect in their waveform?

18. What is the only logical way for the radiographer to modify the exposure intensity while the AEC is engaged?

19. Why does the AEC take longer to shut off the exposure when the patient is rolled up onto his/her side, or when a thicker patient is placed in the beam?

20. When setting the density control, what specific electrical device is being adjusted?

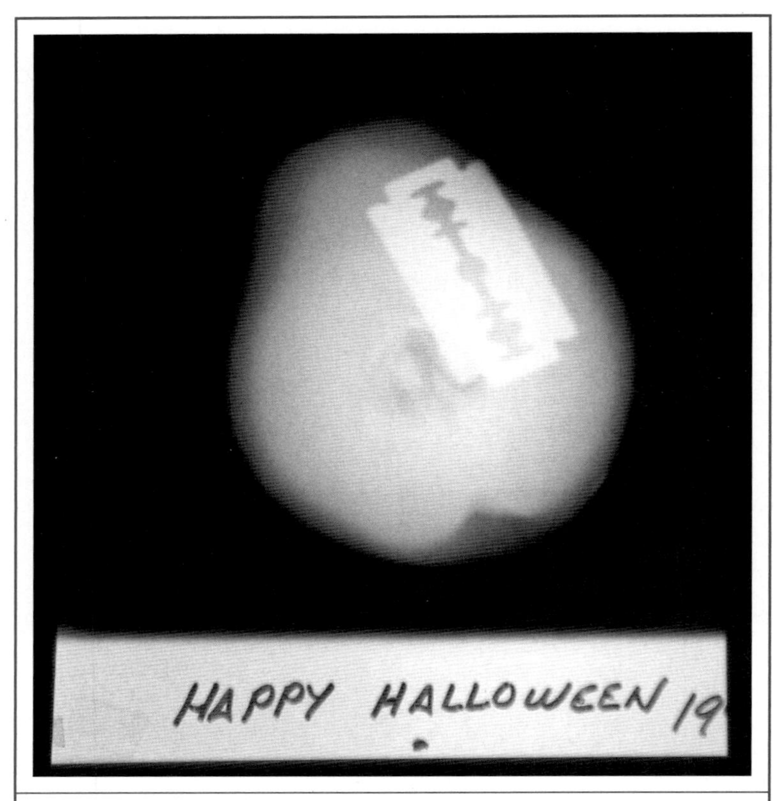

Fruit instead of candy for Halloween. A shaving razor blade had been inserted into this pear, which was brought into the x-ray department for a check.

Chapter 9

THE X-RAY TUBE

Objectives:

Upon completion of this chapter, you should be able to:

1. List the three essential conditions for the production of x-rays, and how each is met.
2. Explain why the creation of a *space charge* around the filament through the process of *thermionic emission* makes the x-ray tube more efficient.
3. Describe the materials and construction of the cathode.
4. Describe *two* ways a repulsive charge on the focusing cup can be used.
5. Distinguish between the purpose of the *filament current* and the *high-voltage* current.
6. Explain why the x-ray tube has 3 wires connected to one end and only one on the other end.
7. Describe the materials and construction of the anode.
8. For the target surface, the anode disc, and the anode stem, state whether its electrical conductivity is high or low, and whether its *heat* conductivity is high or low.
9. Explain the purpose of spinning the anode at high speed.
10. Describe how the copper cylinder of the anode shaft acts as the rotor of the induction motor.
11. Explain the effects of the mA station, the filament selected, and *focal spot blooming* on heat dissipation and image quality.
12. Describe the function of the glass envelope of the x-ray tube.
13. List the causes of x-ray tube failure, both for the filament and for the anode.
14. Calculate the heat unit load for single-phase and 3-phase x-ray machines.
15. Properly interpret x-ray tube rating charts and cooling charts.
16. List ways of extending x-ray tube life.

X-RAY PRODUCTION

There are but three essential conditions for the production of x-rays:

1. There must be a *source* of free electrons.
2. There must be a means of *accelerating* those electrons to extreme speeds.
3. There must be a means of precipitously *decelerating* the electrons.

The source of free electrons is a *filament* wire heated sufficiently to produce *thermionic emission* as described in the previous chapter. A minimum filament temperature of about 3700° F (2000° C) is required. The actual temperature of the filament above this, hence the rate of thermionic emission, is predetermined by the *mA station* selected. When the *rotor* button is depressed, a current sufficient to generate this temperature flows through the filament. With added energy, electrons "jump" from their atoms and right off the wire, forming an *electron cloud* or *space charge* around the filament (Fig. 9-1).

The *space charge* constitutes electrons that are free to move across the x-ray tube to the anode. Some of the electrons fall back into the filament, but are

Figure 9-1

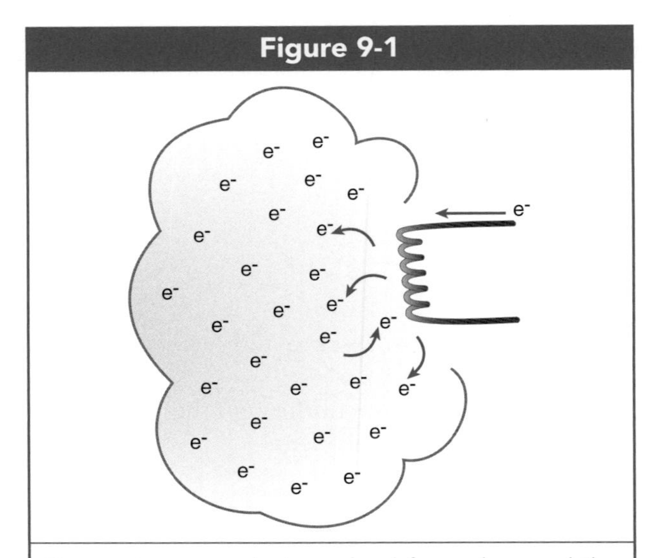

Space charge or electron cloud formed around the filament by *thermionic emission*, each time the rotor switch is depressed.

replaced by other electrons jumping out such that a *constant number* of electrons hover within the cloud. This state of equilibrium is called the *space charge effect*. The number of electrons is predetermined by setting the mA station. All of this process occurs through the *filament circuit* described in the last chapter.

In order to suddenly accelerate the electrons of the space charge in a direction toward the anode, an extremely high-voltage electromotive force (EMF) is applied to the same filament. (The high-voltage circuit of the x-ray machine was described in the last chapter.) The electron cloud "feels" the force of the negative voltage behind it, and is repelled away from the filament. At the same time, the anode of the x-ray tube has acquired a *positive* charge from the same high-voltage circuit, and also *pulls* the electron cloud toward it (Fig. 9-2). This potential difference (in the tens of thousands of volts) is so strong that the electrons can accelerate to more than one-half of the *speed of light* in just one inch of travel before reaching the anode disc!

The anode disc, made of metals with very high atomic-numbers, provides the means of precipitously decelerating these projectile electrons as they "smash" into it (Fig. 9-3). By the law of conservation

Figure 9-2

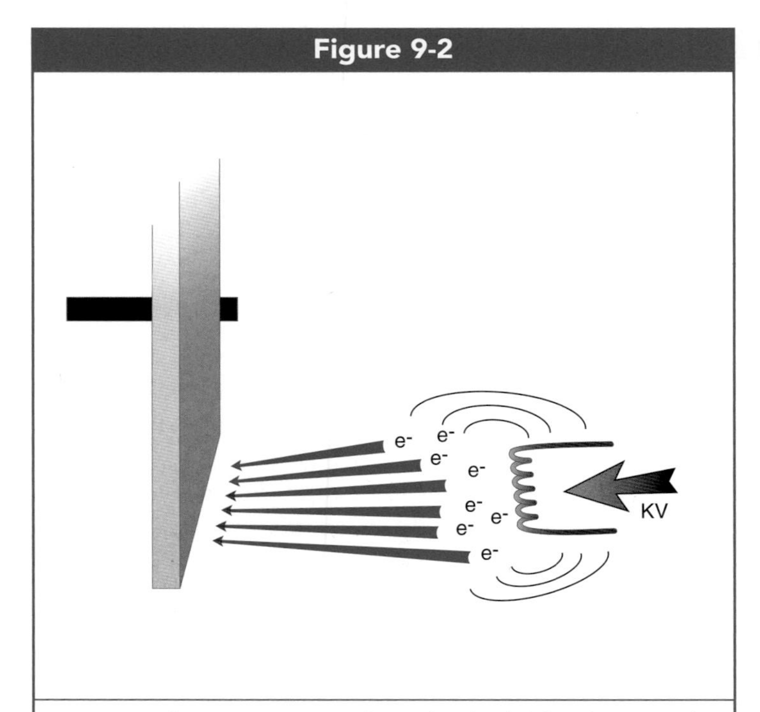

Engaging the exposure switch applies high kilovoltage to the filament and anode, accelerating the electron cloud toward the anode to speeds as high as 80% of the speed of light.

Figure 9-3

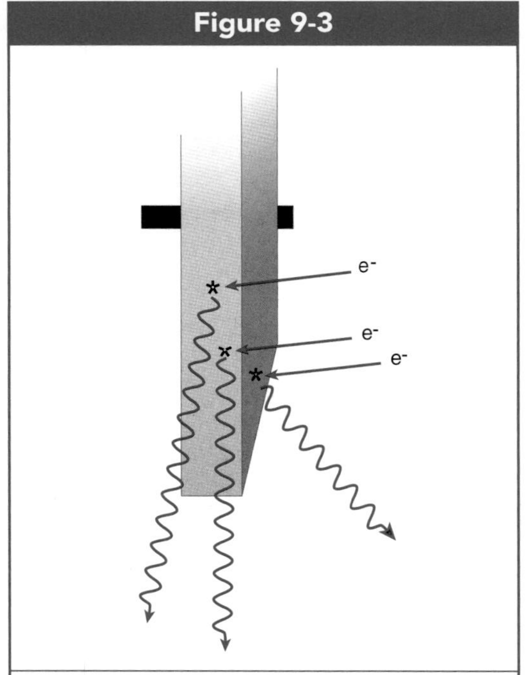

When high-speed projectile electrons are suddenly stopped by atoms in the anode disc, the lost energy is emitted as x-rays.

of energy, the energy lost by the sudden slowing down of the electrons cannot disappear but must be converted into another form. It is emitted from the anode in the form of electromagnetic radiation waves, including infrared, visible light, ultraviolet waves and x-rays.

Unfortunately, the entire process is not very efficient. Only 0.5 percent of the radiation emitted is in the form of useful diagnostic x-rays. The other 99.5 percent is emitted from the x-ray tube and its housing in the form of wasted heat (including infrared radiation), and as visible light—the anode disc, even as thick and dense as it is, reaches "white-hot" temperatures and glows with bright white light.

This is just an outline of what must happen inside the x-ray tube. Of course, there are several other things, electrical and mechanical, that must come together to support these three functions. These might be best presented from the standpoint of the components that make-up the x-ray tube and its housing.

Figure 9-4

Two filaments are embedded in pits (arrow) within the focusing cup of the cathode. These provide for selection of the large and small focal spots at the console.

COMPONENTS OF THE X-RAY TUBE

The Cathode

The x-ray tube is a *diode* tube, containing two charged electrodes. In electrical circuits, a negatively-charged electrode is called a *cathode* and a positively-charged one an *anode*. The cathode end of the x-ray tube consists of two filaments embedded in a *focusing cup* usually made out of molybdenum or nickel, shown in Figure 9-4. Note that the filaments are embedded in pits carved into the focusing cup.

Nearly all x-ray tubes are dual-focus tubes, providing two filaments to choose from, with the larger one ranging from 1.5 to 2 times the length of the smaller one. For standard diagnostic x-ray tubes, the small filament is typically about 1 cm in length. When the radiographer selects the *large* or *small* focal spot setting at the console, it is the filament that is actually being selected.

Because negatively-charged electrons repel each other, the beam of electrons traveling from the filament to the anode tends to spread out. To correct this, the focusing cup has a negative charge placed on it. Within its pit, each filament is thereby surrounded with negative charge. This has the effect of pushing the electrons back toward the middle of the beam as they leave the filament, narrowing and constricting the beam (Fig. 9-5).

In fact, this is so effective that the beam can be *focused* into a spot much smaller than the filament itself by the time the beam reaches the anode (Fig. 9-5). This area on the anode which the electron beam strikes is called the *focal spot*. Typically, the electron beam as it strikes the focal spot is about 1/20th the size of the filament from which the electron beam originated. Therefore, for most standard diagnostic tubes, the *small focal spot* is 0.5 to 0.6 mm

Figure 9-5

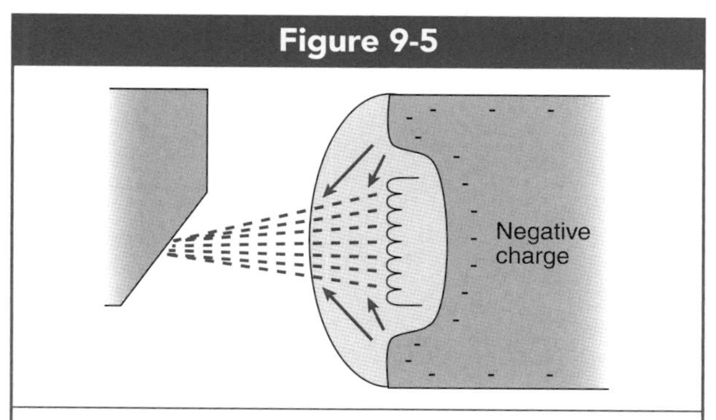

Due to the repulsive charge placed on the focusing cup, the stream of electrons is focused to a much smaller size as it travels toward the anode.

Figure 9-6

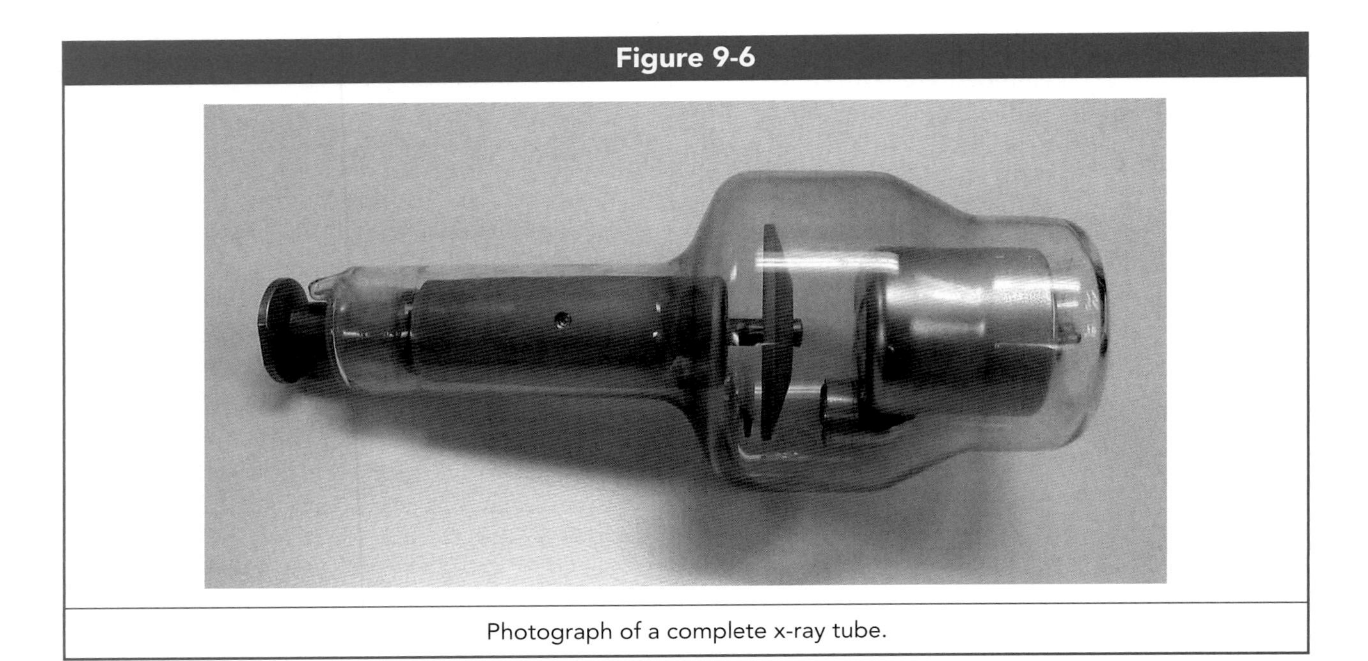

Photograph of a complete x-ray tube.

wide, originating from the 1 cm filament, and the *large focal spot* is 1.0 to 1.2 mm in size, although various customized combinations of focal spot sizes can be designed. For angiography and cardiac catheterization labs, tubes with much smaller focal spots are used. The smallest focal spots that can be engineered are about 0.1 mm. X-ray tubes with this capability are quite expensive.

Figures 9-6 and 9-7 show a photograph and a diagram of a complete x-ray tube. The diagram in Figure 9-7 shows the way in which the focusing cup and filaments are aligned to the anode in the tube.

If the filament is embedded deeply enough into its pit, and the focusing cup is given a strong enough negative charge, the electrons can actually be withheld from leaving the vicinity of the filament by

Figure 9-7

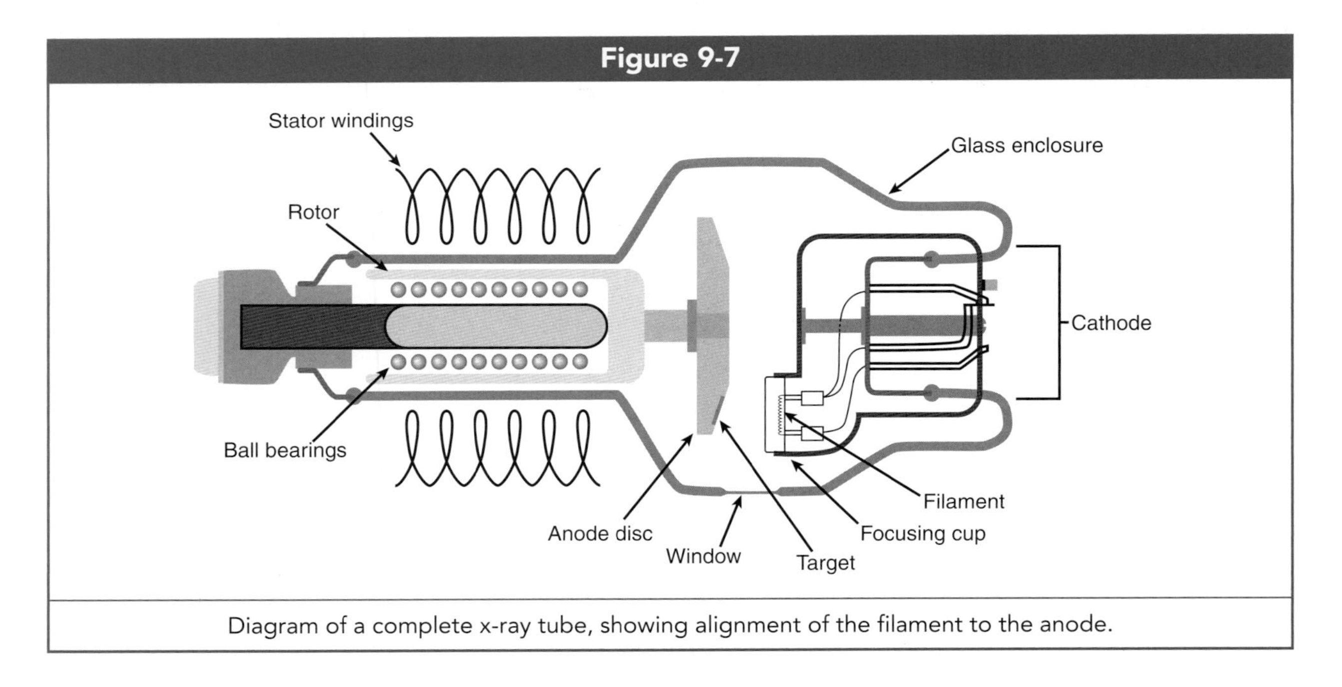

Diagram of a complete x-ray tube, showing alignment of the filament to the anode.

Figure 9-8

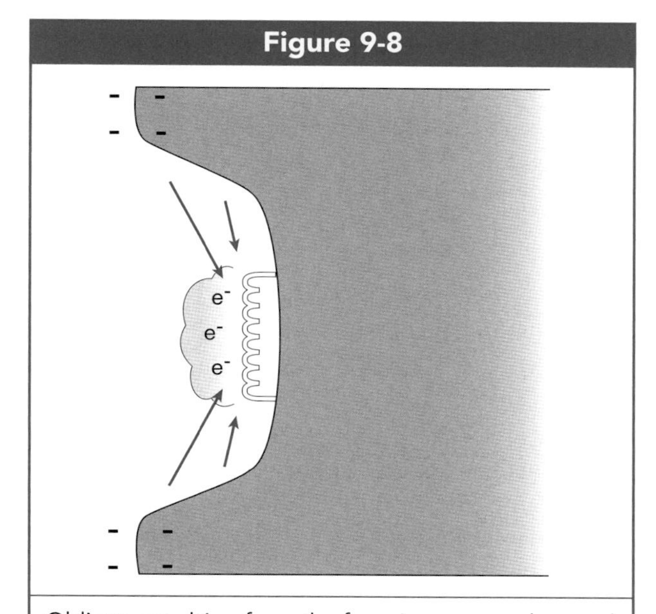

Oblique repulsion from the focusing cup can be used to restrain the electrons within the focusing cup until they are needed.

Figure 9-9

For a *grid-controlled* x-ray tube, a wire mesh can be seen in front of each filament. Rapid-sequence, short exposures are made possible by applying a negative charge on this grid to restrain the electrons until they are needed.

oblique repulsion (Fig. 9-8). In this way, the focusing cup can act as a *switch* for very short exposures. When a rapid sequence of very short exposures is needed, as is typical in angiography and cardiac catheterization labs, a negatively-charged wire mesh or *grid* can actually be placed in front of the filaments (Fig. 9-9) to hold back the electron beam until those precise times when it is needed. These types of x-ray tubes are referred to as *grid-controlled* tubes.

Each filament is made from thorium-impregnated tungsten. Tungsten is used because it is a very good conductor of electricity that also has an extremely high melting point of 6200° F (3400° C). This allows it to endure extreme heat without being destroyed. Tungsten also provides higher thermionic emission than other metals. By adding a small percentage of thorium (atomic number 90), both the efficiency of thermionic emission (which now occurs at 2000° C rather than 2400° C) and the long life of the filament are improved even more.

In the x-ray tube diagram in Figure 9-7, you will note that the filament has two wires going into it from above but only *one* wire leaving it at the bottom. This is because, as described in the last chapter, *two* separate currents with distinct functions merge in the filament. The *filament current* has the sole purpose of heating the filament up to the temperature corresponding to the mA station selected. This current flows into, through, and back out of the filament, accounting for one wire on each end of the filament.

The additional wire entering the filament carries the *tube current*. This is the high-voltage current generated by the main circuit, whose job is to "kick" the electrons boiled off in the space charge across the tube to the anode with extreme force. The energy of this voltage is carried *across the tube to the anode* by the electron beam. The anode is connected to the positive side of the high-voltage circuit by a wire. As these electrons strike the anode, they continue to "feel" the positive charge of that circuit, and they move down the anode shank and out a wire on the anode end of the x-ray tube. This is why a second wire is not seen exiting the filament.

Figure 9-10 shows three terminals (screws) at the cathode end of an x-ray tube. Two of these are for the filament current, one in, one out, and one is for the *tube* current which will cross the x-ray tube and exit through the anode shank at the other end of the tube. In Figure 9-7, two wires can be seen entering the filament and only one leaving it. The odd incoming wire is for the tube current, which jumps across

Figure 9-10

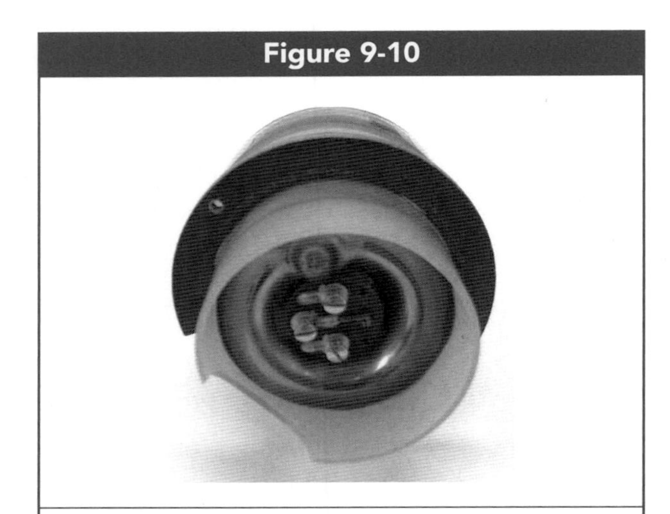

Three terminals (screws) can be seen at the cathode end of the x-ray tube. Two are for the filament current, one for the tube current.

the tube to the anode and exits a wire on the anode end of the tube.

The Anode

The positively-charged anode end of a diagnostic x-ray tube has several components, as can be seen in Figure 9-7. The target surface which the electron beam strikes is mounted on a large, thick disc of molybdenum. High-speed x-ray tubes may have a layer of graphite behind the molybdenum. Molybdenum and graphite are both lightweight, making it easier to rotate the disc. This disc is mounted on a molybdenum shaft. The other end of the shaft is connected to a copper cylinder that sleeves over a support shank. Ball bearings between the cylinder and the shank allow smooth rotation. The end of this rotor system is connected to a wire to conduct electricity away from the anode.

The *focal track* is a ring of tungsten-rhenium alloy embedded within the molybdenum disc, near its outer perimeter (Fig. 9-11). Rhenium is added because it adds mechanical stability for high-speed rotation. A glance at the periodic chart will remind the student that tungsten has an extremely high atomic number of 74, and rhenium can be found next to it at atomic number 75. These are both ideal metals for stopping or "catching" high-speed electrons for the simple fact that their atoms are *crowded* with electrons.

Remember that high-atomic number atoms have only slightly greater diameter than atoms with a low atomic number. This means that tungsten's 74 electrons, and rhenium's 75 electrons, occupy about the *same volume of space* as a "small" atom does. This aspect of an atom is called its *electron density*, the number of electrons per cubic nanometer, for example. A high-speed electron trying to pass through this space has a very high probability of literally colliding with an orbital electron in a tungsten or rhenium atom. This is what we want to happen.

It is also possible for a projectile electron to interact with the atomic nucleus of a tungsten or rhenium atom. These nuclei consist of well over 150 nucleons, so they are physically large and provide a more likely target for the electron to encounter. With 74 and 75 protons, respectively, they create a much greater positive pulling force that extends farther outward from the nucleus. Projectile electrons running into this force will be deviated and slowed down. Any process which slows or stops the projectile electron can produce radiation due to the loss of kinetic energy.

All of the metals mentioned above have very high melting points, and also conduct electricity well. Tungsten conducts heat well, but molybdenum and graphite do not. The shaft supporting the disc is also made of molybdenum because it is a poor conductor of heat but a good conductor of electricity. This forces most of the heat from the focal track to be dissipated by radiation to the glass and surrounding oil, rather than traveling down the shaft and overheating the rotor or ball bearings which are part of the induction motor and must not become warped.

Specialty tubes for mammography use molybdenum or rhodium targets because they produce the lower-kilovoltage x-rays needed for mammograms, but still have high melting points and good electrical conductivity.

Some x-ray machines such as basic dental units do not require high voltages or current. For these units, rather than a rotating anode, stationary anodes consisting of a simple disc of tungsten embedded in a copper shaft are sufficient.

For diagnostic radiology, however, every effort must be made to effectively dissipate the heat created by typical x-ray techniques that bring the entire disc to "white-hot" temperatures. By spinning the anode disc, all of the electrons in the electron beam do not strike the same spot, which would melt the material, but are effectively "strung out" along a track that circles the outer portion of the surface of the anode (Fig. 9-11). In relation to the *patient and image receptor*, the focal spot never changes its position; but, in relation to the spinning anode, it is constantly moving. This spreads the heat over a much greater area so that the heating capacity of the rotating anode is increased by 1000 times that of a stationary anode.

A standard diagnostic x-ray tube spins the anode at 3400 revolutions per minute (rpm). For specialty tubes, or when high techniques are used, the heat-load capacity can be further increased by engaging a *high-speed rotor*, which spins the anode at 10,000 rpm. The anode is rotated by using an *induction motor*, described in Chapter 7. Figure 9-12***A*** shows the x-ray tube separated from the *stator* windings of the motor. Inside the narrow stem of the x-ray tube, the *shank* of the anode can be seen. The outer portion of this shank is a copper cylinder which acts as the *rotor* of the induction motor.

Made of solid copper, this cylinder acts just the same as a coil of copper wire would. The glass stem of the x-ray tube inserted into the *stator* is shown in Figure 9-12***B*** with the supporting brackets all assembled. The stator consists of a series of coils, which are fired in sequence so that the magnetic field around them rotates. This strong, constantly moving magnetic field induces electricity to circulate in the copper cylinder within the glass tube. The induced electrical current magnetizes the cylinder. In accordance with Lenz's law, the induced current flows opposite to the original stator current, therefore, the *north* magnetic pole of the cylinder faces the *north* magnetic pole of the stator. By magnetic repulsion, the cylinder is "pushed" by the circulating magnetic field of the stator, causing the cylinder to rotate.

To allow the smoothest rotation and the longest tube life possible, high-tech ball bearings lubricated with powdered silver are placed between the cylinder and its supporting shank. (Oil or graphite lubricants vaporize in a vacuum, and cannot be used.), The copper cylinder is drilled out in precise spots to carefully balance it (Fig. 9-13), just the opposite approach to adding small weights to the wheels on your car to balance the tires.)

Figure 9-11

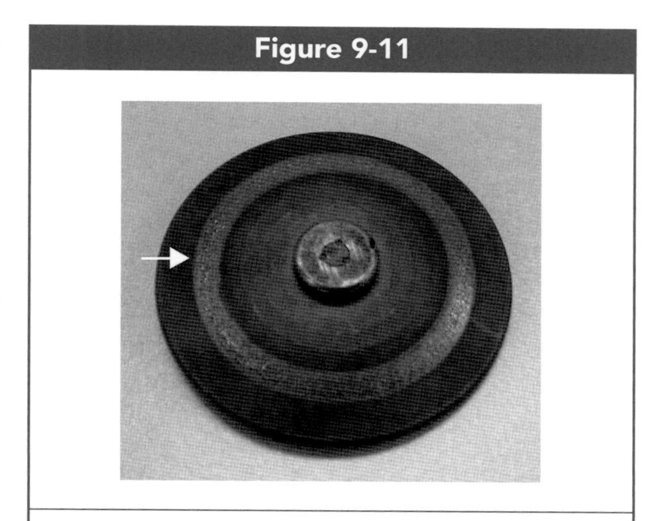

Focal track on a rotating anode disc. Etching of the focal track due to chronic heat trauma over a period of time has given this focal track a glossy appearance with a network of fine cracks in the surface.

After an exposure is completed, one can hear a "shushing" sound from the x-ray tube as the rotor coasts to a stop. The bearings are so perfectly smooth that this would normally take many minutes, but the induction motor is automatically run in reverse to create a "braking" effect. This reduces the coast time to about one minute for a new x-ray tube. As the tube ages and the bearings and anode shank begin to warp, coast time is lengthened and one can hear a rougher sound from the anode as it spins down.

Each time the "rotor" switch is depressed by the radiographer at the console, the induction motor is energized and must bring the spinning anode up to full speed before the tube current is allowed to strike it. This requires a few seconds of delay between rotoring and exposing, which is usually built into the switch system. Recall that during this time, the filament is also being brought up to temperature, which also requires a delay. When both the filament and the anode are prepared, an "exposure ready" light comes on at the console and usually a distinct "click" can be heard.

Most x-ray machines have a two-position switch or two separate buttons for rotoring and exposing, in order to allow the radiographer to "rotor" for some time without committing to exposure. This is

A, stator windings and x-ray tube separated, and ***B***, x-ray tube inserted into stator with supporting brackets.

necessary for situations such as obtaining chest radiographs on pediatric or incoherent patients, when one must carefully time the exposure to a suddenly inspired breath. However, rotoring without exposing must be kept to a practical minimum because of the wear and tear it causes to both the burning filament and the anode shank bearings. Whenever possible, the rotor/exposure switch should be pushed all the way down in one motion to allow the machine to automatically expose as soon as it is ready.

When the small focal spot is selected, the x-ray machine automatically "locks out" higher mA stations (usually those above 300 mA) from being used. Exposure will not occur unless the large focal spot is selected or a lower mA station is set. This is due to limitations in the heat capacity of the anode. A large amount of current focused onto a small enough spot can melt the anode material. This is similar to using a magnifying glass: If enough sunlight is focused onto a small enough spot, it can cause a piece of paper to ignite and burn. Even the tungsten in the anode can only withstand so much concentrated heat energy.

The advantage of using a small focal spot is that it provides for much better sharpness in the resulting image. Fortunately, most procedures requiring maximum sharpness are for extremities which are smaller than the torso of the body and therefore need less radiographic technique. These are frequently exposed with the mA station set at 200 or less. Radiographers should be mindful to take advantage of these lower mA settings by remembering to engage the small focal spot.

The manufacturing of x-ray tubes is very technical and difficult, and the actual sizes of the focal spots vary somewhat from tube to tube. Therefore, regulations allow for the real focal spot to fall within a range of what has been advertised by the manufacturer. The advertised focal spot size is called the *nominal* focal spot (i.e., the "named" focal spot). The actual width is generally allowed to be up to 50 percent larger than the advertised nominal focal spot. The length is given a bit more latitude because it is harder to control. For example, in an x-ray tube whose nominal size for the small focal spot is listed at 0.5 mm, the actual spot may in reality be 0.75 mm wide and 0.9 mm long.

Because electrons repel each other, as more electrons are packed into the space charge cloud produced around the filament, it tends to expand in size, an effect called *focal spot blooming*. Blooming occurs at higher mA stations because more thermionic emission is occurring and there are more electrons in the space charge. Unfortunately, this has an indirect but adverse effect upon resulting image sharpness, because a "swollen" space charge around the filament also means a larger focal spot at the anode when the electron beam reaches it. Not surprisingly then, the *nominal* focal spot advertised by manufacturers is measured at one of the lowest mA stations.

The ideal focal spot would be rectangular in shape with an even distribution of electrons across its area. But, because of the shape of the filament and the repulsive effect of the focusing cup around it, the actual shape of the focal spot is closer to a swollen letter "H," with the electron distribution concentrated at the sidebars, as shown in Figure 9-14.

You will note in Figure 9-7 that the focal track of the anode is on a beveled surface that lies at a steep angle relative to the horizontal electron beam. This results in an *effective* focal spot projected downward from the anode, through the patient's body and to the image receptor, which is substantially *smaller* than the measured area which the electrons actually strike. This *line-focus principle* greatly improves the geometrical sharpness of the image. All of its ramifications and other practical items related to the effective focal spot will be discussed in a later chapter.

Figure 9-13

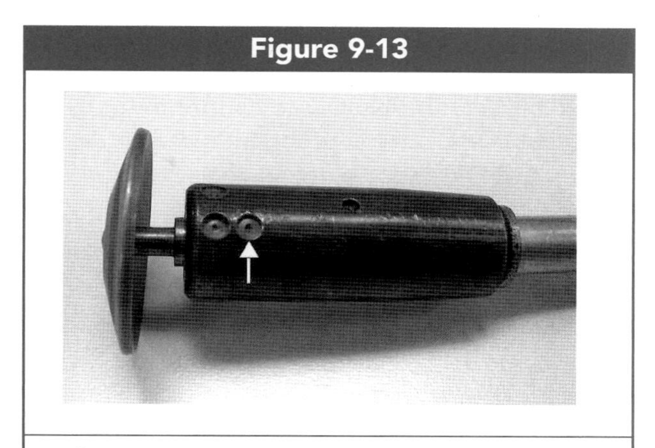

Drill points in copper cylinder of anode shank for balancing.

The Glass Envelope

The entire assembly of the cathode and the anode must be encased within a vacuum. Any molecules of air or other gas within the x-ray tube can impede the flow of projectile electrons from the filament to the anode target. As electrons strike these gas molecules and are deviated, x-ray production at the anode is lost, and radiographic techniques become unreliable.

Figure 9-14

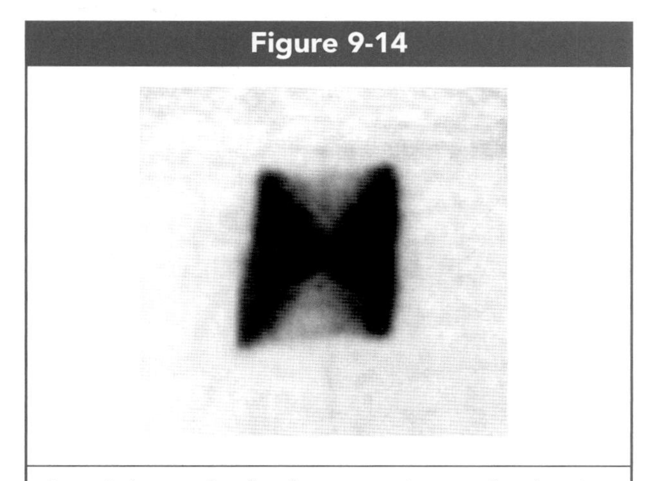

Actual shape of a focal spot on the anode, showing the distribution of projectile electrons striking it.

For standard tubes, an enclosure of thick Pyrex glass is made and carefully vacuumed out. The glass must be airtight and also be able to endure the extreme heat generated from the production of x-rays. The glass is thinner at the *window*, that area through which the effective focal spot is directed.

X-RAY TUBE FAILURE

As an x-ray tube ages, tungsten from both the filaments and the anode is vaporized. It accumulates along the bottom of the tube and begins to coat the window area, acting as filtration and further reducing x-ray output from the tube. When enough of this metal is deposited on the glass, electricity can arc down to it which cracks the glass and causes tube failure. Electrical arcing is the most common cause of tube failure. High-capacity x-ray tubes are often made using a metal enclosure rather than glass. By maintaining a constant negative charge on this metal, electrical arcing is less likely to occur and tube life is extended.

Heat from the anode is conducted down the rotor shaft and to the ball bearings. Prolonged periods of excessive heating can cause these bearings to warp "out of round." As increased friction is generated, rotation becomes even rougher and a spiral effect occurs resulting in rapid deterioration of the bearings and imbalance of the entire rotor. Wobbling of the anode disc causes the projected focal spot to constantly move, destroying the sharpness of images produced. This is another type of tube failure.

Over time, the cumulative effects of heat cause the focal track of the anode to develop a rough surface, as shown in Figure 9-11. Fine cracks and pits form in the anode surface along the focal track. When bearings begin to fail, or if the induction motor fails, so that rotation of the anode is not constant, "pitting" of the anode, shown in Figure 9-15, can result. These are spots where the anode surface has actually melted from the impinging electron beam while the anode was paused in its motion. Any uneven surface destroys the effective focal spot and results in extreme loss of image sharpness.

A third type of tube failure is simply burning out of the filament. Tungsten vaporizes off the filament over time, and it becomes thinner and thinner. Just as with a light bulb, eventually it will fall apart from the heat load, thus breaking the circuit so no further exposures can be made.

(As an x-ray tube ages, the thinning of the filament causes it to reach even higher temperatures than intended when a particular mA is applied. This generates more thermionic emission, more electrons cross to the anode, and ultimately more x-rays are produced. We would say that the *effective mA* is increasing over time. This effect is partially cancelled out by the increased filtration effect of tungsten accumulating on the glass window. However, the overall result is that the *net* output of the x-ray tube increases over time. We say that the tube gets "hotter" over time.)

Rating Charts

To help prevent x-ray tube failure, three types of tube rating charts are provided by manufacturers: The radiographic rating chart, the anode cooling chart, and the housing cooling chart. Modern x-ray machines automatically lock-out and prevent any exposures which might exceed the heat capacity of the anode given its current temperature. Thus, in day-to-day practice, rating charts are rarely needed.

However, these charts also provide a basis for comparing one x-ray machine or x-ray tube to

Figure 9-15

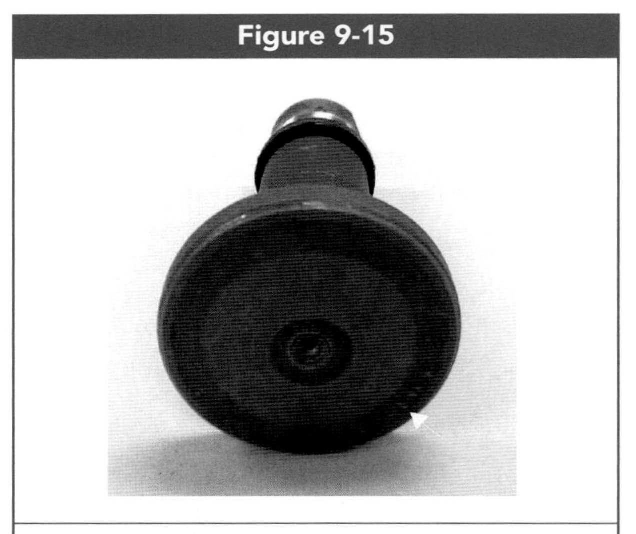

Pitting of the focal track due to failure to keep the anode disc rotating.

another when considering the purchase of equipment. Therefore, radiographers should at least be familiar with these basic graphs and their interpretation, and also develop an appreciation for how radiographic techniques affect the overall heat load in the x-ray tube.

The capacity of the anode or tube housing to withstand and store heat is measured in special *heat units*, defined as the product of the kVp, the mA and the exposure time:

$$HU = kVp \times mA \times s$$

Three-phase and high-frequency x-ray machines generate about 1.4 times more heat than single-phase machines, so this correcting factor is added to the formula:

$$3\Phi\ HU = 1.4 \times kVp \times mA \times s$$

Practice Exercise #1

How many heat units are generated by a three-phase x-ray machine for an exposure using 80 kVp, 300 mA and 0.1 seconds exposure time?

Solution: $1.4 \times 80 \times 300 \times 0.1 = 3,360$

Answer: *There are 3,360 heat units generated by this 3-phase exposure.*

Figure 9-16 is a typical radiographic rating chart for a three–phase x-ray machine, with kVp indicated on the vertical axis, exposure time along the horizontal axis, and mA stations indicated by the curves on the graph. To determine if a particular exposure would be safe, find the intersection of the desired kVp and the desired exposure time on the graph. If this point falls *below* the curve for the desired mA station, it is a safe exposure; if it falls *above* or to the right of the mA curve, it is unsafe and would generate excessive heat on the anode.

For example, using the chart in Figure 9-16, would an exposure of 80 kVp, 500 mA and 1 second be safe? Find the intersection point for 80 kVp and 1 second. Note that this point falls above and to the right of the curve for the 500 mA station. This would be an unsafe exposure.

In comparing x-ray tubes for their heat capacity, assuming that the kVp and time scales are identical on the two graphs, *the farther the mA curves fall upward and to the right, the greater the heat capacity of the x-ray tube.*

Figure 9-16

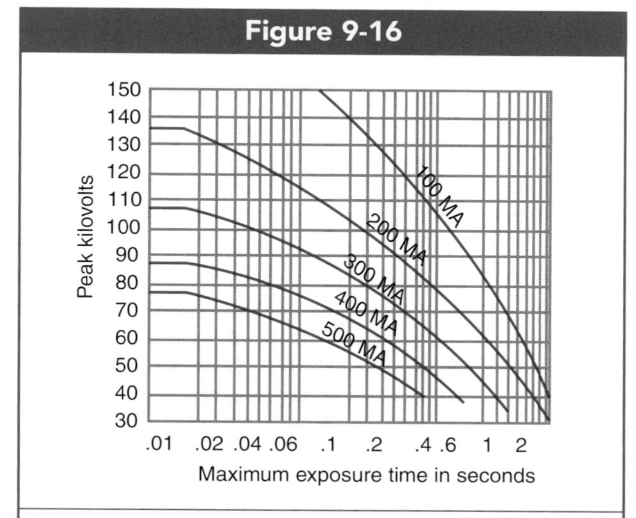

Typical radiographic rating chart for a three-phase machine.

The layout of the chart indicates the *maximum mA* that can be used for a particular kVp and time combination. But, the chart can also identify specific safe levels for kVp and exposure time as follows: For a given mA station and exposure time, find their intersection on the chart; any kVp which lies *above* this point is unsafe. For a given mA station and kVp, find their intersection on the chart; any exposure time which falls *to the right* of this point is unsafe. *Separate rating charts are used for each filament or focal spot size used in the x-ray tube.*

Figure 9-17 is an example of an anode cooling chart. Cooling charts for the housing are very similar. The total heat capacity of the x-ray tube is indicated by the starting point of the curve up the vertical axis. In Figure 9-17, the total heat capacity for this x-ray tube is 350,000 heat units. In comparing the chart of one x-ray tube to another, the one with the curve starting at the highest point has the greatest heat capacity.

To determine if a specific exposure is safe after an initial exposure has been made, you must first calculate the total heat units for each exposure using the heat unit formula above. Find the point along the curve corresponding to the heat units generated by the first exposure. From this point, follow along the curve the amount of time that has elapsed since that

Figure 9-17

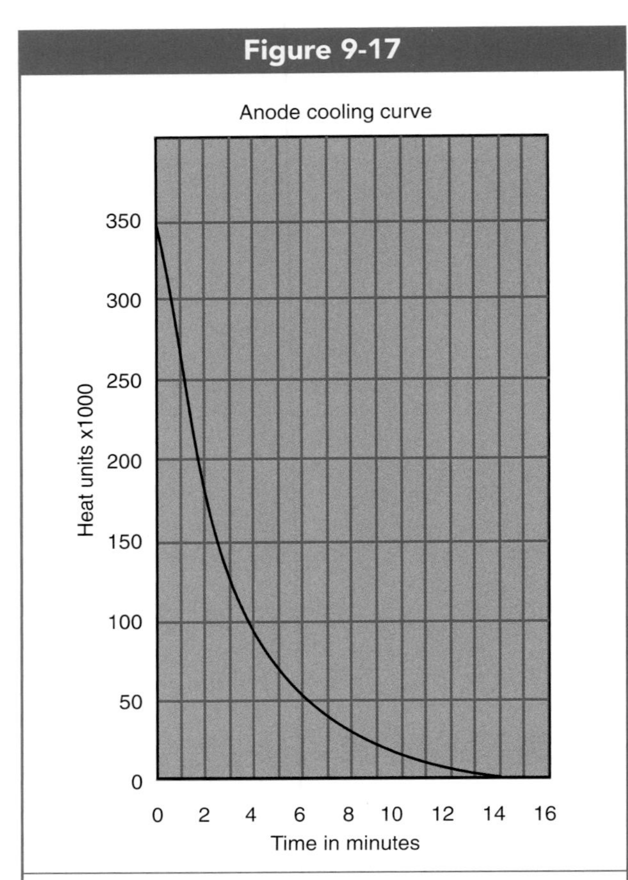

Typical anode cooling chart. The cooling chart for the tube housing is similar.

Figure 9-18

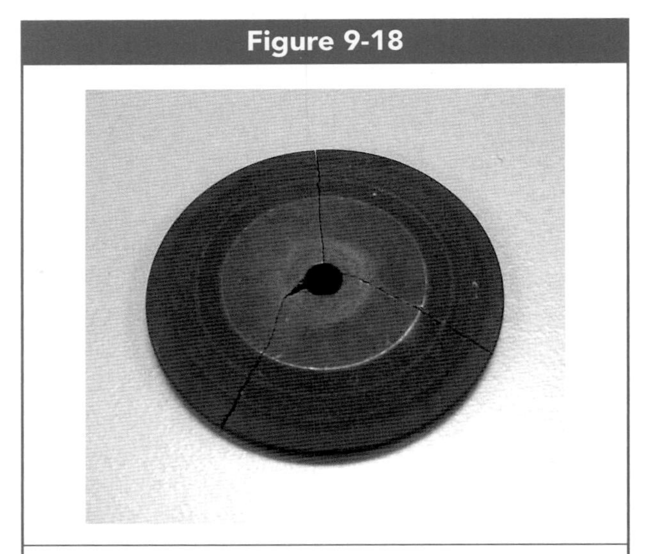

Cracked anode due to sudden high exposure made on a cold anode.

exposure was made. This point indicates the heat remaining on the anode. Add the heat units that will be generated by the second exposure to this amount. If it exceeds the total capacity, indicated by the starting point of the curve, the exposure would be unsafe.

For example, using Figure 9-17, assume that an original exposure generated 125,000 heat units. This point on the curve falls at about 3 minutes. It is desired to take a second exposure exactly 2 minutes after the first. This second exposure will generate 250,000 heat units. Beginning at the 3-minute mark, follow the curve along for two additional minutes. At the 5-minute mark, we can see that there will be 75,000 heat units remaining on the anode from the first exposure. Adding 250,000 heat units to this would generate 325,000 heat units on the anode. This is a safe exposure, since the total capacity is 350,000 heat units.

Note that if these two exposures were taken one right after the other without any wait, 375,000 heat units would be generated, risking damage to the x-ray tube.

Extending X-Ray Tube Life

Thermal shock is caused whenever an object is subjected to a sudden and extreme change in temperature. When a glass is plunged from very hot water into very cold water, or vice versa, the thermal shock can be great enough to cause it to crack. The same phenomenon holds true for x-ray tube anodes and filaments when they are subjected to sudden temperature changes (Fig. 9-18).

To extend the life of the filament, it is kept at a "stand-by" temperature at all times that the x-ray machine power is on. When exposures are made, the filament is boosted from a "warm" temperature to an extreme temperature, rather than from cold (room) temperature to an extreme. In order to extend the life of the anode, radiographers should use a tube warm-up procedure whenever a machine has not been used overnight or when the first exposure to be made after a period of non-use will employ high radiographic techniques. A typical warm-up technique is to make three exposures about 5 seconds apart, using 200 mA, 1 second, and 70 kVp.

As previously discussed, excessive rotoring must be avoided as much as possible to minimize vaporization of the filament and wear of the anode bearings. *Radiographic techniques combining lower mA stations with higher kVp levels not only preserve the x-ray tube by generating less heat, but they also spare radiation exposure to the patient.* These are strongly recommended, especially in this age of digital image processing which allows higher kVp levels to be used without compromising the final image. Consult rating charts when very high techniques, especially in a sequence of multiple exposures, are being considered. The sound of rough rotation of the anode or other unusual noises from the x-ray tube should be reported to managers.

SUMMARY

1. The three essential conditions for the production of x-rays are: (1) a source of free electrons, (2) a means of accelerating the electrons, and (3) a means of suddenly decelerating them.
2. Thermionic emission from one of two filaments in the cathode provides a source of electrons in the form of a constant *space charge* controlled by the filament current set at the console mA station.
3. When the set kVp is applied to the tube current from the high-voltage circuit, the focusing cup uses negative charge to narrow the stream of electrons as they accelerate toward the anode.
4. Tungsten is used in the focal track of the anode as a target material for the electrons to strike and undergo sudden deceleration, resulting in the release of energy as x-rays.
5. The anode disc is spun at high speed to disperse the heat energy of electron collisions across the surface area of the focal track.
6. An induction motor is used to spin the anode, consisting of a copper cylinder around the anode shank which serves as a rotor, and a bundle of stator windings that surround the stem of the x-ray tube.
7. Excessive rotoring increases wear and tear on the rotor bearings from heat, and accelerates evaporation of tungsten from the filament.
8. Metal deposits on the glass envelope from tungsten evaporation of both the anode and the filament accumulate over time, and can cause electrical arcing, the most common cause of x-ray tube failure. Other types of tube failure include warping of the bearings and anode shank, anode etching and pitting, and burning out of the filament.
9. X-ray tube rating charts and tube and housing cooling charts can be consulted to help prevent overheating the system. Heat units are calculated by multiplying the kVp, mA and exposure time ($\times$ 1.4 for 3Φ).

REVIEW QUESTIONS

1. Why must an x-ray tube be vacuumed of all gas?
2. What is the approximate efficiency of x-ray production by an x-ray tube?
3. What type of x-ray tube uses a wire mesh to hold back the space charge until exposure?
4. Name three things that happen during bucky radiography when the rotor switch is depressed:
5. Name the process that provides a source of free electrons in the x-ray tube:
6. What is the focusing cup usually made of?
7. What element is added to the tungsten filament to extend its life?
8. What element is added to the tungsten focal track to help balance the spin of the anode disc?
9. By about what ratio is the electron beam narrowed by the time it reaches the anode?
10. Molybdenum is used for the anode shank because it is a __________ conductor of heat and a __________ conductor of electricity.
11. What are the typical spin rates in rpm for standard rotoring and for high-speed rotoring?
12. As an x-ray tube ages, what is the ever-louder "shushing" sound made after each exposure?
13. The copper anode cylinder in the tube, and the stator windings outside the tube, together form a(n) __________.

(Continued)

REVIEW QUESTIONS *(Continued)*

14. Why are high mA stations not accessible when the small focal spot is engaged?

15. As an x-ray tube ages, the *effective* mA produced at the filament (increases, decreases, or remains constant) __________.

16. What is the cause of *pits* or *melts* on the anode surface?

17. How many heat units are generated by a 3Φ x-ray generator operating at 70 kVp and 200 mA for 1.2 seconds?

18. Using the rating chart in Figure 9-16, is an exposure of 120 kVp, 0.8 seconds and 300 mA safe for the x-ray tube?

19. Why does focal spot blooming occur at high mA stations?

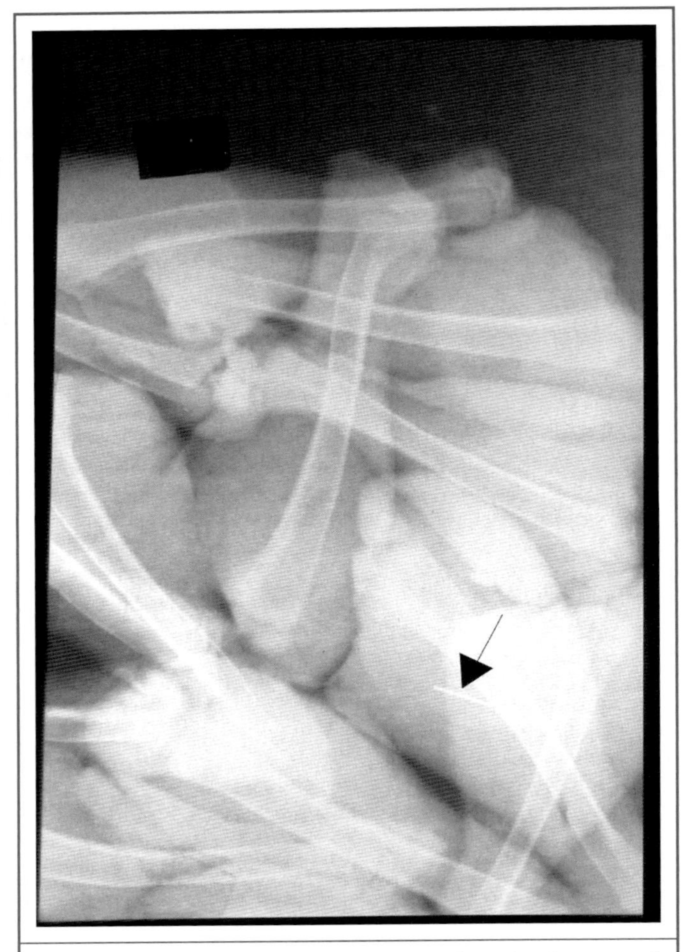

A bucket of fast-food fried chicken was found to have small needles (arrow) inserted into some of the pieces.

Chapter 10

X-RAY PRODUCTION

Objectives:

Upon completion of this chapter, you should be able to:

1. Describe the tremendous kinetic energies and speeds of the electron stream in the x-ray tube.
2. Describe the *Bremsstrahlung* interaction, its effect on the x-ray beam spectrum, and its impact upon the image.
3. Describe the *Characteristic* interaction, its effect on the x-ray beam spectrum, and its impact upon the image.
4. Given the binding energies of various atomic "shells," calculate the energy of characteristic x-rays resulting from different exchanges of electrons between them.
5. Sketch an accurate representation of the x-ray beam spectrum including both bremsstrahlung and typical characteristic x-rays.
6. Describe the poor efficiency of x-ray production.
7. Describe the effects of target material, mAs, filtration and kVp and the type of generator used upon the x-ray beam spectrum.

By definition, x-rays are electromagnetic waves with much higher energies than light and most other forms of electromagnetic radiation, energies in the tens of thousands of volts. In order to generate such energetic waves, electrons emitted from the x-ray tube filament must acquire extreme amounts of kinetic energy by the time they strike the anode disc. The formula for the total kinetic energy that a moving object will be carrying is

$$KE = \frac{1}{2} mv^2$$

where ***m*** is the mass of the object, and ***v*** is the velocity. In the formula we see that velocity is much more important than mass; the energy changes proportionately to the mass, but it increases by the *square* of the velocity. Projectile electrons from the x-ray tube filament are very small objects indeed, with very slight mass, but if we can speed them up to extreme velocities, they will be able to impart enough energy to the anode disc to produce x-rays.

With the tremendous voltage supplied by the step-up transformer of an x-ray machine, the electrons reach speeds from 50% to 80% of the speed of light. At this speed, they would be able to circle the earth more than three times in one second! Furthermore, this incredible speed is achieved in less than one inch of travel from the filament to the anode disc!

As an example, with some unit conversion tables we can readily calculate the maximum energy for an x-ray produced from a projectile electron striking the anode at 56 percent of the speed of light:

Multiplying the speed of light by 56%:

$$3 \times 10^8 \text{ m/s} \times 0.56 = 1.68 \times 10^8 \text{ m/s}$$

The mass of a single electron is 9.1×10^{-31} kg

Using the kinetic energy formula:

$$KE = \frac{1}{2}(9.1 \times 10^{-31} \text{ kg})(1.68 \times 10^8 \text{ m/s})^2$$

$$KE = 1.284 \times 10^{-14} \text{ J}$$

(Units of kilograms and meters per second will yield the kinetic energy in joules.)

There are 1.6×10^{-16} joules of kinetic energy in 1 kV.

Dividing the KE by the joules per kV:

$$\frac{1.284 \times 10^{-14}}{1.6 \times 10^{-16}} = 0.80 \times 10^2 = 80 \text{ kV}$$

Figure 10-1

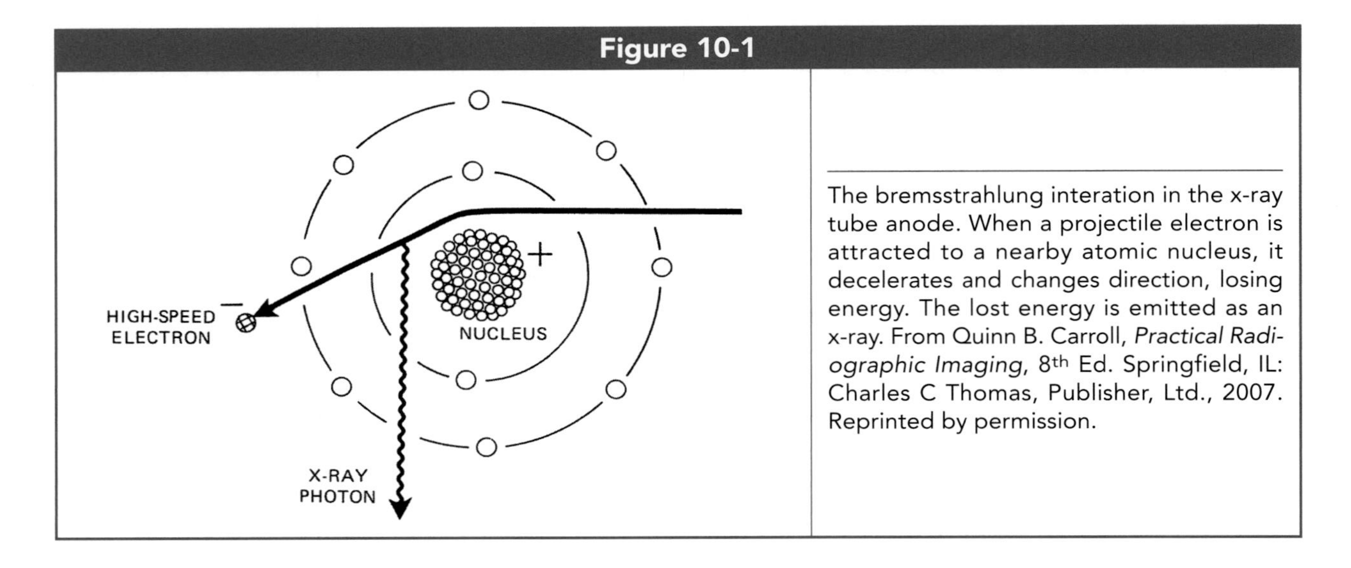

The bremsstrahlung interation in the x-ray tube anode. When a projectile electron is attracted to a nearby atomic nucleus, it decelerates and changes direction, losing energy. The lost energy is emitted as an x-ray. From Quinn B. Carroll, *Practical Radiographic Imaging*, 8th Ed. Springfield, IL: Charles C Thomas, Publisher, Ltd., 2007. Reprinted by permission.

We conclude that if a projectile electron traveling at 56 percent of the speed of light is *completely* stopped by the first atom it strikes, the resulting x-ray emitted will have 80 kV of energy. (Einstein's theory of relativity states that as a particle approaches the speed of light, it gains mass. A more accurate calculation would take this into account, and would give a somewhat higher answer, approximately 84 kV.)

INTERACTIONS IN THE ANODE

A projectile electron entering the anode target material penetrates into its atoms. As described in Chapter 4, atoms are mostly space, so it is possible for the projectile electron to pass through several atoms before it "strikes" anything. When it does, there are two possibilities: The projectile electron may interact with an orbital electron, or it may interact with the nucleus of the atom. Both interactions result in the production of x-rays, but by very different processes.

Bremsstrahlung

If the electron passes near the atomic nucleus, the positive attraction of the nucleus will cause it to *brake* or slow down. This deceleration in the speed of the electron represents a loss of kinetic energy, and that energy which is lost is emitted as an x-ray *photon* (Fig. 10-1). X-rays produced by this interaction are called *bremsstrahlung* (*braking radiation* in German), and they account for the vast majority of the overall x-ray beam.

High-speed electrons may pass by the nucleus at various distances from it. The closer an electron approaches to the nucleus, the greater will be the deceleration of the electron, due to the stronger pulling force of the nucleus. As shown in Figure 10-1, the attractive force of the nucleus also causes the electron to bend in its path of travel toward the nucleus. The greater the deceleration of the electron, the more it deviates from its original direction, and the more kinetic energy is lost. Thus, the closer the electron passes by the nucleus, the higher will be the energy of the emitted x-ray.

Bremsstrahlung, occurring at various distances from the nucleus, produces a wide range of x-ray energies and is thus responsible for the *heterogeneous* or poly-energetic nature of the x-ray beam. Heterogeneity contributes to the differential absorption x-rays within the patient's body by different tissues. It is just this differential absorption which provides *subject contrast* to the remnant x-ray beam and makes the radiographic image possible.

If all of the x-rays were of the same energy, the information reaching the detector plate would essentially be a silhouette image like Figure 10-2. Nearly all of the x-rays would be stopped by dense tissues such as bone, and nearly all of them would penetrate soft tissues. Very little information is contained in such an image. Because bremsstrahlung produces a

Figure 10-2

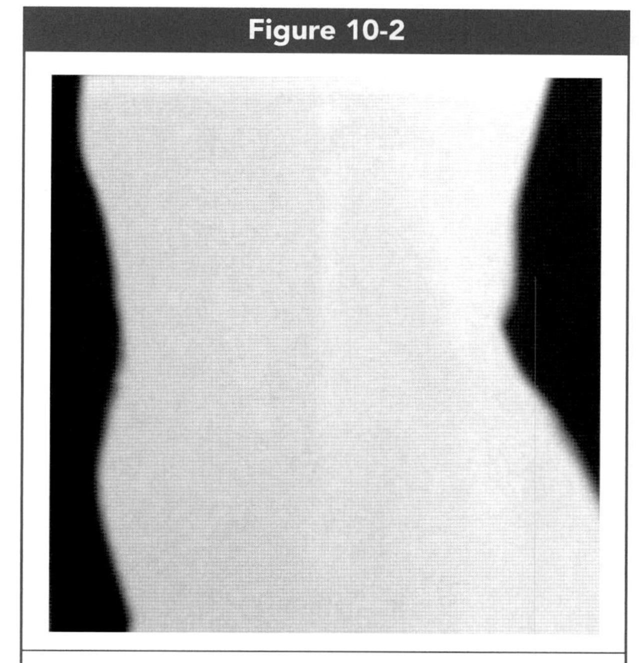

A silhouette image like this would result if the x-ray beam were mono-energetic, all x-rays having the same kV.

Figure 10-3

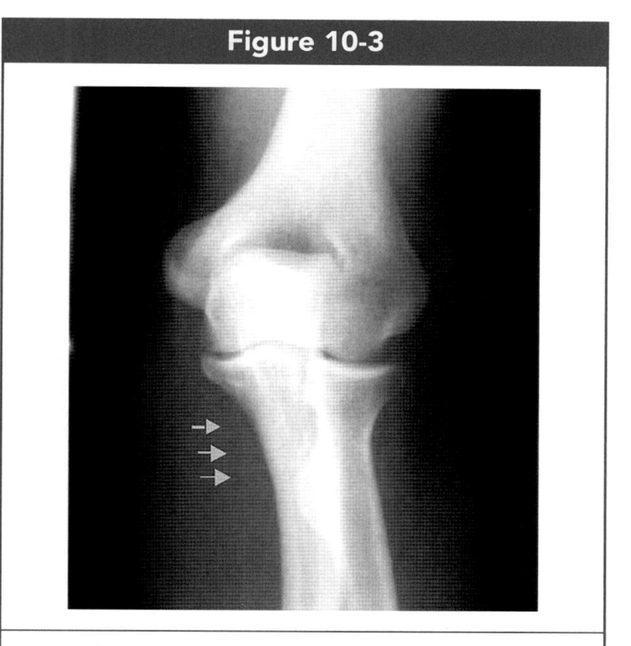

Fat pads (arrows), bone marrow and other intermediate-density tissues are all demonstrated on radiographs because of the heterogeneous, poly-energetic nature of the diagnostic x-ray beam.

whole range of x-rays at different energies, intermediate tissues such as bone marrow and fat pads can be demonstrated because they absorb *portions* of the beam, stopping the lower energies and allowing the higher energies to pass through (Fig. 10-3).

Computers can modify the radiographic image in many ways, but they *cannot create information that was not present in the first place.* The full range of information from different tissues in the body must be represented within the remnant beam that reaches the detectors. Subtle differences between tissues can only be demonstrated when a variety of x-ray energies result in a wide range of radiation intensities reaching the detectors.

The distance at which a projectile electron will pass by the atomic nucleus is a function of statistical probability. Observing Figure 10-4, you can see that at a greater distance from the nucleus, a much larger *volume* is contained within the sphere of that radius surrounding the nucleus. The volume of a sphere will increase by the *cube* of the radius distance (see Chapter 3). As we get further from the nucleus, the volume of space increases exponentially. Therefore, the likelihood that a projectile electron will pass through this volume of space goes up exponentially. It is *much* more likely that an electron will pass further away from the nucleus rather than near it.

In Figure 10-4, we have hypothetically listed the probability that a projectile electron will pass through the inner sphere as 2 percent, and the probability that a projectile electron will pass through the larger, outer sphere as 20 percent. In this example, then, these are the probabilities that an x-ray having 40 kV will be produced (2%), and that an x-ray having only 4 kV will be produced (20%).

Hence, many more x-rays will be produced at lower energies than at higher energies. A plot of the resulting spectrum of x-ray beam energies will look like Figure 10-5. The higher the energy, the fewer bremsstrahlung x-rays are produced, up to the set kVp (kilovolt-peak). Of course, no x-rays are present above the set kVp, because by the law of conservation of energy, the maximum x-ray energy could never be more than the maximum energy of the projectile electrons striking the anode.

Although the graph in Figure 10-5 represents those x-rays *initially produced* within the anode, it

Figure 10-4

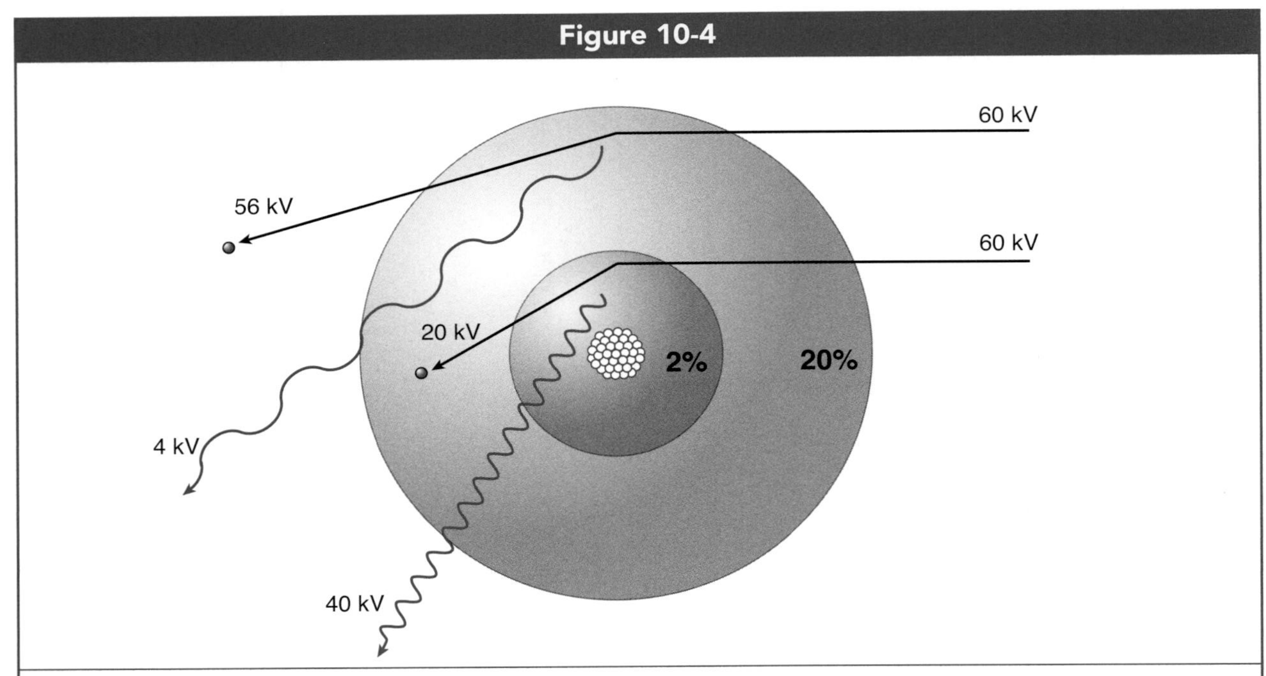

The probability of a projectile electron passing within the inner sphere volume is 2%, whereas the probability of a projectile electron passing within the volume of the larger outer sphere is 20%. To produce higher energy x-rays, the electron must pass closer to the nucleus. In this case, the probability of a 40-kV x-ray being produced is only 2%, whereas the probability of a much lower 4-kV x-ray being produced is 20%.

Figure 10-5

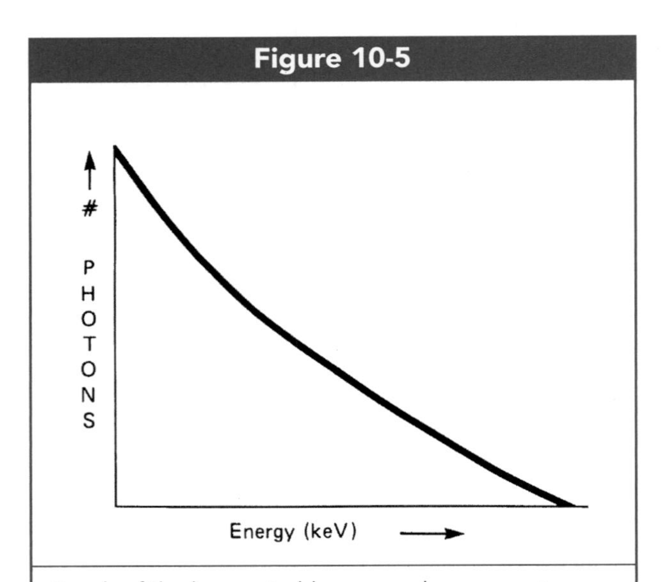

Graph of the bremsstrahlung x-ray beam spectrum as it would appear with no filtration. Most bremsstrahlung are produced at lower energies (kV levels), very few at the highest energies. (From Quinn B. Carroll, *Practical Radiographic Imaging*, 8th ed. Springfield, IL: Charles C Thomas, Publisher, Ltd., 2007. Reprinted by permission.)

does not represent the final product which is the x-ray beam reaching the patient. The x-ray beam must first pass through a number of materials which effectively act as *filters*. These include the anode itself, from which each x-ray must "escape" without being absorbed by another tungsten or rhenium atom. They include the glass window of the x-ray tube and the oil surrounding it, a "beryllium window" filter, an added aluminum filter normally placed between the x-ray tube and the collimator box, the mirror in the collimator (Figure 5-24 in Chapter 5), and other parts of the collimator. All of these filters absorb the x-rays with the lowest energies, so that the remaining bremsstrahlung portion of the emitted x-ray beam is graphed like Figure 10-6.

Reviewing Figure 10-6, fewer x-rays are produced at low energies due to filtration, and fewer x-rays are produced at high energies because of the statistical distribution of bremsstrahlung x-rays produced. This leaves a bell-shaped curve which is somewhat lopsided toward the left, so that the *average* kV within the beam is roughly one-third of the set peak kilovoltage (kVp). The *total* number of x-rays produced is represented by the *total area* under the curve. This area covers the wide range of energies needed to produce subject contrast within the x-ray beam as it passes through the patient's body tissues, rendering a full range of information for the image.

Characteristic Radiation

Now, back to the atoms in the x-ray tube anode: The second possibility for the projectile electron is that it might interact with one of the atoms' orbital electrons. When it passes near an orbital electron, its repulsive negative charge can eject the orbital electron out of its orbit, leaving a vacancy in that electron shell of the atom (Fig. 10-7***A***). The atom, left with a positive charge, will eventually pull in another electron to return to a neutral state. In the meantime, the vacancy created in this specific shell will be filled by any electron available from higher orbits. As the atom attempts to return to its *ground* state, the state with the least energy, electrons from outer orbits will "fall" down into vacancies that are closer to the nucleus.

When an electron falls from an outer orbit down into an inner orbit, there is a loss of *potential* energy (the energy of position). By the law of conservation of energy, this potential energy cannot merely disappear, but must be converted into some other form of energy. It is emitted as a *characteristic x-ray* (Fig. 10-7***B***). Characteristic radiation makes up only a small portion of the overall x-ray beam,

Figure 10-6

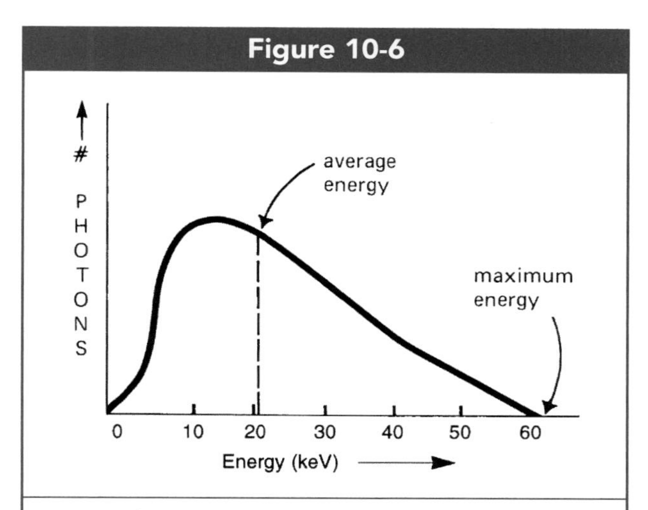

Graph of the actual bremsstrahlung x-ray spectrum with filtration present and the kVp set to 60. The lowest energies have been absorbed by inherent and added filtration. Average energy is about 1/3 of the kVp, or 20 kV. (From Quinn B. Carroll, *Practical Radiographic Imaging*, 8th ed. Springfield, IL: Charles C Thomas, Publisher, Ltd., 2007. Reprinted by permission.)

Figure 10-7

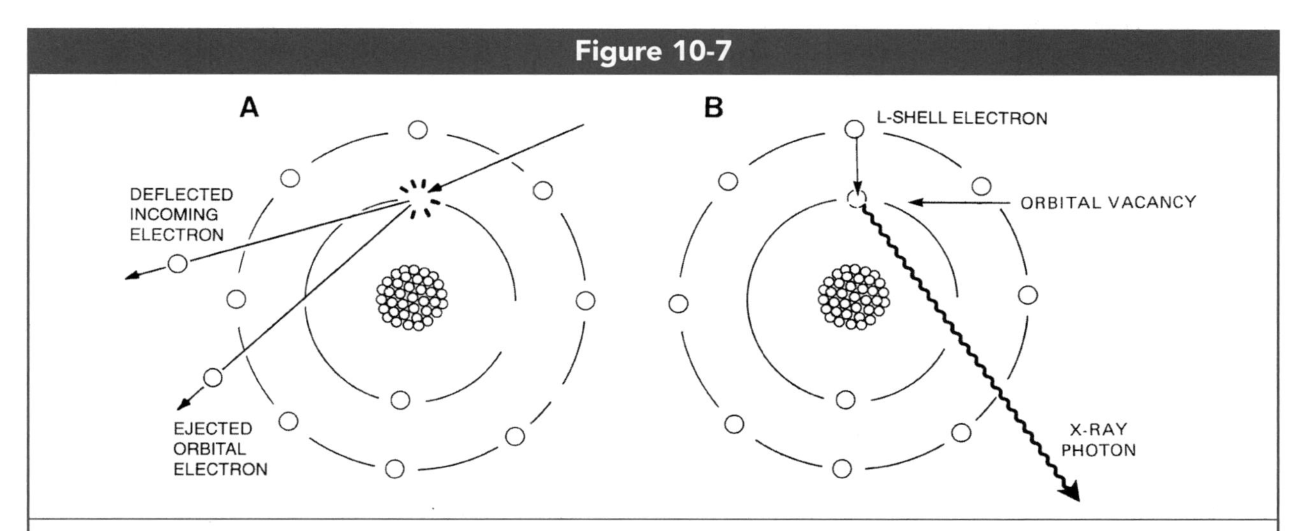

The characteristic interaction in the x-ray tube anode. In step ***A***, a projectile electron collides with and dislodges an orbital electron from the atom. In step ***B***, the atom "pulls" down an electron from a higher shell to fill the vacancy left. As this electron drops into a lower orbit, it loses potential energy which is emitted as an x-ray. (From Quinn B. Carroll, *Practical Radiographic Imaging*, 8th ed. Springfield, IL: Charles C Thomas, Publisher, Ltd., 2007. Reprinted by permission.)

but since it can possess high energies that penetrate through the patient to the detectors, these x-rays are still important in producing a radiographic image.

Characteristic x-rays depend entirely on the difference in energy levels between different orbital shells in the atom. Figure 10-8 lists these energy levels for the various shells of a tungsten atom. While these are given as the *binding energy* for each shell, they also represent the *potential energy* of the electrons in that shell relative to the nucleus. (See SIDEBAR 10-1, on Binding Energies, for a more complete explanation.) Since we know what the energy levels are for each orbit in the atoms of each element, we can accurately *predict* what the characteristic x-ray energies will be for the tungsten and the rhenium in the anode. X-rays will be emitted at discrete energies, rather than over a range of energies like bremsstrahlung, (Fig. 10-10).

For any particular shell, a vacancy may be filled by an electron from the shell immediately above it or from other shells two or three layers higher, Figure 10-9. Thus, we find three or four characteristic x-ray energies that can be produced from each shell. To predict those energies, simply subtract the binding energy for the higher shell from that of the vacant shell (Fig. 10-8). For example, if the vacancy is in the L shell of a tungsten atom, and an electron falls into it from the M shell, the energy of the emitted x-ray will be 12 – 3 = 9 kilovolts. (The actual calculation is –12 –[–3] = –9 kV, but for our purposes, we may simply use the absolute value of these binding energies and ignore minus signs.)

Figure 10-8

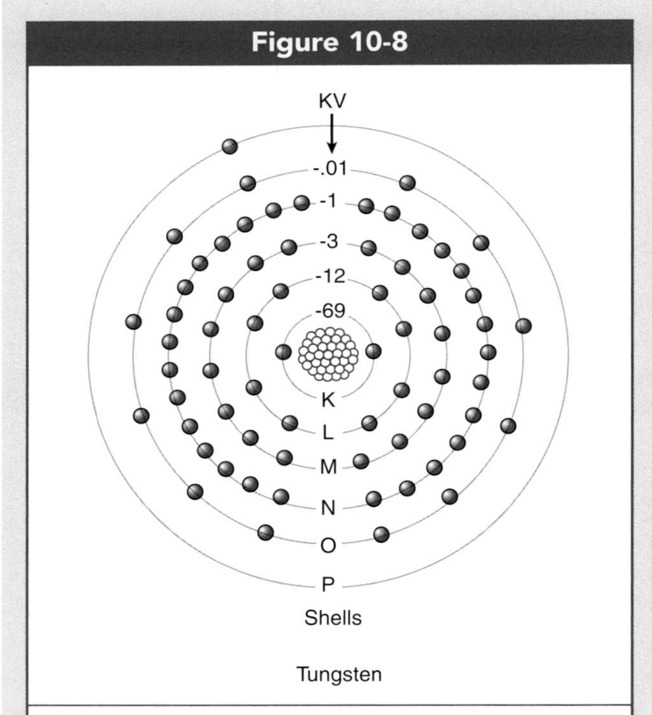

The electron shell binding energies for a tungsten atom, in rounded kilovolts (kV). The energies of characteristic x-rays produced in tungsten can be predicted by subtracting the difference between the various shell binding energies.

SIDEBAR 10-1: The binding energy of a particular shell within the atom may be defined as the amount of energy or *strength* with which its electrons are held in place by the attraction of the nucleus. "Larger" atoms with more shells do not expand much in actual diameter, so as shells are added, the inner shells must collapse in closer to the nucleus. This increases their binding energy. For example, the binding energy for the K-shell of a carbon atom is only 0.3 kilovolts, whereas the binding energy for the K-shell of a tungsten atom is 69 kilovolts. It would take at least this much energy for an incoming projectile electron or x-ray to eject an electron from the K orbit.

Note that in Figure 10-8 the binding energies are listed as *negative* numbers. This is because that much positive energy would have to be put back into the system to *free* an electron from the atom—this is how much energy an incoming projectile electron or photon must have to knock an electron out of this shell. If we take the absolute value of these numbers, (ignoring the negative sign), we might think of these quantities as a measure of the strength with which the nucleus is holding the electrons in place.

Objects falling toward the earth lose potential energy, the energy of position, as they yield to the pull of gravity. So electrons also lose potential energy as they "fall" closer to the nucleus, yielding to its positive electrical attraction.

We might say that for orbital electrons, their potential energies and their binding energies are mirror images of the same thing, positive and negative. As potential energy is lost, binding energy is stronger. However, this strength is measured as a *negative* number. This is consistent. For example, an electron falling from tungsten's L shell to its K shell drops from a potential energy of –12 kV to –69 kV, which is *less*. Potential energy is indeed lost, and is emitted as an x-ray.

Figure 10-9

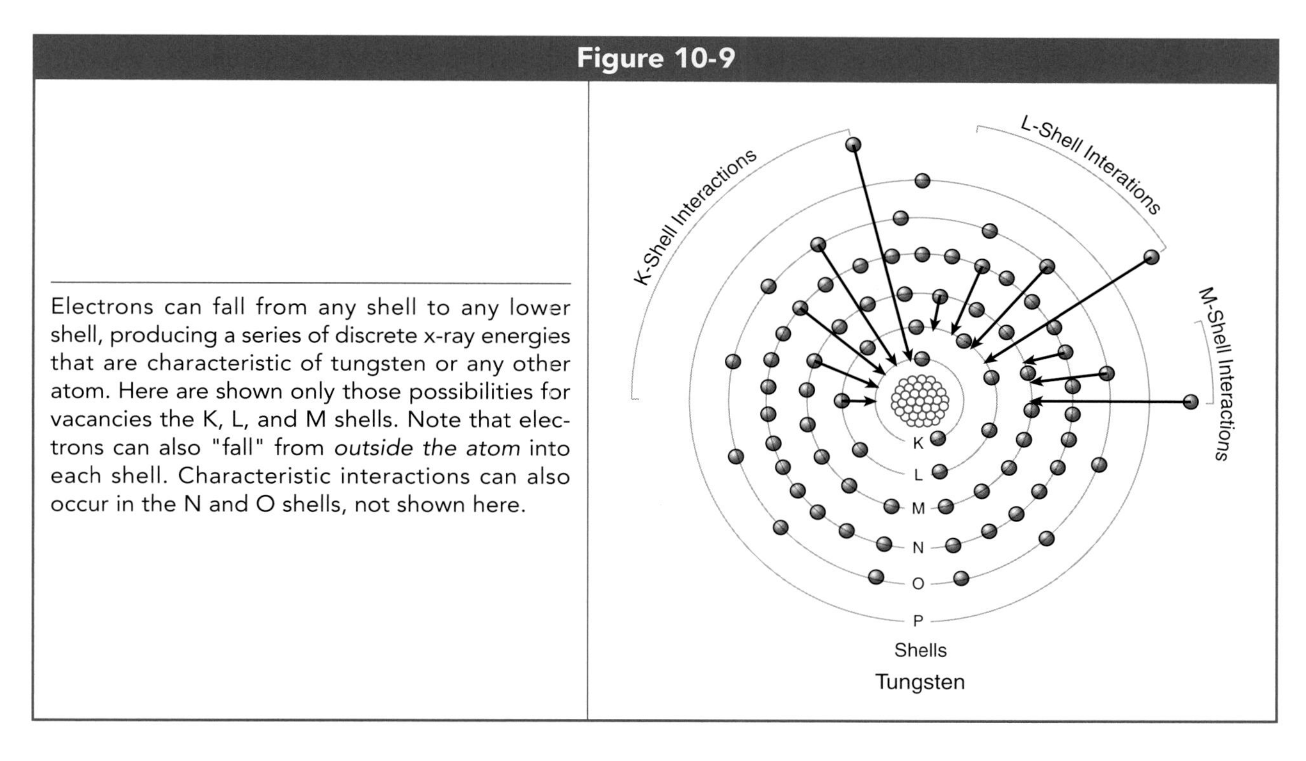

Electrons can fall from any shell to any lower shell, producing a series of discrete x-ray energies that are characteristic of tungsten or any other atom. Here are shown only those possibilities for vacancies the K, L, and M shells. Note that electrons can also "fall" from *outside the atom* into each shell. Characteristic interactions can also occur in the N and O shells, not shown here.

In tungsten atoms, characteristic x-rays will be produced in the innermost electron shell (K), having 57, 66, 68, and 69 kilovolts of energy. A vacancy in the L shell will produce x-rays of 9, 11, and 12 kV, and the M shell will produce x-rays with 2 and 3 kV. (Shells further out will also produce electromagnetic waves, but these will be of such low energy, less than 1 kV, that they would be classified as ultraviolet light rather than as x-rays.)

Figure 10-10 plots the spectrum of characteristic x-rays produced in tungsten. Inherent filtration will remove virtually all of the 2-kV and 3-kV x-rays, so these do not show up on the graph. Filtration also removes most of the 9- and 12-kV x-rays, so the graph plots them but showing a reduced number. Those characteristic x-rays having 57, 66, 68, and 69 kV largely escape the x-ray tube and are considered part of the useful x-ray beam. At each of these energies, a fairly high quantity of characteristic x-rays are produced, so they show up on the graph as tall spikes.

Adding the predominant bremsstrahlung to the characteristic x-rays, a complete graph of the typical spectrum for the x-ray beam finally emitted from the collimator of an x-ray machine appears as in Figure 10-11. Homogeneous radiation (the spikes) combines with heterogeneous radiation (the bell curve) to produce a total filtered x-ray beam which is generally heterogeneous and has an average energy of about one-third of the set kVp. This graph will be used in subsequent sections to help illustrate the effects of changing factors such as mA, kVp and machine phase upon the x-ray beam.

Anode Heat

As previously described, the generation of x-rays in the x-ray tube anode is not a very efficient process. Only about 0.5 percent of the energy deposited by projectile electrons into the anode is converted into x-rays, the other 99.5 percent is lost as heat, infrared radiation and light.

Figure 10-12 shows how a process called *excitation* momentarily raises orbital electrons to a higher energy level. In excitation, the repulsive negative charge of passing projectile electrons can slightly raise the energy of orbital electrons without "knocking them out" of their atoms. These excited electrons immediately fall back into their normal energy levels by emitting their excess energy in the form of infrared radiation and light. In turn, *infrared radiation* is absorbed by the anode's other *molecules as a*

Figure 10-10

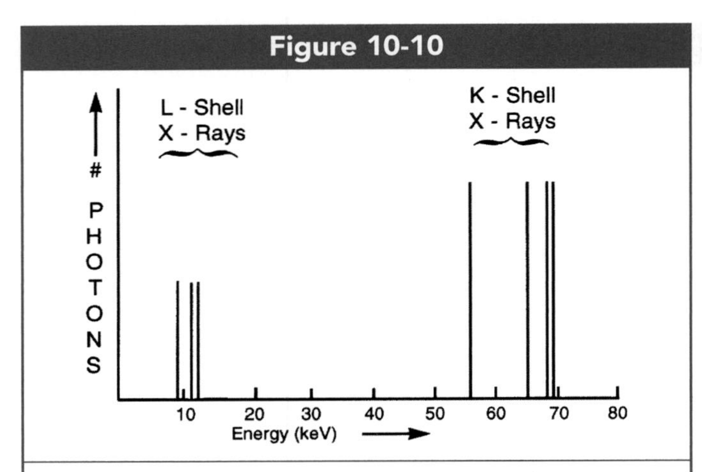

Graph of the characteristic x-ray spectrum for tungsten. X-rays are emitted only at discrete energy levels based on the differences between shell binding energies. Electrons falling from various outer shells into the *K* shell emit x-rays having 57, 66, 68, and 69 kV. The lines to the far left represent *L* shell characteristic x-rays. (From Quinn B. Carroll, *Practical Radiographic Imaging*, 8th ed. Springfield, IL: Charles C Thomas, Publisher, Ltd., 2007. Reprinted by permission.)

Figure 10-11

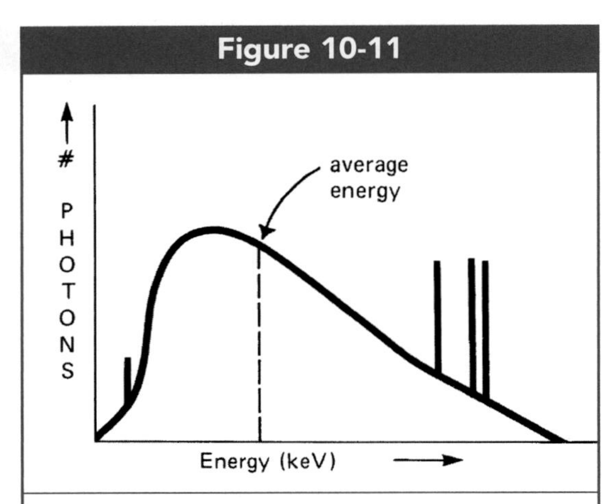

Shape of the complete diagnostic x-ray beam spectrum, representing both bremsstrahlung and characteristic x-ray energies, with filtration present. (From Quinn B. Carroll, *Practical Radiographic Imaging*, 8th ed. Springfield, IL: Charles C Thomas, Publisher, Ltd., 2007. Reprinted by permission.)

whole, causing increased molecular vibration. By definition, increased molecular motion is *heating*—it raises the temperature of the anode. Infrared radiation is responsible for most of the heat generated in x-ray tube anodes.

FACTORS AFFECTING THE X-RAY BEAM SPECTRUM

Target Material

Standard diagnostic x-ray tubes use an alloy of tungsten and rhenium for the target material, embedded in the circular focal track area of the anode disc. With atomic numbers, respectively, of 74 and 75, these elements consist of very "dense" atoms with lots of orbital electrons that the projectile electron from the cathode can strike to create x-rays. The higher the atomic number of the target material, the more efficient it is and more x-rays are produced.

Atoms with higher atomic numbers also improve the quality of the x-ray beam. They emit x-rays with higher energies, both bremsstrahlung and characteristic. The bremsstrahlung x-rays tend to have higher energies because the atomic nuclei in the anode are larger and have more positive charge, so they pull any projectile electrons in the vicinity with greater force, slowing them down more. The amount of kinetic energy lost is greater, and so is the energy of the x-ray emitted. Characteristic x-rays from "larger" atoms also have higher energies because all of the binding energies of the orbital shells are increased, and there is a greater difference between the binding energies of the shells.

Some specialty x-ray tubes use gold (atomic number 79) to achieve this increase in x-ray production and quality. For mammography, an x-ray beam with lower penetration characteristics is needed. By using molybdenum ($Z = 42$) and rhodium ($Z = 45$), x-rays with lower kilovoltages are produced, but a reduced quantity also results. The quantity and quality of characteristic x-rays are affected more than the bremsstrahlung x-rays by changes in target material.

Figure 10-13 graphs the x-ray beam spectrum for different target materials, using the same filtration and set kVp. The bell curve for gold, with its higher atomic number, is slightly higher, indicating more x-rays produced, and also shifted slightly to the right, indicating a somewhat higher average kV for the x-ray beam.

Figure 10-12

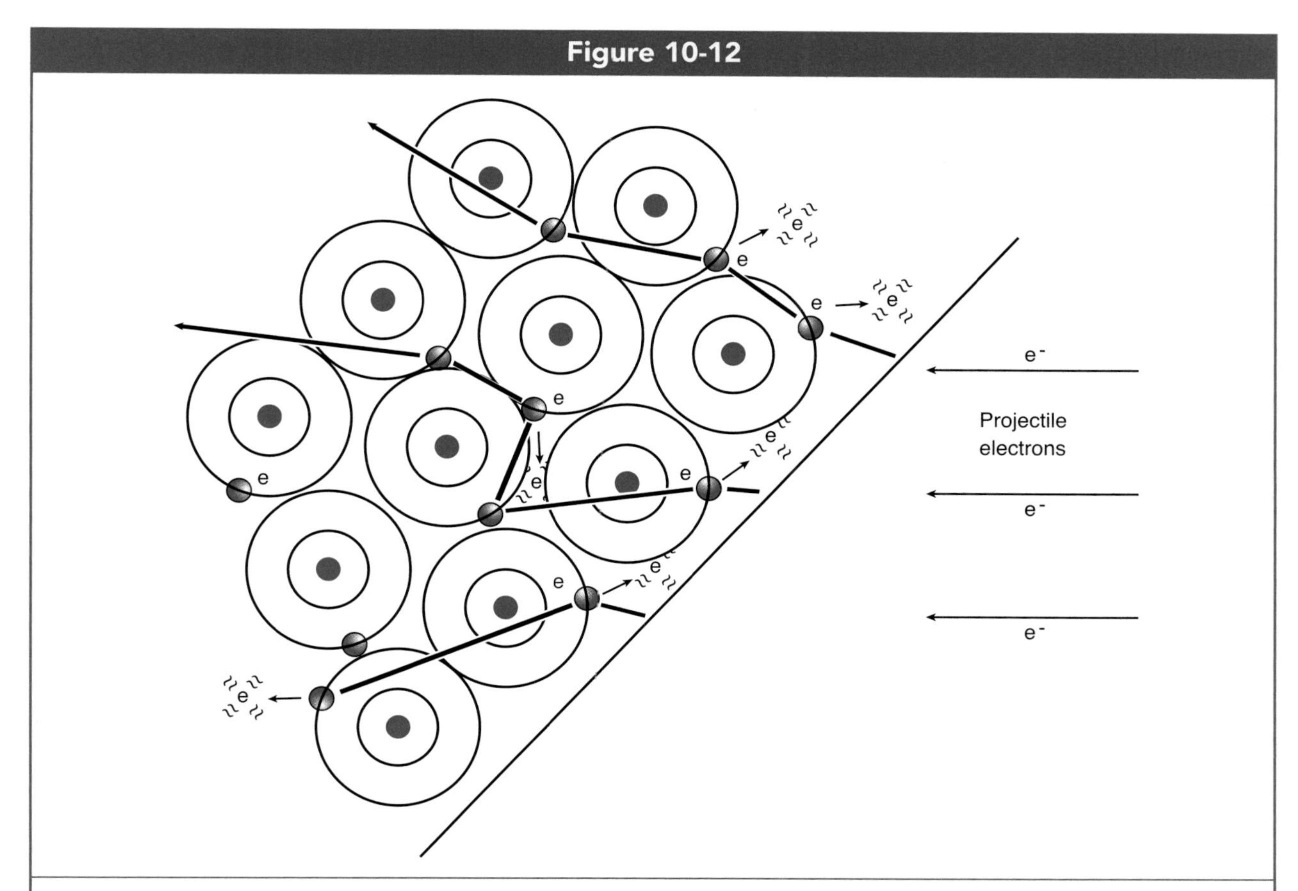

Within the anode, most of the energy of the projectile electrons is spent not on ionization, but rather on *excitation* and on *heating*. Here, *excitation* momentarily raises orbital electrons to a higher energy level; when the excited electrons fall back into their normal positions, *infrared radiation* is produced. *Heating* causes increased vibration of whole molecules within the anode, raising its temperature.

Figure 10-13

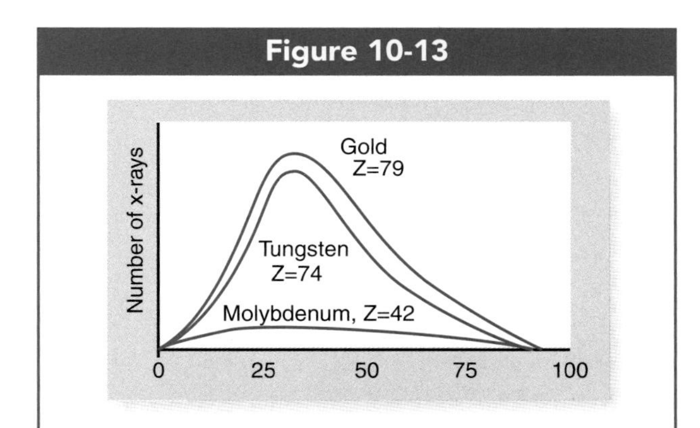

The efficiency of x-ray production also depends on the target material used in the anode. Note that, at the same filtration and set kVp, gold produces more x-rays at slightly higher energies than tungsten, while molybdenum produces less.

Milliampere-Seconds (mAs)

The set milliamperage, the exposure time, and the product of their total mAs, are all directly proportional to x-ray output from the tube. Doubling the mAs results in twice as many x-rays being produced. This is graphed in Figure 10-14, where *each point* along the curve for 200 mA is precisely twice as high as the point below it for the 100 mA station. Therefore, the *total area* under the 200-mA curve is also doubled from that for 100 mA, representing twice the overall intensity of the exposure.

It is important to note, however, that x-ray beam *quality* is not at all affected by changes in mA, time or total mAs. This is notable in Figure 10-14 since both curves begin at the same point on the left, vertically peak at the same kV (about 30), and end at

Figure 10-14

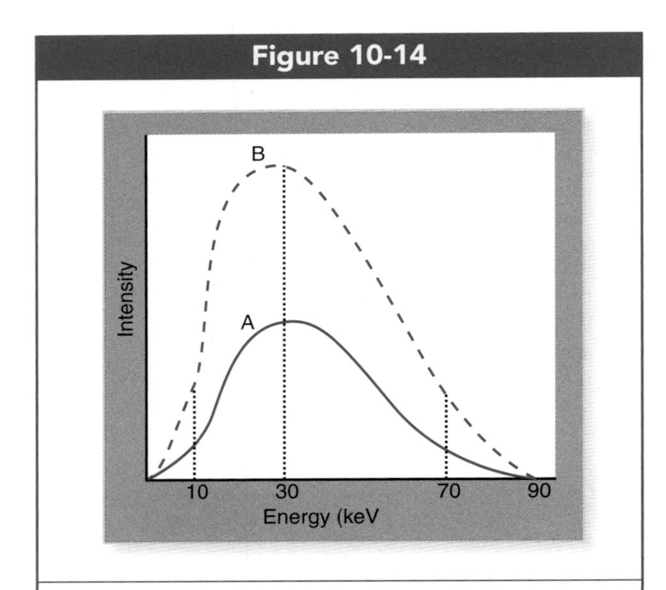

The effect of doubling the mAs on the x-ray beam spectrum, curve ***B***. The number of x-rays at *every* kV level is doubled (dotted lines). Minimum, average, and peak energies remain the same.

the same kV point to the right. Therefore, the *average* kV for these two beams, represented by the dotted line, is unchanged.

When setting radiographic technique, while adjusting the mAs changes the amount of radiation used, it does not alter the percentage of penetration for the x-ray beam.

Added Filtration

Figure 10-15 illustrates the effect of added filtration upon the x-ray beam spectrum. Note that when more filters are added (usually between the x-ray tube housing and the collimator), only the left portion of the curve is shifted. For curve ***A***, representing 2.0 mm of aluminum filtration, the curve starts at 5 kV; this indicates that no x-rays having less than 5 kV of energy have been emitted from the x-ray tube. Curve ***B*** represents the addition of an 0.5 mm filter; the starting point of the curve shifts to the right, indicating that there are now no x-rays in the beam having less than 10 kV.

The ending point of the curve to the far right is only affected by the kVp and has not moved. However, since the starting point shifts to the right, this pushes the *average kV* for this x-ray beam (dotted line) also to the right. If the shortest students are removed from a class, the average height of the remaining students must go up. Likewise, when low energies are removed from the x-ray beam, the average energy must increase. Adding filtration improves the overall penetration characteristics of the x-ray beam, thus enhancing beam quality.

Finally, note that with the loss of those low-energy x-rays represented by the dashed line on the graph, the total area under curve ***B*** has decreased and the peak is slightly lower. This indicates a reduction in the overall quantity of x-rays in the beam, since the added filtration has removed some of them. If one continues to add filters, a point will be reached where the general loss of radiation results in insufficient exposure to the image receptor. This limits us to only moderate amounts of filtration to enhance beam quality without an exaggerated loss in quantity.

Kilovoltage-Peak (kVp)

Higher kVp settings are used by the radiographer to increase the penetration quality of the x-ray beam so that dense or high-atomic number tissues may be fully demonstrated in the final image. In the beam spectrum (Fig. 10-16), increasing kVp is seen as a

Figure 10-15

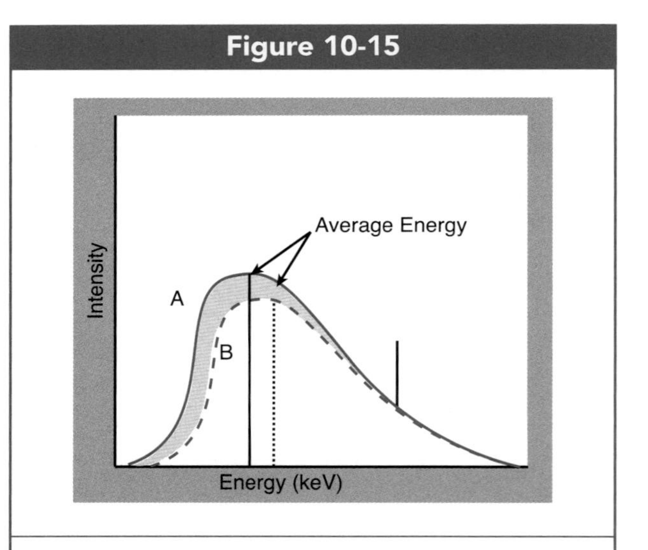

The effect of increased filtration on the x-ray beam spectrum, curve ***B***. Minimum and average energy increase (shift to the right), but maximum energy does not. Overall intensity (height of the curves) is reduced.

Figure 10-16

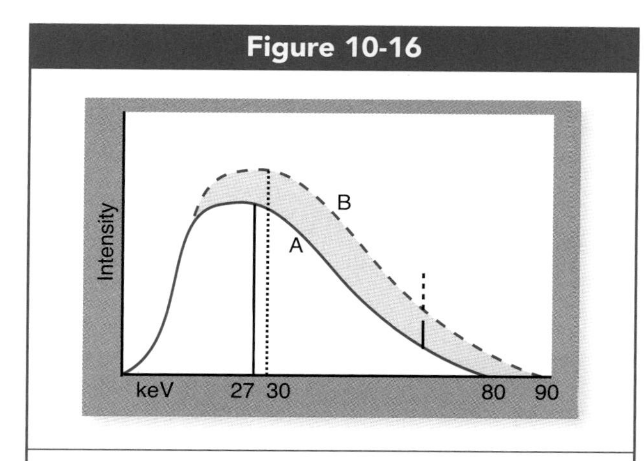

X-ray beam spectrum changes from an increase in kVp, curve ***B***. Peak and average energies of the beam increase, while minimum energy remains the same. The total number of x-rays produced increases by 25-40%. These are all higher-energy x-rays, represented by the shaded area.

shift of the ending point of the curve to the right. This adds high-energy x-rays to the beam, represented by the shaded portion under curve ***B***. If a few tall students are added to a class, the average height of the class must go up. Likewise, the addition of high-energy x-rays to the beam brings up the average kV. This average kV (dotted line) is an appropriate indicator of overall x-ray beam quality.

Higher levels of kVp also affect the efficiency of x-ray production. Within the anode material, a high-speed projectile electron can actually undergo a *series* of interactions. For example, in Figure 10-17, electron ***A*** enters the anode carrying 80 kV of kinetic energy and first undergoes a series of three bremsstrahlung interactions. The first bremsstrahlung slows the electron down by 4 kV, the second interaction by 5 kV, and the third by 2 kV. With 69 kV remaining, it then knocks out an orbital electron from the K shell of the tungsten, in a characteristic interaction, spending every last bit of its kinetic energy and coming to a complete stop.

In Figure 10-17, projectile electron ***B*** enters the anode carrying 90 kV of kinetic energy. It goes through *five* bremsstrahlung interactions before being stopped by a characteristic interaction. In this case, two additional x-rays were produced (for a total of six). The higher the energy of the incoming electron, the more x-rays it can produce before

Figure 10-17

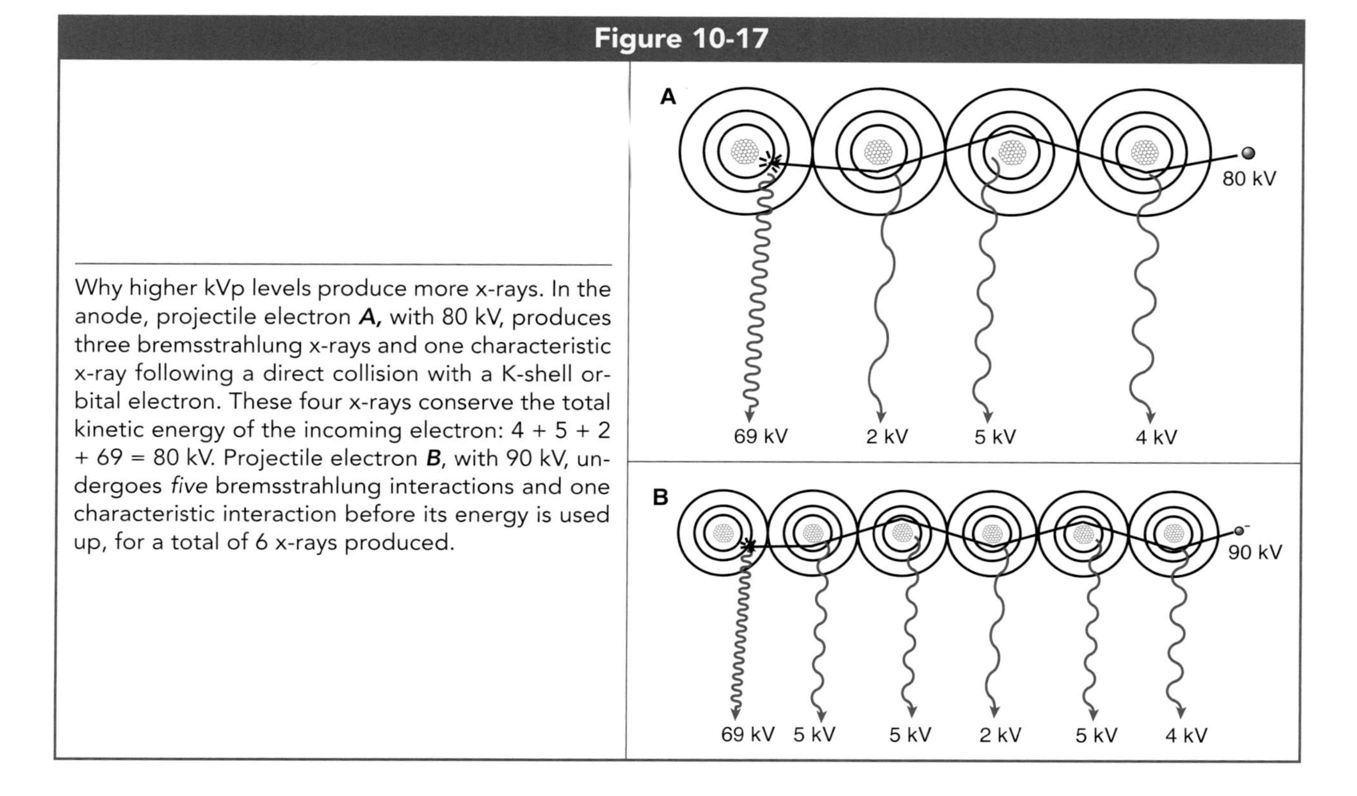

Why higher kVp levels produce more x-rays. In the anode, projectile electron ***A***, with 80 kV, produces three bremsstrahlung x-rays and one characteristic x-ray following a direct collision with a K-shell orbital electron. These four x-rays conserve the total kinetic energy of the incoming electron: 4 + 5 + 2 + 69 = 80 kV. Projectile electron ***B***, with 90 kV, undergoes *five* bremsstrahlung interactions and one characteristic interaction before its energy is used up, for a total of 6 x-rays produced.

Figure 10-18

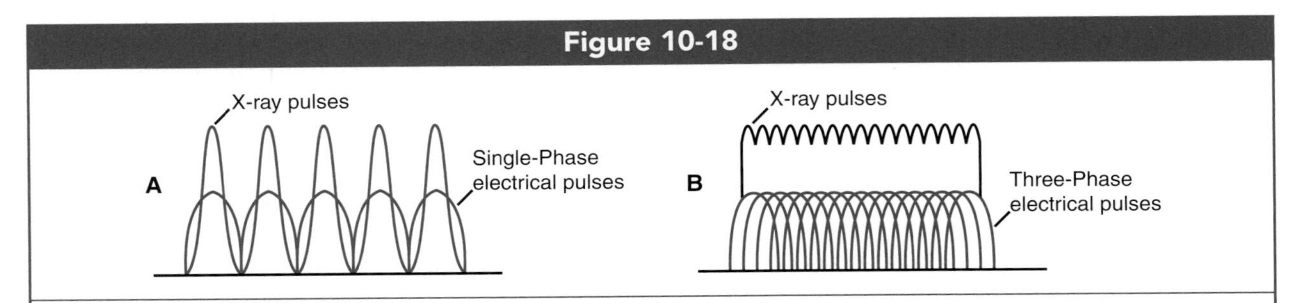

For a single-phase x-ray machine, most x-rays are produced in the middle of the voltage waveform when the voltage peaks, ***A***. Three-phase and high-frequency equipment keeps this x-ray production at its "surge" stage, ***B***, so that many more x-rays are produced at higher average energies.

coming to a complete stop in the anode. The additional quantity of x-rays produced at higher kVp levels is represented in Figure 10-16 by the shaded area under curve ***B***.

In comparing Figures 10-14 and 10-16, note that the increase in the area under the curve for increasing mAs is much greater than for increasing kVp. A 15 percent increase in kVp can add as much as 40 percent to the quantity of x-rays produced, but doubling the mAs, which is an equivalent change for the resulting image, doubles the amount of radiation produced, which is a 100 percent increase. Changes in mAs affect x-ray tube output proportionately. By comparison, changes in kVp affect radiation output much less.

This is important because the tube output represents the quantity of radiation exposure to the patient. When the final exposure to the image receptor can be doubled by using increased penetration (kVp) rather than doubling the original output (mAs), there is a savings in patient exposure. Radiographers should bear this in mind in daily practice.

The actual efficiency of x-ray production in the anode for operation at 60 kVp is about 0.5 percent of the kinetic energy of the incoming projectile electron. At 100 kVp, the overall efficiency doubles to about 1 percent. This leaves more than 99 percent of the original energy which is wasted in the form of heat and light.

Generator Type

As we just discussed in the last section, x-ray production is more efficient at higher voltages. When we look at the voltage waveform from the transformer (Fig. 10-18), we see that the quantity of x-rays produced at the beginning and end of one electrical pulse are very low, but it surges in the middle of the waveform when the voltage is high. In Chapter 8 we illustrated the voltage waveforms for various types of x-ray generators. We noted the "ripple effect" in which three-phase and high-frequency generators maintain an average voltage much higher than low-power generators. At very high power, this average approaches the set peak voltage (kVp). Such a waveform keeps x-ray production at its "surge" stage shown in Figure 10-18.

The x-ray beam spectrum in Figure 10-19 shows how the number of x-rays produced increases for

Figure 10-19

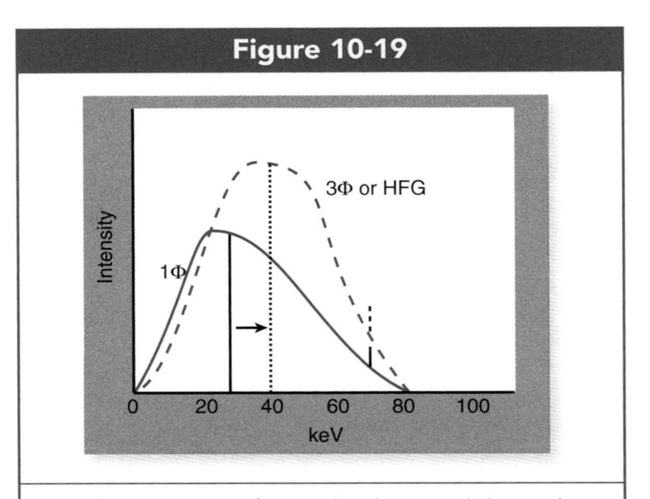

X-ray beam spectra for single-phase and three-phase (or high-frequency) x-ray machines. The higher power generators increase the quantity of x-rays produced, but also produce a higher *average* kV for the beam (shifting the peak of the curve to the right), even though the filtration and set kVp are the same.

higher-power generators. The area under each curve increases as the curves peak at a higher level. However, in this case we also see a shift to the right of the average kVp (dotted line) *even though the filtration and set kVp are the same*. Even though the starting and ending points of the curve are unchanged, the kV at which the curve *peaks* is shifted toward the right. This is because high-power generators increase the average kV for any set kVp.

Hence, we might state that high-power generators increase *both* the quantity and the quality of the x-ray beam. This allows the radiographer to employ either lower mAs settings or lower kVp settings to achieve a particular exposure, which will be discussed in a later chapter.

SUMMARY

1. The tremendous energy required to produce x-rays is acquired by projectile electrons by accelerating them to 50–80 percent of the speed of light.
2. Bremsstrahlung occurs when projectile electrons are slowed by the attraction of atomic nuclei. It accounts for most of the x-rays in the beam, and their heterogeneous energies which make the radiographic image possible.
3. Fewer bremsstrahlung x-rays are produced at higher energies, because it is statistically less likely that projectile electrons will pass very close to an atomic nucleus in the anode.
4. In the final x-ray beam emitted by the tube, inherent and added filtration in the x-ray beam eliminate the lowest-energy x-rays.
5. Characteristic radiation is produced from the refilling of orbital shells after ionization has taken place. The differences between binding energies of the shells determine the discrete energies of the x-rays produced.
6. The vast majority of energy deposited in the anode from projectile electrons is converted into heat and light rather than into x-rays.
7. Target materials used in the anode with higher atomic numbers raise the intensity and the energy of the x-ray beam spectrum produced.
8. Higher mAs raises the intensity of the x-ray beam spectrum produced, but does not change the energy distribution of the beam.
9. Added filtration raises the minimum and average energies of the x-ray beam spectrum, and slightly lowers the intensity.
10. Higher kVp levels raise the maximum and average energies of the x-ray beam spectrum, and slightly increase the intensity as well. The combination of these effects allows only a 15 percent increase in kVp to achieve the same end result exposure at the imaging plate as a 100 percent increase (or doubling) in mAs.
11. Higher-power generators raise the intensity and the average energy of the x-ray beam spectrum produced, such that one-half the mAs can achieve the same end result exposure at the imaging plate for a 3Φ or high-frequency machine as compared to a 1Φ machine.

REVIEW QUESTIONS

1. What does the German term *bremsstrahlung* mean?

2. Why is bremsstrahlung radiation absolutely necessary for the production of subject contrast in radiographic images?

3. The bell-type x-ray beam spectrum curve is "lopsided" toward the left. What is the correct interpretation of this from the graph?

4. Why is it less likely for high-energy bremsstrahlung x-rays to be produced than low-energy x-rays?

5. What type of image would result if the x-ray beam were mono-energetic or homogeneous in its energy?

6. Give an example of *inherent* x-ray beam filtration:

7. In the atoms of the anode target, if an inner-shell electron is ejected by a direct collision from a projectile electron, what x-ray producing interaction will follow?

8. Using the binding energies listed for tungsten in Figure 10-8, how much energy will a characteristic x-ray have when an orbital electron falls from the N shell to the K shell?

9. Using the binding energies listed for tungsten in Figure 10-8, how much energy will a characteristic x-ray have when an orbital electron falls from the O shell to the L shell?

10. When an electron "falls" from *outside the atom* into a particular shell, the characteristic x-ray produced will have a kV equal to the __________ energy for that shell.

(Continued)

REVIEW QUESTIONS *(Continued)*

11. The energies of characteristic x-rays are a characteristic of what?

12. All but 0.5 percent of the energy deposited into the anode from projectile electrons is transformed into heat by what process?

13. Anode target materials with higher __________ will produce more x-rays with higher energies.

14. What is the only variable discussed in this chapter which does *not* affect the average energy of the x-ray beam when it is changed?

15. What is the only variable discussed in this chapter which *reduces* the intensity (quantity) of the x-ray beam as it is increased?

16. When the kVp is increased, do the energies of *characteristic* x-rays increase?

17. Why does a higher-energy projectile electron produce more x-rays?

18. X-ray production is maximum at the _________ of the electrical voltage wave form.

19. Even though the set mA and kVp are equal, high power x-ray machine generators increase both the __________ and the __________ of the x-rays produced.

Part II

PRODUCTION OF THE RADIOGRAPHIC IMAGE

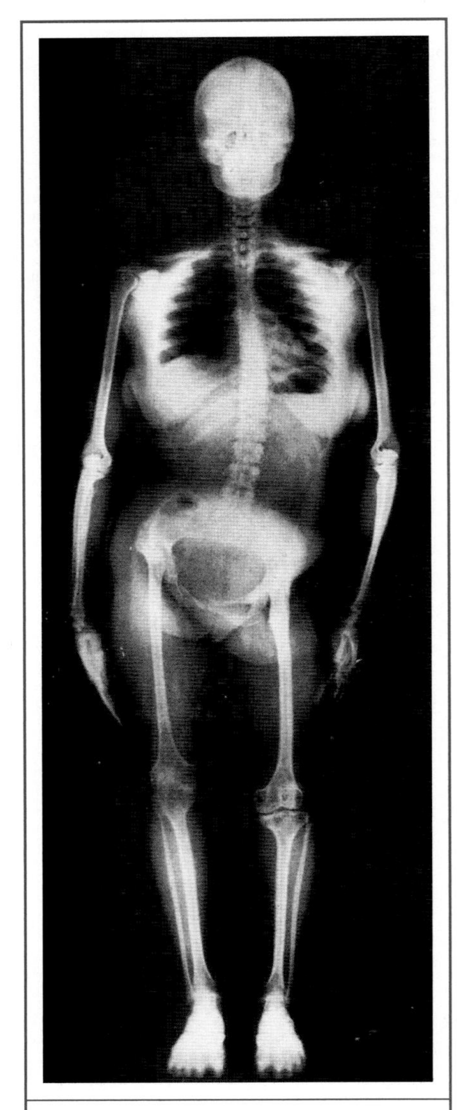

In the early days of x-ray, this whole body radiograph was produced using various filters to balance density.

energy not used in ejecting the electron from its orbit.

The photoelectric interaction cannot occur unless the incoming x-ray photon has energy equal to or greater than the binding energy of the orbital electron. When the incident x-ray does has sufficient energy, the *probability* of the photoelectric effect occurring rapidly decreases as the x-ray energy continues to climb upward. When the x-ray has much more than the binding energy, a photoelectric interaction is possible but *not likely*. Therefore, most photoelectric interactions are produced when the energy of the incoming photons is only slightly higher than the binding energy of the orbital electrons in the inner shells.

For the radiographer, this means that the ideal circumstance for producing high subject contrast is for the *average* kV (which is about one-third of the set kVp) to be slightly higher than the binding energies of the tissues in the body. Table 11-1 gives an overview of the K-shell and L-shell binding energies for those elements common to the body that absorb x-rays effectively, and those commonly used in radiography.

Practice Exercise #1

Within the patient's body, a 30-kV x-ray undergoes a photoelectric interaction with a K-shell electron in a calcium atom. Using Table 11-1, what will be the kinetic energy of the photoelectron speeding away from the atom?

> *Solution: 30 kV (incoming) – 4 kV (binding) = 26 kV kinetic energy*
>
> *Answer: The ejected photoelectron will have 26 kV of kinetic energy.*

The Compton Effect

Orbital electrons in the *outer* shells of atoms, where the influence of the nucleus is weakest, will interact with x-rays throughout the diagnostic range of energies. In this process only a portion of the photon's energy is absorbed, but the atom is still ionized. The substantial photon energy that is left over is reemitted as a new x-ray photon which can be emitted in any random direction or *scattered*. This interaction was discovered by American physicist Arthur Compton and is named after him.

The Compton interaction is also known by two other names, *modified scattering* and *incoherent scattering*. Both terms refer to the state of the scattered x-ray photon, which has been modified in its energy from the original primary photon, and therefore has a new energy which is incoherent with the original. However, these terms can be misleading in that they suggest that the original incoming

Table 11-1

Electron Shell Binding Energies for Radiographically Important Elements

Binding Energy* for:	K-Shell	L-Shell	M-Shell	N-Shell	O-Shell
Carbon	–0.284 kV	–0.022 kV			
Oxygen	–0.543 kV	–0.042 kV			
Calcium	–4.038 kV	–0.438 kV	–0.044 kV		
Iodine	–33 kV	–5 kV	–1 kV	–0.19 kV	
Barium	–37 kV	–6 kV	–1.3 kV	–0.25 kV	–0.03 kV
Tungsten	–69 kV	–12 kV	–2.8 kV	–0.6 kV	–0.075 kV
Lead	–88 kV	–16 kV	–3.8 kV	–0.9 kV	–0.15 kV

*Binding energies are given only for the **s** orbital in each shell.

photon was only changed in some way and continues on. This is not the case. The scattered x-ray should be considered as a *new* x-ray, *created* by an atomic interaction that occurs within the patient, whereas the incident x-ray was created by an interaction in the anode of the x-ray tube. They have different origins. Furthermore, x-rays are often *identified by* the amount of energy they carry, which also determines their frequency and their wavelength. A scattered photon with less energy has a lower frequency and a longer wavelength than the incident x-ray, which makes it "look" different than the primary photon.

Scattered x-rays emerging from this interaction are called *Compton scatter*, while the ejected electron is referred to as a *recoil electron* (Fig. 11-3). Of the x-ray photon's original energy, an amount equal to the binding energy goes into ejecting the electron from its orbit, a small amount also translates into the kinetic energy or speed with which the electron is ejected, and the remainder is reemitted as the Compton scatter x-ray. Mathematically,

$$E_P = E_S + E_B + E_{KE}$$

where E_P is the energy of the incoming x-ray photon, E_S is the energy of the scattered x-ray, E_B is the *absolute value* (dropping the negative sign) of the binding energy for the orbital electron, and E_{KE} is the kinetic energy of the ejected electron.

Figure 11-3

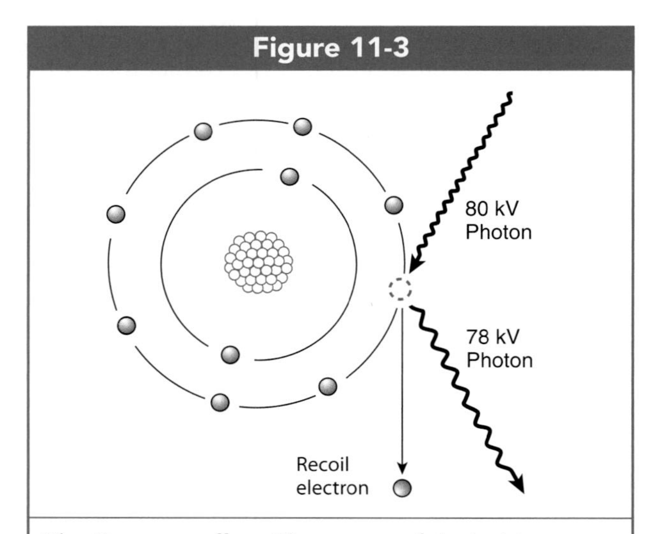

The Compton effect. The energy of the incident x-ray photon is partially absorbed by an outer-shell electron, which is then ejected from the atom as a *recoil electron*. The remaining photon energy is reemitted as a *compton scatter photon* which may penetrate through to reach the image receptor and expose it.

Practice Exercise #2

Within the patient's body, a 40-kV x-ray undergoes a Compton interaction with an L-shell electron in a calcium atom. The absolute value for the binding energy of the L shell is 0.5 kV (rounded). The recoil electron speeds away from the atom with 5 kV of kinetic energy. What is the energy of the Compton scattered photon?

Solution: 40 kV (incoming) = E_S + 0.5 kV (binding) + 5 kV (kinetic energy)

$$E_S = 40 - 0.5 - 5 = 34.5$$

Answer: The scattered Compton photon will have 34.5 kV of energy.

Since binding energies for body tissues are quite low, and only small amounts of energy are likely to be imparted to the ejected electron as kinetic energy, *the scattered photon carries most of the energy from the original x-ray.* This is important to know because it means that scattered radiation, having only slightly less energy than the original x-ray beam, is very likely to penetrate out of the patient's body and reach the image receptor. Therefore, scattered radiation has a substantial impact upon the information reaching the detector.

The recoil electron, on the other hand, will eventually be captured by another ionized atom within the tissue, never making it out of the patient, and has no impact upon image formation.

From a single original x-ray photon, a series of Compton interactions can occur, with each scattered photon having a few kV less energy than the previous one, until that energy is slightly above the binding energies for the tissue. At this point, a photoelectric interaction may take place, finishing off the remaining energy (Fig. 11-4).

There is a relationship between the energy of the Compton scattered photon produced and the angle at which it is scattered from the direction of the original x-ray photon. Scattered rays with higher energy are deflected less from the original direction, so they travel more forward and are more likely to strike the image receptor.

Figure 11-4

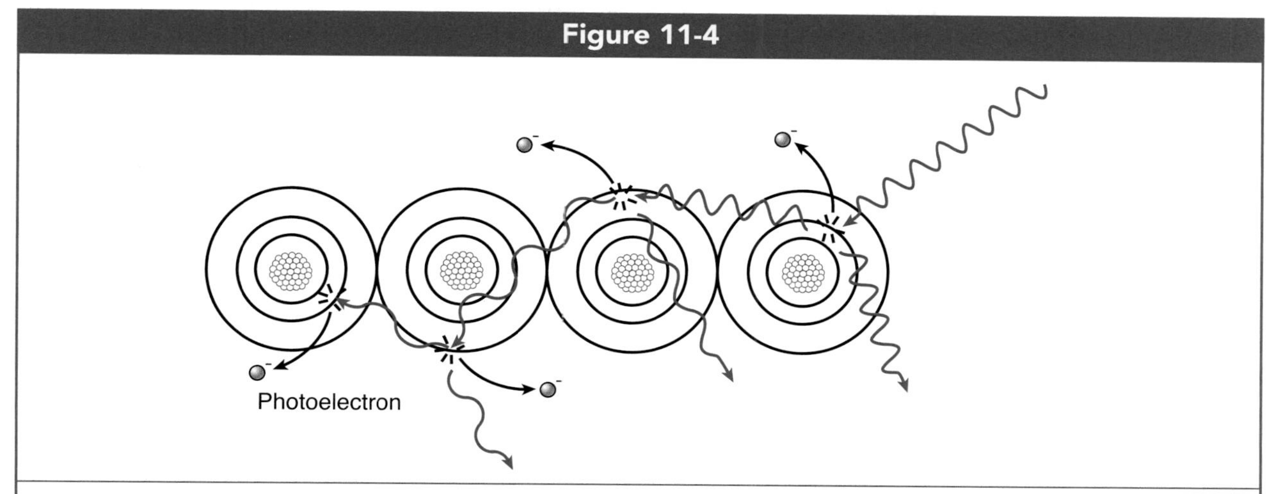

Within the patient, a single x-ray photon can cause a series of compton interactions, each resulting in a photon with less energy, which finally ends in a photoelectric interaction that absorbs all the remaining energy.

Scattered x-rays can be emitted in any direction. As shown in Figure 11-5, scattered x-rays that continue close to the direction of the primary photon have almost 100 percent of the primary photon's original energy. Those deflected at a 45-degree angle possess about 92 percent of the original photon energy. At 90 degrees, they contain 84 percent, and at 135 degrees 76 percent.

Scattered x-rays can even be emitted backwards directly toward the incident beam, at 180 degrees. Radiation emitted from the patient in this general direction is referred to as *backscatter*, and may strike the radiographers if they are standing behind the x-ray tube. Figure 11-5 shows that even these x-rays still retain about 68 percent, or two-thirds of the original photon energy.

The difference between the incident photon energy and the scattered photon energy is imparted to the ejected electron as kinetic energy—the more the energy, the faster the electron streaks out of the atom. For example, suppose a 50-kV x-ray undergoes a Compton interaction, and the scattered x-ray is emitted backward at 180 degrees. How much kinetic energy will the recoil electron have? Referring to Figure 11-5, subtract 68 percent of 50 kV. This is 50 minus 34. The remaining 16 kV goes into the kinetic speed of the recoil electron.

In radiation therapy, where voltages reach into the millions, forward scatter can predominate, but in diagnostic radiography a relatively small percentage of the generated scatter strikes the image receptor (although this is still more than enough to adversely affect the image picked up by the receptor). Much of the scattered radiation veers off at an oblique angle to the central ray (CR). But, *most of it, having energies in the range of ⅔ to ¾ that of the original x-rays, scatters at reversed oblique angles in relation to the CR*. This is *backscatter*.

Since incoming x-ray photons with higher kV tend to produce scattered x-rays with higher energies, the scatter generated at higher levels of kVp is directed more toward the image receptor. The result is that a higher percentage of all scatter produced reaches the image receptor and degrades the radiographic image.

The effects and control of scattered radiation are more fully discussed in a later chapter. Suffice it to say here that scatter radiation *always* degrades the image, because it is random in nature, laying down a "blanket" of unwanted exposure across the image receptor which contains no information. This is image *noise*. It is equivalent to having a bank of fog between yourself and a billboard you are trying to see while driving. The information is still there on the billboard, but its *visibility* is greatly reduced because of a random, yet even, distribution of mist in front of it that effectively reduces the contrast of the image behind it.

Since photoelectric interactions are responsible for the production of subject contrast, and scattered radiation destroys subject contrast, we might say that

Figure 11-5

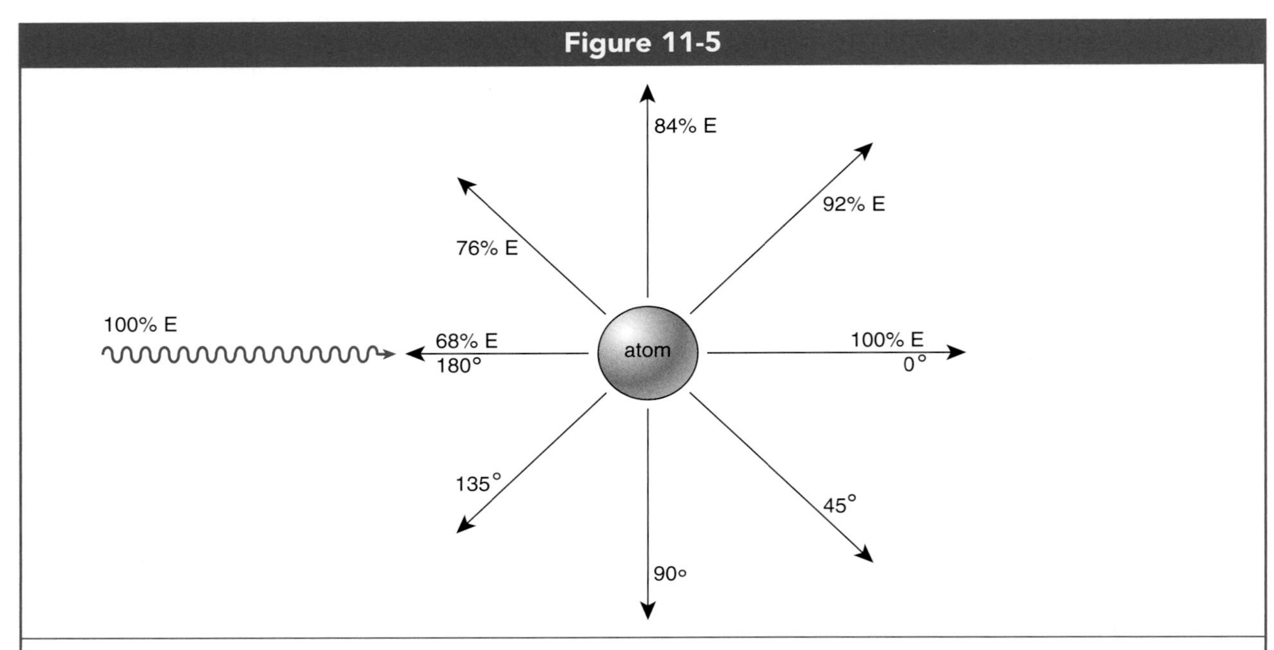

The angle of deflection for scattered x-rays is inversely related to the energy they carry. Scattered x-rays that retain most of the energy of the original incident photon are emitted in a forward direction close to that of the original. Those with 84% of the original energy are deflected at a right angle, and those with about 68% are back-scattered. Since most scattered photons have low energies, most are back-scattered.

Compton interactions work in direct opposition to photoelectric interactions. The Compton effect may be considered synonymous with *scatter*, since approximately 97 percent of all scattered x-rays originate from Compton interactions within the patient. The other 3 percent of scatter comes from coherent scattering.

Coherent Scattering

When the energy of the incoming x-ray photon is substantially less than the binding energy of a strongly bound orbital electron, *coherent scattering* may occur. The British physicist J. J. Thomson discovered that such a low-energy photon can be *momentarily* absorbed by a bound orbital electron, which is raised to a state of *excitation* by the energy, then reemits the entire amount in order to return to its stable state. Shown in Figure 11-6***A***, note that the energy and the wavelength of the incoming and outgoing x-ray photons are identical, hence the terms *coherent* or *unmodified* scattering.

For this interaction, the orbital electron remains in place and the atom is not ionized. Rather, the orbital electron quickly rids itself of the whole amount of energy carried by the incident photon. However, since this energy is *reemitted*, it may emerge in any direction and thus constitutes *scattered radiation*. Also, since any given x-ray travels only in a straight line within a particular medium, an x-ray emitted in a different direction should be considered a new and *different* x-ray from the incident photon.

Another British physicist, John Rayleigh, discovered that the energy of the incident x-ray may also be momentarily absorbed by the entire cloud of electrons, collectively, around an atom (Fig. 11-6***B***). Once again, this energy is immediately reemitted as a scattered photon in a random direction, but retaining the same energy and wavelength as the original. The Rayleigh interaction is identical to the Thompson interaction, except that the entire *atom* is considered to have been raised to a state of *excitation* for a brief moment.

The scattered photons from coherent scattering interactions are of low energy, and only a few penetrate out of the patient's body in the direction of the image receptor, constituting perhaps 3 percent of all scattered x-rays at the image receptor plate. They are therefore of only slight consequence to the final image.

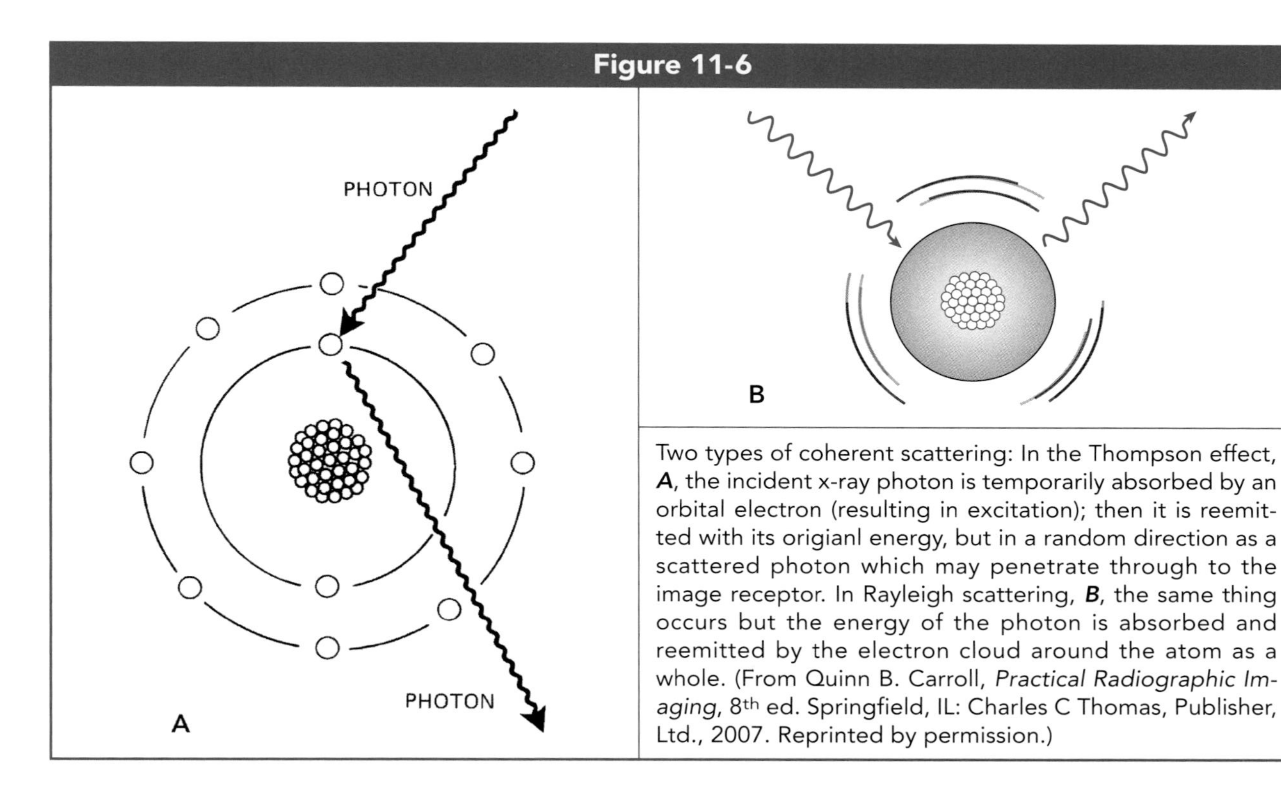

Two types of coherent scattering: In the Thompson effect, ***A***, the incident x-ray photon is temporarily absorbed by an orbital electron (resulting in excitation); then it is reemitted with its origianl energy, but in a random direction as a scattered photon which may penetrate through to the image receptor. In Rayleigh scattering, ***B***, the same thing occurs but the energy of the photon is absorbed and reemitted by the electron cloud around the atom as a whole. (From Quinn B. Carroll, *Practical Radiographic Imaging*, 8th ed. Springfield, IL: Charles C Thomas, Publisher, Ltd., 2007. Reprinted by permission.)

Other types of interactions can occur between x-rays and atoms at energies in the millions of volts, but these are not pertinent to diagnostic radiography and will not be discussed here. However, one final interaction that is common within the patient's body bears mention: The characteristic interaction. This is the same type of interaction which occurs in the x-ray tube anode, but within the patient's body it is caused by an incoming *photon* rather than an incoming electron, and involves much lower energies.

Characteristic Radiation

After any ionizing event within the patient's body, which includes both photoelectric and Compton interactions, an atom is left with an orbital vacancy. It soon pulls another electron into that orbit to fill it, leading to a characteristic interaction. Whether an electron is pulled from a higher orbit to a lower one, or from outside the atom into an orbital shell, potential energy is lost and must be emitted in the form of electromagnetic radiation. (See Figure 10-7B in Chapter 10.)

However, the "size" of the atoms in the soft tissues of the patient is much smaller than those atoms making up the anode target material in the x-ray tube. The binding energies of these atoms are very low (Table 11-1), and so are the subtracted differences between them. Electromagnetic waves emitted from these characteristic interactions that have energies less than about 1 kV would not even be classified as x-rays, but rather as ultraviolet light. Calcium atoms will emit x-rays having a few kV of energy. All of these emitted photons will have such low energies that they do not make it out of the patient to reach the image receptor. Therefore, characteristic interactions from within the patient, even though they always follow Compton and photoelectric interactions, cannot have any effect upon the final radiographic image.

Practice Exercise #3

Following the ionization of a carbon atom, using Table 11-1, what energy will the emitted characteristic photon have when an orbital electron falls from the L shell down into the K shell of the atom?

Solution: $0.284 - 0.022 = 0.262$ *kV*

Answer: The characteristic photon will have 0.262 kV or 262 volts of energy.

ATTENUATION AND SUBJECT CONTRAST

Attenuation is the partial absorption of the x-ray beam, the reduction in intensity that occurs as the x-ray beam traverses a body part. General attenuation of the x-ray beam includes all three of the interactions explained in the previous section, since both the absorption and the scattering of x-rays can prevent them from reaching the image receptor.

Figure 11-7 demonstrates the attenuation of an x-ray beam by a homogeneous object, a "step-wedge" made of pure aluminum. Since the material is of uniform consistency throughout, differences in the remnant radiation beam are entirely due to the changing thickness of the steps in the block of aluminum. Primary radiation striking the first and thinnest step is only slightly attenuated, and the receptor plate behind it receives high radiation exposure. As each step gets thicker, more attenuation occurs and less radiation exposure reaches the receptor plate.

Figure 11-7

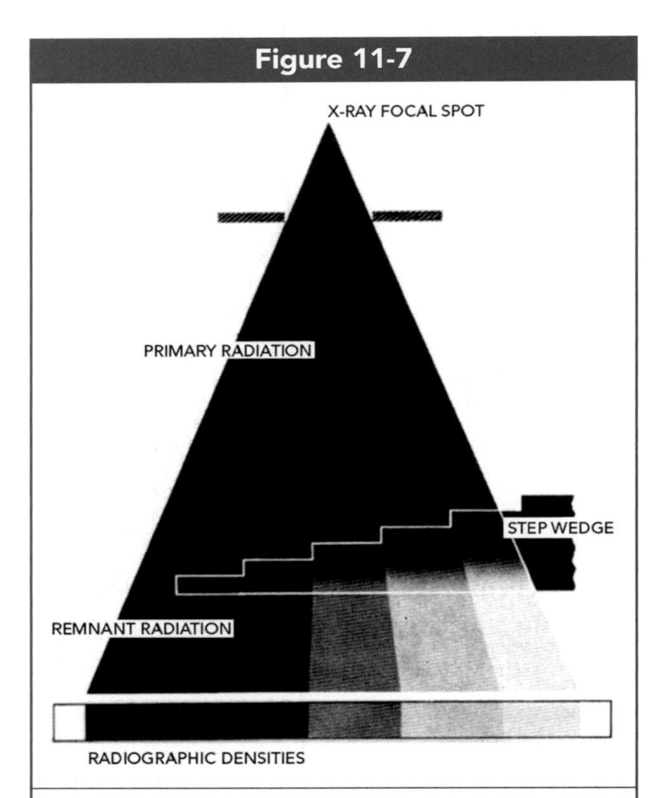

Diagram showing passage of an x-ray beam through a homogeneous material (aluminum) that has steps of varying thickness. (From Quinn B. Carroll, *Practical Radiographic Imaging*, 8th ed. Springfield, IL: Charles C Thomas, Publisher, Ltd., 2007. Reprinted by permission.)

Particulate radiations, such as alpha and beta particles, have a specific range of penetration into the human body before they are *all* stopped. For example, beta particles can only penetrate 1 cm into the body, none make it past this depth. X-rays, in contrast, are *attenuated* exponentially, which means that they are reduced in number by a certain percentage for each incremental thickness of tissue that they pass through. (Hypothetically, exponential attenuation implies that the quantity remaining never reaches zero.)

Figure 11-8 shows this progressive attenuation of the x-ray beam as it passes through a fairly homogeneous body tissue, such as muscle tissue, which possesses close to the same molecular (average) atomic number and physical density as liquid water. The attenuation is about 50 percent (or ½) for every 4 to 5 centimeters of soft tissue thickness. At 4 cm depth, only 500 of the original 1000 x-rays incident upon the body surface remain. At 8 cm, half of these, 250, remain, and so on until, after passing through the full thickness of a 24-centimeter abdomen, only 16 x-rays remain of the original 1000. This is 1.6 percent penetration through the body, which is close to the actual situation.

A related rule-of-thumb used by radiographers to adjust technique for different body thicknesses is to *change the radiographic technique by a factor of 2 for every 4 cm change in part thickness.* For example, if the patient's abdomen is 4 cm thicker than average, double the average mAs or increase kVp by 15 percent; If it is 4 cm thinner, cut the usual mAs setting in half or reduce the kVp 15 percent.

The human body presents additional attenuation factors besides changing thickness; in the body, each tissue presents a different physical density and also a different molecular (average) atomic number. Even though two organs may be of the same thickness, they can present a different percentage of attenuation to the x-ray beam. Less dense or lower-atomic number tissues allow a higher portion of the radiation through than high density or high-atomic number tissues (see Figure 5-25 in Chapter 5). *Differential absorption* refers to the subtle differences in attenuation between all the various tissues and parts of the human body.

Figure 11-8

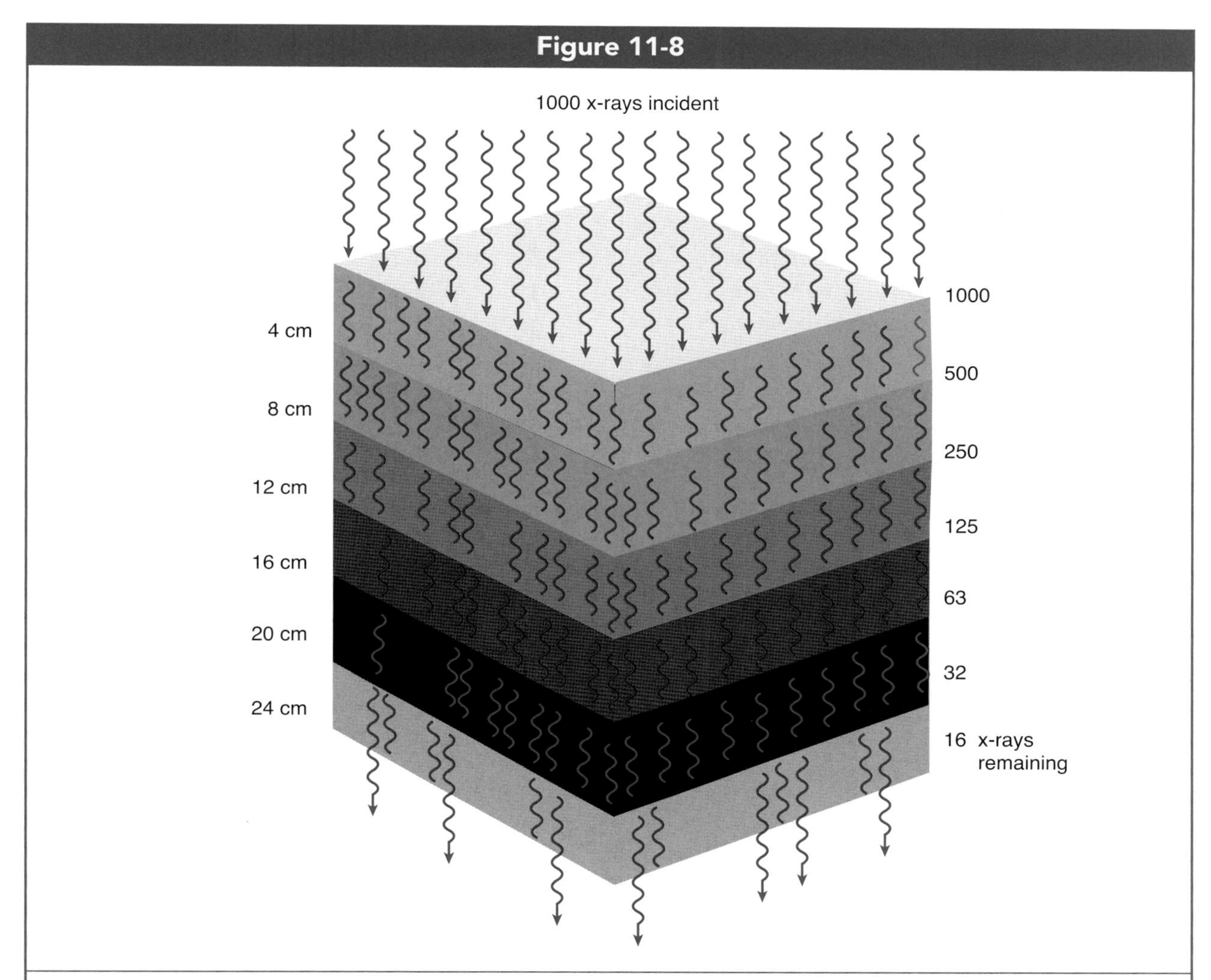

The progressive, exponential attenuation of an x-ray beam as it passes through soft tissues of the body. Each 4 to 5 cm of tissue thickness reduces the x-ray intensity to about one-half. The compensating rule for radiographic technique is to *double technique for every 4 cm increase in body part thickness.*

All these differences in attenuation have a collective effect upon the remnant x-ray beam called *subject contrast.* Subject contrast is most accurately defined as the difference in radiation intensity between the various portions of the remnant x-ray beam behind the patient. These differences represent different tissues and body part thicknesses, and are transmitted to the image receptor as an *unprocessed image.* All of the information which will be processed by the receptor plate and the computer is contained within the subject contrast of the remnant beam. ***It cannot be overemphasized that any information missing from the remnant beam of radiation cannot be recovered at a later stage by computerized post-processing, nor by manipulation of the brightness or contrast of the display monitor screen.***

The greater the x-ray attenuation of a tissue with relation to adjacent tissue, the greater the subject contrast produced. Greater subject contrast is exhibited between bone and soft tissue than between kidney tissue and muscle. For an overall image, the best level of subject contrast is one that results in every anatomical detail being depicted as a particular shade of gray, from light to dark within the anatomical part. To achieve this, *sufficient penetration* of the x-ray beam must match the thickness, physical density and atomic number of the various tissues in such a way that *some* x-rays penetrate through every tissue within the body part.

No anatomical portion of the image should be depicted as "blank white" nor as "pitch black," both of which are areas on the image containing no information. The highest amount of diagnostic information in the image is achieved when all anatomical structures are represented throughout a range of gray shades, something referred to as *long gray scale*. Differential attenuation provides the type of subject contrast that makes this possible.

CAPTURING THE IMAGE

The specifics of how digital image receptors capture the image signal in the remnant x-ray beam are presented in Chapter 34. Image receptors usually employ a *very thin layer* of material which must capture the x-ray photons in order to convert their energy into electrical charges or into light. The only way that such a thin layer of material can effectively absorb x-rays is if the individual *atoms* within it have a high capacity for capturing these high-energy photons. By using elements with a high *atomic number*, many more electrons are made available in a more concentrated *cloud* around the nucleus, making it more likely that an x-ray will "strike" one of them. These elements also have "larger" nuclei with greater positive electrical charge, such that the *binding energies* of their orbital shells are increased. This makes it more likely that photoelectric interactions will occur, releasing electrons with high energy that can reach a positive anode pin or plate and help generate a stored electrical charge.

Any time the goal is to absorb as much x-ray energy as possible, the photoelectric interaction is preferred, because the Compton interaction typically absorbs only a very small fraction of the photon energy. The next chapter discusses the relative occurrence of the photoelectric interaction versus the Compton interaction. All of these principles apply to the image receptor as well as to the patient when we consider how to effectively absorb or capture the energy of the x-ray beam.

SUMMARY

1. The x-ray beam is divided into the primary beam and the remnant beam. Isotropic divergence of the primary beam causes magnification and distortion of structures within the image.
2. The remnant beam, at less than 1 percent the intensity of the primary beam, carries the image-forming *signal* to the image receptor. It includes randomly scattered radiation as well as primary radiation.
3. Six types of variables affect the quality of the final radiographic image. These are technique, geometry, patient condition, the image receptor system, image processing and image viewing conditions.
4. The photoelectric effect, with its "all-or-nothing" absorption of x-rays, is responsible for the production of subject contrast in the radiographic image. Scattered radiation tends to destroy subject contrast in the remnant beam.
5. The photoelectric effect occurs only in the inner atomic shells when the energy of incident x-ray photons is *slightly* higher than the shell binding energies. The Compton effect occurs in the outer shells when the energy of incident x-rays is *much* higher than the binding energies.
6. The Compton effect is responsible for the vast majority of scattered x-rays, which should be considered as newly-created x-rays. Because their energies are altered from the original x-ray photon, the interaction is also known as *modified* or *incoherent scattering*.
7. As the angle of deflection for scattered x-rays increases away from the CR, we find higher amounts of scatter, but at lower energies.
8. *Unmodified* or *coherent scatter* can be produced by the Thompson effect or the Rayleigh effect, in which an x-ray photon temporarily excites an electron or an entire atom and is then reemitted. These account for less than 3 percent of all scatter.
9. X-rays are *attenuated* (partially absorbed) exponentially by body tissues, at a rate of approximately 50 percent for every 4–5 cm of tissue thickness. To compensate, radiographic technique must be roughly doubled for every 4 cm increase in part thickness.

10. The *differential absorption* of x-rays, according to tissue thickness, atomic number and physical density, is responsible for the creation of the *subject contrast* that constitutes a latent image within the remnant x-ray beam.
11. All anatomical structures within a radiographic image should be depicted as a shade of gray due to partial penetration of the x-rays through them. Areas of an image which are blank, due to insufficient penetration of the x-ray beam, represent a loss of information which *cannot* be retrieved by computer processing.
12. Image receptors capture the image by the same atomic interactions that occur within tissue.

REVIEW QUESTIONS

1. The only nondiverging ray in the primary x-ray beam is the __________.

2. A "latent image" or signal is carried to the receptor by the __________ x-ray beam.

3. What does OID stand for?

4. Positioning of the patient actually falls under what category of radiographic variables?

5. Even after the radiographic image is processed and stored, what other type of variables still can alter its quality?

6. Microscopic white spots are produced in the image by which interaction?

7. The ideal conditions for the photoelectric effect to take place are created when the *average* kV of incident x-ray photons is __________ than the inner-shell binding energies of tissue atoms.

8. Within the patient's body, a 35-kV x-ray undergoes a photoelectric interaction with a K-shell electron in a iodine atom. Using Table 11-1, what will be the kinetic energy of the photo-electron speeding away from the atom?

(Continued)

REVIEW QUESTIONS *(Continued)*

9. Within the patient's body, a 30-kV x-ray undergoes a Compton interaction with an L-shell electron in an iodine atom. The recoil electron speeds away from the atom with 5 kV of kinetic energy. Using Table 11-1, what is the energy of the Compton scattered photon?

10. Following the ionization of an oxygen atom, using Table 11-1, what energy will the emitted characteristic x-ray have when an orbital electron falls from the L shell down into the K shell of the atom?

11. Which of the two major interactions can occur in a *series* within the patient from a single x-ray photon?

12. A scattered x-ray is emitted 135 degrees backward from the direction of the original photon. What percentage of the original photon's energy will it have?

13. In terms of the *quantity* of radiation scattered in each direction from the CR, where is the *worst* place for the radiographer to be standing?

14. Within the remnant x-ray beam, scatter radiation forms a type of image __________ which is always destructive.

15. Thompson and Rayleigh scatter photons have the same __________ but a different __________ from the original x-ray photon.

16. Why doesn't characteristic radiation produced within the patient affect the image signal at the receptor plate?

17. Adjust radiographic technique by a factor of 2 for every __________ change in body part thickness.

(Continued)

REVIEW QUESTIONS *(Continued)*

18. What other two aspects of body tissues, besides thickness, affect x-ray attentuation?

19. When x-ray beam penetration is matched properly to the differential absorption of the tissues, every radiographic image detail should be depicted as a __________.

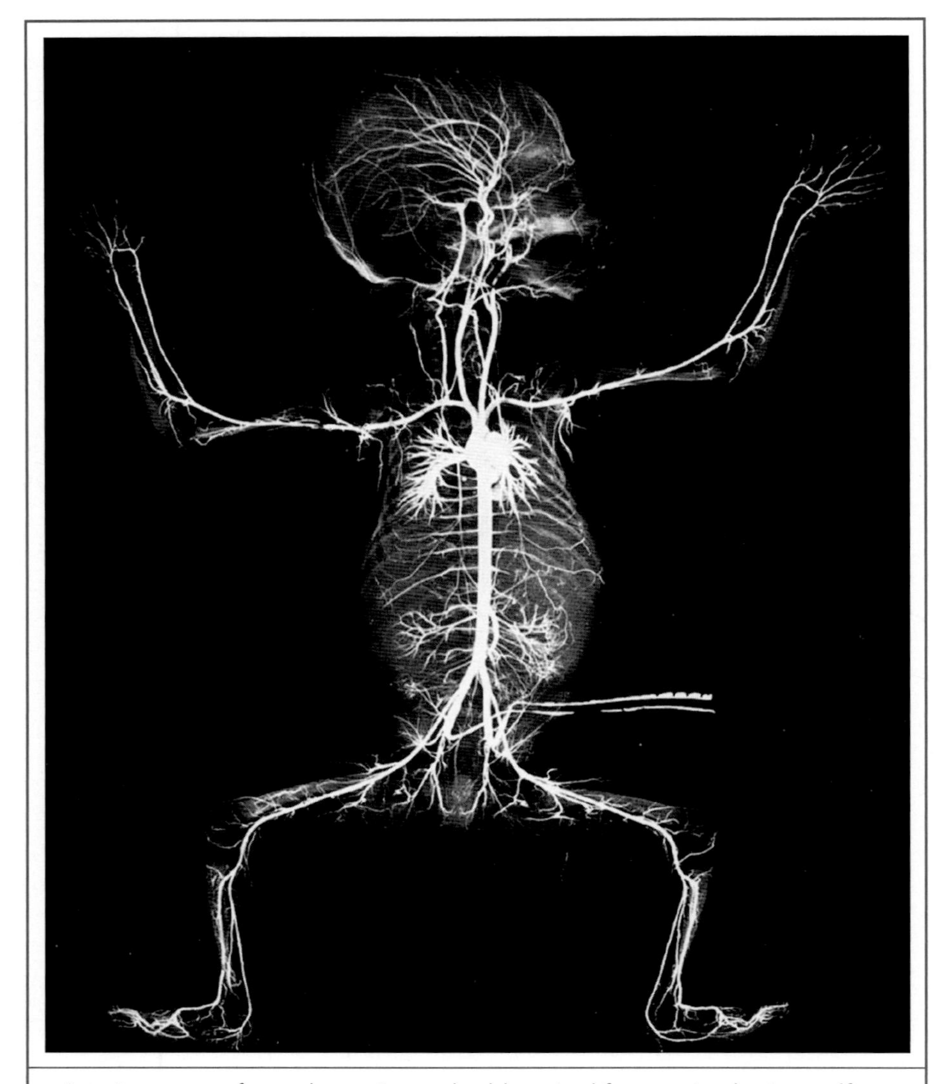

Arteriogram performed on a 5-month-old expired fetus, using barium sulfate.

Chapter 12

PRODUCTION OF SUBJECT CONTRAST

Objectives:

Upon completion of this chapter, you should be able to:

1. Define subject contrast and quantify how it is altered by changes in tissue that result in different *ratios* of interactions.
2. Give a clear mathematical approach for measuring subject contrast.
3. Quantify the effects of tissue thickness, physical tissue density, and tissue atomic number on the production of subject contrast in the remnant beam signal.
4. Specify the effects of scatter radiation on subject contrast in the remnant beam signal.
5. Describe the effect of increasing kVp upon the *predominance* of Compton and photoelectric interactions, and the resulting impact on the image.
6. Describe the effect of increasing the atomic number of tissue or contrast agents upon the *predominance* of Compton and photoelectric interactions, and the resulting impact on the image.
7. Distinguish between the *production* of Compton interactions and the amount of scatter radiation *reaching the image receptor*, and why kVp is a relatively minor influence on subject contrast when compared with field size and part thickness.

Contrast is absolutely essential to the visibility of detail in any radiographic image. The primary concern when discussing the interactions between x-ray beam photons and the atoms within the tissues of the patient is their effect upon *subject contrast* carried to the image receptor by the remnant x-ray beam.

GENERAL ATTENUATION AND SUBJECT CONTRAST

Subject contrast is produced by the *differential absorption* between various tissues of the body. The physical differences between these tissues are already present before the x-ray beam strikes them. Simply put, a tissue such as bone stands out from the "background" of soft tissues because the bone attenuates more x-rays than soft tissue does. This general attenuation of x-rays can be due to either *absorption* of the x-rays by the tissue or to *scattering* of the x-rays by the tissue—either way, the photon may be prevented from reaching the image receptor. *All interactions within the patient, whether photoelectric, Compton, or coherent scattering, represent some degree of absorption of the overall x-ray beam. All interactions attenuate the beam.*

Differential absorption, and hence subject contrast, are a direct consequence of the *percentage* of attenuation by all interactions, versus the unhindered penetration of other x-rays in the beam. In other words, subject contrast is represented by the *ratio* of absorption between two adjacent tissues or anatomical structures. For example, let us examine the subject contrast of bone against soft tissue, and determine how this might change if some disease caused the soft tissue to double its physical density.

Figure 12-1

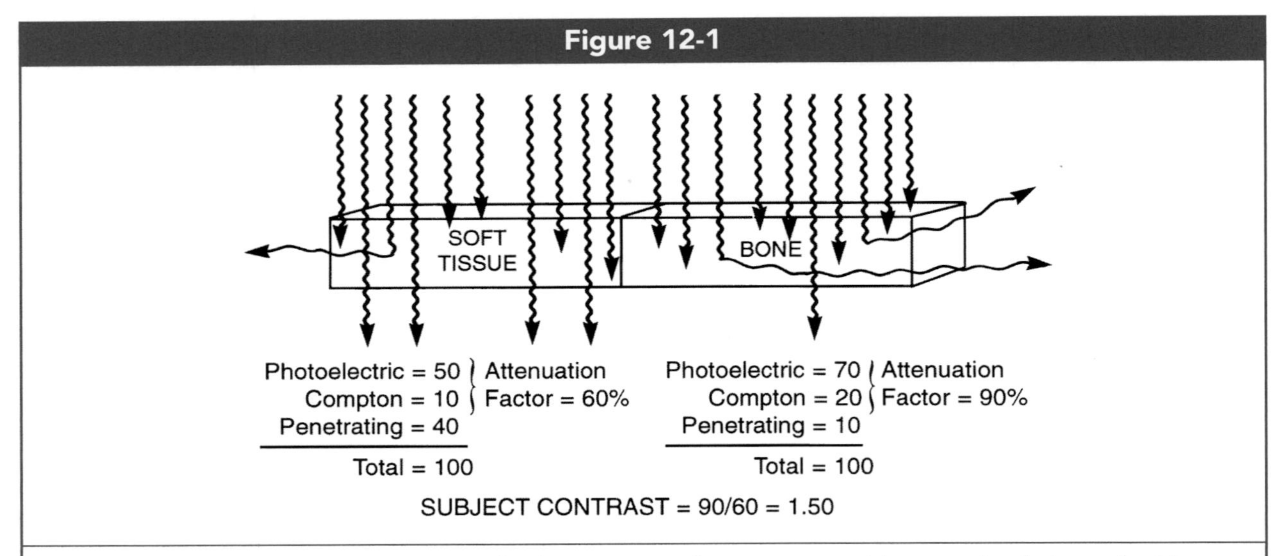

Hypothetical diagram of the production of subject contrast between normal bone and soft tissue. In this case, photoelectric and Compton interactions in soft tissue combine to absorb 60% of the x-rays. In bone, both interactions combine to absorb 90% of the beam. The subject contrast produced is the *ratio* of the two attenuation factors: 90%/60% = 1.50. (From Quinn B. Carroll, *Practical Radiographic Imaging*, 8th ed. Springfield, IL: Charles C Thomas, Publisher, Ltd., 2007. Reprinted by permission.)

Suppose that 100 x-rays per square inch are incident upon the body. Let us assume that, of these 100 x-ray photons, the soft tissue absorbs 50 by photoelectric interaction, and scatters 10 by Compton interaction. This leaves 40 x-rays penetrating all the way through to the image receptor. Figure 12-1 illustrates these ratios. The overall attenuation factor would be 60 percent (consisting of photoelectric absorption plus Compton attenuation). The nearby bone, which is a much denser tissue, will be expected to cause more of both interactions: It absorbs 70 x-rays by photoelectric interaction and scatters 20 by Compton interaction. Of the original 100 incident x-rays, only 10 have penetrated completely through the bone. The overall attenuation for the bone is 90 percent.

The bone has a much higher attenuation (90%) than the soft tissue (60%). Now, the *ratio* between the attenuation factors for the bone and the soft tissue is 90/60 = 9/6 = 1.5. The *subject contrast* between the tissues in Figure 12-1 is 1.5. This is a fairly high ratio; the bone is absorbing 50 percent more radiation than the soft tissue. We expect the bone image to stand out with high contrast against the background of soft tissue.

Now, suppose that the patient has a disease which causes the soft tissues to harden, increase mineral content, and tend to calcify (such as arthritis within a joint space). We would expect this hardening of the soft tissue to increase its overall physical density. There will simply be more atoms per cubic millimeter for the x-rays to "run into." Some of these events will result in photoelectric interactions and some in Compton interactions. Let us assume that both types of interactions increase by some 30 percent in the soft tissue (Fig. 12-2).

There are now occurring 67 photoelectric interactions and 13 Compton interactions. This leaves only 20 x-rays from the original 100 that will penetrate through to the image receptor. The attenuation factor for the diseased soft tissue then increases to a factor of 80 percent, instead of the normal 60 percent. How does this change the *subject contrast* between the bone and the soft tissue? The attenuation factor for the normal bone is still 90 percent as before (Fig. 12-1). The new *ratio* between the two tissues, the new *subject contrast*, is now 90/80 = 9/8 = 1.13. The bone is now absorbing only 13 percent more radiation than the diseased soft tissue. There is much less difference between the two tissues, and much less subject contrast.

In this hypothetical case, the hardening of the soft tissue caused the subject contrast between bone and

Figure 12-2

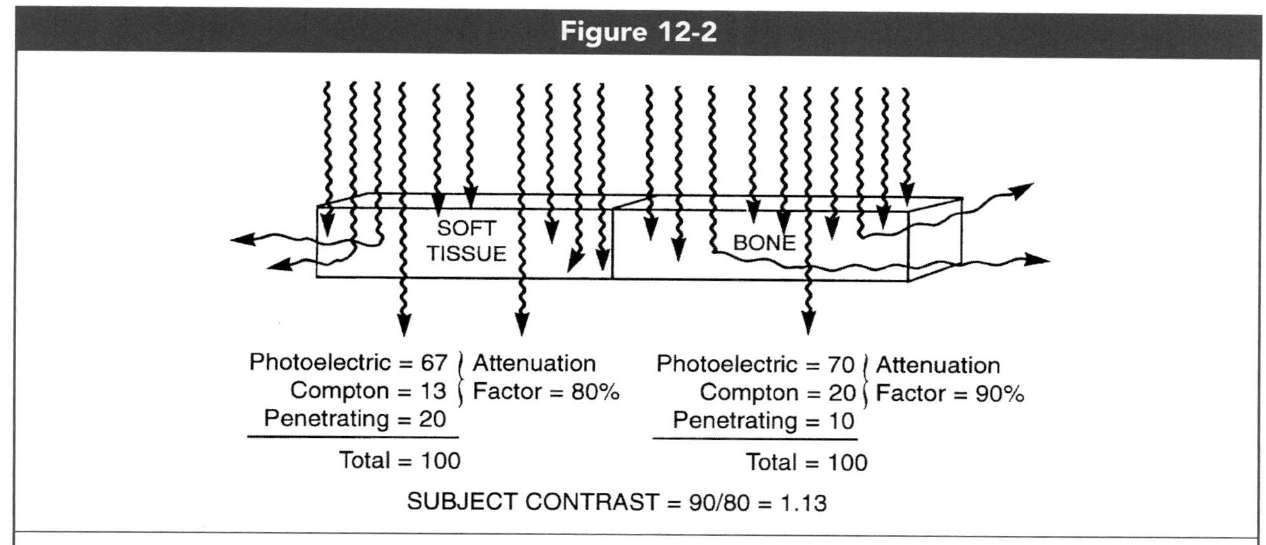

The change in subject contrast from Figure 12-1 if a disease increases the physical density of the soft tissue by some 30%. Both photoelectric and Compton interactions increase in number, and the attenuation factor rises to 80%, with only 2 out of 10 photons penetrating through the soft tissue. The subject contrast between the soft tissue and the adjacent bone is now reduced to 90%/80% = 1.13. (From Quinn B. Carroll, *Practical Radiographic Imaging*, 8th ed. Springfield, IL: Charles C Thomas, Publisher, Ltd., 2007. Reprinted by permission.)

soft tissue to *decrease* from 1.5 to 1.13. Accordingly, there will be less difference between the two tissues in the final radiographic image. Instead of the bone being much lighter than the soft tissue, it will only be slightly lighter. Image contrast will be reduced.

Note that to measure the contrast between two tissues, a *ratio* is used, dividing one factor into the other. In the past, several radiography textbooks have presented this relationship as a *subtracted* difference, rather than as a divided ratio. Dr. Perry Sprawls, in his book *The Physical Principles of Diagnostic Radiology*, gives a clear and scientific definition for subject contrast. He states, "Actually, it is the ratio of the penetration factors, rather than the difference, which determines the amount of contrast." Dr. Sprawls provides the mathematical formula for subject contrast as:

$$C = \left\{1 - \left(\frac{P_o}{P_t}\right)^x\right\} \times 100$$

where P_o and P_t are the penetration factors for the two tissues. Note that these two factors are not subtracted one from the other, but divided to form a ratio. (The other portions of the formula are simply designed to yield an answer which is a percentage figure.)

There are three essential aspects of tissues which determine their attenuation properties and the resulting subject contrast: The thickness of each tissue area, the physical density of each tissue, and the average (or molecular) atomic number of each tissue.

Tissue Thickness

As a tissue area becomes thicker, its attenuation of the x-ray beam is naturally greater. This attenuation increases exponentially and follows a somewhat complicated logarithmic formula which is unnecessary here. A rough rule of thumb for body part thicknesses in general is that for every 4 centimeters of additional thickness, the attenuation of the x-ray beam is doubled, so that the penetration of x-rays through to the receptor plate is cut in half. A body part that is 4 centimeters thicker than some other part will absorb about twice as much of the x-rays, so the exposure to the image receptor will be about one-half as much in this area.

Suppose that body part A has an attenuation factor of 40 percent, and an adjacent body part, B, has a factor of 10 percent. The subject contrast between body parts A and B is 40/10 = 4.0. Now, let body part B increase in thickness by 4 centimeters.

This would double its attenuation factor from 10 percent to 20 percent. The subject contrast between these two body parts will now be 40/20 = 2.0. Clearly in this case, with body part B increasing in thickness, the *subject contrast has been reduced from 4.0 to 2.0.* On the other hand, let body part A increase in thickness by 4 centimeters while part B remains with its original 10 percent attenuation factor. The attenuation of body part A increases from 40 percent to 80 percent, and the ratio between the two will now be 80/10 = 8.0. In this case, the *subject contrast has increased from 4.0 to 8.0.* Hence, changes in body part thickness may cause the subject contrast to either decrease or increase, depending on which part has changed.

Tissue Density

The physical density of a substance refers to the amount of physical mass that is concentrated into a given volume of space, such as grams per cubic centimeter. In the patient, the physical density may be considered as the concentration of atoms or molecules within a tissue. At higher tissue densities, there are more atoms or molecules packed into a given space. At lower densities, atoms or molecules are less concentrated and there is more space between them.

Clearly, if the number of atoms in a particular space is doubled, there will be twice as high a probability that an x-ray photon passing through will actually "hit" one of these atoms. If the tissue density is cut in half, the likelihood for attenuating x-rays is cut in half. This probability applies equally to photoelectric and Compton interactions. Therefore, it may be said that the occurrence of all interactions is *directly proportional* to the physical density of the tissue through which the x-rays pass.

Because of this proportional relationship (as opposed to an exponential one), fairly *extreme differences in physical density between tissues are necessary to result in high subject contrast.* An eminent example is found in chest radiography: The primary difference between air-insufflated lung tissue and soft tissues such as the heart and diaphragm is in their physical density. The average atomic number for air is 7.6 and that for soft tissue is 7.4, nearly equal. But their physical densities are vastly different. Air is a gas and therefore has an *extremely low density* when compared to soft tissue. Soft tissue has a density roughly equal to that of liquid water. The ratio between soft tissue density and lung density is approximately 1000 to 1. Soft tissue will absorb nearly 1000 times more x-rays than the lungs, an extreme enough difference to render subject contrast in the remnant beam reaching the image receptor.

Three types of body tissues may be distinguished from each other primarily on the basis of their differences in physical density: Soft tissues (which generally includes all glandular organs, muscles and connective tissues), gases such as air in the lungs, and fat. Fat is much less dense than soft tissue, so that it shows up visibly darker on a typical radiographic image. For example, on a standard abdomen image (with no contrast agents introduced), it is the quarter-inch thick fat capsules surrounding the kidneys that make them visible against surrounding abdominal muscles. Without these fat layers, it would be difficult indeed to distinguish the kidneys from the muscles. Gasified lung tissues show up much darker than either fat or soft tissue.

Much radiography, however, involves the visualization of bones or of contrast agents. Bones, contrast agents and many metals are not dramatically different from soft tissue in their physical densities. The reason *these* particular materials appear in a radiographic image is because of their high atomic numbers.

Tissue Atomic Number

A high atomic number means that each atom has many more electrons packed within the volume of its shells. The probability of an x-ray striking an electron in these atoms is much higher. We might say that each individual atom is more efficient at absorbing x-rays, rather than the tissue as a whole.

Most organs cannot be distinguished radiographically from the other soft tissues around them, such as muscles and connective tissues, without some form of intervention to artificially provide subject contrast. Contrast agents—substances usually based on the iodine atom or the barium atom—are introduced into cavities within these organs where possible in order to produce an image of them.

Iodine has an atomic number of 53 and barium 56. They not only have this many positively-charged protons in their nuclei, but they also normally have this many electrons packed into their orbital shells. Interestingly, these "large" atoms are not actually much bigger in their overall *diameter* than hydrogen with Z = 1. Rather, their orbital shells are collapsed in closer to the nucleus in order to fit more shells within the atom. The electrons within all of these orbitals are more *concentrated* within the space of each atom. (This concentration is sometimes referred to as the atom's *electron density*.)

A particular tissue is composed of several types of atoms combined into molecules. Thus, the atomic number of a tissue must be expressed as an *average*, which takes into account the number of each different type of atom within these molecules. For example, soft tissue is mostly comprised of water with a molecular Z number of 7.4. Each molecule of water is composed of two atoms of hydrogen with Z# = 1, and one atom of oxygen with Z# = 8. The average atomic number, then, must fall between 1 and 8. It falls closer to 8 because the large oxygen atom has a much greater effect than the very small hydrogen atoms.

The impact of this average atomic number upon the attenuation of x-rays is *exponential*—relatively small differences in atomic number will result in large differences in absorption of the x-ray beam. Specifically, x-ray attenuation is proportional to the *cube* of the atomic number. For example, the average atomic number of bone is around 20, and that of soft tissue (which is mostly water) is 7.4. If each number is cubed (multiplied by itself three times), we obtain the numbers 8000 and 405, respectively. Now, by making a ratio of these two numbers, dividing 8000 by 400, it can be found that bone is approximately 20 times more effective in attenuating x-rays than is soft tissue. This is because of the presence in bone of calcium, phosphorus, and other "larger" atoms with higher atomic numbers than water.

As it turns out, bone also has about twice the physical density of soft tissue. We learned in the last section that this is proportional to x-ray absorption, so bone will also absorb about twice as many x-rays due to its greater density. However, it is 20 times more absorbent due to its atomic number. The atomic number makes 10 times more difference than the density. We might state, then, that bone attenuates x-rays more *primarily* due to its high average atomic number.

SCATTERED X-RAYS AND SUBJECT CONTRAST

The effect of any scattered radiation reaching the image receptor, whether from the Compton interaction, the Thomson interaction, or the Rayleigh interaction, is to reduce the subject contrast carried by the remnant beam. This occurs because scatter radiation is completely random in its direction (Fig. 12-3), so that it lays down a "blanket" of useless exposure across an entire area of imaging plate. Precisely because of its random nature, this "blanket" of exposure is evenly distributed across the area, adding the same amount of exposure to different anatomical portions of the image. It is simple to mathematically demonstrate how this reduces subject contrast.

Let us assume two adjacent tissue structures within the patient, A and B, result in exposures to the image receptor measuring 2 and 4 respectively. Tissue B has allowed twice the number of x-rays to penetrate through it, and the exposure to the image receptor is twice as much in this region as the area under tissue A. The subject contrast between these two portions of the remnant x-ray beam is 4/2 = 2. Now let us add an equal amount of scatter radiation exposure, measuring 1, to *both* areas. Under tissue A the exposure is now 4 + 1 = 5; under tissue B it is 2 + 1 = 3. The subject contrast is now 5/3 = 1.66. Note

Figure 12-3

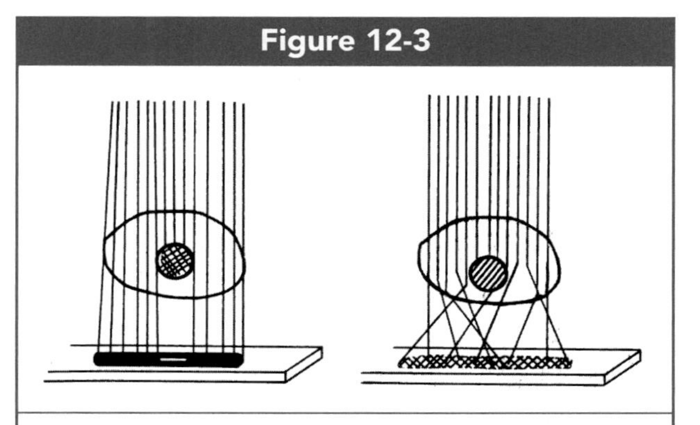

Diagram showing how subject contrast is lost in the remnant beam by scattered x-rays randomly crossing over image boundaries and laying down a "blanket of fog" exposure across the entire area of the image receptor. (From Quinn B. Carroll, *Practical Radiographic Imaging,* 8th ed. Springfield, IL: Charles C Thomas, Publisher, Ltd., 2007. Reprinted by permission.)

Figure 12-4

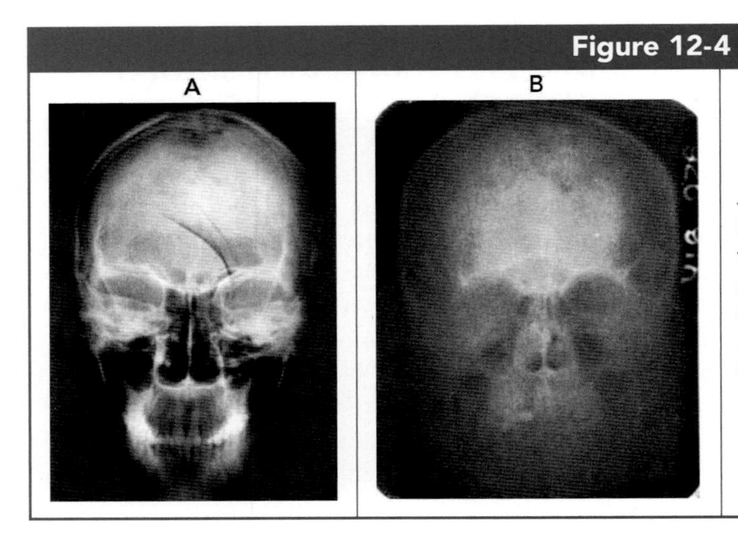

Film radiographs of the skull demonstrating the fogging effect of scattered radiation, ***B***, compared with a properly exposed radiograph, **A**. (From Quinn B. Carroll, *Practical Radiographic Imaging*, 8th ed. Springfield, IL: Charles C Thomas, Publisher, Ltd., 2007. Reprinted by permission.)

that the subject contrast with the scatter radiation has been reduced from 2 to 1.66.

	Normal Remnant Beam	Remnant Beam with Scattered Radiation
Exposure Under Tissue B	4	5
Exposure Under Tissue A:	2	3
Subject Contrast:	**2.0**	**1.66**

Figure 12-4 is a demonstration using film radiographs to illustrate visible fogging of an image from scatter radiation. Although modern digital processing is capable of eliminating most of this fogged appearance, it is important for the radiographer to understand that this is representative of what is going on *at the image receptor during exposure*—scatter radiation is laying down a "blanket" of exposure which constitutes *noise* in the image carried by the remnant x-ray beam.

PREDOMINANCE OF INTERACTIONS AND SUBJECT CONTRAST

X-Ray Beam Energy (kVp)

Changes in the energy levels of the x-ray beam, controlled primarily by the selected kVp, alter the penetration characteristics of the x-rays. Penetration is the opposite of attenuation. As kVp is increased and more penetration is achieved, the subject contrast between different tissues is lessened, but more different types of tissues can also be demonstrated between the extremes of black and white within the image. This is referred to as lengthened *gray scale.* As subject contrast is decreased, gray scale is increased. These effects are due to the penetration of the beam versus the overall attenuation factor, and will hold true regardless of the particular prevalence of the photoelectric effect or the Compton effect.

However, there are implications for the relative prevalence of these two interactions as we study subject contrast. Figure 12-5 is a graph showing the prevalence of the photoelectric interactions and the Compton interactions occurring in soft tissue at increasing levels of kVp. *Occurrence of the photoelectric effect is inversely proportional to the cube of the kVp.* For example, if the kVp is doubled, photoelectric interactions will decline to one-eighth, ($2 \times 2 \times 2 = 8$, inverted $= 1/8$).

The photoelectric effect only occurs at energy levels in which the kilovoltage of the incoming photon is slightly higher than the binding energy of an inner orbital shell. We might say that the inner shell electrons of an atom behave in a very *selective* manner in "choosing" which photons they will interact with.

The outer-shell electrons, on the other hand, are loosely-bound and are not selective either in which photons they will interact with nor in what percentage of those photons' energy they will absorb.

Occurrence of the Compton interaction is only slightly affected by kVp.

In Figure 12-5, upon increasing the kVp, note that although the curve for the photoelectric effect plummets precipitously toward zero, the number of *Compton* interactions decreases only slightly. Since higher kVp levels result in more penetration of the x-ray beam generally throughout the body, there are less interactions of all kinds at higher kVp's. But, since the photoelectric interactions are quickly lost, *the Compton interaction becomes the more prevalent interaction at higher kVp's.*

It is important to distinguish between the *raw number* of Compton interactions occurring within the patient, which goes down with higher kVp, and the prevalence of Compton interactions *as a percentage of all the interactions occuring*, which goes up. It means that even though there is less scatter radiation generated, that scatter reaching the image receptor makes a larger *percentage* contribution to the final image than the photoelectric interactions do. This is the primary significance of Figure 12-5.

To reinforce this concept, let's use the graph in Figure 12-5 to consider what happens more specifically within soft tissue when the set kilovoltage is increased from 40 to 80 kVp. In doing this, we will oversimplify the imaging process a bit by making the following generalizations:

1. Penetrating x-rays produce the "blacks" and darker areas in the image.
2. Photoelectric interactions produce the "whites" or lighter areas in the image.
3. Compton scatter lays down a "blanket" of fog at the image receptor.

By the graph, at 40 kVp the lines for photoelectric and Compton interactions are crossing over each other; there are roughly equal numbers of photoelectric and Compton interactions taking place (0.1 relative number by the scale). Therefore, about one-half of all interactions are photoelectric, producing light areas in the image, and the other half are scatter which lays down a "blanket" of fog at the image receptor. The graph does not include those x-rays that have penetrated all the way to the image receptor without undergoing any interaction, but we must include these in visualizing the final image produced. What we have in the 40-kVp image, then, is light areas and dark areas representing various soft tissue structures, and fog causing some image degradation.

Figure 12-5

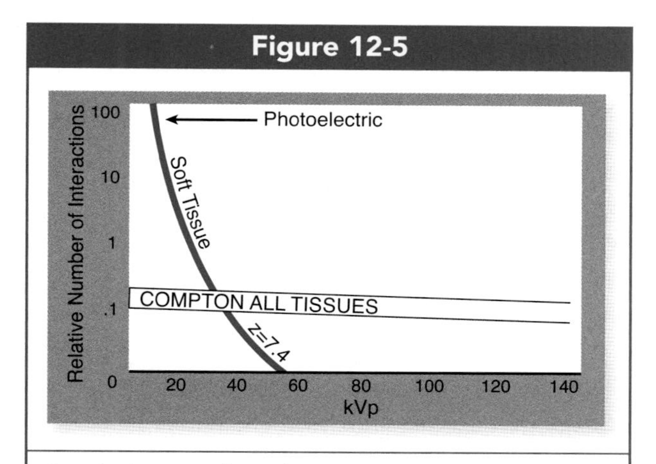

Graph showing the *relative* predominance of photoelectric and Compton interactions within soft tissue at increasing levels of kVp. All interactions decrease in number at higher energies; however, photoelectrics drop precipitously to zero, leaving Compton as the predominant interaction at higher kVp levels.

At 80 kVp this balance of image qualities is altered. By the graph (Fig. 12-5), at 80 kVp there are *no photoelectric interactions* occurring within soft tissue. No portion of the image under soft tissue structures will appear as a light shade. All that remains is dark areas from the penetrating x-rays, and fog from the scatter, which makes these areas darker still. Clearly, information has been lost from this image.

Using this same data for a 40-kVp image and an 80-kVp image, and a hypothetical amount of penetrating x-rays, we could itemize the percentage of the final image which each interaction contributed to, as follows:

Percent of Image A at 40 kVp	
Photoelectric Absorption:	40%
Compton Scatter:	40%
Penetrating X-Rays:	20%
Total:	100%

Percent of Image B at 80 kVp	
Photoelectric Absorption:	0%
Compton Scatter:	60%
Penetrating X-Rays:	40%
Total:	100%

For clarification, using our simplified scheme for translating these interactions into shades of gray in the final image, this table could be rewritten as follows:

Percent of Image A at 40 kVp	
Light Shades:	40%
Fog:	40%
Dark Shades:	20%
Total:	100%

Percent of Image B at 80 kVp	
Light Shades:	0%
Fog:	60%
Dark Shades:	40%
Total:	100%

(Modern digital processing is capable of correcting to an impressive degree for the lack of "lighter" densities in the image, but it is important for the radiographer to understand what is going on *at the image receptor during exposure.*)

Once again, what this analysis shows is that, even though fewer scattering interactions occur within the patient at higher kVp's, nonetheless they constitute a *greater percentage of the image* at higher kVp's; at 40 kVp, only 40 percent of all the information in the image was fog, but at 80 kVp, 60 percent of the image was composed of fog. This has less to do with how much scatter is produced within the patient than it does with the *precipitous decline and disappearance of photoelectric interactions at high kVp's*, whereby photoelectrics are no longer a part of the image information. The end result is that subject contrast is reduced.

It is equally important to appreciate the effects of *lowering* the kVp. When higher subject contrast between the various tissues is desired, lower levels of kVp should be used by the radiographer. As the kVp is reduced, the occurrence of photoelectric interactions within the patient will begin to *increase by a cube relationship*, and lighter areas will quickly be restored to the signal reaching the image receptor.

As will be fully explained in the next chapter, the proper amount of subject contrast is never the maximum nor the minimum achievable, but rather an intermediate level which results in both an adequate *range* of exposures to the image receptor to represent all tissues, and a sufficient difference between these areas of exposure to distinguish them one from another. One of the essential tasks of the radiographer is to find the best balance between these two considerations for each part of the human body, which will result in the most diagnostic information presented in the final image.

Types of Tissue and Contrast Agents

The photoelectric effect is likely to occur only when the energies of the incoming x-ray photons are *slightly* higher than the binding energies of the orbital shells within a particular tissue. As just discussed, one way to accomplish this is to *lower* the set kVp until the average energies in the x-ray beam are just above the binding energies in the tissue. But, another way this can be achieved is to *increase the binding energies* within the tissue until they are just under the average energies of the x-ray beam, without exceeding it.

It may seem impossible at first to do this, since we cannot change what human tissues are made of. What we *can* do in many cases is add contrast agents within the cavities of various organs, which will make them become visible when combined with proper radiographic exposure technique. When this is done, there will be a surge in photoelectric interactions and high subject contrast will be introduced back into the remnant radiation beam.

Figure 12-6 is a version of Figure 12-5 expanded to include bone tissues, and iodine as an example of a contrast agent. Barium, with an atomic number just higher than iodine, would have a similar curve in the graph in Figure 12-6. The student will immediately recognize that bone and iodine produce many more photoelectric interactions than soft tissue does. For example, at 40 kVp, soft tissue shows a relative number of photoelectric interactions measuring 0.1, whereas bone shows its relative number at about 12. This is 120 times more photoelectric interactions in bone than in soft tissue at this same low kVp.

The greatest number of photoelectric interactions is achieved when the kVp is low and the tissue atomic number is high. However, as explained in the next chapter, this must be balanced against the necessity for adequate *penetration* so that sufficient gray scale is also produced in the image.

Figure 12-6

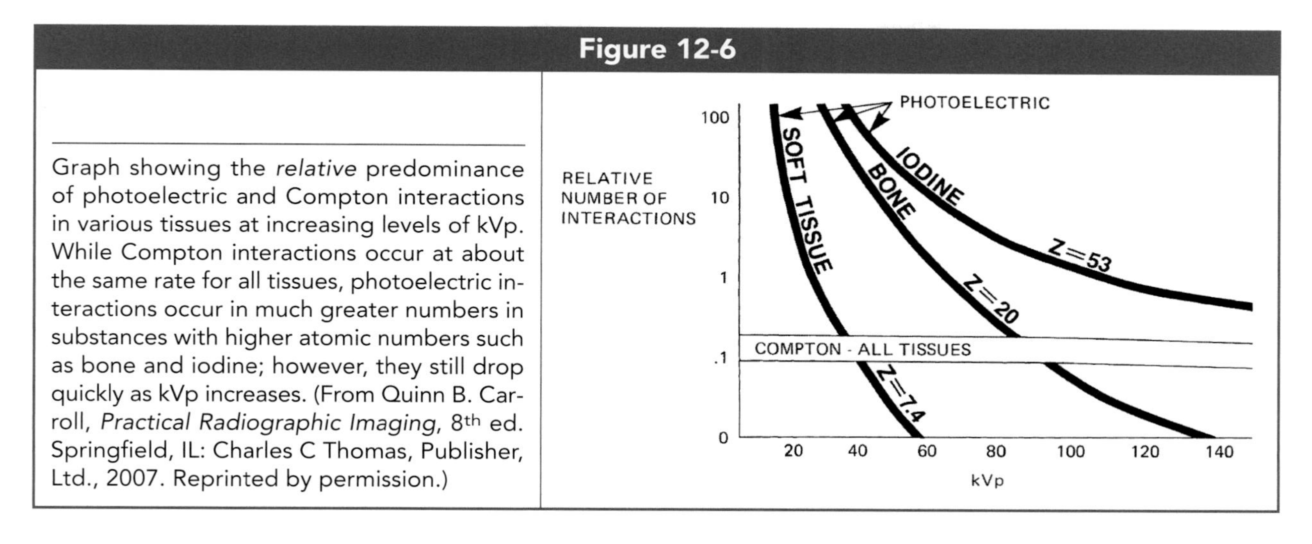

Graph showing the *relative* predominance of photoelectric and Compton interactions in various tissues at increasing levels of kVp. While Compton interactions occur at about the same rate for all tissues, photoelectric interactions occur in much greater numbers in substances with higher atomic numbers such as bone and iodine; however, they still drop quickly as kVp increases. (From Quinn B. Carroll, *Practical Radiographic Imaging*, 8th ed. Springfield, IL: Charles C Thomas, Publisher, Ltd., 2007. Reprinted by permission.)

Observe Figure 12-6 for photoelectric interactions at 80 kVp. As previously described, the soft tissues at this energy are only represented in the image by dark shades, having lost all of their photoelectric interactions. Bone, however, still has plenty of photoelectric interactions occurring at 80 kVp. It will still produce areas of light exposure to the image receptor, producing light shades in the image. Differences in bone tissues, such as that between the compact outer bone and the bone marrow, will be well recorded at 80 kVp, whereas differences between different soft tissue organs will not.

Let us now increase the kVp all the way up to 120, perhaps for a solid-column barium enema examination, and see what fate awaits the bone image at this very high kVp: Checking the graph in Figure 12-6, we see that at 120 kVp, even for bone, the photoelectric interactions are almost gone. Behind the bones, exposure to the image receptor plate now consists almost entirely of penetrating x-rays and scattered x-rays. In the image, bones will be represented only by dark shades unless corrected by computer algorithms. It will be harder to distinguish between cancellous bone and bone marrow, which are now both represented by dark shades in the image.

Contrast agents utilizing iodine or barium still produce lots of photoelectric interactions even at 120 kVp, as shown by the graph (Fig. 12-6). Compton scatter interactions are essentially constant for all tissues. This is because they occur only in the outermost shells of the atoms, which always have extremely low binding energies. Changes in atomic number do not affect these outer shells much because of their distance from the nucleus. Furthermore, their electrons are not "selective" in which x-ray photons they interact with, since they can absorb any percentage of the photon's energy. Thus, the likelihood of a Compton scatter interaction occurring in these outer shells is about the same for all types of tissues.

Relative Importance of kVp in Controlling Subject Contrast

As discussed under *The Compton Effect* (Chapter 11), at higher levels of kVp a greater percentage of generated scatter is directed forward to reach the image receptor. However, this is substantially offset by the fact that less scatter radiation is produced in the first place. Which effect has the larger impact on the end result? The directing of scatter more forward has a slightly greater effect than the drop in the initial production of the scatter. The overall result is that at higher kVp's, the image receptor receives slightly more scatter radiation than at lower kVp's.

It is important to understand that conventional radiography education has tended to overexaggerate the impact of kVp on the final image. That is, high kVp is a relatively minor factor in the production of scatter radiation when compared with other variables such as the collimated field size and the thickness of the patient. (See the Historical Sidebar for further explanation.)

HISTORICAL SIDEBAR 12-1: Scatter Radiation and kVp: With old-fashioned radiographic film, in order to demonstrate a visibly notable increase in fogging of the image due to higher levels of scatter radiation, it was necessary to increase the kVp by as much as 100 percent or double, for example, from 60 kVp to 120 kVp.

But we know by the 15 percent rule that it only takes a 15 percent increase in kVp to double the exposure to the film or imaging plate. This could be done *without any visible increase in image fog* for an average abdomen radiograph, as shown in Figure 12-7, not to mention smaller anatomical parts.

Therefore, kVp could be used to double the density of the image without a visible increase in fogging. This proves that regardless of the traditional teaching that higher kVp's "cause fog," kVp is in fact a minor contributor to scatter at the image receptor. The main causes of scatter are large field sizes, thicker body parts and the types of tissue irradiated.

Figure 12-7

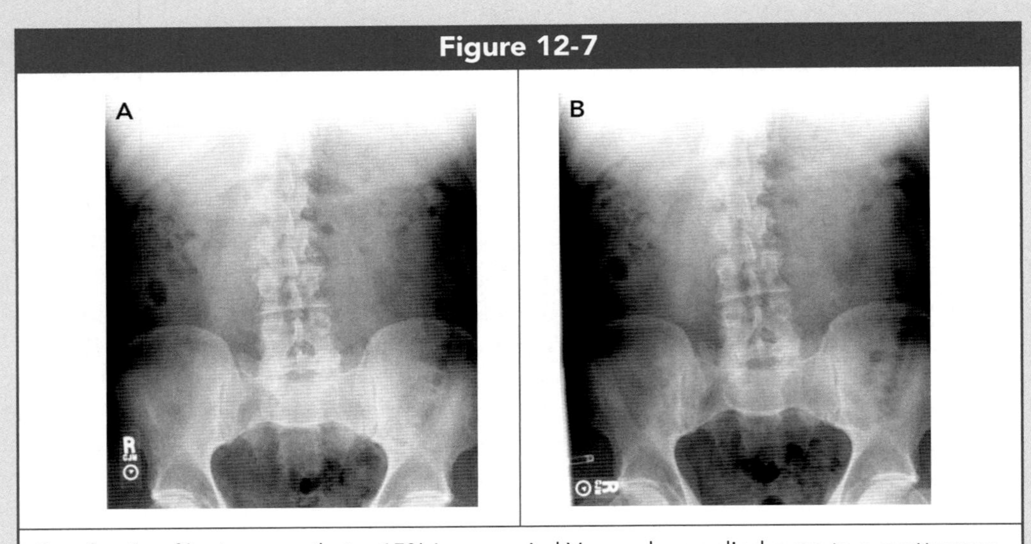

Proof, using film images, that a 15% increase in kVp can be applied even to a scatter-producing abdomen radiograph without visible fogging of the image. Radiograph A was taken at 80 kVp and 40 mAs. Radiograph ***B*** was produced with 92 kVp and on-half the mAs. A slight lengthening of the gray scale is apparent *due to increased penetration*. But, there is no visible fogging of this image (compare to Figure 12- 4). (From Quinn B. Carroll, *Practical Radiographic Imaging*, 8th ed. Springfield, IL: Charles C Thomas, Publisher, Ltd., 2007. Reprinted by permission.)

SUMMARY

1. All interactions attenuate the x-ray beam, and thus contribute to the production of subject contrast.
2. Subject contrast is defined as the *ratio* of x-ray attenuation between two adjacent tissues.
3. For thicker tissues, x-ray attenuation increases exponentially, approximately doubling for every 4–5 cm.
4. X-ray attenuation increases in direct proportion to the physical density of tissue. Therefore, extreme changes in density are required to make a visible difference between tissues in the radiographic image.
5. X-ray attenuation increases by the cube of the molecular (or average) atomic number of the tissue.
6. The subject contrast of the image carried by the remnant x-ray beam is reduced by scatter radiation, which lays down a "blanket" layer of noise across the latent image.
7. Occurrence of the photoelectric effect is inversely proportional to the cube of the kVp. Therefore, at high kVp's photoelectrics are lost, leaving Compton scatter as the prevalent interaction in forming the image, and penetrating primary x-rays.

8. Because of the above, at higher kVp's what would be the *light* densities in the image are lost at the image receptor. This represents a loss of subject contrast. The use of lower kVp's restores the subject contrast.
9. *Positive* contrast agents utilize elements with high atomic numbers, such as iodine and barium, because their high electron densities result in greatly enhanced absorption of x-rays. Negative contrast agents are usually gases and are useful in creating subject contrast primarily because of their extremely low physical density.
10. The greatest subject contrast is achieved when high atomic number tissues are combined with low kVp levels.
11. Although kVp does affect subject contrast, its impact is relatively minor when compared to field size and patient size.

REVIEW QUESTIONS

1. Define subject contrast as it pertains to the remnant x-ray beam reaching the image receptor:

2. Of the three types of x-ray interactions that occur within the patient, which ones contribute to subject contrast in the image carried by the remnant beam?

3. Of the three aspects of a tissue, which one has a *proportional* relationship to x-ray attenuation?

4. Of the three aspects of a tissue, which one must be altered most dramatically to cause a substantive change in subject contrast?

5. Atoms with high atomic numbers are more likely to absorb x-rays because of their high __________ density.

6. Scatter radiation reduces subject contrast because it adds __________ amount of exposure to all local areas of the image.

7. Occurrence of the photoelectric effect is proportional to the __________ of the kVp.

8. As kVp is increased, which way and by how much does the occurrence of the Compton effect change?

9. Ultimately, the subject contrast in the image carried by the remnant beam is dependent, not upon the raw number of the different x-ray interactions, but on their __________ in contributing to the image.

(Continued)

REVIEW QUESTIONS *(Continued)*

10. Which type of interaction is not occurring at all in soft tissues at 90 kVp?

11. Combining the effects of physical density and atomic number on x-ray attenuation, what is the approximate *total subject contrast ratio* between bone and soft tissue?

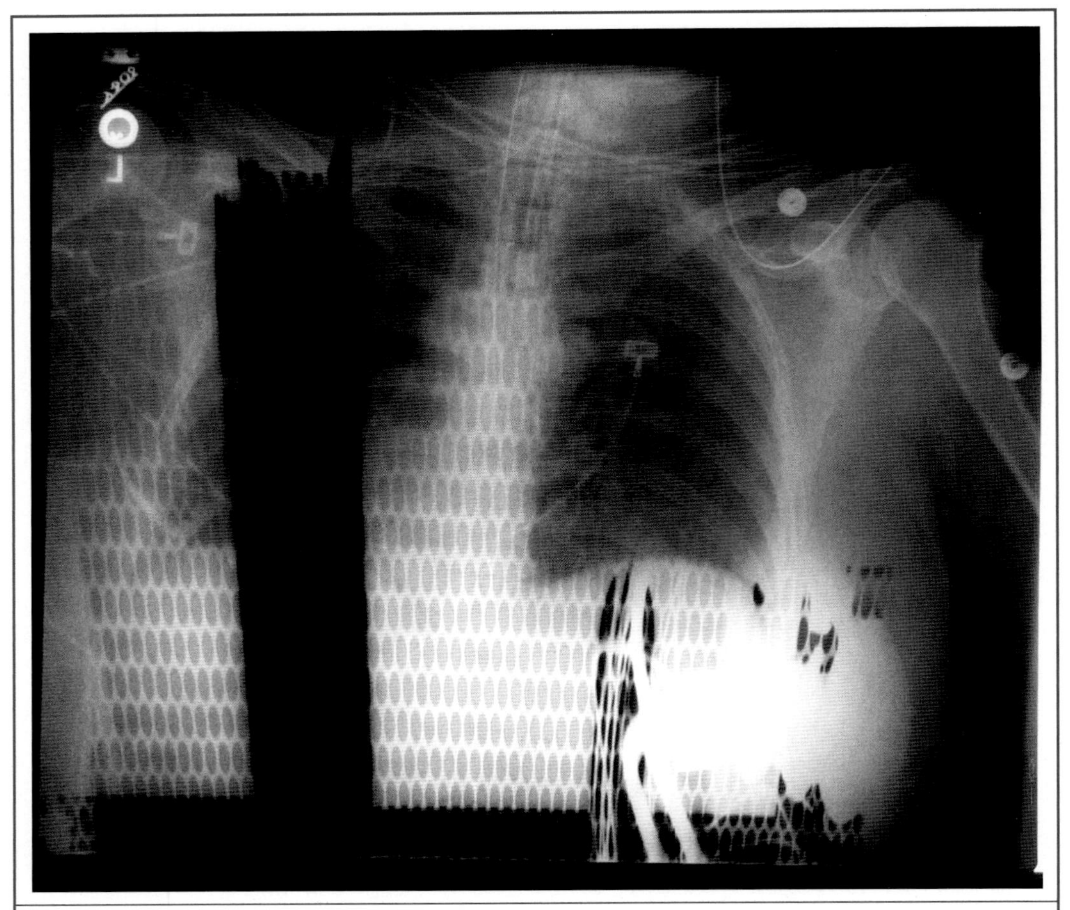

These artifacts were caused by a heating blanket left behind the patient for a chest radiograph.

Chapter 13

VISIBILITY QUALITIES OF THE IMAGE

Objectives:

Upon completion of this chapter, you should be able to:

1. List and define the three *visibility* components of every image.
2. Describe the effects on details for both excessive contrast and insufficient contrast.
3. Identify the best level of brightness (density) for a radiograph, and why sufficient penetration of the x-ray beam is critical to achieve it, especially when contrast agents are used.
4. Define radiographic contrast and gray scale, and how they interrelate.
5. Demonstrate that brightness and contrast are independent of each other as image qualities.
6. Define radiographic noise and give examples.
7. State the two ways that the *signal-to-noise ratio (SNR)* can be improved.
8. Define *artifacts* and give examples.

THE COMPONENTS OF EVERY IMAGE

What are the essential requirements for your eyes to see? They are more complex than you might imagine. Of course, the first necessity is the presence of light. It is noteworthy to consider that light is a form of electromagnetic radiation just as x-rays are. In fact, the only real difference between x-rays and light is the length of their waves and the corresponding energies.

What you actually see is reflected light photons striking the nerve endings in the retina of your eye. When more photons per second strike your eye, the image you see appears brighter. It is more intense. Thus, the rate of the flux of photons is referred to as the *intensity* of the light. When the intensity is very low, you cannot recognize your surroundings because much of the information, although it is present, is not *visible*.

Yet, light alone does not provide any image within your field of vision. Suppose that everything in your field of view was the same white light and all at the same intensity or brightness. You would be just as blind seeing all blank white as you would be seeing only pitch darkness. In order to see any image, there must be *differing shades* of brightness and darkness within your field of view.

The greater the difference between the shade of an object and the shade of the background surrounding it, the easier it is to see the object. A pure black object against a white background, for example, is a most visible combination. It represents almost no light being reflected from the object, and nearly all light being reflected from the background. This difference between the intensities of reflected light is called *contrast.* For contrast to exist, there must be two or more different intensities of light present. If these differences are great, then the image is regarded as having *high contrast.* Some contrast is required for any image to exist. If there is not enough contrast between adjacent objects in an image, one cannot tell that they are separate, distinct objects from each other.

However, it is also possible for an image to have *too much* contrast in it. Figure 13-1 illustrates both extremes using a black-and-white photograph. Photograph ***B***, in the middle, presents a medium range of contrast to which the other two images can be

Figure 13-1

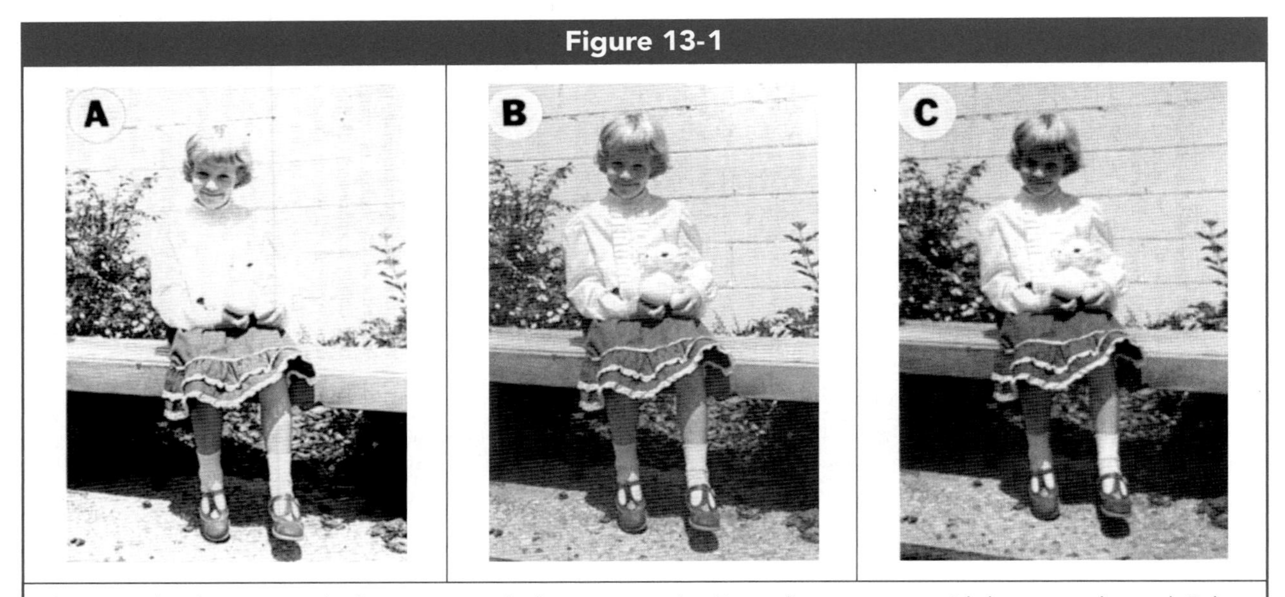

Photographs showing, ***A,*** high contrast with short gray scale, ***B,*** medium contrast with longer scale, and ***C***, low contrast with excessive gray scale. Note that details visible in the stuffed animal, house and sidewalk disappear with excessive contrast (***A***), while details in the shadow under the bench and the plants are lost with excessive gray scale (***C***). Note also that the excessive contrast in ***A*** can mimic underexposure at a glance, when the shadow under the bench is actually darker than that in ***B***. (From Quinn B. Carroll, *Practical Radiographic Imaging*, 8th ed. Springfield, IL: Charles C Thomas, Publisher, Ltd., 2007. Reprinted by permission.)

compared. Photograph ***A*** is a very high contrast image; note that you cannot distinguish between the edges of the girl's blouse and the background wall, the edges between the stuffed animal and the girl's blouse, or the ruffles in the blouse. These details are visible in photograph ***B***. With excessive contrast, as in photograph ***A***, everything tends to be reduced to black or white, with very few intermediate shades of gray. The result is that objects which would have been depicted as a very light gray, such as the ruffles and shadows of the blouse, are recorded instead as white and cannot be discerned against a background of white. Likewise, dark gray objects are depicted as black and will not stand out against a background of black. This means that some details in the image are lost. Note that in Figure 13-1, photograph ***A*** has fewer visible details than photograph ***B***. *Excessive amounts of contrast cause a loss of useful information, a loss of details in the image.*

On the other hand, insufficient contrast also leads to an image in which details are not adequately visible. If there is barely any difference between two adjacent shades of gray, it will be difficult for the human eye to detect that there are indeed two shades there. When these gray shades cannot be distinguished from each other, information is again lost. Photograph ***C*** in Figure 13-1 demonstrates insufficient contrast, with an overall gray appearance. Note that the subtle dark gray shades of leaves under the bench, visible in photograph ***B***, are not as visible in photograph ***C*** because there is little difference between them and the dark shadow of the bench. The different pebbles used in the sidewalk are also less apparent in photograph ***C*** than in photograph ***B***.

To summarize, the ideal amount of contrast in any image lies in an intermediate range. The word *optimum* is used to describe such an intermediate level which is neither too much nor too little. A common misconception for medical imaging students is to assume that the higher the contrast, the better the image. In radiographic imaging, the goal is not to produce *maximum* contrast nor minimum contrast, but *optimum contrast.*

The same principle holds true for the brightness of light in our everyday vision; with too little light, details cannot be seen, yet with too much light we are blinded—our eyes are overwhelmed with the intensity and cannot make out details in this condition either. The ideal amount of light for human vision is an intermediate level, an *optimum intensity*.

Finally, one last factor affects the visibility of an object in your field of view: If it is raining between yourself and the object you are trying to see, you may not be able to see it well in spite of sufficient lighting. The rain represents unwanted "information" which obstructs the wanted information. Any undesirable input that interferes with the visibility of the subject of interest is referred to as *noise*.

Interference, "snow" and static on your TV screen are good examples of noise in an image. These electronic types of noise also affect radiographic images brought up on a TV screen, and are added to any artifacts in the way of the image receptor which obscure the anatomy of interest, and scatter radiation which was discussed extensively in the last chapter. In the fundamental production of any radiographic image, the control and minimizing of scatter radiation is an ever-present concern for the radiographer in eliminating image noise.

These three factors—intensity, contrast, and noise—are the *visibility factors* of an image. Maximum visibility of all image details is attained when intensity and contrast are both optimum and noise is kept to a minimum.

QUALITIES OF THE RADIOGRAPHIC IMAGE

The image on a radiograph has all the same qualities as a visual or photographic image—intensity, contrast, noise, sharpness of detail, magnification and shape distortion. But, since radiography works somewhat differently than your eyes, some of these qualities are given different specific names. Some distinctions must also be made between the electronic image seen on a TV screen and images printed out as "hard copies." An in-depth understanding of these characteristics of the radiographic image is at the heart of the practice of radiography.

Brightness and Density

In the digital age, a diagnostic image is examined first and foremost as an electronic image displayed on a liquid crystal diode (LCD) or cathode ray tube (CRT) monitor. The brightness of this image can be adjusted upward or downward at the display monitor, but the image itself, as it is stored within digital memory, also possesses an inherent brightness level that is based upon both the radiographic technique initially used to produce the image and the computerized processes to which it has been subjected. This inherent brightness of the stored image is independent of the *display* brightness of the observed image when it is brought up on a particular viewing device.

In the previous section we discussed how the intensity of a visual image must neither be too low nor too high, but at an *optimum* level for the maximum amount of information to be conveyed and perceived. In the displayed image, the brightness level of all pixels (picture elements) *within the anatomy of interest* should be neither completely white nor pitch black, but should fall within a broad range of intermediate "gray shades" from a very light gray to a very dark gray.

An intermediate level of brightness indicates that there was proper attenuation of the x-ray beam, with some x-rays penetrating through the tissue and some being absorbed. Where the image is pitch black, virtually all x-rays have penetrated through to the image receptor, and where it is blank white, no x-rays have reached the imaging plate. Under such conditions, only a silhouette image can be produced, like the one in Figure 13-2. The information in a silhouette image is limited to the edges of the anatomy. For a diagnostic image, we wish to see *all* of the anatomy of an organ, including the "front and back," and the inside of it, too, rather than just the edges. To achieve this, a portion of the x-ray beam *must penetrate through* the organ. Such partial penetration always yields some "shade of gray" along the range of image brightness.

High-quality "hard-copies" of diagnostic images can be printed onto transparent film using laser printers. This allows the image to be viewed on an illuminator or "viewbox" which shines light through the image from behind the film. Somewhat lower-quality images can also be printed onto white paper. For these hard-copy images, the overall darkness of the image is often referred to as its optical *density*. Density is the opposite of brightness, but conveys the same overall concept. Individual dark spots within a printed image are also commonly referred to as *densities*. A device called a *densitometer* can be used to

Figure 13-2

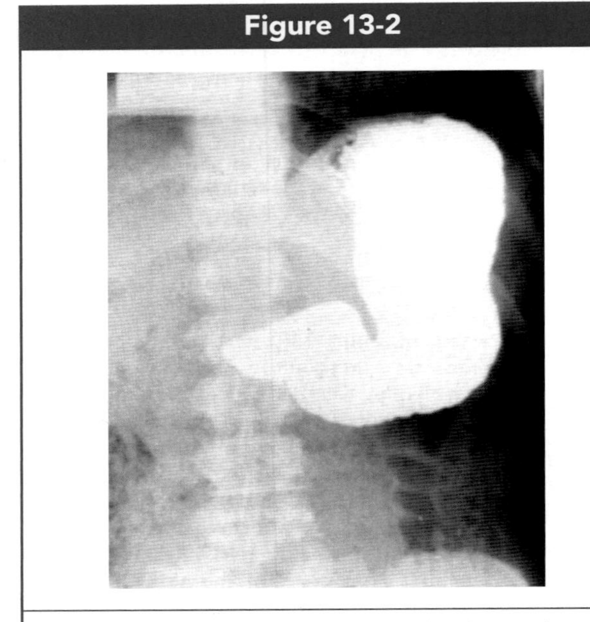

Radiograph of barium-filled stomach taken with only 80 kVp. With inadequate penetration of x-rays through the barium, the only details visible in the stomach are those at the very edges of the barium bolus. Such a "silhouette" image is of little diagnostic value. (From Quinn B. Carroll, *Practical Radiographic Imaging*, 8th ed. Springfield, IL: Charles C Thomas, Publisher, Ltd., 2007. Reprinted by permission.)

measure the darkness of different densities for a hard-copy radiograph that is hung on an illuminator.

Most radiographic images, whether electronic or printed, are *negative images*, meaning that they essentially consist of light details against a dark background. A positive image, such as the print on this page, consists of dark details against a light background. A "black bone" or image reversal feature is available on nearly all computerized imaging systems which allows the radiograph to be displayed as a positive image (Fig. 13-3). Technically, there is no more information present in the image whether it is displayed as a positive or as a negative. Sometimes while "reading" the image, the radiologist can subjectively see a particular detail better when the image is displayed as a positive. However, the convention and the preference continues to be that radiographs are generally presented in the negative format, in which areas having received high x-ray exposure are displayed as dark regions and those tissues which have absorbed x-rays are displayed as lighter areas.

Figure 13-3

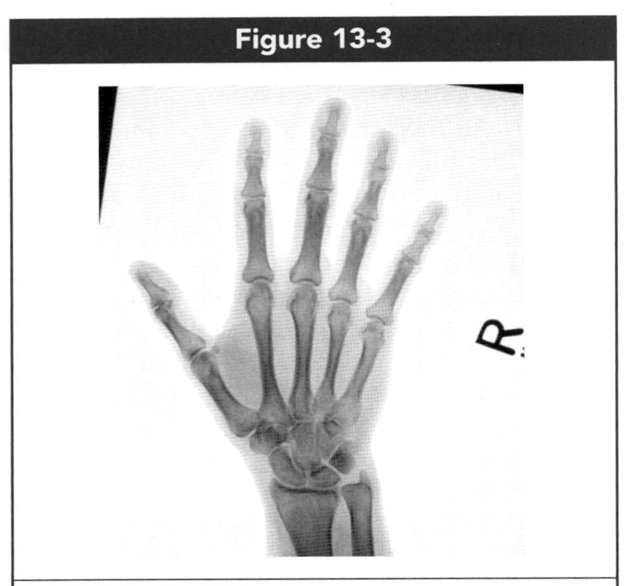

Most radiographic images are *negative*, "white on black" images, but can be reversed as with this positive "black bone" digital processing feature. (From Quinn B. Carroll, *Practical Radiographic Imaging*, 8th ed. Springfield, IL: Charles C Thomas, Publisher, Ltd., 2007. Reprinted by permission.)

Figure 13-4 demonstrates three radiographs of a lateral knee with extreme variations in brightness. In radiograph ***A***, details within the bones are missing due to excessive image brightness. Radiograph ***C*** is also missing image details because it is too dark. The correct optimum brightness in radiograph ***B*** presents a range of gray shades from dark to light within the bones.

Figure 13-4

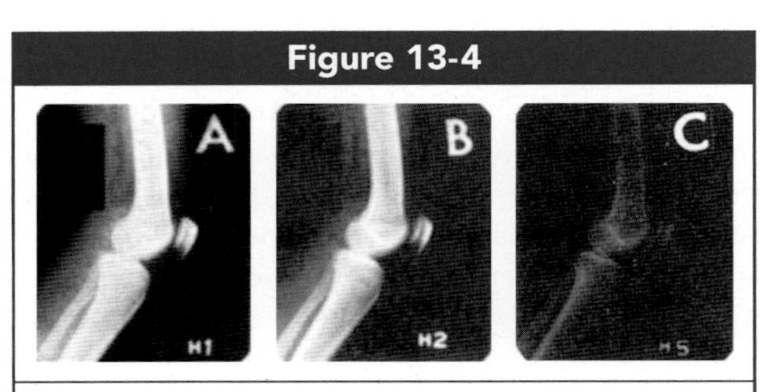

Lateral knee radiographs demonstrating ***A***, excessive brightness; ***B***, *optimum* or intermediate brightness; and ***C***, insufficient brightnesss. (From Quinn B. Carroll, *Practical Radiographic Imaging*, 8th ed. Springfield, IL: Charles C Thomas, Publisher, Ltd., 2007. Reprinted by permission.)

Contrast and Gray Scale

Radiographic contrast is the proportional difference, or ratio, between the brightness of two adjacent details. It is measured as a ratio, dividing the brightness of the lighter detail into that of the darker one (not as a subtracted difference). If one detail appears twice as dark as the one next to it, the contrast is 2/1 = 2.0.

A very contrasty image appears to be more "black and white," whereas a lower contrast image appears somewhat gray overall. Figure 13-5 shows how image contrast is generally *independent* of image brightness: Image ***A*** is a brighter image of two adjacent tissues, for example bone and soft tissue. When these are measured from the hard copy by a densitometer, the soft tissue area measures 2.0, and the bone area reads out at 1.0. The contrast for this image is 2/1 = 2.0. For image ***B***, the brightness is then turned down to a level that appears twice as dark overall, and a hard copy is again measured using a densitometer. All densities in the image have been doubled in their darkness. The soft tissue area now measures 4.0 and the bone area measures 2.0. Note that the original contrast has not changed; it is now 4/2 = 2.0.

A common misconception is that a "darker" image will automatically have less contrast. This false idea originates primarily from comparing densities within the anatomy to the *background* density of the image, outside the anatomy, where there has been no attenuation of the x-ray beam. This background is always pitch black, and has nothing to do with the contrast present between anatomical structures *within the body part*. It is important not to confuse the background with the contrast present within the image itself.

It is true that, as an image is made darker and darker, there will indeed be a point where the darkest areas reach a "pitch black" appearance and cannot become any darker *visually to the human eye*. At this stage, as other densities approach a dark gray, there will obviously be less difference between those densities and the pitch black ones, and contrast will be reduced. This represents a rather extreme scenario, and the important point is that it is not *always* the case.

Within a diagnostic range, images *can* be made brighter or darker without affecting contrast, as demonstrated in Figure 13-5. There are several variables in radiography which affect both the image brightness and its contrast at the same time, but this does not make them *related* qualities. The image contrast itself should be thought of as an image quality separate from the brightness of the image.

Gray scale is the range of different densities (brightnesses) within an image. Gray scale is sometimes called *contrast scale*, but this term will not be used here because it tends to be confused with contrast. *Gray scale is opposite to image contrast.* Where there are many different shades of brightness within an image, it is said to have *long gray scale*. Conversely, an image with only a few levels of brightness present has *short gray scale*. Remember that this term refers to the *range* of densities or brightness.

To better understand the opposing relationship between gray scale and contrast, consider a staircase that leads up 10 feet from one floor to the next. The staircase may be built to have 10 individual stairs, each one being a foot in height. But, it could also be built to have only 5 individual stairs—in this case, each stair must be 2 feet high, (and rather awkward to climb)! There is a greater difference between each stair. The greater the difference from one stair to the next, the fewer stairs there must be.

On a radiographic image, the brightnesses or densities ranging from blank white to pitch black are analogous to the two floors in our staircase example—there is only so much distance to cover between them. When only a few levels of brightness (or densities) are present, we would say that the range is limited and the gray scale is short. In this condition, the difference from one shade to the next will be large—

Figure 13-5

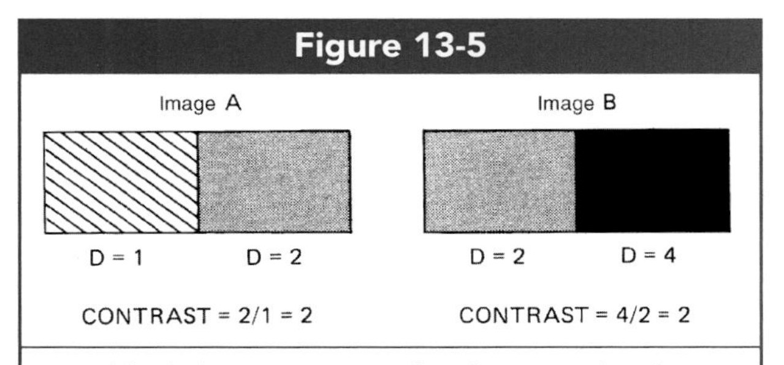

Simplified demonstration of a change in brightness *without* a change in contrast. The ratio of difference between the two adjacent densities remains the same (2) even though image ***B*** is twice as dark overall. An image can be darkened while retaining its contrast. (From Quinn B. Carroll, *Practical Radiographic Imaging*, 8th ed. Springfield, IL: Charles C Thomas, Publisher, Ltd., 2007. Reprinted by permission.)

this constitutes high contrast. When there are many shades present progressing from light to dark, (long gray scale), the differences between them must be small (low contrast).

This is why gray scale and contrast are opposite image qualities. It is very useful, though, to be able to use both terms in discussing radiographs. Furthermore, as we shall find, even though they are opposites, *both* can be simultaneously destroyed by degrading factors such as scattered radiation.

Just as the ideal level of image contrast is an intermediate range (Fig. 13-1), so the gray scale of an image is also optimized at an intermediate level. With excessive gray scale, there will be so little difference between details that it will be difficult to tell them apart. With too short a gray scale, differences between details is exaggerated, reducing them to black or white, such that details are lost from the image.

Short- and long-gray scale radiographs are demonstrated in Figures 13-6 and 13-7, along with step-wedge images that correlate to the staircase analogy used above. In Figure 13-6, long wavelength (low kVp) x-rays were used, producing a short gray scale which can be seen in the image of the aluminum step wedge. The corresponding chest radiograph below, while it demonstrates details in the lungs, forms only a silhouette image of the heart, mediastinal and neck structures. The sternal ends of the clavicles cannot be made out. This is excessive contrast, and very short gray scale.

In Figure 13-7, short wavelength (high kVp) x-rays were used which have penetrated through more of the various tissues within the chest. The step-wedge image shows long gray scale. The chest radiograph below still shows many of the details within the lungs, but in addition to these information is also provided within the radiographic "shadows" of the diaphragms, heart, aorta and bronchial roots; the vertebrae of the neck and upper thorax can now be distinguished along with the air in the trachea running down the middle. The sternal ends of the clavicles (collar bones) can now be identified. (You may note that one clavicle has a displaced fracture.) This is long gray scale, but not excessive. This lower-contrast image has much more information in it than the chest radiograph in Figure 13-6. Figure 13-8 provides some additional examples of short-gray scale images of different body parts, while Figure 13-9 gives several samples of long-gray scale images for comparison.

Figure 13-10 is a series of magnetic resonance images of the spine to further illustrate these important concepts. The contrast for images ***A*** and ***B*** was fixed at the same setting to produce very short gray scale. ***A*** was printed out with a lighter brightness setting and *B* with a much darker setting. Note that *the use of a darker brightness setting does not restore image details that are missing.* In fact, in this case, details were lost. Image ***C*** was printed up with the same brightness setting as ***A***, but with a reduced contrast setting. More details are now visible in the image, including the spinal cord within the spinal canal and distinct muscle and fat tissues behind the spine. Image ***D*** was printed out with the same contrast setting as ***C*** and the same brightness setting as ***B***. It is a darker image than ***C*** but with the same gray-scale. Overall, this image has the most information in it: Although the darkest portions of the vertebral bodies have lost some detail, differences within the vertebral bodies in the darker tissues in front of the spine (left) can still be made out, yet *more* details are clearly seen behind L4 and L5 around the pathology.

The most ideal setting for these MRI images would keep the gray scale of ***C***, but with an intermediate brightness level between ***C*** and ***D***. *Both* ***C*** *and* ***D*** provide vastly more information than either ***A*** or ***B***. *When the gray scale is not correctly adjusted, changes in brightness will not correct for it.*

When these aspects of an image are not already optimum, ***the gray scale and contrast of an image should always be adjusted first, followed by adjustments in the overall brightness***.

Image Noise

Noise may be defined as any undesirable input to the image that *interferes* with the visibility of the anatomy or pathology of interest. The most common form of noise in producing the initial image at the image receptor is radiation *scattered* by interactions discussed in the previous two chapters. X-rays can be scattered from the patient, the tabletop or other objects within the x-ray beam, and are emitted in all directions much like light is dispersed by a mist of fog (Fig. 13-11). The evenly-distributed blanket of unwanted exposure that results at the image receptor reduces the subject contrast of the

Figure 13-6

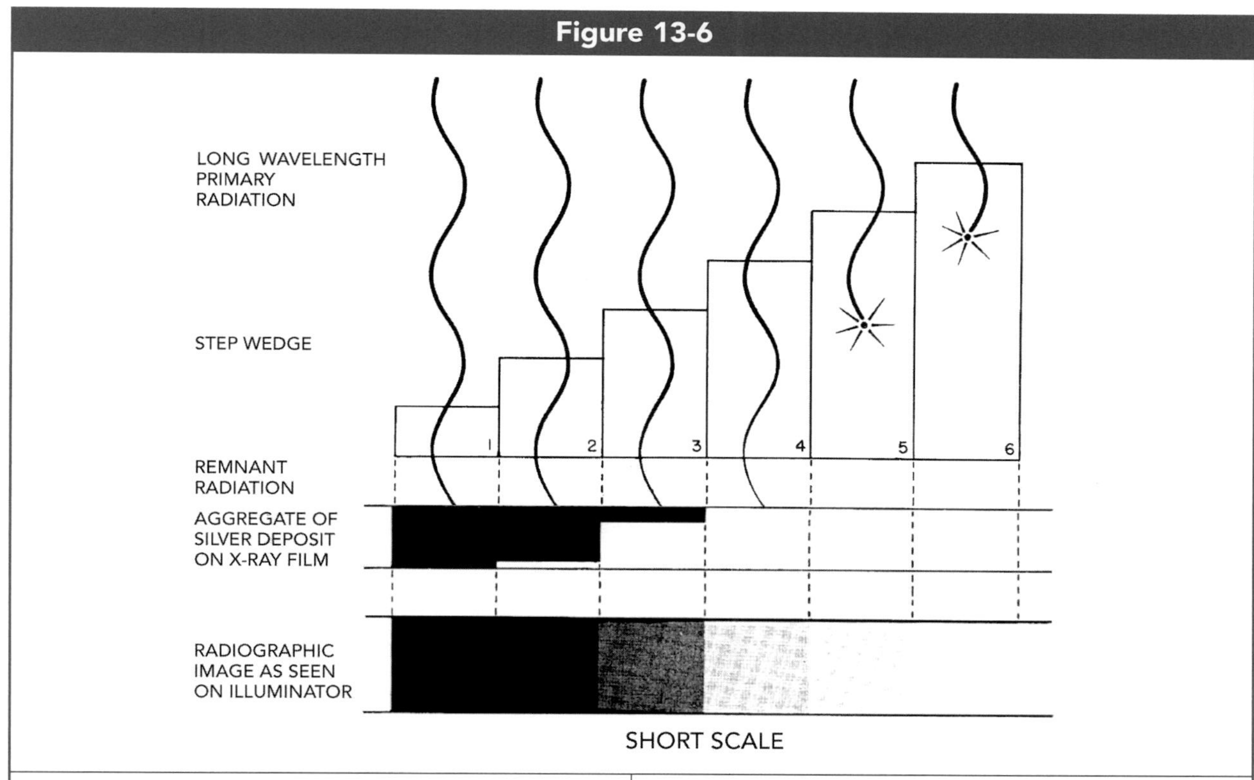

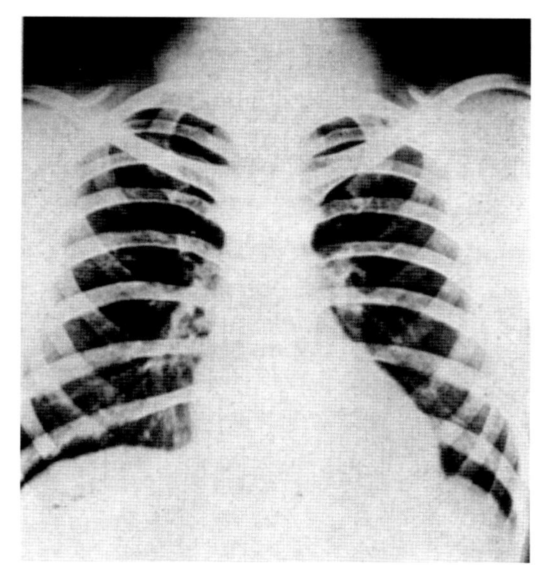

Step wedge diagram and PA chest radiograph illustrating short gray scale from low x-ray penetration. (From Quinn B. Carroll, *Practical Radiographic Imaging*, 8th ed. Springfield, IL: Charles C Thomas, Publisher, Ltd., 2007. Reprinted by permission.)

remnant beam and degrades the final image. (*See* Historical Sidebar 13-1.) Methods for minimizing scatter radiation are discussed in a later chapter.

Another common form of noise for electronic images is "snow" or electrical static that appears on the displayed image. This is caused by "background" fluctuations, small surges and dips, in electrical current which are present in any electronic system. Their control and prevention will be discussed later.

In the digital age, image *mottle* has far exceeded "fog" from scatter radiation as the more common form of noise appearing in the final displayed image.

Figure 13-7

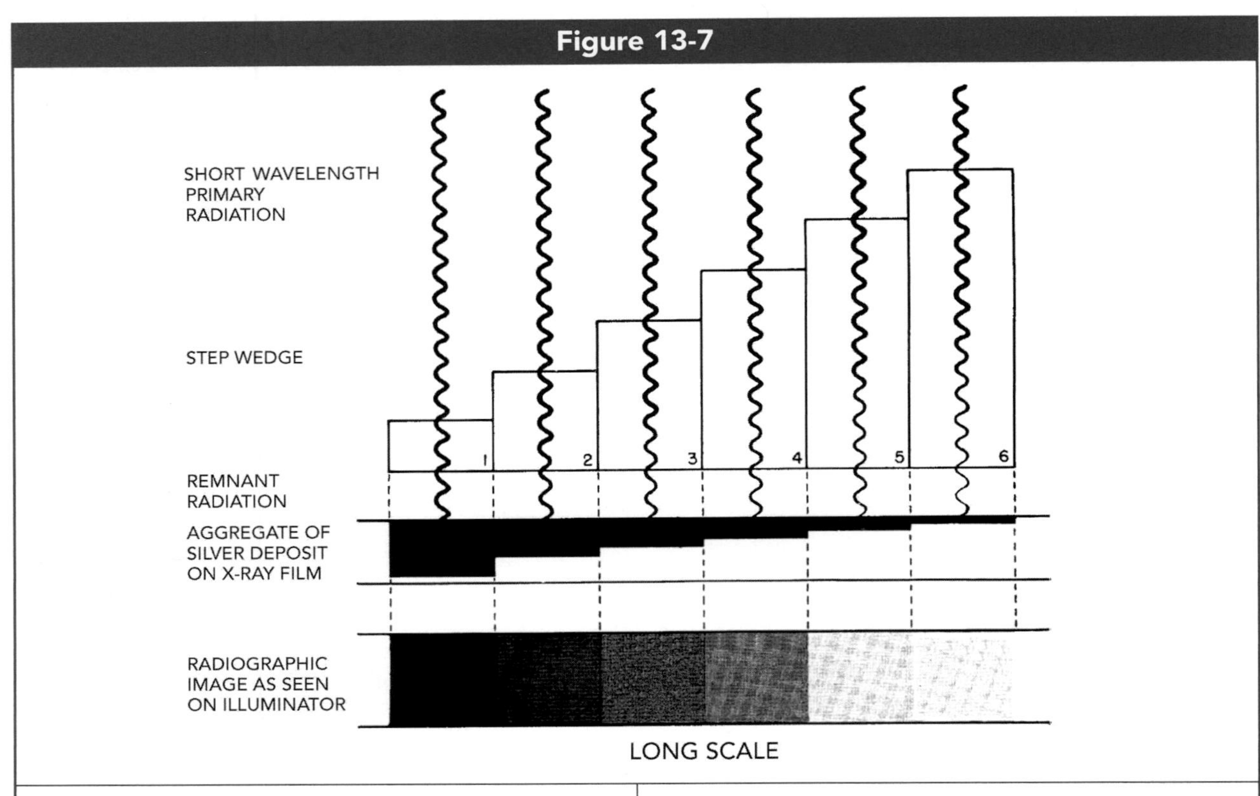

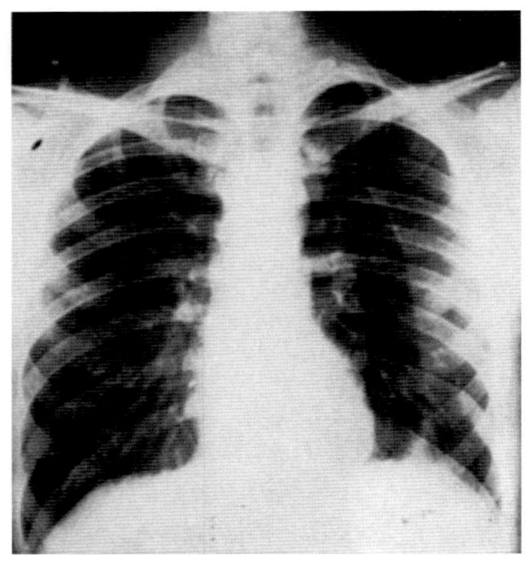

Step wedge diagram and PA chest radiograph illustrating long gray scale from high x-ray penetration. (From Quinn B. Carroll, *Practical Radiographic Imaging*, 8th ed. Springfield, IL: Charles C Thomas, Publisher, Ltd., 2007. Reprinted by permission.)

Mottle appears when any combination of low x-ray beam penetration (kVp), low beam intensity (mAs), and unusually thick or dense anatomy result in an insufficient amount of useful *signal* (discussed next) reaching the image receptor. The types and causes of mottle will be fully explored in subsequent chapters.

Signal-to-Noise Ratio

One of the best methods of measuring the overall visibility of information in an image is the signal-to-noise ratio (SNR). The *signal* refers to all of the desirable information carried by the mechanism of

Figure 13-8

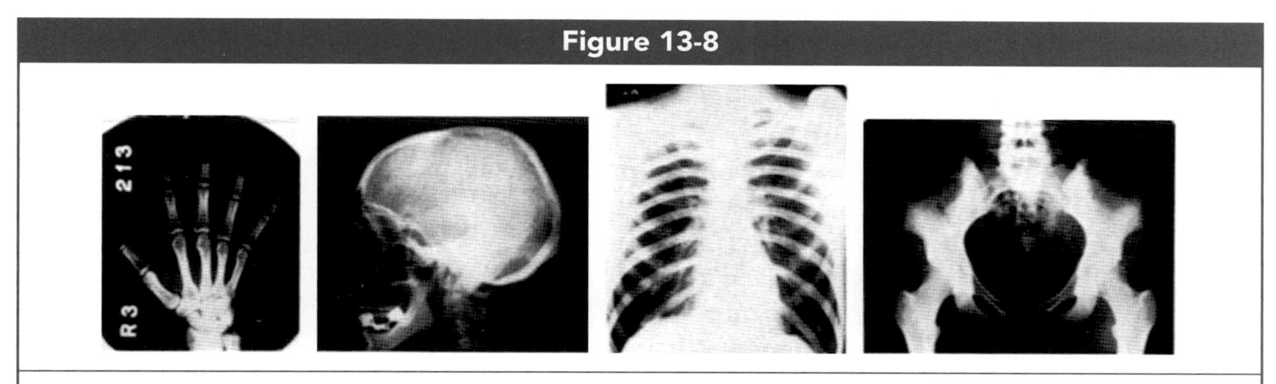

Typical radiographs exhibiting various degrees of short gray scale. (From Quinn B. Carroll, *Practical Radiographic Imaging*, 8th ed. Springfield, IL: Charles C Thomas, Publisher, Ltd., 2007. Reprinted by permission.)

Figure 13-9

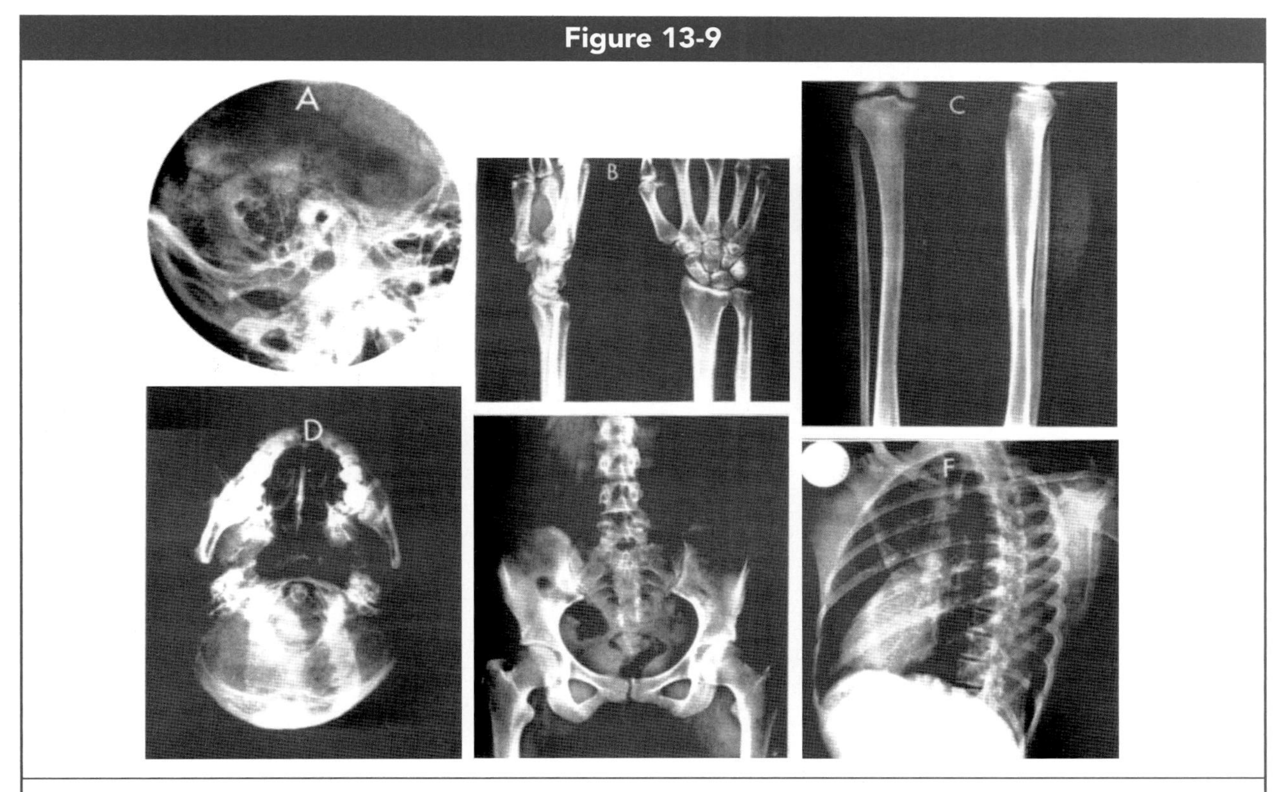

Typical radiographs exhibiting various degrees of long gray scale. An abundance of details is visible because of optimum x-ray penetration through all tissues. (From Quinn B. Carroll, *Practical Radiographic Imaging*, 8th ed. Springfield, IL: Charles C Thomas, Publisher, Ltd., 2007. Reprinted by permission.)

subject contrast within the remnant x-ray beam. (It can actually be measured as the total exposure minus that portion of the exposure which is from scattered radiation.) The signal is represented in the radiographic image by all of the (penetrated) dark shades, the (attenuated) medium shades, and the (absorbed) light shades which constitute useful diagnostic information. Noise includes all of the random scattered x-rays, mottle, random electronic static, and other disinformation in the image.

SNR is a relative number, primarily useful in comparing one exposure to another. An important

Figure 13-10

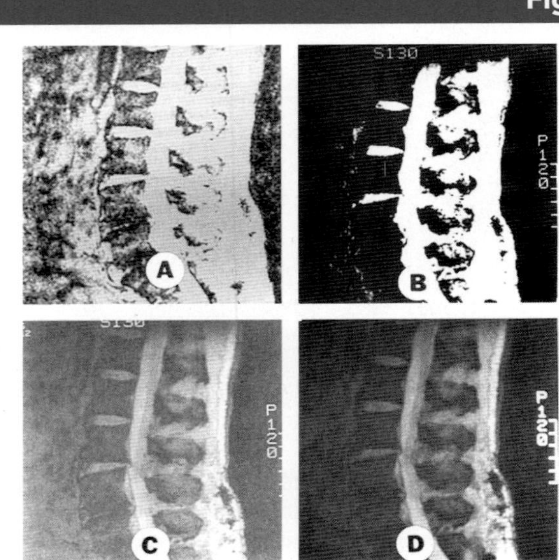

Magnetic resonance images of the lateral lumbar spine showing ***A***, short gray scale and light density (brightness), ***B***, short gray scale and dark density, ***C***, long gray scale and light density, and ***D***, long gray scale and dark density. More details are visible in the long gray scale images. (From Quinn B. Carroll, *Practical Radiographic Imaging*, 8th ed. Springfield, IL: Charles C Thomas, Publisher, Ltd., 2007. Reprinted by permission.)

Figure 13-11

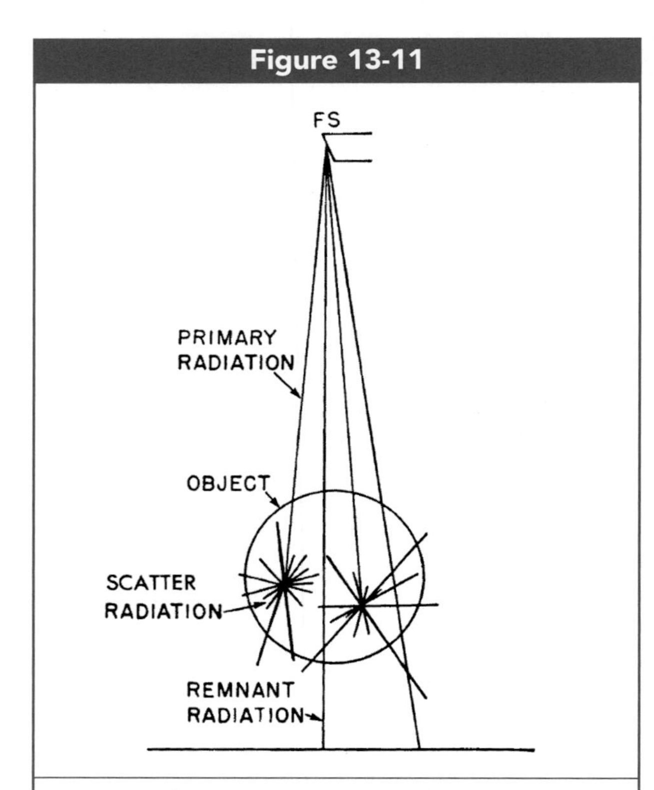

Scatter radiation is generated when x-rays strike atoms in the patient, table or other objects. It is random in direction, and therefore destructive to the image. (From Quinn B. Carroll, *Practical Radiographic Imaging*, 8th ed. Springfield, IL: Charles C Thomas, Publisher, Ltd., 2007. Reprinted by permission.)

aspect of the SNR to understand is its implication that there are *two* ways in which an image may be improved: The obvious way is to reduce the noise (scatter radiation), but the other way is if the signal can be *increased* without adding noise along with it. Following are examples of how these two different approaches increase the measured ratio:

1. With a relative signal of 3, decreasing the level of noise from 2 to 1:

$$\text{Original} \quad \frac{S}{N} = \frac{3}{2} = 1.5$$

$$\text{New} \quad \frac{S}{N} = \frac{3}{1} = 3.0$$

2. With a relative noise of 2, increasing the signal from 3 to 4:

$$\text{Original} \quad \frac{S}{N} = \frac{3}{2} = 1.5$$

$$\text{New} \quad \frac{S}{N} = \frac{4}{2} = 2.0$$

Either approach, reducing the noise or increasing useful signal which transmits subject contrast to the image, will improve the radiograph.

Artifacts

Artifacts are any extraneous images that obscure the desired information, and come from a variety of

HISTORICAL SIDEBAR 13-1: The radiographs in Figure 13-12 are samples of old-fashioned film radiographs demonstrating fog densities from scattered radiation. With film systems, the effects of scattered x-rays in destroying image contrast were immediately apparent. In computer-based systems, much of the resulting loss of contrast can be restored by computer processing algorithms. Noneheless, it is still important that the remnant beam striking the image receptor have as little scatter noise in it as possible, so that the computer will have the best information possible to work with.

It is important to understand that the darker shades in an image produced by better penetration of the x-ray beam through tissues are "good" gray shades. These are desirable in an image, as opposed to the undesirable dark shades that used to be caused by scatter *fogging* such as that illustrated in Figure 13-12. It is also essential to understand the difference between simple overexposure, which overdarkened the film from too many *primary* x-rays, and *fogging*, which darkened the film from too many *secondary scatter* x-rays. The difference in appearance between these two is demonstrated on old-fashioned films in Figure 13-13.

Figure 13-12

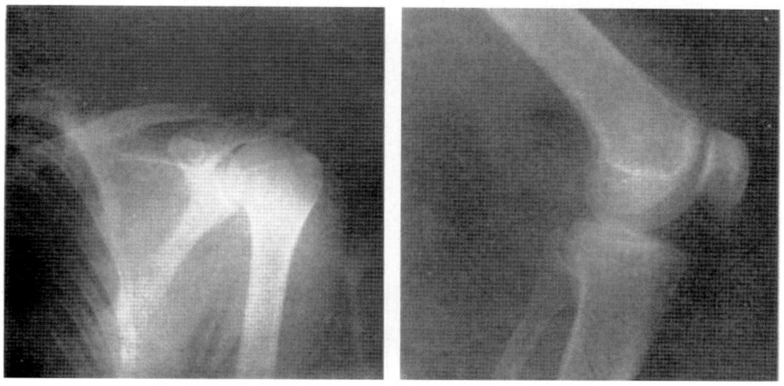

Film radiographs exhibiting evidence of scatter radiation fog. Note the loss of details. (From Quinn B. Carroll, *Practical Radiographic Imaging*, 8th ed. Springfield, IL: Charles C Thomas, Publisher, Ltd., 2007. Reprinted by permission.)

Figure 13-13

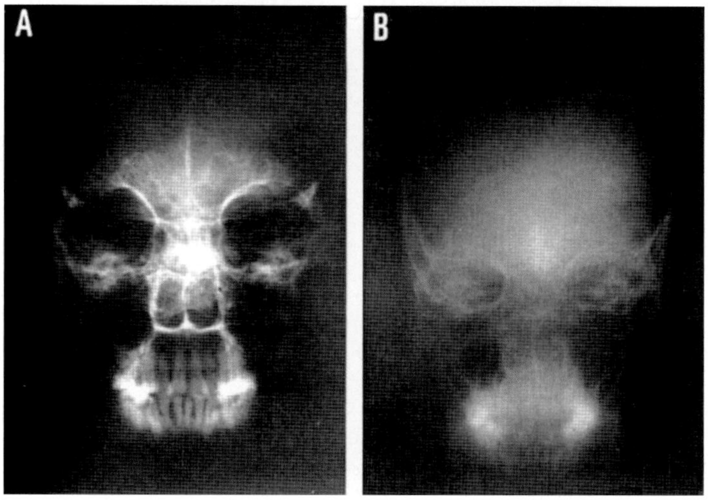

A B

Film radiographs demonstrating the difference in appearance between overexposure, ***A,*** and fogging, ***B***. Note that a fogged image can actually be *lighter* as seen within the orbits and the top area of the skull. (From Quinn B. Carroll, *Practical Radiographic Imaging*, 8th ed. Springfield, IL: Charles C Thomas, Publisher, Ltd., 2007. Reprinted by permission.)

causes too numerous to list. Some examples are iodine or barium spilled onto sponges, removable objects on or in the patient such as hairpins, dentures, jewelry, casts and orthopedic devices, objects in the patient's pockets, snaps on hospital gowns, IV tubes, respiratory equipment, and so on.

Artifacts in the image can also be created by the imaging process itself. Scatter radiation is one example, but others include *false* images which are created by tomographic movement of the x-ray tube, patient motion, or other geometrical anomalies, and static electricity discharges within the imaging plate, computerized processing equipment or display system. (*See* Historical Sidebar 13-2.)

Regardless of their origin or cause, all artifacts obscure the *visibility* of useful image details, are destructive to the image, and are classified as forms of noise.

SUMMARY

1. The components of an image that may affect its visibility are brightness or intensity, contrast, and noise. Optimum levels of brightness and contrast lie in an intermediate range, while noise must always be minimized.
2. Computer-stored images have inherent characteristics based upon the original projection, independent of display qualities. Most radiographic images are inherently *negative* images.
3. Proper balance of brightness and contrast in a radiographic image is characterized by every detail within the anatomy of interest being displayed as a shade of gray, with no "blank" areas or "pitch black" areas present.
4. Radiographic contrast is best defined as the divided *ratio* between the brightness of two adjacent details. Gray scale is the range of brightnesses, and is opposite to contrast.
5. Image brightness and contrast are generally independent of each other (although extreme changes in brightness can affect contrast).
6. Generally, in adjusting an image, the contrast or gray scale should be adjusted first, followed by the brightness.
7. Signal-to-noise ratio (SNR) measures overall *visibility* and should always be maximized in an image. Besides reducing noise, sufficient signal can be ensured by increasing radiographic technique.

Figure 13-14

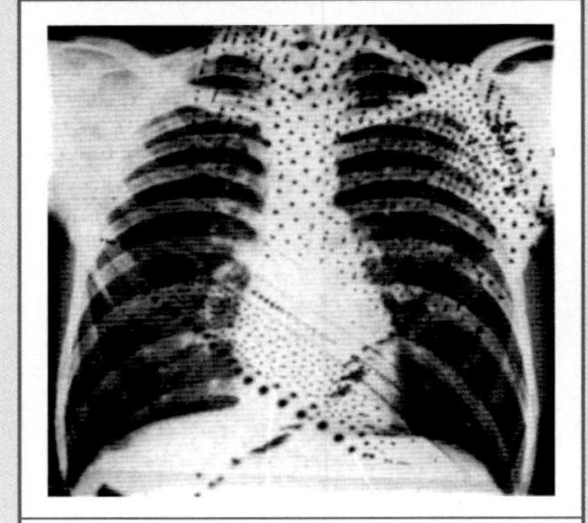

Static electricity artifacts on a film radiograph. (From Quinn B. Carroll, *Practical Radiographic Imaging*, 8th ed. Springfield, IL: Charles C Thomas, Publisher, Ltd., 2007. Reprinted by permission.)

HISTORICAL SIDEBAR 13-2: With film-based imaging systems, discharges of static electricity across the film itself were an ever-present problem, Figure 13-14. Some contraptions, such as conveyor belts that automatically moved films from the chest board into the processor, rubbed against the film, causing static discharges. Humidity in the x-ray department had to be kept at 40-60% to reduce the probability of static from sliding films out of film bins, across counter tops, and into screen cassettes as they were loaded and unloaded by technologists.

Since film processing involved chemicals, artifacts could be caused from chemical spills onto the film or an imbalance or loss of different solutions within the chemical processor. Several artifacts were unique to the roller mechanisms in the processor that moved film from one solution to the next, such as those in Figure 13-15.

Figure 13-15

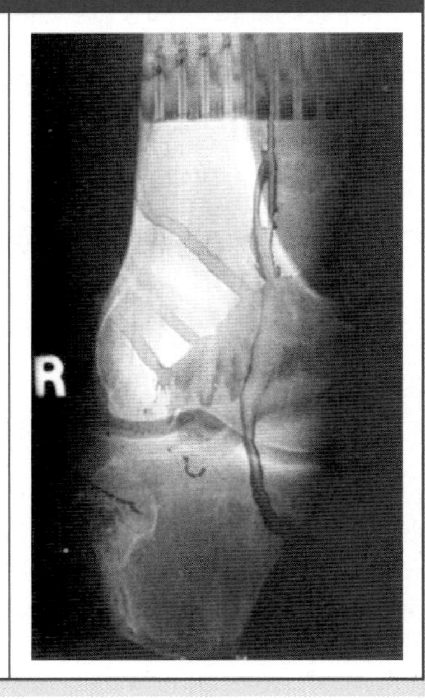

Roller marks caused from splashing developer solution up onto the feed rollers and feed tray of a conventional chemical x-ray film processor prior to running a film through it. (From Quinn B. Carroll, *Practical Radiographic Imaging*, 8th ed. Springfield, IL: Charles C Thomas, Publisher, Ltd., 2007. Reprinted by permission.)

REVIEW QUESTIONS

1. What are the three visibility components of any image?

2. The best amount of brightness or contrast to have in any image is a(n) ________ amount.

3. What is the radiographic opposite of image contrast?

4. List three examples of different types of noise in an image:

5. To ensure that every image detail within the anatomy of interest is depicted as a shade of gray, producing sufficient gray scale in the image, x-ray beam ________ must be sufficient.

6. The optical density of different portions of a hard copy image can be measured using a:

7. What is the radiographic contrast between two adjacent details measuring 3.2 and 1.8?

8. A short gray scale indicates ________ contrast.

9. The use of grids to reduce scatter radiation is an example of changing what portion of the SNR?

10. Increasing the kVp at the console is an example of changing what portion of the SNR?

11. Any extraneous image which obscures the anatomy of interest is classified as a(n):

Fish with scoliosis caught by an orthopedic surgeon.

Chapter 14

GEOMETRICAL QUALITIES OF THE IMAGE

Objectives

Upon completion of this chapter, you should be able to:

1. List and define the three *recognizability* or *geometrical* components of every image.
2. Define image *sharpness* or *spatial resolution*.
3. Define and describe the cause of geometrical penumbra.
4. Calculate radiographic penumbra or *unsharpness* for variable distances and focal spots.
5. Calculate the *relative sharpness* for different distances.
6. Distinguish between poor contrast and poor sharpness.
7. Distinguish between the effects of blur and magnification on a *penumbra diagram*.
8. Calculate the magnification factor for variable distances.
9. Define and quantify shape distortion, as distinct from magnification.
10. Define overall image *resolution*, and list its two primary components.

RECOGNIZABILITY (GEOMETRICAL INTEGRITY)

In addition to the visibility factors of an image, there are what might be called *recognizability factors*. Even though an image is visible, it is worthless if we cannot recognize what it is. We depend upon the geometrical integrity of the image to recognize what real object it represents. If the image is blurry, or if its size or shape are grossly distorted, we may not be able to tell what it is. Recognizability, or geometrical integrity, is made up of three components: sharpness, magnification, and shape distortion.

Sharpness (Spatial Resolution)

The *sharpness* of detail may be described as the *abruptness* with which the edges of a particular image *stop*. To better visualize this principle, imagine driving a microscopic sports car across a black-and-white photograph: You are passing from a white image onto the black background. As you cross over the edge between white and black, if you suddenly find yourself over the black background, then the edge of the white image was *sharp*. On the other hand, if you seem to pass gradually from white into black, then the edge of the white image is *blurred* and *unsharp*.

The American Registry of Radiologic technologists (ARRT) uses the term *spatial resolution* for this quality and defines it as *the sharpness of the structural edges recorded in the image.*

Theoretically, if shadows could be cast from a *point source* of light, such as in Figure 14-1A, there would never be any blur. There would be a single pure, dark shadow projected for each object, with perfectly sharp edges. However, true point sources of light or other radiation are rare indeed in nature. Our shadows from the sun, for example, are projected from a *disk* source of light rather than from a small point. Flashlights, headlights and other artificial sources of light created by man are also generally *area sources* of light rather than small point sources. In all these cases there are *partial shadows* cast

Figure 14-1

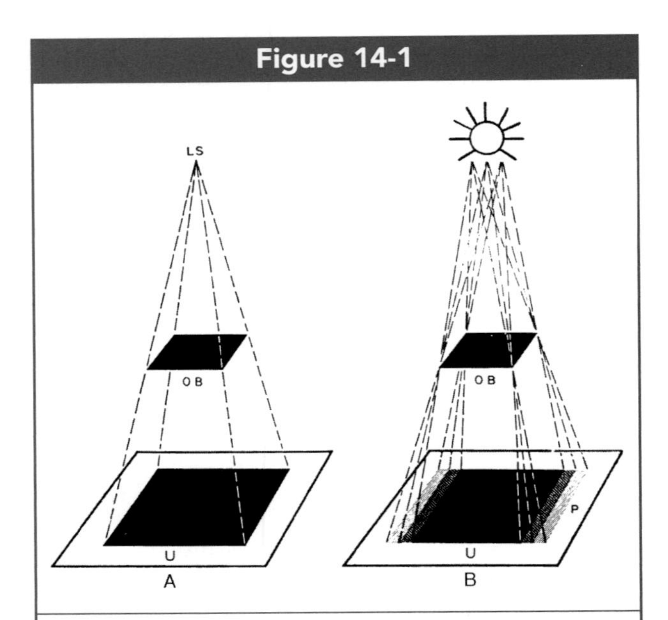

A pure umbra image from a theoretical point source of light, ***A***, would produce an image with perfectly sharp edges. Real sources of light, such as the sun, ***B***, are area sources which produce penumbra (*p*) at the edges of a shadow, from the edges of the object being projected by rays of light from various points. (From Quinn B. Carroll, *Practical Radiographic Imaging*, 8th ed. Springfield, IL: Charles C Thomas Publisher, Ltd., 2007. Reprinted by permission.)

around the edges of the *pure* shadow, as shown in Figure 14-1B. These partial shadows occur because any given edge of the object is actually projected at several different angles from several different points within the light source.

In Figure 14-1B, in which the shadow of a square plate is projected onto the ground by the sun, three different edges are projected from light emitted from three different points across the area of the sun's disk. This is for simplification; in reality, there are thousands of such points of origin for the sun's light, and thousands of partial shadow edges, resulting in shadow edges that appear to gradually fade into their background on all sides.

The inner portion of a shadow has a uniform darkness and is called the *umbra*. The umbra is considered to be the "pure" portion of a shadow. The blurry, fading partial shadow projected at each edge is referred to as the *penumbra*. Penumbra is synonymous with "blurriness" or "unsharpness." *Penumbra* is a scientific term; the width of the penumbra can actually be measured. It can also be predicted geometrically for any given set of circumstances (the diameter of the light source, the distance from the source to the object, and the distance from the object to the ground or screen). The more penumbra is present, the more blurry the edges of the image appear.

Experiment: It is easy to demonstrate umbral and penumbral shadows: Hold your hand, with fingers spread, over an overhead projector or in front of a flashlight and observe the shadow cast upon a wall or screen. Move your hand away from the light source and then back toward it, and you will see the blurry penumbra around the edges of the fingers grow and shrink. Carefully note that when this penumbra grows, it not only expands outward but also inward, actually *invading the umbra* shadow. The umbra actually shrinks from this effect. This is an important point to remember about the production of penumbral blur.

At an extreme distance from the screen, you can make the pure umbra portion of the shadow actually disappear, so that the remaining image is *all* penumbra. All that is left is a nebulous, blurry density in the area of the image. When enough penumbra is present, the shadow can no longer be recognized as the shadow of a hand. This is a prime example of the fact that even though an image is visible, it may not be recognizable if its geometrical integrity is not preserved. From a diagnostic standpoint, this renders the image useless.

Now consider the problems caused with recognizability when observing the very blurry images of two objects closely adjacent to each other. (Repeat the above experiment observing two fingers held close together.) As penumbra grows, the edges of the two shadow images actually begin to overlap and blur into each other. When two shadows overlap in this way, they may deceptively appear as the shadow of one object. Severe penumbra has the effect of making it impossible to distinguish two adjacent objects in an image as being distinct and separate objects.

Sharpness is destroyed not only by *geometrical penumbra* as described above, but also by *motion penumbra*. Movement of the source of light, the object, or the recording surface spreads penumbra laterally such that its width is expanded. In radiography, movement of either the x-ray tube, the patient, or the image receptor plate during an exposure will likewise result in a blurred image. Sharpness of recorded detail has a strong impact upon our ability to recognize an image.

Magnification (Size Distortion)

Excessive magnification of the size of an image can make it difficult to recognize what real object it represents. Imagine standing just one inch away from the side of a rhinoceros! With just its skin in your field of view, how sure could you be that it was not a hippopotamus or an elephant? When something is grossly magnified, we may literally lose the ability to recognize what it is.

In radiography, we define *magnification* as the difference between the size of the real object and the size of its projected image. Magnification is often referred to as *size distortion.* This is the term adopted by the ARRT (American Registry of Radiologic Technologists). To avoid confusion, the term *distortion* will only be used in this textbook when referring to changes in the *shape* of the image. *Magnification* shall be consistently used to describe any change in the size of the image.

In most radiographic applications, our goal is to minimize any magnification present in the image. Projecting the true size of an object can be important to correct diagnosis. For example, chest radiographs are routinely checked by the radiologist for any enlargement of the patient's heart. If the heart size is magnified because of the way the radiographer positioned the patient, it can simulate this pathological condition. Magnification can lead to misinformation.

In contrast, at times important anatomical or pathological details are just too small to recognize with certainty in an image. In such cases, magnification can be desirable to make these small details recognizable. Therefore, magnification techniques are sometimes intentionally used in radiography, especially in angiography where blood clots or other anomalies in very small arteries or veins must be diagnosed.

Shape Distortion

Shape distortion is defined as the difference between the shape of the real object and the shape of its projected image. In a given axis of direction (lengthwise or crosswise), shape distortion will consist of either a foreshortening of the image or elongation of the image. An example is given in Figure 14-2, in which the shadow of a square object is projected onto a surface from an angled direction. The shadow is elongated into a rectangular image which does not represent the true square shape of the object. A spherical object projected by such an angle will be distorted into an oval shadow or image.

In radiography, similar distortion can be produced not only by angling the x-ray tube, but also by angling the body part being radiographed or the image receptor plate. *Misalignment* of any one of these things, which includes off-angling or off-centering, can distort the shape of the image in such a way that diagnostic information is misleading or misrepresented.

Measuring Unsharpness

The most important component of the recognizability of an image is the *sharpness of detail*, previously described as the abruptness of the image edges. Sharpness may also be described as the *lack* of penumbra (blur) at the edges of an image. Figure 14-3 is an exaggerated example of blurred edges and the subsequent loss of details in the bones of a hand caused by movement.

How can the sharpness of a radiographic image actually be measured? Only indirectly. As described in the last section of this chapter, there are ways to measure the *resolution* of an image, and if the

Figure 14-2

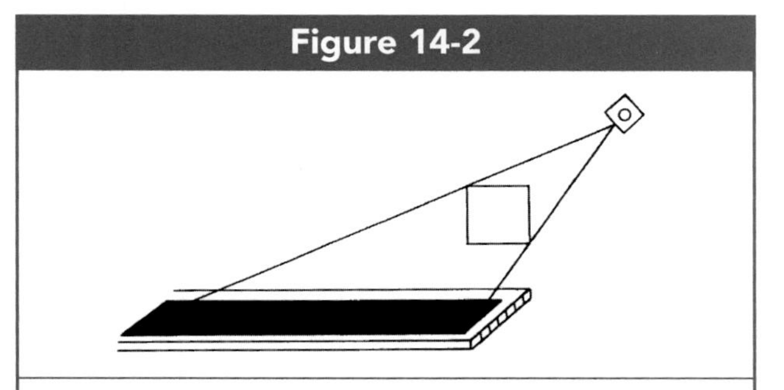

Misalignment or angling of a beam of light will project the shadow of a square (cubical) object as an elongated rectangle. This is shape distortion. (From Quinn B. Carroll, *Practical Radiographic Imaging*, 8th ed. Springfield, IL: Charles C Thomas Publisher, Ltd., 2007. Reprinted by permission.)

Figure 14-3

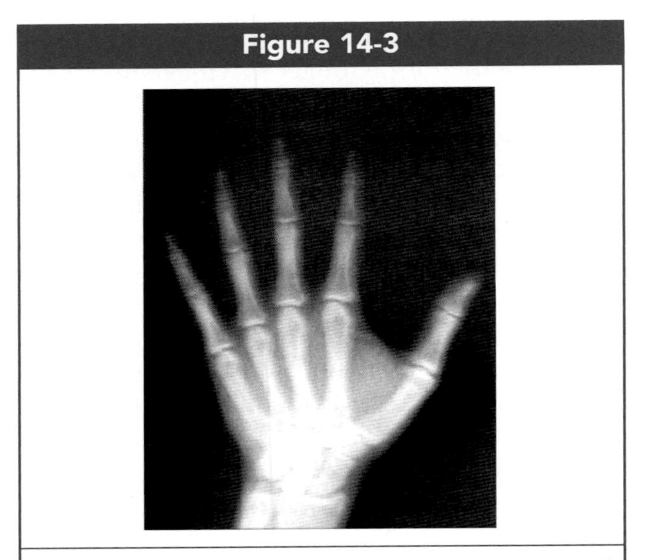

Radiograph of the hand showing blurred edges of bones and loss of details within the bones due to motion. (From Quinn B. Carroll, *Practical Radiographic Imaging*, 8th ed. Springfield, IL: Charles C Thomas Publisher, Ltd., 2007. Reprinted by permission.)

visibility factors are known to be constant, any change in that resolution can be extrapolated as a change in sharpness. Or, we can directly measure *unsharpness*, and take the relative sharpness as the *inverse* of that measurement. Mathematical expressions of sharpness are always *relative* numbers, indirectly derived from these methods.

Unsharpness, on the other hand, is an objective quantity that can be directly measured, geometrically predicted and mathematically calculated. Because it is easier to quantify, we will first spend some time diagraming and calculating unsharpness. Afterwards, it should be easier for the student to visualize *sharpness* as the *inverse* of these calculations.

In the geometry of any radiographic projection, unsharpness is controlled by three things; they are the source-to-object distance (SOD), the object-image receptor distance (OID), and the size of the focal spot being used in the x-ray tube. It is only because the focal spot size can be known that we are able to proceed with calculations. Figure 14-4 is a *penumbra diagram* showing how the actual extent of the penumbral shadow can be plotted. This diagram is not to scale—the distances are shortened a great deal (so we don't need 40 inches of paper!), and then the focal spot sizes are exaggerated also to make the effects more obvious. It is possible to draw a penumbra diagram precisely to scale. (As long as the SOD/OID ratio is maintained, a drawn 4" SOD and 1" OID will precisely yield the same results as an actual 40" SOD and 10" OID, but the focal spot would need to be drawn exactly as measured.) Penumbra diagrams will be used extensively in later chapters.

To plot a penumbra diagram, simply extend a projected line *from each end* of the focal spot area through each end of the object being radiographed

Figure 14-4

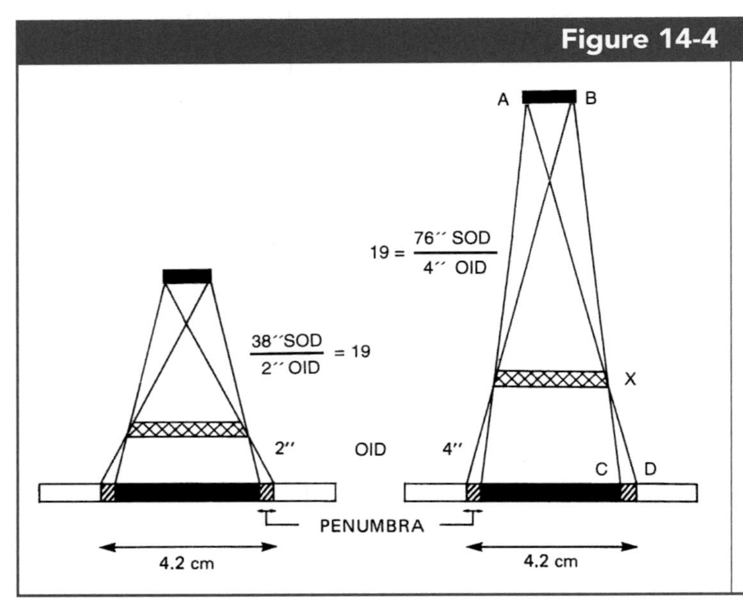

A penumbra diagram. When the focal spot size is known, the actual extent of the penumbral shadow (blur) can be both predicted and measured for any set of distances. (From Quinn B. Carroll, *Practical Radiographic Imaging*, 8th ed. Springfield, IL: Charles C Thomas Publisher, Ltd., 2007. Reprinted by permission.)

to the imaging plate surface. Since x-rays can be emitted from any point across the focal spot area between these two lines, *the spread of penumbra is precisely indicated by the spread of these lines projected to the imaging plate.* In other words, a particular edge of the object is being projected at all points *between* these two lines, by x-rays emanating from different portions of the focal spot.

In terms of plane geometry, what this diagram shows *at each edge* of the object is two similar triangles that are *inverted* and *reversed* to each other (Fig. 14-5). We know that for similar triangles, the ratios of their height to their base must always be equal. In this case, the base of the larger triangle is the focal spot and its height the SOD, while the base of the smaller triangle is the penumbral spread and its height the OID.

$$\frac{\text{Base}}{\text{Height}} = \frac{\text{FS}}{\text{SOD}} = \frac{\text{Penumbra}}{\text{OID}}$$

Cross-multiplying the last two entries, we find that image penumbra is directly proportional to the focal spot size, and also to the ratio of OID/SOD. (Each specific relationship between these factors will be demonstrated in later chapters.) After cross-multiplying, we find the entire formula derived for the spread of penumbra, or *unsharpness*, is

$$\textit{Unsharpness} = \text{Penumbra} = \frac{\text{FS} \times \text{OID}}{\text{SOD}}$$

Practice Exercise #1

Given a focal spot size of 1.5 mm, an OID of 30 cm and an SOD of 60 cm, what will the spread of penumbra, or unsharpness, measure?

Solution: $\frac{FS \times OID}{SOD} = \frac{1.5 \times 30}{60} = \frac{45}{60} = 0.75$

Answer: *The penumbra spreads across an area of .75 mm.*
Unsharpness is .75 mm.

Try the following exercise, and check your answers from Appendix #1.

EXERCISE #14-1

Given the following factors, calculate the image unsharpness for each set:

Figure 14-5

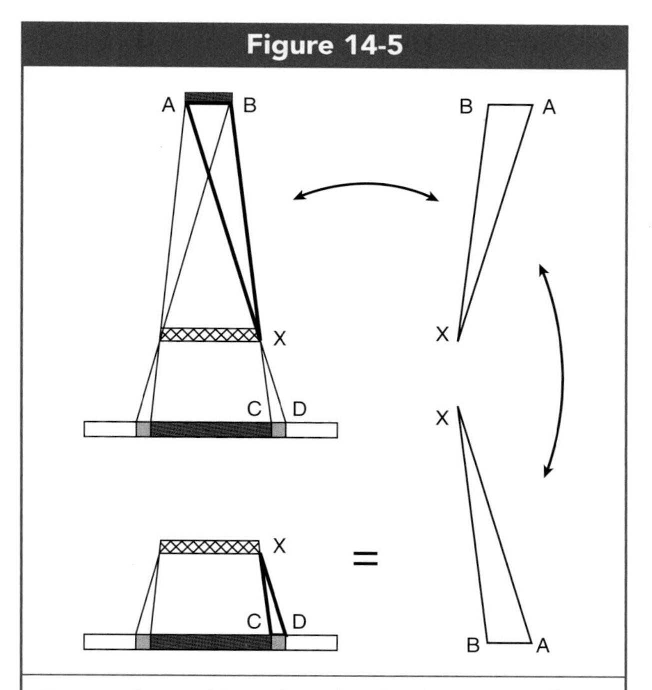

By reversing and inverting triangle ABX, we see that it forms a similar triangle to CDX. Therefore, all corresponding angles are equal and all ratios between lengths are equal.

Focal Spot Size	OID	SOD	*Unsharpness*
1. 2 mm	30 cm	100 cm	________
2. 0.5 mm	5"	35"	________
3. 0.3 mm	2"	38"	________
4. 0.3 mm	4 cm	80 cm	________
5. 0.5 mm	15 cm	150 cm	________

These results should reflect what is visually seen in a radiographic image; when the penumbra is calculated to be twice as great, the image will appear twice as blurry to the eye.

Radiographic Sharpness

By *inverting* the formula for unsharpness, we should be able to derive a formula for *sharpness* or *spatial resolution.* The problem is that we cannot express the concept of sharpness *in millimeters* or any other distance unit. Therefore, the focal spot portion of the formula must be left out because the units are inappropriate. We can, however, take the ratio that remains, the SOD/OID ratio, as an indicator of *relative sharpness*, meaning that it can still be readily used for purposes of comparison.

For example, if the SOD/OID ratio for one particular exposure is twice that of another, we can say that the first radiograph will have twice the sharpness of the second, even though there are no particular units to measure that sharpness with. The following practice exercise provides another example.

Practice Exercise #2

Exposure ***A*** employed an SOD of 90 cm and an OID of 5 cm. If the SOD is changed to 180 cm *and* the OID is reduced to 2.5 cm for exposure ***B***, how will image sharpness be affected?

Solution: *For Exposure A:* $\frac{SOD}{OID} = \frac{90}{5} = 18$

For Exposure B: $\frac{SOD}{OID} = \frac{180}{2.5} = 72$

Ratio of B to A: $\frac{72}{18} = 4$

Answer: Relative Sharpness for Exposure B is 72, that for A is 18.
Exposure B will be 4 times sharper than Exposure A.

For reinforcement, try the following exercise and check your answers from Appendix #1.

EXERCISE #14-2

1. What is the relative sharpness for an SOD of 100 cm and an OID of 30 cm?
2. What is the relative sharpness for an SOD of 35" and an OID of 5"?
3. An original exposure used an SOD of 36" and an OID of 2". If the SOD is increased to 54", by what factor much will the sharpness be improved?
4. An original exposure used an SOD of 80 cm and an OID of 4 cm. If the OID is increased to 6 cm, to what fraction will the sharpness be reduced?
5. An original exposure used an SOD of 150 cm and an OID of 15 cm. If the SOD is increased to 180 cm *and* the OID is reduced to 10 cm, by what factor will the sharpness be improved?

It can be quite difficult to visually evaluate sharpness in a radiograph, because factors that make the anatomy more *visible* can mimic improved sharpness to the human eye at first glance. For example, it is common in radiography to mistake *high contrast* for improved sharpness. In Chapter 13, compare the two chest radiographs in Figures 13-6 and 13-7, and see if you can determine which one has sharper edges between the ribs and the background density of the lungs. In reality, they are both equally sharp, since they were taken with identical geometrical conditions. The high contrast present in the chest radiograph in Figure 13-6 is deceptively appealing to the human eye. The ribs are *brighter*, and this makes their edges stand out more. The edges themselves are more visible. But this does not constitute better sharpness. Remember that sharpness is measured by how "quickly" the transition from light to dark changes as one scans *laterally* across the image. The edges of these ribs may be bright and yet still fade off gradually from light to dark across the same distance as those in Figure 13-7.

The converse is also true, in that low contrast can be mistakenly perceived as poor sharpness. Scatter radiation can reduce subject contrast in the remnant beam, hindering the visibility present in the initial image at the image receptor plate. As will be demonstrated later, the scattering process is *unrelated* to the formation of penumbra at the edges of an image, so it is impossible for scatter to have anything to do with real sharpness. The edge of an image may be less *apparent*, yet this does not necessarily constitute *blur*.

To reinforce this distinction between poor contrast and poor sharpness, Figure 14-6 presents three simplified diagrams of two adjacent densities. In ***A***, there is no blur present and high contrast for comparison. In ***B***, high contrast remains, but there is blurriness or penumbra at the edge between the two brightnesses or densities. In ***C***, there is low contrast between the two, which makes the edge between them less *visible*, but the edge is still sharp. Image ***C*** is an example that low contrast does not constitute blur, and image ***B*** is an example that high contrast does not constitute sharpness.

There are devices and methods to directly measure the sharpness produced with different geometrical factors (Fig. 14-7), but as demonstrated above, sharpness can be tricky to evaluate by the subjective visual inspection of actual anatomy in a radiograph.

Figure 14-6

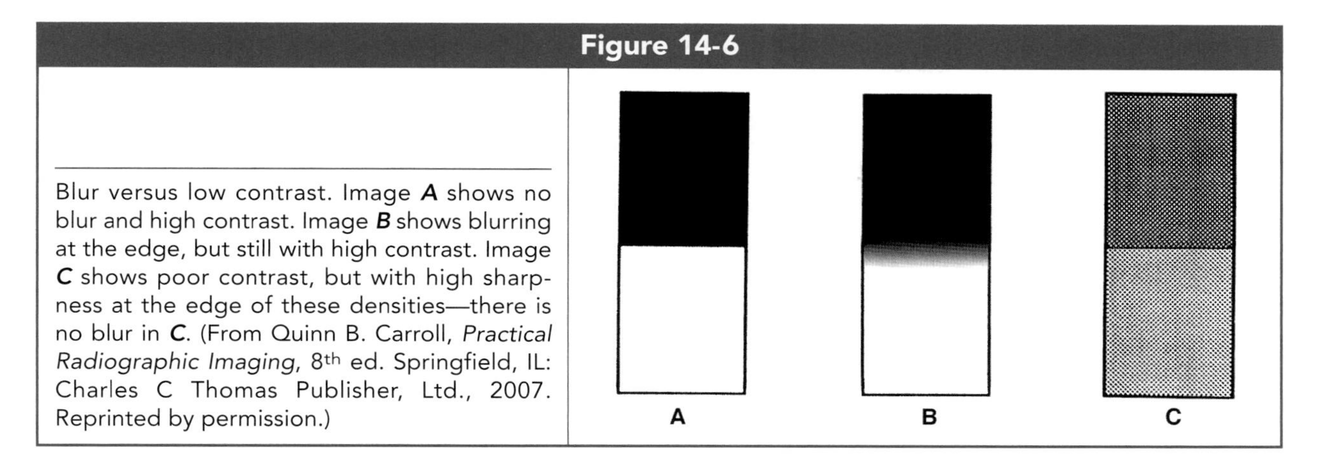

Blur versus low contrast. Image ***A*** shows no blur and high contrast. Image ***B*** shows blurring at the edge, but still with high contrast. Image ***C*** shows poor contrast, but with high sharpness at the edge of these densities—there is no blur in ***C***. (From Quinn B. Carroll, *Practical Radiographic Imaging*, 8th ed. Springfield, IL: Charles C Thomas Publisher, Ltd., 2007. Reprinted by permission.)

Just like the rest of an image, the *edges* of anatomy also have *both* visibility and recognizability functions. Just because they are more visible does not automatically mean they are also more sharp.

Two factors primarily control sharpness within the image carried by the remnant x-ray beam to the imaging plate: These are the prevention of *motion*, which should be considered the arch-enemy of sharpness, and the use of optimum *geometry* in the projection beam, including all distances involved (SID, SOD, and OID), and the size of the focal spot used. In computer processing of the image, the size of the detector elements (dels) in the image receptor plate play a critical role, and in presenting the electronic image at the display monitor, the size of pixels (picture elements) in the display monitor screen itself and the electronic processes used can affect sharpness. All these considerations must be monitored to ensure the best diagnostic value of radiographic images.

Radiographic Magnification

The *magnification* of a radiographic image can be quantitatively measured by determining the difference between its size and the size of the actual object it represents *in both axes, lengthwise and crosswise*. If both the length and the width of the image are identical to those of the object, no magnification is present. In a magnified image, *both the length and the width of the image will measure larger than the real object by equal proportions*. For example, if the image is both twice as long and twice as wide as the object, this effect is due to magnification.

For magnification to be present, *the pure shadow, the umbra of the image must be larger*. It is possible for the penumbra to spread outward due to blurring, yet if it expands *without the umbra being also enlarged*, this does not constitute magnification, only blurring. This is illustrated with a penumbra diagram in Figure 14-8. Note that in projection ***B***, where a larger focal spot is used, more blurring occurs as expected and witnessed by the expansion of penumbra. However, the *umbra* actually *shrinks* in projection ***B***. When this "clear" portion of the shadow is smaller, it is impossible for real magnification to

Figure 14-7

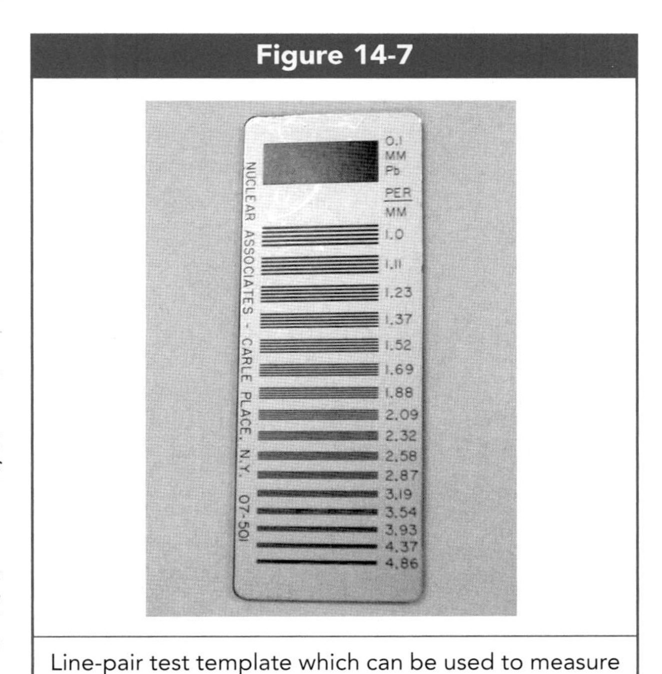

Line-pair test template which can be used to measure sharpness and resolution.

Figure 14-8

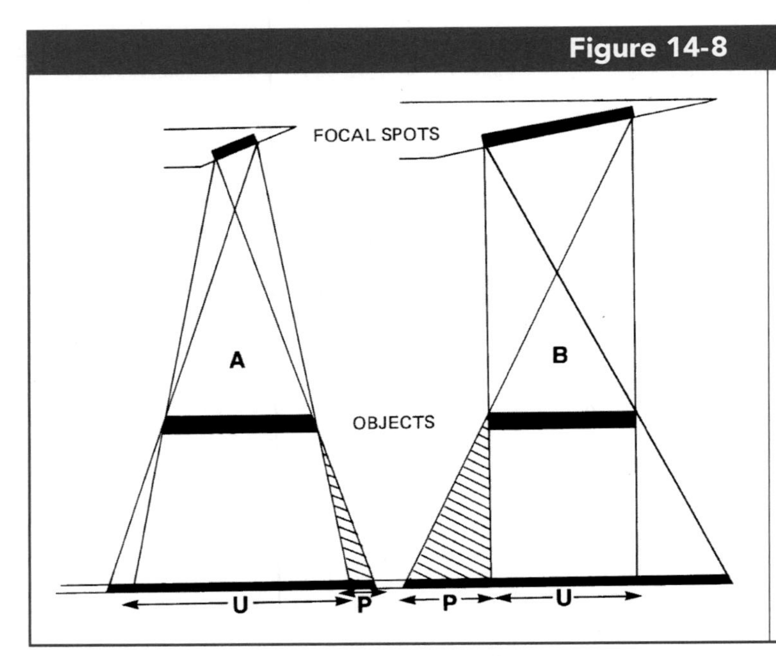

Demonstration that for magnification to be present, the clear *umbra* portion of an image must expand in size. Compared to ***A***, the shadow in projection ***B*** may appear larger with the penumbra included, but this *cannot be magnification* since the clear umbra actually shrinks in size. (This is due to the spread of penumbral blur *inward* as well as outward.) (From Quinn B. Carroll, *Practical Radiographic Imaging*, 8th ed. Springfield, IL: Charles C Thomas Publisher, Ltd., 2007. Reprinted by permission.)

have occurred, even though at a glance it appears that the entire image with the penumbra included has expanded.

This can be further demonstrated by considering where the human eye locates the visual edge of the image. If this were an image of a bone, for example, very close examination of a single edge would reveal more blurriness in projection ***B***; however, the bone would appear to be of just the same *size* on projection ***B*** as on projection ***A***, because the human eye will tend to locate the visual edge of the bone in the middle of the penumbral shadow, indicated by the dotted line in Figure 14-9. This phenomenon is so effective that even upon lining up a ruler to measure the bone images, one will obtain the same measurements because the marks on the ruler will be aligned with the center-point of each penumbral shadow.

Magnification Formula

As with unsharpness, the magnification of an image can be both predicted by geometric diagraming and calculated mathematically. On a penumbra diagram, one can simply measure the *umbra*. Magnification is *controlled* by the ratio of the SID to the SOD, and is affected by no other variables. This makes it quite simple to predict the degree by which any projected image will be magnified, as long as the distances are known, by the formula:

$$\text{Magnification} = \frac{\text{SID}}{\text{SOD}}$$

As we have defined magnification as the ratio between the size of the image and the size of the real object it represents, we can restate the formula as follows:

$$\frac{\text{Image Size}}{\text{Object Size}} = \frac{\text{SID}}{\text{SOD}}$$

Also like sharpness, magnification is based solidly on the geometry of similar triangles. Only this time, in diagraming the triangles, the penumbra is left out and the focal spot is treated as a single point as in Figure 14-10. In this diagram, the height of the large triangle is formed by the SID, and the height of the small triangle by the SOD with the object as its base. These are completely different triangles than the ones used for penumbra diagrams, with the large triangle encompassing the smaller one within it.

The *ratios* formed between any two corresponding parts of similar triangles must always be equal. Thereby, we can make a ratio between their heights to predict the ratio between their bases where magnification is represented. That is, the *ratio between the distances controls the ratio of magnification* between the image and the object.

When the ratio of SID/SOD is calculated, the result is the *factor* of linear magnification; that is, it represents the multiplying factor by which the *length* and the *width* of the image will each be increased (not the area). For ratios up to a doubling, a simple operation can be used to convert this factor into a *percentage*: Simply subtract 1 and multiply by 100. For example, a magnification of 1.75 would also be expressed as 75 percent magnification. Following are a few examples of these calculations:

Practice Exercise #3

An object measuring 40 cm in width is radiographed using an SID of 100 cm and an SOD of 60 cm. How wide is the projected image of the object?

Solution: $\frac{SID}{SOD} = \frac{Image}{Object}$

$$\frac{100}{60} = \frac{X}{40}$$

Cross-multiplying $60X = 4000$

$$X = \frac{4000}{60} = 66.7$$

Answer: The image width is 66.7 cm.

Practice Exercise #4

A radiograph is taken of an object 6 cm in width and the resulting image measures 18 cm in width. If a 180 cm SID was used, what was the SOD?

Solution: $\frac{SID}{SOD} = \frac{Image}{Object}$

$$\frac{180}{X} = \frac{18}{6}$$

Cross-multiplying $18X = 1080$

$$X = \frac{1080}{18} = 60$$

Answer: The SOD was 60 cm.

Practice Exercise #5

An object is radiographed at 200 cm SID and 150 cm SOD. What is the *percentage* magnification?

Solution: $\frac{200}{150} = \frac{X}{1}$

Figure 14-9

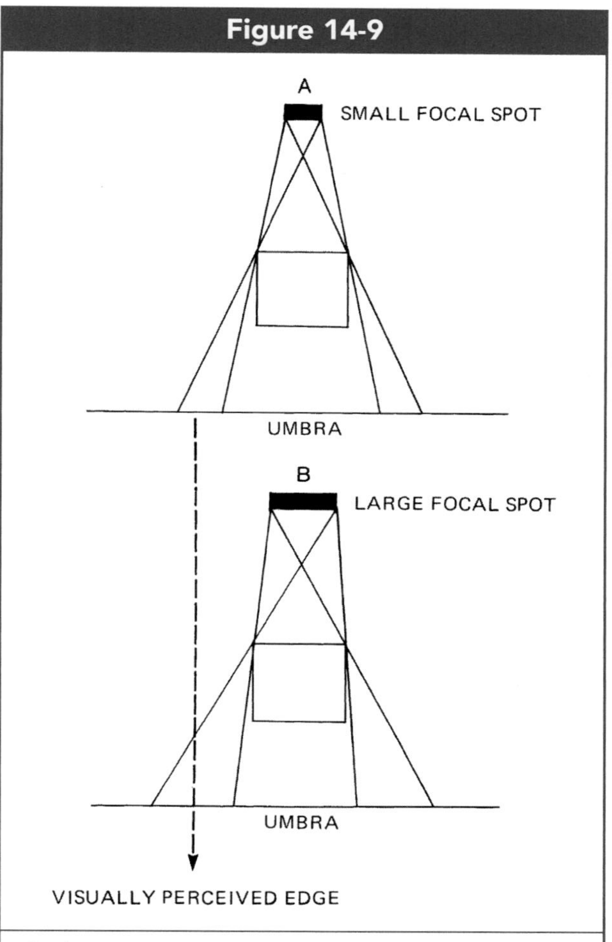

The human eye always locates the edge of an image at the center of the penumbra. Again, even though the penumbra in B is spread out more, an observer will conclude that there is no magnification upon measuring the image with a ruler. (From Quinn B. Carroll, *Practical Radiographic Imaging*, 8th ed. Springfield, IL: Charles C Thomas Publisher, Ltd., 2007. Reprinted by permission.)

Cross-multiplying $150X = 200$

$$X = \frac{200}{150} = 1.33$$

$$1.33 - 1 = .33$$

$$.33 \times 100 = 33$$

Answer: The factor of magnification is 1.33 times. The percentage is 33% magnification.

Finally, if the *OID* is given rather than the SOD, don't forget to subtract the OID from the SID to obtain the SOD before proceeding. Complete the

Figure 14-10

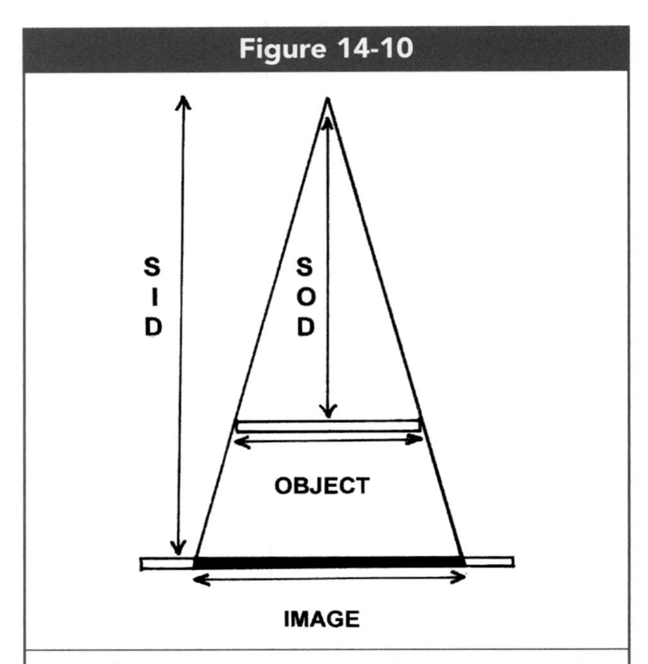

Magnification diagram, using the similar triangles formed by the SID and SOD from the focal spot as a common apex. The SID/SOD ratio controls the magnification ratio of the image (base of the large triangle) to the actual object being radiographed (base of the small triangle).

following exercise and check your answers from Appendix #1:

EXERCISE 14-3

Fill in the blanks for the following sets of factors:

	Object Width	*SID*	*SOD*	*Image Width*
1.	40 cm	100 cm	60 cm	______
2.	30"	200 cm	50 cm	______
3.	15"	200 cm	150 cm	______
4.	50 cm	100 cm	______	75 cm
5.	_____	180 cm	135 cm	20 cm
6.	12 cm	______	60 cm	20 cm

7. A 25 cm object is projected using an SID of 100 cm and an OID of 40 cm. What will be the length of its projected image?
8. An object is radiographed at 180 cm SID and 160 cm SOD. What is the *percentage* magnification?

Radiographic Shape Distortion

The "length" of an image is defined radiographically as its measurement *in the direction of any angling or off-centering of the x-ray beam.* If that length measures longer than that of the real object, while its width remains unchanged, *elongation distortion* is present. If the length measures shorter than that of the real object, then *foreshortening distortion* is present in the image. Elongation and foreshortening are forms of *shape distortion,* which is defined as any difference between the shape of the projected image and the shape of the object it represents.

The primary distinction between shape distortion and magnification is that for shape distortion to be present, *the length and the width of the image must have been altered by different degrees.* As described in the previous section, if they are both increased equally, only magnification is present. It is possible for an image to be *both* magnified and distorted. To better understand how these two effects could be sorted out when both are present, we will use the concept of a *shape ratio.*

The *shape ratio* of an object is simply defined as its length *divided by* its width. The shape ratio of a cube or sphere will be 1:1, or 1.0, since they have equal measurements in each dimension. The shape ratio of the circular shadow ***A*** in Figure 14-11 is also 1.0. Now suppose that a change in our projection geometry (distances, angles or centering) results in the projected shadow ***B*** in Figure 14-11; if the image is both 50 percent longer and 50 percent wider, its shape ratio will be calculated as 1.5 divided by 1.5, still equal to 1.0. Therefore, the shape has not changed, only the size, and we may state that there is 50 percent magnification with no distortion.

Comparing shadow ***A*** to shadow ***C*** in Figure 14-11, we see that the length of ***C*** has been elongated to a doubled amount, yet the *width* is still equal. In this case, the new shape ratio is 2:1 or 2.0. Shape distortion is present, but since the *width* has not increased at all, magnification is *not* present.

For shape distortion to be present, it is necessary that the length and width change by *different amounts.* For a 1-centimeter object, suppose that both the length and the width of its projected shadow are greater than those of the object, but the length is 3 times longer, while the width is only 2 times wider. We can sort out precisely how much distortion *and* magnification are present. The magnification is simply the *lesser* of the two amounts, in this case, the increase in the width, which was twofold. Magnification is 2:1 or 100 percent. This would account for the length of the shadow increasing from 1 cm to 2 cm. The remaining increase in length, from 2 cm to 3 cm, must be due to shape distortion (elongation). The new shape ratio for this image is 3 divided by 2, or 1.5, indicating that the shape has been distorted by 50 percent (or 1.5 times). The image has been magnified by 100 percent and elongated by 50 percent.

In radiography, it is essential to minimize all forms of distortion as much as possible, and in general practice, to also keep magnification at the achievable minimum.

Figure 14-11

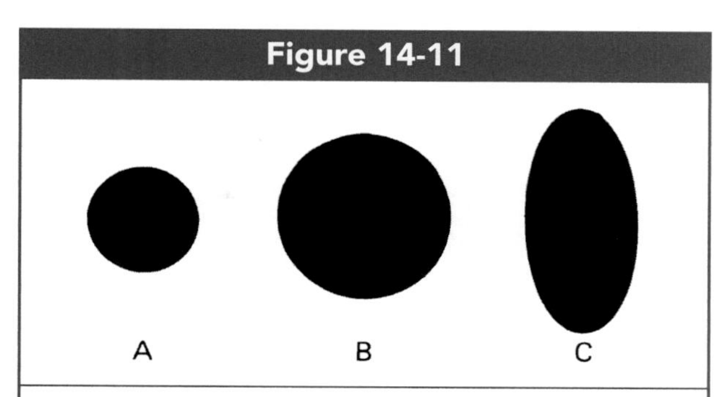

The difference between magnification and shape distortion: If an image measurement changes equally in *both* axes (length and width), a magnified but still circular image results, ***B***. Only when one axis changes by a different ratio than the other is shape distortion present, producing in this case an oval shadow, ***C***. (From Quinn B. Carroll, *Practical Radiographic Imaging*, 8th ed. Springfield, IL: Charles C Thomas Publisher, Ltd., 2007. Reprinted by permission.)

RESOLUTION

Finally, it is important to understand the concept of image resolution. In a broad sense, resolution can be thought of as *the total amount of useful information present in an image*. But, more specifically, *resolution is defined as the ability to distinguish any two adjacent details in the image as being separate and distinct from each other*. A well-resolved image requires *both* high visibility and optimum recognizability. *All* of the image qualities affect its overall resolution. However, the two most important aspects in the resolution of a particular image detail are its *contrast* compared to other details nearby, and its *sharpness* against background details.

As emphasized in the section on *Sharpness (Spatial Resolution)*, sharpness and contrast are not the same thing, and must not be confused. (In fact, they fall under completely different classifications, with sharpness being a *recognizability* or *geometrical* factor, and contrast being a *visibility* factor in the image.) Nonetheless, sharpness and contrast work hand-in-hand in producing good resolution of details. This is illustrated in Figure 14-12, a diagrammatic image projected from a test template that consists of alternating lines of lead foil and open slits. X-ray exposure of such a template produces in the image pairs of alternating black and white lines.

In Figure 14-12, the lower portion of image ***A*** represents high contrast but low sharpness. The recorded lines are black and white, but penumbral blur at their edges causes a gradual transition from one to the next. (In print, this is simply "bleeding" of the ink at the edges of the dark lines.) Resolution is

Figure 14-12

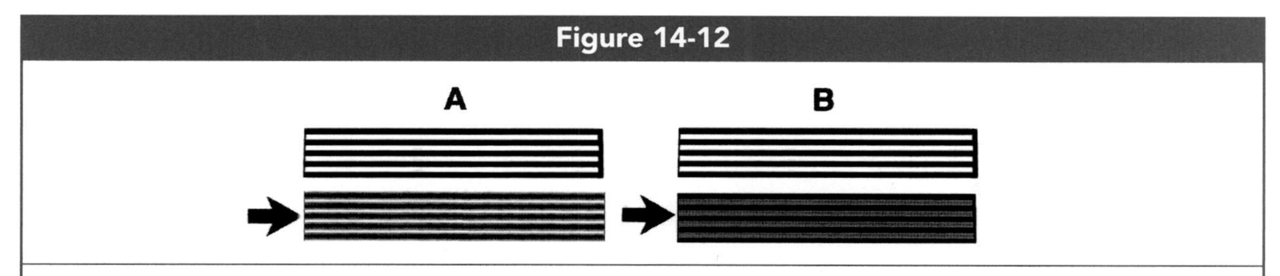

Resolution template images showing that overall image resolution can be lost by either (A, arrow) blur resulting in ragged edges that begin to run into each other, or (B, arrow) a loss of contrast even though the edges are sharp. In both cases individual lines are more difficult to distinguish. (From Quinn B. Carroll, *Practical Radiographic Imaging*, 8th ed. Springfield, IL: Charles C Thomas Publisher, Ltd., 2007. Reprinted by permission.)

Figure 14-13

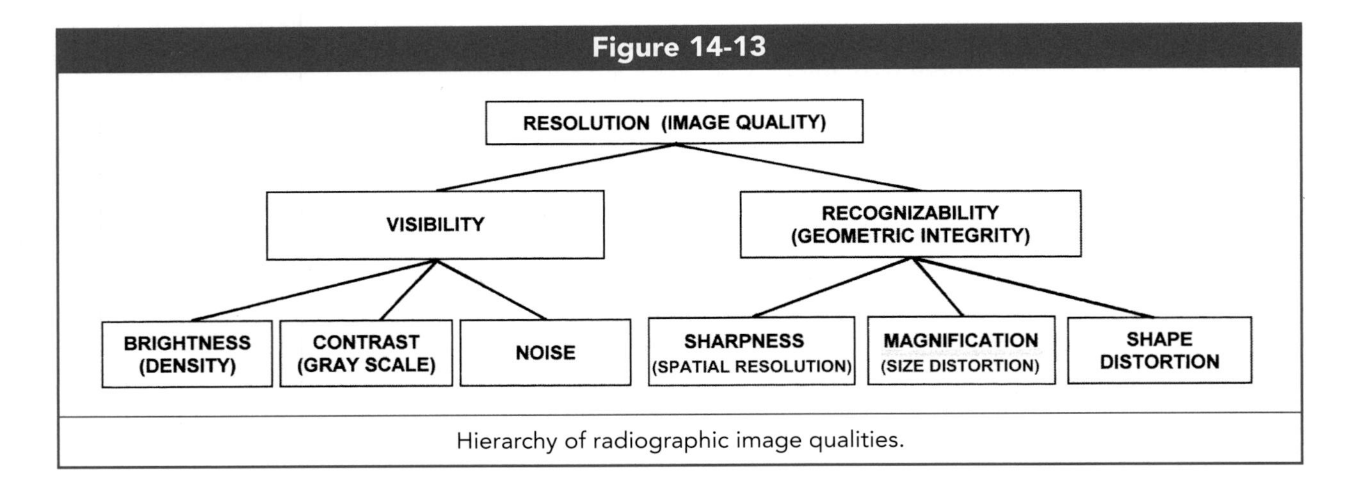

Hierarchy of radiographic image qualities.

hindered because of the crossing-over of the edges of image details. The lower portion of image ***B*** preserves the sharpness of the image above it, but has reduced contrast. Even though good sharpness is present, it is hard to *distinguish these lines apart* because of poor contrast, so resolution is again hindered. Both poor sharpness and poor contrast can degrade overall image resolution.

HIERARCHY OF IMAGE QUALITIES

By way of summary, Figure 14-13 presents a graphic hierarchy of all the essential image qualities for a radiograph. There are three factors affecting image visibility, and three factors affecting the recognizability or geometrical integrity of the image. All of these factors combine to produce good *resolution* in any image.

Standard definitions for digital radiography used by the American Registry of Radiologic Technologists (AART) are found in Appendix 2.

SUMMARY

1. The geometrical components of an image that affect its recognizability are sharpness of recorded detail, magnification and shape distortion. Sharpness should be maximized, and shape distortion minimized. Magnification is generally undesirable, except when a detail is too small to be recognized.
2. Every projected image consists of a central, clear *umbra* component and a blurry *penumbra* component at the edges. Penumbra can be caused by the projection geometry or by motion. As it grows outward, it also spreads inward, invading and shrinking the umbra.
3. *Sharpness* is defined as the abruptness with which the edges of a particular detail "stop" as one scans across an image.
4. Radiographic penumbra or unsharpness can be predicted and measured by penumbra diagrams or by the formula $FS \times OID/SOD$.
5. Relative sharpness in a radiographic image is proportional to the ratio SOD/OID.
6. To the human eye, high contrast can mimic improved sharpness, and poor contrast can be mistaken as blur. They are unrelated image characteristics and must not be confused.
7. Radiographic magnification is proportional to the SID/SOD ratio. For magnification to be present, the *umbra* portion of the image must expand (not just the penumbra).
8. Shape distortion is present only when the shape ratio changes because one dimension of the image (length or width) is changed by a different amount than the other.
9. Resolution indicates the overall quality of an image, and is specifically defined as the ability to distinguish adjacent details as being separate and distinct from each other. Although all image characteristics bear upon resolution, its two most important components are contrast and sharpness of detail.

REVIEW QUESTIONS

1. The recognizability factors in an image depend upon the ______ integrity of the projection.

2. Why is penumbra present to some degree in all practical images?

3. The umbra of very small details in an image can completely disappear by what geometrical process?

4. Misalignment of the x-ray beam, part, and image receptor results in what undesirable image quality?

5. Given a 2 mm focal spot size, and SOD of 90 cm and an OID of 30 cm, what is the geometrical unsharpness of the image (include units)?

6. Given an 0.5 mm focal spot, and *SID* of 150 cm and an OID of 15 cm, what is the geometrical unsharpness of the image (include units)?

7. What is the relative sharpness for an SOD of 90 cm and an OID of 15 cm?

8. An original exposure used an SOD of 120 cm and an OID of 12 cm. If the SOD is increased to 150 cm *and* the OID is reduced to 10 cm, by what factor will the sharpness be improved?

9. With an SID of 30 inches and an *OID* of 20 inches, by what factor will the image be magnified?

10. An image measures 8 inches across. A 30-inch SID was used with an SOD of 20 inches. What is the size of the original object?

11. What is the percentage magnification for an image with a magnification factor of 1. 44?

12. The radiographic image of a particular bone turns out 4 times longer than the real bone, but only 2 times wider. What is the factor of magnification? What is the factor of shape distortion?

13. The resolution of line pairs from a test template can be reduced by either of which two processes?

14. The ability to distinguish any two adjacent details as separate and distinct details is the definition for what aspect of an image?

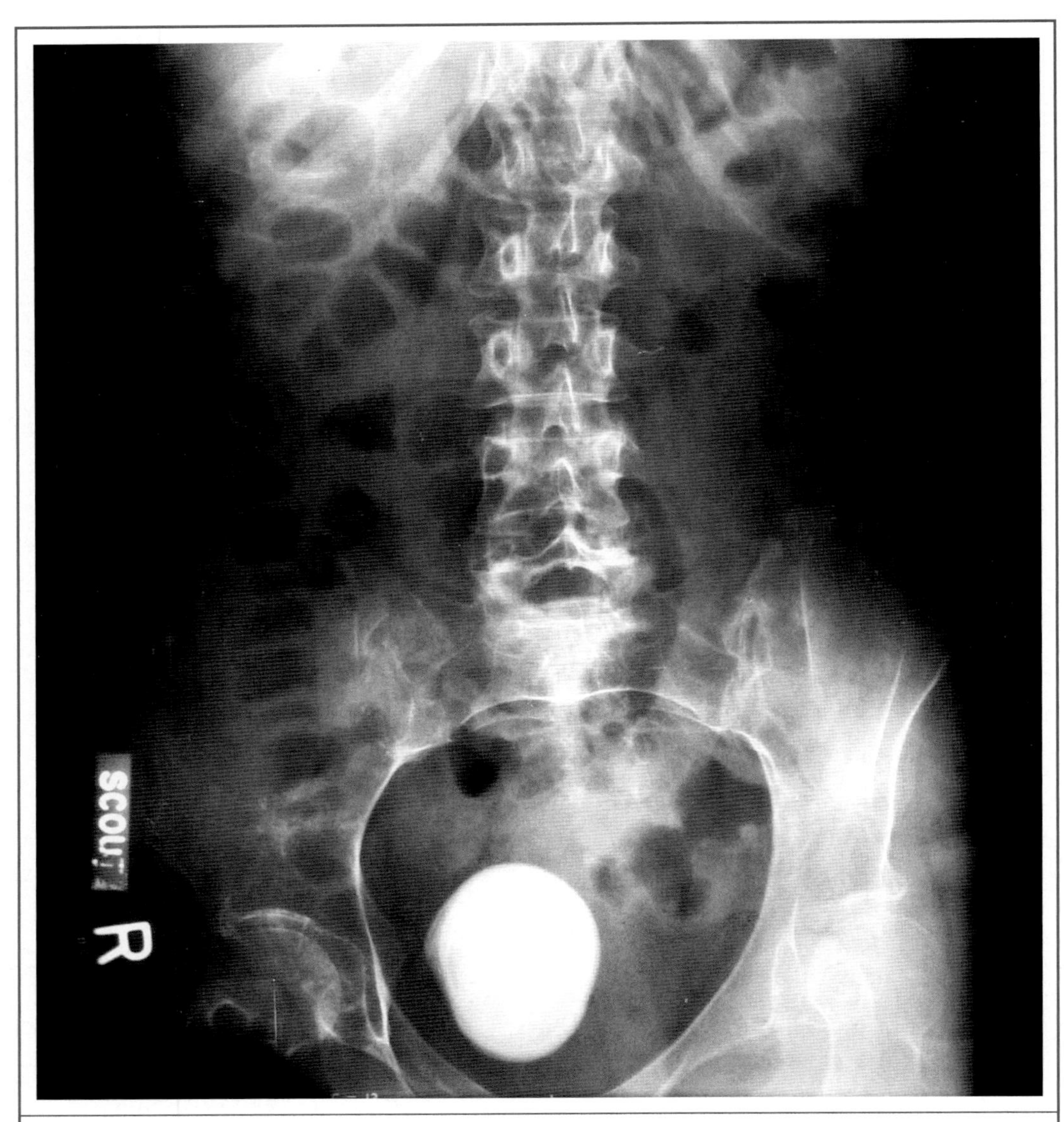

A giant bladder stone. Often kidney and bladder stones are small enough to urinate out. Not this one.

Chapter 15

MILLIAMPERE-SECONDS (mAs)

Objectives:

Upon completion of this chapter, you should be able to:

1. Describe what mAs controls in the x-ray beam, and distinguish between the proper units for x-ray beam intensity and electrical current intensity.
2. Describe why mAs is considered the *primary control* for exposure.
3. Calculate the total mAs for various combinations of mA and exposure time, and vice versa.
4. Develop the ability to do simple mAs calculations mentally.
5. Explain the root causes of quantum mottle or scintillation, and how to correct it.
6. List those image qualities which are *not* directly affected by mAs.
7. Accurately explain the relationship between exposure time and motion unsharpness.

Milliamperage, abbreviated *mA*, is a measure of the quantity of electrical current flowing through a circuit. As described in Chapter 7, it is a *rate* representing the number of electrons passing down a wire per second. The mA control on the x-ray machine console is actually a selector which taps off one of a series of different *resistors* in the filament circuit. The less the resistance, the more *intense* the flow rate of electricity to continue through the circuit toward the x-ray tube.

When the radiographer sets a higher mA setting, the greater is the flow rate of electricity passing through the filament in the x-ray tube cathode each time the *rotor* switch is depressed. As more current is forced through the thin filament wire, more friction results and the hotter the filament burns. Because of the high temperature of the filament, more electrons are *boiled off* of it by the process of *thermionic emission* (see Chapter 8).

These liberated electrons form a *space charge* or electron cloud around the filament, which reaches an equilibrium number of electrons based on the original set mA. This is the number of electrons available to be accelerated across the x-ray tube to strike the anode when the exposure switch is fully depressed. As the exposure progresses, the filament continues to replenish the electron cloud. Thereby, the rate of electrons per second striking the anode is steadily maintained throughout the exposure. This, in turn, determines the rate of x-rays produced each second. *The intensity rate of the x-ray beam is directly controlled by the mA station set at the console.* If the mA station is doubled, the "flow" of x-rays emitted in the beam is twice as much per second.

Exposure time, abbreviated *s* for *seconds*, is simply the amount of time during which the beam is activated and x-ray exposure is occurring. Since mA is a rate, the mA multiplied by the exposure time *s* will give an indication of the total intensity of the entire x-ray exposure made. In a similar way, the speed at which you drive your car (set on *cruise control*) multiplied by the amount of time you drive will yield the total miles driven, that is, the end product. (For example, 40 miles per hour × 3 hours = 120 miles covered.) For electricity, the mA is analogous to your car's speed, and the exposure time analogous to the amount of time you have driven.

The radiographic term *mAs* derives from this product of multiplying the *mA* × *s*, and is commonly pronounced as "mass." Strictly speaking, *mA* and *mAs* are electrical terms. They refer, respectively, to the rate of electrical current and the total amount of electricity used during an exposure. (And, strictly speaking, the proper terms that should be applied to the actual x-ray beam exposure are *R/s* (*Roentgens per second*) for the exposure rate, and *R* (*Roentgens*) for the total exposure.) However, the electrical factors *directly control* these exposure factors, so it is common for radiographers to use these terms interchangeably, or to use the electrical terms to describe the x-ray beam, as, for example, having produced "so much *mAs* of x-rays."

The effect of doubling the set mA upon the x-ray beam spectrum was described in Chapter 10. The occurrence of both bremsstrahlung and characteristic x-rays is doubled, making the spectrum curves on a graph twice as high. Doubling the exposure time has the same effect in terms of the total exposure achieved.

CONTROL OF X-RAY EXPOSURE

In radiography, milliampere-seconds, the product of the mA and the exposure time set at the x-ray machine console, is the primary electric *control* over x-ray exposure. (Many factors *affect* x-ray exposure, but when we state in this textbook that a particular variable is a "primary *control*," we mean that it is the *preferred* way to manipulate the x-ray beam.) One reason that the mAs is the preferred way to control the quantity of exposure is that it *only* affects the quantity of exposure, whereas the use of other variables may have an undesirable effect on things we do not wish to alter. (For example, kVp can be used to increase or decrease the exposure, but it also changes the penetration characteristics of the x-ray beam and the subject contrast present in the remnant beam, things we may not wish to tamper with.)

If *either* the mA station or the set exposure time is doubled, the total radiation exposure is doubled to both the image receptor plate and to the patient. (See Historical Sidebar 15-1.) Since this total is the product of the mA × time, we can further state that for a particular desired amount of exposure, *mA and exposure time are inversely proportional to each other*. That is, while maintaining a certain amount of exposure, if the mA is doubled, the time must be

HISTORICAL SIDEBAR 15-1: Figure 15-1 shows a series of old-fashioned film radiographs with the total mAs used indicated on each exposure. For each doubling of the mAs, the radiographs turned out twice as dark. This gave immediate feedback to the radiographer as to whether too much or too little exposure had been used.

In computer-based systems, underexposure or overexposure is compensated for by computer algorithms in the presentation of the final image on a display screen. One disadvantage of digital radiography systems is the lack of immediate and obvious feedback to the radiographer regarding the exposure level used.

Figure 15-1

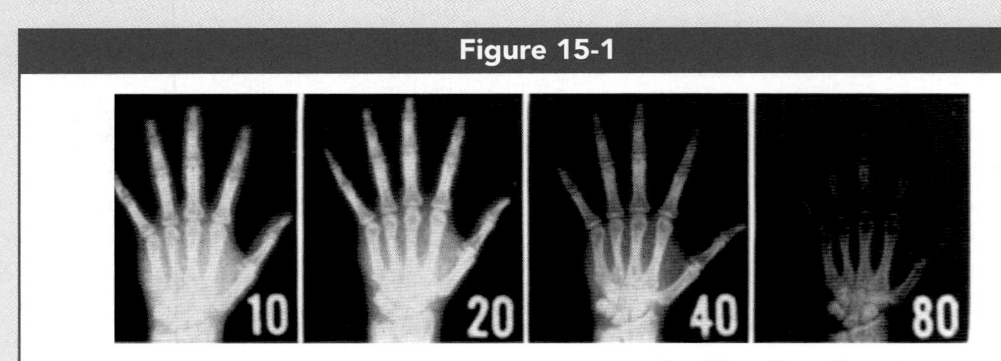

For conventional film radiographs, each doubling of the mAs resulted in a doubling of the overall image density, shown here with the mAs listed on each radiograph, all taken at 40 kVp. Note that digital systems can restore the 80-mAs image, but *cannot* restore information missing from the 10-mAs image. (From Quinn B. Carroll, *Practical Radiographic Imaging*, 8th ed. Springfield, IL: Charles C Thomas, Publisher, Ltd., 2007. Reprinted by permission.)

cut in half, and vice versa. This inverse relationship is mathematically expressed in the formula

$$mA_o \times s_o = mA_n \times s_n$$

where ***o*** is the old mA and exposure time and ***n*** is the new mA and exposure time. The product of mA times the exposure time must always yield the same total mAs to maintain a particular level of exposure.

Practice Exercise #1

Assume that 10 mA and an exposure time of 0.5 seconds was employed and resulted in a proper exposure. To control possible motion, it is desired to reduce the exposure time to 0.05 seconds. What mA should be used to assure a comparable overall exposure?

Solution:

$$10(0.5) = X(0.05)$$
$$5 = X(0.05)$$
$$\frac{5}{0.05} = X$$
$$100 = X$$

Answer: 100 mA should be used.

Practice Exercise #2

Assume that 500 mA and an exposure time of 0.5 seconds was employed and resulted in a proper exposure. To allow a tomographic breathing technique, it is desired to reduce the mA station to 100 mA. What new exposure time should be used to assure a comparable overall exposure?

Solution:

$$500(0.5) = 100(X)$$
$$250 = 100(X)$$
$$\frac{250}{100} = X$$
$$2.5 = X$$

Answer: An exposure time of 2.5 seconds should be used.

Doing the Mental Math

Time can be saved by doing the above two problems *mentally* if you recognize that for Practice Exercise #1, the new exposure time was 1/10th the original, so the new mA needs to be 10 times the original; and, for Practice Exercise #2, the new mA was one-fifth the original, so the new exposure time needs to be 5 times the original.

Chapter 3 includes instruction and practice exercises for converting fractions to decimals and vice versa, and Chapter 2 explains how to convert units such as seconds into milliseconds. These are strongly recommended for your review if you struggle with any of the math in this chapter.

In calculating the total mAs, one helpful thing for radiographers to remember is that most mA stations are listed in multiples of 100, so that if you *move the decimal place in the exposure time to the right two places*, you can *move the decimal place in the mA to the left two places*, so that the time is simply multiplied by the *first* number of the mA setting. For example, the total mAs for

300 mA at 0.05 seconds = 3 mA at 5 seconds = $3 \times 5 = 15$ mAs

Think of 300 mA at 0.08 seconds as simply $3 \times 8 = 24$ mAs, and 100 mA at 0.025 seconds as $1 \times 2.5 = 2.5$ mAs. As an example of an exposure time with only one digit after the decimal, think of 200 mA at 0.4 seconds as $2 \times 40 = 80$ mAs. As an example of an mA station with only two digits, think of 50 mA at 0.08 seconds as $0.5 \times 8 = 4$ mAs. *Milliseconds* must be converted to seconds before doing this.

For practice, complete the following two mAs calculation exercises. The answers are found in Appendix I.

EXERCISE #15-1

mA X decimal time = total mAs

1. 100 @ .05 =
2. 200 @ .125 =
3. 300 @ .7 =
4. 300 @ .025 =
5. 400 @ .33 =
6. 500 @ .03 =
7. 600 @ .25 =
8. 600 @ .008 =

EXERCISE #15-2

mA X milliseconds = total mAs

1. 50 @ 50 ms =
2. 100 @ 35 ms =
3. 200 @ 125 ms =
4. 300 @ 33 ms =
5. 300 @ 6 ms =
6. 400 @ 80 ms =
7. 500 @ 5 ms =
8. 600 @ 150 ms =

One of the most common math problems a radiographer faces on an everyday basis is having a desired total mAs in mind, and needing to mentally determine a good mA-time combination that will yield that total. The first step is to decide upon an appropriate mA station. Generally, the *highest* mA station available for the desired focal spot should be used, in order to minimize exposure time and the probability of patient movement. If high sharpness is more important than patient exposure, as is generally the case for the distal extremities, the *small focal spot* should be used, and the highest mA station allowed for the small focal spot (usually 200 or 300 mA) should be engaged.

When patient exposure is more important than fine detail, generally the case with all chest and abdomen procedures, the large focal spot should be employed and mA stations as high as 500 or 600 can be engaged. It is possible for the mA station to be too high to obtain a particularly low desired total mAs—use the highest mA at which this total mAs can be obtained. For tomography procedures and "breathing techniques," the situation is reversed: Long exposure times are desired, and very low mA stations are recommended to achieve the longest practical times.

As an example of this type of mAs calculation, let us assume that you need to obtain 80 mAs for a total exposure, high sharpness of detail is desired, and the 200 mA station is the highest mA available with the small focal spot. Simply divide the mA station *into* the desired mAs:

$$mAs = mA \text{ X } time$$

$$80 = 200(\text{X})$$

$$\frac{80}{200} = \text{X}$$

$$0.4 = \text{X}$$

The appropriate exposure time is 0.4 seconds, or 400 milliseconds. Try the following exercise for practice, and check your answers in Appendix #1.

EXERCISE #15-3

Total mAs = mA X seconds

1. 2.5 mAs = 100 mA X _______
2. 40 mAs = 100 mA X _______
3. 1.25 mAs = 50 mA X _______
4. 5 mAs = 200 mA X _______
5. 14 mAs = 200 mA X _______
6. 50 mAs = 300 mA X _______
7. 6 mAs = 300 mA X _______
8. 21 mAs = 300 mA X _______
9. 180 mAs = 300 mA X _______
10. 240 mAs = 400 mA X _______

UNDEREXPOSURE AND QUANTUM MOTTLE

During a light rain shower, you can see the individual raindrops on a sidewalk. If you count the raindrops on each square of cement, you will see that they are not evenly distributed; more raindrops fall in some areas than others. It is a random phenomenon and a matter of statistical probability which squares of cement will be exposed to more raindrops. When a very *heavy* rainfall comes, the same uneven distribution of raindrops is still there, but you can no longer tell, because the sidewalk is now saturated with water, leaving no dry spots between the wet areas.

The x-ray beam is a *shower* of x-rays which have a random distribution just like the drops in a rain shower. When very low mAs values are used and there are few photons striking the image receptor, one can see the uneven distribution of the exposure. The radiographic image appears *grainy*, that is, very small "freckles" or blotches of dark and light are seen across the image when closely examined. An extreme example is presented in the knee radiograph in Figure 15-2. These small blotches are called *quantum mottle*, a mottled appearance of the image caused by the *quanta* or photons in the x-ray beam. Visible mottling of the image indicates that an insufficient amount of x-rays have reached the image receptor plate.

Like the rain shower on the sidewalk, only by delivering plenty of exposure to the image receptor can the variations in intensity be subdued. Quantum mottle, like scatter radiation, is a form of *image noise*, and is a factor in the *signal-to-noise ratio* (SNR). But, the randomness of the x-ray beam distribution is not something we can have much influence over—it is an ever-present aspect of the x-ray

Figure 15-2

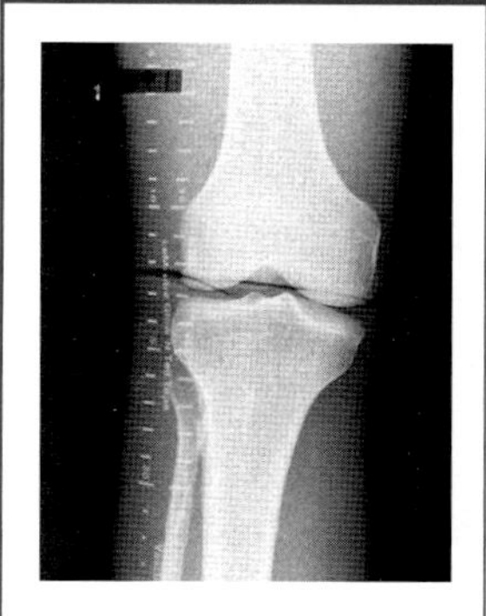

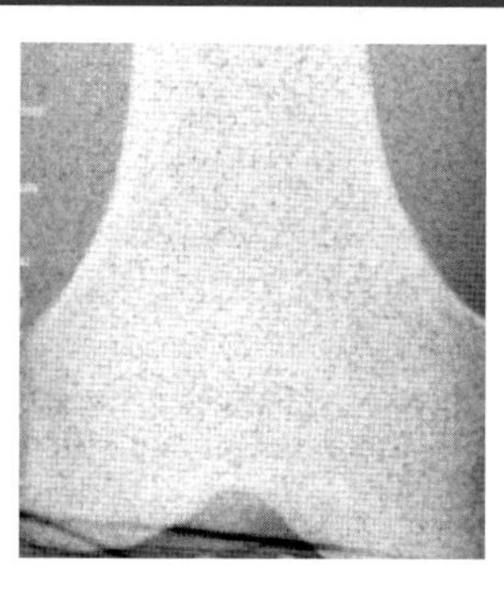

A knee radiograph demonstrates quantum mottle from an insufficient exposure level at the image receptor plate.

beam. Therefore, the best way to subdue it is to "overwhelm it with good exposure," that is, with information-carrying exposure or *signal*. By doing this, the mottle does not go away but becomes less apparent in the image; it makes a lesser *percentage contribution* to the overall image.

Figure 15-3 is a series of fluoroscopic images of some catheters, displayed on a CRT monitor screen. The mottle in image ***A*** is pronounced. As the intensity rate of the fluoroscopic x-ray beam (the mA) is turned up, the mottle becomes less and less apparent. The same phenomenon affects "overhead" radiographs or "still" images.

A common cause of visible quantum mottle in radiographs is an insufficient *mAs* setting. However, anything which results in an insufficient amount of exposure at the image receptor plate can lead to mottle, such as inadequate kVp to penetrate the part, or rectifier failure in the x-ray machine.

Computer-based systems essentially amplify the input from each pixel of the image receptor, which exaggerates the level of image noise. This makes all digital systems highly sensitive to quantum mottle. Whereas extreme underexposure was required to reveal mottle in film imaging systems, it shows up in digital imaging systems *with only slight underexposure* at the image receptor plate. This makes it absolutely critical that the radiographer employ radiographic techniques that ensure sufficient exposure will reach the image receptor. What is worse, for all electronic display systems such as LCDs, *electronic mottle* constitutes additional noise in the image. These will be fully discussed in the section on digital radiography.

Generally, to correct for excessive quantum mottle in digital images, one must first ensure that a level of kVp sufficient to fully penetrate the body part is used, taking into account extreme body part size, disease conditions or any other factor that will reduce the flux of the x-ray beam. Then, be sure to set sufficient mAs so that the initial quantity of x-rays in the beam is intense enough that the small percentage (about 1% on average), which reaches the receptor plate is still adequate to provide plenty of signal full of useful information.

For digital images, overexposure due to the use of excessive mAs is *not* apparent in the quality of the image. The only means for the radiographer to become aware of overexposures is to monitor the

Figure 15-3

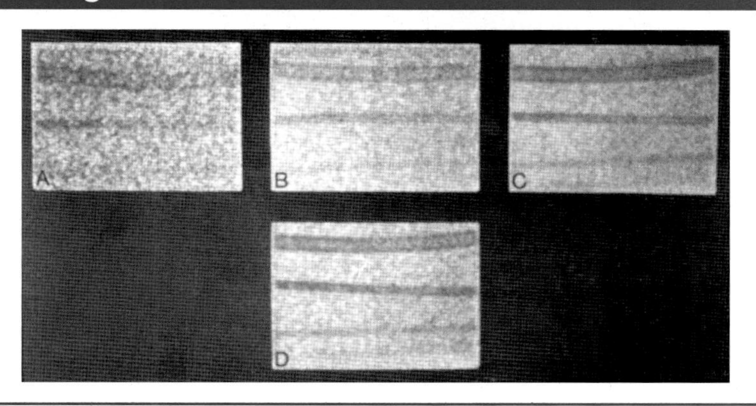

TV monitor images of catheters demonstrating mottle or scintillation. As the signal is increased by turning up the fluoroscopic mA, the signal-to-noise ratio (SNR) is improved, mottle becomes less apparent and the image appears much smoother. (Reprinted with permission, Lea & Febiger, *Christensen's Physics of Diagnostic Radiology*.)

exposure indicator readout which may or may not be annotated on the image.

SUBJECT CONTRAST AND OTHER IMAGE QUALITIES

As demonstrated in the x-ray beam spectrum graphs in Chapter 10, changes in mAs affect the intensity of the x-ray beam, or the quantity of x-rays, but *not* the *energies* in the beam, whether peak, minimum, or average. For example, when the mAs is doubled, there will be twice as many x-rays *at every energy level*. Since the *average energy* remains unchanged, the penetration characteristics of the beam are the same. Within the patient, the number of penetrating x-rays doubles, as does the number of photoelectric interactions and Compton scatter interactions. Since all interactions increase by an equal amount, their *percentages* of occurrence relative to each other remains the same.

In Chapter 12 the production of *subject contrast* within the remnant beam was fully discussed. Remember that the subject contrast is determined by the *percentage* contribution from each type of interaction to the total information reaching the image receptor. Changes in mAs do not alter these percentages in any way, nor the penetration of the x-ray beam, which itself is a *percentage*. Therefore, mAs cannot have any direct bearing upon subject contrast except in the most extreme cases.

Furthermore, in Chapter 13, we proved that image contrast is *independent* of overall image brightness (or density). Changes in the one do *not* automatically alter the other. *The mAs is not a factor in determining contrast.*

The sharpness of recorded detail in the image, magnification and shape distortion are all *geometrical* factors in the image, whereas the mAs is an *electrical* factor. The mAs can have no relation whatever to any of these recognizability functions in the image.

To summarize what mAs *does* affect, it controls the intensity of the exposure at the image receptor plate, which bears upon image *brightness*, and although it has no relationship to scatter radiation, it does have a great deal to do with image *noise in the form of quantum mottle.*

EXPOSURE TIME AND MOTION

Long exposure times are not the *direct* cause of motion—it is the patient who moves. When the patient is fully cooperative, holds still and does not breath during the short time it takes to make an exposure, no motion occurs. However, neither the patient nor the radiographer has control over *all* types of motion. In chest radiography, the movement of the heart is unavoidable. Its effects in blurring the image can only be eliminated by the use of very short exposure times that effectively "freeze" the motion, just as high-speed photographic exposures must be used to freeze the motion of athletes or race cars.

Also, it is clear that the longer the exposure time, the greater *chance* there is for motion to occur. Peristaltic motion in the stomach or intestines may surge, a child or an intoxicated patient might move or breathe. Therefore, exposure time is generally considered to be a *contributing factor* for motion blur during radiographic exposures.

(A fascinating sidelight is that in this regard, and *only* in this regard, the factor of exposure time "crosses over" from the visibility factors in the *hierarchy of image qualities* to the geometrical factors affecting the image. Einstein proved that time *is* indeed a form of geometry, constituting a fourth dimension.)

Motion is the greatest enemy to sharpness of recorded detail in the image. Shorter exposure times cannot *guarantee* that motion will not occur, but since some forms of movement are beyond the control of the radiographer, *it is generally assumed that the shorter the exposure time, the sharper the images are likely to be.*

SUMMARY

1. Milliamperage (mA) measures the rate of electricity flowing through the x-ray tube, and controls the rate of x-rays emitted from it.
2. The mA stations at the console actually select from different resistors to control the amount of amperage flowing through the filament to

maintain a steady space charge boiled off by thermionic emission.

3. Milliampere-seconds (mAs) are the product of mA and exposure time, and control the total amount of x-rays delivered from the x-ray tube during an exposure. In producing a given exposure, mA and exposure time are inversely related to each other.
4. The mA, s, and mAs are all directly proportional to delivered exposure. The mAs is the preferred *controlling factor* for total exposure.
5. Generally, the highest mA station available for a particular focal spot should be used to minimize the chance of patient motion. For distal extremities which do not require higher mA stations, the small focal spot should be used to maximize image sharpness.
6. Underexposure, which can result from insufficient mAs, results in the appearance of quantum mottle in the image, especially in digital systems.
7. Overexposure from excessive mAs is not apparent in digital images, and can only be monitored by checking the *exposure indicator readout.*
8. The mAs does not affect the average energy or penetration characteristics of the x-ray beam, and has no impact on the relative percentages of different interactions contributing to the image. Therefore, mAs is not considered a factor in controlling subject contrast.
9. The mAs has no direct relationship with recognizability factors in the image (sharpness, magnifcation and distortion). However, shorter exposure times make unsharpness due to motion less likely to occur.

REVIEW QUESTIONS

1. Strictly speaking, mA is a unit which measures ___________.

2. Give two reasons why mAs should be considered as the prime factor in controlling x-ray exposure:

3. The mAs does not alter subject contrast in the remnant beam because it does not change the ___________ of different types of interactions occurring within the patient.

4. For what type of radiographic procedure would a low mA and long exposure time be needed?

5. Insufficient mAs can cause the appearance of ___________ in the image.

6. Generally, short exposure times are desirable to minimize the probability of __________.

7. In maintaining an overall exposure, if the mA is tripled, the exposure time should be changed to ___________.

Calculate the following total mAs values from the mA and exposure time combination listed:

8. 200 mA @ .035 sec =

9. 300 mA @ .006 sec =

10. 500 mA @ .124 sec =

11. 50 mA @ 300 ms =

12. 600 mA @ 12 ms =

(Continued)

REVIEW QUESTIONS *(Continued)*

Give the decimal exposure time required to complete each of the following:

13. 16 mAs = 100 mA @:

14. 2.5 mAs = 50 mA @:

15. 120 mAs = 200 mA @:

16. 75 mAs = 300 mA @:

17. 80 mAs = 400 mA @:

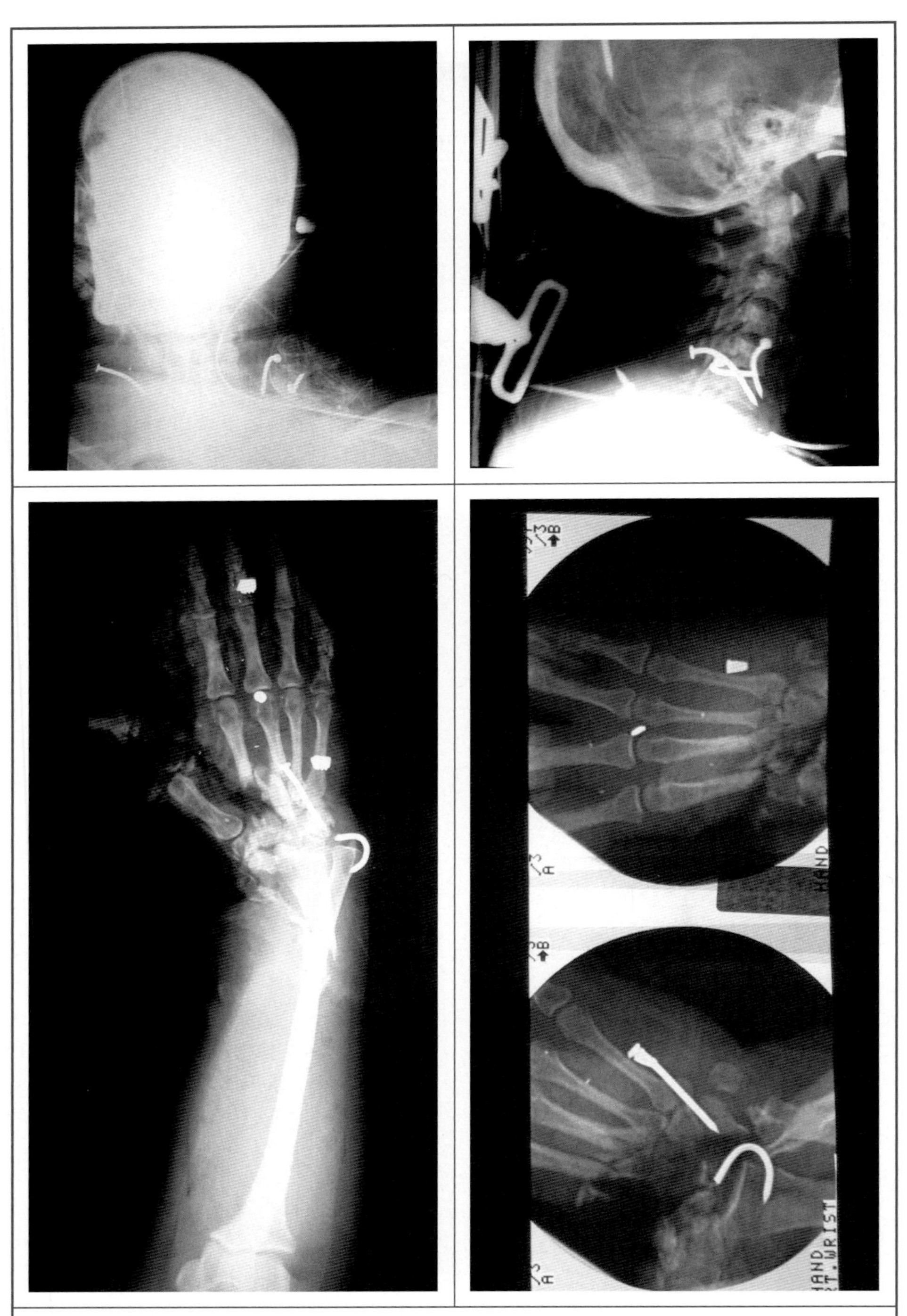

Unfortunate victim of a homemade pipe bomb. Shown here are various nails embedded in her head, neck and upper extremities.

Chapter 16

KILOVOLTAGE-PEAK (kVp)

Objectives:

Upon completion of this chapter, you should be able to:

1. Describe what kVp controls in the x-ray beam, and distinguish between the proper units for x-ray beam penetration and electrical force.
2. Describe the relationship between increasing beam penetration and subject contrast in the remnant beam signal.
3. Define the *minimum kVp* for a body part, and what determines it.
4. Explain why increases in x-ray intensity (mAs) cannot compensate for insufficient penetration (kVp).
5. Distinguish between the effects of increased kVp on x-ray tube output and on the remnant x-ray beam reaching the image receptor.
6. Accurately calculate proper mAs compensations for various kVp changes using the 15% rule.
7. Define *optimum kVp* for digital imaging, and how it differs from *minimum kVp*.
8. Describe how the 15% rule can be used to advantage in lowering patient exposure.
9. Rank kVP against body part thickness and field size as a contributor to *scatter* radiation.
10. Explain why much higher kVp's can be used with digital imaging than used to be used for film radiography.
11. List those image qualities which are *not* directly affected by kVp.

Kilovoltage, abbreviated *keV* or *kV*, is a measure of the electrical force or *pressure* behind a current of electricity, which causes it to flow. It is a measure of electrical *energy*. In the x-ray machine, the kilovoltage control at the console is actually the *autotransformer* in the high-voltage circuit. Whenever a potential difference exists between two points in a conductor, one end having a relative negative charge and the other a relative positive charge, electrons will flow through the conductor toward the positive charge. Extreme positive and negative charges are applied across an x-ray tube.

The greater the potential difference, the more "pressure" is exerted on the electrons to flow, the greater the *energy* pushing the current, and the higher kV will be measured. Whereas *mAs* has been described as a measure of electrical quantity, kV is a measure of electrical *quality*.

Due to the rotation of magnetic fields in the AC electrical generator that powers most x-ray equipment, the actual kilovoltage of the current supplied to the x-ray tube varies up and down in a sine-wave pattern (Chapter 8), rising to a peak and then falling back to zero repeatedly. Since the kilovoltage is constantly changing, it is necessary to measure it in terms of either the *average* value or the *peak* value attained during this repeating cycle (Fig. 16-1). Hence, the term *kilvoltage-peak*, or *kVp*.

The effects of higher kVp upon the x-ray beam spectrum were fully discussed in Chapter 10. The main purpose for adjusting the kVp is to set the *penetration* level of the x-ray beam. The percentage of x-ray penetration through any particular tissue or patient is a direct function of the *average energy* of those x-rays. This average energy is pulled upward when the *peak energy* is increased.

Figure 16-1

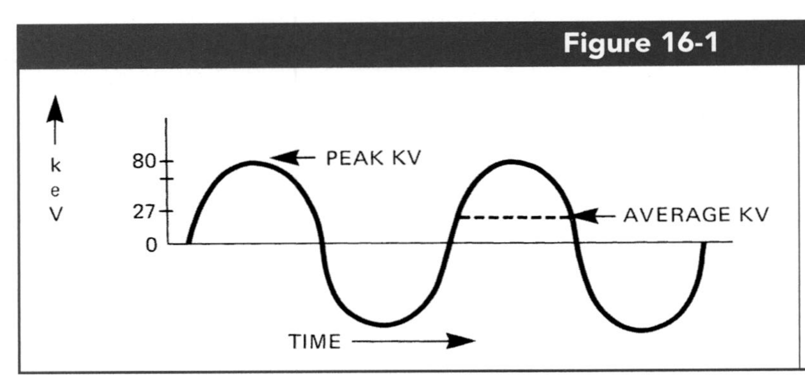

The sine-wave graph for AC current shows that the kilovoltage is constantly changing. Therefore, kV must be measured as either the *average kV* or the *peak kV, (kVp)*. For single-phase current, the average kV is about one-third of the kVp. (From Quinn B. Carroll, *Practical Radiographic Imaging*, 8th ed. Springfield, IL: Charles C Thomas, Publisher, Ltd., 2007. Reproduced by permission.)

But, an increase in kVp also results in more bremsstrahlung x-rays being produced within the x-ray tube anode, so that the *quantity* as well as the quality of the x-rays is increased. This is a somewhat undesirable side-effect for the radiogapher when the intent is mainly to adjust penetration. Generally, the radiographer prefers to use *mAs* to control x-ray quantity because mAs is directly proportional to output and does not affect other aspects of the beam. But, increased output from the x-ray tube cannot be avoided at higher kVp's, and must be taken into account when considering the total exposure that penetrates through the patient to the image receptor.

SUFFICIENT PENETRATION AND SUBJECT CONTRAST

An x-ray beam with higher average energy will be capable of penetrating through more different types of tissue. *This is the most important function of the kVp, to provide at least partial penetration through all tissues to be recorded.*

As described in Chapter 11, the ideal degree of subject contrast within the remnant x-ray beam is an *intermediate* level. At the proper level, *all* tissues will be represented at the image receptor by some degree of x-ray exposure that has penetrated through them.

Imagine beginning at an extremely low kVp and gradually increasing it to observe the effects upon the image. When the set kVp is too low, it is possible for two adjacent soft tissue organs to both absorb nearly all of the x-rays incident upon them. They would both be represented as essentially blank or "white" areas on the image. There is little contrast between them, so they would not be distinguishable from each other. Information is lost from the image. This is dramatically shown in Figure 16-2, where *all* of the series of film radiographs ***A*** through ***D*** present silhouette images with no detail at all visible *within the bones* of the palm because only 30 kVp was used.

As the kVp is gradually increased, tissues with lower atomic numbers are recorded as different shades of gray, but bones, barium and iodine may still be recorded as "white" silhouette images. Bone marrow and other osseous details may not be visible, nor any anatomy *through* the contrast agents. Continuing to still higher kVp, these details within the bone and through the contrast agents become visible, recorded as light shades of gray while other tissues are seen as medium and dark shades.

Figure 16-2

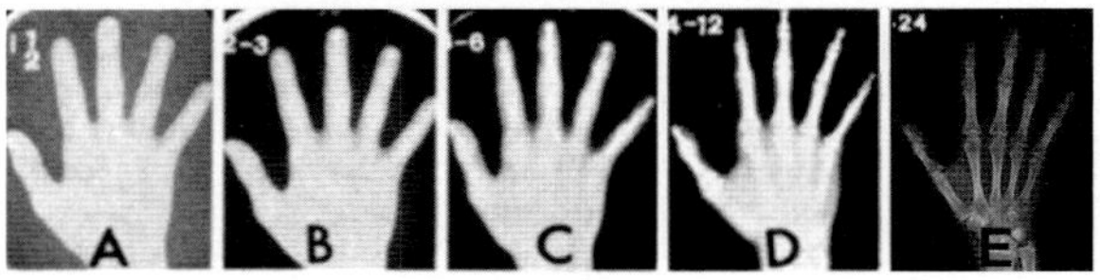

Film-based hand radiographs showing a whole series of "silhouette" type images, ***A*** through ***D***, taken at 30 kVp which provided insufficient penetration. Even at 360 mAs, ***D***, the image is still missing anatomical details within the bones. *No amount of radiation (mAs) can compensate for insufficient penetration (kVp).* Using direct exposure film, the mAs values used were ***A***, 45; ***B***, 90; ***C***, 180; and ***D***, 360. Radiograph ***E*** was taken using 54 kVp and 50 mAs, for comparison. (From Quinn B. Carroll, *Practical Radiographic Imaging*, 8th ed. Springfield, IL: Charles C Thomas, Publisher, Ltd., 2007. Reproduced by permission.)

Since more information is present, it is generally *desirable* to have long gray scale in an image. Figure 16-3 presents a series of film-based chest images to illustrate this essential point. Note that the high contrast chest image (#1) has the least amount of information in it. One can literally count the visible details for comparison.

The minimum kVp for a particular body part is defined as the lowest kVp that still provides some degree of penetration through all tissues of interest. Table 16-1 is a recommended list of minimum kVp's for various landmark portions of the body.

(Note that higher *optimum kVp's* are recommended in Table 16-3 for digital imaging. Here, we are only addressing the question of *minimum* kVp to produce sufficient signal in the remnant x-ray beam, regardless of the type of image receptor system used, be it film, phosphor plates, or direct-capture DR detectors.)

What, then, determines the *minimum kVp for any given body part*? The answer is that, in addition to part thickness, the *predominant type of tissue* that makes up the body part must be taken into consideration. The typical lumbar spine series provides an excellent example of this concept: When comparing the AP, oblique, and lateral projections, the predominant type of tissue in the abdomen is *soft tissue* for all of these projections. The obliques and lateral present a greater thickness of tissue, so *some* aspect of technique must be increased, but it does not have to be the kVp. Either the mAs or the kVp could be increased to restore the proper exposure to the image plate behind the abdomen. (In terms of saving *patient exposure*, the kVp is actually preferred, but in terms of maintaining exposure at the image receptor, either one will do.)

However, when changing from the full lateral projection to the "coned-down" L5–S1 spot, *it is essential to increase the kVp*, rather than the mAs. This may seem odd at first glance because particularly on a male patient *the lateral thickness of the waist and the hips will likely be the same.* The key difference is that within the lateral pelvis, *bone tissue becomes much*

Figure 16-3

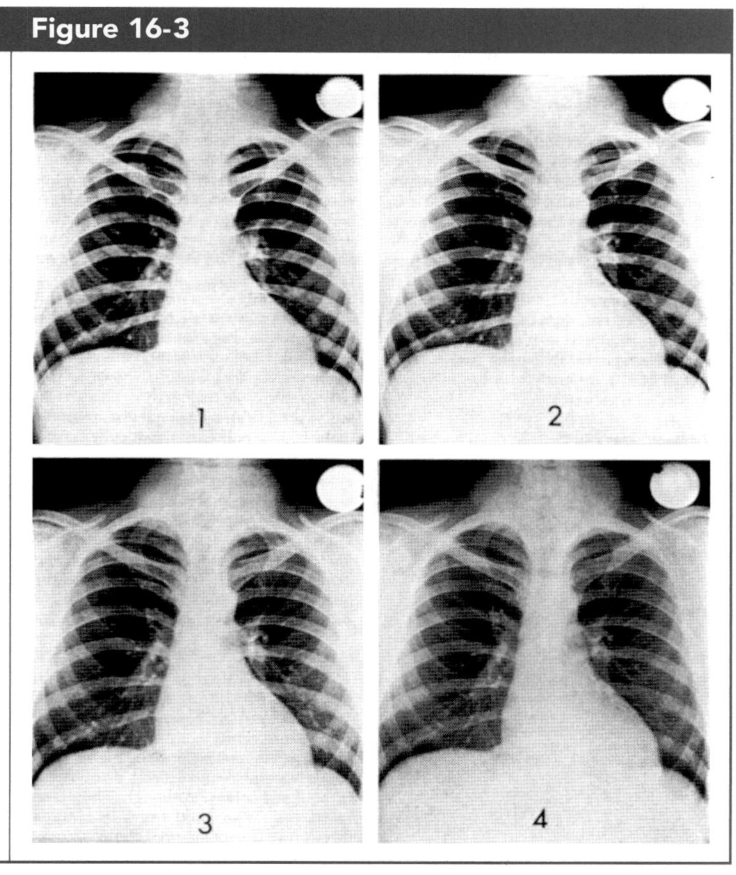

Film-based chest radiographs demonstrating the desirability of long gray scale in the image. The kVp was increased from 50 in image #1 to 100 in image #4, allowing the x-ray beam to progressively penetrate through more and more different types of tissue, literally adding to the number of details visible in the longer gray scale images. (From Quinn B. Carroll, *Practical Radiographic Imaging*, 8th ed. Springfield, IL: Charles C Thomas, Publisher, Ltd., 2007. Reproduced by permission.)

Table 16-1

Minimum kVp for Sufficient Penetration

Procedure	Minimum kVP
Hand/Wrist/Digits	54
Elbow/Forearm/Foot	62
Ankle/Leg	66
Knee/Humerus	70
Mandible, Tangential Skull	70
Femur/Shoulder/Sinus/Ribs	76
Cervical and Thoracic Spines	76
Iodine Procedures (IVP, cystog)	76
Abdomen/Pelvis/Lumbar Spine	80
Non-Grid Chest (fixed unit)	80
Skull	80
Air Contrast Barium Studies	92
Esophagram	92
Solid-Column Barium Studies	110
Grid Chest (fixed unit)	120

more predominant as a percentage of the overall tissue thickness; the sacrum is as wide as three vertebrae combined, and this is overlapped by both right and left iliac bones as well. As a proportion of the total tissue in the body part, the percentage of bone in the lateral pelvis is approximately *4 times* the percentage of bone in the waist portion of the abdomen. This requires an increase specifically in the *penetration power* of the x-ray beam, not just an increase in quantity. Most radiographers increase 8–12 kVp for this view over the routine lateral L-spine.

Likewise, when a large "solid-column" bolus of barium is introduced into an organ such as the stomach or colon, the kVp must be raised to the range of 110–120 in order to penetrate the contrast agent. For "air-contrast" studies which include a coating of thick barium around the lumen of the stomach or colon mixed with air, kVp should be set in the range of 90–100. For intravenous urograms (IVU's or IVP's), 76–80 kVp is sufficient because the column of iodine is very thin as it passes through the renal pelvis and narrow ureters. A larger bolus of iodine in the stomach, such as *Gastrografin*™, would require higher kVp, in the range of 86–90.

To appreciate the importance of kVp, the radiographer must understand that *no amount of radiation intensity can ever compensate for insufficient penetration* of all tissues within the body part. Suppose that the kVp is set too low for any substantial penetration through bone. An example might be setting 50 kVp for a lumbar spine series. The bones will be recorded at the receptor plate as blank areas where almost no exposure to the image receptor is measured. Now if the mAs is doubled, twice as many x-rays will be incident upon the bone; nonetheless, with insufficient energy, almost none of them are able to penetrate through the bone. The bones will *still* be recorded as essentially blank areas. This is dramatically illustrated with film radiographs in Figure 16-3. *No amount of mAs can ever compensate for insufficient kVp.*

With the same scenario of a lumbar spine series, but with the kVp set at 90, let us assume that penetration through the vertebral bone is now 5 percent. If the mAs is then doubled, the bone image will receive twice the previous exposure along with all the other tissues, but the penetration through the bone will still be 5 percent, and the subject contrast, determined by the *ratio* between the penetration of the bone and the penetration of the adjacent soft tissue, will be unchanged. Penetration and subject contrast are both controlled primarily by kVp and are both independent of mAs.

THE FIFTEEN PERCENT RULE

The relationship between kVp and actual exposure at the image receptor is not a linear one, but an *exponential* one. Remember that higher kVp not only increases the penetrating power of the x-ray beam, but also results in a higher quantity of x-rays being produced within the anode. Thus, both the quantity and the quality of the x-ray beam are affected. The actual output of the x-ray tube goes up approximately by the square of the kVp.

However, this formula only applies to the intensity of the *primary* beam, not the end result at the image receptor. The radiographer is not very concerned with the initial output of the x-ray tube, but rather with the remnant beam exposure *at the imaging plate* which determines the qualities of the final image. Here, the exposure increases approximately by the *fourth power* of the kVp. This is a much greater increase than the initial tube output, for the fundamental reason that the remnant x-ray beam is the product not only of the initial intensity of the radiation, but also of the *penetration* factor for that radiation after it has passed through the patient.

In other words, to estimate the effect of a higher kVp on exposure to the imaging plate, we must take into account *both* the increased output of the x-ray tube and the increased penetration through the patient. When this is done, we find that the final exposure rate to the image receptor goes up by about the fourth power of the kVp.

A simpler rule-of-thumb conventionally used by radiographers is called the *15 percent rule*. It states that *a 15 percent change in kVp will change the exposure of the image receptor to the remnant beam by a factor of 2*. This is equivalent to the fourth-power rule, but easier to calculate mentally and to apply in daily practice. If the kVp is increased by 15 percent, from 80 to 92 kVp for example, the total exposure delivered to the image receptor will double. If the kVp is *reduced* 15 percent, from 70 down to 60 kVp for example, the total exposure to the imaging plate will be cut to about one-half.

This effect on total remnant exposure is due to *the combination of both* the change in penetration and the change in intensity of the primary beam. Note that a doubling of the exposure is expressed in percentage as a 100 percent increase. That is, a 15 percent increase in kVp results in a 100 percent increase in net exposure at the image receptor. Of this 100 percent, about 25 percent comes from the original increase in x-ray tube output, and the remaining 75 percent from increased penetration through the patient.

There are procedures for which it is desired to alter the penetration characteristics of the beam or the subject contrast *while maintaining the original overall exposure* at the image receptor. For example, a solid-column barium enema requires the same overall exposure as a routine abdomen radiograph with no contrast agent in the abdomen, but must have increased beam penetration for the x-rays to get through the barium column. To calculate an appropriate technique, *the 15 percent rule can be used to adjust the mAs.*

For example, let us assume that for a particular three-phase x-ray room a technique of 80 kVp and 20 mAs produces good results for a routine AP projection of the abdomen. For a solid-column barium enema, approximately 110 kVp is desired to assure full penetration through the barium column. What combination of mAs should be used with the 110 kVp? To obtain the answer, use the 15 percent rule *in steps*, cutting the mAs in half for each 15 percent step increase, as follows:

Starting technique = 80 kVp @ 20 mAs

15% of 80 is +12 kVp

1st 15% increase = 92 kVp @ 10 mAs

15% of 92 is +14 kVp

2nd 15% increase = 106 kVp @ 5 mAs

This approximates our goal of 110 kVp—slightly less mAs can be used with slightly more kVp to get this number right on. Clinical experience confirms that in the range of 106 to 110 kVp, 4 or 5 mAs gives the correct end result exposure at the imaging plate for a solid-column barium study. (For air-contrast barium studies, a kVp in the range of 90–94 is often desired; using only the first-step increase above, we can see that 8 to 10 mAs would be appropriate for this range.)

Thus, the 15 percent rule can be re-stated as a *technique adjustment* rule as follows: *To maintain exposure at the image receptor, for every 15 percent change in kVp, adjust the mAs by a factor of 2.*

Doing the Mental Math

To find 15 percent of any number, take 10 percent of it and then add one-half that much again. For example, 10 percent of 80 would be 8, plus one-half of 8 for a total of 12 kVp. In the range of 80 kVp, a 12-kVp change is required to double or to halve the exposure at the image receptor. Examine the three pairs of techniques listed below. Using the 15 percent rule, you should be able to surmise that each pair consists of two equivalent techniques in terms of the final remnant beam exposure produced at the imaging plate.

A	*B*
60 mAs at 40 kVp	25 mAs at 92 kVp
30 mAs at 46 kVp	50 mAs at 80 kVp

C
15 mAs at 110 kVp
7 mAs at 126 kVp

For accuracy when adjusting kVp in steps, it is important to recalculate 15 percent *for each individual step*. In the above example using a barium enema, two step-increases were made. If the 15 percent rule is simply doubled and applied to the whole problem in one step, we would take 30 percent of 80 kVp, which is 24, and add it for a total of 104 kVp. But note that by taking 15 percent of 80 in the first step, and *then 15 percent of 92* for the second step, we get a higher result of 106 kVp. This is more accurate. Each 15 percent step change should be applied *to the result* of the previous calculation.

Astute radiographers can also learn to apply the 15 percent rule *in portions*. For example, suppose that it is desired to increase exposure to the image receptor by 50 percent. This is one and a half times the original, or half-way to doubling it. Since a 15 percent increase will double the exposure, *one-half of a 15 percent increase* should approximate our target amount. This would be calculated as follows for a starting point of 80 kVp:

15% of 80 kVp = 12

½ of 12 = 6

80 + 6 = 86 kVp

Solution: 86 kVp will produce a 50% increase in exposure over 80 kVp.

When considering a *reduction* in exposure using kVp, *factors* can always be applied consistently, but *percentages* work a little differently and can be tricky. (For example, when we consider a factor of 2, we find that a doubling is reported in percentage as a 100% increase, but a *halving* is reported in percentages not as a 100% reduction, but as a *50%* reduction.)

With this in mind, if we repeat the above problem, but in terms of *going down 15 percent in kVp*, how would we report the results? Using factors, the language is consistent: Adding 6 kVp increased the exposure half-way to double, subtracting 6 kVp would decrease the exposure half-way to one-half. But, be careful reporting this in percentages—it comes out to 75 percent of the original exposure (*down* half-way from 100% to 50%). A reduction from 80 kVp to 74 kVp reduces exposure to about three-fourths.

To reinforce the use of the 15 percent rule in steps and in portions, complete the following practice exercise, without pencil and paper if possible, and then check your answers from Appendix #1.

EXERCISE #16-1

1. What is *one-half* of 15% of 120?
2. Starting at 120 kVp, what new kVp would result in one-half the exposure?
3. Starting at 60 kVp, what new kVP would result in 50% more (1½ times) exposure?
4. Starting at 80 kVp, what new kVp would result in 1/4 the original exposure?

For the following pairs of techniques, fill in the kVp that would maintain equal exposure to the original technique:

5. 400 mA @ 0.05 s and 90 kVp = 300 mA, 0.0167 s and ________ kVp
6. 50 mA @ 0.0167 s and 50 kVp = 400 mA, 0.0083 s and ________ kVp
7. 300 mA @ 0.033 s and 120 kVp = 400 mA, 0.1 s and ________ kVp
8. 300 mA @ 0.05 s and 70 kVp = 150 mA, 0.15 s and ________ kVp

Some radiographers oversimplify the 15 percent rule into a "10-kVp" rule, stating that every 10 kVP changes exposure by a factor of 2. (This stems from the fact that 15% of the *average* kVp used in diagnostic radiography, about 70 kVp, is 10.) But, this rule is too inaccurate, since in the range of 40 kVp 15 percent is only 6, and at 100 kVp, 15 percent is 15 kVp. One must consider the range of kVp used for a particular procedure and take the trouble to make the 15 percent calculation.

The 15 percent is useful not only for practical adjustments in daily practice, but for the development of technique charts as well, as will be demonstrated in Chapter 26.

OPTIMUM kVp

With modern computer-based imaging systems, there is considerable flexibility with how *high* a kVp setting can be employed for a particular body part; as long as the kVp is above the minimum required for adequate penetration, the computer can resolve a good image. The kVp range was more limited with conventional film radiography. (See Historical Sidebar 16-1.) Since the use of high kVp/low mAs techniques reduces *patient exposure*, it is difficult to find any justification or argument for lower kVp settings that barely meet the minimums listed in Table 16-1. The advent of computer-based systems has forced radiographers to rethink their definition of what constitutes an *optimum kVp*, or recommended kVp, for each body part.

We will define the *optimum kVp* as a level of kVp *well above the minimum* required for sufficient penetration, which strikes an appropriate balance between saving patient exposure and preventing excessive scatter radiation at the image receptor plate. The issues involved are as follows:

Patient Exposure and the 15 Percent Rule

Patient exposure is always an important consideration. There are two approaches to increasing exposure to the image receptor when it is needed: If the mAs is doubled, then patient exposure, which is proportional to the mAs, is also doubled. This is a 100 percent increase in patient exposure. If the kVp is increased by 15 percent instead, there is only a 25–40 percent increase in patient exposure.

Table 16-2 lists several measurements of the contribution of kVp to the patient's entrance skin exposure (ESE) when a 15% step increase in kVp is made, based on dosimeter readings taken at the typical distance to the patient's upper surface. The amounts range from 14 to 19 percentage points *above* the "expected" 50% mark from cutting the mAs in half. Any percentage *above* this amount can be attributed to the kVp. (As explained in Chapter 10, this is due to an increased number of *bremsstrahlung* x-rays produced in the anode at higher kVp's.) When this difference is taken as a ratio of the expected 50%, it ranges from 28% to 38%, with a *median* average of 33%.

We conclude that these 15% steps in kVp increased patient exposure by about one-third. The important point for the student to remember is that, since halving the mAs cuts exposure to 50% first, **the *net* result for applying the 15% rule is exposure averaging about 67% of the original, a net savings in patient dose of about 1/3.** *Patient exposure is saved whenever kVp can be increased instead of mAs.*

HISTORICAL SIDEBAR 16-1: With film-based radiography, once some degree of x-ray penetration had been achieved in *all* tissues present, further increases in the kVp could lead to overexposure if the mAs was not compensated. In the series of hand radiographs in Figure 16-4, this result is apparent in the last radiograph where 70 kVp was used—the radiograph is too dark overall, and differences between cortical bone and bone marrow in the phalanges are less apparent than in the images taken at lower kVp levels.

The *range* of acceptable exposure levels was limited between underexposure and overexposure. With modern digital imaging systems, there is effectively no upper limit placed on this range in this regard, because the computer is able to adjust brightness levels of the tissues.

There *is*, however, a lower limit to the needed kVp, and always will be, because *the computer cannot restore information which is not present in at the image receptor due to insufficient x-ray beam penetration.*

Figure 16-4

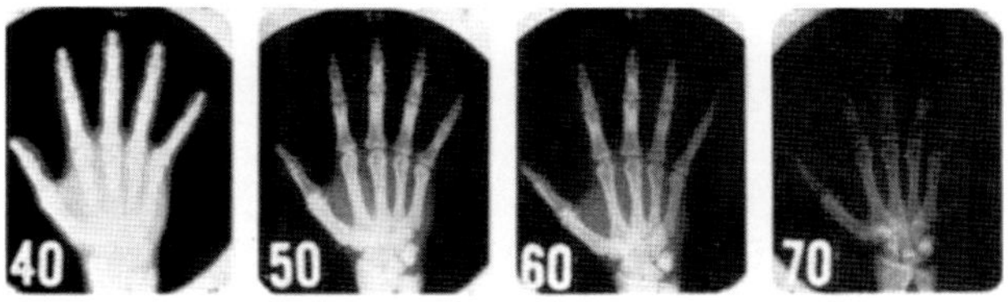

40 50 60 70

A series of 10-kVp increases on images of the hand demonstrate the limited range of exposure factors that could be used for film-based radiography, with an *upper limit* imposed when *without compensating mAs* overexposure caused too dark an image to result. With digital imaging, there is a lower limit to the exposure range but *no practical upper limit.* (From Quinn B. Carroll, *Practical Radiographic Imaging*, 8th ed. Springfield, IL: Charles C Thomas, Publisher, Ltd., 2007. Reproduced by permission.)

Table 16-2

Percentage Increases in Entrance Skin Exposure Due to 15% Step Increases in kVp

Dosimeter Measurements from Lab, UNC, May 2013

Exp #	kVp	mAs	mR Measured	% of Previous Exposure	% Increase Due to kVp
1	70	16	146.0	–	–
2	81	8	98.1	67%	17 = **34%** of 50
3	93	4	64.5	66%	16 = **32%** of 50
4	105	2	41.4	64%	14 = **28%** of 50

Courtesy of University of North Carolina, Chapel Hill, NC.

Dosimeter Measurements from Clinical Unit, 2010

Exp #	kVp	mAs	mR Measured	% of Previous Exposure	% Increase Due to kVp
1	70	20	221	–	–
2	81	10	153	69%	19 = **38%** of 50
3	96	5	104	68%	18 = **36%** of 50

Courtesy of Digital Radiography Solutions, Marina California.

Impact of Scatter Radiation on the Image

In the past, concern over the effects of scatter radiation in fogging the film restrained radiographers from using kVp to increase exposure at the imaging plate. There are two very substantial reasons why this restraint no longer applies: First, the fear of scatter production from high kVp's was somewhat exaggerated in the first place. Proof of this is presented in Figures 16-5 and 16-6.

In Figure 16-5, radiographs ***A*** and ***B*** are an identical elbow radiographed with conventional film, with the kVp raised all the way from 64 to *94 kVp* on radiograph ***B*** with mAs adjusted downward. A lengthening of the gray scale can be seen, demonstrating more details in the thickest bones, but *no fog is present from scatter radiation.* This demonstrates *that high kVp does not bring with it any significant increase in scatter radiation when the body part is not large enough to produce substantial scatter in the first place.* Scatter radiation **originates** primarily from large body parts and large field sizes, *not* from kVp.

Second, even when scatter radiation *is* being produced by the tissues in substantial quantities, kVp is only a *minor* contributing factor. This allows some flexibility in adjusting kVp upward even on scatter-producing body parts such as the abdomen, as illustrated in Figure 16-6. Here, an average abdomen measuring 22 cm was radiographed at 80 kVp and 92 kVp for comparison. The 15 percent rule was used to adjust the mAs, cutting it to one-half for the 92-kVp radiograph, and thus maintaining overall exposure at the film. As expected, the 92-kVp radiograph shows better penetration of x-rays through the tissues, resulting in a lengthened gray scale and more information, all of which is desirable. But, this radiograph *is not visibly fogged.* Even with large scatter-producing anatomy, kVp can be adjusted upward by

Figure 16-5

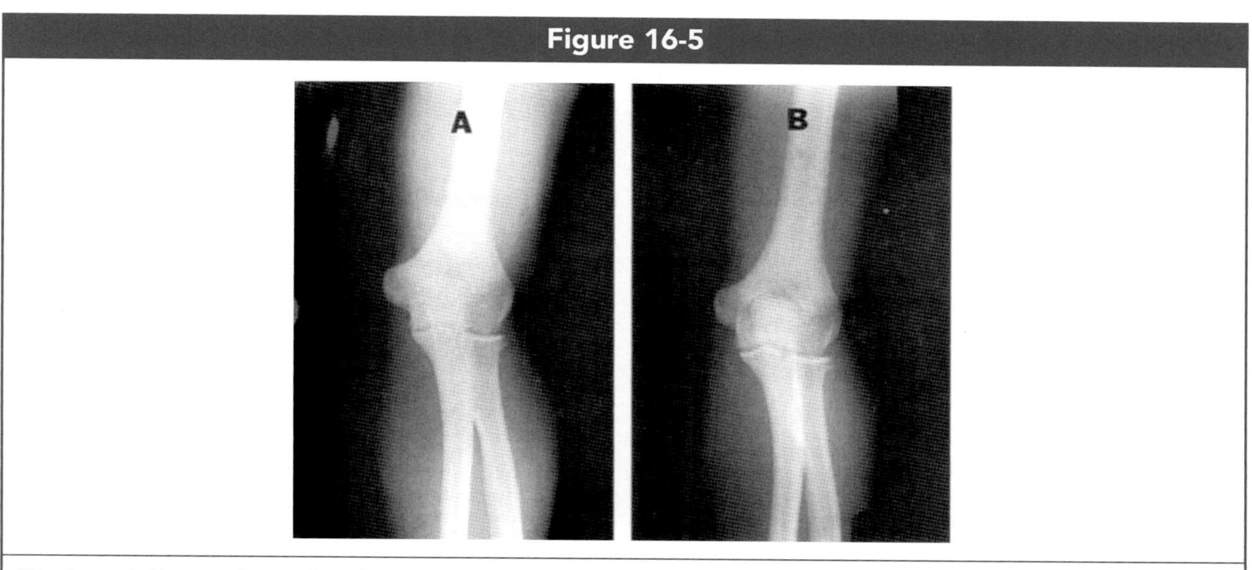

Film-based elbow radiographs taken at ***A***, 64 kVp and ***B***, 94 kVp, with mAs adjusted, demonstrate that even a 30-kVp increase does not produce visible fog when the part itself is too small to generate significant levels of scatter radiation. Increased gray scale is apparent in ***B*** from enhanced *penetration* (not from scatter). (From Quinn B. Carroll, *Practical Radiographic Imaging,* 8th ed. Springfield, IL: Charles C Thomas, Publisher, Ltd., 2007. Reproduced by permission.)

Figure 16-6

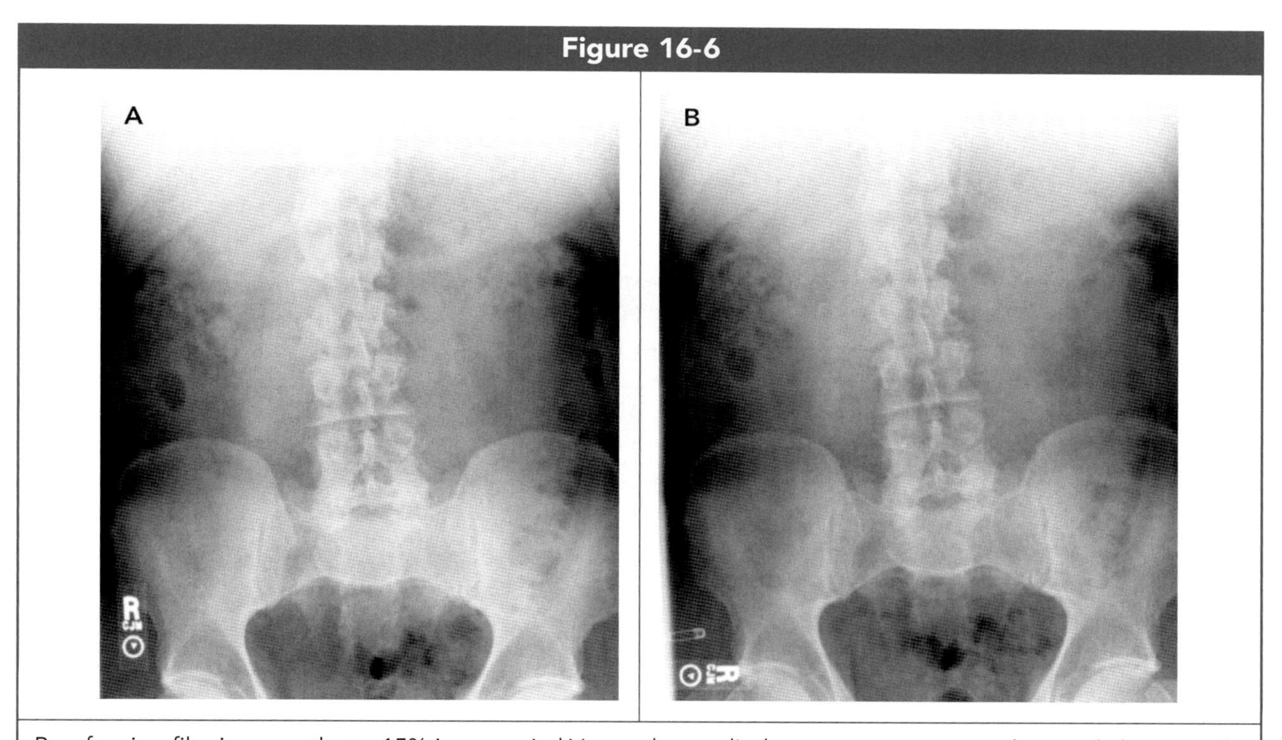

Proof, using film images, that a 15% increase in kVp can be applied even to a scatter-producing abdomen radiograph without visible fogging of the image. Radiograph ***A*** was taken at 80 kVp and 40 mAs. Radiograph ***B*** was produced with 92 kVp and one-half the mAs. A lengthening of the gray scale is apparent *due to increased penetration*, but there is no visible fogging of this image. (From Quinn B. Carroll, *Practical Radiographic Imaging,* 8th ed. Springfield, IL: Charles C Thomas, Publisher, Ltd., 2007. Reproduced by permission.)

a single 15 percent step without causing a substantial increase in scatter production—it takes a large increase in kVp to see actual degradation of the film image from the impact of scatter.

To top all this off, modern digital processing can easily correct for moderate losses of subject contrast in the remnant beam striking the image receptor. Computer algorithms adjust each image for optimum presentation at the display screen. Only in the case of *extreme* loss of subject contrast from scatter radiation will the computer not be able to correct the image.

Conclusion

Let us summarize these implications:

1. Higher kVp does not generate significant scatter radiation in smaller body parts.
2. Even in large body parts, kVp can be increased by one step of 15 percent *without* a substantial surge in scatter production.
3. Whatever very small loss of subject contrast *does* occur from a single 15 percent increase in kVp is easily corrected by computer-based imaging systems.
4. Higher kVp can ensure adequate exposure to the image receptor *with minimum exposure to the patient.*

Even with film-based systems, it was possible to make an across-the-board 15 percent increase in kVp and cut all mAs values in half for a standard technique chart, and spare radiation exposure to patients. *With digital image processing, all barriers to the use of high kVp techniques, real or imagined, have been removed. There is simply no compelling reason to keep kVp at previous levels. As a profession, radiographers have an opportunity here to lower radiation dose to the public, and for professional reasons, they are obligated to do so.*

Therefore, we can now define the *optimum kVp* as that kVp *well above the minimum required for sufficient penetration, which strikes an appropriate balance between saving patient exposure and preventing excessive scatter radiation at the image receptor plate.* Based upon this definition, Table 16-2 presents a listing of new optimum kVp's for various landmark body parts. *These optimum kVp's are strongly recommended for daily clinical use in the digital age.*

OTHER IMAGE QUALITIES

Like mAs, kVp is an *electrical* factor and therefore has no bearing at all upon any of the *recognizability* or *geometrical* functions in an image. It has no

Table 16-3

Recommended Optimum kVp for Digital Imaging

ProcProcedure	Optimum kVP	Procedure	Optimum kVP
Hand/Wrist/Digits	64	Iodine Procedures (IVP, cystog)	80
Elbow/Forearm/Foot	72	Abdomen/Pelvis/Lumbar Spine	90
Ankle/Leg	76	Non-Grid Chest (fixed unit)	86
Knee/Humerus Tabletop	80	Skull	90
Knee in Bucky	84	Air Contrast Barium Studies	100
Femur/Shoulder/Sinus/Ribs	86	Esophagram	92
Mandible, Tangential Skull	76	Solid-Column Barium Studies	120
Cervical and Thoracic Spines	86	Grid Chest (fixed unit)	120
Pediatric Extremities	60–70	Pediatric Chest	70–80

impact upon the sharpness of recorded detail, magnification or shape distortion.

SUMMARY

1. In the x-ray tube, higher kVp results in both higher average energies and higher quantities for the x-rays produced.
2. The most important function of the set kVp is to provide at least partial penetration through all tissues to be recorded in the image. The minimum kVp for a body part is the lowest kVp that still provides some degree of penetration through each of its tissues.
3. The predominance of different tissues within a body part determines the minimum kVp that should be used.
4. No amount of radiation intensity can ever compensate for insufficient penetration of the x-ray beam. No amount of mAs can ever compensate for insufficient kVp.
5. Subject contrast in the remnant x-ray beam depends on kVp, but is independent of mAs.
6. A 15 percent change in kVp, up or down, will alter the intensity of radiation reaching the image receptor by a factor of 2. This is due to the combination of changes in penetration and x-ray output produced at different kVp levels.
7. For modern digital imaging, we define *optimum kVp* as a level well above the minimum needed for sufficient penetration, which strikes a balance between saving patient exposure and preventing excessive scatter radiation.
8. Using the 15% rule to increase kVp and reduce mAs, a net savings in patient exposure is always achieved because the reduction in mAs has a much greater impact tha the increase in kVp.
9. The set kVp is a relatively minor contributing factor in the production of scatter radiation, when compared with the thickness of the body part and the field size.
10. The kVp has no direct impact on the geometrical aspects of the image: sharpness, magnification and distortion.

REVIEW QUESTIONS

1. The most important function of kVp is to produce sufficient __________ of the x-ray beam.

2. What electronic device is controlled by the kVp settings at the x-ray machine console?

3. Why is higher kVp, rather than higher mAs, particularly recommended when changing from the AP projection of the lumbar spine to the lateral L5–S1 "spot" view?

4. No amount of __________ can ever compensate for insufficient kVp.

5. A technique of 100 mA, ½ second and 80 kVp results in a radiograph with motion blurring. Using the 15 percent rule in one step, what new kVp and exposure time would improve it?

6. What is 15 percent of 60?

7. Starting at 120 kVp, what new kVp would result in 50 percent more (1½ times) the original exposure?

8. Using steps of 15 percent, if 40 mAs at 80 kVp produced the correct exposure for an abdomen, and a "solid column" barium enema required the same overall technique but with the mAs adjusted all the way down to 5 mAs, what new kVp would be indicated?

9. If a particular 15% increase in kVp brought patient exposure up by 20%, but the accompanying halving of the mAs had first cut it to 50%, what is the *net* patient exposure?

10. Suppose you reduced the kVp by *3 steps* of 15 percent each. Even though you doubled the mAs three times, the radiograph may still turn out underexposed because of:

(Continued)

REVIEW QUESTIONS *(Continued)*

11. For a small body part such as the elbow, could 90 kVp be used without producing a substantial amount of scattered radiation?

11. Does scatter radiation cause blurring of the image?

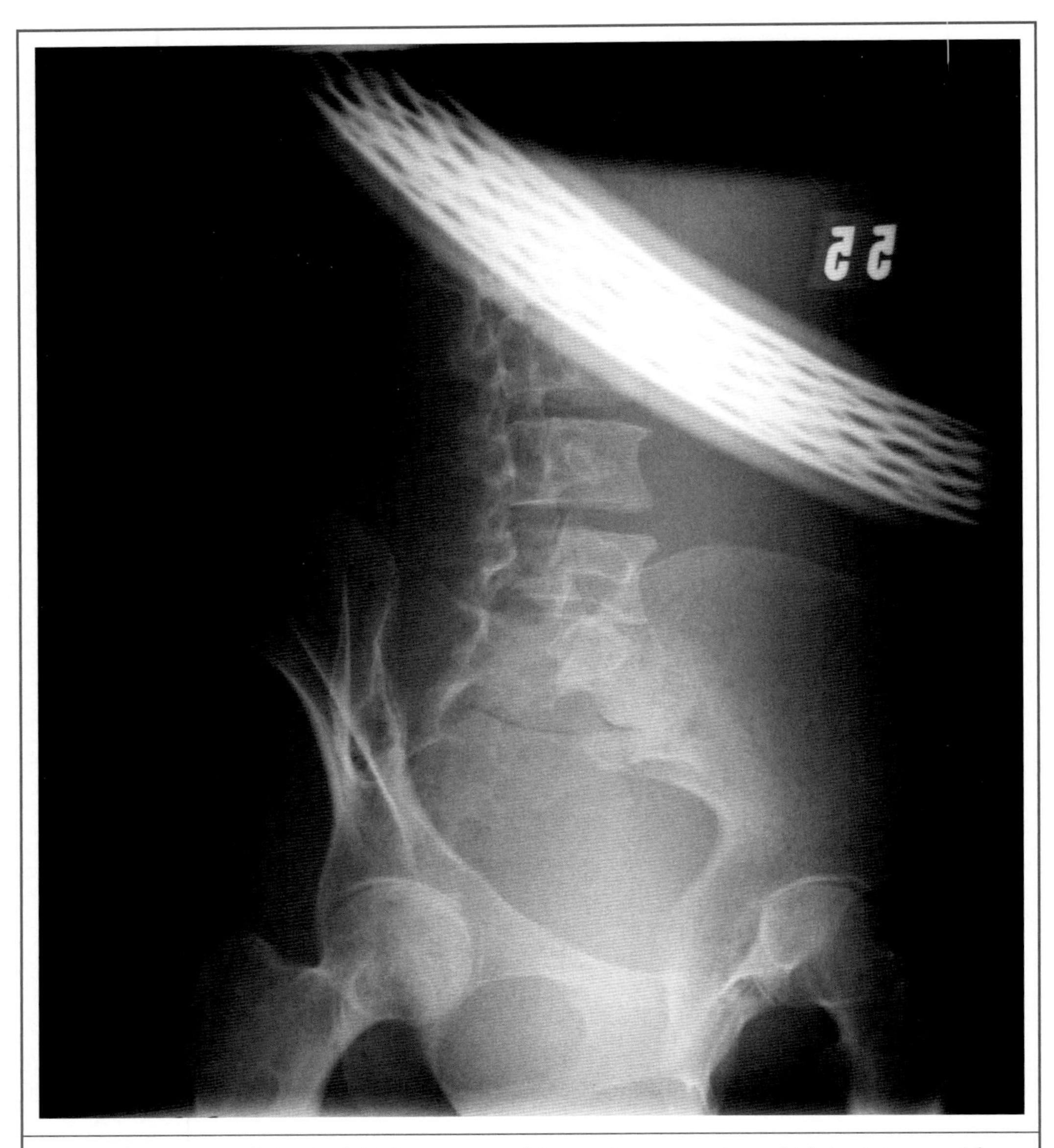

This artifact is one of the thick high-voltage cables to the x-ray tube, which had dropped from its ceiling suspension cable into a position under the collimator.

Chapter 17

GENERATORS AND FILTRATION

Objectives:

Upon completion of this chapter, you should be able to:

1. Explain the effects of higher power generators upon effective mA and effective kV.
2. State the appropriate reductions in radiographic technique for higher power generators.
3. List the image qualities that are affected by generator power, and those not affected.
4. State the minimum filtration requirement for x-ray machines operating above 70 kVp.
5. Describe the purpose of protective filtration, and what limits the amount used.
6. List the two types of protective filtration and give examples.
7. Define *half-value layer (HVL)* and how it is measured.
8. List the three determining factors for x-ray beam penetration and "hardening of the beam."
9. Describe the purpose of compensating filters and where they might still be of value with digital procedures.

GENERATOR TYPE

The various types of electrical rectification and x-ray machine generators, along with the waveforms they produce in the electrical current, were described in Chapter 8. Our purpose in this chapter is to examine their effect on the setting of radiographic techniques. To do this, a brief review of the impact they have upon the electrical current that flows through the x-ray tube is in order.

Effect of Rectification and Generators on Exposure

As the rectification becomes more sophisticated, and as the power of the generator increases, *both the quantity and the quality* of the electrical current flowing across the x-ray tube are affected. Let us review the quantity of electrical current produced first, using a fully rectified, single-phase generator as a reference point for comparison. This generator produces 120 pulses of electricity per second crossing the x-ray tube to the anode and producing x-rays. This is 60-cycle AC current, with those pulses that would have flowed "backwards" across the tube corrected in their direction by the rectifier bridge so that all 120 pulses can be taken advantage of.

Recall that, in self-rectified or half-rectified equipment, these reversed pulses are merely *blocked* from reaching the x-ray tube, rather than corrected. Only 60 pulses cross the tube per second. One-half as many x-rays are produced in the anode each second—the *rate* of x-ray production is *one-half* as great as a fully-rectified machine. For radiographic technique, this means that *the effective mA* is one-half that of a fully-rectified machine. To achieve a particular mAs, exposure times must be set twice as long. Half-rectified x-ray machines are rare in modern diagnostic radiology.

Three-phase machines *overlap* the pulses of electricity, so that gaps in x-ray production that would have been present in a single-phase machine are *filled* over the time of the exposure. This is shown in Figure 8-11 in Chapter 8. The *effective mA* is increased accordingly, and the *rate* of x-ray production goes up. More x-rays per second are produced for a particular mA set at the console.

As we move to more and more powerful generators, the *ripple* of the waveform lessens and these gaps in the waveform are filled even better, yielding slightly more x-ray production for 12-pulse generators, and still a bit more for high-frequency generators (HFGs). All of these improvements result in a corresponding increase in x-ray production within the anode, so that the *exposure rate of the primary x-ray beam is increased* with each step in generator power.

In addition to increasing the rate of x-ray production, high-power generators also enhance the *average* or *effective kV* at each step. As a percentage of the set peak kilovoltage (kVp), the effective kV for each generator type is as follows:

Single-phase (self-, half-, or full rectified):	Appx. 33% of the set kVp
Three-phase, 6-pulse:	Appx. 91% of the set kVp
Three-phase, 12-pulse:	Appx. 97% of the set kVp
High-Frequency Generator:	Appx. 99% of the set kVp

In the x-ray beam produced, the energies of the x-rays correspond with those energies of the projectile electrons striking the anode in the x-ray tube. *Each increase in the average or effective kV represents a more penetrating x-ray beam. Thus, each step in generator power raises beam penetration.*

The radiographer is primarily interested in the end-result exposure level (signal) at the image receptor, which controls the adequacy of information fed into the computer system. This net information is carried by the remnant x-ray beam, which is dependent upon *penetration* of the radiation through the patient as well as the original exposure rate of the primary beam. In terms of this final exposure to the image receptor plate, *three-phase x-ray machines are about twice as efficient overall than single-phase machines, with high-frequency machines slightly more effective than three-phase machines. When moving from a single-phase x-ray room into a three-phase or high-frequency room, mAs values may be cut in half.* This is demonstrated in Figure 17-1 with lateral skull radiographs taken on old-fashioned film to show the density effect.

Conversely, when changing from a three-phase room or HFG room to a single-phase room, the mAs for each type of procedure and projection will need to be doubled. Technique charts in these rooms should reflect this adjustment.

In terms of practical technique, the distinction between a three-phase machine and a high-frequency generator is not so great that a different technique rule must be adopted for the HFG machine. Both are generally considered to be about "twice as hot" as a single-phase machine.

Note that to reduce the mAs, if exposure time is cut in half rather than the mA station, there is an additional benefit in helping prevent motion.

Figure 17-1

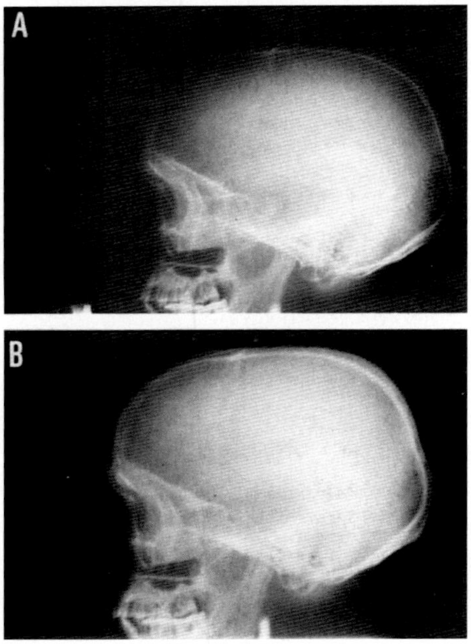

A

B

Film radiographs of the lateral skull demonstrate the maintenance of exposure by cutting the mAs to one-half the original when changing from a single-phase x-ray machine (***A***), to a three-phase machine (***B***). The mAs would also be cut nearly in half when changing from a single-phase machine to a high-frequency generator. (From Quinn B. Carroll, *Practical Radiographic Imaging*, 8th ed. Springfield, IL: Charles C Thomas, Publisher, Ltd., 2007. Reprinted by permission.)

Other Image Qualities

With a higher average kV produced for its x-ray beam, a higher-powered generator will provide better penetration resulting in a slight drop in subject contrast at the image receptor plate. This small change was barely visible on old-fashioned film radiographs as slightly lengthened gray-scale (Fig. 17-1***B***). In digital systems, it is easily compensated for and the image contrast presented at the display screen can be adjusted by postprocessing algorithms. Therefore, the effects of generator type upon the contrast in the final digital image are completely negligible.

High-power generators allow much shorter exposure times to be used, which is particularly helpful in minimizing the effects of patient motion. For pediatric radiography, x-ray rooms equipped with high-power generators are recommended. These are also helpful with inebriated or mentally-retarded patients, or any other situation in which patient cooperation may be compromised.

Because it affects the ability to reduce exposure times, we might state that generator power has an *indirect* effect upon the sharpness in the radiographic image. It should not be thought of, however, as a controlling or a causative factor for image sharpness. It is an electrical variable, and should be generally considered to be unrelated to any of the geometrical functions in the image: sharpness, magnification, and shape distortion.

Nor should the generator be considered a significant contributor to image noise. Slight increases in scatter radiation are compensated for by the computer, enhanced penetration reduces quantum noise, and *electronic* noise in a modern image originates from the circuits of the *computer and the electronic display system*, not from the original exposure generated.

Battery-Operated Mobile Units

Most modern "portable" or mobile x-ray units use a bank of powerful batteries as the power source for exposures. The circuitry for these machines is sometimes referred to as *constant potential generators* (*CPGs*). The batteries are charged between exposures by plugging the machine into a wall outlet. Electrical charge stored in the batteries is used both for driving the mobile unit and for making exposures. It is essential to keep these machines fully charged between uses so that the full mAs and kVp values set by the radiographer will be supplied during exposure.

Batteries normally supply DC (one-way) current. This electrical current must be changed into alternating current so it can be manipulated and controlled at the console. Remember that transforming electricity can only be done when *moving magnetic fields* are used to induce a secondary current. Before it reaches the x-ray tube, this AC current must be converted *back* into DC current so it will always flow from cathode to anode. Because of these transformations, the current reaching the x-ray tube has slight fluctuations in it, but for all practical purposes is considered "straight-line DC" current rather than pulsed DC.

The importance of this waveform is that the *average* kilovoltage is almost *equal* to the peak kV set at the console, just like a high-frequency generator produces. The average kV is generally considered to be 100 percent of the kVp. This explains why, in practice, substantially lower kVp settings are frequently used on mobile equipment than on stationary x-ray machines in the x-ray department. For example, for a *non-grid* AP chest (such as a *wheelchair* chest) taken in a typical x-ray room, 76–82 kVp is likely to be used. Yet, the same view taken with a mobile x-ray unit is often done using 68–72 kVp. This is 8 to 10 kVp less. This can be done because the *average* kV on the mobile unit is higher for the set kVp.

Lower mAs values are also an option for mobile units. Battery-powered units are more efficient than even three-phase x-ray machines. They are approximately equivalent to high-frequency generators. A workable rule-of-thumb is that *battery-powered mobile units require about 8 kVp less than three-phase x-ray rooms.*

BEAM FILTRATION

Protective Filters

The primary purpose of filtering the primary x-ray beam before it reaches the collimator is to *eliminate unnecessary patient exposure*. Without filtration, there would be millions of very low-energy x-rays which would practically have no chance of penetrating through the patient to the image receptor. (Statistically, there is *always* a chance of a particular x-ray, regardless of its energy, making it through the patient; but, at the extremely low energies we are discussing here, that probability is so infinitesimally small as to defy measurement. In practical terms, these x-rays should be considered as *nonpenetrating*.) Since they cannot contribute to the image in any way, they are radiographically useless. In the spirit of the *ALARA* concept (*As Low As Reasonably Achievable* dose), they must therefore be disposed of. If they are not filtered out before they reach the patient, then the patient absorbs virtually *all* of them. Such exposure to the patient would be completely unnecessary.

There are two types of filtration in any x-ray unit: Inherent filtration and added filtration. *Inherent filtration* consists of various components of the x-ray tube and its housing which are essential to the operation of the x-ray tube and permanently fixed in place. These include the glass of the x-ray tube itself, which absorbs some radiation, the oil in the tube housing, and a *beryllium window* (Fig. 17-2), which provides preliminary filtration of the very lowest energy x-rays.

Technically, the mirror and the plastic windows in the collimator, through which the x-ray beam must pass, may also be considered as part of the inherent filtration. In the x-ray tube, the anode itself acts as a filter; this phenomenon, known as the *anode heel effect*, will be fully discussed in a later chapter.

Added filtration normally consists of thin slabs of pure aluminum which may be installed between the x-ray tube and the collimator, or found in a slot in the upper portion of the collimator as shown in Figure 17-3. These slabs are easily removed or reinserted.

Because filtration is such an important issue in protecting the general public from unnecessary radiation exposure, minimum amounts are set by government regulation. All x-ray machines capable of operation above 70 kVp must have a minimum total filtration that is *equivalent to 2.5 mm of aluminum*. In modern equipment, the *inherent* filtration of everything within the tube housing normally totals 1.5 mm aluminum equivalency, so that a 1 mm slab of added aluminum at the collimator will satisfy the overall total requirement.

The added filter should never be removed by a radiographer unless a certified radiation physicist has verified that sufficient penetration will be maintained for those procedures the x-ray machine is used for, and great care is taken to ensure that the filter is replaced for routine procedures.

Figure 17-2

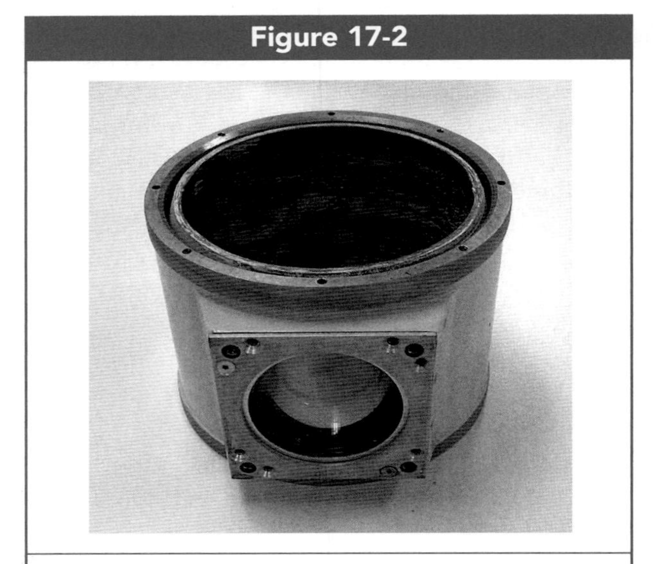

The beryllium window from an x-ray tube housing. This is part of the inherent filtration.

Figure 17-3

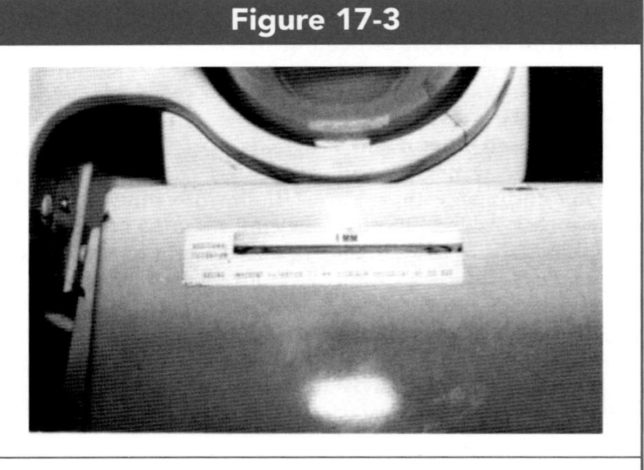

Photograph of the added 1 mm slab of aluminum filtration inserted into a slot in the upper portion of the collimator of a mobile x-ray unit. This filtration should never be removed. (From Quinn B. Carroll, *Practical Radiographic Imaging*, 8th ed. Springfield, IL: Charles C Thomas, Publisher, Ltd., 2007. Reprinted by permission.)

Half-Value Layer

In Chapter 10 we illustrated how, as filters remove the low-energy x-rays from the beam, the *average energy of the beam is increased.* This is referred to as *hardening the x-ray beam*, because as a whole it is more penetrating. Of those x-rays that remain in the beam after filtration, a higher percentage will make it through the patient to the image receptor.

Since penetration is the opposite of radiation absorption, *hardening the beam lowers patient exposure.* A special unit, the *half-value layer*, directly measures the actual penetration capability of the x-ray beam. Half-value layer, abbreviated *HVL*, is defined as that amount of a specified absorbing material needed to reduce the intensity of the beam to precisely one-half the original x-ray output from the tube.

To this point, we have discussed *three* determining factors for beam penetration: kVp, the power of the generator, and filtration. All three of these bring up the *average kV* of the x-ray beam as they are increased. All three have some effect in hardening the beam. But none of them directly *measure* beam penetration. We have stated that government regulations set a minimum amount of filtration to be used, but this alone does not guarantee sufficient penetration, because the kVp settings could be out of calibration or the generator could be malfunctioning. Therefore, government regulations also designate *minimum HVL requirements* in published tables.

A separate table of required HVL's must be published for each type of generator, be it single-phase, three-phase, high-frequency or another configuration. HVL's are listed for the whole range of kVp settings in multiples of ten. For example, the HVL required for a three-phase machine operating at 80 kVp is 2.34 mm of aluminum. (Copper, tin, or lead can also be used to measure HVL's.) *If the measured HVL comes out less than this amount, and calibration of the kVP or generator cannot correct it, added filtration must be increased until the HVL requirement is met. Even if the machine meets the minimum filtration requirement, (normally 2.5 mm aluminum equivalency), more filters must be added when the HVL test fails.*

HVL requirements trump filtration requirements because HVL is an actual measurement of the end-result beam penetration *after* all three controlling factors (the generator, the kVp and the filtration) have impacted upon it. Beam penetration is what we are really concerned about in minimizing patient exposure.

Although a radiation physicist will do a thorough HVL check at all kVp levels once a year on each x-ray machine, a radiographer can easily do a quick check at 80 kVp to see if it exceeds 2.34 mm aluminum if an ion chamber is available. The procedure is in Chapter 39.

Effects on Exposure and Beam Spectrum

When an additional sheet of aluminum is placed in the x-ray beam at the portal of the x-ray tube, both the quantity and the quality of the primary x-ray beam are altered. These effects show up in the beam spectrum graph as shown in Chapter 10 (Fig. 10-15). As a filter screens out low-energy x-rays from the beam, the overall quantity of x-rays is decreased, reducing the area under the spectrum curve. Yet, at the same time, the quality of the beam is *increased* as it becomes more penetrating. This shows up in the spectrum curve as an overall shift to the right, with an attendant increase in the *average* kV.

It is important to understand that, even though protective filters reduce the intensity of the *primary* beam, *they do not reduce the intensity of the remnant beam reaching the image receptor.* In fact, if this were the case, it would indicate that too much protective filtration was being used, because when exposure at the imaging plate is lost, radiographic techniques have to be increased to compensate. This defeats the whole purpose of the filtration, adding exposure back again *to the patient.* The correct amount of protective filtration is defined as the maximum possible *without affecting final exposure at the imaging receptor.* This amount has generally been determined to be around 2.5 mm of aluminum equivalency.

Hypothetically, because filtration hardens the beam, an excessive amount of protective filters would not only lower the exposure rate in the remnant beam, but also lower subject contrast and increase gray scale. (These effects are sometimes taught in the classroom, but they confuse the effects on the *primary beam* with effects on the *remnant beam.*) This is just not a practical, realistic scenario when a real patient undergoes general radiography.

If any of these effects actually occurred within the remnant beam reaching the image receptor, it would be an indication that far too much protective filtration was being used and that some of it should be removed. *In daily practice, protective filtration should not be considered as a factor affecting any of the radiographic image qualities for general radiography.*

Compensating Filtration

Very thick filters, over one-half inch or 12 mm thick, can be specifically shaped to even out the intensity of the remnant beam when unusual-shaped body parts are radiographed. As shown in Historical Sidebar 17-1, filters of such extreme thickness *did* affect the appearance of the final image for film-based radiography. At the thickest portion, the filter both reduced the density and slightly lengthened the gray scale in the image.

With digital imaging systems, computer algorithms make corrections that largely balance the brightness across the image, so that many of the traditional applications of compensating filters, such as for the AP thoracic spine, are done away with and no longer in general use. In fact, as shown with the lateral cervical spine view in Figure 17-6, computer software can now target specific portions of the image for brightness correction.

However, in some situations the differences in exposure at the IR from one anatomical area to another can be so extreme that some brands of digital processing software are unable to fully compensate. These include such procedures as the groin lateral hip shown in Figure 17-8, or the "swimmer's" type projection through the cervicothoracic area. Special contoured filters (Figure 17-7) have regained popularity in many departments for these types of applications, and can be effective.

With digital imaging systems, computer algorithms make these corrections to balance the brightness across the image, and compensating filters are no longer found in general use. However, there may be

HISTORICAL SIDEBAR 17-1: Figure 17-4 shows an aluminum "wedge" filter used on AP thoracic spine and AP foot projections to even out exposure to the film at the thinnest portions of the anatomy. An example of the resulting balance in overall density is demonstrated in the AP foot images in Figure 17-5. Technique had to be compensated for the thick part of the filter by adding 2 kVp for each mm of additional thickness.

Figure 17-4

In film-based radiography, an aluminum "wedge" filter could be attached to the collimator to balance the image density for wedge-shaped anatomy such as the foot or AP thoracic spine. (From Quinn B. Carroll, *Practical Radiographic Imaging*, 8th ed. Springfield, IL: Charles C Thomas, Publisher, Ltd., 2007. Reprinted by permission.)

Figure 17-5

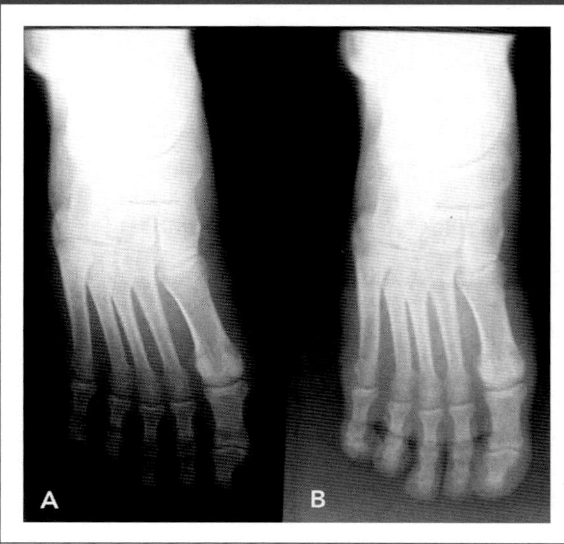

Film images of an AP foot, ***A,*** without a wedge filter, and ***B,*** with a wedge filter. Note in particular the more balanced density in the area of the toes.

Figure 17-6

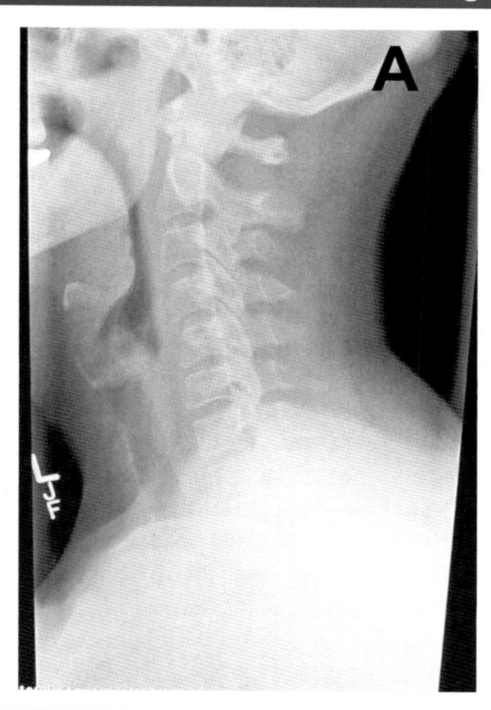

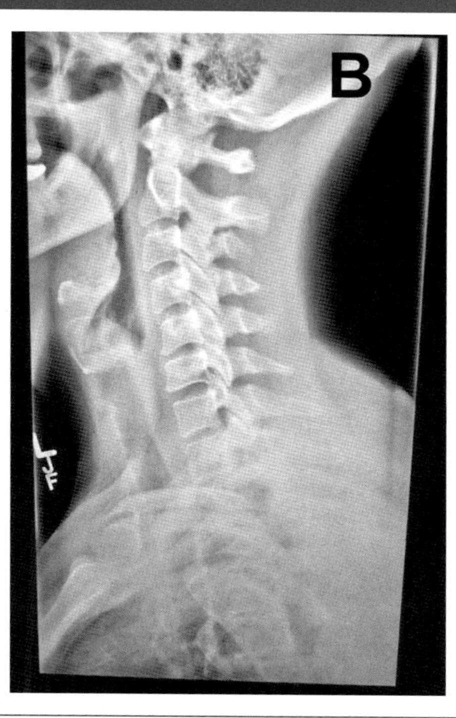

Computer software can now *target* specific portions of the image for brightness correction. In this lateral C-spine, General Electric's *underpenetrated* function first identifies the underexposed portion within the shoulders, ***A***, then darkens the "density" (reducing brightness) in only this portion of the image, resulting in a balanced view of all vertebrae of interest, ***B***.

unique, extreme situations for which the computer algorithms are unable to completely compensate, when a compensating filter might assist by balancing the remnant exposure at the receptor plate.

Figure 17-7

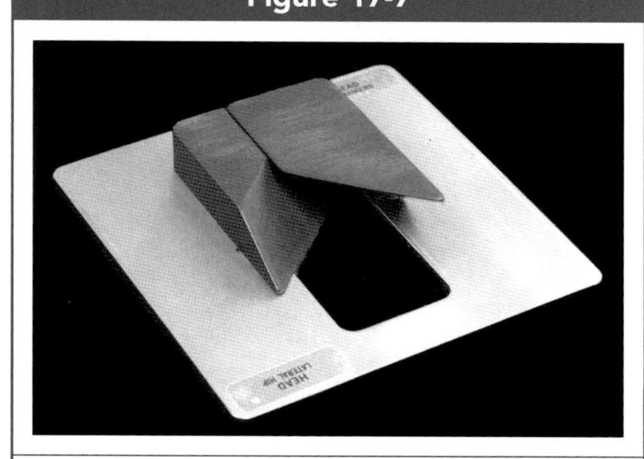

This *Ferlic*™ filter is specifically crafted to balance the intensity of the remnant x-ray beam having passed through a cross-table "groin" lateral projection of the hip, where extreme differences in absorption occur between the upper femur and pelvic bones. (Courtesy, Digital Radiography Solutions, Marina, CA.)

SUMMARY

1. Higher-power x-ray generators produce both a higher rate of x-rays per second and higher average x-ray beam energy, that is, both a higher *effective* mA and a higher *effective* kV. The end result is much higher exposure by the remnant beam at the image receptor plate.
2. Radiographic techniques may be cut to one-half when changing from half-rectified to fully-rectified equipment, and when changing from single-phase equipment to three-phase or high-frequency generators.

Figure 17-8

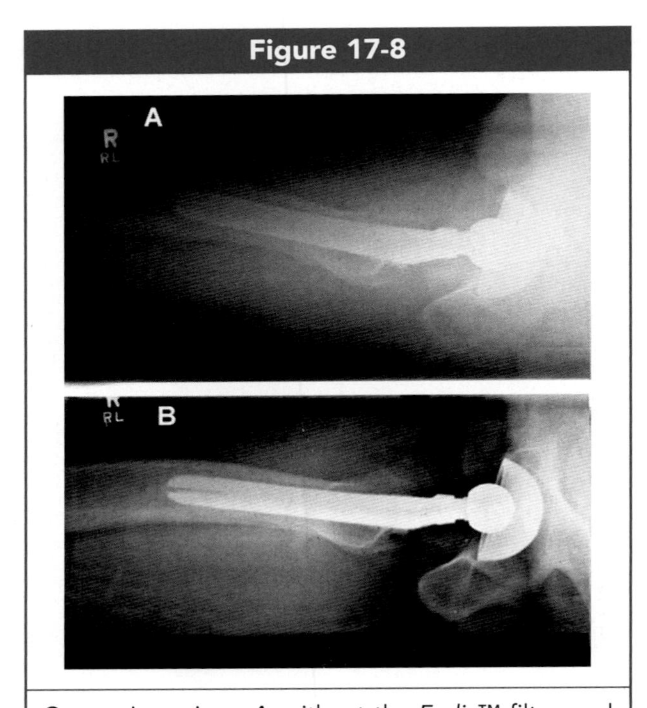

Comparison views ***A***, without the *Ferlic*™ filter, and ***B***, with the *Ferlic*™ filter in place, for a cross-table "groin" lateral projection of the hip with a hip pin. In such extreme cases, filters can still be useful when a particular digital processing suite is unable to fully correct the image balance. (Courtesy, Digital Radiography Solutions, Marina, CA.)

3. Although the subject contrast in the remnant x-ray beam is slightly reduced with higher-power generators, no image qualities are visibly affected.
4. The main impact of higher-power generators is a reduction in the needed radiographic technique. By using exposure time for this reduction, high-power generators can be used to advantage for pediatric, mentally retarded, or inebriated patients to help prevent motion blur.
5. Battery-operated mobile units can be operated at about 8 kV less than a 3Φ fixed unit. High-frequency generators are also slightly more efficient than 3Φ machines.
6. X-ray machines capable of operating at greater than 70 kVp must have a minimum of 2.5 aluminum equivalency total filtration. Added filters must not be removed from the beam.
7. The sole purpose of protective filtration is to spare unnecessary patient exposure from the *primary* x-ray beam. If protective filtration affects the remnant beam exposure at the receptor plate in any way, an excessive amount has been used.
8. The actual measurement of x-ray beam penetration is made using HVL. Because of their beam-hardening effect, increasing either the kV or the amount of filtration raises the HVL of the x-ray beam.
9. A minimum HVL is designated for each kVp level (in intervals of 10). If this amount is not achieved, calibration of the kVp should be checked, and filtration may need to be added.
10. Compensating filters can be used to balance remnant exposure to the image receptor plate, but with the increasing power of digital image processing, they are only needed in extreme cases.

REVIEW QUESTIONS

1. Why would a self-rectified x-ray unit only produce one-half the exposure of a single-phase x-ray machine?

2. A three-phase x-ray machine produces more remnant exposure at the receptor plate both because of increased __________ and increased __________.

3. For a high-frequency x-ray machine, what is the average kV as a percentage of the kVp?

4. If 80 kVp were used in a fixed 3Φ x-ray room for an AP chest, what kVp would be indicated for the same procedure using a battery-powered (CPG) mobile x-ray unit?

5. For digital imaging systems, how is the image contrast affected by switching from a 1Φ x-ray machine to a high-frequency (HFG) machine?

6. The recommended change in radiographic technique when changing from a 1Φ x-ray machine to a 3Φ machine is to __________ the __________ to __________ the original.

7. Protective filtration should *only* affect the __________ x-ray beam.

8. The glass of the x-ray tube envelope, the oil surrounding it, and the beryllium window are all examples of:

9. If a 1 mm slab of aluminum filtration were removed from the collimator, what would be the immediate effect upon the subsequent images?

10. What, besides insufficient filtration, could cause thé measured HVL for an x-ray machine to be inadequate?

11. HVL is defined as the amount of absorber required to:

12. The correct total amount of protective filtration is defined as the __________ possible without affecting the _________ exposure at the __________.

13. Give an example of a projction where compensating filters might still be useful, even with the power of digital processing.

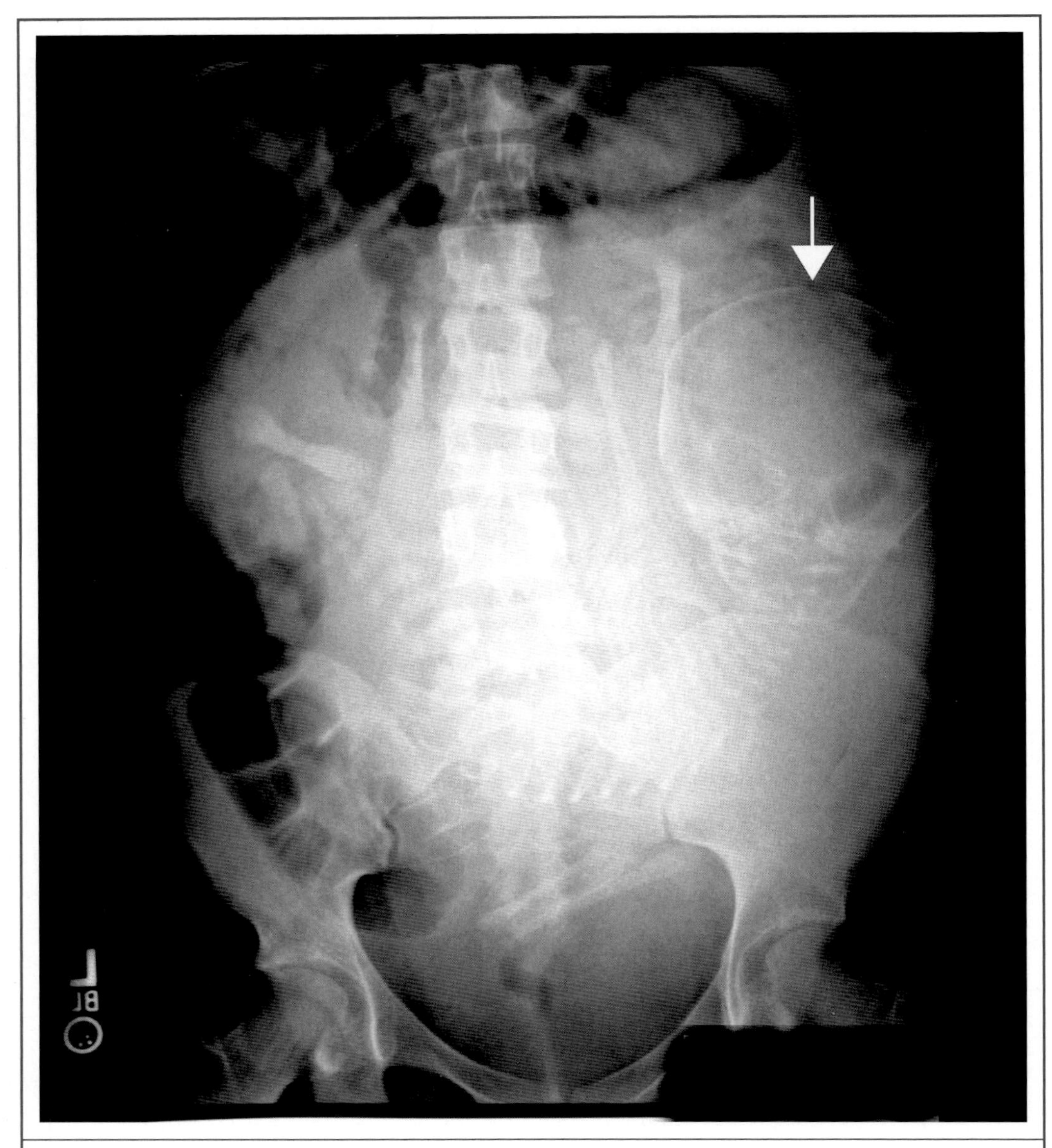

For this fetogram, the full-term fetus faced directly "into the camera" (arrow).

Chapter 18

FIELD SIZE LIMITATION

Objectives:

Upon completion of this chapter, you should be able to:

1. State the major guidelines for determining proper field size.
2. Quantify how effective field size limitation can be in reducing organ dose.
3. Define off-focus radiation and which collimating devices are most effective at reducing it.
4. Describe the history and proper use of *positive beam limitation (PBL).*
5. State the risks associated with over-collimation, and the guideline to avoid clipping anatomy.
6. Explain how a larger field size contributes to scatter production, and why subject contrast is reduced.
7. Explain how a larger field size contributes to overall exposure level at the IR.
8. List those image qualities that are *not* affected by field size.
9. For various distances and apertures, calculate the resulting field size, and vice versa.

The purposes of beam size limitation (collimation) are twofold: (1) to minimize radiation exposure to the patient, and (2) to preserve subject contrast in the remnant beam. The minimizing of patient exposure should be of paramount concern to every radiographer, and limiting the size of the x-ray beam is one of the most effective ways to do this. X-ray field size must never be larger than necessary to include the anatomy of interest, and must never be larger than the size of the receptor plate. By controlling field size, organs with critical sensitivity to radiation, such as the gonads, thyroid gland, and lenses of the eyes can be kept outside the *primary* x-ray beam. The resulting reduction in dose to these organs can be as much as a hundredfold. A variety of devices is available for this purpose.

COLLIMATION DEVICES

In 1899, just four years after the discovery of x-rays, a dentist named William Rollins developed the first x-ray beam filters as well as the first devices for limiting the size of the x-ray beam. Both developments significantly reduced unnecessary radiation exposure to patients. Rollins called his earliest "collimation" or beam-limiting devices *diaphragms*. These were simple lead plates with different sizes of aperture openings cut into the middle to constrict the area of the projected x-ray beam, Figure 18-1. These were soon followed by lead cones, cylinders, and extension cylinders, also shown in Figure 18-1.

All modern x-ray machines have a *collimator box* mounted to the x-ray tube housing. The collimator may be manual only, with dials that the radiographer must adjust, or it may use *positive beam limitation* (PBL), in which sensors in the bucky tray detect the size of a cassette plate and send an electronic signal to activate small motors in the collimator to adjust the shutters. In addition to the collimator, other ancillary devices such as lead cones and cylinders are still used.

The most effective of these devices in providing a sharp border at the edges of the x-ray field are those which cause the beam to pass through *two* apertures (openings) rather than one (Fig. 18-2). Collimators and cylinders (commonly misnamed "cones") provide

Figure 18-1

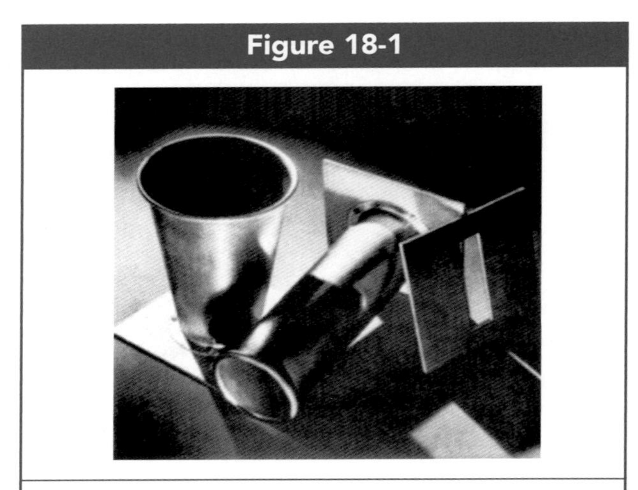

A cone, extension cylinder and aperture diaphragm, devices used to restrict field size. (From Quinn B. Carroll, *Practical Radiographic Imaging*, 8th ed. Springfield, IL: Charles C Thomas, Publisher, Ltd., 2007. Reprinted by permission.)

Figure 18-2

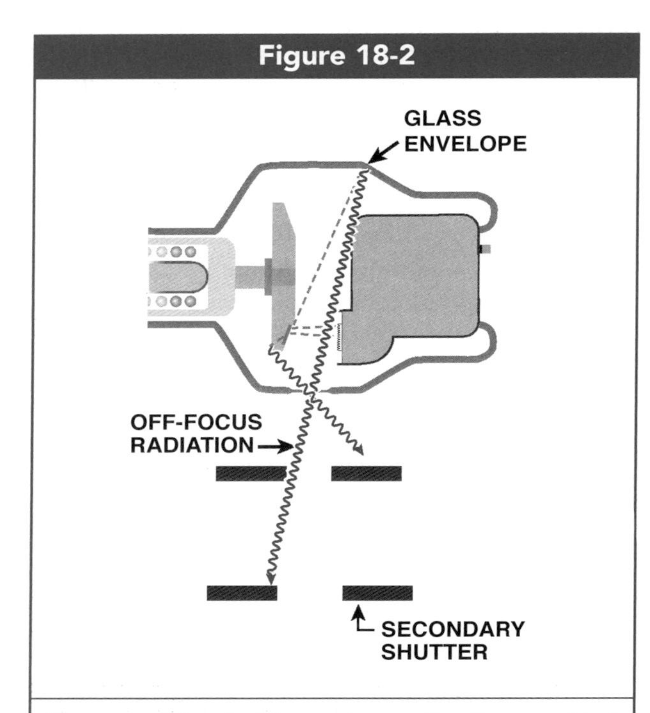

Diagram of the double-aperture collimation provided by extension cylinders and collimators. The second set of shutters, or lower aperture, absorbs off-focus radiation (**A**), and provides a sharper edge to the field (***E***) than the upper aperture would. (From Quinn B. Carroll, *Practical Radiographic Imaging*, 8th ed. Springfield, IL: Charles C Thomas, Publisher, Ltd., 2007. Reprinted by permission.)

such double-apertures. (A true cone expands in diameter with the beam, creating only one effective aperture.) The main purpose of the secondary aperture is actually to reduce *off-focus* or *extrafocal radiation*. These are x-rays produced anywhere outside of the focal spot.

In the x-ray tube during an exposure, some projectile electrons striking the focal spot are *scattered* from the anode and strike other portions of the x-ray tube. Low-energy x-rays can thus be produced from the glass envelope, cathode assembly, or other parts of the tube (Fig. 18-2). In addition, remember that x-rays are initially emitted from the focal spot in all directions—these can interact with atoms outside of the focal spot to produce scattered x-rays. From these two sources, some x-rays will escape through the window of the x-ray tube that are not aligned with the regular, geometrically-controlled beam. Off-focus x-rays can constitute as much as 25% of the beam and are responsible for the "ghost" images that are often seen around the edges of the field.

Figure 18-2 shows how an off-focus x-ray that has made it through the first set of collimator shutters (or the upper aperture of a cylinder) is prevented from reaching the image receptor by the lower set of shutters. In a cylinder, these rays will run into the casing wall before reaching the lower opening.

Even though off-focus x-rays are produced from what might be called "scattering" events, they are properly considered as part of the *primary beam* originating from the x-ray tube, and should not be confused with scatter radiation from the patient or table. Yet, since off-focus x-rays travel in a crooked direction relative to the rest of the beam, those few that make it through the patient to the image receptor add destructive noise to the signal.

Since a secondary aperture is closer to the patient and image receptor, it also provides a smaller projected field with sharper edges (Fig. 18-2). Sharper edges on the projected light field and x-ray beam must not be confused with *sharpness of details within the image field*, which are *not* affected by field size limitation.

Positive Beam Limitation

From the mid-1970s to the mid 1990s, all new x-ray machines manufactured were required to have positive beam limitation or "automatic collimators."

Most modern equipment provides PBL, but with an optional switch to override it when manual collimation is desired. Electronic detectors are connected to the cassette clamps in the bucky tray, which measure the size of each plate when the clamps are closed down on it. This information is electronically relayed to the collimator, where the shutters are moved by small motors to match the plate size. This ensures that the field size will never be larger than the cassette or plate size.

The original idea was to protect the patient from radiation exposure to unnecessary portions of the body. However, many would say that PBL turned out to be more a curse than a blessing; there are many projections which should be taken with the field size collimated *smaller* than the plate size. Taking this automation for granted, many radiographers fell out of the habit of ever checking the collimation at all, and stopped collimating smaller than the plate when they should. Consequently, after studying the effects on actual radiographic practice, the requirement to install PBL was rescinded in the 1990s.

PBL is still available on many x-ray machines. Generally, it should not be completely disengaged; but, whenever a "coned-down" or collimated view is in order, radiographers should take the trouble to override the PBL and adjust the field size manually. Failure to do so results in unnecessary x-ray exposure to the patient.

Figure 18-3

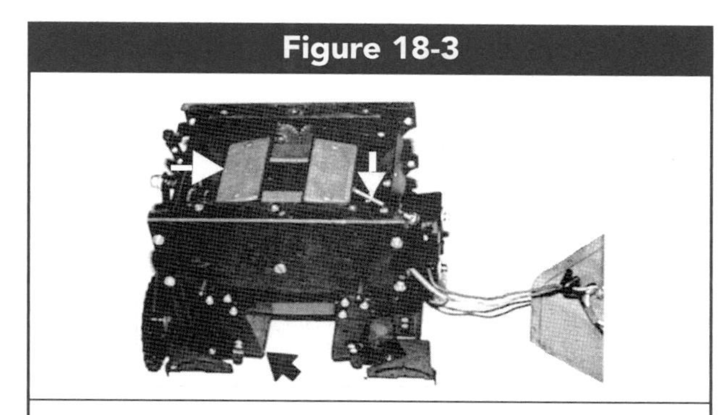

Photograph of a collimator mechanism, showing the upper shutters (horizontal white arrow) and lower shutters (black arrow). A screw for adjusting the exact angle of the light field mirror can be seen, (vertical white arrow). (From Quinn B. Carroll, *Practical Radiographic Imaging*, 8th ed. Springfield, IL: Charles C Thomas, Publisher, Ltd., 2007. Reprinted by permission.)

OVERCOLLIMATION

The *anatomy of interest* should always be included within the x-ray field. Overzealous collimation can result in "clipping" essential anatomy of interest from view and necessitate a repeated exposure. Repeats more than defeat the main purpose of collimation in saving patient exposure, since a repeat *doubles* the exposure to the patient in obtaining a particular view.

With experience, the student will come to appreciate that the edges of the x-ray beam are not always perfectly aligned with the edges of the projected light field. This is due to the fact that the light field is projected from a light bulb mounted in the side of the collimator and reflected downward from a mirror, while the x-ray beam itself passes through the mirror from above. The angle of the mirror can be adjusted by a screw inside the collimator (Fig. 18-3). It is not uncommon for the light field to be projected as much as one-half inch off of the actual x-ray beam. In fact, each edge of the light field is only required to fall within about one-half inch accuracy to the edges of the actual x-ray field.

With this in mind, a word to the wise would be to *always allow at least one-half inch (1 cm) of light beyond each edge of the anatomy of interest*, as long as it does not extend beyond the edge of the cassette plate. For example, when radiographing the hand, allow at least one-half inch (or 1 cm) of light beyond the fingertips and to each side of the hand shadow. This will ensure that the fingertips are not clipped off by the edge of the actual x-ray beam, which may lie somewhat *inside* the edges of the light field. This will help prevent repeated exposures due to clipping off anatomy.

SCATTER RADIATION AND SUBJECT CONTRAST

The amount of scatter radiation produced in the remnant x-ray beam is a direct function of the *amount of tissue exposed*. Two variables control the amount of exposed tissue: The size of the patient

Figure 18-4

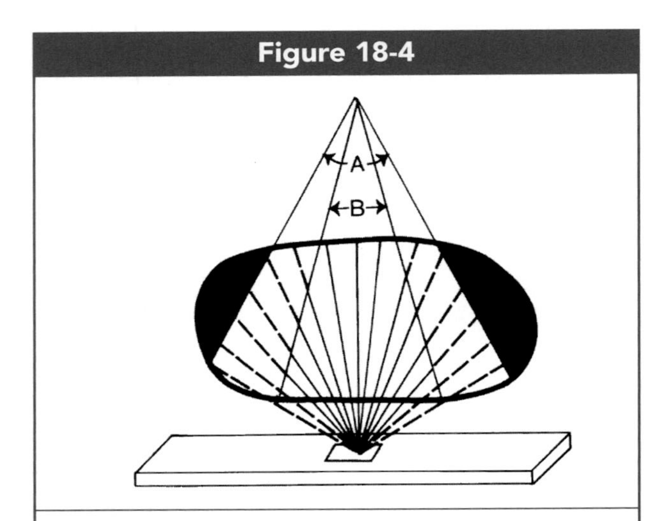

Diagram illustrating how smaller field sizes (***B***) expose a lesser volume of tissue and a narrower range of angles (solid lines) from which scattered radiation can affect a given area at the image receptor. This reduction in scatter increases the subject contrast of the remnant beam (along with some loss of overall exposure) reaching the receptor plate. (From Quinn B. Carroll, *Practical Radiographic Imaging*, 8th ed. Springfield, IL: Charles C Thomas, Publisher, Ltd., 2007. Reprinted by permission.)

and the size of the x-ray field. Therefore, collimation of the x-ray beam, besides saving patient exposure, is also very important in minimizing scatter radiation. By limiting the size of the x-ray beam, less tissue within the patient is available to produce scatter radiation. Further, the number of different *angles* from which any given point on the image receptor may receive scattered x-rays is reduced, as shown in Figure 18-4.

Examining Figure 18-4, we see that with a wide beam, ***A***, scatter may be produced from a large volume of tissue including all of the solid-line and dashed-line areas. This scatter can reach any point on the image receptor from various points within the body part and from many different angles. If we examine a small chosen area on the image receptor plate, represented by the small square, we see that both the volume of tissue producing scatter and the number of angles from which that scatter may strike this spot are limited to only those solid-line portions by collimating the field to a smaller size. (Scattered rays are in fact emitted in all directions and can strike any spot on the image receptor, but we examine only one spot on the plate to illustrate this concept.)

We learned in Chapter 12 that the production of good subject contrast in the remnant beam is dependent upon the *proportions of different interactions within the patient relative to each other*, that is, the *percentage* of each interaction's contribution to the final image. We have established here that with narrower x-ray beams, less scatter will strike a particular area on the image receptor; but, for subject contrast to be altered, we must also show that the number of photoelectric absorptions and penetrating x-rays striking the same area have *not* decreased by the same amount. If this were so, the effects of decreased scatter rays and decreased penetrating rays would cancel each other out.

Figure 18-5 provides this demonstration, using diagrams of penetrating and absorbed portions of the beam. Once again, we focus on a small chosen area on the image receptor, represented by a small square of black to represent penetrated exposure, with a white spot in the middle to represent an area of photoelectric absorption. Note that the number of penetrating primary rays *striking the black square* is equal (4 in each case), regardless of whether the collimated beam is wide, ***A***, or narrow, ***B***. The white spot in the middle is also unchanged, indicating a single photoelectric interaction above it. The *concentration* of both penetrating and absorbed x-rays in this area is not affected by the width of the overall x-ray beam. This is due to the fact that these are geometrically controlled portions of the beam, emanating from the *x-ray tube* and not the patient. They follow straight, predictable paths which do not change just because the overall beam is wider or narrower.

Scatter, on the other hand, is produced from tissues *within the patient*; it is geometrically unpredictable and random in direction, so it can strike the small square area on the image receptor from *anywhere* in the patient's body where tissue has been exposed to radiation, as shown in Figure 18-4. If scatter is decreased from collimation, while photoelectric absorption and penetrating x-rays remain unchanged, then the subject contrast will be improved. The Compton/photoelectric ratio has decreased, and so has the Compton/penetration ratio. *The smaller the collimated field size, the higher the subject contrast present in the remnant beam.*

EFFECT ON EXPOSURE

The *total exposure* reaching the image receptor from the remnant beam of x-rays consists of the sum of penetrating primary x-rays and scattered secondary x-rays. In Figure 18-5, we saw that the *concentration* of primary x-rays remains the same regardless of the field size. However, in Figure 18-4, we saw that the *concentration of scattered x-rays is reduced* for any given area of the image receptor when the beam is collimated to a smaller field size. Therefore, the *total intensity* of radiation reaching any point on the image receptor is reduced by collimation, due to the loss of some of the scatter exposure. This is true for all other small areas on the image receptor plate, so that the overall exposure of the whole plate is decreased.

With the old film-based systems, additional "coned-down" views were frequently ordered by the radiologist for the express purpose of obtaining an enhanced-contrast view of specified anatomy. (See Historical Sidebar 18-1.) Due to the reduction in *total* exposure, these views also turned out lighter, enough so that technique had to be increased to compensate. With computer-based systems, no compensation in technique is needed.

With digital imaging systems, both the contrast and the brightness of the final image are corrected by computer algorithms and can be further adjusted at the digital workstation. It is still important, nonetheless, for the radiography student to understand what is going on at the intersection of the remnant radiation beam with the image receptor, as part of the overall process of image formation.

OTHER IMAGE QUALITIES

None of the recognizability functions of the image can be affected by the field size. The sharpness of recorded detail, magnification, and shape distortion present in the image are strictly due to the geometry formed between the *focal spot and the edges of those details*, not the edges of the field

Figure 18-5

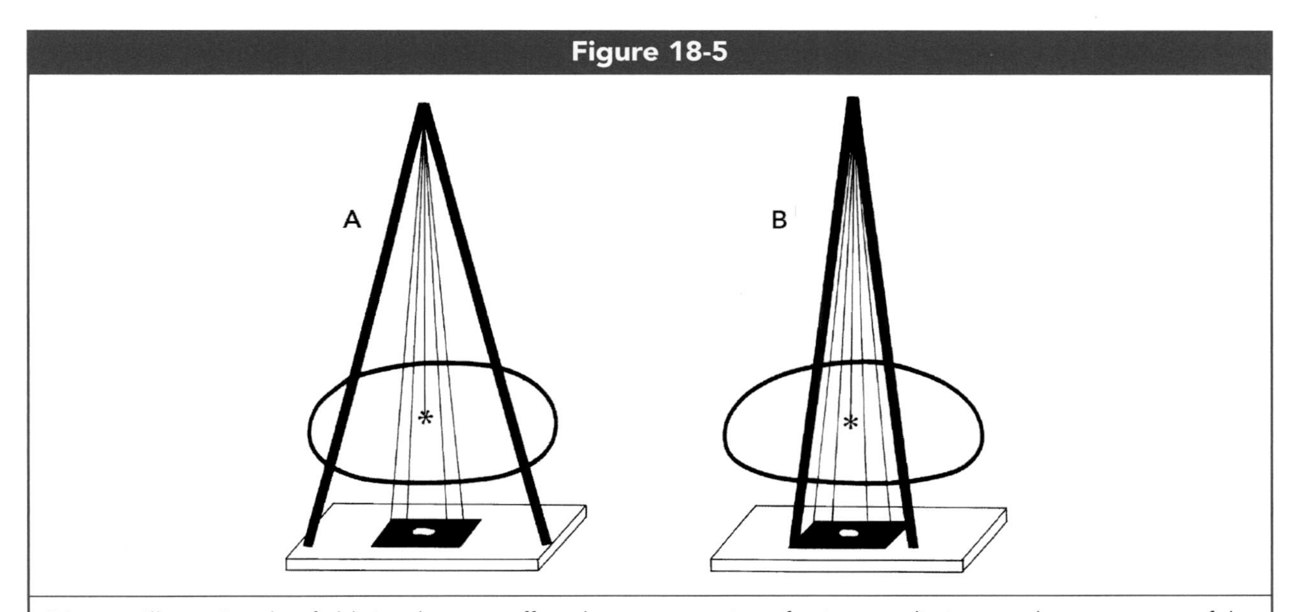

Diagram illustrating that field size does not affect the concentration of *primary* radiation nor the occurrence of the photoelectric effect (asterisk). Field size is indicated by the bold lines. In ***A***, four x-rays reach the square area at the image receptor, and one is absorbed by photoelectric interaction, leaving a "blank" spot in the remnant image. In the more collimated projection, ***B***, the square area still receives four x-rays with one photoelectric absorption in the middle. The effects of changing field size are *entirely* due to alterations in the scatter radiation, not to any change in the primary radiation. (From Quinn B. Carroll, *Practical Radiographic Imaging*, 8th ed. Springfield, IL: Charles C Thomas, Publisher, Ltd., 2007. Reprinted by permission.)

HISTORICAL SIDEBAR 18-1: Figure 18-6 is a series of sinus radiographs taken on film. They manifest progressively increasing contrast as well as an overall lightening of image density. Both effects are due to the reduction of scatter exposure reaching the film as the field size was collimated smaller and smaller, using extension cylinders.

Figure 18-6

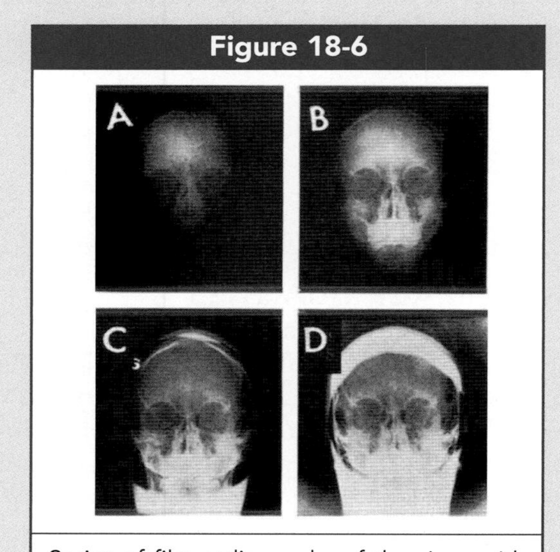

Series of film radiographs of the sinus with progressively reduced field size, using extension cylinders. Note the dramatic improvement in contrast and slight loss of density due to scatter reduction in the tightly-collimated view, **C.** (From Quinn B. Carroll, *Practical Radiographic Imaging*, 8th ed. Springfield, IL: Charles C Thomas, Publisher, Ltd., 2007. Reprinted by permission.)

itself. Chapter 19 will give a full explanation as to why scatter radiation is unrelated to sharpness of detail.

To the extent that scatter radiation is presented at the image receptor, it may be said that field size has some impact upon image *noise.* Field size only affects the visibility functions of image formation.

CALCULATING FIELD SIZE COVERAGE

The total area of field size coverage is directly proportional to the square of the distance from the x-ray tube to the image receptor plate. Any given side of a rectangular field, or the diameter of a circular field, is directly proportional to the SID. As illustrated in Figure 18-7, this relationship follows the law of similar triangles, where the distances involved are represented by the height of each triangle formed by the beam, and the sides or diameter of the fields are represented by the bases of the triangles.

The apex of both triangles is always defined as the focal spot in the x-ray tube. The exact location of the focal spot is normally marked on anode end of the tube housing with a small, red *plus* sign or "x". It can be generally estimated to be about 1½ inches below the center of the x-ray tube housing.

A large triangle is formed by the SID with the field size at the image receptor as its base. A smaller triangle is formed by the SOD (source-to-object distance) with its base representing the *lower aperture* of any collimating device. The *ratio* formed between any two similar portions of these triangles, including their heights, will always be equal to the ratio between their bases. We can see in Figure 18-7, then, that *magnification of the field size is always proportional to the ratio of SID/SOD.* Either of the following proportion formulas can be used:

$$\frac{SID}{F_{IR}} = \frac{SOD}{F_A} \quad \text{OR} \quad \frac{SID}{SOD} = \frac{F_{IR}}{F_A}$$

where F_{IR} is the field size at the image receptor and F_A is the field size at the lower aperture of a cone or cylinder, or the lower shutters of a collimator.

These formulas are simple ratios and can be expressed "in English" as follows:

1. The SID is to the final field size as the distance to the lowest aperture is to its diameter, or
2. The SID is to the SOD as the final field size is to the lowest aperture.

Because these are simple ratios, in practice the radiographer can make fairly simple calculations mentally. For example, from the x-ray tube, if the distance from the end of a cylinder "cone" is ¼ of the distance to the x-ray table, the SID/SOD ratio is 4 to 1, and the field size at the tabletop will have 4 times the diameter of the cylinder. To find the dimensions of a rectangular field, the problem would have to be worked twice, once for the length and once for the width of the field.

Another way of stating this relationship is that the SID/SOD ratio determines the *magnification* of the field. The very same ratio is used to determine the magnification of any object or anatomical part within the projection, which was explained in Chapter 13 and will be further discussed in Chapter 23.

After reviewing the following examples, try Exercise 18-1. Your answers may be checked in Appendix #1.

Practice Exercise #1

Suppose that the end of a cylinder "cone" is about 15 inches from the focal spot, and the SID being used is 45 inches. The bottom opening, or aperture, of the cylinder is 3 inches in diameter. What will be the diameter of the field at the image receptor plate?

Solution:

Using the first formula, $\frac{SID}{F_{IR}} = \frac{SOD}{F_A}$

$$\frac{15}{3} = \frac{45}{X}$$

Cross-multiplying: $15(X) = 3 \times 45$

$$X = \frac{135}{15}$$

$$X = 9$$

Answer: The diameter of the field at the image receptor will be 9 inches.

Practice Exercise #2

Suppose that the end of a cylinder "cone" is about 10 inches from the focal spot, and the bottom opening, or aperture, has a diameter of 4 inches. What SID should be used to obtain a field diameter of 16 inches at the image receptor?

Solution:

Using the first formula, $\frac{SID}{SOD} = \frac{F_{IR}}{F_A}$

$$\frac{X}{10} = \frac{16}{4}$$

Cross-multiplying: $4(X) = 16 \times 10$

$$X = \frac{160}{4}$$

$$X = 40$$

Answer: The SID that should be used is 40 inches.

Figure 18-7

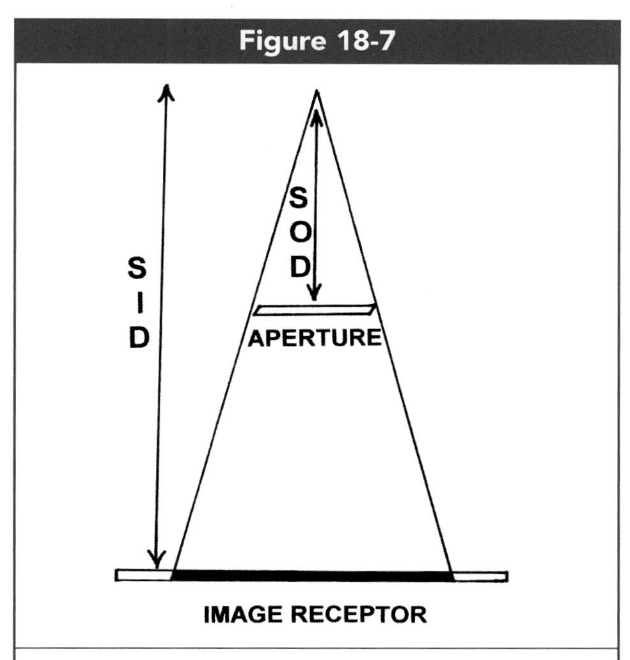

Diagram of similar triangle geometry formed by collimation of the x-ray beam. The field size is proportional to the SID/SOD ratio, where SOD is the distance to the lowest aperture of any collimating device. (From Quinn B. Carroll, *Practical Radiographic Imaging*, 8th ed. Springfield, IL: Charles C Thomas, Publisher, Ltd., 2007. Reprinted by permission.)

EXERCISE #18-1

In the following problems, the "aperture" refers to the width of the bottom opening of a cylinder or cone, or of the bottom shutters in a collimator. Using similar triangle proportions, fill in the missing size or distance.

	Distance from FS to aperture	*SID*	*Size of Aperture*	*Field Size at Image Receptor*
1.	10"	40"	6"	_____
2.	12"	36"	5"	_____
3.	40 cm	100 cm	8 cm	_____
4.	30 cm	60 cm	_____	32 cm
5.	18"	36"	_____	24 cm
6.	18"	45"	_____	25"
7.	15"	_____	5"	10"
8.	40 cm	_____	8 cm	24 cm
9.	_____	80 cm	12 cm	32 cm
10.	_____	72"	3"	12"

SUMMARY

1. X-ray field size must never be larger than necessary to include the anatomy of interest, and must never be larger than the size of the receptor plate.
2. Radiation exposure to a particular organ can be reduced to approximately 1/100th solely by collimating it outside of the primary x-ray beam.
3. Collimation devices with a secondary aperture reduce off-focus radiation.
4. Automatic collimation devices should be overridden in order to collimate smaller than the receptor plate size when the anatomy is substantially smaller than the plate.
5. To allow for inaccuracies in field light alignment, always allow one-half inch (1 cm) of field light beyond each edge of the anatomy of interest.
6. Excessive field size reduces subject contrast in the remnant beam, because the production of scatter radiation within the patient is increased such that it contributes a higher percentage contribution to the formation of the image at the receptor plate.
7. By reducing the amount of scatter reaching a particular area of the receptor plate, the total exposure to that area is also reduced.
8. As a form of noise, scatter radiation reduces the visibility functions of image formation—it has no relationship to the geometrical image qualities: sharpness, magnification or distortion.
9. Field size follows the projection geometry of similar triangles and obeys the magnification formula of SID/SOD (where the SOD is the distance to the lowest aperture of the collimating device.

REVIEW QUESTIONS

1. What are the *two* purposes of reducing field size?

2. How does collimation affect the concentration or intensity of the *primary* x-ray beam?

3. Why is a secondary set of shutters or second aperture in a collimating device desirable?

4. Where is off-focus radiation produced?

5. What does *PBL* stand for?

(Continued)

REVIEW QUESTIONS *(Continued)*

6. It is acceptable to override an automatic collimator in order to:

7. If anatomy of interest is clipped from a view due to overzealous collimation, how much is patient exposure increased for that view upon repeating it?

8. The lower shutters of a collimator are 30 cm from the focal spot. If the shutters are opened to a 10-cm square, how big will the x-ray beam be at 105 cm from the focal spot?

9. The end of a 6-inch diameter extension cylinder is 15 inches from the focal spot. At what distance will the circular field have an 18-inch diameter?

In the following problems, the "aperture" refers to the width of the bottom opening of a cylinder or cone, or of the bottom shutters in a collimator. Using similar triangle proportions, fill in the missing size or distance.

	Distance from FS to aperture	*SID*	*Size of Aperture*	*Field Size at Image Receptor*
10.	12"	36"	4"	________
11.	15"	60"	________	12"
12.	30 cm	________	12.5 cm	25 cm
13.	________	100 cm	8 cm	32 cm

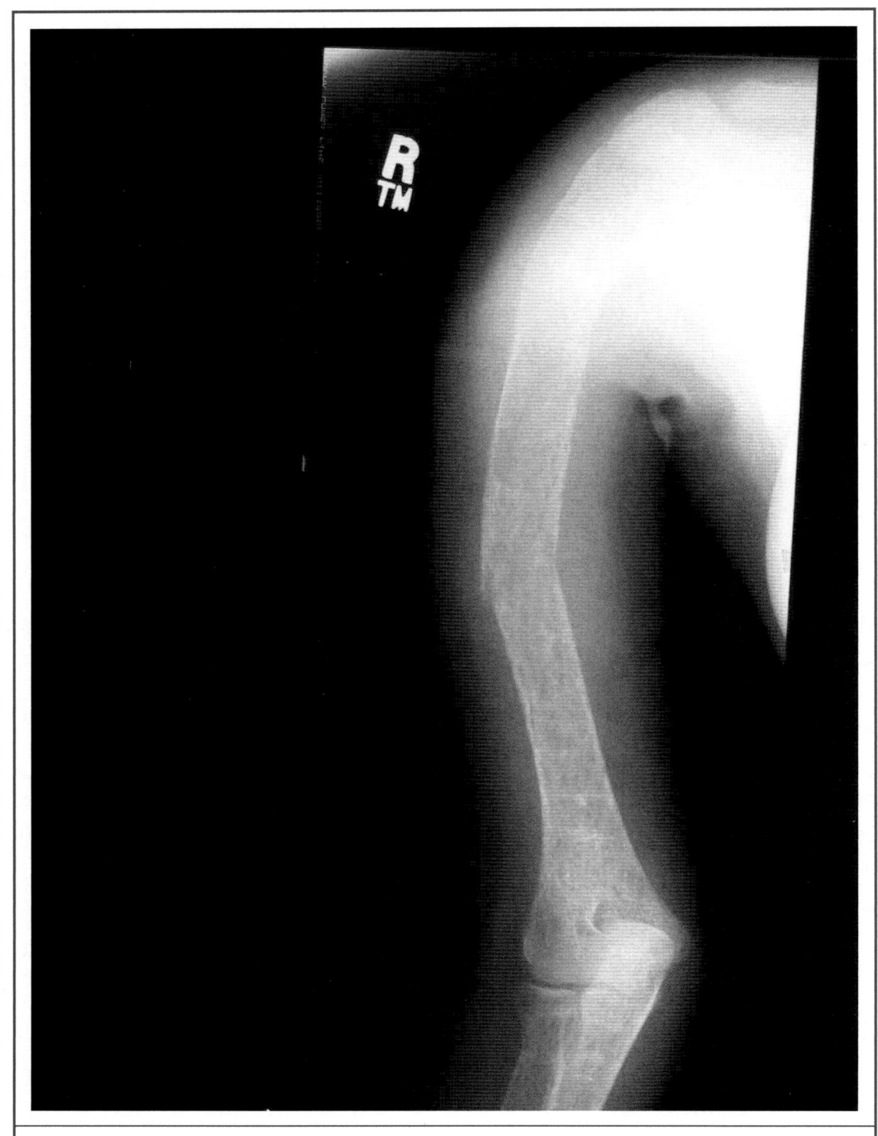

This humeral fracture is secondary to the disease *rickets*, due to a deficiency of vitamin D.

Chapter 19

PATIENT CONDITION, PATHOLOGY, AND CONTRAST AGENTS

Objectives:

Upon completion of this chapter, you should be able to:

1. Describe the proper locations and methods for measuring body part thicknesses.
2. Explain why radiographic technique can be approached in a systematic and scientific way.
3. State the average thicknesses for the adult torso in AP and lateral projection.
4. Use the 4-centimeter rule to derive appropriate techniques for variable part thicknesses.
5. State the rule for minimum change of exposure factors.
6. Describe each of the five types of body habitus and the associated adjustments for technique.
7. Describe the influence of age and anthropological factors on radiographic technique.
8. List the five general types of materials that are radiographically demonstrable, in order from most radiolucent to most radiopaque.
9. Describe the impact of the molecular (average) atomic number and physical density of body tissues upon subject contrast in the image.
10. Describe the impact of the molecular (average) atomic number and physical density of contrast agents upon subject contrast in the image.
11. Describe the effects of patient cooperation and stage of respiration on chest radiographs.
12. Define radiographically additive and destructive diseases.
13. For the most common additive and destructive diseases, describe the appropriate types of adjustments in radiographic technique.
14. State the technique guidelines for post-mortem radiography.
15. State the technique guidelines for soft-tissue radiography.
16. State the technique guidelines for cast and splint radiography.

GENERAL PATIENT CONDITION

Radiographic image quality, regardless of the type of processing used, is always first dependent upon the production of good subject contrast in the remnant x-ray beam. We have discussed the impact of mAs, kVp, generator power, filtration, and field size on the remnant beam. These are all variables that are under our control. The condition of the patient is something we have almost no control over, but must be aware of and take into account when selecting the technical factors for an exposure.

Thickness of the Part

Despite the many natural variations in body type and shape among humans, it is possible to establish certain patterns of thickness measurement for radiographic purposes that may conform to what may be termed *average* thickness ranges. Such measurements can be standardized and are an important means of standardization for radiographic exposures.

The type of calipers to be used for measuring different body parts is shown in Figure 19-1. *Generally, measurements should be taken along the course of the projected central ray; however, for wedge-shaped body parts that differ substantially in thickness from one end to the other (including the foot, the thigh and often the chest), it is best to take the measurement slightly toward the thicker portion of the anatomy from the CR, at about the thickest one-third.*

Figure 19-1

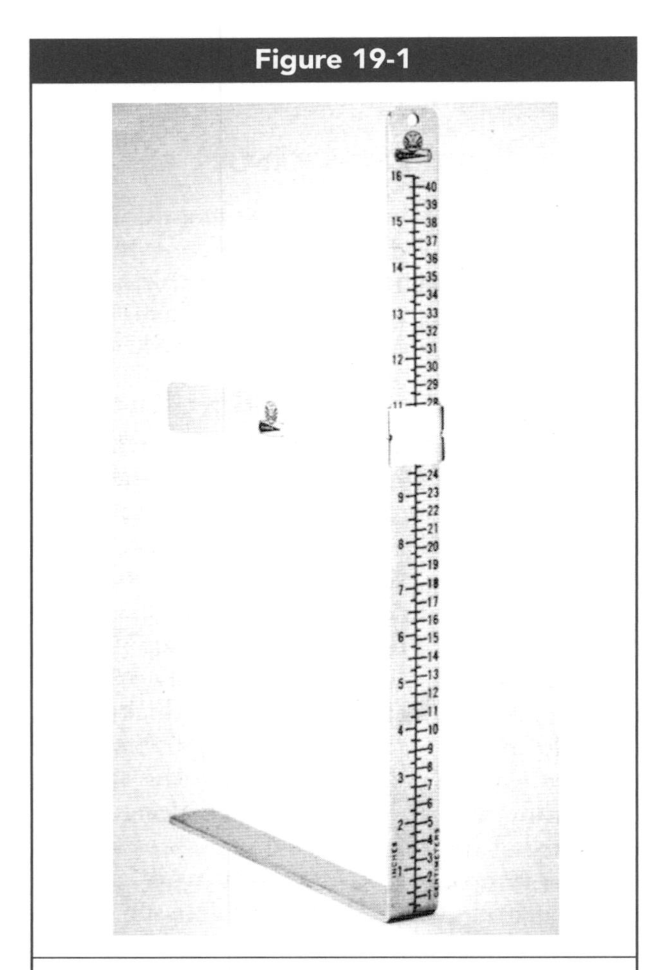

The type of calipers that should be used for measuring body parts. (From Quinn B. Carroll, *Practical Radiographic Imaging*, 8th ed. Springfield, IL: Charles C Thomas, Publisher, Ltd., 2007. Reprinted by permission.)

Measurements for the PA or AP chest, for example, are generally taken at about the nipple-line for a male, and for a female patient should include the upper portion of the breast thickness. Chests measurements should be made with the patient standing or sitting as they will be radiographed, under normal respiration.

Usually, the non-movable leg of the calipers is placed opposite the body entrance point for the CR and the movable leg is closed until it *just* makes contact with the skin. For an accurate measurement, the calipers should not be squeezed such that they displace any tissue or the movable leg is in a bent position. When measuring a supine abdomen, it is important not to simply place the non-movable leg of the calipers on the tabletop, but rather to *lift* it up into contact with the patient's back. When the patient is standing or sitting erect, abdominal organs gravitate downward, packing together and increasing the overall density and thickness to be penetrated. An increase in technique will be expected when compared with the supine abdomen.

Patient weight or height are not reliable indices for the formulation of radiographic techniques. Reliable techniques may be determined only by using the *measured part thickness* as a fundamental guide, with the patient's condition then taken into account.

Thickness Ranges

Statistically, the frequency with which adult patients conform to an average thickness range for a given projection is relatively high. Table 19-1 presents the data from thousands of accurate measurements of body part thicknesses. The frequency with which various thicknesses appeared for each given projection was then tabulated. The higher thickness frequencies were then grouped into an *average thickness range*, to which could be applied a common set of exposure factors with the assurance that satisfactory exposure would be obtained in more than 85 percent of the cases.

The implications of Table 19-1 for radiographic technique are very significant indeed. First, these statistics show that techniques can be usefully standardized, because a relatively high percentage of patients fall within the "average range" of thickness for most types of projections. Note that for most distal extremities, and for head anatomy, more than *90 percent* of all adult patients will conform to a single "average"

Table 19-1

Table Listing Average Thickness Ranges for Various Projections and the Frequency with Which the Adult Patient Dimensions Fall into each Average Range

<table>
<tr><th colspan="2">Region</th><th colspan="3">Average Thickness/Adult—Cms.</th><th>Percent Frequency</th></tr>
<tr><th colspan="2"></th><th>AP</th><th>PA</th><th>LAT</th><th></th></tr>
<tr><td colspan="2">Thumb, fingers, toes</td><td></td><td>1.5–4</td><td></td><td>99</td></tr>
<tr><td colspan="2" rowspan="2">Hand</td><td colspan="2">3–5</td><td></td><td>99</td></tr>
<tr><td colspan="2"></td><td></td><td>93</td></tr>
<tr><td colspan="2" rowspan="2">Wrist</td><td colspan="2">3–6</td><td></td><td>99</td></tr>
<tr><td colspan="2"></td><td>5–8</td><td>98</td></tr>
<tr><td colspan="2" rowspan="2">Forearm</td><td colspan="2">6–8</td><td></td><td>94</td></tr>
<tr><td colspan="2"></td><td>7–9</td><td>92</td></tr>
<tr><td colspan="2" rowspan="2">Elbow</td><td>6–8</td><td></td><td></td><td>96</td></tr>
<tr><td colspan="2"></td><td>7–9</td><td>87</td></tr>
<tr><td colspan="2" rowspan="2">Arm</td><td>7–10</td><td></td><td></td><td>95</td></tr>
<tr><td colspan="2"></td><td>7–10</td><td>94</td></tr>
<tr><td colspan="2">Shoulder</td><td>12–16</td><td></td><td></td><td>79</td></tr>
<tr><td colspan="2">Clavicle</td><td></td><td>13–17</td><td></td><td>82</td></tr>
<tr><td colspan="2" rowspan="2">Foot</td><td colspan="2">6–8</td><td></td><td>92</td></tr>
<tr><td colspan="2"></td><td>7–9</td><td>91</td></tr>
<tr><td colspan="2" rowspan="2">Ankle</td><td>8–10</td><td></td><td></td><td>86</td></tr>
<tr><td colspan="2"></td><td>6–9</td><td>96</td></tr>
<tr><td colspan="2" rowspan="2">Leg</td><td>10–12</td><td></td><td></td><td>85</td></tr>
<tr><td colspan="2"></td><td>9–11</td><td>89</td></tr>
<tr><td colspan="2" rowspan="2">Knee</td><td>10–13</td><td></td><td></td><td>92</td></tr>
<tr><td colspan="2"></td><td>9–12</td><td>92</td></tr>
<tr><td colspan="2" rowspan="2">Thigh</td><td>14–17</td><td></td><td></td><td>77</td></tr>
<tr><td colspan="2"></td><td>13–16</td><td>76</td></tr>
<tr><td colspan="2">Hip</td><td>17–21</td><td></td><td></td><td>76</td></tr>
<tr><td rowspan="3">Cervical Vertebrae</td><td>C1–3</td><td>12–14</td><td></td><td></td><td>77</td></tr>
<tr><td>C4–7</td><td>11–14</td><td></td><td></td><td>98</td></tr>
<tr><td>C1–7</td><td colspan="2"></td><td>10–13</td><td>90</td></tr>
<tr><td colspan="2" rowspan="2">Thoracic vertebrae</td><td>20–24</td><td></td><td></td><td>76</td></tr>
<tr><td colspan="2"></td><td>28–32</td><td>81</td></tr>
<tr><td colspan="2" rowspan="2">Lumbar Vertebrae</td><td>18–22</td><td></td><td></td><td>69</td></tr>
<tr><td colspan="2"></td><td>27–32</td><td>77</td></tr>
<tr><td colspan="2">Pelvis</td><td>19–23</td><td></td><td></td><td>78</td></tr>
<tr><td colspan="2" rowspan="2">Skull</td><td></td><td>18-21</td><td></td><td>96</td></tr>
<tr><td colspan="2"></td><td>14–17</td><td>88</td></tr>
<tr><td rowspan="3">Sinuses</td><td>Frontal</td><td></td><td>18–21</td><td></td><td>97</td></tr>
<tr><td>Maxillary</td><td></td><td>18–22</td><td></td><td>88</td></tr>
<tr><td></td><td colspan="2"></td><td>13–17</td><td>96</td></tr>
<tr><td colspan="2">Mandible</td><td></td><td></td><td>10–12</td><td>82</td></tr>
<tr><td colspan="2" rowspan="3">Chest</td><td></td><td>20–25</td><td></td><td>82</td></tr>
<tr><td>OBL</td><td></td><td>27–32</td><td>84</td></tr>
<tr><td>24–30</td><td></td><td></td><td>83</td></tr>
</table>

From Quinn B. Carroll, *Practical Radiographic Imaging*, 8th Ed. Springfield, IL: Charles C Thomas, Publisher, Ltd., 2007. Reprinted by permission.

technique. In fact, the lowest percentage in the table is for the AP projection of the lumbar spine, at 69 percent. The AP lumbar spine may be considered equivalent to the AP abdomen projection. Therefore, it might be safely stated that even for abdominal radiography, where, not surprisingly, the greatest variation between patients occurs, roughly two out of three patients will still conform to a single "average" technique. *Radiographic technique can be confidently approached in a systematic and scientific way.*

Second, it is useful to memorize the average thicknesses from Table 19-1. These provide an essential starting point from which to derive techniques for various thicknesses of body parts. Note that, combining the figures for the chest, thoracic spine, lumbar spine and pelvis, we conclude that *average thicknesses for the adult torso in general are 22 centimeters (about 8½ inches) in AP projection, and 30 centimeters (about 12 inches) in lateral projection.*

The Four Centimeter Rule

As described in Chapter 12, part thickness affects the absorption of x-ray photons in an exponential fashion, that is, small changes in thickness cause relatively large changes in absorption of the x-ray beam. Radiographers must become adept at evaluating the patient and adjusting radiographic technique accordingly.

The *four centimeter rule* states that *for every 4 centimeters change in part thickness, adjust the technique by a factor of 2.* For example, when the patient's abdomen measures 26 cm, this is 4 cm greater than the average 22 cm for the abdomen, and overall technique should be doubled. When the patient is 4 cm thinner than average, cut the technique in half.

The rule may be applied in steps; a 30 cm abdomen is thicker than the average 22 cm by *two sets of 4 cm.* To compensate, double the technique *twice.* It can also be adapted for 2-centimeter differences by adjusting the technique *half-way to double* (a 50% increase) for a patient who is 2 cm thicker, or *half-way to one-half,* (a 25% decrease) when the patient is 2 cm thinner.

The above proportional and percentage changes would be directly applied to adjustments in *mAs.* If it is desired to change the *kVp* rather than the mAs, the four centimeter rule can be adapted using the 15 percent rule as follows: *For every 4 centimeters change in part thickness, adjust the kVp by 15 percent.* (*For every 2 cm change, adjust it by 8%.*)

Try the following brief exercise to practice applying the 4-centimeter rule, then check your answers using Appendix #1.

EXERCISE #19-1

In the technique charts below, the recommended mAs or kVp is given for a particular thickness of the body part being radiographed. For each problem, list the actual mAs or kVp you would use after adapting for the actual thickness of the part. You can check your answers in Appendix #1.

PART A: Adapting mAs

	Technique Chart		*Actual Part*	
Part	*mAs*	*Thickness*	*Thickness*	*mAs Used*
1. Elbow	5	6 cm	10 cm	________
2. Femur	10	17 cm	13 cm	________
3. UGI AP	7.5	20 cm	25 cm	________
4. L-Spine	30	24 cm	16 cm	________
5. Chest	10	22 cm	24 cm	________

PART B: Adapting kVp

	Technique Chart		*Actual Part*	
Part	*mAs*	*Thickness*	*Thickness*	*mAs Used*
1. Elbow	60	6 cm	10 cm	________
2. Femur	74	17 cm	13 cm	________
3. AC UGI	100	20 cm	25 cm	________
4. L-Spine	80	24 cm	16 cm	________
5. Chest	110	22 cm	24 cm	________

Minimum Change Rule

Any time an increase in exposure to the image receptor is needed, it is wasted effort to bring the overall radiographic technique up by less than one-third. A 35 percent increase in the mAs should be considered the *minimum change* to bring about any *significant* alteration in the final exposure. (See Historical Sidebar 19-1.)

It is an easy matter to adapt the 15 percent rule for kVp to this purpose; the desired result in remnant beam exposure is an increase at least one-third of the way to 100 percent (or a doubling). To adapt this for kVp, we would make a minimum increase one-third of the way to 15 percent, or 5 percent.

When exposure needs to be adjusted, the minimum change that should ever be made in radiographic

HISTORICAL SIDEBAR 19-1: Figure 19-2 is a series of film-based radiographs of the knee which clearly demonstrate the minimum change rule for technique. In the first radiograph labeled ***N*** and ***B***, the mAs was increased by precisely 25 percent from ***N*** to ***B***. Observe the middle portion of the upper tibia to confirm that radiograph ***B*** is not visibly darker. In the second set, the mAs was increased by precisely 50 percent from ***N*** to ***C***. Here, it can be discerned that radiograph ***C*** is *just visibly darker*.

This principle is applicable to exposure at the image receptor for digital radiography systems as well: To make any *significant* change in exposure, overall technique must be changed by at least one-third, or about 35 percent.

Figure 19-2

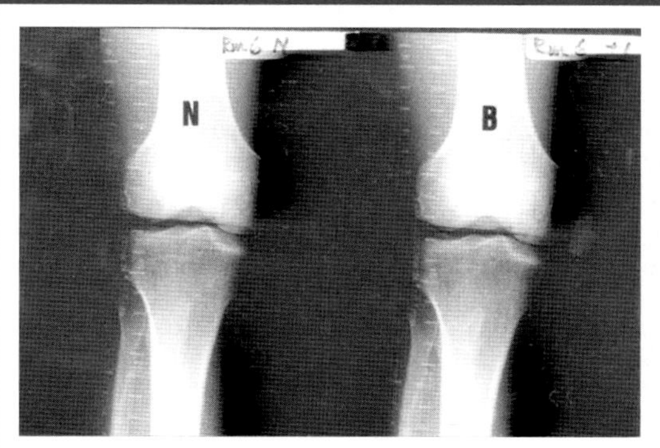

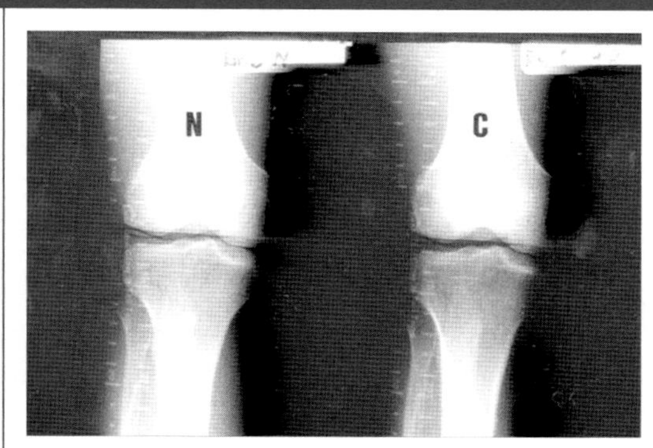

Film radiographs of the knee prove that a 35% change in x-ray exposure is the minimum required to make a substantial difference. For comparison, observe the upper middle portion of the tibia: In the first set, the mAs was increased by precisely 25% from ***N*** to ***B,*** and there is not a visible increase in density. In the second set, mAs was increased 50% from ***N*** to ***C,*** and the radiograph becomes just visibly darker than ***N***. (From Quinn B. Carroll, *Practical Radiographic Imaging*, 8th ed. Springfield, IL: Charles C Thomas, Publisher, Ltd., 2007. Reprinted by permission.)

technique is a 35 percent change in mAs or a 5 percent change in kVp.

Body Habitus

No two human physiques are identical. Not only do they vary in their general family and racial characteristics, but individual parts vary widely is size, shape and tissue density. For radiographic purposes, we traditionally classify each body into one of four *somatotypes*, or types of *body habitus*. These are *sthenic*, *hyposthenic*, *asthenic* and *hypersthenic*. However, *five* types will be described here, adding a category for *large muscular* patients.

Sthenic

The *sthenic* body type is taken as a healthy average, strong and active. The bones are large, with full musculature, and the skin is thick with an abundance of underlying tissue. These persons are physically well-balanced, with the chest larger than the abdomen.

For typical adults with sthenic habitus, an AP caliper measurement of the abdomen will average 22 cm, but considering all racial types and ages, it will fall within a fairly broad range of plus-or-minus 4 cm from this average (18–26 cm). What will be more consistent is the *ratio* between the AP and the lateral abdominal measurements. The lateral measurement is typically 8 cm greater than the AP. This gives the patient's torso an oval shape in cross-section, diagrammed in Figure 19-3.

Hyposthenic

In terms of adjusting radiographic techniques, the hyposthenic body habitus is not an extreme variation from sthenic or average, in that this person is strong, active and generally healthy, but unusually thin. In spite of their small size, hyposthenic patients

Figure 19-3

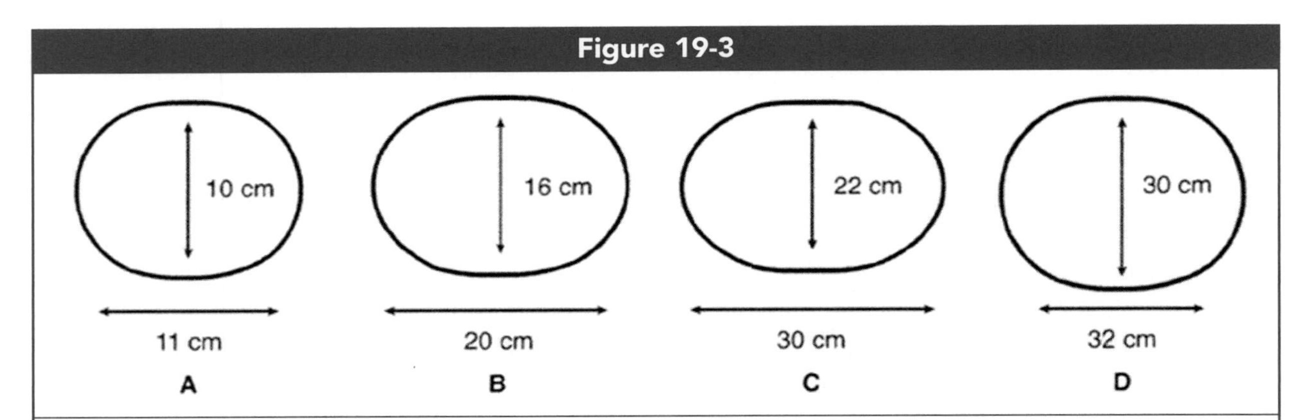

Cross-sectional diagrams of body torso shapes and dimensions, of A, a premature infant presenting a nearly round torso, B, a healthy toddler with a more oval-shaped torso, C, an average adult with an oval cross-section measuring on average 22 X 30 cm, and D, a fluid-distended or hypersthenic adult with AP and lateral dimensions nearly equal. (From Quinn B. Carroll, *Practical Radiographic Imaging*, 8th ed. Springfield, IL: Charles C Thomas, Publisher, Ltd., 2007. Reprinted by permission.)

have a body composition with muscle/fat radios similar to sthenic patients. Therefore, technique may be adjusted from the average entirely on the basis of the patient's measured size without futher consideration as to condition. A technique chart with small enough measurements represented on it may be followed.

In cross-section (Fig. 19-3), the oval shape of the torso is elongated more laterally, so that the difference between the AP and lateral measurements is greater than 8 cm. A somewhat greater adjustment in radiographic technique is likely to be necessary when changing from frontal to lateral projections.

Asthenic

The asthenic body type may best be described as overly thin *due to ill health*. Because of advanced age or disease, this patient is physically weak and emaciated in appearance. The body is delicate and characterized by small bones and stringy muscles. The ribs and shoulder blades protrude, and the abdomen is flat and shallow in depth.

In cross-section, the asthenic patient's torso is similar to the elongated oval shape of a hyposthenic patient. However, due primarily to the loss of cortical bone density, radiographic techniques may need to be reduced for asthenic patients more than their measurement alone would indicate.

Hypersthenic

The hypersthenic body type is characterized by roundness and softness throughout the various body regions, with an excess of subcutaneous fat. The torso is barrel-like and dominant compared to the head and extremities. The chest is relatively short and wide at the base. The neck is short and thick and the waistline is high. The back is well-padded with fat. This type of patient tends toward obesity, hypertension and other degenerative diseases of the circulatory system and kidneys.

In cross-section, the hypersthenic patient's torso would be described as *round* rather than oval (Fig. 19-3). This makes the AP and lateral thicknesses comparable when measured with calipers. Both measurements are likely to be 30 cm or greater. For a 30 cm hypersthenic patient, all views of the torso, whether frontal, oblique, or lateral, would require a radiographic technique equivalent to an average *lateral* projection on a technique chart. This applies to all chest and abdomen techniques, abdominal procedures using contrast agents, and in many cases also to pelvis techniques.

It is not uncommon for hypersthenic patients to measure considerably larger than 30 cm, and they may be larger than any of the thicknesses listed on a particular technique chart. In such cases, the *four-centimeter rule* may be used to further adjust the technique. The four-centimeter rule is calibrated on the assumption that much larger patients are generally of the hypersthenic type. In other words, it brings techniques up at a rate that anticipates increased adipose tissue (fat) rather than all muscle. This is appropriate, since the vast majority of patients with large measurements are, indeed, hypersthenic.

Large Muscular

A fifth category will be suggested here, one that certainly requires an adjustment in technique and is not uncommon in modern times. To date, this body habitus has not been given a proper medical name, perhaps because it was *not* so commonplace in the first half of the twentieth century. Some patients, predominantly but not exclusively male, can only be described as both very large *and* very muscular. This body habitus does not properly fit into any of the four conventional categories listed above. These patients are typically very *tall*, so that their height-to-weight ratio is not very far out of line. Although their body shape fits that of a sthenic adult, (*C* in Figure 19-3), they are large-boned, with a triangular chest and the solid muscle mass of an athlete who works out regularly.

Experienced radiographers know that when they are presented with a patient of this type, radiographic techniques must be increased quite beyond what a technique chart would indicate even for the measured thickness. Nor does the four-centimeter rule bring technique up enough to fully compensate. To ensure adequate exposure at the image receptor plate, *after* increasing technique in accordance with the measured thickness of the body part, *an additional* 35–50 percent increase in technique is needed.

Large muscular patients must be individually evaluated to decide if *more* than this is required. In extreme cases, a 50 percent increase or even more may be called for. Remember that this amount must be added to increases already made strictly for larger thickness of the body part.

Generally, muscle tissue has the approximate density of water, which is already denser than fat tissue. But, for athletic patients who regularly work out, the density of *their* muscle tissue is greater still as it is well-engorged with circulatory and interstitial fluids. For this reason, as well as in the interest of minimizing patient dose, the radiographer may wish to effect the needed increase in technique using kVp rather than mAs so that it is achieved through enhanced x-ray penetration. Adapting the 15 percent rule for kVp to this purpose, we can state that a 5–8 percent increase in kVp would achieve the same result.

Influence of Age

Throughout life there is a constant flux in bone mineralization. In infants and very young children, calcification of bones is very slight, and reduced kVp will provide proper penetration. During the growth period, bone formation is more rapid than resorption. During middle age, bone formation and bone resorption are in balance. In later life, resorption exceeds bone growth and mineralization decreases. There is also some atrophy, or loss of water and minerals, from soft-tissue organs. Lower kVp levels are required to avoid overpenetration.

Pediatric techniques are often derived by modifying adult techniques according to the proportionate thickness and composition of the anatomy. For example, an adult knee technique might be adapted for a skull projection on a toddler—the skull may be slightly larger but has a lesser proportion of bone within that thickness. It is well to note the proportionate differences in the growth of various body parts: The thigh (femur) represents the extreme in growth—it is over five times longer and thicker in an adult than in an infant. On the other hand, the human skull expands less than three times its original diameter from infancy. This means that the pediatric technique for a skull will be roughly one-third to one-half that of an adult skull technique, while the pediatric technique for a femur will be as little as one-fifth that of an adult.

Anthropological Factors

The radiographer must continually train his or her eyes in evaluating the patient for radiographic purposes. There are physical differences in the mass, density and proportionate shape of various bones for different human races which can be substantial enough to require adjustments in technique. For example, these differences can be particularly notable in skull radiography, where some types of skulls are of rounder-cross section and tend toward much thicker prominences than others. By comparison, such skulls will require more technique overall, but with less difference between the AP and lateral projections, than a more oval skull.

MOLECULAR COMPOSITION OF TISSUES

Apart from the thickness of a body part, the attenuation of x-rays as they pass through it is a function of both the molecular or average *atomic number* of the tissues and the average *physical density* of the tissues. X-ray beam absorption due to both part thickness and atomic number is *exponential*, while absorption due to physical density is only *proportional* and thus requires huge discrepancies to make a difference. These relationships were thoroughly discussed in Chapter 11.

There are basically only *five* types of materials discernible by the differences in their attenuation of an x-ray beam (subject contrast). These are *gas*, *fat*, *fluids*, *bone* and *metals*. (The term *fluids* is a misnomer but is commonly used in this context—a more correct term would be *liquids*.) Soft tissues include most visceral organs, muscles and connective tissues which are normally difficult to distinguish from one another in a radiographic image. When a very large difference in the thickness exists, such as for the *psoas muscles* in the abdomen or the liver, these organs can just be made out against other adjacent soft tissues (Fig. 19-4). Another exception occurs when the bladder is full of pure liquid, setting it apart from the surrounding *mixture* of liquids and solids that make up soft tissue. The postprocessing features of digital imaging systems can be used to enhance these differences. Even so, they are slight.

Figure 19-4

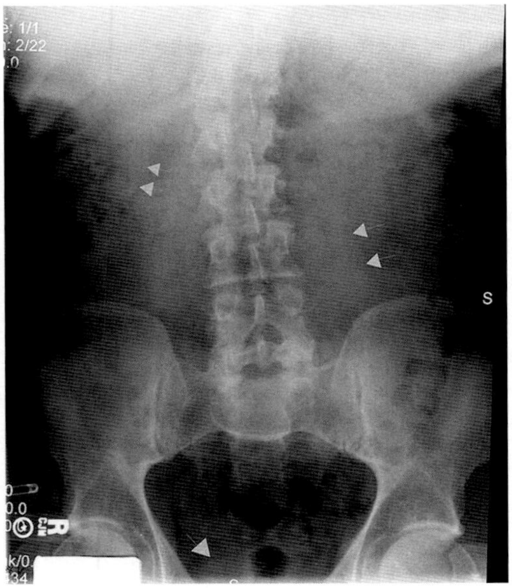

In a typical AP projection of the abdomen, the margins of the psoas major muscles and gall bladder (arrows) can just be made out against surrounding soft tissues. When the bladder is fully distended with fluid, it can also be recognized.

Soft tissues should generally be depicted in a radiographic image as a mid-level shade of gray, and can be used as a standard for comparison with the other four types of materials. Fat shows up slightly darker than soft tissue, and gases such as air in the lungs show up much darker, both on account of substantial differences in their physical density. Bone shows up well against soft tissue mostly because of its average atomic number, which makes it 8 times more attenuating to x-rays, but also because of its density which doubles the absorption again. It is also primarily the atomic number of most metallic objects such as bullets or orthopedic devices which makes them so impenetrable to x-rays.

CONTRAST AGENTS

Precisely because it is so difficult to demonstrate soft tissue visceral and circulatory organs against the background of similar soft tissues, various *contrast agents* have been developed, non-toxic chemical preparations that have high x-ray absorption properties and can be either injected or ingested into these organs (Fig. 19-5).

Contrast agents are broadly classified as *positive* or *negative* agents based on the radiopacity in comparison to typical soft tissues. *Positive* contrast agents are those which are *more* absorbent to the x-ray beam than surrounding soft tissues, so that they are presented on radiographic images as *radiopaque*

Figure 19-5

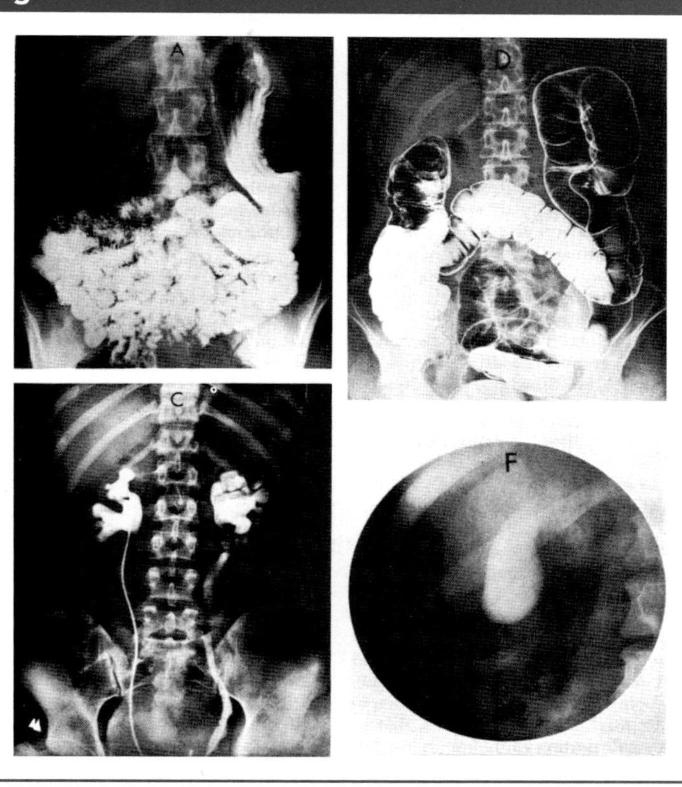

Radiographs in which contrast agents have been employed to delineate soft tissue cavities. ***A***, stomach and small intestine with barium, ***C***, kidneys and ureters with iodine, ***D***, colon with barium and air, and ***F***, gall bladder with iodine. (From Quinn B. Carroll, *Practical Radiographic Imaging*, 8th ed. Springfield, IL: Charles C Thomas, Publisher, Ltd., 2007. Reprinted by permission.)

areas that are lighter against the gray soft tissue background. *Negative* contrast agents are all gases. These include normal room air which may be drawn up into a syringe or trapped in a bag, carbon dioxide, and nitrous oxide, all of which are more *radiolucent* than soft tissue because of their extremely low physical density as gases.

Positive contrast agents are all based upon either the element *iodine* or the element *barium* bound into molecules with organic salts to make them nontoxic. Both iodine and barium are used on account of their very high atomic number; for iodine Z = 53, and for barium Z = 56. These "large" atoms have lots of orbital electrons packed within their diameter, with which x-rays may "collide" in the absorption process. Since x-ray absorption is proportional to the *cube* of the atomic number, these elements in their pure form are literally *hundreds of times* more absorbing than the surrounding soft tissue with its average atomic number of just 7.6. Iodine, for example, absorbs approximately 531 times more x-rays than soft tissue. (The calculation is $[53 / 7.6]^3$.) This effect is somewhat diluted by the mixture with other organic salts and water that make up the medium, but it still falls well over a factor of 100.

In fact, this very effectiveness can work against us when the bolus of contrast agent becomes very thick, as frequently occurs within the lumen of larger organs such as the stomach and colon. This can be seen in the barium enema radiograph ***D*** in Figure 19-5; even though this is an air-contrast study, the bolus of barium in the patient's cecum (to the viewer's lower left) is a thick, "solid column" of barium and demonstrates a blank white image of little diagnostic value because the only information is at the edges of the white silhouette.

For maximum radiographic information, it is essential to be able to see at least partially *through* the bolus of a positive contrast agent. This is well demonstrated in radiograph ***A*** in Figure 19-5, an upper GI radiograph—note that in the mid-portion of the stomach, the folds of *rugae* in the walls of the stomach can be made out as tortuous lines of gray *within* the lighter barium. In other words, anatomy in the *front and back* of the organ is demonstrated through the contrast agent, not just anatomy at the edges of the organ.

This need to see *through* the contrast agent explains the origin of *air-contrast studies* for the gastrointestinal system; by using less barium, mixing it to a thicker viscosity so it will *coat* the walls of the organ, and mixing it with air within the organ, more details of the anatomy can generally be visualized radiographically. This coating effect is seen in the barium enema radiograph in Figure 19-5, in both colonic flexures, the descending colon and the sigmoid portion, where the folds between the *haustral compartments* of the colon are demonstrated against the darker air. These folds would not be seen through a "solid-column" of barium, nor would some pathology in these areas such as small polyps or diverticuli. In radiograph ***A*** of the stomach, the same positive result is obtained naturally from the bubble of air that normally resides in the upper stomach.

Also, to see *through* a bolus of positive contrast agent, the x-rays in the beam must be of sufficient energy to penetrate through the agent and record information at the image receptor. Compare the UGI in Figure 19-5, radiograph ***A***, with Figure 13-2 in Chapter 13, where insufficient kVp was used to penetrate a stomach full of barium. Observe the iodine studies, radiographs ***C*** and ***F*** in Figure 19-5; most of the iodinated areas are demonstrated as light shades of gray rather than as blank white (an exception being the catheter and renal pelvis of the patient's right kidney). This is the proper balance between the amount of contrast agent present and the kVp needed to penetrate it. In the gall bladder (***F***) and the patient's left ureter and kidney (***C***), the presence of some density behind the bolus of contrast assures us that we are seeing the entire organ.

With this in mind, the *optimum kVp* recommended for these types of studies is 80 for the urinary system, 120 for "solid-column" studies of the stomach, small intestine and colon, and 100 kVp for all air-contrast studies of the digestive system. Digital imaging allows us to use the higher end of these kVp ranges.

STAGE OF RESPIRATION AND PATIENT COOPERATION

In radiography of the thorax, the stage of respiration has a great impact on the permeability of the air-containing tissues to x-rays. Figures 19-6 and 19-7 are film radiographs which amply demonstrate the effects of poor inspiration; failure by the patient to take a large, full inspiration before exposure can mimic underexposure through the lung fields. Many a poor inspiration has been confused with insufficient technique. To distinguish the two, the radiographer must get in the habit of carefully observing the position of the patient's diaphragm, which will lie above the 10th rib when a poor breath was obtained.

Expiration radiographs are intentionally ordered at times to rule out specific pathologies such as a

Figure 19-6

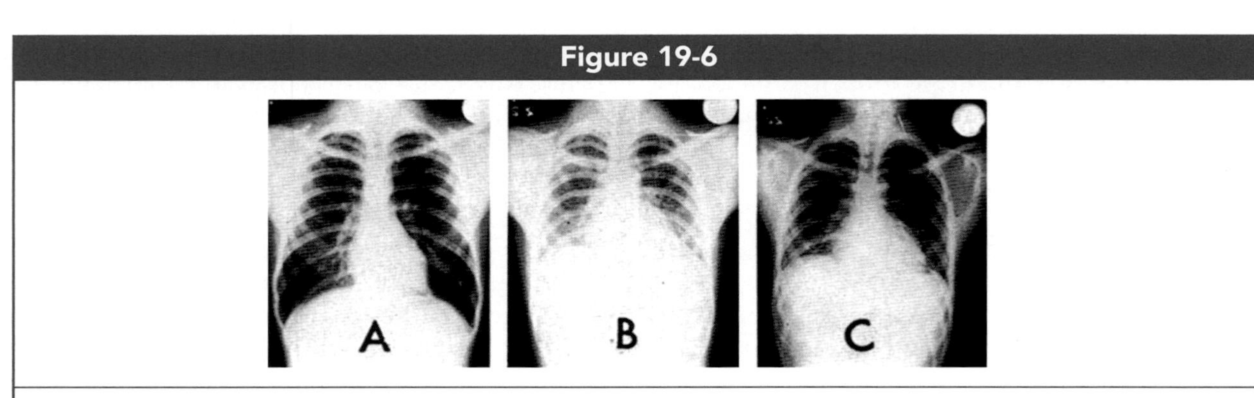

Film radiographs of the PA chest made on ***A***, inspiration, and ***B and C***, expiration, all at 80 kVp. Both ***A*** and ***B*** were exposed using 2.5 mAs. In C, the mAs was increased to 3.4 to compensate for the expected increase in overall tissue density upon expiration. (From Quinn B. Carroll, *Practical Radiographic Imaging*, 8th ed. Springfield, IL: Charles C Thomas, Publisher, Ltd., 2007. Reprinted by permission.)

Figure 19-7

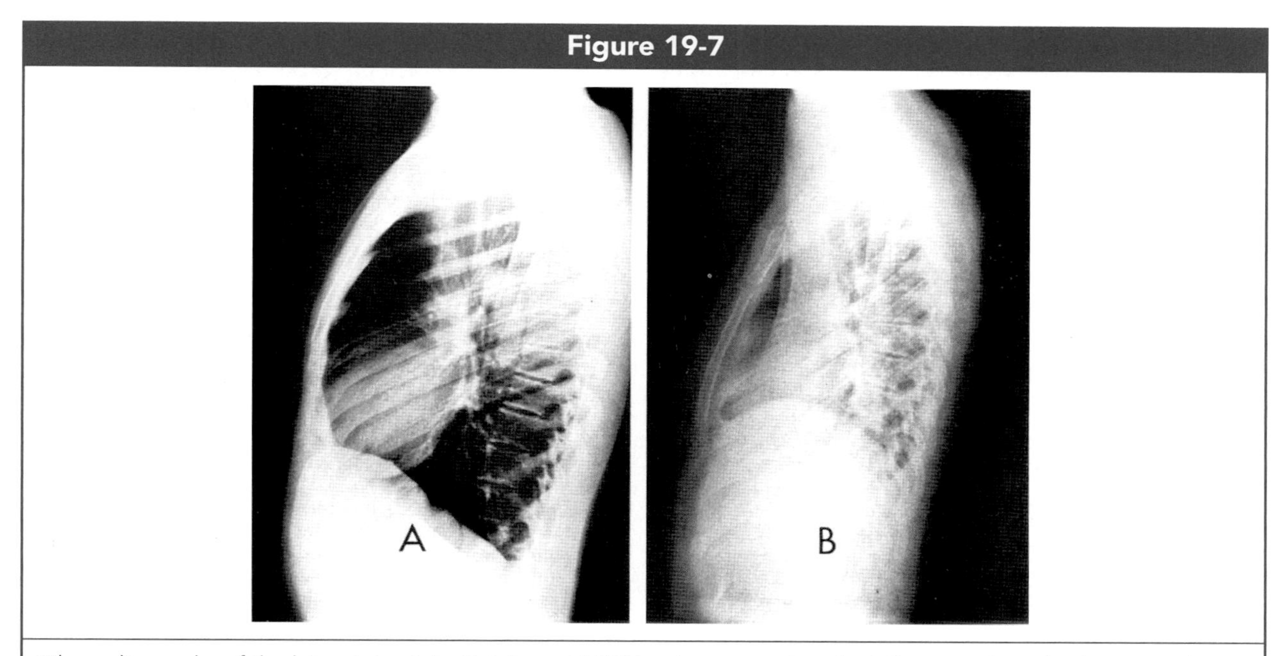

Film radiographs of the lateral chest, both taken at 85 kVp, demonstrating the influence of respiration upon the inherent density of the image; ***A***, upon inspiration at 10 mAs, and ***B***, upon expiration using 13.3 mAs. (From Quinn B. Carroll, *Practical Radiographic Imaging*, 8th ed. Springfield, IL: Charles C Thomas, Publisher, Ltd., 2007. Reprinted by permission.)

small pneumothorax. When an exposure is made on expiration of the breath, the technique should be increased by at least 35 percent to compensate for the increased density of lung tissues. Radiograph ***C*** in Figure 19-7 is an expiration chest as witnessed by the position of the heart and diaphragms, but the mAs has been increased by 35 percent to restore the proper exposure.

With digital imaging systems, even though these effects may be compensated for by corrective computer algorithms, it is better to properly compensate radiographic technique in the first place to assure sufficient exposure at the image receptor.

Poor cooperation from the patient can result in motion blur, and radiographic positioning itself is essentially a geometrical process. All of these will be examined in-depth in later chapters. As regards the final image qualities, these variables affect the recognizability functions in the image indirectly and to a much lesser degree than the visibility functions that bear upon subject contrast in the remnant beam. As can be surmised by the amount of text in this chapter dedicated to the subject, our paramount concern with patient condition is its effect upon subject contrast.

PATHOLOGY

The condition of the patient is the greatest variable the radiographer faces in producing quality radiographs. In addition to the general condition of the patient thus far discussed, the radiographer must also be conscious of abnormal changes due to disease, trauma, or medical intervention. The radiographer should ascertain the reason that each procedure was ordered, because this knowledge often includes information that bears directly upon technique selection and can prevent repeated exposures. To obtain this information, one should review the x-ray requisition prior to every procedure, glance at recent entries for the inpatient's charts when available, and obtain a brief verbal history from the patient whenever possible. In situations that are unclear, a quick call to the referring physician's office can often prevent additional hassle for the patient as well as for the radiographer.

It should be emphasized that it is the *radiographer's responsibility*, not that of the referring physician nor the radiologist, to obtain all information pertinent to adapting radiographic techniques and then applying it. Not all disease conditions are radiographically visible. Many do not appreciably alter the exposure reaching the imaging plate. For a pathological condition to require an adjustment in radiographic technique, it must substantially alter one of the five radiographically demonstrable materials: air, fat, fluid, bone or metal.

Additive Diseases

Abnormal conditions which lead to an increase in *fluid*, *bone*, or *metal* are, for radiographic purposes, considered as *additive conditions*. They require increased technique factors in order to attain proper exposure at the image receptor. In the case of excessive bone tissue, calcification of joints, or presence of metals, increased technique is necessitated primarily because of their high atomic numbers. For fluid accumulation in the lungs (Fig. 19-8), an increase in factors is required because these fluids have a density nearly 1000 times that of normal air in the lungs. Fluid distention of the abdomen requires an increase in technique mostly due to the increase in part thickness; however, because it constitutes a bolus of nearly pure liquid (rather than fluid mixed with tissue), it requires somewhat more technique than the *four-centimeter* rule would dictate. (Abdominal fluid distention is easily distinguished clinically by palpation of the belly: Instead of loose, flaccid tissue, fluid distention makes the skin tight and rather hard to the touch.)

Figure 19-8

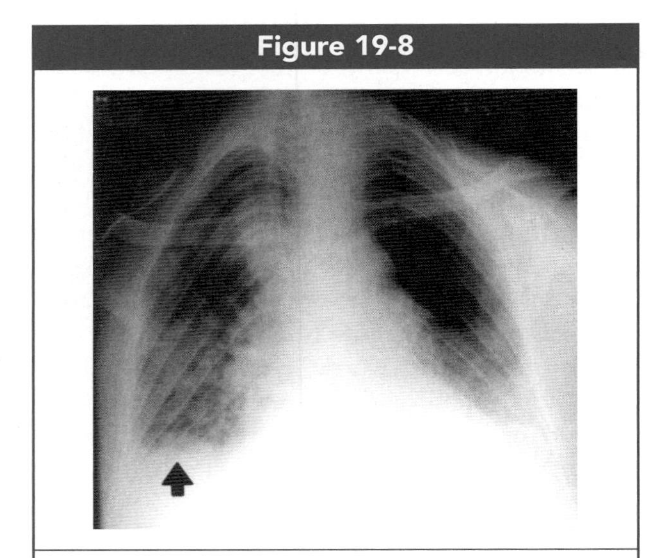

Film radiograph using increased technique for a patient presenting with cardiomegaly (enlarged heart) as well as a pleural effusion (note fluid level at base of right lung). (From Quinn B. Carroll, *Practical Radiographic Imaging*, 8th ed. Springfield, IL: Charles C Thomas, Publisher, Ltd., 2007. Reprinted by permission.)

For additive diseases, technique increases may range from 35 percent (one-third) to over 100 percent (a doubling) when a disease is in its advanced stage. *As a rule-of-thumb, increase overall technique by 50 percent for additive diseases in an advanced stage*; this will be in addition to any increase made from measured part thickness. Table 19-2 lists commonly encountered diseases that require a substantial increase in technique. It assumes the disease to be in an advanced stage. Many of these diseases change the mineral content of tissue, such that an increase in penetration is indicated. In these cases, an 8 percent increase in kVp will substitute for a 50 percent increase in mAs. Naturally, the radiographer must modify this recommended change upward or downward upon careful observation and assessment of each individual patient.

Acromegaly, osteoarthritis, osteochondroma, osteopetrosis, osteomyelitis, Paget's disease, and advanced syphilis involve either excessive bone growth or the replacement of cartilagenous tissues with bone. In the case of osteoarthritis, even though there is degeneration within the bones themselves, bony spurs grow into the joints and bone tissue replaces normal cartilage and fluid in the joint spaces, necessitating a net increase in technique. In the case of actinomycosis, fibrous carcinomas, cardiomegally (Fig. 19-8), cirrhosis, pneumoconiosis, and pulmonary tuberculosis, the growth or overexpansion of dense fibrous tissues replaces normal tissue. For ascites, pulmonary edema, hydrocephalus, hydropneumothorax, pleural effusion, and pneumonia, the accumulation of abnormal amounts of body fluids or the displacement of aerated tissues with fluid requires greater radiographic techniques.

Destructive Diseases

Abnormal conditions which lead to an increase in *air* or *fat*, or to a decrease in normal body fluid or bone, are radiographically considered as *destructive conditions*. These require a reduction from typical

Table 19-2

Additive Diseases Requiring Increased Technique

Acromegaly: 8–10% kVp	Hydrocephalus: 50–75% mAs	Paget's Disease: 8% kVp
Actinomycosis: 50% mAs	Hydropneumothorax: 50% mAs	Pleural Effusion: 35% mAs
Ascites: 50–75% mAs	Osteoarthritis (Degenerative Joint Disease): 8% kVp	Pneumoconiosis: 50% mAs
Carcinomas, fibrous: 50% mAs	Osteochondroma: 8% kVp	Pneumonia: 50% mAs
Cardiomegaly: 50% mAs	Osteopetrosis: 8–12% kVp	Syphilis: 50% mAs
Cirrhosis: 50% mAs		Tuberculosis, pulmonary: 50% mAs
Edema, pulmonary: 50% mAs		

From Quinn B. Carroll, *Practical Radiographic Imaging*, 8th Ed. Springfield, IL: Charles C Thomas, Publisher, Ltd., 2007. Reprinted by permission.

exposure techniques to assure proper exposure at the image receptor. Both air and fat are significantly less dense than soft tissue, and absorb fewer x-ray photons.

For destructive diseases, overall technique may need to be reduced by 30–50 percent. *As a rule-of-thumb, reduce technique by 35 percent in mAs, or by 8 percent in kVp, for destructive diseases in their advanced stage.* Table 19-3 lists destructive diseases commonly found in radiography, along with the recommended technique reduction for each.

Aseptic necrosis, various bone carcinogens, Ewing's tumor, exostosis, gout, Hodgkin's disease, hyperparathyroidism, osteitis fibrosa cystica, osteoporosis, osteomalacia, osteomyelitis and rheumatoid arthritis result in either a demineralization of bone or an invasive destruction of bone tissue. Less penetration is required to secure optimum subject contrast, so generally the kVp should be reduced-for these cases. In blastomycosis, yeast-like fungi produce gas pockets within the tissues, while bowel obstructions, emphysema (Fig. 19-9), and pneumothorax are processes which essentially "trap" air or gas within body cavities. A reduction in technique is indicated.

TRAUMA

Trauma to the body can certainly produce some of the pathological conditions just discussed, including excessive aeration which would require a reduction in technique, or internal bleeding in the abdomen and hematomas in the brain that may require a 35 percent increase in mAs. Blood pooling in the lungs can necessitate increases up to 100 percent from the usual.

Postmortem Radiography

Experienced radiographers learn to expect considerable pooling of blood and fluids in a dead body,

Table 19-3

Destructive Diseases Requiring Decreased Technique

Aseptic Necrosis: 8% kVp	Exostosis: 8% kVp	Osteomalacia: 8% kVp
Blastomycosis: 8% kVp	Gout: 8% kVp	Osteomyelitis: 8% kVp
Bowel Obstruction: 8% kVp	Hodgkin's Disease: 8% kVp	Osteoporosis: 8% kVp
Cancers, osteolytic: 8% kVp	Hyperparathyroidism: 8% kVp	Pneumothorax: 8% kVP
Emphysema: 8% kVp	Osteitis Fibrosa Cystica: 8% kVp	Rheumatoid Arthritis: 8% kVp
Ewing's Tumor: 8% kVp		

From Quinn B. Carroll, *Practical Radiographic Imaging*, 8th ed. Springfield, IL: Charles C Thomas, Publisher, Ltd., 2007. Reprinted by permission.

Figure 19-9

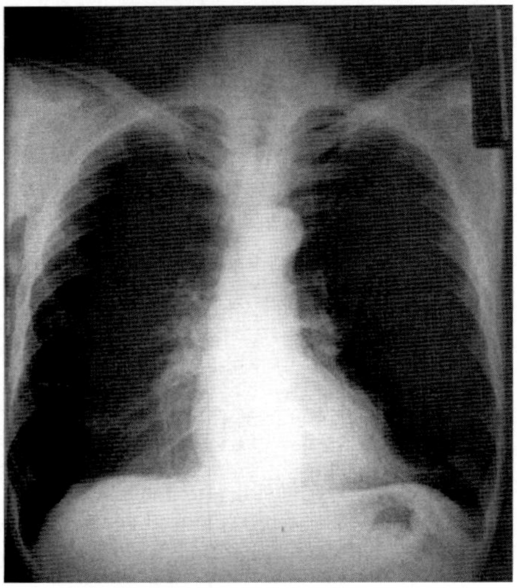

Film radiograph using decreased technique for a patient with emphysema, (note over-distention of lungs from trapped air). (From Quinn B. Carroll, *Practical Radiographic Imaging*, 8th ed. Springfield, IL: Charles C Thomas, Publisher, Ltd., 2007. Reprinted by permission.)

particularly in the head, thorax, and abdomen. An increase in technique, typically 35–50 percent, should be *anticipated* and applied on the very first exposure. Even shortly after death, blood and fluids begin to pool immediately in the lungs as they follow gravity. Since postmortem radiographs are normally taken with the body recumbent, these fluids will pool across the entire lung field.

In addition, bear in mind that by definition, postmortem chest radiography requires an *expiration* technique (no pun intended). Without the normal air insufflation of the lungs, another increase in technique is indicated. Thus, there are *two* distinct reasons and a compelling case for expecting an increase in technique. *Postmortem chest techniques should be increased by at least 35 percent immediately after death, and by 50 percent if a half-hour or more has elapsed.* These adjustments will ensure adequate exposure at the image receptor.

Soft-Tissue Technique

Normally, for the radiographic localization of bullets and other metallic foreign bodies, it is *not* necessary nor productive to increase technique from the normal. Such an increase only reduces subject contrast of the surrounding tissues without necessarily penetrating the metal object. On the other hand, many small foreign bodies such as slivers of wood and glass, or swallowed bones, are better visualized by using a reduced-kVp *soft-tissue technique.*

The experience of most radiographers in the age of digital imaging is that even though a regular-technique image may be simply windowed at the workstation or display monitor, better results are obtained when the original radiographic technique is adjusted, so that the actual *subject contrast* reaching the image receptor plate is enhanced. To achieve this, a very practical rule-of-thumb is—*for soft-tissue techniques, reduce the kVp by 20 percent, without making any compensation in the usual mAs.*

Many foreign bodies, such as small chicken bones lodged in the laryngopharynx, have only slight inherent subject contrast rendering their radiographic appearance very subtle and making them extremely difficult to detect. Wood splinters are particularly troublesome; because of ubiquitous air pockets they contain, *immediately* after their introduction they may show up slightly *darker* than the soft tissue around them. Yet, they quickly absorb body fluids which makes them blend in with surrounding tissue. Even metal and glass, which are normally highly radiopaque, can be hard to demonstrate in the form of very small slivers.

It is essential to maximize the subject contrast within soft tissue structures in all these cases. This requires not only a reduction in x-ray beam penetration, but *also* an overall lightening of the overall exposure. The 20% reduction in kVp recommended above accomplishes *both* of these objectives, as shown in Figure 19-10.

Soft tissue visualization is most frequently called for in radiography of the hands and neck. It is not only useful for demonstrating foreign bodies, but for traumatic damage to the soft tissues themselves. A common traumatic condition from automobile accidents is *padded dash syndrome*, in which there is soft tissue damage in the anterior neck from a frontal collision where a passenger lurches forward striking his or her neck against the padded dashboard of a car. Soft tissue radiographs can be combined with

CT scans for full diagnostic capability in these types of cases.

Casts and Splints

As with soft tissue techniques, most radiographers have found that digital imaging systems, while they make some compensation for the artifacts created in an image by casts or splints overlying the anatomy of interest, just cannot replace the need for adjusting the original radiographic technique. The penetration and quantitative characteristics of the x-ray beam itself as it passes through these materials and tissues must be set appropriately at the x-ray machine console. Fine adjustments in the resulting image may then be made at the digital workstation to perfect the final image.

It must be allowed, at the same time, that *the magnitude* of technique adjustments needed has been lessened somewhat by the advent of digital imaging, because of the tremendous exposure latitude proffered by these systems.

The proper increase in technique factors is dependent upon the type of material used for a cast or splint, how thick it is, and whether it is still wet or has completely dried. The radiographer must be cognizant of all these factors.

Radiographs taken after a broken limb has been set and casted are called *post-reduction radiographs*. Generally, *most full plaster casts require a doubling in mAs, or a 12–15 percent increase in kVp*. If the cast is small, such as for a child's forearm, or if it is a half-cast held in place with an Ace™ bandage, a 50 percent increase in mAs or 8 percent increase in kVp will usually suffice. If the cast is unusually large, such as a thick cast on the thigh, *or* if the plaster has been just applied and is still wet, more than a doubling, sometimes up to three times the technique, may be required.

Fiberglass mesh has become popular as a casting medium. Ironically, it is so radiolucent that normal techniques may be used in most cases. Fiberglass is frequently mixed half-and-half with plaster. In this case, the general rule would be to *increase technique by 50 percent* (Fig. 19-11).

A splint made of a single thick piece of wood or aluminum will likely require a 50 percent increase in technique. Plastic splints and thinner aluminum splints may not require any increase at all. Table 18-4 summarizes these recommendations.

Figure 19-10

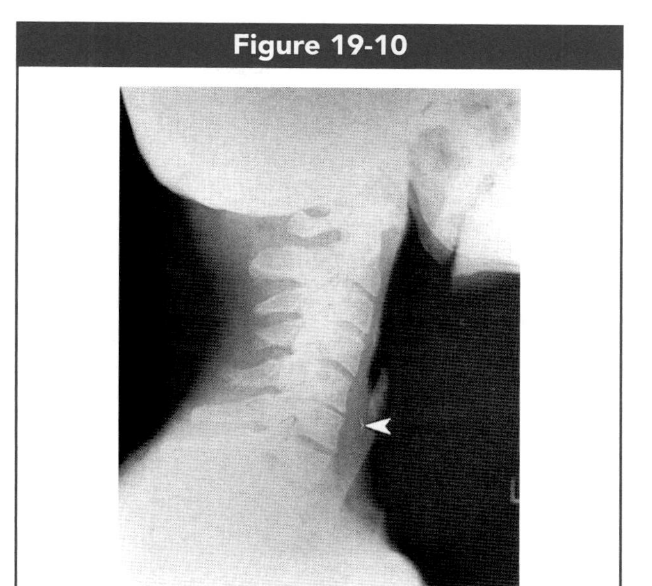

Film radiograph showing soft tissue technique to demonstrate a sliver of glass (arrow) in the neck. From a cervical spine technique, the kVp was decreased from the listed 76 kVp to 64 kVp, all other factors unchanged. (From Quinn B. Carroll, *Practical Radiographic Imaging*, 8th ed. Springfield, IL: Charles C Thomas, Publisher, Ltd., 2007. Reprinted by permission.)

Figure 19-11

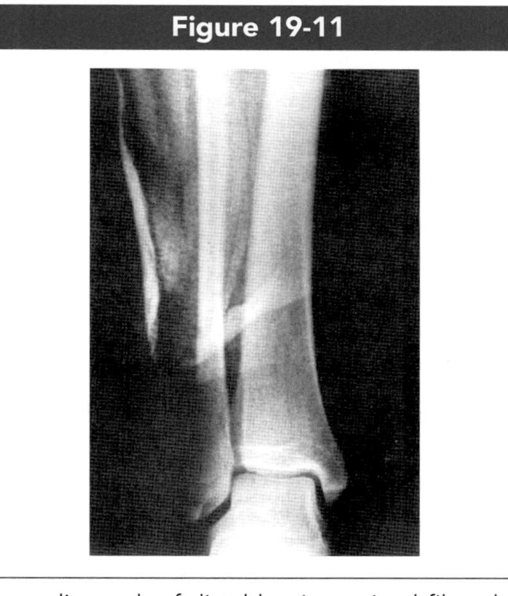

Film radiograph of distal leg in a mixed fiberglass/plaster cast, using 50% increase in mAs, all other factors equal. (From Quinn B. Carroll, *Practical Radiographic Imaging*, 8th ed. Springfield, IL: Charles C Thomas, Publisher, Ltd., 2007. Reprinted by permission.)

Table 19-4

Technique Increases for Casts and Splints

4 × mAs or 30% kVp	Plaster, thick and wet	50% mAs or 8% kVp	Plaster half-cast, dry Fiberglass/plaster full cast Plastic splint Wood splint
3 × mAs or 22% kVp	Plaster, medium and wet Plaster, thick and dry		
2 × mAs or 15% kVp	Plaster, medium and dry	No increase needed	Fiberglass, pure Inflatable air splint

SUMMARY

1. Average thickness ranges have been established which apply to more than 85 percent of all patients for all body parts. Standardized *technique charts* are therefore applicable to a high percentage of patients and should be regularly used by radiographers.
2. Radiographic techniques must be further refined by evaluating the body habitus, age, and condition of the patient. Large muscular patients require a greater increase in technique than obese patients. Appropriate information should be obtained from the patient's chart, x-ray requisition, and by good verbal communication.
3. The average torso thickness is 22 cm in AP projection and 30 cm laterally. Technique should be adjusted by a factor of 2 for every 4 cm deviation in part thickness from the average. Body part measurements should be taken carefully using calipers.
4. The five materials demonstrated on radiographs are, from most radiopaque to most radiolucent: metals, bone, fluid (soft tissues), fat and gas (air).
5. Positive contrast agents, metals and bone are demonstrated radiographically primarily due to their atomic numbers. Negative contrast agents, fat and gases are demonstrated primarily due to their extreme differences in physical density.
6. An increase in radiographic technique is indicated for expiration chests, advanced additive diseases, post-mortem radiography, and casts. A decrease in technique is required for advanced destructive diseases and for "soft tissue" techniques.
7. For soft tissue technique, reduce the kVp 20 percent from the usual without any compensation in mAs. This also demonstrates small embedded foreign bodies.
8. The minimum change in radiographic technique to bring about any significant change in the final exposure is 35 percent.

REVIEW QUESTIONS

1. How does the *asthenic* body habitus differ from the hyposthenic habitus?

2. How might a radiographer obtain an exaggerated measurement of the AP abdomen when using proper calipers on a patient lying supine on the x-ray table?

(Continued)

REVIEW QUESTIONS *(Continued)*

What percentage of adult patients fall within the "average" thickness range for the following:

3. Lumbar spine AP:
4. Sinuses PA:
5. Foot AP:

6. A technique chart lists 50 mAs for a 24-cm AP lumbar spine projection. Your patient measures only 16 cm in AP. What total mAs should you use?

7. If the above patient measures 24 cm in *lateral* projection, what mAs should be used?

8. List the five general types of materials demonstrated on conventional radiographs:

9. What part of the body changes *least* in size from infant to adult?

10. Measurements of body parts should generally be taken along the path of the __________.

11. The skin of a fluid-distended abdomen feels __________.

12. Body fat is demonstrated radiographically as a darker density than surrounding soft tissues, primarily because of its very different __________.

13. What percentage change in kVp is equivalent to a 50 percent increase in mAs?

14. For a soft tissue technique, what change in kVp should be made, and what change, if any, in mAs should be made?

15. A thick, wet, pure plaster cast on the femur may require up to __________ increase in technique.

16. Fresh wood slivers appear as a __________ density against the background of soft tissue.

17. For a mixed fiberglass/plaster full cast on the leg, increase technique by about __________ percent.

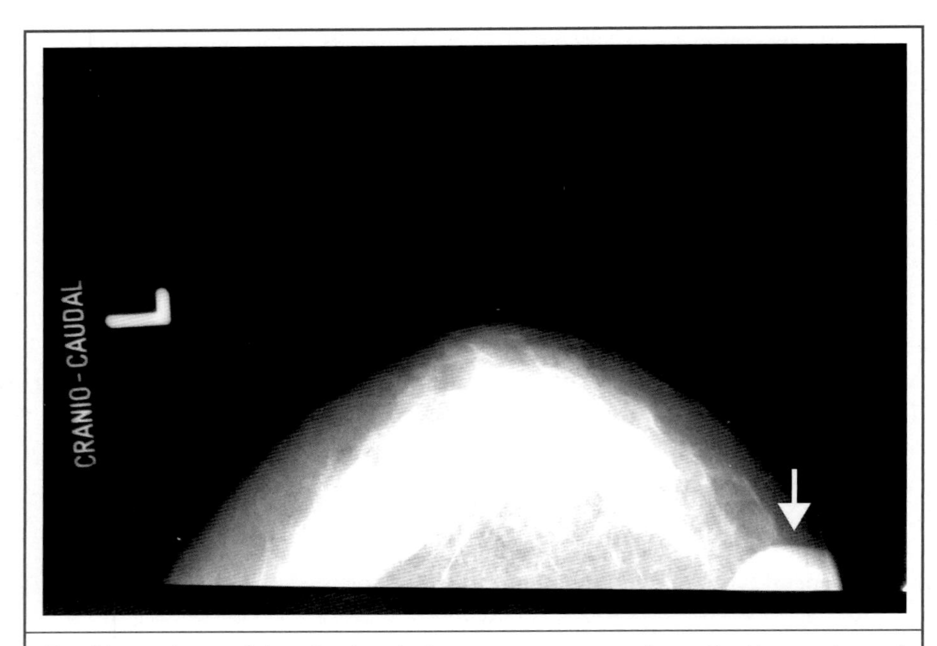

For this cranio-caudal projection during a mammogram, the patient's nose (arrow) was caught in the view.

Chapter 20

SCATTERED RADIATION AND GRIDS

Objectives:

Upon completion of this chapter, you should be able to:

1. Prioritize the causes of scatter radiation by their degree of influence.
2. Understand the nominal role of high kVp levels in image degradation, especially for digitally-processed images.
3. Distinguish between the *visibility* effects of scatter radiation and the *geometrical* effects of blur, and that they are not directly related to each other.
4. Describe the geometry of how grids eliminate scatter radiation from the remnant beam.
5. Define *grid ratio*, describe how it improves subject contrast in the remnant beam, and explain why it is the only reliable method of indicating the effectiveness of a grid.
6. Calculate the grid ratio from grid dimensions.
7. Define *grid frequency* and *lead content*, and explain why they are *not* reliable ways of indicating grid effectiveness.
8. Describe the role of grids for digital imaging.
9. Prioritize part thickness, field size and set kVp in determining grid use and appropriate grid ratio.
10. Define the *bucky factor* and *selectivity* of a grid.
11. State technique compensations for landmark grid ratios and from tabletop to bucky.
12. List those image qualities *not* affected by grids.
13. Distinguish between *grid lines* and *grid cut-off.*
14. State the cut-off effects of distances outside the grid radius for focused and parallel grids.
15. State the cut-off effects of off-centering or off-angling the x-ray beam and grid crosswise to the grid strips.
16. Describe grid line patterns and the directions one *can* off-center or angle the x-ray beam without causing cut-off effects.

The effects of scattered radiation are among the most misunderstood concepts in radiography. This chapter will attempt to consolidate all of the aspects of this topic covered in previous chapters, clarify common misconceptions, and then summarize those methods that can be used to control scatter radiation. A careful review of Chapters 12 and 13 ("Production of Subject Contrast," and "Qualities of the Radiographic Image"), may be helpful.

Scatter radiation lays down a "blanket" of exposure, evenly distributed across an area of the image receptor, thus adding a fixed amount of exposure to *every* tissue area within that area. In Chapter 12 we demonstrated mathematically how this reduces subject contrast. If all other factors are kept constant, the addition of scatter radiation also adds to the total exposure reaching an area of the imaging plate.

However, it is important to understand that an image with excessive scatter radiation present in it can be *underexposed* at the same time. That is, the *proportion* of the total exposure that is constituted by scatter radiation can go up, and yet the *total exposure* itself can still be insufficient. Proof of this is visually presented in the old film-based radiograph in Figure 20-1, an AP abdomen projection which is both *too light* overall and at the same time *excessively fogged* from scatter.

Scatter is *completely* destructive to the image. We have defined "contrast" and "gray scale" as opposite image qualities. Yet, it is intriguing to find by experimentation that *both* contrast and gray scale are destroyed by scatter radiation, as shown in Figure 20-2: Even though the difference between different steps in this image is lessened by scatter, at the same time the *number* of steps discernable is reduced also—this represents a loss of gray scale along with the loss of contrast.

Figure 20-1

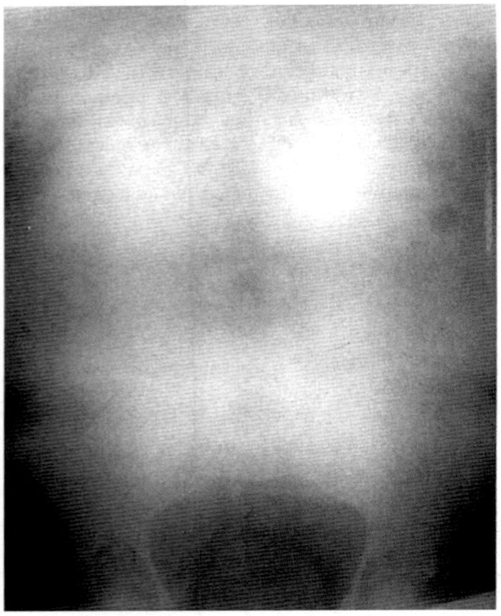

Film radiograph of a large patient demonstrating both fog from scatter radiation and underexposure overall. (From Quinn B. Carroll, *Practical Radiographic Imaging*, 8th ed. Springfield, IL: Charles C Thomas, Publisher, Ltd., 2007. Reprinted by permission.)

THE CAUSES OF SCATTER

Almost all scatter radiation is caused by the Compton interaction as described in Chapter 11. Let us briefly consider each of the three factors that increase the amount of scatter radiation produced. In the order they have been covered, they are:

1. High levels of kVp (Chapter 16)
2. Large field sizes (Chapter 18)
3. Large body part thicknesses of soft tissue (Chapter 19)

High kVp Levels

At high kilovoltages, slightly *less* scatter interactions actually occur within the *patient* (Chapter 12). However, that scatter radiation which *is* produced has higher energy to penetrate through to the image receptor, *and* is emitted in a more forward direction toward the image receptor. Furthermore, photoelectric interactions are rapidly lost at higher kVp's, (Chapter 12). The *net* result of all these factors is that a higher *percentage* of the total exposure reaching the image receptor consists of scatter noise.

When large amounts of exposed tissue are *already* creating a great deal of scatter, kVp must be also taken into consideration as a contributing factor. *Compared to the effects of field size and patient size, however, the impact of this increased percentage of scatter is relatively minor.*

Higher kVp settings are *desirable* for several reasons; they provide adequate *penetration* in order to produce good subject contrast in the remnant beam. This is especially critical for studies using contrast agents and for body parts with high proportions of bone tissue. Radiographic techniques that combine high kVp with low mAs save patient exposure, an important consideration. Also, the long gray scale produced by high kVp's is *desirable* because it represents more information in the final image. These benefits *far outweigh* the small increase in scatter radiation reaching the image receptor. Therefore, the selection of kVp should be based primarily upon the penetration and subject contrast needed, with scatter radiation only as a secondary consideration.

Large Field Sizes

Large field sizes allow greater amounts of exposed tissue to generate scatter radiation. Since the concentration of the primary beam remains unaffected by field size, the *net contribution* of scatter as a percentage of the total radiation reaching the image receptor is increased. Additional scatter is also produced from the tabletop or other objects any time the light field is allowed to extend well beyond the anatomy. As long as the field is adequate to include all anatomy of interest, there is *no benefit* to further increases in field size.

Large Soft-Tissue Part Thicknesses

Larger patients or larger body parts present more exposed tissue to generate scatter radiation, even while the primary beam is further attenuated. The loss of useful rays combined with the increased scatter results in a dramatic increase in the contribution of scatter as a percentage of the total radiation reaching the image receptor.

For some procedures, a *compression paddle* can be used by the radiologist during fluoroscopy, or patients may be rolled over onto their stomach, both of which compress the tissue thickness within the beam and effect some reduction in the scatter generated.

(It must be conceded that the *type* of tissue also makes a difference; note that the *chest* presents an exception to part thickness as a major cause of scatter radiation. With fully insufflated lungs, so much of the chest cavity is composed of gaseous air that very little scatter radiation is produced. This is primarily due to the extremely low physical density of gases, which offer only a few molecules per unit volume for x-rays to interact with. All other portions of the human body are composed mostly of soft tissues and bone which produce ample scattering as thicknesses increase.)

Conclusion

Of these three causes of scatter radiation, we see that high kVp levels have advantages which outweigh the impact of scatter, and that part thickness is largely outside the control of the radiographer. This leaves *field size limitation (collimation) as the primary method of preventing scatter radiation from being produced.*

Figure 20-2

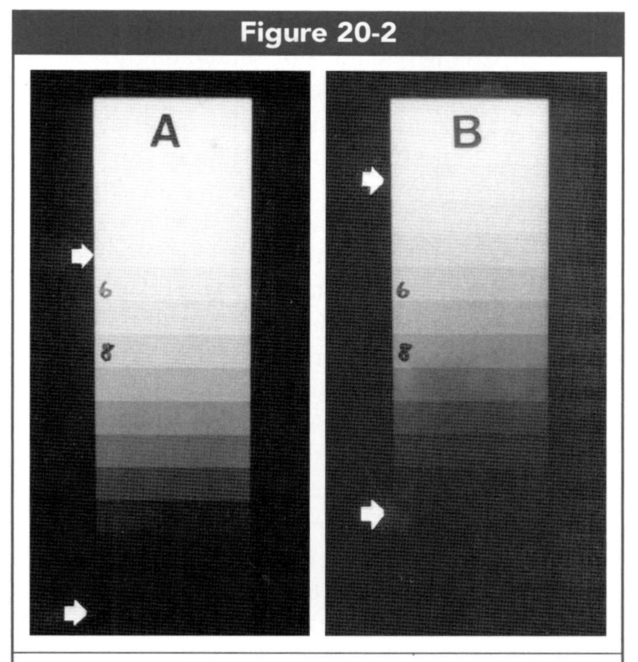

Film images of a step wedge showing that even slight fogging in image ***B*** can reduce gray scale even as it reduces contrast. One can count 11 steps from pitch black to blank white in ***A,*** but only 10 steps in ***B.*** The darkest gray shades are lost as they turn to pitch black from the added scatter radiation. (From Quinn B. Carroll, *Practical Radiographic Imaging*, 8th ed. Springfield, IL: Charles C Thomas, Publisher, Ltd., 2007. Reprinted by permission.)

After scatter radiation has been produced, the only practical method available to help reduce it before it reaches the image receptor is the use of grids, to be discussed shortly. (The *air-gap technique* is briefly discussed in Chapter 22. In conventional radiography, it was very effective at reducing scatter but also magnified and blurred the image. The advantages of digital postprocessing over this approach has made it obsolete in practice.)

SCATTER VERSUS BLUR

There is a common misconception that scattered x-rays "undercut the edge" of an image, causing it to become blurred. It has even been mistakenly asserted that *grids* improve sharpness of detail by eliminating scatter. These statements are false and misleading. As described in Chapter 13, a loss of image contrast can reduce the overall *resolution* of details, but this is due to less *visibility*, not to a lack of geometrical integrity. The sharpness of recorded detail at the edges of an image depends upon the width of the *penumbra*. Anything which does not change the spread of penumbra cannot be said to affect sharpness.

Scatter is completely random. It does not "select" the edges of the image to affect, but simply lays down a blanket of exposure over an *entire* area of the image. Blurring, on the other hand, is geometrically predictable through penumbra diagrams (see Chapter 14), and relates specifically to the edges of the image as projected from the x-ray tube. Scatter, emanating from the patient, cannot and does not affect the alignment of the primary rays projected from the x-ray tube.

Figure 20-3 illustrates that a scattering event is separate from the geometry of penumbra formation: In both ***A*** and ***B***, the same geometrical factors (distances and focal spot) are maintained, so the spread of penumbra *has to be equal* as shown. Note that the overlaying of scatter exposure from a nearby object in ***B*** *does not alter the measured width of the penumbra.* The only effects of scatter radiation at the image receptor are to add to overall exposure and reduce subject contrast. Figure 14-6 in Chapter 14 demonstrated that an image can be blurred without any loss of contrast, or lose contrast without any blurring.

To summarize, scatter radiation affects *all three* of the visibility functions in an image—the exposure, the subject contrast and the level of noise. Scatter, being random in nature and direction, does *not* affect any of the recognizability functions in an image—the sharpness of detail, magnification or distortion.

REDUCING SCATTER WITH GRIDS

Gustav Bucky invented the radiographic grid in 1913, just eighteen years after the discovery of x-rays. The function of the grid was to absorb scatter radiation that has already been produced in the patient's body before it reaches the image receptor. This required that the grid be placed *between* the patient and the imaging plate. The grid is a flat rectangular plate containing alternating strips of lead foil separated by a radiotransparent interspacer material.

Figure 20-4 shows a cross-section cut-away view of a grid plate; it can be seen that when the grid plate is lying flat or horizontal, the lead strips point vertically toward the x-ray tube. The lead foil strips

Figure 20-3

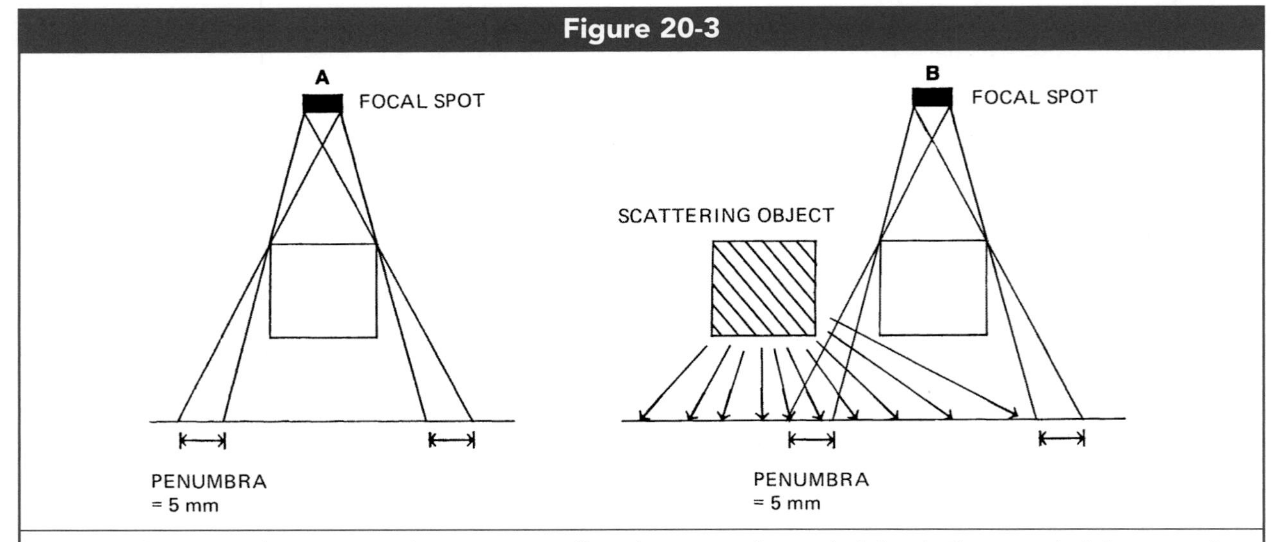

Diagram illustrating that scatter radiation cannot affect sharpness of recorded detail. The spread of the penumbra shadow is 5 mm prior to fogging (***A***), and still measures 5 mm across *after fogging*, (***B***). Although the *visibility* of the edge is decreased because scatter reduces subject contrast, the sharpness of the image is unchanged. (From Quinn B. Carroll, *Practical Radiographic Imaging*, 8th ed. Springfield, IL: Charles C Thomas, Publisher, Ltd., 2007. Reprinted by permission.)

Figure 20-4

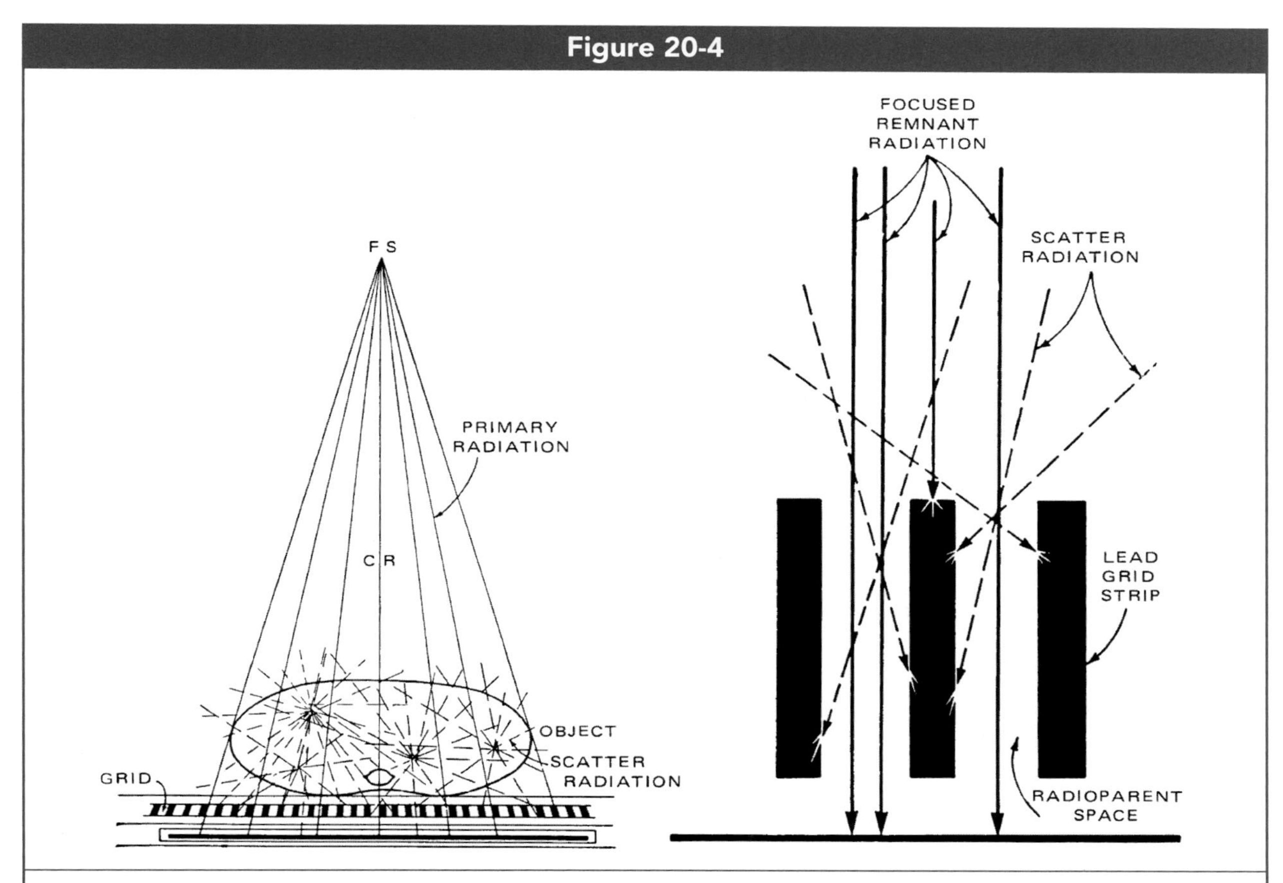

Diagrams illustrating the function of the Potter-Bucky diaphragm in absorbing scattered radiation before it reaches the image receptor plate, with a close-up view (right). (From Quinn B. Carroll, *Practical Radiographic Imaging*, 8th ed. Springfield, IL: Charles C Thomas, Publisher, Ltd., 2007. Reprinted by permission.)

are approximately 0.005 inch thick. These strips are vertically aligned to the fanning primary beam of x-rays so that, in use, most of the focused remnant beam of x-rays will pass *between* these strips to reach the imaging plate, while most of the unfocused, randomly-directed scatter radiation will strike the lead strips broadside and be absorbed. For any x-rays to pass through the interspaces between the lead strips, they must be traveling *nearly parallel* to the strips.

The grid as Dr. Bucky invented it was not too practical at first because it was used in a stationary position and the pattern of the lead strips was apparent on the resulting image as a series of thin white lines. These lines not only represented a form of noise, but also a loss of some radiographic information. If you carefully examine Figure 20-9 in the latter part of this chapter, you can make out these white lines that occur as artifacts from a stationary grid. It was for Dr. Hollis Potter to solve the problem by *moving* the grid during the x-ray exposure. In so doing, all *grid lines* were blurred out of the image. The first commercial moving grid was announced in 1920 and became known as the Potter-Bucky diaphragm. Bucky's name was also attached to the tray under some x-ray tabletops for holding cassettes or imaging plates, which to this day is still called the *bucky tray*.

The effectiveness of the Potter-Bucky diaphragm in reducing scatter radiation and enhancing image contrast was scientifically established over the next decade. Modern grids attenuate 80% to 90% of scattered radiation. With digital imaging systems, both Potter mechanisms and stationary grids are still useful in controlling the amount of noise-generating scatter that reaches the image receptor plate from the patient and surrounding objects, such as the tabletop, that fall within the field of the x-ray beam.

Improvements in grids continued over the years. More radiolucent interspacer materials were found. The technology in cutting lead foil allows strips so thin that 110 lines of foil per inch could be packed into the grid. At 60–80 lines per inch, the strips were easily seen on a stationary grid exposure such as the one in Figure 20-9. At 110–120 lines per inch they are barely perceptible at normal viewing distances. The stationary grid is still frequently used for mobile radiography and cross-table exposures.

In the Potter-Bucky mechanism, small motors move the grid plate back and forth in a *reciprocal* movement, or in an *oscillating* pattern that is circular. Either way, the grid lines are effectively blurred out of the image, reducing noise and enhancing subject contrast.

Grid Ratio and Effectiveness

Lead (Z# = 82) is so effective at absorbing x-rays that even a thin foil of the material will stop a substantial percentage of those having energies in the diagnostic range (further reduced by scattering). The obliqueness with which most scattered x-rays strike the lead strips also increases the effective thickness of the foil through which they must pass.

This leaves the *dimensions of the interspaces between the lead strips* as the primary factor controlling the grid's efficiency in cleaning up scatter radiation. The mechanism by which more of the scatter is removed lies in *reducing the angle at which scattered rays can pass through these interspaces.* We shall call this the *effective angle of penetration*. As shown in Figure 20-5, this angle is formed simply by the *diagonal* drawn across an upper corner and a lower corner of any interspace. Comparing grids **A** and **B** in Figure 20-5, we see that the narrower interspaces of grid **B** have reduced the effective angle of penetration from 25 degrees to 18 degrees. For any scattered x-ray to make it through this space, it must be traveling at an angle *less than 18 degrees*. This grid will be more *selective* in the scatter that is allowed through it. Grid **A** allows a much broader range of scattered rays to pass through, up to 25 degrees from the vertical.

It is the dimensions of these radiotransparent interspaces that determines the effective angle of penetration. The *grid ratio* is defined as the relationship between the *height* and the *width* of the interspaces between the lead foil strips. This is expressed in the simple formula

$$\text{Grid Ratio} = \frac{H}{D}$$

where H is the height of the strips of lead and D is the distance between them, which is to say, the height and width of the spaces. An 8:1 grid ratio, for example, means that these interspaces are 8 times taller than they are wide. The *height* of these spaces, and of the lead strips themselves, actually corresponds to the overall *thickness of the whole grid as a plate.*

Practice Exercise #1

What is the grid ratio for a grid that is 3 millimeters thick as a plate, and has spaces between the lead strips that are 0.5 millimeters in width?

Solution: $\frac{H}{D} = \frac{3\ mm}{0.5\ mm} = 6$

Answer: The grid ratio is 6:1.

Obviously, the grid ratio may be increased by either increasing *H* or by decreasing *D*. When the height of the strips is increased, however, this results

Figure 20-5

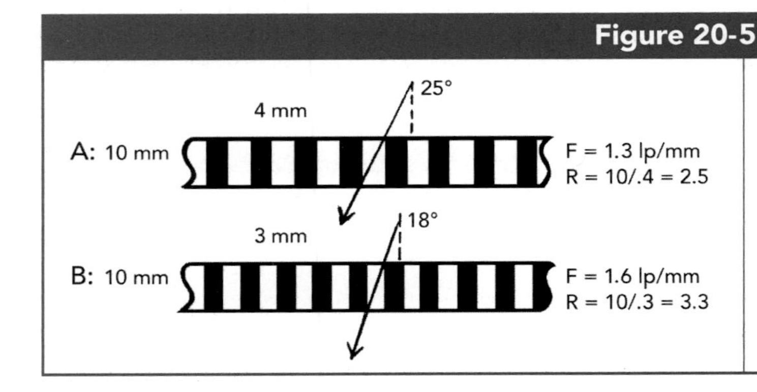

Due to narrower interspaces, Grid ***B*** has a higher grid ratio (3.3) than Grid ***A*** (2.5). The selectivity is increased, indicated by a reduction in the maximum *angle of penetration* at which the scatter can pass through. Grid efficiency is improved. *Grid ratio* is the key to grid efficiency. (From Quinn B. Carroll, *Practical Radiographic Imaging*, 8th ed. Springfield, IL: Charles C Thomas, Publisher, Ltd., 2007. Reprinted by permission.)

in a thicker grid plate which might not fit into the Potter-Bucky mechanism or other holding devices. Therefore, the primary challenge in the technology of grid manufacturing has been to make ever thinner interspaces and lead strips. When this is done, more lead strips can be packed into each inch of the grid's overall width as a plate. This number of lead lines per inch is known as the *grid frequency*.

Grid Frequency and Lead Content

Grid frequency and the total lead content in a grid are both directly related to the grid ratio. *Grid frequency* is the number of lead strips counted per inch when scanning transversely across the grid. Grid frequencies can range from 60 lines per inch up to 200 lines per inch (25–80 lines/cm). The most commonly used grids fall in the frequency range of 85–103 lines per inch (33–41 lines/cm).

It is generally assumed that the more lead strips there are, the more *concentrated* they are, and the more efficient the grid will be. This is true as long as the lead strips *themselves* are of a standard thickness, but this is not always the case and should not be *assumed.* If the lead strips themselves are thinner, more could be packed into each inch with the interspaces *unchanged.* The grid ratio remains unchanged, and we find that the actual effectiveness of the grid is the same.

Let us assume that the interspaces of a grid are made narrower, but the *height* of the lead strips and interspaces is also reduced proportionately by making the overall grid plate thinner; as shown in Figure 20-6, these two changes will cancel each other out and the grid ratio will remain the same. No change in efficiency will occur. In other words, if the grid plate is made thinner, it is possible to change the grid frequency *without* changing the grid ratio. As shown in Figure 20-5, *it is the grid ratio which is the critical factor in efficiency.*

On the other hand, if the thickness of the grid plate is maintained, preserving the height of the interspaces, as in Figure 20-5, then an increase in grid frequency would have to bring with it a reduction in the effective angle of penetration, an increase in grid ratio, and more effectiveness.

The *lead content* of a grid, which can be measured as grams per square inch or as total grams for the entire grid, is also sometimes cited as an indication of a grid's efficiency. Clearly, at a higher grid frequency with more lead strips per inch, there will be more lead content. But, a cheaply manufactured grid might also simply have thicker lead strips in it, in which case it could be claimed that the lead content was high. This would not necessarily change the grid ratio, as can be seen by examining Figure 20-6 *backwards*, comparing ***A*** to ***B*** rather than ***B*** to ***A***. In this case, again, there is no gain in real efficiency.

Although it can be true that more lead content or a higher grid frequency improves the performance of a grid, these assumptions are not *always true. The critical factor in a grid's efficiency is the grid ratio, not the frequency nor the lead content.* The grid ratio is at the heart of actual efficiency.

Effect on Subject Contrast

The sole purpose for which the grid was invented was to *restore contrast* in the final image which otherwise would have been destroyed by scatter-producing fog. The actual improvement in the image can be seen in the film-based chest radiographs in Figure 20-7. (See Historical Sidebar 20-1, page 310, for more samples

Figure 20-6

Due to narrow interspaces, Grid ***B*** has a higher grid frequency than Grid ***A***; nonetheless, since the height of the interspaces was also reduced proportionately by making the grid plate thinner, there is no difference in *grid ratio*, and selectivity is unchanged since x-rays scattered up to 25 degrees can make it through both grids. (From Quinn B. Carroll, *Practical Radiographic Imaging*, 8th ed. Springfield, IL: Charles C Thomas, Publisher, Ltd., 2007. Reprinted by permission.)

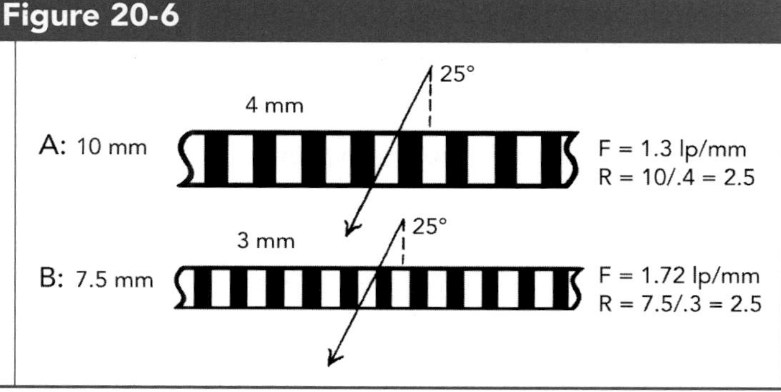

Figure 20-7

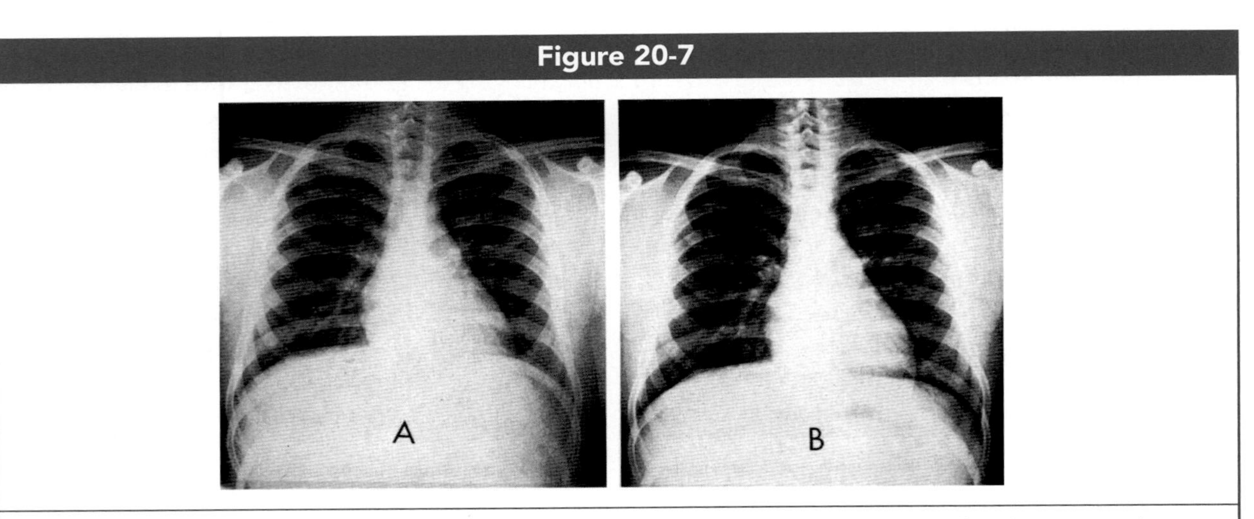

Film radiographs of the PA chest demonstrating the improvement of contrast due to the elimination of scatter radiation with a grid, ***B***. (From Quinn B. Carroll, *Practical Radiographic Imaging*, 8th ed. Springfield, IL: Charles C Thomas, Publisher, Ltd., 2007. Reprinted by permission.)

combined with the effects of reduced field size.) The higher the grid ratio, the greater the improvement in contrast.

With digital imaging systems, the final contrast of the image can be adjusted not only by the computer algorithms employed, but also at the display monitor window control. But, even though the effects of scatter radiation are not immediately apparent in computerized images, *it is important that the amount of noise in the original signal reaching the image receptor be minimized as a component of the information that is fed into the computer system.*

Most digital systems do a remarkable job of improving a scatter-fogged image from the remnant x-ray beam signal and can now even correct for an image with *grid lines* caused by a stationary grid. But, if a computer is fed an *excess* of poorly composed data, it cannot produce outstanding results with it. The quality of the remnant x-ray beam *must* be attended to, so that the computer will have both full and good information to work with. Therefore, grids still play an important role in reducing scatter radiation, thus minimizing noise in the digital imaging process.

Indications for Grid Use

Radiographers often speak of "doing a procedure *bucky*," which means that the imaging plate will be placed in the bucky tray of the table or vertical bucky. For the purposes of our discussion here, this is synonymous with *using a grid*. Remember that each time we speak of using a grid, it can mean placing a CR cassette in the bucky tray, using a DR imaging board with a grid already installed in it, or using a stationary plate "wafer" grid placed in front of the imaging receptor.

The previous section of this chapter listed the prime causes of scatter radiation. These serve perfectly as guidelines here for when a grid should be used. But, the order of *priority* is important to appreciate: *The most critical consideration is the thickness of the soft-tissue part being radiographed, followed by the size of the field, and finally the kVp employed.*

Part Thickness

By far, the *first* consideration for grid use, and the main indicator for it, is the thickness of the soft-tissue part. The general rule is that *body parts thicker than about 13 cm produce enough scatter radiation to require a grid*, especially if the field size is also large. An 8:1 ratio grid is ideal at this thickness of tissue, with a 10:1 grid more appropriate for an 18 cm part, and a 12:1 grid for a part measuring more than 22 cm. With this in mind, most radiology departments install grids within this range, 8:1 to 12:1, in their x-ray tables and vertical buckies.

We have noted that the *chest* presents an exception to thickness as a guide, because of its unusual tissue composition which includes so much air. This explains why clinically, mobile chest radiographs and "wheelchair" chests in the department can be done without a grid, if kVp is reduced to compensate.

The author has even obtained completely satisfactory mobile *abdomen* views without a grid when the patient was very thin (saving patient exposure because of the reduced techniques). Certainly the same principle would apply for pediatric abdomens and other torso and head projections when the anatomy is small. One should not get locked into the mentality that grids are used for certain *projections* rather than for certain *thicknesses*. It is the thickness that really matters.

Field Size

Projections using a 14" × 17" (35 × 43 cm) field size produce substantially more scatter radiation than those with smaller fields. As it turns out, most procedures that involve thicker anatomy such as the torso *also* employ larger fields, adding scatter to scatter. The connection between large fields and the need for grids has not been taught or written about much in the past, and should be brought more to the fore in radiography instruction. Field size has a greater impact upon scatter production than high kVp. When large field sizes are used, a grid is more likely to be needed.

Kilovoltage

Consulting Table 16-1 on *Minimum kVp* in Chapter 16, one can readily surmise that those anatomical body parts exceeding 13 cm in thickness in an adult *all* use kVp's higher than 70. The shoulder and neck, for example, require minimum kVp's in the mid-70s. It has been conventionally taught that grids should generally be used for procedures employing more than 70 kVp, but this has more to do with the coincidence just mentioned than with the impact of kVp itself. In other words, all anatomy requiring more than 70 kVp meets the 13 cm thickness requirement. It is the thickness, more so than the kVp increase, that necessitates a grid.

Chapter 16 firmly established that kVp generally plays a minor role in scatter production, and that consequently a radiographer should not hesitate to make a 15 percent kVp increase when other conditions favor it. However, *extreme* increases in kVp can create enough scatter to need a grid. What we mean by *extreme* is, for example, an increase from 70 to 120 kVp, which represents almost *four* 15 percent step increases, a change one would almost never see in compensating technique during a particular procedure. We mention it here to explain the use of grids for high-kVp *chest* radiographs.

Recall that air in the lungs produces almost no scatter. This allows us to employ extremely high voltages in the range of 110 to 120 kVp to achieve outstanding penetration and gray scale through the heart and mediastinal structures which are terribly important to demonstrate, while still resolving lung details. Note that mobile non-grid chests are typically performed with 70–74 kVp. *This change, from 74 to 120 kVp for the same anatomy, represents just the kind of extreme increase where the kVp itself becomes a significant factor in scatter production.* It is the price we pay for the excellent penetration obtained, that 120 kVp contributes enough to scatter production to require a grid.

A minor factor, if changed by an extreme amount, can then cause a substantial effect.

To summarize, the three questions that should be asked to determine when a grid should be used are, *in order of priority*:

1. Does the body part consist of more than 13 cm thickness of *soft tissue*? If so, use a grid.
2. Will a large field size be used? If so, a grid may need to be applied even though the body part may not be quite 13 cm thick.
3. Will *extremely* high kVp be used? If so, a grid may need to be applied even though the body part may not be quite 13 cm thick.

Measuring Grid Effectiveness

Naturally, the most direct way to measure the actual effectiveness of a grid would be to measure the change in image contrast when it is used and compare this to the image contrast without the grid, all other factors kept equal. (See Historical Sidebar 20-1.)

HISTORICAL SIDEBAR 20-1: With film radiography, density measurements could be taken directly off the film image using a densitometer, Figure 20-8. The image contrast could then be calculated using these measurements. The most direct and scientific way to measure the actual effectiveness of a grid was the contrast improvement factor, (CIF), formulated as:

$$\text{CIF} = \frac{\text{Measured contrast with the grid}}{\text{Measured contrast without the grid}}$$

Figure 20-8

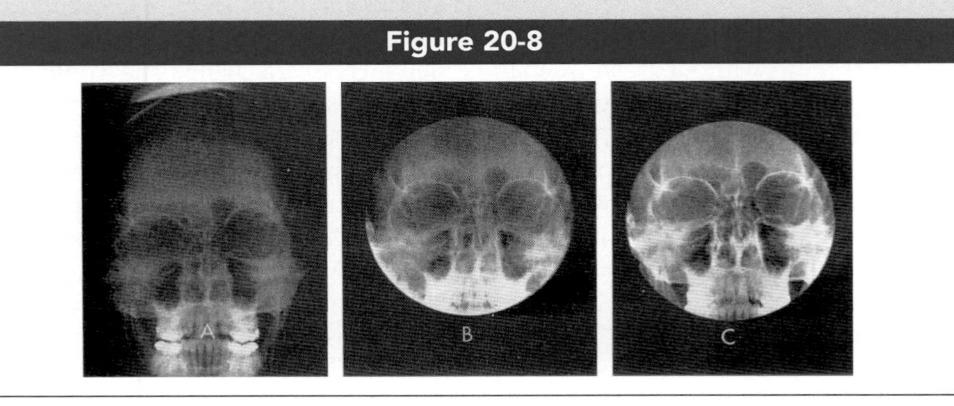

Film radiographs of the frontal sinuses demonstrating progressive contrast enhancement by the use of ***B***, an extension cylinder to reduce field size, then combined with ***C***, a grid. Technical factors were equal for all three views. The *contrast improvement factor (CIF)* could be measured using a densitometer on corresponding light and dark areas in each radiograph. (From Quinn B. Carroll, *Practical Radiographic Imaging*, 8th ed. Springfield, IL: Charles C Thomas, Publisher, Ltd., 2007. Reprinted by permission.)

Unfortunately, with digital imaging systems, the final image has *always* been "washed" through various software *filters* and computer algorithms, so that its contrast and other characteristics are the product of postprocessing the original data. Digital images cannot be validly used in comparing grids.

Bucky Factor

The *bucky factor* is defined as the ratio of mAs required with the grid to the mAs required without the grid in order to maintain the exposure in the remnant beam. Remnant beam exposure is best measured by using a standardized "phantom" as a beam absorber and taking x-ray measurements behind it with an ion chamber. Some have advocated its use as an indicator of grid effectiveness. But, as explained above, a grid can have more lead in it without necessarily being more efficient. The fact that a particular grid requires more technique is *not* a reliable indication of its effectiveness, and can even be due to poor construction. The bucky factor has been useful, however, in determining technique compensations for the various grid ratios that will be discussed later in this chapter.

Selectivity

The *selectivity* of a grid is mathematically defined as the ratio of primary radiation transmitted through the grid to scatter radiation transmitted through the grid. This ratio can be extrapolated from direct measurements taken with an ion chamber. Since it takes into account that the objective of the grid is to attenuate scatter with *minimal attenuation of the primary beam*, the selectivity ratio is an excellent measure of actual grid effectiveness.

Both selectivity and the bucky factor require ion chamber measurements to be taken, something more suited to a medical imaging physicist than a radiographer. Radiographers should be familiar with these terms and concepts, however, so they can understand tables or other formats of data for comparison when considering the purchase of a grid.

Technique Compensation for Grids

Overall exposure at the image receptor includes both scatter radiation exposure and exposure from primary beam x-rays that have penetrated through the body. Grids are designed to eliminate scatter radiation, and in so doing they lessen the exposure at the image receptor. They also have the undesirable effect of absorbing some of the *primary* beam x-rays which strike the lead strips in the grid end-on. This is one reason why the lead strips must be made as thin as feasible in manufacturing the grid. The thicker the lead strips, the more of the primary beam is absorbed, the thicker are the resulting white grid lines from a stationary grid, and the lighter the exposure at the image receptor.

This combination of reduced scatter and primary radiation causes the overall, total exposure at the image receptor to plummet, and must be compensated for by increasing technique factors at the console. The recommended technique adjustments for each grid ratio have been worked out many times by manufacturers, governmental and scientific organizations, and authors. Most of these studies are based on the *bucky factor* concept described earlier, but some did not use absorbing phantoms to simulate real clinical applications, others used different types of absorbing phantoms, different approaches and different biases for interpretation that have led to a plethora of confusing results. We shall attempt here to both simplify the entire topic and condense it into a few easily-memorized rules-of-thumb, by using the following rules of engagement:

1. The raw data used will be that from Report No. 102 of the National Council on Radiation Protection and Measurements (NCRP).
2. We will limit this data to results for grid ratios from 8:1 through 12:1, those grid ratios commonly found in general-use x-ray tables.
3. We shall make a reasonable attempt to round the numbers to whole integers.

Table 20-1 Lists the NCRP bucky factors for 8:1, 10:1 and 12:1 grids at 70 kVp and 95 kVp. Based upon those factors boxed in, Table 20-2 presents rounded-out mAs conversion factors that can easily be committed to memory. These are the multiplication factors by which mAs should be increased when changing from a non-grid technique to a grid with the listed ratio.

The most important of these is the general conversion factor for changing from a tabletop technique to a bucky technique or vice versa—the conversion factor is 4. For example, when changing from a tabletop non-grid knee to a bucky knee, multiply the non-grid mAs by 4 times. When moving from the bucky tray to the tabletop, reduce mAs to ¼.

Note from Table 20-1 that the technique compensation factors (bucky factors) increase at higher kVp levels as well as for higher grid ratios. Across all of the possible ratios and kVp's used in diagnostic radiography, these factors range from 2 (for a 5:1 grid at 70 kVp) to as much as 6 (for a 16:1 grid at 120 kVp). These two extremes bear mention: A 5:1 or 6:1 grid, having lower *selectivity*, is generally recommended for mobile radiography to allow for more positioning latitude with these more difficult cases. Depending on the anatomy, the technique compensation

Table 20-1

Technique Conversion Factors (Bucky Factors) for Various Ratio Grids at Different Ranges of kVp*

Grid Ratio	8:1	10:1	12:1
70 kVp	3.5	3.8	4.0
95 kVp	3.8	4.0	4.3

*From Report #102 of the National Council on Radiation Protection and Measurements (NCRP).

Table 20-2

Rounded Technique Conversion Factors for Common Bucky Grid Ratios

Grid Ratio	mAs Conversion Factor
Non-Grid	1
8:1	3½
10:1	3¾
12:1	4.0
Bucky Grid Average	**4**

would then be 2 to 2.5 times the non-grid mAs. Grid ratios of 15:1 or 16:1 are so selective that they can only be used with dedicated "chest" units that have the x-ray tube permanently locked into a perfectly centered and perpendicular relationship to the bucky. On these units, technique compensation would be 5 to 6 times the non-grid mAs.

Other Image Qualities

Earlier in this chapter, we established that scatter radiation has nothing to do with image sharpness. By the same reasoning, neither can grids have any direct relationship to image sharpness. An indirect problem can be related to the use of the *bucky tray*, in that a small distance of 2 to 4 inches is created between the patient and the image receptor. But, any geometric effects of this are caused by a change in the *OID*, not simply because a grid is in place or is not in place. By the same token, a stationary grid can be placed on an imaging plate without any change in the geometry of the x-ray beam at all.

To summarize, grids have nothing to do with any of the recognizability factors in the image: sharpness, magnification or distortion. They do affect all three visibility functions: Exposure is lessened, subject contrast is improved and noise in the form of scatter is reduced.

Grid Cut-Off

The lead strips of a grid absorb some of the primary x-rays along with many of the scattered rays. With a stationary grid, this absorption will leave blank lines in the image called *grid lines* (Fig. 20-9), where a loss of information occurs. Grid lines are classified as a form of *noise* in the image. Although these grid lines can be minimized by the use of proper alignment and distance, they can never be completely eliminated when *stationary* grids are used.

The employment of the Potter-Bucky diaphragm blurs out the grid lines from the final image, but there is still a loss of exposure called *grid cut-off*, caused by the absorption of remnant beam radiation. Any time improper distances are used, or when the x-ray beam is not properly aligned with a grid, primary-beam x-rays strike the lead grid strips broadside or at an oblique angle, rather than passing through the grid parallel to these strips (Fig. 20-10). Grid cut-off becomes much worse, taking large amounts of information out of the image.

Grid Radius

Due to the fact that x-ray beams diverge or spread out as they approach the image receptor, most grids in modern use are *focused*, that is, the lead strips are tilted more as they move away from the center of the grid, such that they are always pointing toward the source of x-rays—the x-ray tube focal spot. This tilt of the lead strips, called *canting*, is illustrated in Figure 20-10, *right*.

With a focused grid, the *grid radius* is defined as that distance from the grid at which lines drawn from the canted lead strips, with their various degrees of tilt, *converge* to a focal point. This is the distance at which the focal spot, or x-ray tube, should be positioned. Since there is some margin for error, actual grid radii are given by manufacturers as *ranges* rather than as a single number. The two most common ranges are 36–42 inches (91–107 cm), and

Figure 20-9

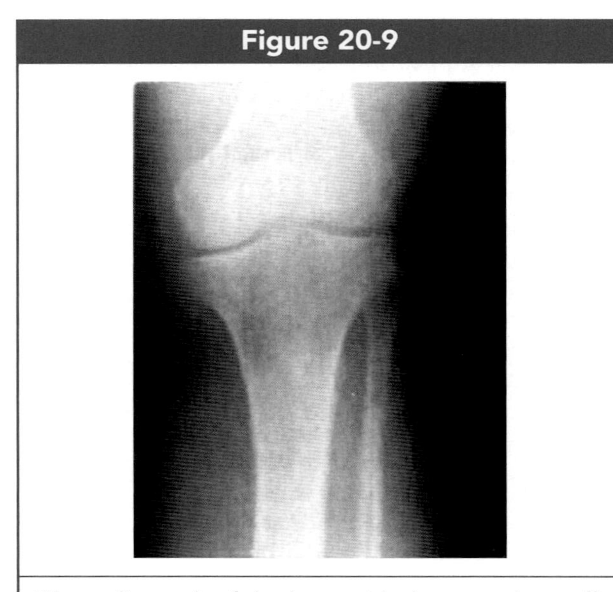

Film radiograph of the knee with the central ray off-centered 2 inches to a 12:1 ratio stationary grid. Thin, vertical white grid lines can be best seen toward the middle. (From Quinn B. Carroll, *Practical Radiographic Imaging*, 8th ed. Springfield, IL: Charles C Thomas, Publisher, Ltd., 2007. Reprinted by permission.)

Figure 20-10

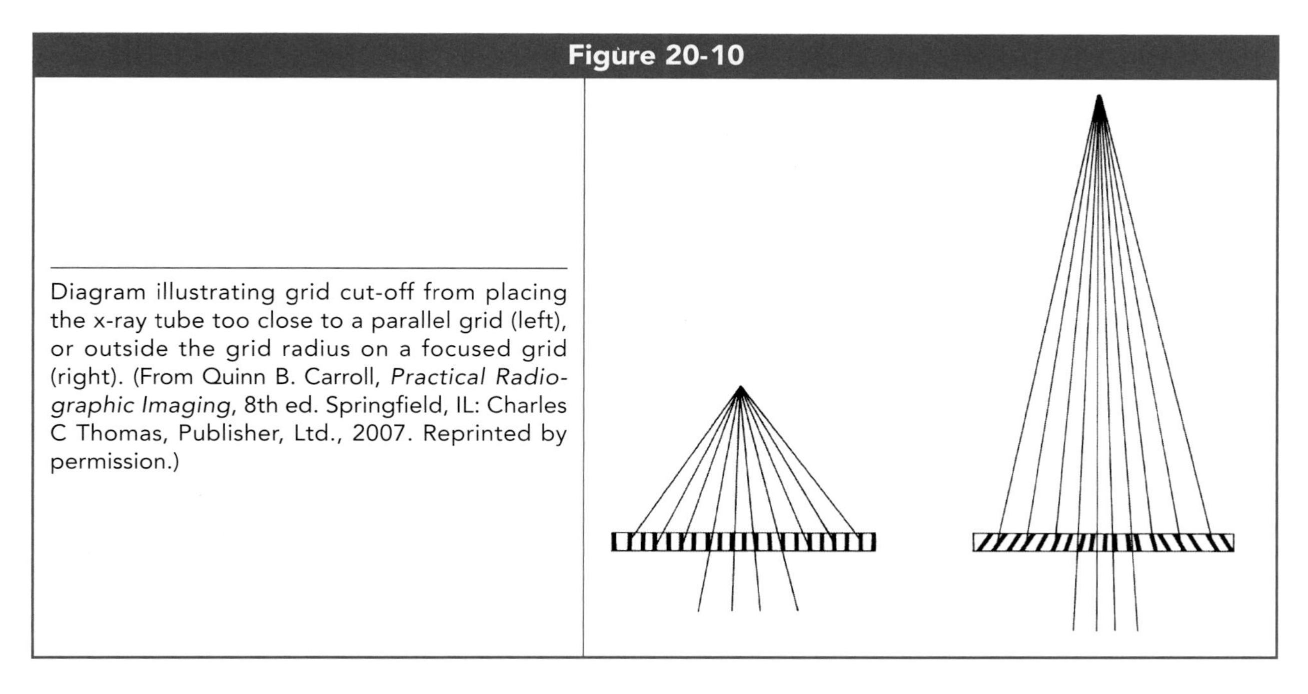

Diagram illustrating grid cut-off from placing the x-ray tube too close to a parallel grid (left), or outside the grid radius on a focused grid (right). (From Quinn B. Carroll, *Practical Radiographic Imaging*, 8th ed. Springfield, IL: Charles C Thomas, Publisher, Ltd., 2007. Reprinted by permission.)

66–74 inches (168–188 cm), obviously designed for the two common SID's used.

If the x-ray tube is not positioned at a distance from the grid within the grid radius designated, excessive grid cut-off will occur. The lead strips will absorb much more of the primary radiation than desired, in a *symmetrical pattern toward each side of the image*, because peripheral x-rays are striking the grid strips "broadside" at an increasing angle, as shown in Figure 20-10. Whether the x-ray tube is placed *too close* to the grid or *too far away*, the same symmetrical pattern of grid cut-off occurs, with a loss of exposure toward both lateral edges of the image.

Parallel grids have no canting of the lead strips. Each strip points straight up, as shown in Figure 20-10, *left*. Therefore, parallel grids have a grid radius of infinity—the further the x-ray tube from the grid, the more parallel are the primary rays to the lead strips in the grid. Therefore, grid cut-off problems never occur from *excessive* SID for a *parallel* grid. Too short an SID, however, causes greatly increased beam divergence. An inordinate amount of primary radiation is cut off toward the periphery, and severe grid cut-off occurs with a loss of information, Figure 20-10, *left*.

Figure 20-10, *right*, shows how too long an SID with a *focused* grid results in just the same sort of pattern for grid cut-off. Using a conventional film radiograph in Figure 20-11, the actual result of *off-radius* grid cut-off can be seen in the exposure pattern. Upon close examination, white grid lines themselves can be seen in the transition areas between black and white densities, indicating that a

Figure 20-11

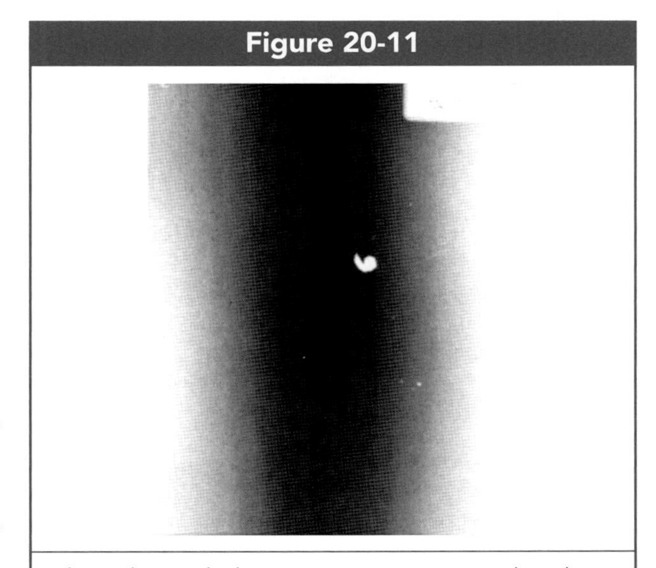

Film radiograph demonstrating symmetrical grid cut-off from placing the x-ray tube at 20" SID with a 12:1 40-inch radius focused grid. (From Quinn B. Carroll, *Practical Radiographic Imaging*, 8th ed. Springfield, IL: Charles C Thomas, Publisher, Ltd., 2007. Reprinted by permission.)

Figure 20-12

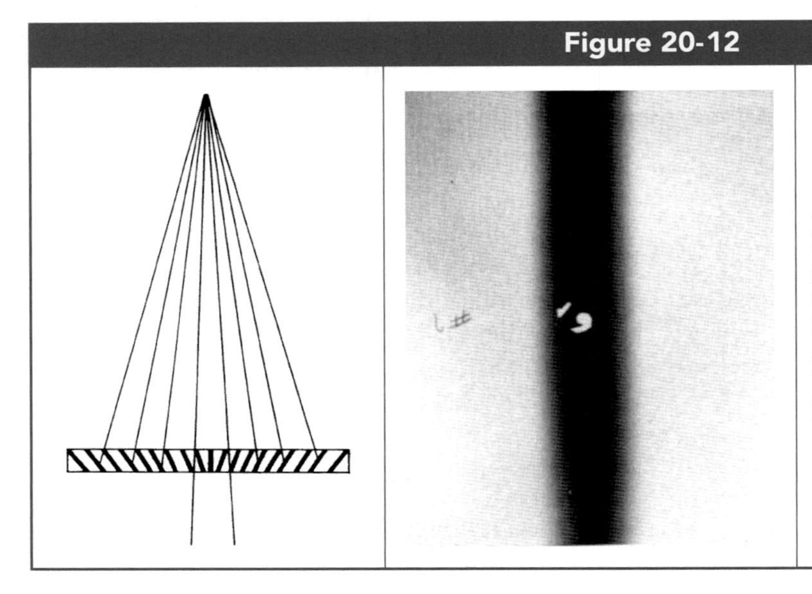

Diagram and radiograph showing extreme grid cut-off caused by placing a focused stationary grid upside-down. (From Quinn B. Carroll, *Practical Radiographic Imaging*, 8th ed. Springfield, IL: Charles C Thomas, Publisher, Ltd., 2007. Reprinted by permission.)

stationary grid was used. All white areas represent a loss of information.

Figure 20-13

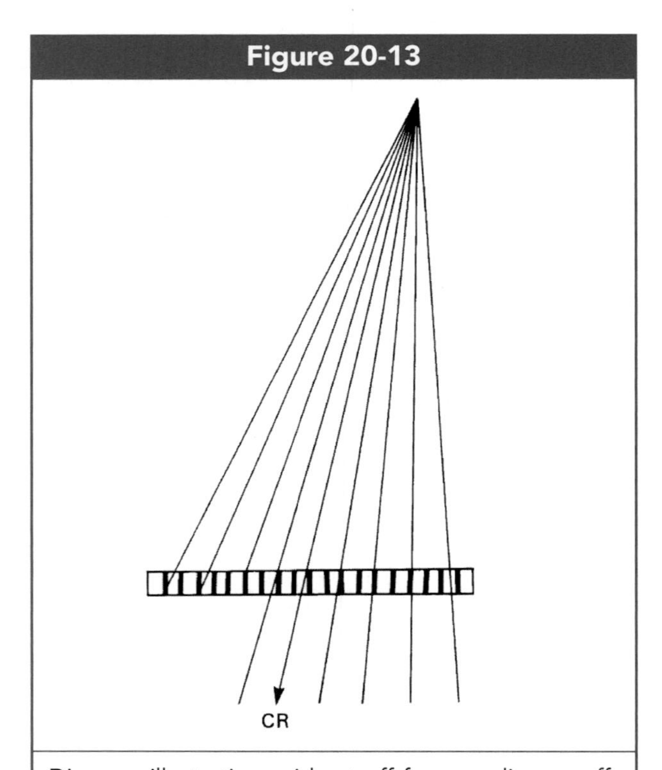

Diagram illustrating grid cut-off from angling or off-centering the x-ray beam across the lead strips, or by tilting the grid crosswise. (From Quinn B. Carroll, *Practical Radiographic Imaging*, 8th ed. Springfield, IL: Charles C Thomas, Publisher, Ltd., 2007. Reprinted by permission.)

More selective grids (with higher grid ratios) have shorter ranges for their radii, within which the SID must fall. There is less margin for error, and the same types of mistakes result in more severe grid cut-off. With *wafer* grids which are separate from the imaging plate rather than built into it, a common mistake is to lay the grid upside down over the image receptor. With a focused grid, the effect is to *obliterate* the peripheral exposure, causing an even more exaggerated effect (Fig. 20-12).

Alignment of the Beam and Grid

When performing mobile radiography, two other problems commonly occur. First, the imaging plate and grid may be tilted, so that they are not perpendicular to the CR. This is equivalent to angling *across* the lead strips laterally (Fig. 20-13). Second, the imaging plate may be off-centered transversely to the CR. Whether off-centering or off-angling laterally *across the lead strips*, the effect is the same, demonstrated using a film radiograph in Figure 20-14. There is an *asymmetrical* loss of exposure that is much greater toward one side of the image than the other. Again, these effects are more pronounced when working with higher-ratio grids.

Note that on *digital* images all types of grid cut-off are much less apparent, due to the computer's ability to *partially* compensate for the loss of exposure in these areas. They appear slightly lighter and

Figure 20-14

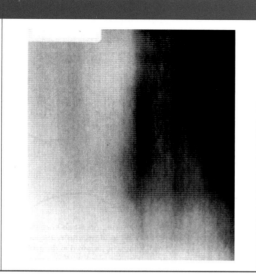

Film radiograph demonstrating asymmetrical grid cut-off caused by off-centering the CR 2" across the strips of a 15:1 stationary grid. The effect of angling the beam or tilting the grid crosswise is similar. (From Quinn B. Carroll, *Practical Radiographic Imaging,* 8th ed. Springfield, IL: Charles C Thomas, Publisher, Ltd., 2007. Reprinted by permission.)

present significant mottle. For this chapter, the more dramatic effects on film radiographs have been retained for the purposes of illustration.

Although special grids can be designed with cross-hatch patterns, most grids are *linear*, with all of the lead strips running parallel to each other along the *length* of the grid. These are designed for use in imaging plates and bucky trays so that *longitudinal tube angles* (cephalic and caudal) can be used without grid cut-off. When using a standard x-ray table bucky or chest board bucky, the x-ray tube must never be angled *transversely* across the table, or grid cut-off will occur.

There are times, particularly in mobile radiography, when intentional off-centering relative to the grid is needed. Figure 20-15 illustrates such a scenario, in which a trauma patient is supine and a "cross-table" lateral skull projection must be obtained with a horizontal beam. If a grid is placed in the typical fashion, crosswise to the skull, the lead strips will run vertically. Note that superior-to-inferior centering must be aligned with the grid, even though it off-centers the CR to the skull itself, to avoid grid cut-off. Yet, in the *vertical* dimension the CR may be off-centered to the grid in order to center to the skull. If superior-to-inferior centering to the skull is more critical than anterior-to-posterior centering, the grid should then be placed *lengthwise* to the skull, so that the lead strips run horizontally to allow it.

The more selective a grid is (the higher the grid ratio), the more critical centering and alignment become in preventing grid cut-off. Grids with a ratio of 15:1 or 16:1 are only used in *dedicated* chest units where the x-ray tube is permanently locked into a perfectly centered and perpendicular position in relation to the image receptor plate. On the other

Figure 20-15

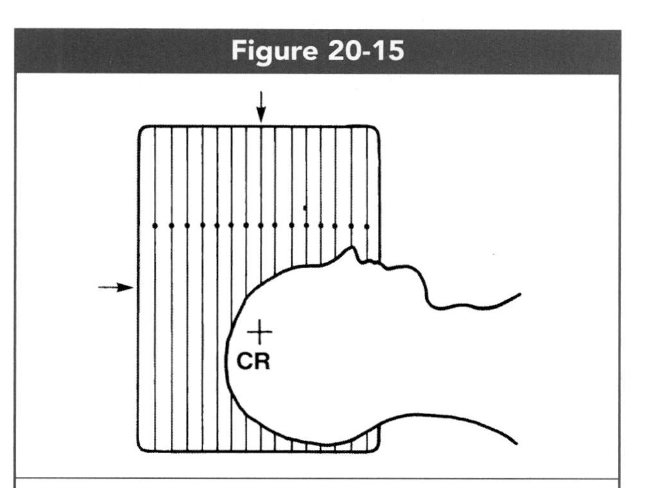

Mobile radiography scenario in which the CR must be centered to the grid crosswise even though this may place it off-centered supero-inferiorly to the skull. The CR may be centered to the skull in an antero-posterior direction even though doing so places it off-centered to the grid *lengthwise.* (From Quinn B. Carroll, *Practical Radiographic Imaging,* 8th ed. Springfield, IL: Charles C Thomas, Publisher, Ltd., 2007. Reprinted by permission.)

extreme, some departments purchase 6:1 ratio grids for use only on mobile procedures, allowing wide flexibility in positioning and margin for error in centering.

With careful attention to these considerations, grids can be properly applied in radiography of anatomy thicker than 13 cm, and they continue to be a valuable aid in the digital age for minimizing scatter noise at the image receptor.

SUMMARY

1. Overexposure must not be assumed when excessive scatter radiation is produced—it is possible for *under*exposure to occur along with an excessive proportion of scatter radiation in the remnant x-ray beam.
2. Scatter radiation destroys *both* subject contrast and gray scale in the remnant beam image.
3. The major causes of scatter radiation are large patients and large field sizes, both of which increase the volume of soft tissue exposed to the primary x-ray beam. The most important tool available to the radiographer for preventing scatter production is *collimation* (field size restriction).
4. Although high kVp becomes a contributing factor for scatter production, its impact is relatively minor and must be weighed against the substantial benefits of full penetration and reduced patient exposure.
5. Scatter radiation has no impact on sharpness or other geometrical qualities of the image.
6. Scatter radiation impacts *all* visibility functions of the image: It increases exposure intensity, reduces subject contrast, and contributes to image noise.
7. Grids are still useful for digital imaging in reducing the percentage of scatter noise in the remnant beam. Their entire purpose is to restore subject contrast in the signal at the image receptor.
8. The controlling factor for grid selectivity and effectiveness is the *grid ratio*. Grid frequency and lead content are not good indicators of grid quality.
9. Generally, grids should be used whenever the size of the body part measures greater than 13 cm. Large field sizes and very high kVp levels can contribute to the need for grids.
10. *Selectivity* is an indication of the effectiveness of a grid, and is defined as the ratio of primary radiation to scatter radiation transmitted through a grid. The *bucky factor* is the ratio of technique adjustment required to maintain the remnant exposure when a grid is used.
11. Grids have no direct impact upon any of the recognizability (geometrical) functions in the image.
12. Grid cut-off is underexposure in a portion of the image due to misalignment of the grid or using an SID outside the grid radius. Grid *lines* can be visible when stationary grids are used, and constitute noise in the image.
13. Off-centering and angulation of the CR can be employed in a lengthwise direction with a linear grid, but not in the crosswise dimension. The higher the grid ratio, the more severe is the grid cut-off from misalignment or incorrect SID.

REVIEW QUESTIONS

1. What is the most important tool available to the radiographer in preventing the *production* of scatter radiation?

2. High kVp levels actually result in *less* scatter radiation being produced within the patient. Give *two* reasons why increased proportions of scatter reach the receptor plate:

3. Although extreme kVp increases can contribute to scatter, give *two* advantages of using higher kVp:

4. As scatter radiation increases, what happens to the degree of *penumbra* or blur in the image?

5. Unless *all* tissues of interest are penetrated to some degree by the x-ray beam, gray scale in the remnant beam image is too __________.

6. The distance at which lines extended from the lead strips of a focused grid converge defines the:

7. The primary purpose of a grid is to restore __________ in the remnant beam image or signal.

8. A grid has a frequency of 40 lines per cm. The plate is 1 mm (0.1 cm) thick. What is the grid ratio?

(Continued)

REVIEW QUESTIONS *(Continued)*

9. What is the purpose of the mechanism invented by Hollis Potter?

10. What are the *two* types of motion utilized by Potter-Bucky diaphragms?

11. Using a stationary grid, the resulting radiograph shows grid cut-off and thicker grid lines toward one side of the image. Name *two* possible causes:

12. As a general rule, what bucky factor or technique change should be used when changing from the tabletop into a typical table bucky?

13. A mobile cross-table skull is performed using a grid cassette with the lines running vertically. The x-ray beam is centered horizontally but not vertically because the patient's head is raised on a sponge. The grid has a radius is 40 inches and a ratio of 8:1. To maneuver around ER equipment, an SID of 55 inches is employed. What *single* problem will appear on the finished radiograph?

14. As the effective angle of penetration is reduced, the *ratio* of primary to scatter radiation passing through the grid is increased. This is the definition for grid _________.

15. Indirectly, what is the impact of higher-ratio grids upon patient exposure?

16. Grid frequency and lead content are not reliable indicators of improved efficiency because they do not necessarily increase the _________.

(Continued)

REVIEW QUESTIONS *(Continued)*

17. What grid ratio is recommended for body parts measuring 18 to 22 cm in thickness?

18. What is the *only* recommended application for grids with a ratio of 15:1 or 16:1?

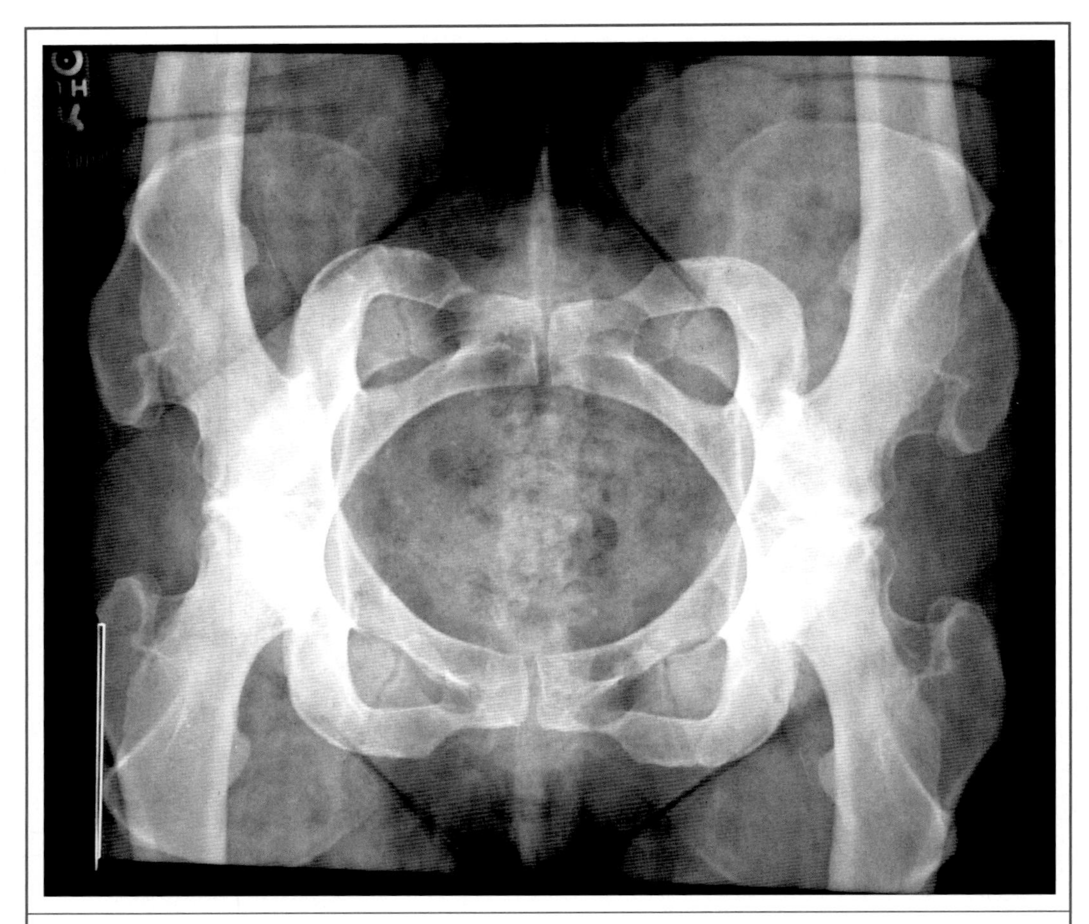

For an AP pelvis projection, the x-ray film got folded crosswise in the cassette, recording this mirror-image of the lower pelvis and femurs.

Chapter 21

THE ANODE BEVEL AND FOCAL SPOT

Objectives:

Upon completion of this chapter, you should be able to:

1. Describe how the line-focus principle is used to allow for high image sharpness while providing for adequate heat dispersion at the anode surface.
2. State how the line-focus principle causes the image to be sharper at one end of the field than the other.
3. Describe the cause of the anode heel effect, and the factors that make it worse.
4. For different body parts of variable thickness, state the best end of the x-ray table to position the patient's head end, in order to minimize the anode heel effect.
5. Quantify the relationship between focal spot size and image penumbra.
6. Explain why geometrical penumbra occurs.
7. Explain how the umbra portion of the image can be made to completely disappear.
8. Describe why magnification is not affected by the focal spot size.
9. Describe why the focal spot is uniquely considered as the controlling factor for sharpness.

The *anode bevel* refers to the angle of the target surface of the anode in relationship to a vertical line drawn perpendicular to the long axis of the x-ray tube (Fig. 21-1). The angle of this surface affects both the size of the *effective* focal spot (line-focus principle), and the distribution of x-ray intensity within the beam (anode heel effect).

LINE-FOCUS PRINCIPLE

The size of the projected or effective focal spot is crucial to the sharpness of any radiographic image. It is controlled by both the width of the beam of electrons striking the anode (which is determined by the size of filament used), and the angle of the anode bevel.

The smaller the effective focal spot, the greater the sharpness in the image. In accordance with the line-focus principle, the *effective* focal spot can be made much smaller than the actual stream of electrons. If the anode bevel were exactly 45 degrees, the resulting effective focal spot would be the same width as the beam of electrons impinging upon the anode surface (Fig. 21-2). By beveling the anode at a *lesser* angle, the target surface is made *steeper*, or more vertical. Line-focus geometry shows that this change causes the effective focal spot to be projected at a smaller size than the *actual focal spot*. (The true actual focal spot is measured along the beveled anode surface, as illustrated in Figure 21-2.)

The area of the *actual* focal spot is the area available for the dispersion of *heat* generated in the anode from the colliding electrons. As previously described, the anode reaches tremendous temperatures during exposure, so much so that the entire disc can glow "white hot" like a light bulb in spite of its thickness and size. The goal at hand is to *achieve maximum sharpness while maintaining good heat dispersion*. If

Figure 21-1

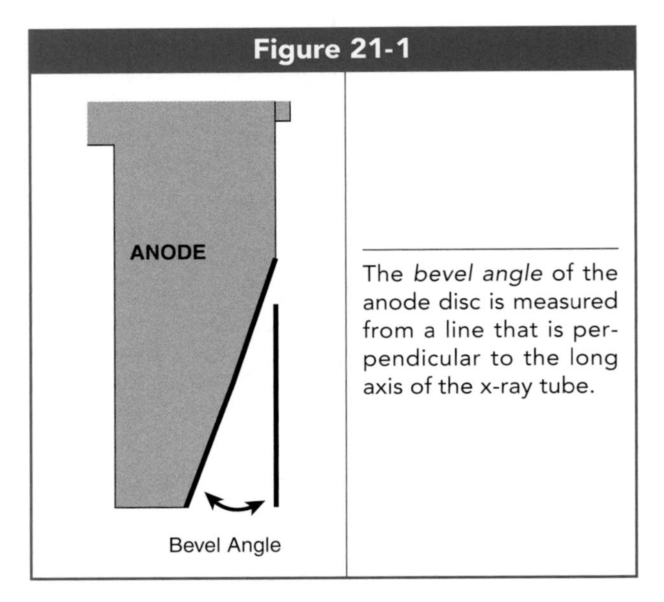

The *bevel angle* of the anode disc is measured from a line that is perpendicular to the long axis of the x-ray tube.

the width of electron beam were narrowed to achieve the desired focal spot, heat would be so concentrated that it might melt the anode surface.

The line-focus principle makes it possible to achieve this goal of a very small *effective focal spot* while allowing at the same time the anode heat to dispersed over a sufficient area. Most x-ray tubes for standard diagnostic use have anode bevels of 15–17 degrees. Special procedures tubes such as those used for angiography and cardiac catheterization can have anode bevels as shallow as 7–10 degrees in order to achieve extra-small effective focal spots.

Typically, the *large* focal spot setting on the console refers to an effective focal spot ranging from 1 to 2 mm in size, while the *small* FS ranges from 0.5 to 1 millimeter. *Fractional* focal spots used for magnification techniques in special procedure tubes can be as small as 0.2 mm.

Since the size of the *effecive* focal spot is dependent upon the relative angles formed between the anode bevel, the electron beam, and the image receptor below, one might suspect that it changes according to the *angle of projection* toward portions of the imaging plate other than the center-point. This is the case, and is demonstrated graphically in Figure 21-3. Perhaps the best way to visualize this effect is to imagine oneself *looking at the focal spot* from the perspective of the imaging plate; from the *cathode* end of the plate, away from the anode, the focal spot appears as a larger area because it is "seen" more face-on. Indeed, it is *projected* from the anode to this end of imaging plate as a *larger focal spot.* At the center-point of the imaging plate, where the x-ray beam CR is perpendicular to the plate, we find the *effective size* of the focal spot as defined by the line-focus principle, which is smaller. At the *anode* end of

Figure 21-2

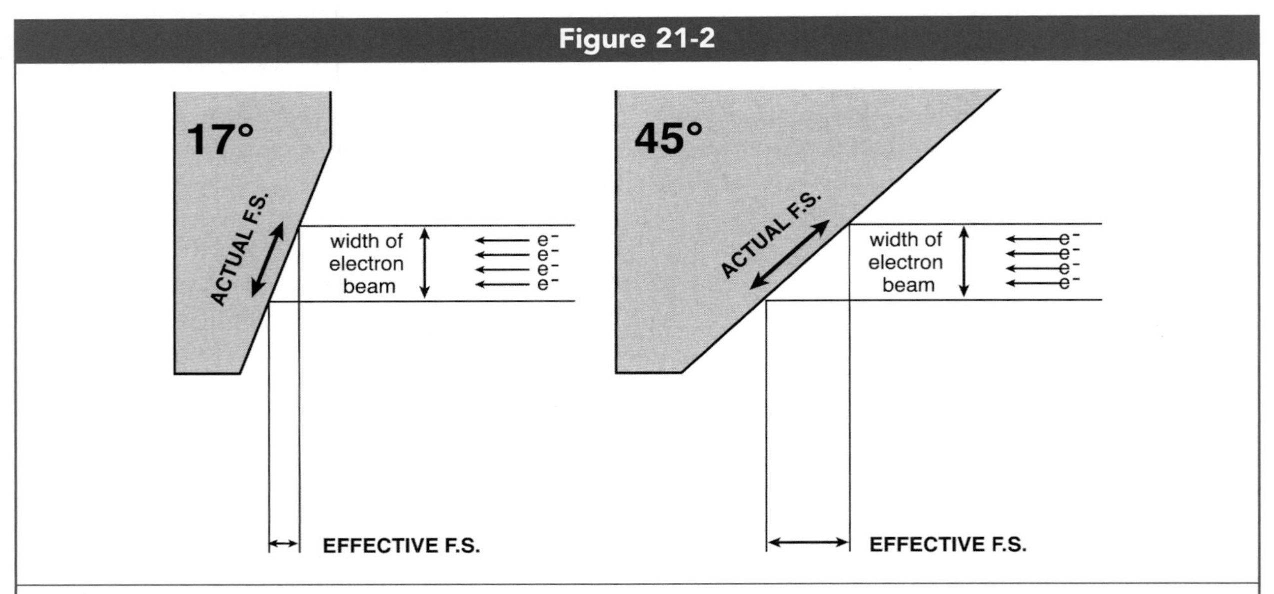

Diagram of the line-focus principle. By using an anode bevel angle much steeper than 45 degrees (*left*), effective focal spots much smaller than the actual focal spot can be obtained.

the plate (usually to the *left* as one approaches an x-ray table), the focal spot would be "seen" from the plate at a very steep obliquity and would appear foreshortened. Indeed, it is *projected* toward this end of the plate as a smaller spot.

This phenomenon has an interesting implication: It means that the anode end of the resulting image is actually *sharper* than the cathode end. This is true for all radiographs. With routine SID's and smaller sizes of imaging plates this variation in sharpness may be too small to measure, but when large plates (14" × 17" or 35 × 43 cm) are used at 40" (100 cm) SID or less, it can become exaggerated enough to measure and even see.

(With digital radiography, image sharpness is additionally limited by the pixel size, fully discussed in later chapters. If the pixel size is larger than the effective focal spot size, the effects of the line-focus principle are pre-empted in the final image produced, *but are still present in the remnant beam image at the image* receptor plate.)

Note the precise use of the terms *effective focal spot* and *projected focal spot*. There is only one *effective focal spot*, defined as the apparent size of the FS as perceived in a line *perpendicular* to the long axis of the x-ray tube, or perpendicular to the electron beam shown in Figure 21-2. There are any number of *projected focal spots*, which vary with the angle at which they are perceived from any point along the length of the image receptor as illustrated in Figure 21-3.

Figure 21-3

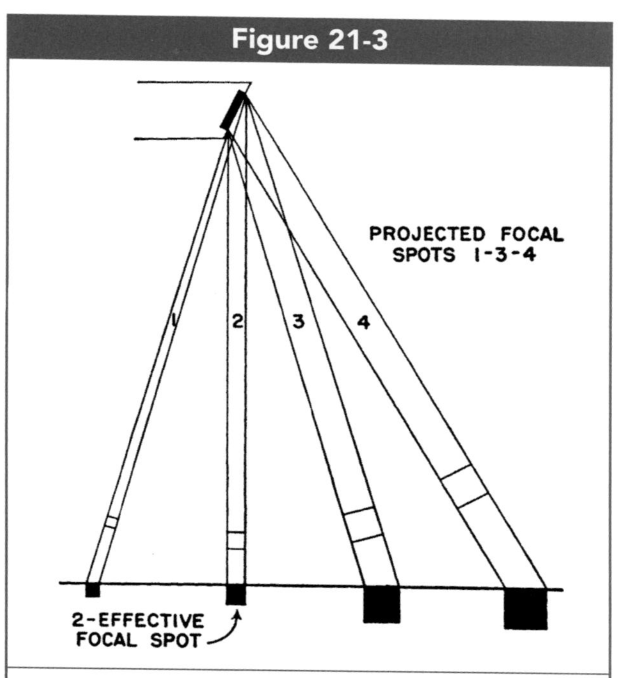

The size of the *projected* focal spot varies with the angle at which it is projected from the target to different points along the length of the image receptor. It is smallest toward the *anode end* of the x-ray tube; consequently, this end of the image has the greatest sharpness. Note that only one projected FS is properly labeled the *effective focal spot*—the one that lies perpendicular to the long axis of the x-ray tube. (From Quinn B. Carroll, *Practical Radiographic Imaging*, 8th ed. Springfield, IL: Charles C Thomas, Publisher, Ltd., 2007. Reprinted by permission.)

ANODE HEEL EFFECT

The *anode heel effect* is a variation in the x-ray intensity along the longitudinal tube axis. The intensity of the x-ray beam diminishes fairly rapidly from the central ray toward the *anode* side of the x-ray beam. We have previously mentioned that the anode acts as a form of *inherent filtration*; x-rays produced at any given point inside the anode must first *escape* the anode itself by passing through other atoms of tungsten and rhenium. Those that exit in a direction perpendicular to the anode bevel surface have *less distance* to travel to escape the anode than those that exit straight downward or toward the *heel* of the anode (Fig. 21-4). The *heel* is defined as the lower back corner of the anode disc. Therefore, as we progress along the long axis of the tube from the cathode toward the anode, the x-rays produced have effectively more and more filtration to pass through just to be emitted from the tube.

The impact of the anode heel on actual exposure is dramatically demonstrated by the flash exposure of a film placed vertically in the x-ray beam in Figure 21-5. The loss of exposure toward the anode can be readily seen. With more acute anode bevel angles such as those used for special procedure x-ray tubes, the thickness of the anode increases much more quickly across the length of a particular projected focal spot, and the heel effect can become so severe with large field sizes that field sizes must be limited to 14" (35 cm) in length (rather than the usual 17" or 43 cm). The heel effect is also more pronounced

Figure 21-4

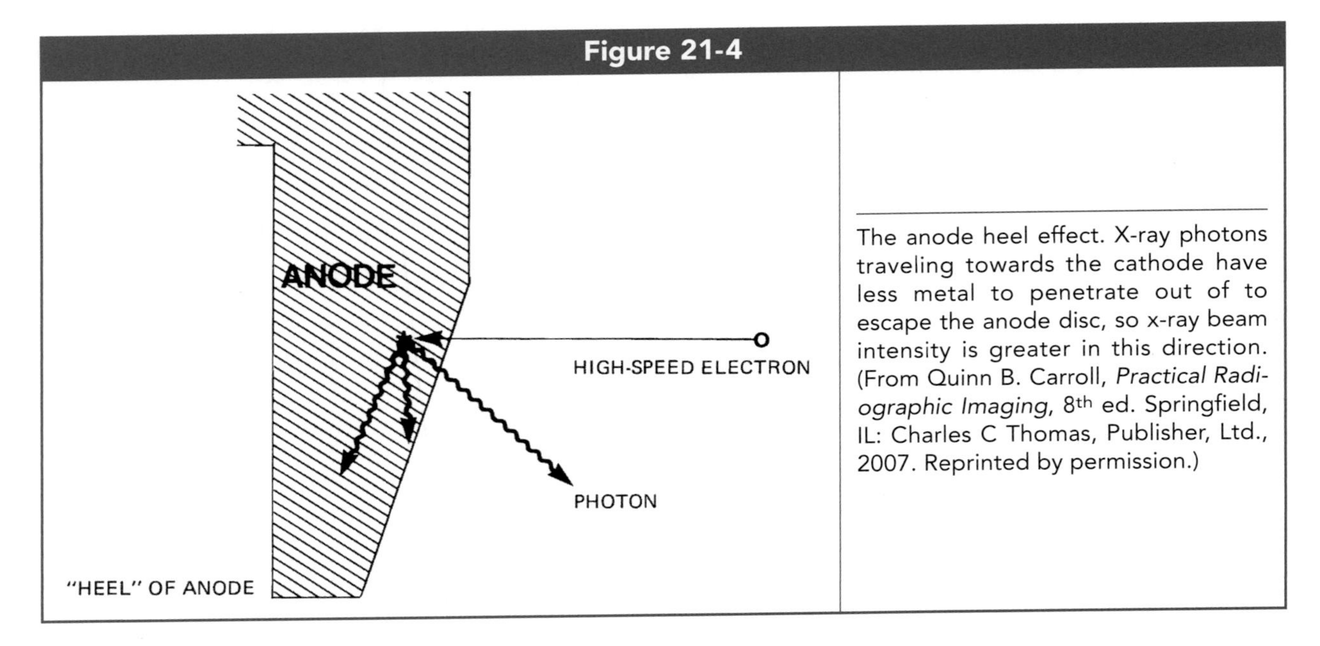

The anode heel effect. X-ray photons traveling towards the cathode have less metal to penetrate out of to escape the anode disc, so x-ray beam intensity is greater in this direction. (From Quinn B. Carroll, *Practical Radiographic Imaging*, 8th ed. Springfield, IL: Charles C Thomas, Publisher, Ltd., 2007. Reprinted by permission.)

Figure 21-5

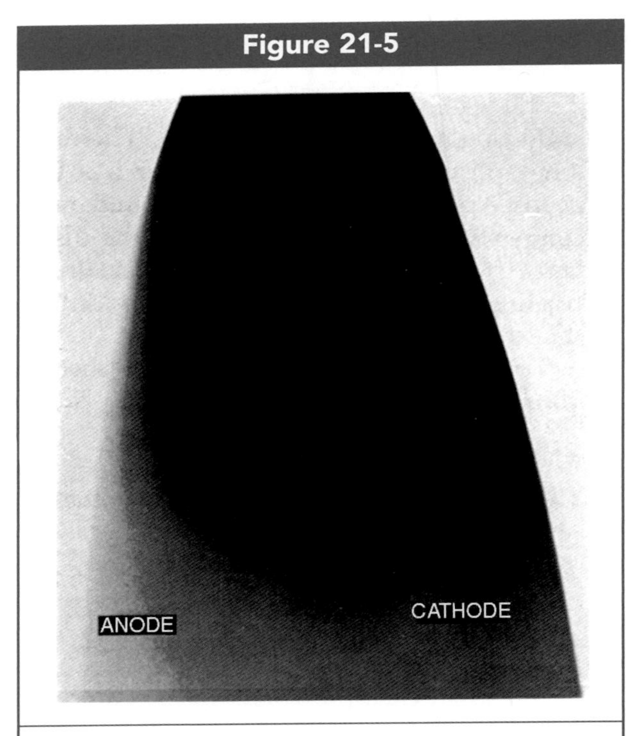

A "flash" x-ray exposure taken with a film placed vertically in the beam shows dramatic evidence of **A**, the anode heel effect toward the left of the exposure, and ***B***, the inverse square law as the exposure decreases at greater distance from the source. (From Quinn B. Carroll, *Practical Radiographic Imaging*, 8th ed. Springfield, IL: Charles C Thomas, Publisher, Ltd., 2007. Reprinted by permission.)

with larger focal spots than with small ones, since the difference in anode material thickness is greater from one end of the focal spot to the other.

The importance of the heel effect to clinical practice is that when body parts of graduating thickness are radiographed, it is always best to *place the thinnest end of the anatomy toward the anode end of the x-ray tube*, where beam intensity is the least. Examples of such body parts include the thorax (which is much thinner at the shoulder end), the femur, the humerus, and the foot. Failure to follow this guideline can result in a surprising degree of underexposure at one end of the image receptor when the thicker anatomy is *combined* with the heel effect, as well as overexposure at the other end. (See Historical Sidebar 21-1 for a dramatic example using a film exposure of the humerus in Figure 21-6.)

Figure 21-7 presents the data obtained from many dozens of experimental trials with meticulous measurements taken to determine the actual percentage distribution of the exposure rate across the anode-cathode axis. These measurements are listed at the bottom of the diagram as percentages of the mean exposure at the central ray, and are given in intervals of every 4 degrees of deviation from the CR. Moving to the left from the CR, we see the expected decline in percentage exposure toward the anode. Moving to the right, we see a gradual increase up to plus-12 degrees, then a decline at the extreme cathode end.

HISTORICAL SIDEBAR 21-1: Conventional film radiographs of the AP humerus illustrate the dramatic effect of the anode heel effect when combined with a body part of increasing thickness toward one end. Radiograph **A** was taken with the patient's thicker shoulder toward the anode, resulting in insufficient density at this end as well as overexposure at the elbow end. For radiograph *B*, the patient was turned with his shoulder toward the cathode, and a balanced density was achieved across the image. These images were exposed with a 30" SID, so the collimator had to be opened more, exaggerating the anode heel effect.

Figure 21-6

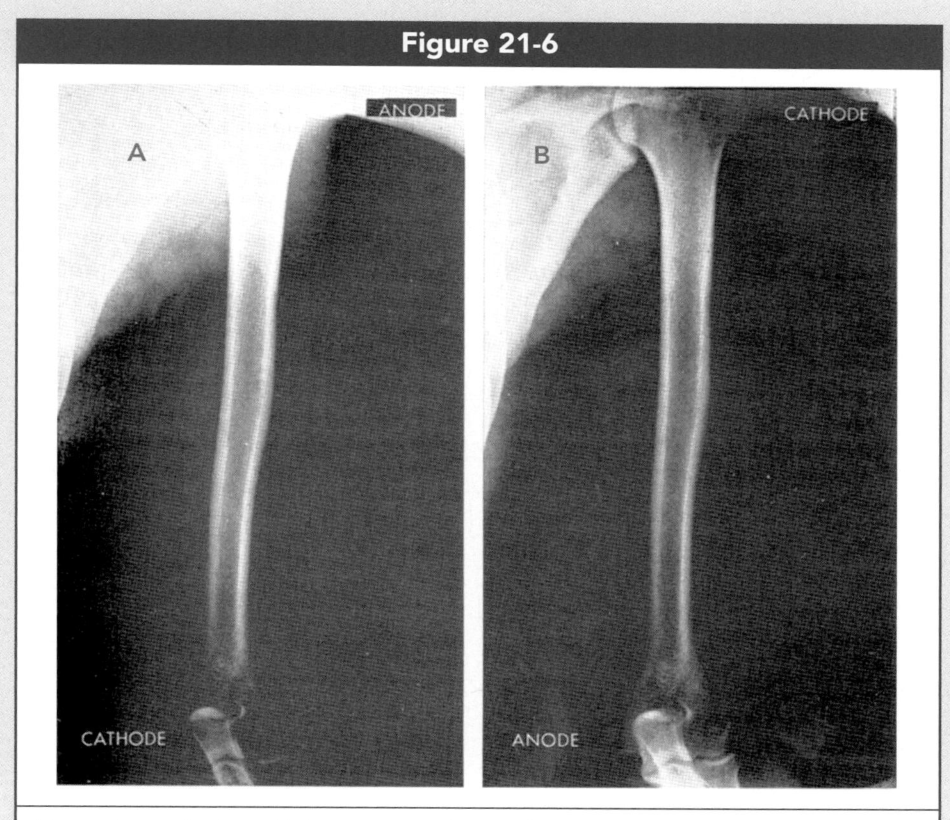

(From Quinn B. Carroll, *Practical Radiographic Imaging*, 8th ed. Springfield, IL: Charles C Thomas, Publisher, Ltd., 2007. Reprinted by permission.)

Two other important aspects of the heel effect are also found diagrammed in Figure 21-7. First, note that various SID's are listed vertically on the left side. We see that as distance to the image receptor increases, the effect is more spread out, and at shorter distances it is more concentrated. Short SID's amplify the impact of the anode heel effect on the distribution of exposure. Second, at the top of the diagram just under the x-ray tube, various sizes of image receptors are listed. By following the dotted lines downward from these sizes, we can find the degree of change in the percentage exposure across the length of each size of image receptor plate or cassette.

The change is most extreme for the 36" orthopedic cassette, moving from about the 68 percent mark on the left to about 104 percent on the right. Subtracting these two numbers, we conclude that for a 36" cassette there will be a 36 percent difference in exposure from one end to the other. This is certainly a visible difference. How much difference is there for a standard 17" (43 cm) length cassette or plate? The dotted lines extend from about 89 percent past the 103 percent mark. This is a difference of 14 percent in exposure from one end of the plate to the other. By itself, it is not a very substantial change, but remember that when combined with thicker anatomy

Figure 21-7

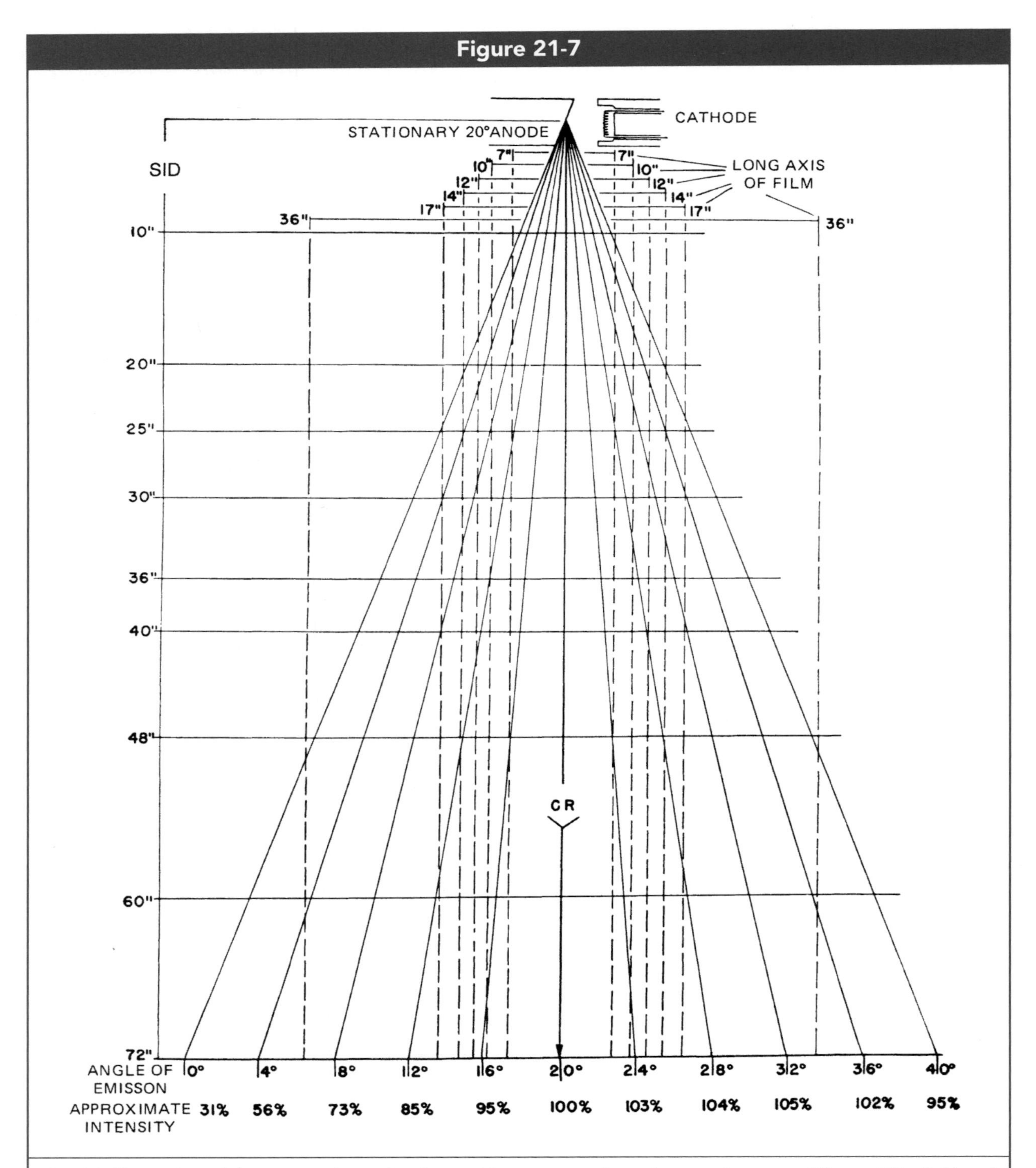

Graphic illustration of the mean values of radiographic exposure along the x-ray beam at different SIDs (left) and for different field sizes (top), based on experimental data using various x-ray tubes. The ratios of exposure at the bottom are percentages of the exposure obtained at the central ray, and illustrate the anode heel effect at different angles from it. (From Quinn B. Carroll, *Practical Radiographic Imaging*, 8th ed. Springfield, IL: Charles C Thomas, Publisher, Ltd., 2007. Reprinted by permission.)

being placed at the wrong end of the plate, it can make a substantial difference (as shown in Historical Sidebar 21-1). The heel effect is negligible for small image receptor plate sizes.

With digital imaging systems, the heel effect is not generally compensated for because of its variance at different distances. Therefore, radiographers must still be cognizant that, if the thicker end of the anatomy is placed at the anode end of the tube axis, it is possible for the combined effect to be enough to make a difference in the final image, resulting in mottle at this end of the image.

It is the anode-heel effect which explains why x-ray tubes are always installed with the anode *to the left* as the radiographer approaches the x-ray table. The convention in positioning is to normally lie patients down with their *head* to the left as well. Note that for chest and thoracic spine radiographs, this places the thinner shoulders and neck toward the anode of the x-ray tube. For a humerus, femur or a foot, however, it makes sense to reverse the patient's orientation. Even though the foot is done on a small plate, it presents the most dramatic change in part thickness, so there is no sense in adding the anode heel effect to this.

FOCAL SPOT SIZE

The term *focal spot* derives from the *focusing* of the electron beam in the x-ray tube down onto a small area on the anode target surface, but the term is used by radiographers on a daily basis in referring to the *effective* focal spot as "seen" from the viewpoint of the image receptor. The *effective* focal spot is generally about the size of a pinhead (1 mm), and all useful x-rays are emitted from this small area. Image quality is determined primarily by the size of the *effective* focal spot. For the purposes of the rest of this chapter, all references to the focal spot are *only* in regard to the effective focal spot.

Effect Upon Sharpness

The focal spot of the x-ray tube is comparable to the light source employed in shadow formation. Both follow *projection geometry* based upon similar triangles. The influence of the focal spot on image detail is confined to sharpness. With all other factors constant, *the smaller the focal spot, the sharper the recorded detail* in the image. *The focal spot is inversely proportional to image sharpness.* A large focal spot, though capable of withstanding the heat generated by high electrical energies, does not produce the sharpness of detail that is characteristic of a small focal spot. The high heat dispersion of the *rotating* anode makes it possible to use these smaller sizes and maximize image sharpness.

The visible effect of the focal spot size upon the sharpness of recorded detail is demonstrated in Figure 21-8 using a line-pair test pattern made of lead foil. Note that smaller line-pairs can be resolved and recognized when using the smaller focal spot. The geometry of this relationship is illustrated in Figure 21-9, where it can be seen that a larger focal spot results in increased spread of *penumbra* at the edges of image details. With more penumbra, a blurry image results.

In Figure 21-9, note that as penumbra grows, it spreads *inward* as well as outward, invading the umbral shadow so that the "good" part of the shadow actually shrinks in size. In angiographic studies, increased object-image receptor distance (OID) is often used, which further expands the penumbra. If an excessively large focal spot is used in conjunction with these distances, it is possible for the umbra to shrink to the point of disappearing entirely, so that a blurry patch of nebulous exposure is all that remains to represent some important small detail, as shown in Figure 21-10. Information is lost.

This effect is of great significance in angiography, because vascular embolisms (clots) and other pathology of interest may be smaller than the projected focal spot used, and would not be visibly resolved in the image, leading to a misdiagnosis.

In Chapter 13, the *formulas* for calculating unsharpness and relative sharpness were presented. Recall that these formulas include as variables the focal spot size, the SOD, and the OID. These are the only three variables that control the *projected, geometrical sharpness* of details in the image.

Penumbra

X-rays can be emited at various different angles and from different points within the area of the focal spot, and yet record the same edge of an object at the

Figure 21-8

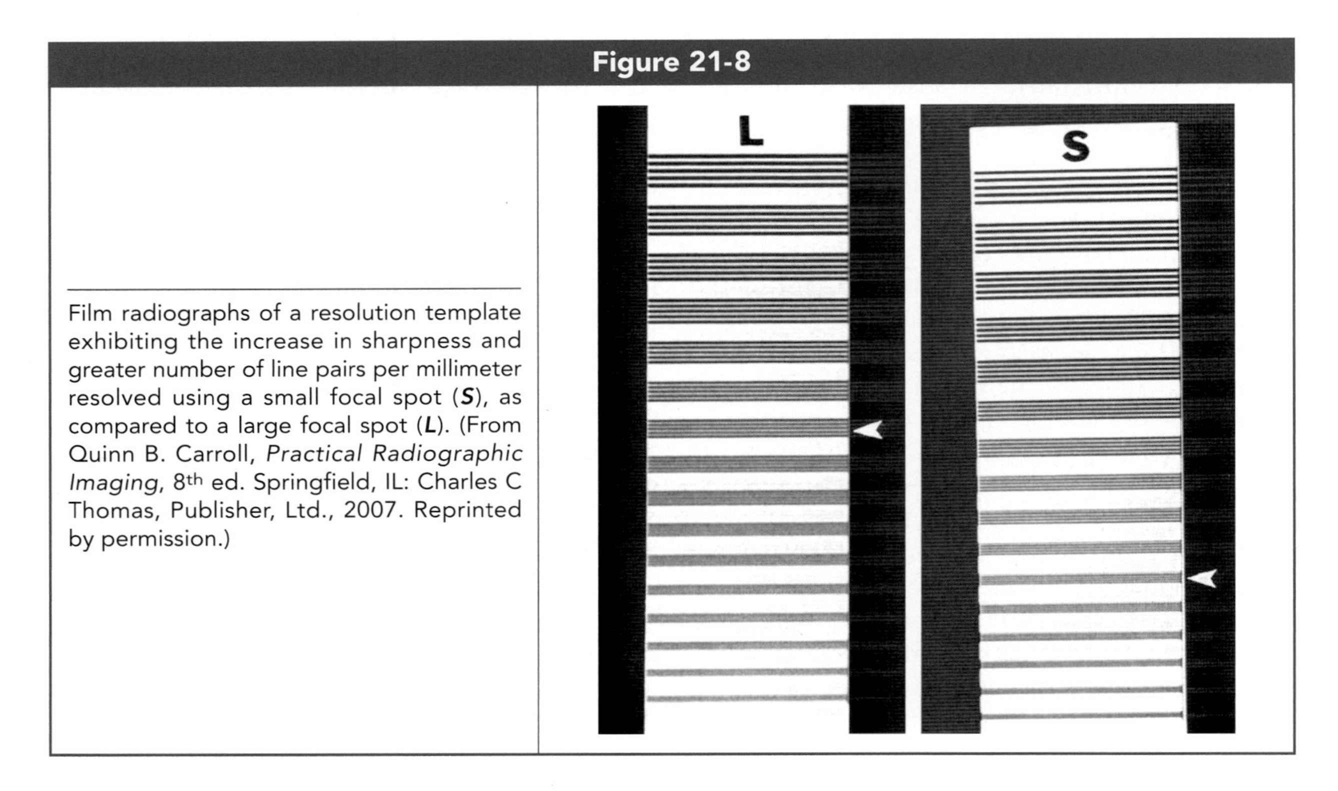

Film radiographs of a resolution template exhibiting the increase in sharpness and greater number of line pairs per millimeter resolved using a small focal spot (***S***), as compared to a large focal spot (***L***). (From Quinn B. Carroll, *Practical Radiographic Imaging*, 8th ed. Springfield, IL: Charles C Thomas, Publisher, Ltd., 2007. Reprinted by permission.)

image receptor. This means that the same edge of the object will actually be recorded several times in various different locations (Fig. 21-11). This "spreading" of the edge constitutes blur or *penumbra*.

Figure 21-11 is a diagrammatic analysis of why penumbra occurs specifically in radiographic projections. For simplification, we will assume an object which completely absorbs any x-rays that strike it;

Figure 21-9

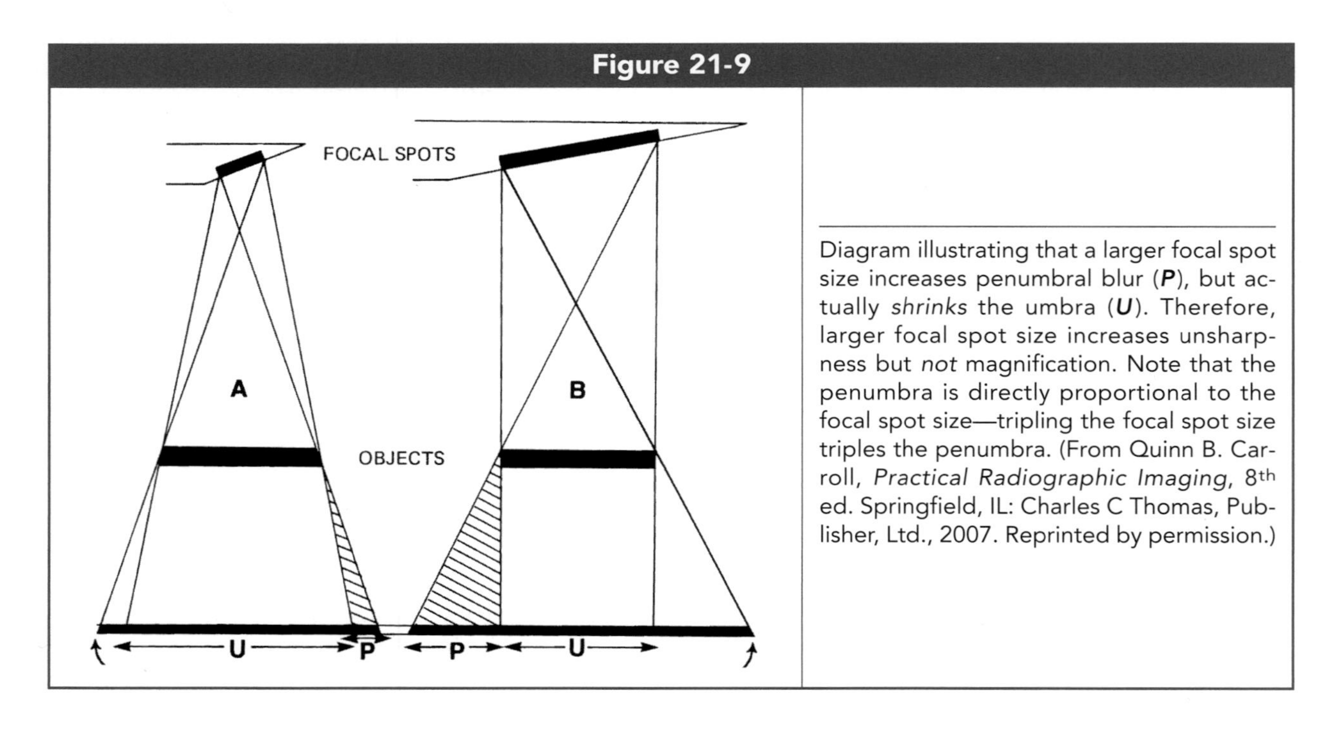

Diagram illustrating that a larger focal spot size increases penumbral blur (***P***), but actually *shrinks* the umbra (***U***). Therefore, larger focal spot size increases unsharpness but *not* magnification. Note that the penumbra is directly proportional to the focal spot size—tripling the focal spot size triples the penumbra. (From Quinn B. Carroll, *Practical Radiographic Imaging*, 8th ed. Springfield, IL: Charles C Thomas, Publisher, Ltd., 2007. Reprinted by permission.)

Figure 21-10

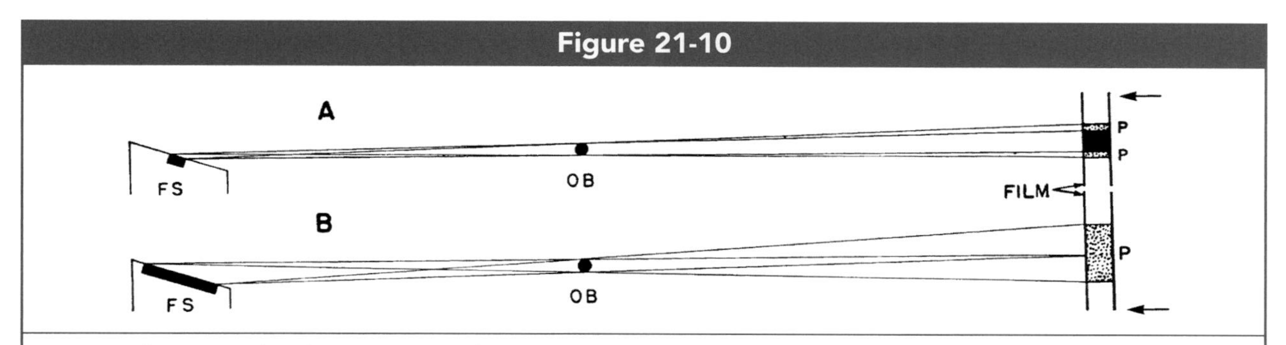

Diagram illustrating the *disappearance* of the umbral image when the object of interest is significantly smaller than the size of the focal spot used. (From Quinn B. Carroll, *Practical Radiographic Imaging*, 8th ed. Springfield, IL: Charles C Thomas, Publisher, Ltd., 2007. Reprinted by permission.)

note that all x-rays emitted at the same angle as ***A*** but from other regions of the focal spot (dashed lines), are subject to absorption by the object. All x-rays emitted at the same angle as ***B*** but from other regions of the focal spot (dotted lines), reach the imaging plate unattenuated.

There is a *transition* from total absorption to total penetration that occurs in between beams ***A*** and ***B***. In this region, different graduating amounts of x-rays are absorbed by the object *depending on their angle and point of origin within the focal spot.* In other words, between ***A*** and ***B*** absorption of the x-ray beam is *partial*, increasing in nearer proximity to the object. This causes the edge gradient known as *geometrical penumbra.*

Thick metal objects, such as an intact bullet, realistically fit this scenario. Penumbra will be evident at the edges of their image entirely due to the geometry described here. Of course, most objects and anatomy do not fit this category, and are partially penetrated by the x-ray beam to one degree or another across their breadth. (Indeed, the particular *shape* of the

Figure 21-11

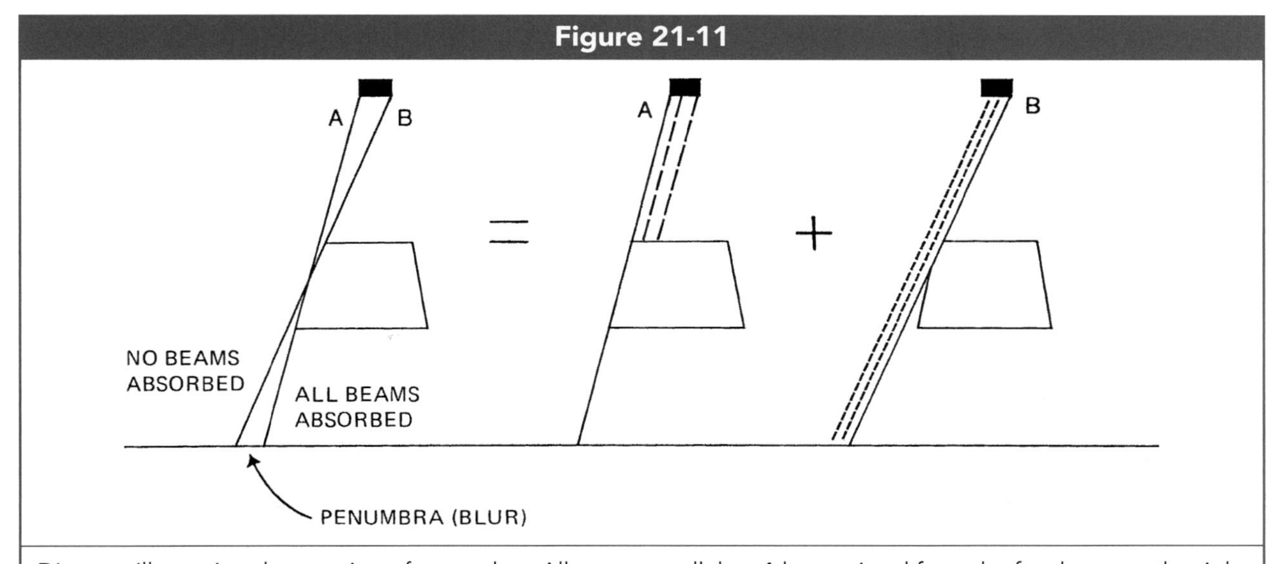

Diagram illustrating the creation of penumbra. All x-rays parallel to ***A*** but emitted from the focal spot to the right of it (dashed lines) are subject to absorption by the object. None of the beams parallel to ***B*** but emitted to the left of it (dotted lines) are subject to absorption by the object. Between ***A*** and ***B*** some beams are subject to absorption and others are not, depending on their point of origin and angle from within the focal spot. This *partial* absorption causes the partial shadow called *penumbra.* (From Quinn B. Carroll, *Practical Radiographic Imaging*, 8th ed. Springfield, IL: Charles C Thomas, Publisher, Ltd., 2007. Reprinted by permission.)

Figure 21-12

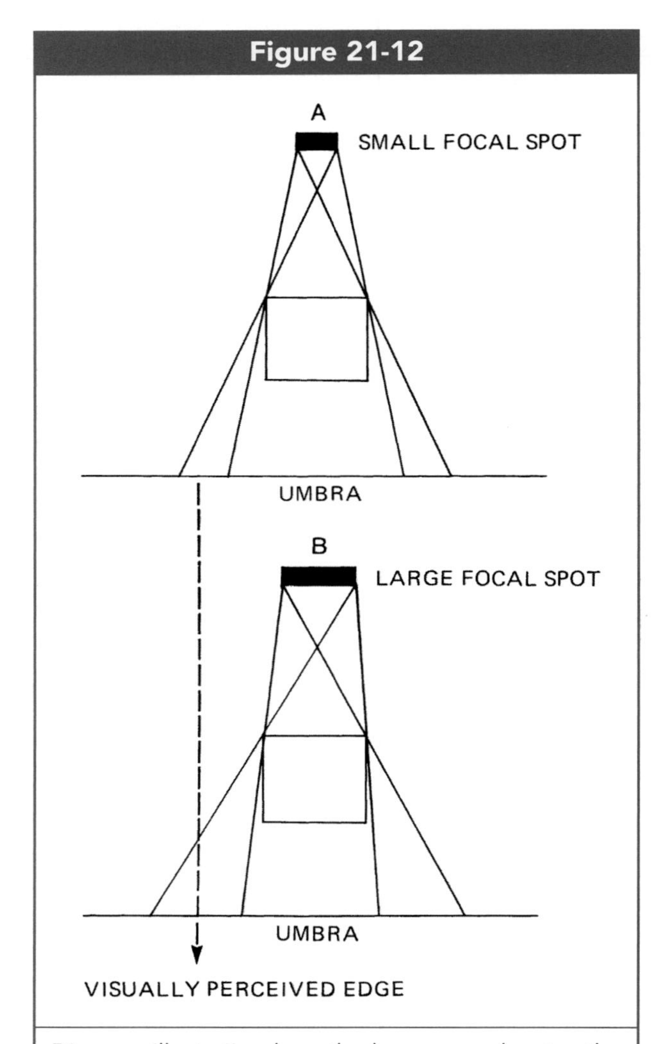

Diagram illustrating how the human eye locates the edge of an image at the mid-point of the penumbra regardless of the amount of penumbral spread. Hence, the apparent size of the gross image does not change with differences in focal spot. (From Quinn B. Carroll, *Practical Radiographic Imaging*, 8th ed. Springfield, IL: Charles C Thomas, Publisher, Ltd., 2007. Reprinted by permission.)

object can result in *absorption penumbra* which shall be described in a later chapter. Here, we wish to focus strictly on geometrical processes.) Let us assume that, because of partial penetration, a homogeneous object appears in the image as an overall medium gray shade; geometrical penumbra is still present at the edges of this image, and is defined as the gradual transition from zero absorption to that amount of absorption *the object is capable of.* If the background shade were pitch black, the penumbra for this object would transition from black to medium gray.

Focal spot size is directly proportional to penumbra. Doubling the size of the FS doubles the spread of penumbra.

Magnification

As shown in the penumbra diagrams (Fig. 21-12), when the penumbral shadow is included, a cursory glance at the change from ***A*** to ***B*** may give the impression that the entire image has spread out larger in ***B***. This has led to considerable confusion among both radiographers and educators regarding focal spot size and the question of *magnification.* A more careful examination of Figure 21-12 reveals that the *umbra* in ***B*** has actually *shrunk in size.* How can magnification be present if the "good" part of the shadow, the umbra, is smaller?

A practical experiment with changing focal spot sizes will demonstrate that the measured size of the *gross* image of a bone or other anatomical structure does *not* change with focal spots. This is because as a ruler is lined up, the human eye will tend to locate the edge of the object *in the middle of the penumbra*, regardless of how wide the penumbra has spread. This is indicated by the dashed line in Figure 21-12. Since penumbra grows *inward* as well as outward as it spreads, its midpoint does not change apparent location. This squares the theory with the practical results.

Magnification must be defined as an increase in the size of the *gross* image as a whole. It *must* include an expansion of the *umbra* portion of the image to be present. This has not occurred in Figure 21-12 from ***A*** to ***B*** when the focal spot size at the top of the diagram was increased in size. If the umbra is not larger, magnification has not occurred.

We conclude that focal spot size *does not affect true magnification* of the image, a result that can be documented by simple practical experiments. The focal spot is unrelated to image magnification.

Other Image Qualities

Focal spot size has no relation to shape distortion, because it is not a factor in *alignment* of the beam, part and image receptor. Nor does it have any direct impact upon the visibility functions of the image: exposure, contrast or noise. Another common misconception about focal spots is that larger focal spots allow more x-ray output, since the small focal spot is only available with the lower mA settings at the console. But, this limitation is due to the risk of overconcentrating the *heat load* on the anode with smaller focal spots if too high a current is allowed in the electron beam.

The number of x-rays produced depends only upon the quantity and energy of these electrons which strike the anode. Whether this production occurs within a half-millimeter space or within a one-millimeter space, *the same number of x-rays is produced.* (It only begs the question to argue about any difference in the distribution of the produced x-rays within a large focal spot versus a small focal spot, because any differences between 0.5 mm and 1.0 mm, both smaller than a pinhead, become absolutely negligible when this distribution is spread out over a 14-inch field that falls 40 inches away from the anode.) Remember that essentially, the focal spot is a geometrical factor, whereas the quantity of x-ray production is an *electrical* factor.

Conclusion

In summary, it is worthy of note that the size of the focal spot affects *only* the sharpness of detail in a radiographic image. It does not affect any other aspect of the radiographic exposure or of the final image. This makes the focal spot *the primary controlling factor for sharpness of recorded detail* for the projected image reaching the image receptor. The various distances involved in aligning the x-ray tube, patient and image receptor (to be discussed in the following chapters), do have an effect upon sharpness of detail, but they also affect *all* of the other image qualities except shape distortion; they impact upon exposure, subject contrast, noise, and magnification. *The focal spot is only technical factor which exclusively affects sharpness of detail in the image reaching the IR.*

SUMMARY

1. The line-focus principle uses steep anode bevel angles to project an effective focal spot that is much smaller than the actual focal spot. This allows for higher image sharpness, while providing for adequate heat dispersion at the surface of the anode.
2. Because of the line-focus principle, the projected focal spot is smaller toward the *anode* end of the x-ray tube, and produces a sharper image at that end of the receptor plate.
3. The anode heel effect results in less radiation being emitted at the anode end of the x-ray tube. Therefore, anatomy of variable thickness should be positioned with the thinnest portion toward the anode end of the x-ray tube.
4. The anode heel effect is more pronounced with steeper anode bevels, larger focal spots, shorter SIDs, and longer field sizes.
5. The size of the projected focal spot is directly proportional to penumbra, and inversely proportional to the sharpness produced in the remnant beam image.
6. Penumbra is produced by rays from different portions of the focal spot recording the same detail edge in different locations at the receptor. As penumbra spreads, it also invades the umbral shadow of the image detail, causing the umbra to shrink.
7. If an anatomical detail of interest is significantly smaller than the focal spot used, it may not be resolved in the image, disappearing from view.
8. Focal spot size has no effect upon image magnification, distortion, or any visibility function of the image.
9. *The focal spot affects image sharpness exclusively.* It is the only radiographic variable that does so, and therefore should be considered the *controlling factor* for image sharpness for the projected image at the detector. The small focal spot should always be engaged for smaller anatomical parts.

REVIEW QUESTIONS

1. What is the typical range of anode bevel angles for a general purpose x-ray tube?

2. Rather than using the line-focus principle, why don't we just focus down the beam of electrons so that the *actual* focal spot is 0.5 mm?

3. The projected focal spot is largest at the end of the _________ image receptor.

4. Explain why the anode heel effect occurs:

5. X-ray tubes are conventionally installed with the anode to your __________ (left or right) as you approach the x-ray table.

6. For a radiograph of the AP foot or femur, it is best to position the patient on the x-ray table with his/her head to your _________ (left or right).

7. At a shorter SID, the collimator must be opened up to cover the same field size, so the anode heel effect _________ (increases, decreases, or is not affected).

8. What is the most compelling reason to consider focal spot size as the *controlling* factor for sharpness in the remnant beam image?

9. Why is the small focal spot not available when high mA stations are engaged?

10. The *actual* focal spot is best measured along the:

11. Explain how a focal spot which is much larger than an anatomical detail can cause it to disappear from the resolved image:

(Continued)

REVIEW QUESTIONS *(Continued)*

12. If the size of the focal spot is cut exactly in half, the sharpness of the image carried by the remnant x-ray beam will change which way and by precisely how much?

13. Where within the penumbra does the human eye locate the edge of an image detail?

14. Geometrical penumbra is caused by the _________ absorption of x-rays as a function of the total absorption which the anatomical structure is capable of.

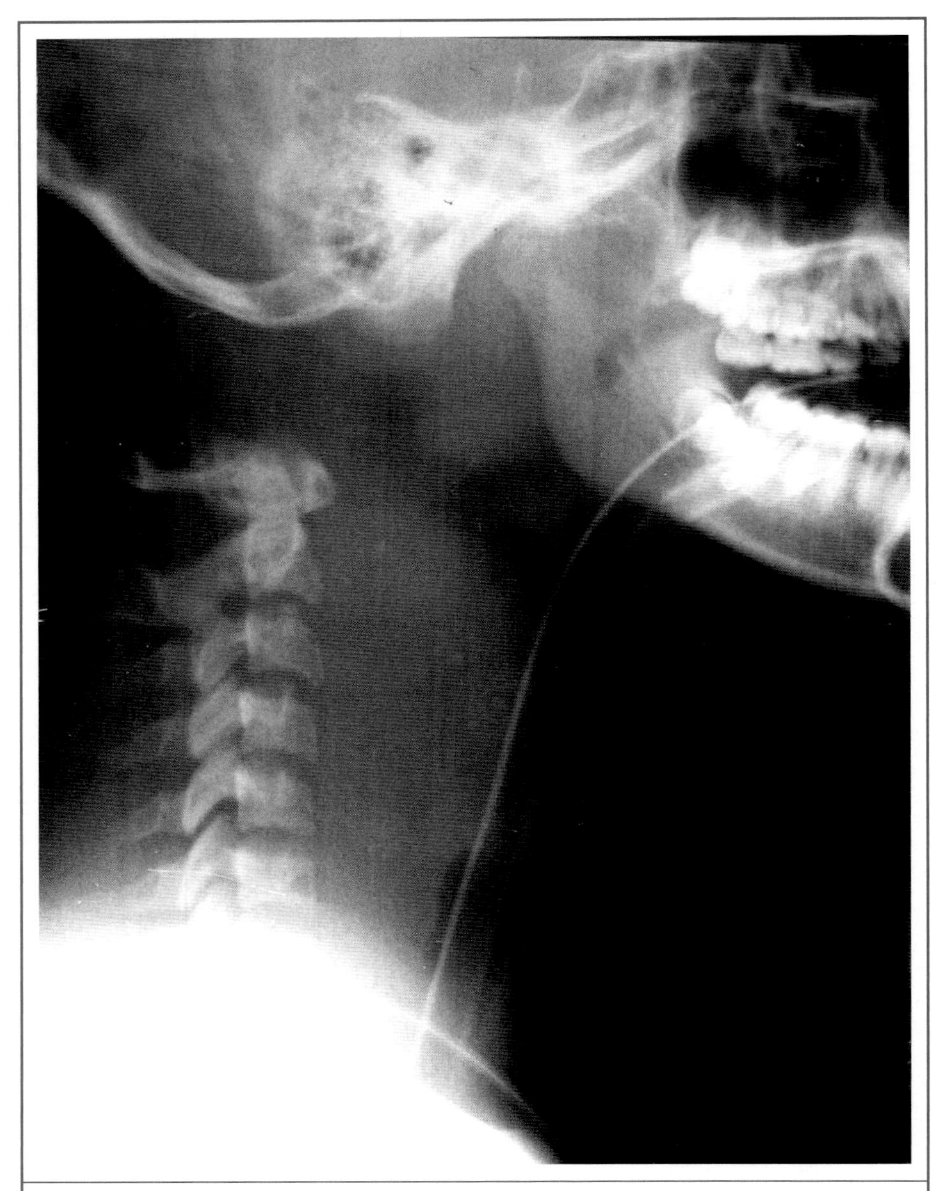

This unfortunate victim, ejected from a car during an accident, had her spinal cord severed. A large gap can be seen between cervical vertebra #1 and the occipital bone.

Chapter 22

SOURCE-TO-IMAGE RECEPTOR DISTANCE (SID)

Objectives:

Upon completion of this chapter, you should be able to:

1. Describe both the preferred methods and back-up methods for measuring or estimating SID.
2. Explain the effect of SID on image penumbra, and on image sharpness.
3. Explain the effect of SID on image magnification.
4. Describe procedures where the intentional use of short SID can be beneficial for diagnosis.
5. Explain why the inverse square law applies to changes in the SID.
6. Explain why any change greater than 10% in the SID should be compensated for with technique, especially for digital images.
7. State the radiographic formula for the inverse square law.
8. Calculate exposure changes using the inverse square law.
9. State the formula for the *square law.*
10. Calculate various technique adjustments using the square law.
11. State the rules of thumb for landmark SID changes, especially between 40" (100 cm) and 72" (180 cm).
12. List those image qualities *not* directly affected by SID.

The distance from the x-ray tube to the image receptor is very important because it has substantial influence on three image qualities—exposure level, magnification and sharpness of recorded detail. Since the source-to-image distance (SID) is also easily manipulated by the radiographer, it has come to be conventionally known as the fourth of the "prime factors of radiography," along with mA, exposure time and kVp.

A red plus-sign or "x" is engraved in the anode end of the x-ray tube housing to mark the exact vertical location of the focal spot. This is the precise point from which SID and SOD should be measured. Most fixed x-ray machines have electronic indicators or locks that click in at 40" (100 cm) and at 72" (180 cm) from the x-ray table and chest board bucky trays. Still, when performing a wheelchair chest or stretcher (gurney) chest, another method will be needed to accurately measure distance.

Some collimators have a tape measure built into them which is already calibrated to the location of the focal spot. Others may have a tape measure connected to the side. When these are available, they should be used for the sake of accuracy. *This is especially true for mobile radiography.* When a tape measure is *not* available, there are some clever ways to *estimate* the SID.

A human's "wingspan" from fingertip to fingertip with the arms fully outstretched is very close to the person's height. For example, if you happen to be close to six feet tall, you can estimate a 72-inch (180 cm) SID by stretching your fingertips from the imaging plate to the middle of the x-ray tube housing. A 40-inch (100 cm) SID extends roughly from the fingertips of *one* outstretched arm to the *opposite* side of the chest. For those of us who are shorter, it is easy to adapt these rules-of-thumb to just about any height by subtracting the difference and finding a part of the collimator that is this distance away from the focal spot. You will simply extend your fingertips from the imaging plate to that identified part of the collimator to estimate 72 inches.

In terms of the geometrical effects of distance on the image, it is generally desirable to use the *maximum feasible SID*. The reason that a 40-inch (100-cm) SID became the convention for x-ray table procedures was simply a matter of how high the average radiographer could comfortably *reach* and see the various indicators, buttons and knobs on the collimator. (The 100-cm SID also makes *one meter*, which is the scientific standard unit for measuring lengths.) Some authorities have advocated long and hard for increasing the standard SID to 44 inches (112 cm), but this very excellent idea has simply never caught on with practicing radiographers.

Figure 22-1

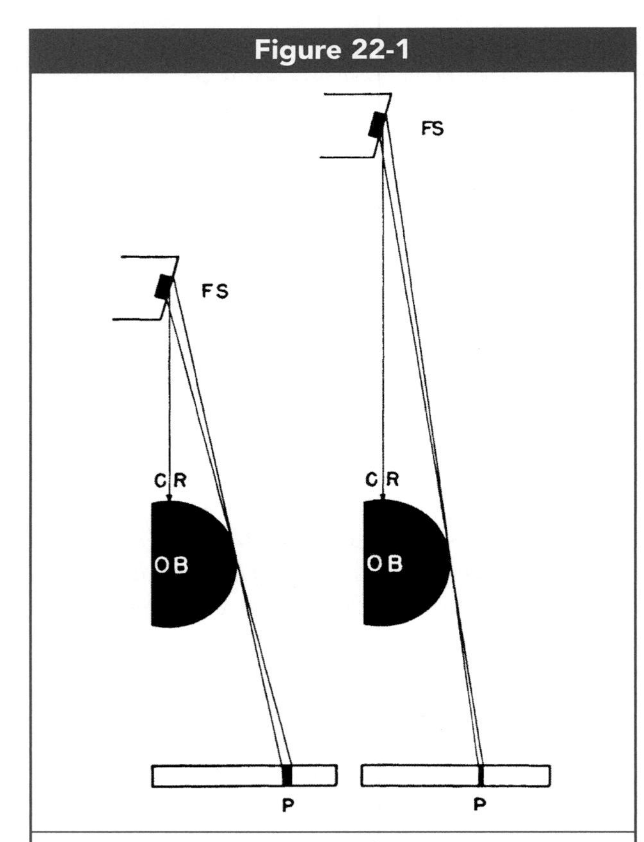

Diagram illustrating that, by projecting the edges of an object with beams more parallel to the CR, a longer SID minimizes penumbral shadows and thus improves sharpness of detail. (From Quinn B. Carroll, *Practical Radiographic Imaging*, 8th ed. Springfield, IL: Charles C Thomas, Publisher, Ltd., 2007. Reprinted by permission.)

As for upright procedures, the 72-inch (180-cm) SID was settled upon because, at this distance, magnification of the heart (at about 3%) becomes statistically negligible. Dedicated chest units, however, can have a locked-in SID of 96 inches (8 feet, or 244 cm), because chest techniques tend to use very low mAs values.

EFFECT ON SHARPNESS

When the x-ray tube is moved farther away from the patient and image receptor, the amount of blur or penumbra produced at the edges of the image diminishes. As illustrated in Figure 22-1, this occurs because those beams recording the edges of the image are nearer to and more parallel to the central ray. Thus, the SID has a direct effect upon image sharpness: *The longer the SID, the sharper the recorded detail.*

Visual proof is provided in Figure 22-2, using a lead-foil line pattern template; as we scan from sets of thicker lines to thinner lines, the first point where resolution of the lines is lost has been marked for each distance. With the longer SID used for image ***B***, it is clear that much smaller line-pairs have been resolved, due to the improvement in sharpness at their edges.

As a rule then, in order to optimize image sharpness, the maximum feasible SID should be used.

It is important to understand that, in reality, it is the *SOD* that is directly responsible for these sharpness effects. As presented in Chapter 13, sharpness is actually determined by the ratio of SOD over OID. Any *calculations* of sharpness or unsharpness must use the SOD rather than the SID. All other factors equal, any increase in SID will always result in an increase in the SOD. The increased SOD, in turn, improves sharpness of recorded detail as demonstrated in Figure 22-2.

EFFECT ON MAGNIFICATION

All other factors equal, *increasing the SID reduces magnification of the image size.* We have defined magnification as an increase in the *umbra* of an

Figure 22-2

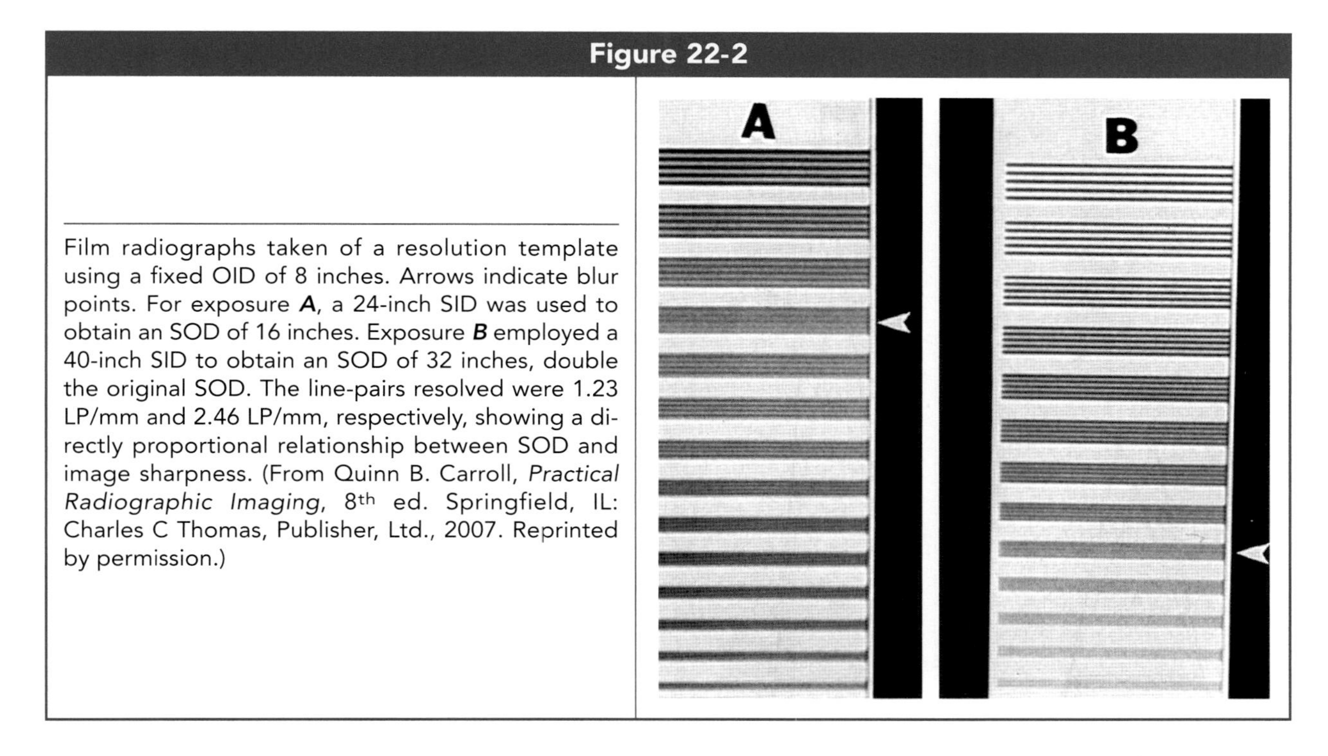

Film radiographs taken of a resolution template using a fixed OID of 8 inches. Arrows indicate blur points. For exposure ***A***, a 24-inch SID was used to obtain an SOD of 16 inches. Exposure ***B*** employed a 40-inch SID to obtain an SOD of 32 inches, double the original SOD. The line-pairs resolved were 1.23 LP/mm and 2.46 LP/mm, respectively, showing a directly proportional relationship between SOD and image sharpness. (From Quinn B. Carroll, *Practical Radiographic Imaging*, 8th ed. Springfield, IL: Charles C Thomas, Publisher, Ltd., 2007. Reprinted by permission.)

image. The penumbra diagram in Figure 22-3 clearly shows that when a longer SID is employed, the *umbral portion* of the image is reduced in size (along with the reduction in penumbra). For all objects which lie at any significant distance from the image receptor plate, magnification will be noticeable (Fig. 22-4). For most radiographic applications, magnification is undesirable. This is especially true for chest radiography where it could falsely simulate an enlarged heart, and for any procedure in which orthopedic *measurements* must be made. For accuracy in measuring *any* anatomy, magnification must be kept at a minimum. Therefore, in this regard also, the goal is generally to use the *maximum feasible SID*.

If a *very thin* object, such as a coin, is placed directly on the imaging plate, we find that the degree of magnification is insignificant *at virtually any SID*. But, clinically, this would be a rare scenario indeed. Most organs and bones of interest lie in the middle of a body part possessing some thickness; these always present some degree of distance to the imaging plate, even when the body part is placed directly on the plate. Also, note that the use of a bucky tray in any x-ray table or chest board places the imaging plate a few inches from the tabletop and the anatomy. In all of these cases, any change in the SID will significantly impact magnification.

Intentional Use of Short SID

It is never desirable for the *anatomy of interest* to be blurred, and generally not desirable for it to be magnified. However, there are times when the blurring and magnification effects of shorter SID's *on structures superimposing the anatomy of interest* can be used to advantage. Short-SID techniques have become something of a lost art in radiography, due partly to exaggerated concerns over the effects on patient exposure, but mostly due to apathy. There are several examples of contrasty anatomy, such as a bone, with considerable *separation* from the anatomy of interest but *superimposing* it in the course of the projection, and thus obscuring it from full view. Several projections of the *skull* fit this description, where a TMJ, mandibular ramus, mastoid tip or air cells (to name a few examples), are partially obscured from bones on the *opposite side* of the skull (even with an angled CR).

In those cases where there is considerable separation between the overlying anatomy and the anatomy of interest, and the anatomy of interest can be placed

Figure 22-3

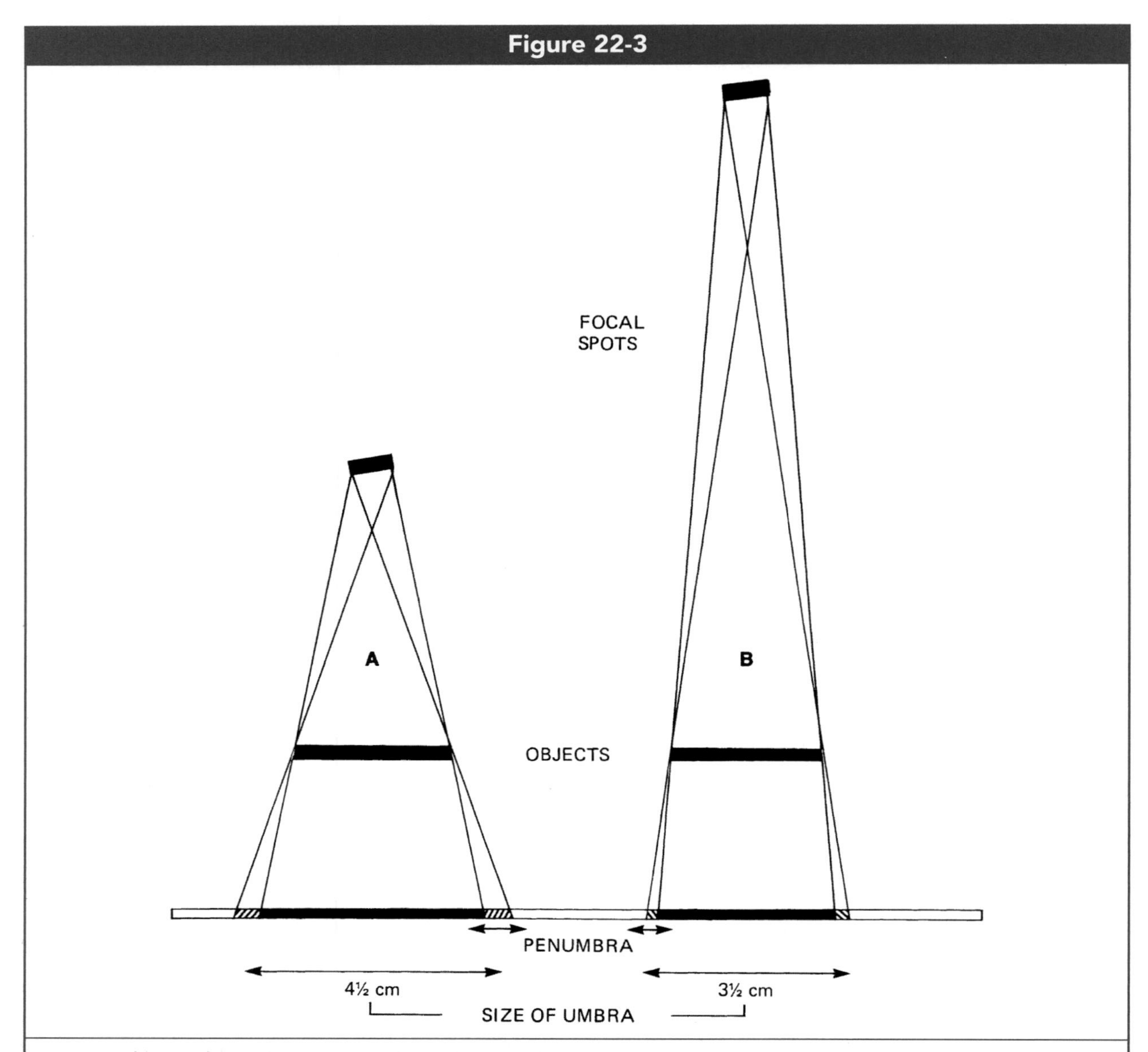

Diagram showing how a longer SID reduces the size of the umbra, (as well as the penumbra), reducing the magnification of the image. ***A*** represents 40 inches SID, ***B,*** 72 inches. (From Quinn B. Carroll, *Practical Radiographic Imaging,* 8th ed. Springfield, IL: Charles C Thomas, Publisher, Ltd., 2007. Reprinted by permission.)

very close to the imaging plate, a short SID of 30 inches (76 cm) will both blur and magnify the obscuring anatomy. This makes it possible to better see *through it* to the anatomy of interest.

There are other examples of practical value for a short SID. In the case of the sternoclavicular joints, the contrasty spine and proximal ribs superimpose the PA projection. Short SID can be *added* to beam angle or part rotation in demonstrating the PA sternum—superimposing posterior ribs will be magnified and blurred, while the sternum itself, close to the imaging plate, is scarcely affected. On the angled projection of the *sigmoid colon,* short SID increases beam divergence, helping desuperimpose loops of bowel.

When a 30-inch (76-cm) SID is used, technique should be compensated by cutting the mAs to approximately one-half, as compared to a 40-inch technique. Studies have shown that even with the

Figure 22-4

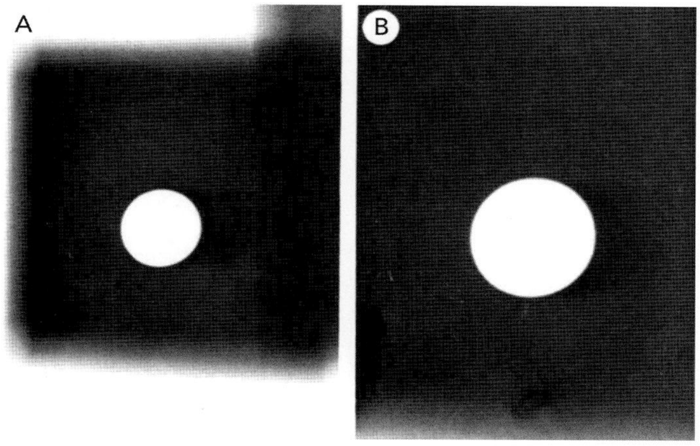

Film radiographs demonstrating magnification of a flat coin with a fixed OID of 4 inches. ***A*** was projected with 40 inches SID, ***B***, with 20 inches. A shorter SID magnifies the image. (From Quinn B. Carroll, *Practical Radiographic Imaging*, 8th ed. Springfield, IL: Charles C Thomas, Publisher, Ltd., 2007. Reprinted by permission.)

compensated technique, a *slight* increase in patient skin exposure may result. As with all other radiologic considerations, this should be weighed against the benefits of obtaining the modified view. There are certainly cases where the diagnostic value of the view obtained with a shortened SID could justify such a small increase (5 or 6%) in patient exposure. The short SID method should be kept in our arsenal of positioning tools, but used only in cases where the anatomical or pathological justification is clear.

Shape Distortion

Shape distortion is a function of the alignment and angling of the x-ray beam and the object being radiographed, as well as the shape of the object itself, and is thoroughly discussed in Chapter 24. Figure 22-4 demonstrates that at a reduced SID, a circular coin is magnified *but maintains its circular shape.* The *shape* of images is not altered as long as the centering and perpendicularity of the beam, part and imaging plate are maintained.

In Chapter 13, shape distortion was defined as a change in the *shape ratio*. Some confusion about distances and distortion stems from the fact that an image which is *already distorted* is more obvious when it is *also magnified*. Consider a spherical object which is being projected onto the imaging plate by an angled beam as an *oval* image. Reducing the SID will magnify the overall image. Both the width and the length of the oval shape will be multiplied by the same proportionate amount. The shape ratio is not changed. It may be more *apparent* to the observer that this image is distorted because of the added magnification. But, the SID change was not the *cause* of the distortion. *There is no direct relationship between SID and shape distortion.*

EFFECT ON EXPOSURE

X-rays diverge as they are emitted from the focal spot, and, proceeding in straight paths, cover an increasingly larger area with *lessened intensity* as they as they travel from their source. This principle is illustrated in Figure 22-5: Note that at an SID of 12 inches, the x-rays have spread out from the focal spot over an area of 4 square inches in plane ***C***. When the distance is doubled to 24 inches to plane ***D***, the x-rays now cover 16 square inches—an area 4 times as great as that at ***C***. It follows, therefore, that the *intensity* of the radiation per square inch at plane ***D*** is only one-quarter that at the level ***C***.

You may recognize this as the *inverse square law* described in Chapter 3, the same law that governs the spreading out of forces such as gravity (Chapter 2), magnetic and electrical forces (Chapter 6), and all kinds of radiation including light, radio waves, etc. (Chapter 5). In all these cases, the *area* over which the force or radiation spreads out increases as the *square* of the distance. Thus, in Figure 22-5,

Figure 22-5

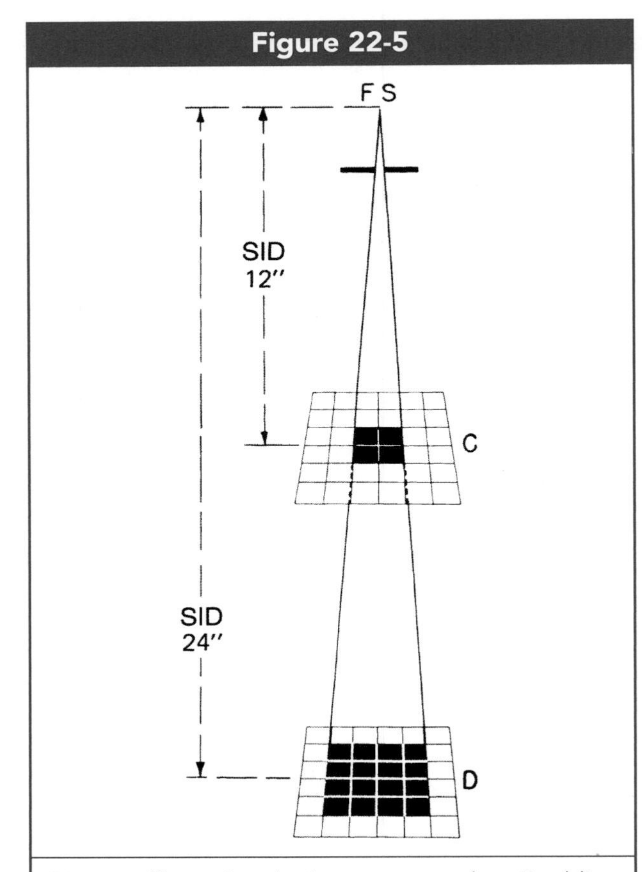

Diagram illustrating the inverse square law. Doubling the distance allows the x-ray beam to spread out over a fourfold area, reducing exposure intensity to one-quarter the original. For further clarification, see Figure 3-5 in Chapter 3. (From Quinn B. Carroll, *Practical Radiographic Imaging*, 8th ed. Springfield, IL: Charles C Thomas, Publisher, Ltd., 2007. Reprinted by permission.)

we see that when the SID increases by a factor of 2, the area increases by 2 *squared* or 4 times. Therefore, the intensity of the radiation has spread out such that its *concentration* is now 1/4 as much per square inch.

The actual effect of the inverse square law on the x-ray beam can be seen in the flash x-ray exposure of a film placed *vertically* in the beam in Figure 21-5 in the previous chapter. As the radiation coursed downward along the film, the intensity of exposure can be seen to diminish as the SID increased. (The difference in density along the horizontal plane from cathode end to anode end can also be seen as the anode heel effect.) Historical Sidebar 22-1 (page 343) gives a visual demonstration of the effect of increasing SID upon exposure at the image receptor, in the form of a progressive lightening of a hand image using conventional film.

With digital image processing, this loss of exposure is not immediately apparent in the brightness of the final image, yet it *does* still represent a loss of original data that can be fed into the computer, and *can quickly lead to visible image noise in the form of quantum mottle.* Remember that at low levels of exposure, the signal-to-noise ratio (SNR) is lowered and mottle becomes visible in the image. For digital imaging systems, *it does not take a very great loss of exposure for mottle to become apparent in the image.*

Therefore, the radiographer must be mindful to compensate the mAs for any significant alteration in the SID. What constitutes a *significant* alteration? Generally, *any change greater than 10 percent in the SID should be compensated for by adjusting the mAs.*

Radiographic Formula for the Inverse Square Law

For radiographic applications, the inverse square law formula can be written in a format that expresses the effect of any change in the SID upon the intensity of x-ray exposure, as follows:

$$\frac{E_o}{E_n} = \frac{(D_n)^2}{(D_o)^2}$$

where E is the x-ray exposure intensity, D is the SID, "o" refers to the *old* or *original* value, and "n" to the *new* value. Thus, this formula states that the ratio of the *old* exposure to the *new* exposure is proportional to the ratio of the *new* distance squared to the *old* distance squared.

Let us apply this formula to the distance change illustrated in Figure 22-5 as an example. In this case, the original distance was 12 inches and the new distance is 24 inches. The original exposure is not given in *R* (roentgen units), so we cannot calculate the actual exposure—but we *can* calculate *relative* exposure for comparison with the original; this is done by assigning the original exposure a value of unity or 1.0. Whenever this is done, the result will be the *factor* to which the original exposure is changed; that is, a result of 3.0 would mean the original exposure, whatever it was, had tripled; a result of 0.75 would mean that the new exposure was three-quarters of the original, and so on. For Figure 22-5, then, the formula would be set up as follows:

Practice Exercise #1

Calculate the *relative* new exposure when changing the SID from 12" to 24", as illustrated in Figure 22-5.

Solution: $$\frac{1}{X} = \frac{(24)^2}{(12)^2}$$

Squaring first: $$\frac{1}{X} = \frac{576}{144}$$

Cross multiplying: $$576(X) = 144$$

$$X = \frac{144}{576}$$

$$X = 0.25$$

Answer: The new exposure will be 25 percent, or one-quarter, of the original. This agrees with Figure 22-5, for there are four times as many squares across which the x-ray beam has spread. Each square must have one-fourth the original concentration of exposure.

Practice Exercise #2

At an SID of 80 cm, the measured exposure for a particular technique is 50 mR. Calculate the new exposure if the SID is changed to 140 cm.

Solution: $$\frac{50}{X} = \frac{(140)^2}{(80)^2}$$

Squaring first: $$\frac{50}{X} = \frac{19{,}600}{6400}$$

Cross multiplying: $$19600(X) = 320{,}000$$

$$X = \frac{320{,}000}{19{,}600} = \frac{3200}{196}$$

$$X = 16.3$$

Answer: The new exposure will be 16.3 mR.

Simple inverse square law problems can be solved mentally by simply applying the *name of the law* to the *proportional change in distance: Invert the change, then square it.* For example, if the distance is to be reduced to one-third the original, what would the resulting change in exposure be? Mentally invert ⅓—this is 3. Now square 3 × 3 for a product of 9. The exposure will be *9 times* greater than the original if the distance is reduced to one-third, and no compensation in technique is made. Try the following exercise in applying the formula, and check your answers from Appendix #1.

EXERCISE #22-1

Use the inverse square law formula to solve for the missing factor: (Note that in #4, you will need to use a calculator to take a square root. See Chapter 3 if you need help setting this up.)

	From	To
1.	40" SID Exposure = 1	72" SID Exposure = ________
2.	128 cm SID Exposure = 1	92 cm SID Exposure = ________
3.	60" SID Exposure = 20 mR	72" SID Exposure = ________
4.	72" SID Exposure = 20 mR	New SID = ________ Exposure = 11.2 mR

COMPENSATING TECHNIQUE: THE SQUARE LAW

Distance changes are normally compensated for by using mAs, although the kVp can be adapted to make the correction by applying the 15 percent rule. Changes in the SID alter only the intensity of the x-ray beam—energy levels and penetrability are not affected. Adjusting the mAs restores the original intensity of the exposure, without changing its quality.

Whereas the inverse square law predicts the resulting exposure from a change in SID, the *square law* is used to compensate the mAs in order to maintain the original exposure level when changes in SID are made. *This* is the problem radiographers face in day-to-day practice, so radiographers have much more occasion to utilize the *square law* than the inverse square law. As indicated by its name, the only difference in the formula between the two is that for the *square law, the distances are not inverted. This only simplifies the math* for the radiographer. The formula for the square law, then, is:

$$\frac{mAs_o}{mAs_n} = \frac{(D_o)^2}{(D_n)^2}$$

It states that the ratio of the original mAs to the new mAs is proportional to the ratio of the squares of the original SID to the new SID. In other words, *the needed change in technique is simply the change in SID squared*. Using Figure 22-5 again as an example, when the SID is doubled, the mAs should be increased by two-squared, or four times. If the original technique had used 20 mAs, 80 mAs would be needed to compensate and maintain the original exposure.

Practice Exercise #3

Using an SID of 40", a radiographic technique of 72 kVp and 10 mAs results in a satisfactory exposure. If the SID is changed to 96", what new mAs should be used to maintain the exposure level?

Solution: $$\frac{10}{X} = \frac{(40)^2}{(96)^2}$$

Squaring first: $$\frac{10}{X} = \frac{1600}{9216}$$

Cross multiplying: $$1600(X) = 92,160$$

$$X = \frac{92,160}{1600}$$

$$X = 57.6$$

Answer: Exposure will be maintained by using 57.6 mAs.

Historical Sidebar 22-1 demonstrates how well the square law works in maintaining the darkness of a conventional radiograph of the hand. Table 22-1 was derived using the square law formula. This table covers a variety of distance changes, and would make a handy pocket-reference for radiographers, especially for mobile radiography.

The square law formula applies to each individual component of mAs—the mA and the exposure time—as well. The formula to solve for a new mA station is:

$$\frac{mA_o}{mA_n} = \frac{(D_o)^2}{(D_n)^2}$$

Practice Exercise #4

Using an SID of 180 cm, a radiographic technique of 62 kVp, 300 mA and 0.05 seconds results in a satisfactory exposure. If the SID is changed to 100 cm, what new mA should be used to maintain the exposure level?

Solution: $$\frac{300}{X} = \frac{(180)^2}{(100)^2}$$

Squaring first: $$\frac{300}{X} = \frac{32,400}{10,000}$$

Cross multiplying: $$32,400(X) = 3,000,000$$

$$X = \frac{3,000,000}{32,400} = \frac{30,000}{324}$$

$$X = 92.6$$

Answer: Exposure will be maintained by using 92.6 mA, which can be rounded up to the 100 mA station.

The formula to solve for a new exposure time when changing the SID, using the letter T to represent the exposure time, is:

$$\frac{T_o}{T_n} = \frac{(D_o)^2}{(D_n)^2}$$

Practice Exercise #5

Using an SID of 40", a radiographic technique of 62 kVp, 100 mA and 0.5 seconds results in a satisfactory exposure. If the SID is changed to 30", what new exposure time should be used to maintain the exposure level?

Solution: $$\frac{0.5}{X} = \frac{(40)^2}{(30)^2}$$

Squaring first: $$\frac{0.5}{X} = \frac{1600}{900}$$

Cross multiplying: $$1600(X) = 450$$

$$X = \frac{450}{1600}$$

$$X = 0.28$$

Answer: Exposure will be maintained by using 0.28 seconds exposure time.

The *factor* by which the overall technique must be multiplied can still be calculated even if the specific mAs is unknown, by using unity, 1.0, as the *old* mAs, set up as follows:

$$\frac{1}{X} = \frac{(D_o)^2}{(D_n)^2}$$

Practice Exercise #6

Suppose the SID is increased from 45 inches to 60 inches. By what factor should the overall technique be

HISTORICAL SIDEBAR 22-1

Figure 22-6

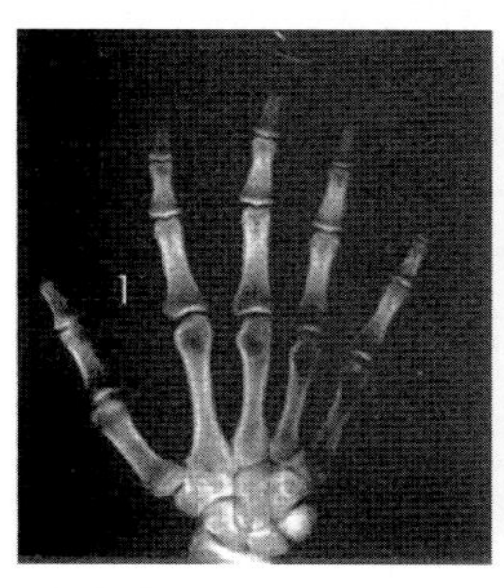

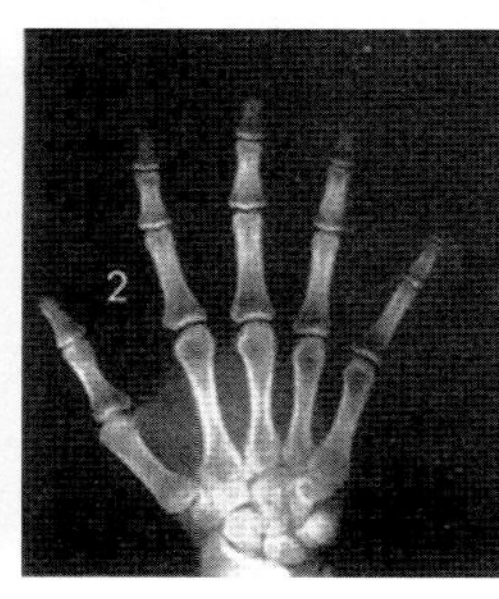

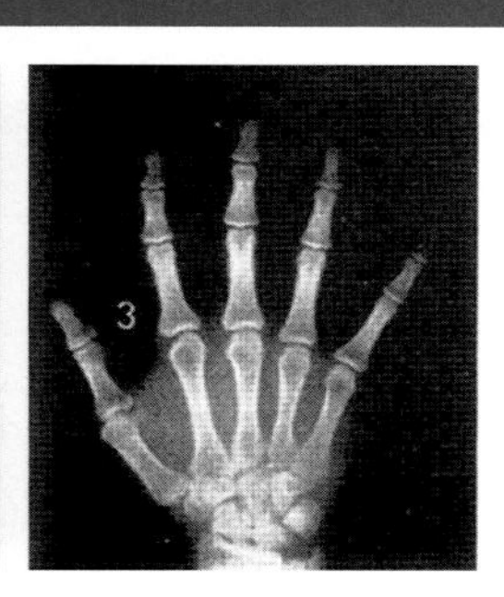

Demonstration of the effect of the *inverse square law* on film radiographs of the hand. ***A*** was taken using 25 inches SID, ***B***, 36 inches, and ***C***, 48 inches, all with the same mAs and kVp. With each increase in distance, the radiographic exposure becomes lighter. (From Quinn B. Carroll, *Practical Radiographic Imaging*, 8th ed. Springfield, IL: Charles C Thomas, Publisher, Ltd., 2007. Reprinted by permission.)

Figure 22-7

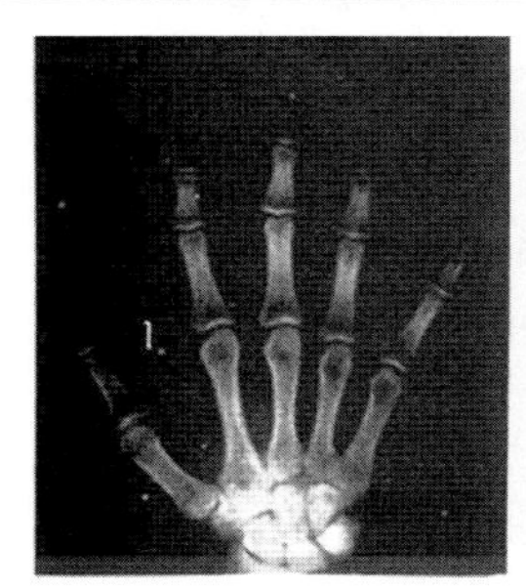

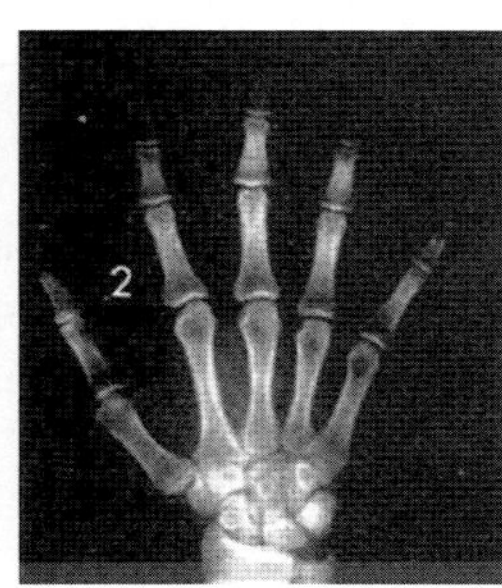

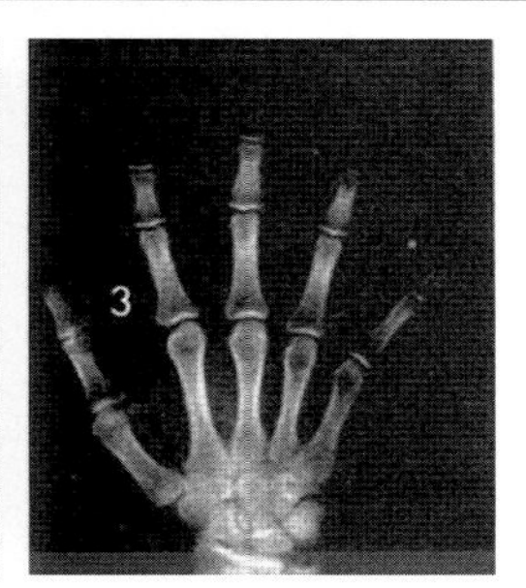

Demonstration of the *square law* maintaining density on film radiographs of the hand. ***A*** was taken using 25 inches SID and 25 mAs, ***B***, 36 inches SID with 50 mAs, and ***C***, 48 inches with 90 mAs, in accordance with the factors listed in Table 21-1, to maintain the same exposure level at the film (image receptor). (From Quinn B. Carroll, *Practical Radiographic Imaging*, 8th ed. Springfield, IL: Charles C Thomas, Publisher, Ltd., 2007. Reprinted by permission.)

increased to compensate and maintain the exposure level?

Solution: $\frac{1}{X} = \frac{(45)^2}{(60)^2}$

Squaring first: $\frac{1}{X} = \frac{2025}{3600}$

Cross multiplying: $2025(X) = 3600$

$X = \frac{3600}{2025}$

$X = 1.78$

Answer: Exposure will be maintained using 1.78 times the original technique.

Try the following practice exercise and check your answers from Appendix #1.

EXERCISE #22-2

Use the square law formula to solve for the missing factor: (Note that in #5, you will need to use a calculator to take a square root. See Chapter 3 if you need help setting this up.)

Table 22-1

mAs Multiplying Factors for Use When SID Is Changed

Original SID	*New SID*								
	20"	25"	30"	36"	40"	48"	60"	72"	
20"	1.00	1.56	2.25	3.22	4.00	5.75	9.00	12.96	*mAs Multiplying Factors*
25"	.64	1.00	1.44	2.07	2/56	3.68	5.76	8.29	
30"	.44	.69	1.00	1.44	1.77	2.56	4.00	5.76	
36"	.31	.48	.69	1.00	1.23	1.77	2.77	4.00	
40"	.25	.39	.56	.81	1.00	1.44	2.25	3.24	
48"	.17	.27	.39	.59	.69	1.00	1.56	2.25	
60"	.11	.17	.25	.36	.44	.64	1.00	1.44	
72"	.08	.12	.17	.25	.31	.44	.69	1.00	
	mAs Multiplying Factors								

From Quinn B. Carroll, *Practical Radiographic Imaging*, 8th ed. Springfield, IL: Charles C Thomas, Publisher, Ltd., 2007. Reprinted by permission.

	From	To
1.	50" SID mAs = 5	60" SID mAs = __________
2.	150 cm SID mAs = 2.5	200 cm SID mAs = __________
3.	96" SID T = 1.2 sec	30" SID T = __________
4.	120 cm SID mA = 200	80 cm SID mA = __________
5.	90 cm SID Exposure = 80 mR	New SID = __________ Exposure = 45 mR

RULES OF THUMB FOR SID CHANGES

Although the formulas for the square law and inverse square law are important to understand, in daily practice, radiographers are rarely seen to pull out a calculator or pencil for this purpose—rather, when doing a mobile procedure at an estimated 60 inches SID, for example, they will likely make a mental estimate of the technique required when compared to a 40-inch SID. Surprisingly accurate rules-of-thumb can be developed to make this process more systematic. They are particularly helpful in mobile radiography.

We shall select a handful of landmark distances (30", 50", 60" and 72"), and compare them to a standard of 40" SID. It helps to think of think of distance changes in factors of two, that is, *sets* of doubling or halving the original SID. We know from the square law, for example, that a doubling from 40" to 80" would require double-squared, or 4 times, the original mAs. By thinking of this factor, 4 times, as *two doublings*, rules-of-thumb can be derived for other distances.

Think of increasing the SID from 40" to 60" as going *half-way to doubling* the SID (half-way from 40" to 80"). Since the 80" distance would require two doublings in technique, a 60-inch SID will require one-half of that increase, or *one doubling* of the mAs. A 60-inch SID requires double the mAs used at 40 inches.

Continuing this extrapolation process, we find all of the other values needed: 30 inches is *half-way to cutting 40" in half*, and requires one *halving* of the 40-inch technique. Fifty inches is *half-way* to 60 inches, and requires *half-way* to a doubling, or 1.5

times, a 50 percent increase. The standard chest board distance of 72" is approximately *half-way* between 60", which required 2 times the technique, and 80", which would require 4 times the technique. The rounded factor is 3 times. Table 22-2 summarizes these rounded-out rules-of-thumb next to the actual calculated value using the square law formula, to show how surprisingly close the rules-of-thumb fall.

The primary advantage of the rule-of-thumb approach is that it facilitates easy memorization. It is well to emphasize in particular the rounded-out ratio between 40 inches (100 cm) and 72 inches (180 cm) which is the standard distance for chest projections. *The relationship between a 40-inch technique and a 72-inch technique is a factor of 3.* For example, when an upright abdomen is ordered in connection with a chest x-ray, it is *not* necessary to change the SID between the two views if the rule-of-thumb has been memorized. If the PA chest projection was first obtained at 72" SID, the upright AP abdomen can be taken afterward also at 72 inches, which geometrically only *improves* both sharpness and magnification. Simply take the recommended technique for a standard 40-inch abdomen and *triple the mAs.* An example going in the opposite direction is provided when a 72-inch chest technique must be adapted for a supine trauma patient on a stretcher—if the x-ray tube can be raised to 40 inches above the gurney, use *one-third* the mAs from a 72-inch technique.

Note that these factors in Table 22-2 can also be applied to any other *proportionate* change between two distances. For example, changing from 30" to 45" is the same proportionate change as moving from 40" to 60" (both a 50% increase in SID), the same factor of doubling the mAs can be applied.

The factors in Table 22-2 can also be adapted for changing between *any two SID's listed*, by forming a simple ratio between the factors listed. For example, the factor for 60" is 2 and that for 72" is 3. A change from 72" to 60" required 2/3 the original technique, while increasing from 60" to 72" requires three-halves, or a 50 percent increase. Complete the following exercise *mentally, using the rules-of-thumb.* Your answers may be checked using Appendix #1. The answers listed are based on Table 22-2, *not* on mathematically solving for the square law formula. The exercise is intended for *mental* practice only.

EXERCISE #22-3

Use *only* the rules-of-thumb from Table 22-1 to solve for the missing factor:

	From	To
1.	40" SID mAs = 25	72" SID mAs = ________
2.	40" SID mAs = 15	60" SID mAs = ________
3.	40" SID mAs = 30	30" SID mAs = ________
4.	60" SID mAs = 20	72" SID mAs = ________

Table 22-2

Rules of Thumb for Adjusting Technique for Changes in SID from 40 Inches

New Distance	Technique Change Computed by the Square Law	Rule of Thumb Technique Change
30" (76 cm)	0.56	½
40" (100 cm)	1 (standard)	1
50" (127 cm)	1.56	1½ × (50% incr.)
60" (152 cm)	2.25	2 ×
72" (180 cm)	3.24	3 ×
80" (200 cm)	4	4 ×

OTHER IMAGE QUALITIES

The intensity of *both* the primary and the secondary radiation in the remnant beam changes as the inverse square of the distance used, so they *both change by the same proportion*. No matter what SID is employed, all other factors equal, the *ratio* of scattered x-rays to primary x-rays will remain the same. Therefore, subject contrast cannot be affected by the SID.

Source-image receptor distance, as a controlling factor of radiation exposure, is similar in its affects to the mAs—both simply control the *quantity* of radiation or intensity of the beam. As with mAs, very *extreme* changes can be forced to a point of affecting subject contrast, but within the ranges of distance changes employed in real practice, SID is just not a consideration for subject contrast.

SUMMARY

1. Electronic locks or a tape measure are provided on most equipment for accuracy in setting the SID. When these are not available, the human "wingspan" can be adapted to make a close estimation.
2. A longer SID reduces penumbra and therefore enhances the sharpness of recorded detail. Increasing the SID extends the SOD, which is directly proportional to sharpness.
3. A longer SID reduces magnification of the image.
4. The intentional use of short (30-inch or 76-cm) SID can be beneficial in desuperimposing contrasty anatomy that obscures the anatomy of interest. This should only be done, however, when the anatomy of interest can be placed very close to the image receptor.
5. Increasing the SID reduces exposure intensity at the image receptor plate by the *inverse square* of the distance, because x-rays spread out isotropically.
6. Any change greater than 10 percent in SID should be compensated for by adjusting technique.
7. The adjustment of radiographic technique for changes in SID follows the *square law*—adjust the mAs according to the distance change ratio squared, or use the 15 percent rule to make an equivalent adjustment in kVp.
8. The rules-of-thumb for distance changes in Table 21-2 should be memorized. When changing between an SID of 40 inches (100 cm) and 72 inches (180 cm), the adjustment factor for technique is 3.
9. To nullify significant magnification of the heart for chest radiography, at least a 72-inch (180-cm) SID should be used whenever possible. Dedicated chest units may be locked into a permanent SID of 96 inches.
10. SID has no direct impact upon image contrast or shape distortion.

REVIEW QUESTIONS

1. Decreasing the SID __________ (increases, decreases) penumbra.

2. How does a short SID benefit the visualization of the sternoclavicular joints?

3. If a change in the SID has resulted in the exposure increasing by *4 times*, all other factors equal, the SID must have been changed by what amount?

4. A technique chart lists 0.05 second for a procedure using 40 inches (100 cm) SID. At a new SID of 80 inches (200 cm), what new time would be required to maintain the exposure at the image receptor?

5. For any change greater than __________ percent in the SID, radiographic technique factors should be adjusted to compensate and maintain exposure.

6. When changing from 72 inches (180 cm) to 40 inches (100 cm), change the mAs to __________ the original.

7. Image sharpness is directly proportional to which radiographic distance?

8. In the interest of image sharpness, generally the __________ feasible SID should be used.

9. As a rule-of-thumb, when changing from a 40-inch (100 cm) SID to a 50-inch (127 cm) SID, increase the mAs by __________.

10. An SID of 128 cm results in an exposure at the receptor plate of 10 mR. Using the inverse square law, if the SID is changed to 92 cm, what will the new exposure be?

11. At an SID of 96 inches, 120 mA results in a satisfactory exposure. If the SID is changed to 30 inches, what new mA must be used to maintain the exposure at the receptor plate?

12. At an SID of 80 inches, 25 mAs results in a satisfactory exposure. If the SID is changed to 36 inches, what new mAs must be used to maintain the exposure at the receptor plate?

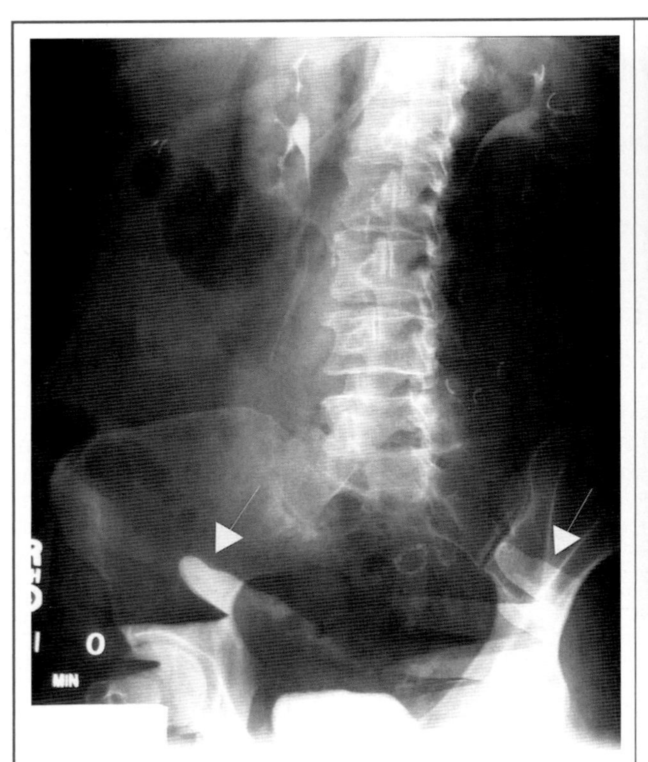

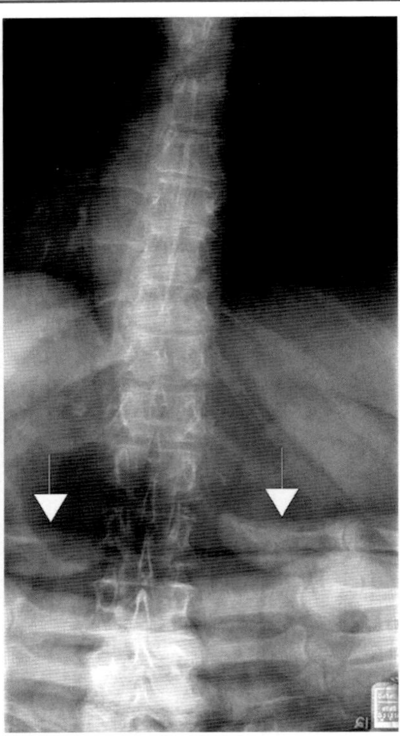

Sets of hands seen superimposing these two radiographs (arrows) are from different causes. The first, (*left*) belong to a dark room technician who fogged the film by leaving it on a counter too long with his hands resting on it before loading it into a cassette for use. The second (*right*) are the patient's hands folded on her abdomen during exposure.

Chapter 23

OID AND DISTANCE RATIOS

Objectives:

Upon completion of this chapter, you should be able to:

1. Describe the effect of OID upon subject contrast in the remnant x-ray beam.
2. Explain how OID affects overall exposure level at the image receptor.
3. Describe how OID impacts the spread of penumbra and image sharpness.
4. Describe the influence of OID on image magnification.
5. Describe how the intentional use of long OID can be beneficial for some procedures.
6. Calculate the magnification for various distance combinations using the SID/SOD ratio.
7. Calculate the relative sharpness for various distance combinations using the SOD/OID ratio.
8. Explain why distance ratios, rather than any single distance, must be used to accurately describe their effects on penumbra, sharpness and magnification.

OBJECT-IMAGE RECEPTOR DISTANCE

The object-image receptor distance (OID) is unique among radiographic variables in that it affects *so many* functions of the image; in fact, the *OID has an impact upon every aspect of the image except shape distortion.* In organizing this chapter then, we shall set aside penumbra diagrams for a moment to examine the effects of OID upon the *visibility* functions in the image: exposure, subject contrast, and noise. We will then return to imaging geometry and discover how the *relationship* of OID to SID and SOD controls both sharpness and magnification in the image.

Effect on Subject Contrast

We first consider subject contrast because it is the one thing that increasing OID has a *positive* impact upon. (See Historical Sidebar 23-1.) The emission of scatter radiation and its influence on the image may be simulated radiographically by the use of a large block of paraffin—a material that generates quantities of scatter radiation similar to soft tissue, and a coin to simulate the *anatomy of interest* to be radiographed. The effects are diagrammed in Figure 23-1 and demonstrated in film radiographs in Figure 23-2.

When the paraffin block is placed directly on the imaging plate with the coin on top (Fig. 23-1, ***A***), we see that scatter radiation produced within the paraffin *undercuts* the object and reaches the image receptor in quantities great enough to constitute significant *noise* in the image. In the corresponding image in Figure 23-2, ***A***, we see this noise made visible as *fog* by the use of film as an image receptor. Modern digital imaging systems are able to "clean up" much of this noise through postprocessing algorithms, but it is still essential in the initial imaging process to *minimize* the amount of noise reaching the image receptor in the first place, so that the computer has the best possible SNR (signal-to-noise ratio) to work with.

Now, in part ***B*** of Figure 23-1, the scatter-producing material along with the object are moved upward away from the image receptor, creating an *OID* gap, and we see that at this distance the scatter radiation is allowed to *spread out* a great deal more before it reaches the image receptor. The result is enhanced contrast of the object (anatomy of interest) as

Figure 23-1

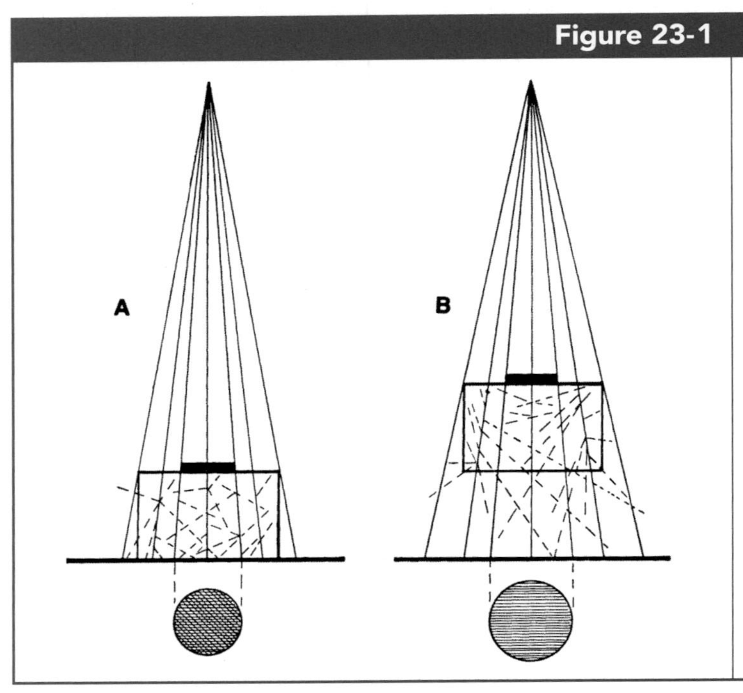

Diagram of experimental set-up for the film radiographs in Figure 23-2. A coin was placed atop a block of paraffin as a scatter emitter, and exposed to x-rays. In ***B***, when the block and coin are moved away from the image receptor, increasing the OID, scatter radiation (dashed lines) is allowed to spread out more, whereas the primary beam geometry remains the same. The ratio of scatter to primary radiation reaching the receptor plate is reduced. (From Quinn B. Carroll, *Practical Radiographic Imaging*, 8th ed. Springfield, IL: Charles C Thomas, Publisher, Ltd., 2007. Reprinted by permission.)

demonstrated in Figure 23-2, ***B***. *By increasing the OID, the intensity of scatter radiation reaching the image receptor diminishes.*

If we propose to prove that the actual *subject contrast* in the remnant x-ray beam is improved by this process, we must show that the *concentration* of penetrating primary beam x-rays has *not* been also spread out by the same proportion—otherwise, the two effects would cancel each other out. For Figures 23-3 and 23-4, we have designated a single square inch of area (shaded black) in the center of the image receptor plate to examine the concentration of radiation striking it.

Figure 23-3 illustrates that, all other factors equal, simply raising the object higher from the imaging plate *does not alter primary beam geometry*—the

Figure 23-2

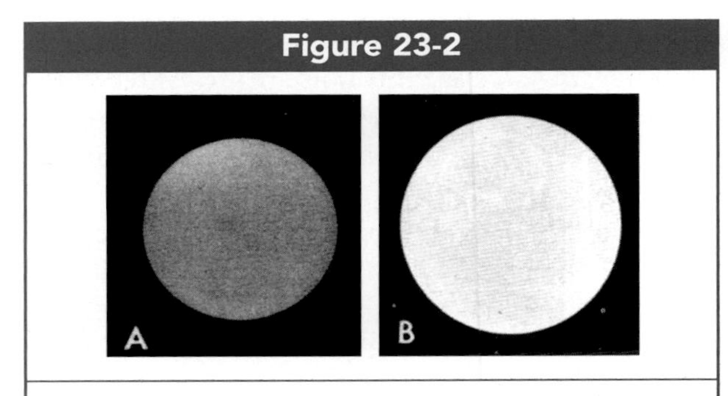

Film radiographs of a coin on a paraffin block as diagramed in Figure 23-1. When the OID is increased from zero, ***A***, to 5 inches in ***B***, scatter fogging of the coin image is seen to diminish, enhancing contrast and reducing density in the coin image. (From Quinn B. Carroll, *Practical Radiographic Imaging*, 8th ed. Springfield, IL: Charles C Thomas, Publisher, Ltd., 2007. Reprinted by permission.)

Figure 23-3

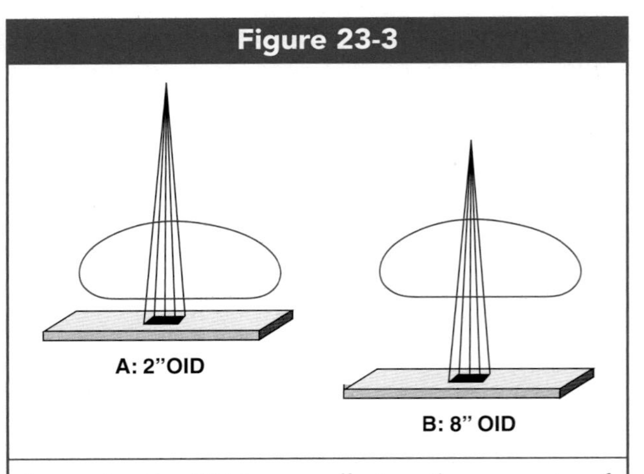

Increasing the OID has no effect on the geometry of the primary beam of x-rays, so their concentration at a selected square inch area, (5 x-rays), remains unchanged.

Figure 23-4

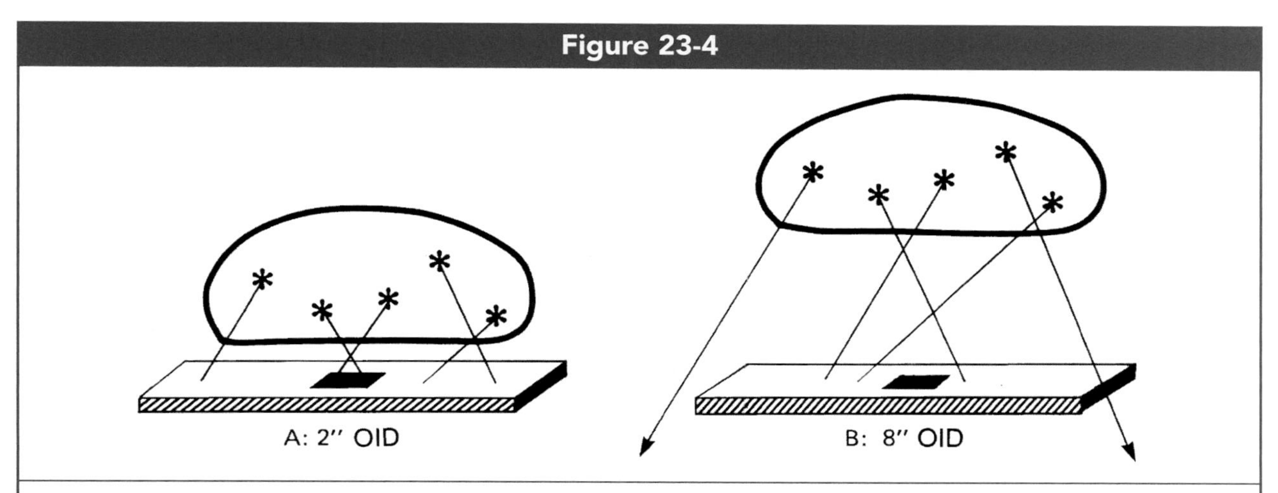

From ***A*** to ***B***, increasing the OID from 2 inches to 8 inches reduces the intensity of *scattered* photons from 2 to zero at the selected square inch area, and from 5 to 3 across the whole area of the receptor plate. The scatter-to-primary x-ray ratio is reduced, so subject contrast in the remnant beam image is enhanced. (From Quinn B. Carroll, *Practical Radiographic Imaging*, 8th ed. Springfield, IL: Charles C Thomas, Publisher, Ltd., 2007. Reprinted by permission.)

HISTORICAL SIDEBAR 23-1: The *air-gap technique* was a method for restoring as much image contrast as possible when *extremely large* patients defeated even the effectiveness of grids because of the amount of scatter radiation produced. Figure 23-5 shows the marginal improvement in a severely-fogged abdomen from increasing the OID to 12 inches. The unfortunate side effects included reduced sharpness and increased magnification in the image—a severely-fogged image was traded for a somewhat fogged and more blurry image!

The postprocessing features of digital imaging systems do a far superior job of restoring image contrast for cases of extreme body size without causing magnification or blurring of the image. The advantages of digital radiography have thus rendered the air-gap technique obsolete.

Figure 23-5

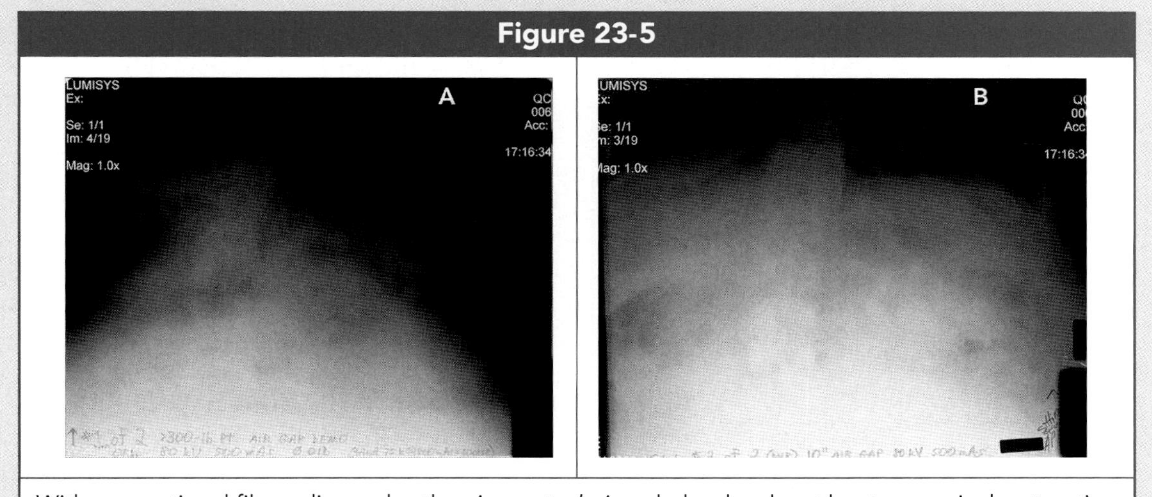

With conventional film radiography, the *air gap technique* helped make at least a marginal restoration of some image contrast, ***A*** to ***B***, when the patient was so large that scatter fogging rendered the image of almost no diagnostic value. ***A*** was taken with zero OID, ***B*** with 10 inches (25 cm) OID by sliding the patient forward seated on a gurney.

concentration of penetrating x-rays at the selected square inch is still *5 primary x-rays per square inch in either case.* Figure 23-4 shows, however, that for scatter radiation, which is produced *in the object* rather than in the x-ray tube, the same change in OID *does* reduce the number of *scatter photons* striking our designated square inch from 2 to 0. (Indeed, when the imaging plate *as a whole* is examined in Figure 22-4, we see that some scattered rays that originally struck it now miss the plate entirely.) The concentration of scatter noise has been reduced, while the concentration of penetrating primary rays has been preserved. *The ratio of primary-to-scatter radiation has been improved. Subject contrast in the remnant beam is increased.*

Effect on Exposure

Recall that the *total* exposure reaching the image receptor is the sum of both primary radiation and scatter radiation. Any reduction in scatter radiation, such as is achieved by increasing the OID, will also result in a reduction in the total exposure. This will not be readily apparent on digitally-processed images, but is clearly demonstrated using film images in Figure 23-2, where the increased-OID image in ***B*** is lighter.

An interesting point about scatter radiation is that even though it is emitted in *random* directions, it still follows the *inverse square law* just as the primary x-ray beam does. We can state, then, that as the patient is moved farther away from the image receptor plate, the intensity of scatter radiation reaching the plate is inversely proportional to the square of the *OID*. Each time the OID is doubled, scatter intensity falls to one-quarter the original.

Generally, for geometrical reasons, the OID should always be kept at a minimum. Sometimes in clinical situations, particularly with trauma radiography, a substantive OID cannot be avoided. In order to maintain the full original exposure to the image receptor, a small increase in mAs (35%) is recommended.

Figure 23-6

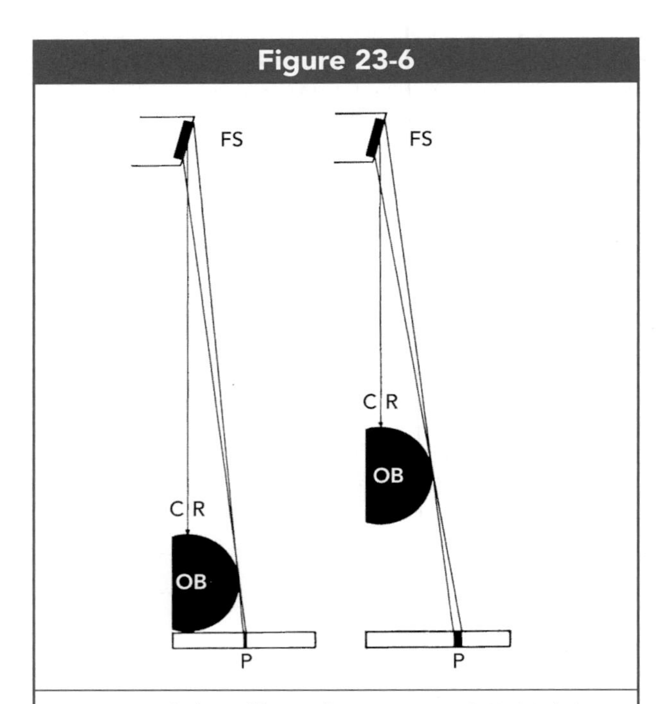

Diagram of the effect of increasing OID (right) on spreading the penumbra, thus reducing sharpness. (From Quinn B. Carroll, *Practical Radiographic Imaging*, 8th ed. Springfield, IL: Charles C Thomas, Publisher, Ltd., 2007. Reprinted by permission.)

Effect on Sharpness

Figure 23-6 is a penumbra diagram showing the effect of OID on the sharpness of recorded detail: As long as the object is placed directly upon the imaging plate, the spread of penumbra is minimal (*left*). When an OID gap is introduced between the object and the plate, the penumbral shadow is spread out, and blur at the edges of the object increases. *The larger the OID, the less the sharpness of recorded detail.*

Observe the series of radiographs of a hand in Figure 23-7; these were taken at increasing OID with the hand in ***A*** placed directly on the plate. The sharpness of the edges of the finger and metacarpal bones in radiograph ***A*** is striking compared to ***D*** where 8 inches of OID was introduced. The use of a bucky tray in the x-ray table or chest board always causes a 2–4" OID and introduces a small amount of image unsharpness.

Effect on Magnification

Some degree of magnification is present in nearly all radiographic projections, because it is rare that the anatomy of interest is so peripheral in the body, and so thin itself, that it can be so positioned as to be considered directly in contact with the imaging plate. Figure 23-8 is a penumbra diagram with somewhat

Figure 23-7

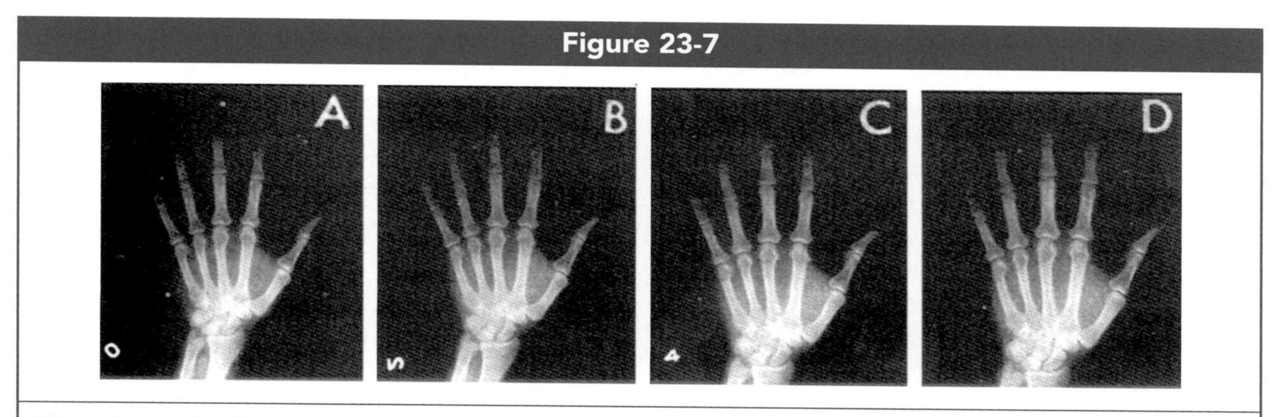

Film radiographs of the hand demonstrate that as the OID increases, magnification and unsharpness at the edges of the bones become apparent. ***A***, zero OID, ***B***, 2 inches OID, ***C***, 6 inches OID, and ***D***, 8 inches OID. (From Quinn B. Carroll, *Practical Radiographic Imaging*, 8th ed. Springfield, IL: Charles C Thomas, Publisher, Ltd., 2007. Reprinted by permission.)

exaggerated OID's to demonstrate the magnification effects on the *umbra* of the image. Remember that for true magnification to be present, the *umbra* must increase in size, not only the penumbra.

A substantial OID is already present in ***A***, Figure 23-8, and a 2.9-cm object is magnified to an umbra that measures 3.25 cm, representing about 14 percent magnification. In ***B***, the OID is doubled, while

Figure 23-8

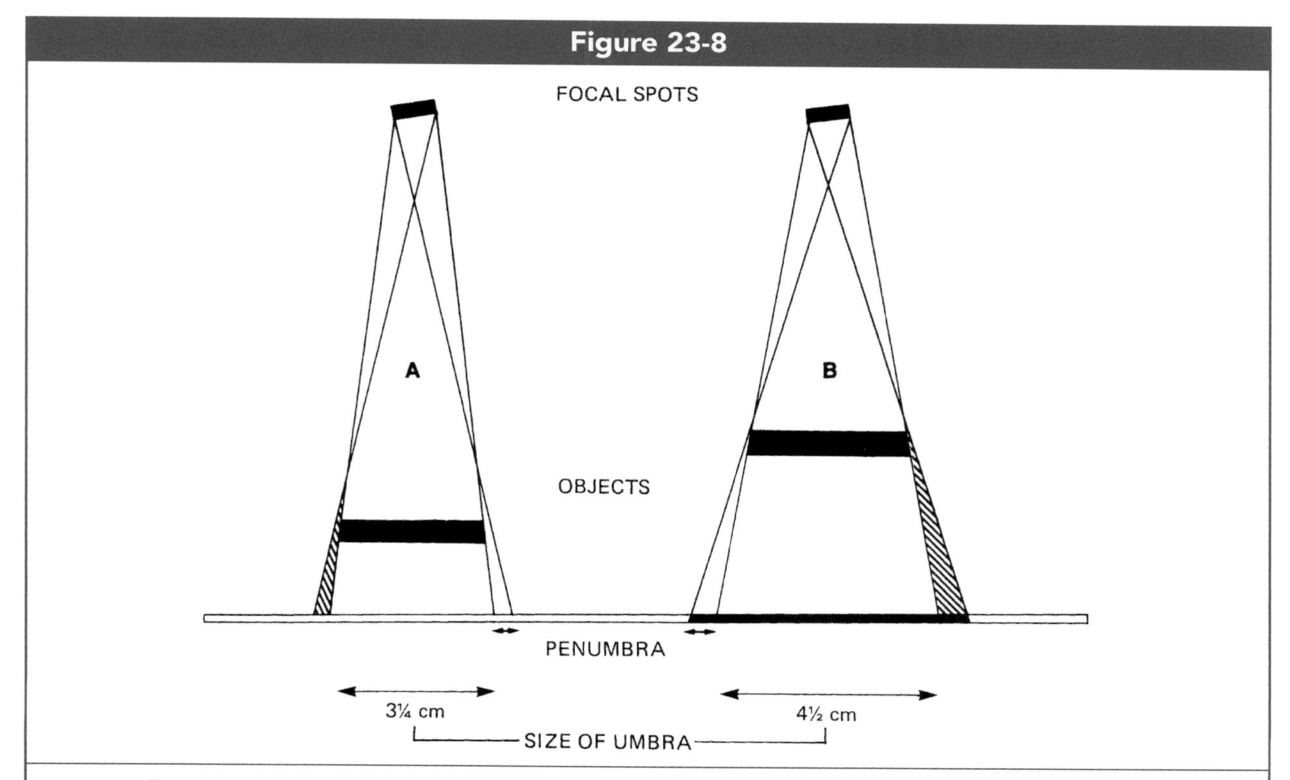

Diagram illustrating the effect of doubling the OID upon image magnification. In ***A***, the *umbra* is magnified by 14%, in ***B*** the umbra is magnified by 54%. (From Quinn B. Carroll, *Practical Radiographic Imaging*, 8th ed. Springfield, IL: Charles C Thomas, Publisher, Ltd., 2007. Reprinted by permission.)

keeping all other factors (focal spot size and SID) equal. The projected umbra of the 2.9-cm object is now nearly 4.5 cm, representing *54 percent* magnification. The series of hand radiographs in Figure 23-7, in order of increasing OID, clearly demonstrate the graduating magnification of the hand image along with the loss of sharpness. *The greater the OID, the greater the magnification of the image.*

The student may have noticed a trend by now that when any of the distances are modified, magnification of the image is accompanied by a loss of sharpness. This pair of effects are so persistently found together that some have assumed the one *causes* the other (magnification causes unsharpness). This is false. Although they are both affected by many of the same geometric variables in a projection, they are *not*, perforce, directly connected with each other. An example follows in the next section.

Intentional Use of Long OID

In angiographic procedures, magnification of small arteries and veins, especially in the head, is often desirable and can be *intentionally* created using OID. Referred to as *magnification radiography* or *macroradiography*, the OID in these procedures can be as great as one-half the SID, resulting in a *doubling* of the image size. Just as with using a magnifying glass, small details are made more *visible* by this process.

As described above, this use of an extended OID also introduces undesirable *blur* into the image. The sharpness can be restored by using a *fractional focal spot size*, which does *not* reintroduce magnification back into the mix. The fact that the focal spot affects sharpness *without* affecting magnification proves that the two, while related, are not the same thing. We might say that, while magnification and unsharpness *usually* go together, they do not *have* to.

Shape Distortion

As with SID, changes in OID do not affect shape distortion. Distortion occurs from improper alignment of the beam, part and imaging plate independent of the distances used. An image can be both shape-distorted and magnified, and the magnification might make the distortion more *obvious* or apparent, but these are different processes. The measured *shape ratio* will not be altered by changes in any distance.

DISTANCE RATIOS FOR MAGNIFICATION AND SHARPNESS

When discussing the geometrical effects of distance on a projected image, the results just cannot be accurately described by considering the any one distance—the SID, the SOD, or the OID—as a separate entity. These three types of distance must be considered *in relation to each other*. This is because they follow the dictates of *similar triangle geometry*, which is entirely based upon the *ratios* formed between corresponding parts of similar triangles, Figures 23-9 and 23-10. (Similar triangles are overviewed in Chapter 13 in connection with Figure 13-25.)

Magnification: The SID/SOD Ratio

Image *magnification* was defined in Chapter 13 as *the ratio between the size of the image and the size of the object* it represents. In the projection diagram, Figure 23-9, the size of the image is represented by M_1, the base of the triangle *XCD*, and the object being projected is represented by M_2, the base of the triangle *XAB*. The ratio M_1 to M_2 should equal the ratio between the distances D_1 and D_2, (the heights of the triangles):

$$\frac{M_1}{M_2} = \frac{D_1}{D_2}$$

By definition, the ratio M_1 / M_2 is the factor of magnification, and may be abbreviated *M*. For radiography, D_1 is the SID and D_2 is the SOD. Making these substitutions, the formula for *radiographic magnification* is derived as:

$$M = \frac{\text{SID}}{\text{SOD}}$$

The controlling factor for magnification in radiography is the SID/SOD ratio. If the SID/SOD ratio is 2, it means that the projected image is twice as large as the projected object. (This is 100% magnification.) If the ratio is 1.5, the image is half-again as large so there is 50 percent magnification.

Note that *OID* is not used in calculating magnification. In Figure 23-9, the OID does not represent any triangle but rather a trapezoid formed by the points *A*, *B*, *D* and *C*. It cannot be used for calculations. Whenever the OID is given in a problem and the SOD is not, you must first subtract the OID from the SID to obtain the SOD, then work the formula.

Practice Exercise #1

With an SID of 30 inches and the object placed 20 inches above the imaging plate, how much will the projected image be magnified?

Solution: First, find the SOD as:

$$SID - OID = 30 - 20 = 10$$

$$\frac{SID}{SOD} = \frac{30}{10} = \frac{3}{1} = 3.0$$

Answer: The magnification factor is 3.0 The image will be magnified 3 times as big as the real object.

Practice Exercise #2

An image measures 8 inches across its width. A 30-inch SID was used along with an SOD of 20 inches. What is the *size of the original object*?

Solution: $$\frac{SID}{SOD} = \frac{30}{20} = \frac{3}{2} = 1.5$$

►*The magnification factor is 1.5. The image size, given as 8 inches, is 1.5 times larger than the real object. Designating the real object as X, then 8 = 1.5 times X:*

$$1.5(X) = 8$$

$$X = \frac{8}{1.5} = 5.3$$

Answer: The real object must be 5.3 inches.

Note that any change in SID, SOD, or OID will be exactly offset by a *proportionate* change in one of the other distances. This underscores the importance of thinking of magnification in terms of the *ratio*, and never in terms of the influence of one particular distance.

For example, suppose that OID is doubled, but the SID is *also* doubled. As shown in Figure 23-11, *this results in the SOD also being doubled.* The SID/SOD ratios for both projections is equal, (40/38 = 80/76). The magnification will be identical as long as the distances are changed proportionately one to another, so that they offset each other.

Figure 23-9

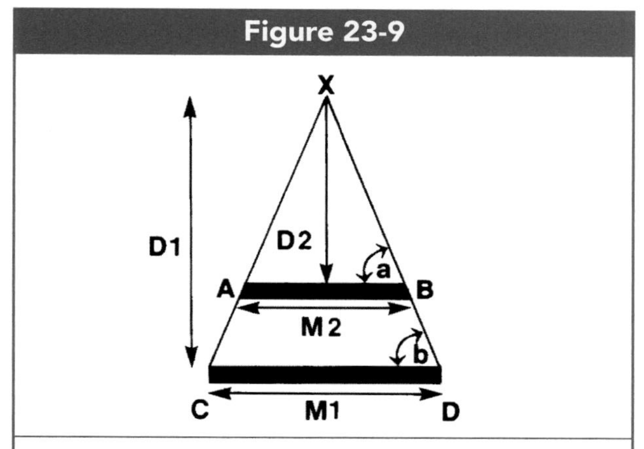

Diagram of similar triangle geometry for magnification: Triangles XAB and XCD are similar, (angle a = angle b). AB represents an object or the aperture of a collimating device. CD is the image or field size at the receptor plate. The ratio D_1/D_2, or SID/SOD, will determine the factor of magnfication M_1/M_2. (From Quinn B. Carroll, *Practical Radiographic Imaging*, 8th ed. Springfield, IL: Charles C Thomas, Publisher, Ltd., 2007. Reprinted by permission.)

Try the following exercise for practice, then check your answers from Appendix #1.

EXERCISE #23-1

Use the SID/SOD formula to solve for the missing factor:

	Width of Actual Object	*Width of Image*	SID	SOD
1.	10 cm	______	80 inches	20 inches
2.	6 cm	______	100 cm	80 cm
3.	______	24 cm	96 inches	72 inches
4.	8 cm	12 cm	60 inches	______
5.	12 cm	18 cm	______	60 inches
6.	8 inches	______	60 inches	OID = 20"

Sharpness: The SOD/OID Ratio

The sharpness of recorded detail in the image is also determined by similar triangle geometry, but in this

Figure 23-10

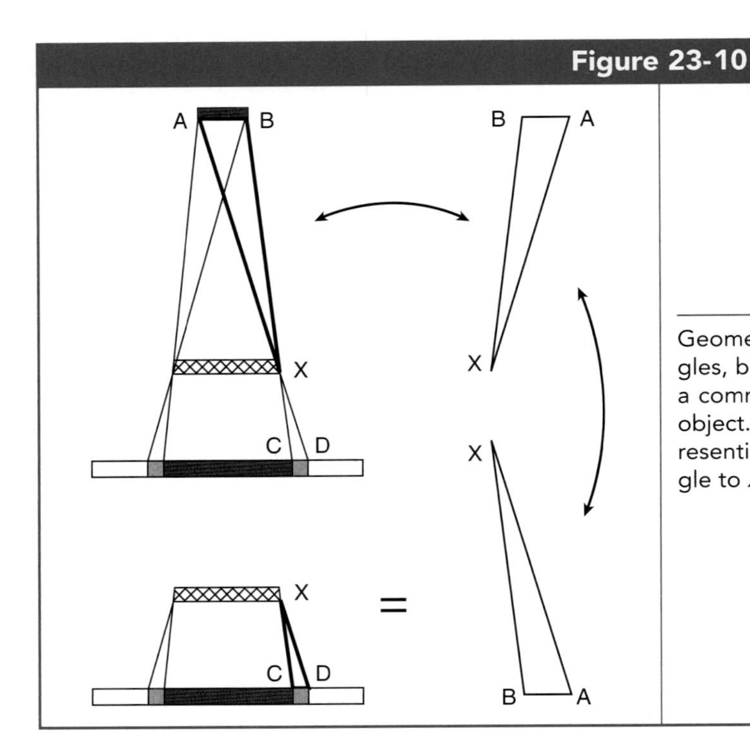

Geometrical sharpness is also based upon similar triangles, but in this case the two triangles are formed with a common apex ***X*** which is the edge of the projected object. By reversing and inverting the triangle ***XAB*** representing the SOD, we can see that it is a similar triangle to ***XCD*** formed by the OID.

Figure 23-11

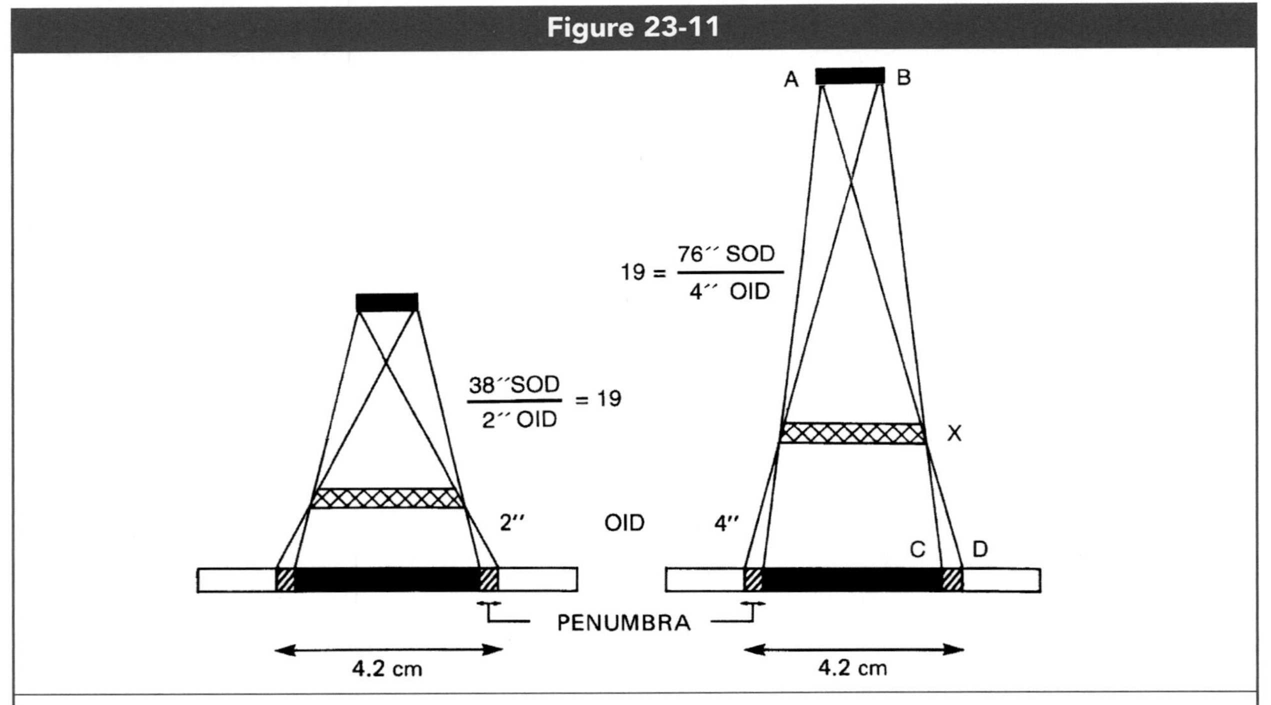

Diagram showing a doubling of the SID, SOD and OID. As long as these distances are all changed by the same proportion, the ratios between them remain unchanged, and there is no change in either sharpness or magnification of the image. Relative sharpness (SOD/OID) = 38/2 = 76/4 = 19 for both images. Magnification (SID/SOD) = 40/38 = 80/76 = 1.05 for both images. (From Quinn B. Carroll, *Practical Radiographic Imaging*, 8th ed. Springfield, IL: Charles C Thomas, Publisher, Ltd., 2007. Reprinted by permission.)

case the pair of triangles is formed by the penumbra diagram at *each edge* of the object being projected; as shown in Figure 23-10, triangle *XAB* and triangle *XCD* are inverted to each other with their common apex, *X*, being the edge of the projected object. The height of triangle *XAB* represents the source-to-object distance (SOD). The height of triangle *XCD* represents the object-image receptor distance (OID).

Sharpness of recorded detail is proportional to, and controlled by, the ratio of SOD/OID.

Observe Figure 23-11; if the OID *alone* were doubled, the spread of penumbra at the imaging plate at the bottom of the diagram would also double. Doing the math from the formulas for unsharpness and for relative sharpness (Chapter 13), we would find that *unsharpness* or *blur* had doubled, and conclude that *sharpness of detail* had been cut in half. However, in Figure 23-11 (*right*), we see that the *SID* has also been doubled, resulting in a doubling also of the SOD. On the *left*, the SOD/OID ratio is 38"/2" = 19. On the *right*, the SOD/OID ratio is 76"/4" = 19. The distances, having been changed proportionately, offset each other exactly. The sharpness of recorded detail is therefore maintained equal.

In Chapter 13, we explained how to calculate for actual unsharpness, or penumbral spread, in millimeters or other small units of length. We also introduced the concept of *relative sharpness* values, which can be used in comparing different configurations of the distances. *The relative sharpness of an image is dependent upon the SOD/OID ratio.* The following practice exercise illustrates how relative sharpness values can be applied and interpreted.

Practice Exercise #3

Both using the same focal spot, two techniques are compared for their relative sharpness. For technique *A*, an 80 cm SOD was used with a 20 cm OID. For *B*, an SOD of 100 cm was combined with an OID of 40 cm. Which combination will produce the greater sharpness, and by how much?

Solution:

$$\textit{Relative Sharpness for A} = \frac{80}{20} = 4.0$$

$$\textit{Relative Sharpness for B} = \frac{100}{40} = 2.5$$

$$\textit{Ratio between them} = \frac{4.0}{2.5} = 1.6$$

Answer: Technique A will be 1.6 times sharper than Technique B.

For practice, calculate the relative sharpness for each of the following, and check your answers in Appendix #1.

EXERCISE #23-2

Use the SOD/OID formula to solve for the value of relative sharpness:

	SOD	OID
1.	60"	15"
2.	40"	2"
3.	80 cm	15 cm
4.	72"	4"

Visibility Functions and Distance Ratios

Distance ratios are not directly related to the visibility functions in the image. However, it is interesting and useful to note the additive or canceling effects of increasing or decreasing both the SID and the OID together.

Regarding overall exposure to the image receptor, since increasing SID reduces exposure, and increasing OID *also* reduces overall exposure (due to the lessened *scatter* component), an exaggerated loss of exposure occurs when both distances are increased. When we examine subject contrast, we find that increasing the SID has no affect on it, whereas lengthening the OID increases it; therefore, increasing both distances will have the *net* effect of increasing subject contrast.

SUMMARY

1. The OID impacts every image quality except shape distortion.
2. Increasing the OID enhances subject contrast in the remnant beam image, because scattered x-rays are allowed to spread out while the primary beam retains its intensity.

3. Increasing OID lessens the total exposure at the image receptor.
4. Increasing the OID results in the spread of penumbra and thus reduces the sharpness of recorded detail in the image.
5. Increasing the OID also magnifies the image.
6. Although magnification and unsharpness often share the same causes, they do *not* always occur together.
7. In magnification radiography or macroradiography, an extended OID may be intentionally used, in combination with a fractional focal spot to restore most of the lost sharpness.
8. Magnification of the image is proportional to the SID/SOD ratio, which is its controlling factor.
9. Sharpness of recorded detail in the image is proportional to the SOD/OID ratio, which is its controlling factor.
10. Because both magnification and sharpness are controlled by the *ratios* between distances, changing one distance proportionately to another can cancel out its effects.

REVIEW QUESTIONS

1. In general radiography, the ___________ possible OID should always be used.

2. The greater the OID the ___________ the subject contrast in the remnant beam image.

3. The greater the OID, the ___________ the sharpness of detail in the image.

4. At greater OID, ___________ radiation is allowed to spread out more, while ___________ radiation remains at the same concentration.

5. Magnification is directly proportional to the ___________ ratio.

6. Image sharpness is directly proportional to the ___________ ratio.

7. If the SID is 40 inches and the SOD is 20 inches, the projected image of the object will be magnified by a factor of ___________.

(Continued)

REVIEW QUESTIONS *(Continued)*

8. An original technique uses an SID of 40 inches and an OID of 2 inches. If the OID were increased to 3 inches, what new SID would be required in order to completely eliminate the blurring effects of the OID change?

9. An original magnification ratio is 50 percent. If the SID and OID are both tripled, the new magnification ratio will be ___________.

10. If both SID and OID are increased by 6 inches, what will be the net effect on the exposure intensity at the image receptor plate (increase, decrease, or no change)?

11. If both SID and OID are increased by 10 inches, what will be the net effect on the subject contrast in the image at the receptor plate (increase, decrease, or no change)?

Use the SID/SOD formula for magnification to solve for the missing factor; *note that the OID is given, not the SOD*:

	Width of Actual Object	*Width of Image*	SID	*OID
12.	10 inches	___________	48 inches	8 inches
13.	4 cm	5 cm	___________	6 cm
14.	___________	12 inches	96 inches	8 inches
15.	4 cm	6 cm	96 inches	___________

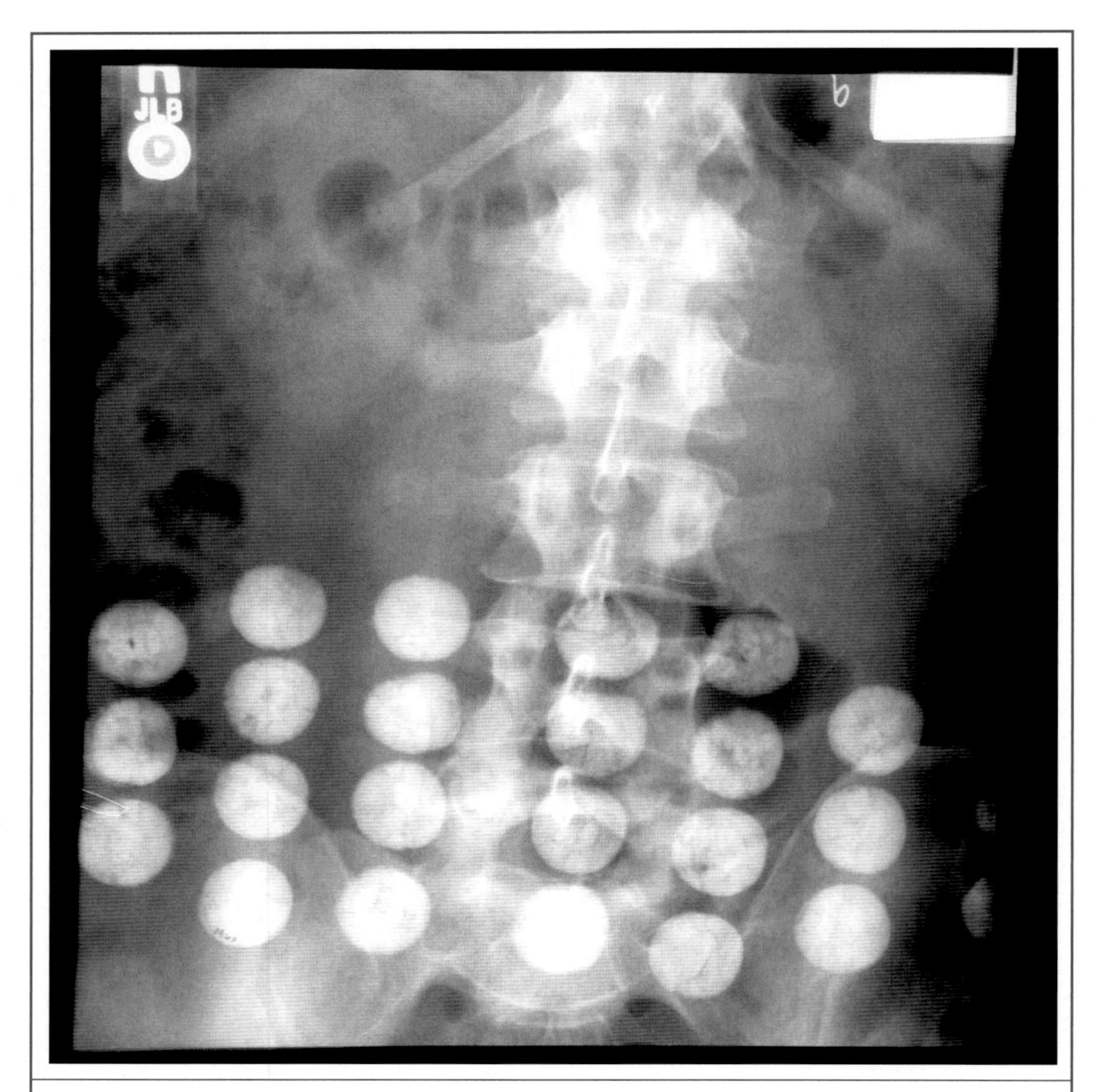

This interesting artifact was caused by a commercial chemical heating pad left around the patient's waist during abdominal radiography.

Chapter 24

ALIGNMENT AND MOTION

Objectives:

Upon completion of this chapter, you should be able to:

1. Explain why off-centering of the beam causes the same types of effects as off-angling.
2. Describe how the shape and size of the object affect the degree of distortion that occurs.
3. Describe the extremes of distortion caused by placing the CR perpendicular to either the IR or the object when the object is tilted in relation to the IR.
4. Apply Ceiszynski's law of isometry in minimizing distortion for a tilted object.
5. Explain why thick objects will always undergo elongation distortion with an angled beam.
6. Give examples of how beam divergence can be taken advantage of to better demonstrate the anatomy for some procedures and situations.
7. Quantify the degrees of beam divergence per inch or per cm away from the CR for standard SIDs.
8. Describe how the beam divergence rule can be used to determine the angle for an L5/S1 "spot" projection from observation of the routine lateral L-spine view.
9. Explain how shorter SID becomes a contributing factor for distortion once it is already occurring due to misalignment.
10. State the rule for compensating tabletop-tube distance according to degrees of angulation, in order to maintain exposure level at the IR.
11. List those image qualities *not* affected by the SID.
12. State the four geometrical objectives of radiographic positioning.
13. Describe why a minimum of two projections must be taken as a general rule for diagnosis.
14. List the three means of controlling the effects of motion.
15. List the types of motion, both for the patient and for equipment.
16. Describe the effect of motion on penumbra and image sharpness.
17. Describe the effects of moderate and extreme motion on image contrast.
18. Define *false images*, and distinguish between the effects of motion and true distortion.

ALIGNMENT AND SHAPE DISTORTION

Alignment of tdahe x-ray beam, the part being radiographed, and the image receptor plate refers both to centering these three in relation to each other, and to any angulation between them outside of the ideal perpendicular relationship. We may say that *off-centering* and *off-angling*, that is, *incorrect* centering and angles, cause shape distortion in the image. However, the effects of beam-part-film alignment on shape distortion are somewhat complex, and there are situations where, for example, an angle on the CR is needed to *reduce distortion*. The extent of distortion depends not only on the degree of off-angling or off-centering, but also on the size and shape of the object itself.

Off-Centering Versus Angling

Off-centering of either the central ray or the part in relation to each other places the part in the *diverging* peripheral rays of the x-ray beam. As shown in Figure 24-1, these peripheral rays angle away from the central ray. The further they are from the CR, the more angled they are in relation to it. Therefore, *off-centering has identical types of effects to angling the beam or part.* In clinical practice, the *degree* of distortion we experience is generally less for off-centering than for off-angling, but this is only because typical off-centering is limited to a few inches, whereas angles are frequently much more extreme by comparison. Although off-angulation is used for all of the following demonstrations, remember that the same effects result from off-centering as well.

Position, Shape, and Size of the Anatomical Part

The traditional rule of keeping the part parallel to the imaging receptor plate and the central ray perpendicular to both the part and the plate will minimize distortion in the image. But, this ideal situation is not always possible. Some cases consist of anatomy which cannot be placed parallel to the receptor plate. Many cases require angles in order to desuperimpose structures overlying the anatomy of interest. For these common situations, shape distortion must be still be minimized, and keeping the beam perpendicular to the imaging plate or to the part is not always applicable.

The thickness of the anatomical part or object being radiographed affects the degree to which it will be distorted by off-centering or off-angling. *The thicker the object, the greater the resulting distortion.* The shape of the object is also critical; objects that are spherical or cubical in general shape will have their projected image distorted under more circumstances, and to a greater degree, than flat, tubular, or wedge-shaped objects. Examples of spherical or cubical anatomy include the cranium, the femoral condyles, the heads of the femur and humerus, the vertebral bodies, and the tarsal bones.

Objects which are flat (such as the sternum and the blades of the scapulae), wedge-shaped (such as the sacrum or anterior teeth), or tubular (such as the shafts of the long bones and the ureters), will often not distort visibly, even with tube angles, providing their long axis is kept parallel to the plane of the image receptor plate.

Objects with a Distinct Long Axis

When a relatively flat, wedge-shaped, or tubular object is kept parallel to the receptor plate, angulation of the central ray may not lead to appreciable distortion (Fig. 24-2). This is because at *any* given ray in the beam, the angle will have caused the SOD to increase by the same proportion as the SID. The SID/SOD ratio is preserved *across the beam,* so there is no magnification of any particular portion of the part, which might cause the *shape ratio* to change.

When a flat, wedge-shaped or tubular object is tilted in relation to the receptor plate, *foreshortening*

Figure 24-1

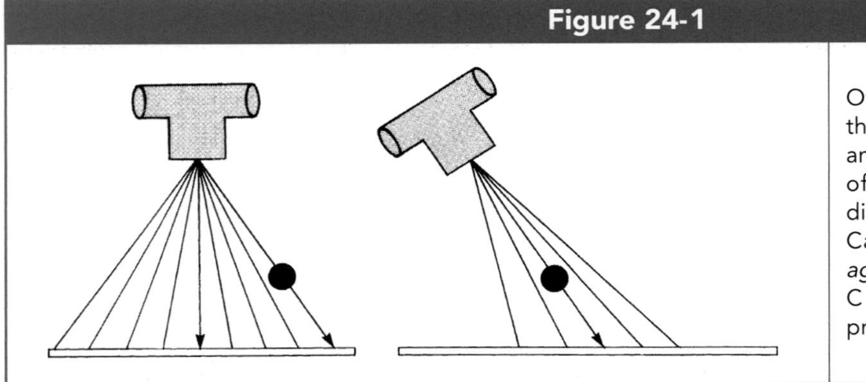

Off-centering of the x-ray beam to the part has an equivalent effect to angling the x-ray beam, (right) since off-centering places the object in diverging rays (left). (From Quinn B. Carroll, *Practical Radiographic Imaging,* 8th ed. Springfield, IL: Charles C Thomas, Publisher, Ltd., 2007. Reprinted by permission.)

Figure 24-2

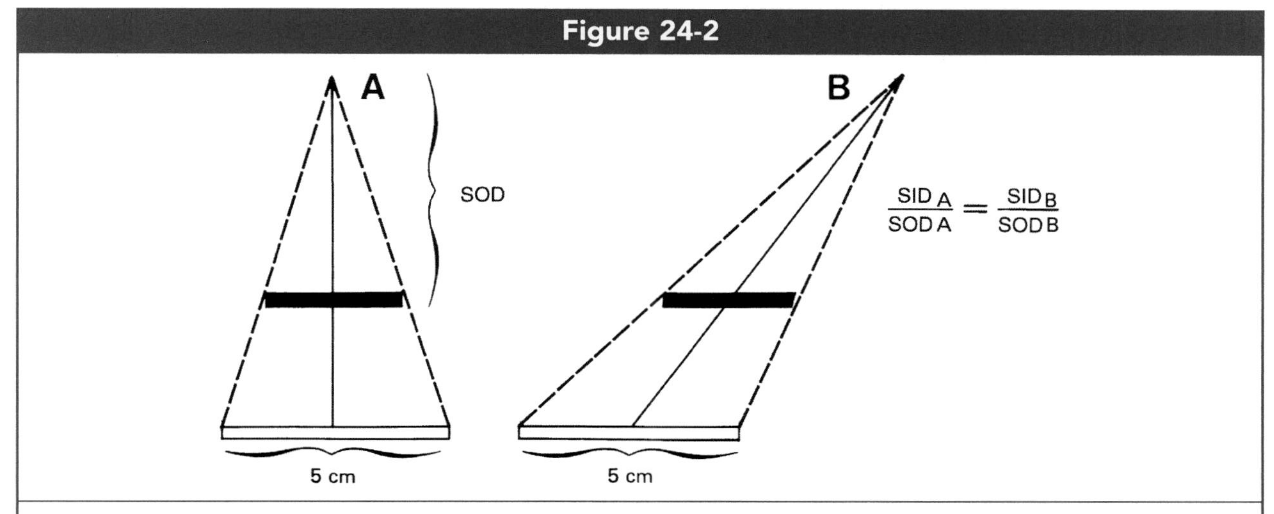

Angling or off-centering the x-ray beam against a thin, flat object may not cause visible shape distortion if the object is kept parallel to the image receptor, since the ratios of SID/SOD *across the image* are maintained. Only relatively thick objects exhibit shape distortion. (From Quinn B. Carroll, *Practical Radiographic Imaging*, 8th ed. Springfield, IL: Charles C Thomas, Publisher, Ltd., 2007. Reprinted by permission.)

distortion will occur, as shown in Figure 24-3. The image recorded is shorter in the axis of the tilt than the real object is. This can cause misleading information on the radiograph. (Since a spherical or cubical object has no single, identifiable long axis, its orientation in relation to the receptor plate is inconsequential.)

Now, let us assume that a flat, wedge-shaped or tubular object is tilted in relation to the receptor plate, and the central ray is angled also to place it perpendicular to the long axis of the object (Fig. 24-4). *Elongation distortion* will occur in the axis of tilt and angulation. Misinformation is again recorded in the radiographic image.

Ceiszynski's Law of Isometry

Since tilted objects present distortion regardless of whether the central ray is placed perpendicular to the object itself or perpendicular to the image receptor plate, foreshortening on the one hand and elongation on the other, minimal distortion of the image must occur somewhere between these two opposite effects, that is, at some angle between these two extremes. *Ceiszynski's Law of Isometry* states that an *isometric* angle, equal to one-half the angle formed between the long axis of the object and the plane of the receptor plate, will eliminate or minimize distortion effects.

This rule is widely known in *dental radiography*, as it must be used for all *bitewing* types of projections of the incisor and cuspid teeth. The bitewing film

Figure 24-3

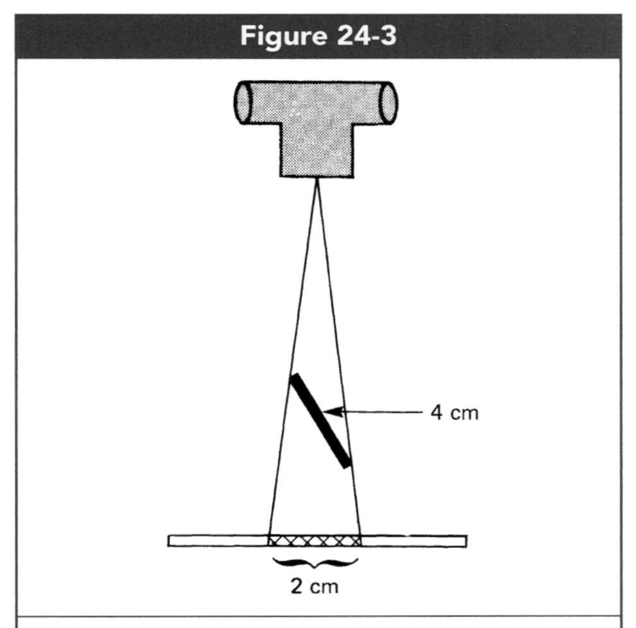

Foreshortening distortion is caused by tilting the object while the CR and image receptor are kept perpendicular to each other. (From Quinn B. Carroll, *Practical Radiographic Imaging*, 8th ed. Springfield, IL: Charles C Thomas, Publisher, Ltd., 2007. Reprinted by permission.)

Figure 24-4

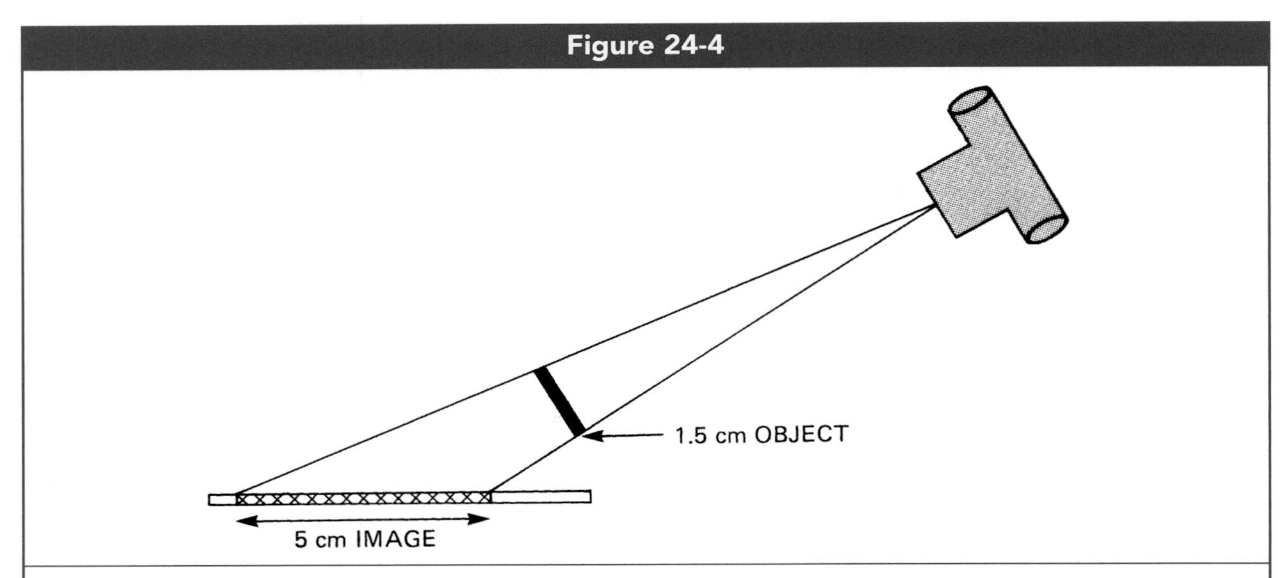

Elongation distortion is caused when the CR is directed perpendicular to an object that is tilted in relation to the image receptor. (From Quinn B. Carroll, *Practical Radiographic Imaging*, 8th ed. Springfield, IL: Charles C Thomas, Publisher, Ltd., 2007. Reprinted by permission.)

Figure 24-5

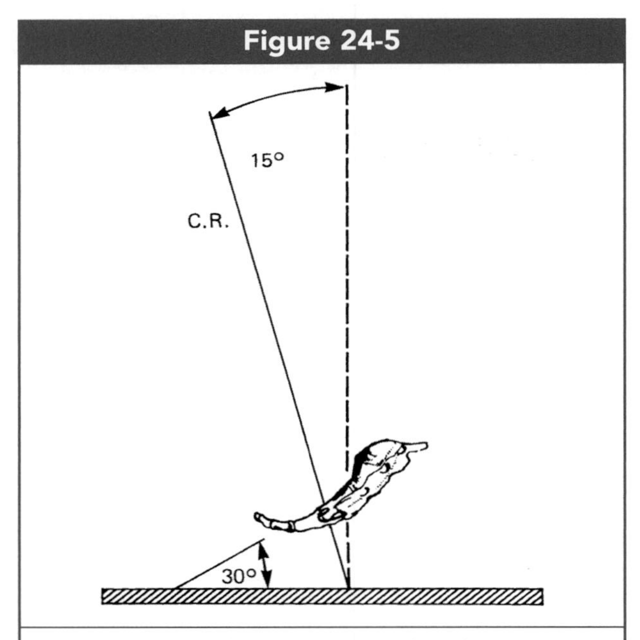

Ceiszynski's law of isometry applied to the AP projection of the sacrum: To minimize shape distortion of an object that cannot be placed parallel to the image receptor, angle the CR one-half of the part/receptor angle. Since the sacrum lies at a 30-degree angle to the receptor plate, the CR should be angled 15 degrees. (From Quinn B. Carroll, *Practical Radiographic Imaging*, 8th ed. Springfield, IL: Charles C Thomas, Publisher, Ltd., 2007. Reprinted by permission.)

cannot be placed parallel to the long axis of the tooth because it runs into the gums and roof of the mouth. For the upper teeth, the dental x-ray tube is pointed downward at *one-half the angle* formed between the tooth and the film.

In diagnostic radiography, a classical example of this same situation is found in the AP projection of the *sacrum*. The sacrum is a relatively flat, wedge-shaped object which normally lies at an angle of 30–35 degrees from the imaging plate when the patient is lying supine. No feasible repositioning of the patient can correct this angle. Note that positioning atlases *do not recommend a 30–35 degree angle for the AP sacrum, but rather a 15-degree angle*, as illustrated in Figure 24-5. This *isometric* angle, at one-half the anatomical angle of the sacrum, minimizes shape distortion in the image. (It also avoids projecting the pubic symphysis over the sacrum, so there are *two* compelling reasons to use 15 degrees.)

Actual results from all of these projection scenarios for a relatively flat object are demonstrated in Figure 24-7, ***A*** through ***E***, using a coin for a flat object. In ***A***, the object is parallel and the CR perpendicular to the image receptor. ***B*** shows foreshortening due to angling the object only, and ***C*** shows elongation due to angling the CR perpendicular to the tilted coin. ***D*** is of particular note: The coin was angled 45 degrees,

At 40 inches (100 cm) SID, moving away from the central ray in any direction results in approximately 2 degrees of beam divergence per inch at the tabletop.

(This divergence is equivalent to a 2-degree angle for those x-rays passing through joints which lie one inch away from the CR, a 4-degree angle for anatomy two inches away, and so on.)

At 72 inches (180 cm) SID, moving away from the central ray in any direction results in approximately 1 degree of beam divergence per inch at the tabletop. Note, for easy memorization, that this is one-half of the rule for 40 inches (100 cm).

Using the metric system, the measured divergence comes out to about 0.7 degrees per centimeter using a 100 cm SID, and 0.4 degrees per cm using a 180 cm SID. Although it requires a little more stretch of the imagination to round these values up, a functional and helpful rule may be derived: *At 100 cm SID, beam divergence is very roughly 1 degree per centimeter from the CR; at 180 cm SID, beam divergence is about 1 degree for every 2 cm.*

A beautiful example of the practical value of the divergence rule is found again in lumbar spine radiography. A common positioning rule-of-thumb states that if the L5-S1 joint space is nicely opened on the *routine lateral L-spine view*, a caudal angle should be employed when centering directly over this joint for the L5–S1 "spot" view. This idea is based on beam divergence, and works well; yet, the amount of angle indicated is frequently *underestimated.* By using the beam divergence rule, this amount can be quantified with a fair degree of confidence as illustrated in Figure 24-10.

For a 10 × 12 inch (25 × 30 cm) view, centering for the routine lateral L-spine is the mid-body of L3, about one inch above the iliac crest. On average, the L5–S1 joint lies about 3½ inches below this point. The actual amount of beam divergence at this point on the lateral view can be found by multiplying this distance by 2 degrees per inch. This comes to about 7 degrees of beam divergence. That is, when the CR is perpendicular and centered over L3, the rays passing through the L5–S1 joint are *angled caudally about 7 degrees.* If the joint space is nicely opened on this view, the indicated angle for a "spot" lateral projection centered over L5-S1 itself is 7 degrees. (Note

Figure 24-9

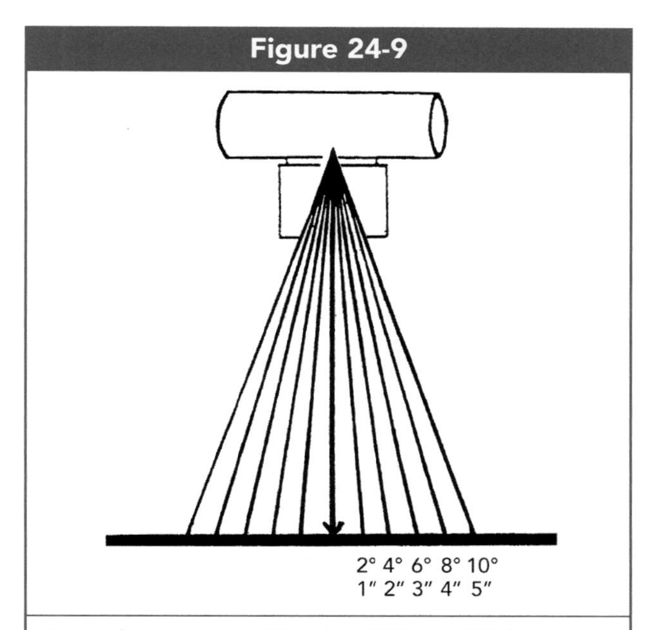

Beam divergence at 40 inches (100 cm) SID is approximately 2 degrees per inch (1 degree per cm) in any direction away from the CR. At 72 inches (180 cm) SID, it is approximately one-half this amount.

Figure 24-10

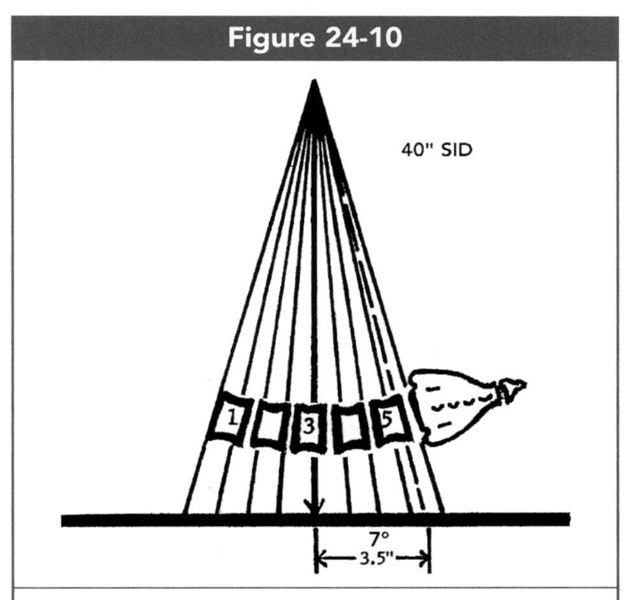

Example of applying the divergence rule: If the L5-S1 joint space is open on the *routine lateral* projection of the L-spine, then the indicated angle for the *spot lateral* projection is approximately 7 degrees caudal. (The common "5-degree" angle would be underestimated.) (From Quinn B. Carroll, *Practical Radiographic Imaging*, 8th ed. Springfield, IL: Charles C Thomas, Publisher, Ltd., 2007. Reprinted by permission.)

that this is *more* than the 5 degrees some radiographers routinely use. As we have stated, the tendency is to *underestimate* the needed angle to open these intervertebral joints spaces.)

SID as a Contributing Factor

In the case of off-centering objects in the beam, using a shorter *SID* can exaggerate distortion effects which are already present in the projection. When a shorter SID is used the radiographer must often open the collimator or otherwise increase the field size to avoid cutting off anatomy of interest. In doing so, objects that are placed at a given distance away form the central ray will be recorded at the receptor plate by more peripheral beams that are more divergent, or *more angled away* from the CR (Fig. 24-11). In effect, these objects have been angled against more, and if they are relatively thick objects, they will be distorted more as illustrated by a set of mugs in Figures 24-12.

Therefore, because the degree of *beam divergence* increases at greater distances from the CR, the *off-centering of objects causes more severe distortion effects at short SID's than at long distances.* Nonetheless, it must be remembered that distances are never the *cause* of distortion. Any object centered to the CR will merely be magnified, not distorted, by a shorter SID. The cause of the distorted mugs in Figure 24-12 is misalignment in the form of *off-centering.* Once distortion is being generated, the SID can be an exacerbating factor.

Maintaining Exposure: Compensating Tube-to-Tabletop Distance

Angling the central ray, part or receptor plate does not directly affect the intensity of exposure. However, angling the CR *without* compensating the tube-to-tabletop distance (TTD) results in an increased SID (Fig. 24-13). The longer SID then leads to a loss of exposure intensity in accordance with the inverse square law (Chapter 22). It is this change in *distance,* not the angle itself, that causes the lessened exposure.

Of course, compensation is not necessary for very minor angles. A good rule is to *compensate the tube-to-tabletop distance (TTD) any time that the CR is angle more than 15 degrees.* An angle of 15 degrees increases a 40 inch (100 cm) SID to just under 42 inches (about 104 cm). CR angles greater than 15 degrees will alter the actual SID by more than 5 percent if they are not compensated for.

The exact amount of compensation in the TTD for increasing degrees of CR angulation can be geometrically calculated. This has been done, and the results rounded to formulate the following conventional wisdom:

> *For every 5 degrees of CR angulation, reduce the tube-to-tabletop distance (TTD) by 1 inch (2.5 cm).*

By following this rule, the actual SID from the focal spot to the center of the image receptor will be maintained at 40 inches (100 cm). Any unexpected loss of exposure intensity will be prevented.

It is worth noting that, due to the same geometry, angling the CR may extend the distance the x-ray beam has to penetrate through the *body part,* effectively making it thicker, which further reduces the remnant exposure reaching the receptor plate. However, this is not a factor we have any control over.

Other Image Qualities

The student will note that all of the diagrams for distortion in this chapter demonstrate the effects of alignment changes in *one axis only*—the axis of the angle or off-centering. In the opposite axis, crosswise to the shift, no change occurs. This is confirmed by comparing the *crosswise* width of the femoral head in Figure 24-7, ***F*** and ***G,*** which does not change even while elongation distortion occurs. Since we have defined magnification as in increase in the *total size* of the image across *both* axes, it may be stated that alignment of the beam, part and image receptor is *not* directly related to magnification.

Close examination of these same images, ***F*** and ***G*** in Figure 24-7, reveals no visible change in the sharpness of recorded details such as bone marrow trabeculae. Alignment changes result in a proportionate change for both SID and SOD, such that the SID/SOD ratio and the SOD/OID ratio are both maintained. As long as this ratio remains equal, and there is no change in the focal spot, neither sharpness nor magnification in the image will be affected.

Alignment of the beam, part, and image receptor is strictly geometrical in nature. It has no bearing

Figure 24-11

When shape distortion is *already occurring* due to off-centering, the SID becomes a contributing factor for shape distortion, because a reduced SID (left) places the object in more divergent beams. However, SID is not a *cause* of shape distortion. (From Quinn B. Carroll, *Practical Radiographic Imaging*, 8th ed. Springfield, IL: Charles C Thomas, Publisher, Ltd., 2007. Reprinted by permission.)

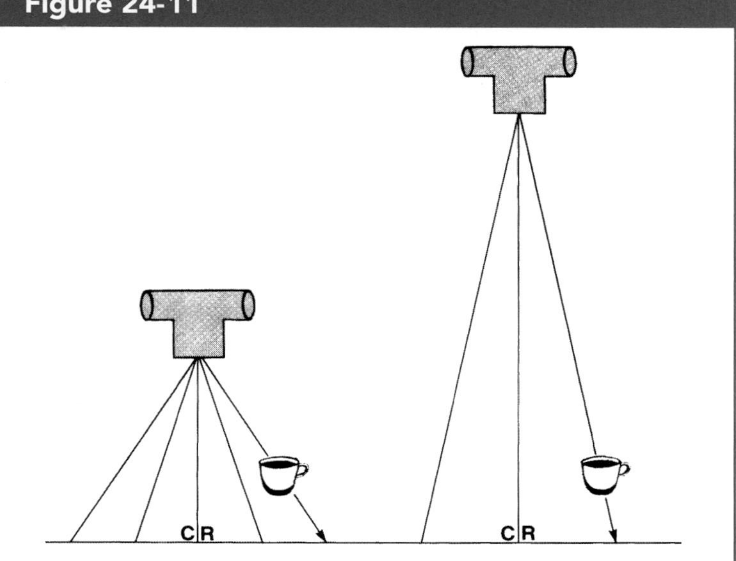

Figure 24-12

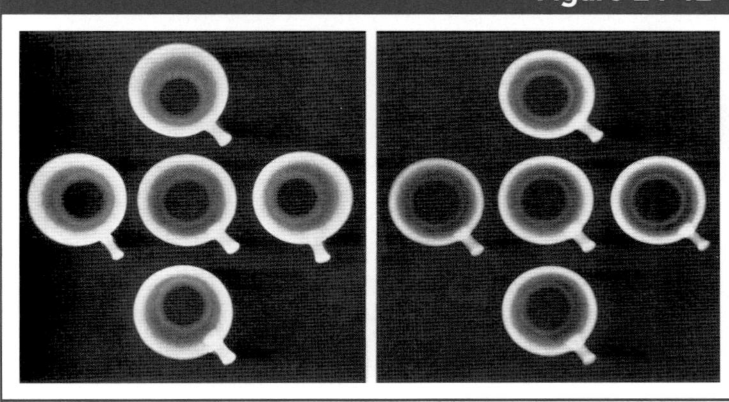

Radiographs of several mugs demonstrating the principle from Figure 24-11. At reduced SID of 30 inches (left), note that the darker area within each peripheral cup is off-centered, indicating shape distortion, while this effect is reduced at 72 inches SID (*right*). The *center cup* is not significantly distorted at either distance, because it is not placed in substantially diverging beams. (From Quinn B. Carroll, *Practical Radiographic Imaging*, 8th ed. Springfield, IL: Charles C Thomas, Publisher, Ltd., 2007. Reprinted by permission.)

Figure 24-13

Angling the x-ray tube without compensating the tube-to-table-top distance (TTD) causes an increase in the SID, ***B***, which can lead to a loss of exposure at the image receptor.

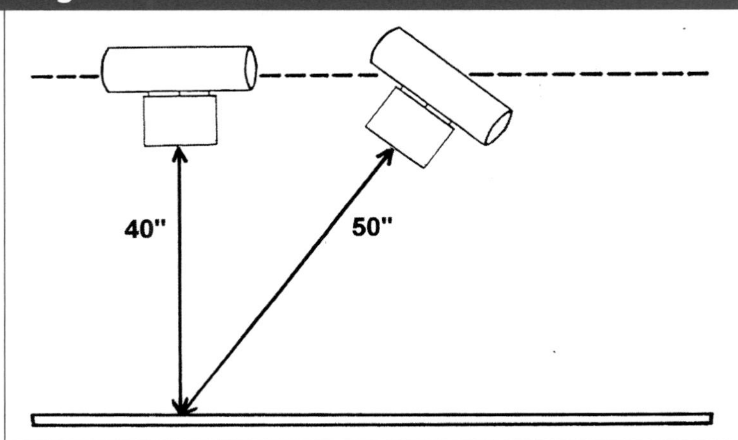

upon the proportions of scatter radiation generated within the patient or upon the penetration characteristics of the x-ray beam. It cannot, therefore, have any relationship to subject contrast.

GEOMETRIC FUNCTIONS OF POSITIONING

Radiographic positioning, at its most fundamental level, is a geometrical factor affecting all of the recognizability functions in an image: sharpness of detail, magnification, and shape distortion. It also bears upon the one visibility function of *noise*, in that the unwanted radiographic "shadows" of overlying anatomical structures are desuperimposed from the anatomy of interest by body part rotation, flexion, extension, abduction, adduction or tilt, and by manipulation of the CR angle and centering point (alignment).

Every specific position has as its purpose one of the following objectives:

Figure 24-14

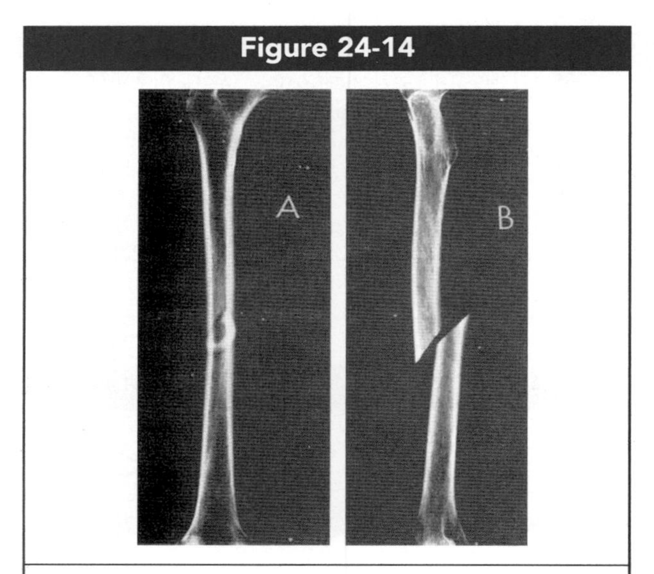

Radiographs of a femur with a simulated displaced fracture dramatically emphasize the necessity of obtaining a second projection at right angles to the first, ***B***, such that the displacement can be fully evaluated in three dimensions. The displacement is not demonstrated in radiograph ***A***. (From Quinn B. Carroll, *Practical Radiographic Imaging*, 8th ed. Springfield, IL: Charles C Thomas, Publisher, Ltd., 2007. Reprinted by permission.)

1. Increase sharpness of recorded detail by placing the anatomy of interest closer to the image receptor.
2. Reduce magnification by placing the anatomy of interest closer to the image receptor (or intentionally cause magnification by the opposite course of action).
3. Minimize distortion of the shape of the anatomy of interest by optimizing the alignment of the anatomy with the x-ray beam and receptor plate.
4. Increase the visibility of the anatomy of interest by desuperimposing other overlying, contrasty anatomical structures which are not of interest and therefore constitute a form of *noise*.

Projection routines for each type of diagnostic study help standardize the quality of images and consistency of results, and are strongly recommended. Groups of radiologists should strive for consensus in developing these routines for the radiographers to follow, rather than each having different customized procedures.

To accurately ascertain the size, shape and location of foreign bodies, the displacement of fractures and the condition and extent of pathological processes, radiographic projections should generally be made from different directions. The need for multiple views is clearly demonstrated in Figures 24-14 and 24-15. In routine radiography, a minimum of two views at right angles to each other are made.

Figure 24-14 graphically demonstrates how the displacement of a fractured bone cannot be ascertained by the single frontal projection, ***A***, where the alignment looks normal. In Figure 24-15, a single frontal projection, ***A***, results in the demonstration of a foreign body only as a circular metallic object, whereas projection ***B*** brings out its profile shape as a regular bullet.

MOTION

Motion of a body part during a radiographic exposure blurs the sharpness of image details (Fig. 24-16). Movement can be voluntary or involuntary on the part of the patient. The only means for controlling the effects of motion are (1) patient cooperation, (2) immobilization of the part being examined, and (3) short exposure times.

Voluntary motion includes not only general movement, but breathing motion which can normally be controlled by the patient, and esophageal peristalsis which occurs during swallowing.

Involuntary motion is mostly associated with the physiologic activity of body organs. Blurring from movement of the heart and great vessels can only be subdued by very short exposures. The same is true for peristaltic movement throughout the gastrointestinal tract. A peristaltic wave in the stomach usually lasts 15 seconds or more. Peristalsis in the small intestine is about 10 centimeters per second. The gall bladder and bile duct exhibit rhythmic contractions that can last from 5 to 30 minutes. In the urinary tract, peristaltic waves move down the ureters from each kidney to the bladder at a frequency of 3 to 6 contractions per minute.

Nor is motion limited to the patient; when the x-ray tube has not been properly locked into place after centering for a projection, it is not unheard-of to find it drifting during the exposure. Image plates can sometimes be shaken by the vibration of the motors or grid in the Potter-Bucky diaphragm. The effects on the image are the same regardless of the source of movement.

Figure 24-15

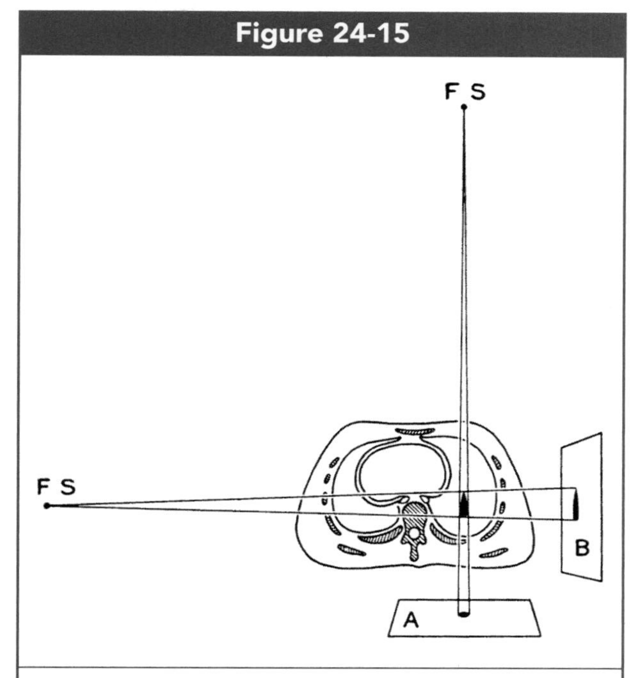

Projection ***A*** demonstrates only a circular metallic object in the image. The true nature of the foreign body (as a bullet) is only ascertained by an additional second projection taken at right angles to the first. (From Quinn B. Carroll, *Practical Radiographic Imaging*, 8th ed. Springfield, IL: Charles C Thomas, Publisher, Ltd., 2007. Reprinted by permission.)

Effect on Sharpness

Motion is the greatest enemy of image sharpness. Besides being the most common cause of blurring, only small amounts of movement wreak havoc with image sharpness, whereas fairly substantial changes in the focal spot and distances are required to do the same. The influence of motion in producing image unsharpness is diagrammatically shown in Figure

Figure 24-16

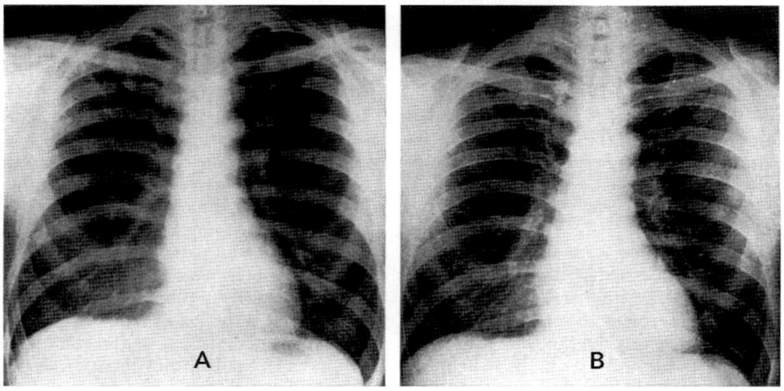

A
B

Radiographs of the PA chest demonstrate ***A***, blurring of pulmonary and mediastinal anatomy due to movement, compared to a properly exposed image, ***B***. (From Quinn B. Carroll, *Practical Radiographic Imaging*, 8th ed. Springfield, IL: Charles C Thomas, Publisher, Ltd., 2007. Reprinted by permission.)

24-17. In this illustration, it can be seen that the movement of the object during exposure extends the spread of *penumbra* at the margins of the image. The resulting appearance for radiographs of the hand is featured in Figure 24-18, ***B*** and ***C***.

Immobilization is imperative in radiography. The radiographer is responsible for using good communication skills to secure full cooperation from each patient, and for exercising good professional judgment as to when immobilizing devices should be used and which type of device is most appropriate for each specific situation. Available devices include special clamps, compression bands, sponges and sandbags.

After every effort is made at immobilization, some movement can still be expected with inebriated patients and many pediatric cases, where it becomes imperative to minimize the exposure time. As a rule, *to freeze motion, exposure time should not exceed 0.033 (1/30) second.*

Figure 24-17

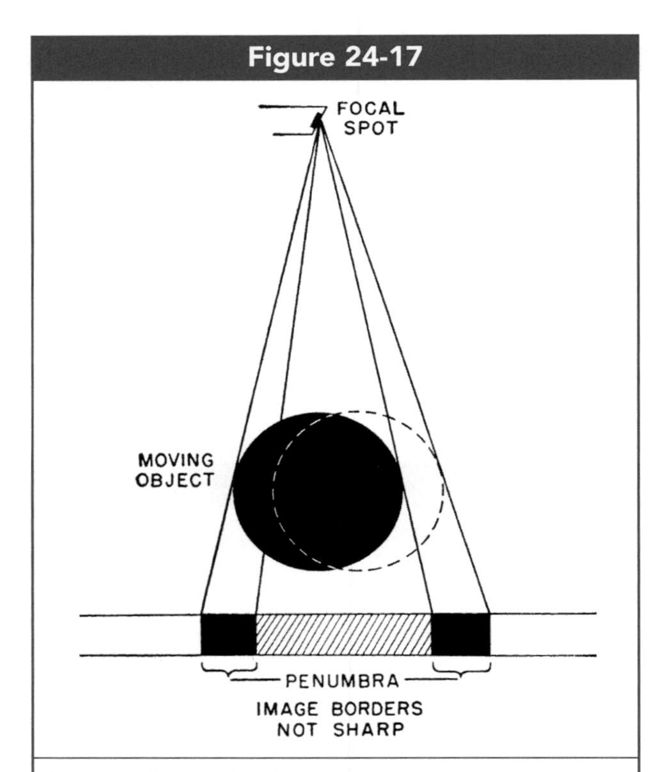

Diagram illustrating how motion causes spreading of the penumbra, producing unsharpness in the image. (From Quinn B. Carroll, *Practical Radiographic Imaging*, 8th ed. Springfield, IL: Charles C Thomas, Publisher, Ltd., 2007. Reprinted by permission.)

Effect on Image Contrast

Motion of the x-ray tube, the patient, or the receptor plate during the exposure causes extraneous densities from other structures to be superimposed over the anatomy of interest. If this overlapping of adjacent densities is severe, the differences between them becomes less apparent and image contrast is reduced. If the movement is slight, the loss of contrast only occurs at the peripheral portions of each tissue type, but is nonetheless sufficiently destructive to render an unacceptable image. Note in Figure 24-16 that in radiograph ***A*** where motion occurred, the contrast in the peripheral portions of the heart shadow against the density of the surrounding lung tissue is not as great as that for radiograph ***B***.

Motion is unique in that it is the only variable which destroys subject contrast *without* doing so through the mechanism of scattered radiation. In digital radiography, the effects of scatter radiation during an exposure can generally be compensated for by correcting algorithms. But, there is little that can be done by digital processing for a radiograph blurred from severe motion during the original image capture process. The loss of subject contrast *caused by severe motion* will not be corrected for, but will be carried through to the final image displayed at the display screen.

The hand radiographs in Figure 24-18 demonstrate how the effect of motion on contrast is dependent on the severity of the motion. When radiograph ***B*** is compared to ***A***, it is clearly blurred, but a reduction in the overall contrast is arguable. In radiograph ***C***, however, severe motion has resulted in a clear loss of contrast: The finger bones are a "washed-out" darker gray, while the soft tissues of the palm are a "washed-out" lighter gray, with less difference between the two.

Other Image Qualities

The effects of some types of motion, such as peristalsis, can be *confined* to only a portion of the radiograph. In discussing exposure, we normally refer to the overall resulting brightness or density of the image. In the chest radiographs in Figure 24-16, note that although ***A*** demonstrates substantial blurring, the overall density is comparable to that of radiograph ***B***. Motion should not be considered as a factor directly related to image brightness or density.

Figure 24-18

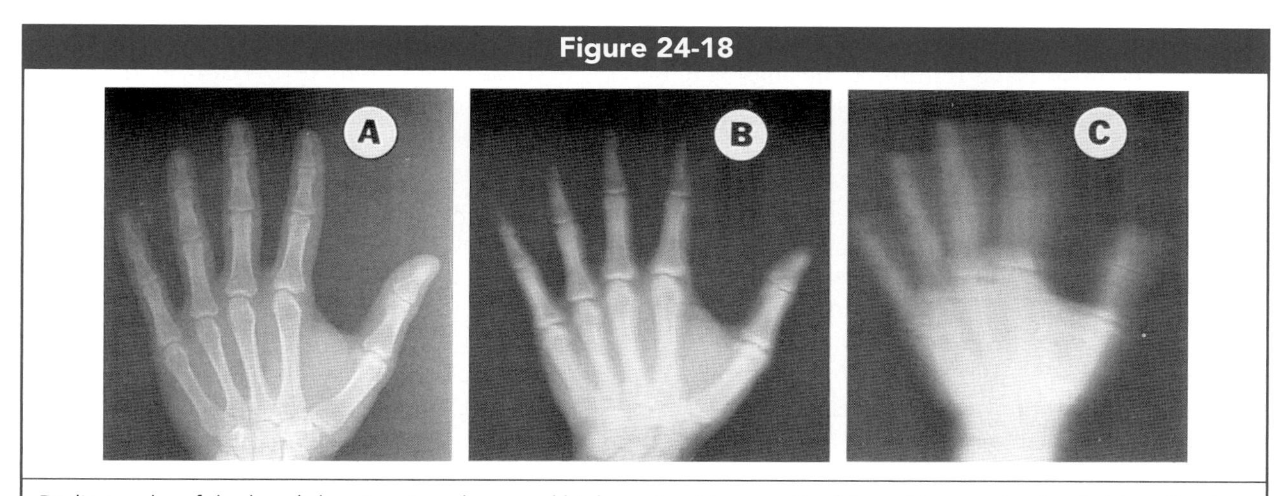

Radiographs of the hand showing ***A***, a sharp and high-contrast image, ***B***, a blurred image due to slight motion but still possessing high contrast, and ***C***, a severely blurred image in which contrast as well as sharpness has been destroyed. (From Quinn B. Carroll, *Practical Radiographic Imaging*, 8th ed. Springfield, IL: Charles C Thomas, Publisher, Ltd., 2007. Reprinted by permission.)

A common misconception is that motion causes *shape distortion* in an image. True elongation and foreshortening of the image were explained in a previous section, and are caused *only* by changes in alignment. Yet, we know that motion can cause a change in the apparent shape of the image, and need to find a more descriptive term for this effect that does not imply the same type of cause. The proper term for an image altered in shape by the effects of movement is a *false image*. The distinction between a false image and a distorted image is that the false image is *a new image created by the interaction between the anatomy present and the motion itself.* It is not a distortion of the image of the real object, but a *new* image which does *not represent the real object at all.*

The most common example of a false image is the familiar *streaks* seen in linear tomographs, for example, of the urinary system; the ureters are long, tubular, contrast-filled structures that happen to fall in line with the longitudinal movement of the x-ray tube during the tomographic exposure. This *relationship* between the anatomy and the movement causes white streaks to appear, which should no longer be considered to represent the ureters at all. Ring-shaped false images can be produced when circular tomographs are taken of tubular structures, and fall into the same category.

False images are a type image *noise*, under the general classification of *visibility* functions in the image, whereas truly distorted images fall under the general category of *recognizability* functions. For the same reasons that motion does not affect distortion, neither does it affect magnification.

To summarize, then, the effects of motion are limited to the destruction of sharpness, a loss of contrast in severe cases, and the possible interposition of *noise* in the form of false images.

SUMMARY

1. Shape distortion in the image is caused only by misalignment of the CR, part and receptor, and is absent when these are maintained respectively perpendicular, parallel, and centered.
2. Off-centering results in identical types of effects as angling the beam, although usually to a lesser degree.
3. Angling or off-centering the beam against spherical or cubical objects causes elongation distortion.
4. Flat, tubular or wedge-shaped objects tilted in relation to the CR and receptor will undergo foreshortening distortion.
5. Angling the CR perpendicular to an object which is tilted in relation to the receptor will cause elongation distortion.

6. When an object cannot be placed parallel to the receptor, distortion is minimized by angling the CR one-half of the angle between the two, according to Ceiszynski's law of isometry.
7. Flat or linear objects kept parallel to the receptor may not be visibly distorted by CR angles.
8. Unless the tube-to-tabletop distance (TTD) is adjusted one inch for every 5 degrees of CR angulation, the SID is effectively increased by an angle, and exposure therefore reduced.
9. CR angulation has no effect upon subject contrast, magnification, or sharpness of detail.
10. Beam divergence for a 40-inch (100 cm) SID is approximately 2 degrees per inch (1 degree per cm). At 72 inches (180 cm) SID, it is about one-half this. This rule can be effectively applied to improve positioning accuracy.
11. Radiographic positioning is essentially a geometrical factor used to maximize sharpness, minimize magnification, minimize shape distortion, or desuperimpose overlying structures.
12. Standardized positioning routines including at least two projections at right angles to each other are generally needed for radiographic diagnosis.
13. Motion is the most destructive factor for sharpness of recorded detail in the remnant beam image.
14. Severe motion can destroy subject contrast by superimposing various densities.
15. Motion can create false images.
16. Motion has no direct relationship to image brightness, magnification or shape distortion.
17. To freeze motion, exposure times should not exceed 0.033 second.

REVIEW QUESTIONS

1. Why is off-centering of the CR identical in its effects to angling the x-ray beam?

2. Why do positioning atlases recommend a 15-degree angle for the AP projection of the sacrum, when the sacrum actually lies at a 30–35 degree angle?

3. Which of the following would be most distorted by a 30-degree CR angle: The head of the femur, the shaft of the femur, or the sternum?

4. If a flat object is tilted in relation to the receptor, and the CR is angle perpendicular to the object, what type of distortion, if any, occurs?

5. The distorting effects of off-centering are indirectly worsened when the SID is __________.

6. If the CR is angled 20 degrees, the TTD should be changed from 40 inches (100 cm) to __________ in order to maintain the exposure level.

(Continued)

REVIEW QUESTIONS *(Continued)*

7. List the four aspects of beam-part-receptor alignment which impact upon shape distortion:

8. Why does the cranium distort more than the head of the femur with an angled projection?

9. What is the angle of x-ray beam divergence at a point 3 inches cephalic to the CR?

10. For a lateral projection of the lumbar spine, why is it not desirable to build the spine up all the way to a horizontal position?

11. What are the four geometrical objectives of radiographic positioning?

12. Why are multiple projections at different angles necessary for fracture radiography?

13. Contrasty anatomy that obscures the anatomy of interest is a form of image __________.

14. Penumbra in the image is __________ by motion of the part, x-ray tube, or receptor.

15. How might the x-ray tube accidentally be moving during an exposure?

16. By superimposing various image densities, severe motion can destroy image __________ as well as sharpness.

17. Streaks or circular artifacts caused by tomographic movement are classified as __________ images.

18. List the three methods of minimizing motion during radiographic exposures:

19. To freeze motion, exposure time should not exceed __________ second.

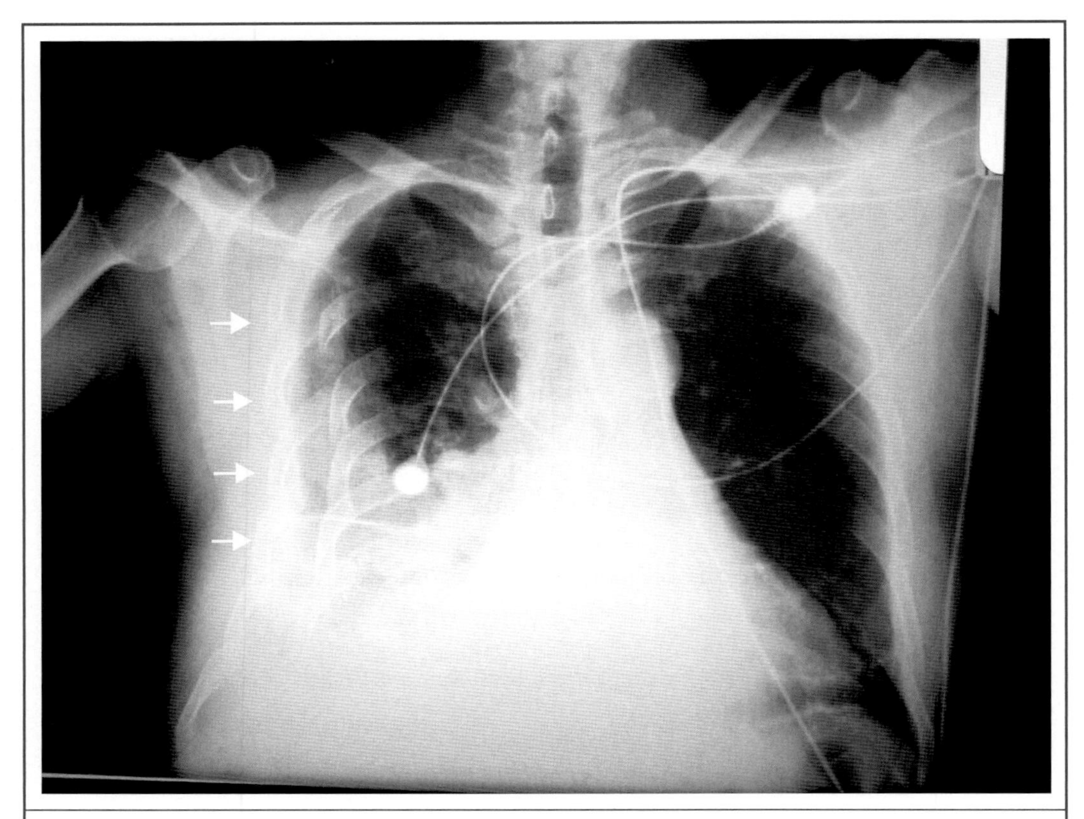

Flail chest from trauma. An entire series of right ribs show displaced fractures, (arrows).

Chapter 25

ANALYZING THE LATENT RADIOGRAPHIC IMAGE

Objectives:

Upon completion of this chapter, you should be able to:

1. List all of the variables affecting exposure level at the image receptor.
2. List all of the variables affecting subject contrast at the image receptor.
3. List all of the variables affecting image noise at the image receptor.
4. List all of the variables affecting sharpness of detail at the image receptor.
5. List all of the variables affecting magnification at the image receptor.
6. List all of the variables affecting shape distortion at the image receptor.
7. Describe the production of absorption penumbra.
8. Interpret *exposure trace diagrams*.
9. Describe how object shape affects the degree of absorption penumbra.
10. Explain how absorption penumbra and geometrical penumbra combine to form the total penumbra.
11. Define spatial resolution and the unit for spatial frequency.
12. Calculate the smallest object size that can be resolved at a given spatial frequency.
13. Define contrast resolution and modulation transfer function (MTF)
14. Describe how, as image details become smaller and smaller, the effects of penumbra eventually affect contrast resolution on a microscopic scale.
15. Describe the limiting factors for image resolution with modern digital equipment.

We have now covered some twenty types of variables which affect the projected radiographic image prior to processing it. While we have treated each one separately, they frequently interrelate to each other in complex ways in determining the quality of the final image. The overall ability of an imaging system to bring out the maximum number of details in an image, or the most visible information, is called its *resolution*.

In the digital age, the resolution of the *final* image is dependent upon many additional processing and display factors. Digital processing includes several computer algorithms that alter both the contrast resolution and the spatial resolution of the image that was initially acquired at the image receptor. We shall refer to this unprocessed image as the latent image.

(Historically, the term *latent image* was used to refer to the barely-visible chemical information contained within a film emulsion after it had been exposed to x-rays but *prior* to processing. Just as chemical processing "brought out" this image into a fully visible form on the film, so computer processing now "brings out" the electronic information acquired at the image receptor to a fully visible image. Thus, the term *latent image* is perfectly appropriate in referring to a digital image that has not yet been computer-processed.)

Nearly all radiographic images are now displayed on *electronic* devices such as LCDs. These display monitors have their own characteristics and pixel sizes which further affect the contrast resolution and spatial resolution of the final image. All of these digital processing and electronic display factors will be fully discussed in later chapters. At this point, however, it would be well to summarize those variables that impact upon each particular quality of the *latent* image as it has been acquired at the image receptor, because these qualities are critical to the *data set* from which the image will be developed.

It would be good practice, as a student, to make your own list of variables which affect each characteristic of the remnant beam image first, then see how your list compares with the following. The *six* variables for the remnant beam as it reaches the receptor plate are: (1) exposure, (2) subject contrast, (3) noise, (4) sharpness of detail, (5) magnification, and (6) shape distortion. Also, where applicable, note those relationships which are directly proportional, inversely proportional, or exponential.

Variables Affecting Exposure at the Image Receptor

1. Milliamperage (mA) affects it in direct proportion.
2. Exposure time (s) affects it in direct proportion.
3. The total mAs affects it in direct proportion, and is the *prime factor* for controlling it.
4. Kilovoltage-Peak (kVp) affects it in an exponential manner.
5. The type of generator and rectification used affect it.
6. In clinical application, protective filtration does not affect it, but *compensating* filtration is thick enough to affect it.
7. Field size (collimation) affects it.
8. Part thickness affects it in an exponential fashion.
9. Patient condition, pathology and contrast agents all affect it.
10. Scatter radiation levels affect it.
11. Grids affect it.
12. The anode heel effect modifies it.
13. Source-to-image receptor distance (SID) affects it by the inverse square law.
14. Object-to-image receptor distance (OID) affects it.

Variables Affecting Subject Contrast at the Image Receptor

1. Kilovotlage-Peak (kVp) controls sufficient *penetration*, which is critical to it.
2. The type of generator used affects it.
3. In clinical application, protective filtration does not affect it, but *compensating* filtration is thick enough to affect it.
4. Field Size affects it very substantially.
5. Part thickness affects it profoundly.
6. Patient condition, pathology and contrast agents all affect it.
7. Scatter radiation levels affect it profoundly.
8. Grids affect it very substantially.
9. Object-to-image receptor distance (OID) affects it substantially.
10. Severe motion can reduce it.

Variables Affecting Image Noise at the Image Receptor

1. Insufficient mAs (mA or exposure time) makes quantum mottle apparent.
2. Insufficient kVp makes quantum mottle apparent. Very excessive kVp contributes to the scatter ratio in the remnant beam.
3. Large field sizes contribute to scatter production.
4. Part thickness is the major source of scatter production.
5. Patient condition may include artifacts.
6. Grids reduce scatter noise, but can cause grid lines or grid cut-off.
7. Increased OID reduces scatter.
8. Motion can generate false images.
9. Positioning can affect it insomuch as artifacts or obscuring anatomy are superimposed over the anatomy of interest.

Variables Affecting Sharpness at the Image Receptor

1. Focal spot size is the *prime factor* in controlling sharpness.
2. The anode bevel controls the *projected* focal spot size at various angles.
3. The SOD/OID ratio is a *controlling factor* for sharpness.

4. SID, SOD and OID all affect it insomuch as they alter the SOD/OID ratio.
5. Positioning can affect it insomuch as it alters the SOD/OID ratio.
6. Motion is the *prime enemy* of sharpness.
7. Exposure time can affect it by allowing substantial motion to occur.

Variables Affecting Magnification at the Image Receptor

1. The SID/SOD ratio is the *prime factor* for controlling magnification.
2. SID, SOD and OID all affect it insomuch as the alter the SID/SOD ratio.
3. Positioning can affect it insomuch as it alters the SID/SOD ratio.

Variables Affecting Shape Distortion at the Image Receptor

1. Alignment of the beam, part, and image receptor is the *controlling factor* for shape distortion. Alignment includes both centering and angling.
2. Positioning can affect it insomuch as it alters alignment.

One can see at a glance from this list that the last three factors, which are geometrical in nature, are much more straightforward than the first three. All these projection variables can sometimes interrelate in complex ways that begin to blur the distinction between visibility and recognizability. One such example is the phenomenon of *absorption penumbra.*

Absorption Penumbra

For the following discussion, *exposure trace* diagrams like the one in Figure 25-1 will be used. At first glance, this looks like the penumbra diagram that has been used in previous chapters, but at the bottom of the diagram is a shaded area whose *depth* represents the exposure received by the image receptor. The higher the top "surface" line of this shaded area, the greater the exposure.

Figure 25-2 provides a closer examination of an *exposure trace*; the normal height of the trace line represents a dark background density surrounding a particular image detail. Note that the *contrast* of this detail is indicated by the reduction in the exposure trace in the center of the image, such as a bone, where the resulting image would be the lightest. This appears as a kind of "pit" in the exposure trace, the depth of which represents the contrast produced. Penumbra at the edges of the image is represented by a gradient slope which indicates that the background density gets lighter and lighter toward the center of the image. The *horizontal* measurement of this slope is the extent of the penumbra. The steeper the slope (the less the horizontal spread), the sharper the image.

Figure 25-1

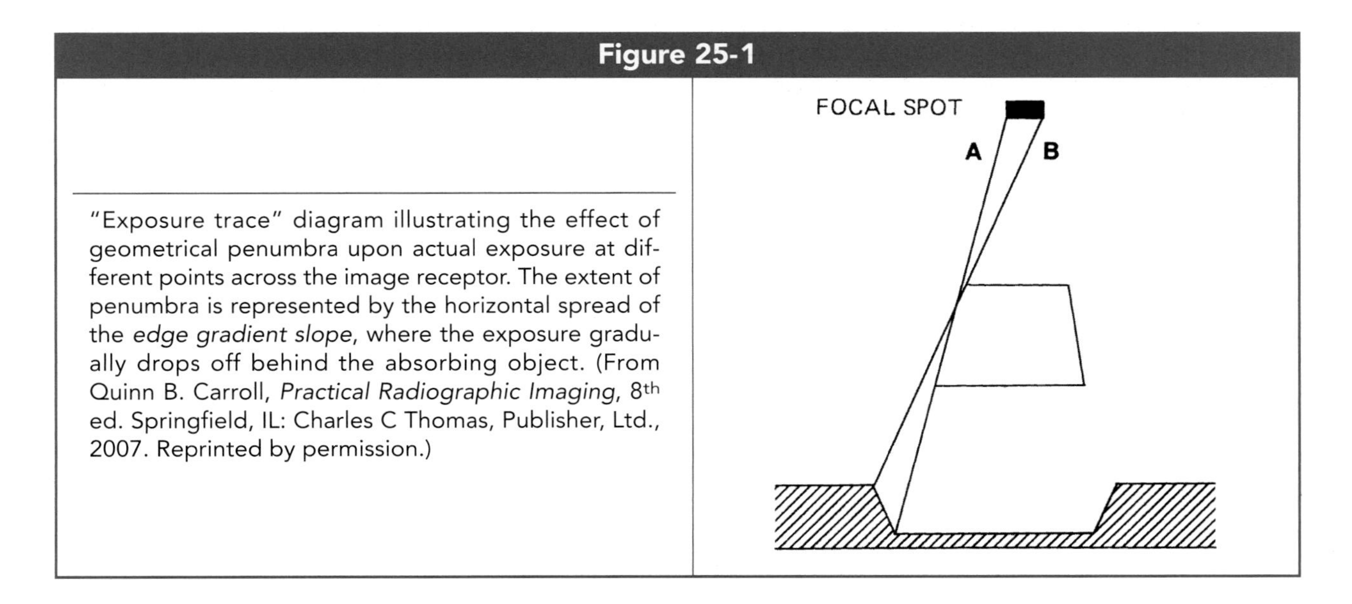

"Exposure trace" diagram illustrating the effect of geometrical penumbra upon actual exposure at different points across the image receptor. The extent of penumbra is represented by the horizontal spread of the *edge gradient slope*, where the exposure gradually drops off behind the absorbing object. (From Quinn B. Carroll, *Practical Radiographic Imaging*, 8th ed. Springfield, IL: Charles C Thomas, Publisher, Ltd., 2007. Reprinted by permission.)

Figure 25-2

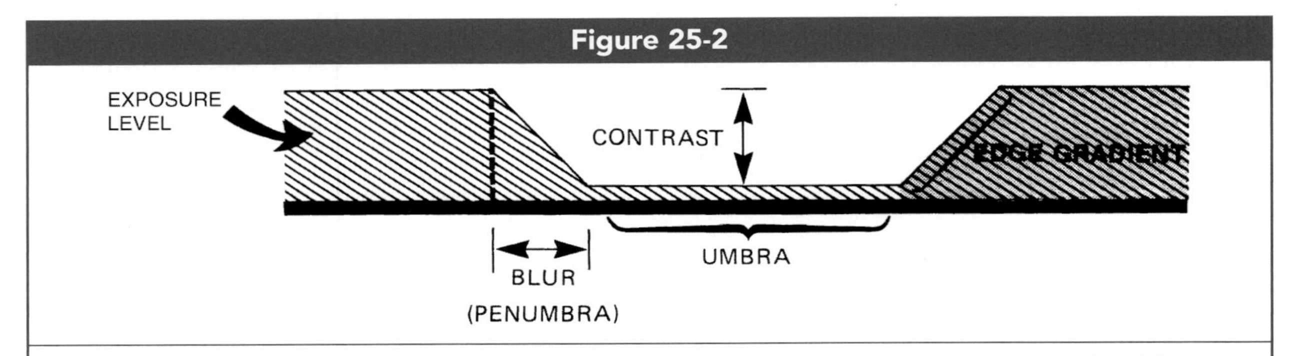

Interpretation of an "exposure trace" diagram: Background exposure is represented by the height of the exposure trace—the higher this thickness, the *darker* the resulting image. *Subject contrast* is represented by the vertical depth in the middle of the umbra shadow. The *edge gradient* is the slope at the edge of the image—the steeper the slope, the more "quickly" the exposure drops off toward the umbral shadow of the image. Penumbra (blur) is indicated by the *horizontal* spread of the edge gradient. (From Quinn B. Carroll, *Practical Radiographic Imaging*, 8th ed. Springfield, IL: Charles C Thomas, Publisher, Ltd., 2007. Reprinted by permission.)

Geometrical penumbra was fully explained in Chapters 13, 21, and 23 as penumbra created by the relationship between the focal spot size and the various distances involved with the projected x-ray beam. We described it as a *partial shadow* resulting from *partial absorption* of the x-rays which increases gradually toward the object. However, the use of the term *partial* here referred to a portion of the total amount of x-rays which the object is *capable* of absorbing, and therefore assumed an object of uniform *thickness* across its breadth (Fig. 25-1).

To complete the picture, we must acknowledge that most objects are of *varying thickness*, and there is also a process of *partial absorption* that graduates from the *thinnest to the thickest portions of an object.* The thickness of the part is normally thought of as bearing primarily upon the visibility functions of the image, but when the object's thickness tapers at its edges, the visible effects of partial absorption can be *indistinguishable from geometrical penumbra.*

Let us examine this effect by considering three objects having different shapes, but made out of the same homogeneous material, about to be radiographed; the "ideal shape" for an object to be radiographed would be a trapezoid whose slanted sides coincide exactly with the angles of the diverging x-rays (Fig. 25-3, ***A***). All portions of the x-ray beam striking such an object will be attenuated by the same thickness of material. The *darkness* of the resulting image is represented by the thickness of the exposure trace at the bottom of the diagram, with the thickest areas representing a uniform, black background.

Note that at the edges of projection ***A***, this exposure or darkness drops *vertically* off, indicating a perfectly sharp edge to the lighter image under the object. In ***B***, a cubical object absorbs only a few of the x-rays striking it at the upper left corner, then increasingly more x-rays as the beam passes through thicker portions until the full thickness of the objects is reached passing through the *lower left* corner. At the image, we see the exposure trace slope downward, indicating a lighter and lighter density to this point—this is *absorption penumbra*. Then a uniformly light exposure extends across the middle portion of the image where the object thickness is consistent. We might say that absorption penumbra is caused by the decrease in *projected* thickness of an object toward its edges.

In projection ***C*** of a *spherical* object, we see that, due to such gradual changes in part thickness, not only does this slope in the density trace extend *all the way to the middle* of the image, but that it also begins much more subtly at the edge, almost blending with the background. A spherical object presents the worst case scenario in producing absorption penumbra.

A representation of total blur or penumbra may be obtained by combining the diagram for geometric blur (Fig. 25-2) with that for absorption blur (Fig. 25-3, ***B***). The result is Figure 25-4; in this case, the maximum absorption occurs inside the dashed absorption penumbra line. Within the geometric

Figure 25-3

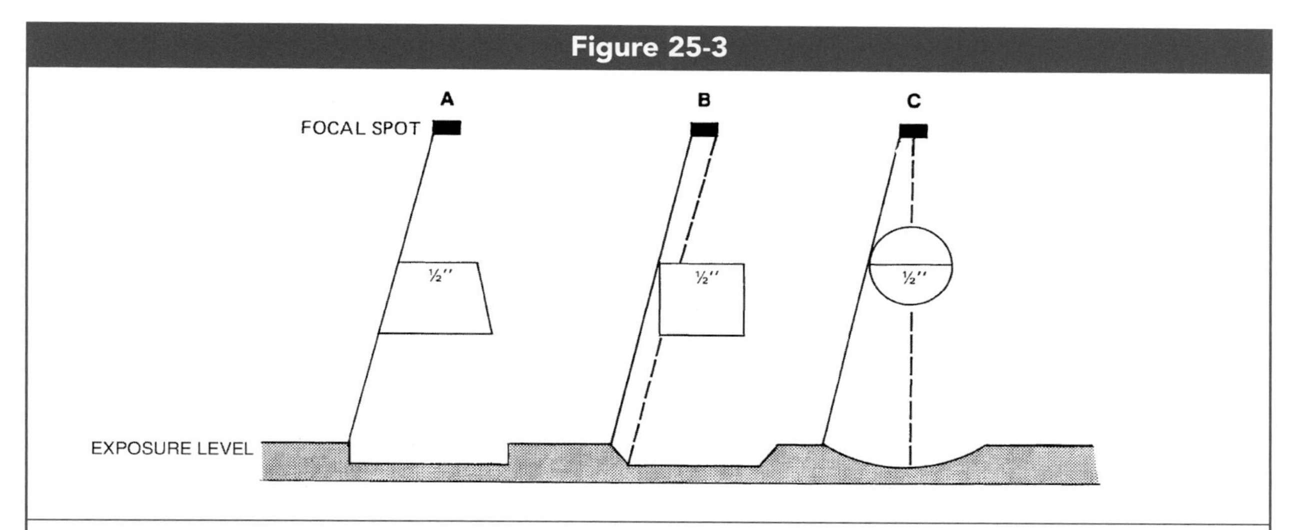

Exposure trace diagrams for *absorption penumbra*, using objects of equal thickness but different shapes. The trapezoid object, ***A***, if perfectly aligned with the diverging x-ray beam, will produce no absorption penumbra as shown here (but *will* produce *geometrical penumbra* as shown in Figure 24-1). A cubical object, ***B***, absorbs more radiation at the thickest portion of its projected corner, causing a penumbral edge gradient. For a spherical object, ***C***, the penumbra extends all the way from the edge to the middle of the image—This represents the most severe form of absorption penumbra. (From Quinn B. Carroll, *Practical Radiographic Imaging*, 8th ed. Springfield, IL: Charles C Thomas, Publisher, Ltd., 2007. Reprinted by permission.)

penumbra lines, absorption varies because beams originate at different points within the focal spot. Within the absorption penumbra lines, it continues to vary, but strictly because of changing object thickness. *Total penumbra is comprised of geometric penumbra plus absorption penumbra.*

RESOLUTION

In evaluating the overall quality of any image, the concept of *resolution* is of essence. A dictionary defines resolution as the "ability to distinguish the individual

Figure 25-4

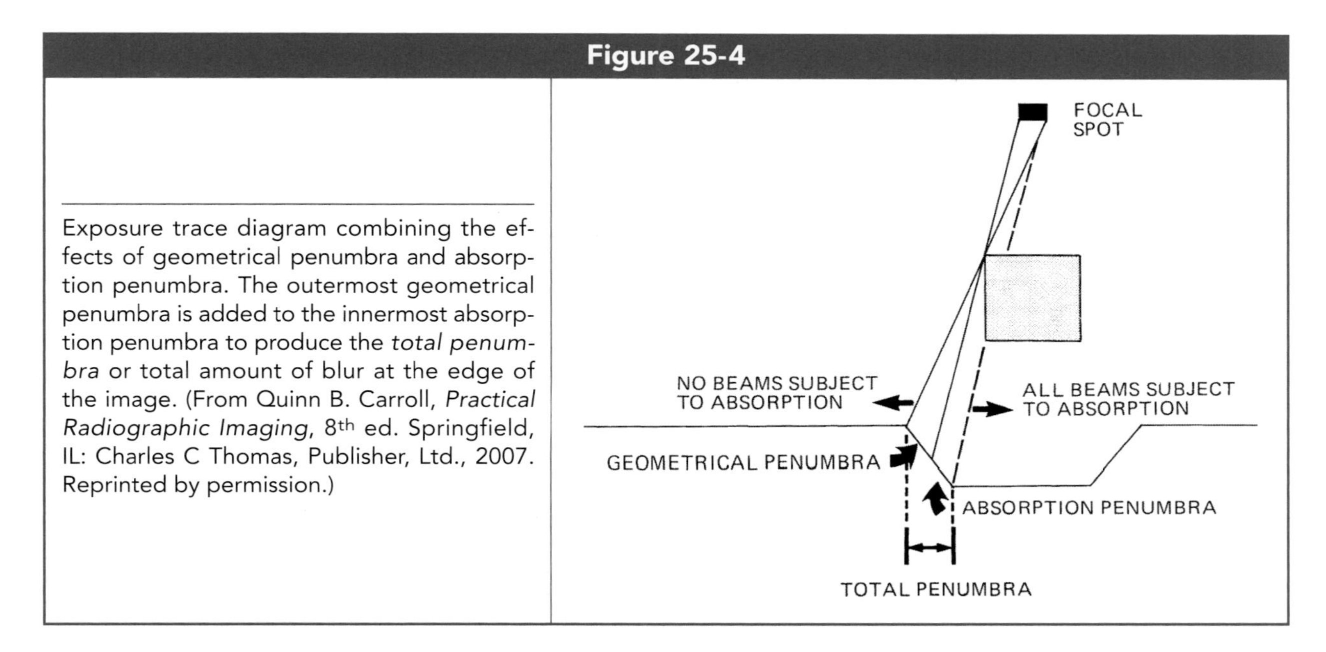

Exposure trace diagram combining the effects of geometrical penumbra and absorption penumbra. The outermost geometrical penumbra is added to the innermost absorption penumbra to produce the *total penumbra* or total amount of blur at the edge of the image. (From Quinn B. Carroll, *Practical Radiographic Imaging*, 8th ed. Springfield, IL: Charles C Thomas, Publisher, Ltd., 2007. Reprinted by permission.)

parts of an object or closely adjacent images." On any image, in order for two small, closely adjacent details to be recognized as distinct and separate from each other, *all* of the image qualities must be optimized. Of particular influence, however, are *image contrast* from the visibility functions, and *sharpness of detail* from the recognizability functions.

When two details are adjacent to each other, contrast is essential to be able to make out their *difference*, and sharpness is essential to visually *separate* them. High contrast images can still have poor resolution if their edges are blurred; sharp images are still poorly resolved if their visibility is impaired.

Both aspects of an image can be measured. The most common way of measuring spatial resolution is called the *spatial frequency*, defined as the number of details that can be "fit" into a given amount of space. Contrast resolution is measured by *modulation transfer function* or *MTF*.

Spatial Resolution: Spatial Frequency

Spatial resolution can be tested by taking the *spatial frequency* of the image from a lead foil *line-pair* template (Fig. 25-5); it has a series of fine slits cut into lead foil, which, when radiographed, produce alternating black and white lines on the image. The unit of measurement is *line-pairs per millimeter (LP/mm)*. The corresponding LP/mm is etched into the template alongside the slits. On the resulting radiographic image, the observer scans the image from the thicker sets of lines to the thinner sets until a point is determined where the black and white line pairs can no longer be clearly distinguished from each other (Fig. 25-6). The last set of *resolved* lines, just prior to these, is where the corresponding LP/mm should be read out. The thinner the lines that can be resolved, the more line pairs fit within a millimeter.

Figure 25-5

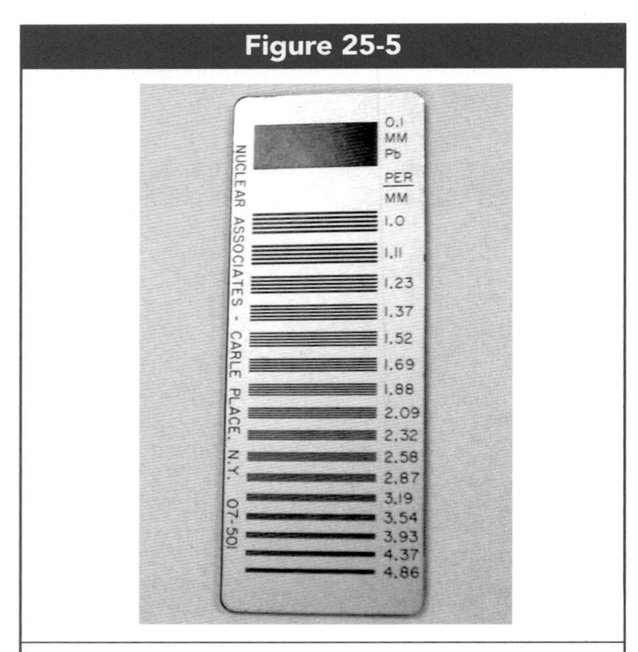

Line-pair resolution test template made of lead foil with fine slits cut in line patterns, used to measure spatial resolution.

The spatial resolution determines the size of object that can be reproduced as an image by the system. *The smallest absolute object size that can be reproduced is inversely proportional to one-half the spatial frequency.* This is just a fancy way of saying that the smallest object that can be imaged is the width of *one* of the resolved black or white lines in the resolution template pattern, (as opposed to a *pair* of lines). It is expressed by the formula:

$$\text{Minimum Object Size} = \tfrac{1}{2}\left(\tfrac{1}{SF}\right)$$

where *SF* is the spatial frequency in line-pairs per millimeter (LP/mm).

Practice Exercise #1

For example, what is the smallest object that can be resolved by an x-ray machine with a spatial frequency of 3 line-pairs per millimeter?

Solution: $OS = \tfrac{1}{2}\left(\tfrac{1}{3}\right)$

$OS = \tfrac{1}{6} = 0.17$

Answer: The smallest resolvable object would be $\tfrac{1}{6}$ *mm, or 0.17 mm in size.*

For digital imaging systems, the size of a *pixel* becomes an additional limiting factor. No object smaller than a single pixel could be imaged. In terms of line pairs on a spatial resolution template, one line and its interspace would require at least two rows of pixels to image.

Contrast Resolution: MTF

Physicists use a more complex measurement of the *contrast resolution* capacity of a whole imaging system

Figure 25-6

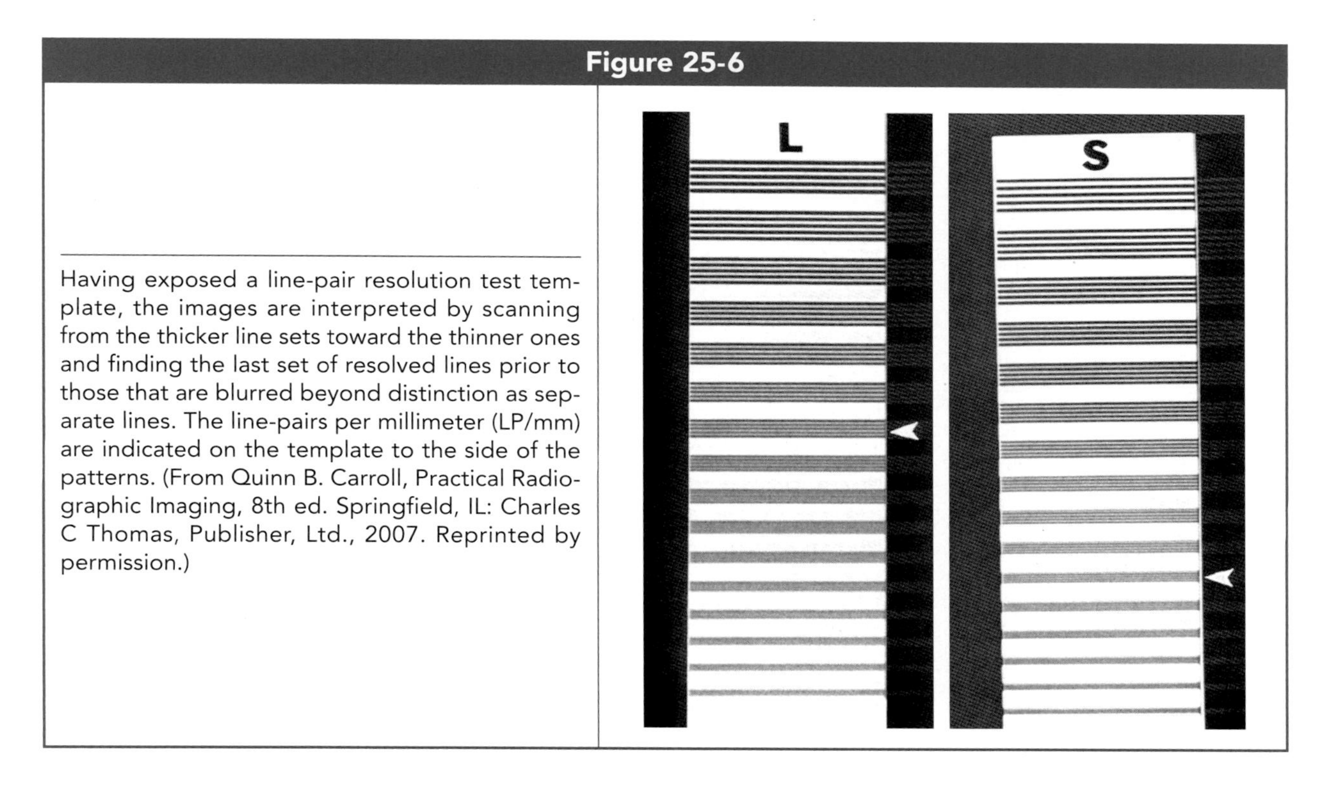

Having exposed a line-pair resolution test template, the images are interpreted by scanning from the thicker line sets toward the thinner ones and finding the last set of resolved lines prior to those that are blurred beyond distinction as separate lines. The line-pairs per millimeter (LP/mm) are indicated on the template to the side of the patterns. (From Quinn B. Carroll, Practical Radiographic Imaging, 8th ed. Springfield, IL: Charles C Thomas, Publisher, Ltd., 2007. Reprinted by permission.)

called the *modulation transfer function* (*MTF*). In the line-pair pattern produced by the template in Figure 25-6, note that the highest level of *contrast* would be for each dark line to be pitch black and each lighter line to be a blank white. To graphically illustrate the concept of MTF, we begin with a simple *exposure trace diagram* like the one in Figure 25-1. When this type of diagram is constructed for the series of alternating radiolucent and radiopaque strips presented by a resolution template (Fig. 25-5), the effects of penumbra *round off* the corners of the resulting trace as shown in Figure 25-7.

This degree of penumbra, or rounding-off, remains constant no matter how *small* the lines themselves become or how *close together* they are. This is because the degree of penumbra is a function of the imaging system, not the template. In the next Figure 25-8, carefully observe what happens to the image trace as the

Figure 25-7

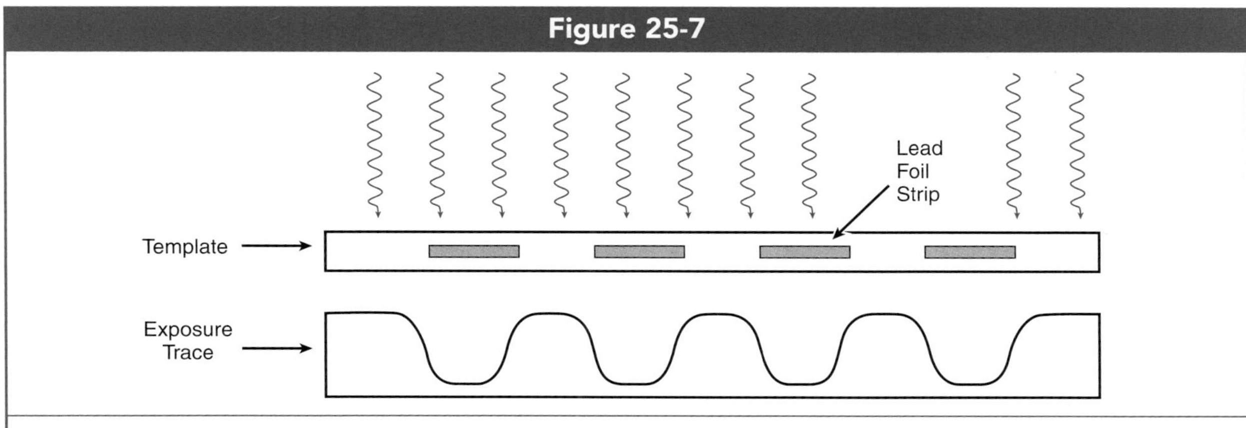

An exposure trace diagram for a line-pair resolution test template image would show rounded corners rather than sharp ones, due to the edge gradient or penumbral blur at the edges of the lines.

Figure 25-8

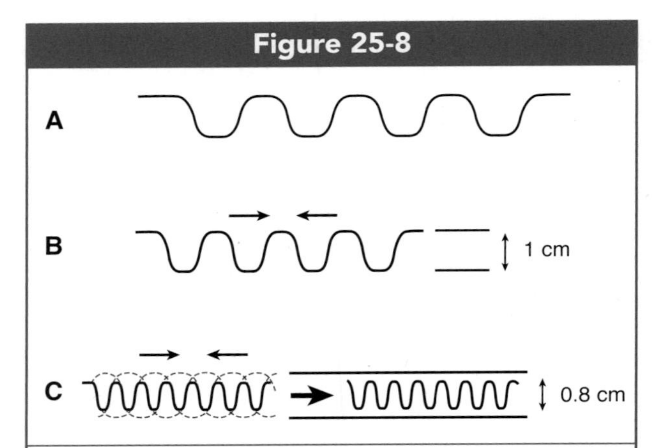

As line-pairs become smaller and closer together, ***A*** to ***B***, their rounded penumbral edges begin to overlap, such that the exposure trace diagram begins to look like a *sine wave*. When they become small enough and close enough, ***C***, the *depth* of the troughs between the sine waves begins to shallow, indicating that, at a microscopic scale, *contrast* begins to be reduced between the lines. This depth of the sine wave, as a percentage of the full *subject contrast* of the real object, is measured by physicists as the *modulation transfer function* (*MTF*).

lines then are moved closer and closer together: First, shown in ***B***, when they get so close that their rounded penumbral edges begin to overlap, the resulting trace begins to look like a *sine wave*. Second, as they are brought closer still in ***C***, we see that the *depth* of the "troughs" traced between them begins to shallow. This depth, against a background density, represents *contrast*. It means that, where there were blank whites lines (representing the lead foil) in between the black lines before, there are now only *gray* lines separating the black ones. *This is a loss of image contrast, caused on a microscopic scale, from the effects of penumbra.*

(As with absorption penumbra, this blurs the distinction between visibility and recognizability functions in the image, but only *at the microscopic level*.)

Figure 25-9

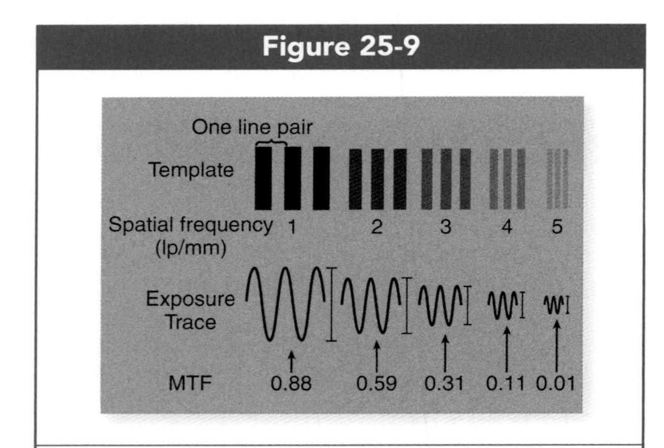

The resulting MTF values (bottom) as resolution template line pairs become smaller and closer together. The exposure trace in the middle shows graphically the loss of contrast, which is also indicated in the simulated image of the lines at the top.

In the simplest terms, modulation transfer function (MTF) is the *ratio of the recorded contrast of an image to the real object's subject contrast*. If the image has one-half the contrast of the real anatomy, the MTF is $\frac{1}{2} = 0.5$. If the depth of the trace diagram represented 100 percent of the original contrast in the object (the contrast between the lead foil strips and the slits between them, in this case), the resulting perfect MTF would be 1 to 1, or 1.0. Since no imaging system is perfect, real modulation transfer functions always fall somewhere between zero and one.

These values are illustrated in Figure 25-9 for the gradually thinning lines of the test template. Note that as the lines get thinner and closer, the corresponding sine-waves below become *shorter* vertically with the loss of depth at the troughs representing lost contrast. The resulting contrast *ratios* are listed below. These are the MTF's. They can also be expressed as *percentages* by simply moving the decimal points to the *right* two places, 0.97 being *97 percent* of the actual object's contrast, and so on.

When the MTF, or *contrast resolution*, is plotted against the spatial frequency of the lines, we obtain a graph such as the one in Figure 25-10. The most ideal improvement in an imaging system would be to obtain both higher contrast resolution (MTF) *and* higher spatial resolution, represented by curve ***B*** in Figure 25-10. Unfortunately, a trade-off often occurs, as shown in Figure 25-11, where higher spatial resolution is obtained at the expense of contrast resolution and vice versa.

This precisely describes the historical situation in having moved from film-based radiography to digital radiography: As mentioned in the previous section, no object smaller than a single pixel can be resolved by computer-based systems. The pixel size is limited by how *small* individual electronic detector elements in the receptor plate can be made. On

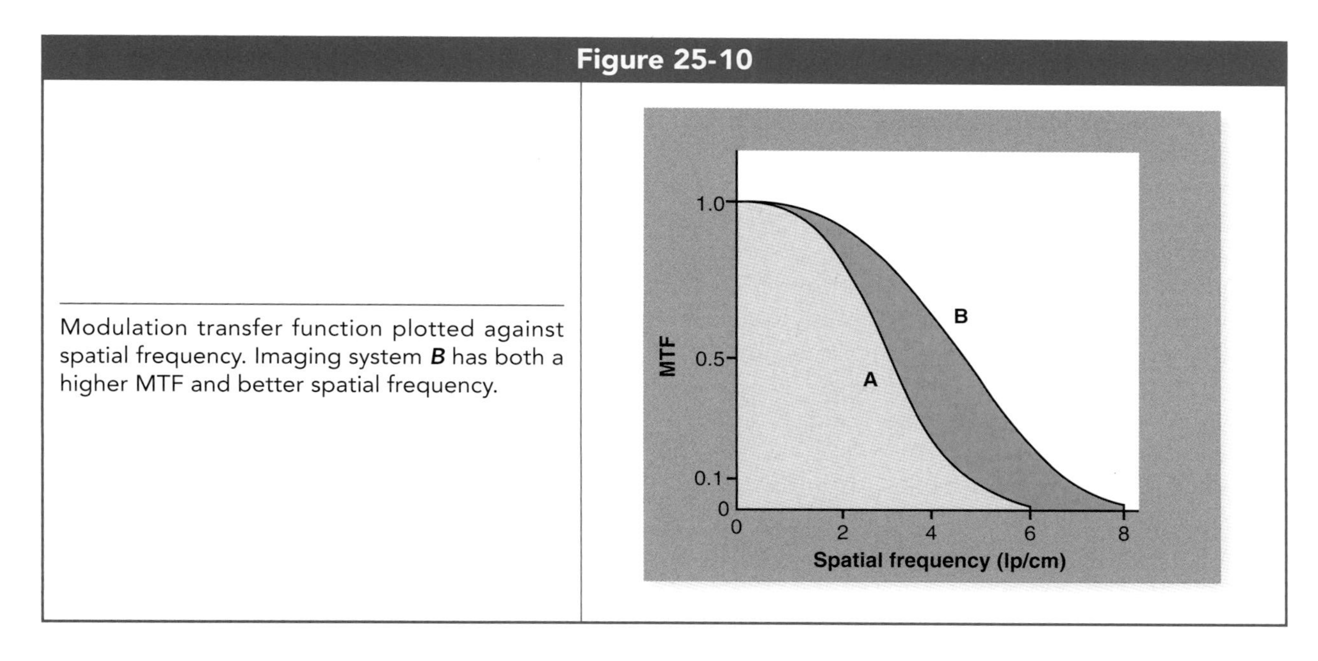

Figure 25-10

Modulation transfer function plotted against spatial frequency. Imaging system ***B*** has both a higher MTF and better spatial frequency.

radiographic *film*, the limiting factor was the size of a single *crystal of silver bromide*, much smaller than an electronic detector can be made to date. Therefore, in terms of the image receptor, digital systems would have lower spatial resolution, which was traded off by greatly enhanced *contrast resolution* that software has made possible.

However, recall that if an object is much smaller than the *focal spot*, it also cannot be resolved because of geometrical penumbra (Chapter 21). This limiting factor affects *both* film-based and computer-based systems. The individual *dels* or *detector elements* used in direct-capture digital radiography now have a typical square size of just 0.1 mm, much smaller than the routine focal spot for an x-ray tube, (0.5–1.2 mm). This makes the *focal spot* the primary limiting factor once again for spatial resolution, rendering any smaller pixel technology a moot point for the time being.

These issues will be fully discussed in the following chapters. It is still important to appreciate the need for maximizing the quantity and quality of data fed into the computer system. This is all dependent upon the initial aspects of the radiographic *projection* itself, and the resulting characteristics carried by the remnant x-ray beam. This "latent image," the image inherent to the remnant beam as it approaches the receptor plate, possesses its own spatial resolution and contrast resolution. These can be measured by intercepting the remnant x-ray beam with *film* or with *ion chambers*, prior to reaching the receptor plate.

SUMMARY

1. The student should be able to accurately list the variables affecting the six characteristics of the

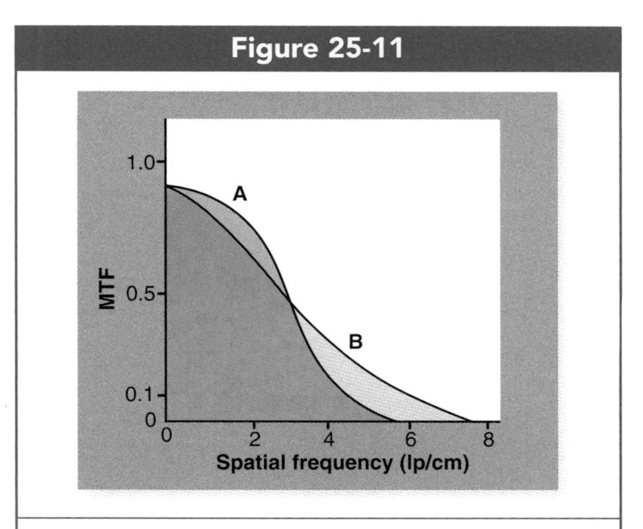

Figure 25-11

For most imaging systems, there is a trade-off between MTF and spatial resolution, as shown here, where system ***B*** has higher spatial resolution but lower MTF (contrast).

remnant beam image as it reaches the receptor plate.

2. Absorption penumbra is caused by the decrease in *projected* thickness of an object toward its edges, and is visually indistinguishable from geometrical penumbra. The total penumbra produced in an image is the sum of the absorption penumbra and the geometrical penumbra.
3. Overall image resolution is the ability to distinguish adjacent details as being separate and distinct. It is affected by all aspects of the image, but of particular importance are the contributions of image contrast and sharpness of detail.
4. Spatial resolution, or sharpness, is measured by spatial frequency with the unit LP/mm. The minimum object size that can be resolved by an imaging system is inversely proportional to one-half of the spatial frequency.
5. Contrast resolution is measured by modulation transfer function (MTF), defined as the ratio of a recorded image's contrast to that of the real object on a microscopic scale.
6. With most imaging systems, there is a trade-off between higher contrast resolution and higher spatial resolution. It is difficult to achieve both.
7. For digital radiography, with the development of hardware pixels smaller than one millimeter in size, the *focal spot* of the x-ray tube has once again become the primary limiting factor for spatial resolution.

REVIEW QUESTIONS

1. What is the primary *controlling factor* for the following:

   a. Exposure at the image receptor:

   b. Penetration of the x-ray beam:

   c. Sharpness of detail:

   d. Magnfication of the image:

   e. Shape distortion:

2. What are the two types of image penumbra?

3. As demonstrated by the exposure trace diagram, what are the two main image qualities that combine to determine overall image resolution?

(Continued)

REVIEW QUESTIONS *(Continued)*

4. What is the unit for spatial frequency? (Do not abbreviate.)

5. If the smallest object an imaging system can resolve is 0.2 mm in size, what is the spatial frequency for this system?

6. An MTF of 88 percent means that the __________ present in the image is 88 percent of that of the real object being projected.

7. As the line pairs of a resolution test template become smaller and closer together, the *contrast* between them begins to diminish because the __________ of the line images begins to overlap.

8. If the spatial frequency of the remnant x-ray beam image is 4 LP/mm, and the hardware pixels of a digital imaging receptor plate are 0.2 mm in size, what is the smallest object that can be resolved by this system?

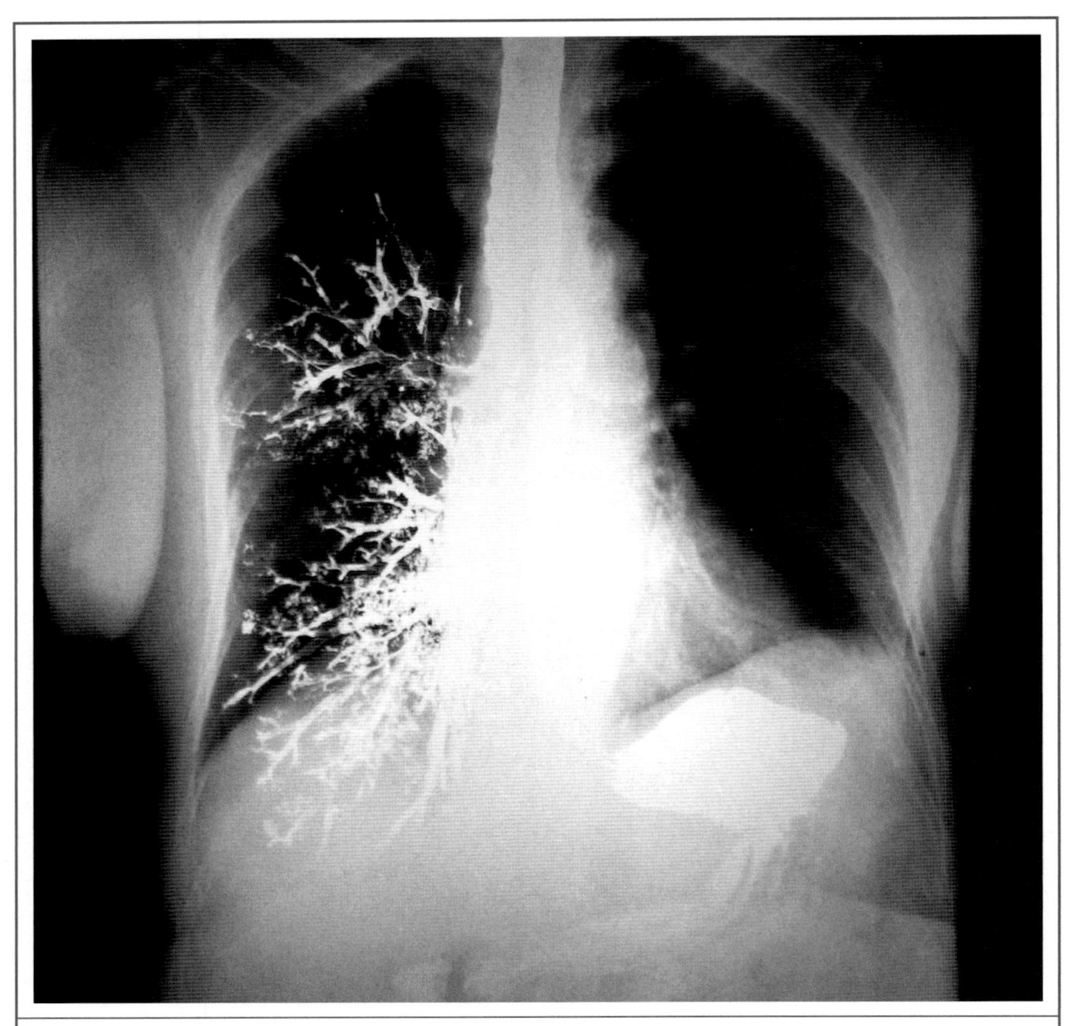

Accidental barium bronchogram. The barium was aspirated into the patient's right bronchial tree through a tracheoesophageal fistula during an attempted upper GI study.

Chapter 26

SIMPLIFYING AND STANDARDIZING TECHNIQUE

Objectives:

Upon completion of this chapter, you should be able to:

1. List the three general areas of standardization within a medical imaging department.
2. Describe the advantages and disadvantages of the variable-kVp and fixed-kVp approaches to radiographic technique.
3. Using the base-60 method, apply the variable-kVp approach to developing techniques.
4. List the objectives and advantages of the proportional anatomy approach to technique.
5. List those procedures in the major proportional anatomy groupings which share equivalent overall techniques.
6. Give landmark examples of proportional anatomy derivations from one body part to another.
7. Describe essential considerations for torso *shape* in adjusting technique.
8. Explain the advantages, proper development, and proper use of technique charts.
9. Describe the basic steps in developing a technique chart "from scratch."

The standardization of procedures in radiography produces benefits on several levels. The common lack of unanimity on the subject of radiographic techniques has been aptly demonstrated by numerous recipes advanced in the form of complicated exposure tables and systems. The lack of a systematic, straightforward and workable method becomes readily apparent when it is necessary to train a student in radiographic exposure.

In the quest for optimal image results, using a standardized approach certainly contributes by providing *consistency* from one image to the next, but it also does much more—it *narrows* the range of potential causes when some failure in image quality must be investigated. It facilitates not only the training of students, but the learning curve and adaptation period for an experienced, but newly-hired radiographer in the department. Simplification reduces the probability of errors occurring in the first place.

Standardization within a radiography department resolves itself into three phases, all of which should be addressed for the best level of patient care:

1. **Standardization of Projections:** A routine series of projections to be taken for each radiographic procedure should be established, which does not vary substantially from doctor to doctor. Optional views can always be added as needed for particular patient conditions, but it is incumbent upon radiologists and managers to provide basic routines that are based on typical diagnostic necessities agreed upon by consensus, rather than on individual whim in the guise of "personal preferences."

   Further, for each projection the basic guidelines for positioning the part, including centering, collimation and the receptor plate size (for CR) should be standardized within the department. This is not meant to preclude adaptations as needed in any way, but to provide a routine which can be applied consistently for the typical majority of cases. A written *routines* manual is recommended for every department.

2. **Standardization of Exposure Factors:** The provision of *technique charts* in each radiographic suite and on all mobile units is not only strongly recommended, but often required by law. Staff radiographers should be allowed to provide input and feedback on the accuracy of charts, but should also then be given the incentive to use them regularly.

   Technique charts should be of assistance not only in standardizing, but in *simplifying* radiographic exposure factors. It is entirely possible to establish the full range of needed exposures for any given series of projections within a radiographic procedure, including all needed adaptations of exposure for different-sized patients, *while adjusting only one variable.* For example, the kVp can be standardized for the entire procedure (with a few exceptions), and the mA station as well, such that only the exposure time is adjusted between views. The typical intervals between one exposure time and the next still allow sufficient flexibility to obtain an appropriate technique for each view and for every size of patient.

3. **Standardization of Image Processing:** The consistency in quality and appearance of the final image depends as much on how it is *processed* as how it was initially projected to the receptor plate. With digital imaging systems, standardization of processing occurs automatically *for a particular algorithm* or view.

   However, radiographers can manually change the algorithm used to process each image. For example, a cervical spine can be processed using the algorithm for an *abdomen* view to give it a different appearance, especially in regards to the image contrast. Within a department, consensus should be established as to when and how these changes are are to be used, and guidance should be employed, such as in a *routines* manual, so that their application is not perceived as completely arbitrary on the part of individual radiographers.

VARIABLE kVp VS. FIXED kVp APPROACHES

In Chapter 16, tables were presented recommending a *minimum* kVp (Table 16-1), and an *optimum* kVp (Table 16-2) for the various body parts. Historically, there have been two general approaches to the setting of radiographic techniques: The *fixed-kVp* philosophy and the *variable-kVp* approach. It is essential to understand that whichever approach is used, *the minimum kVp required for sufficient penetration of the anatomy, as listed in Table 16-1 for each body part, always applies.*

The *fixed-kVp* approach follows the philosophy of standardizing exposure factors just as described above, such that only one variable is changed from projection to projection whenever possible. For film-based images, there was a more clear distinction as to what *optimum kVp* should be employed: It was the *minimum* kVp that achieved full penetration of all anatomy of interest. This provided the maximum image contrast that still allowed for full penetration. Further, in order to avoid changes in contrast from one view to the next within a series, an attempt was made to generally keep this kVp level the same throughout a procedure. The primary goals of the fixed kVp method, then, were to (1) assure sufficient penetration, (2) keep image contrast consistent, and (3) simplify and standardize for accuracy.

Proponents of the *variable-kVp* approach, on the other hand, had as their best and primary argument the issue of *patient exposure.* As we have shown, an increase in kVp has less impact upon patient exposure than a corresponding increase in mAs. This is a valid consideration. In practice, nearly all variable-kVp systems actually adjusted *both* kVp *and* mAs for each projection, making the whole approach more complicated. In fact, for very *thin* patients, mAs had to be reduced rather than kVp because of the risk of inadequate penetration through the tissues of that particular body part. Since kVp is not *proportional* to exposure as is mAs, all adjustments were essentially based upon adaptations of the 15 percent rule and were more a matter of *estimation* than of calculation.

Which philosophy is more suitable for *digital imaging*? Since the contrast of the image is ultimately determined by postprocessing algorithms and kept automatically at a very constant level, this key argument for the fixed-kVp approach is taken away. Furthermore, the priority of minimizing patient exposure has been brought to the fore by the requirement of many CR systems for increased exposure factors over those that were previously used for film-screen systems. Unwisely, many radiology departments have accepted an across-the-board doubling of the mAs to make this conversion over to CR technology, when a 15 percent increase in kVp would have accomplished the same objective. Some techniques, such as for a PA chest, have seen as much as an eightfold increase in mAs. As a profession, this should be unacceptable to radiographers.

Therefore, the variable-kVp approach, as a general philosophy for radiographic technique, is embraced in this textbook as the most conducive approach for digital imaging. This is not done without at least two caveats, however: First, the minimum kVp's listed in Table 16-1 *must* always be met in order to assure sufficient penetration of the particular tissues within the body part. When *reductions* in technique from the average are called for, they should generally be made using mAs. When *increases* from the average are called for, they should normally be made using kVp. Second, the variable-kVp method used must be one of *systematic calculation*, achieving as much simplification and standardization as possible. Guesswork must be avoided. The following section provides guidelines in constructing such a systematic approach.

APPLYING THE VARIABLE kVp APPROACH

Over the years, three or four systems have been developed for varying the kVp according to body part thickness. One of the more common systems consisted of using a "base" kVp of 50, to which an additional value equal to *two times the measured part thickness* is added. Since we have advocated a high-kVp approach for digital imaging (Chapter 16, Table 16-3), we will modify this system to use a "base" kVp of *60* rather than 50. As an example, for a projection of the cervical spine measuring 13 centimeters in thickness, the calculation would be:

$$60 + (2 \times 13) = 86 \text{ kVp}$$

A proper *fixed mAs* value must still be established for each *general* body area, at least one for each of the following:

1. Upper extremities, non-grid
2. Lower extremities, non-grid
3. Pelvis and femurs
4. Head procedures
5. General abdomen and urinary procedures
6. Gastrointestinal barium procedures
7. Thorax and shoulder procedures

Grid and non-grid procedures must be grouped together; for example, if the knee is done with a grid, it should be in the same group as the gridded femur projection rather than with the non-grid lower extremities.

Practice Exercise #1

Using the "base-60" approach to variable kVp, a base mAs of 6.4 is established for all non-grid lower extremities. List the kVp and mAs that should be used for each of the following:

1. AP Foot measuring 7 cm:
2. AP Ankle measuring 9 cm:
3. AP Knee measuring 12 cm:

Solutions for kVp:

1. *AP Foot: $7 \times 2 = 14 + 60 = 74$*
2. *AP Ankle: $9 \times 2 = 18 + 60 = 78$*
3. *AP Knee: $12 \times 2 = 24 + 60 = 84$*

Answers:
1. *74 kVp at 6.4 mAs for the AP foot*
2. *78 kVp at 6.4 mAs for the AP ankle*
3. *84 kVp at 6.4 mAs for the AP knee*

As this system is applied to all procedures, you will recognize the levels of kVp fall very close to those in Table 16-3, *Recommended Optimum kVp for Digital Imaging* presented in Chapter 16.

For grid chests, a minimum of 110 kVp is recommended for the PA and 120 kVp is recommended for the lateral projection. When performing an AP chest non-grid with a *mobile unit*, 82 kVp may be used because of the high power of the battery-supplied constant-potential generator. This formula seems to work very well indeed across the range of radiographic procedures.

THE PROPORTIONAL ANATOMY APPROACH

A *proportional anatomy* system was first proposed by John Cullinan, RT, in the 1960s. Further developed by the author in the 1980s, it is founded on the basic proposition that, for the average body habitus, a proper radiographic technique for one body part can be derived from a known technique for another body part. A central pattern emerges that simplifies technique, because it allows the formulation of *anatomical groupings* that share the same overall technique. By *overall technique*, we mean the total combination of mAs and kVp. Using the 15 percent rule to adjust for any differences in kVp, the resulting mAs values can be compared for their proportionality.

For example, the following three techniques for different body parts, taken from a chart, are equal in terms of their *overall* impact upon remnant beam exposure to the image receptor plate:

AP Abdomen = 80 kVp at 30 mAs
AP Pelvis = 92 kVp at 15 mAs
Townes Skull = 86 kVp at 20 mAs

The *pelvis* technique is equal to the abdomen by a single-step application of the 15 percent rule. The Townes view of the skull is also equal to the abdomen, as follows: There is a 6-kVp increase which represents *one-half* of a 15 percent step; for the resulting exposure, this is equivalent to increasing the mAs half-way to a doubling, or three-halves, $3/2$, of the abdomen technique. The adjusted mAs, then, is $2/3$.

Proportional anatomy tables are presented in this chapter in two formats; the first one, Table 26-1, lists *groupings* of anatomy that share equivalent overall techniques. As with the examples given above, the radiographer may prefer different specific combinations of kVp and mAs, yet it is important to note that an identical technique *can* be used for all anatomy listed in a grouping. In the above example, the abdomen technique, 80 kVp at 30 mAs, could be used also for the AP pelvis and for the Townes skull views. In each of the groupings listed in Table 26-1, the same overall technique can be applied for all projections listed. This facilitates memorization.

Table 26-2 is formatted to show how to *derive* techniques from one body part to another. This is not only useful when a technique chart is not available or is incomplete, but also serves as a tool in *developing technique charts*, as demonstrated in the next section. Perhaps the most useful application of memorizing proportional anatomy derivations is that, when one inquires after a suggested technique for a particular view, one can then derive all the remaining techniques for a series without returning to ask or look these up. For example, having been given a technique for an AP skull, the radiographer armed with proportional anatomy will already know just how to adjust that technique for the lateral, Townes, or submentovertex projections.

Proportional anatomy is a tool that assists with *any* specific system for setting radiographic techniques—it does not compete with, but rather complements the variable kVp approach, as well as the fixed kVp approach. Nor does the proportional anatomy system designate whether to use kVp, mAs or any particular combination of the two; it only lists the desired end result in terms of adjusting the

Table 26-1

Proportional Anatomy Groupings

Group 1 (Reference: AP Abdomen, average 22 cm):		
AP Abdomen	Lat Hip	
AP Pelvis	(AP Dorsal Spine)	
AP Lumbar Spine	Townes (Grashey) Skull	
AP Hip		
Group 2 (Reference: AP Cervical Spine, Average 13 cm):		
AP Cervical Spine	Lateral Skull	
All C-Spine Views including Odontoid (if at same distance and grid)	AP & Lat Femur	
	(AP & PA Chest)	
	(AP Sacrum/Coccyx)	
AP Shoulder/Clavicle/Scapula)		
Group 3 Barium Studies (Reference: AP Upper GI solid column):		
Solid Column:	**Air Contrast:**	
UGI AP = BE AP	UGI AP = BE AP	
UGI Obl = BE Obl	UBI Obl = BE Obl	
UGI Lat = BE Lat	UGI Lat = BE Lat	
Group 4 Extremity Groupings:		
A. AP Foot	B. Lat Foot	C. AP Leg
AP Elbow	Lat Ankle	Lat Leg
	AP Humerus	Lat Humerus

Table 26-2

Technique by Proportional Anatomy

A. TRUNK:

AP Abdomen
AP Pelvis
AP Hip
Lat Hip
AP Lumbar Spine
AP Dorsal Spine
} all are roughly equal to each other and equivalent to a Townes (Grashey) Skull

30° Obliques on All Above = 1½ × AP

45° Obliques on all Above = 2 × AP (average thickness increase is about 4 cm)

Laterals on All Above = 4 × AP (average thickness increase is about 8 cm)

All Barium Studies = about ¼ mAs and up about 30 percent kVP

B. SKULL:

PA/Caldwell = ⅔ Abdomen, = ⅔ Townes, = 2 × Lateral, = 2 × C-Spine

Lateral = ½ C-Spine, = Shoulder = Femur = PA or AP grid Chest

Lateral Sinus = ⅓ PA

Townes = 1½ × PA, = 3 × Lateral, = AP Abdomen

Submentovertex = Townes (1½ × PA)

Waters = PA + 4–6 kVp

Mandibles, Mastoids, etc. = roughly ⅓ PA

C. CHEST:

PA CXR = AP Shoulder, = Cervical Spine, OR = AP Knee + 10 kVp IF BOTH are done non-grid or both in the bucky (if one is grid and one non-grid, technique must be adjusted by a factor of 4)

Lateral = 3–4 × PA (the common "double mAs and up 10 kVp" is equivalent to two doublings, or 4×)

D. PEDIATRICS:

Skull:	Newborn = ¼ Adult
	1 Year = ½ Adult
	5 Years = ¾ Adult
Torso:	Newborn = ¼ Adult
	1 Year = ½ Adult
	5 Years = ¾ Adult
Extremities:	Newborn = ⅙ Adult
	2 Years = ¼ Adult
	8 Years = ½ Adult
	12 Years = ¾ Adult

E. EXTREMITIES, ETC.:

Cervical Spine: AP = Shoulder grid, = Lateral Skull, = ½ PA Skull, = Femur

Obliques
Lateral*
Odontoid
} All equal to AP

*If lateral is done at 72" SID, but non-grid, technique will equal AP at 40" grid. Otherwise, adjustment must be made for distance or grid changes.

Femur = ⅔ AP Abdomen, = 2 × Knee grid, = Shoulder, = Cervical Spine; Lateral = AP

Knee = 2 × Ankle, = PA Chest – 10 kVp if BOTH are grid or both are non-grid, = AP Shoulder – 8 kVP or AP C-Spine – 8 kVP if grid
Lateral = AP

Leg = ½-way between AP Ankle and AP Knee
Lateral = AP

Ankle = 2 × AP Foot, = ½ AP Knee, = Lateral Foot
Lateral = AP

Foot = ½ AP Ankle (usually down 8 kVp, NOT less mAs), = ¼ AP Knee, = AP Elbow
Lateral Foot = 2 × AP, = Ankle

Shoulder = C-Spine, = Knee + 8 kVp, = ½ PA Skull, = PA Chest IF gird, = Lat Skull, = Femur

Humerus = ½-way between elbow and shoulder
Lateral = AP + 6 kVp

Elbow = 2 × Wrist, = ⅔ Humerus, = AP Foot
Lateral = AP + 4 kVp

Forearm = ½-way between wrist and elbow

Wrist = ½ Elbow, = ½ Foot, = 1.5 × Hand (usually up 8 kVp, not mAs)
Oblique = 1½ × PA; Lateral = 2 × PA

Hand = ⅓ Elbow, = ⅔ Wrist
Oblique = 1½ × PA; Lateral = 2 × PA

Digit = PA Hand

F. CASTS AND SPLINTS:

Plaster Casts: Dry plaster, small extremity = 2 × non-cast technique

Wet plaster, small extremity, OR
Dry plaster, large extremity (femur)
} = 3 × cast

Wet plaster, large extremity (femur = 4 × non-cast

1" Wood Splint, two ½-inch wood splints, or plaster half-cast = 1½ × non-cast technique

Pure Fiberglass cast or Air Splint = No Change in technique

Fiberglass Plaster Cast = 1½ × non-cast technique

Wet Fiberglass Cast = 1½ × non-cast technique

From Quinn B. Carroll, *Practical Radiographic Imaging*, 8th Ed. Springfield, IL: Charles C Thomas, Publisher, Ltd., 2007. Reprinted by permission.

overall technique—the radiographer must then decide whether to use kVp or mAs to make the adjustment. If kVp is used, it must be done in accordance with the 15 percent rule.

For example, Table 26-2 states that a knee technique may be derived by doubling an ankle technique. However, if this is done by doubling the mAs only, the optimum kVp for the knee listed in Table 16-2 might not be met. A combination of increasing the kVp by 8 percent and increasing the mAs by 50 percent would result in a doubling of the *overall* technique.

In applying these tables, bear in mind that they are applicable only to *average, adult patients*. In assessing a patient to derive an appropriate technique, the radiographer must be conscientious enough to consider the *shape* of the patient as well as the body habitus and size. Figure 26-1 illustrates how body torso shape, in cross-section, deviates from that of the average adult, ***C***, for premature babies, healthy infants, and fluid-distended or hypersthenic patients.

The average adult torso is oval in shape, measuring 22 cm in AP by 30 cm laterally. Table 26-2 recommends 4 times the AP technique for lateral projections of all torso anatomy, including the chest and abdomen. Note that this also agrees with the 4-centimeter rule, since two doublings in technique would be required where the anatomy is 8 centimeters thicker. This very workable rule-of-thumb fails, however, for very young children or for fluid-distended abdomens. Figure 26-1, ***B*** shows the cross-sectional torso shape for a healthy newborn, which is much rounder than an adult. A good rule-of-thumb for the chest or other torso anatomy of such a young child is to double the overall technique from the AP to the lateral view. The popular increase of 8 to 10 kVp accomplishes this overall doubling.

Experienced radiographers know that the same technique can be used for both the AP and the lateral projections on newborn intensive care babies that are well below normal birth weight—their bodies are nearly circular in shape (Fig. 26-1, ***A***). Many hypersthenic patients, and patients with fluid-distended abdomens acquire a very round torso shape (Fig. 26-1, ***D***). After increasing the technique according to the AP thickness of the patient, the radiographer often uses about the same technique for a lateral view. Radiographers must be vigilant in assessing this aspect of the patient's body habitus.

The human neck is nearly circular in cross-section, so it might be assumed that the same overall technique could be used for the lateral and oblique projections as for the AP projection. This is, in fact, true when all these projections are taken at the same distance (SID) and using the same grid ratio. Note that the technique adjustments listed in Table 26-2 do not take into account any changes in SID or grids, but assume all other factors to remain equal.

In a cervical spine series, for example, when changing from an AP to a lateral view, one may need to adjust for the difference of using 40-inch SID at the x-ray table for the AP to a 72-inch SID at the chest

Figure 26-1

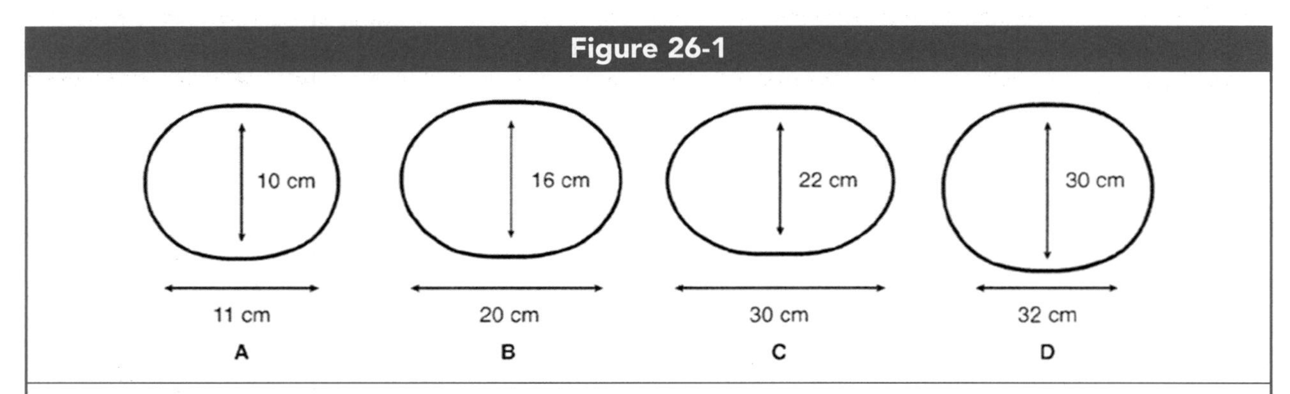

Cross-sectional diagrams of body torso shapes and dimensions (not to scale) of ***A***, a premature infant presenting a nearly round torso, ***B***, a healthy newborn with a more oval shape, ***C***, an average adult measuring 8 cm more in the lateral than in the AP dimension, and ***D***, a fluid-distended hypersthenic adult (or pregnant patient) with the AP and lateral dimensions roughly equal. (From Quinn B. Carroll, *Practical Radiographic Imaging*, 8th ed. Springfield, IL: Charles C Thomas, Publisher, Ltd., 2007. Reprinted by permission.)

board for the lateral. This adjustment is *not* listed in Table 26-2, but the radiographer must be mindful to make it as a correction for changing the SID. If all cervical spine projections were taken with the same distance and grid ratio, the proportional anatomy guidelines should be accurate.

Figure 26-2 presents a series of film-based radiographs demonstrating the general accuracy of the proportional anatomy approach. A note of caution regarding the derivation of oblique torso techniques: Note that in Table 26-2 oblique projections of the torso are broken down into 30-degree and 45-degree obliques. There is great variation in the amount of actual obliquity used in practice for lumbar spines, urograms and upper and lower GI series. The oblique for the lumbar spine is supposed to be a true 45 degrees which would require a doubling of technique (being 4 cm thicker as measured through the CR). However, most radiographers position this much closer to 30 degrees, which only requires a 50 percent increase in technique. Be sure to use the 30-degree guideline for these shallower obliques.

Figure 26-2

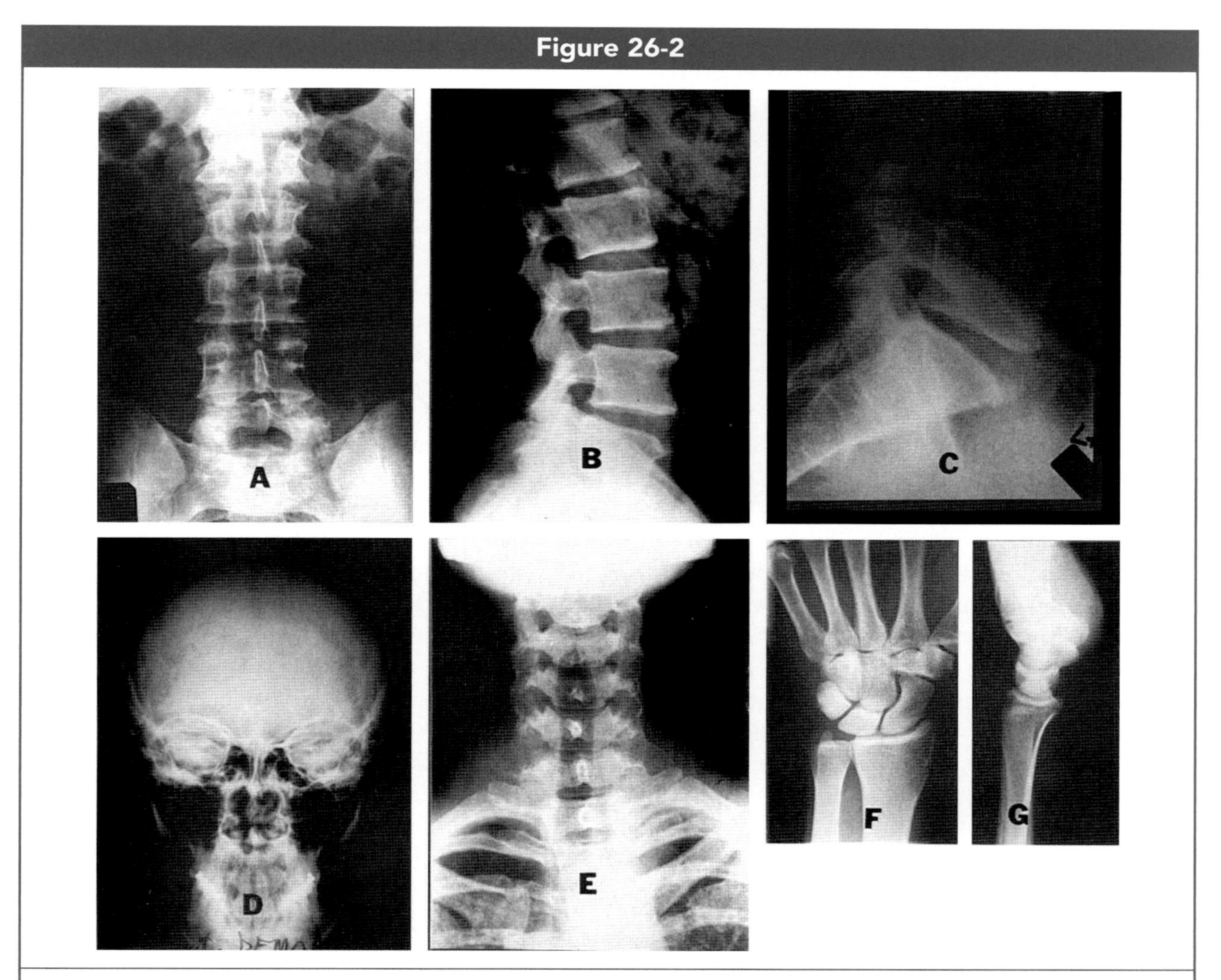

Film radiographs demonstrating the usefulness of proportional anatomy derivations on an average patient. On the lumbar spine series, the mAs was increased 4 times from ***A*** to ***B***, and the kVp increased by 10 from ***B*** to ***C***. The mAs used for the PA skull in ***D*** was 2/3 that for the L-spine in ***A***. From the PA skull in ***D*** to the AP C-spine in ***E***, mAs was cut in half. For the wrist radiographs, mAs was doubled from ***F*** to ***G***. No other changes were made. These adjustments are all in exact accordance with Table 26-2. (From Quinn B. Carroll, *Practical Radiographic Imaging*, 8th ed. Springfield, IL: Charles C Thomas, Publisher, Ltd., 2007. Reprinted by permission.)

The following practice exercise helps reinforce a number of key or landmark derivations from the proportional anatomy system. The answers are in Appendix #1.

EXERCISE #26-1

For each of the following changes in projection, write the proportional change in *overall* technique that would be indicated:

1. PA skull to lateral skull:
2. Lateral facial bones to PA facial bones:
3. AP cervical spine to oblique cervical spine:
4. Barium enema AP projection to 35-degree sigmoid projection:
5. PA chest to lateral chest:
6. Townes skull to AP abdomen:
7. PA wrist to AP elbow:
8. AP knee to AP shoulder:
9. AP leg to AP leg with dry plaster cast:
10. PA adult skull to AP infant skull:
11. AP foot to lateral foot:
12. AP ankle to lateral ankle:

USING TECHNIQUE CHARTS

Every radiology department should develop technique charts for each machine. In the final analysis, the philosophy of using technique chart amounts to a simple willingness to *write down what works*, so that all may benefit. The technique chart has been proven to reduce exposure errors considerably, and it is indispensable as a starting point and as a reference in setting radiographic techniques. It has been empirically demonstrated that the use of technique charts both reduces patient radiation exposure and helps control costs for the radiology department. In short, it is professionally unethical to ignore this tool for simplifying and standardizing technique.

Table 19-1 in Chapter 19 demonstrated from thousands of measurements that many body parts are so consistent in thickness (for adults) that a single listed technique may suffice as a guide. To adapt a technique chart for those body areas, primarily in the torso, which present substantial variation from one patient to the next, the chart should be broken down into columns according to the *measured thickness* of the part as illustrated in Table 26-3. This is a fixed-kVp chart, varying only the mAs for different thicknesses. (Specifically, the kVp's listed on this chart are based on the *optimum kVp* recommendations made in Table 16-2 in Chapter 16 for digital technology, considerably higher than used to be used for film/screen radiography.) Note that the techniques listed there follow the *four-centimeter rule*, doubling for each increase in thickness of four centimeters, and cutting it in half for each reduction of four centimeters.

Once a chart such as this is developed, having been tested for thorough accuracy, *the entire chart should be able to be adjusted up or down by a certain ratio, making the same change "across the board" for all listed techniques, to adapt it to different x-ray machines*. This is because there *is* consistency in the ratio of thickness between different body parts for the average body habitus (the basis for the *proportional anatomy approach* to technique).

The chart in Table 26-3 is presented primarily to illustrate the usefulness of its *format*, which demonstrates economy of space (using only two sheets of paper for the entire chart), while at the same time providing thorough coverage of all body parts including pediatric landmarks. The torso section is organized according to variable part thickness; each column represents a change in part thickness of 2 centimeters, a fine enough distinction to cover all needed adjustments in technique, and facilitating application of the four-centimeter rule.

The extremity section lists a single technique for the average-measured body part, but can be organized with three or four columns in order to list a technique for "small," "average," and "large" patients. In some cases it may be desirable to add a "very large" category. It is helpful to standardize what is meant by these terms (something easy to do yet rarely done). An excellent suggested guideline is to define them as follows:

- SMALL = ¾ average thickness for the part
- LARGE = 1¼ to 1½ times average thickness for the part (up to 50% larger than average)
- VERY LARGE = 1½ to 2 times average thickness for the part

Figure 26-3 illustrates other formats that allow the mA and time to be listed separately if desired, rather than the total mAs. This overall format in

Table 26-3

Digital Optimum kVp Technique Chart

Abbreviations: GD = Grid, NG = Non-Grid ▸ **FOR CR, DOUBLE ALL mAs's**			**TORSO / SKULL** Total MAS by PART SIZE										
PROCEDURE	**Notes**	**VIEW**	**kVp**	Ave CM	–6cm	–4cm	–2cm	**AVE**	+2cm	+4cm	+6cm	+8cm	+10c
GRID CHEST	72"	PA/AP	120	22	0.5	0.6	0.9	1.2	1.8	2.4	3.6	4.8	7.2
		LAT	120	30	1.5	2	3	4	6	8	12	16	24
NON-GRID CHEST	72"	AP	86	22	0.6	0.9	1.2	1.8	2.4	3.6	4.8	7.2	9.6
		LAT	96	30	1.2	1.8	2.4	3.6	4.8	7.2	9.6	14	18
SUPINE CHEST 40" NG		AP	86	22	0.3	0.4	0.6	0.8	1.2	1.6	2.4	3.2	4.8
RIBS / Sternum	72"	AP↑Diap	70	22	3.6	5	7.5	10	15	20	30	40	60
		OBL↑Diap	70	24	6	8	12	16	24	32	48	64	96
	*72"	AP↑Diap	86	22	3.6	5	7.5	10	15	20	30	40	60
ABDOMEN / IVP	GD	AP/PA	90	22	3.6	5	7.5	10	15	20	30	40	60
		30°OBL	90	24	5	7.5	10	15	20	30	40	60	80
PELVIS / HIP	GD	AP	90	22	3.6	5	7.5	10	15	20	30	40	60
HIP (Unil or Groin)	GRID	LAT	84	22	3.6	5	7.5	10	15	20	30	40	60
SACRUM	AP/Coccyx All		84	20	3.6	5	7.5	10	15	20	30	40	60
	GD	LAT	90	28	7.5	10	15	20	30	40	60	80	120
LUMBAR SPINE	GD	AP	90	22	3.6	5	7.5	10	15	20	30	40	60
		45°OBL	90	26	7.5	10	15	20	30	40	60	80	120
		LAT	90	30	15	20	30	40	60	80	120	160	240
		L5/S1 SP	**102**	30	15	20	30	40	60	80	120	160	240
THORACIC SPINE	GD	AP	84	22	3.6	5	7.5	10	15	20	30	40	60
	Breathing LAT		68	30	12 mA / 2 s		12 mA / 4 sec			25 mA / 4 s		50 mA / 4 s	
TWINING C/T	GD	LAT	86	28	7.5	10	15	20	30	40	60	80	120
CERVICAL SPINE	40" GD	AP/Odon	86	14	1.3	1.8	2.6	3.5	5.2	7	11	15	22
	72" **GD**	OBL/LAT	86	14	3.7	5	7.5	10	15	20	30	40	60
	72" **GD**	OBL/LAT	86	14	1.3	1.8	2.6	3.5	5.2	7	11	15	22
SKULL	GD	PA/Cald	90	19	2.2	3	4.5	6	9	12	18	24	36
		LAT	86	15	1.2	1.8	2.2	3	4.5	6	9	12	18
		Townes	90	22	3	4.5	6	9	13	18	26	36	52
SINUSES / FACIAL BONES	GD	PA/Cald	84	19	2.2	3	4.5	6	9	12	18	24	36
		Waters	78	20	2.2	3	4.5	6	9	12	18	24	36
		Lat	72	15	1.2	1.5	2.2	3	4.5	6	9	12	18
AIR CONTRAST U.G.I. / B.E.	GD	AP/PA	92	22	1.8	2.5	3.7	5	7.5	10	15	20	30
		OBL/SIG	92	24	2.6	3.8	5.2	7.5	11	15	22	30	44
		LAT	92	30	7.5	10	15	20	30	40	60	80	120
SOLID-COLUMN U.G.I. / B.E.	GD	AP/PA	120	22	1.5	2	3	4	6	8	12	16	24
		OBL/SIG	120	24	1.8	2.5	3.7	5	7.5	10	15	20	30
		LAT	120	30	5.2	7.5	11	15	20	30	40	60	90

(Continued)

Table 26-3 (Continued)

Digital Optimum kVp Technique Chart

Abbreviations: GD = Grid, NG = Non-Grid, SFS = Small Focal Spot,
PIGOST = Pig-O-Stat$_{TM}$ ▸ **FOR CR, DOUBLE ALL mAs's**

EXTREMITIES

PROCEDURE	Notes	VIEW	kVp	mAs
HAND	PA / All Fingers		64	0.3
	SFS	OBL	64	0.4
		Fanned Lat	64	0.6
WRIST	SFS	PA	64	0.5
		OBL	64	0.7
		LAT	64	1
FOREARM	SFS	AP	72	0.3
		LAT	72	0.5
ELBOW	SFS	AP	72	0.4
		OBL	72	0.6
		LAT	72	0.8
HUMERUS	SFS	AP	80	0.4
		LAT	80	0.35
TOES	SFS	All	66	0.3
FOOT	SFS	AP	72	0.35
		OBL	72	0.5
		LAT	72	0.7
CALCANEUS	SFS	PD	76	1
		LAT	76	0.5
ANKLE	SFS	AP/OBL	76	0.5
		LAT	76	0.5
LEG	SFS	AP	76	0.7
		LAT	76	0.7
TABLETOP KNEE	TBLTOP	AP	80	0.7
		LAT	80	0.7
BUCKY KNEE	GD	AP/OBL	84	2.5
		LAT	84	2.5
FEMUR	GD	AP/LAT	86	5
HIP	GRID	FRG/GROIN	86	10
SHOULDER	GD	AP/Transax	86	3.5
		Transthoracic	90	15
CLAVICLE	GD	AP/PA	86	3
SCAPULA	GD	AP	96	3
		LAT	86	5

FACIAL

PROCEDURE	Notes	VIEW	kVp	mAs
MANDIBLE	GD	PA	76	8
		OBL LAT	66	8
	NON-GRID OBL LAT		60	5
ORBIT Rheese	GD	PA OBL	76	5
SINUSES	GD	SMV	80	10
ZYG. ARCH	TBLTOP	SMV	62	0.5
NASAL BONE	TBLTOP	LAT	54	0.5

PEDIATRIC

PROCEDURE	Notes	VIEW	KVp	mAs
"PREMIE" CHEST	40" NG	AP	64	0.3
		LAT	64	0.3
INFANT CHEST	40" NG	AP	68	0.5
		LAT	76	0/5
2-YEAR CHEST	72" PIGOST	PA	74	0.8
		LAT	84	0.8
6-YEAR CHEST	72" NG	PA	80	1
		LAT	90	1.5
ABDOMEN / IVP / PELVIS SPINES	INFANT **NG**		70	0.8
	8-YEAR GD		76	2
	6-YEAR GRID		80	5
CERVIC. SP	2-YEAR	All	70	0.8
SKULL (PA/AP)	INFANT **NG**		70	0.8
	2-YEAR GD		76	2
	6-YEAR GD		80	4
UPPER EXTREMITY	INFANT		58	0.08
	2-YEAR		62	0.1
	6-YEAR		66	0.3
LOWER EXTREMITY	INFANT		62	0.08
	2-YEAR		66	0.15
	6-YEAR		70	0.3

From Quinn B. Carroll, *Practical Radiographic Imaging*, 8th Ed. Springfield, IL: Charles C Thomas, Publisher, Ltd., 2007. Reprinted by permission.

Table 26-3 should also be used for *variable kVp charts*, that is, the two-centimeter breakdown for thicknesses is still recommended, along with the above definitions for small, average, large and very large body parts. For a variable kVp chart, a single average *mAs* should generally replace the kVp column to the left, as shown in Table 26-4, and each measurement column then presents the kVp to be used, consistent with the 15 percent rule.

The student will note, upon examining Table 26-4, that several entries at both the lowest and highest ranges of technique are shaded with green and have an "m" written next to the number—these are necessary because of two important limitations to the variable kVp approach: First, remember from our previous discussion that care must be taken to never drop below the *minimum kVp* listed in Table 16-1 in Chapter 16. (Insufficient penetration of the tissue of interest will result, which cannot be compensated for.) Therefore, the shaded entries on the left of this chart indicate reductions in the listed *mAs* because the kVp should not be decreased any further. Second, most x-ray machines do not allow kVp settings much higher than 130. Therefore, for very large patients which fall to the right portion of this technique chart, once 130 kVp has been reached, only by changing the mAs can further increases in overall technique be effected.

The use of calipers to measure the part thickness is a must for torso procedures and for pediatric radiography. Radiographers can become very adept at estimating body part thicknesses after acquiring some experience, and may only require calipers for exceptional cases. The student must not take this practice in the wrong light—the less the experience, the greater the need for the calipers. In order to *develop* this skill, it is essential that calipers be used to check one's estimates over a considerable period of time (several months), using persistent practice to develop accuracy.

The above statement should not be taken to excuse any radiographer from using calipers where it is legally required, *which is the case in most states*. Actual measurements will always be more reliable than any estimation, as long as the calipers are used properly as described in Chapter 19.

The use of technique charts has its limitations; charts are designed to provide consistency in *routine* situations. No device can replace the need for the radiographer's careful assessment of each individual patient and situation, adapting exposure factors as needed for pathology and unusual conditions.

Figure 26-3

VARIABLE TIME CHART

Procedure	kVp	mA	Time for Each View		
WRIST	60	200	PA	OBL	LAT
			.025	.035	.05

VARIABLE MA /TIME CHART

Procedure	kVp	mA / Time for Each View		
WRIST	60	PA	OBL	LAT
		200 / .025	200 / .035	200 / .05

DEVELOPING A CHART FROM SCRATCH

The proper construction and use of technique charts must include the following:

1. A quality control program for calibration of x-ray machines and standardization of image processing.
2. Input from all radiographers using a given machine as the chart is developed and corrected for it.
3. Strict enforcement by administrators that individual radiographers not be allowed to alter a chart in any way once it has been developed and tested. Frequently, when individuals make such alterations, they are using incorrect distances, receptors, or positioning. The next radiographer to use the room may use correct factors with poor results because of rewritten techniques.
4. Encouragement by administrators that all radiographers use the system.
5. Periodic checks (every 6 months) and updates of all technique charts allowing full input from staff radiographers.

Table 26-4

Variable kVp Technique Chart

Abbreviations: GD = Grid, NG = Non-Grid ▸ **FOR CR, DOUBLE ALL mAs's**			**TORSO / SKULL** KVP by PART SIZE										
PROCEDURE	**Notes**	**VIEW**	**mAs**	Ave CM	–6cm	–4cm	–2cm	**AVE**	+2cm	+4cm	+6cm	+8cm	+10cm
GRID CHEST	72"	PA/AP	2.5	22	86	94	102	110	118	126	134	4mAs	5mAs
		LAT	8	30	94	102	114	120	130	12m	16m	24m	32m
NON-GRID CHEST	72"	AP	1.8	22	1mAs	1.4m	76	80	86	92	98	104	112
		LAT	3.6	30	2.7m	78	84	90	96	104	112	120	134
SUPINE CHEST 40" NG		AP	0.8	22	0.4m	0.6m	76	80	86	92	98	106	114
RIBS / Sternum	72"	AP**8**Diap	6	22	4.5m	68	74	80	86	92	98	106	114
		OBL**8**Diap	10	25	8mAs	68	74	80	86	92	98	106	114
	***72"**	AP**9**Diap	10	22	76	82	88	94	100	108	116	126	134
ABDOMEN / IVP	GD	AP/PA	10	22	76	82	88	94	100	108	116	126	134
		30EOBL	10	24	80	86	92	98	106	114	122	130	15m
PELVIS / HIP	GD	AP	10	22	76	82	88	94	100	108	116	132	15m
HIP (Unil or Groin)	GRID	LAT	8	22	76	82	88	94	100	108	116	132	12m
SACRUM	AP/Coccyx All		10	20	72	78	84	90	98	104	110	120	128
	GD	LAT	10	28	84	90	98	106	112	122	130	15m	20m
LUMBAR SPINE	GD	AP	10	22	76	82	88	94	100	108	116	132	15m
		45EOBL	10	26	80	86	94	102	110	118	127	136	15m
		LAT	10	30	88	94	102	110	118	126	134	15m	20m
		L5/S1 SP	10	30	88	94	102	110	118	126	134	15m	20m
THORACIC SPINE	GD	AP	8	22	66	72	78	84	90	96	104	112	120
	Breathing LAT		68kV	30	12 mA / 2 s		12 mA / 4 sec			25 mA / 4 s		50 mA / 4 s	
TWINING C/T	GD	LAT	12	28	84	82	88	96	104	112	120	130	18m
CERVICAL SPINE	40" GD	AP/Odon	4	14	3mAs	74	80	86	94	100	108	116	126
	72" **GD**	OBL/LAT	12	14	9mAs	74	80	86	94	100	108	116	126
	72" **GD**	OBL/LAT	4	14	3mAs	74	80	86	94	100	108	116	126
SKULL	GD	PA/Cald	6	19	4.5m	78	84	90	98	104	112	120	130
		LAT	4	15	3mAs	74	80	86	94	100	108	116	126
		Townes	6	22	4.5m	82	88	96	102	110	118	126	9m
SINUSES / FACIAL BONES	GD	PA/Cald	6	19	3mAs	74	80	86	94	102	110	118	126
		Waters	6	20	3mAs	78	84	90	98	104	112	120	130
		Lat	3	15	3mAs	70	76	82	88	94	100	108	115
AIR CONTRAST U.G.I. / B.E.	GD	AP/PA	5	22	74	80	88	94	100	106	115	122	132
		OBL/SIG	5	24	80	85	92	98	112	120	130	7.5m	10m
		LAT	5	30	88	94	102	110	120	128	136	7.5m	10m
SOLID-COLUMN U.G.I. / B.E.	GD	AP/PA	120k	22	1.2m	1.5m	2.2m	3mAs	4.5m	6m	9m	12m	18m
		OBL/SIG	120k	24	1.5m	2m	3m	4mAs	6m	8m	12m	16m	24m
		LAT	120k	30	4.5m	6m	9m	12m	18m	24m	36m	48m	72m

(Continued)

Table 26-4 *(Continued)*

Variable kVp Technique Chart

Abbreviations: GD = Grid, NG = Non-Grid,
PIGOST = Pig-O-Stat$_{TM}$ ► **FOR CR, DOUBLE ALL mAs's**

EXTREMITIES

PROCEDURE	Notes	VIEW	mAs	kVp
HAND	PA / All Fingers		0.3	56
	SFS	OBL	0.3	60
		Fanned Lat	0.3	66
WRIST	SFS	PA	0.5	56
		OBL	0.5	60
		LAT	0.5	64
FOREARM	SFS	AP	0.4	62
		LAT	0.4	68
ELBOW	SFS	AP	0.5	64
		OBL	0.5	66
		LAT	0.5	70
HUMERUS	SFS	AP	0.5	68
		LAT	0.5	66
TOES	SFS	All	0.4	54
FOOT	SFS	AP	0.5	64
		OBL	0.5	66
		LAT	0.5	72
CALCANEUS	SFS	PD	0.5	74
		LAT	0.5	66
ANKLE	SFS	AP/OBL	0.5	70
		LAT	0.5	68
LEG	SFS	AP	0.5	74
		LAT	0.5	72
TABLETOP KNEE	TBLTOP	AP	0.5	76
		LAT	0.5	74

EXTREMITIES *(Continued)*

PROCEDURE	Notes	VIEW	mAs	kVp
BUCKY KNEE	GD	AP/OBL	2	76
		LAT	2	74
FEMUR	GD	AP/LAT	3	84
HIP	GRID	FRG/GROIN	6	90
SHOULDER	GD	AP/Transax	3	78
		Transthoracic	15	90
CLAVICLE	GD	AP/PA	2.5	78
SCAPULA	GD	AP	2.5	78
		LAT	2.5	78

FACIAL

PROCEDURE	Notes	VIEW	mAs	kVp
MANDIBLE	GD	PA	5	86
		OBL LAT	5	78
	NON-GRID OBL LAT		3	74
ORBIT Rheese	GD	PA OBL	3	86
SINUSES	.GD	SMV	5	94
ZYG. ARCH	TBLTOP	SMV	0.6	60
NASAL BONE	TBLTOP	LAT	0.5	56

Suppose you are in charge of a newly-constructed radiology clinic and must devise some technique charts "from scratch." Using the rules and relationships discussed in Chapters 15, 16, 17, 19 and 26, you can write a reasonably accurate preliminary technique chart completely from just a few test exposures taken on plexiglass "phantoms" of the abdomen, skull and one extremity. (A chest phantom is also recommended when available.) With digital imaging systems, the only way to ensure that excessive exposure has not been used for a particular projection is to read out the *exposure index* number on the image. These exposure indices are fully explained in Chapter 32. For these few test exposures on phantoms, the technical factors must be adjusted and the exposures repeated until an exposure index very close to (within 10% of) the ideal index given by the manufacturer of the equipment is achieved. Once the ideal exposures are determined for these few test objects, an *entire preliminary chart* can be developed as demonstrated in Exercise 26-2, by using proportional anatomy, the 4-centimeter rule, and the 15 percent rule.

For full reinforcement of how this is done, run through the entire Exercise 26-2 for a fixed kVp chart, and Exercise 26-3 for a variable kVp chart.

Formats for technique charts seem to be as varied as the individuals who use them. Charts can be made using a box of index cards with a separate card for each procedure or projection, by using a drawing of the AP and lateral body with techniques written near each body part, using a loose-leaf "flip chart," or using a table or grid lined out on a couple of pieces of paper that are hung on the wall in the control booth.

Whichever format is used, it is important to standardize all other factors, such that only one factor is varied from column to column. Tables can only accommodate two variables, one for the rows and one for the columns, and since one (usually the columns) must be used for differing part thickness measurements, the other (the rows) can be used for one variable but not all three. Exercise 26-4 illustrates all three resulting general formats that might be employed: variable mAs, variable time, and variable kVp. (Two examples of variable *exposure time* charts are presented, one with standardized kVp *and mA* listed to the left, the other with only the standardized kVp to the left, and the mA/time combination indicated for each position.) Try these exercises and check if your answers come reasonably close to those in the Appendix #1.

Any chart should include at least one column for "notes," such as the use of a grid, the bucky tray versus table-top technique, the small focal spot, special receptor plate sizes or other non-routine information. Also, at least for all procedures in the human

EXERCISE #26-2

Constructing a Variable mAs Technique Chart from Scratch

EXERCISE: Assuming that the listed technique was good, complete the rest of the chart using proportional anatomy and the 4 cm rule of thumb.

Procedure	Projection	KVP	Ave CM	MAS						
Lumbar Spine	AP	80	22	18 cm	20 cm	**22 cm** **24**	24 cm	26 cm	28 cm	30 cm
	Oblique	80	26	22 cm	24 cm	**26 cm**	28 cm	30 cm	32 cm	34 cm
	Lateral	80	30	26 cm	28 cm	**30 cm**	32 cm	34 cm	36 cm	38 cm
Skull	PA	76	19	15 cm	17 cm	**19 cm**	21 cm			
	Lateral	76	15	11 cm	13 cm	**15 cm**	17 cm			
Shoulder	AP	76	14	10 cm	12 cm	**14 cm**	16 cm			
Knee	AP	70	12	8 cm	10 cm	**12 cm**	14 cm			

EXERCISE 26-3

Constructing a Variable kVP Technique Chart from Scratch

EXERCISE: Assuming that the listed technique was good, complete the rest of the chart using proportional anatomy, the 4 cm rule, and the 15% rule in steps.

Procedure	Projection	mAs	Ave CM	KVP						
Lumbar Spine	AP	12	22	18 cm	20 cm	**22 cm 96**	24 cm	26 cm	28 cm	30 cm
	Oblique	12	26	22 cm	24 cm	**26 cm**	28 cm	30 cm	32 cm ↑ mAs	34 cm ↑ mAs
	Lateral	12	30	26 cm	28 cm	**30 cm**	32 cm ↑ mAs	34 cm ↑ mAs	36 cm	38 cm
Skull	PA	10	19	15 cm	17 cm	**19 cm**	21 cm			
	Lateral	10	15	11 cm	13 cm	**15 cm**	17 cm			
Shoulder	AP	10	14	10 cm	12 cm	**14 cm**	16 cm			
Knee	AP	8	12	8 cm	10 cm	**12 cm**	14 cm			

EXERCISE 26-4

Variable Time, Variable mA/Time, and Variable kVP Technique Chart Formats

Directions: Assuming that the listed technique was good, complete the rest of the chart using proportional anatomy, the 4 cm rule, and the 15% rule in steps.

Procedure	KVP	mAs for Each View		
		PA	OBL	LAT
WRIST	60	4	6	8

VARIABLE TIME CHART

Procedure	KVP	mA	Time for Each View		
			PA	OBL	LAT
WRIST	60	200			

(Continued)

EXERCISE 26-4 *(Continued)*

VARIABLE MA/TIME CHART

Procedure	KVP	mA/Time for Each View		
WRIST	60	PA	OBL	LAT

VARIABLE KVP CHART

Procedure	mAs	kVp for Each View		
WRIST	4	PA	OBL	LAT

torso, the chart should be broken down for different part thicknesses listed in columns. It is suggested that thickness columns be made in intervals of 2 centimeters above and below the average, with twice as many columns on the "thicker than average" side. For a table format, the result should look similar to Table 26-3.

Finally, although every mobile x-ray machine should have a technique chart attached to it, this does not always happen, and when it does charts are not always complete. *Every radiographer should carry a "portable" technique reference* in the form of a telephone/address booklet or similar pocket-sized notepad. Simply list procedures alphabetically, and make a note of techniques that are proven to work by experience. Be sure to always make careful note of the *thickness* of the part for which the technique applied, using at the least the categories described above of *small*, *average*, *large* and *very large*. Mobile equipment often requires techniques that are very different from those in the radiology department. If it works, write it down!

SUMMARY

1. By standardizing routine projections, exposure factors, and image processing, the probability of errors is reduced, and the range of potential causes for errors is narrowed.
2. Technique charts should be provided for every x-ray unit, and can be simplified to the point where only a single variable is changed from one view to the next within a radiographic series. The use of charts and calipers helps reduce both patient exposure and departmental costs.
3. With digital imaging systems, the *variable kVp* approach to radiographic technique is generally advocated, because of its potential for reducing patient exposure, but it must be implemented so as to ensure adequate kVp for sufficient penetration at all times. To do this, increases should generally be made using kVp, while decreases in technique should usually be made using the mAs.
4. When converting to CR equipment, any needed increases in radiographic technique should be made using kVp rather than mAs. A 15 percent across-the-board increase in kVp for all techniques is recommended.
5. The "base-60" approach to variable kVp works extremely well for developing variable kVp techniques.
6. The proportional anatomy system for deriving radiographic techniques is useful and applicable regardless of the specific approach adopted, because it is based upon consistent ratios of thickness between average body parts.
7. Radiographers must be conscientious of the shape of the torso in addition to its general thickness.

8. A properly developed technique chart can have all of its listed techniques adjusted upward or downward by the same ratio to adapt it to different x-ray machines.
9. A technique chart can be developed "from scratch" using proportional anatomy, the 4 cm rule, and landmark exposures using phantoms, then refined with practice.
10. Proper implementation of technique charts includes quality control, full input from staff, incentives and enforcement, and periodic follow-up checks for accuracy.
11. "Pocket" technique booklets are recommended as a back-up for mobile radiography and other situations where a regular technique chart may not be immediately accessible.

REVIEW QUESTIONS

1. What are the three general areas in which standardization should be sought?

2. A well-developed technique chart will have thickness measurement columns for every __________ cm and alter only __________ variable from view to the next within a radiographic series.

3. When converting from a film/screen system to CR, instead of an across-the-board doubling of mAs, what technique adjustment would be preferred?

4. Define "optimum mA":

5. Compared to the average adult, the torso of an infant requires __________ (less, more, or the same) adjustment in technique from a frontal view to a lateral view.

(Continued)

REVIEW QUESTIONS *(Continued)*

Using proportional anatomy, what is the change in overall technique needed for each of the following:

6. AP to lateral lumbar spine:

7. Barium enema AP to 30-degree oblique projection:

8. AP ankle to AP knee:

9. PA wrist to lateral wrist:

10. PA skull to lateral skull:

11. AP cervical spine to AP shoulder:

12. Using the "base-60" variable kVp system, what kVp should be used for a body part measuring 9 cm thickness?

13. Using the "base-60" variable kVp system, what kVp should be used for a body part measuring 24 cm thickness?

(Continued)

REVIEW QUESTIONS *(Continued)*

14. Especially for use with "pocket" technique booklets, how have we defined the typical thicknesses for the following classifications of patients, as a ratio of the average:

    A. Small:

    B. Large:

    C. Very large:

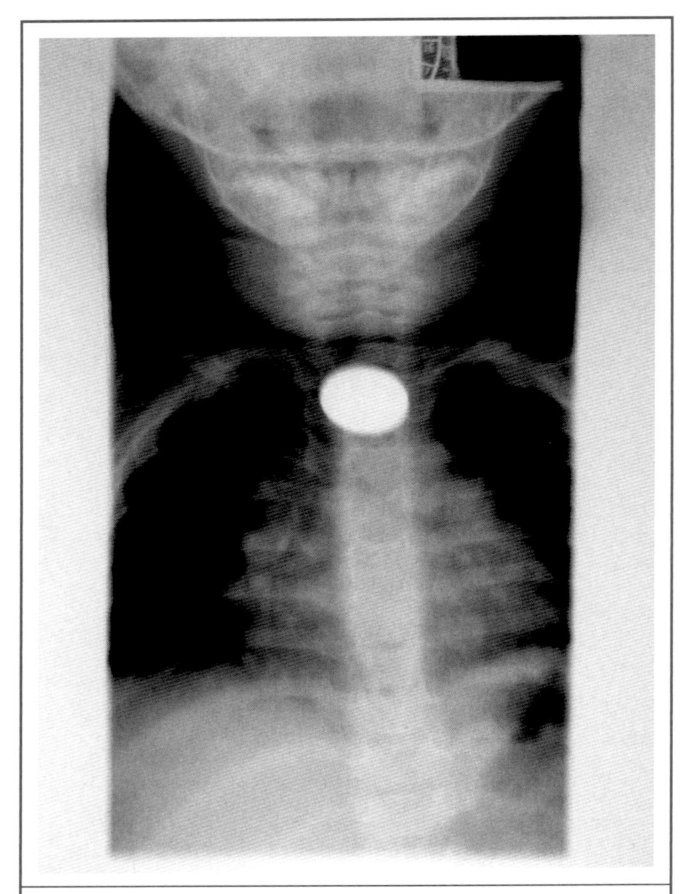

This infant swallowed a penny which became lodged in the esophagus. Coins aspirated within the trachea tend to lodge end-on because the horseshoe shape of the cartilages leaves a "soft" spot posteriorly formed by ligaments. The esophagus is flattened between the trachea and the spine, so coins tend to lodge crosswise.

Chapter 27

USING AUTOMATIC EXPOSURE CONTROLS (AEC)

Objectives:

Upon completion of this chapter, you should be able to:

1. List those technique factors which must still be "manually" set by the radiographer even when the AEC is engaged.
2. Describe the limitations imposed by the minimum response time of the AEC circuit, and how technique can be compensated to correct for them.
3. State the general formula for appropriate back-up mAs or time, and why it is so important.
4. Explain why back-up times pre-set by the manufacturer must be checked and often overridden.
5. Describe typical formats for the intensity (density) control.
6. Describe the proper use of the intensity (density) control setting.
7. Explain five limitations for using the AEC, situations for which "manual" technique may be more appropriate.
8. Describe the major factors that determine the proper AEC detector cell configuration for a particular projection.
9. List AEC detector cell configurations in order, from the least resulting exposure to the most.
10. List the common causes of overexposure using AEC.
11. List the common causes of underexposure using AEC.
12. List essential components for an AEC technique chart.
13. Describe the limitations of programmed exposure controls.

The first automatic exposure control device was demonstrated by Russell H. Morgan in 1942. Automatic exposure controls or *phototimers* were developed for the purpose of achieving more consistent exposures, reducing retakes, and ultimately saving radiation exposure to the patient. All automatic exposure controls work on the same physical principles, based upon the ability of radiation detection devices to convert radiant energy into an electrical current.

The term *phototimer* originates from the fact that early AEC devices employed a fluorescent screen which emitted light *photons* when struck by x-rays. This light was then absorbed by a photomultiplier tube that converted its energy into electrical current.

Nearly all modern AEC's use simpler *gas ion chambers* that normally consist of a flat, rectangular double-plate of very thin aluminum or plexiglass with a layer of gas encased in it. The ion chamber induces an electrical current when the gas atoms are ionized by impinging radiation, freeing electrons from the gas atoms. These electrons are then attracted to and strike a positively-charged anode plate at one end of the chamber. Continuing to be attracted toward the positive terminal within a circuit, they flow out of the anode plate and down a very thin wire, thus becoming an *electrical current.*

The basic circuit for an AEC is illustrated and described in Chapter 8 and diagrammd in Figure 8-16. Electrical charge from the induced current is stored up on an electrical *capacitor* until it reaches the preset threshold amount that corresponds to an ideal amount of radiation exposure. The *thyratron* in the circuit then releases the charge in a surge of electricity that is used to activate an electromagnet. The electromagnet pulls open the exposure switch, terminating the exposure.

When a patient is turned sideways, or when larger patients are radiographed, more radiation is absorbed within their bodies so that there is less radiation *per second* striking the ion chambers. It therefore takes a *longer time* for the capacitor to reach the preset amount of charge, so that the radiation exposure is lengthened until the desired exposure is attained.

It should be emphasized that an AEC only controls the exposure time and consequently the total mAs used for an exposure. *Optimum kVp and optimum mA must still be determined and set by the radiographer when using the AEC,* in accordance with all of the principles discussed in Chapters 15 and 16. If the set kVp is insufficient to achieve proper penetration of the body part, sensors will detect a reduction in the exposure rate and the AEC will allow a longer exposure to try to compensate. As explained in Chapter 16, *no amount of exposure can compensate for inadequate penetration.* Some areas of the image will still be too light, regardless of the increased exposure time. Nor will increasing the *density control* setting properly correct this problem because it has no effect on x-ray beam penetration.

Optimum mA was defined in Chapter 15 as the maximum mA available for a given focal spot size, which does not overload the x-ray tube heat capacity. Defining the optimum mA is a bit more complicated for automatic exposure control, because there is the additional consideration of *minimum response time* for the AEC circuit to properly operate.

MINIMUM RESPONSE TIME

All electronic devices require a minimum amount of time and signal (input) in order to operate. The automatic exposure control is no exception. It takes time, albeit thousandths of a second, for the circuit to detect and react to the radiation received.

Minimum response times vary greatly from one radiographic unit to another and between manufacturers. When a new unit is installed, it is a good idea for the quality control technologist to post or otherwise ensure that staff radiographers are made aware of its minimum response time (MRT), especially if the unit will be frequently used for pediatric radiography. Typical MRT's range from 0.002 seconds for state-of-the-art equipment to 0.02 seconds for older units.

High power generators are often employed to reduce exposure times for pediatric radiography, sometimes in combination with increased digital processing speeds. It is possible for the actual exposure time to be reduced to such an extent that it is too short for the AEC circuit to respond to. When the machine does not shut off until it reaches the MRT, overexposure to the patient results. In such a circumstance, the best alternative is to *decrease the mA station* until sufficient exposure times are produced.

For example, let us assume an MRT of 0.005 seconds for a new high-power generator. To keep exposure times short, the radiographers are in the habit of using the 300 mA station. At this mA, the *minimum total mAs* that the machine can produce is:

$$300 \text{ mA} \times 0.005 \text{ seconds} = 1.5 \text{ mAs}$$

Suppose that an ideal technique for a PA chest on a child is 65 kVp and 0.8 mAs. If the radiographer uses the AEC and the 300 mA station on this child, an overexposure of almost twice too much radiation will be delivered to the child. *Note that in this situation, adjusting the density knob to a "minus" setting will not help.* The machine is not capable of making a shorter exposure. The proper solution is to reduce the mA to *100*, since 200 mA would still produce a minimum mAs of 1.0.

Optimum mA takes on a new meaning, then, when applied to AEC exposures: It is defined as an mA high enough at a given focal spot size to minimize motion, but *not* so high that resulting exposure times are shorter than the MRT.

BACK-UP mAs OR TIME

Although it is a rare occurrence, it is possible for the AEC circuit to fail. A "back-up time" or "back-up mAs" must be set to prevent excessive duration of the exposure in this event. There are two important reasons for taking this precaution: One is to prevent excessive heat overload of the x-ray tube which may damage the anode, but more importantly, excessive and unnecessary radiation exposure to the patient must be prevented.

On older equipment, the regular electronic timer should be set as a "back-up timer." Most newer x-ray units set a *total back-up mAs* rather than a back-up time. Some authors recommend a back-up time of 50% more, or 1½ times the expected exposure time for the anatomy. Generally, the back-up time should never be set to more than *2 times the expected exposure time or mAs for a particular projection.* For example, if a typical manual technique for an AP abdomen were 15 mAs, what would be an appropriate back-up time when using the AEC at 300 mA? The manual exposure time at this mA would be:

$$15 \text{ mAs} / 300 \text{ mA} = 0.05 \text{ second}$$

Multiply this exposure time by 2 for a resulting back-up time of 0.1 second. Using the same rule for total mAs, an appropriate back-up mAs for this projection would be (2 × 15 =) 30 mAs. The four-centimeter rule (Chapter 19) can be used for different thicknesses of patients in estimating adjusted back-up times.

An appreciation for the importance of back-up time or mAs is gained by examining the extremely short exposures required for frontal chest projections: Let us assume that a particular chest projection requires an exposure time of 1/40 (0.025) second. The AEC fails, and a very quick radiographer realizes the exposure is continuing beyond normal and releases the manual exposure switch after just 2 seconds. The patient will have received the equivalent of *80 chest x-rays!* Clearly, radiographers should be *certain* that a back-up time or mAs is always preset. An old practice was to always set the back-up time at 1 or 2 seconds for all AEC exposures. This is unacceptable, since in the scenario just given even 1 second would result in 40 times too much exposure.

A common error while using the AEC is to forget to activate the correct *bucky* mechanism, such as when performing a chest radiograph at the vertical chest board but leaving the *table* bucky on. The bucky selection button also activates the AEC detectors for that bucky. In this case, the exposure would continue indefinitely at the chest board, while the detectors at the table "wait" for an adequate exposure level to be reached. Once again, an appropriate *back-up time or mAs* is the only way to prevent excessive exposure to the patient. Always check all stations at the console before making an AEC exposure, including the bucky selection, and the density control to make sure it has not been left on a *plus* or *minus* setting from the previous patient.

Preset Automatic Back-up mAs or Time

Most modern units have all of the back-up mAs values preset by the manufacturer upon installation, but *many modern x-ray machines are preset to excessive back-up times or mAs values.* A department survey recently conducted by the author discovered two dramatic examples:

Example #1: A DR unit of *Brand A* displays the back-up mAs at the "manual" *mAs* knob when the AEC is engaged. Setting AEC for a PA chest projection, the back-up mAs displayed is *80 mAs.* The average mAs for the PA chest listed on the technique chart for this unit (also provided by the manufacturer upon installation) is 4 mAs. The back-up mAs is *20 times* the average mAs for a PA chest. If the wrong bucky were activated, the patient would receive the equivalent of 20 chest x-rays before the exposure was terminated.

Example #2: In the same department, a CR unit of *Brand B* displays the back-up mAs on a touch-screen read-out under the heading *max* whenever the AEC is engaged. *Maximum* or *back-up* mAs values listed include:

- 500 mAs for all barium procedures
- 1000 mAs for the abdomen and IVP
- 1000 mAs for the PA chest, 1250 for the lateral
- 800 mAs for the C-spine, shoulder girdle, and L-spine
- 2000 mAs for the oblique sternum, 1600 for AC joints
- 800 mAs for the PA skull, but 2000 mAs for the lateral

> If we take 50 mAs (intentionally overstated) as an average for an abdomen projection, with other body parts following the ratios of *proportional anatomy* from that value, we can characterize the above amounts as generally *20 times* the average or more. What is worse, some appear to have not been calibrated even in the *right direction*: Note that the skull settings (listed last) *double* for the lateral projection over the PA, when the lateral skull is *thinner* anatomy. Perhaps the installer of this equipment was focused on heat-load to the x-ray tube rather than on patient exposure—what is clear is that *these settings are not consistent with the anatomy.*

These trends point up a serious need for radiographers to get involved in the interest of their patients. First, on an individual level, it is important for each radiographer to appreciate the magnitude of overexposure to the patient that failure of the AEC system or engaging the wrong bucky can cause. Many of these preset back-up times or mAs values *can be overridden* and reduced to a more appropriate level at the touch of a knob or button. *Back-up values of 2 to 4 times the expected exposure should cover nearly all contingencies that might arise.*

Second, upon installation of new equipment, quality control technologists and managers should get personally involved with the manufacturer in determining appropriate preset values for back-up times or mAs. We should take ownership of the issue as a profession, and insist upon having input into this process.

THE AEC INTENSITY (DENSITY) CONTROL

A knob or series of buttons may be found on the console labeled *density* which applies to the AEC circuit. This control increases or decreases the preset sensitivity of the thyratron by specific percentages, so that the exposure time will automatically be extended or shortened by those amounts. With digital imaging, a more appropriate term for this device would be the *intensity control*, since it actually determines the intensity of exposure at the receptor plate rather than the end result density of the postprocessed image. So far, however, the label "density" control has continued to be used by manufacturers.

There are various formats for this control; some have only three settings, for *small*, *average*, and *large* patients. In this case the *small* setting will usually cut the exposure time to one-half, and the *large* setting will double it. Others are labeled as ½, ¾, N, 1½ and 2, in which the *N* represents *normal* or average and the other numbers are fractions or factors of that amount. (For example, the 1½ setting is 50% more than average, or 1½ times average.) Many have seven settings labeled as –3, –2, –1, N, +1, +2 and +3 or even a range from –5 to +5. In some cases there is no labeling but only symbols showing a bar or light that becomes wider to indicate an increase and narrower to indicate a decrease. In this format, unless otherwise specifically labeled, each of the seven stations usually represents a 25 percent change. (For example, the +2 setting would *not* be a doubling of the exposure time but rather *2 sets* of 25% increases for a total increase of 50% from *N*.)

Remember from Chapter 19 *the minimum change rule* which states that overall technique must be increased by at least 35 percent (or more than ⅓) in order to make a significant difference. This means that a density setting of +1, if it translates to a 25 percent increase, is not likely to result in any substantial improvement in exposure to the image receptor—this was dramatically demonstrated in Historical Sidebar 19-1 (page 287) in Chapter 19 *using AEC exposures*. We conclude that *a minimum intensity control setting of +2 is recommended when more exposure is needed for any reason.*

(The –1 setting usually *does* result in a significant decrease because –25% results in ¾ exposure, the exact inverse of 4⁄3 or a 33% increase.)

As long as the AEC circuit is functioning properly, the intensity (density) control should only need to be used infrequently, for special projections as noted in this chapter, and for special circumstances. When constant adjustments are made for routine positions, it is indicative of either a calibration problem with the equipment or poor positioning by the radiographer. For example, Figure 27-1 illustrates how poor centering over the lumbar spine can result in a significant portion of the detector cell having soft tissue rather than *bone tissue* over it. Since more radiation penetrates through the soft tissue, this can result in the AEC shutting off the exposure somewhat early, and underexposure of the tissue of interest which is the bony spine.

Figure 27-2 is an example of how setting the intensity control can be an integral part of automatic exposure control for some procedures. In this unilateral "frog" position for the hip, you can see that the detector, centered over the neck portion of the femur, is doing a perfect job of maintaining image density there, but that in doing so the *acetabulum* and *head of the femur* (black arrow) are much too light. These are essential anatomy of interest for a hip series, but are in a corner of the view where the detector cannot properly measure and adjust for the exposure. One solution is to maintain proper centering over the neck of the femur as shown, but to use a +2 or higher setting on the intensity control to darken this area.

At this point, we must emphasize that digital imaging systems can generally compensate for overexposure at the receptor plate, but cannot compensate for information which is simply missing due to underexposure at the receptor plate. If an AEC exposure is too light in some portions of the remnant beam image, insufficient data or skewed data is provided to the computer regarding the overall image; rather than just coming out too light in this one area, as an old film radiograph would, different histogram analysis and post-processing errors can occur that result in a bizarre image that is too light or too dark, has too low or too high contrast, or which manifests extensive mottle.

To prevent processing errors, sufficient exposure must be provided to the receptor plate across all areas of the image field. It is therefore *critical* to identify those projections, such as this "frog" lateral hip, where a *plus* setting at the density control is indicated for routine use.

When equipment is out of calibration, the intensity control setting can provide a *temporary* coping tool while waiting for service. The intensity control *itself* can be out of calibration, and there is a simple way for a radiographer to check it: Most AEC units include a mAs indicator on the console which reads out the actual total mAs used *after* the exposure is

Figure 27-1

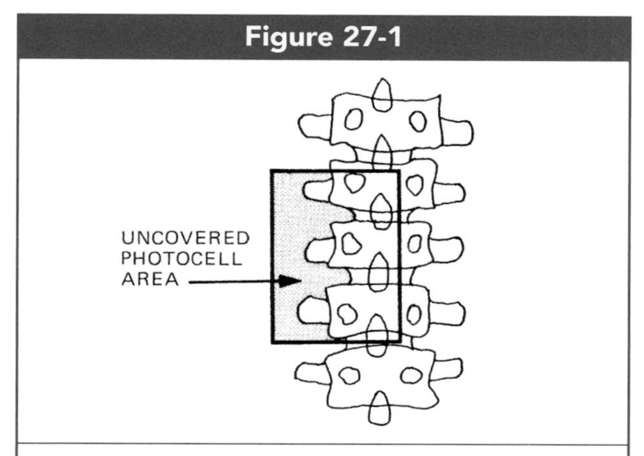

Off-centering of the oblique spine by only 3/4 inch (2 cm) places about one-fourth of the detector cell area outside the tissue of interest (bone). Underexposure of the spine will result from the AEC shutting off too soon. (From Quinn B. Carroll, *Practical Radiographic Imaging*, 8th ed. Springfield, IL: Charles C Thomas, ,Publisher, Ltd., 2007. Reprinted by permission.)

Figure 27-2

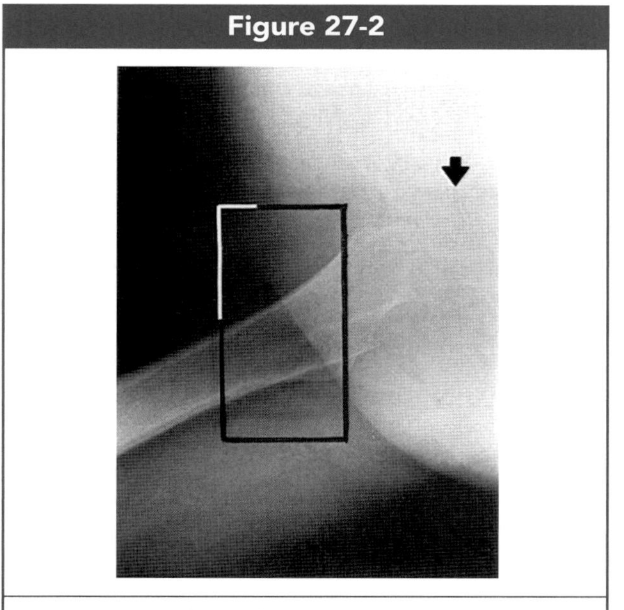

Using the AEC for a "frog" lateral hip projection, the anatomy over the detector cell is properly exposed on this film radiograph, but anatomy of interest includes the head of the femur and acetabulum (arrow), which are underexposed, because they are in the corner of the field and no detector cell is activated there. Digital systems *cannot* correct for this lack of information in the image. (From Quinn B. Carroll, *Practical Radiographic Imaging*, 8th ed. Springfield, IL: Charles C Thomas, Publisher, Ltd., 2007. Reprinted by permission.)

completed. If a plexiglass "phantom" of the knee, skull or torso is available, or some other absorber to simulate a body part, the post-exposure readouts can be written down for the different intensity control settings, starting with the *N (normal), average or zero* setting. Ensure that no other variables are changed between exposures, and write down the mAs readout immediately after each exposure. Now, for each *plus* or *minus* setting, make the following calculation:

$$\frac{N - A}{N} \times 100$$

where *N* is the mAs readout for the *N* or *average* setting, and *A* is the mAs readout for each plus or minus setting. This calculation simply yields the *percentage* by which the technique was changed up or down from "N."

Observe the following results for one x-ray machine tested by the author:

+1 = +17%	–1 = –14%
+2 = +42%	–2 = –51%
+3 = +75%	–3 = –67%

It seems clear that this machine was designed for each setting to be an increment of 25 percent; the –2 and +3 settings fit almost exactly. The +2 and –3 settings each fall about 8 percent short of a 25 percent increment. Most interesting are the +1 and –1 settings: The +1 setting is only increasing technique by 17 percent. We have stated that even a 25 percent increase is not likely to make a substantial difference in exposure intensity, so a 17 percent increase would *certainly* fit this description. The –1 setting is *44 percent short* of the intended incremental decrease. Radiographers using this machine would do well to go right to the +2 or –2 settings when needed.

On another machine tested by the author, it was discovered that both the –2 and -3 settings actually *increased* the mAs readouts by nearly double! This should serve as a wake-up call to radiographers and quality control technologists, that intensity controls can be and often are seriously out of calibration. The procedure to check them is easy to perform and well worth the short time it takes. Stations that are far out of calibration should be marked to avoid until a service technician corrects the problem.

LIMITATIONS OF AEC

The AEC circuit is automatically engaged when the machine is set for spot-filming during fluoroscopic procedures. For overhead radiographs, AEC can be used at the radiographer's discretion. While repeat rates have been reduced by the use of AEC for many procedures, it was never intended to be used on *all* procedures. AEC does have its limitations. Radiographers who use AEC as a "crutch," as an escape from the mental work needed to set manual techniques where appropriate, may *cause* repeats rather than prevent them. The major constraints on the use of AEC are as follows:

1. AEC should never be used on anatomy that its too small or narrow to completely cover at least one detector cell. This includes most distal extremities, and extremities in general on small children. Detectors measure the *average* amount of radiation striking the area they cover. Portions of the detector cell not covered will receive too much radiation and the AEC will shut off too soon, resulting in underexposure. In digital x-ray imaging, this can be a major cause of mottle in the final image.

2. Care must be taken when radiographing anatomy that is *peripheral*, that is, close to the edge of the body, such as the clavicle (Fig. 27-3), the mandible, and lateral projections of the sternum or scapula. In each of these situations the CR is normally centered close to the edge surface of the body part, so that portions of the detector cell may extend beyond the part into the raw x-ray beam. As it averages the measured exposure, the detector cell will terminate the exposure early, resulting in underexposure and mottling of the digital image.

Radiographers have been known to adapt a position to the point where the CR is centered one or two inches away from the "textbook" centering point, just to be sure the detector cell is covered so that the AEC can be used and "manual technique" avoided. *Proper positioning should never be compromised merely to allow use of the AEC.*

3. Even when AEC is used for proper applications, positioning and centering must be perfected. For an ideal exposure, *the tissue of interest, not just any anatomy, must cover most of the detector cells used.* For some procedures, the demands of nearly

Figure 27-3

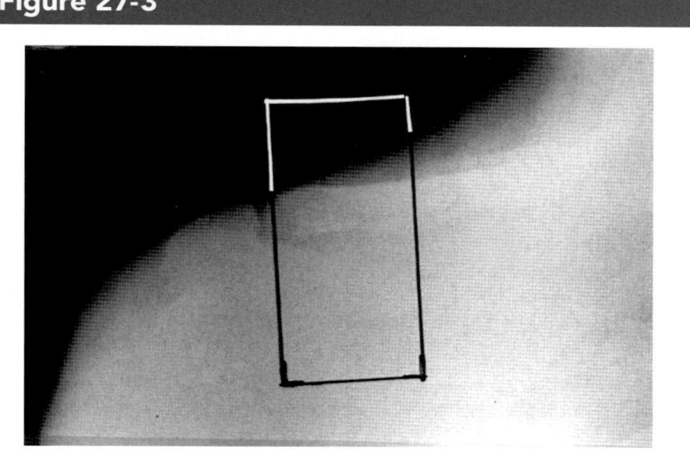

Film radiograph using AEC for an AP clavicle projection, in which this peripheral anatomy resulted in the detector cell extending above the anatomy and receiving "raw" x-ray exposure. This resulted in the AEC shutting off too soon, underexposing the image. (From Quinn B. Carroll, *Practical Radiographic Imaging*, 8th ed. Springfield, IL: Charles C Thomas, Publisher, Ltd., 2007. Reprinted by permission.)

perfect positioning may not be within reasonable limits. For example, as shown in Figure 27-1, the lumbar spine is about the same width as a detector cell. Therefore, in order to cover the cell with *bone* tissue rather than soft tissue, the centering must be nearly perfect. This is not difficult to achieve for the AP projection, but is much more difficult for the *oblique* positions, especially on large patients.

4. *The x-ray field must be well-collimated to the anatomy of interest.* When the field is too large, excessive amounts of scatter radiation from the table and body will cause the AEC to shut off prematurely. *Digital imaging is particularly vulnerable to this problem.* A frequent result is a "washed-out" appearance to the digital radiograph in which it appears light overall and begins to manifest mottle. Side-to-side collimation for cervical spine projections, "swimmers" views, and "groin lateral" (cross-table lateral) hips must be *tighter than the plate* size to prevent these effects when using the AEC.

5. *The AEC should never be used when there is any type of radiopaque surgical apparatus, orthopedic corrective devices, extensive orthodontic dental work, or other large metal artifacts which cannot be readily removed from the area of interest.* Such artifacts leave large areas over the detector cell where almost no exposure is being received (see Fig. 27-4 in Historical Sidebar 27-1). The detector cell *averages* the exposure rate across its entire area. The effect of large radiopaque objects is to lower this measured rate, such that the AEC will stay on much longer in order to reach its preset cut-off value.

Historical Sidebar 27-1 gives two demonstrations of how this resulted in gross overexposure of the surrounding anatomy of interest for film-based radiography. Digital imaging systems are able to compensate for *general* overexposure and restore an image to diagnostic quality, but this scenario with AEC presents a very special case in which only a portion of the image is grossly overexposed, while another section of the image is nearly devoid of all data. This huge discrepancy can lead to a number of different histogram analysis and postprocessing errors that result in an image that is too light or too dark, or has too low or too high contrast.

In addition there is the important issue of *excessive patient exposure* when the AEC overextends the exposure time while trying to compensate for a large radiopaque artifact. For example, when attempting automatic exposure for "frog" or "groin" lateral hip projections with a hip prosthesis present, exposure times have been known to run all the way to the back-up time or mAs. Overexposure to the patient is likely to be from 4 to 10 times the necessary radiation, and, as discussed under *Back Up mAs or Time*, it can be as much as 20 times the average. The unacceptability of such levels of radiation goes without saying.

To prevent any risk of this happening to the patient, *"manual" technique should be set for all such situations. AEC should never be used when large radiopaque artifacts are present.* Radiographers must be conscientious enough to screen patient's charts and x-ray requisitions, and to communicate verbally

HISTORICAL SIDEBAR 27-1: Figures 27-4 and 27-5 are both examples of large radiopaque artifacts within the patient that caused an AEC exposure too stay on much too long, overexposing the surrounding anatomy of interest. Since these are both film radiographs, the predictable result was much too dark an image. For modern digital imaging, the effects upon the resulting final image are *not* predictable, because they are the result of various possible mathematical misinterpretations by the computer processing algorithms. The best way to avoid these errors is to employ *manual technique*.

In Figure 27-4, a radiographer unwisely attempted an AEC exposure for a "frog" lateral hip projection on a patient with a large hip prosthesis.

In Figure 27-5, a radiographer attempted to use AEC on an AP projection for the odontoid process through the open mouth. The area of the detector cell is outlined, which included an entire metal tooth and extensive orthodontic bridges. These still appear blank white even after the AEC grossly extended the time in an attempt to average the overall exposure. For a film-based radiograph, the result was nearly pitch-black density of the cervical spine and odontoid, rendering the image diagnostically useless.

Figure 27-4

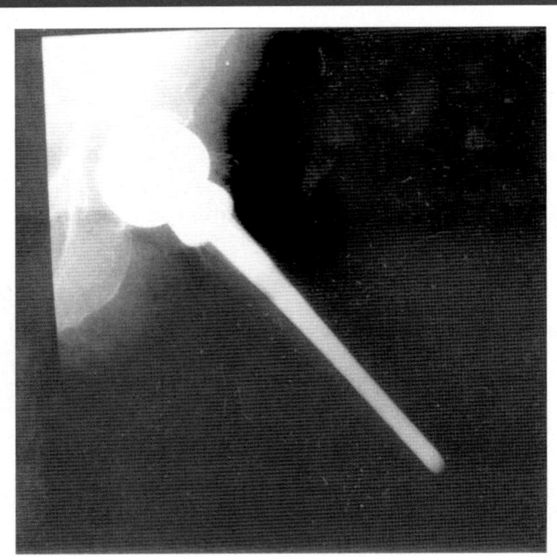

Large radiopaque artifacts such as this hip pin leave large areas over the AEC detector cell in which almost no exposure is being received. Because the detector cell *averages* the measured exposure across its entire area, the exposure time will be extended and overexposure will result at the receptor plate and to the patient. (From Quinn B. Carroll, *Practical Radiographic Imaging*, 8th ed. Springfield, IL: Charles C Thomas, Publisher, Ltd., 2007. Reprinted by permission.)

Figure 27-5

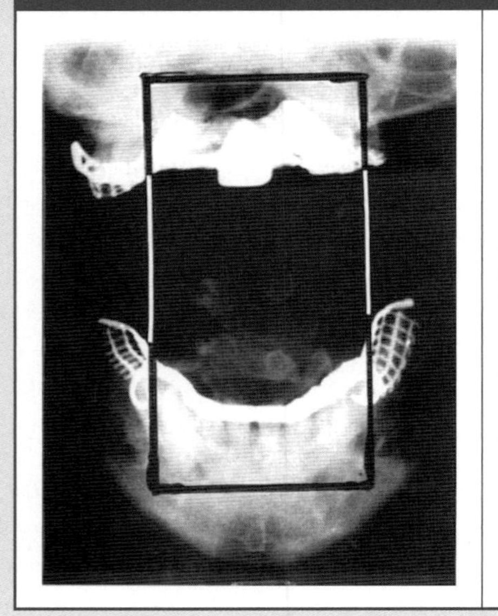

A film radiograph of the odontoid process in which x-ray absorption by dental hardware resulted in the exposure time being extended by the AEC, resulting in extreme overexposure for the anatomy of interest (the upper cervical spine). (From Quinn B. Carroll, *Practical Radiographic Imaging*, 8th ed. Springfield, IL: Charles C Thomas, Publisher, Ltd., 2007. Reprinted by permission.)

with the patient so that the presence of any hardware can be anticipated prior to radiographing the patient.

DETECTOR CELL CONFIGURATION

The location of the detector cells is normally demarcated by a triad of rectangles drawn on the radiographic tabletop or on a wall-mounted "chest board" (Fig. 27-6). Sometimes the detectors are demarcated by dark lines within the field light, projected from a plastic insert in the collimator. For this type, the indicated size of the cells will be accurate *only* at a specified distance (usually 72 inches or 180 cm for chest units).

The actual detectors are made of very thin aluminum and located behind the tabletop, but in front of the grid, imaging plate and bucky tray. At extremely low kVp, these chambers and the very thin wires leading from them can show up on the finished image, but at most radiographic techniques with anatomy in the image they are not normally visible.

Positioning should be such that the detector cells are covered as much as possible with the *tissue of interest*, and *the thickest portions of the anatomy are over an energized cell*. This will assure adequate exposure to the image receptor plate prior to the termination of exposure. When using AEC, *selection of the best configuration of the three detector cells to energize is part and parcel of the positioning effort.* When a radiographer has developed a keen sense of the relative densities of the tissues and is able to visualize the location of the internal anatomy well, he or she can be more creative in the selection of the detector cells used, and to great advantage.

Figure 27-6

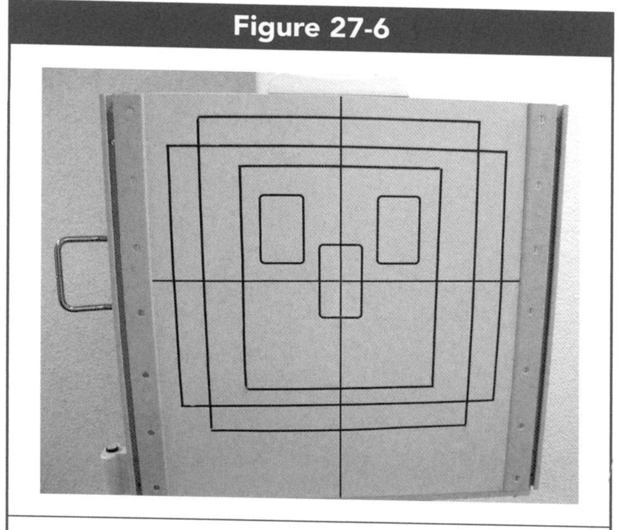

The typical location of the triad of AEC detector cells for a vertical bucky or x-ray table (small rectangles).

A classic example of this principle is provided by the "frog" lateral hip radiograph, Figure 27-2 (page 413). In order to produce sufficient exposure over the acetabulum and head of the femur, the *side* detector cell which lies under the *thicker, medial portion of the hip area* can be energized, either alone, or in combination with the center detector, as shown in Figure 27-7. (Even with this configuration, an intensity control setting of +2 is also recommended.)

Figure 27-7

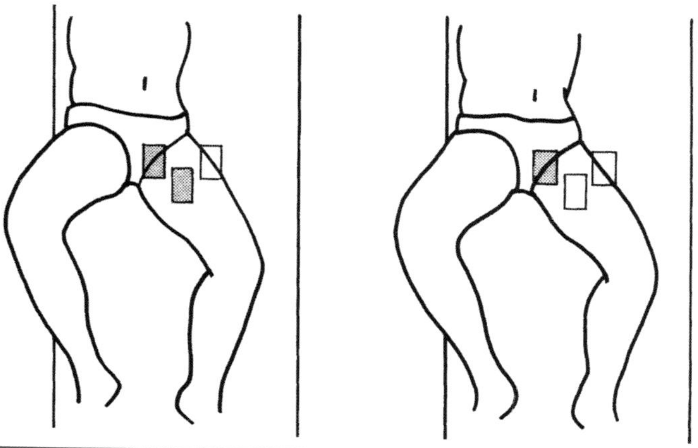

Two options for a more proper configuration of the energized AEC cells for a "frog" lateral hip (See Figure 26-2), placing the cells over the thickest portion of the anatomy to ensure adequate exposure at the receptor plate. (From Quinn B. Carroll, *Practical Radiographic Imaging*, 8th ed. Springfield, IL: Charles C Thomas, Publisher, Ltd., 2007. Reprinted by permission.)

Chest radiography with digital systems presents a unique situation in selecting the best configuration of AEC detector cells. For the PA projection, many radiographers are in the habit of using only the two side cells, a practice held over from the days of film-based radiography. Remembering that two-thirds of all chest x-rays are done primarily for evaluation of the *heart*, it is essential that the heart and mediastinum be penetrated and demonstrated in light shades of gray rather than as a blank white silhouette image. For digital images, such underexposed portions can also cause mottle to appear. Remember that digital systems *cannot* compensate where no information penetrates through to the receptor plate in the first place. It is simply better to *assure adequate exposure through the heart and mediastinum*, and let the digital processor then lighten the lung areas.

As illustrated in Figure 27-8, the detector cell ***A*** to the patient's *right*, lies primarily over lung tissue, whereas the *left* cell, ***B*** typically has a considerable portion of the heart (dotted area) overlying it. The center cell, ***C***, is over the densest anatomy, overlying the spine as well as heart and some abdominal tissue (shaded area). The lined area over detector cell ***A*** represents the additional breast tissue for female patients which overlies the lower portions of both cells ***A*** and ***C***. With all this in mind, we can formulate the following list of detector cell configurations with the relative exposures that result. Each step results in only slightly more radiation exposure reaching the receptor plate than the previous step.

Least exposure	=	Right cell only
Increased exposure	=	Right and left cells
Increased exposure	=	All three cells
Increased exposure	=	Left cell only
Increased exposure	=	Left and center cells
Greatest exposure	=	Center cell only

With digital systems, it is recommended that the exposure level achieved by using *all three cells* be considered a *minimum*, and that the use of the two side cells only be abandoned. In fact, many radiographers have come to prefer using the *center cell only*, which you will find produces consistent quality images while maintaining the exposure index number within perfectly acceptable limits.

CHECKLIST OF AEC PRECAUTIONS

Digital imaging systems do such a great job of covering up errors that in most cases it is not apparent from the image itself that the AEC has been used incorrectly. Where film-based images once turned out obviously too light or too dark, the only thing radiographers really have to go on now is the exposure

Figure 27-8

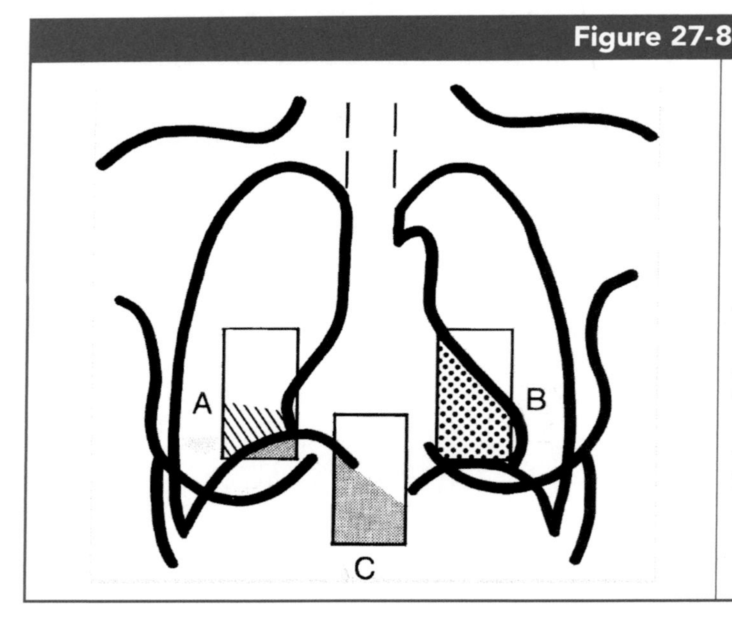

For digital radiography of the chest, activation of only the two side cells will result in underpenetration through the mediastinum, because approximately ¾ of the cell area is over lung tissue. To ensure adequate exposure through the mediastinum, adding the center detector cell, at a minimum, is recommended. (Many prefer the center cell *only*.) (From Quinn B. Carroll, *Practical Radiographic Imaging*, 8th ed. Springfield, IL: Charles C Thomas Publisher, Ltd., 2007. Reprinted by permission.)

indicator (index) number. Yet, since this is an indication of the exposure to the patient, *it has become a critical ethical and professional issue for radiographers to develop a habit of monitoring the exposure index.*

One staff radiographer made the comment that, "I don't care what the exposure index is, as long as that image looks right." Such a remark is tantamount to saying, "I don't care how much radiation the patient receives." Our very mission as a profession is to obtain the highest quality images *with the least radiation exposure to the patient.* The day that we lose sight of this will be the day when people with no training can be hired to operate x-ray equipment. Each individual student and radiographer must be willing to make a personal commitment to minimize patient exposure, then act on that commitment in daily practice.

High exposure index numbers indicate overexposure to the patient for a particular projection. However, it is just as important to monitor for *low* index numbers, because these can indicate insufficient penetration of the x-ray beam or underexposure that can lead to a mottled image, to name just two examples that might require repeating the projection. Since every repeated exposure essentially *doubles* the amount of radiation to the patient for that view, underexposures are just as important to monitor as overexposures in the overall scheme of minimizing radiation to the patient.

Tables 27-1 and 27-2 form a complete checklist of errors that would explain high or low exposure index numbers. Whenever the exposure indicator is unusually high or low, the radiographer should go over this checklist mentally, in a process of elimination, to isolate the probable cause of the incorrect exposure.

Table 27-1

Causes of Overexposure Using AEC

- Wrong bucky activated
- Needed exposure time *less* than minimum response time (small anatomy, high mA)
- Density control left on *plus* setting from previous patient
- Electronic malfunction of the AEC (Back-up buzzer sounds)
- Incorrect detector cell configuration, such that activated cell(s) lie under tissue *denser* or *thicker* than the tissue of interest
- Presence of radiopaque artifacts or appliances within the anatomy (hip or knee prosthesis)
- Presence of external radiopaque artifacts such as lead sheets or sandbags over the sensor

AEC TECHNIQUE CHARTS

Use of the AEC does not obviate the need for the radiographer to set technique factors. The only thing that an AEC controls automatically is the *exposure time*, and therefore the resulting total mAs. An appropriate kVp must still be set "manually" to ensure adequate penetration. An *optimum mA* station must be selected by considering the minimum response time for the machine, the associated focal spot size, the probability of motion, and heat load to the x-ray tube. The small focal spot should be selected for extremity procedures. An intelligent choice for the configuration of the detector cells must be made. An appropriate back-up time should be assured.

For all these reasons, it makes good sense to construct and use *AEC technique charts* for all AEC units. An example is provided in Table 27-3, showing

Table 27-2

Causes of Underexposure Using AEC

- Backup time shorter than needed exposure time (esp. on large patient)
- Density control left on *minus* setting from previous patient
- Inadequate collimation (excessive scatter radiation reaching sensors
- Incorrect detector cell configuration, such that activated cell(s) lie under tissue *less dense* or *thinner* than the tissue of interest
- Detector cells not fully covered by the tissue of interest:
  - Anatomy too peripheral
  - Anatomical part too small
  - Specific tissue area too small
  - Specific tissue area not centered over selected detector cells

Table 27-3

AEC Technique Chart

Procedure	View	kVp	mA	Backup Time(s)	Detector Selection	Density Setting	Notes
CHEST	PA/AP	106	200	0.1	■□■	N	72"
	LAT	116	200	0.2	□■□	N	72"
6-YR CHEST	PA	90	200	0.05	□■□	–2	72"
RIBS ↑ DIAPH	AP/OBL	60	400	0.2	□■□	–1	72"
RIBS ↓ DIAPH	AP	76	400	0.2	□■□	N	72"
ABDOMEN	AP	80	400	0.2	■■■	N	
IVP	AP/OBL	80	400	0.2	■□■	N	
G.B. / Coned	AP/OBL	80	400	0.2	□■□	N	
PELVIS	AP	80	400	0.2	□■□	N	
HIP Unilateral	AP/LAT	80	400	0.2	■□□	+2	*medial cell
LUMBAR SPINE	AP	80	400	0.2	□■□	N	
	PB:	80	400	0.4	□■□	N	
	LAT	80	400	1.0	□■□	N	
	L5 / S1	90	400	1.0	□■□	N	
THORACIC SPINE	AP	74	400	0.2	□■□	N	
	LAT	60	10	5	□■□	N	breathing
TWINING C/T	LAT	76	400	0.4	□■□	N	
CERVICAL SPINE	AP	76	200	0.2	□■□	N	40"
	Odontd	76	200	0.2	□■□	N	40"
	Lat/Obl	76	200	0.5	□■□	N	72"
SKULL (SINUS) (FACIAL)	PA/Cald	80	400	0.2	□■□	N	
	Waters	80	400	0.2	□■□	N	
	Townes	80	400	0.3	□■□	N	
	LAT	76	400	0.1	□■□	N	
Mastoid/Coned	LAT	76	400	0.1	□■□	N	
SHOULDER	AP	76	200	0.2	□■□	N	
FEMUR	AP/LAT	76	200	0.2	□■□	N	
KNEE / LEG	All	70	200	0.1	□■□	N	
U.G.I.-A.C.	All	90	400	0.2	□■□	N	air con
B.E.-A.C.	AP/OBL	90	400	0.2	□■□	N	air con
	Sigm/Lat	90	400	0.2	□■□	N	air con

From Quinn B. Carroll, *Practical Radiographic Imaging*, 8th Ed. Springfield, IL: Charles C Thomas Publisher, Ltd., 2007. Reprinted by permission.

columns for all this information and additional notes. By providing technique charts like this, AEC practices within a department can be more standardized, and the learning curve for students and new radiographers shortened.

Modern radiographers must learn *both AEC and "manual" technique skills* thoroughly, and maintain clinical proficiency in both. Because of the advent of AEC, there is *more*, not less, to learn about radiographic technique. For radiography of the distal extremities, cross-table projections, adaptations for trauma situations, and the presence of unavoidable artifacts in the projection, "manual" techniques must be used. There will always be situations in which "manual" technique is more appropriate than using the AEC. These skills must not be lost.

PROGRAMMED EXPOSURE CONTROLS

Many modern x-ray machines have been designed to simplify technique manipulation for the radiographer by programming preselected technical factors into computer memory for each type of procedure. At the control console, each procedure is listed with a drop-down menu for each specific projection. The radiographer simply selects the projection, and a kVp and mAs are displayed. Additional settings for large, average, and small patients may be available, which modify the kVp and mAs up or down by preset percentages.

The important thing for the student to understand is that these listed techniques are not the "final word" in setting technique. They may each be overridden, and *should be* when appropriate, by manually turning the kVp or mAs up or down to refine a final technique setting. These techniques should be thought of only as *suggested starting points* or basic guides. Preprogrammed techniques cannot take into account all of the many variations of body habitus and conditions that occur from patient to patient. They lack flexibility. The patient size indicators are often limited to *large*, *average*, and *small* which certainly do not cover all possibilities for patient size. While programmed techniques may expedite or simplify the setting of technique, *they do not and never can replace the need for the independent judgment of the radiographer* in refining those techniques to obtain the best possible exposure results.

SUMMARY

1. AECs were developed to improve the consistency of radiographic exposures, *but they are not suited for every procedure.* Some procedures, particularly those with complex anatomy, large artifacts present, anatomy that is peripheral or too small to fully cover the detector cells are better radiographed using "manual" technique.
2. Proper positioning should never be compromised merely to allow the use of AEC.
3. When using AEC, the engaged detector cells should be fully covered with the *tissue of interest.*
4. Especially for digital image, when using AEC, the x-ray field must be well-collimated.
5. Optimum kVp and mA must still be set by the radiographer when using AEC.
6. Adequate penetration must be assured and ample remnant beam signal must reach all portions of the receptor plate to provide sufficient data to correctly process a digital image.
7. High-power generators, high-speed digital processing, and high mA stations can combine to bring needed exposure times down to less than the *minimum response time* of the AEC. In this case, overexposures will occur.
8. The *back-up time or mAs* should be set at 2–4 times the expected exposure time. Many automatic settings are excessive, and should be overridden and reduced by the radiographer. Managers and quality control technologists should be involved in determining appropriate automatic settings upon installation of equipment.
9. It is appropriate to adjust the intensity (density) control for special projections and situations. When increasing exposure with the intensity (density) control, change to at least +2 to make a significant difference.
10. The intensity (density) control can frequently be out of calibration, and can be easily checked by a radiographer.

11. The correct configuration of AEC detector cells is essential for proper exposure. The thickest and densest portion of the anatomy should be positioned over the correctly engaged cells. For the PA chest, the middle cell should be engaged to ensure adequate exposure through the mediastinum.
12. Radiographers should be adept at both "manual" technique and AEC technique.
13. AEC technique charts are just as important as "manual" technique charts and should be provided for every x-ray unit capable of AEC.
14. Preprogrammed exposure controls can and should be over-ridden for unusual circumstances.

REVIEW QUESTIONS

1. What is the disadvantage of preprogrammed exposure controls?

2. It takes longer for the AEC to shut off the exposure on a thicker body part because the __________ is reduced.

3. A radiographer uses a 72-inch (180 cm) SID for an AP cervical spine at the vertical "chest board." If the AEC is used, why will an underexposure *not* result?

4. How can extremely high-speed imaging systems lead to overexposure when the AEC is engaged?

5. As a rule, the back-up time should be set at __________ the expected exposure.

6. List four items that should be included on an AEC technique chart.

7. The needed total mAs for a particular procedure is 0.8 mAs. The minimum response time for the x-ray machine is 0.006 seconds. Using the AEC at the 200 mA station, will the exposure time be too long, too short, or correct?

8. When increased exposure is needed, why is the +1 setting at the intensity (density) control never recommended?

(Continued)

REVIEW QUESTIONS *(Continued)*

9. What is the formula for finding the percentage change from *N* for each station of the intensity (density) control?

10. What is the change in resulting exposure when the *wrong bucky* is activated during an AEC exposure?

11. Many x-ray units are preset to __________ back-up times, and should be overridden.

12. During an AEC exposure, those detector cells activated should be covered by the __________ (thinnest, average, thickest) portion of the anatomy.

13. During an AEC exposure, the detector cell should be covered not just by the anatomical area of interest, but by the __________ of interest.

14. A low-contrast and light digital image can result from insufficient __________ during an AEC exposure.

15. For *digital imaging systems*, extreme overexposure of parts of the image combined with extreme underexposure of other portions of the image will cause __________ (predictable or unpredictable) results in the final image.

16. What are two reasons why back-up times are still very important even if the digital image were unaffected by them?

17. A *low* exposure indicator (index number) is important to note because it may be an indication of insufficient __________ of the x-ray beam.

18. If the image turns out OK, why is a high exposure indicator (index number) still important?

Part III

DIGITAL RADIOGRAPHY

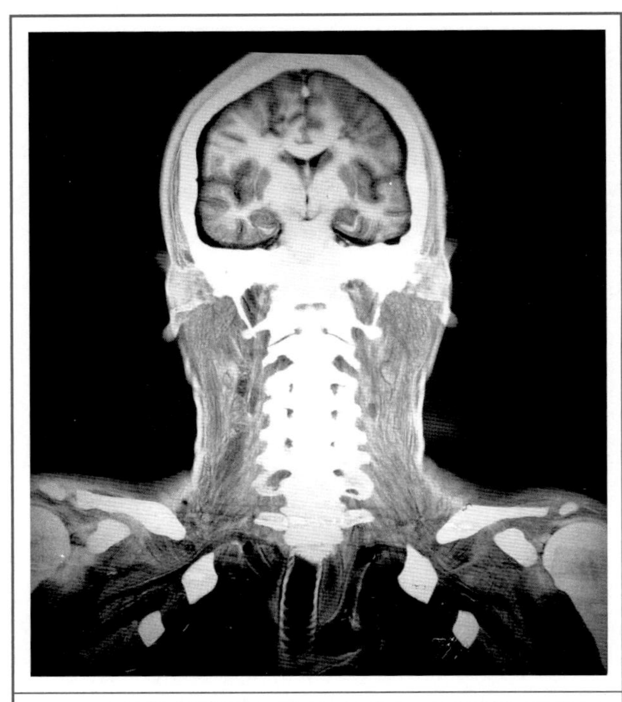

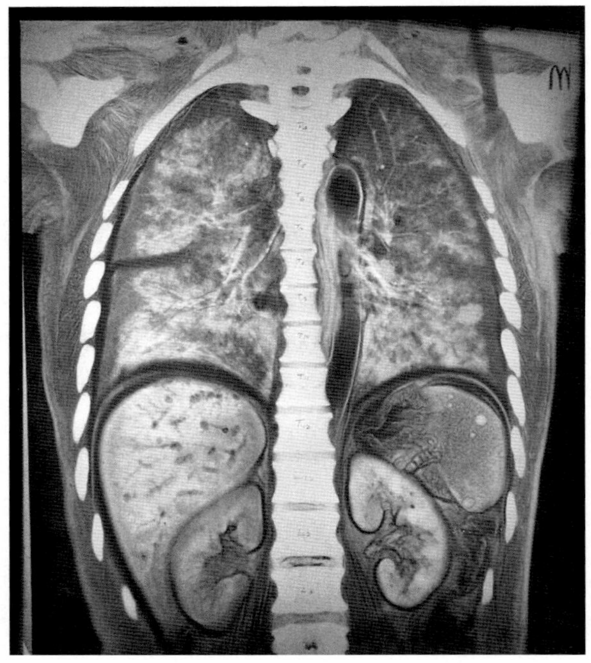

Conventional radiographs of autopsied coronal slices through the chest and head of a human corpse, appearing somewhat like MRI images.

Chapter 28

COMPUTER BASICS

Objectives:

Upon completion of this chapter, you should be able to:

1. Overview how computer hardware and software interact to perform tasks at high speed.
2. List the types of computers and terminals, and how they relate to radiography.
3. Overview the history and development of computers and micro-circuitry.
4. Describe how peripherals integrate with the central processing unit.
5. Describe the types of storage and main components in the CPU.
6. Describe the types of storage and major components of a typical PC.
7. Distinguish between the various characteristics of modern digital memory.
8. Analyze the differences between analog and digital data and how they relate to radiographic images.
9. Understand the basic aspects of binary code and ASCII code.
10. Overview the general types of software and levels of machine language.
11. Define the four levels of data processing.
12. Overview the hardware components and compatibility of digital communications systems.

A computer is any machine that can perform mathematical computations, manipulate information, make decisions and interact accurately and quickly. All of these functions are based upon the fundamental ability of the machine to follow preprogrammed instructions known as *algorithms*. Each algorithm is a concise set of instructions for a *single, specific task*, such as how to subtract two numbers that are inputted into the computer by the user. A *computer program* is a collection of many hundreds or even thousands of interrelated algorithms which allow the user to perform a general application such as calculating taxes, word processing, or organizing a data base.

To avoid repetitious programming and wasteful duplication, algorithms that will be used repeatedly within a program, called *subroutines*, are written only once and stored apart from the overall instructions, where they can be accessed as often as needed by a "go to" command.

Artifical intelligence (AI) describes the ability of a machine to make decisions based on *logic functions* such as "do," "if then," and "if else." An example of an algorithm for an "if else" statement might be as follows:

1. Store number ***A*** inputted from keyboard at memory address
2. Retrieve permanently saved number ***B*** from memory to calculator
3. Retrieve inputted number ***A*** from memory to calculator
4. Subtract ***B*** minus ***A***
5. IF the result of step 4 is positive, (if ***B*** is greater than ***A***), go to line 7
6. ELSE, (if ***B*** is NOT greater than ***A***), go to subroutine starting at line 11
7. $C = [A \times 0.5]$
8. Print out at monitor screen: ***C*** "will be deducted from your tax"
9. Count for 5 seconds
10. Go to (next section of tax instructions)
11. Print out at monitor screen: "You cannot deduct this from your taxes"

Figure 28-1

A typical microprocessor for a personal computer (PC). This is the CPU.

12. Wait for "ENTER" command
13. Go to (next section of tax instructions)

The part of a computer that interprets and executes instructions is called the *central processing unit*, or CPU. A CPU that is contained on a single integrated circuit chip is called a *microprocessor* (Fig. 28-1). The microprocessor is the heart of the computer. We think of the power of a computer in terms of how much data it can input, process and output in a given amount of time. The unit for this is *millions of instructions per second*, or *MIPS*. Actual processing speeds range from hundreds of MIPS for microcomputers to thousands of MIPS for mainframe computers.

This overall power is determined primarily by the *speed* of the microprocessor. This speed is determined, in turn, by an internal clock. The faster the clock, the faster the processing. Recall from Chapters 5 and 7 that the unit for frequency is the *hertz*, defined as one cycle per second. For an analog clock, one cycle represents the completion of one *circle* around its face by the clock's hand. The speed of a microprocessor is expressed as the *rate* of cycles the clock can complete or *count* each second. As with all other aspects of computers, we have seen this rate increase exponentially over time: Once measured in kilohertz and then megahertz, we now talk of the speed of microprocessors in common PC's in units of *gigahertz*, or *billions of cycles per second*. Speeds measured in terahertz will likely be achieved in the very near future.

Perhaps the most common way to classify computers is by their size. We generally think of a computer as the "PC" (personal computer) that fits on our desk at home. Only a few decades ago, the computing power of a modern PC required a computer as large as an entire room. All of the computing power of the lunar module which landed on the moon is now contained within a small hand-held calculator. As miniaturization in electronics continues to progress, it becomes more difficult to make clear distinctions between sizes of computers, and the "size" of the computational power is more pertinent than the physical size in application. With the understanding that some overlapping of terms is unavoidable, we can broadly categorize the sizes of computers as follows:

1. *Microcomputers* usually have one single microprocessor, and generally fit on a desktop such as a PC (personal computer) or "notebook" computer.
2. *Minicomputers* contain many microprocessors that work in tandem, and are too large and heavy to be placed on a desktop. The smallest minicomputers occupy a single cabinet ranging in sizes comparable to various refrigerators, placed on the floor. Larger minicomputers can occupy three or four large cabinets taking up a portion of a room. *CT and MRI computers are examples of minicomputers.*
3. *Mainframe computers* and *supercomputers* consist of microprocessors numbering in the hundreds or even thousands, and can support thousands of users. They require the space of an entire room or even a whole floor of a building. They are used in telecommunications companies, military and government organizations, airlines, and weather forecasting applications, to name a few.

The operating console of a standard diagnostic x-ray machine is essentially a *microcomputer*, with about the same overall processing power as a PC, but with all that power dedicated to the selection of proper radiographic technique while compensating for electronic and other variables.

THE DEVELOPMENT OF COMPUTERS

Tools for performing mathematical calculations date back thousands of years to the *abacus*, invented in

Figure 28-2

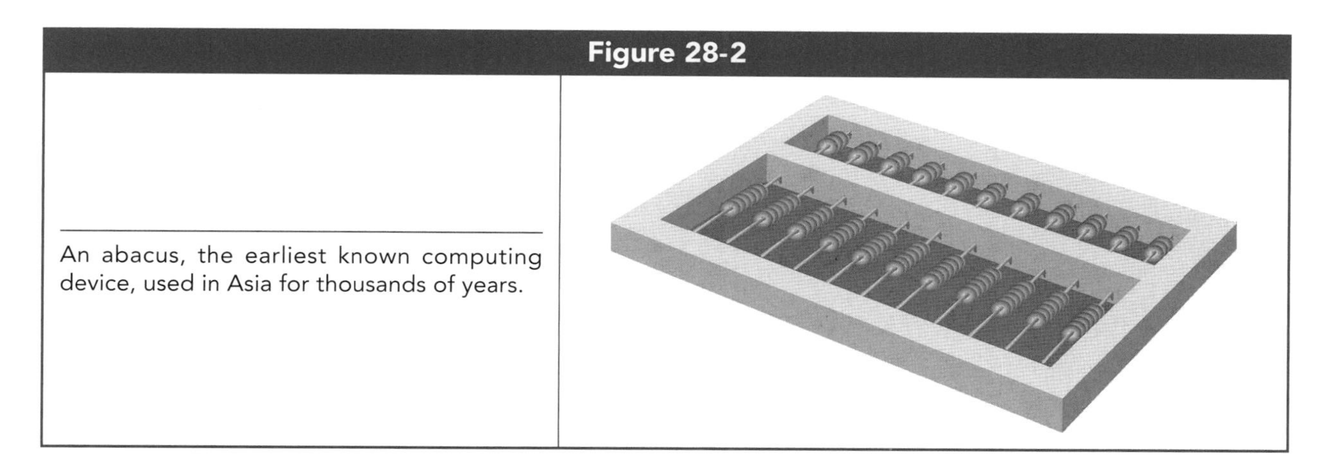

An abacus, the earliest known computing device, used in Asia for thousands of years.

China. The abacus consisted of a frame containing columns of beads separated by a crossbar (Fig. 28-2). Each column held five beads below the crossbar, representing *ones*, and two above the crossbar representing *fives*. Each whole column represented a power of 10 above the column to its right, such that 13 columns could represent numbers reaching into the *trillions*. Equally impressive, the abacus could be used not only for all four standard mathematical operations, but also to calculate square roots and cube roots.

The first major step in the evolution of a completely automatic, general purpose, digital computer was taken by an English mathematician, Charles Babbage, in 1830 when he began to build his *analytical engine*. One hundred years ahead of his time, the limitations of technology prevented Babbage from completing the machine in his lifetime. Meanwhile, another English mathematician, George Boole, devised a system of formulating logical statements symbolically which led to the design of switching circuits in the arithmetic/logic units of electronic computers. After Babbage's death in 1871, no significant progress was made in automatic computation until 1937 when American professor Howard Aiken began building his *Mark I* digital computer. Completed in 1944, it was the realization of Babbage's dream, but the *Mark I* still contained some components that were mechanical rather than electronic. It could perform up to five arithmetic operations per second.

The first fully *electronic digital computer* was completed at the University of Pennsylvania in 1946 by J. Presper Eckert and John Mauchly. Called the *Electronic Numerical Integrator and Calculator (ENIAC)*, it consisted of 18,000 vacuum tubes (Figs. 28-4 & 28-6), weighed 30 tons, and took up 1500 square feet of floor space (Fig. 28-3). It could perform 5000 arithmetic operations per second. This same year, John Von Neumann, a Hungarian-born American mathematician, published an article proposing that entire programs could be coded as numbers and stored with the data in a computer's memory. Almost everything he suggested was incorporated into the *EDVAC* (Electronic Discrete Variable Automatic Computer) designed by Eckert and Mauchly's new company. This was the first *stored-program digital computer*, completed in 1949.

In the meantime, a breakthrough in computer hardware took place in 1948 with the development

Figure 28-3

The first electronic digital computer, the ENIAC, took 1500 square feet of floor space and weighed 30 tons. (U.S. Army photo.)

Figure 28-4

A technician replacing a burned-out vacuum tube, one of 18,000 such tubes in the ENIAC. (U.S. Army photo.)

of the first *transistor* at Bell Telephone Laboratories. The transistor (Fig. 28-7), is a very small electronic (rather than mechanical) switch, which alternately allows or does not allow electrical current to pass through it. Eckert and Mauchly quickly integrated the transistor with their basic *EDVAC* design to produce the much more advanced *UNIVAC I* (Universal Automatic Computer), completed in 1951. *The UNIVAC was mass-produced within a few years and became the first commercially available computer.* Unlike earlier computers, it handled numbers and alphabetical characters equally well, and was the first computer to separate input and output operations from the central computing unit (Fig. 28-5).

The UNIVAC I used both vacuum tubes (Fig. 28-6), and transistors (Fig. 28-7). Both the vacuum tube and the transistor are able to represent binary digits, or *bits* of computer language, by simply allowing the two states of being switched on or off. (The "on" condition indicates a "yes" or the number 1, and the "off" state indicates a "no" or the number 0.) But, vacuum tubes were bulky, and the heated filaments would often burn out just as light bulb filaments do, making them very unreliable indeed.

The transistor allowed two critical developments to evolve: First, by the miniaturization of memory components, the size and weight of computers dropped dramatically, facilitating their mass production, their portability, and their use. More importantly, memory components were now *solid state*, based on small crystals rather than on heated wire filaments—this lengthened their life span as much as 100 times, and also dramatically reduced the electrical power needed to run the computer. The economy and efficiency of computing skyrocketed. *Therefore, the solid state transistor is perhaps the single most important invention in history for the development of computer hardware.*

Since 1951, computers are considered to have evolved through at least four *generations* based on continued radical improvements in technology. These

Figure 28-5

The UNIVAC was the first mass-marketed computer, and the first to separate input/output modules from the main computer. (U.S. Navy photo.)

Figure 28-6

Vacuum tubes, with cathode pins and anode plates (arrows). Tubes like these were the earliest switching elements in computers.

Figure 28-7

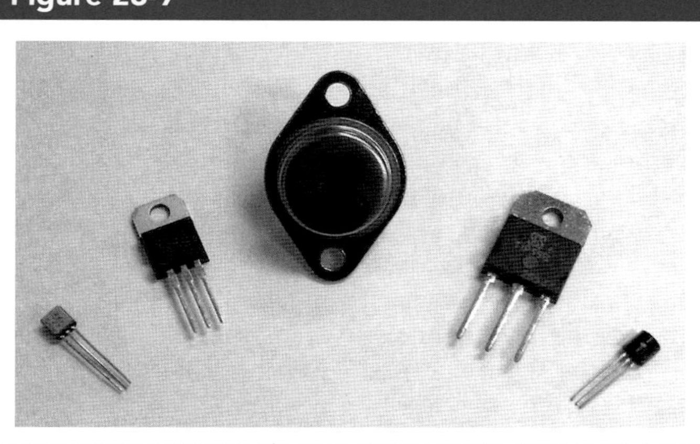

Various sizes of solid-state transistors. The transistor, used as a switching element, was perhaps the single most important development in the evolution of computers. (Courtesy, Tom O'Hara, PhD.)

generations are briefly defined in Table 28-1. Since the invention of the transistor, most advancements have been made in the area of miniaturization. In the mid-1960s a method was developed in which hundreds of miniaturized components could be chemically fused onto a small silicon *chip*, typically about 1 cm in size, to form microscopic circuits. These came to be known as *integrated circuits.*

Silicon is a *semiconductor*—it can be *doped* by other chemicals to make it conduct, resist, or block the flow of electricity. By introducing chemical *impurities* such as aluminum or boron in specific arrangements, microscopic *capacitors, diodes, resistors,* and *transistors* can be created. Specific areas of the chip are treated with various chemicals to serve these functions. With these areas in mind, the particular circuit is first mapped out on a large board. Special *photography* is used to reduced the pattern to microscopic size, form a photographic negative and project the pattern onto the silicon chip. More chemical impurities are baked into specified portions of the wafer to complete the circuit.

Further advancements in this miniaturization process have led to microprocessors which now contain millions of circuit elements within a square centimeter of silicon.

COMPUTER HARDWARE COMPONENTS

The *hardware* of the computer consists of all the physical components, including input devices, the

Table 28-1

Generations of Computers

Generation	Logic and Memory Circuit Components	Generally Available
1st:	*Vacuum Tubes for both*: Conducting = filament heated = "on"	1951
2nd:	*Transistors for logic*: Conduction = silicon charged = "on" *Magnetic cores* for memory	1958
3rd:	*Integrated Circuits*: Miniaturized components chemically fused onto a small silicon chip in microscopic circuits	1965
4th:	*Microchips*: Enhanced miniaturization of integrated circuits Large-Scale Integration (LSI) = thousands of elements Very Large Scale Integration (VLSI) = millions of circuit elements onto a 1 cm chip	 1970s 1990s

processing system, memory and storage devices, output devices and systems for communication. These physical components are connected as shown in Figure 28-8. From this diagram, it is clear that there is a flow of information from input, output, and memory storage devices to the *central processing unit* or *CPU*. This flow of data is carried by a multi-wire line called a *bus*. The connections of bus lines to each of the devices are called *ports*. *Serial ports* transmit data sequentially one bit at a time. *Parallel ports* have multiple channels to transmit data in batches, so the jacks that fit them typically have more than a dozen prongs.

Input/output or *I/O* devices, also called *peripherals*, transmit data to and from the computer. Input devices include the keyboard, the mouse, the trackball, the joystick, the touchpad, and the light pen. Most of these are pointing devices which control the location of the *cursor* (usually an arrow), which indicates the insertion point on the screen where data may be entered. These devices all require the user to enter information one character or menu selection at a time, and are somewhat slow. In order to more quickly *copy* information directly from a document, or from an audio or visual scene, *source-data entry devices* were developed. These include bar code readers, scanners and fax machines, sensors, microphones, and digital cameras and camcorders.

Output devices include printers, display screens and speaker systems. The display screen or monitor is typically a liquid crystal display (LCD)—two plates of glass with a substance between them that can be activated in different ways to make the crystals appear lighter or darker. To create smooth-looking letters and numbers on a monitor screen, a *character generator* is used to illuminate selected dots in a 7 × 9 matrix for each character.

A *video display terminal* (*VDT*) uses a keyboard and mouse or trackball for input, and a display screen

Figure 28-8

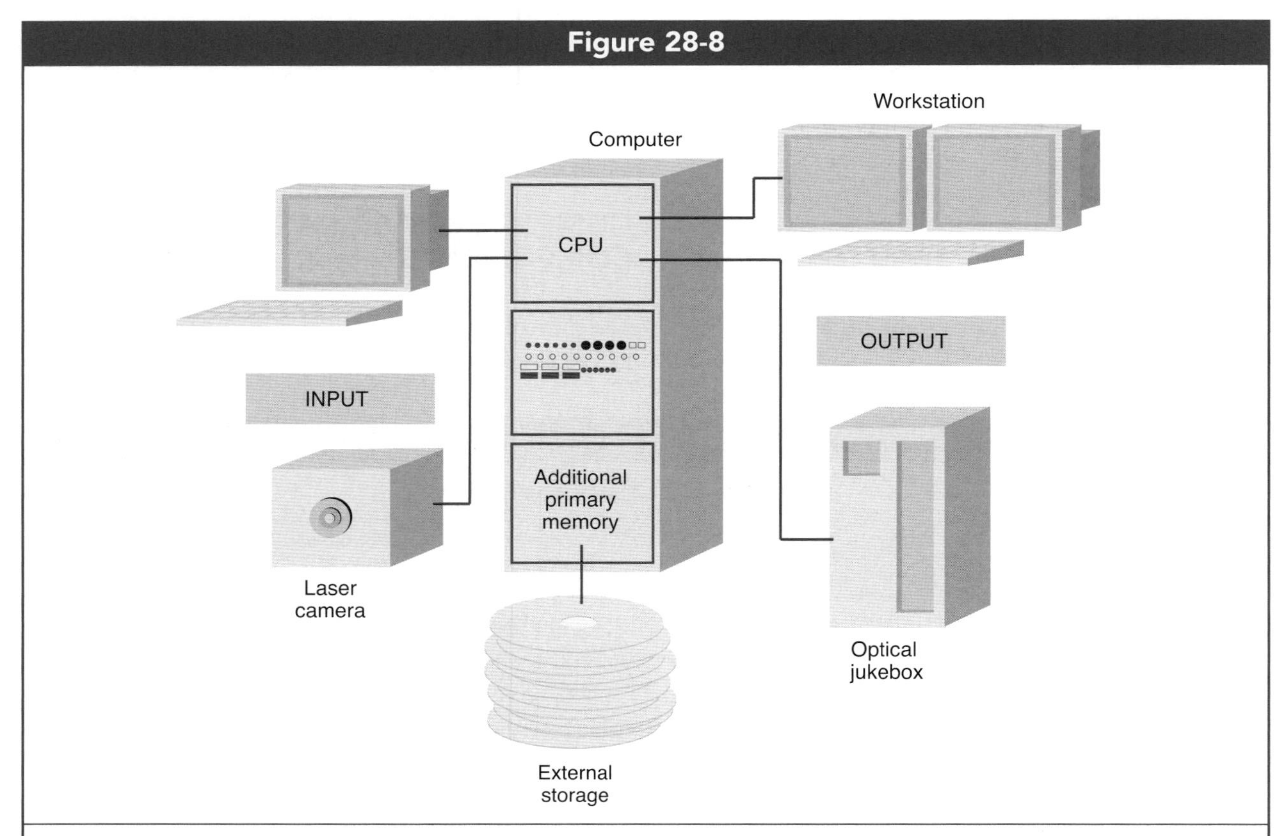

The central processing unit directs data flow from input devices, between primary and secondary memory and the arithmetic/ logic unit, and to output devices.

for output. A *dumb* terminal cannot do any processing on its own, but is used only to input or receive data from a host computer, such as is done at airport check-in counters. An *intelligent* terminal has built-in processing capability and memory, but does not have its own substantial storage capacity. *Most x-ray machine consoles would be categorized as intelligent terminals.*

Most modern printers are either ink-jet printers or laser printers. Ink-jet printers place an electric charge onto small drops of ink that are then sprayed onto the page. Laser printers form an image on a drum which is then treated with a magnetically charged ink-like substance called *toner*, and then transferred from the drum to paper. While ink-jet printers are quieter and less expensive, they can print only 10 to 20 pages per minute.

Laser printers have their own memory to store such information as fonts separate from the computer, and their own limited data processor. They provide sharper resolution in the image (up to 1200 dots per inch), and can print from 32 to 120 pages per minute depending on the power of the computer they are connected to.

Most radiographic images are viewed as *soft copies* on the LCD monitor screen. Sometimes it is desirable to print them out on transparent plastic film which can be hung on an *illuminator* or *view-box* for examination, or physically carried from place to place. Images or text that have been printed onto paper or plastic film are referred to as *hard copies.*

The *multiformat camera* is the most commonly used output device in radiology to produce hard copies of images, because it is capable of reproducing several small radiographs in series on a single sheet of film. The radiographer selects the desired images and the format of rows and columns (3 by 4 or 4 by 5, for example) in which to arrange them on the film. Inside its cabinet, the machine brings up each individual image on a monitor screen and uses a camera to take a snapshot of the screen. The photograph of the first monitor image is projected onto a specified area in the corner of the film. The film is then moved from projection to projection such that the radiographs are arranged in columns and rows. The film is then processed and treated for permanence.

The Central Processing Unit

The *central processing unit* (CPU) performs data manipulation in the computer. It tells the computer how to carry out software instructions. The CPU for a mainframe computer may be large enough to occupy its own separate cabinet, while the CPU for a typical PC is usually a single microprocessor. All CPU's may be divided into two basic components: The control unit, and the arithmetic/logic unit. These two operate on information and data retrieved from a *primary memory* storage system.

The *control unit* directs the flow of data between the primary memory and the arithmetic/logic unit, as well as between input devices, the CPU, and output devices. The control unit is analogous to a traffic cop directing the flow of traffic through an intersection. It tells input devices when to start and stop transferring data to the primary memory. It also tells the primary memory unit when to start and stop transferring data to output devices.

The control unit coordinates the operations of the entire computer according to instructions in the primary memory. It is programmed to select these instructions in proper order, interpret them, and relay commands between the primary memory and the arithmetic/logic unit. Each set of instructions is expressed through an *operation code* that specifies exactly what must be done to complete each task. The operation code also provides *addresses* that tell where the data for each processing operation are stored in the memory.

Somewhat like a very sophisticated hand-held calculator, the *arithmetic/logic unit* (*ALU*) performs all the arithmetic calculations and logic functions required to solve a problem. Data to be operated upon must be retrieved from addresses in memory, and are temporarily held in the ALU's own storage devices called *registers.* These registers are connected to circuits containing transistors and other switching devices.

To perform arithmetic and logic operations, electrical signals must pass through three basic circuits called the *AND-gate*, the *OR-gate*, and the *NOT-gate*, used in different combinations. One combination of these gates results in subtraction, another selects the larger of two numbers, and so on. The result of a calculation is first stored in the ALU's main register called the *accumulator*. Results may then be exported from the accumulator to internal or external memory, or directly to an output device such as a display screen.

Primary memory is also referred to as *main memory* or *internal memory*, mostly stored on chips. Four sectors of primary memory space are reserved for distinct functions as follows:

1. The *program storage area* retains program statements for a specific application, transferred from an input device or secondary storage. Upon the request of the control unit, these instructions are "read" and executed one at a time to perform the operations of a saved program.
2. The *working storage* or *scratch-pad storage* area temporarily holds data that is being processed by the arithmetic/logic unit, and intermediate results.
3. There is a designated temporary storage area for data received from input devices which is waiting to be processed.
4. There is a designated temporary storage area for processed data waiting to be sent to output devices.

The unit for measuring storage capacity is one *byte*, consisting of eight *bits* (*binary digits*) of information. The significance of this number is that eight bits are sufficient to create a single *character* which can represent almost any alphabetical letter, number, other value or symbol needed to communicate. The *bit*, an acronym for *binary digit*, is the smallest unit of storage, consisting of a 0 or 1.

An *address* is assigned to each permanent character stored within the memory. Therefore, each address consists of eight storage units, whether all of them are needed or not to contain a particular character. Just as the number of a particular mail box at the post office has nothing to do with what is contained therein, the addresses within computer memory are *only designated locations* where bytes are stored, and have nothing to do with the particular character stored there. They are necessary for the control unit to locate each character when it is needed.

Physically, most primary memory is contained in RAM (random access memory) and ROM (read-only memory) *chips* mounted on boards and connected directly to the CPU. Most computers have slots for additional boards of RAM chips to be inserted (Fig. 28-9) which generally speeds up the computer's response time.

The *motherboard* or *system board* is the main circuit board for a computer, usually the largest board within the casing (Fig. 28-10). It anchors the microprocessor (CPU), RAM and ROM chips and other types of memory, and expansion slots for additional circuit boards such as video and audio cards that enhance specific capabilities of the computer.

The *power supply* for a computer must be carefully controlled. Most computer circuits are designed to operate at 5 volts or 12 volts. A power supply box (Fig. 28-11), includes a step-down transformer (Chapter 7) and resistors used to reduce the voltage of incoming electricity to levels that will not burn out delicate computer components. Additional resistors leading into specific devices may be found on the motherboard.

Computer components also require a steady, reliable supply of power that will not be immediately affected by split-second interruptions, reductions or surges in the incoming electricity supply. For this purpose, numerous *capacitors* may be found on the

Figure 28-9

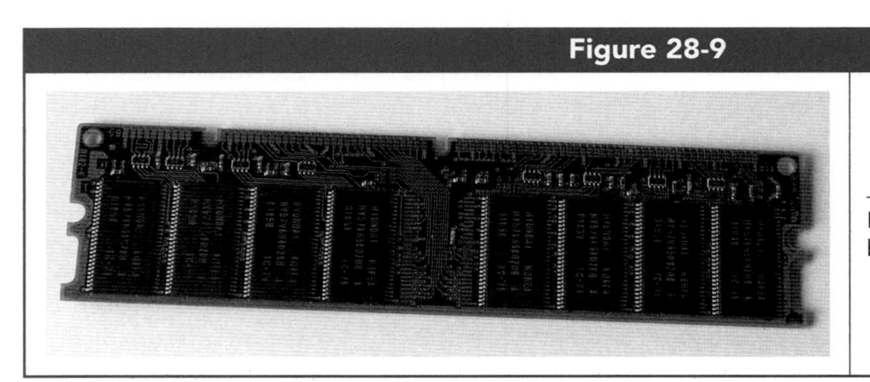

RAM chips mounted on a removable board.

Figure 28-10

The motherboard from a PC, showing ***A***, the microprocessor (CPU) with a cooling fan over it, ***B***, banks of RAM, and ***C***, slots for additional circuit cards.

motherboard, which store up incoming electrical charge and then release it in a controlled, constant stream.

Figure 28-11 gives a broad overview of the major components one will see upon opening the processor casing for a typical PC. These include the power supply, optical disc drives (CD and DVD) and flash memory drive, and the motherboard with the CPU (microprocessor), banks of RAM chips and slots for additional memory, banks of ROM chips, and various attached *cards* containing audio, video, and modem circuits.

Secondary Storage Devices

Several physical formats are available for the storage of secondary memory. *Hard disc drives* (Fig. 28-12), include one or more thin, rigid discs of glass or metal. Both sides of each platter are coated with a very thin layer of *ferromagnetic* material, (see Chapter 6). A small, button-like *read/write head* is suspended by an arm just over each surface of each platter (Fig. 28-12). With the disc spinning, when electrical current is passed through this head, magnetic fields are generated around it which magnetize the microscopic fibers on the surface of the disc. As the electrical current varies, the magnetic field around the read/write head changes shape and orientation. This results in the north and south poles of the magnetic elements or fibers on the disc being "pointed" in different fixed directions, such that they are arranged in distinct patterns representing the data.

For a disc to be *read* back, electrical current being fed to the read/write head is shut off so that it is in a passive "listening" mode. As the disc spins past it, by *electromagnetic induction* (Chapter 7), the magnetized elements passing by the read/write head induce a small electrical current flowing back into the system,

Figure 28-11

Inside of a typical PC, showing ***A*** the power supply, and ***B***, brackets to hold disc drives. The motherboard can be seen at the lower right.

Figure 28-12

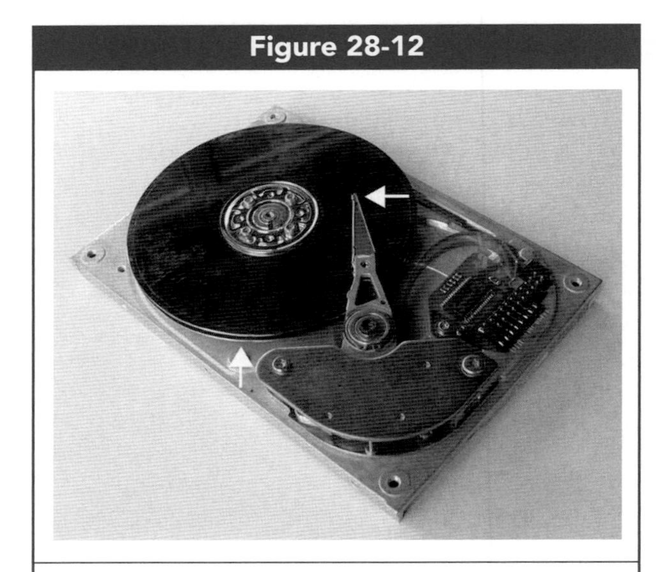

Inside of a hard drive unit, showing one of three magnetic read-write heads (horizontal arrow), and a double-disc (vertical arrow).

whose patterns precisely mirror those of the original recorded data.

Data is recorded onto discs in individual circular tracks (rather than a spiral track), forming a series of closed, concentric rings. When the read/write head completes reading one track, it must "jump" to the next one. As with a CD music player, a slight microsecond delay in outputting data allows these jumps to be made while the output flows continuously and seamlessly. Hard discs can squeeze thousands of tracks per inch within their radius. The tracks are organized in up to 64 invisible sections called *sectors* for storage reference. Figure 28-13 shows how sectors of data and their addresses are arranged in a circular track.

As shown in Figure 28-14, multiple hard discs can be stacked within a disc drive, with several read/write heads suspended between them on different arms. When they are stacked this way, the reading speed can be enhanced by using the *cylinder method* to locate data; this involves reading one circular track, then electronically switching to the *same track on the next disc below*, where the read/write head is already in position, rather than waiting for the read/write arm to mechanically move to the next outer track on the same disc. When information is recorded, it is placed *vertically* on all of the corresponding tracks throughout the stack of discs before moving the read/write heads to the next outer track. One can visualize the data stored on virtual *cylinders* that are arranged concentrically (Fig. 28-14).

Hard disc drives for a typical PC can hold up to 10 gigabytes (GB) of memory, and spin at high speeds, making them suitable for recording radiographic images. Larger computers use removable fixed disc drives with stacks of up to 20 hard discs, reaching memory capacities that are measured in *terabytes* (trillions of bytes). A mainframe computer may have as many as 100 stacked disc drives, each sealed within its own cabinet, attached to it.

A *Redundant Array of Independent Discs* (*RAID*) is a storage system with two or more hard drives that duplicate storage of the same information. In this way, if one disc drive fails or is damaged, other drives which may have their own independent power supplies and connections to input and output devices will preserve the information. These are used in medical imaging departments to ensure that patient records and images are not lost, and have obvious applications for the government and military.

The *recording density* refers to the number of bits that can be written on a disc per inch of radius. An extended-density (ED) disc can generally hold twice as many megabytes as a high-density (HD) disc, and allows more sectors to be organized. As of 2013, the typical storage capacity for hard discs is *10 gigabytes.*

Large spools of *magnetic tape* are still used with some larger computers for back-up and archiving. Magnetic tape employs the same basic technology as magnetic discs, in which fibers of iron oxide coated onto the tape take on magnetized patterns to represent data, and upon being read, induce small electrical currents in a read/write head.

Invented in 1958, *optical discs* have a light-reflective surface into which pits are etched by a laser beam. The most familiar form of optical disc is the compact disc (CD) used for recording and playing back music. Supported by a clear polycarbonate plastic base, the reading surface of an optical disc is an extremely thin layer of shiny aluminum, into which a microscopic spiral groove has been cut extending from the innermost track to the outermost. Seen from different angles, this spiral groove reflects light in a diffused "rainbow" pattern, creating an iridescent appearance to the disc.

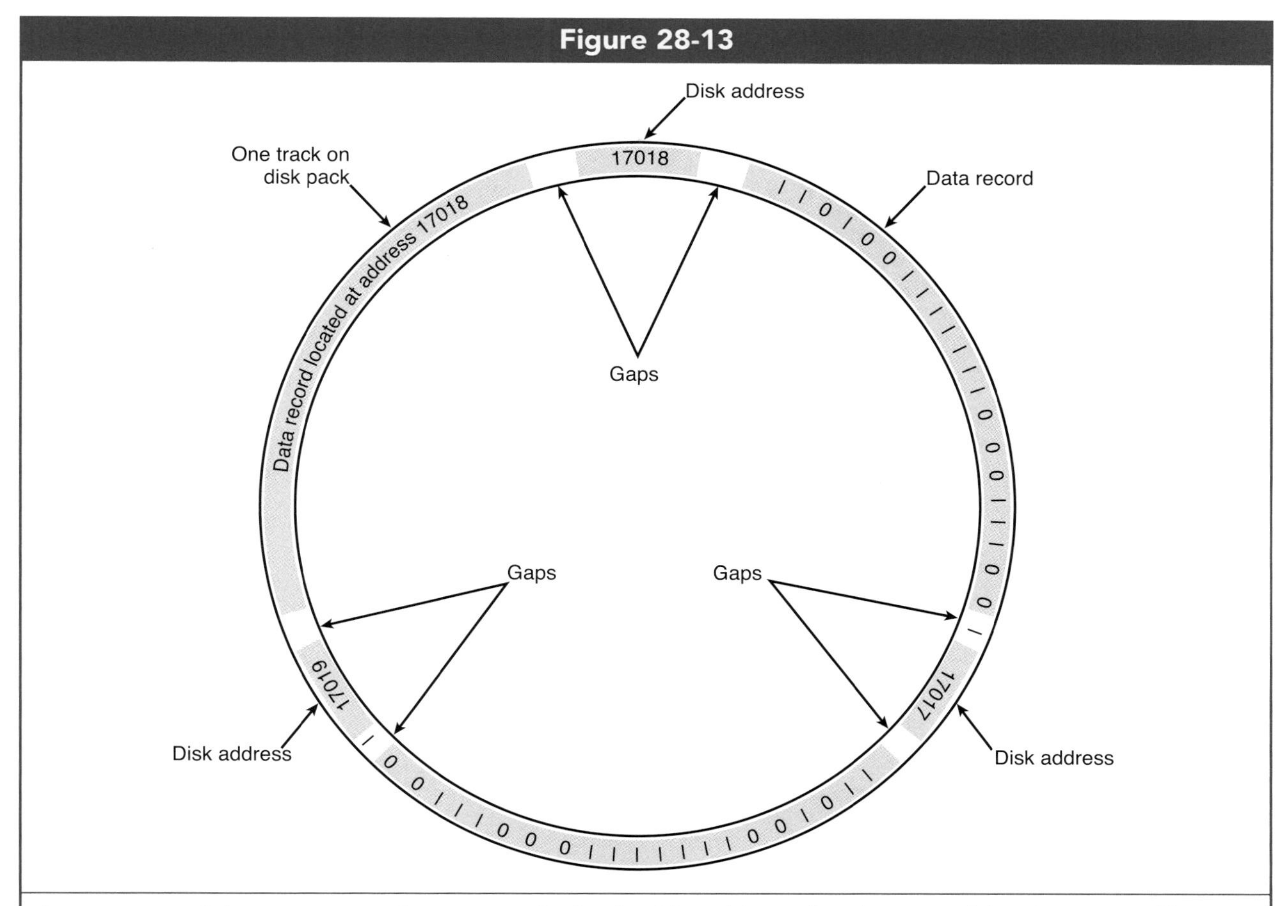

Arrangement of three sectors on the outer track of a disc. The address of each sector of data is separated by gaps between the sectors. Up to 64 sectors can be configured.

Upon recording, an ultra-thin beam of laser light is used to cut a series of microscopic *pits* into the grooved track, leaving flat spaces of equal size, called *lands*, between the pits (Fig. 28-15). Each pit represents the binary number 0 or an "off" condition, and each land represents a 1 or an "on" condition. To read the disc back, a less intense laser beam is reflected off the surface of the track and picked up by a light detector. Lands reflect the laser light for a positive read-out, while pits diffuse the light rather than reflect the intact beam directly to the detector.

Optical discs come in various sizes from 3 to 12 inches in diameter, and are typically 1.2 mm (0.05 inches) in thickness. In the mid 1990s, the second generation of optical disc, the *digital versatile disc* or *digital video disc* (*DVD*) was developed. Thinner tracks, with a *pitch* (distance from the center of one groove to the center of the next) of 0.74 microns versus 1.6 microns, made it possible to store more data in the same diameter, and allowed use of a shorter wavelength of laser light. The increased storage capacity was sufficient to support large video applications. Storage capacity went from 700 megabytes for a typical CD to nearly 5 gigabytes for a typical DVD.

A third generation, developed by 2006, employed a blue-violet laser, with a wavelength of 405 nanometers, rather than 650-nanometer red light. This shorter wavelength made it possible to focus the laser spot with even greater precision. Combined with a smaller light aperture, this made it possible to store up to 25 gigabytes of memory, enabling the recording of high-definition (HD) video. Since then, multiple layering of discs has been developed, with up to 20 reflective layers stacked on a single disc pushing storage capacities to 500 gigabytes.

Figure 28-14

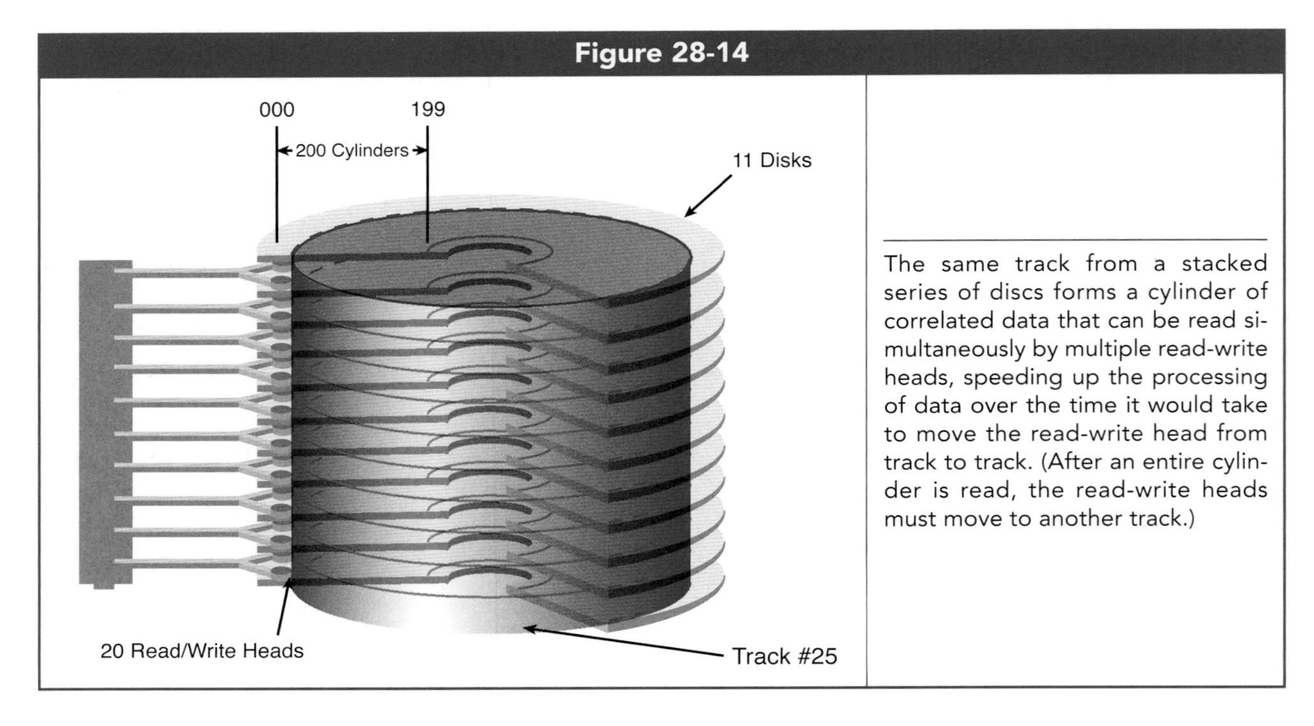

The same track from a stacked series of discs forms a cylinder of correlated data that can be read simultaneously by multiple read-write heads, speeding up the processing of data over the time it would take to move the read-write head from track to track. (After an entire cylinder is read, the read-write heads must move to another track.)

Dual layer discs have several reflective surfaces at different depths within the plate. The laser beam, upon writing or reading, can be *focused* to reflect sharply from only the indicated depth within the disc, and is thus able to single out each layer.

Standardized suffixes apply to all types of optical discs alike: A DVD-*ROM* is *read only memory* and cannot be written onto to record new data. A DVD-*R* (recordable) can be written onto only once and then played back as a DVD-ROM. A DVD-*RW* (re - writable) or DVD-*RAM* (random access memory) can be erased and recorded onto multiple times. Rewritable discs include a layer of metallic *phase-change* material that allows the surface to be completely smoothed out for erasing. The DVD+R uses a different format than the DVD-R, and the *plus* or *minus* sign must match that of the playback device being used.

Figure 28-15

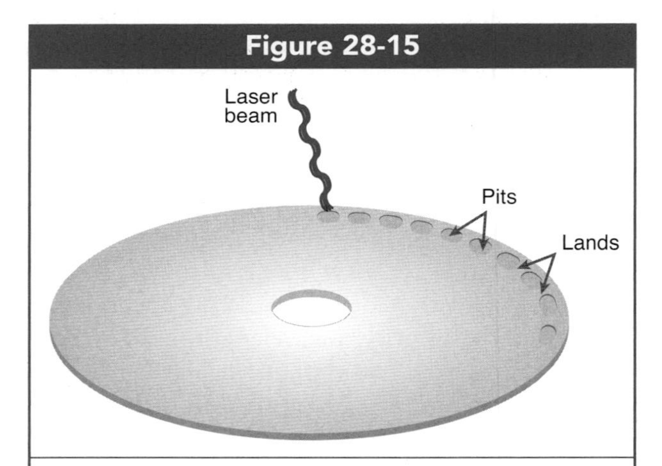

A high-intensity laser beam is used to melt pits into the aluminum reflective surface of an optical disc. To read the disc, a low-intensity laser beam is reflected off of the *lands* between the pits and intercepted by a detector, while the pits diffuse the light, to represent ones and zeros respectively.

Flash memory, developed in the early1980s, stores data in the form of electrical charges, but does so in such an effective way that the charge can be maintained for very long periods of time before "bleeding off." It is a type of *EEPROM* chip, which stands for *Electronically Erasable Programmable Read Only Memory*, and got its name because the "flash" of electrical current used to erase it reminded its developers of the flash of a camera. Your home computer's *BIOS (Basic Input/Output System)* chip is an example of a common application for flash memory.

Each functional memory cell of a flash drive consists of two electronic gates, the *control gate* and the *floating gate*, separated by a thin oxide layer (Fig. 28-16). Because the oxide layer completely surrounds the floating gate, it is electrically insulated,

Figure 28-16

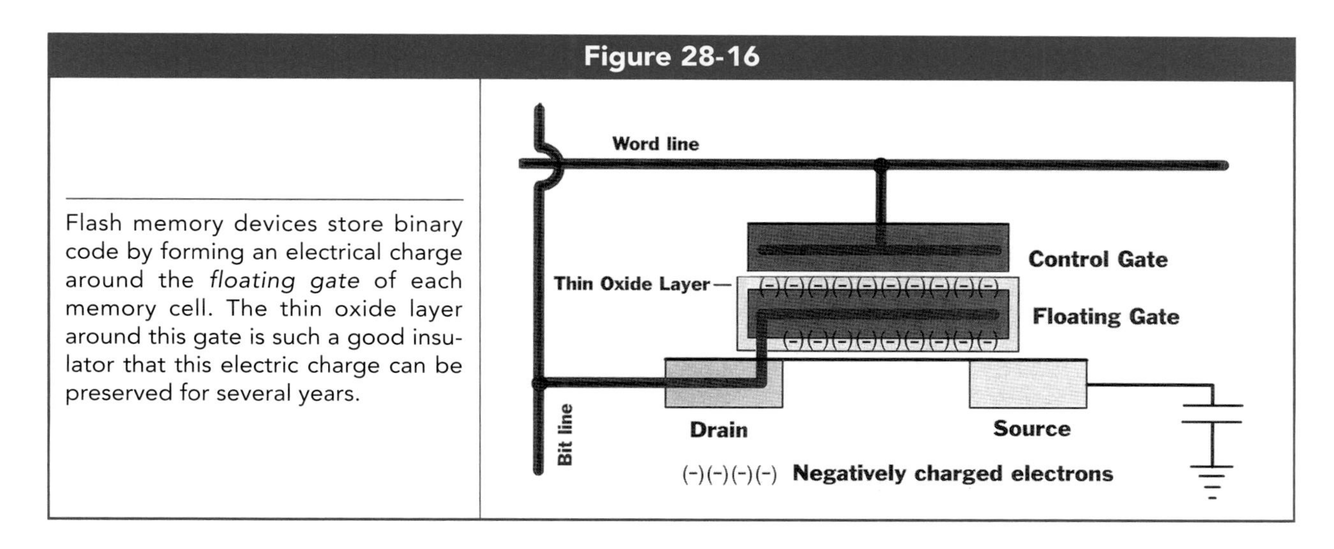

Flash memory devices store binary code by forming an electrical charge around the *floating gate* of each memory cell. The thin oxide layer around this gate is such a good insulator that this electric charge can be preserved for several years.

and any electrons trapped there will not discharge for several years. When enough charge is held by the floating gate, the memory cell as a whole becomes more resistant to the flow of electricity through it. This is its "on" state. When a small voltage is used to test a series of cells, their "on" and "off" states form a binary code.

Flash drives had a history of data corruption problems due to electronic bleed-off, but have now reached a level of reliability similar to hard disks. Flash memory "sticks" (Fig. 28-17) have become more popular than hard disk drives for use in portable devices because of their high resistance to mechanical shocks or jolts. When compared to hard disk drives which require moving, mechanical devices, solid-state drives such as flash memory have higher speed, make less noise, consume less power, and provide greater reliability. They are beginning to be used in high-performance computers and servers with RAID architectures. (A new type of memory called *phase-change random access memory* or *PRAM*, developed in 2006, appears to have 30 times the speed and 10 times the lifespan and may eventually replace flash memory.)

However, two significant advantages remain for magnetic hard drives: They are drastically cheaper per gigabyte of memory, and they have substantially greater capacity. For the purposes of medical imaging, flash drives can provide great convenience in moving image files from one place to another, but due to cost and capacity, a RAID system using hard disk drives will continue to be the preferred method for long-term storage of medical images for the near future. For extremely long-term backup storage, optical disks are best, provided they are properly stored in protective cases. Disc technology itself continues to advance; the *holographic versatile disc* (*HVD*) uses collinear holography to record data in three dimensions. Although its current maximum storage capacity is 500 gigabytes, it may one day hold up to 3.9 *terabytes* of memory.

Types of Memory

There are several ways in which memory can be categorized into one of two types. These methods of

Figure 28-17

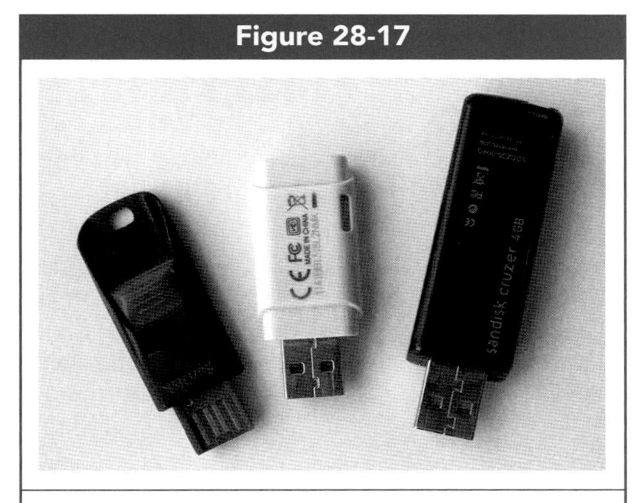

Three examples of "memory sticks" based on flash drive memory.

typifying memory are not directly connected to each other. That is, one categorization does not necessarily determine another. For example, *internal* memory is not necessarily always *primary* memory, and internal memory can be either ROM or RAM. For a particular device, one or the other description applies in *each* of the following approaches to categorizing it:

Internal Vs. External Memory: *Internal memory* physically resides within the processor casing of the computer and is *addressed* (each memory location is assigned a label to denote its position for the control unit. *External memory* includes flash memory sticks, CDs, etc. stored outside the processor casing of the computer. External hard drives can be attached to a computer, so even a hard drive is not necessarily *internal.*

Primary Vs. Secondary Memory: *Primary memory* is that memory which is *necessary for the computer to function generally*, regardless of which operating system or particular program is being used. An example is the *bootstrap* program, so named because it "pulls the computer up by its own bootstraps," to use an old adage, whenever the computer is turned on. From the time that electrical power begins to be supplied to the computer, it needs instructions from the CPU in order to seek out the operating system that has been installed and bring up its particular screen or "desktop" format to prompt the user to interact with it, and also provide corrective options should the operating system fail to initiate properly.

Secondary memory is specific to the operating system and the application being used at any given time. It is essential to the *program*, but not to the *computer.*

Volatile Vs. Nonvolatile Memory: *Volatile memory* is erasable. While this is desirable and often essential for different applications, it also means that the memory is susceptible to *accidental erasure* by the user striking the wrong key or clicking on the wrong menu item.

Nonvolatile memory is physically not erasable and is thus protected from accidental loss. It is absolutely essential that the *primary memory* for the computer be *nonvolatile* so that the user cannot accidentally render the computer itself nonfunctional.

RAM Vs. ROM: *Random Access Memory (RAM)* gets its name from the fact that it can be accessed from anywhere on the disc or other medium in approximately equal amounts of time, regardless of where the data is specifically located. This is in contrast to *taperecorded* data, such as songs on an audio cassette tape or movies on a videotape. With tape-based media, in order to get to the *fourth song* in the album or the *second part of a movie* on the videotape, the user has no choice but to "fast-forward" *through* all of the tracks preceding it, in sequence.

Random access means that the user can go more or less directly to the desired track. (Ironically, old-fashioned *records*, which preceded audiotapes, provided random access, since the user could drop the needle of the record player anywhere on the disc. The invention of audiotapes was a step *backward* in this regard, but the tapes were less vulnerable to damage.)

The importance of random access is that it vastly improves the *speed* with which different portions of a program can be brought to the video screen or speakers and then manipulated by the user. Such speed is essential to video gaming and critical to military applications, but has come to be expected by users for all types of computer applications that are interaction-intensive such as wordprocessing. (An example of an application that is *not* interaction-intensive is batch-processing of data.)

Although its name does not indicate it, *RAM quickly* came to be associated with *temporary memory* because most data that required high speed access was also data intended for the user to be able to change at will. Therefore, RAM is generally synonymous with "read-write" memory which may be altered and is volatile or erasable. *Static RAM (SRAM)* retains its memory when power to the computer is turned off. An example of this type of application is when the user *saves* the location within a game where he or she left off, in order to pick up at the same point later. *Dynamic RAM (DRAM)* is lost when power to the computer is shut off, but because it is cheaper and requires less space, it is the more predominant form of RAM in the computer.

Physically, the term *RAM* in actual usage refers to banks of computer *chips* arranged on cards, which serve the above purposes. Most computers have slots on the motherboard to insert additional cards of RAM chips in order to upgrade the RAM capacity. RAM capacities vary widely between computers, and are generally expressed in megabytes (MB), gigabytes (GB) or terabytes (TB).

Read only memory (*ROM*) also generally refers to banks of chips, only these contain instructions from the manufacturer, known as *firmware*, that cannot be erased or written over. These instructions can only be read and followed. An example is the "bootstrap" program mentioned under *Primary Memory* above. *ROM BIOS* is the ROM for the *Basic Input/Output System* which directs the flow of information between the keyboard, mouse, monitor screen, printer, and other I/O devices. (*EPROM, erasable programmable ROM* chips have been developed which can be changed with special equipment, thus blurring the distinction between ROM and RAM.)

Figure 28-18

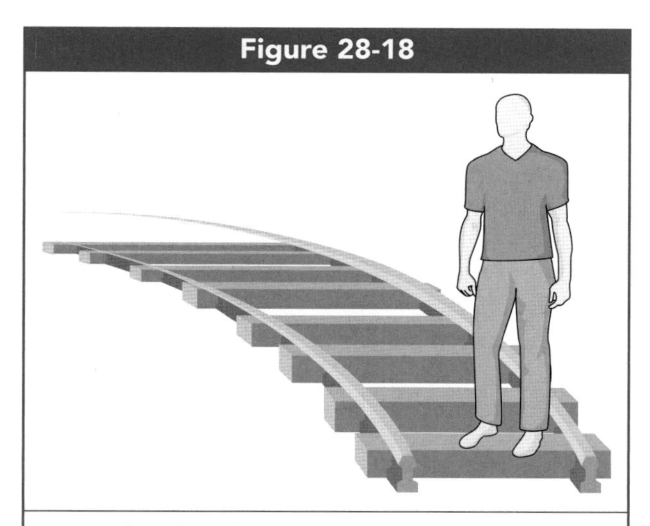

On a railroad track, the steel rails are continuous and can be infinitely subdivided, representing *analog* information. The wooden ties, on the other hand, represent *discrete* or *digital* information, since they cannot be divided into fractions as one steps across them.

MANAGING DATA

Analog vs. Digital Data

Imagine that you are running along a railroad track (preferably with no trains coming). There are two ways you can measure your progress: by measuring the distance (in meters, for example) that you have come along the *rails*, or by counting the number of wooden railroad *ties* you have passed (Fig. 28-18). The rails are *continuous*, consisting of smooth lines. The measurement of your distance along them can include fractions of a meter. The ties, on the other hand, are *discrete* or separated. They cannot be measured in fractions because of the spaces between them. You must count them in whole integers. This is precisely the difference between analog and digital information.

Data transmission can be in *analog* or *digital* form. Mathematically, the term *analog* means *precisely proportional. Analog data* is data presented in *continuous* form, such that its presentation is precisely proportional to its actual magnitude. This means that, in effect, its *units* are infinitely divisible.

An example is an old-fashioned mercury thermometer, in which a column of liquid rises within a glass tube as the temperature gets hotter. (Older-style barometers and blood-pressure cuffs use the same type of system.) This column of liquid mercury rises and falls in a smooth, continuous movement that can place its top surface at *any conceivable location* between the degrees marked on the glass tube. Conceptually, it can indicate a temperature of 70.004 degrees or 70.005 degrees—the number of *decimal places* can be extended as far as one wishes for accuracy, that is, the data is being presented in units that can be infinitely subdivided.

Digital data, on the other hand, is presented on a *discrete* scale, a scale made up of separated, distinct parts. How small these parts are limits the degree to which measurements can be subdivided. The units are defined such that the number of decimal places is limited. (For railroad ties, no decimal places past the zero are allowed. If you are standing in a space between them, you must state that you have traveled past 153 ties or 154 ties, no fractions are allowed.) Because the number of allowed decimal places in a digital system is preset, when analog information comes into it the measured values must be *rounded* to the nearest discrete value allowed by the system.

In a computer system, the magnitude of measured incoming data can be represented by the *voltage* of electrical charge accumulated on a capacitor. Let us connect an *analog* computer to the old-fashioned liquid thermometer mentioned above. When the temperature is 70.004 degrees, the analog computer can store 70.004 millivolts to record it; when it is 70.005, the computer can store this voltage as well, or any other fraction. Now, let us connect a *digital* computer to the thermometer, a computer whose discrete units are limited to *hundredths* of a millivolt. When a temperature measurement of 70.004 degrees is fed into it, it must round this number *down* to 70.00 millivolts in order to record it. When a temperature of 70.005 degrees is fed into the digital computer, it must round this number *up* to 70.01 millivolts, the next available unit in hundredths.

This rounding-out process may seem at first to be a disadvantage for digital computers. Strictly speaking, it is less accurate. Yet, when we take into consideration the limitations of the human eye, we find that it can actually be *more* accurate in *reading out* the measurement; the human eye is not likely to detect the difference between 70.00 degrees and 70.01 degrees in the height of the mercury column on a liquid thermometer, but a digital read-out can make this fine distinction. *As long as the discrete units for a digital computer are smaller than a human can detect, digitizing the information improves read-out accuracy.*

An everyday example of this principle is found in clocks and watches. For an *analog* clock, the hands sweep out a continuous circular motion. Since the second-hand is continuously moving, even though it is technically accurate, it is difficult for a human to look at it and determine how many *tenths* of a second have passed by when timing some event. A digital read-out clock can be stopped at a space *between* two discrete values and read out to the tenths or even to the hundredths of a second. Even though it is effectively rounding these measurements out to the nearest hundredth, this is a much finer distinction than the human eye can make from watching an analog clock.

When a photograph is taken, the information coming into the camera lens consists of light in *analog* form, in various colors and intensities of all imaginable shades, values than can be infinitely subdivided. A digital camera must round these values out to discrete units it can process. If these units are smaller than the human eye can detect, the resulting digital picture will appear to have the same quality as an analog photograph.

The same holds true for radiography. *The various intensities of x-rays that strike the image receptor can have any value and therefore constitute analog information* (Fig. 28-19***A***). For a digital imaging system, these values must be rounded out to the nearest allowable discrete unit so that the computer can manage them (Fig. 28-19***B***). This is the function of a device called the *analog-to-digital converter, or ADC* (Fig. 28-20). All image data must be converted into digital form by the ADC before being passed along to any computerized portion of the equipment.

Figure 28-19

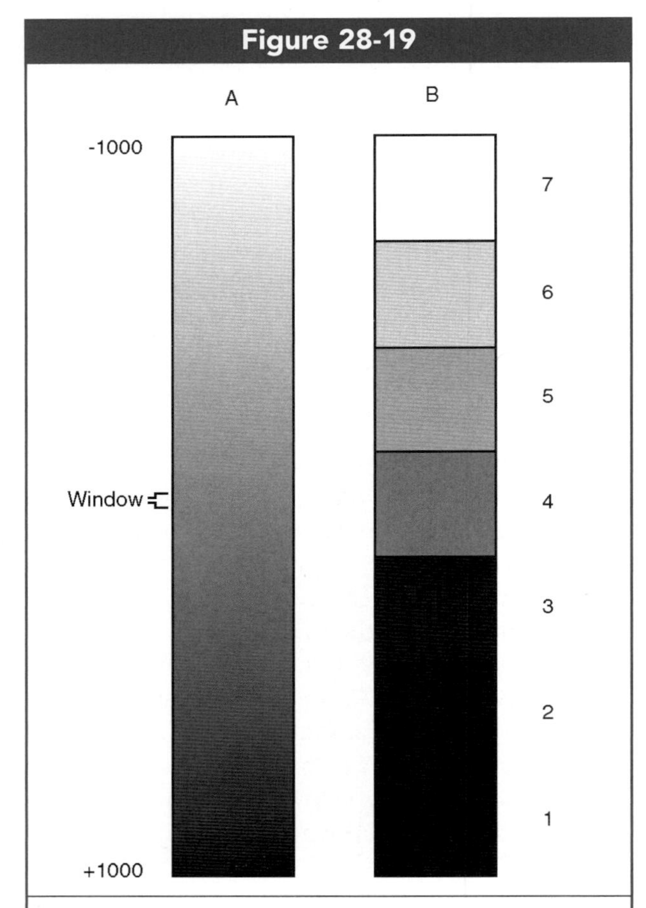

The x-ray beam that strikes the image receptor carries analog information. Its various intensities can have any value along a continuous spectrum as shown in ***A***. For all digital imaging systems, these values must be "rounded" by an analog-to-digital converter (ADC) into discrete pixel values as shown in ***B***. This is necessary because the computer cannot manage an infinite range of numbers. The range of numbers it *can* handle is called the dynamic range.

Binary Code

In the CPU, the *operation code*, which provides step-by-step instructions for every task, is in binary form (*bi-* referring to *two* states only). Much more complex

Figure 28-20

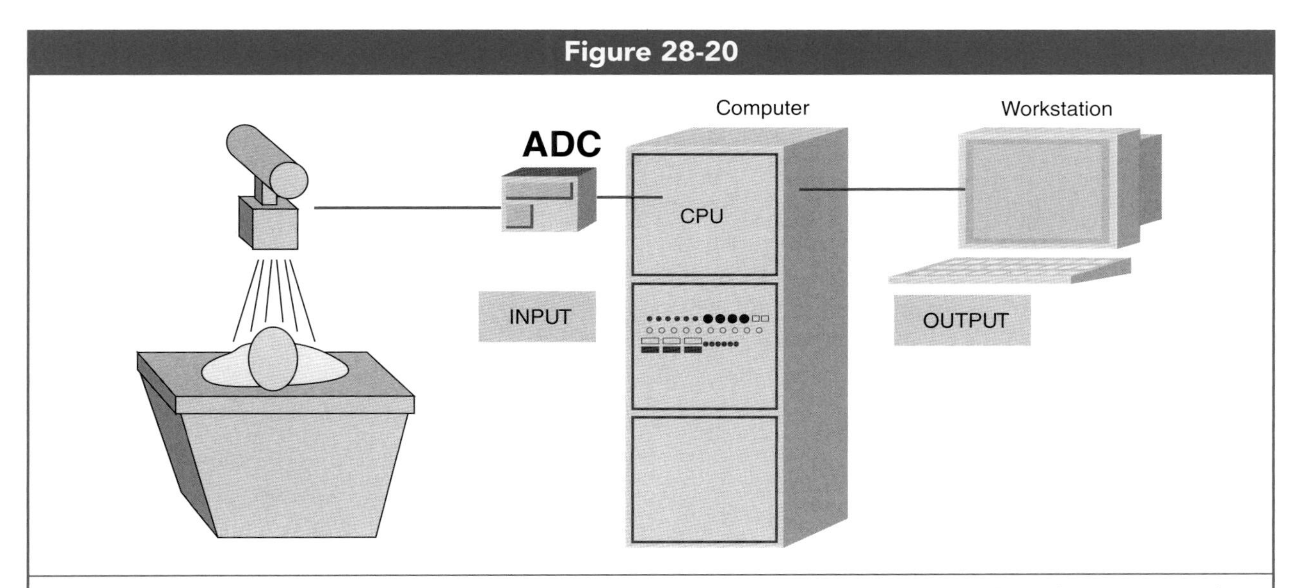

Every digital imaging system must pass incoming data from the image receptor system through an *ADC (analog-to-digital converter)* before it enters the computer. The ADC effectively *rounds out* x-ray exposure measurements into discrete values that the digital computer can cope with.

computer languages are used for operating systems software and for various applications, but these languages are all based upon the basic binary code because the *hardware* of the computer requires this format. Ultimately, every *bit* of information within a computer must be able to be represented as a transistor in the condition of either being turned on or turned off. A basic understanding of the *binary number system* is important because it shows how all possible numbers can be reduced to an expression using only these two states of *on* or *off, yes* or *no*, 1 or 0.

For radiographers, it is also important to understand power of 2 notation, because not only is image storage capacity expressed in powers of 2, but so are the dynamic range (gray scale) and the matrix sizes of the images themselves. For example, typical image sizes are 256 by 256 *pixels* (picture elements), 512 × 512 pixels, and 1024 × 1024 pixels, all binary numbers based on powers of 2.

The unit for the binary number system is one "bit," an acronym for **bi**-nary digi-**t**. Table 28-2 compares the way the familiar *decimal* system of numbers is organized to the way the *binary* system is organized. For the decimal system, the value of the number's *place* position to the right or left of the decimal point is based upon the exponent of the base 10. For the

Table 28-2

Decimal vs. Binary Number System

Decimal System					**Binary System**				
Places to Left		Exponent of 10		Value	Places to Left		Exponent of 2		Value
1st place	=	10^0	=	1's	1st place	=	2^0	=	1's
2nd place	=	10^1	=	10's	2nd place	=	2^1	=	2's
3rd place	=	10^2	=	100's	3rd place	=	2^2	=	4's
4th place	=	10^3	=	1000's	4th place	=	2^3	=	8's
5th place	=	10^4	=	10,000's	5th place	=	2^4	=	16's

binary number system, the value of this place from right to left of is based upon the exponent of the base 2. Examine the layout of the numbers in Table 28-2 to understand this placement concept.

For example, in the decimal system, a "1" positioned in the third place to the left of the decimal point would indicate *hundreds*, or groupings of 10^2. But, in the binary system, a "1" positioned in the third place to the left would indicate *fours*, or groupings of 2^2. Table 28-3 lists several examples of how the placement of a single "1" in binary translates into decimal numbers.

To read a binary number, the number *1* indicates a "yes" that the number *represented by that place of position* is a component of the whole number being represented. A *0* indicates that it is not. For example, to interpret the binary number 1011, begin at the right-most place and ask the question, "Is there a *1* in this number?" If the value there is one, there is a *1* in the number. Move to the *left* one place and ask if there are any *2's* in the number. In this case, the value there is one, indicating a "yes" to the question. A zero in the next place to the left indicates that there are *no 4's*, and a one in the next indicates that yes, there is an *8*. Finally, *sum* all of the numbers for which a "yes" was indicated. In this case, an 8 *plus* a 2 *plus* a 1 indicates the final value of 11. To better illustrate:

8's	4's	2's	1's
1 = yes	0 = no	1 = yes	1 = yes

8 + 2 + 1 = 11

To reinforce the binary concept, try the following exercise, and check your answers from Appendix #1.

Table 28-3

Resulting Decimal Values

Binary Number	Decimal Equivalent
1	1
10	2
100	4
1000	8
10000	16
100000	32
1000000	64

EXERCISE #28-1:

PART A: Convert the following binary numbers into decimal numbers:

1101 = __________

110010 = __________

11111011 = __________

PART B: Write the following numbers in binary:

7 = __________

19 = __________

63 = __________

There are only 10 kinds of people in the world—Those who understand binary, and those who don't.

The next obvious question is, "How can alphabetic characters and other symbols, rather than just numbers, be represented in binary code?" Several different schemes have been developed. What most of them have in common is that they require no more than *8 bits* to represent all the characters needed to communicate. This explains the origin of the *byte* unit for memory. One *byte* equals eight *bits*, and these sets of eight bits are separated by a space. One byte is sufficient to represent any single character from a keyboard. Therefore, stating that a particular storage medium, such as a compact disc, can hold 700 megabytes, or 700 million bytes, is tantamount to saying that it can store 700 million alphanumeric characters.

To provide an example of why eight bits is more than sufficient to any alphanumeric character, we shall take a brief look at the *American Standard Code for Information Interchange* (*ASCII code*). This was the first binary code developed through the *collaboration* of several different computer manufacturers in order to standardize computer language. Before ASCII was developed, programs written for one brand of computer could not be run on any other brandname.

ASCII code is actually a 7-bit code in which the first three digits were called *zone bits* and gave an indication whether the four digits following represented a number or a letter. Table 28-4 lists the codes for the ten decimal digits and all 26 letters of the English alphabet. Note that the codes for all of the decimal numbers begin with *011*—these are the zone bits indicating that these are numerical values. The remaining four digit places are sufficient to represent the numbers 0 through 9, with 9 being coded as *1001* (8 + 1).

Note that at this point in the list (Table 28-4), the zone bits change to the code *100*, indicating that the character will be a letter rather than a number. The remaining four digits simply begin with the value *1* for the letter "A," *2* for a "B," and so on until these four digit places are exhausted upon reaching *1111* at the letter "O." At this point, the zone bits change to *101*, also indicating letters, and the remaining four bits begin at *0* all over again.

Since $2^7 = 128$, 7 bits can be combined in 128 different ways to represent characters, the sum total of all characters needed for the English alphabet and the decimal digits is only 26 + 10 = 36, leaving *92* additional characters that can be coded to cover punctuation marks, letters from other languages, scientific, mathematical and iconic characters that might be entered at a keyboard.

For ASCII code, the eighth bit in each byte is used as a *parity bit*; it is coded as a *1* or a *0* to ensure that the number of *on* bits in each byte is either even or odd. Each microprocessor is designed to work on the basis of *odd* or *even* parity. This helps the computer catch coding errors, since a mistake would throw off the evenness or oddness of *on* bits within a byte. The parity bits are not shown in Table 28-4.

The capacity of computer memory is often expressed in units of *kilobytes*, *megabytes* and *gigabytes*. Note that when applied to computer memory, these prefixes, *kilo-*, *mega-*, and *giga-*, are *not metric but binary* expressions. They are based upon increasing the *exponent* by which the number 2 is raised in increments of ten, as illustrated in Table 28-5 (as opposed to raising the exponent by which the number 10 is raised in increments of 3 for the decimal system).

You will note that these binary numbers actually come out very close to the decimal equivalents, with a kilobyte being slightly more than one thousand bytes, a megabyte being slightly more than one million bytes, and a gigabyte being slightly more than one billion bytes. To convert kilobytes, megabytes, or gigabytes into *bits*, the correct number under the *binary system* in Table 28-5 would have to be multiplied by 8. Taking the kilobyte as an example:

Table 28-4

American Standard Code for Information Interchange

Character	ASCII Bit Representation
0	0 1 1 0 0 0 0
1	0 1 1 0 0 0 1
2	0 1 1 0 0 1 0
3	0 1 1 0 0 1 1
4	0 1 1 0 1 0 0
5	0 1 1 0 1 0 1
6	0 1 1 0 1 1 0
7	0 1 1 0 1 1 1
8	0 1 1 1 0 0 0
9	0 1 1 1 0 0 1
A	1 0 0 0 0 0 1
B	1 0 0 0 0 1 0
C	1 0 0 0 0 1 1
D	1 0 0 0 1 0 0
E	1 0 0 0 1 0 1
F	1 0 0 0 1 1 0
G	1 0 0 0 1 1 1
H	1 0 0 1 0 0 0
I	1 0 0 1 0 0 1
J	1 0 0 1 0 1 0
K	1 0 0 1 0 1 1
L	1 0 0 1 1 0 0
M	1 0 0 1 1 0 1
N	1 0 0 1 1 1 0
O	1 0 0 1 1 1 1
P	1 0 1 0 0 0 0
Q	1 0 1 0 0 0 1
R	1 0 1 0 0 1 0
S	1 0 1 0 0 1 1
T	1 0 1 0 1 0 0
U	1 0 1 0 1 0 1
V	1 0 1 0 1 1 0
W	1 0 1 0 1 1 1
X	1 0 1 1 0 0 0
Y	1 0 1 1 0 0 1
Z	1 0 1 1 0 1 0

$$1 \text{ Kilobyte} = 2^{10} \text{ bytes} =$$
$$1024 \text{ bytes} \times 8 = 8192 \text{ bits}$$

Some microprocessors work with groups of 16 consecutive bits rather than 8. Each group of 16 bits constitute a *word*, and a space is left between words.

Table 28-5

Decimal vs. Binary Number System

Decimal System	Binary System
Kilo = 10^3 = 1000	2^{10} = 1024
Mega = 10^6 = 1,000,000	2^{20} = 1,048,576
Giga = 10^9 = 1,000,000,000	2^{30} = 1,073,741,824

A word, then, is equivalent to two bytes. Within the memory, each word is assigned its own address, a physical location within the microscopic hardware.

COMPUTER SOFTWARE

Computer *software* refers to all the instructions given to the hardware of the computer in order to carry out tasks, which is written in higher-level codes called *computer languages*. All languages are ultimately reduced to binary or *hexadecimal* code which can be understood by the CPU. Hexadecimal code (*hex* = 6, *deci* = 10), consists of 16 characters including the numbers 0 through 9 and the letters A through F. Each of these characters represents a string of four binary numbers, therefore *two hexadecimal characters can be used to represent a byte or 8 bits of binary code*. Hexadecimal notation becomes a kind of *shorthand* for binary code, and serves as an intermediary coding system between high-level languages and binary.

Systems software includes assemblers, compilers, interpreters and operating systems designed to make the computer easier for the user to operate in general, that is, to make the entire system more *user-friendly*. These programs bridge the gap between *machine language* which only the computer understands and high-level languages that imitate human communication.

The assembly of programs using machine language is tedious, time-consuming and costly. Mid-level computer languages were developed which use commands in the form of symbolic names, acronyms and abbreviations to carry out repetitive functions. Examples are *READ* for "read file," *ADD*, *SUB* for "subtract," *LD* for "load file," and *PT* for "print." An *assembler* is a program that translates these symbolic commands into a binary or hexadecimal form which the *machines* (the printer, the modem, and the CPU, for example) will understand.

Interpreters and *compilers* translate the highest-level language of specific applications software into a form suitable for the assembler. From a description by the user of what task must be completed, the *compiler* or *generator* actually generates whole instructions and commands as needed in mid-level machine language, and organizes (compiles) them in proper order. The high-level instructions inputted into the computer are sometimes referred to as the *source code*, while its translation into low-level machine language is called the *object code*.

An *operating system* determines the general format of operation for a computer, based on the broadest sense in which it is intended to be used (home, business, or scientific use), and presents an appropriate interactive interface (or "desktop") at the display screen for the user in connection with the most appropriate input devices (keyboard, mouse, trackball, etc.). Operating systems are often written by the computer manufacturer and stored in ROM in the CPU. Examples of operating systems are Windows, Unix, Linux and MAC-OS. Typical commands for an operating system include such basic functions as *run file*, *save file*, *minimize* or *exit/escape*.

Specific user applications, the types of software one commonly buys at a store, are written in the highest-level programming languages such as Visual BASIC, C++, Pascal, VisiCalc (for spreadsheets), COBOL (for business) or FORTRAN (for scientific applications), and LOGO (for children). *Applications software* describes programs written in these languages to carry out specific types of user tasks such as word-processing, communications, spreadsheets, graphics and database management. Examples of some specific applications software packages include Microsoft Word, Quicken, Lotus and Excel.

When using an applications program, particular sets of instructions generated by the user may be found to be needed repeatedly in different projects. It is more efficient to write them once and store them as a separated module that can be accessed with a single command or key-stroke. *Macros* carry out these user-defined functions at the stroke of a key. *Function* keys serve a similar purpose, but macros can be defined to use any letter or character on the keyboard. (Macros serve exactly the same purpose as *subroutines* within a program, but macros are *created by the user*.)

Files created by the user from various applications are generally stored on the hard drive, not in the RAM memory. Each software program includes some instructions that are critical to its proper function and which must not be tampered with or accidentally changed by the user. These instructions are technically *volatile* since they can be changed or erased, but are made inaccessible to the user by placing the files in memory locations that are hard to get at or require passwords which only a specially-trained service representative would know. This is even more important for operating systems.

PROCESSING METHODS

There are four general approaches to processing data on a computer. For *on-line processing*, transactions are processed immediately upon entering a command, and the user must be present at the terminal to execute the command. Many functions entered at the console of an x-ray machine would fit this category. *Batch processing* refers to the method used when large amounts of data must be processed and only a few operations need to be executed on it. After the program, data and control statements are entered, the user may leave while the computer performs these operations. For *real-time processing*, an *array* of processors work in parallel to perform a complex computation on a large amount of data at high speed. This creates the illusion of *instantaneous* feedback or image display. Radiographic imaging systems must use real-time processing to display images with quick access and manipulation capability.

Time-sharing refers to the use of a large central computer that creates the illusion of serving several terminals simultaneously. This type of processing is also common in medical imaging, particularly in the form of *Picture Archiving and Communication Systems* (*PACS*) which allow centralized patient files to be brought up at a number of different terminals.

COMMUNICATIONS

An *interface* describes the connection between a computer or imaging machine and any of its peripherals, other computers or devices. For communication to take place between all of these machines, both hardware and software components must be *compatible*, that is, they must operate on the same physical principles and use the same basic languages and codes. Compatibility may be divided into two broad categories: *Internal compatibility* is the ability of computer's own components and software to work together, including graphics and sound cards, modems, printers, and software programs. *External compatibility* is the ability of different computer systems to communicate with each other.

The use of telephone lines to transfer data between computers was made available by the development of the *modem*. The word *modem* is an acronym for **Mo**dulator-**Dem**odulator. Musically, "modulation" means adjusting the pitch of a musical note or key signature upward or downward. A modem receives digital information from the computer in the form of electronic signals of differing voltages. It converts these into analog audio signals, or distinct tones, for transmission over phone lines. These are just the same types of tones one hears while dialing a telephone, with each tone or pitch representing a different number, only on a more sophisticated scale. At the other end of the telephone line, another modem converts these audio tones back into voltages that represent the data. Collectively, these signals can be reassembled to formulate an entire photograph or radiographic image, or a complete musical composition.

A similar process can be used with optical fiber bundles to transmit different wavelengths of light along a cable from one computer to another. This process still requires a form of *modem* at each end of the transmission, to code the electronic signals into different light frequencies and decode these at the other end of the line.

Teleradiology refers to any system which allows the remote transmission and viewing of radiographic images via modems over phone or cable lines. The images transmitted may come directly from computer storage, or they may be *scanned* off of a hard-copy radiograph using an optical scanner. The details of how a scanner works will be covered later.

The *baud rate* is the speed of transmission in bits per second (bps) or *kilobits per second* (*K*). Baud rates for more and more powerful modems are generally described in multiples of 14 *kilobytes,* such as *28K, 56K,* and so on, numbers which have been rounded out. For example, a *28K* modem actually transmits 28,800 bps.

Teleradiology makes it possible for images to be sent great distances for a specialist to collaborate with a radiologist, and for images stored at a hospital to be accessed almost instantly by doctors at their individual clinics. A common use of teleradiology is to transmit images to a radiologist's *home* during off-hours. For these types of access, it is often not necessary for any specific data operations to be performed on the image—the only immediate need is for the image to be *displayed,* so that the doctor can phone in or e-mail a reading. In such cases, it is not even necessary for the image data to pass through the CPU of the computer, which only slows down its arrival at the display screen. *Direct memory access* (*DMA*) controllers were developed for this purpose. Transmissions intended for direct delivery to the monitor screen are coded. The DMA controller detects this signature, and allows the transmission to bypass the CPU, speeding up delivery to the display screen or other output device.

Each individual point within a communications network where data may originate or be accessed is called a *node.* When a transmission is sent from a smaller computer or less important node to a larger centralized computer, a more important node within the network, or a satellite, we refer to this process as *uploading* data. When a transmission flows from a satellite, a central computer, or a central node within a network to a less important or smaller computer, we call it *downloading* the data.

A *local area network (LAN)* is a computerized communications network generally contained within a single building or business. The devices in a LAN share one server, and, typically, the system is privately owned. A *WAN,* or *wide area nework,* extends to other businesses or locations that may be at great distances. A WAN is usually publicly or commercially owned and uses transmission services provided by common carriers such as phone or cable companies.

Both LANs and WANs are widely used in medical imaging. There are at least three types of *LAN*'s with which radiographers should be familiar: the *PACS (Picture Archiving and Communication System),* the RIS *(Radiology Information System),* and the *HIS (Hospital Information System).* The picture archiving and communication system (PACS) is used within a medical imaging department to make radiographs, CT and MRI scans, ultrasound and nuclear medicine images for a particular patient available at any *node* within the network. This allows radiologists and radiographers to access these images from various locations, improving the efficiency of communication.

Every computer within a network has a *unique internet protocol* or IP address. Expressed in "dotted-quad" format, this number always has four components separated by periods, such as: 172.8110.3.1. The first number set, before the first period, identifies the network, and the remaining sets of digits indicate the specific computer, device, or host. To set up a network, a network interface card with accompanying software must be installed in each computer or device. The card is a small circuit board which may be installed inside the computer or connected on the outside. If the network is wireless, the interface cards will include an antenna for radio transmission.

A network *switch* connects various nodes within a network, and is considered *smart* in that it "knows" where a particular type of data needs to go without always searching the entire network. A *router* connects two or more networks. Routers can have "firewall" hardware or software that filters access to the connected networks. Wireless routers now allow "point-of-care" access to a network for physicians and other caregivers via the personal digital assistant (PDA) they may carry in their pocket, a "tablet," or a laptop computer.

The *radiology information system* (*RIS*) performs just the same function, but for a data base of written records and files on patients, making them accessible from different locations within the radiology department. The *hospital information system* (*HIS*), does the same for all of a patient's general medical files throughout the hospital. The greatest efficiency of communication is achieved when these systems, the PACS, the RIS and the HIS are compatible and fully integrated (Fig. 28-21).

Figure 28-21

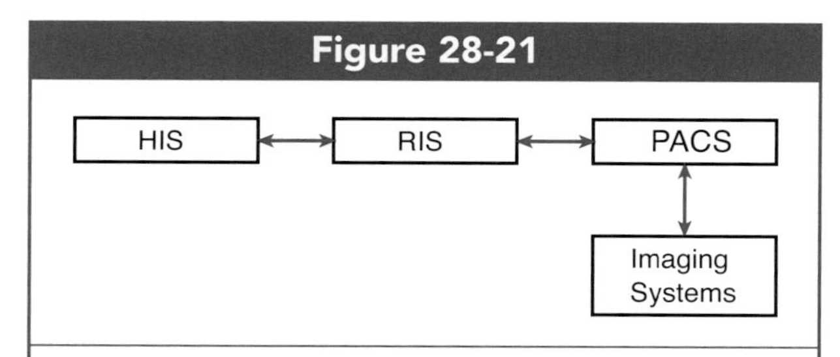

A PACS must be fully integrated and compatible with all imaging systems in the department, with the radiology information system and with the hospital information system.

SUMMARY

1. A computer program is a collection of hundreds or thousands of algorithms, each of which instructs the computer how to perform a single, specific task.
2. The power of a computer is measured by how many millions of instructions per second (MIPS) it can process, largely determined by the speed of the microprocessors which is measured in gigahertz or terahertz.
3. Most x-ray machine consoles are microcomputers. MRI and CT scanners use minicomputers.
4. The first electronic digital computer was developed in the year 1946, and by 1951 the first mass-marketed computer was available, made possible by the invention of the transistor. Since that time, computers have evolved through four generations.
5. Photographic and chemical processes are now used to miniaturize and fuse millions of circuit elements into an integrated circuit on a silicon chip about 1 cm in size.
6. All peripherals and storage devices communicate with the CPU via bus lines that are connected through ports. The CPU consists of the control unit and the arithmetic/logic unit, which work in tandem to manage all data.
7. The VDT consists of a display screen and input devices (keyboard and mouse), and can be intelligent if it has its own processing capability and memory. Most x-ray machine consoles are intelligent terminals.
8. The multiformat camera is the most widely-used system for processing hard copies in medical imaging.
9. Operation code from primary memory directs the activities of the control unit and provides addresses for locating data storage. In the ALU, data for calculations are temporarily stored in registers, and intermediate results of calculations are stored in the accumulator.
10. A byte consists of eight bits and is sufficient to create a single character. Each address in computer memory stores one byte of data.
11. The motherboard supports all of the main circuits, which generally operate at 5-volt or 12-volt electrical current that has been stepped down from the incoming power supply.
12. Hard discs use magnetized surfaces to store data, and electromagnetic induction to read and write data. By using the cylinder method to locate data within a stack of discs, the reading process is accelerated.
13. The RAID system, widely used in medical imaging, prevents the accidental loss of information by multiple, independent back-up storage.
14. Optical discs use the reflection of a laser beam from a pitted mirror surface to read data. A higher intensity laser beam is used to melt these pits into the surface in the writing process.
15. Flash memory drives store electric charges in their cells for years, forming a binary code. They are more reliable than magnetic hard drives, but for the purposes of medical imaging, the lower cost and higher capacity of hard disk drives makes them the preferred method for long-term storage of medical images.
16. Memory can be internal or external, primary or secondary, volatile or nonvolatile, and RAM or ROM.

17. Analog information is on a continuous spectrum, whereas digital information is discrete. Mathematically, the ADC essentially rounds out numbers to discrete values, thus reducing the volume of data to a dynamic range which the computer can manage.
18. Although digitized information is inherently less accurate than analog information, as long as the discrete units are smaller than a human can detect, *read-out* accuracy is improved.
19. By using base 2 notation rather than a base 10 numbering system, binary code allows all data to be reduced to two values or bits, 1 and zero.
20. Machine languages, based on hexadecimal code, are intermediate languages that form a kind of "shorthand" notation which facilitates repetitive functions such as "read," "load," and "print" for assemblers, interpreters and compilers. Since ASCII code was established, most of these languages also provide compatibility between different manufacturers.
21. Systems software includes the operating system which determines the general format for data input and display, and all of the machine code for a computer system. Applications software uses high-level language to carry out specific types of user tasks in user-friendly format. It provides *source code* to the computer system.
22. Data processing can be executed on-line, in batches, in real-time or in a time-sharing format.
23. Modems provide the ability to transmit images and other data over phone lines or fiber optic lines. Direct memory access speeds up the display process through bypassing the CPU.
24. In medical imaging the PACS is a local area network which facilitates access to and management of images for all the nodes in the RIS and HIS systems.

REVIEW QUESTIONS

1. Artificial intelligence is defined as the ability to perform __________ functions such as "if then," and "if else."

2. List the three general size categories of computers:

3. A PC or other microcomputer usually has a single __________.

4. What was the name of the first computer, completed in 1949, that incorporated John Von Neumann's theories to provide *stored programs*?

5. In what year was the transistor developed?

6. A multiwire line is called a(n) __________, and if its plug includes 24 prongs, it must be connected through a(n) __________ port.

7. A bar-code reader is an example of a __________ entry device.

8. The combination of a display monitor screen with a keyboard and mouse makes up a __________.

9. Which portion of the CPU directs the flow of data between the ALU, primary memory, and input and output devices?

10. To perform arithmetic and logic operations in the ALU, electrical signals must pass through which three types of basic circuits or "gates" in different combinations?

11. List the four main sectors of primary memory:

12. Each address in primary memory consists of how many bit storage units?

(Continued)

REVIEW QUESTIONS *(Continued)*

13. What type of transformer must be used for regular electrical power coming into a computer?

14. When reading data from a hard disc, patterns of magnetized elements on the surface of the disc induce __________ in the read/write head.

15. When the cylinder method is used to locate data on a stack of magnetic hard discs, reading speed is increased because multiple __________ can be used to simultaneously read the data.

16. What does *RAID* stand for?

17. The number of bits that can be written to a magnetic disc per inch of radius is known as its recording __________.

18. During the reading of an optical disc, what does a *land* do to the laser beam which strikes it?

19. DVD discs can hold much more information that CD discs because they have less __________ from the center of one groove to the center of the next.

20. Comparing flash memory with magnetic hard drives, which of the two provides more reliable storage?

21. Memory which can be easily erased is termed:

22. Memory which can be accessed without indexing through previous files or recordings is termed:

23. *Firmware*, such as the bootstrap program, is written as __________ memory.

(Continued)

REVIEW QUESTIONS *(Continued)*

24. Data which can have any value, without limitation on its number of decimal places is __________ data.

25. The term *bit* is an acronym for:

26. What decimal number is represented by the binary number 110110?

27. What is the binary code for the number 24?

28. In ASCII code, we know when the last four of seven bits represent a *letter* rather than a number because of the first three digits, called __________ bits.

29. How many bits are there in 2 megabytes?

30. Interpreters and compilers translate source code inputted from applications software into machine language or __________ code.

31. The ability of a single computer's peripherals and components to all work together is termed its __________ compatibility.

32. The speed, in kilobits per second, with which data can be transmitted between modems is called the __________ rate.

33. Any single access point within a WAN or LAN is called a(n) __________.

34. What code was the first standardization of intermediate computer languages which provided compatibility between different manufacturers?

35. The specific type of LAN used for managing images within a radiology department is called a:

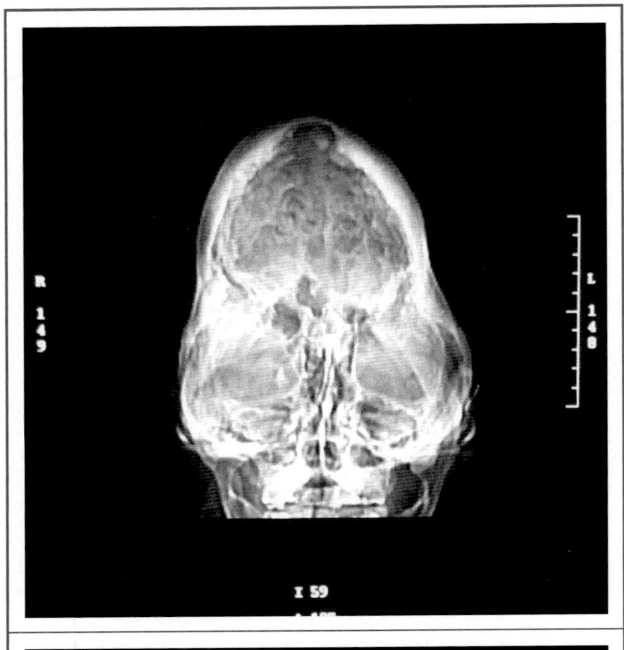

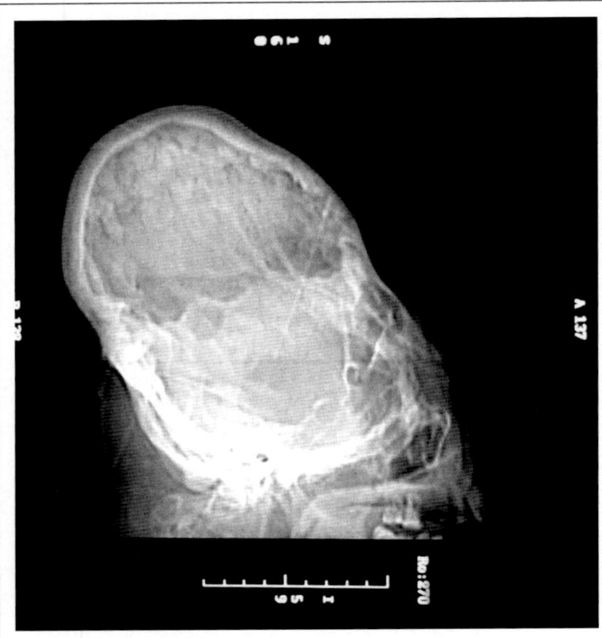

Microcephaly. This unfortunate patient was born with an underdeveloped cerebrum and cranium.

Chapter 29

CREATING THE DIGITAL IMAGE

Objectives:

Upon completion of this chapter, you should be able to:

1. Describe the aspects of a digital image matrix and how it impacts image resolution.
2. Relate pixel size to the displayed field of view and matrix size.
3. Define the three steps in digitizing any analog image.
4. Explain the relationships between bit depth, dynamic range and image gray scale in providing image resolution.
5. Describe the nature of *voxels* for CT, CR and DR imaging and how the x-ray attenuation coefficient for each is translated into the gray levels of pixels.
6. Describe the development and limitations of contrast resolution and spatial resolution for digitized radiographic images.
7. Explain how the selection of specific procedural algorithms impacts the displayed image.
8. Fully define window level and window width and how they translate into displayed image brightness and gray scale.
9. Describe the components and function of the PACS, RIS and HIS and the DICOM standard.
10. Define the types, characteristics and proper use of workstations and display stations.
11. Overview how lasers work, and their four types of applications in the medical imaging department.

THE NATURE OF DIGITAL IMAGES

All digital images, whether photographic, radiographic, or fluoroscopic, consist of a *matrix* of numeric values that can be stored in computer memory. The matrix is a pattern of cells laid out in rows and columns that cover the entire area of the image. As shown in Figure 29-1, each cell can be identified by its column and row designations and corresponds to a specific location within the image. For radiographic images, the numerical value stored for each cell represents the *brightness* (or *density*) assigned to that location. This brightness level is taken from a range of values stored in the computer which represent different shades from "pitch black" to "blank white."

Figure 29-2 is a visual trick to illustrate how an image of different tissues within the body can be represented by a matrix of numbers. In this case, the bone tissue of the femur, which should be represented on the display screen as a very light gray shade, nearly white, has been assigned a numerical value of 555. The soft tissue of the thigh surrounding the bone has been assigned a value of 11 which will bring up a dark gray shade on the monitor. The background of the image, which will be pitch black, has been assigned a pixel value of 0. Observing this pattern, you can just make out how denser tissues can be represented by higher numbers to build up an image of the bone within the thigh.

In Figure 29-3, both matrices can be found to have higher numbers around the center of the matrix and extending downward and somewhat to the right.

Figure 29-1

COLUMNS

	1	2	3	4	5	6
1	1-1	2-1	3-1	4-1	5-1	6-1
R 2	1-2	2-2	3-2	4-2	5-2	6-2
O 3	1-3	2-3	3-3	4-3	5-3	6-3
W 4	1-4	2-4	3-4	4-4	5-4	6-4
S 5	1-5	2-5	3-5	4-5	5-5	6-5
6	1-6	2-6	3-6	4-6	5-6	6-6

A digital image matrix is made up of individual picture elements, each designated by its column and row number.

Figure 29-2

In this simplified representation of a digital image as it is stored in the computer, the background density is assigned a numerical value of 0, the soft tissues of the thigh are given a value of 11, and the bone of the femur a value of 555. Although one can make out the pattern of the anatomy visually here, in the computer the image is purely numerical in nature. (From Quinn B. Carroll, *Practical Radiographic Imaging*, 8th ed. Springfield, IL: Charles C Thomas, Publisher, Ltd., 2007. Reprinted by permission.)

0	0	0	0	0	0	0	0	0	0
0	0	0	11	11	11	11	0	0	0
0	0	11	11	11	11	11	11	0	0
0	11	11	11	555	555	11	11	11	0
0	11	11	555	555	555	555	11	11	0
0	11	11	555	555	555	555	11	11	0
0	11	11	11	555	555	11	11	11	0
0	0	11	11	11	11	11	11	0	0
0	0	0	11	11	11	11	0	0	0
0	0	0	0	0	0	0	0	0	0

Figure 29-3

A

9	18	20	35	23	7
20	25	77	76	61	11
22	36	88	92	56	8
26	47	78	77	50	34
19	51	60	75	74	43
3	42	58	72	72	53

B

9	2	19	7	4	3	5	12	6	15	8	7
11	14	8	5	2	4	7	9	13	10	4	3
13	10	6	11	18	20	35	88	92	5	3	2
11	7	17	25	22	39	44	61	84	9	6	5
12	14	19	41	36	77	76	56	61	13	8	6
15	18	21	54	47	78	77	50	55	16	11	9
20	24	30	57	60	51	75	74	53	19	14	12
26	31	28	66	62	58	72	72	43	22	18	15
29	27	33	45	56	47	61	56	52	30	24	21
19	22	27	35	44	50	54	48	38	26	19	17
9	12	15	23	28	33	33	31	27	20	10	8
3	2	5	8	13	15	18	14	12	7	4	1

Two digital images with higher pixel values toward the center and lower right, but using different size matrices. Image ***A*** is a 6×6 matrix and image ***B*** a 12×12 matrix. Covering the same physical area, image ***B*** with 144 pixels must have smaller pixels than image ***A*** with 36 pixels. Since the smaller pixels produce sharper resolution, it is easier to make out the pattern of larger numbers in ***B***.

These are not as apparent as the pattern in Figure 29-2, but upon close examination one can make out what might represent a distinct anatomical part in this region on both digital images.

Each cell within a digital image is called a *pixel* (from "**pic**ture-**el**ement"). In Figure 29-3, ***A*** is a matrix that is 6 pixels in height and 6 pixels in width, for a total of 36 pixels, while ***B*** is a matrix of the same overall area, but with 12 columns and 12 rows of pixels for a total of 144 pixels. The *size* of the matrix is expressed in terms of this total number of pixels (not the actual area of the image). ***B*** is a larger matrix than ***A*** and has many more pixels. What becomes immediately apparent is that *for a larger matrix, the pixels must be of smaller size.*

Figure 29-4 illustrates a progression of increasing matrix sizes for the same image. As the matrix size grows, the individual pixels become smaller, so that smaller details can be resolved in the image. The result is an image with *sharper resolution of details.*

Larger matrix = Smaller pixels =
Improved sharpness

For digital images, this *pixel size becomes a limiting factor for the spatial resolution of the image.* As described in Chapter 25, spatial resolution or *sharpness* can be measured in terms of the spatial frequency which has units of *line-pairs per millimeter (LP/mm)* (Fig. 25-6). At least two pixels are required to record a line pair with one line having a brighter shade and one having a darker shade of density. With pixels measuring 0.4 mm, no more than 1.25 line pairs per millimeter can be resolved from a standard resolution phantom (Fig. 25-5 in Chapter 25). When pixel size is reduced to 0.1 mm, spatial resolution increases to about 5 LP/mm.

By the 1990s, improvements in both computers and monitor screens had improved the resolution of digital imaging systems to 6–8 LP/mm. Just prior to the conversion of diagnostic radiology departments to digital systems, high-speed film/screen systems were being used that had a spatial resolution of 8–10 LP/mm. Modern digital systems approach this value, but still cannot compete with the 10–12 LP/mm once achieved by slow-speed film systems. It is important to understand that *digital radiographic images could only achieve the resolution of* analog *images by reducing pixels to the size of a single silver bromide crystal (several molecules).* Generally, digital images have *poorer* spatial resolution than analog images, but this is offset by vast improvements in *contrast resolution.*

When considering different imaging modalities, the obvious differences in spatial resolution can be directly correlated to the matrix sizes employed. The images appearing most *blurry* are those in nuclear medicine where image matrices are about 64 × 64 pixels. Sonograms, with a matrix size approximating 128 × 128, are still quite blurry to the human eye. Computed tomography (CT) and magnetic resonance imaging (MRI) appear much sharper. They

Figure 29-4

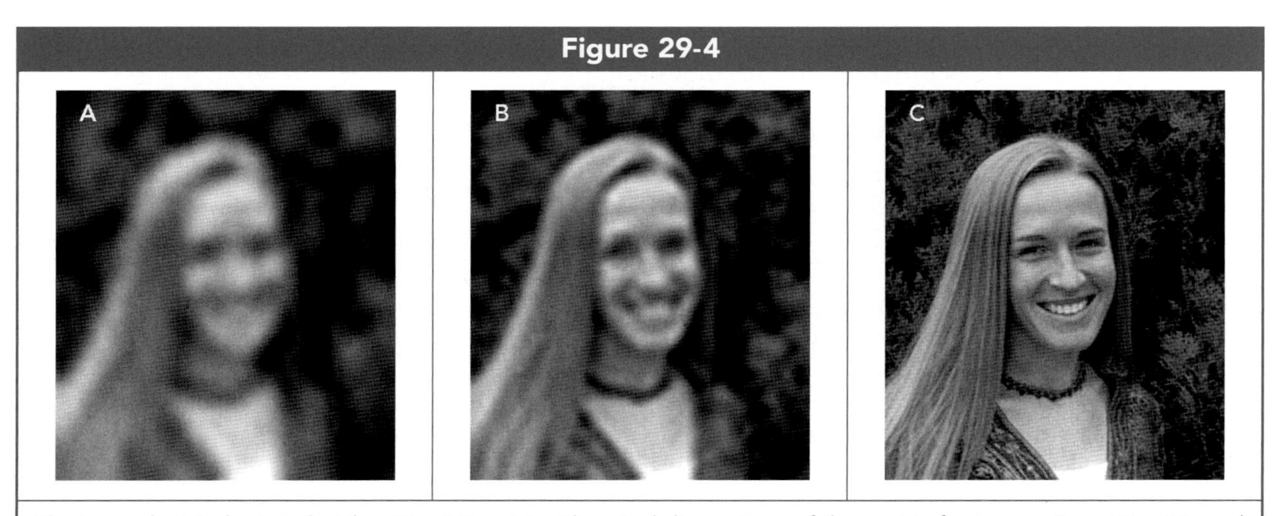

Photograph ***A*** is depicted with a 26 × 32 matrix. The pixel dimensions of the matrix for image ***B*** are 51 × 64, and those of image ***C*** are 200 × 251. The larger the matrix, the sharper the image.

generally use matrices of 512 × 512 pixels, with some applications at 256 or 1024 pixels. Sharper still are direct-capture digital radiography (DR) and computed radiography (CR), which use matrices of 1024 × 1024 with some applications reaching 2045 pixels.

From Chapter 25, remember that the spatial frequency in LP/mm can also be derived from the minimum object size that can be imaged, in this case a single pixel. The formula for this relationship would be rewritten as:

$$SF = \frac{1}{2(PS)}$$

where *SF* is the spatial frequency in line-pairs per unit length, (usually millimeters), and *PS* is the pixel size in the same units. The spatial frequency is a measure of image *resolution*. Following is a practice exercise applying the formula:

Practice Exercise #1

For a pixel size of 0.3 mm, what is the associated resolution in line-pairs per millimeter?

Solution:

$$SF = \frac{1}{2(PS)}$$

$$SF = \frac{1}{2(0.3)} = \frac{1}{0.6} = 1.6$$

Answer: The resolution of this image is a spatial frequency of 1.6 LP/mm.

Repeating the same calculation for a smaller pixel size of 0.2 mm, we see that the spatial resolution increases to 2.5 LP/mm. As the pixel size becomes smaller, spatial resolution is improved.

More specifically, we can state that the size of the image matrix, by pixel count, is *inversely proportional* to pixel size, and *directly proportional* to spatial resolution.

Displayed Field of View and Pixel Size

The size of a pixel is not only related to the size of the matrix, but also to the displayed *field of view* (*FOV*) presented on the display screen. In conventional radiography, the field of view was the collimated portion of the x-ray beam projected onto the film and containing the anatomic structures of interest. For digital radiography, the FOV is that portion of the imaging plate that contains relevant anatomic information and is displayed at the monitor screen. For example, the field of view for an adult hand would be considerably smaller than that for the chest. *Since the images of both the hand and the chest are displayed at the monitor screen using the same image matrix, the smaller field of view will consist of more pixels in a given display area than the larger FOV used for the chest.*

The relationship of pixel size to the size of the matrix and the field of view is summed up in the formula:

$$\text{Pixel Size} = \frac{\text{FOV}}{\text{Matrix}}$$

This formula states that the size of the pixels in an image is directly proportional to the size of the displayed field of view and inversely proportional to the size of the matrix. The following exercises are two examples of how this formula may be applied to calculations.

Practice Exercise #2

What is the pixel size for a 12 × 12 inch digital image reconstructed at the display screen on a 1024 × 1024 matrix?

Solution: First, convert the field of view from inches into millimeters:

12 inches × 25.4 millimeters per inch = 305 mm

$$\frac{FOV}{Matrix} = \frac{305}{1024} = 0.298 = 0.3$$

Answer: The size of the pixels in the displayed image is 0.3 mm.

Practice Exercise #3

A 41 × 41 cm field of view is projected onto a monitor screen with 0.8 mm pixels. What is the *size of the matrix on the monitor screen*?

Solution: First, convert the field of view from cm into millimeters:

41 cm = 410 mm

$$Pixel\ Size = \frac{FOV}{Matrix}$$

$$0.8 = \frac{410}{Matrix}$$

Cross-multiplying: $\quad Matrix\,(0.8) = 410$

$$Matrix = \frac{410}{0.8} = 512.5$$

Answer: *The size of the matrix at the monitor screen is 512 × 512 pixels.*

Figure 29-5 is a CT scan of the head reconstructed using 7 mm pixels for the upper half of the image, and 3 mm pixels for the lower half. A step-stair appearance can be seen in both portions of the image (arrows), but it is very much diminished in the lower portion where the spatial resolution is improved.

DIGITIZING AN ANALOG IMAGE

Light images enter a camera lens in analog form—the various intensities of light can have any value. Likewise, x-rays from a radiographic projection enter the image receptor plate in analog form (as do radio waves during an MRI scan or sound waves during a sonogram procedure). All of these forms of input must be converted into digital form in order to allow computerized processing.

There are three fundamental steps to *digitizing* an image which are relevant to all forms of imaging. In the first step, the field of the image is divided up into an array (a matrix) of small cells by a process called *scanning*. Each cell becomes a *pixel* or picture element in the final image. Based on which column and row it falls into, each pixel is assigned a designator for its location, as shown in Figure 29-1 at the beginning of the chapter. Here, in Figure 29-6, the scanning process results in a 9 × 11 matrix composed of 99 pixels.

A standard photocopying machine, a radiographic film scanner, or the scanner connected to your home PC can all be heard completing a precopying sweep across an image, which performs this function of determining matrix size and pixel allocation. In computed radiography (CR), the *reader* (processor) is set to scan the imaging plate in a designated number of lines which are divided into individual sectional measurements corresponding to pixels.

In direct-capture radiography (DR), the number of *available* pixels is the number of thin-film transistors (TFT's) physically embedded in the imaging plate, but *collimation* of the x-ray beam effectively selects which of these will comprise the image and so is analogous to the *scanning* function. For digital fluoroscopy (DF), as well as for video cameras in general, a charge-coupled device (CCD) which picks up the light image is composed of a preset number of charge-collecting electrodes that constitute pixels.

Figure 29-5

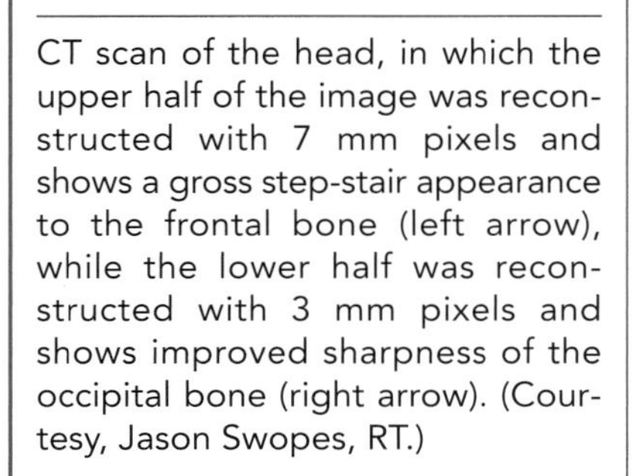

CT scan of the head, in which the upper half of the image was reconstructed with 7 mm pixels and shows a gross step-stair appearance to the frontal bone (left arrow), while the lower half was reconstructed with 3 mm pixels and shows improved sharpness of the occipital bone (right arrow). (Courtesy, Jason Swopes, RT.)

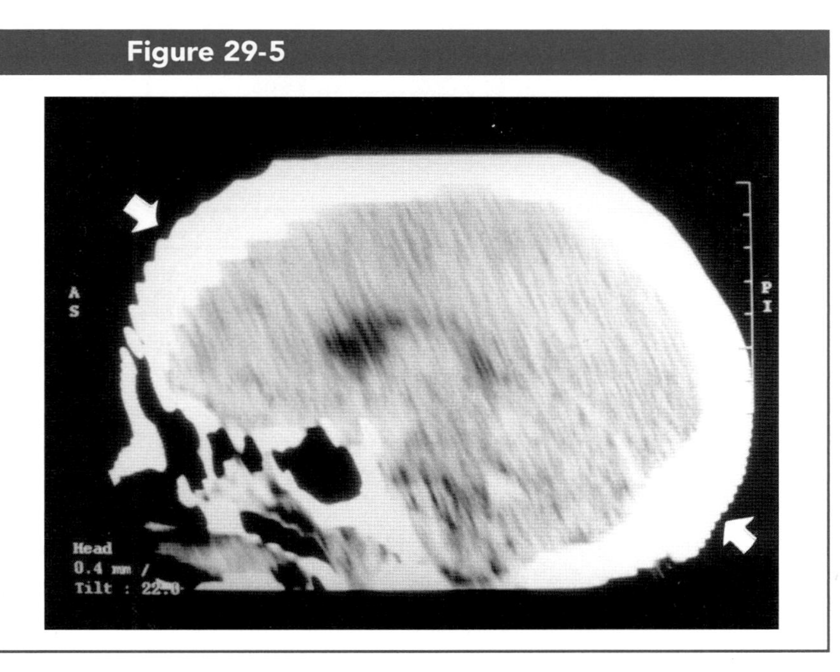

Figure 29-6

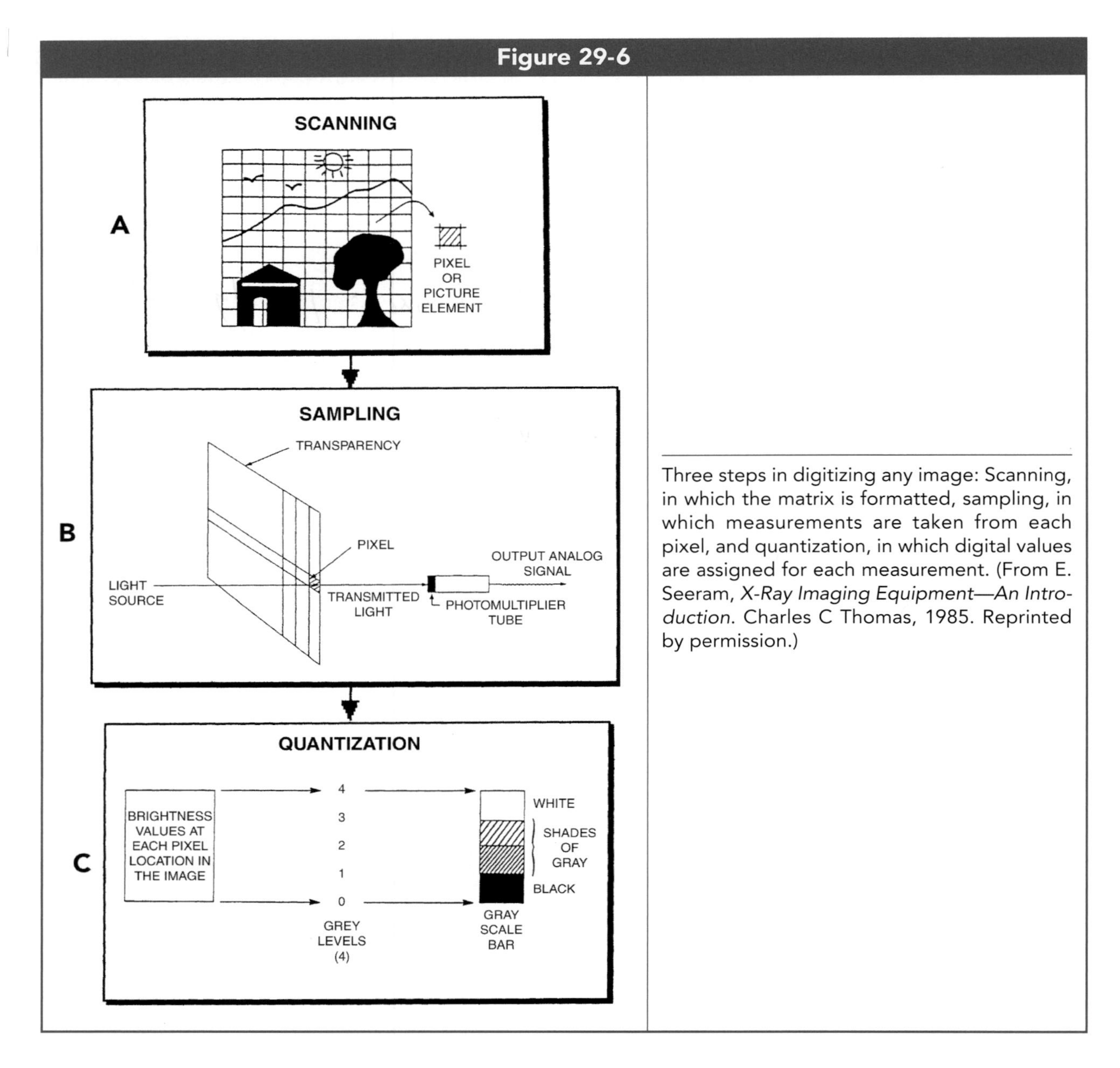

Three steps in digitizing any image: Scanning, in which the matrix is formatted, sampling, in which measurements are taken from each pixel, and quantization, in which digital values are assigned for each measurement. (From E. Seeram, *X-Ray Imaging Equipment—An Introduction.* Charles C Thomas, 1985. Reprinted by permission.)

Regardless of which particular method is used, all forms of digital imaging require the *formatting of a matrix with a designated pixel size.* The term *scanning* may be broadly applied to all the different approaches to achieving this preliminary step.

The second step in digitizing an image is known as *sampling.* In sampling, the *intensity* of light or radiation from each designated pixel area is *measured* by a detector. For a photographic or radiographic scanner, the light reflected from a page or transmitted through a radiograph is detected by a *photomultiplier* tube (Fig. 29-6***B***), which converts the light into electricity and amplifies the signal. For CR, DR, DF, CT, MRI and all other forms of medical imaging, the *sampling* stage may be considered as the function of the specific imaging machine itself, that is, the detection and measurement of various forms of radiation which occurs at the imaging plate, at an array of detectors, or at a radio antenna (for MRI).

Instruments used to sample the pixels in an image can have different sizes and shapes for their *sampling aperture* or opening through which the pixel value is measured. An interesting difference between DR and CR is that for DR, the sampling aperture is *square* and the samplings are adjacent to each other, since the detector is a square detector element, whereas for CR the aperture is *round* and the samplings overlap each other (Fig. 29-7), because the *laser beam* which strikes the phosphor plate to stimulate it is round. The specific methods of how detection and measurement are accomplished for CR, DR and DF will be discussed in following chapters.

The final step in digitizing an image is *quantization*. The end result of the quantizing process must be a value assigned to each pixel representing a discrete, predesignated *gray level*, a number which the computer can understand and manipulate. This gray level can only be selected from a predetermined *range* of gray levels called the *dynamic range*. In Figure 29-6***C***, there are only four shades of gray to choose from—the *dynamic range* is 4. Actual values of brightness that fall *between* these four shades must be *rounded up or down to the nearest available gray level*.

Recall from the previous chapter that digital computers can only handle *discrete* numbers which have a limited number of places beyond the decimal point. Analog numbers coming into the system which fall between these values must be rounded up or down to the nearest available *digital* number so the computer can understand it. This is the function of the *analog-to-digital converter* (*ADC*), to essentially round out all inputted data into digits allowed by the computer system.

(A digital-to-analog converter, or DAC, may be used for signals flowing *out* of the computer to display screens in order to speed up transmission and make the signals compatible for the electronics in the device to process. The actual values of the data, however, are not changed, since a number cannot be "de-rounded" once the initial analog value is lost.)

The maximum range of pixel values the computer or other *hardware* device can store is expressed as the "bit depth" of the pixels. *Bit depth* is the exponent of the base 2 that yields the corresponding binary number. We say that the pixels are *6 bits deep* for a range of $2^6 = 64$, *7 bits deep* for a range of $2^7 = 128$, and *8 bits deep* for a range of $2^8 = 256$. All of

Figure 29-7

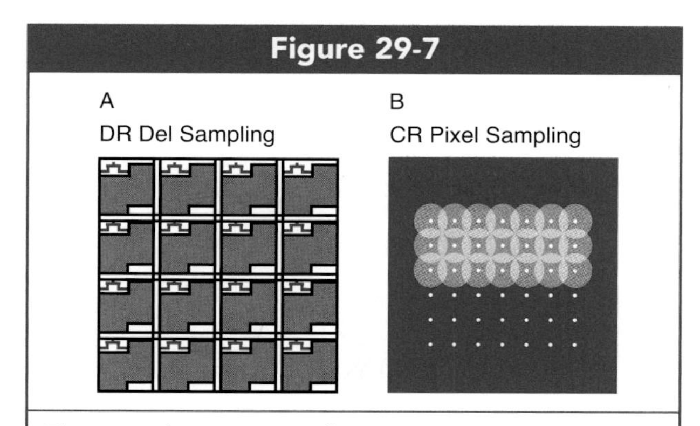

The *sampling aperture* for DR equipment is roughly square, ***A***, but misses some information between actual detection surfaces. The aperture for CR is round, ***B***, and must overlap adjacent samplings in order to fill square pixels in the constructed digital image.

these bit depths are beyond the capability of the human eye, which can only discern approximately 2^5 or 32 different levels of brightness; such bit depths result in images that are indistinguishable from analog images to the human eye. Therefore, it is not necessary to utilize the *full* capacity of the computer in presenting images to the human eye. That is, the full bit-depth need not be used in presenting images at a display screen. By *selecting* a smaller range of pixel values from the bit depth, which will be made available to build up images, the processing time for images can be accelerated.

The *range* of gray levels made available to construct images is called the *dynamic range* of the imaging system. *The dynamic range set by the system software determines the gray scale available for the image to be displayed.*

As with bit depth, the dynamic range is always a binary number—therefore, the image can be represented in 2, 4, 8, 16, 32, 64, 128, 256, 512 or 1024 gray levels. This is the number of gray shades with which *each pixel* can be represented. The brightness level for each pixel in the image must be "selected" from this scale.

Figure 29-8 illustrates a series of images with increasing dynamic range and the resulting lengthened gray scale. It becomes readily apparent that when the dynamic range is too low and the gray scale is too short, as in ***A***, details are actually lost to the image. As the gray scale increases in this series, more and more details of the image are discerned. *The greater*

Figure 29-8

Photographs of the face of a moth. ***A*** has a bit depth of only 1, generating a dynamic range or gray scale of 2^1 = 2 shades, black and white. ***B*** has a bit depth of 2, generating a dynamic range of 2^2 = 4 shades of gray. ***C*** has a bit depth of 3, generating a dynamic range of 2^3 = 8 shades of gray, and ***D*** has a bit depth of 8, generating 2^8 = 256 shades of gray. The greater the dynamic range, the longer the gray scale, and the more details can be resolved. (Courtesy, Brandon Carroll.)

the dynamic range and the longer the gray scale, the more details can be represented in an image.

What, then, constitutes an ideal dynamic range to be selected from the bit depth capability of a digital imaging system? An excessive dynamic range can slow down processing time unnecessarily, while too short a range causes image details to be lost. An important third factor is that the range chosen must allow postprocessing *manipulation of the image*, such as adjusting the brightness or the contrast, to meet all reasonable contingencies. For example, a dynamic range of 256 (2^8) is *eight times the capability of the human eye* (32). This would allow for the overall brightness of any image to be doubled or cut in half two or three times without running out of available gray levels. Visually, then, a dynamic range 8 bits deep would seem to be more than sufficient for most applications. However, for special processing features such as *subtraction* this may still not be sufficient.

The dynamic range of the remnant x-ray beam as it exits the patient is approximately 2^{10}. Furthermore, the *main advantage* of digital imaging is its enhanced *contrast resolution*, which depends entirely upon an extended dynamic range and the processing latitude it affords. The enhanced contrast resolution and processing features of CT and MRI systems require a 12-bit dynamic range. Overall, then, most digital imaging systems have their dynamic ranges set at 2^8 (256), 2^{10} (1024), or 2^{12} (4096).

Even though the storage capacity of modern computers and recording media is very impressive, the large computer *file size* of medical images can become an important issue when many thousands of images are stored. The file size of an image is the product of its *matrix size multiplied by its bit depth.*

$$\text{File size} = \text{Matrix size} \times \text{Bit depth}$$

Although medical images require both high spatial resolution and high dynamic range, the PACS administrator or informatics technologist must make prudent decisions regarding studies which can be stored with larger pixel sizes (such as digital fluoroscopy) or with less bit depth and still retain adequate diagnostic quality.

ROLE OF X-RAY ATTENUATION IN FORMING THE DIGITAL IMAGE

Conventional film-based radiography, direct-capture digital radiography (DR), computed radiography (CR), and computed tomography (CT) all work on the basis of measuring the *attenuation of x-rays* as the x-ray beam passes through the patient. The ratio or percentage of the original x-ray beam intensity that is absorbed by a particular tissue area within the patient is the tissue's *attenuation coefficient.* In tissues that have a greater thickness or a higher physical density, a smaller proportion of the incident radiation reaches the image receptor. In such areas where the attenuation coefficient is higher, a lighter gray level is assigned by the computer to the corresponding pixel in the image.

To determine the attenuation coefficient for various tissues, data are acquired from three-dimensional volumes of tissue within the patient called *voxels* (from "**vo**lume-**el**ements"). For radiographic images, *each pixel in the image represents a voxel within the patient.* As shown in Figure 29-9, each voxel from a CT scan is in the shape of a *cube* representing a small portion of the three-dimensional "slice" that is being sampled. The CT scanner is capable of isolating this cube of tissue because it uses *multiple projections* to acquire data from hundreds of angles through the patient. *Within* each voxel, the attenuation coefficients for various tissues are averaged to obtain a single number representing the entire voxel, which will be translated into a single gray level to by displayed in the corresponding *pixel* in the final image.

By comparison, direct-capture digital radiography (DR) and computed radiography (CR) (as well as conventional film-based radiography) all produce images from a *single projection*, meaning that the *voxels* of tissue that are sampled take on the shape of *square tubes* that extend all the way from the front to

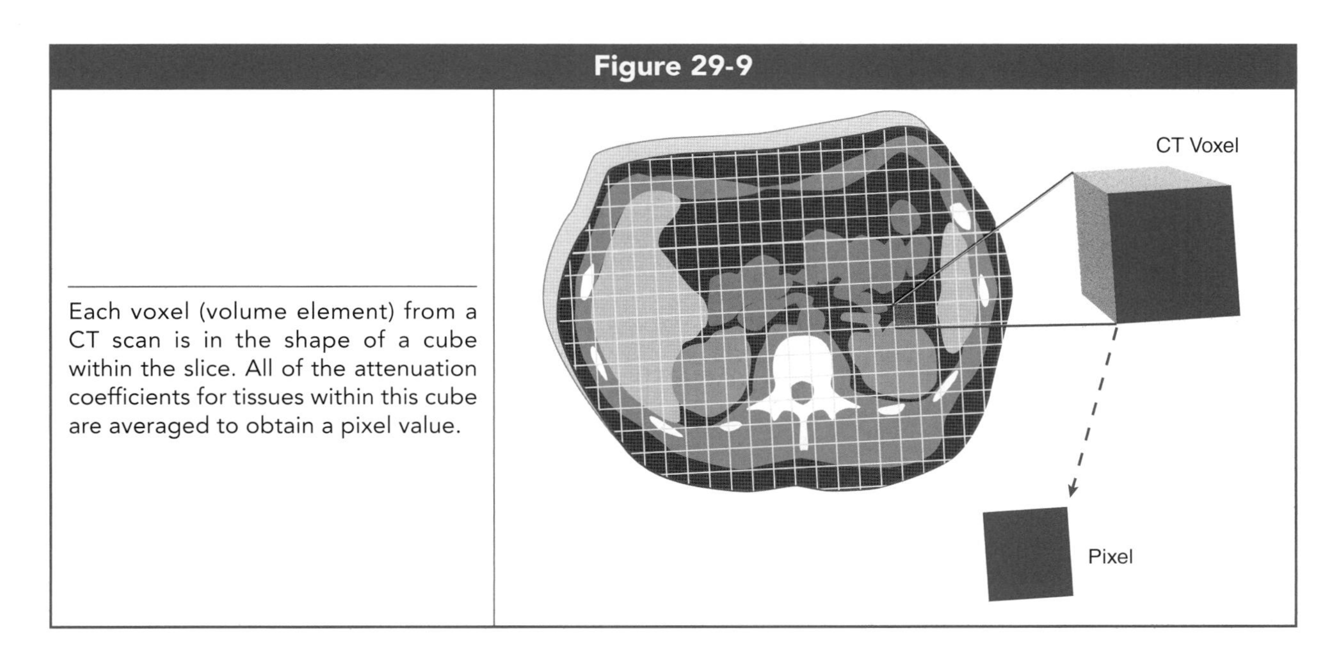

Figure 29-9

Each voxel (volume element) from a CT scan is in the shape of a cube within the slice. All of the attenuation coefficients for tissues within this cube are averaged to obtain a pixel value.

the back of the patient, as shown in Figure 29-10. This is because the x-ray beam passes clear through the whole thickness of the patient and records an attenuation coefficient for that entire thickness for each pixel. As with a CT scan, the attenuation coefficient measured from each voxel must be *averaged* for all of the tissues within that tube-shaped volume, so that a single gray level can be assigned to the corresponding pixel in the final image.

These attenuation coefficients must first be rounded out by an analog-to-digital converter (ADC) to discrete values the computer can interpret, then the computer selects from the dynamic range a corresponding gray level to assign to each pixel. These gray level values are stored in digital memory and collectively constitute the virtual image. Whenever the image is brought up on a display screen, the *brightness* of each pixel in the displayed image is controlled by the amount of electrical voltage applied to it, which, in turn, is a function of the gray level number. In other words, the brightnesses of all of the individual pixels that make up an electronic image are ultimately derived from the averaged attenuation coefficients of voxels within the patient.

ENHANCEMENT OF CONTRAST RESOLUTION

A main advantage of digital imaging is its ability to manipulate the gray scale values of the pixels after the image is acquired, thus allowing alteration of the appearance of the image *without* reexposing the patient. Special software and processing functions enable the selection and assignment of amplified gray scale values to low subject-contrast tissues in the image.

Figure 29-11***A*** shows a set of four adjacent pixels, three gray and one black with their corresponding attenuation coefficients indicated in bar graph form. There is low inherent subject contrast between these tissues, as indicated by the slight difference in the depth of the attenuation coefficient bars. The resulting gray levels of the pixels themselves also show low contrast in setting apart the black pixel against the gray ones. The application of software programs makes it possible to use a different formula in producing the pixel gray levels, from the same attenuation coefficients. As demonstrated in ***B***,

Figure 29-10

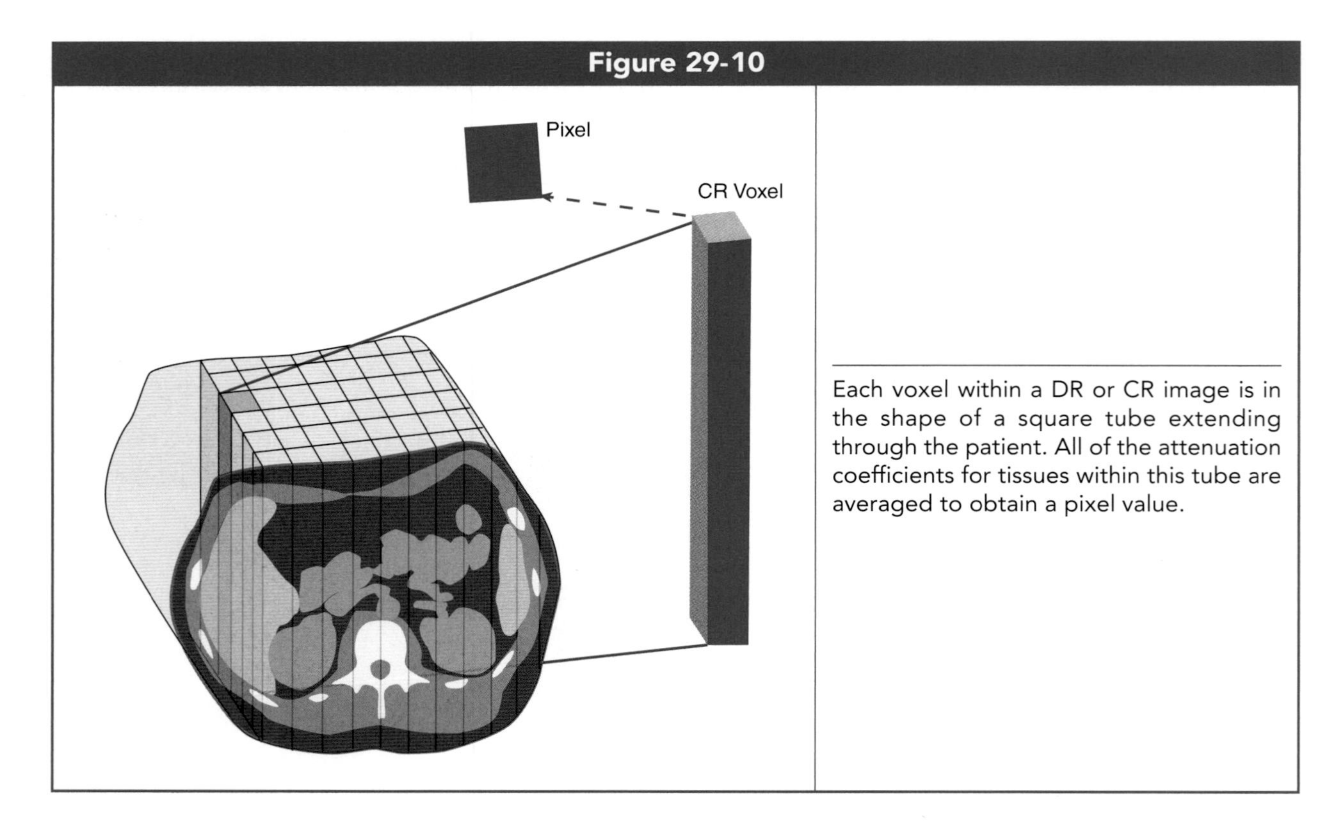

Each voxel within a DR or CR image is in the shape of a square tube extending through the patient. All of the attenuation coefficients for tissues within this tube are averaged to obtain a pixel value.

this mathematical adjustment has resulted in three of the pixels being assigned a lighter gray value, such that the contrast between them and the black pixel is enhanced.

Because of its poor contrast resolution capability, film-based radiography required a *subject contrast* difference of at least 10 percent between adjacent tissues to enable the perception of adjacent structures. Because of the contrast-enhancing capability of digital imaging software as shown in Figure 29-11, the perception of adjacent structures with a *subject contrast* as low as 1 percent is made possible. In the head, for example, digital images are capable of portraying the difference between blood, cerebrospinal fluid and brain tissue, none of which can be distinguished from each other on film-based radiographs.

The graph in Figure 29-12 serves as summary comparison between film-based analog images and digital images. When the minimum 10 percent subject contrast is provided for an analog image, it provides superior spatial resolution; as witnessed by the vertical portion of its curve being placed farther to the *left* than the digital curve, we see that smaller objects with less separation between them can be imaged by the analog system. However, the analog system does not resolve adjacent objects at all that have less than 10 percent subject contrast. For a digital image, the enhanced detectability of low-subject contrast structures enables perception of structures with very small differences in physical density. The trade-off for the digital image is that extremely *small* details with slight separation between them cannot be resolved.

Figure 29-11

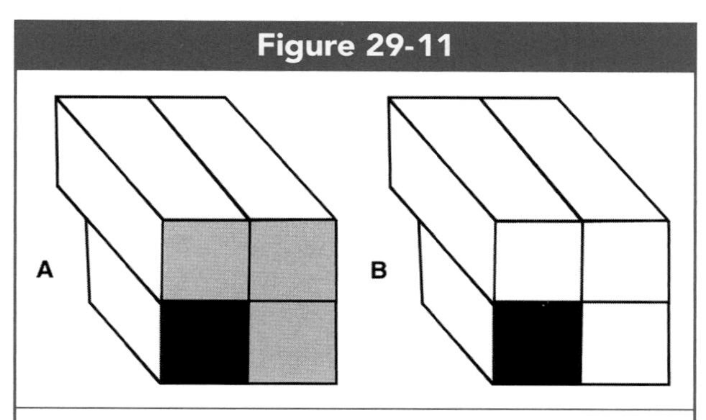

The relative gray scale values for four pixels shown in ***A*** were obtained from the true attenuation coefficients of the tissues. This low subject contrast results in low image contrast between the lower left pixel and the others. In ***B***, digital postprocessing has resulted in different gray scale values being assigned to the lighter three pixels, resulting in enhanced contrast and greater visibility of the lower left pixel. (From Quinn B. Carroll, *Practical Radiographic Imaging*, 8th ed. Springfield, IL: Charles C Thomas, Publisher, Ltd., 2007. Reprinted by permission.)

Figure 29-12

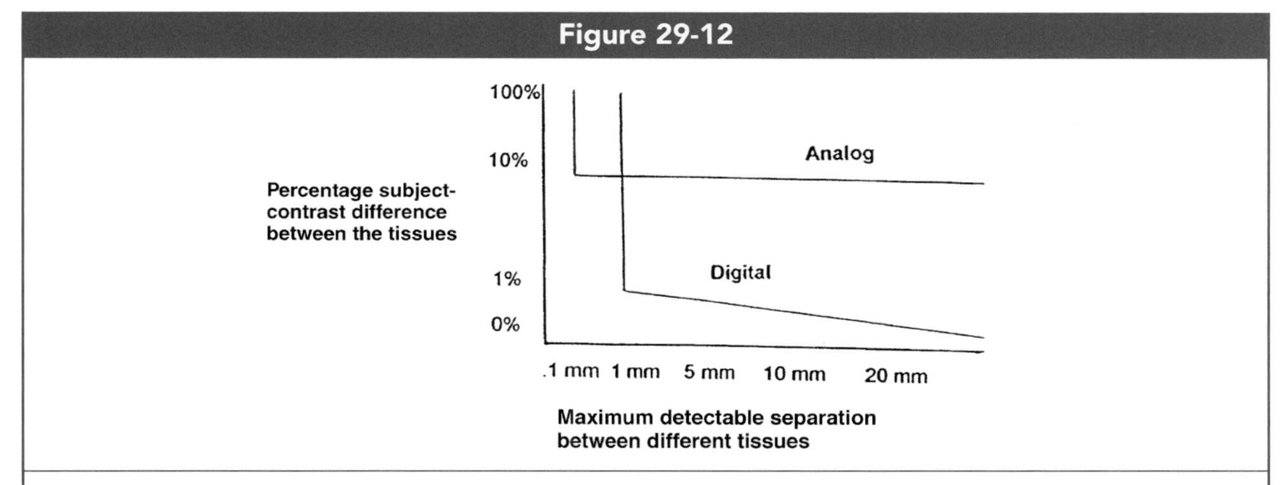

Graph comparing the contrast detectability and spatial resolution between analog and digital images. For analog images, higher spatial resolution is possible *when* at least a 10% subject contrast is present. There is a nearly complete loss of image perception as the subject contrast falls below 10%. For a digital image, the enhanced detectability of low contrast structures enables perception of structures with only a 1% difference in physical density. (From Quinn B. Carroll, *Practical Radiographic Imaging*, 8th ed. Springfield, IL: Charles C Thomas, Publisher, Ltd., 2007. Reprinted by permission.)

Table 29-1

Imaging System	Low Subject-Contrast Detectability	Maximum Subject-Contrast Detectability	Spatial Resolution
Analog (general film-screen)	Fair	10%	8–12 lp/mm
Analog (fluoroscopy)	Fair	10%	4–8 lp/mm
Digital Fluoroscopy (DSA)	Excellent	1%	4–5 lp/mm
Computerized Radiography	Excellent	1%	6–8 lp/mm

From Quinn B. Carroll, *Practical Radiographic Imaging*, 8th Ed. Springfield, IL: Charles C Thomas, Publisher, Ltd., 2007. Reprinted by permission.

Table 29-1 summarizes these same points in written form. As described in Chapter 14, the *overall resolution* of an image is an indicator of *total image quality*, and is dependent upon both spatial resolution and contrast resolution. The initial adoption of digital imaging systems in medical practice occurred when the power of computers, which had increased *exponentially* over time, reached a point where *contrast resolution* made such an overwhelming impact upon the overall image as to outweigh the loss of *spatial resolution*. Since that time, improvements in the miniaturization of pixel detectors and other technology has continued to narrow the difference in *spatial resolution* between digital and analog images.

PROCEDURAL ALGORITHMS

We have just given an example (Fig. 29-11) of how the use of a different formula to reset the dynamic range parameters can be used to enhance image contrast. Other types of alterations are desirable for specific anatomical procedures, including the *reduction* of contrast in some cases, or the enhancement of contrast by *different degrees* in selected portions of the anatomy. The set dynamic range directly determines the gray scale of the displayed image.

Also, for each specific anatomical area of interest, there is an "ideal" overall level of *brightness* which brings out details in their best visual presentation. This is controlled by adjusting the *average gray level* (or average brightness) up or down. In the thorax, for example, a lighter gray level may be used when interest is primarily in the lungs, and a much darker level would be used to bring out details in the thoracic spine.

All digital radiographic systems are designed with *preselected* gray scale and average gray level (brightness) settings to optimize the visualization of specific anatomical regions. These protocols are programmed into the operating system. They are customized for each type of radiographic procedure by anatomy, and are automatically engaged when that anatomical procedure is selected at the processing console of the digital system. Radiographers have come to refer to these settings as procedural *algorithms*; a "chest algorithm," a "foot algorithm," an "abdomen algorithm," and so on. This is proper terminology since the different formulas setting parameters on the gray scale and gray level are indeed computer algorithms.

WINDOWING

The brightness and the contrast of the image displayed at the monitor screen can both be adjusted upward or downward by controls on the console *as the image is being viewed*. This is generally referred to as *windowing*. By visually examining the image *as* the brightness and contrast are adjusted up and down, and by going back and forth between these two controls, the image can be *fine-tuned* to the precise results desired.

The fine points and limitations of windowing will be discussed in the following chapters, but we will define the two windowing functions here in broad terms that apply to all digital imaging systems.

The *window level* (*WL*), sometimes referred to only as the *level* (*L*), controls the overall *brightness* of the image. In some modalities (digital angiography), the window level may be referred to as the *center*, because it sets the center-point of the entire gray scale at a selected gray level. This center-point of the gray scale also represents the *average gray level* for the image. As this average is moved up and down the scale, the entire *window* of displayed gray shades moves with it.

To illustrate, imagine a literal window of fixed size on a wall—outside this window is a complete scale of gray shades arranged in ascending steps from white to black (Fig. 29-13). We measure the height of this window from the floor by using its *center-point*. In ***A***, the center-point of the window is 3 feet above the floor, placing it level with the *fourth step* of the gray scale, a medium shade of gray. An observer from inside the building can see five steps of the gray scale through the window, but would say that the *average* brightness of the entire image being observed is 4.0 (the fourth step, which is centered in the window).

Now, let the wall be remodeled with the window placed higher, 6 feet above the floor. In ***B***, we see that the view through the window still includes a range of five steps on the gray scale, but appears darker overall. The *average* brightness, centered in the window, is now 7.0 or the seventh step. The *window level has been increased, resulting in a darker average brightness.* A darker average brightness translates into a darker *overall* appearance.

The *window width* (*WW*), sometimes referred to only as the *width* (*W*), controls the length of gray scale in the presented image. Gray scale refers to the number of different shades of gray (or brightness) presented, so *longer scale* means *more shades of gray*. Observe Figure 29-14, a continuation of our analogy using a literal window on a wall. In

Figure 29-13

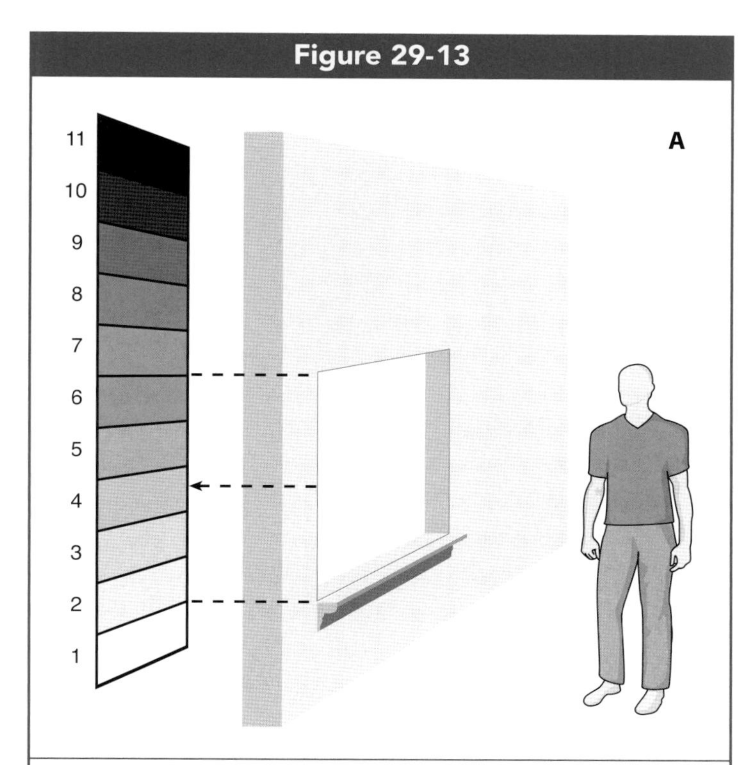

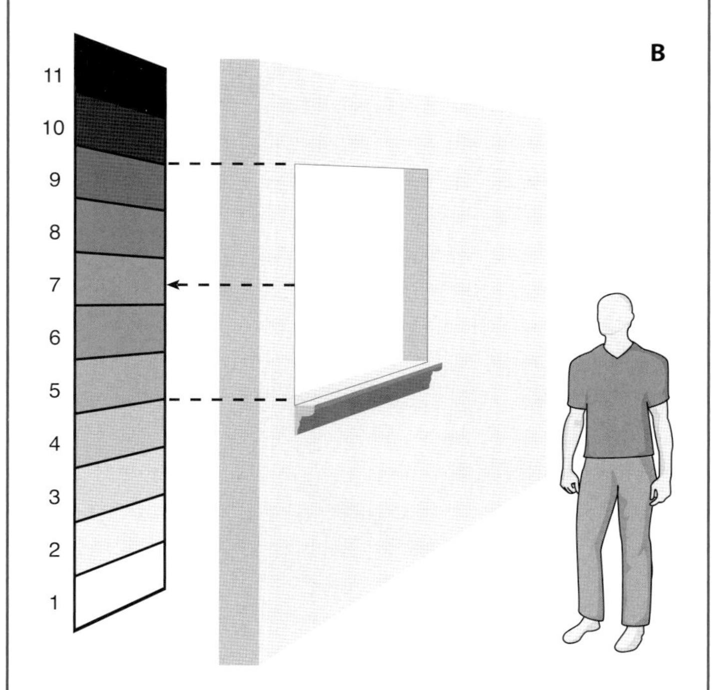

From ***A*** to ***B***, when the window level is raised, we see an overall darker image, but the *range* of shades within each image (between the dashed lines) remains equal, at 5. The value of the window level represents the average darkness, or *center* (dashed arrow) of the gray scale.

Figure 29-14

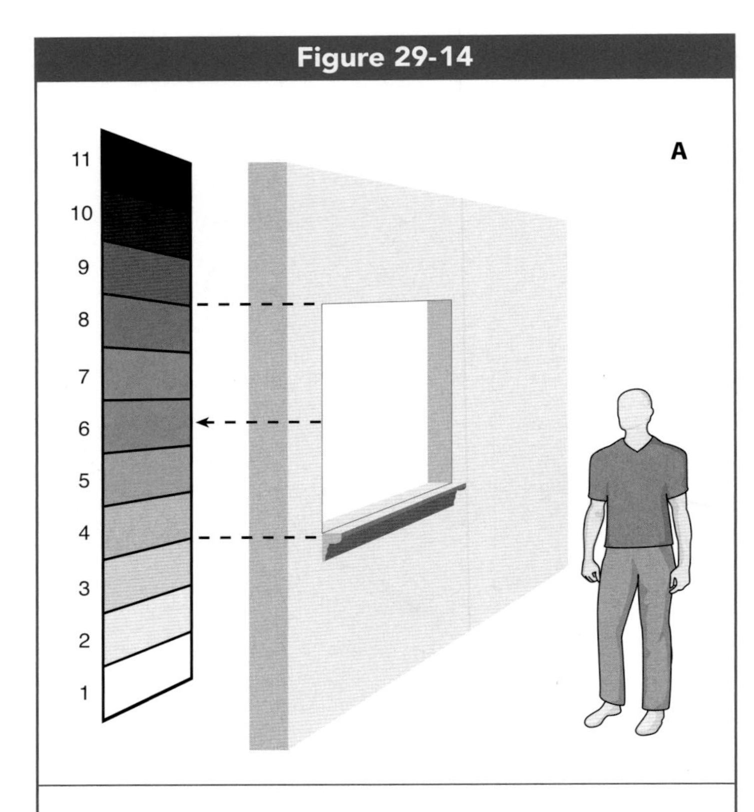

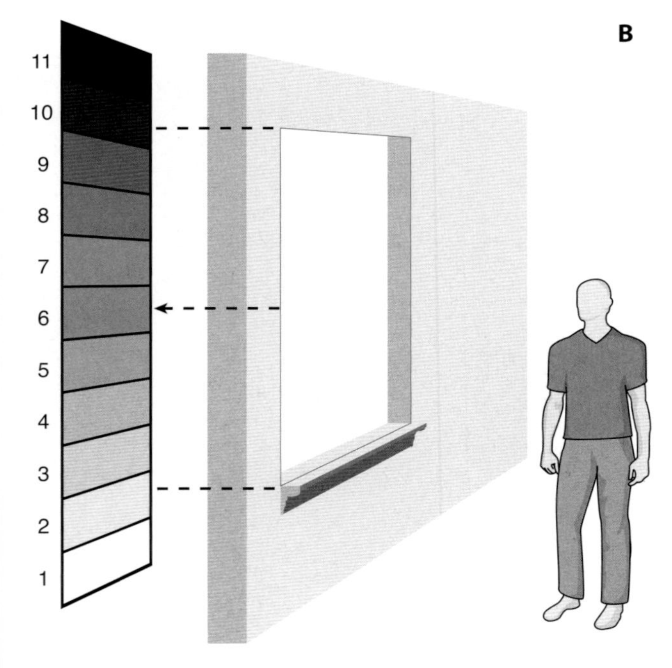

Without changing the center or window level (dashed arrow), the window width can be expanded as shown from ***A*** to ***B***, increasing the *range* of visible shades from 5 to 9. This is the gray scale.

this case, both windows, ***A*** and ***B***, are centered vertically on the wall. Measured from the *center* of each window, they are both at step six. Window ***B***, however, is opened *vertically wider*, that is, it is a "taller" window. Through window ***B***, the observer inside the building can count nine steps on the gray scale, compared to only five through window ***A***. For the observer, the gray scale in the image has been increased, even though the *average* gray level is equal and the overall brightness of the image is the same. This is an example of *increasing the window width without changing the window level.*

The actual results of these windowing changes on a radiographic image are illustrated in Figure 29-15 using a CT scan of the head. CT images were selected because in CT scanning these terms, *window level* and *window width*, are used precisely as defined here. Images ***A*** and ***B*** demonstrate the effect of increasing window *level* without changing window width. Images ***C*** and ***D*** show the effect of increasing window *width* without changing the window level.

Some issues relating to terminology must be clarified. Strictly speaking, *increasing the window level makes the image on the monitor screen* ***darker***, as shown in the CT scan images ***A*** and ***B*** in Figure 29-15. Most DR and CR systems now have a *brightness* control rather than a window level setting. This is a more user-friendly format, but remember that *brightness is the opposite of true window level.* Increasing the brightness setting is equivalent to *decreasing* the window level.

Likewise, the terminology relating to image contrast can be confusing. Strictly speaking, *increasing the window width lengthens the gray scale of the image. This reduces contrast*, as demonstrated with the CT scan images ***C*** and ***D*** in Figure 29-15. Many DR and CR systems have a more user-friendly *contrast* control rather than a *window width* setting. Again, this is the *opposite* of true window width, but accomplishes the same function.

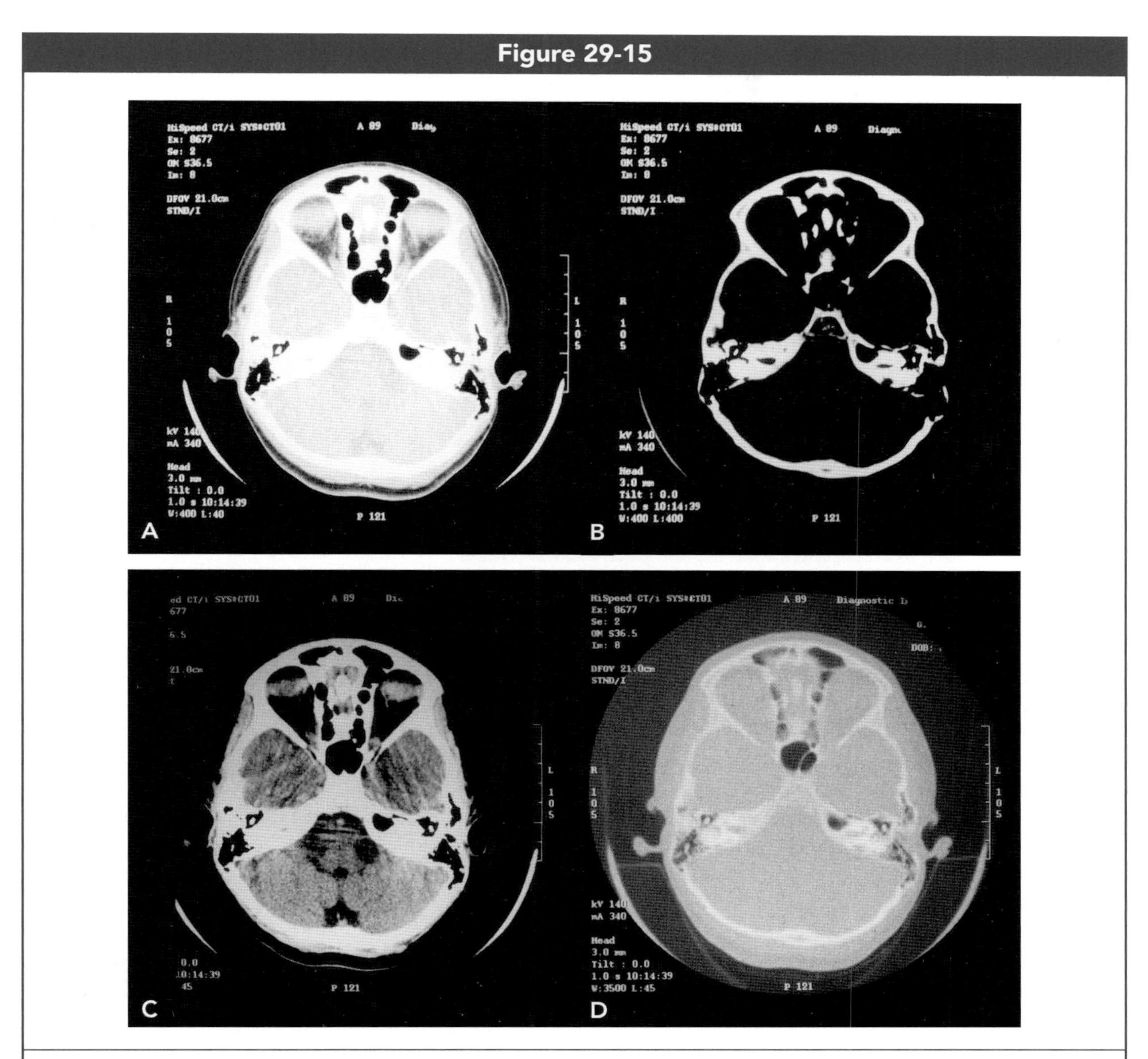

Figure 29-15

Window level vs. window width using axial CT scans of the head. From ***A*** to ***B*** (top), without changing the width (fixed at 400), the window level was increased from 40 to 400, dramatically darkening the overall brightness. From ***C*** to ***D*** (bottom), with the level fixed at 45, the window width was increased from 97 to 3500, expanding the gray scale but with an equal overall brightness. (From Quinn B. Carroll, *Practical Radiographic Imaging*, 8th ed. Springfield, IL: Charles C Thomas, Publisher, Ltd., 2007. Reprinted by permission.)

Workstations and Display Stations

A *workstation* is defined as a fully-equipped computer terminal that cannot only access images, but can be used to manipulate image quality and permanently *save* changes made to the image in the PAC system. Patient information attached to each image can also be added or deleted. The control console of a CT or MRI machine can be classified as a workstation.

Generally, all of the terminals within a radiologists' reading room are high-quality workstations with two monitors large enough to present 14 × 17" (35 × 42 cm) radiographs to their original scale, as

Figure 29-16

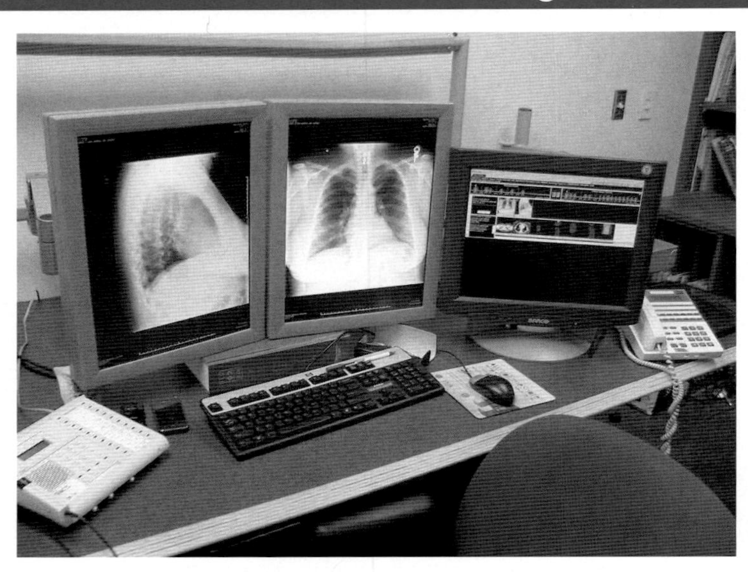

A radiologist's workstation with a pair of high-resolution monochrome monitors. (Courtesy, Patricia Hughes, M.D.)

shown in Figure 29-16. These are high-resolution (2000 × 2000 pixel) monitor screens, connected to a control monitor, keyboard and pointing devices to enable numerous post-processing functions.

Low-resolution (1000 × 1000 pixel) workstations are commonly found in the technologists' image review area. These are also capable of making permanent modifications to images and saving the changed images to the PAC system for storage, but they usually have only a single low-resolution monitor screen. Most workstations are also connected to a multiformat camera or other device with the capacity to print out hard copies of images.

A *display station* is defined as a computer terminal that is limited to the display of stored images—there is no ability to permanently manipulate or change the image, nor generally to print out hard copies. Since display stations are rarely used to obtain a primary diagnosis on a patient, monitor screens with lower resolution than that of a workstation can be used, along with a simple keypad (Fig. 29-17). This change lowers the cost of display stations to a fraction of the cost of a workstation, making the system more accessible to off-site centers within the network.

One of the most beneficial aspects of a PACS involves the strategic placement of display stations in key locations within a medical facility, which allows clinicians ready access to medical information and reports at the touch of a keyboard. By placing these display terminals in the ER, OR, ICU and CCU, doctors have nearly instant access to images and information that could otherwise take critical minutes or hours to obtain without the system. Display stations are also used in wide area network (WAN) systems so individual doctor's offices and clinics affiliated with a particular hospital can access images.

Figure 29-17

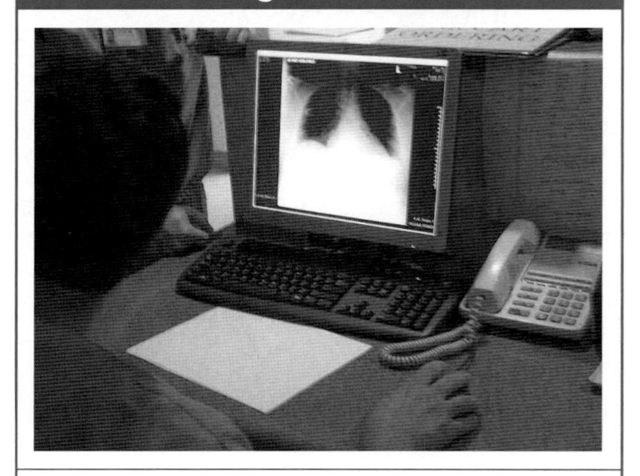

A typical display station with a low-resolution monitor.

Display stations, as well as workstations, should always be located in an area where the ambient lighting can be controlled. Too much room light is detrimental to the contrast that can be perceived in the image, further discussed in Chapter 39. In order to reduce the effects of extraneous light from the room, all display monitors should be provided with anti-reflection coatings. There is a growing body of evidence that when these considerations for viewing conditions are properly addressed, the use of soft-copy images does not compromise but often enhances the diagnosis of medical images.

SUMMARY

1. The larger the digital image matrix size, the smaller the pixel size (by inverse proportion), and the higher the spatial resolution (by direct proportion).
2. Pixel size for a displayed image is determined by the ratio of the measured field of view to the matrix size in pixels.
3. The three steps for digitizing an image are scanning, sampling, and quantization by an ADC.
4. As the gray scale of an image lengthens, more details can be resolved. The gray scale is dependent upon the dynamic range selected by system software from the bit depth capability of the computer system.
5. While the "bit depth" capability of the human eye is about 2^5 shades of gray, the dynamic range set for most digital imaging systems is 2^8, 2^{10}, or 2^{12}.
6. For CR, DR and CT, the brightness (gray) level assigned to each pixel is based on an attenuation coefficient for x-rays that is averaged throughout the corresponding voxel within the patient.
7. Although analog images inherently possess higher spatial resolution, a 10 percent difference in subject contrast was required to resolve details at all, whereas the ability of digital systems to amplify subject contrast allows adjacent structures with only a 1 percent difference to be imaged. This is the main image quality advantage for digital systems.
8. An "ideal" average gray level (brightness) and gray scale is preset in system software for each specific anatomical procedure. A particular image may be processed under a different anatomical algorithm.
9. Windowing controls allow fine-tuning of the image brightness and contrast as it is being viewed. *Level* controls the average or center of the image's gray scale, while *width* controls the length of the displayed gray scale.
10. Workstations allow permanent manipulations of the image, whereas display stations allow only viewing. Radiologists' workstations use very high-resolution monitors.

REVIEW QUESTIONS

1. The matrix size of an image (by pixel count) is __________ proportional to the spatial resolution or sharpness of the image.

2. What is the spatial frequency for an image with 0.1 mm pixels?

3. What is the pixel size for a 10 × 10 inch digital image reconstructed at the monitor screen on a 512 × 512 matrix?

4. Since DR uses hardware pixels, __________ of the x-ray beam becomes analogous to the *scanning* function in digitizing an image.

5. For CR, the scanning aperture is __________ in shape.

6. The initial measurement of pixel values within an image field is called __________.

7. Why may the actual dynamic range set for image reconstruction be less than the bit depth of the computer system?

8. The set dynamic range for an imaging system must be much greater than the capability of the human eye in order to allow for __________ - __________ of the image.

9. For a particular image pixel, the higher the attenuation coefficient for x-rays in the corresponding voxel, the __________ the gray level that will be assigned to the pixel by the imaging system.

10. For CR and DR, the sampled voxels within the patient take on the shape of __________.

(Continued)

REVIEW QUESTIONS *(Continued)*

11. The brightness of a pixel in the displayed electronic image is a function of the electric __________ applied to that segment of the monitor screen.

12. Digital radiographic imaging became feasible when the exponential increase in the power of __________ brought contrast enhancement capabilities up to an overwhelming level.

13. For each anatomical procedure, what two image qualities are preset within the procedural algorithms?

14. What are the three different general ways that postprocessing parameters can be modified?

15. Changing the window level from 80 to 300 causes the image to become __________.

16. The user-friendly "contrast" adjustment is exactly inverse to window __________.

17. What is the formula for the computer file size required for a medical image?

18. What type of image viewing station is likely to be placed in an emergency room?

19. Workstations should be located in an area where ambient __________ is subdued.

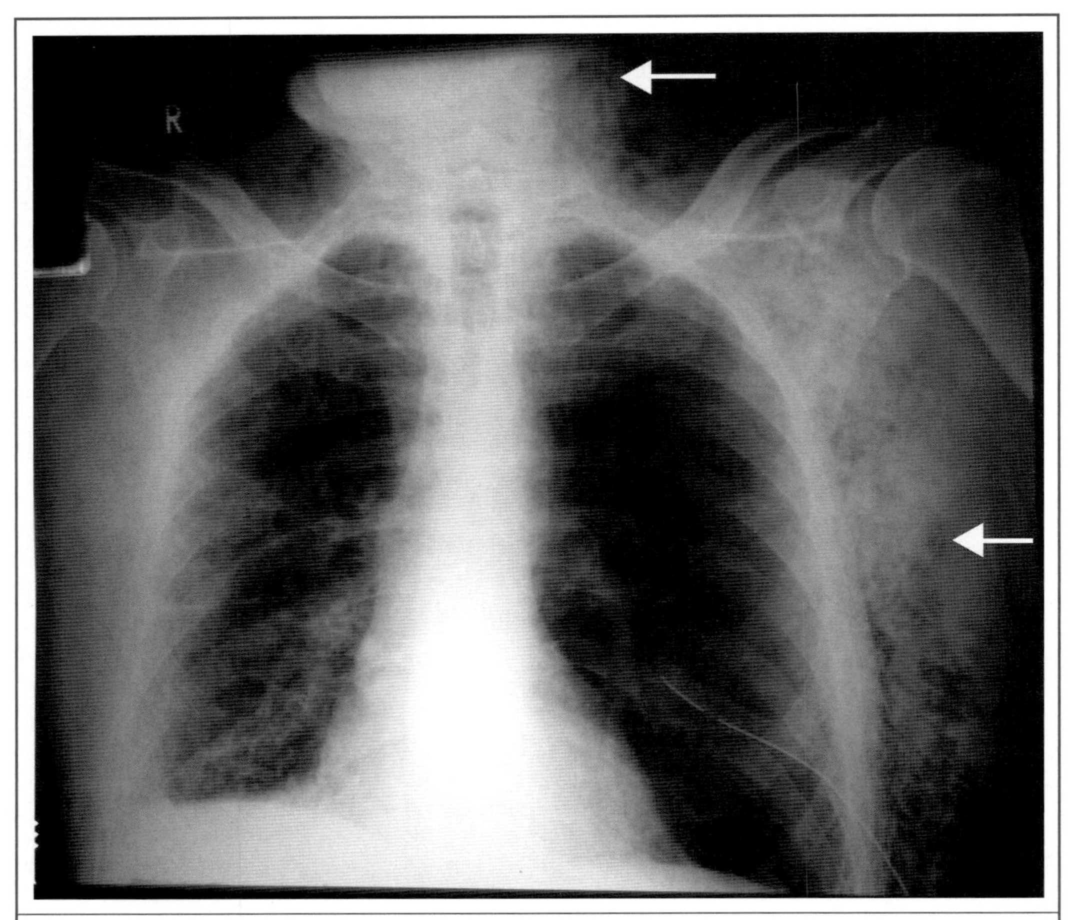

This patient suffered massive subcutaneous pneumothorax when a chest tube, seen in the lower left lung, was not fully inserted. Air pockets can be seen within the muscles under the skin throughout the chest and neck (arrows).

Chapter 30

DIGITAL IMAGE PREPROCESSING

Objectives:

Upon completion of this chapter, you should be able to:

1. Define *preprocessing* and *postprocessing.*
2. Describe five types of corrections made to the acquired image to establish field uniformity.
3. Explain how spatial kernels can be used to correct for *del* drop-out.
4. Define partitioned pattern recognition and exposure field recognition.
5. Explain the construction of the image histogram.
6. Give two examples of how the computer can identify key landmarks within the image histogram.
7. Define the characteristics of three general types of histogram analysis, and why they must be matched to the actual acquired histogram.
8. Define the *values* or *volume of interest.*
9. Interpret the gray scale curve.
10. Explain how bizarre histogram shapes can cause errors in rescaling.
11. Describe how pixel values in the image are re-mapped to rescale brightness and correct for scatter radiation.
12. Define the limitations on what the computer can do with the data set from an image.

INTRODUCTION

Throughout this book we have identified the capability for *postprocessing* as the major advantage of digital imaging over conventional imaging, and we have broadly defined post-processing as the ability to manipulate the image any number of times without re-exposing the patient to radiation. The term *preprocessing* has now come into common use in describing the computerized operations that must be executed on "raw" digital images to correct for flaws in digital image acquisition.

Physicists Ulrich Neitzel and Mathias and Cornelia Prokop have stated,

> The term "post-processing" is commonly associated with the processing option available for the user and is often distinguished from the default processing that all digital radiographs are subjected to. In reality, this distinction is arbitrary since available processing options are generally identical. In an ideal environment, the default processing should be chosen so that no additional "postprocessing" is necessary.[1]

For example, brightness and contrast adjustments may be made by the radiographer after the image is first displayed, but are also part of the default processing that every digital image is subjected to before it is displayed. The internal computer functions that take place are identical. Furthermore, if the correct algorithm is selected and the default processing is completely successful, no further adjustments would be needed by the radiographer after the image is first displayed. Therefore, using the timing of the first

[1] Prokop, Neitzel and Schafer-Prokop. "Principles of Image Processing in Digital Chest Radiography," *Journal of Thoracic Imaging,* Vol 18, No 3, 2003.

displayed image as a distinction between *processing* and *postprocessing* seems very arbitrary.

A clearer distinction between preprocessing and postprocessing can be made by defining them in *functional* terms rather than by the time they occur. There are at least two reasons for this: First, most operator adjustments to the image, such as windowing, are adjustments to operations that have already taken place once in the default processing of the image, so the operator's adjustment is not the first time the actual algorithm has been applied. (This is what Neitzel and Prokop are pointing out in the above quote.) Second, many manufacturers will repeat a particular step, such as noise reduction or gradation processing, twice, once to initially normalize the image from raw data, and again after detail processing has resulted in an alien-looking image, in effect re-normalizing the image just prior to display.

We suggest here a *functional* delineation of the terms *preprocessing* and *postprocessing* which should encompass all processing operations that may be executed on any digital image, as follows:

Preprocessing: All *corrections* that are made to the "raw" digital image data due to physical flaws in *image acquisition* that are inherent to the elements and circuitry of the particular image receptor system, or the physical elements and circuitry of the processor. Preprocessing may also be termed *acquisition processing*.

Postprocessing: All manipulation and adjustments of the digital image (whether by default settings in the processor or by an operator) made *after corrections have been made for data acquisition.* These operations are targeted at *refinement* of the image, and although they may be performed as part of the default processing of the image, are also somewhat subject to personal preference.

We might summarize by stating that preprocessing is directed toward image corrections, while postprocessing is intended for image refinement.

Some time is still needed for a common vocabulary to solidify for digital radiographic processing. Beside "preprocessing," other terms such as "unsharp masking" and "spatial frequency processing" are inherently confusing to the student. In this textbook, we will attempt to provide clear definitions and consistent usage of terminology. We will organize next two chapters into preprocessing and postprocessing functions as defined above, and begin with preprocessing.

PREPROCESSING I: FIELD UNIFORMITY

To prepare the image for postprocessing, physical limitations that are inherent in every image acquisition system must be corrected for. Several flaws are found in the electronics and optics of the receptor system and processor.

In direct-capture DR systems, the detector elements (dels) constituting individual pixels can suffer from various electronic faults that are not found in the reading process for CR plates. These flaws introduce noise into the image or cause a loss of pixels. Additional software is configured to compensate for these electronic problems. Therefore, direct-capture DR systems typically undergo more pre-display processes than those required in computed radiography (CR).

Dark noise can include accumulated background exposure to a CR phosphor plate and *dark current* in a DR detector system, a small amount of current that may be flowing through the system when no exposure is taking place. *Del drop-out* or faulty detector elements in a DR system constitutes another form of noise. Noise reduction is discussed in the next section, *Preprocessing II.*

Flat-Field Uniformity Corrections

Flat-field corrections are made for the purpose of evening out the overall signal or brightness across the entire area of the imaging field. A flat-field uniformity test is simple to obtain by making a low exposure to the image receptor without any phantom or other object in the x-ray beam. The pixel values from areas in the center and four corners of the image are then compared, Figure 30-1. These must be aligned to each other within a narrow percentage range (Chapter 39). Several variables affect the uniformity of the flat field.

The anode heel effect (Chapter 21) is familiar to most radiographers. It results in less x-ray intensity reaching the anode-end of the image receptor plate than the cathode end of the plate. The "flat-panel" detector arrays used in DR are particularly sensitive to the heel effect. Electronic amplification or computer software could be configured to compensate for this deviation from one end of the field to the other; however, as described in Chapter 21, the degree of anode heel effect is dependent on the SID used—the shorter the distance, the more the collimator must be opened to accommodate a particular field size, and the more severe the anode heel effect becomes. Since radiographic projections require different SIDs, there is no feasible way to apply this capability. Therefore, digital imaging systems do *not* compensate for the anode heel effect.

Electronic Response and Gain Offsets

The active matrix array in a DR system consists of hundreds of individual detector elements or *dels*, and due to unavoidable variations in their manufacture, some are bound to have slightly less or slightly greater *response* to the x-rays or light striking them. In a thin-film transistor (TFT), variable sensitivity of charge conversion can occur from electric charge becoming trapped within the semiconductor. In some systems, across the array, electronic amplification may be applied by multiple amplifiers which may not be perfectly aligned with each other in the degree to which they boost the signal. From dels within the array, signals that must pass down longer wires encounter greater electrical resistance and are slightly affected. These flaws and others within the electronic hardware of the image receptor and amplification systems collectively affect the uniformity of the detected field. This non-uniformity can be tested for and corrected (Fig. 30-1).

Figure 30-1

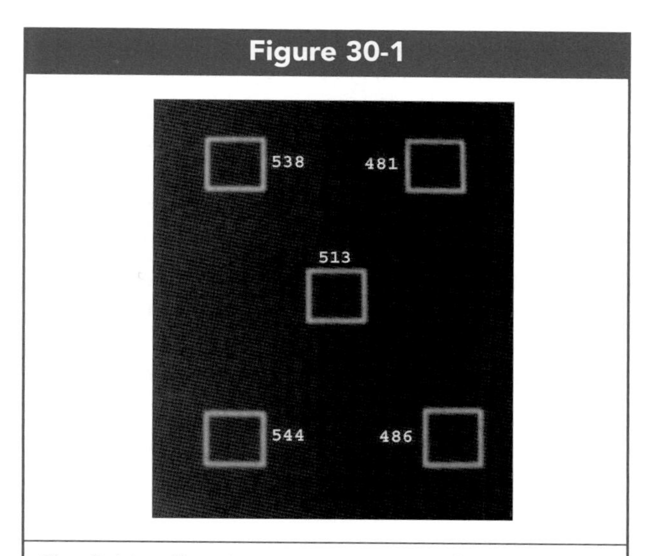

Flat-field uniformity test. Five areas of the field, the center and four corners, are sampled and compared for consistency, using a low exposure with no object in the field.

Variable Scintillator Thickness

In computed radiography (CR), the phosphor layer of the PSP (photostimulable phosphor plate) may have very slight variations in its thickness from manufacture. Note that for indirect-capture DR a phosphor layer is also used to convert x-rays into light which then strikes the active matrix array. These phosphor layers are also subject to slight variations in thickness. Thicker portions of the phosphor layers will have slightly greater absorption efficiency, and will produce slightly more light output than other areas of the phosphor.

Light Guide Variations in CR

The CR processor uses fiber-optic light guides to direct the light emitted from the CR plate phosphor to the photomultiplier tube. Light intensity variation can be caused by differing efficiency between light guides, due to variable length, quality, or flaws. These light fluctuations can be detected, and adjustments can be made in amplification of the signal over specific spots on the image receptor, either by hardware or by software.

PREPROCESSING II: NOISE REDUCTION FOR DEL DROP-OUT

In DR, individual *dels* or detector elements can drop completely out of the image due to TFT failure or switching transistor malfunction. This is commonly called "pixel drop-out," but should not be confused with the functioning of actual pixels in an LCD monitor. Here, we are talking about flaws in the *image receptor* that result in missing information in the data set sent into the computer for processing. So, this is properly categorized as a *preprocessing* correction. But, it is a misnomer to label it "pixel drop-out" when it should properly be called *del drop-out.*

A *kernel* is defined as a sub-matrix that is passed over the larger matrix of the image executing some mathematical function (Fig. 30-2). (Kernels are more fully described in the following chapter on *postprocessing.*) A kernel can be configured to average the pixels surrounding the malfunctioning component and fill this "dead" space in the image by interpolation. The pixels above, below, to either side and at every corner of the dead pixel are sampled. The values for these eight pixels are simply summed and averaged, and then this value is inserted into the *software pixel* corresponding to the dead detector element or *del.*

Entire rows or columns of dels can also fail. Algorithms can be configured to interpolate pixel values from the rows above and below, or from the columns to either side, to correct for very mild cases of short, single lines in the image, but many cases involving entire rows or multiple rows of malfunctioning dels are severe enough, as shown in Figure 30-3, that the only solution is to replace an expensive detector plate. Normal "wear and tear" of image receptors over long periods of time naturally includes the accumulation of dead dels, and eventually necessitates replacement of the plate itself.

Left uncorrected, malfunctioning dels would result in pixels or small pixel groups within the display monitor that appear "dead." Since any image can be reversed (white-on-black or black-on-white), it is impertinent to state whether these areas of the image would be specifically black or white on the monitor, only that they represent gaps of missing information. They can affect diagnosis and so must be corrected.

PREPROCESSING III: IMAGE ANALYSIS

Before any manipulation of the actual acquired image can take place, its components and characteristics must be identified and analyzed. The processes for accomplishing this include partitioned pattern

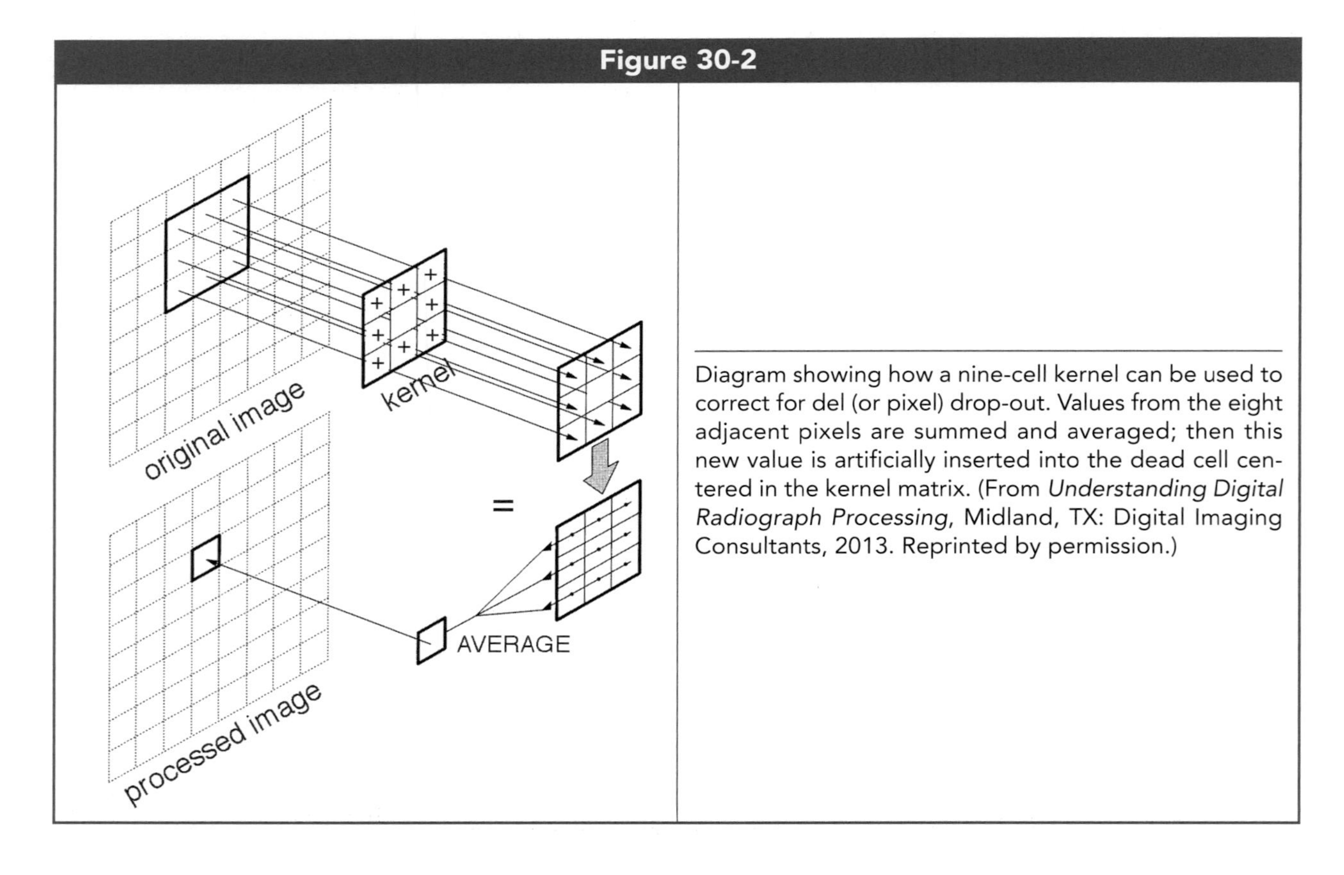

Figure 30-2

Diagram showing how a nine-cell kernel can be used to correct for del (or pixel) drop-out. Values from the eight adjacent pixels are summed and averaged; then this new value is artificially inserted into the dead cell centered in the kernel matrix. (From *Understanding Digital Radiograph Processing*, Midland, TX: Digital Imaging Consultants, 2013. Reprinted by permission.)

recognition or *segmentation*, exposure field recognition, construction of the initial histogram, and histogram analysis.

Segmentation and Exposure Field Recognition

Before an exposure indicator can be calculated and the image can be correctly rescaled, two particular types of "noise" must be eliminated from the data set of the acquired image: Densities outside of the collimated fields, such as might occur from background exposure to a CR plate or from scatter produced within the body during exposure, and the pitch-black background densities typical of exposure to the "raw" x-ray beam outside the anatomy.

Segmentation or *partitioned pattern recognition* software is used in CR to identify and count the number of views taken on a single PSP plate, so that each exposed area can be treated separately by the reader/processor. Partitioned pattern recognition (PPR) software identifies each edge of an exposed field by the sudden dramatic change in density, almost clear to very dark. If areas outside the collimated fields have accumulated much density from background and scatter radiation, the software may effectively mistake them for lighter anatomical areas *within* an exposed field. Two or more exposed fields are treated as one image and these relatively light areas are included in constructing the histogram. The computer will calculate the average density for the image much lower and overcompensate during the rescaling of the image, making the displayed image too dark (Fig. 30-4). (Figure 30-9, which shows the effect of large amounts of metal or barium within the image, also serves as a graphical representation of the type of effects this has on the histogram.) Partitioned pattern recognition is only used in CR, which scans across the PSP plate row by row, whereas DR scans any exposed field from the center outward until the edges are encountered.

Exposure field recognition (EFR) is used in both CR and DR systems primarily to identify the pitch-black background densities from "raw" x-ray beam exposure outside of the anatomy. CR systems are more prone to EFR errors, so they are further discussed in Chapter 33. EFR is accomplished by recognizing landmark changes in the histogram, so it is technically part of histogram analysis, in which the "volume of interest" (VOI) to be analyzed from the original data set is defined. This is described in a following section, but before the histogram can be analyzed, it must be constructed from the acquired data.

Figure 30-3

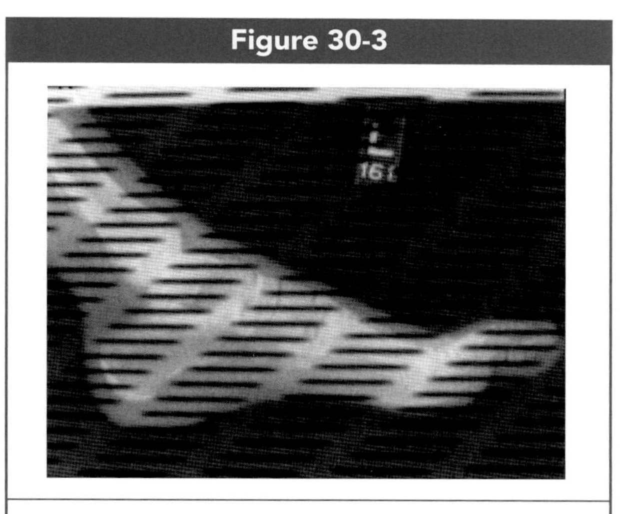

Example of pixel drop-out for entire rows of pixels. Software can compensate for moderate cases by interpolating and filling these pixels. (Courtesy, Fujifilm Medical Systems, Stamford, CT.)

Figure 30-4

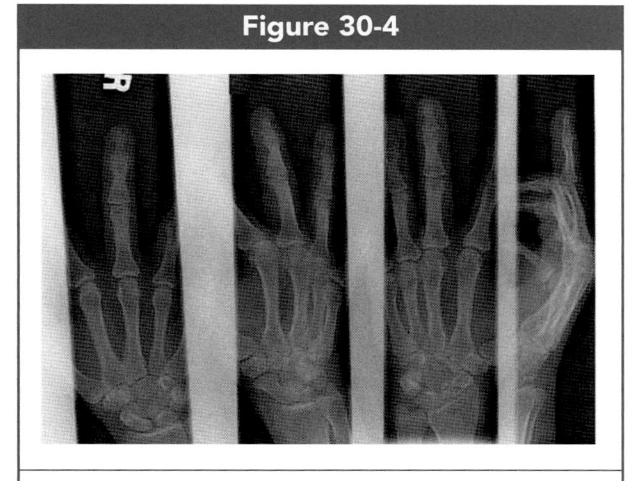

Segmentation error resulted in the entire plate being treated as a single exposure, so that scatter radiation densities *between* collimated fields were included in constructing the histogram. With large areas of relatively light density added to the data, the computer compensated the rescaled image by overall darkening. (Copyright by the America Society of Radiologic Technologists. All rights reserved. Reprinted by permission.)

Figure 30-5

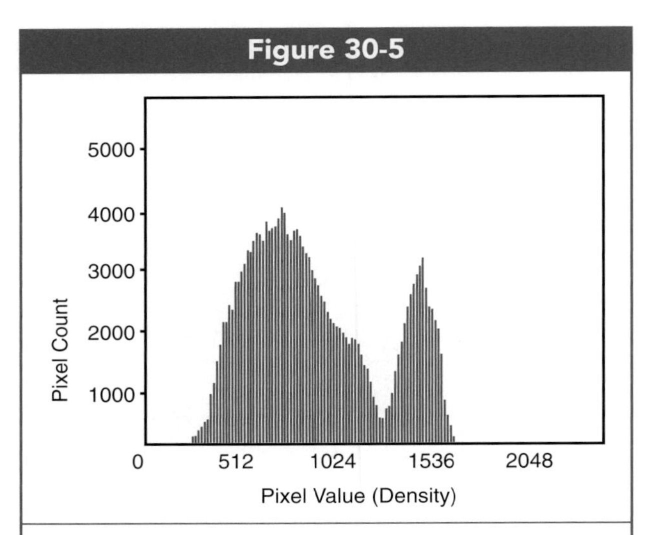

A histogram is actually a "bar" graph constructed by counting the number of pixels that contain each gray level indicated along the bottom axis; from left to right, levels increase from white to black. (From Quinn B. Carroll, *Practical Radiographic Imaging*, 8th Ed. Springfield, IL: Charles C Thomas, Publisher, Ltd., 2007. Reprinted by permission.)

Constructing the Histogram

The histogram representing an image is constructed by simply counting the number of pixels within the image at each density or brightness level as the computer scans across the entire image. A histogram showing the actual data bars is shown in Figure 30-5. Each vertical bar indicates the number of pixels that hold the gray level indicated along the bottom axis of the graph. The "peaks and valleys" of the graph represent variations in the anatomical structures in the image. For radiographs, these are normally presented on a simple scale of brightness (or "density") from white to black, as read from left to right. Thus, metallic objects or contrast agents would be represented on the far left of the graph, followed by bone, then soft tissues near the center, fat and finally gaseous or air densities to the far right (Figs. 30-6 and 30-7).

Although the original histogram is actually a bar graph, the displayed histogram connects the tops of all these bars with a "best fit" line. As shown in Figure 30-6, this forms a curved graph with dips and peaks in the curve. (See Chapter 3 if you need to review how graphs such as histograms are generally interpreted. We will rely heavily upon graphs to illustrate the concepts being discussed, but remember that the computer is actually looking at sets of *numbers*, not graphs. Graphs are used for our benefit to help visualize what is going on with the math in the computer.)

Different general portions of the body each produce a histogram shape with characteristic "peaks

Figure 30-6

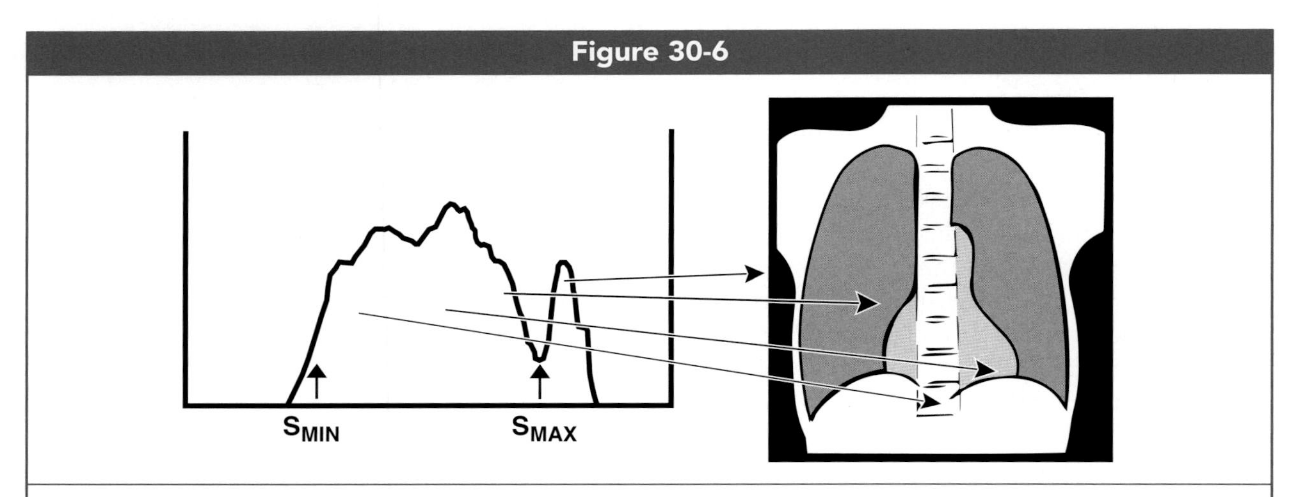

Construction of a histogram for a chest image. The left-most portion of the bell-curve represents bone densities, the mid-portion represents soft tissues, the right portion aerated lung, and the spiked "tail" represents the "pitch-black" background densities. Also see labeled areas in Figure 30-7. For most applications, the computer must recognize the area from S_{MIN} to S_{MAX} and analyze only this information in processing the image. (From Quinn B. Carroll, *Practical Radiographic Imaging*, 8th Ed. Springfield, IL: Charles C Thomas, Publisher, Ltd., 2007. Reprinted by permission.)

Figure 30-7

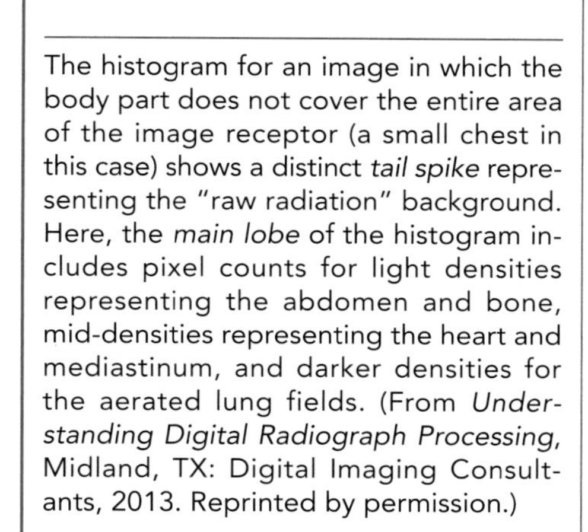
The histogram for an image in which the body part does not cover the entire area of the image receptor (a small chest in this case) shows a distinct *tail spike* representing the "raw radiation" background. Here, the *main lobe* of the histogram includes pixel counts for light densities representing the abdomen and bone, mid-densities representing the heart and mediastinum, and darker densities for the aerated lung fields. (From *Understanding Digital Radiograph Processing*, Midland, TX: Digital Imaging Consultants, 2013. Reprinted by permission.)

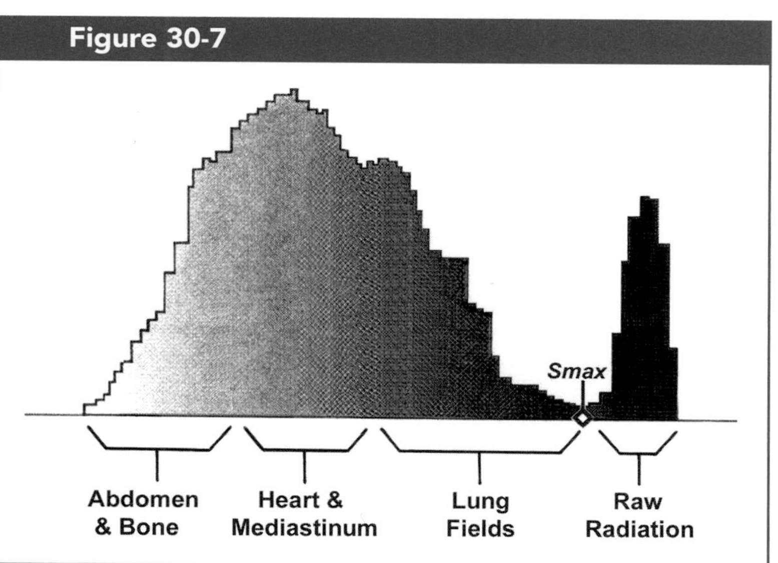

and valleys" that allow key data landmarks to be identified. Histogram analysis consists of identifying these landmarks within the numerical data, so that only *values of interest* are included when the image is rescaled and manipulated.

Figure 30-7 illustrates the most common type of histogram, obtained for body parts that do *not* cover the entire area of the image receptor, leaving some areas of "raw" radiation background. This histogram is typical for an extremity or small chest, and consists of a roughly bell-shaped curve called the "main lobe," followed by a distinct "spike" in the graph toward the right side. The bell-shaped portion is due to the fact that *within the actual anatomy* there are a few very light densities (to the left of the curve) and a few very dark densities (to the right), while most tissues are displayed as intermediate shades of gray (the mid-peak of the curve). The "spike" portion represents the *background* density of the image, outside of the anatomy, which is usually pitch dark. This is referred to as the "tail" of the graph. (It is absent on images of the torso that cover the entire cassette with anatomy, since no pitch dark background densities are present in the image—see Fig. 30-8.) The *area* under this tail curve is representative of the amount of the image consisting of background density.

For most applications, the computer must identify this "tail" portion and eliminate it from the histogram

Figure 30-8

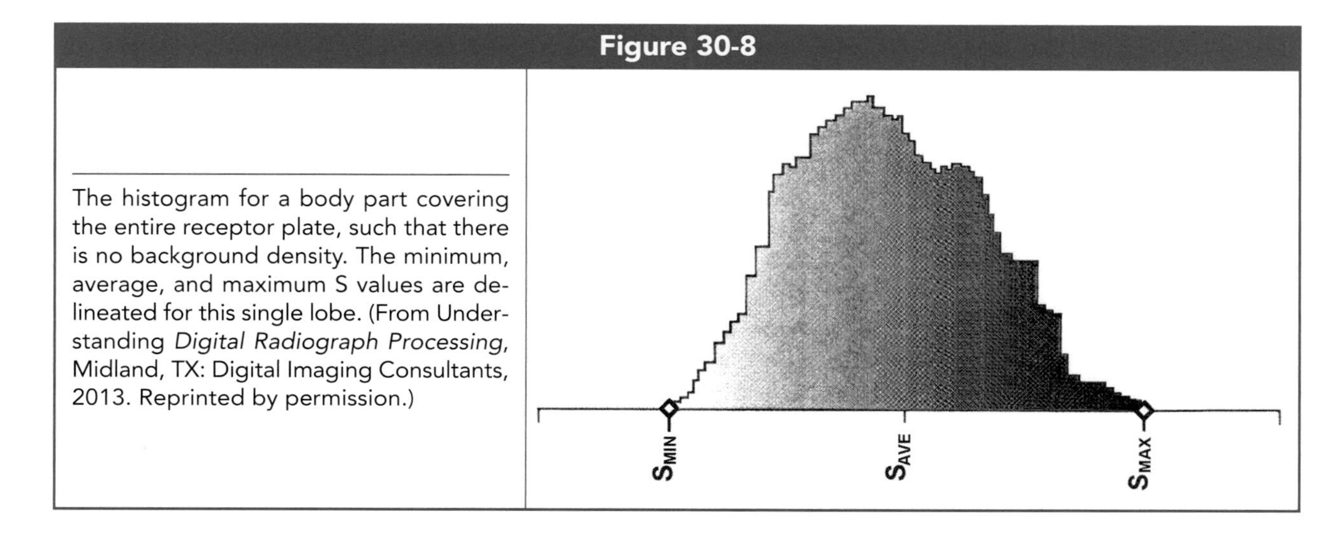

The histogram for a body part covering the entire receptor plate, such that there is no background density. The minimum, average, and maximum S values are delineated for this single lobe. (From Understanding *Digital Radiograph Processing*, Midland, TX: Digital Imaging Consultants, 2013. Reprinted by permission.)

analysis performed. If these pitch black background densities are included in the computer's calculations, the final image produced will be skewed away from the desired brightness levels. Only data from within the anatomy must be included during processing. The process of excluding undesirable portions of the histogram from the desired data set is called *exposure field recognition* (EFR) or *exposure data recognition* (EDR). Other variables that can throw off the histogram analysis include positioning and collimation errors, unusual pathological conditions, removed or added anatomy, and artifacts such as prostheses that are present during the initial exposure. In Figure 30-7, the landmark point labeled S_{MAX} must be identified by the computer so that all data to the right of this point can be excluded.

Figure 30-8 shows the simplest type of histogram which is produced when a large body part such as the abdomen completely covers the image receptor area, or collimated field area, with body tissue, such that there is no background density. A different type of histogram analysis algorithm must be used for these types of images, which does not seek for an S_{MAX} point.

Where a large radiopaque prosthesis is present, or for solid-column barium procedures, large areas of the image are expected to consist of nearly white densities where very few x-rays have penetrated through to the image receptor. The resulting type of histogram is demonstrated in Figure 30-9. Just as large areas that are pitch black can skew the data for averaging, so can large areas that are nearly blank white, so it is desirable to also eliminate this data from the rescaling process and from calculations for the exposure indicator. To do so, the landmark *S minimum* must be identified (shown to the left), and all data to the left of this point must be excluded from calculations.

Figure 30-10 illustrates the rather unique histogram that is typical for a mammogram procedure. It is desirable to eliminate data from the chest wall for rescaling and exposure indicator calculations, so a similar S_{MIN} point, (shown to the left), must be identified. Thus, for both barium procedures and mammograms, an entirely different type of histogram analysis algorithm must be used, which seeks for both S_{MIN} and S_{MAX} points.

When the radiographer is at the control console, selecting the *procedure* algorithm assigns the type of histogram analysis to be used, based on the shape the acquired histogram will be *expected* to have.

Data is scanned *inward* from each end of the histogram to identify the expected landmarks. If we think of computer files as "bins" holding pixel counts for each gray level, as shown in Figure 30-11, the computer "hops" from bin to bin to locate the first bin containing pixels. Here, at the left, this first bin containing the lightest non-zero pixel count is designated as the S_{MIN} point. Continuing inward, any telltale, sudden changes in the number of pixels registered can be used as an identifiable landmark.

Figure 30-9

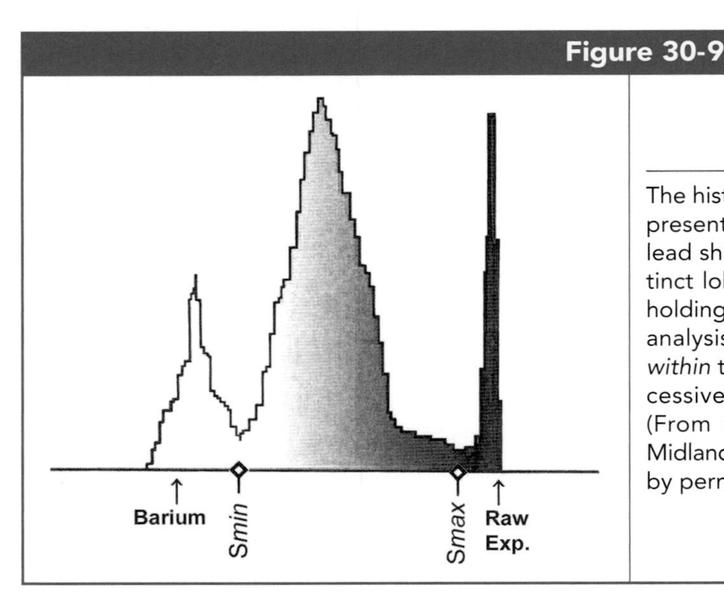

The histogram for an image with a large radiopaque area present, typical of a "solid-column" bolus of barium, a lead shield, or a large metallic prosthesis presents a distinct lobe (left) representing the large number of pixels holding these light densities. To avoid errors in histogram analysis, the S_{MIN} and S_{MAX} points must be identified *within* the data set so that both excessively light and excessively dark values are excluded from calculations. (From *Understanding Digital Radiograph Processing*, Midland, TX: Digital Imaging Consultants, 2013. Reprinted by permission.)

Figure 30-10

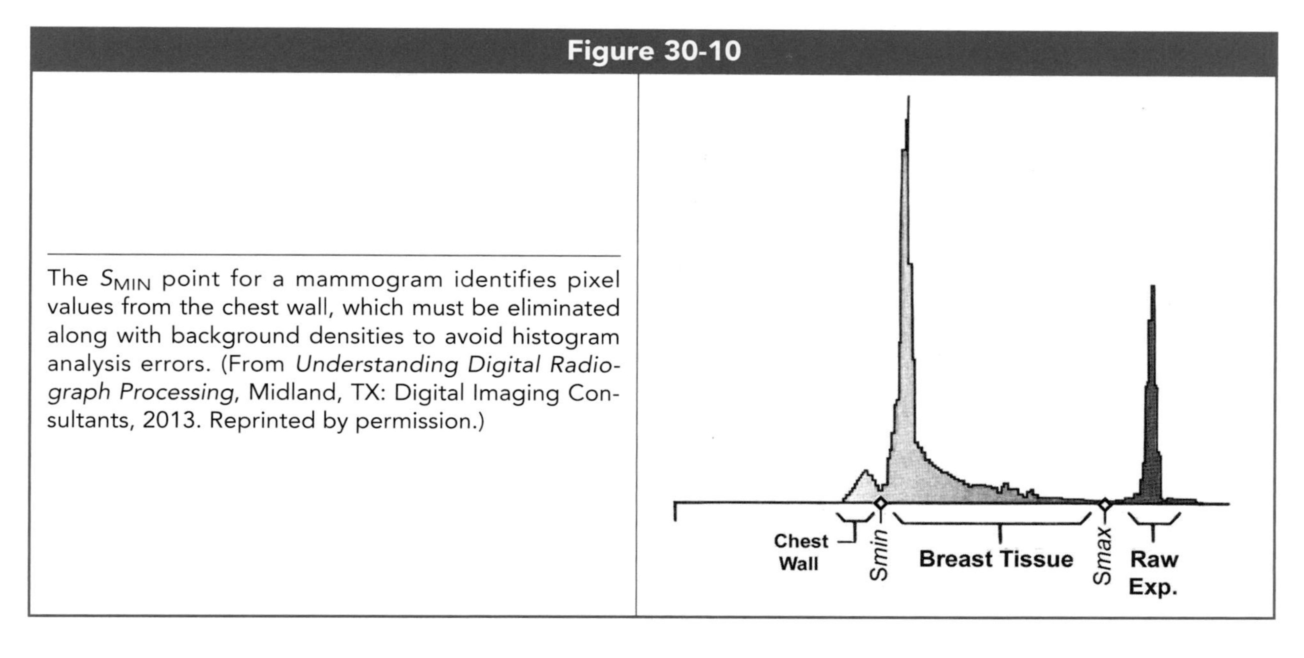

The S_{MIN} point for a mammogram identifies pixel values from the chest wall, which must be eliminated along with background densities to avoid histogram analysis errors. (From *Understanding Digital Radiograph Processing*, Midland, TX: Digital Imaging Consultants, 2013. Reprinted by permission.)

On this histogram for a typical extremity that is smaller than the detector area, the computer must also identify the S_{MAX} point *within* the data curve. This represents the darkest density (highest pixel value) *within the anatomical body part*, with only the "raw x-ray exposure" or "background density" to the right of it.

The S_{AVE} is the average density *within* the anatomy, or average pixel value for the *main lobe* of the histogram, between the S_{MIN} and the S_{MAX}, and is particularly important for calculating the exposure indicator discussed in Chapter 32.

How does the computer identify these landmarks within the data set? Several methods can be used.

Figure 30-11

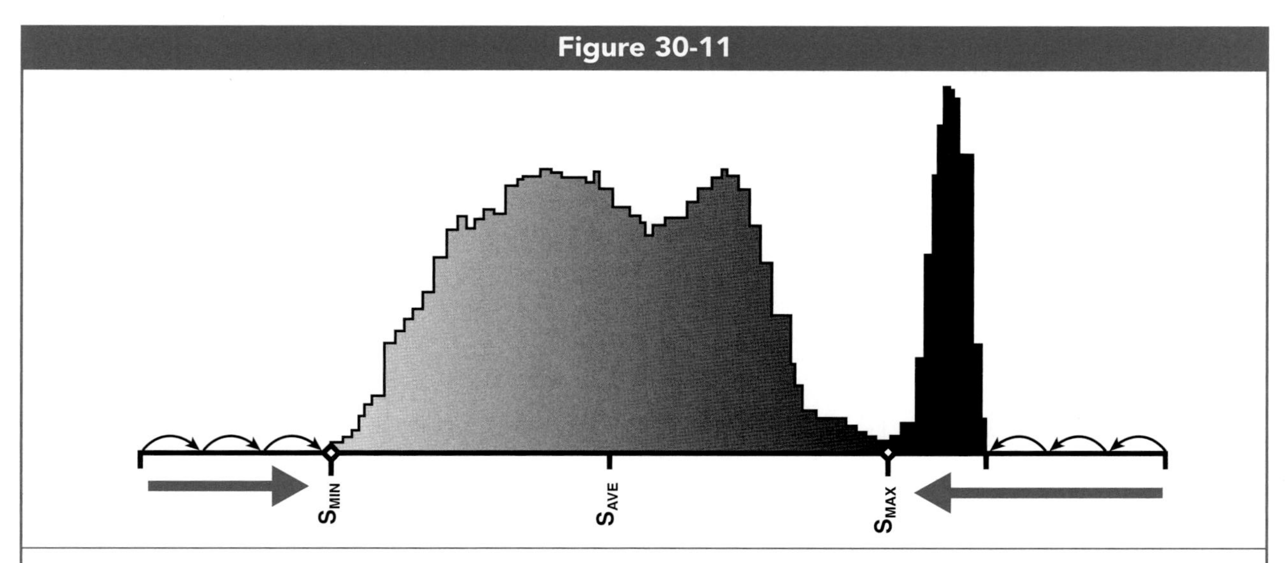

To identify the S_{MIN} to S_{MAX} points, the computer scans inward from each end of the histogram, hopping from bin to bin, until a non-zero pixel count or a threshold count of pixels is found. Various methods for identifying landmark points can be used, but a simple example is an algorithm that searches for changes up or down in the count of pixels holding each value. (From *Understanding Digital Radiograph Processing*, Midland, TX: Digital Imaging Consultants, 2013. Reprinted by permission.)

Most manufacturers apply "threshold" algorithms to both the right and left ends of the main lobe. These thresholds only begin storing values when a certain number of pixels holding these values is reached, illustrated in Figure 30-12. This approach results in a tight, specific range for calculations from the main lobe.

However, just to demonstrate how simple it can be to write an algorithm to identify data landmarks, the pixel count in each bin could simply be *subtracted* from the count for the previous bin. From *right to left* in Figure 30-11, note that this process will first yield negative numbers as the computer "climbs" the right side of the tail spike, then positive numbers as it progresses down the left side of the tail, then negative numbers again at the right slope of the main lobe. The S_{MAX} can be identified as the point where these subtractions begin to result in negative values *for the second time*, a simple instruction for the computer to execute. For the histogram in Figure 30-12, S_{MAX} can be defined as the *second time the threshold pixel count is reached*, scanning right-to-left, *while negative values are resulting from the subtractions.*

Types of Histogram Analysis

The type of Histogram analysis to be applied is set when the radiographer selects a *procedure* algorithm at the control console. There are actually three general types of histogram analysis that are each based upon certain assumptions about the histogram that will be available for data input. In effect, each type of histogram analysis "expects" a particular shape of histogram and performs its calculations based on this. The important point here is that these three types of analysis must be properly matched with the actual types of histograms acquired in order to avoid processing errors.

Type 1 histogram analysis is designed with the expectation that a direct exposure area (the spike at the right or "tail" in the graph) will be present, as we see in the graphs in Figure 30-7. Data from this extraneous area must not be fed into the LUT, so the computer must recognize the location of S_{MAX}. This is one source of the myth that CR projections must not be collimated smaller than the receptor size. This is not true. However, if *over-collimation* on an extremity resulted in a complete lack of "background density," and Type 1 histogram analysis was improperly applied, this would lead to histogram errors.

Type 2 analysis operates on the assumption that there will be no "tail" in the histogram, or background density "spike." See Figure 30-8 for an example. The histogram analysis must attempt to localize

Figure 30-12

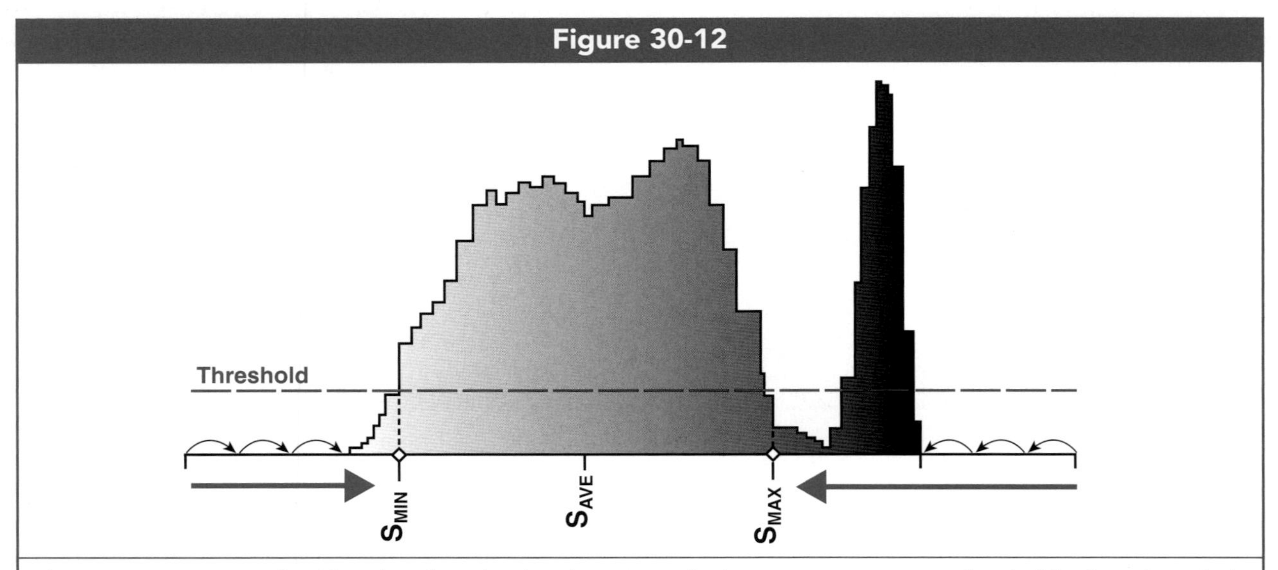

The most common method for identifying landmark points in the histogram is to use a "threshold" algorithm which only begins storing values when a certain number of pixels holding these values is reached. (From *Understanding Digital Radiograph Processing*, Midland, TX: Digital Imaging Consultants, 2013. Reprinted by permission.)

S_{MAX} without expecting any direct exposure area to be included in the curve. (In our "simple subtraction" example, rather than S_{MAX} being identified as the "2nd point where negative values result," it would simply be identified as the first bin with a non-zero pixel count, scanning from right-to-left.) This type of analysis tends to result in some oversaturation in the least dense soft tissue areas, such as the skin, but it is essential for projections of large body parts, such as the abdomen, which may leave no "background density" to work with.

Type 3 histogram analysis is designed for a large bolus of positive contrast agent (e.g. barium) to be included in the projection. It can also compensate for the presence of a lead shield or radiopaque hardware such as a metal prosthesis. As shown in Figure 30-9, in preparation for type 3 analysis to be applied, S_{MIN} must be identified as the point between the metal densities and the normal tissue curve. S_{MAX} must also be identified. All data outside of these two points must be eliminated from the calculations used to produce the final image.

An averaged gray scale curve is usually included in the histogram; This is the "S"-shaped curve in Figure 30-13, a copy of an actual histogram. As with the "H & D" (Hurter and Driffield) curves that were used to analyze the response of film to x-rays, this gray scale curve is built up by plotting the output "densities" (the inverse of brightness) in the final processed image against the *input exposure* that they were generated from (Fig. 30-14).

As shown in Figure 30-14, a *steep* gray scale curve indicates *high contrast* (or short gray scale) in the image, since, as exposure increases the darkness of the pixel goes up rapidly. *Long gray scale* (or low contrast) is indicated by a curve with a more shallow slope that gradually ascends. *The steeper the gray scale curve, the higher the image contrast.*

We can select the range of pixel values from the histogram that represent the anatomy of interest, and send only that range to the LUTs for gradation processing. This range is designated as the *values* or *volume of interest (VOI)* or *range* or *region of interest (ROI)*. Figure 30-15 illustrates how bony anatomy or soft tissue might be accentuated in the final image in this way. In histogram ***A***, the values of interest were selected to accentuate bony anatomy, closely aligning the left portion of the actual histogram and

Figure 30-13

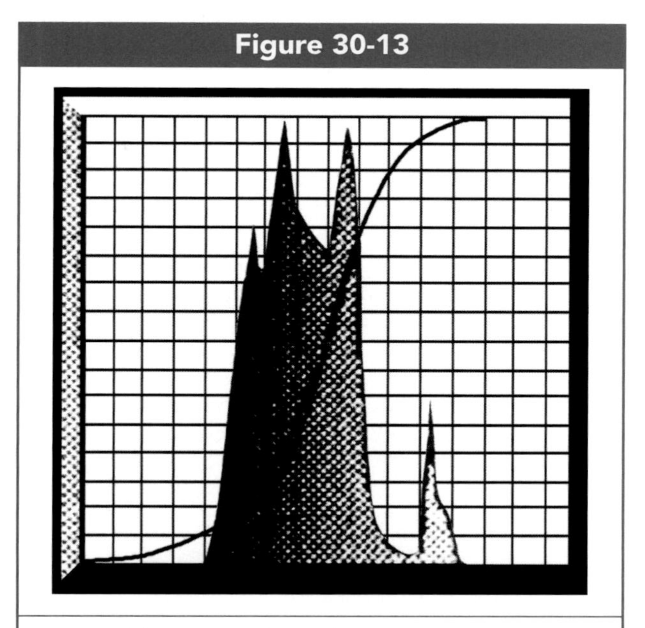

A typical histogram from the image review screen, with superimposed gray scale curve. (From Quinn B. Carroll, *Practical Radiographic Imaging*, 8th ed. Springfield, IL: Charles C Thomas, Publisher, Ltd., 2007. Reprinted by permission.)

Figure 30-14

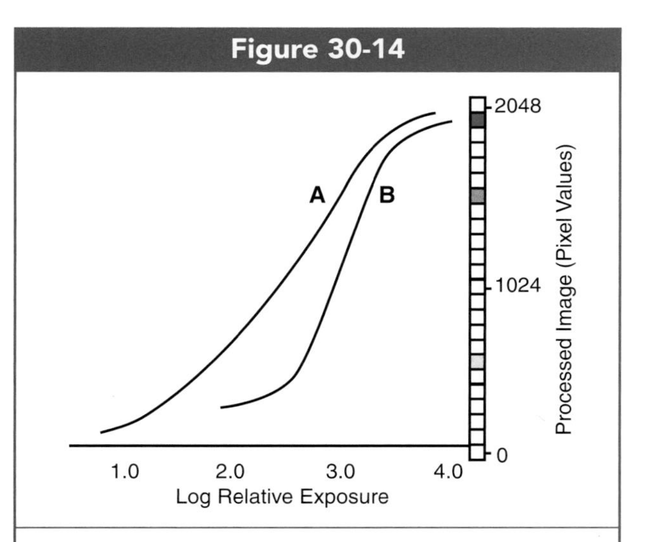

The gray-scale curve is plotted as the output pixel values against the original x-ray *exposure* at the image receptor. When the "body" portion of the curve is *shallow*, ***A***, it indicates a long gray-scale (low contrast) output image. A *steep* curve, ***B***, represents a high contrast image. (From Quinn B. Carroll, *Practical Radiographic Imaging*, 8th ed. Springfield, IL: Charles C Thomas, Publisher, Ltd., 2007. Reprinted by permission.)

Figure 30-15

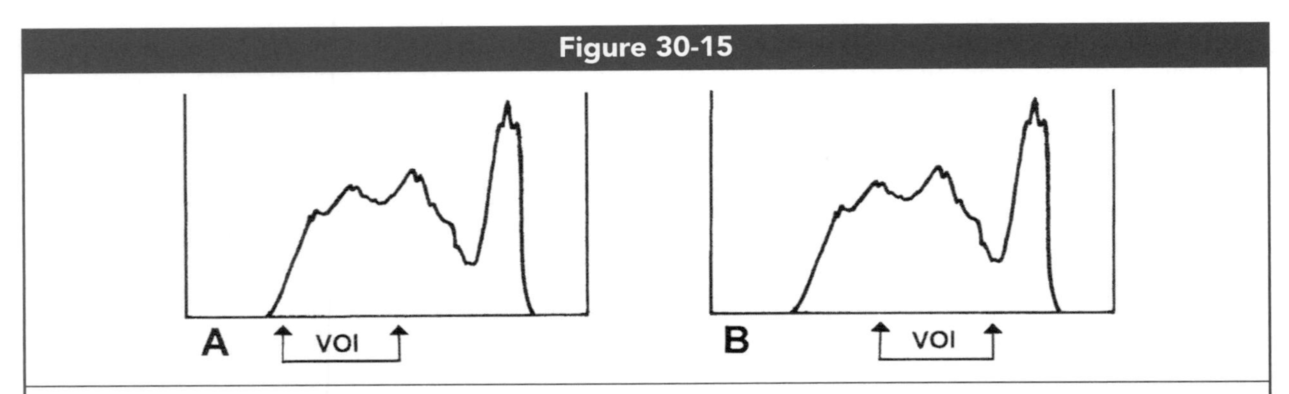

Histogram **A** shows the selection, from the original data set, of a *volume of interest (VOI)* designed to accentuate bony anatomy, by relocating the range S_{MIN} to S_{MAX} farther to the left. Only that portion of the data will be fed to the LUT for gradation processing. Histogram B shows a VOI selected for accentuating soft tissues, farther to the right. (From Quinn B. Carroll, *Practical Radiographic Imaging*, 8th ed. Springfield, IL: Charles C Thomas, Publisher, Ltd., 2007. Reprinted by permission.)

reference histogram curves, and these values are processed at the expense of the soft tissue densities to the right. If the same radiograph is processed using a soft-tissue algorithm, the values of interest will lie in the right region of the bell curve representing tissues from dense organs to skin (histogram ***B***). This area will be aligned at the expense of the bone densities.

For some manufacturers, once the data is accumulated from 50 histograms of actual procedures performed, the ideal reference histogram for the procedure is "updated" by averaging this data in, and then continues to use the most recent 50 histograms for that particular anatomy to update this reference histogram with each new procedure performed. Abnormalities in the histogram for a particular procedure are identified by comparison with this ideal reference histogram. This is referred to as a *neural* histogram (as opposed to a fixed *a priori* histogram).

Histogram Analysis Processing Errors

Many digital processing errors relate to *histogram analysis*. Particularly in the case of CR, segmentation and exposure field recognition errors contribute to failures in histogram analysis. For both DR and CR, errors occur when the type of histogram analysis used is mismatched to the shape of the actual histogram acquired.

Analysis fails, too, when the histogram of the acquired image attains any bizarre, unexpected shape from very extreme exposure conditions or unique situations where multiple exposure errors are combined, making it difficult or impossible for the computer algorithms to identify key landmarks and to distinguish between the different lobes of the histogram.

For example, imagine a situation in which a patient has a very large, thick prosthetic or trauma device made of dense plastic or aluminum which absorbs much of the x-ray beam but not as much as lead. The corner of a lead apron is also draped over the gonads. The resulting histogram might look like Figure 30-16, with *two* left spikes, the first representing the white lead area, and the second representing the large device. The radiographer correctly selects a processing menu that accounts for a leaded area and uses *type 3* histogram analysis. As the computer scans into the data curve from the left, it identifies the S_{MIN} point between the spike for the lead and the spike for the device. This is shifted much too far to the left from the real S_{MIN} point, which should exclude the large device as well. The average is skewed to the left, the "lighter" side, the exposure indicator is corrupted to read lower than it should, and upon rescaling, the entire image is over-corrected toward darker densities.

The type of histogram analysis used must match the shape of the actual histogram obtained. To do so, we must be able to *anticipate* the general shape to be expected for histograms of different anatomy. Anything that presents a bizarre variation from the

Figure 30-16

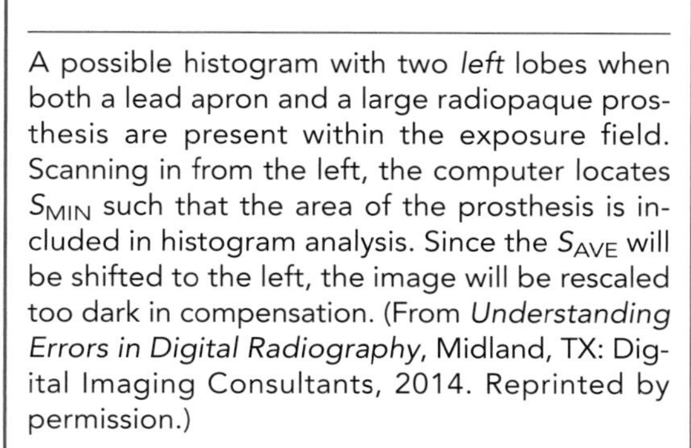

A possible histogram with two *left* lobes when both a lead apron and a large radiopaque prosthesis are present within the exposure field. Scanning in from the left, the computer locates S_{MIN} such that the area of the prosthesis is included in histogram analysis. Since the S_{AVE} will be shifted to the left, the image will be rescaled too dark in compensation. (From *Understanding Errors in Digital Radiography*, Midland, TX: Digital Imaging Consultants, 2014. Reprinted by permission.)

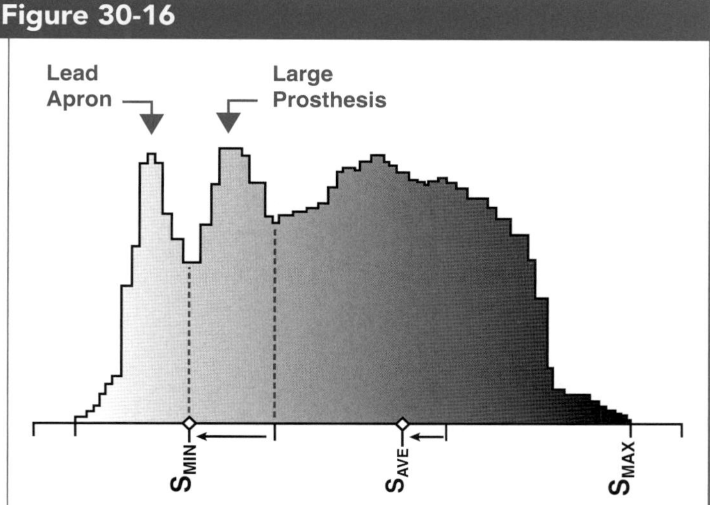

normal anatomy can throw off histogram analysis. Note that for a particular situation that continues to arise in a department, *custom histogram analysis protocols can be developed* by the manufacturer for a particular processing system.

Various histogram analysis and rescaling errors can result in an image that is too light (Fig. 30-17), too dark, with excessive contrast or with excessive gray scale. It is important to understand that these errors occur in mathematical data processing, so *the resulting poor appearance of the image is not directly related to the original radiographic technique used*, but rather to the computer's ability to interpret and adjust incoming data produced from that technique.

The algorithms for histogram analysis have become increasingly robust, and nowadays it is quite difficult to throw the computer off, so to speak. For a digital processing error to occur, it requires either a very extreme condition such as very severe amounts of scatter radiation, or a very bizarre situation in which several different problems (such as the presence of a large prosthesis combined with excessive amounts of scatter) come together to create a histogram that is far from the normal shape expected by the type of histogram analysis algorithm used.

Histogram analysis errors are less common with DR systems, because they include only the exposed pixels in the image data base, whereas CR systems begin by scanning the entire plate and then try to sort out exposed fields from unexposed portions of the plate.

RESCALING (NORMALIZING) THE IMAGE

If the "raw" image captured by the detector system were directly transformed into gray levels on a display monitor, it would appear as extremely "washed out" ghost-like image with almost no contrast, as

Figure 30-17

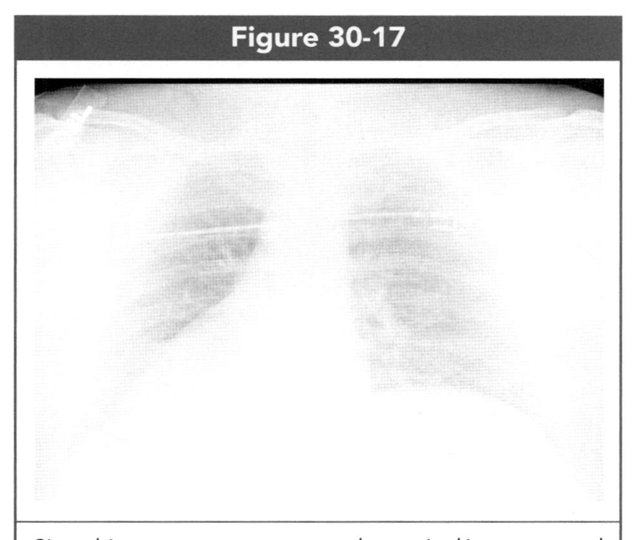

Since histogram errors are *mathematical* in nature and not directly controlled by the original exposure, they can just as easily result in a light digital image, shown here, as in a dark image. (From Quinn B. Carroll, *Practical Radiographic Imaging*, 8th ed. Springfield, IL: Charles C Thomas, Publisher, Ltd., 2007. Reprinted by permission.)

Figure 30-18

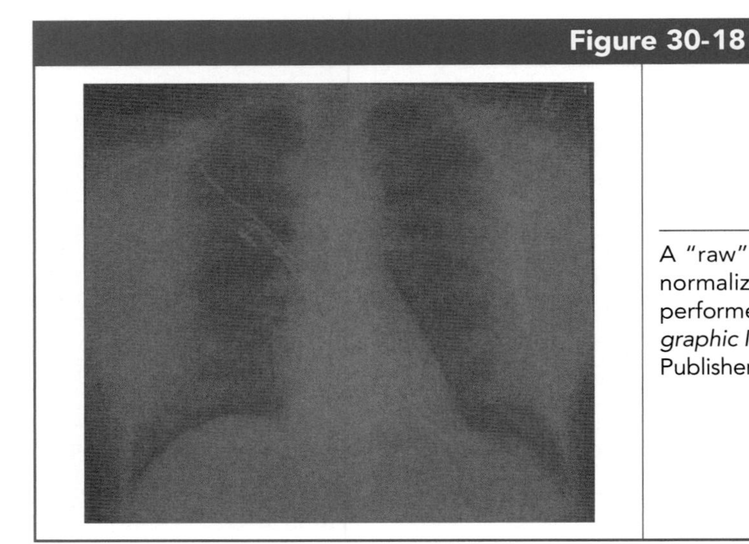

A "raw" digital image from the image receptor before normalization or any other gradient processing has been performed on it. (From Quinn B. Carroll, *Practical Radiographic Imaging*, 8th ed. Springfield, IL: Charles C Thomas, Publisher, Ltd., 2007. Reprinted by permission.)

Figure 30-19

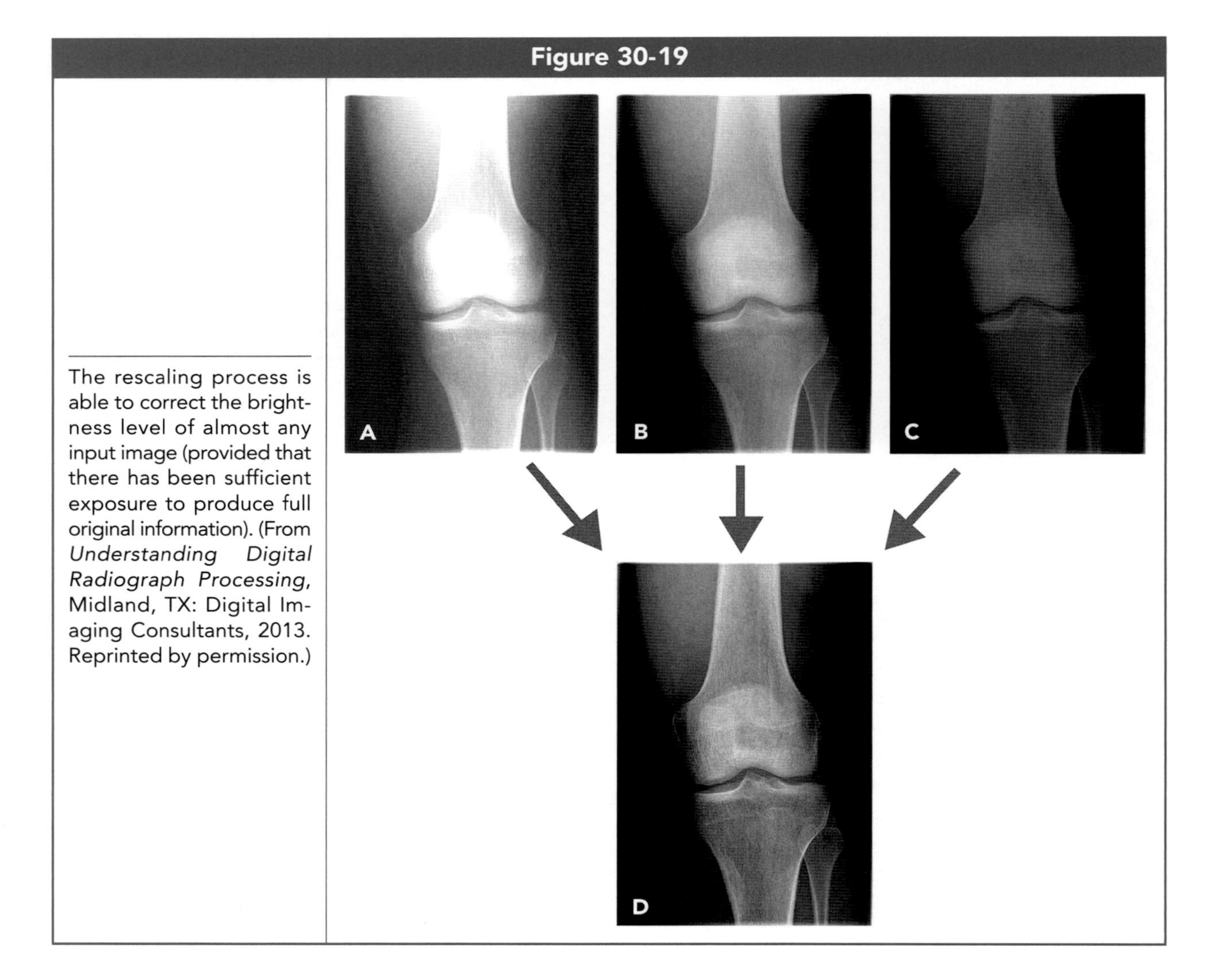

The rescaling process is able to correct the brightness level of almost any input image (provided that there has been sufficient exposure to produce full original information). (From *Understanding Digital Radiograph Processing*, Midland, TX: Digital Imaging Consultants, 2013. Reprinted by permission.)

shown in Figure 30-18 (*facing page*). The term *normalization* of the image refers to the initial round of default processes designed to make this image appear like a "normal" or conventional radiograph. They involve relatively simple alignments of the brightness of the image (and to a certain degree the gray scale range) and are referred to as rescaling.

It is the rescaling step that results in final displayed digital images that nearly always have the correct brightness level (Fig. 30-19, *facing page*). For such a profound result, the explanation actually involves concepts that are very easy to understand. As with many other digital processing operations, rescaling can be accomplished using either an electronics approach or a software approach, but due to its several advantages, software programming is universally used.

Rescaling, as we will see, has to do with algebraically applying the same labels to incoming data regardless of what that data actually is.

For the purpose of rescaling the image brightness and gray scale, the "bins" of data, (actually computer files), from the S_{MIN} to the S_{MAX} are given algebraic labels. For example, in Figure 30-20, they are designated as *S2*, *S3*, *S4*, and so on up to *S1022*, *S1023*, and finally S_{MAX}.

We begin with a pre-set look-up-table, such as the one in Table 30-1, that is permanently stored in the computer. This table lists a set of labels we will designate as Q values, from Q_{MIN} to Q_{MAX}. This example is based on a dynamic range for the software from zero to 4095. For the gray scale in the final image, this table always sets the output for Q_{MIN} at 511 and the output for Q_{MAX} at 1534, with 1532 values in between labeled as *Q2*, *Q3*, *Q4* and so on up to *Q1023*. *Q2* is always output at a gray level or pixel value of 512, *Q3* at 513, and so on up to the Q_{MAX} at agray level of 1534.

Now, in order for input from the acquired image to be aligned with this output scale, the range of gray levels to be sampled from the acquired image must always match the number of entries in this table. In other words, the number of S values (Figure 30-20) must match the number of Q values stored in the permanent LUT (Table 30-1). Regular digitization of the incoming image facilitates this matching process. Depending on the degree to which incoming pixel values are rounded up or down, they can be made to fit a pre-set *range* of S values (Fig. 30-21). We simply match the Q range to this S range.

Since measured gray levels are rounded out by digitization, there is only a limited number of levels.

Figure 30-20

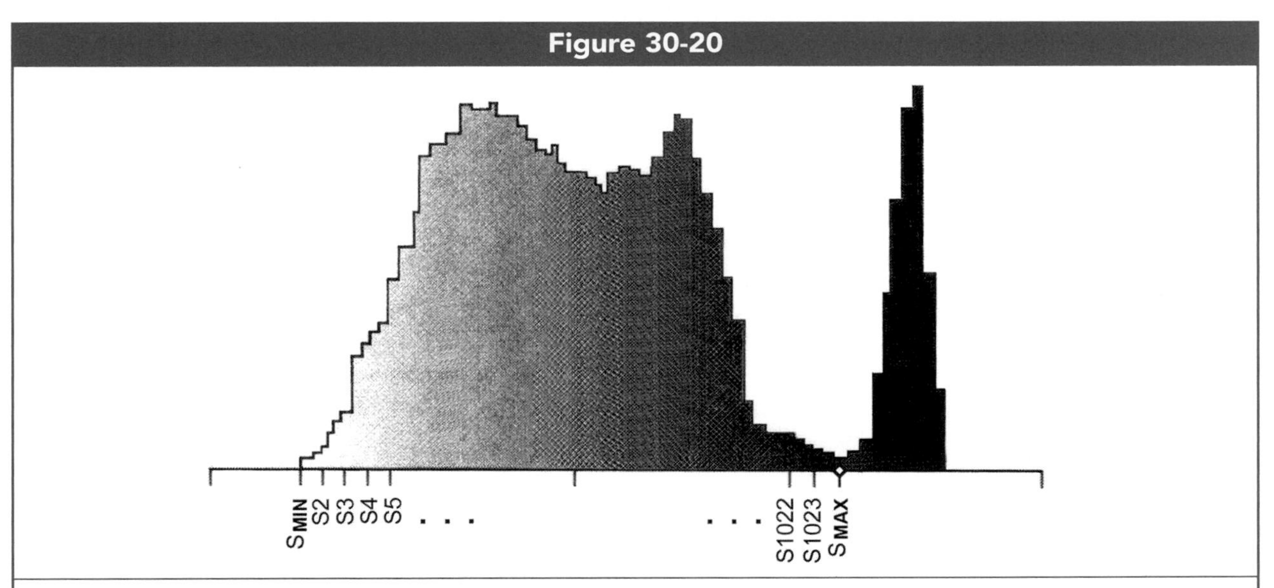

For rescaling, each pixel count is treated as a separate "bin" of data (computer file). Each bin is given an algebraic label from the S_{MIN} to the S_{MAX} such as *S1*, *S2*, *S3* and so on here up to *S1023*. These are the designations that will be used for rescaling as shown in Tables 30-1 and 30-2. (From *Understanding Digital Radiograph Processing*, Midland, TX: Digital Imaging Consultants, 2013. Reprinted by permission.)

Table 30-1

Q Values Stored in the Permanent LUT

Q_{MIN}	511
Q2	512
Q3	513
Q4	514
Q5	515
—	—
—	—
Q1022	1532
Q1023	1533
Q_{MAX}	1534

Adapted from *Understanding Digital Radiograph Processing*, Midland, TX: Digital Imaging Consultants, 2013. Reprinted by permission.

These gray levels can be assigned to different "bins" or individual computer files. As an example, for a *gray scale* of 1535, 1535 individual computer files are needed. An available dynamic range of 4096 is far beyond the discernment of the human eye and allows for the image to be windowed up and down along the available dynamic range.

Table 30-2

Reassigning S Values as Q Values

Algorithm:		
Set S_{MIN}	=	Q_{MIN}
Set S2	=	Q2
Set S3	=	Q3
Set S4	=	Q4
Set S5	=	Q5
	—	
	—	
Set S1022	=	Q1022
Set S1023	=	Q1023
Set S_{MAX}	=	Q_{MAX}

Adapted from *Understanding Digital Radiograph Processing*, Midland, TX: Digital Imaging Consultants, 2013. Reprinted by permission.

We now write a computer algorithm that simply reassigns incoming S values as Q values shown in Table 30-2: *Set S_{MIN} to Q_{MIN}. Set S2 to Q2, Set S3 to Q3* and so on to the Q_{MAX}. Regardless of what S_{AVE} is, it is re-mapped to Q_{AVE} which always has the same set value (1023 here). S_{MIN} is always re-mapped to Q_{MIN} (511 here), and S_{MAX} is always re-mapped to Q_{MAX} (1534 here).

The term *re-mapping* means that the computer simply re-assigns all pixels in a particular "bin" or computer file to this pre-selected number, *no matter what the original input value is for these pixels.* Regardless of what measurement *S2* is, it is always labeled as *Q2*, *S3* is labeled as *Q3*, *S4* is labeled as *Q4* and so on. Thus, *regardless of the input values, the output values are always the same.*

It is important to understand what the computer can and cannot do with the digital image. Figure 30-22 is designed to illustrate that there is no magic here—for all the computer can do, there are limitations. In Figure 30-22***A***, the computer compares the S_{AVE} of the acquired histogram to the S_{AVE} for the permanently-stored reference histogram for a particular body part, and by simple subtraction determines the difference to be 0.3. In line ***B***, the entire acquired data set is adjusted by this fixed amount, simply subtracting 0.3 from all values. The effect is to shift the acquired histogram to the left into alignment with the ideal histogram at its mid-point. The average brightness level (window center) is now aligned.

However, note in Figure 30-22***B*** that the gray scale *range*, represented by the width of the histogram at the bottom, may still not be aligned from the minimum to the maximum pixel values of the ideal Q range. By progressively decreasing the pixel values (in Table 30-1) as they spread from the average, both upward and downward, the resulting histogram can be narrowed as shown in Figures 30-21 and 30-22***C***. All this is achieved by the algorithms in Table 30-2, and the end result is that the average, high point, and low point of the gray scale can be aligned with the ideal histogram as illustrated in Figure 30-22***C***. The gray scale of the image is thus adjusted *to a degree*,

Figure 30-21

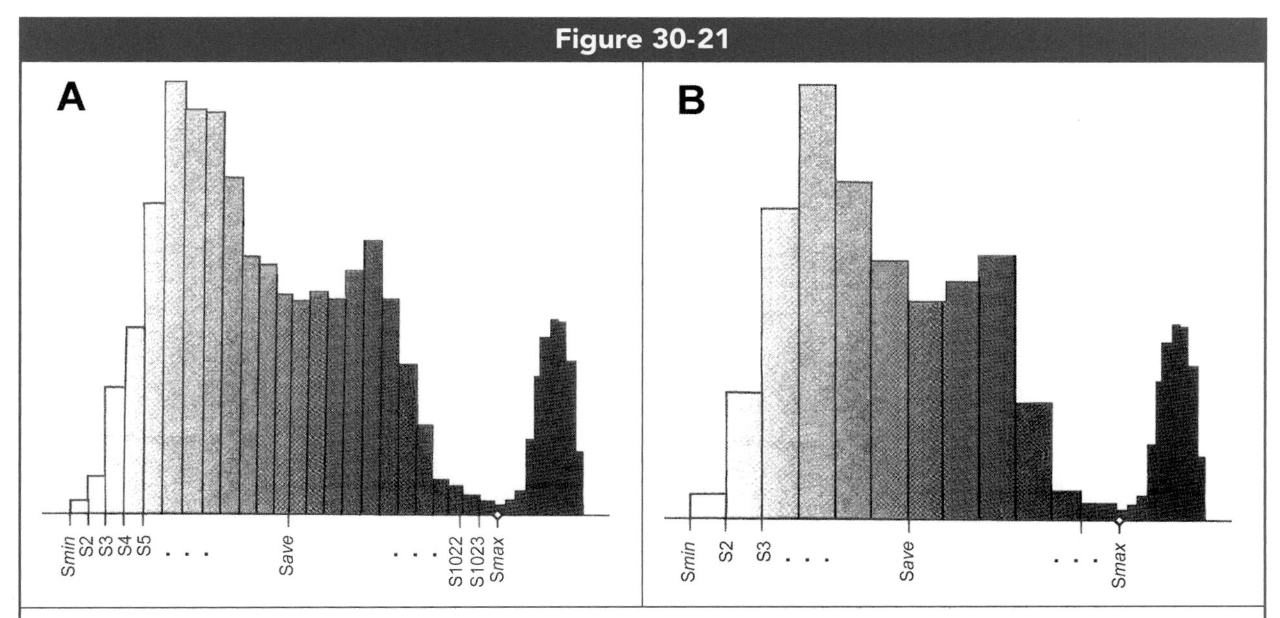

Depending on the degree to which digitization rounds incoming pixel values, they can be made to fit a preset *range* of S values. In ***B***, more extreme rounding of pixel values results in fewer values from one end of the histogram to the other.

Figure 30-22

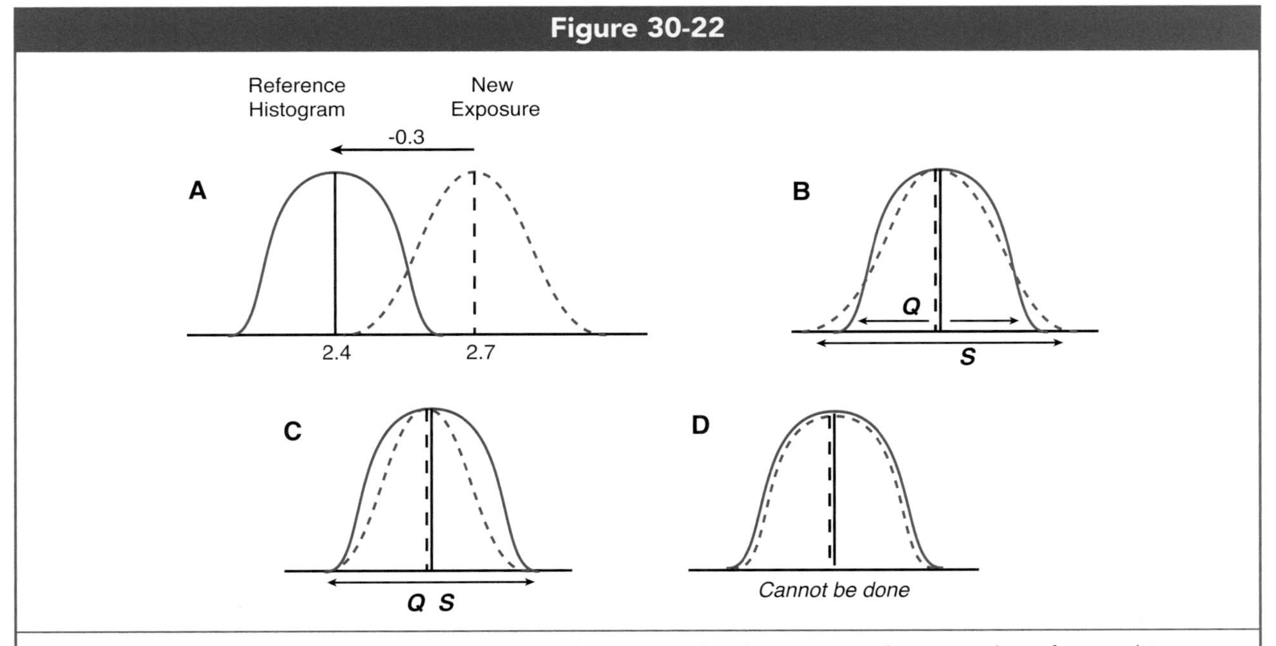

From ***A*** to ***B***, the computer finds the difference in the *average brightness* (S_{AVE}) between the reference histogram and the acquired histogram to be 0.3. By simply subtracting this amount from all pixel values for the new exposure, its average brightness is aligned with the ideal histogram. From ***B*** to ***C***, the *range* of pixel values for the new exposure (***S***) can also be rescaled to match the low, mid, and high points (**Q *values***) of the reference histogram (see Figure 30-21). This roughly corrects the gray scale. What the computer *cannot do* is change the pixel counts themselves in each bin, (which would perfectly match the *shape* of the histogram as shown in ***D***). (From *Understanding Errors in Digital Radiography*, Midland, TX: Digital Imaging Consultants, 2014. Reprinted by permission.)

Figure 30-23

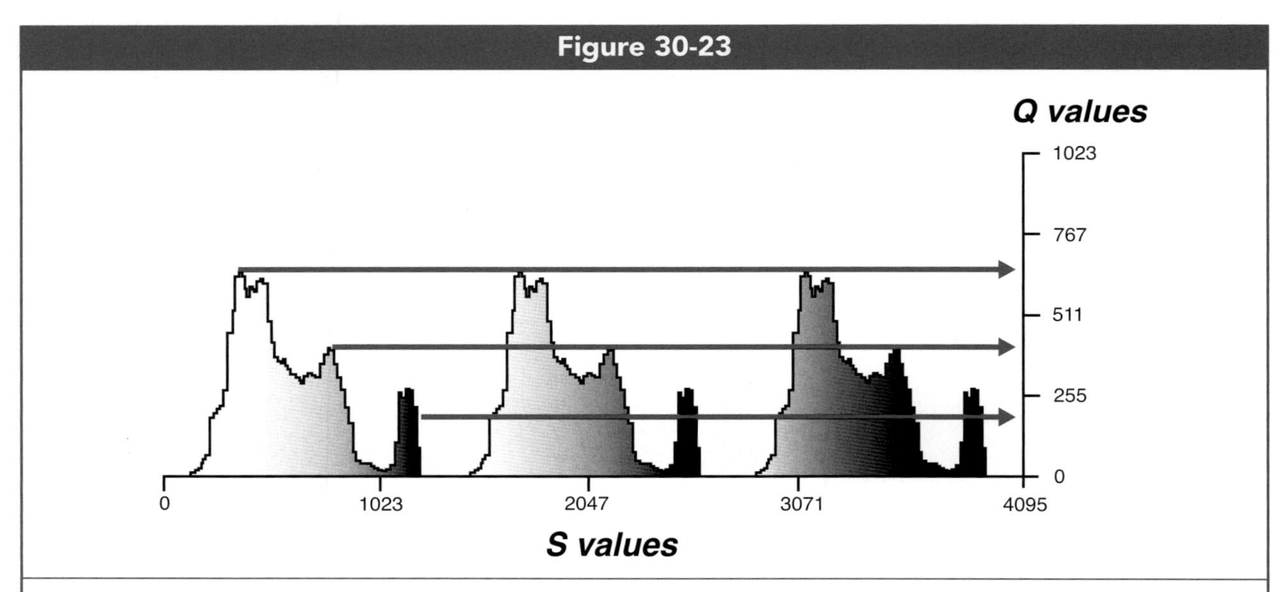

Graphic illustration showing that no matter where the histogram of the acquired image lies, it can always be re-aligned by rescaling to the **Q** range of the ideal histogram (right). These are the "for processing" values that will be used for gradation processing. (From *Understanding Digital Radiograph Processing,* Midland, TX: Digital Imaging Consultants, 2013. Reprinted by permission.)

but will be further refined during later gradation processing as explained in the next chapter.

What cannot be done is to force the *shape* of the acquired histogram to precisely fit that of the reference histogram as shown in Figure 30-22***D***. We *cannot* vertically align the number of pixels in each bin, that is, we cannot change the *pixel count* for a particular gray level. This is impossible because we cannot change *which pixels across the image hold a particular initial value*—we can only alter the values themselves.

Rescaling has the power to align image brightness perfectly (Fig. 30-19), but can only align image gray scale or contrast *partially* by aligning the high, low and average values.

Figure 30-23 shows graphically how the brightness level of the image is always standardized, because no matter where the histogram of the incoming image lies, it is always repositioned to the center of the Q range. In effect, rescaling brightness re-aligns the histogram for the new exposure to the "ideal" histogram of the Q range. It is this adjusted data set that is then fed into a procedure-specific LUT for fine-tuning based on the anatomy to be demonstrated. This anatomical LUT is set by the operator when the procedure is selected from a menu. LUTs are fully discussed under *gradation processing* in the next chapter.

Physicists' Terminology

For clarity in this chapter, we have broadly used the terms *S* values and *Q* values when referring to the incoming and the outgoing data sets, respectively. Physicists have more recently adopted the following definitions:[2]

- *Q values:* The "for processing" values of pixels that have undergone field uniformity and noise corrections.
- Q_K *values:* The normalized (rescaled) "for processing" pixel values that have been converted to relate to exposure levels in *kerma.*
- Q_P *values:* The final "for presentation" pixel values after gradation processing and all other postprocessing operations are complete and the image is ready for display. (This is covered in the next chapter.)

[2] Shepard et al.: Exposure Indicator for DR: TG116 (Executive Summary) in *Medical Physics*, Vol. 36, No. 7, July 2009.

While it is not important for the radiography student to understand all the nuances of these definitions, the concepts of "for processing" and "for presentation" pixel values should be understood, and the terms can be translated into those used in this chapter. A few manufacturers continue to use the *S value* when referring to data that has not yet been rescaled, as we do here.

SUMMARY

1. Preprocessing consists of corrections for flaws in the *acquisition* of the original image, image analysis and normalization.
2. Field uniformity corrections include adjustments for flaws in digital acquisition hardware and del drop-out effects. Software can be configured to fill in pixel values for dead hardware pixels, and to adjust for flat-field non-uniformity, by interpolation.
3. Image analysis includes partitioned pattern recognition or segmentation, exposure field recognition, construction of the initial histogram, and histogram analysis. Each of these steps must be correctly executed for the following procedures to result in a proper image.
4. To analyze a histogram, the computer scans in from each end of the data set, searching for identifiable landmarks created by changes or trends in the pixel counts in each "bin" or file.
5. The proper type of histogram analysis must be matched with the acquired histogram for images that contain a background density, radiopaque prostheses or large bolus of barium. For an acquired histogram with a bizarre shape, failure to properly locate the expected landmarks is a common cause of digital processing errors.
6. The *volume* or *values of interest (VOI)* is a range within the histogram that can be relocated to accentuate different tissue types in the image.
7. A steeper gray scale curve superimposed over the histogram indicates higher image contrast.
8. Most *rescaling* (normalization) methods use software that algebraically re-labels input pixel values such that the output values are always the same. To accomplish this, the range of *S* values acquired must match the range of *Q* values in a permanent output look-up table (LUT).
9. The computer can adjust the levels and ranges of pixel values, but cannot *change* the pixel count for any particular value.
10. Physicists use the term *Q value* to describe *for processing* pixel values that have not yet been rescaled, Q_K *values* when referring to *for processing* pixel values that have been rescaled, and Q_P *values* to describe *for presentation* pixel values that have been postprocessed by gradation processing.

REVIEW QUESTIONS

1. Why does DR require more preprocessing than CR?

2. Of the variables listed which affect flat field uniformity, which are the only *two* that apply to direct-capture DR?

3. What is the mathematical process called by which a kernel averages surrounding pixel values in order to fill a dead pixel from a faulty del?

4. In addition to exposure field recognition, CR units must also employ ____________ or ___________ ____________ ____________ software that can count and separate multiple views taken on a single PSP plate.

5. Exposure field recognition (EFR) prevents black densities __________ the anatomy from being included in histogram analysis.

6. On a typical histogram, as we move from *left to right* the ____________ of the pixels is changing from ____________ to ____________.

7. For histogram analysis, the desired portion of the main lobe to be used, between points S_{MIN} and S_{MAX}, is referred to as the ______________ _______ ________________.

8. On the displayed histogram for an image, a steeper gray scale curve indicates higher ____________ in the image.

9. To avoid processing errors, we must be able to anticipate the general ____________ of the acquired histogram to be expected for different types of anatomy.

(Continued)

REVIEW QUESTIONS *(Continued)*

10. Which type of histogram analysis should be applied when a large bolus of barium or an area of lead shielding is expected within the field of view?

11. Normalizing the raw digital image, so that it has the typical overall brightness and appearance of a conventional radiograph, is also called ____________.

12. The computer *re-maps* all pixels in a particular bin (or file) to a pre-selected *Q* value, no matter what the original ____________ value was for these pixels.

13. During rescaling, is the computer system capable of matching the low, mid-, and high points of the *S value range* to the ideal *Q* range, roughly correcting the gray scale of the image?

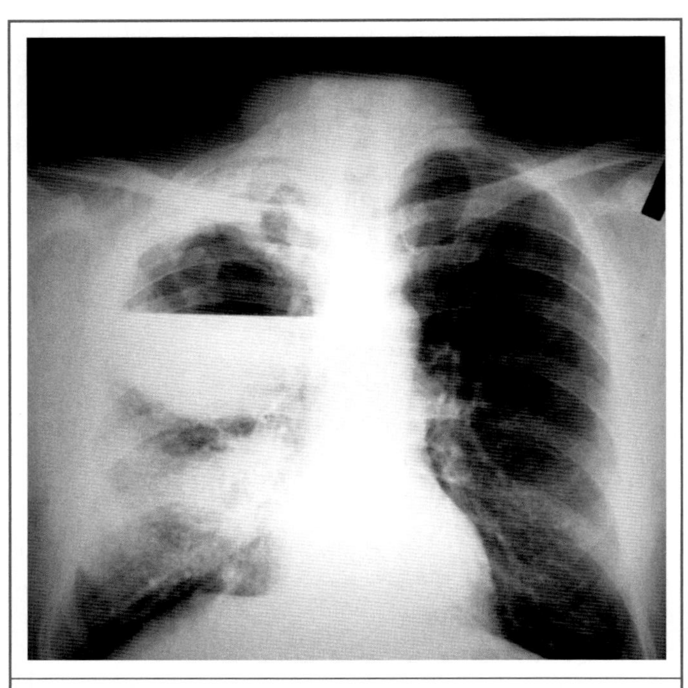

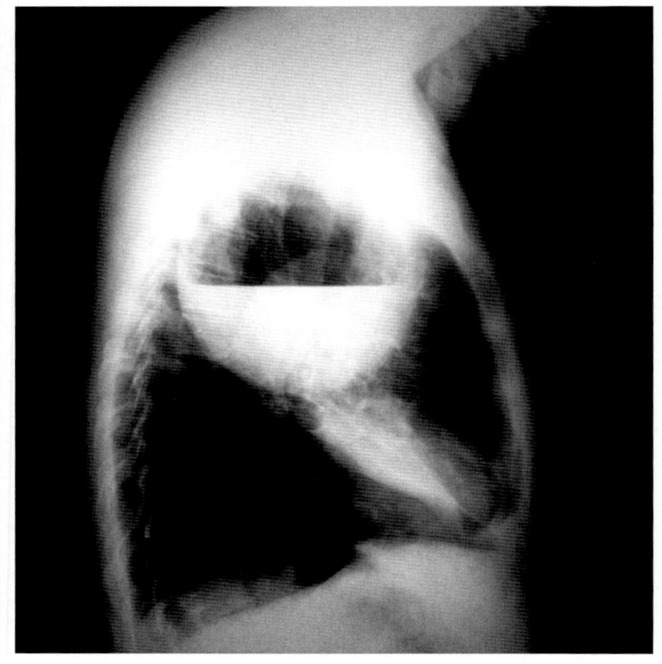

A 5-inch cyst in the upper right lung has filled half-way with pus, creating a classic air-fluid level on this radiograph.

Chapter 31

DIGITAL IMAGE POSTPROCESSING

Objectives:

Upon completion of this chapter, you should be able to:

1. Define *spatial domain*, *intensity domain*, and *frequency domain* processing of image details.
2. Explain how an image can be represented in the form of wavelengths and frequencies.
3. Define gradient processing and describe how LUTs are used to customize image brightness and gray scale according to the anatomical procedure.
4. Explain the proper graphical interpretation of the function curves from which different LUTs are generated.
5. Define data clipping, dynamic range compression and tissue equalization.
6. Describe point processing using kernels, and how kernels are used to enhance image detail.
7. Describe the steps in *unsharp mask filtering*, and how *subtraction* is used in this process.
8. List the risks of over-applying smoothing or edge-enhancement functions.
9. For the electronically-displayed image, define the relationships between object size, frequency, and image sharpness.
10. Define *Fourier transformation*, *high band-pass filtering*, and *low band-pass filtering*.
11. Define *pyramidal decomposition* and the steps of *multiscale processing*.
12. Define the parameters of frequency processing and how they are used to enhance detail.
13. Discriminate between random noise and periodic noise, and state the best methods for reducing each.
14. Explain the necessity for additional gradation processing before image display.
15. Overview processing suites used by manufacturers.
16. Describe operator adjustments to the image, including windowing, edge enhancement, smoothing and tomographic artifact suppression.
17. Explain how dual-energy subtraction works and the different approaches to achieving it.
18. State the concept of targeted artifact suppression such as is used for tomographic artifacts.

DIGITAL PROCESSING DOMAINS

There are three general approaches to processing any digital image. These are such completely different ways of looking at the image, as if from different universes, that we refer to them as "domains." We speak of processing the image in the spatial domain, in the intensity domain, or in the frequency domain. These domains are really different ways of *sorting* the information contained in the image.

The spatial domain is defined as a sorting of the image by the *location* of the pixels in space, that is, according to their coordinates within the image matrix—which column and which row they are in. Any operation executed on the image identifies the pixels by these spatial matrix coordinates.

Spatial domain operations can be further divided into three broad classifications: (1) point processing, (2) area processing, and (3) global processing. Point-processing operations perform a specific algorithm on each individual pixel in sequence, pixel by pixel or "point by point." Subtraction is a good example of a point-processing operation: The value contained in a specific pixel is subtracted from the value contained in the very same pixel from another image, but these pixels are *identified* by the computer according to their location in the matrix.

Area processing operations execute some function on a local group of pixels or a subsection of the image, and are also call "local operations." A good example of area processing is the zoom or magnification feature: Here, a subsection of the image is selected. Within that local area, the value contained in each pixel can be reassigned to a group of four pixels or nine pixels. The effect is that they combine to form a larger pixel containing a single value. The selected area expands to fill the monitor or "field of view." This is all done according to the locations of the pixels in the matrix.

Global processing carries out some massive spatial function across the entire image. A great example of a global processing operation is image *translation*, or "flipping" the image left-to-right. Figure 31-1 illustrates this concept using a small matrix with only eleven columns, ***A*** through ***I***; For a particular row of pixels such as row ***5***, the value in ***A5*** is reassigned to ***I5*** and vice versa, the value in ***B5*** is reassigned to ***H5*** and vice versa, the values in ***C5*** and ***G5*** are exchanged, and those in ***D5*** and ***F5***, progressively toward the middle column. To invert an image, the same process is executed row-by-row, top to bottom.

The application of a "kernel," which is a submatrix that is passed over the original pixels of the image executing a mathematical function on them, is also a spatial domain operation. A convolution kernel can be applied for edge enhancement or smoothing, which derives the final value of a pixel from a surrounding group of pixels, according to their locations. Kernels are fully discussed later in this chapter.

Figure 31-1

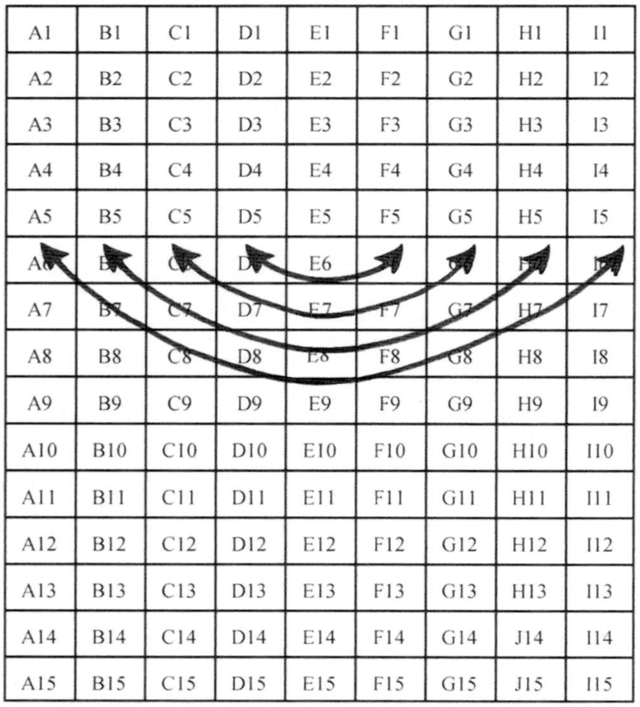

A1	B1	C1	D1	E1	F1	G1	H1	I1
A2	B2	C2	D2	E2	F2	G2	H2	I2
A3	B3	C3	D3	E3	F3	G3	H3	I3
A4	B4	C4	D4	E4	F4	G4	H4	I4
A5	B5	C5	D5	E5	F5	G5	H5	I5
A6	B6	C6	D6	E6	F6	G6	H6	I6
A7	B7	C7	D7	E7	F7	G7	H7	I7
A8	B8	C8	D8	E8	F8	G8	H8	I8
A9	B9	C9	D9	E9	F9	G9	H9	I9
A10	B10	C10	D10	E10	F10	G10	H10	I10
A11	B11	C11	D11	E11	F11	G11	H11	I11
A12	B12	C12	D12	E12	F12	G12	H12	I12
A13	B13	C13	D13	E13	F13	G13	H13	I13
A14	B14	C14	D14	E14	F14	G14	J14	I14
A15	B15	C15	D15	E15	F15	G15	J15	I15

Image *translation*, an example of a global processing operation. The pixel values contained in cells ***A5*** and ***I5*** are exchanged, the values in ***B5*** and ***H5*** are exchanged, those in ***C5*** and ***G5*** are exchanged, and those in ***D5*** and ***F5***, progressively toward the middle column. The result "flips" the image left-to-right.

We define the intensity domain as a sorting of the image strictly by the *values* of the pixels, that is, the gray level stored in each pixel, regardless of their location. Examples of operations in the intensity domain include the construction of the original histogram and histogram analysis covered in the previous chapter, where pixels from various locations are grouped together in "bins" or computer files that share the same value (Fig. 30-5). In this chapter, gradation processing with its application of anatomical LUTs, and subsequent windowing of brightness and contrast by the operator are all intensity domain functions.

As the pixels are sorted into their bins according to their intensity value, the computer keeps a record of each pixel's original location. After intensity domain operations are completed, the image transitions back into the spatial domain when the pixels are "taken out of their bins" and re-assigned to their locations in the matrix.

We define the frequency domain as a sorting of the image according to the *sizes of the objects or details* in the image. This is very different from the first two approaches, which refer to the pixels of the image rather than objects in the image. This concept of sorting structures in the image by their size will be fully explained in a following section on *detail processing*. Examples of operations in the frequency domain include edge enhancement and smoothing, which are often applied during the default processing of the image but can also be applied later at the touch of a button by the operator.

These three general approaches to processing (or sorting out) the information in a digital image—by location, by intensity, or by structure size—each result in a different product: As illustrated in Figure 31-2, sorting the image by spatial location results in a matrix, sorting the image by pixel intensity results in a histogram, and sorting the image by the size of objects or structures results in a frequency distribution.

Figure 31-2

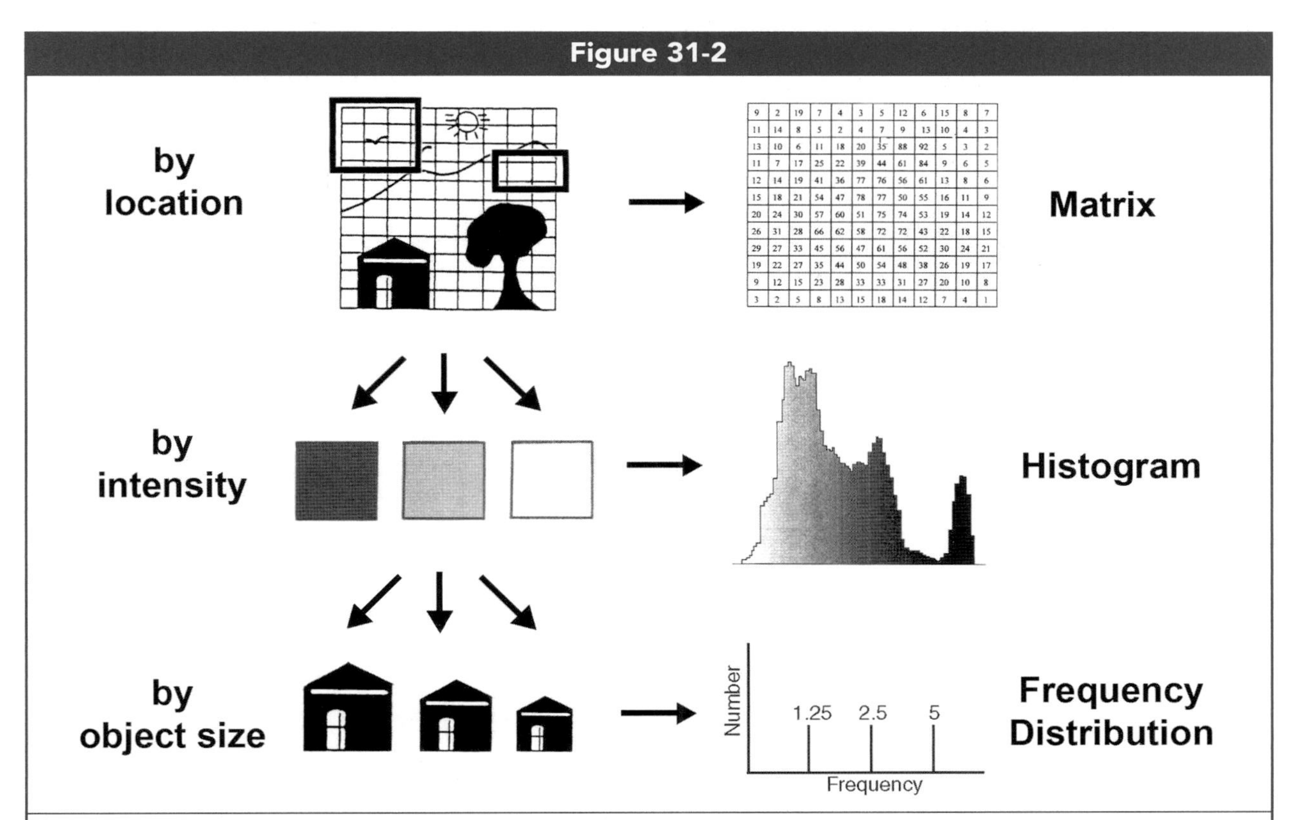

Three approaches to sorting (processing) any digital image and what they produce: Sorting the image by spatial location results in a matrix. Sorting the image by pixel intensity results in a histogram. Sorting the image by the size of objects or structures results in a frequency distribution.

POSTPROCESSING I: GRADATION PROCESSING

Initial Gradation Processing

Gradation is defined as a gradual passing from one tint or shade to another, so for radiography it refers to the gray scale of the image. A "gradient" is defined in the dictionary as the rate of change in an ascending or descending scale, or the steepness of a slope. *Gradient* processing is named after the *gradient curve* of an image, shown in Figure 31-3. The *left-to-right* position of the curve indicates image *brightness*, while the *steepness* of the curve represents image *contrast*.

When a gradient curve is plotted on a graph showing the various densities in an image, we can think of the average brightness of the image in terms of where this curve is *centered* (left-to-right), and the contrast as how *steep* the slope (or gradient) of this curve is. (Older technologists will recognize this concept as the *H & D* curve.) So, gradient processing is a good term to describe the manipulation of both the centering and the steepness of this gradient curve.

Figure 31-3

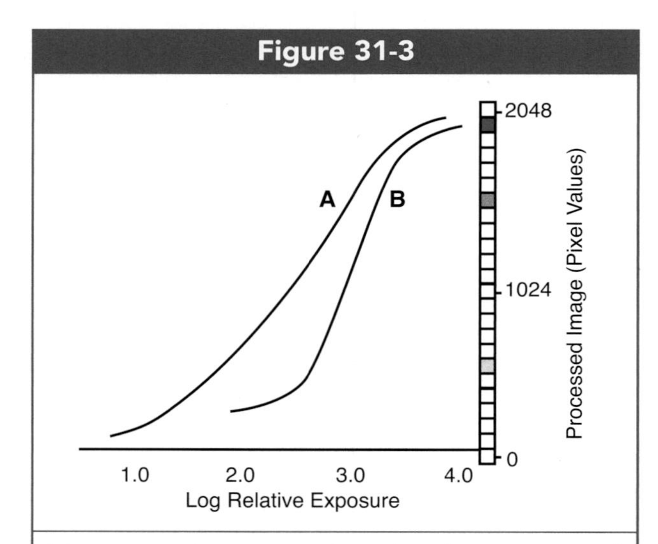

The gray-scale curve is plotted as the output pixel values against the original x-ray *exposure* at the image receptor. When the "body" portion of the curve is *shallow*, **A**, it indicates a long gray-scale (low contrast) output image. A *steep* curve, **B**, represents a high contrast image.

In gradient processing, the rescaled data set is fed into an anatomical LUT. The anatomical LUT is selected by the operator when a procedure is entered at the console. This LUT *customizes* the gray scale and brightness according to the specific anatomy to be demonstrated.

Look-Up Tables really are *tables* (not graphs). Computers work with numbers, not graphs, but *graphs* are often valuable for *human* benefit in understanding LUTs. An LUT is a simple table of two columns of numbers representing input and output values. Examples of two actual LUTs are presented in Table 31-1. Note that the input for both tables is identical, because the number set has already been rescaled. Look-up Table *A* produces medium contrast; whenever the computer reads an input value of 32 (in the middle of the table), it outputs the value 256. When it reads 34, it outputs 264. For every increase of 2 in input, the output is increased by 8. Table *B* produces high contrast from the *same* data. Note that in the output column, the *average* gray level or *center (GC)* is the same for both tables. But for Table B, each time the input increases by 2, the output is now increased by 32. For Table B, the top values are higher than for Table A, but the bottom values are also lower than for Table A. This is high contrast.

The conversion of input values to output values is *generated* by a mathematical formula (function) that can be represented on a graph as a *function curve*. The curve represents a particular computer algorithm with algebraic formulas that the input numbers are being subjected to. The graph generating a typical LUT plots the input pixel values along the bottom axis, the function curve (Fx) representing the mathematical formula that is acting upon these input values, and the converted output pixel values, reflected off the function curve, on the vertical (abscissa).

Figure 31-4 shows a function curve for enhancing image contrast. When we choose 2 input values such as 12 and 10 and calculate their contrast ratio, we find that 12 divided by 10 is 1.2. However, after these values have been changed by the formula (represented by the curve), we have a new contrast of 18 divided by 7 or 2.6. Thus, the original contrast of 1.2 has been converted to an output contrast of 2.6

All gradation processing takes place in the intensity domain. In computer graphics, these types of operations are called *intensity transformations*. The radiographer certainly does not need to know any of these formulas themselves, but understanding a little bit about how they work in general helps to understand not only gradation processing, but also the daily practice of windowing the image.

Below are three samples of Intensity transformation formulas for window width, controlling the displayed gray scale and contrast.

Gamma Transformation: $s = cr^{\gamma}$

Log Transformation: $s = clog(1 + r)$

Image Negative: $s = L - 1 - r$

You will note that the only 2 elements common to all 3 formulas are the terms "s" and "r." In each case, r is the input pixel value and s is the output pixel value. *Windowing* directly applies the formula converting input to output. For an example, we will use the gamma transformation formula,

$$s = cr^{\gamma}$$

In this formula, the term "c" is a constant that affects the type of change made to the image. To control window width, c is normally set at 1.0, so we can ignore it here.

The exponent, *gamma*, represents the window width setting that the radiographer can turn up or down at the console. As this exponent changes, its effect on "r" is manifested in the output pixel value "s".

Table 31-1

Actual Look-Up Table Format: Medium and High Contrast for Same Input

Look-Up Table A Medium Contrast			Look-Up Table B High Contrast	
INPUT	OUTPUT		INPUT	OUTPUT
44	304		44	448
42	296		42	416
40	288		40	384
38	280		38	352
36	272		36	320
34	264		34	288
32	256	⇐GC⇒	32	256
30	248		30	224
28	240		28	192
26	232		26	160
24	224		24	128
22	216		22	96
20	208		20	64
18	200		18	32
16	192		16	0

Figure 31-4

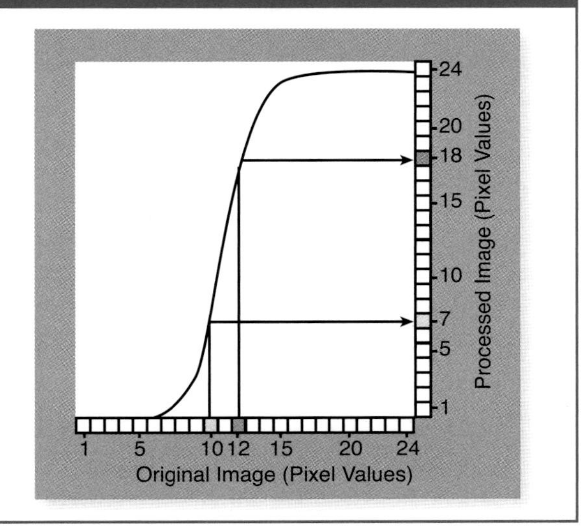

LUT curve for enhancing image contrast. Input contrast is the ratio of steps 12/10 = 1.2. After applying the algorithm to these pixel values, output image contrast is the ratio of steps 18/7 = 2.6, more than double the original contrast.

Figure 31-5

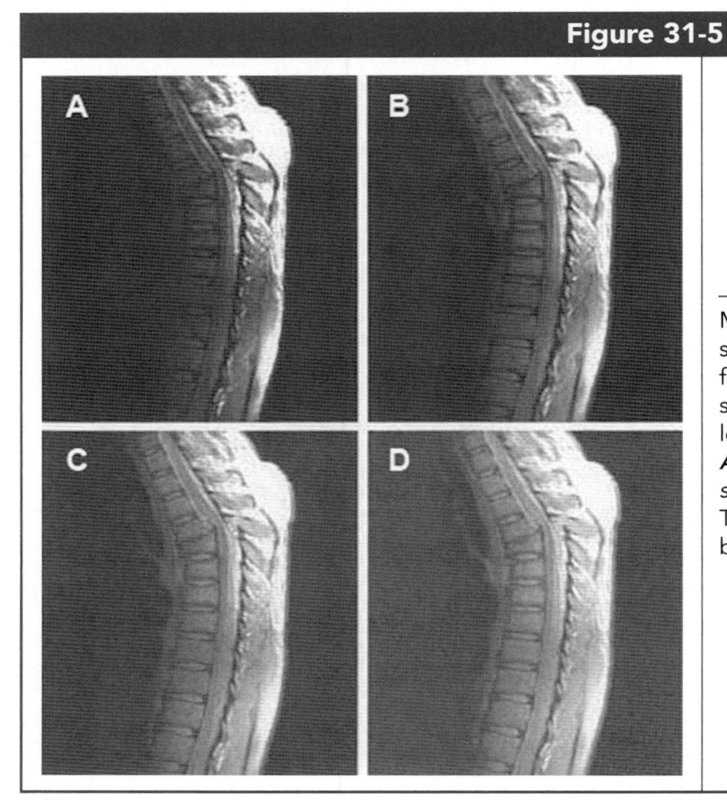

MR images of the lateral thoracic spine demonstrate application of the gamma transformation formula: As the gamma value is decreased to smaller and smaller fractions of 1, we see a lengthening of the gray scale. Gamma values are ***A*** = 1.0, ***B*** = 0.6, ***C*** = 0.4 and ***D*** = 0.3. (From *Understanding Digital Radiograph Processing*, Midland, TX: Digital Imaging Consultants, 2013. Reprinted by permission.)

Figure 31-6

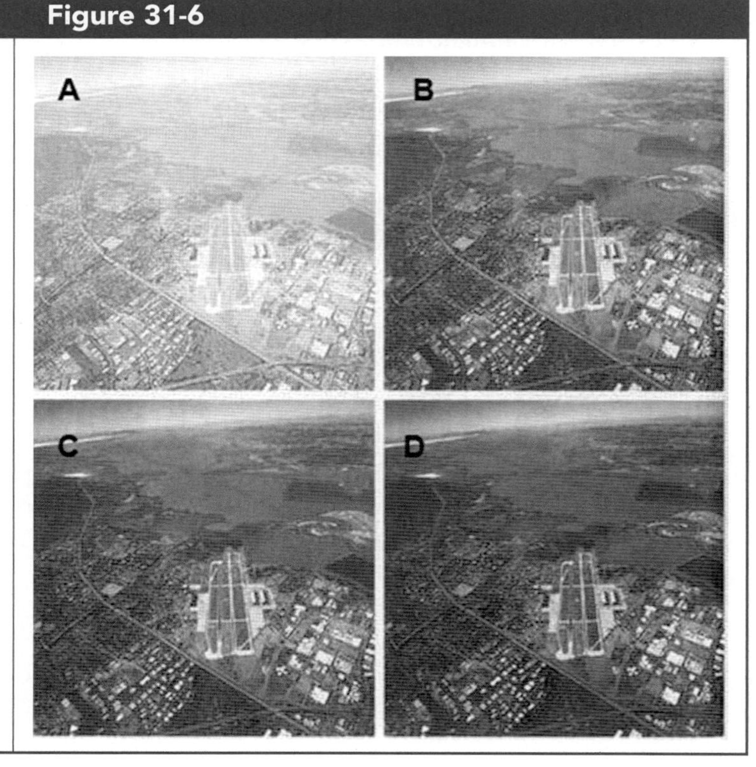

Aerial photographs of an airport show the effect of increasing the gamma value: ***A*** = 1.0, ***B*** = 3, ***C*** = 4, and ***D*** = 5. As gamma increases, we see an increase in image contrast or shortening of the gray scale. (From *Understanding Digital Radiograph Processing*, Midland, TX: Digital Imaging Consultants, 2013. Reprinted by permission.)

Figure 31-5 (*facing page*) shows a series of MR images of the lateral thoracic spine for C = 1: As the gamma value is decreased to smaller and smaller fractions of 1, we see a lengthening of the gray scale. Figure 31-6 (*facing page*) is an aerial photograph of an airport for C = 1: As the gamma values are increased from 1 upward, we see an increase in image contrast or shortening of the gray scale.

For c = 1, the effects of changing the gamma value in this formula can be graphed as function curves shown in Figure 31-7. This is the origin of the LUT "processing curves" used by different manufacturers. The type, shape and slope of the curves used by these manufacturers are set by the formula (math function) used for gradation processing. Since many different mathematical approaches can be used, the particular algorithms employed by different manufacturers result in different levels of effectiveness, and this drives competition.

Table 31-2 lists the parameters for Gradient Processing used by FujiMed. These parameters apply to the *formula* used, not to the actual input numbers. Each one changes a single factor in the formula in a specific way. On the manufacturer's LUT graph, these mathematical changes in the formula alter the position, slope or shape of the "processing curve." For example, curves with a *slope* steeper than 45 degrees in the *body portion* of the curve represent formulas that *increase contrast*. Curves with a *slope* shallower than 45 degrees represent formulas that *decrease contrast*. The *right* to *left* position of the curve represents changes to the image *brightness*.

Figure 31-8 shows just four of Fuji's 26 *Gradient Type (GT)* Curves. Each curve represents a different *formula* acting on the input. For example, Note that curve M reverses the image to black on white rather than white on black (often called a *"black bone" feature*).

Note that when windowing, the formulas and parameters used first set up the LUT, then output is simply read out from the LUT as input is fed into it. All these calculations take place in just seconds. Remember for *rescaling, an already-generated permanent LUT* was used, with output designated as mathematical *labels—Q values*. In that case, there is no real calculating done at the time the image is processed, rather, the input values are simply assigned to the pre-set labels and the output values are read out from the LUT.

Figure 31-7

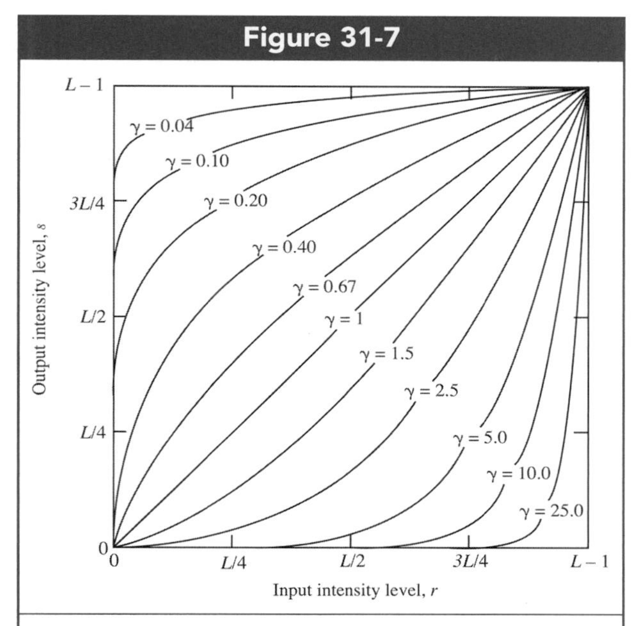

For c = 1 in the gamma transformation formula, these are the actual graphed function curves for different values of gamma, decreasing from 1.0 at the upper-left, increasing from 1.0 at the lower-right. (From *Understanding Digital Radiograph Processing*, Midland, TX: Digital Imaging Consultants, 2013. Reprinted by permission.)

So, gradation processing is used both during *default processing* when anatomical LUTs are applied, and again any time the displayed image is windowed by the operator at the console. Since gradation processing is aimed at *refinement* of the image according to the specific anatomy involved, it is best classified as a post-processing operation, whereas *rescaling* of the image was considered as a *preprocessing* operation that compensates for flaws in image acquisition.

Parameters for Gradient Processing

Table 31-2 itemizes for each type of radiographic procedure the pre-set parameters used by Fuji for *gradient processing*, which control the shape and position of the curve in the look-up table (LUT) that will be applied. Four parameters are listed, *GT*, *GS*, *GA* and *GC*.

GT stands for *gradation type*, and represents the shape of the curve which the algebraic formula for the LUT generates, four of which are illustrated in Figure 31-8.

Table 31-2

Fuji* Parameters for Gradation Processing

Anatomical Region	Gradient Curve Parameters			
	GA	GT	GC	GS
General chest (Lat)	1.0	B	1.6	–0.2
General chest (PA)	0.6	D	1.6	–0.5
Portable chest (Grid)	0.8	F	1.8	–0.05
Portable chest (No Grid)	1.0	D	1.6	–0.15
Pediatric chest (NICU/PICU)	1.1	D	1.6	–0.2
Finger	0.9	O	0.6	0.3
Wrist	0.8	O	0.6	0.2
Forearm	0.8	O	0.6	0.3
Plaster cast (arm)	0.8	O	0.6	0.4
Elbow	0.8	O	0.6	0.4
Upper Ribs	0.8	O	1.6	0.0
Pelvis	0.9	O	0.6	0.2
Pelvis Portable	0.9	O	0.6	0.2
Tibia / Fibula	0.9	N	0.6	0.25
Foot	0.8	O	0.6	0.3
Foot	1.2	N	0.6	–0.05
Os calcis	0.8	O	0.6	0.4
Foot cast	0.8	O	0.6	0.5
C-spine	1.1	F	0.6	0.5
T-spine	0.8	F	1.8	–0.05
Swimmers	1.2	J	0.9	0.3
Lumbar spine	1.0	N	0.9	0.4
Breast specimen	2.5	D	0.6	0.35

*Courtesy, Fujimed, Inc.

Figure 31-8

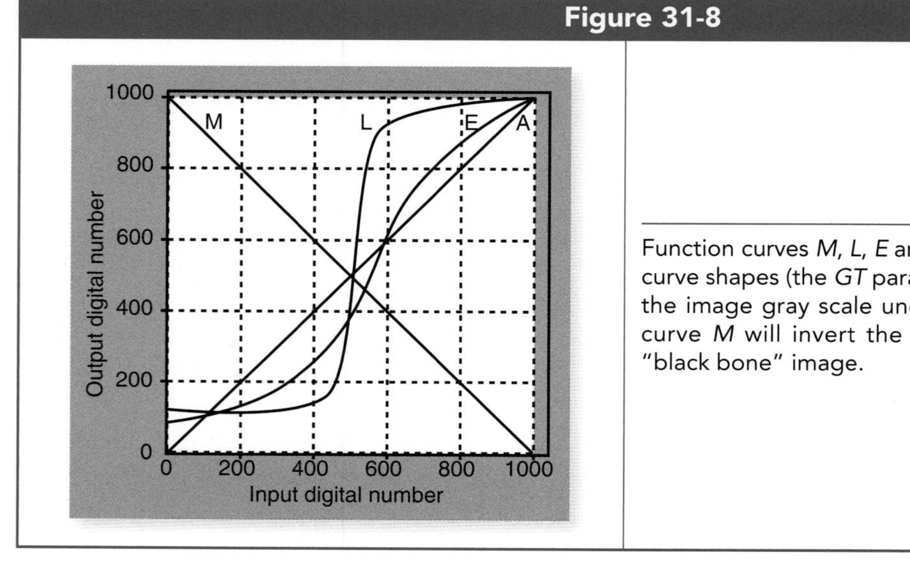

Function curves *M*, *L*, *E* and *A*, just four of the 26 types of curve shapes (the *GT* parameter) available for processing the image gray scale under the Fuji system. Note that curve *M* will invert the gray scale to form a positive "black bone" image.

GS stands for *gradation shift* and is a *sensitivity* or *speed* setting which slightly adjusts the overall *brightness* of the entire range of pixels. Figure 31-9 illustrates how changes in the GS setting move the entire LUT curve to the right or left.

GA stands for *gradation amount* or *angle*, and determines the *slope* or angle of the body portion of the LUT curve. As was shown in Figure 31-3, whenever this slope is greater than 45 degrees, the result is an increase in *contrast* of the output image compared to the original image. In Figure 31-10, this is the case for the curve marked *GA >1*, which will *shorten the gray scale* of the image. The steeper the LUT curve, the shorter the resulting gray scale. In Figure 31-10, *GC* stands for the *gradation center*, and is a specific point in the X and Y axes of the graph around which the LUT curve rotates when it is changing its slope or *GA*. Notice in Table 31-1 (p. 501) that for these two actual look-up tables, *GC* has the same output value for both scales, medium contrast or high contrast. This is because the *average brightness*, which is controlled by *GS*, has *not* been changed in this example. This is the effect of "rotating the gradation curve without shifting it."

Continuing from Figure 31-10, if the curve labeled *GA < 1* is used, which has a slope *less* than 45 degrees, the gray scale from the original image will be lengthened.

Each particular LUT is constructed based upon these four parameters. Then, in effect, the graph is "read out" by the computer by effectively taking each input pixel value and reflecting it off of the curve to obtain the output values that will be used in constructing the new image. Remember, however, that these graphs are visual aids. Within a computer, everything is reduced to numbers, not graphs. In reality, *GT*, *GS*, and *GA* are mathematical *factors* inserted into an algebraic formula that results in look-up tables such as those in Table 31-1.

Figure 31-9

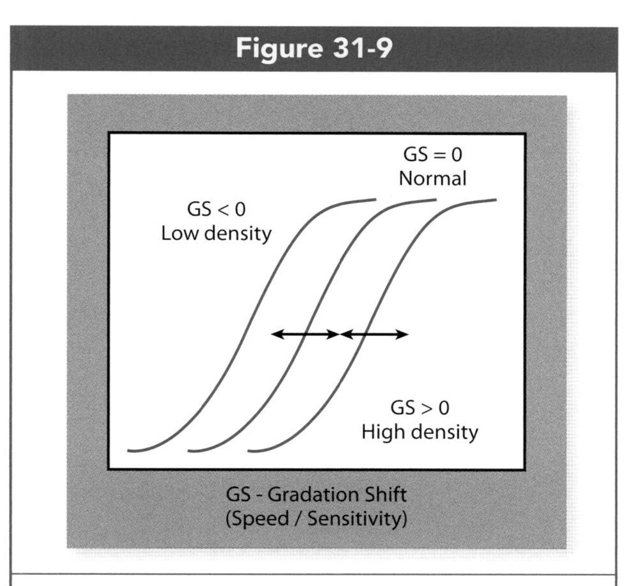

The *gradient shift* (*GS*) simply moves the curve left or right and is equivalent to fine adjustments in the overall brightness control. This shifts the image histogram to the left or right.

Figure 31-10

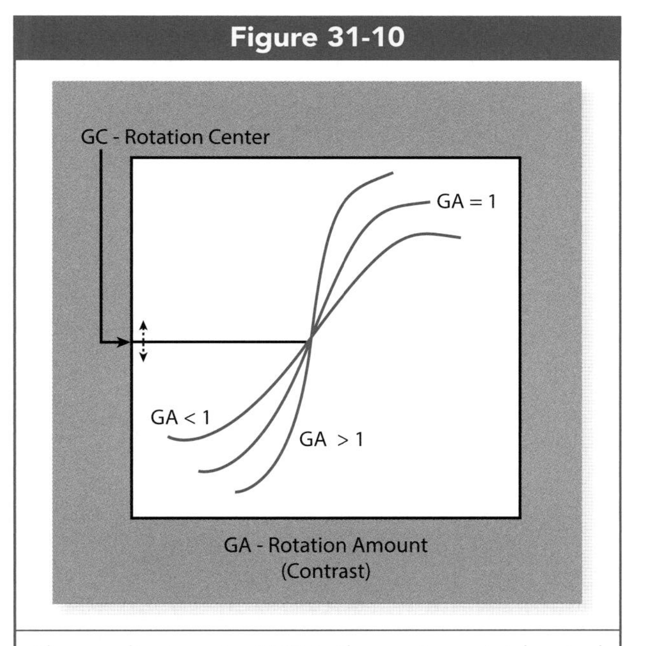

The *gradient center* (*GC*) is the center point located in the *x* and *y* axes around which the *gradient amount (GA) or angle* of the slope of the function curve rotates for the Fuji system. A *GA* less than 1 (positioning the function curve shallower than 45 degrees) lengthens the image gray scale. A *GA* greater than 1 (positioning the curve steeper than 45 degrees) increases image contrast.

Data Clipping

If the dynamic range or *bit depth* of a digital processing system is limited, it is possible for *data clipping* to occur when either brightness or contrast adjustments are made. Figure 31-11 illustrates graphically how this might occur for a system with an 8-bit pixel depth; in ***A***, a brightness increase results in the data

Figure 31-11

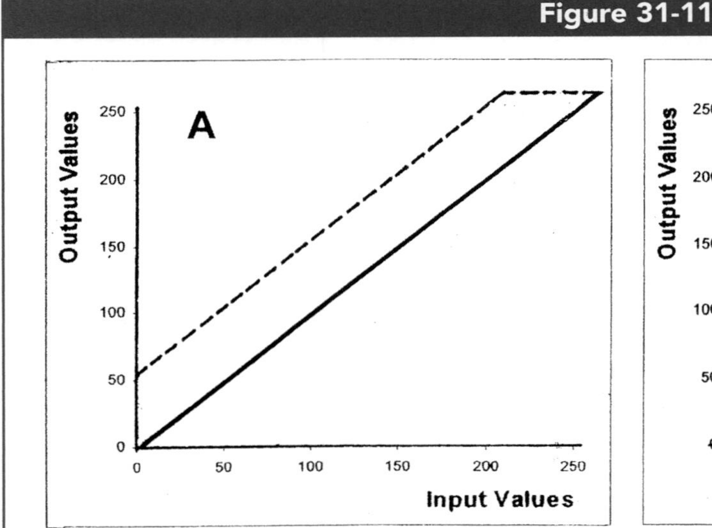

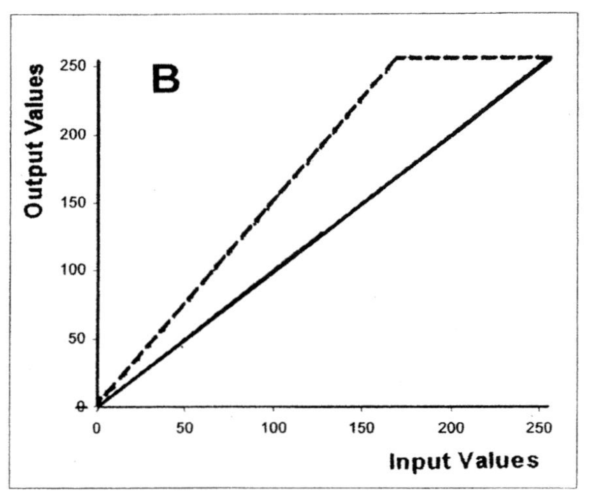

Two examples of data clipping. Using a dynamic range 8 bits deep, only 256 pixel values are possible. In ***A***, increasing the window level (reducing brightness) resulted in the pixel values in the image being truncated at the darkest value of 255 (dotted line). In ***B***, increasing the window width (contrast) resulted in the same effect as the slope of the curve (dotted line) became steeper. (From Quinn B. Carroll, *Practical Radiographic Imaging*, 8th ed. Springfield, IL: Charles C Thomas Publisher, Ltd., 2007. Reprinted by permission.)

curve (dotted line) running into the 256-pixel value limit before the entire curve is plotted. After multiplication, new pixel values that exceed 256 are lost from the image because the computer cannot process values above this number. In ***B*** we see the same effect from an increase in contrast, where the steeper slope of the curve (dotted line) results in its truncation at 256 pixels before the data are all plotted.

The dynamic range of the software, supported by the computer hardware, must extend sufficiently above and below typical input values to allow for all probable adjustments that might be made in diagnostic radiology. By constricting the data set available for an image, data clipping limits the radiologist's ability to window the image and can affect diagnosis.

Dynamic Range Compression (DRC) or Equalization

The *dynamic range* has been defined as the number of different gray levels or brightness levels that can be represented in the displayed digital image. This is the range made available by the entire computer system including its installed software. The range of the computer itself, or any other *hardware device* such as an LCD monitor, is referred to as its *bit depth.* Since the bit depth of a computer can far exceed the range of human vision, computer storage space can be saved by not using the entire bit depth of the system. The use of dynamic range compression for this purpose is shown in the graph in Figure 31-12***A***.

The dynamic range can be compressed to such a degree that its effects become visible in the image as a loss of the lightest light shades and the darkest darks—that is, the most extreme shades are removed from the actual gray scale of the image. This is graphically shown in Figure 31-12***B***. Note that this compressed image can still be darkened or lightened as a whole, or *windowed*, sliding the actual gray scale up or down the actual available dynamic range.

Without postprocessing manipulation of the image, most radiographic techniques designed to demonstrate *bony* anatomy will result in soft tissue areas that are too dark in the image to be of diagnostic value. Conventionally, *soft tissue techniques* were required to produce separate images of these areas, which then depicted the bones too light.

Visibly compressing the dynamic range clips the gray scale curve so that it is shorter on the graph,

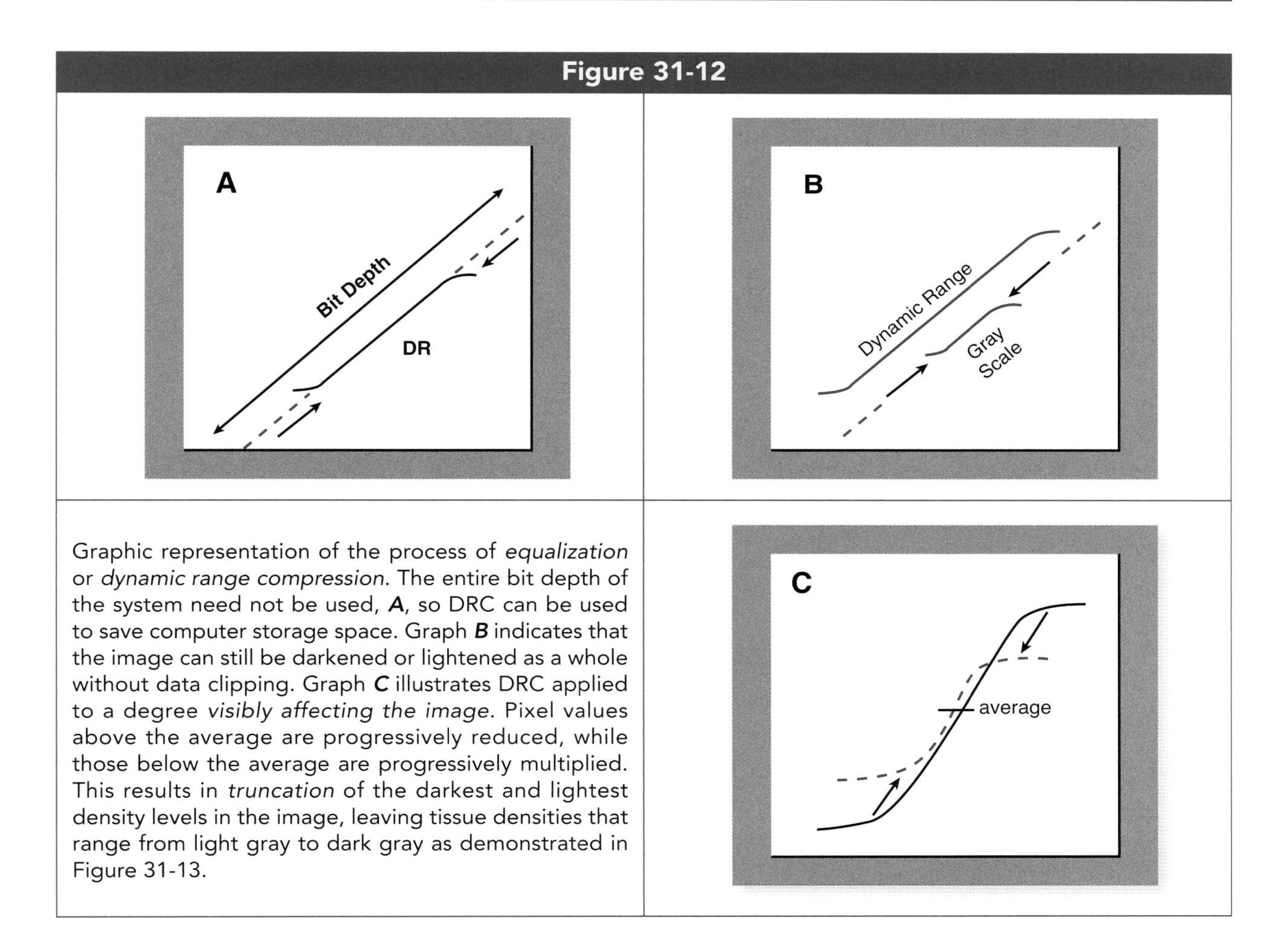

Graphic representation of the process of *equalization* or *dynamic range compression*. The entire bit depth of the system need not be used, ***A***, so DRC can be used to save computer storage space. Graph ***B*** indicates that the image can still be darkened or lightened as a whole without data clipping. Graph **C** illustrates DRC applied to a degree *visibly affecting the image*. Pixel values above the average are progressively reduced, while those below the average are progressively multiplied. This results in *truncation* of the darkest and lightest density levels in the image, leaving tissue densities that range from light gray to dark gray as demonstrated in Figure 31-13.

Figure 31-12***C***. The toe (bottom) of the original curve represents extremely light densities in the image. Compressing the curve brings these extremely light densities up to a darker level. At the same time, extremely dark areas of the image, represented by the shoulder (top) of the curve are made lighter. The result is that more types of tissue can be seen. This compression is done mathematically by the computer; after defining a mid-point average brightness or density level, the computer progressively reduces pixel values above this point, and progressively increases pixel values below it.

Both blank white portions of the image and pitch dark portions are diagnostically useless. While darkening up the light areas can bring some details into better visibility, it must be remembered that *details which are absent from the original image due to underpenetration of the x-ray beam can never be recovered*. On the other hand, DRC is especially useful in bringing out image details that are present within the image but obscured by excessive density. Compression of the gray scale is also known by the term *tissue equalization* (used by GE and others) or *contrast equalization*. Figure 31-13 provides a demonstration of the effectiveness of DRC in improving the image. A significant improvement over conventional radiographs is that soft tissue areas are well demonstrated while maintaining proper density for bone detail.

The concept of dynamic range compression can be very difficult for the radiographer to understand because it seems to contradict the conventional axiom that "shortening of the gray scale increases image contrast." On the contrary, DRC results in a *grayer*-looking image. It may help to think of DRC as gray scale *truncation* rather than compression. Conventional shortening of the gray scale might be described as "squeezing" a set range of densities that

Figure 31-13

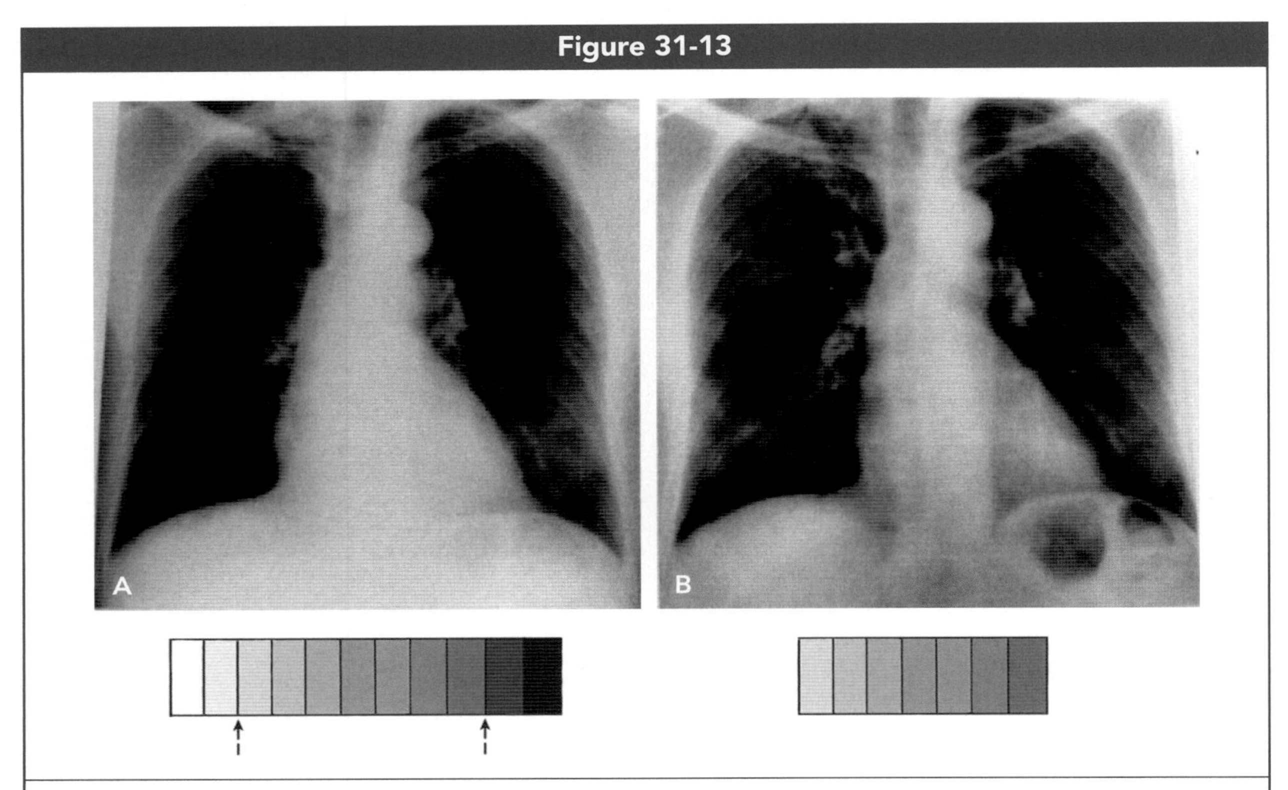

Example of improvement of the radiograph using *Fuji's dynamic range control (DRC)*, a tissue equalization algorithm. In image ***B,*** structures through the heart shadow are now visible due to darkening, while in the lung fields more vascular and bronchial structures are now visible due to lightening. Gray scales under each image illustrate *truncation* of the darkest and lightest densities from the gray scale, leaving a grayer-appearing image. (Courtesy, Fujifilm Medical Systems, Stamford, CT.)

still extends from white to black, whereas DRC *cuts* or *clips off* the ends of the range of densities. This is what is meant by the term *gray scale truncation*—the darkest and lightest densities are cut out and actually lost from the image. The resulting image appears grayer, Figure 31-13, because its entire gray scale only ranges from light gray to dark gray. More tissues can be seen because those that were *excessively dark* have been lightened, while those that were *excessively light* have been darkened.

Fuji's *Dynamic Range Control* is somewhat more involved than simply compressing the gray scale curve; a *DRR* function sets the *range* of the new curve (to which the original will be compressed), a *DRT* function selects one of 20 *types* of software filters to apply to this set of data, and a *DRE* function sets the level of *enhancement* to be applied, or strength of the filter, customizing the final gray scale to the anatomy to be demonstrated.

POSTPROCESSING II: DETAIL PROCESSING

Detail processing is set apart from all other forms of processing by its ability to treat fine details *as a separate component of the image*, without changing the overall contrast or brightness of the image as a whole. For example, the visibility of small details can be enhanced, while the general contrast of the overall image remains relatively unchanged so that the image retains its gray scale for larger structures. This aspect is often referred to as *decoupling local contrast from global contrast*, and it results in operations that seem almost magical in their ability to alter the image. Detail processing can be executed in either the frequency domain or in the spatial domain, and the choice used by a particular manufacturer is based on the final results.

Applying Kernels in the Spatial Domain

A *kernel* is a *submatrix* or smaller "core" matrix that is passed over the larger matrix of the image, executing some mathematical function. In the previous chapter it was shown how a kernel can be used to sum surrounding pixel values to "fill" a dead pixel and thus reduce noise in the image. By placing *multiplication factors* in the cells of a kernel, it can be used to enhance the edges of structures, smooth the "hardness" of those edges, or suppress larger background structures, all forms of detail processing.

In these operations, for each pixel in the image, pixel values in the local neighborhood are multiplied by the values in the kernel. These products are then all summed to form a new pixel value, which is assigned to the pixel over which the kernel is *centered*, as illustrated in Figure 31-14. This process is usually applied starting at the top left corner of the image, after which the kernel is moved one pixel to the right and the operation is repeated using the original pixel values. The kernel indexes left to right, repeating these calculations column by column until the entire image is covered, then the whole process is repeated vertically, row by row.

Different functions can be executed on the image according to the interrelationships between these "core" values in the kernel. Figure 31-15 shows three examples: In ***A***, all core values are equal and greater than zero—this results in *smoothing* of the image, in which noise is reduced with an attended slight loss of edge contrast. The kernel in ***B*** contains both negative and positive values which sum up to 1—this results in edge enhancement, an increase in contrast only at the small detail level. The kernel in ***C*** contains negative and positive values that sum up to zero—this results in background suppression. This is a broad reduction in the contrast of the larger, gross structures in the image that constitute the background, leaving image details in the "foreground" more visible.

(The general, *overall* contrast of the image can also be changed using kernels. Overall contrast is increased when negative and positive core values sum up to a result greater than one, overall contrast is decreased when negative and positive values sum to less than one. If the sum is 1, as in Figure 31-15***B***, the overall contrast of the processed image will be

Figure 31-14

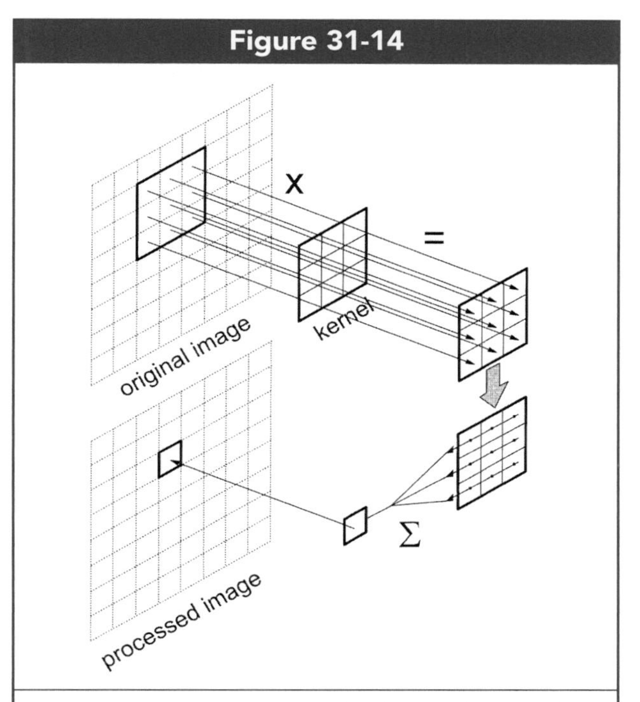

Spatial enhancement using a 3 × 3 *kernel*. Pixel values from the original image are multiplied by preset values in the kernel, then summed to reassign the value for the center pixel in the region. (From Bushberg et al., *The Essential Physics of Medical Imaging,* 2nd ed., Baltimore: Williams & Wilkins. Reproduced by permission.)

Figure 31-15

A

0.5	0.5	0.5
0.5	0.5	0.5
0.5	0.5	0.5

B

-1	-1	-1
-1	9	-1
-1	-1	-1

C

-1	-1	-1
-1	8	-1
-1	-1	-1

Examples of types of kernels: ***A***, with all cells holding the same value, results in noise reduction or *smoothing* of the image. ***B***, whose cells sum to 1, results in *edge enhancement*, and ***C***, which sums to zero, results in suppression of the background.

similar to that of the original image. Although general image contrast is better controlled by dynamic range control processing using look-up tables, these examples show that kernels can be used for a variety of applications.)

Unsharp Mask Filtering

Unsharp mask filtering, or unsharp masking, is a form of edge enhancement that is best performed in the spatial domain using kernels. It normally consists of 3 key steps. In step one, an "unsharp mask" is created that contains only the larger, gross structures from the original image. This is accomplished by passing a kernel over the original image, which has the effect of averaging local pixel values. Structures smaller than the kernel size are suppressed and no longer visible. Figure 31-16 demonstrates the result of passing the kernel over the original image—an "unsharp mask" containing only gross structures in the image, without any fine details present.

Note that the unsharp mask is not truly blurred in any geometrical sense. Rather it *appears* blurry or *unsharp*, because the finer details have been removed through averaging, and only larger, gross structures, such as broad tissue areas, remain. In fact, it would be more accurate to refer to unsharp masking as "gross structure masking," but this is a cumbersome term. The larger the *size* of the kernel matrix, and the *wider* the region it uses for averaging, the more "blurred" the mask image appears.

In the second step, shown in Figure 31-17, a *positive* of the unsharp mask is createdby image reversal. Image reversal is an intensity operation in which the scale of pixel values throughout the image is mathematically inverted such that the highest numbers are re-mapped as the lowest numbers, numbers slightly higher than the mid-range are re-mapped slightly lower than the mid-range, and so on. The result is that the darkest densities become the lightest densities while densities slightly darker than the mid-range become slightly lighter than the mid-range, and so forth.

In step 3 this positive mask is subtracted from the original image. Only structures present on both the original image and the mask image are subtracted.

Figure 31-16

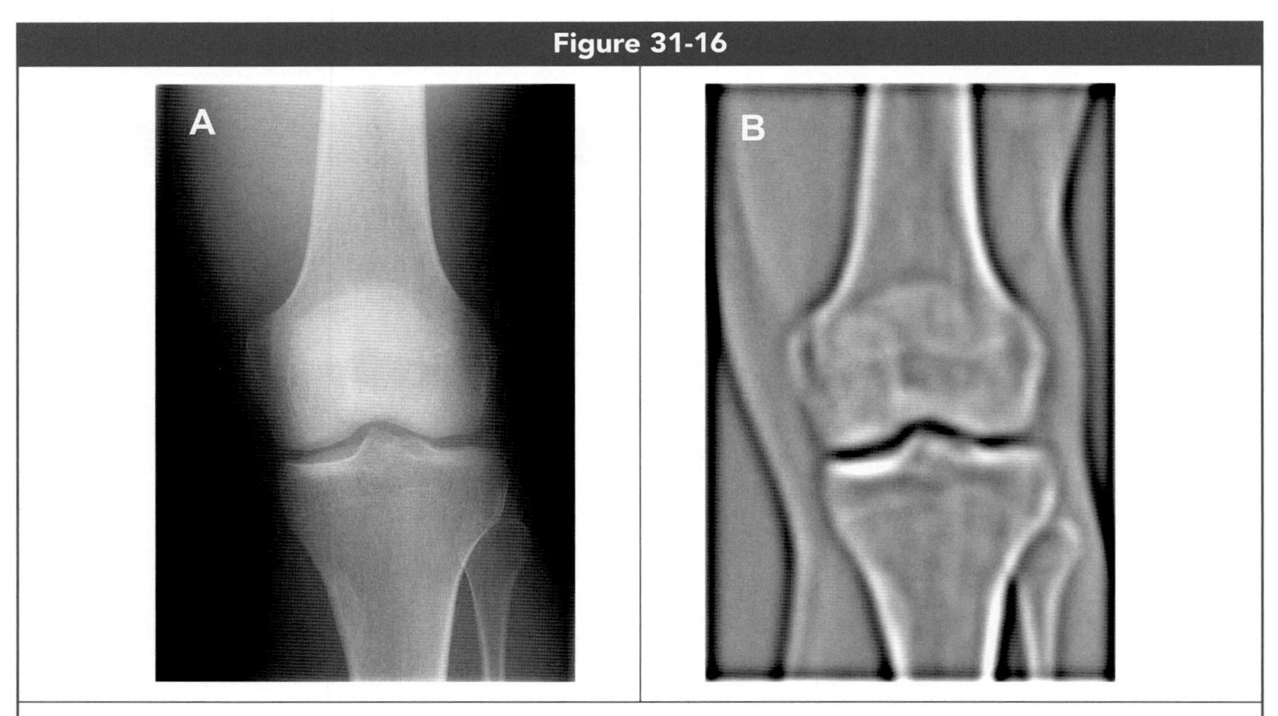

In step 1 of unsharp mask filtering, a large kernel is passed over the original image, ***A***, which removes the smaller, fine details, leaving only the larger, gross structures in image component ***B***, which appears more "blurry." (Courtesy, Philips Healthcare, Bothell, WA.)

Figure 31-17

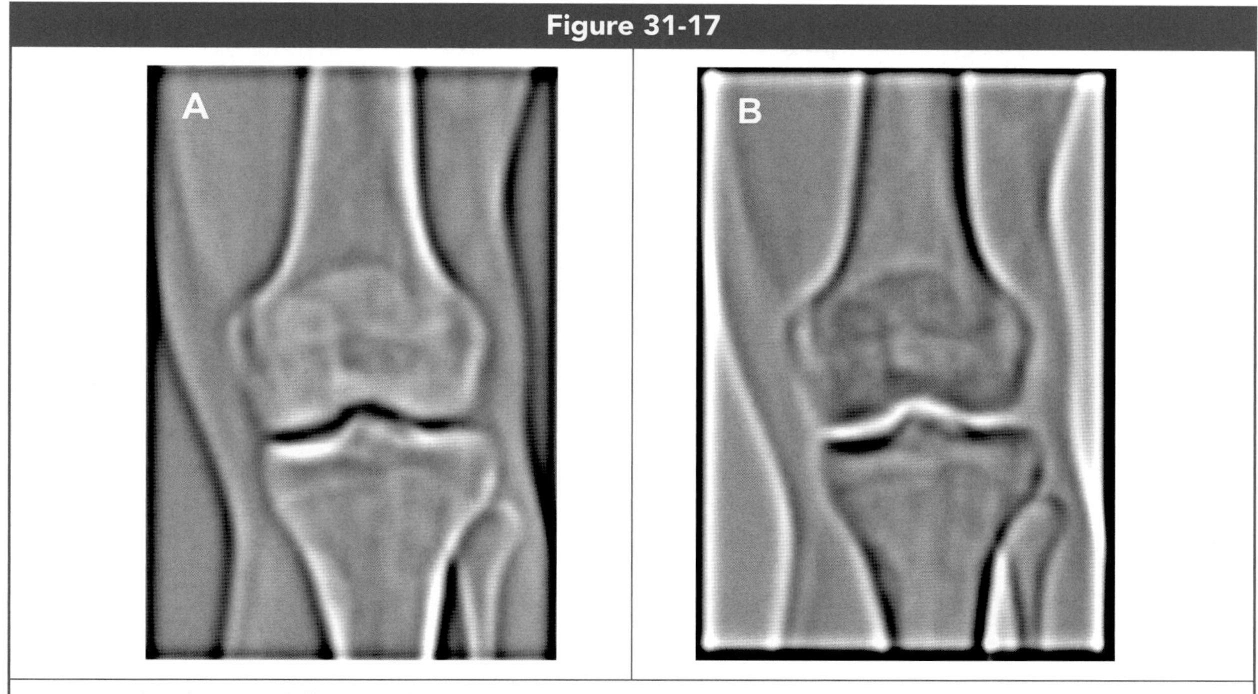

In step 2 of unsharp mask filtering, the gross-structure component of the image (***A***) is reversed to create a *positive* mask (***B***). (Courtesy, Philips Healthcare, Bothell, WA.)

For example, Figure 31-18 illustrates the older, conventional method of subtraction using films with different chemical characteristics. The reversed film, B, was laid over the original to the left and a new film was exposed to light shining through the 2 overlaid films. Structures present on both films would be cancelled out, dark over light, to form nebulous gray shades.

However, note that on the original there was an iodinated contrast agent introduced into the arteries of the brain (shown by the arrow) and that this is not present on the reversed mask, because the mask image was created from an exposure taken before the iodine was injected. Thus, when the mask was subtracted from the original, the iodinated arteries are not subtracted out with the bones and tissues.

The same rule applies to these digital images during unsharp masking, Figure 31-19; the mask image created contains only the larger, gross structures in the image. These are present on both images, so when the mask is effectively superimposed over the original, only large, gross structures will be cancelled out, leaving mid-size and small structures in the final image.

The final effect is to suppress the visibility of gross structures, thus leaving the fine details more visible. Taken as a whole, the process of unsharp mask subtraction results in an *edge-enhanced* image. "Before" and "after" images in Figure 31-20 demonstrate the effectiveness of unsharp masking. A moderate improvement in the visibility of small details can be seen in the image at the right.

The "detail" image left over from the kernel filtering (Fig. 31-19***C***) can also be utilized to produce the final image—different combinations of the original image, the "blurred" or gross-structure image, and the detail image can be combined in a weighted fashion. For the processing parameters listed by Fuji in Table 31-3, the weighting factor is designated as *RE, the "enhancement factor,"* and typically ranges from zero to 1.0 On this particular scale, enhancement factors above 2.0 are likely to over-enhance the image such that the level of noise is increased to unacceptable degree.

Kernels can be used to target a specific threshold *size* of detail to be eliminated throughout the image. This is controlled by the size of the kernel in pixels. The larger the kernel (the more pixels in it),

Figure 31-18

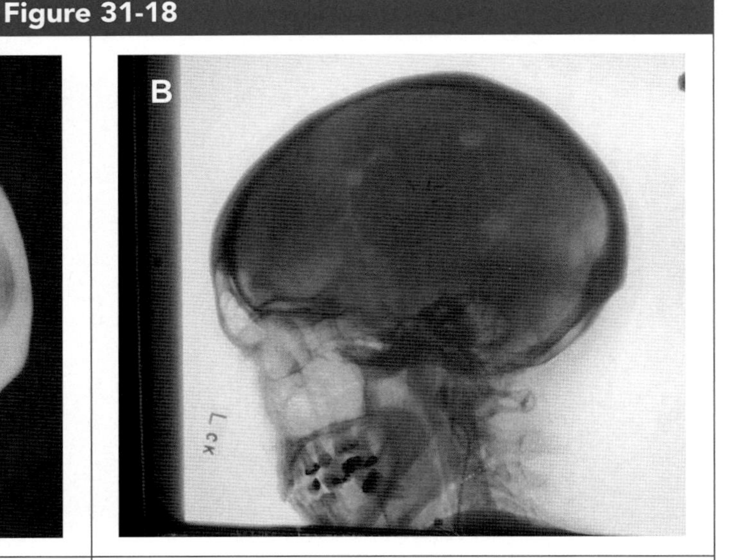

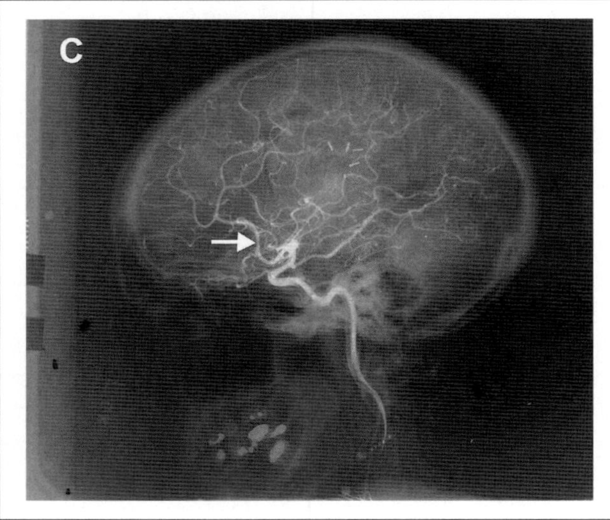

Conventional method of subtraction using films. The reversed film, ***B***, was laid over the original (***A***) to the left and a new film (***C***) was exposed to light shining through the 2 overlaid films. Bones and tissues present on both films would be cancelled out, dark over light, into a gray shade. However, iodinated contrast agent in the arteries of the brain (arrow) was *not* subtracted out, because it was not present on the reversed mask, only on the original. (From *Understanding Digital Radiograph Processing*, Midland, TX: Digital Imaging Consultants, 2013. Reprinted by permission.)

the larger the structures suppressed. The kernel size used must be carefully selected and matched to the type of anatomy being radiographed, since too *small* a kernel may remove details from the image such that diagnostic information is lost.

It is tempting to overuse edge enhancement features or even to apply them automatically to all images that come up for review. But it should be kept in mind that edge-enhancing algorithms also enhance *noise* levels in the image, to such an extent that only a 30 per-cent reduction in exposure can result in unacceptable image mottle. A sharp eye should be kept out for this. As edge enhancement increases *local* contrast of smaller details, the edges of details may acquire a "halo effect," in which the darker density side is further darkened while the lighter side is lightened up more. This is also a form of image noise. Edge enhancement should not be used blindly, but with careful evaluation of each study.

Using Kernels for Noise Reduction and Smoothing

Just opposite to the effects of unsharp masking are the effects of *smoothing* on the image. From ***A*** to ***B*** in Figure 31-21, CT images of the elbow demonstrate how applying a smoothing function can reduce noise in the image, pointed out by the arrows. Since

Figure 31-19

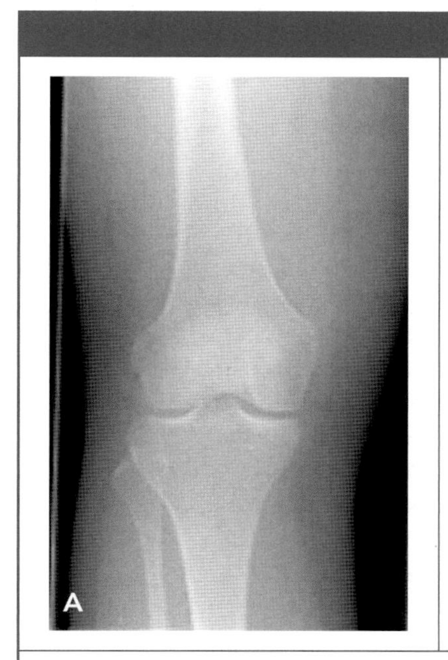

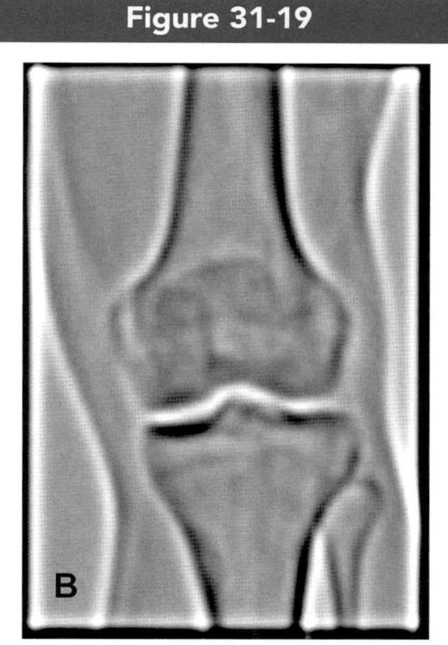

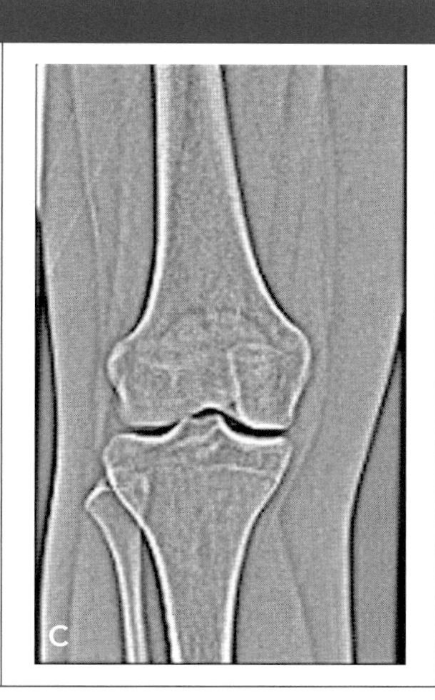

In the final step of unsharp mask filtering, the positive mask (***B***) is subtracted from the original image (***A***). (The remaining image component, ***C***, forms a "high-detail" image which can be used for additional operations.) (Courtesy, Philips Healthcare, Bothell, WA.)

most types of noise consist of very small artifacts, smoothing can be used to effectively reduce noise in the image, and also to soften the edges of details.

However, *in an image that already has long gray scale or low contrast, applying a smoothing function can lead to a loss of details in the image.*

Figure 31-20

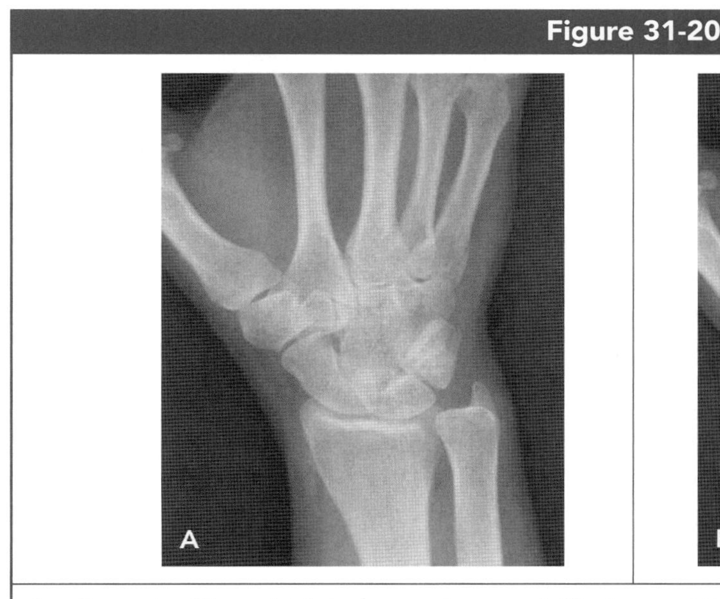

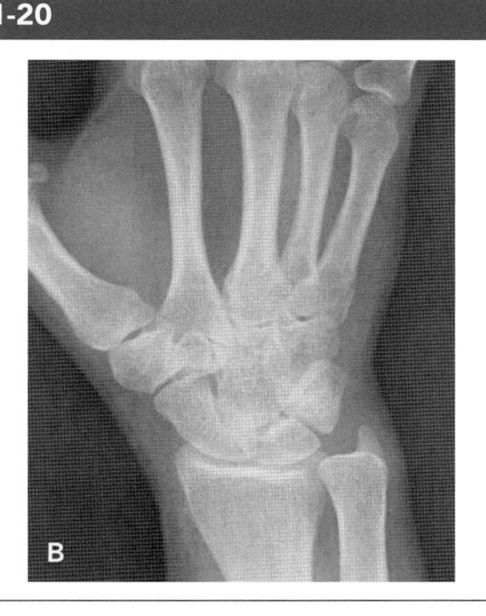

A radiograph of the wrist ***A***, before unsharp mask filtering, and ***B***, after unsharp mask filtering which demonstrates increased local contrast around fine details due to the subtraction of the blurrier components of the image. (Courtesy, Jason Swopes, R.T.)

Table 31-3

Fuji* Parameters for Frequency Processing

Anatomical Region	Frequency Enhancement Parameters		
	RN	RT	RE
General chest (Lat)	4.0	R	0.2
General chest (PA)	4.0	R	0.2
Portable chest (Grid)	4.0	T	0.2
Portable chest (No Grid)	4.0	R	0.5
Pediatric chest (NICU/PICU)	3.0	R	0.5
Finger	5.0	T	0.5
Wrist	5.0	T	0.5
Forearm	5.0	T	0.5
Plaster cast (arm)	5.0	T	0.5
Elbow	7.0	T	1.0
Upper Ribs	5.0		1.0
Pelvis	6.0	T	1.0
Pelvis Portable	4.0	T	0.5
Tibia / Fibula	5.0	F	0.5
Foot	5.0	T	0.5
Foot	7.0	T	0.5
Os calcis	5.0	F	1.0
Foot cast	5.0	F	0.5
C-spine	5.0	P	0.5
T-spine	4.0	T	0.2
Swimmers	5.0	T	0.5
Lumbar spine	5.0	T	1.0
Breast specimen	9.0	P	1.0

*Courtesy, Fujimed, Inc.

Both kernel operations (in the spatial domain) and frequency operations can be used to suppress noise, but kernels are particularly well-suited for suppressing *random* noise (whereas frequency processing works better for suppressing periodic noise such as electronic noise). One of the best examples of *random* noise is quantum mottle. Since quantum mottle appears as specks of various sizes and occurs in an irregular pattern in the image, frequency filtering is not as effective as kernels in this case.

Understanding the Frequency Domain

Breaking an image into its *spatial* components, which are its pixel locations, or its *intensity* components, which are its pixel values, are both intuitive to us and easy to understand. Frequency processing breaks the image into its *wave* components. This is not intuitive and will likely require some effort for the student to grasp.

In this connection, it is interesting to recall from Chapter 5 that, at the subatomic level, what we think of as particles also have *wave functions* and share the behavior of waves. In a similar way, we can think of an image as a collection of pixels in space, *or* as a collection of *waves with different frequencies.* To understand this concept, it is useful to refer back to the *density trace diagrams* used in Chapter 25 to illustrate geometrical penumbra and absorption penumbra in the image. See in particular Figure 25-7. It is an easy leap of imagination to convert a density trace diagram into *graph* of the image, in which higher points represent darker densities and lower points represent lighter densities. With no penumbra or blur at all present in the image, or with large pixels, such a graph would appear *squared* as in Figure 31-22***A***. *Either* by mathematically averaging the values in this graph, or by adding penumbral blur into the image (which all images have to some degree) and making the pixel size small enough, the effect on the plotted curve would be to round it such that it begins to appear more and more like a sine-wave graph, Figure 31-22***B***.

Note that in Figure 31-22, as we move from graph ***A*** to graph ***B***, the brightness, "density" or gray level of each pixel becomes the *amplitude* (height) of the sine waves, and the number of pixels occupied by each object in the image becomes the *frequency* (shortness in length) of the waves.

Now, imagine Figure 31-22 as representing a *single row* of pixels in an image. As we scan across the graph from left to right, we begin to surmise that *shorter waves represent smaller details, and longer waves represent larger objects* within the image. Recall from our discussion of waves in Chapter 5 that shorter wavelengths also represent higher frequencies. We conclude that, in any image, all of the extremely small details can be *collectively* represented as those parts of the image possessing very short wavelengths and very high frequencies. Mid-sized objects are collectively represented by middle-range frequencies. Very large objects in the image may be collectively described as portions of the image with long wavelengths or low frequencies.

Figure 31-21

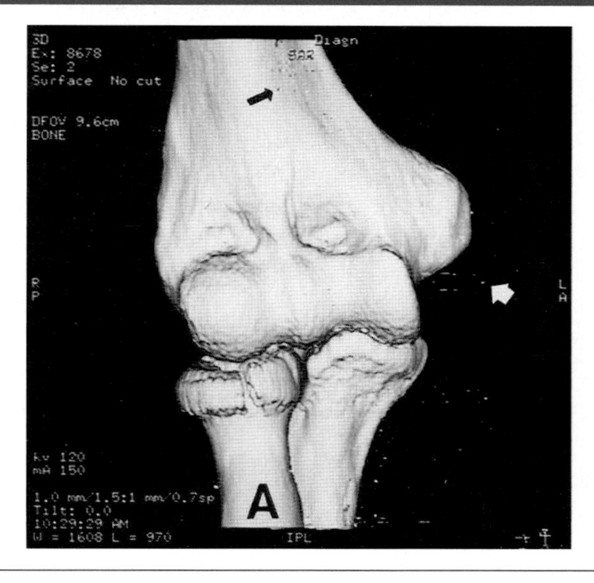

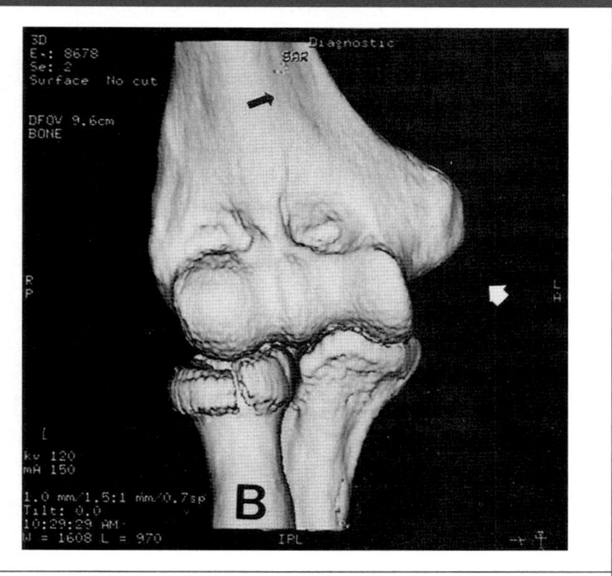

Smoothing from low-pass filtering of a 3D CT image of the elbow. Note that in the high-contrast image, ***A***, high-frequency noise appears in both the background (white arrow) and over the anatomy (black arrow). A smoothing algorithm was used to remove these artifacts in ***B***. (Courtesy, Jason Swopes, R.T.)

Figure 31-22

A pixel-by-pixel graph of one row in a digital image would appear much like a density trace diagram (Chapter 25), but with squared steps representing discrete gray level values, ***A***. By mathematical averaging, or by adding image blur and reducing the size of the pixels in the graph, we obtain a sine-wave type graph, ***B***. Thus, each row of the image can be treated as a series of wave components or frequencies rather than a row of spatial pixels. *Note that the amplitude (height) of the sine waves represents the gray level or brightness of the pixels, while the frequency or wavelengths correspond to the size of the image detail, or how many pixels it occupies.*

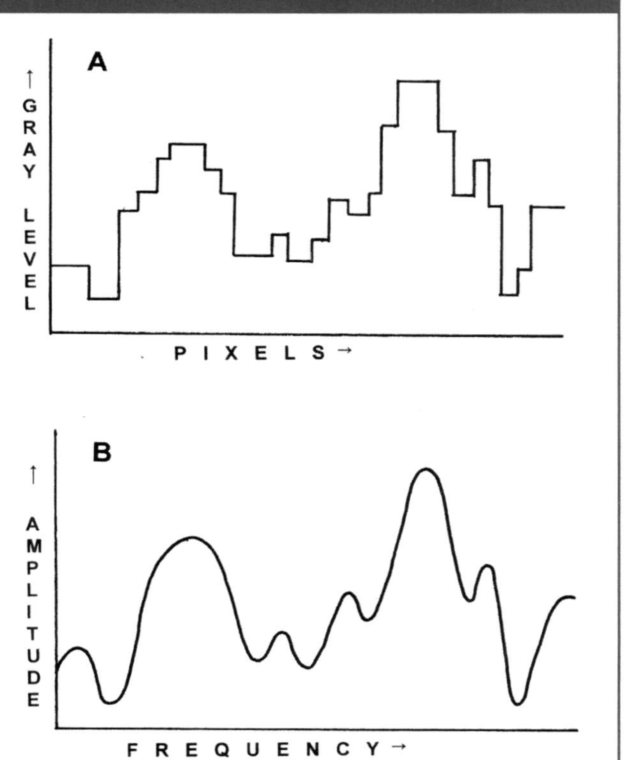

To concisely summarize how object size is translated into frequencies, *large objects produce long waves,* and so are classified as low frequency objects. *Small objects produce short waves,* so they are classified as high-frequency objects. Frequency processing breaks the image down into sine-wave functions in order to manipulate them mathematically. When this is completed, these sine-waves are then re-assembled back into an image.

Processing in the Frequency Domain

Although frequency processing is more difficult to understand than spatial location processing, MRI, which is based on radio waves, provides an everyday example of frequency or *wavelength* processing.

Figure 31-23 illustrates how any particular *line* or row of pixels within an image can be sampled such that the values of the pixels across its length form a sine-wave *graph.* The simplest scenario would be for these pixels to consist of alternating black and white densities. This pattern can be represented as a mathematical function resulting in a sine-wave, ***A*** in Figure 31-23, in which the peaks represent a pitch black density and the troughs represent blank white. If the pattern were alternating *gray* and white densities instead of black and white, ***B***, the *amplitude* or height of the waves would be *shorter*; however, the *frequency* of these waves in both ***A*** and ***B*** is equal, because their *wavelength* is equal. (See Chapter 5 for the relationship between frequency and wavelength.)

Figure31-23

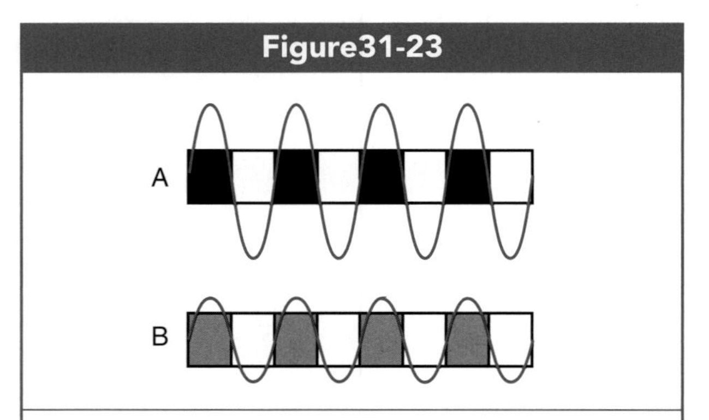

The wave functions for pixels of two alternating values in *one row* of an image matrix. ***A*** is a high-amplitude wave representing black and white pixels; ***B*** is a wave with lower amplitude to represent alternating *gray* and white pixels.

Upon closer examination of Figure 31-23, we see that the zero-point of the *wavelength* crosses through the transition border between each pair of pixels. Therefore, the wavelength is representative of the *size* (specifically the *width*) of the pixels. (This is also equivalent to the pixel *pitch*, which is the distance from the center of each pixel to the center of the next pixel.) The shorter the wavelengths, the smaller the pixels. Remember from Chapter 5 that frequency and wavelength are inversely proportional, so shorter wavelengths represent higher frequencies. In other words, the smaller the pixels, the more pixels per line in the image. High frequency also means there are fewer pixels *per detail* in the image.

$$\text{Frequency} = \frac{1}{\#\text{ pixels per detail}}$$

Fewer pixels per detail translates into higher *sharpness* at detail edges, or a "harder" edge.

$$\begin{aligned}\text{High Frequency} &= \text{Fewer Pixels Per Detail}\\ &= \text{High Sharpness of Detail}\end{aligned}$$

In Figure 31-24 we depict the *highest* and the *lowest* possible frequencies for a display monitor screen that is ten pixels across. Remember that the unit for frequency, the *hertz*, represents one full *cycle* per second. In this case, however, one hertz represents one full cycle *per matrix width*, or one cycle completed across the entire display screen. A full cycle must consist of a positive pulse and a negative pulse, here representing a *pair* of densities, one black and one white, after which the next cycle starts over. To simplify this discussion, each *density* pair will represent one detail (or the *edge* of a structure). The lowest possible frequency for an image matrix that is 10 pixels wide would be one cycle, consisting of a pair of densities, one black and one white, with each density being 5 pixels wide, ***A*** in Figure 31-24. For this image, at a frequency of 1 hertz, there are 5 pixels per density, or 10 pixels per detail. Only one detail can be imaged across the width of the matrix.

The highest possible frequency for a 10-pixel image is shown in ***B*** in Figure 31-24. Here, the smallest number of pixels to record a detail (or *edge* of a structure) is 2, one black and one white. We can fit no more than 5 full cycles across the width of the image. The frequency is 5 hertz. (Five details can be imaged.)

The graph in Figure 31-25*A* represents a single row (or column) in a real image. We see that this is a complex wave, not only with different *heights* of waves representing different gray levels, but also with different *widths* of waves representing larger details that take up more than one pair of pixels and smaller details that require only a single pair of pixels to record.

Fourier transformation is a mathematical process which allows a complex waveform such as that in Figure 31-25 to be broken down into the individual sine-waves (or functions) that make it up. A complex wave such as ***A*** in Figure 31-25 may be constructed by adding individual waves of different frequencies, ***B through D***, together. In places where the *peaks* of the shorter waves coincide with those of the longer wave, an amplified *spike* in the resulting complex wave will be seen. In other places, the peaks of the shorter waves coincide with *troughs* of the larger waves, having the effect of *cancelling out* these pulses, and causing a negative "dip" in the pulse seen in the resulting complex wave.

Let us assume that, having performed Fourier analysis on the image represented by ***A*** in Figure 31-25, we find that it is composed of only three distinct frequencies. As illustrated, these are frequencies at 1.25 hertz, 2.5 hertz, and 5 hertz. The *large objects* in line ***B*** consume 4 pixel-pairs each. Their frequency, noted at the left, is 1.25 Hertz—this is simply 4 pixel pairs, or 8 pixels, divided into the 10 available pixels for this row. In line ***C***, mid-sized structures require 2 pixel-pairs each, and therefore have a frequency of 2.5 Hz (10 divided by 4). Line ***D*** represents very small details that only consume 1 pixel-pair each, so their frequency is 5 Hz (10/2). The smaller the object or detail, the higher the frequency associated with it.

Figure 31-24

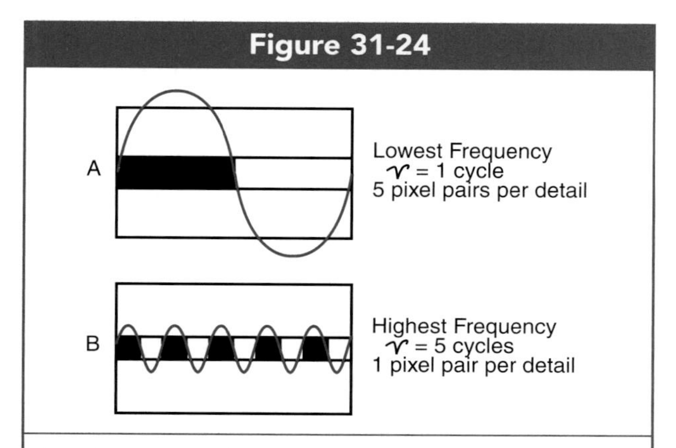

The lowest possible frequency in an image would be one cycle across its width, ***A***. This represents one detail edge, which requires a *pair* of densities to demonstrate. If 10 pixels are available in the row, each density consumes 5 pixels. The highest possible frequency for a row of 10 pixels is 5 cycles, shown in ***B***, with each cycle or each detail consuming two pixels.

Figure 31-25

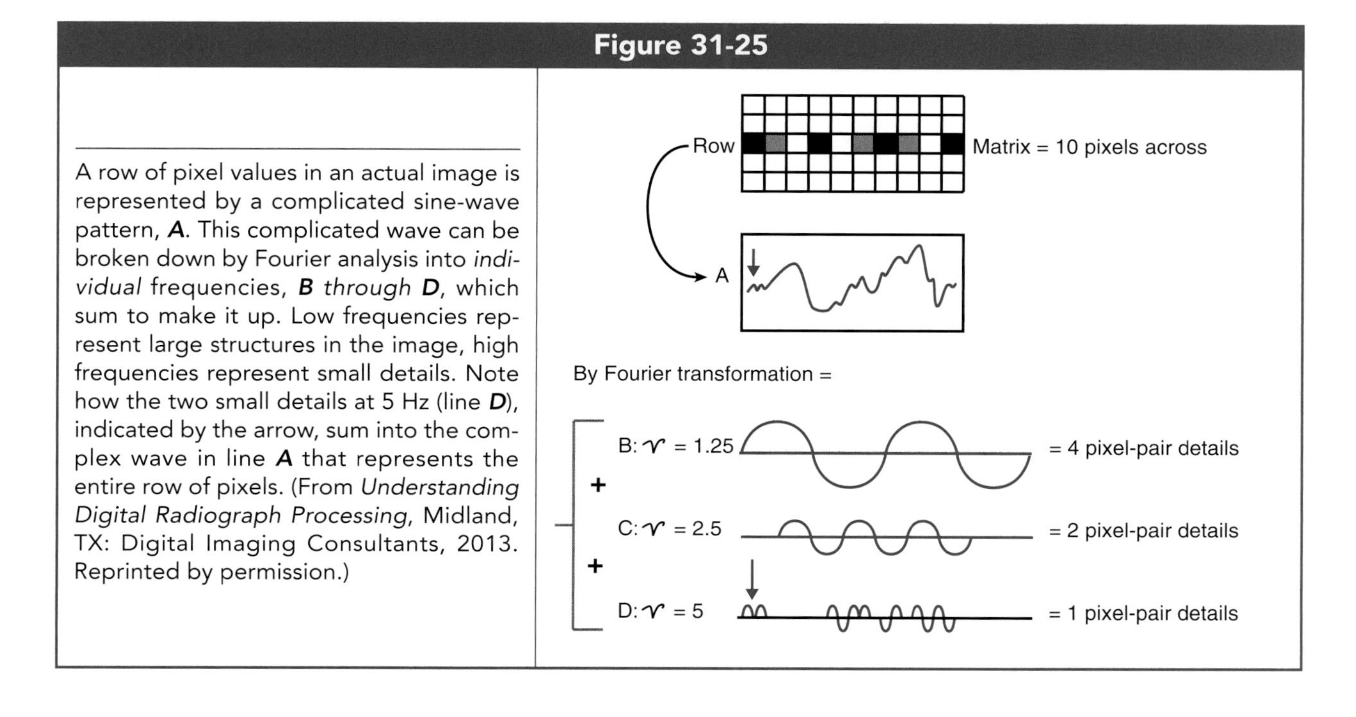

A row of pixel values in an actual image is represented by a complicated sine-wave pattern, ***A***. This complicated wave can be broken down by Fourier analysis into *individual* frequencies, ***B through D***, which sum to make it up. Low frequencies represent large structures in the image, high frequencies represent small details. Note how the two small details at 5 Hz (line ***D***), indicated by the arrow, sum into the complex wave in line ***A*** that represents the entire row of pixels. (From *Understanding Digital Radiograph Processing*, Midland, TX: Digital Imaging Consultants, 2013. Reprinted by permission.)

Figure 31-26 displays the actual "layers" of a hand image split into eight separate frequencies. Each image represents a separate computer file which holds only details at a particular frequency (or wavelength) that has been separated out by Fourier transformation. At the highest frequencies (B, C and D), we see only the smallest details from the image, such as trabecular markings in the bone marrow. At mid-frequencies (E and F), the thicker cortical bone margins and similar-sized structures can be made out. At very low frequencies (G, H and I), all bone detail is lost, but the gross soft tissue structures of the hand are now demonstrated. Note that the background density of the image only shows up in the last three layers (G, H and I). Because it covers a large portion of the field area, it is treated as a large "structure" by Fourier transformation.

While these image "layers" are separated, different operations (such as gradation processing) can be executed *on any individual layer*. These will be described in the next section on multiscale processing. Graphs such as the ones in Figures 31-25 and 31-27 help us to visualize what Fourier transformation does with the image, but remember that the actual processes are all *mathematical*, not graphical, in nature.

Once all the layers of the image have been treated for a particular result, a mathematical process called *inverse Fourier transformation* effectively adds all of the separated wave forms from their different computer files back together, Figure 31-27, to form a complex wave. Such a reconstituted wave-form will represent each row of the re-assembled image. On the display monitor, the voltage applied to each row of pixels is controlled by the peaks and valleys, highs and lows of this wave, as shown in Figure 31-28. The end result is a varying *brightness* level for each pixel displayed, which constitutes the final image.

Figure 31-26

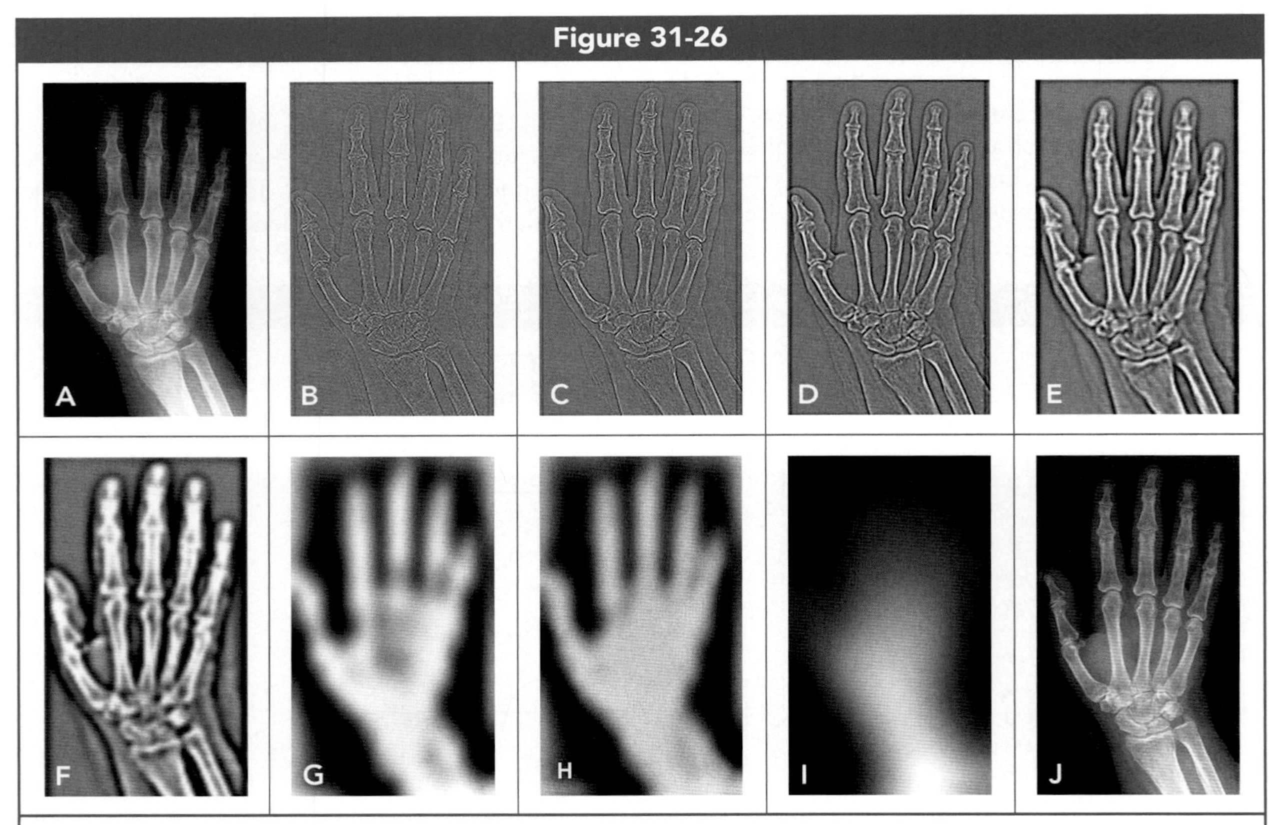

Decomposition of an original image of the hand, **A**, into eight separate bands of decreasing frequency, ***B through I***, and the final composite image which has been edge enhanced, ***J***. (Courtesy Philips Healthcare, Bothell, WA).

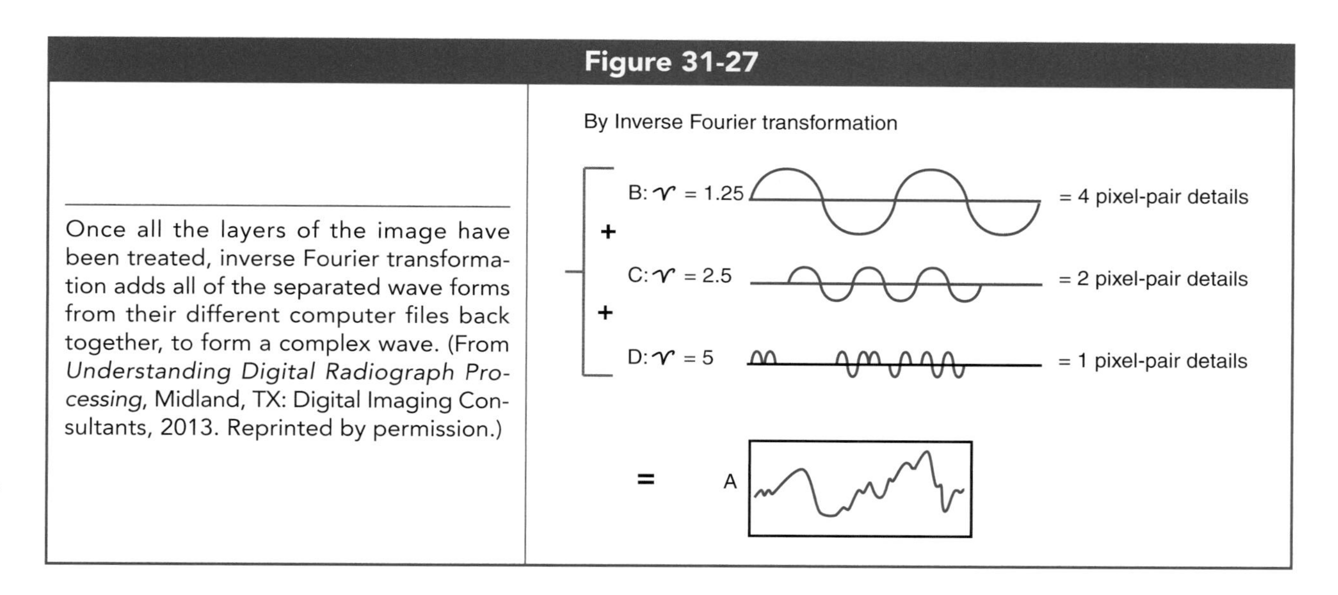

Once all the layers of the image have been treated, inverse Fourier transformation adds all of the separated wave forms from their different computer files back together, to form a complex wave. (From *Understanding Digital Radiograph Processing*, Midland, TX: Digital Imaging Consultants, 2013. Reprinted by permission.)

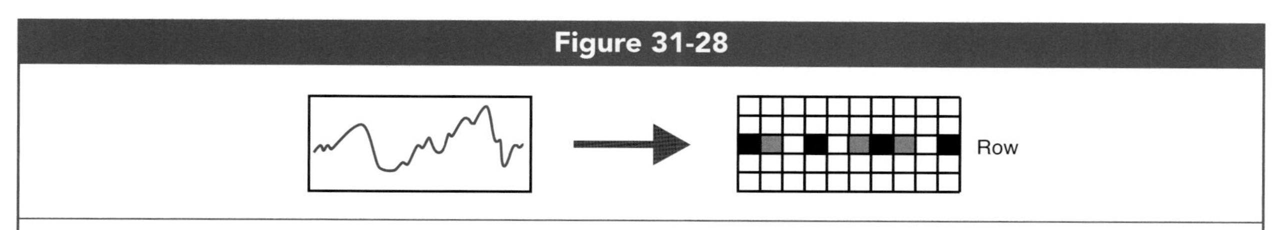

On the display monitor, the voltage applied to each row of pixels is controlled by the peaks and valleys, highs and lows of the complex wave form. The end result is a varying *brightness* level for each pixel displayed, which constitutes one row of the final image. (From *Understanding Digital Radiograph Processing*, Midland, TX: Digital Imaging Consultants, 2013. Reprinted by permission.)

Multiscale Processing and Band-Pass Filtering

Multiscale processing decomposes the original image into eight or more separate frequency bands, each of which contains only information from a particular structural size. Each of these sub-bands can be separately subjected to different filtering methods and parameters before they are all added back together to form the final image. This yields a wide variety of processing options.

Multiscale processing was pioneered by Philips Healthcare as *UNIQUE (Unified Image Quality Enhancement)*, and by Agfa as *MUSICA (Multi-Scale Image Contrast Amplification)*. Fuji followed with *MFP (Multi-Objective Frequency Processing)*, as did several other manufacturers under various acronyms.

To decompose the original image into multiple frequency bands, it is repeatedly split into a high-frequency component and a low-frequency component. The high-frequency image is set aside while the low-frequency image is subjected to the next division. This creates a kind of "3D" stack of image layers, Figure 31-29. Going downward in the stack, each layer is at a lower frequency and therefore is more blurry.

Note in Figure 31-29 that the bottom images in the stack are depicted as having a smaller matrix size. This is because the last images to be split have the lowest frequencies, which represent only very large structures from the original image. The size of the original matrix is simply not needed to record these very large structures, so it is a waste of computer storage space to maintain a large image matrix with thousands of pixels to record a blurry image of only gross structures. Later, when the layers are reconstructed, these smaller images must all be placed into a larger matrix matching the original, by interpolation, so that they all "line up."

Figure 31-29

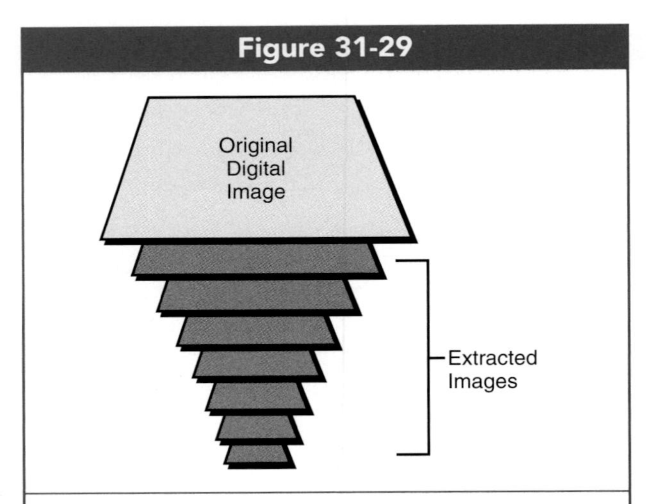

Multiscale processing uses *pyramidal decomposition* to break the original image down into a "stack" of images at different frequency bands. Each image can then be subjected to a variety of processing procedures before summing them to reconstruct the output image. Each descending image in the stack is at lower-frequency, requiring fewer pixels in its matrix. It may therefore be considered as a "smaller" image file which consumes less computer memory.

To help visualize these images, note that the concept is often used when downloading images over the internet with a slow computer system. First, a blurry, blocky version of the image appears because low-resolution images are smaller in data storage size and transmit faster. This image is followed by successively sharper detail layers until the final image is built up. The blocky image is the lowest-frequency image in the stack, and the last layer to be added is the highest-frequency image with the smallest details present. Decomposition of the digital image is just this process *in reverse*. It is referred to as *pyramidal decomposition* because the smaller and smaller files create an inverse *pyramid* shape to the stack as shown in Figure 31-29.

While the image is separated into its various frequency bands, Figure 31-26, we can literally "target" the *size of structures* we wish to enhance or suppress. The amplitude of the signal for any specific layer can be *boosted* (Figure 31-30), enhancing the visibility of low-contrast details at that frequency relative to the other frequency layers. If the gross "background" anatomical structures of an image are impeding the visibility of smaller details, low-frequency bands can also be *suppressed* before adding them back into the image. Note that *each* of the decomposed image layers can also be separately targeted for additional gradient processing, or the entire layer may be left out upon image reconstruction. To summarize, *any* single layer of the image can be:

1. boosted (amplified)
2. suppressed (de-amplified)
3. removed from the stack upon recomposition
4. subjected to gradient processing, noise reduction, or other treatments

Figure 31-30

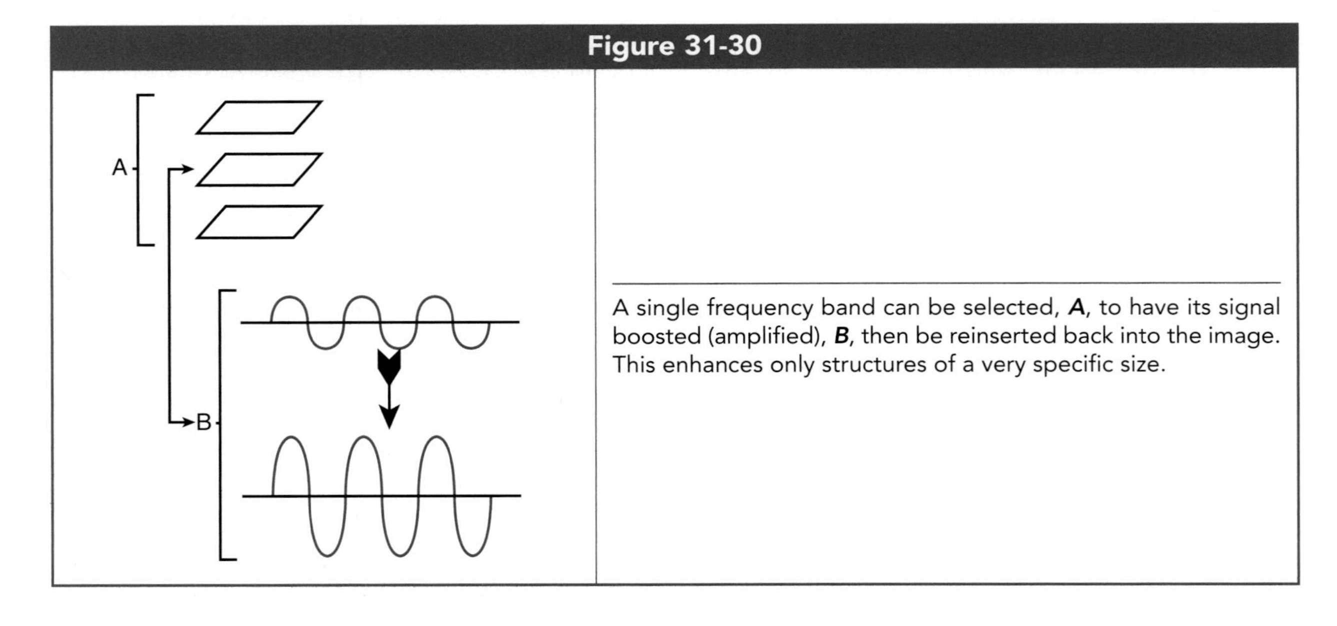

A single frequency band can be selected, **A**, to have its signal boosted (amplified), **B**, then be reinserted back into the image. This enhances only structures of a very specific size.

Many combinations and complicated series of processing operations can be imagined, which drives competition among manufacturers.

Suppose there is a white spot in the middle of a row of pixels, consisting of two adjacent white pixels, which represents image *noise* due to del drop-out. We desire to filter out this noise, this white spot, in the image. The frequency processing method that will accomplish this is called *low-pass filtering*.

The *low-pass filtering* function tells the computer to *keep the low frequencies in the image*. That is, *high frequencies will be filtered out, low frequencies will be allowed to pass*. The computer selects the highest-frequency band, band ***D*** in Figure 31-27, which has been separated by the Fourier transform, and *does not add this band back in when the image is reconstructed*. Small details taking up only one pixel-pair each (including the white noise spot) will not be demonstrated in this image. For radiographers, this function at the console is usually referred to as *smoothing*. An example was provided in Figure 31-21 using a CT scan of the elbow.

The price we pay for this is that the resulting image will also appear more *blurry*. Only the larger 2-pixel-pair and 4-pixel-pair details will be visible, and fine edges will no longer be present. The elimination of image noise requires that we accept some loss of detail along with it.

High-pass filtering allows high frequencies to pass through to the final image, but filters out the *low frequencies, or longer wavelengths*, that represent larger-detail areas of the image such as background densities. In Figure 31-27, waveform ***B*** would be filtered out. The result is that the fine edges of structures stand out better against background densities. At the console, this function is typically labeled as *edge enhancement*. An example was provided in Figure 31-20 (hand).

After this filtering process is completed, an *inverse Fourier transform* is used to re-assemble the *remaining* component wavelengths back into a complex wave that will form each line of the final image.

The method used by Carestream Health (Kodak), called *enhanced visualization processing (EVP)* is similar to steps #3 and #4 in MUSICA (see *Processing Suites*, p. 524), but decomposes the original image into only two component images, one at low frequency and on at high frequency. The contrast of the low frequency layer is suppressed, and that of the high frequency layer with its fine details is amplified. The two layers are then recombined. EVP is applied after edge enhancement and analysis for tone scaling has taken place, but before tone scaling is applied. Figure 31-31 demonstrates an EVP-processed image.

Kernels as a Form of Band-Pass Processing

We have described the larger structures within an image as those having a lower spatial frequency. We could state, then, that during unsharp masking, the kernel filtering process used to create the mask leaves only the gross, *low-frequency* structures in the image, and has removed the high-frequency structures or fine details. This is a form of *low-pass filtering*, but it is performed in the spatial domain rather than in the frequency domain.

The size of the kernel used defines a transitional frequency between the high-pass and the low-pass filtered images; all spatial frequencies above this point are enhanced, while all spatial frequencies lower than this are suppressed. The kernel size used must be carefully selected and matched to the type of anatomy being radiographed, since too *small* a kernel may remove details from the image such that diagnostic information is lost.

The *end result* of unsharp masking is a form of edge enhancement achieved by suppressing or removing gross structures, so, taken as a whole, unsharp masking may be considered as a spatial form of *high-pass filtering*.

Parameters for Frequency Processing

The table of parameters presented in Table 31-3 (p. 514) are for Fuji digital image processing and are applied in connection with the gradient processing parameters found in Table 31-2. These are *edge-enhancement parameters* specified for each type of anatomical procedure, using the frequency processing methods we have just described. *RN* stands for *frequency number* or *rank*. It selects one of ten particular frequencies in the image to be enhanced. Frequencies 0 through 3 represent large objects in the image with a low spatial resolution of .09–.25 LP/mm (line pairs per millimeter). Frequencies 4 and 5 represent moderate-sized structures such as pulmonary vessels and bone contours that have a spatial resolution of .35–.5 LP/mm. Frequencies 6 through 9 represent very small objects in the image such as bone marrow details and pulmonary trabeculae with a resolution of .71–2.0 LP/mm.

RT specifies the *type* of enhancement function that will be executed on the selected frequency, that is, the format and amount by which it will be amplified or de-amplified. There are ten types of operations labeled *F* and *P* through *X*. The *RE* function sets the amount or degree of enhancement that will be applied to the frequency on a scale from 0.1 to 9.9.

In summarizing detail processing for digital images, we might say that unsharp masking and other kernel operations attack the *pixels* to filter the image, while multiscale processing attacks the *frequency bands* to filter the image. Spatial operations are based upon the location of each pixel, while frequency processing operations are based upon the *size* of structures in the image.

POSTPROCESSING III: PREPARATION FOR DISPLAY

Manufacturers universally include some type of edge enhancement (high-pass filtering) as part of their detail processing. As described in the previous section, this has the ill side-effect of increasing noise in the image. Therefore, even though noise inherent in the incoming data set was corrected during preprocessing, noise reduction must be applied again at this stage. It is here that we will address the general subject.

Noise Reduction

There are two general types of noise: periodic noise and random noise. *Periodic* noise consists of small artifacts that are all of a roughly consistent size and occur in a regular pattern across the area of the image. These artifacts all occur at the same frequency, so they tend to all be in the same image frequency layer. Electronic mottle such as the "snow" artifact often seen in TV images (Fig. 38-6, p. 659) is an example of periodic noise.

Random noise consists of small artifacts of variable sizes and occurs in an irregular, chaotic pattern across the image. It can therefore be found in several image detail layers. The best example of random noise is quantum mottle (Fig. 32-10, p. 552).

Normal anatomy varies in size, so it occurs at various frequencies and is spread across several image layers. This makes frequency processing ideal for removing electronic noise and other forms of periodic noise without significantly affecting the normal anatomy. Frequency processing usually targets one specific image layer, while normal anatomy is spread across multiple layers. Still, some normal anatomy will be removed along with the noisy layer.

By removing the single image layer corresponding to the highest amount of noise identified, a large portion of the noise in the image can be eliminated. When the final image is reconstructed by inverse Fourier transformation, this layer is simply left out.

Sophisticated methods for identifying noise in the image are available, such as the CNR image concept used by Agfa. By then removing the single layer corresponding to the highest amount of noise identified, a large proportion of image noise can be eliminated with only a slight loss of anatomical details.

Frequency processing is generally poor at separating *random* noise from the image. Random image noise, such as *quantum mottle, is best suppressed using spatial processing*, (kernels) rather than frequency processing.

Most manufacturers include at least one step within their default processing suite designed to suppress image noise. Additional noise reduction can be applied later by the radiographer as a *smoothing* function.

Additional Gradation Processing

After noise reduction and detail processing operations are completed, the image possesses extremely high visibility and delineation of details, but typically has a foreign, almost *alien* appearance to it. In order to restore the overall appearance to that of a conventional radiograph for display, an additional round of gradation processing will be applied prior to display. The processes used are precisely those described early in this chapter under "Postprocessing I: Gradation processing."

Perceptual Tone Scaling

The method used by Carestream Health (Kodak) for final preparation of the image, called *perceptual tone scaling (PTS)*, is somewhat unique and bears mention. Perceptual tone scaling is based upon the original H & D curve (Hurter and Driffield curve) used for film-based radiography, in which a single characteristic curve for a particular type of film plotted the density response of the film against the logarithm of the actual x-ray exposure received. Kodak conducted "psychophysical studies" to identify three *brightness models* describing human perception of radiographic images. From these models, formulas were developed which generate H & D type curves as shown in Figure 31-3.

Each type of radiographic procedure is assigned to be processed using one of these curves. A key difference from other processing approaches is that these curves do not convert input pixel values into output *pixel values*. Rather, as can be seen at the bottom of the graph in Figure 31-3, they map input *log exposure values* into output pixel brightness on the displayed image. Since digital image receptors do not measure the actual x-ray intensity in *exposure* units, calibration tables or analog log amplifiers must be used to extrapolate the estimated log exposure from original digital pixel values.

The stated goal of PTS is to select the speed, contrast, toe and shoulder of the tone scale curve such that equal log exposure differences are reproduced as equal perceived brightness differences on the image display. While the overall approach mimics film-based radiography, in application the tone scale curves have been "optimized" in such a way as to effectively *equalize* images, lightening up soft tissue areas that were often portrayed too dark on film-screen images.

Figure 31-31 demonstrates the digital processing sequence used by Carestream Health (Kodak), in

Figure 31-31

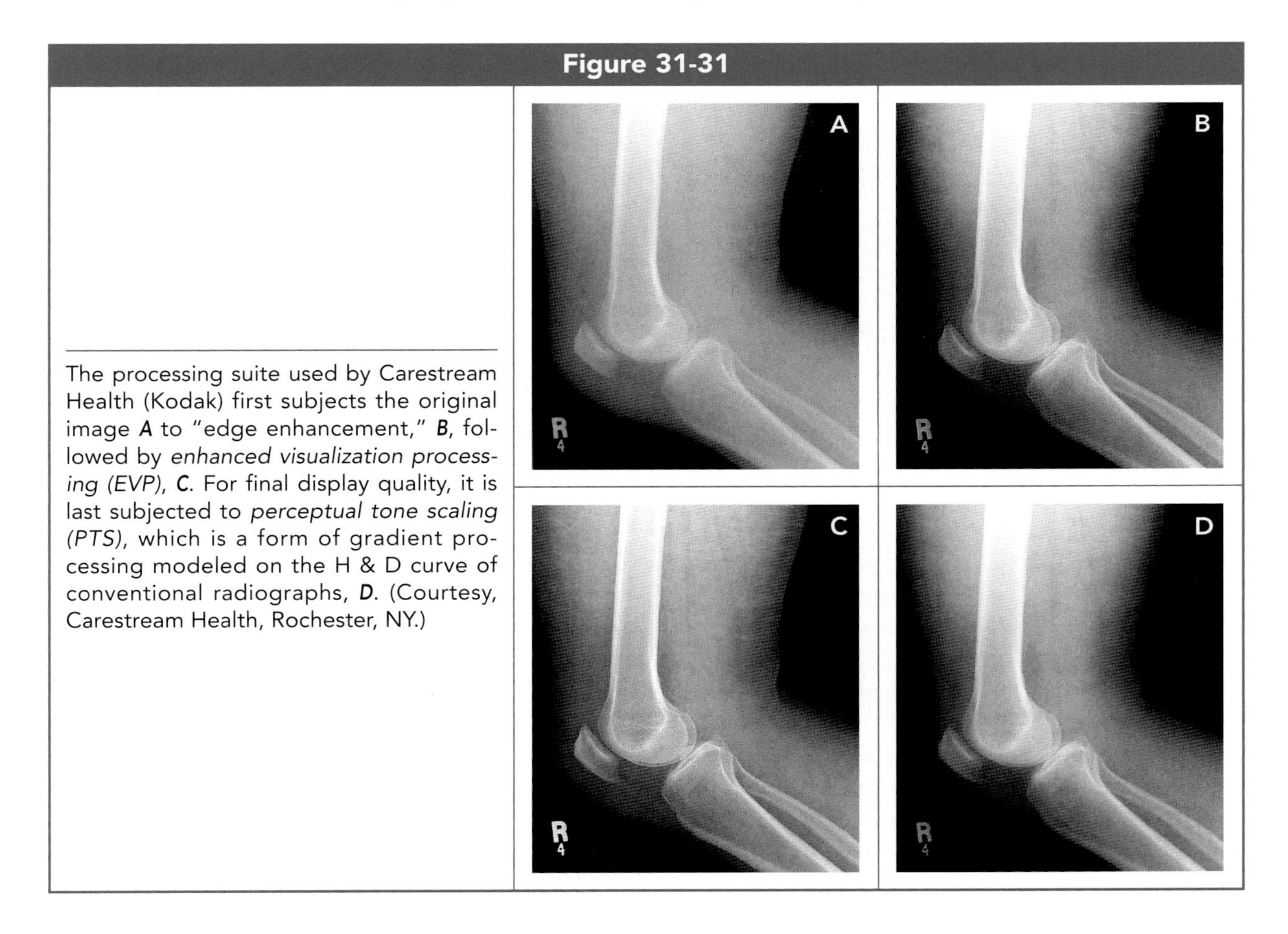

The processing suite used by Carestream Health (Kodak) first subjects the original image ***A*** to "edge enhancement," ***B***, followed by *enhanced visualization processing (EVP)*, ***C***. For final display quality, it is last subjected to *perceptual tone scaling (PTS)*, which is a form of gradient processing modeled on the H & D curve of conventional radiographs, ***D***. (Courtesy, Carestream Health, Rochester, NY.)

which edge enhancement and enhanced visualization processes are followed by perceptual tone scaling in final preparation for image display.

Formatting for Display

The dynamic range and other parameters of the radiographic image acquisition system can be quite different, usually much more powerful, than the capabilities of image display systems, which are discussed in Chapter 35. The display system is usually the weakest link in the radiographic imaging chain. All data from the image acquisition stage is stored by the PAC system in the standardized DICOM format so that it will be generally compatible with and able to be "read by" different display systems. But, when an image comes up for display on a particular monitor screen, the number of pixels in the matrix of the image must match the number of pixels available on that monitor screen so that it "fits" on the screen. Also, the scaling of brightness and contrast between the stored image and the display monitor controls must be aligned so that, for example, the adjustment available at the monitor does not only cover a truncated small portion of the dynamic range of the image. These computerized corrections are typically made within the monitor itself.

DIGITAL PROCESSING SUITES

We can now summarize the general processes of digital radiograph preprocessing as (1) field uniformity corrections, (2) noise and del drop-out corrections, (3) image and histogram analysis, and (4) rescaling (normalization); and the steps for postprocessing as (1) gradation processing, (2) detail processing, (3) preparation for display, (4) operator adjustments and (5) application of special features. Operator adjustments and special features are discussed in the following chapters. Each of these general processes can be accomplished by a variety of specific software operations that can be arranged in different sequences. This rather expansive set of options (along with the quest to develop new techniques), is just what continues to drive competition between manufacturers of digital radiograph processing equipment.

To provide an example of one of the more complex processing suites, we turn to Agfa's *MUSICA (Multi-scale Image Contrast Amplification)*. *MUSICA* consists of eight steps as follows:

1. Gain adjustment
2. Image decomposition using multi-scale transform
3. Excess contrast reduction
4. Subtle contrast enhancement
5. Edge enhancement
6. Noise reduction using contrast-noise ratio (CNR)
7. Image reconstruction using multi-scale transform
8. Gradation processing

Armed with the information in these last two chapters, the reader might venture to translate these proprietary terms as follows: The gain adjustment is a form of *rescaling of brightness*. Multi-scale transform refers to pyramidal decomposition by Fourier transformation as was shown in Figure 31-29. MUSICA calculates from the finest detail (highest frequency) image an estimate of the noise level and derives a *contrast-to-noise ratio (CNR)*. It computes a "CNR image" which is later used to help distinguish between actual noise and clinically relevant local contrasts. A conversion function is then applied to the coarser image layers where larger high contrast structures predominate to suppress their contrast. Higher frequency fine detail layers are subjected to a contrast amplification function. Thus, contrast enhancement of smaller image details is combined with contrast suppression of larger structures.

Frequency processing as described above is used to execute an additional edge-enhancement process. Following this, the CNR image is used to help identify noise. Periodic noise which occurs *at the same frequency*, and therefore *in the same detail layer*, is eliminated by removing the single layer corresponding to the highest noise frequency identified in the CNR image.

At this point, the remaining detail layers are added back together to reconstruct the image. Finally, the composite image is subjected to the gradation processing described earlier in the chapter. *MUSICA*^{2} is an upgraded "intelligent" version of this software which does not even require the type of exam to be entered, or the presence of contrast agents or metal implants, in order to analyze the image and proceed with its processing functions.

Using the *UNIQUE* software developed by Philips Healthcare, "before and after" images demonstrating the power of multiscale processing are demonstrated in Figures 31-32 and 31-33.

POSTPROCESSING IV: OPERATOR ADJUSTMENTS

Operator adjustments to the image are made at the console by the technologist or radiologist. They consist of windowing the brightness and contrast, applying edge enhancement or other features to the image that are available under proprietary names unique to each manufacturer, adding annotations, and making global geometrical changes such as re-orienting the image right-to-left, or zooming in.

While these operations go by various labels at the console, within the computer most of them are identical processes to those that have been used in *default* processing of the initially displayed image. For example, adjusting the *window width* is likely to be simply a re-calibration of the LUT curve initially used for gradation processing.

Added to these are a number of geometrical operations such as zooming (magnification), rotating, inverting, or translating the image (that is, flipping it left to right), all of which are spatial domain operations.

All of these operator adjustments are fully discussed in subsequent (and previous) chapters. The operator may also choose to apply *special* postprocessing features described in the next section.

POSTPROCESSING V: SPECIAL POSTPROCESSING

Dual-Energy Subtraction

Impressive separation of a digital image into a *tissue only image* and a bone only image, illustrated in Figure 31-34, can be achieved through dual energy subtraction. In more complex portions of the body such as the chest, this is of special diagnostic value in discriminating whether a particular abnormal density belongs to the bony or the soft tissue anatomy. For example, a lesion in the lung may be superimposed

Figure 31-32

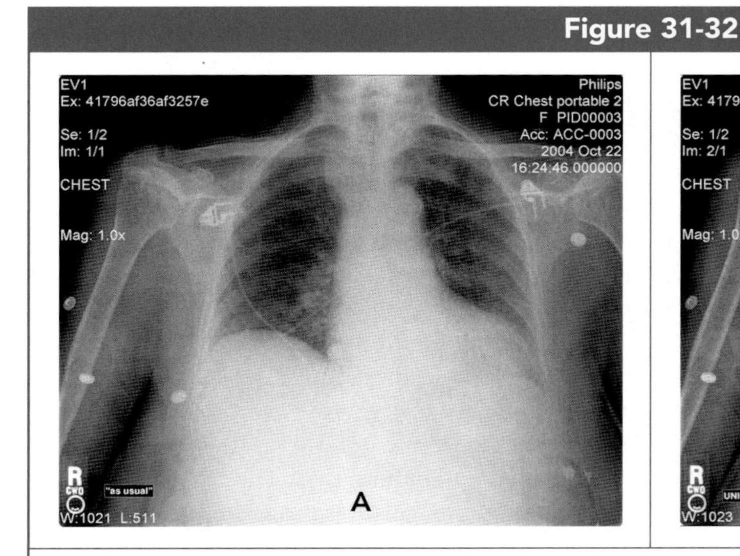

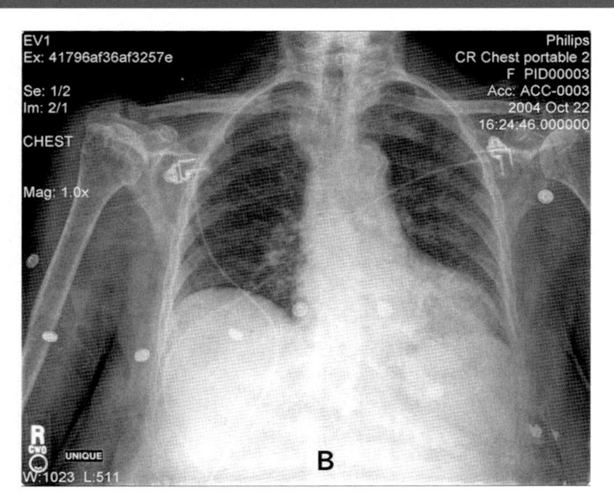

Chest radiograph ***A***, before, and ***B***, after processing with Philips' *UNIQUE* software. Note the dramatic improvement in visualization of soft tissue details in the neck, mediastinum, and abdomen. (Courtesy, Philips Healthcare, Bothell, WA.)

Figure 31-33

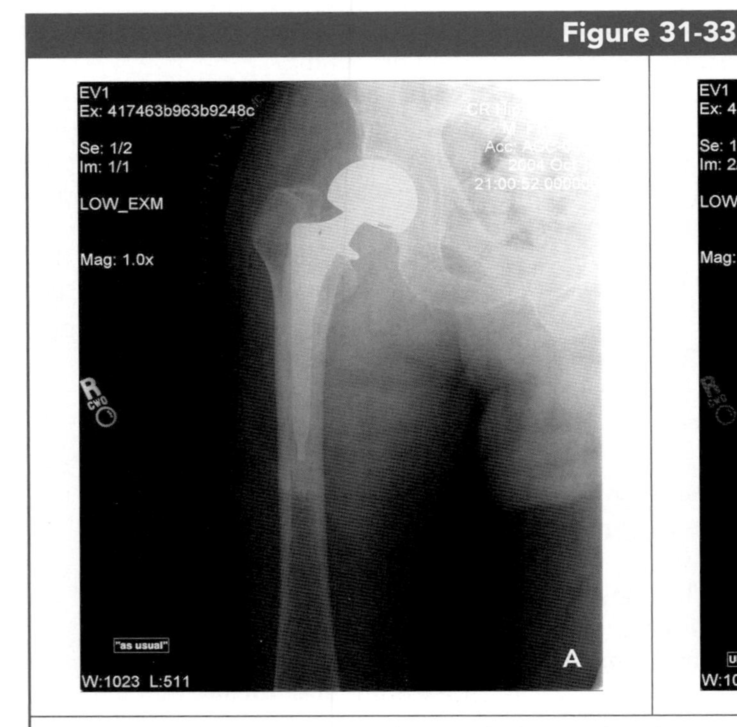

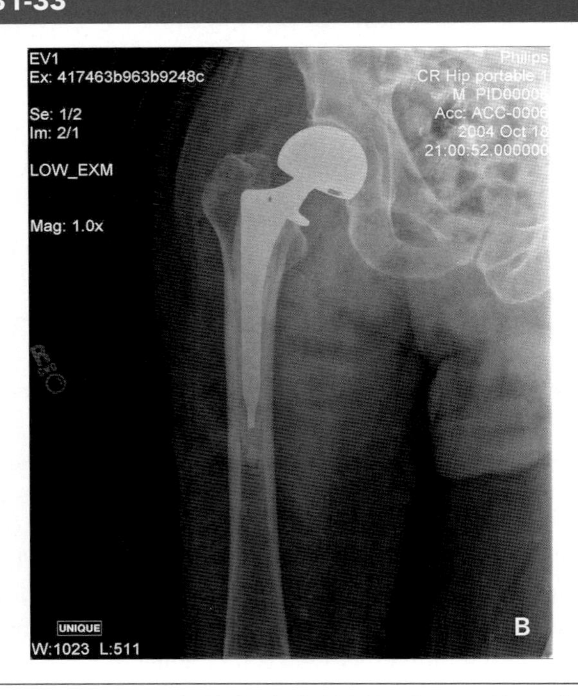

Radiograph of a hip with a prosthesis ***A***, before, and ***B*** after processing with Philips' *UNIQUE* software. Note the enhancement of detail in both bone marrow and cortical bone structures. (Courtesy, Philips Healthcare, Bothell, WA.)

by a rib. Separation of soft tissue and bone images allows the distinction of whether it is a growth on the rib or a lesion in the lung tissue behind the rib. Dual energy subtraction has also proven of considerable value in mammography in determining whether lesions have any calcific content.

In order to produce separate soft tissue and bone images, a *high energy image* and a *low energy image* must first be obtained (Fig. 31-35). There are two general approaches to obtaining these: The first requires a double exposure in order to obtain two images taken at different kVp settings. While the patient holds his/her breath, the x-ray unit must make a high-kVp exposure and then quickly switch to a low kVp setting and re-expose the detector plate. This method is used by GE, whose system requires about 200 milliseconds (1/5 second) for switching between the exposures.

The second approach places a filter between two or more image receptor plates, Figure 31-36. The front plate records the low energy image. The remnant x-ray beam passing through the filter is hardened prior to reaching the back plate(s), since the filter removes lower energy x-rays. With the right filter material, the average kV of the x-rays will be substantially raised. This beam records the high energy image. Fuji places two image receptor plates behind the filter and combines data from these two plates to improve the signal-noise ratio (SNR) in the high energy image.

The double-exposure technique produces a greater difference between the two average kV levels, allowing the computer to make an easier distinction between the two tissue types, but has the disadvantage of requiring more time to obtain the exposure which increases the probability of motion blurring. Manufacturers are trying to develop detectors that can effectively make a separate but simulataneous count of high-energy and low-energy x-ray photons during a single exposure.

The principle behind dual-energy subtraction is that, between the low-energy and high-energy exposures, there will be a greater change in absorption for soft tissues than for bone tissue. In Chapter 12 we learned that there is a radical difference in the photoelectric absorption characteristics between bone

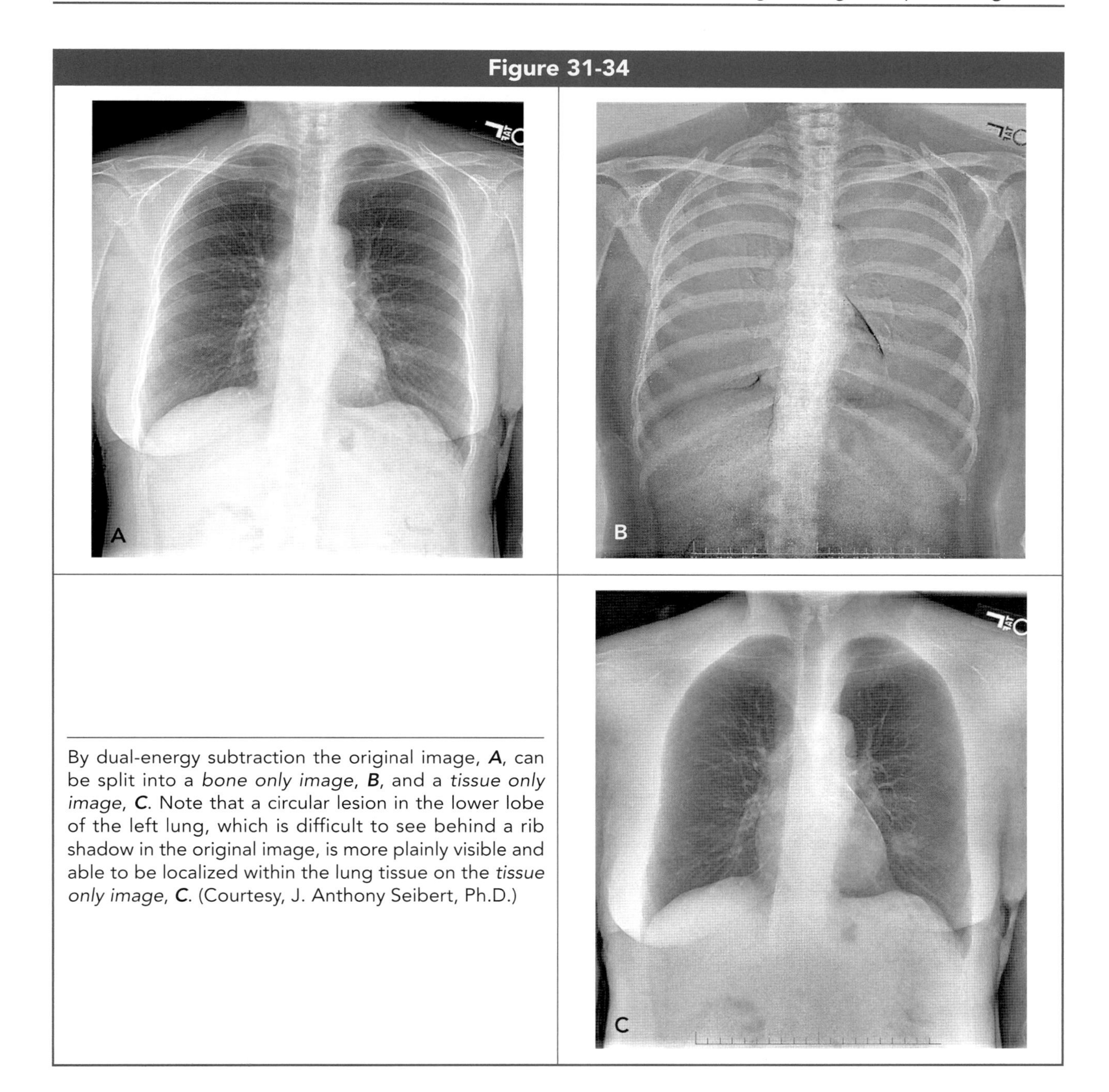

Figure 31-34

By dual-energy subtraction the original image, ***A***, can be split into a *bone only image*, ***B***, and a *tissue only image*, ***C***. Note that a circular lesion in the lower lobe of the left lung, which is difficult to see behind a rib shadow in the original image, is more plainly visible and able to be localized within the lung tissue on the *tissue only image*, ***C***. (Courtesy, J. Anthony Seibert, Ph.D.)

and soft tissue; Shown in Figure 31-37, as the average kV increases, photoelectric interactions drop more dramatically for soft tissues than for bone tissue. The computer is able to compare the high-energy and low-energy exposures, and identify those areas where the absorption dropped more rapidly as soft tissues. This allows the computer to reconstruct images using only the data for soft tissue, or only the data for bone tissue.

Tomographic Artifact Suppression

Artifacts that possess any type of consistent characteristics can be identified by computer software. By frequency layer filtering, these artifacts can then be removed from the image. A great example is the linear streaking artifacts typically present in tomography images: Linear streaking artifacts are unique because they occur in a *single axis* in the image

Figure 31-35

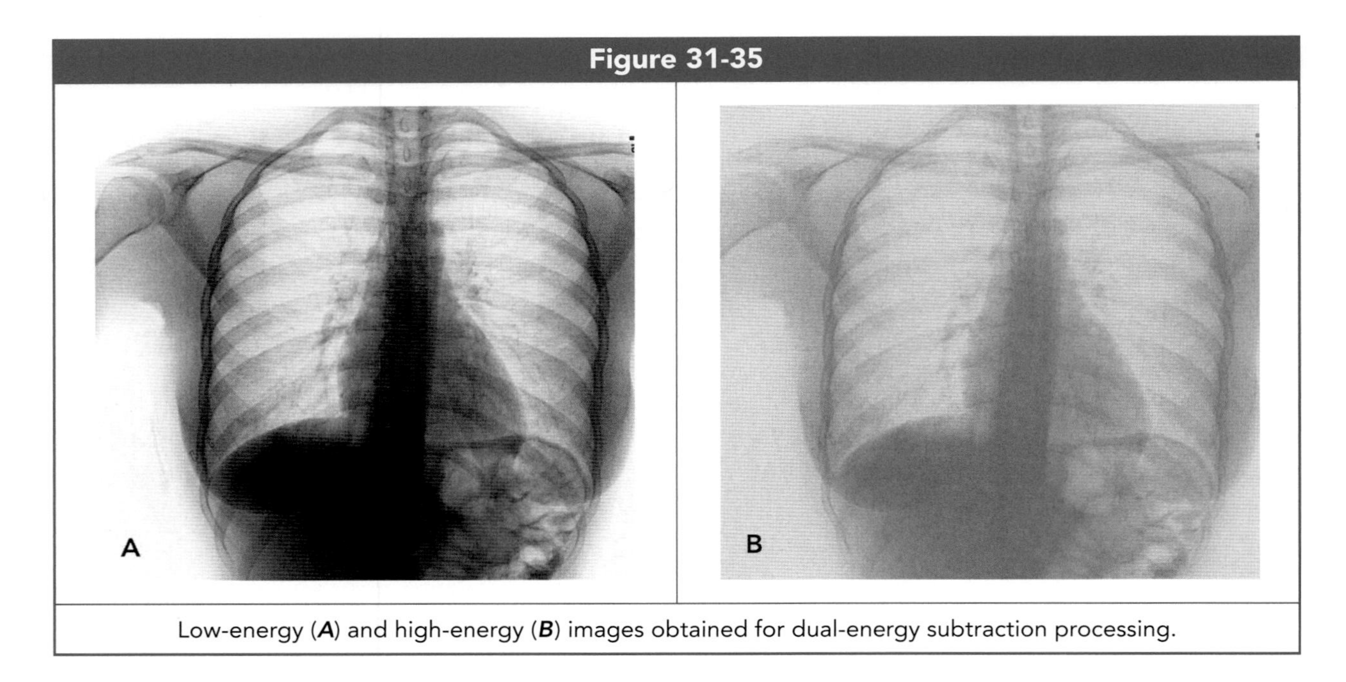

Low-energy (*A*) and high-energy (*B*) images obtained for dual-energy subtraction processing.

(either lengthwise or crosswise only). They can further be classified as a *low frequency* phenomenon because they are usually large (long). It is easy to write a software program to search only low-frequency layers of the image for artifacts that occur only in one direction. By filtering out one or two of these low-frequency layers upon image reconstruction, streaks are effectively removed from the image. GE has pioneered this process and calls it *tomographic artifact suppression* or *TAS*.

Fuji's TAS algorithms identify and suppress these artifacts with an *ORR* function that targets the expected *size* of the artifact based upon the tomographic angle used, an *ORE* function that sets the level of

Figure 31-36

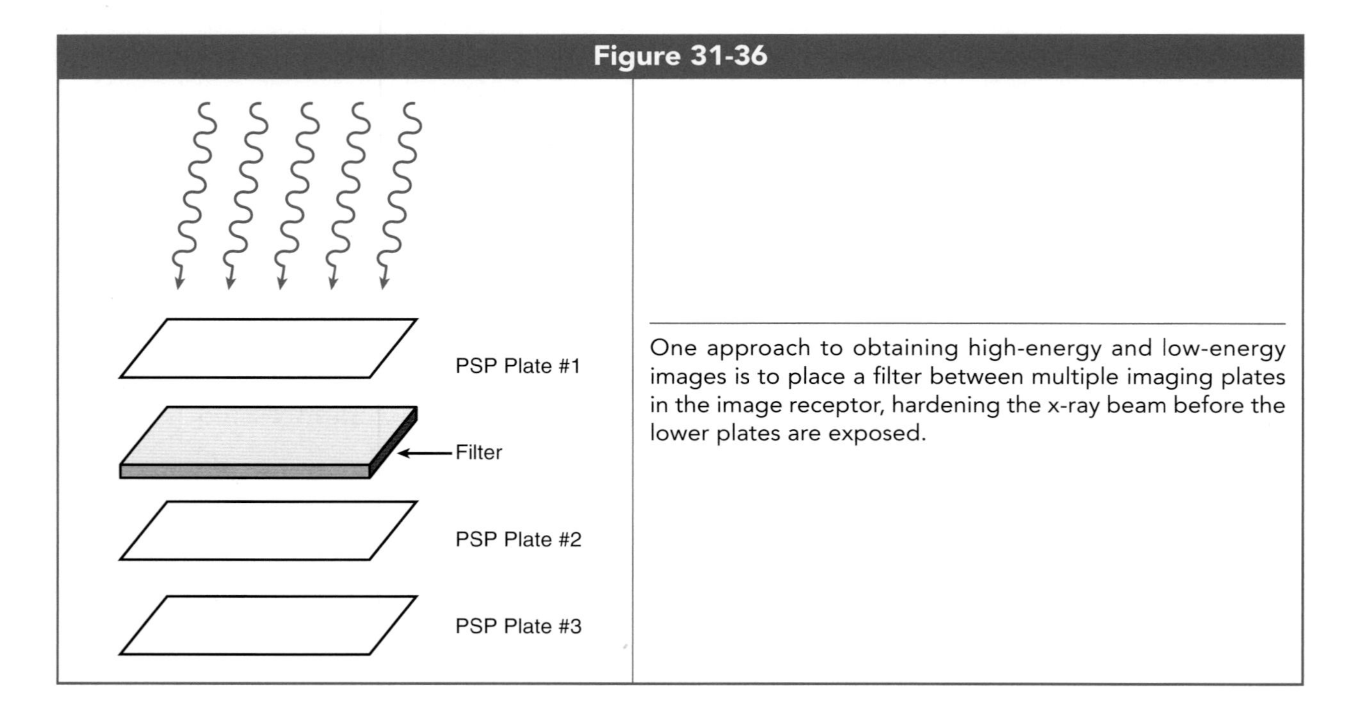

One approach to obtaining high-energy and low-energy images is to place a filter between multiple imaging plates in the image receptor, hardening the x-ray beam before the lower plates are exposed.

Figure 31-37

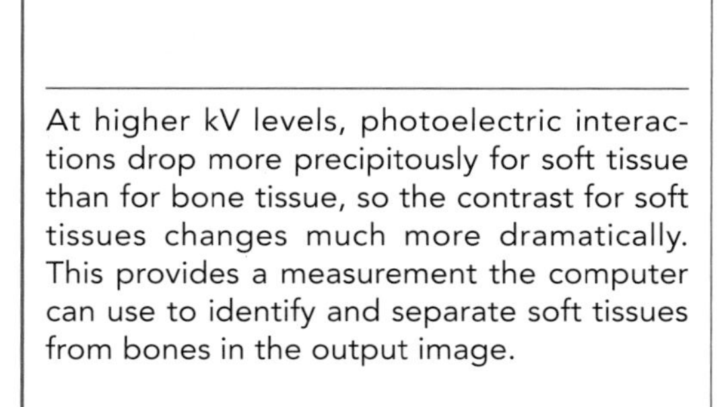

At higher kV levels, photoelectric interactions drop more precipitously for soft tissue than for bone tissue, so the contrast for soft tissues changes much more dramatically. This provides a measurement the computer can use to identify and separate soft tissues from bones in the output image.

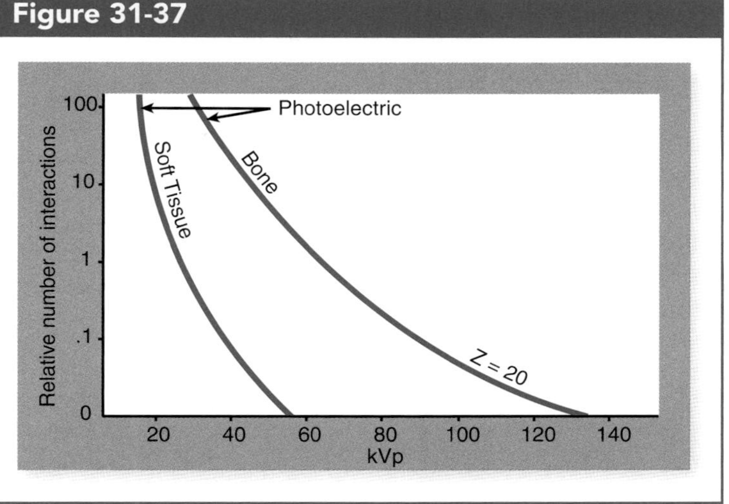

suppression to be applied, and an *ORD* function which identifies the expected *direction* of the artifact. The image is then subjected to regular gradation processing and *multi-objective frequency processing (MFP)*.

CONCLUSION

By way of review, preprocessing or acquisition processing consists of (1) field uniformity corrections, (2) noise reduction including del drop-out corrections, (3) image and histogram analysis, and (4) rescaling or normalization of the acquired raw image in order to formulate an initial digital image that is "normal" in appearance to the human eye. Preprocessing *makes corrections to the acquired image data* in order to compose a typical radiograph.

Postprocessing includes (1) gradation processing, (2)detail processing, (3) preparation of the image for final display, (4) operator adjustments, such as windowing, after the image is initially displayed, and (5) special features *applied by the operator* to the final image or image series. As shown in Table 31-4, the

Table 31-4

Summary of Digital Processing Steps

Steps	
Preprocessing	**Default Processing**
1. Field Uniformity Corrections	
2. Noise and Del Drop-out Corrections	
3. Image and Histogram Analysis	
4. Rescaling (Normalization)	
Postprocessing	
5. Detail Processing	
6. Gradation Processing (LUTs)	
7. Formatting for Display	
Image Displayed	
8. Operator Adjustments	
9. Application of Special Features	

From *Understanding Digital Radiograph Processing.* Midland, TX: Digital Imaging Consultants, 2013. Reprinted by permission.

first seven of these nine steps constitute the *default processing* that produces the image initially displayed.

All digital processing operations can be broadly classified as falling into one of the three domains described at the beginnng of this chapter. From the sequential list of processing operations just discussed, we can state that field uniformity and del drop-out corrections are performed within the *spatial domain*; histogram analysis, rescaling, gradation processing, and windowing are performed within the *intensity domain*; and detail processing and noise reduction can be performed either in the *frequency domain* or in the *spatial domain* using kernels.

SUMMARY

1. When processing an image in the spatial domain, it is sorted according to pixel *location*; in the intensity domain, the image is sorted by pixel *value*, and in the frequency domain it is sorted by *object size.*
2. Image postprocessing includes gradation processing, detail processing, preparation for display, and adjustments and features *applied by the operator* to the final displayed image.
3. Gradation or *gradient* processing fine-tunes the image brightness and gray scale by feeding rescaled pixel values into a look-up table (LUT) that is customized for the particular anatomy of interest.
4. Digital processing *function curves* graph the mathematical formulas used to generate different LUTs.
5. Gradation processing is used both as a default processing operation and again any time the displayed image is windowed by the operator.
6. Data clipping can occur when the dynamic range or bit depth of a system does not extend sufficiently above and below typical input values to allow for windowing adjustments.
7. Dynamic range compression (DRC) truncates the gray scale curve, cutting out the darkest darks and lightest light shades in the image. Structures that were depicted too dark are lightened, while those that were depicted too light are darkened, making more tissues visible.
8. A spatial *kernel* can be used for detail processing by placing multiplication factors in its cells and passing it over the image column-by-column and row-by-row. Depending on the factors by which pixel values are multiplied, the kernel may execute smoothing, edge-enhancement, or background suppression on the image, and can also affect overall image contrast.
9. High pass filtering leaves only high-frequency or small details in the image. Low pass filtering leaves only low-frequency or large structures in the image. *Unsharp mask filtering* is a poorly-named process which subtracts a low pass-filtered imaged from the original to enhance the visibility of details.
10. The reduction of random image noise such as quantum mottle is best achieved with spatial (kernel) processing, whereas the suppression of periodic noise, such as might be caused by hardware, is best achieved with frequency processing.
11. For an image that already has long gray scale, applying a smoothing function can lead to loss of details; for an image that leady has high contrast, applying an edge-enhancement function can lead to noise.
12. In the frequency domain, the amplitude (height) of a wave represents a brightness level or pixel value, whereas the frequency or length of the wave represents the size of a structure or the number of pixels occupied by it along a particular row.
13. For a digital image, high-frequency translates to high sharpness of detail, because each detail occupies fewer pixels.
14. A particular advantage of multiscale frequency processing is its ability to *decouple* or separate the enhancement of the contrast of local details (edge enhancement) from the enhancement of general image contrast. This and other improvements are made possible by breaking the image down into numerous frequency bands that can each be subjected to many different processing procedures.
15. Because detail processing often results in an alien-looking image, additional noise-reduction and/or gradation processing may be applied prior to image display.
16. Digital processing suites include many combinations of processing procedures in different

sequences which distinguish one manufacturer from another.

17. Dual energy subtraction allows the separation of *soft tissue* only images from *bone only* images, of particular utility in chest imaging.
18. Tomographic artifact suppression (*TAS*) is an example of how artifacts with any type of consistent identifying characteristics can be specifically targeted for elimination by software.

REVIEW QUESTIONS

1. List the three *domains* in which a radiographic image can be processed:

2. List the three general types of *spatial* processing:

3. List the five general types of operations categorized as *postprocessing* of the image:

4. In computer graphics, the types of formulas that are used to generate LUTs for gradation processing are collectively called *intensity* ____________.

5. The formula used to generate an LUTs can be represented on a graph as a ____________ curve.

6. The LUT itself is *not* a graph, but an actual ____________ with input and output values listed.

7. For Fuji's gradation processing parameters, the particular *shape* of the gradient curve used for rescaling the contrast is abbreviated as:

8. For Fuji's gradation processing parameters, if *GA* is less than 1.0, how is the gray scale of the original input image altered?

(Continued)

REVIEW QUESTIONS *(Continued)*

9. Gradation processing, which operates on each pixel in the image, is categorized into which image domain?

10. When the dynamic range or bit depth of a system does not extend sufficiently above and below typical input values, ____________ ____________ can occur when the image is windowed.

11. Mathematically truncating the gray scale curve, such that extremely light densities are darkened, and extremely dark densities are lightened, is referred to as ____________ _________ ____________.

12. For detail processing a spatial kernel may be passed over the image which multiplies each pixel value by a core factor, then ____________ the surrounding cell values to arrive at a new value for the centered cell.

13. A kernel with both negative and positive core values that sum to 1 will result in __________.

14. Overall image contrast is increased when both positive and negative core values in a kernel sum to a result ____________ than one.

15. What is the name of the mathematical procedure by which complex waves can be broken down into the series of individual wavelengths or frequencies that make them up?

16. A particular frequency band which has been separated from an image represents a particular ____________ of structures within the image.

17. In unsharp mask filtering, structures that are ____________ than the size of the kernel are suppressed.

(Continued)

REVIEW QUESTIONS *(Continued)*

18. In unsharp mask filtering, the creation of the gross image mask or "blurred" mask is a form of ____________-pass filtering.

19. In multiscale processing, if the amplitude of the highest frequency band is boosted before composing it back into the image, the ____________ details in the image will be enhanced.

20. Is spatial filtering or frequency filtering best suited for suppressing random noise such as quantum mottle?

21. Since the parameters for CareStream's gradation processing are based partially on perceptual surveys of radiologists, they call it ____________ _________ ____________.

22. In order to more effectively distinguish between actual noise and clinically relevant local contrasts, Agfa's *MUSICA* computes a ____________ image from the highest frequency band.

23. Dual energy subtraction is made possible by the great difference in what type of absorption (interaction) between high-energy and low-energy images.

24. For dual energy subtraction, what are the two optional methods for creating high-energy and low-energy images?

25. Tomographic artifact suppression is made possible by the fact that linear tomographic streaks can be identified as a ____________-frequency phenomenon occurring in a ____________ axis.

26. In frequency processing to enhance image detail, what is the term for the mathematical process which is applied to re-assemble individual wavelengths in order to formulate the final displayed image?

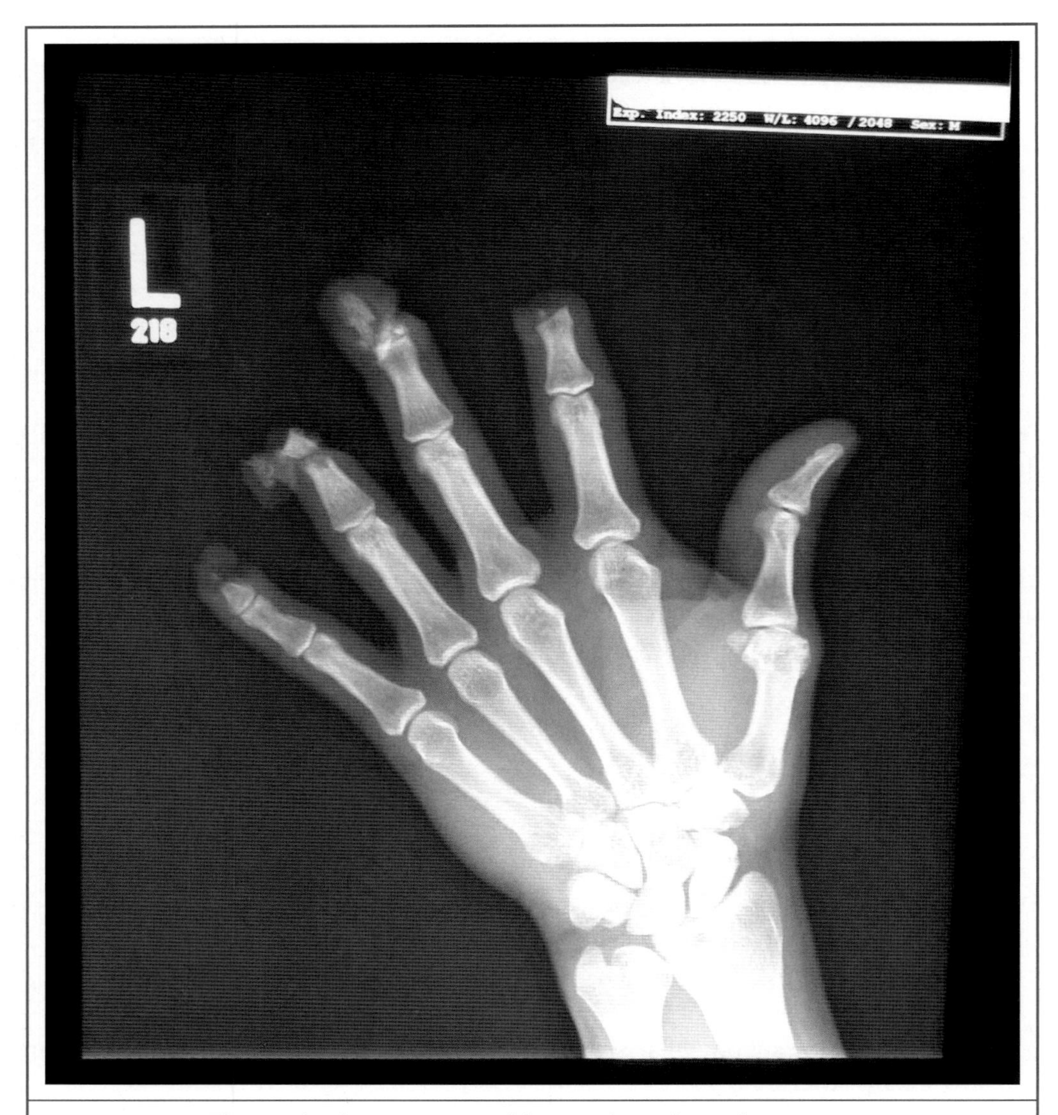

The result of running several fingers through an electric saw.

Chapter 32

POSTPROCESSING OPERATIONS IN PRACTICE

Objectives:

Upon completion of this chapter, you should be able to:

1. Overview the typical menu screen functions for operator adjustments to the image.
2. Define *speed class* and how it is best used to minimize patient exposure.
3. List the three rules for proper use of the *exposure indicator*.
4. Describe the general approaches to formulating the exposure indicator and their accuracy.
5. Define the three general types of scales used for exposure indicators (logarithmic, proportional, and inversely proportional) and overview how they are used by different manufacturers.
6. Recite and interpret the recommended exposure indicator control limits for clinical images published by the AAPM.
7. State the typical parameters for acceptable exposure based on different types of indicators.
8. List typical causes of exposure indicator error.
9. Explain the proper use and limitations of alternative procedural algorithms in manipulating the image.
10. Describe the proper use of windowing, edge enhancement and smoothing in manipulating the image.
11. Define various features for adjusting the image, (dark masking, image reversal, resizing, image stitching, etc.)
12. Clearly define the seven criteria for digital radiography image quality.

NAVIGATING THE MENU SCREENS

At the user terminal, menu screens are presented in an extremely user-friendly format, usually replete with icons that can be understood internationally. Anyone familiar with consumer photography software will quickly adapt to the CR and DR menu screens. Although every brand name has its own layout and specific terminology, some generic observations can be made about the typical functions available to the user.

A "main menu" screen usually includes options for entering patient and study data and for reviewing images, accessible to all users. "Key operator functions" are for setting default procedure codes to be used in processing images, and other generic settings for the equipment, and are accessible only to quality control technologists with a password. Consultant and service functions are accessible only to designees of the manufacturer.

For CR systems, there will be an "erase cassette" button on the main menu. CR cassettes are very vulnerable to both background radiation and scatter radiation. Any plate that has not been used for several days will have accumulated sufficient background fog to affect processing, and should be erased. Any cassette left in the exposure area of a radiographic room should be erased prior to use. Whenever the condition of a cassette is in question at all, it should be erased!

On a "study data" or patient data screen, patient identification and demographic information are typed into the computer, as well as information on the radiographic examination being conducted. The "submit" button then sends this file to the PACS system for storage. Various images can later be "assigned" to this file.

On a typical "image review" screen, diagrammed in Figure 32-1, each acquired radiograph is displayed using a default or preset brightness and contrast level. Windowing is accomplished by changing the default brightness and contrast values displayed. Some systems refer to contrast as the "gamma." Some present a "density" adjustment rather than a "brightness" control. Finally, the histogram and the exposure indicator for that radiographic exposure are displayed.

Most systems also allow for adjustment of the image brightness, contrast, or magnification by cursor movement: Typically, side-to-side movement of the cursor while depressing the mouse button will increase brightness from left to right; contrast is increased by moving the cursor up and decreased by moving it down (extreme reduction will reverse the image). By clicking on a magnifying glass icon, either cursor movement will magnify or minify the image.

By touching the histogram on the image review screen, an image reprocessing screen is displayed. Selecting the procedure name allows alternative processing algorithms to be applied to the image. Options are provided to add a text box for patient information, to add markers or arrows to the image, mask it with a black border, reverse it into a positive,

Figure 32-1

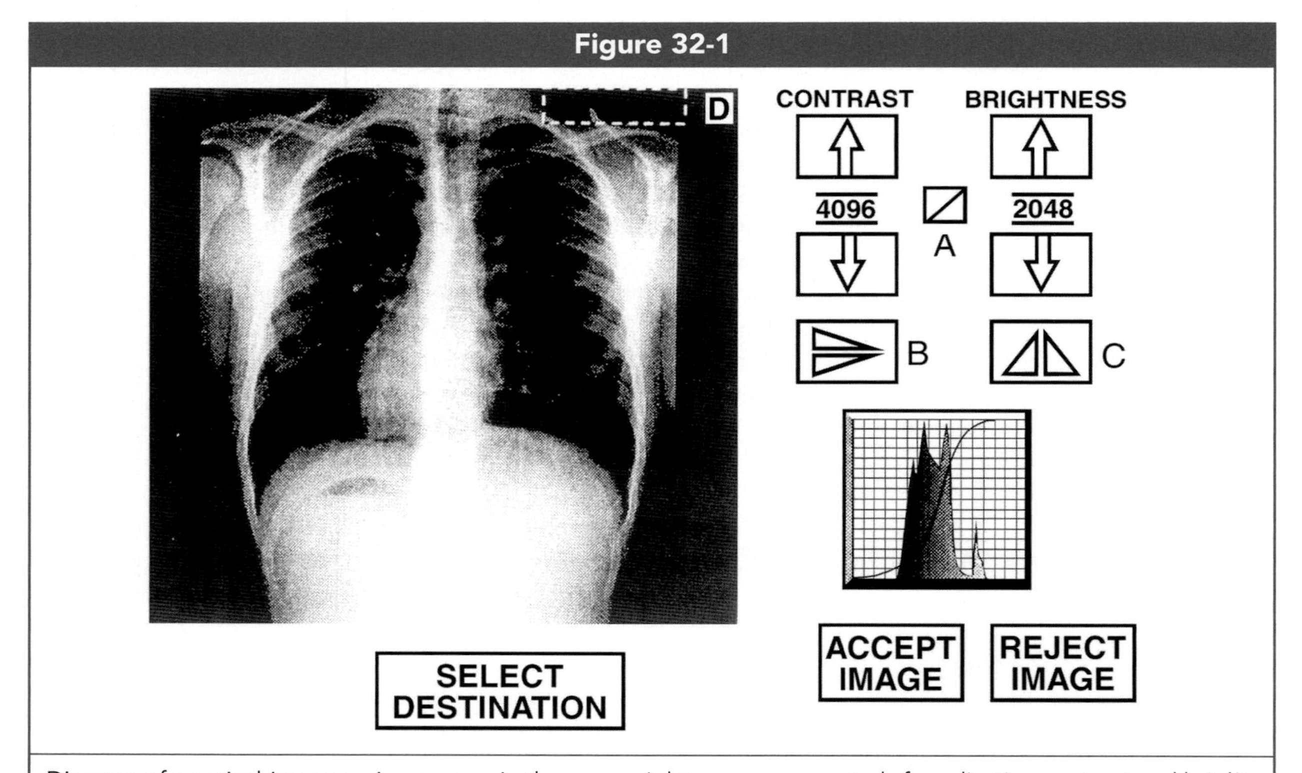

Diagram of a typical *image review* screen. In the upper right corner are controls for adjusting contrast and brightness. Touching the button labeled ***A*** restores these settings to their original values. Buttons ***B*** and ***C*** flip the image horizontally and vertically. Touching the image itself allows the placement of blockers (***D***). Touching the histogram allows for alternate algorithms, markers, masks, edge enhancement, image reversal, and other features to be applied. "Accept image" sends the image to a file in the PAC system, directly to a printer, or to a remote site as designated by the "select destination" button. (From Quinn B. Carroll, *Practical Radiographic Imaging*, 8th ed. Springfield, IL: Charles C Thomas, Publisher, Ltd., 2007. Reprinted by permission.)

black-on-white image (Fig. 32-16), and rotate or flip the image. Other features available include edge enhancement, smoothing and equalization, all discussed in the previous chapter.

On the image review screen, a button labeled "assign," or "stamp view" dedicates the specific image to a study or radiographic series. With CR this can be done prior to beginning the actual exam if the "study data" have been created: The cassettes to be used are "assigned" by scanning their bar codes into the system using a laser bar code reader. Finally, the "accept image" or "deliver" button sends the image, with all its current settings, into the PAC system to the patient's file. When the entire study is displayed, the "accept all images" or "store" button can be used to send the entire study into PACS storage. As an alternative, images can be sent directly to a printer without being saved in the PAC system, by using the "select destination" option and changing the default setting.

While right and left markers can be generated by the computer, it is still strongly recommended that physical markers be used to reduce errors. Not all vendors include the exposure indicator in the DICOM header so that it is included for the radiologist's viewing. Some allow the exposure indicator to be turned off. On the Fuji and Konika systems, it is lost if the brightness is adjusted and the image is saved with the change. It is strongly recommended that all images viewable from the PAC system include the associated exposure indicator, and that this information be kept as part of all patient records stored within the PAC system.

SPEED CLASS

The *speed* of any imaging system is an expression of its sensitivity to radiation, and is always *inverse* to the amount of exposure required to produce an adequate signal at the image receptor:

$$\text{Speed} = \frac{1}{\text{Exposure Required}}$$

For traditional film-based radiography, speed was very clearly defined because the target outcome was to produce an average density on a standard film that measured 2.5 on a densitometer device. This would appear to the eye as a medium gray. When twice as much radiation was required to reach this density, the speed of the imaging system was said to be *one-half* as fast, in accordance with the above formula. Using the most common type of film/intensifying screen combination at the time, a *standard* speed was defined at a value of *100*, which required an exposure of 2 mR to produce an average density of 2.5. All other systems were compared to this standard.

With modern digital imaging systems, the brightness of digital images is *always* adjusted by the computer prior to being displayed, so there is no longer a set "density" or brightness which can be referenced in defining speed. Nonetheless, the traditional values still work as *relative measures that can be compared to one another*. Therefore, the speeds of digital imaging systems are still based upon this historical standard and expressed in multiples of 100.

Before the advent of digital imaging systems, the most commonly used film/ screen combinations in radiography had reached a speed of 400. Early CR systems were typically installed to operate at a speed of *200*, which is the inherent speed of the photostimulable phosphor plate used as an image receptor for CR. (The experiment illustrated in Figure 32-2 confirms this speed.) Thus, the initial change from film/screen radiography to computed radiography actually required a *doubling* of radiographic techniques—this was a step *backward* for patient dose, resulting in considerable controversy.

A novel aspect of any computed digital radiography system is that the *speed* at which it operates can actually be *selected* without any physical change of the receptor plates used. This is because the operating speed of the system is based upon the computer processing rather than on the cassette or plate. *The quality control supervisors and managers of a radiology department make the decision as to what speed the digital system will be operated at, upon installation.* The installers then select default settings for the computer based on this desired "speed class."

A *speed class* of 100 assumes that an average exposure of 2 mR will penetrate through the patient to reach the receptor plate. A speed of 200 assumes an average plate exposure of 1 mR, and a speed of 400 assumes a plate exposure of 0.5 mR. These speed classes are aligned with traditional film/screen speeds, so operation of a DR or CR system in the "200" speed

Figure 32-2

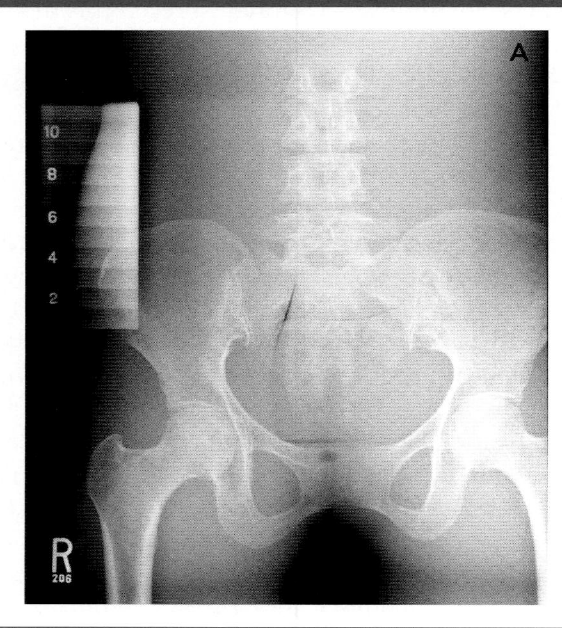

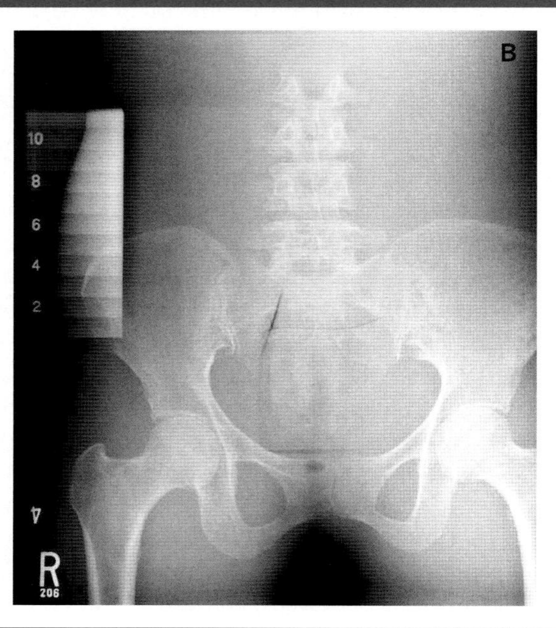

Experimental proof that a CR processing system operated at the 200-speed class is equivalent to a conventional 200-speed film/screen combination. This department had been using a 400-speed film/screen system for which the average technique had been 76 kVp and 25 mAs. When this same technique was used with the new CR system, ***A***, it resulted in a low exposure index of 1790 in a system with a target index of 2000. When the kVp was increased 15%, effectively doubling the exposure to account for an assumed speed of 200, ***B***, the exposure index read out at 2040, very close to the "ideal" index. These results were confirmed when the experiment was repeated for tabletop radiographic techniques on an elbow. (From Quinn B. Carroll, *Practical Radiographic Imaging*, 8th ed. Springfield, IL: Charles C Thomas, Publisher, Ltd., 2007. Reprinted by permission.)

class would require radiographic techniques (and, consequently, patient exposure levels) very close to those of a 200-speed film/screen combination.

The image receptor plate itself has an *inherent* speed. For CR plates this inherent speed is typically 200, and, partly for this reason, many CR units continue to be installed with the processor or *reader* matching this speed. But, *the reader can be adjusted to amplify or reduce the speed for the whole system.* When the reader is set for the 400 speed class, the plate need only receive half as much exposure (0.5 mR) in order for the whole system to produce the proper image.

This reduction in exposure can often be achieved without unacceptable mottle appearing in the images, because manufacturers have tended to set the default speed at 200 in the first place in an effort to double-ensure against the appearance of mottle. Yet, *nearly all CR and DR systems can be operated at a speed class of 300 or 350, and many at 400, without the appearance of substantial mottle in the image.* The ability to set the speed class of a digital processor to 400 without the appearance of substantial mottle should be a primary consideration in purchasing a particular brand name of digital imaging equipment. Radiologists can be involved in determining the levels of mottle that can be allowed for diagnosis. It cannot be overemphasized that quality control supervisors and managers of diagnostic imaging departments *have a choice* in determining the overall speed class at which every DR processor or CR *reader* will be operated.

EXPOSURE INDICATORS

A final CR or DR image has *always* been manipulated by the computer. Since the final image produced is a result of the operation of computer algorithms upon the original data acquired, *neither the brightness nor the contrast of the image can be attributed entirely to the original radiographic technique.* In fact, images taken at an increased mAs sometimes turn out brighter, or *lighter* in appearance, than those at lower mAs. To the student, it may seem that all of the conventional rules for technique become irrelevant with CR and DR.

Within normal ranges of radiographic technique, the *only* digital image quality which is directly affected by the set technique is *noise* in the form of mottle, and this is indicative *only* of underexposure. Aside from this, any immediate reinforcement as to whether the best combination of kVp and mAs was selected is now absent, since the image nearly always "turns out right." The only remaining motivation for being thoughtful and careful in setting technique is something much more abstract—namely, cumulative radiation exposure to the public. While patient exposure is generally acknowledged as an important issue, in practice it is easy for the "out of sight, out of mind" attitude to prevail.

Therefore, we must emphasize as strongly as possible the original mission of every radiographer, to produce images of the highest diagnostic quality possible *while keeping patient exposure to a minimum.* It has always been the case that untrained personnel can be brought in "off the street" and sufficiently trained in which buttons to push, since many combinations of kVp and mAs can produce a *passable* image. With CR and DR, this is more true than ever. It might be said, then, that *minimizing public exposure* has become the primary benefit of requiring certification for radiographers.

Exposure indicators are provided for this purpose. With digital equipment, they are the *only* means for determining whether a "correct" technique was used for the original exposure. It is important to understand that exposure indicators are *not* related to the brightness of the image on the display screen nor to the density of a printed hard copy. They are only an indication of the original x-ray exposure to the receptor and, by implication, to the patient.

The following three rules should be used as guidelines for CR and DR technique:

1. Insufficient techniques resulting in low exposure indicators can cause an unacceptable level of *mottle* noise in the image (Fig. 33-8***D***).
2. Very high exposure indicators reflect an unacceptable level of exposure to the patient.
3. In achieving exposure indicators within the correct recommended range, it is essential *that high kVp, low mAs technique combinations,* be generally utilized.

The combination of high kVp's with lower mAs values is not different than the conventional recommendations for screen-film radiography, and serves the same two essential purposes: Ensuring adequate penetration of the signal through to the detector system, and minimizing patient exposure. This is fully discussed in Chapter 33.

For most manufacturers, the exposure indicator is read by the computer from the midway point on the same image histogram generated for processing (Fig. 32-3). This mid-way point, however, is defined as the *median* rather than the *mean* value of the histogram. This is important because a median value is not unduly skewed by small errors in segmenting the image for measurement. For example, including a

Figure 32-3

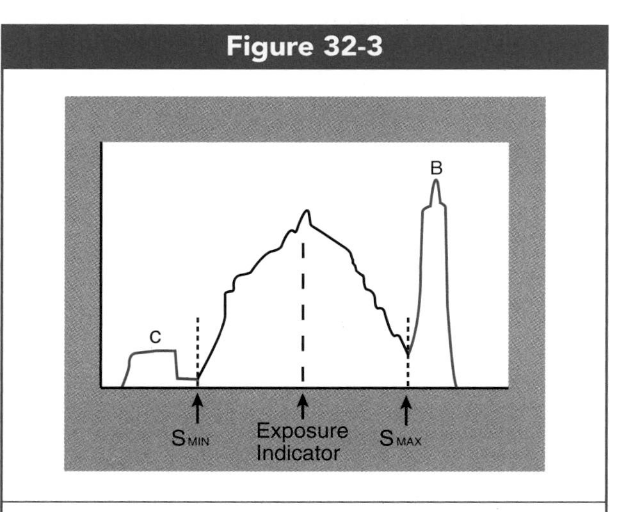

The exposure indicator is derived from the median pixel value of the *main lobe* of the generated image histogram, midway between the S_{MIN} and S_{MAX}, after the "tail" lobe for background radiation ***B*** has been identified and eliminated from the data set. Because DR scans the entire plate, a DR histogram includes any areas outside the collimated fields (which have accumulated scatter, off-focus, and background radiation), represented here by lobe ***C*** to the left. This lobe of data must also be identified and eliminated from the data set to avoid skewing the calculation of the EI.

small amount of direct exposure area will cause a significant shift in the mean, but the median is affected only slightly.

The median is the *counting average*, rather than the *arithmetic average*, of the pixel values. A counting average is found by simply counting the number of pixels, adding one, and dividing by 2, then reporting the value stored by this pixel number by counting up from the bottom. The formula for finding the median is

$$\text{Mdn} = \frac{N + 1}{2} \text{ count up}$$

where *N* is the number of pixels. This is opposed to the arithmetic average, which uses the formula *sum of all pixel values/N*.

For an example, let us use test scores for a group of 24 students as listed in Table 30-1. In the left-hand column, *X*, the actual scores are listed. The second column, *f*, is the *frequency* with which this score occurred, that is, the number of students that received this score. Finally, in the right-hand column, the product *f* times *X* is listed so all of the points can be summed.

The arithmetic average or *mean* score for this test is found as the sum of all points at the bottom of the right-hand column divided by the number of students, 1764/24 = 73.5. However, the counting average or *median* score is located as follows: *N + 1 = 25, divided by 2* = 12.5. We now go back to the frequency column, *f*, and count up from the bottom to 12.5. We find that 12 students achieved a score of 73 or less, and 12 achieved a score of 77 or more. Therefore, the score correlating to the 12.5th place is midway between 73 and 77, or 75.

The *mean* score for this test was 73.5, but the *median* score was 75. This illustrates the difference between a mean and a median value. By using the median rather than the mean in calculating an exposure indicator, the indicator is less severely affected by segmentation and other preprocessing errors.

For each manufacturer, a *target* median pixel value, which represents the "ideal" amount of x-ray exposure at the detectors, is set as the center point of the exposure indicator scale. This then becomes the target exposure indicator, and a range of acceptable exposure indicator values is framed around this center point for guidance to radiographers as to whether the radiographic technique fell within an acceptable range of exposures. *The target exposure indicator and the range of acceptable exposure indicators are all based (generally) upon the median pixel value in the image*, determined by each manufacturer's particular method.

Table 32-2 lists the names, symbols and units used by 14 different manufacturers for their exposure indicators. The many different terms used and display formats have been a source of confusion for radiographers. In 2009, Task Group 116 of the American Association of Physicists in Medicine concluded that, "A standardized indicator . . . that is consistent from manufacturer to manufacturer and model to model is needed." In their report, "An Exposure Indicator for Digital Radiography," they proposed a generic *deviation index (DI)* that can be used by all manufacturers regardless of their specific method of calculating the EI, and recommended that the DI "should be prominently displayed to the operator of the digital radiography system immediately after every exposure."

This standardized deviation readout, along with an indicator of the actual exposure (air kerma) in microGray (μGy) delivered to the image detector, should also be included in the DICOM header, the "metadata" record for every image that can be

Table 32-1

Data from Student Test Scores to Illustrate *Mean* and *Median* Averaging

X	f	fX
99	1	99
93	1	93
87	1	87
84	4	336
77	5	385
73	3	219
67	3	201
62	3	186
57	1	57
52	1	52
49	1	49
		1764

X = score, f = frequency; fX = product of score and frequency.

Table 32-2

Manufacturer	Exposure Indicator Name	Symbol	Units
Agfa	Log of Median	LgM	bels*
Alara CR	Exposure Indicator Value	EIV	mbels*
Canon	Reached Exposure Value	REX	*unitless*
Fuji	S Value (speed or sensitivity)	S	*unitless*
General Electric	Detector Exposure Index	DEI	*unitless*
Hologic	Dose Area Product	DAP	*unitless*
iCReo	Exposure Index	–	*unitless*
Imaging Dynamics Co.	Accutech	f#	*unitless*
CareStream/Kodak	Exposure Index	EI	mbels*
Konica	Sensitivity Number	S	*unitless*
Phillips	Exposure Index	EI	*unitless*
Shimadzu	Reached Exposure Value	REX	*unitless*
Siemens	Exposure Index	EXI	μGy
Swissray	Dose Indicator	DI	*unitless*

*The *bel* is a logarithmic unit for which every change of 0.3 bels represents a change in actual exposure by a factor of 2 (a doubling or a halving of exposure). The unit *mbel* (millibel) is one-thousandth the size of the unit *bel*, so each change of 300 mbels is equivalent to a factor of 2 in actual exposure. These units were derived from the *decibels* used to describe the loudness of sound or intensity of ultrasound waves.

accessed within the PACS system. Manufacturers are adopting the *deviation index* on all new equipment and trying to retrofit older units with the system. Table 32-4 in the next section shows the deviation index criteria which define when a radiograph may or must be repeated and when exposure guidelines have been exceeded. This will be of great help to radiographers in practice, but will take some time to become universally implemented. In the meantime, we present Table 32-2 as a summary of the older terms used by different manufacturers and continue to provide the following overview of their different approaches to presenting an exposure indicator.

There are three broad approaches to constructing the scale for an exposure indicator; these are logarithmic scales, proportional scales, and inversely proportional scales. In any logarithmic scale, we find that changes of 0.3, or some multiple thereof, represent a change in exposure by a *factor of 2*; that is, when the exposure indicator increases by 0.3 or its multiple, the actual exposure has doubled; when the EI decreases by 0.3, the exposure was one-half the original.

Logarithmic Scales

CareStream/Kodak: The "exposure index" used by CareStream/Kodak, abbreviated "EI," is derived from the *average pixel value* of the region of interest (the mid-point of the histogram). The EI value is calculated by the formula $1000 \times \log(\text{exposure in mR}) + 2000$. For a processing speed class of 200, the center of the scale is set at 2000, which indicates that 1 mR of exposure was received at the receptor plate using 80 kVp with 1.5 mm of aluminum filtration and 0.5 mm of copper as added filtration.

This is a logarithmic scale, and every change of plus or minus 300 EI represents a change in actual exposure by a factor of 2. (This standard logarithmic relationship allows the EI to be expressed in units called *mbels* as defined under Table 32-2.) Thus, an EI of 2300 would indicate twice the "ideal" exposure level, and an EI of 1700 would be one-half the ideal exposure. Special high-resolution (HR) or "detail" cassettes are available from this manufacturer, which, it is important to note, use a different target EI value.

ALARA CR: The *EIV* or "exposure indicator value" used by Alara CR is nearly identical to the system used by Kodak/CareStream, calculated by a slightly different formula: EIV = 1000 × log(SC X mR)/2, where *SC* is the speed class at which the image processor is being operated. After the anatomical region of the image is identified, the *mean* (not the median) pixel value for this area is calculated and converted to mR. Measurements are calibrated to 70 kV and 21 mm aluminum filtration. The EIV can be expressed in units of mbels, defined under Table 32-2. As with the CareStream/Kodak system, the target EIV is 2000, and with the logarithmic scale used, every change of 300 in the EIV corresponds to a change by a factor of 2 in the actual exposure. A feature of this system is that the EIV is also graphically displayed at the bottom of thumbnail views of each image as a horizontal bar.

IMAGING DYNAMICS CORPORTION AND ICRO: The *Accutech* system used by Imaging Dynamics Corporation abbreviates the exposure index as the *f#* (reminiscent of the *f-stop* used in photography, which correlates to the "speed" of the lens in gathering light). IDC set as its goal the production of a *user-friendly* index with a scale familiar to the average radiographer. The scale is centered on a target value based on the typical exposure required to produce a mid-range optical density on a 200-speed film/screen system (1 mR). This is calibrated for each radiographic system using 1 mm of copper filtration and 80 kVp, to obtain the ratio of mR per digital number. This ratio is then stored for all calculations.

The readout is a rounded out value presented in a scale radiographers will recognize as similar to that used on many AEC density controls: +2, +1, 0, –1 and –2. *It must be read, however, as exponents of the base 2*, that is, +1 indicates $2^1 = 2$ times the target exposure level, +2 indicates $2^2 = 4$ times the target exposure level, and so on. A –1 reading indicates 50 percent of the target exposure, and –2 indicates 25 percent. The target exposure range is defined between –1 and +1. This implies that the acceptable exposure range lies between one-half and two times the target exposure.

Digital systems from iCRco use an nearly identical approach, with the feature that readings outside the –2 to +2 range are displayed in red rather than green.

AGFA: Agfa takes their exposure indicator from the central lobe of the image histogram, identified as the region of anatomical interest after eliminating the spike from background radiation. That is, the values from most of the pixels in the main lobe are used to find the median value. The logarithm of the median exposure value is then calculated. It is abbreviated *LgM* for "log of median," and is calibrated on an exposure of twenty micro-Grays, 20 µGy, being received at the receptor plate using 75 kVp and 1.5 mm of added copper filtration.

For general torso procedures, the center of this scale is set at 2.5. In this logarithmic scale, every change of plus or minus 0.3 represents a change in exposure by a factor of 2. So, an LgM of 2.8 would indicate twice the "ideal" exposure (2.5), and an LgM of 2.2 represents one-half of the ideal exposure.

You may note that these units are 1000 times larger in size than those used by Kodak and Alara, so they can be expressed in *bels* as defined under Table 32-2. The speed class at which each image will be processed is selected by the user. This complicates the practical use of the LgM somewhat because there are different target exposure ranges for each speed class.

Agfa systems provide for the monitoring of dose history per exam over time. The LgM values for 50 consecutive images in every category are stored and averaged. Any image which substantially deviates from this average is flagged, and the extent of under- or overexposure is indicated with an attendant bar graph.

Proportional Scales

SIEMENS: The exposure index used by Siemens, abbreviated *EXI*, is also independent of the anatomical menu selection made by the user for processing and other image processing parameters. With the exposure field divided into a 3 × 3 matrix, the EXI is derived as the average pixel value in the central segment. It is then multiplied by a calibration factor (also adjusting for 70 kV and 0.6 mm of copper filtration) to obtain the estimated exposure in units of microgray (µGy). It is an advantage that the EXI readout is directly proportional to the actual exposure, and that the microgray is a standard unit for radiation dose. Indeed, this unit has been recommended by a task force of medical physicists for adoption by all manufacturers.

GENERAL ELECTRIC: For earlier GE flat-panel DR systems, no exposure indicator was presented upon display, but images other than chests were darkened or lightened upon display to simulate film/screen radiographs as a form of exposure level feedback to the radiographer. Newer units display a "detector exposure index," or *DEI*, along with lower and upper limits which can be user-defined. A target pixel value range is defined using AEC exposures of acrylic phantoms. After the anatomical area of the image is defined, the DEI is derived from the *median anatomy value*, a median of the pixel values in the "raw" image after preprocessing functions such as bad pixel and gain corrections have been applied.

SHIMADZU AND CANON: Canon developed cassette-based DR systems for retrofitting existing x-ray generators, which employ an exposure indicator called the "reached exposure value" or *REX*. The numerical value of REX is about 100 mR, but the value is a function of "brightness" and "contrast" settings chosen by the operator at the x-ray machine console. This is also the system used by Shimadzu. A possible advantage of this unusual arrangement is that it facilitates oversight of the exposure factors used by radiographers.

HOLOGIC AND SWISSRAY: Most exposure indicators are ultimately *derived* from pixel values, so they are *estimates* that suffer from questionable validity. Hologic Corporation and SwissRay use generic radiation units based on the *dose area product (DAP)*, which is explained in Chapter 41. Although this system is based on actual radiation units, the measurements are typically taken by a "DAP meter" connected to the *collimator* of the x-ray machine. Therefore, the final readout is an *extrapolated* estimate of the receptor exposure derived from measurements taken of the x-ray beam *before it passes through the patient's body*. This approach introduces its own set of problems for validity.

Inversely Proportional Scales

FUJI AND KONICA: These manufacturers use a scale called the *sensitivity number*, or "*S*" number. Fuji's S number is the oldest exposure indicator developed, and is taken from the image histogram *after* it is normalized electronically. From the main histogram lobe identified between S_{MIN} and S_{MAX} (Fig. 32-3), the median input pixel value is found and mapped to a digital output value of 511 in a 10-bit range. The sensitivity number is then calculated as $S = 4 \times 10^{(4 - K)}$, where *K* is the median input value. Konika also states that their S value is determined "following gradation processing," and although it is useful as a relative exposure index, it is not determined directly from pixel values relating to the original x-ray exposure.

The center of this scale actually indicates the speed class at which the system is being operated to achieve an exposure of 1 mR to the receptor plate using 80 kVp. Thus, if the CR or DR system is processing images at the 200 speed class, then an S number of 200 indicates proper exposure.

This is a linear scale, but is *inverted relative to the actual exposure*, which creates some confusion. For example, operating at 200 speed, an S number of 100 indicates that the exposure was twice too high. An S number of 400 would indicate the exposure was ½ of the ideal. (This may be thought of as the "speed class" at which the actual exposure was taken, that is, an S number of 100 means that the exposure actually used would have been correct for a 100-speed system, but *not* when operating the system at 200-speed.) In this scale, high numbers risk image mottle, while very low numbers represent *excessive* patient exposure. The manufacturer states, "S numbers under 75 should be considered overexposed, even though they may look normal in appearance."

On both the Fuji and the Konika systems, the S number actually serves two functions. The initial S number displayed on the monitor screen by the computer is the exposure indicator. However, once the image is displayed on the monitor screen, the S number becomes the image ***brightness control****, and can be adjusted up or down. If the brightness is altered and the image saved with the change, the original exposure indicator displayed by the computer can be lost unless it is recorded.* (The original S number may be permanently stored as *metadata* in the PAC system, but the *displayed* number no longer serves as an exposure indicator.) *Remember, once the S number is changed to adjust brightness, it no longer can be used as an exposure indicator.*

PHILIPS: An important improvement in the reliability of the "exposure index" was incorporated into Philips DR systems when the EI calculation was decoupled from the process of determining the region of interest (ROI) in the histogram. Remember that this determination is affected by the anatomical menu selection made by the user for processing the image. The use of alternative processing algorithms can change the way the histogram is read and analyzed, and thus affect the allocation of the mid-point or average.

Rather than use this approach, newer Philips units derive an estimated x-ray exposure by averaging the original pixel values (1) in the areas of the activated detector cells of the AEC when it is engaged, or (2) from the central 25 percent area of the image (called the "quarter field") when "manual" technique is used. This average is then adjusted according to the kVp used, since this affects the sensitivity of the DR detectors. In this way, the validity and accuracy of the derived EI are rendered more reliable and consistent.

The EI readout for Philips is intentionally confined to rounded values (100, 125, 160, 200, 250, 320, 400, and so on) in steps of about 25 percent. Increments of 25 percent were selected because they correspond to scales conventionally used for screen-film speeds, grading x-ray generators, and the "exposure point" system known to many radiographers. (Remember, too, that changes in the optical density of an image of much less than 35 percent are not visible to the human eye.) From a practical standpoint, Philips understandably considered it a waste of computer storage to use smaller increments.

On the other hand, this scale is *inverted* from the actual exposure in order to represent a relative "speed class" that would accommodate that exposure level in producing a medium optical density using conventional screens and film. Thus, an EI readout of 50 means that *twice the exposure was received* compared to an EI of 100 (thus, an imaging system half as "fast" would have been appropriate for this exposure level).

Limitations for Exposure Indicators

All exposure indicators are subject to skewing from "preprocessing" errors, such as segmentation failure, exposure field recognition failure, the presence of large prostheses or shielding, and other unusual circumstances, and most are also somewhat susceptible to the effects of extreme changes in scatter radiation, collimation and kVp level.

Many exposure indicators are derived from the image histogram, so anything which can lead to histogram analysis errors can cause the exposure indicator to be inaccurate. Several factors can cause errors in analyzing the histogram, which are discussed in the following sections.

ACCEPTABLE PARAMETERS FOR EXPOSURE

As discussed above, the primary function of exposure indicators is to help the radiographer avoid *overexposure to the patient.* A lower limit to exposure is imposed by the appearance of image mottle. To avoid either extreme, it is useful to define a target *range* for the exposure indicator, within which radiographers should strive to keep all exposures. *The acceptable ranges of exposures is ultimately decided by the management of each department, preferably in consultation with quality control technologists and radiologists.*

The common approach to determining ranges of acceptable exposure on older units was to *allow from one-half to 2 times the "ideal" exposure,* delivering an average exposure to the receptor plate of 0.5 to 2 mR. For operation at the 200 speed class, this formula yielded the following acceptable ranges of exposure indicators:

For 200 Speed Class:	Fuji = 100 – 400
Low exp to High exp:	CareStream/Kodak = 1700 – 2300
	Agfa = 2.2 – 2.8
	Phillips = 55 – 220

Radiography of the distal extremities presented a special problem for CR systems: Due to the part thickness, the exposures required were inherently low, and mottle became apparent in the digital images. In order to provide sufficient signal to the receptor plate, higher kVp's than traditionally used for distal extremities were recommended, as can be seen in Table 33-1 (Chapter 33) which is a manufacturer's recommended technique chart for a CR unit. Also, the acceptable range for the exposure indicators had to be adjusted upward, and the common approach was to base these upon operation at the 100 speed class. The following ranges were recommended:

For Distal Extremities: (100 speed class)	Fuji = 75 – 200
	Carestream/Kodak = 2200 – 2400
Low exp to High exp:	Agfa = 2.1 – 2.2 *set at 100 speed class*
	Phillips = 40 – 70

Note that DR systems differ from CR when it comes to imaging the distal extremities. On a DR system, all procedures are done with a grid in place, including all extremities. The grid requires higher techniques to be used, and generally sufficient signal reaches the receptor to obviate the need to specify a different range of exposure parameters. Thus, on DR systems the regular ranges listed above for the 200 speed class will apply to all procedures.

To help clarify the preceding discussion, the practical parameters of five exposure indicator systems are summarized in Table 32-3, showing for each the target indicator number for an ideal exposure level, what the indicator would read if the exposure were doubled, and what the indicator would read if the exposure level were cut in half.

Note that, as illustrated by comparing the *Distal Extremities* section above to the previous *200 Speed Class* section, we see that *any time the speed class for processing is changed, the range of acceptable exposure indices changes with it.* Radiographers must adjust to a new "ideal exposure" level or *target EI* (EI_T). In Chapter 33 we will a recommend a high-kVp approach to digital radiography techniques. If a dramatic enough change is made (two or more 15% steps), it may be necessary to establish a new EI_T. This is perfectly allowable; as target EI's are based upon the desired outcomes for the department, they are *not* inherent to a particular unit or processor. Managers, in consultation with manufacturers, radiologists and physicists, establish the desired image quality and patient dose outcomes for a department. The "speed class" for processing different examinations is thus determined, and an appropriate EI_T is derived for that speed class.

As mentioned in the previous section, in 2009 a task group of the medical physicists' association (AAPM) established a standardized *deviation index (DI)* that manufacturers are implementing to be displayed immediately after every exposure. The formula for the deviation index is

$$DI = 10\log_{10}(EI/EI_T)$$

where *EI* is the exposure indicator read-out and EI_T is the target EI for "ideal exposure."

Table 32-4 presents the deviation index criteria, with an added column to clarify the *actual exposure deviation* amounts as a percentage of the ideal exposure level.

In the physicists' description of the table, they note that "The index changes by +1.0 for each +25% (increase in exposure, and by) –1.0 for each –20% change." However, these changes are *multiplicative, not additive.* That is, each step increase multiplies the *previous* amount (not the original amount) by 1.259, and each step decrease multiplies the previous amount

Table 32-3

Exposure Indicators for CR Systems Operating at 200-Speed Class

System	Exposure Indicator	Ideal Exposure	Double Exposure	½ Exposure
Kodak	Exposure Index	2000	2300	1700
Fuji	S Number	200	100	400
Phillips	Exposure Index	110	55	220
Agfa	Log Median Value	2.5	2.8	2.2

From Quinn B. Carroll, *Practical Radiographic Imaging*, 8th Ed. Springfield, IL: Charles C Thomas, Publisher, Ltd., 2007. Reprinted by permission.

Table 32-4

Recommended Exposure Indicator Control Limits for Clinical Images

Deviation Index	Exposure Deviation	Descriptor	Recommended Action
> +3.0	> 100% high	Excessive patient exposure	**No repeat** unless image burn-out occurs from saturation. Management follow-up
+1 to +3	25% to 100% high	Overexposure	Repeat only if burn-out occurs
–0.5 to +0.5	–20% to +25%	Target range	–
–1 to –3	20% to 50% low	Underexposed	Consult radiologist for repeat
< –3.0	< 50% low	Excessive underexposure	**Repeat** (Excessive mottle is certain)

by 0.794. For example, an index of –2 is calculated as 0.794 × 0.794 = 0.63 or 63% of the target exposure. An index of +3 is calculated as 1.259 × 1.259 × 1.259 = 1.995 or, rounded up, two times the target exposure. As shown in Table 32-4, the +3 and –3 indices are roughly equivalent to a *factor of 2* in exposure. A –3 index means one-half the target exposure.

Interpreting Table 32-4, we can state that with an exposure indicator reading less than 80% of the target exposure index (EI_T or K_T), the image is considered underexposed but ***should not be repeated unless a radiologist finds the level of noise in the image unacceptable.***

Any image displaying an exposure index less than 50% of EI_T *should be repeated.* Clinical experience establishes that such a low level of exposure can be expected to result in unacceptable levels of noise (mottle) regardless of which brand-name of equipment is used, CR or DR.

An overexposed digital image should ***not*** *be repeated, no matter the E_I, unless saturation has occurred. Saturation* is a condition in which the detector elements (dels) in a particular area of the image receptor have reached the maximum electrical charge that they can store. Since any further increase in exposure cannot be measured, the data becomes meaningless. All pixels in these tissue areas are displayed as pitch black on a standard radiograph, representing a complete loss of data in these areas, as shown in the lung fields in Figure 32-4. *Do not mistake a very dark image for a saturated image. If any details at all can be made out in the dark portion of the image, it is overprocessed but not saturated. True saturation presents a flat black area with absolutely no details present.*

Even the most sensitive digital units require *at least* 8–10 times (three or more doublings of) the typical radiographic technique for saturation to occur, and this would be with an average patient rather than a large patient. Such an extreme condition for exposure can happen from time to time, but is quite rare in practice.

Figure 32-4

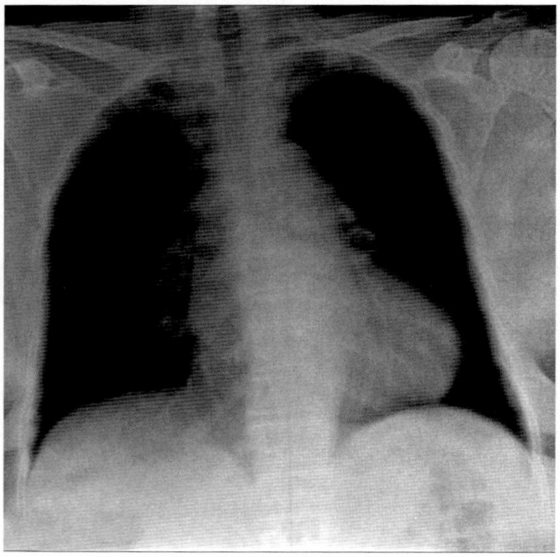

Saturation of the system from extreme overexposure in the lung areas on this chest radiograph resulted in a complete loss of data, shown as flat black areas in the image.

Inappropriate Clinical Use of the Deviation Index (DI)

The following is directly quoted from the report of AAPM Task Group 16 in the journal *Medical Physics*, Vol. 36, No. 7, the July 2009 issue:

> Even if images being produced clinically have corresponding DI's well within the target range, the clinical techniques used may still not be appropriate. One can just as readily achieve an acceptable DI for an AP L-spine view with 65 kVp as with 85 kVp; evidence of underpenetration and concomitant excess patient exposure with the lower kVp may be... windowed and leveled out in a digital image. Similarly, poor collimation, unusual patient body habitus, the presence of prosthetic devices, or the presence of gonadal shielding in the image may raise or lower DI's (depending on the exam and projection) and perhaps hide an inappropriate technique.

Further, "The deviation index is intended to indicate the acceptability of SNR (signal-to-noise ratio) conditions" from selected values of interest (VOIs) in the image histogram. It is *one indicator*, not *the indicator*, of image quality. Furthermore, since it is derived from pixel values, it can be used as an *indicator*, but *not as a measurement*, of patient exposure. Strictly speaking, it is taken from exposure data at *the image receptor, not at the patient's surface nor within the patient even in any simulated sense.* The EI_T and DI are guides to exposure and should not be misconstrued to other purposes.

EXPOSURE INDICATOR ERRORS

Since most exposure indicators are derived from the image histogram, anything which can lead to histogram analysis errors (discussed above) can cause the exposure indicator to be inaccurate. These problems include:

- extraneous exposure information, including scatter
- exposure field recognition error
- unexpected material in field
- collimation margins not detected
- extreme underexposure or overexposure
- delay in processing

USING ALTERNATIVE PROCESSING ALGORITHMS

One of the greatest advantages of digital imaging is that an unsatisfactory image can be reprocessed to change its appearance without repeating radiographic exposures to the patient. This is true for an image which turns out too light, too dark, or with poor contrast, but remember that a mottled image due to *gross underexposure* may be correctable only by a repeated exposure.

After the advent of CR imaging, radiographers quickly realized that an image could be modified by selecting a different preset algorithm for processing. That is, the procedure entered into the computer is changed. For example, if a chest image comes out too dark or too "gray," it might be re-entered into the computer as a knee procedure. The computer then processes the chest using algorithms that were designed for an "ideal" knee image, algorithms that will result in a lighter and higher-contrast image.

Figure 32-5 is a lateral lumbar spine that was reprocessed as a lateral T spine; due to the large amount of lung field in this view from high centering, the T spine algorithm resulted in an improved level of contrast and brightness for these vertebrae.

To change the algorithm applied, on the image review screen, touch the histogram (graph) display. A list of body parts or procedures appears on the next screen. Touch the procedure desired. The computer will apply the algorithm for that procedure to reprocess the image, aligning the image characteristics with a different reference histogram or look-up table.

Manufacturers sternly warn radiographers against altering the window level and window width to correct the brightness and gray scale of a poor image, and then saving that image into the PAC system under those new window settings. Remember, once an image is "saved" or sent to the image storage system, the data from the original image histogram is lost and cannot be retrieved. The radiologist, upon examining the image or diagnosis, must be able to adjust the window settings to his preference. He must be allowed the full range of settings to choose from. When a technologist alters these window settings and saves them into the storage system, it results in a

Figure 32-5

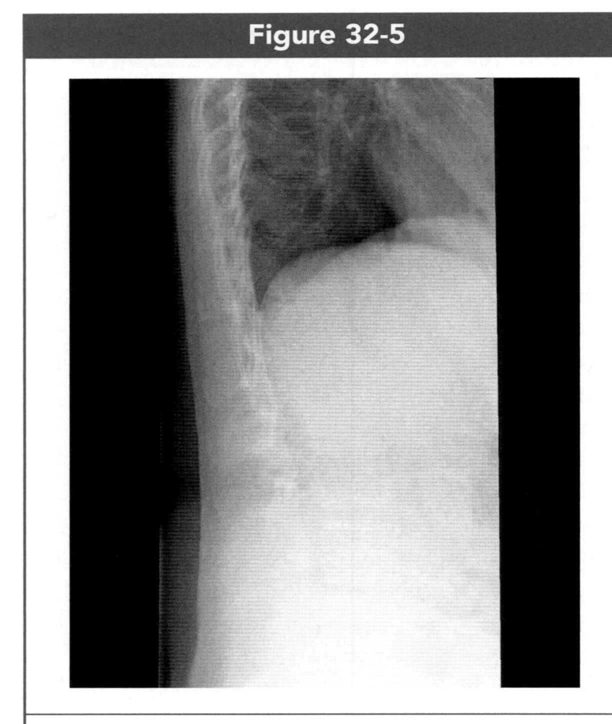

Lateral lumbar spine with improved brightness and contrast for those vertebrae within the lung field, by reprocessing as a thoracic spine algorithm. (From Quinn B. Carroll, *Practical Radiographic Imaging*, 8th ed. Springfield, IL: Charles C Thomas, Publisher, Ltd., 2007. Reprinted by permission.)

narrower range of values for the radiologist to choose from. It limits his ability to adjust the image precisely the way he wants it to appear.

A similar problem occurs when alternate algorithms are applied: Although the original data set may not always be *compressed*, it must be stressed that *if the image is saved to the PAC system* under the new algorithm, the original data set is permanently lost.

Even more problematic is the fact that the altered image is saved into the PAC system under a changed DICOM header as a different procedure. This complicates record keeping and retrieval of radiographic studies and can cause legal problems. Such an action would never be prudent without the collaboration and explicit approval of a radiologist or supervisor.

Medical malpractice lawsuits have been decided on the basis that the information in a radiograph had been tampered with when a radiographer reprocessed the image under a different algorithm and *saved* the changed image into the PAC system. Imaging departments should develop clear policies and provide stringent oversight of this practice.

A workable compromise is to make a copy, reprocess the copy under the new algorithm with annotation of the change added to it, and then save it to the PAC system as an additional image.

The application of alternate algorithms can be of value, as shown in Figure 32-6, if used objectively, scientifically, and conservatively. On the other hand, changes in image contrast and brightness are not always even appreciable for some alternate algorithms. Others result in a worse image, as demonstrated in Figure 32-7. These changed images must be carefully examined: Did the change truly result in

Figure 32-6

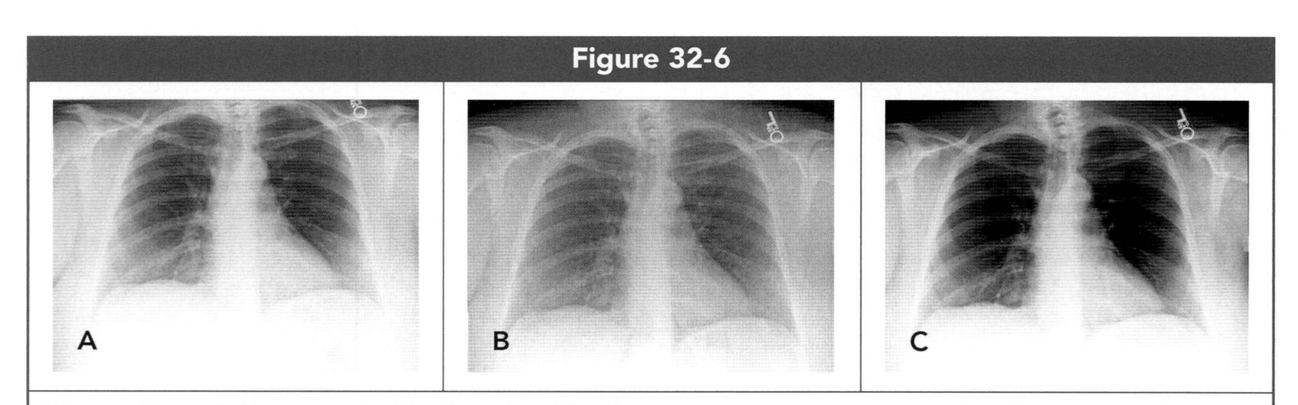

Chest radiograph ***B*** processed with a foot algorithm demonstrates increased gray scale and an improved quantity of radiographic details compared to radiograph ***A*** processed with a normal chest algorithm. Radiograph ***C*** was processed with an abdomen algorithm and demonstrates enhanced contrast but no added detail. (From Quinn B. Carroll, *Practical Radiographic Imaging*, 8th ed. Springfield, IL: Charles C Thomas, Publisher, Ltd., 2007. Reprinted by permission.)

Figure 32-7

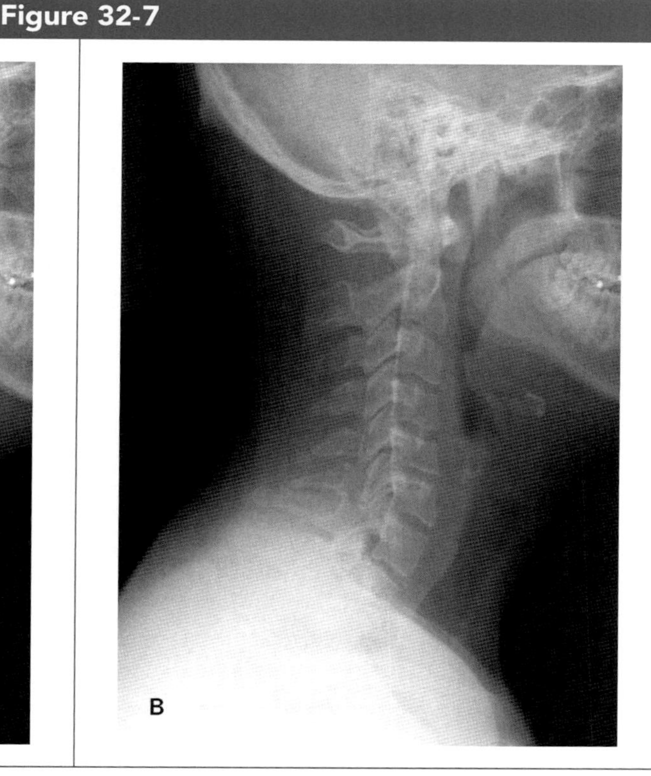

Cervical spine radiographs processed with ***A*** the normal cervical spine algorithm, and ***B***, an abdomen algorithm. In this case, no real improvement is achieved by the alternative algorithm. (From Quinn B. Carroll, *Practical Radiographic Imaging*, 8th ed. Springfield, IL: Charles C Thomas Publisher, Ltd., 2007. Reprinted by permission.)

any significant visual difference in image contrast or brightness? And, was that change indeed in the desirable direction? Discretion must be used.

When digital imaging systems are functioning properly and used correctly, the great majority of produced images should have ideal quality using the default processing settings. Therefore, the use of alternate algorithms should be the exception rather than the rule. *Routine use of alternate processing algorithms indicates a problem with equipment or with technical practice.*

Examples of Alternative Processing Algorithms

Figure 32-6 presents three chest radiographs processed under different algorithms for comparison. Radiograph ***A*** was processed with the normal chest algorithm. Radiograph ***B*** was re-processed under a "foot" algorithm. The image contrast dropped from 2.3 to 1.5, and is apparent. Note that structures in the mediastinum are indeed more visible: The spine can be better seen through the heart, and structures within the heart and aorta are not obliterated by high contrast. This is a significant difference. In the authors' opinion, more details are also visible within the lung fields and this is in line with the convention that chest radiographs are best presented with relatively long gray scale. However, some radiologists insist upon higher contrast in the lung fields and so this comparison is largely a matter of personal preference.

Chest radiograph ***C*** in Figure 32-6 was reprocessed under an "abdomen" algorithm. It demonstrates markedly increased contrast. This is excessive contrast for a chest image. Ironically, this is the opposite effect that the "abdomen" algorithm had on the cervical spine in the following experiment with a cervical spine.

Figure 32-7 presents two cervical spine radiographs for comparison. Radiograph ***A*** was processed with the original C-spine algorithm. Radiograph ***B*** was processed using the algorithm for "abdomen." This made the image turn out slightly darker and also having lower contrast between the vertebral bodies and the adjacent soft tissue. While both images may be within a diagnostically acceptable range of brightness and contrast and are not dramatically different, the changed algorithm resulted in a poorer image.

WINDOWING

Image brightness and contrast can be adjusted from the image review screen shown in Figure 32-1. An example of a proper application of windowing is presented in Figure 32-8, where, from window settings set to demonstrate fluid in the lungs in ***A***, the window level was greatly reduced (by tenfold) and the window width was more than tripled in order to better demonstrate the soft tissues of the lung parenchyma in ***B***.

On the CareStream/Kodak CR system, brightness can be set from 1 to 4094, and usually has a default setting of 2048. The contrast can be adjusted from 2 to 8191, and normally has a default setting of 4096. On the Fuji system, remember that the "S" number which comes up when the image is processed is the *exposure indicator* and therefore has no "default setting," but it can then be adjusted as the *brightness control.* Once it is changed, the exposure indication is lost unless you record it. The "S" number can be set from 1 to 20,045. An "L" number is also presented on the Fuji image review screen, indicating the *latitude* or gray scale level of the image. The default setting is 1.6 to 1.8. This gray scale can be adjusted from 0.5 up to 4.0. Since gray scale is opposite to contrast, in order to increase contrast in these Fuji images, turn the "L" number *down.*

Remember that when windowing of the brightness and contrast is applied to the image and then it

Figure 32-8

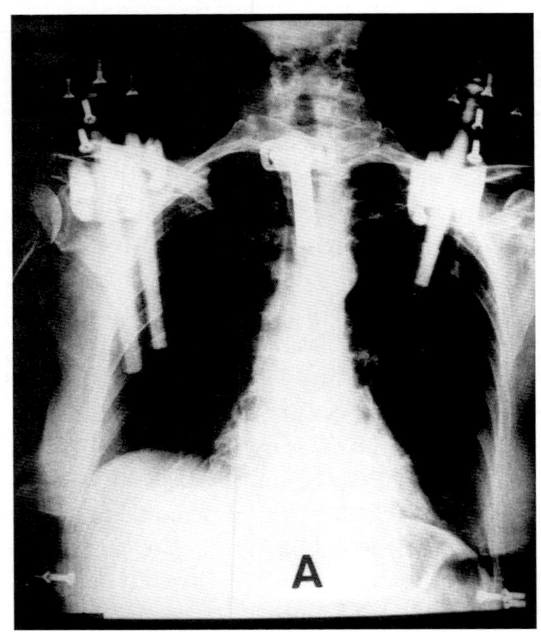

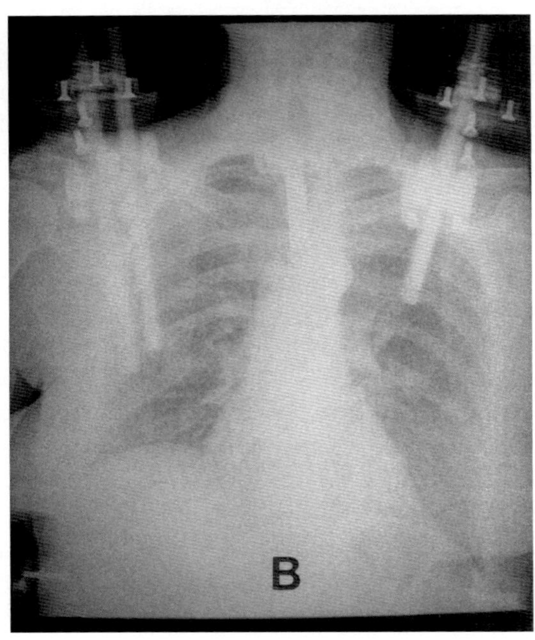

An example of proper windowing of a digital image of the chest, in which ***A*** was processed with a window level of 50 and window width of 500 to show fluid levels, whereas ***B*** was reprocessed with a greatly reduced window level (-500) and tripled window width (1600) to improve conspicuity of the air-filled soft tissues (parenchyma) of the lungs. (Courtesy, Robert DeAngelis, R.T.) (From Quinn B. Carroll, *Practical Radiographic Imaging*, 8th ed. Springfield, IL: Charles C Thomas Publisher, Ltd., 2007. Reprinted by permission.)

Figure 32-9

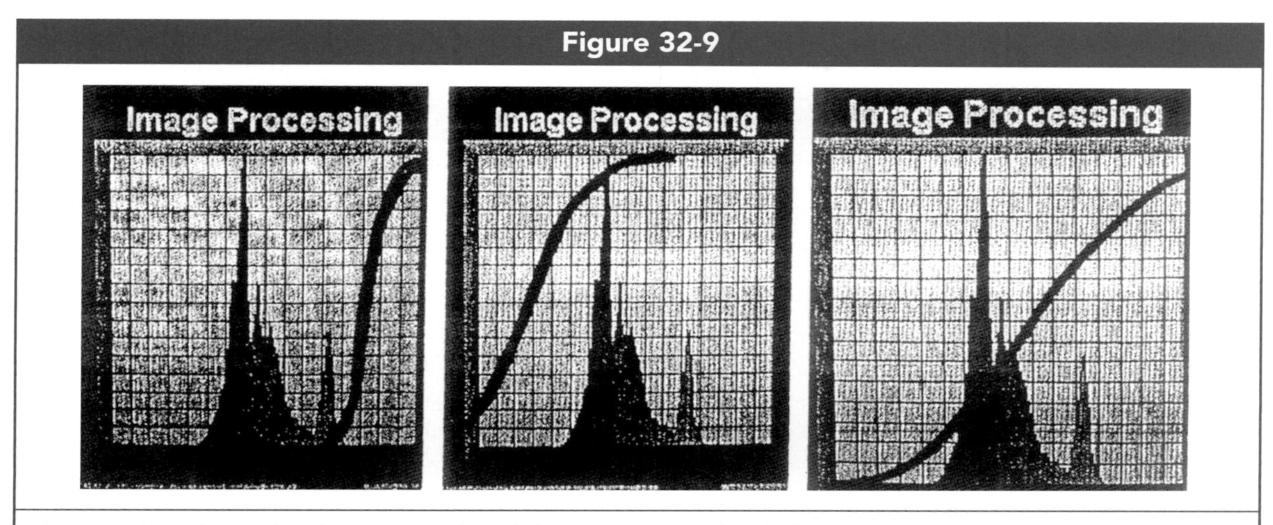

Continued windowing by the radiographer, if the image is saved with the changes, results in a loss of raw image data, reflected in the histogram as a progressively shallower slope on the gray scale curve (*left* to *right*). This means that the radiologist has less information from which to adjust the image and make a diagnosis. (From Quinn B. Carroll, *Practical Radiographic Imaging*, 8th ed. Springfield, IL: Charles C Thomas Publisher, Ltd., 2007. Reprinted by permission.)

is saved into the PACS system, the range of windowing available to the radiologist is always narrowed. Information is permanently lost from the original set of data. This loss of information is reflected in the histogram by a shallower slope on the characteristic curve as shown in Figure 32-9. Manufacturers recommend that saving windowed changes by the technologist generally be avoided. Minor windowing adjustments are allowed in some departments.

Some clarification of windowing terminology is helpful: Remember from Chapter 29 that the two generic terms for adjusting all computerized images are *window level* and *window width*. These terms are widely used in CT, MRI and angiographic imaging. They do apply to CR, DR and DF imaging, but manufacturers use many variations of terminology. Increasing the *window level* is usually understood as making the image *darker*. For CR, DR and DF, increasing the brightness or the "S" number is the **opposite** to increasing window level, but of course serves the same function visually.

Increasing the *window width* was described in Chapter 29 as lengthening the gray scale of the image. For the Fuji and Phillips systems the "L" number refers to image *latitude*, a synonym for gray scale, so these terms are consistent: Increasing the "L" number corresponds with increasing the window width. CareStream/Kodak and other systems have the more user-friendly adjustment for *contrast*. Increasing contrast is **opposite** to increasing window width, but again, serves the same function visually. Radiographers cross training from CR to CT, for example, will need to remember these distinctions.

SMOOTHING AND EDGE ENHANCEMENT

Smoothing algorithms, sometimes called "noise compensation," suppress the visible mottle in an image (Fig. 32-10). These are very useful when applied to relatively small amounts of mottle. It must be kept in mind, however, that severe mottle also indicates underexposure which represents a loss of small detail information within the image. The use of a smoothing algorithm *will not recover this lost information*. Radiographs presenting severe mottle from underexposure should be repeated.

As described in Chapter 31, either kernels or low-pass frequency filtering algorithms can be used to smooth the image. These also result in some loss of image contrast, so it is important that the original image not already possess low contrast which would

Figure 32-10

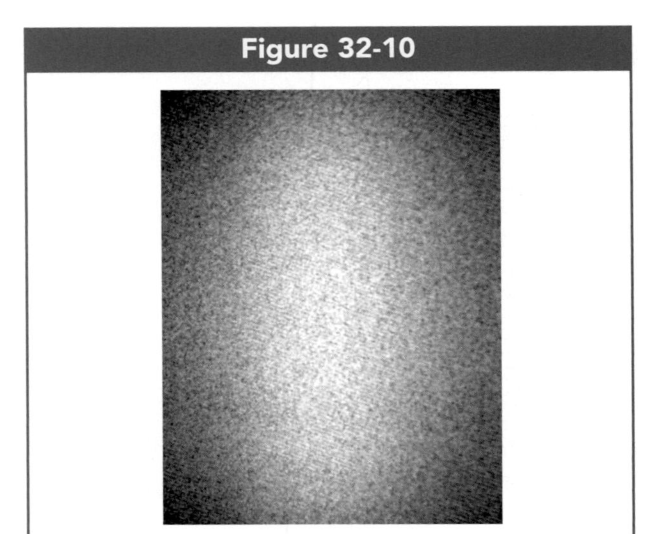

Demonstration of quantum mottle without a superimposed image for better visibility. Smoothing algorithms suppress mottle somewhat.

be made worse. The smoothing feature should be used with discretion. A beautiful example of image smoothing in a CT image was presented in Figure 31-21 in the previous chapter.

As with smoothing, the edge enhancement feature should also be used with discretion. Using edge enhancement on an already contrasty image can result in visible noise, including quantum mottle and *halo* artifacts. In the halo effect, the darker density side of detail edges is further darkened, while the light side is lightened even more. Figure 32-11 is a pair of veterinary images of an animal's knee with an orthopedic plate installed. In image ***A*** to the left, the halo effect is apparent at the condyles of the femur and around the metal plate (arrows). In the image ***B*** the halo artifacts were eliminated by simply *disengaging the edge-enhancement feature.*

Different manufacturers use various terms for edge enhancement and smoothing functions. In Figure 32-12 for example, a DR unit manufactured by GE has a feature called the "look" of the image, that can be set to *normal, hard,* or *soft.* The "hard" look setting actually applies edge enhancement to the image, while the "soft" look setting applies a smoothing algorithm which has the opposite effect. Figure 32-13 presents the resulting images for a lateral view of the C spine: ***A*** is the edge-enhanced or "hard" rendering of the image, ***B*** is the original or normal image for comparison, and ***C*** is the smoothed or "soft" rendering of the image.

It is important to remember that applying edge enhancement to an image that already has adequate

Figure 32-11

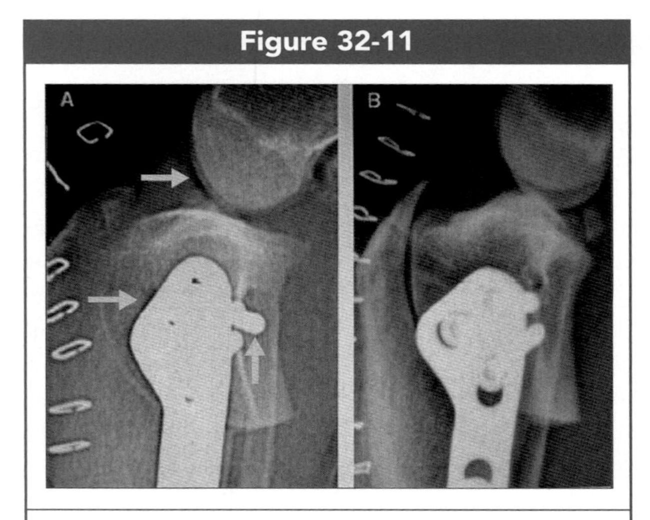

On a lateral view of an animal's knee, the halo effect can be seen around the femoral condyles and orthopedic plate (arrows) in image ***A*** when edge enhancement was used. In the image ***B*** the halo artifacts were eliminated by simply disengaging the edge-enhancement feature.

Figure 32-12

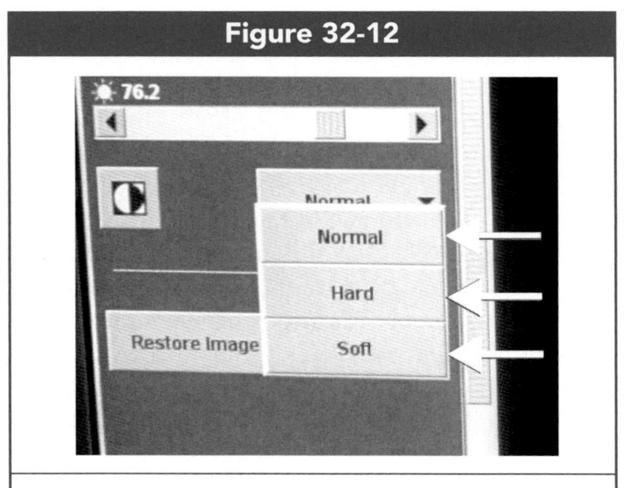

Selections from the "look" feature on a DR unit manufactured by GE include *normal, hard,* or *soft.* The "hard" look setting applies edge enhancement, while the "soft" look setting applies a smoothing algorithm. (From *Understanding Digital Radiograph Processing,* Midland, TX: Digital Imaging Consultants, 2013. Reprinted by permission.)

Figure 32-13

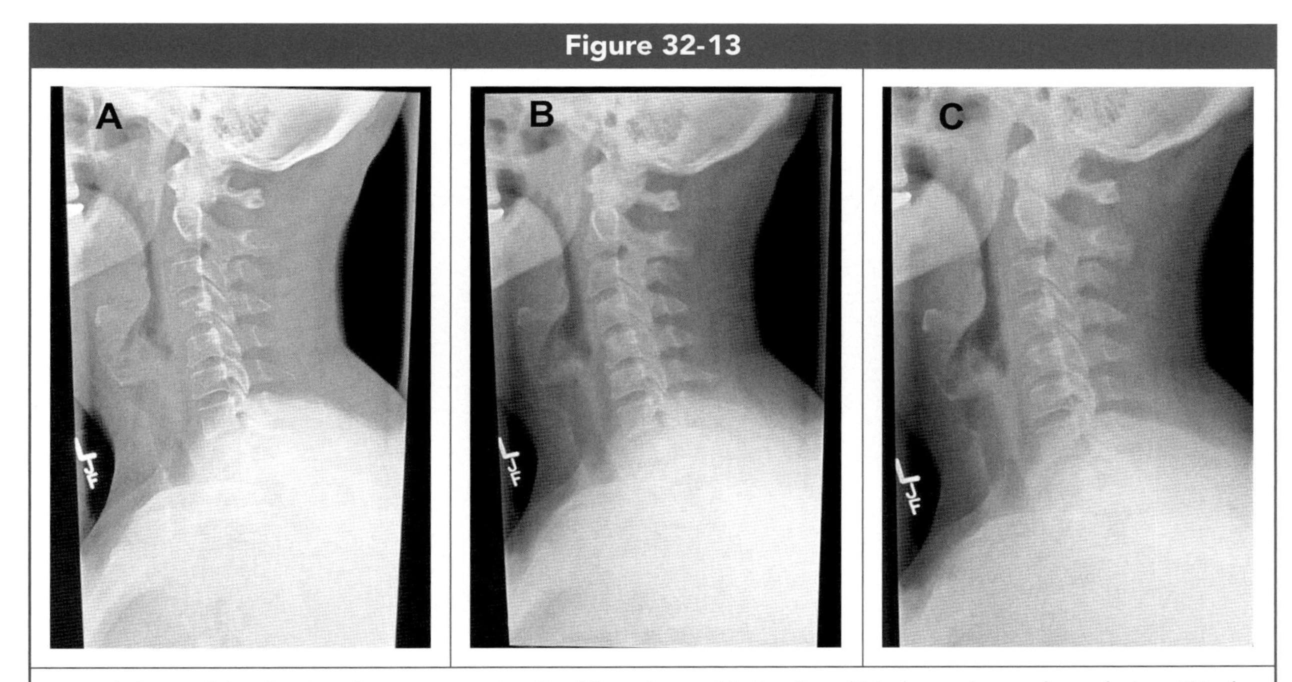

Lateral views of the C spine demonstrate the "look" settings: ***A*** is the "hard" (edge-enhanced) rendering, ***B*** is the normal setting for comparison, and ***C*** is the "soft" (smoothed) rendering of the image. (From *Understanding Digital Radiograph Processing*, Midland, TX: Digital Imaging Consultants, 2013. Reprinted by permission.)

contrast can actually cause some details to be obscured from visibility, possibly leading to misdiagnosis. In Figure 32-14, observe the bony area on C4 indicated by the arrow and compare the edge-enhanced radiograph ***B*** to the normal radiograph ***A***. In an actual clinical case, a hairline fracture in an area like this was found by the radiologist to be obscured from visibility when the *factory default* edge-enhancement feature was applied and this area became too bright as shown in image ***B***. It was determined that the factory default setting for this application over-applied the feature.

It was decided to develop a customized level of edge enhancement to be activated whenever the "hard" look setting was selected, that would use a more subtle application. Figure 32-15 shows the customization screen for the "look" of the image on this GE unit. *Most manufacturers provide the option for customizing applications of edge enhancement or smoothing features, available to the QC technologist in password-protected menus*. When a new unit is installed, always consult closely with the radiologists in the use of default or factory settings or developing customized settings for edge enhancement and smoothing features.

Smoothing and edge enhancement act as opposites: Excessive smoothing can lead to a loss of detail, excessive edge enhancement to noise.

MISCELLANEOUS PROCESSING FEATURES

Dark Masking

It is a simple matter for the computer to apply an offset pixel value in order to reverse blank collimation areas around the image into a black density, having already performed segmentation in the preprocessing stage. By reducing extraneous glare, black masking *always* improves the apparent visual contrast in the radiographic image, and is recommended to be applied to all images generally.

Image Reversal (Black Bone)

The image reversal or "black bone" feature is demonstrated in Figure 32-16. All of the pixel values

Figure 32-14

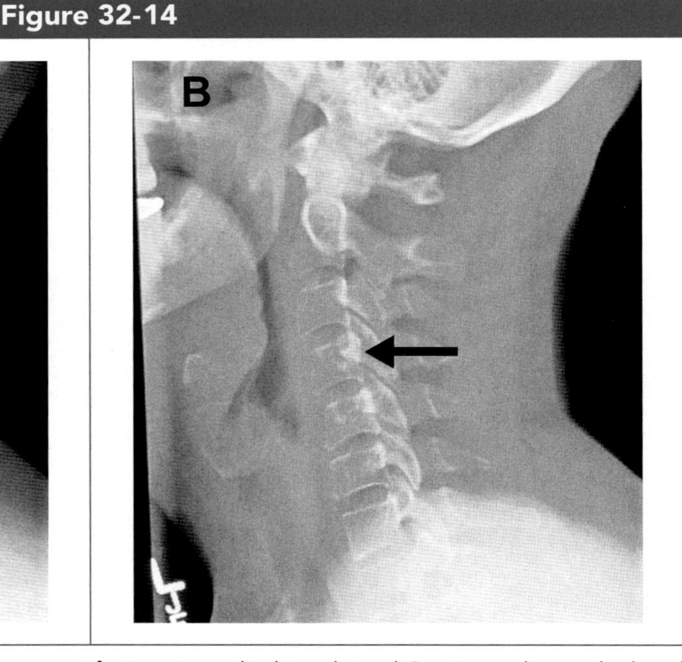

In image ***B***, when the *factory default* edge-enhancement feature is applied to a lateral C-spine radiograph already possessing high contrast, details are lost from the bony area on C4 indicated by the arrow (compare to normal radiograph ***A***). Clinical cases have shown that hairline fractures can be misdiagnosed this way. The settings for edge enhancement and smoothing features may need to be customized in collaboration with radiologists. (From *Understanding Digital Radiograph Processing*, Midland, TX: Digital Imaging Consultants, 2013. Reprinted by permission.)

within the image are simply inverted, such that a positive image (as opposed to the usual negative form of a radiograph) results. It is important for the student to understand that image reversal *produces no new information or details in the image.* However, it can sometimes make certain details more apparent to the eye. Even though this is a subjective effect, it can be helpful in diagnosis and is readily available at the touch of a button with CR and DR imaging.

Resizing

Resizing is used not only for *zoom* or *reduce/expand* functions which the operator may execute at the workstation, but to initially adapt the image size to the display device which may have much lower resolution than the image itself. To resize an image, the pixels are *mapped* onto a smaller or larger image matrix. This can be done by several different interpolation and convolution methods which are beyond the scope of this text.

Image Stitching

For scoliosis series or other body-length procedures, three CR plates can be exposed with a single exposure, by using a special wire mesh alignment grid. "Image stitching" software is available which uses the imaged alignment grid lines to accurately align and crop the 3 images to form a single body-length image. DR has the same type of stitching software available, but uses a moving detector and only requires 2 exposures to produce the image, shown in Figure 32-17.

CRITERIA FOR DIGITAL RADIOGRAPHIC IMAGE QUALITY

Having presented various methods by which the radiographer can adjust the digital image in this chapter, this would be a good place to review in concise terms the qualities we seek in the digital image. The student

Figure 32-15

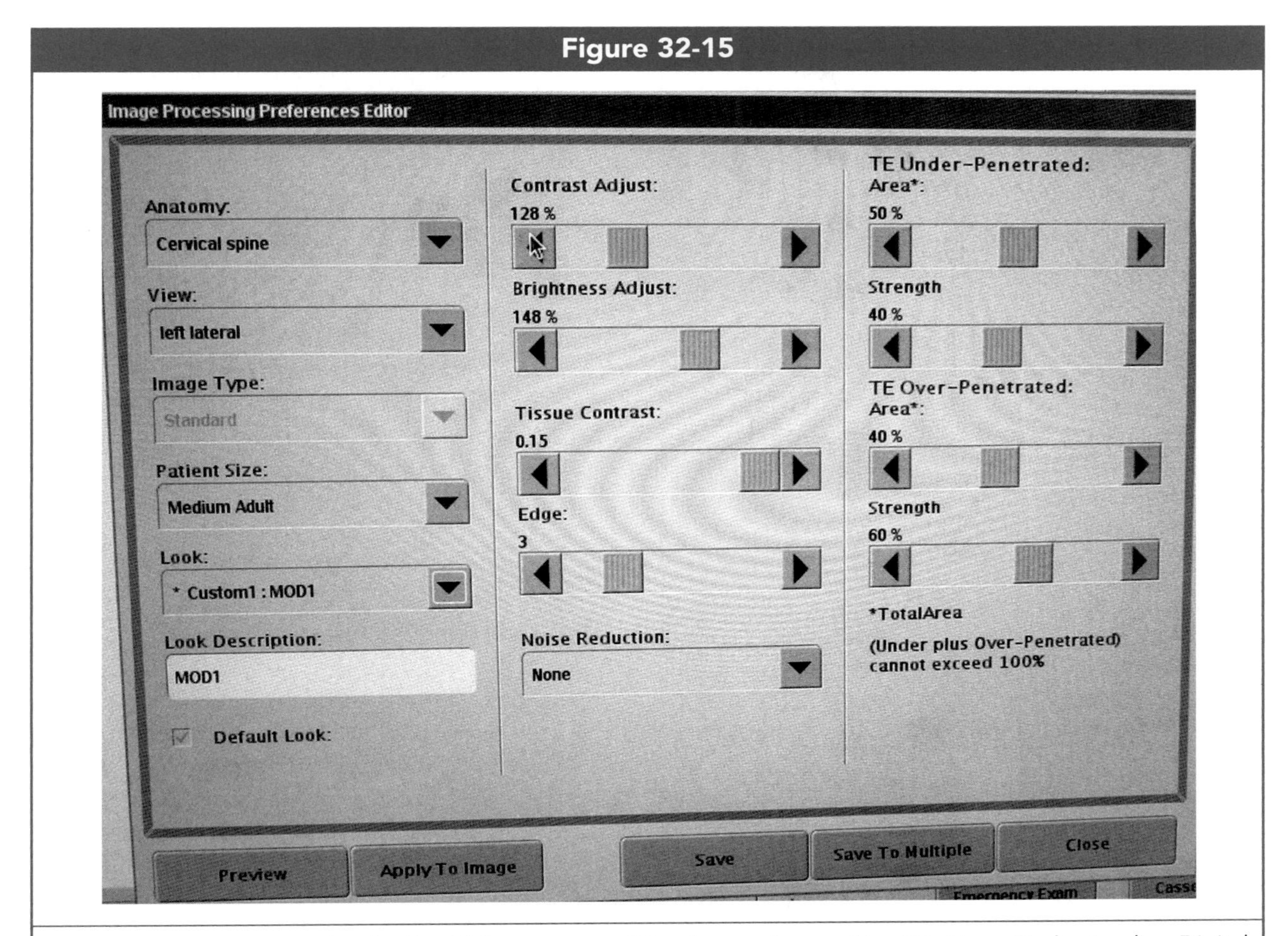

The customization screen for the "look" features on a DR unit manufactured by GE. (From *Understanding Digital Radiograph Processing*, Midland, TX: Digital Imaging Consultants, 2013. Reprinted by permission.)

should commit these concepts to memory, as they effectively constitute the *objectives* of radiographic practice. Eight essential criteria for the quality of digital images may be summarized as follows:

1. All pixel **brightness** levels within the anatomy of interest should be neither completely white nor completely black, but an intermediate shade along a broad scale ranging from a very light gray to a very dark gray. Brightness level may need to be adjusted to achieve this.
2. Image **contrast and gray scale** should be balanced such that the number of anatomical details present in the image is maximized, while preserving the visual differentiation between details. Contrast or gray scale settings may need to be adjusted to achieve this.
3. A maximum **signal-to-noise ratio (SNR)** must be achieved in every image. To ensure maximum penetration of signal through the patient to the detectors or imaging plate, *sufficient kVp must be used in the initial radiographic technique*. High kVp techniques not only ensure this level of penetration, but also spare radiation dose to the patient as opposed to the use of higher mAs techniques. Sufficient SNR may be visually evaluated by the lack of mottle or electronic noise in the displayed image. To check for image mottle, it is helpful to use the *magnification mode* in displaying the image.
4. Maximum **spatial resolution (sharpness)** should be apparent in the electronic image. For digitized static images, resolution should be at least 8 line-pairs per millimeter (LP/mm); for digital

Figure 32-16

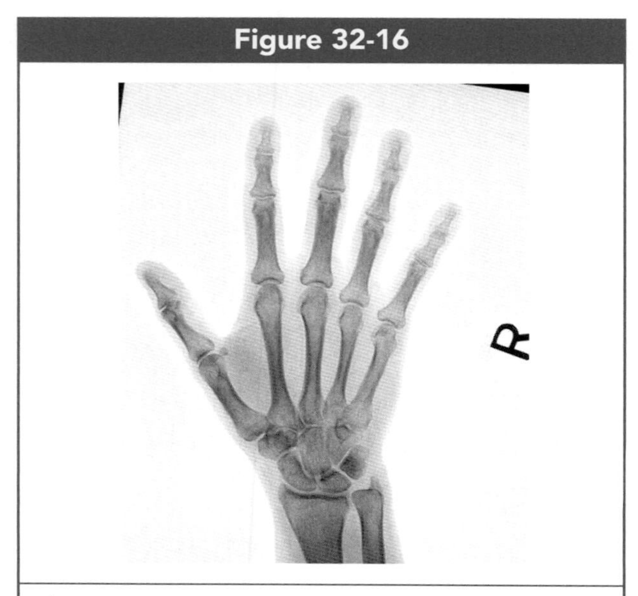

The image reversal or "black bone" feature produces a black-on-white, positive image. (From Quinn B. Carroll, *Practical Radiographic Imaging*, 8th ed. Springfield, IL: Charles C Thomas, Publisher, Ltd., 2007. Reprinted by permission.)

Figure 32-17

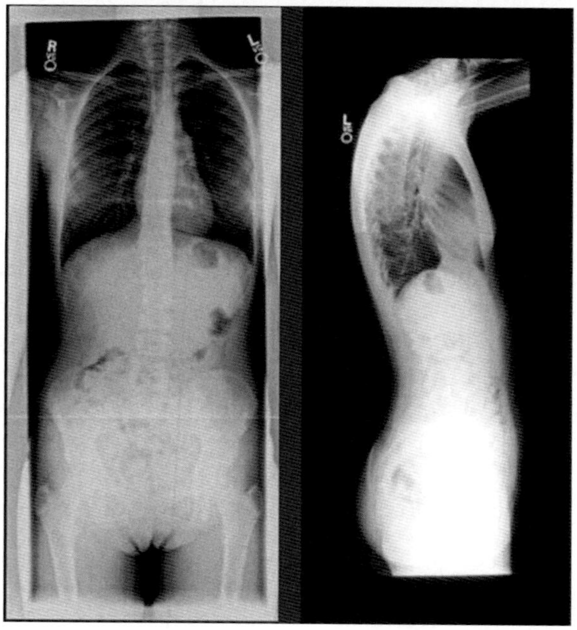

Full-spine scoliosis radiographs using the *image stitching* feature. Note the two "seam" artifacts where the three images were pieced together by the computer.

fluoroscopy, it should be 6 LP/mm. This is achieved not only by the initial geometrical factors employed in the radiographic projection, but also by the vertical and horizontal resolution capabilities of the display monitor. (For a digital image, resolution is affected by magnification, see #7.) Sharpness can also be affected by the focal spot size.

5. **Artifacts** of all kinds must be absent from the image to the degree possible. These include all types of removable objects between the x-ray tube and the image receptor, unnecessary superimposition of contrasty anatomy, mottle and all other image noise as discussed in #3.
6. **Shape distortion** must be minimized, such that the accurate representation of the anatomy of interest is achieved. This is primarily a function of positioning.
7. **Geometric Magnification (size distortion)** of the original image is generally undesirable (except for macroradiography). Structures in the image should be as close to the real size of the objects they represent as possible. Geometrical magnification is strictly controlled by the SID/SOD ratio and related positioning.
8. **Display Magnification** of the digital image is relative to the image matrix size and the physical display screen size, and is subjective in evaluation. Nonetheless, display magnification should generally not be so extreme that the image becomes "pixelly," where individual pixels become apparent and the sharpness of image details is lost.

   At certain levels of magnification or "zoom," depending on the matrix size of the original image, a moire or aliasing pattern becomes apparent in the displayed image. This is a very distracting form of image noise that can only be eliminated by changing to a higher or lower zoom level.

GLOSSARY AND ARRT STANDARD DEFINITIONS

By now, the reader will recognize that digital radiography is fraught with misnomers and misleading terminology. To assist, a glossary of digital imaging nomenclature and other key radiographic terms is presented at the back of the book. Also, the most current version of the *Standard Definitions* from the *American Registry of Radiologic Technologists (ARRT)* relating to digital radiography is presented after the glossary, as *Appendix I*.

"CONTROLLING" FACTORS FOR DISPLAYED IMAGE QUALITIES

It has been a tradition in radiography education to designate a primary *controlling* factor for each image quality, one variable that stands above other mere *contributing* factors for its impact on the image. Table 32-5 shows what these factors were for film-based radiography, compared side-by-side with what must now be acknowledged as the primary factors controlling the same qualities in the *displayed digital image*. The reader will note that *only one of the five image qualities listed* (shape distortion) *can still be attributed primarily to the same "controlling factor."* This table bears testimony to the deep impact that the advent of image digitization has had upon our thinking and paradigms in radiography.

SUMMARY

1. User-friendly touch screens on most digital radiography equipment allow radiographers to add or modify patient information, manipulate images, and deliver them to printers or storage devices with changes made.
2. Since the brightness and contrast of digital radiographic images is not directly related to the radiographic technique used, a permanently stored original exposure indicator which cannot be altered should be included on all digital images. Not all manufacturers meet this criteria.
3. The speed class of a digital image processing system can be selected. Although manufacturers tend to install equipment at the 200 speed class, it is strongly recommended that the 300 or 400 speed class be used for general radiography, and that the ability of the system to process images at 400 speed without substantial mottle become a standard for purchasing decisions.
4. As a guideline to acceptable exposure levels, the new *deviation index* published by the AAPM is being implemented by all manufacturers. The index ranges from +3 (indicating twice the ideal exposure) to –3 (indicating ½ ideal exposure), in 25% *multiplicative* increments. Digital exposures should not be repeated unless saturation or mottle are *certain*.
5. To minimize preprocessing errors, most manufacturers sample pixels only from a central portion of the image to calculate the exposure indicator.
6. *Exposure indicator errors* may result from scatter radiation, extreme over- or underexposure, unexpected materials within the field, histogram errors or delays in processing.
7. The reprocessing of images under *alternative algorithms* has become widespread and can be used to advantage, but should be done with discretion and in consultation with the quality control technologist.

Table 32-5

Comparison of Primary "Controlling" Factors for *Displayed* Image Qualities

Image Quality	Film-Based Radiography	Digitized Radiography
Brightness/Density	mAs	Rescaling
Gray Scale/Contrast	kVp	LUTs
Sharpness	Focal Spot	Pixel Size
Magnification	Distances (SID/SOD Ratio)	Matrix Size
Shape Distortion	Alignment	Alignment

8. *Windowing* of the brightness and contrast of images can be done, but results in a loss of data from which the radiologist can manipulate the image if it is saved with the changes. Minor adjustments may be allowable.
9. The image *smoothing* feature suppresses noise in the image, including moderate levels of quantum mottle. It should be used with discretion. Radiographs with *severe* mottle should be repeated. Edge enhancement should also be used with discretion, as it can re-introduce noise into the visible image or cause *halo* artifacts.
10. Dark masking is *always* recommended for digital radiographs, as it enhances visual contrast. Image reversal may subjectively make some details more apparent, but adds no actual information to the image.
11. Radiographers should commit to memory the eight *Criteria for Digital Radiographic Image Quality*, page 555.

REVIEW QUESTIONS

1. On most digital image displays, the cursor can be moved left and right while holding the mouse button down to adjust __________.

2. For digital images, the only indication the radiographer has that the set radiographic technique has resulted in an overexposure is the __________.

3. The __________ method of averaging pixel values within the region of interest is the most resistant to skewing by preprocessing errors.

4. For a logarithmic scaled exposure indicator, if the EI reads 0.3 (or multiple thereof) below the target value, the actual exposure to the detectors was what ratio (fraction or multiple) of the target exposure level?

5. Upon resetting the speed class of a digital processor, if twice the previous exposure is required to obtain the target exposure indicator, the speed class must have been changed to __________ the original.

6. To minimize preprocessing errors, most manufacturers sample pixels only from a __________ portion of the image to calculate the exposure indicator.

7. According to the AAPM __________ guidelines, the *only* time the radiographer may assume that a repeat exposure is indicated for digital radiography is when the deviation index falls:

(Continued)

REVIEW QUESTIONS *(Continued)*

8. Generally, the parameters for an acceptable range of exposures are set from __________ to __________ of the target level for actual exposure.

9. On the Fuji and Konika systems, once the "S" number is changed, it becomes the control for image _________.

10. For distal extremities, many manufacturers base their exposure indicator recommendations on operation at the 100-speed class. Compared to the 200-speed class, this means that the normal range of exposure readings is expected to be (higher, lower, or equal):

11. Which manufacturer refers to its exposure indicator as the *log median value*?

12. When a digital exposure is 68% too high, the deviation index will read:

13. List three possible causes of exposure indicator errors:

14. What is the significance of the characteristic curve on the histogram developing a more shallow slope after windowing the image?

15. The image smoothing feature should not be applied to an image which already has very low __________.

16. For proper brightness in digital image, no portion of the anatomy of interest should appear __________ or __________.

17. For digital radiographs, spatial resolution should be at least ________ LP/mm.

18. For digital radiographs, excessive magnification is characterized by a ________ appearance to the image.

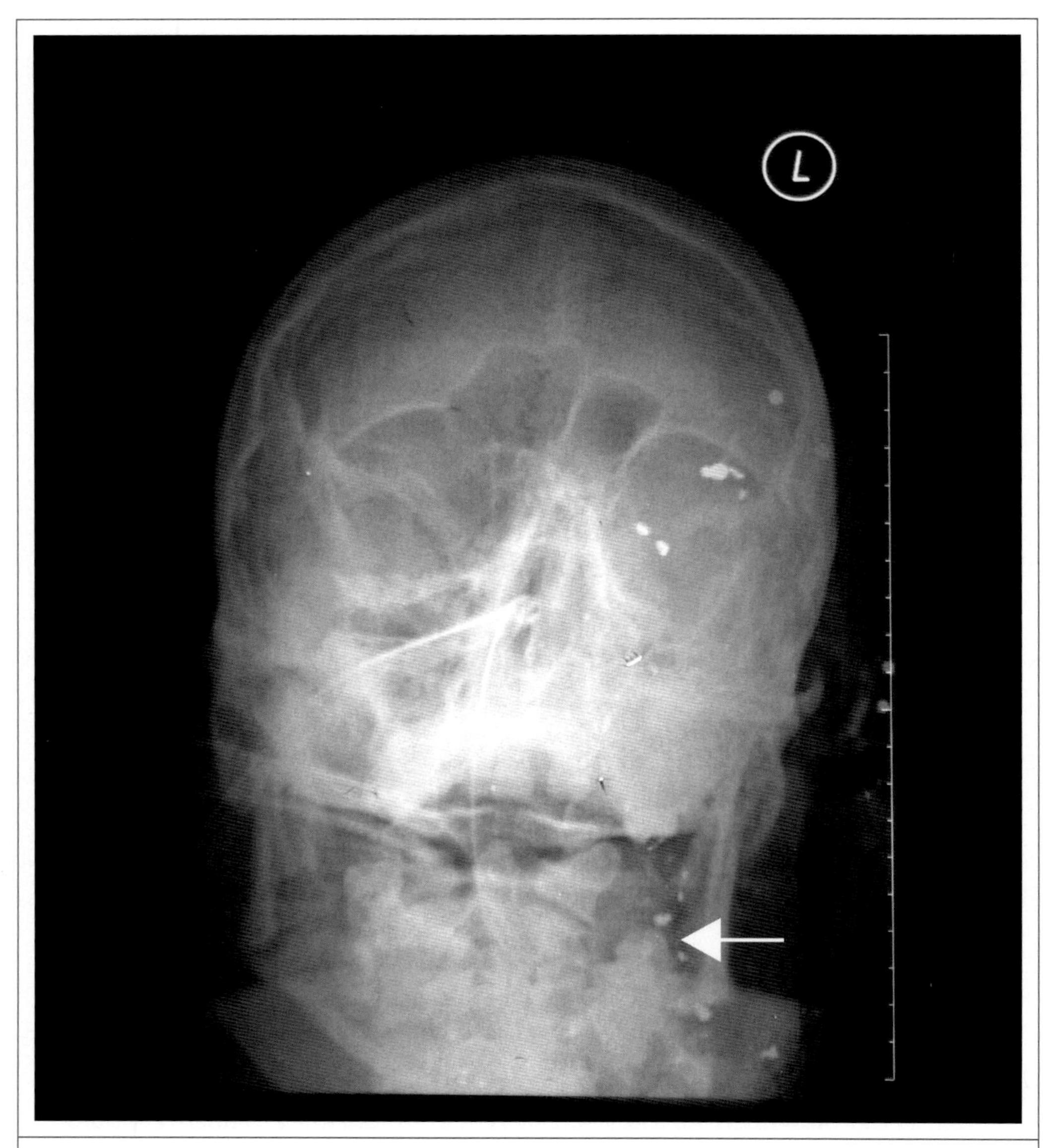

An attempted suicide by gunshot to the right side of the patient's head shows the devastation of tissues from the *left* side exit wounds where the bullet has fragmented into numerous pieces. Extensive soft tissue damage is evident along with bone destruction. A fragment of the left lower jaw with molars evident (arrow) is displaced more than an inch below its normal position.

Chapter 33

APPLYING RADIOGRAPHIC TECHNIQUE TO DIGITAL IMAGING

Objectives:

Upon completion of this chapter, you should be able to:

1. Explain the rationale for allowing much higher kVp levels in digital imaging than were used in conventional film-based radiography and why a 15% "across-the-board" increase in kVp is recommended for reducing patient exposure.
2. Describe the exposure latitude for digital images, what factors set its upper and lower limits, how it has impacted patient exposure, and what we in the profession should do about it.
3. Explain the importance of sufficient penetration and a high signal-to-noise ratio (SNR) for digital imaging.
4. Explain the effects and ranges of underexposure and overexposure for both mAs and kVp in digital imaging.
5. Dispel common myths about technique settings for digital imaging.
6. Clearly describe the resilience of digital imaging to localized fog patterns from scatter radiation caused during exposure versus "pre-fogging" of a CR cassette.
7. Define the continuing role of "manual" technique skills and proportional anatomy in the age of digital imaging.
8. Describe the proper use of AEC for digital imaging and preventing mottle in AEC images.
9. Describe the impact of centering on mottle in DR images.
10. Describe the effects of centering of the anatomy, alignment of multiple fields on the IR, and overcollimation on processing errors in CR.
11. State important considerations for horizontal projections and bilateral views to avoid CR processing errors.
12. Describe the retention of the image on CR phosphor plates and its implications for practice.

The advent of computer processing for radiographic images has introduced not only new concepts, but whole new paradigms for the radiography community. A *paradigm* is an underlying assumption, something so taken for granted to be true that it forms a foundation for a whole structure of beliefs and behaviors that are built up upon it. It is no surprise, then, that the *decoupling* or disconnection between the qualities of the final digital image and the original technique factors used to acquire the image created a conceptual challenge for radiographers living through the transition from screen/film technology to the age of digital imaging.

In its 2009 report, Task Group 116 of the American Association of Physicists in Medicine states, "Unlike screen-film imaging, image display in digital radiography is independent of image acquisition."[1] In the

[1] Shepard et al.: Exposure Indicator for DR: TG116 (Executive Summary) in *Medical Physics*, Vol. 36 No. 7, July 2009.

final displayed image, the *only* quality that has typically *not* been altered by digital processing is typically shape distortion, which is primarily determined by positioning. Brightness, contrast, noise, sharpness and magnification have all been "tampered with" upon final display. What role, then, *does* the original radiographic technique play in determining final image quality?

The answer is that technique plays one, and only one, very critical role, and that is to *ensure that adequate signal reaches the detector system such that computer algorithms can be successful in making corrections and refinements to the image.* In more familiar terms, the one objective for setting radiographic technique on a digital unit is to get plenty of exposure to the image receptor without unnecessary exposure to the patient. This can be directly measured by the signal-to-noise ratio (SNR) at the detector.

The mAs setting and the SID combine to determine the *quantity* or intensity of radiation incident upon the patient, but since kVp controls the percentage penetration of that radiation *through* the patient, it also has a profound effect on the final intensity of the *remnant beam reaching the detector* behind the patient. To achieve adequate exposure to the image receptor and a good SNR, mAs *and* kVp in relation to each other (along with distances, filtration and generator type) must all be taken into consideration.

We will find in this chapter that all of the physics and technique concepts related to this goal (of achieving sufficient exposure at the detector) remain critical to the practice of radiography, and have not changed. For example, the concept that *no amount of mAs can compensate for insufficient kVp* still holds true. On the other hand, particularly when discussing the qualities of the final displayed image, we find that many old concepts must be completely discarded in order to avoid confusion. This chapter will attempt to sort out which concepts belong to the "still true" group and which to the "discard completely" group.

MINIMIZING PATIENT EXPOSURE

Many CR systems are installed and being operated at a speed class of 200. This is only one-half the speed of the "regular" rare earth screens (400) that were popular over the last quarter of the twenthieth century. In making the change from rare earth screen systems to CR, many radiology departments have at least doubled the mAs values used for most Bucky procedures, with some adjustments being more than this. This resulted in an undesirable doubling of x-ray exposure to patients undergoing pelvic, abdominal and head procedures, just where the most radiosensitive organs are located.

As described in Chapter 32, operation of a CR or DR system at the 400-speed class assumes an average exposure reaching the imaging plate of 0.5 mR. It is possible for this level of exposure to be insufficient in some cases, based on the *assumption* of using previously popular kVp levels. But, by increasing *kVp* rather than mAs, penetration of the x-ray beam through to the imaging plate *does* result in sufficient exposure to the receptor elements, and allows operation at the 400-speed class.

High kVp and Scatter Radiation

Figures 33-1 and 33-2 use conventional radiographs to demonstrate how the effects of higher kVp levels upon the production of scatter radiation have traditionally been over-emphasized. The primary causes of scatter radiation are *patient size* and *collimation*, both of which bear upon the *volume of exposed tissue.* The effects of kilovoltage, while important to understand, are *secondary* when compared to these issues of tissue volume. Figure 33-1 shows a pair of AP elbow exposures taken at 65 kVp and 90 kVp for comparison. While desirable penetration and gray scale are achieved in radiograph ***B***, no significant *fogging* is visible *even when 25 more kVp than usual is used.* This is because the anatomy has too small a volume of tissue to generate much scatter radiation at *any* kVp. Figure 33-2 demonstrates two abdomen radiographs of the same patient taken at 80 kVp and 92 kVp for comparison. Both were taken using the Bucky grid to attenuate scatter radiation. Again, while Radiograph ***B*** shows increased gray scale and penetration as expected, it is *not visibly fogged*—this result in spite of the fact that the abdomen is the portion of the body most prone to generate scatter radiation. While there is no question that kVp as high as 120 would generate unacceptable levels of scatter radiation during an

Figure 33-1

Proof that high kVp can be used on small anatomy without substantial generation of scatter radiation. Using an AP projection of an elbow, kVp was increased from ***A*** to ***B*** by *25 kV*, from 65 to 90. Using conventional film radiographs, we can see an increase in gray scale *due to enhanced penetration* of the x-ray beam, but no evidence at all of fog. (From Quinn B. Carroll, *Practical Radiographic Imaging*, 8th ed. Springfield, IL: Charles C Thomas, Publisher, Ltd., 2007. Reprinted by permission.)

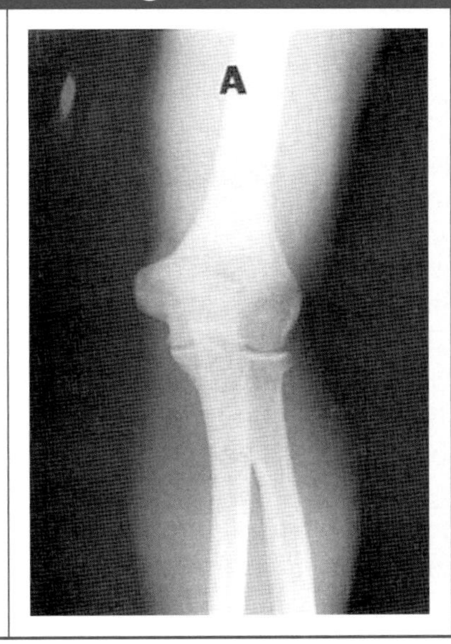

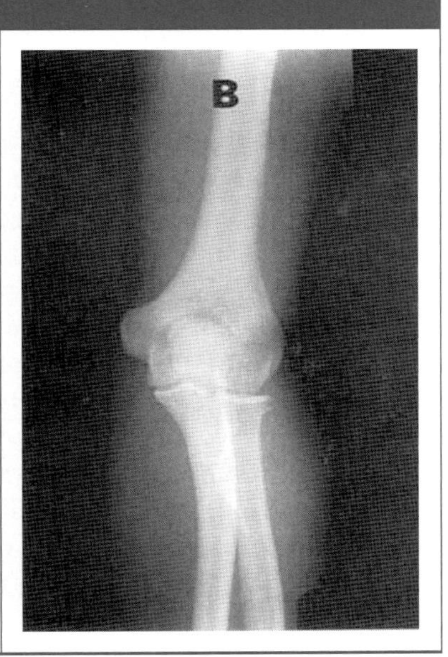

Figure 33-2

A

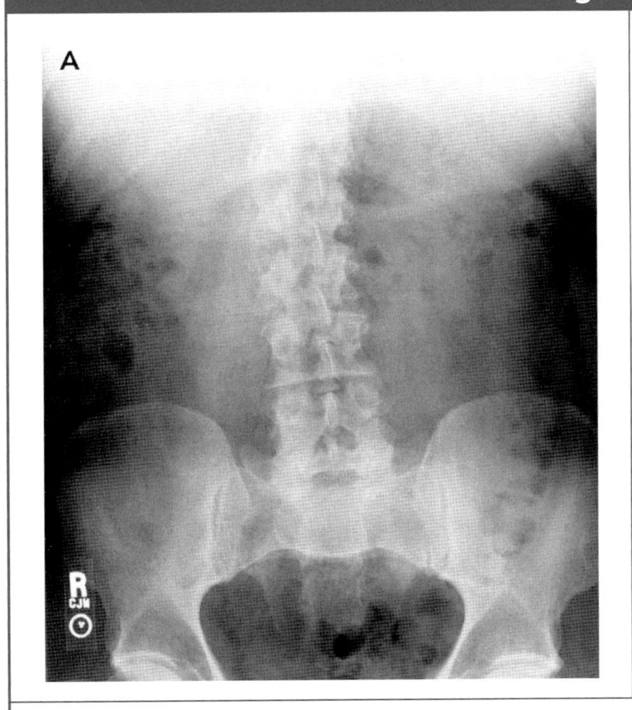

B

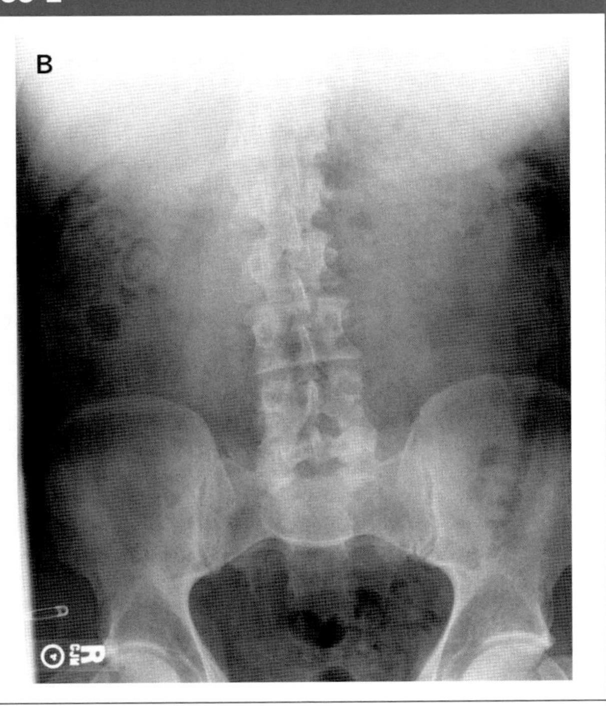

Further proof that "across-the-board" technique increases of 15 percent in kVp can be applied to all techniques without a substantial increase in scatter radiation. The abdomen is the body part expected to produce the most scatter radiation. Using screens and film, radiograph ***A*** was taken using 80 kVp and 40 mAs. Radiograph ***B*** was produced with 92 kVp and one-half the mAs. A slight lengthening of gray scale is apparent *due to increased penetration*. There is no visible fogging of this image. (From Quinn B. Carroll, *Practical Radiographic Imaging*, 8th ed. Springfield, IL: Charles C Thomas, Publisher, Ltd., 2007. Reprinted by permission.)

abdomen exposure, this experiment demonstrates that ***kVp levels can be increased by 15 per cent "across the board" for technique charts without substantial generation of scatter radiation.***

Using such an approach would allow all mAs values to be cut in half, bringing patient exposures closer to the previous levels using rare earth screen technology. Some would use the fact that digital detectors are highly sensitive to scatter radiation to argue against using high kVp's with digital equipment. This may be true in discussing the physics of the detector, but it does not really address the outcome of the final image after processing, which is what really matters. As will be shown later in this chapter (see Figure 33-9), digital processing of the image includes the noise reduction features described in Chapter 31, which are able to effectively subdue the effects of scatter radiation *caused during the radiographic exposure.* In fact, manufacturers themselves have recommended the general approach of using higher kVp's with CR and DR systems, as demonstrated by the manufacturer's technique chart in Table 33-1. They discourage employing less than 60 kVp even on distal extremities, (hand radiographs were routinely exposed using 52 kVp with film/screen radiography).

To test the application of this idea to digital radiography, an experiment was conducted by several colleagues on *nine different brands* of digital processing units, using a series of 15% step increases in kVp with compensating reductions in mAs (1/2 mAs for each step). The study was repeated on abdomen, chest and knee phantoms, totaling 33 series of images (165 images) for analysis. Several representative examples of these series are illustrated in Figure 33-3. Two key questions were addressed: (1) Was final displayed image contrast affected in any significant, diagnostically destructive way, and (2) At what point, if any, did image mottle reach a diagnostically destructive level?

Of the 33 series, 4 showed *completely indiscernible or barely visible changes* in contrast from one 15% step to the next, 21 displayed *slight but noticeable* reductions in contrast, and 5 demonstrated *moderate to significant* contrast loss. Of the nine manufacturers, one (Fuji) showed contrast changes that could be generally described as *indiscernible to very slight*, 7 as *slight to moderate*, and 1 as *moderate to significant.* At the end of each row of images in Figure 33-3, the first and last image are placed side-by-side for comparison. These represent a *52-kVp* difference. With conventional film/screen radiography, the image to the right would have been seriously fogged rather than just demonstrating reduced gray scale.

We may conclude that increasing kVp still generally reduces contrast in the final displayed image just as it did in conventional radiography, but that the impact is less (and in some cases may not even be discernible for a single 15% increase in kVp).

Later in this chapter we demonstrate that digital processing is very effective at "cleaning up" the effects of scatter radiation. This makes sense because scatter can be identified as a low-frequency (large area) phenomenon in the image. It also means that the reduction of contrast at higher kVp levels shown in Figure 33-3 must be due primarily to *increased penetration* rather than scatter radiation. Higher beam penetration reduces subject contrast in the "latent" image picked up at the detector, and it appears that digital processing preserves this general relationship, although lessening its impact, through to the final displayed image.

On a more practical note, in the digital age it has become a universal practice for radiologists to immediately window the images they examine, increasing contrast or gray scale according to the anatomy or the particular condition to be ruled out. This renders the contrast differences illustrated in Figure 33-3 not only minor from step-to-step, *it renders them practically irrelevant.* In daily clinical practice, what difference does it make that a 15% kVp increase results in a slightly lower-contrast image initially displayed, when the radiologists immediately window the contrast to their personal preference anyway?

High kVp and Mottle

Using 200% magnification for evaluation of mottle, Figures 33-4, 33-5 and 33-6 demonstrate the effects of a single 15% step increase on a non-grid extremity procedure, a grid abdomen exposure for maximum scatter production, and a grid chest exposure for minimal mAs values. These pairs of images are taken from 15 of the 33 series studied and include samples from all nine manufacturers (Agfa, Fuji, General Electric, Kodak [now CareStream], Konica, Philips, Shimadzu, Siemens and Toshiba).

Table 33-1

Computed Radiograph

(Addendum #1)		**Radiographic Exposure Recommendations**				**MAS**		
Anatomical Region	**Measurement (cm) (Medium)**	**Exam**	**kVp**	**SID**	**GRID**	**Small**	**Medium**	**Large**
Skull	**18** to **21**	Skull PA/AP	80	40	Yes	12	20	30
	14 to **17**	Skull Lateral	80	40	Yes	6	12	18
	18 to **21**	Skull Townes, Waters	85	40	Yes	12	24	40
	14 to **17**	**Facial Bones** Lat.(Bucky)	80	40	Yes	6	10	12
	14 to **17**	Facial Bones Lat (Non-Bucky	60	40	**No**	2	4	6
	18 to **21**	**Nasal Bones** (Non-Bucky)	60	40	**No**	2	3	4
Spine	**11** to **14**	**Cervical** AP/OBL	80	40	Yes	6	12	18
	11 to **14**	Cervical Lateral	80	72	Yes	18	28	45
	11 to **14**	Cervical Odontoid	80	40	Yes	8	14	20
	21 to **25**	C-7/T-1 Swimmers	85	40	Yes	24	40	64
	20 to **24**	**Thoracic** AP/OBL	80	40	Yes	18	28	40
	28 to **32**	Thoracic Lateral	85	40	Yes	25	35	50
	18 to **22**	**Lumbar** Spine AP/OBL	80	40	Yes	24	40	64
	27 to **32**	Lumbar Spine Lateral	85	40	Yes	30	64	100
	27 to **32**	Lumbar L-5/S-1 Spot	90	40	Yes	30	64	100
Chest	**20** to **25**	**Chest** PA	100	72	Yes	4	6	10
	27 to **32**	Chest Lateral	110	72	Yes	8	12	20
	20 to **25**	Chest **Portable** **(GRID)**	100		**Yes**			
	20 to **25**	Chest **Portable** (Non-GRID)	<85		**No**			
Thorax	**20** to **25**	**Sternum** RAO	80	40	Yes	20	30	40
	27 to **32**	Sternum Lateral	85	40	Yes	30	40	50
	20 to **25**	**Ribs** AP/PA/OBL Upper	70	40	Yes	15	25	40
	20 to **25**	Ribs AP/PA/OBL Lower	80	40	Yes	20	36	50
Shoulder	**12** to **16**	**Shoulder** AP	80	40	Yes	6	12	15
	4 to **6**	Shoulder Axillary (non-Bucky)	70	40	**No**	3	5	8
	12 to **16**	**Scapula** AP	80	40	Yes	10	20	36
	13 to **17**	Scapula Lateral	80	40	Yes	12	24	40
Abdomen	**18** to **22**	**Abdomen**—KUB	80	40	Yes	24	40	64
	18 to **22**	Abdomen—Upright Decubitus	85	40	Yes	28	48	64
	18 to **22**	**Barium Studies** (GI, BE)	100	40	Yes			
	18 to **22**	**Contrast Studies** (IVP, GB)	80	40	Yes			
Pelvis	**19** to **23**	**Pelvis** AP	80	40	Yes	24	40	64
	17 to **21**	**Hip** AP	80	40	Yes	20	35	50
	17 to **21**	Hip X-Table Lateral	85	40	Yes	25	40	60
Upper Extremity	**1.5** to **4**	**Fingers**	60	40	No		1	
	3 to **5**	**Hand** AP/OBL	60	40	No		2	
	3 to **5**	Hand Lateral	60	40	No		3	
	3 to **6**	**Wrist** AP/OBL	60	40	No		2	
	3 to **6**	Wrist Lateral	60	40	No		3	
	6 to **8**	**Forearm** AP	60	40	No		3	
	6 to **8**	Forearm Lateral	60	40	No		3	
	6 to **8**	**Elbow**	60	40	No		3	
	7 to **10**	**Humerus** (Bucky)	75	40	**Yes**	6	10	15
Lower Extremity	**1.5** to **4**	**Toes**	60	40	No		1	
	6 to **8**	**Foot** AP/OBL	60	40	No		3	
	6 to **8**	Foot Lateral	60	40	No		3	
	8 to **10**	**Oscalsis**	60	40	No		6	
	8 to **10**	**Ankle** AP—Mortise	60	40	No		3	
	8 to **10**	Ankle Lateral	60	40	No		3	
	10 to **12**	**Tib-Fib** AP	60	40	No		3	
	10 to **12**	Tib-Fib Lateral	60	40	No		3	
	10 to **13**	**Knee** AP—Lateral(Bucky)	80	40	**Yes**	6	12	20

From Quinn B. Carroll, *Practical Radiographic Imaging*, 8th Ed. Springfield, IL: Charles C Thomas, Publisher, Ltd., 2007. Reprinted by permission.

Figure 33-3

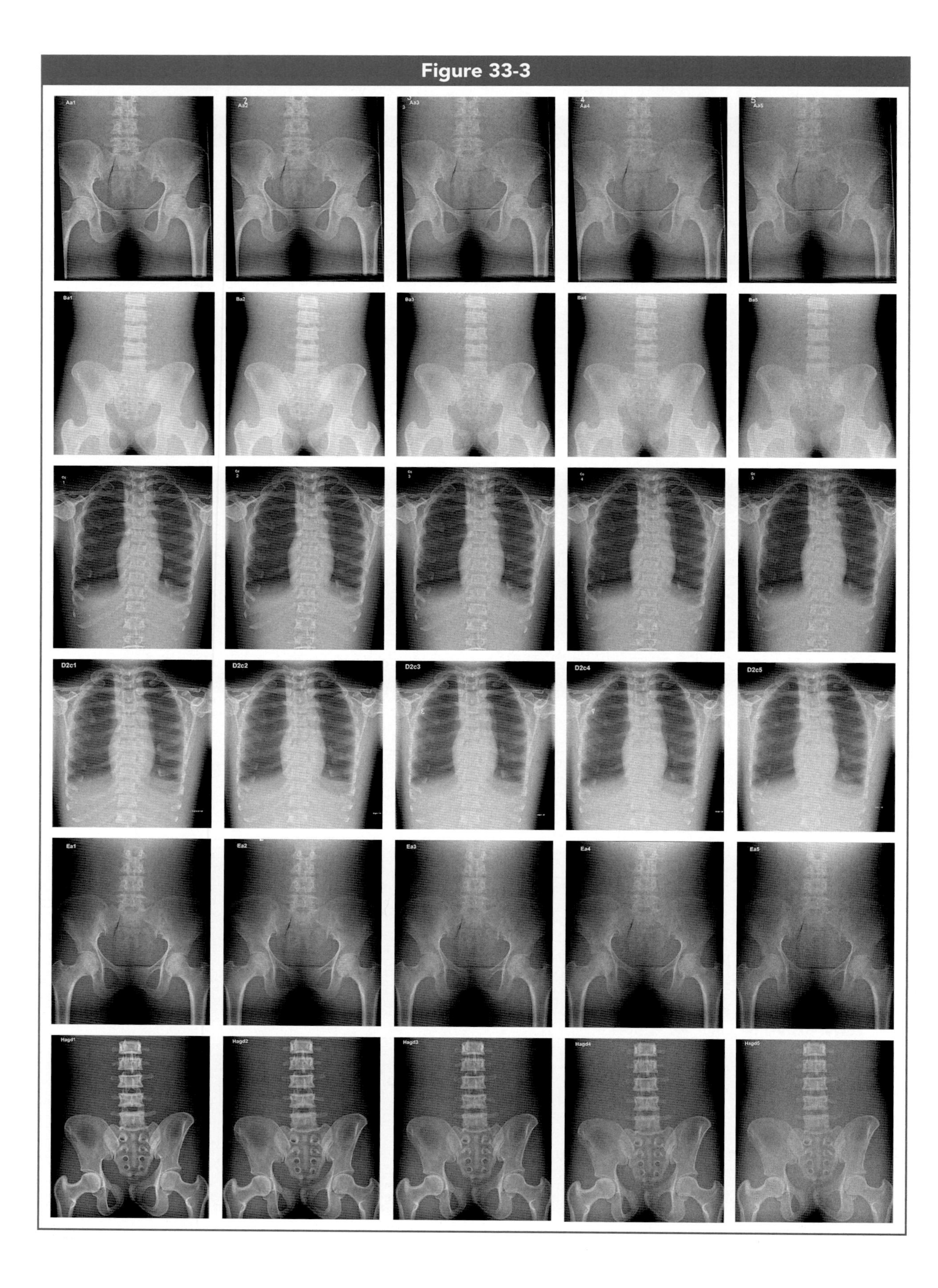
Aa1
Aa2
Aa3
Aa4
Aa5
Ba1
Ba2
Ba3
Ba4
Ba5
D2c1
D2c2
D2c3
D2c4
D2c5
Ea1
Ea2
Ea3
Ea4
Ea5
Hagd1
Hagd2
Hagd3
Hagd4
Hagd5

Figure 33-3 *(Columns Continued)*

Six examples of the 33 series of 15% step increases in kVp taken across nine different manufacturers. Each series consisted of *four* step increases, with each step compensated by a halving of the mAs. At the right, the minimum and maximum kVp's from each series are compared alongside each other, representing an average *52-kVp* spread.

On the facing page, the first three rows show that these manufacturers' processing algorithms were robust enough that image contrast is almost indistinguishable from one 15% step increase in kVp to the next, and even after four step increases show only a moderate loss of contrast.

The bottom three rows are examples of manufacturers whose algorithms show visible reductions in image contrast from one 15% step increase to the next, with a substantial lengthening of gray scale brought about by a 52-kVp increase. Note that even these images do *not* show conventional fogging patterns, only reduced overall contrast.

Row A: Grid exposures of abdomen phantom processed with a Shimadzu *Rad Speed* DR unit. Techniques: (1) 70 kVp at 40 mAs, (2) 80 kVp at 20 mAs, (3) 92 kVp at 10 mAs, (4) 106 kVp at 5 mAs, (5) 122 kVp at 2.5 mAs.

Row B: Grid exposures of abdomen phantom processed with an Agfa CR unit. Techniques: (1) 70 kVp at 40 mAs, (2) 80 kVp at 20 mAs, (3) 92 kVp at 10 mAs, (4) 106 kVp at 5 mAs, (5) 122 kVp at 2.5 mAs.

Row C: Grid exposures of chest phantom processed with a Fuji CR unit. Techniques: (1) 70 kVp at 32 mAs, (2) 80 kVp at 16 mAs, (3) 92 kVp at 8 mAs, (4) 106 kVp at 4 mAs, (5) 122 kVp at 2 mAs.

Row D: Grid exposures of chest phantom processed with a General Electric *Definium 8000* DR unit. Techniques: (1) 80 kVp at 16 mAs, (2) 92 kVp at 8 mAs, (3) 106 kVp at 4 mAs, (4) 122 kVp at 2 mAs, (5) 140 kVp at 1 mAs.

Row E: Grid exposures of abdomen phantom processed with a Kodak (CareStream) *CR-950* unit. Techniques: (1) 70 kVp at 40 mAs,(2) 80 kVp at 20 mAs, (3) 92 kVp at 10 mAs, (4) 106 kVp at 5 mAs, (5) 122 kVp at 2.5 mAs.

Row H: Grid exposures of abdomen phantom processed with a Siemens *Axiom Aristos* DR unit. Techniques: (1) 70 kVp at 40.25 mAs, (2) 81 kVp at 20.24 mAs, (3) 93 kVp at 10.25 mAs, (4) 105 kVp at 5.26 mAs, (5) 121 kVp at 2.74 mAs.

(Courtesy of Philip Heintz, PhD, Quinn Carroll, MEd, RT, and Dennis Bowman, AS, RT. Reprinted by permission.)

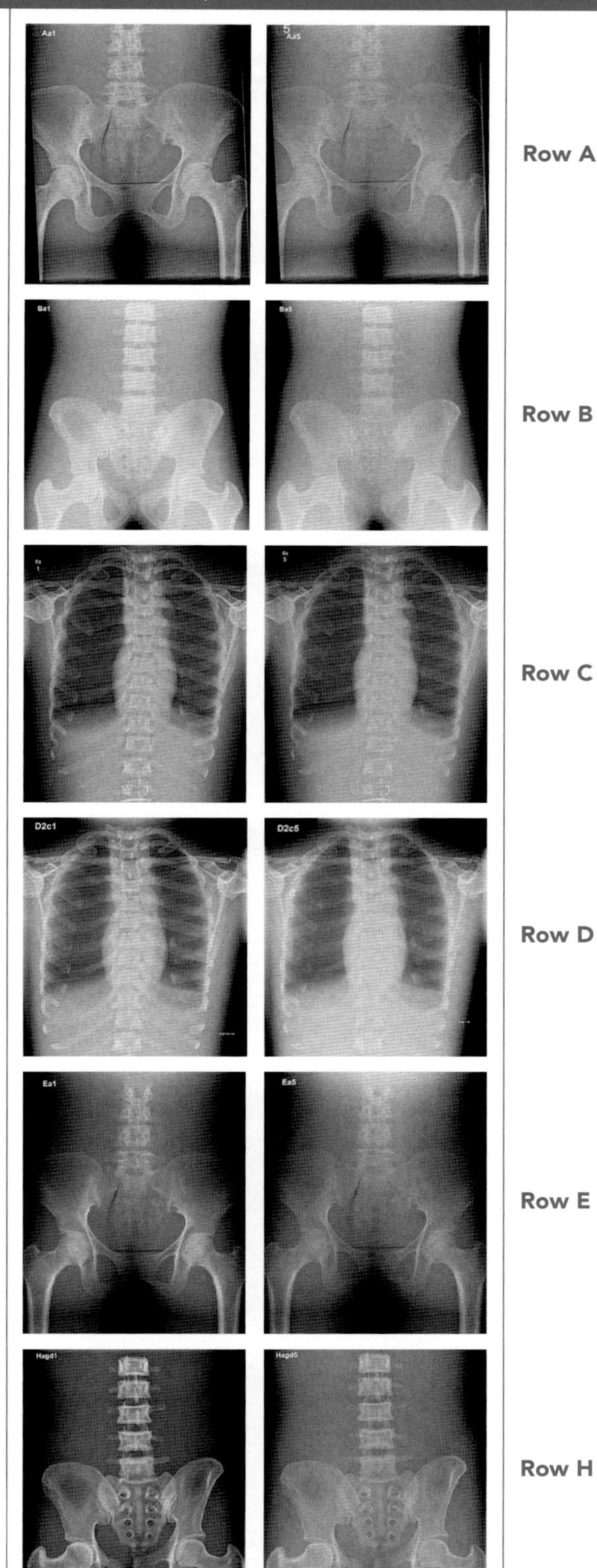

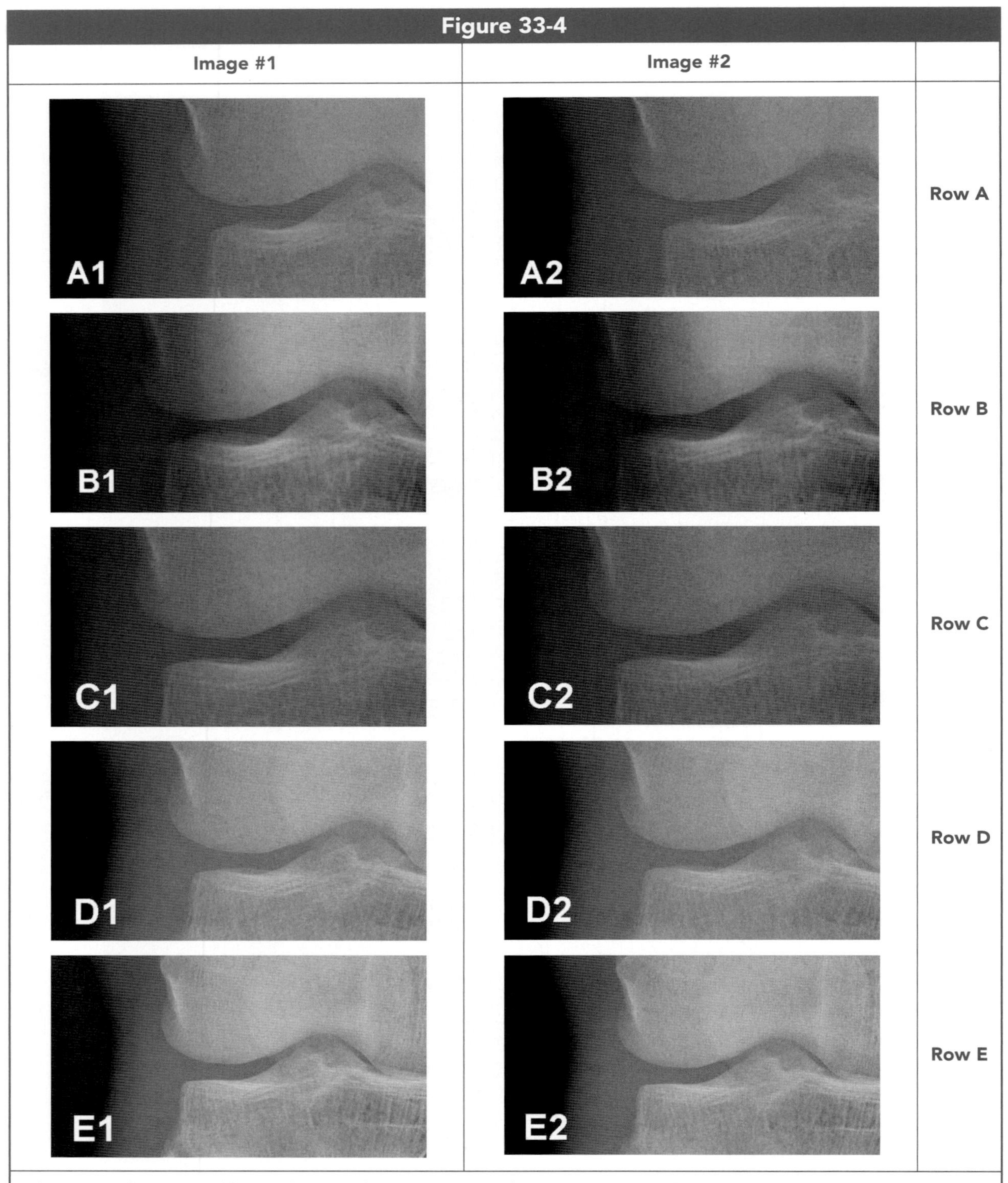

Close-ups of a non-grid knee phantom for comparison of mottle with a single 15% step increase in kVp from left to right, and with the mAs cut in half, for five different manufacturers. All five sets show *no visible change in mottle* from left to right, findings confirmed by a radiologists on high-definition *class-1* monitors. (Techniques: For all series, image **#1** was exposed with 70 kVp at 8 mAs, image **#2** with 80 kVp at 4 mAs; Processors: **Row A:** Fuji CR unit; **Row B:** GE *Definium 8000* DR unit; **Row C:** Kodak *DirectView 850* CR unit; **Row D:** Philips DR unit; **Row E:** Shimadzu *RadSpeed* DR unit. (Courtesy of Philip Heintz, PhD, Quinn Carroll, MEd, RT, and Dennis Bowman, AS, RT. Reprinted by permission.)

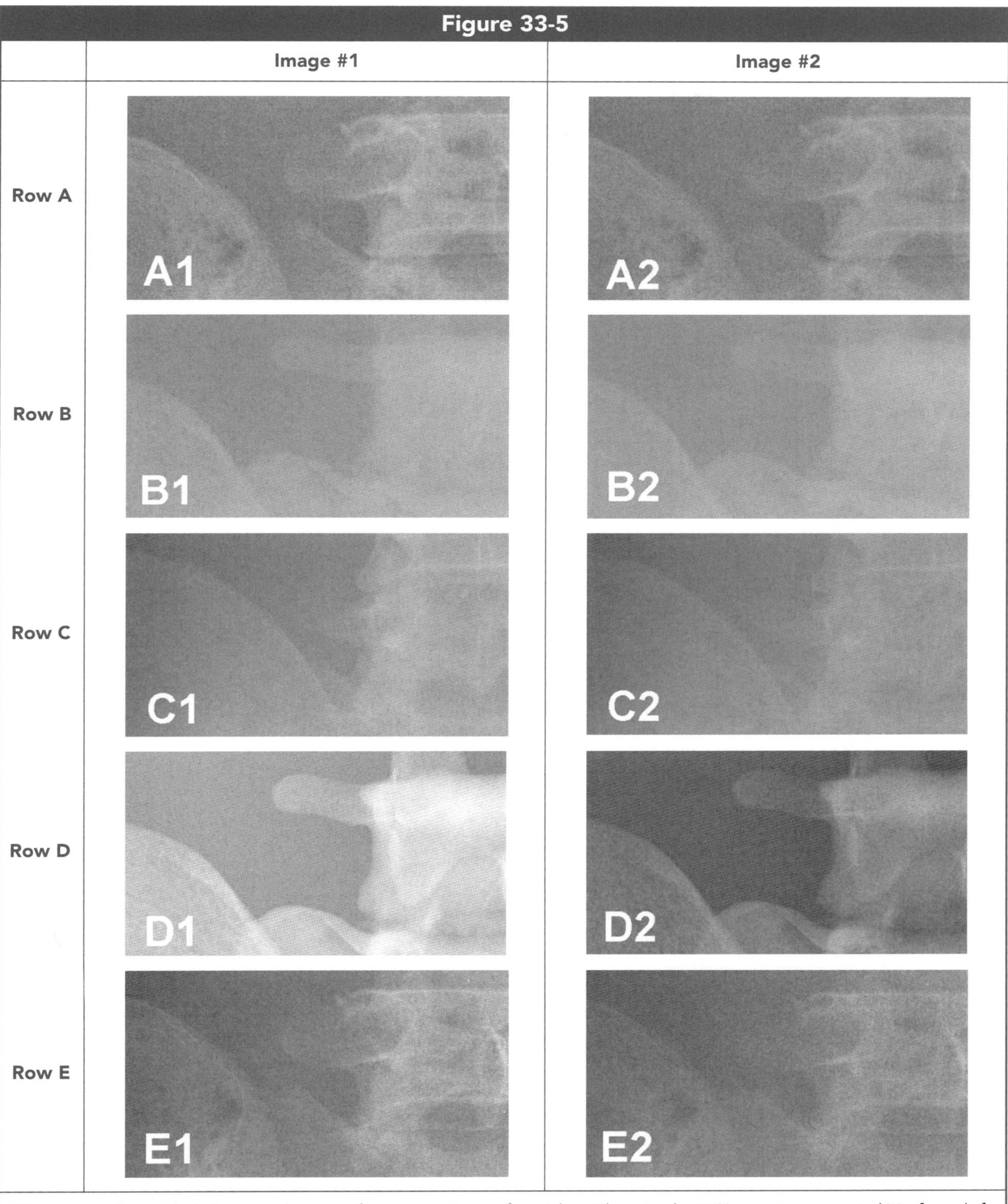

Close-ups of a grid abdomen phantom for comparison of mottle with a single 15% step increase in kVp from left to right, and with the mAs cut in half, for five different manufacturers. A barely visible increase in mottle may be apparent in row **E**, while all four other rows show *no visible change in mottle* from left to right, findings confirmed by a radiologists on high-definition *class-1* monitors. (Techniques: For all series, image **#1** was exposed with 80 kVp at 20 mAs, image **#2** with 92 kVp at 10 mAs; Processors: **Row A:** Shimadzu *Rad Speed* DR unit; **Row B:** Agfa CR unit; **Row C:** Konica *Regius 190* DR unit; **Row D:** Siemens *Axiom Aristos* DR unit; **Row E:** Kodak *CR-950* unit. (Courtesy of Philip Heintz, PhD, Quinn Carroll, MEd, RT, and Dennis Bowman, AS, RT. Reprinted by permission.)

Figure 33-6

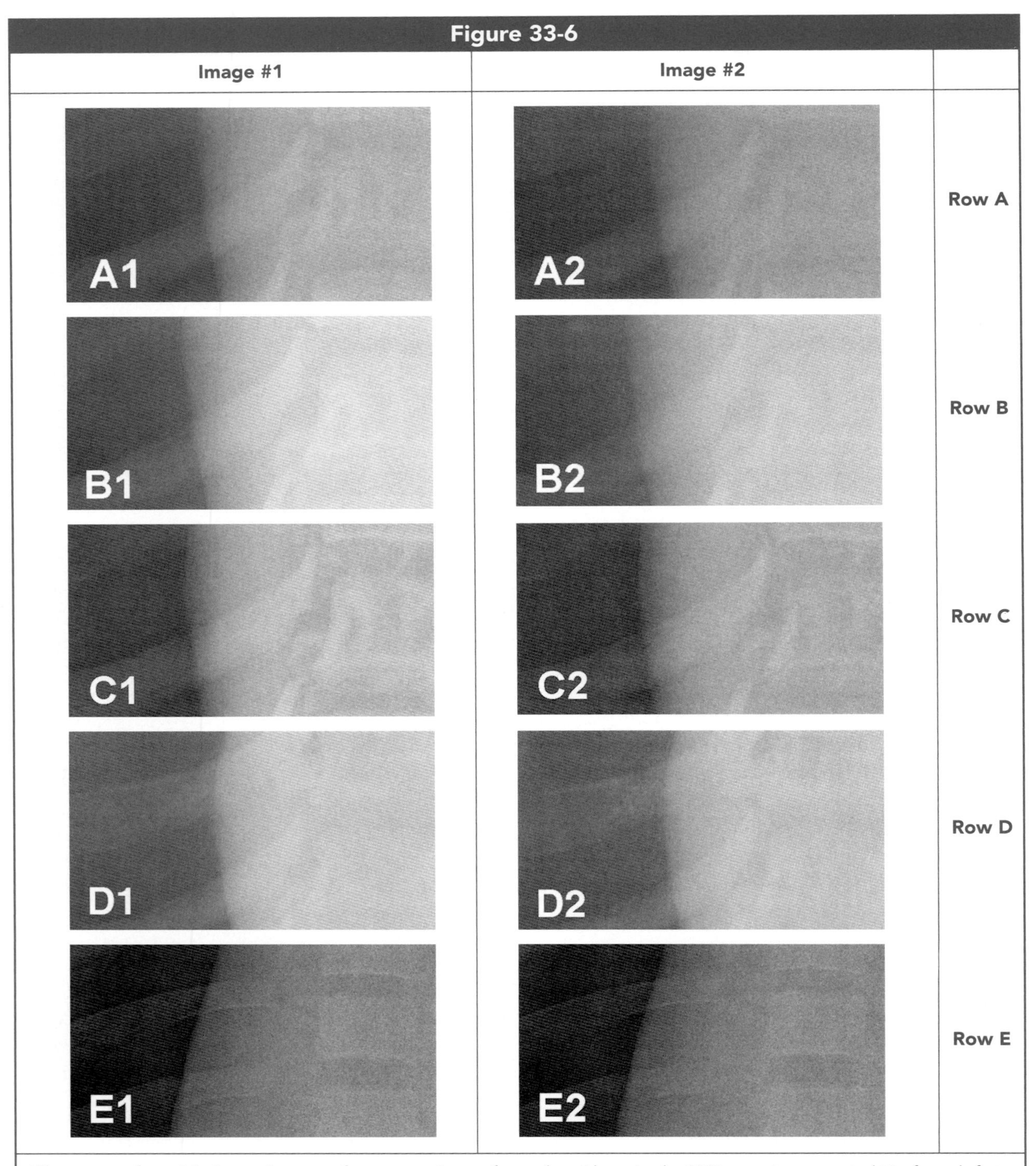

Close-ups of a grid chest phantom for comparison of mottle with a single 15% step increase in kVp from left to right, and with the mAs cut in half, for five different manufacturers. Although Row **E** shows significant mottle in both views, rows **B**, **D** and **E** show no visible difference in mottle between the two techniques. A barely visible increase in mottle may be apparent in rows **A** and **C**. Findings were confirmed by a radiologists on high-definition *class-1* monitors. (Techniques: For rows A, B, D and E image **#1** was exposed with 122 kVp at 4 mAs, image **#2** with 140 kVp at 2 mAs; For row C, image **#1** was exposed with 106 kVp at 4 mAs, image **#2** with 122 kVp at 2 mAs; Processors: **Row A:** Kodak *CR-950* unit; **Row B:** Philips DR unit; **Row C:** Fuji CR unit; **Row D:** GE *Definium 8000* DR unit; **Row E:** Toshiba *Radrex-I* DR unit. (Courtesy of Philip Heintz, PhD, Quinn Carroll, MEd, RT, and Dennis Bowman, AS, RT. Reprinted by permission.)

As you observe these images, mottle is best evaluated by examining a relatively homogeneous tissue or "smooth" density area, (something actually facilitated by the use of phantoms). For the knee in Figure 33-4, note in particular the soft-tissue area within the joint space. For the abdomen (Fig. 33-5), note the soft-tissue area between the transverse process of L5 and the sacral ala. For the chest (Fig. 33-6), the soft tissues between the ribs, both the lighter heart tissue and the darker lung tissue, are recommended.

The student may note the presence of mottle even in the first image of some of these sets. When evaluating these image pairs, it is important to understand that we are not assessing whether mottle is present, but strictly whether an *increase* in mottle from image A to image B can be determined, since this represents the 15% kVp change. Preliminary evaluation by radiologists using Class-1 monitors supports the following conclusions:

For the non-grid knee in Figure 33-4, all five rows show *no obvious* increase in mottle. For the grid abdomen in Figure 33-5, again, four of the five rows show *no obvious* increase in mottle, while row ***E*** may show a barely visible increase. For the grid chest in Figure 33-6, rows ***B***, ***C*** and ***D*** show no visible difference in mottle between the two techniques. A barely visible increase in mottle may be apparent in row ***A***, and a significant increase in mottle can be seen in Row ***E***.

It is significant that of the 15 rows of images, those 5 demonstrating any visible change at all in mottle represent a cross-section of 5 different manufacturers, so there appears to be little correlation to the manufacturer. Only the chest series in Figure 33-6 had more than one row showing a visible difference, and also has the worst case of mottle (Row ***E***). This may indicate that chest exams tend to be more vulnerable to mottle than other procedures, likely due to the very low mAs values associated with typical chest techniques.

With 11 of the 15 series demonstrating no visible change, and 3 with barely visible changes, we may conclude that a single *15% step increase in kVp with an accompanying halving of the mAs may be generally applied for digital radiography without creating significant mottle, with the possible exception of chest radiography which already uses very high kVp and low mAs values.*

Recommendations for Reducing Patient Exposure

As described in Chapter 16, higher kVp levels result in more bremsstrahlung production within the x-ray tube anode. For a 15 percent step increase, this results in a 30–35 percent increase in entrance skin exposure (ESE) to the patient. Since the mAs is cut in half, we see a halving of the patient exposure to 50 percent, accompanied by an increase of 17 percent (35% of 50) due to bremsstrahlung, for a net reduction in patient skin exposure to about 67 percent. Much more importantly, however, the *absorbed dose* to the patient is reduced by an even more significant amount, due to increased penetration of the x-rays.

By applying the 15 percent rule in a single step, surface exposure can be reduced by one-third, to 65–70 percent, and absorbed dose to the patient can be reduced even more. As we have just demonstrated, this can be generally accomplished without significant mottle appearing in the images, and without substantial loss of image contrast.

(Note that if the 15 percent rule is applied in *two steps*, the net result in patient ESE would be about 67% of 67%, or 44 percent of the original exposure, a literal halving of patient surface exposure. Preliminary results indicate that this may be achievable *with some brands of equipment* without mottle levels reaching diagnostically destructive levels. Future improvements in processing may also contribute to this possibility.)

Given all of this data, it is strongly recommended that medical imaging departments adopt a policy of implementing ***at least one*** *15 percent step increase in kVp from the conventional kVp levels used for film/screen radiography, accompanied by a halving of the mAs, and that this approach be applied "across the board" for all techniques, with the possible exception of chest procedures.*

Any reduction in radiation dose to the public is worth pursuing. We as a profession have an opportunity to restore public exposure much closer to previous levels, by simply establishing new "optimum kVp's" as recommended in Table 16-2.

Even departments electing not to make an *across-the-board* adjustment in kVp's used should consider special circumstances for which there is compelling reason to do so. One example is scoliosis imaging in adolescents: Scoliosis imaging involves repeated, whole-spine exposures to children who have an increased risk of breast and bone cancer from these exposures. The image quality required in follow-up scoliosis views need not always be equivalent to a diagnostic bone radiograph—this imaging is done to measure angles and to see hardware. Some departments have been able to cut mAs values in half or more by using high kVp techniques. All departments should consult with their radiologists about procedures that may be targeted for exposure reduction by the use of high kVp, especially high cumulative dose procedures (such as running scoliosis series) that affect adolescents and children.

Note that when a substantially different approach to technique is used, *target exposure indicator* values, discussed in the last chapter, may need to be adjusted. The single 15-percent step change recommended here may not be radical enough to require adjustment of the EI.

Quality control technologists can consult with manufacturers as needed to confirm their selection of appropriate target exposure indicators.

DOES kVp STILL *CONTROL* IMAGE CONTRAST?

We can say categorically that kVp still controls the *subject contrast* of the remnant x-ray beam signal that reaches the image detector, as fully explained in Chapter 12 and Chapter 15. As described in the introduction to this chapter, the percentage penetration of the x-ray beam is a critical part of ensuring an adequate signal for the computer to work with. However, at the beginning of this chapter we also characterized the final displayed image as being "independent of image acquisition."[1] *Final digital image qualities are primarily the result of computer operations on the acquired data.*

Rescaling modifies the contrast of the incoming data set. In the digital age, the greatest impact on the contrast of the final displayed image is the *LUT* applied during gradation processing. In addition to all this, the contrast can be further adjusted by *windowing*. In light of all these considerations, the word "control" is too strong to describe the relationship between kVp and digital image contrast; the notion that kVp is *the controlling factor* for contrast (a popular concept with conventional radiography) should be avoided. Rather, in the digital age, kVp should be thought of as one of several factors simply *affecting* the contrast of the initially displayed digital image.

EXPOSURE LATITUDE, OVEREXPOSURE, AND PUBLIC EXPOSURE

Exposure latitude is the margin for error in setting technique that a system will allow. While film/screen systems had an exposure latitude from -30 percent to +50 percent, CR and DR are said to have a latitude from -50 percent to more than +400 percent, a remarkable range. A "passable" image can vary visually in density up or down by a factor of 2, but there is no question that with the linear response of digital imaging, a far wider range of error is allowed, especially in the direction of overexposure.

The previous chapter described *saturation* from extreme overexposure. It requires as much as eight to ten times the normal exposure to reach saturation, corresponding to an exposure index number higher than *3000* on the Kodak system or an "S" number less than 25 on the Fuji system. In the series of knee radiographs in Figure 33-7, note that even at four times the normal mAs, the CR system is able to perfectly compensate so that the final processed image is unchanged.

A lower limit for exposure is easy to discern since mottle becomes quickly apparent as techniques are reduced (see Figure 33-8, radiograph ***C***.) But, in terms of typical technique adjustments, a *practical* upper limit eludes us.

As radiographers gain experience with digital images, they soon realize how slight an underexposure is required to cause visible mottle. The fear of this, added to the lack of any immediate consequences to overexposure, has led to an unfortunate tendency toward overexposure. Recall from the previous section

[1] Shepard et al.: Exposure Indicator for DR: TG116 (Executive Summary) in *Medical Physics*, Vol. 36 No. 7, July 2009.

Figure 33-7

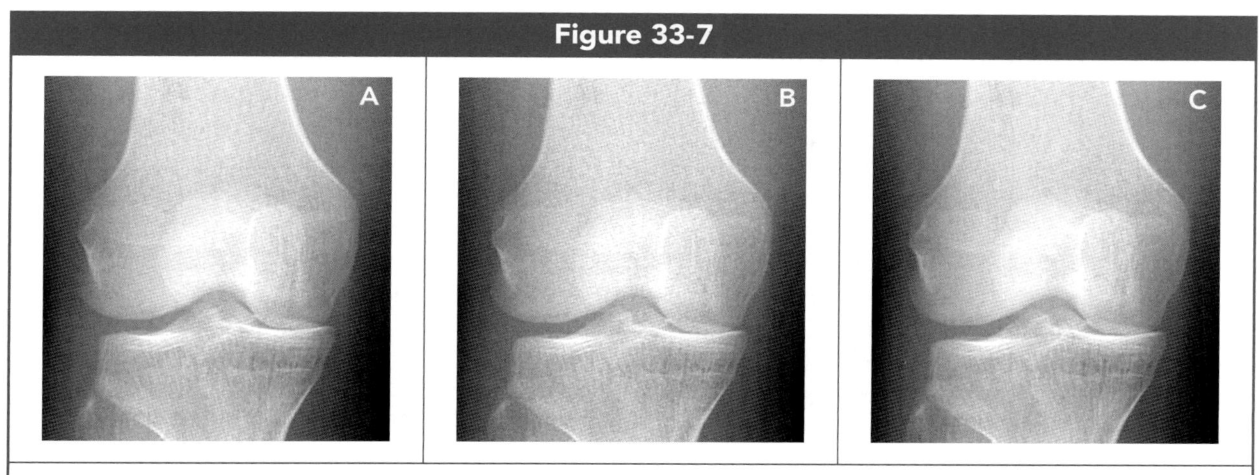

Magnified sections of CR radiographs of the knee taken with very low and very high mAs while maintaining kVp and all other factors constant. From a standard technique, ***A***, radiograph ***B*** was taken using one-fourth the mAs, and mottle is apparent. However, radiograph ***C***, taken at four times the mAs, shows no image degradation at all. Generally, overexposure is not apparent in the quality of DR and CR images. (From Quinn B. Carroll, *Practical Radiographic Imaging*, 8th ed. Springfield, IL: Charles C Thomas, Publisher, Ltd., 2007. Reprinted by permission.)

on *Minimizing Patient Exposure* that there has already been a doubling of typical exposures with many CR units due to the selection of the 200-speed class of operation while leaving kVp's at traditional levels. Combining all of these issues, it can be seen that increased exposure to the public is becoming a serious issue with digital imaging.

Even though repeat rates have been brought down from approximately 5 percent to 3 percent, an increase in radiation delivered per exam has been documented at approximately 40 to 50 percent. The net change is an increase in public exposure. In one regional study, it was revealed that 33 percent of mobile chest radiographs had doubled in average exposure, 36 percent of pediatric mobile abdomens had doubled, and exposures for pediatric trunk exams had increased by 4 times the previous film/screen levels! This is an issue we *must* deal with as a profession.

To summarize, with CR and DR imaging there are at least four factors combining to inexorably push us toward unacceptable levels of patient exposure. They are:

1. The popular selection of the 200 speed class rather than 300 or 400 which are available
2. The irrational resistance to using higher kVp's across the board
3. The legitimate fear of image mottle resulting from underexposure
4. The amazing inherent ability of CR and DR to "correct" for overexposure

Due to the speed and convenience with which digital radiography units bring an image up for review right at the acquisition workstation, another very disturbing trend has begun among radiographers—namely, discarding an original exposure and repeating the exposure to the patient in order to make minor positioning corrections or to "hide" the positioning error. Images are displayed so quickly after exposure that the radiographer can leave the patient in position while checking the image in the control booth, and repeat the exposure to the patient for errors that may have been embarrassing but were nonetheless "passable" for diagnosis. The original can be discarded from the study submitted into the PAC system, so that the fact that the exposure was repeated may only be discovered by accessing the metadata at the acquisition unit itself. When this is done for relatively minor corrections to the position, the increased exposure to the patient is unwarranted, and this practice should be avoided. More than ever, educators, managers and leaders in the field must strive to instill a strong sense of ethical duty in students and employees toward minimizing radiation exposure to patients and public.

SUFFICIENT PENETRATION AND SIGNAL-TO-NOISE RATIO

In Chapter 6 we learned that *no amount of mAs will ever compensate for inadequate kVp*. Or re-stated, *no amount of radiation will ever compensate for inadequate penetration*. This is still true for CR and DR. Sufficient signal must reach the radiation detector, whether that detector is a CCD, a TFT, or an old-fashioned film. There must be enough good signal to overwhelm the presence of noise, such as quantum mottle, within the exposure process. This signal must *reach* the detector by penetrating through the anatomy in the first place. A minimum kVp level recommended for each anatomical part is still applicable. It cannot be over-stressed that even though computerized systems may seem to be able to compensate the brightness of the image for reduced kVp, the use of kVp much below the minimum levels recommended in Table 16-1 (Chapter 16) in practice is unethical, resulting in excessive patient exposure as mAs values are brought up in an attempt to restore index numbers to their recommended levels.

Recall from Chapter 16 also that if one had to choose, it is always better to slightly over-penetrate than to under-penetrate. The algorithms of a computer-based system can easily adjust an image which has been slightly over-penetrated. Adequate signal is present. The information is there to adjust. But, just as with screen-film radiography, risking under-penetration means risking the *loss* of information. Even a CR or DR system cannot replace or adjust information which is simply absent.

EFFECTS OF kVp CHANGES ON THE IMAGE

Figure 33-8 demonstrates the effects of moderate and extreme changes in kVp upon the CR image. Extreme under-penetration results in obvious quantum mottle in the image. Even slight under-penetration, or slight underexposure resulting in index numbers that are *low but within the published acceptable range*, can produce some visible mottle in the image. Over-penetration, on the other hand, does not result in obvious fog or any other apparent decline in image quality. Even at extreme increases in kVp, the computer algorithms are able to produce a good image with no apparent fog. As indicated previously, this means that patient exposure can be reduced using the 15-per cent rule as a department develops technique charts for CR or DR imaging, without fear of any reduction in image quality.

EFFECTS OF SCATTER RADIATION ON DIGITAL IMAGES

When it comes to the quality of the final digital image, there is an intriguing difference between the effects of scatter radiation which occurs *during a radiographic exposure* and the effects of "pre-fogging" a CR plate to scatter or background radiation *prior* to using it for a radiographic exposure.

Let us first examine the effects of scatter radiation during the radiographic exposure. These effects are compensated for by digital processing to an impressive degree, as shown in Figure 33-9. The top row of images are film radiographs that demonstrate a dramatic increase in fog from ***A*** (at 80 kVp) to ***C*** at 120 kVp. Radiograph ***D*** was exposed at 120 kVp using a CR system. Even when 120 kVp was used, the CR system was able to compensate for the effects of scatter radiation so well that the exposure appears much like the *80 kVp* exposure (***A***) using film.

Even localized areas of scatter radiation exposure are at least partially corrected by digital processing algorithms. Figure 33-10 illustrates two series of lateral lumbar spine radiographs. With conventional film/screen radiography these views were persistently degraded by a fog density that obscured the spinous processes posteriorly. The border of this fog density is pointed out with arrows on image ***A1***. This fog was caused by scatter radiation generated primarily from the tabletop being exposed to the "raw" x-ray beam just behind the patient. Image ***A4*** shows one rare exception to this trend that is apparent in the other three images in *Series A*.

Series B in Figure 33-10 proves that digital processing largely corrects for this scatter radiation, demonstrating most of the posterior spinous processes with

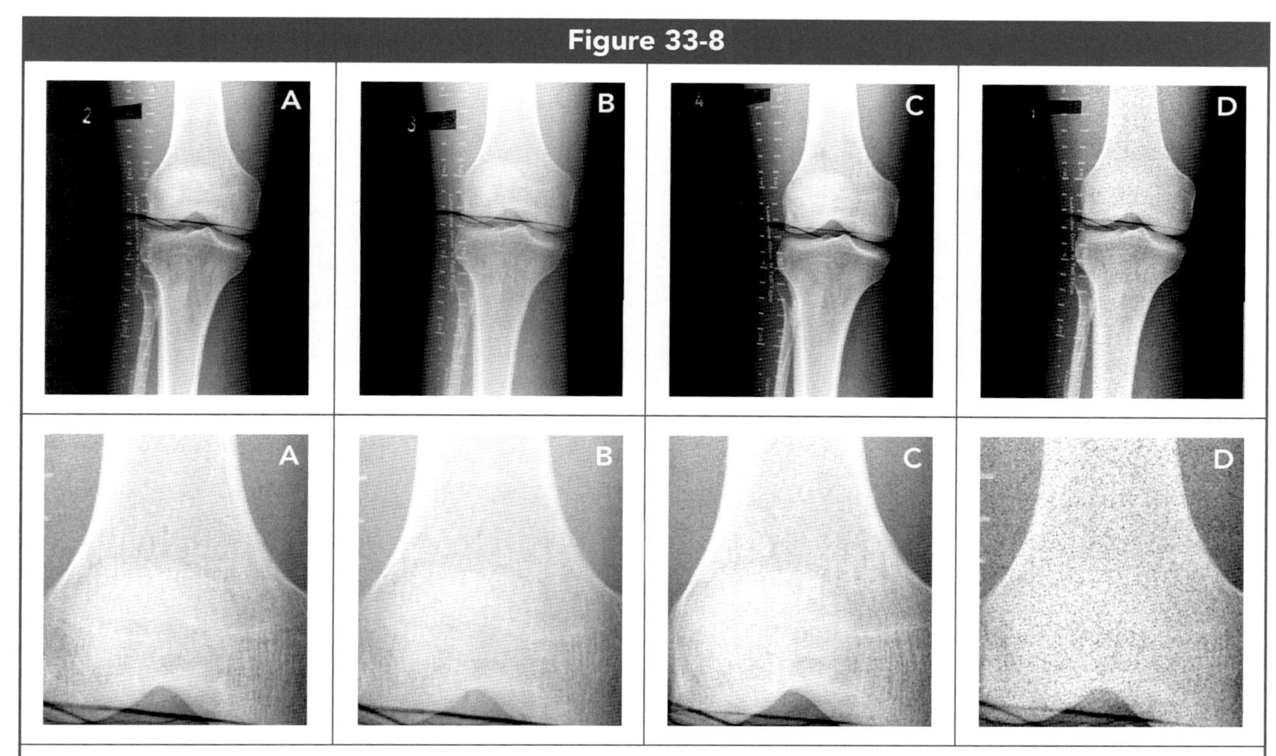

Figure 33-8

Effects of extreme changes in kVp on CR images of the knee are shown with a second series of magnified sections below to closely observe mottle. From a standard technique, ***A***, radiograph ***B*** was taken at a 30% increase in kVp, and shows no degradation at all due to overpenetration. Radiograph ***C***, taken with a 15% reduction in kVp, demonstrates that even with a modest reduction, mottle begins to appear in the image. Radiograph ***D*** was taken at minus 30% kVp and presents severe mottle. (From Quinn B. Carroll, *Practical Radiographic Imaging*, 8th ed. Springfield, IL: Charles C Thomas, Publisher, Ltd., 2007. Reprinted by permission.)

remarkable clarity, and even the soft tissue of the patient's back. In cases where an extreme amount of scatter radiation is produced, the fog pattern may not be fully corrected by digital processing and may be apparent in the displayed image. Even so, these poorer images will still represent a considerable improvement over their conventional film/screen counterparts.

One likely explanation for the ability of digital processing to correct for fog patterns within the exposed field that are caused during exposure is illustrated by the histograms in Figure 33-11. From histogram ***A*** to histogram ***B***, you can see where an added fog pattern has widened the black "tail spike" and raised the lowest point between the two lobes. However, as long as this low-point remains below the set threshold, note that the computer is *still able to properly identify the* S_{MAX} *landmark just as in histogram* ***A*** (scanning from right to left, as the second time the threshold is exceeded while pixel count subtractions are resulting in negative values). This means that all of the undesirable data in the tail portion can still be eliminated from calculations for rescaling and for the exposure index, and these functions will work properly.

Digital equipment is remarkably resilient to the effects of scatter radiation caused during an exposure. This means there is considerable latitude in choosing exposure factors, including high kVp levels and opting to perform more procedures non-grid.

On the other hand, experiments confirm that if a CR plate has been fogged by either background radiation or scatter radiation *prior* to using it for a radiographic exposure (shown in Figure 34-16 in Chapter 34), a fogged digital image can result. The effects of pre-fogging can also be explained by examining the way it changes the histogram. In Figure 33-12, from histogram ***A*** to histogram ***B***, note that

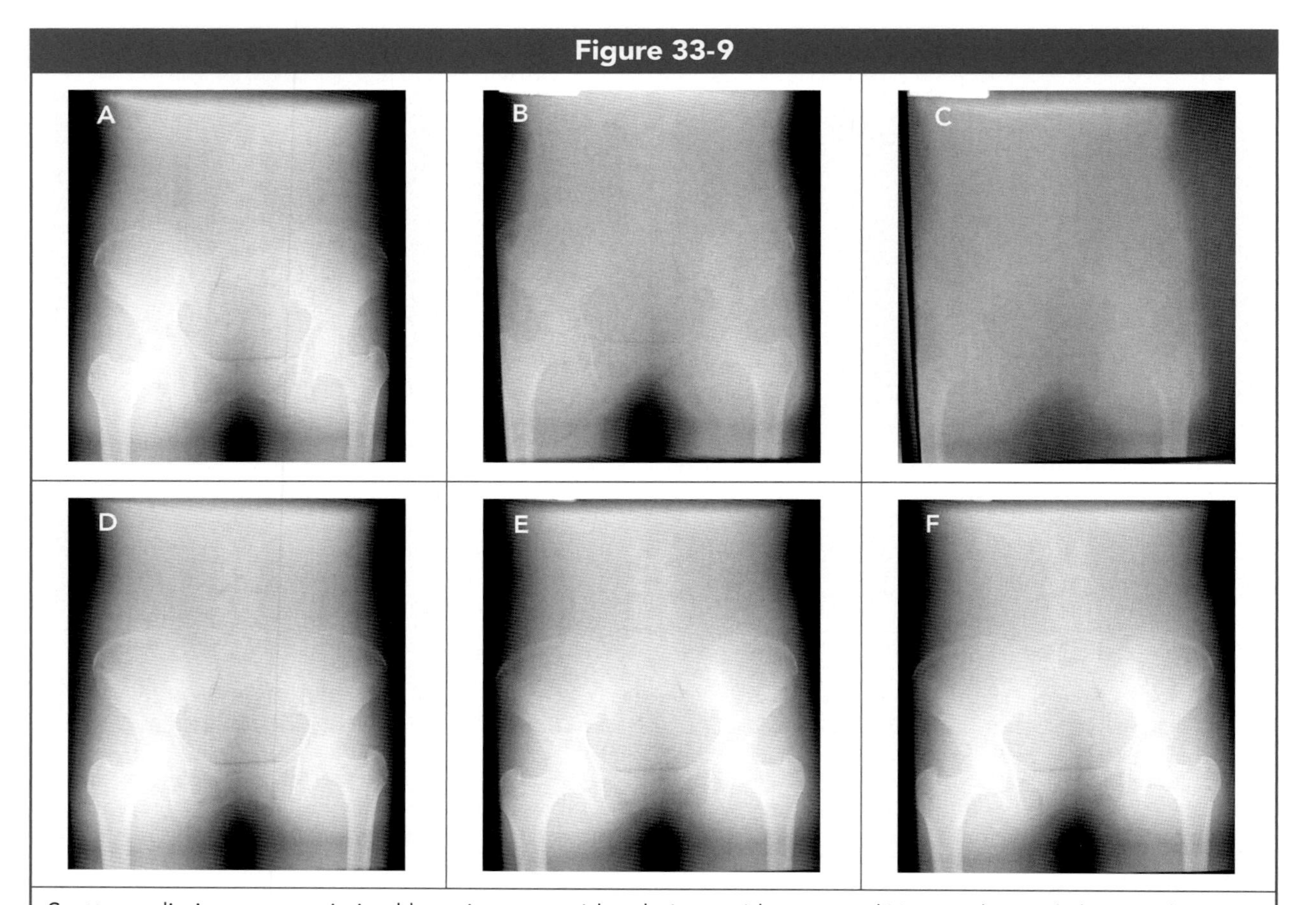

Figure 33-9

Scatter radiation was maximized by using non-grid technique with extreme kVp on a large abdomen phantom. The top row of images are film radiographs demonstrating a dramatic progression of fog from ***A*** to ***C***, while the bottom row employed similar techniques using a CR system. Even doubling the mAs at 120 kVp in image ***F***, the CR system was able to completely compensate for the effects of scatter radiation, producing digital images with no fog at all. (Techniques for film radiographs: ***A***, 5 mAs at 80 kVp for comparison; ***B***, 1 mAs at 120 kVp; ***C***, 2.5 mAs at 120 kVp. Techniques for digital radiographs: ***D***, 1 mAs at 120 kVp (compared to ***B***); ***E***, 2.5 mAs at 120 kVp (compared to ***C***); and ***F***, 5 mAs at 120 kVp.)

"pre-fogging" the PSP plate has added a "spike" of light *gray* densities at the *left* end of the histogram and that there are no more "blank" white pixels present. This is quite different from the exposure fog illustrated in Figure 33-11. Here, the S_{MIN} point is shifted to the right, pulling the S_{AVE} point also to the right. This skews both rescaling of the image and calculations for the exposure index. In effect, the fog goes "undetected" and is kept in the image data set all the way through to display. This typ of fog can remain uncorrected.

This underscores again the importance of erasing CR plates prior to use if there is any question of their condition. *Note that DR image receptors are not vulnerable to this problem of pre-fogging, because they are automatically "erased" between exposures.*

TECHNIQUE MYTHS

Several myths abound regarding CR and DR radiography: "Never use less than 70 kVp"; "Never use more than 80 kVp"; "This system is mAs-driven, adjust only mAs"; "This system is kVp-driven, adjust only kVp"; "You can't collimate with CR"; "You can't use grids with CR or DR." These are all inaccurate statements.

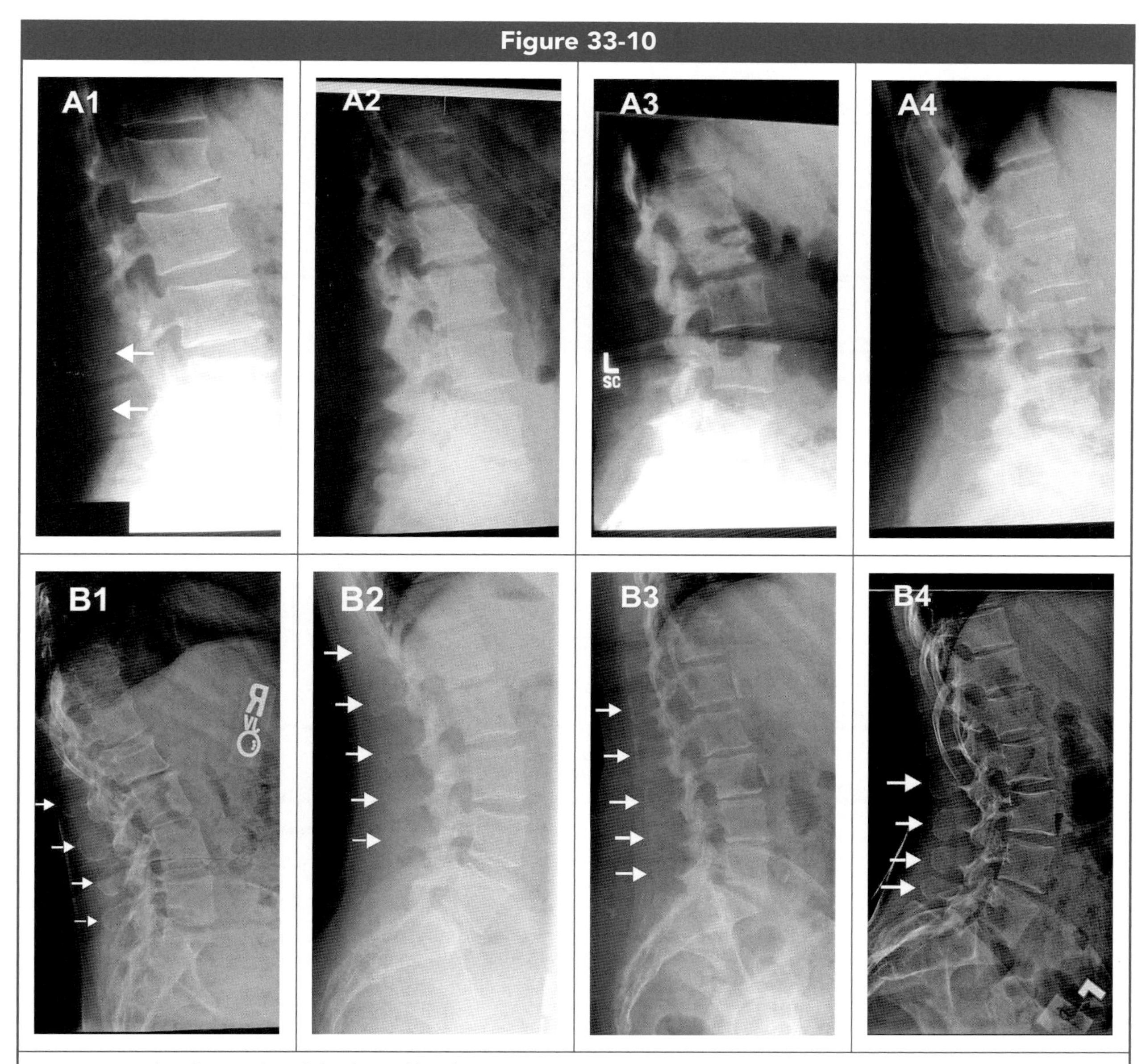

Conventional radiographs of the lateral lumbar spine, ***Series A***, were generally plagued by a fog density (arrows, ***A1***) caused by scatter radiation from the tabletop, that obscured the spinous processes posteriorly. ***A4*** shows one rare exception. ***Series B*** proves that digital processing largely corrects for this scatter radiation, demonstrating the posterior spinous processes and even the soft tissue of the back extremely well, except for the most extreme cases. (Courtesy, Dennis M. Bowman, AS, RT.)

The use of appropriate collimation and grids is discussed in the following sections. The selection of kVp and mAs levels is firmly based upon the same principles as conventional radiography, all of which have been taught in previous chapters. Technologists should beware of unscientific methods in applying radiographic technique. The recommended range of kVp for adults is 60–120, for pediatric patients weighing less than 100 lbs, 50–90. Higher kVp levels display more anatomical data and lower patient dose, and the radiologists can window higher image contrast as desired using computer display software. Technique charts like the one in Table 33-1 are still highly recommended.

Figure 33-11

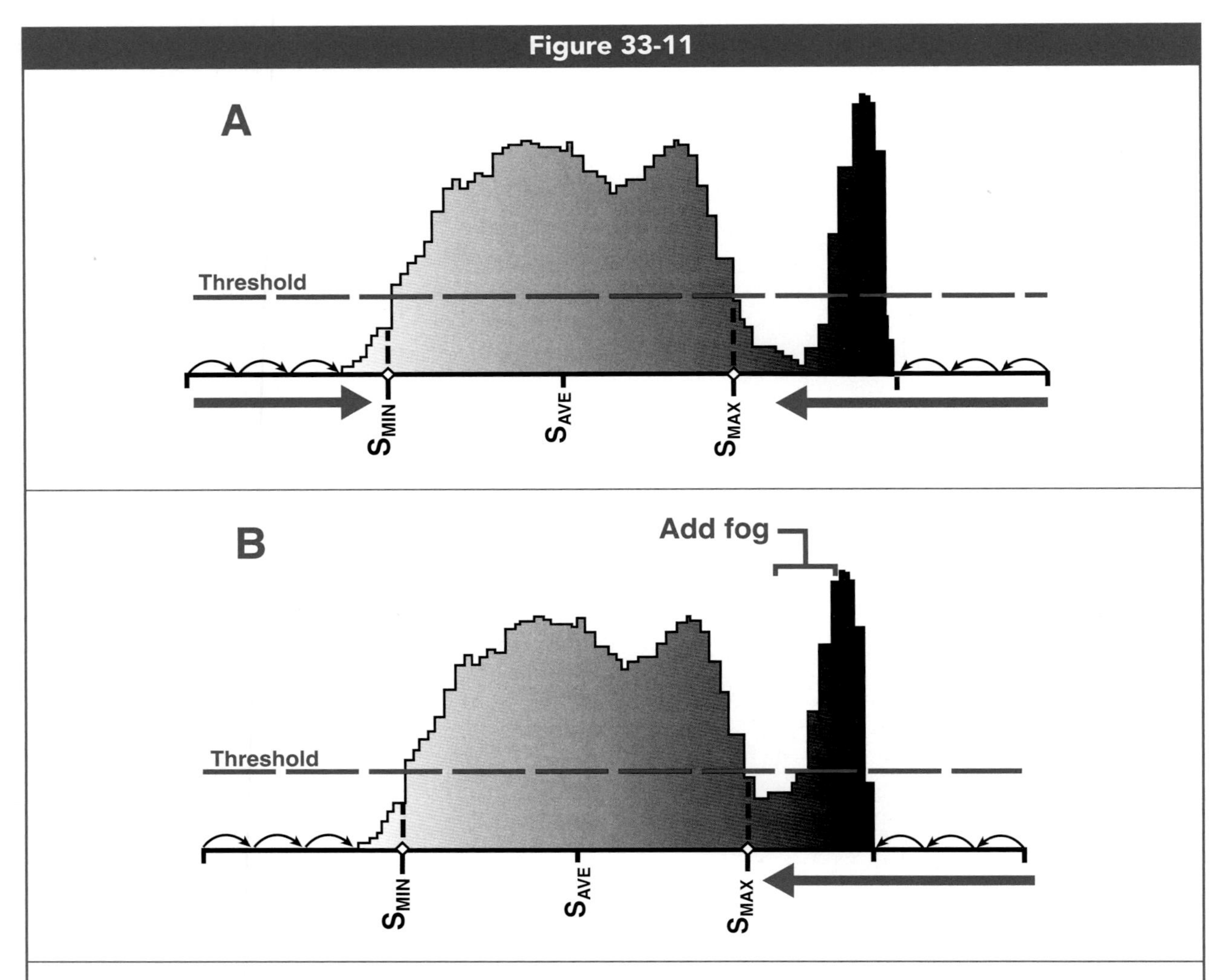

In histogram ***B***, the addition of a localized, dark fog pattern (such as behind a lateral lumbar spine projection) shows widening of the "tail spike" and raising of the lowest point between the main lobe and tail lobe. Nonetheless, in this case, the computer is still able to locate the threshold for the S_{MAX} landmark, so no errors result in histogram analysis.

PROPORTIONAL ANATOMY AND MANUAL TECHNIQUE RULES

Chapter 26 presented a "proportional anatomy" system for deriving techniques from one body part to another. The advent of CR and DR introduces some intriguing possibilities for this approach to radiographic technique: In most applications, there is no longer an "extremity cassette" for which techniques must be adjusted. All procedures are processed at the same speed, usually 200. This makes the use of a proportional anatomy approach much easier. Experience shows that although some of these rules of thumb as presented in Chapter 26 might need refinement, they are generally still accurate enough that the system continues to have practical value.

Examine the manufacturer's recommended CR technique chart in Table 33-1 (p. 565). Note that the lateral skull, AP cervical spine, and AP shoulder techniques are all equal to each other and about one-half of the PA skull technique, just as the proportional anatomy system predicts. The abdomen, AP lumbar spine, and AP pelvis all share the same technique. The AP elbow is equal to the lateral wrist, and so on. Approximately two-thirds of this chart,

Figure 33-12

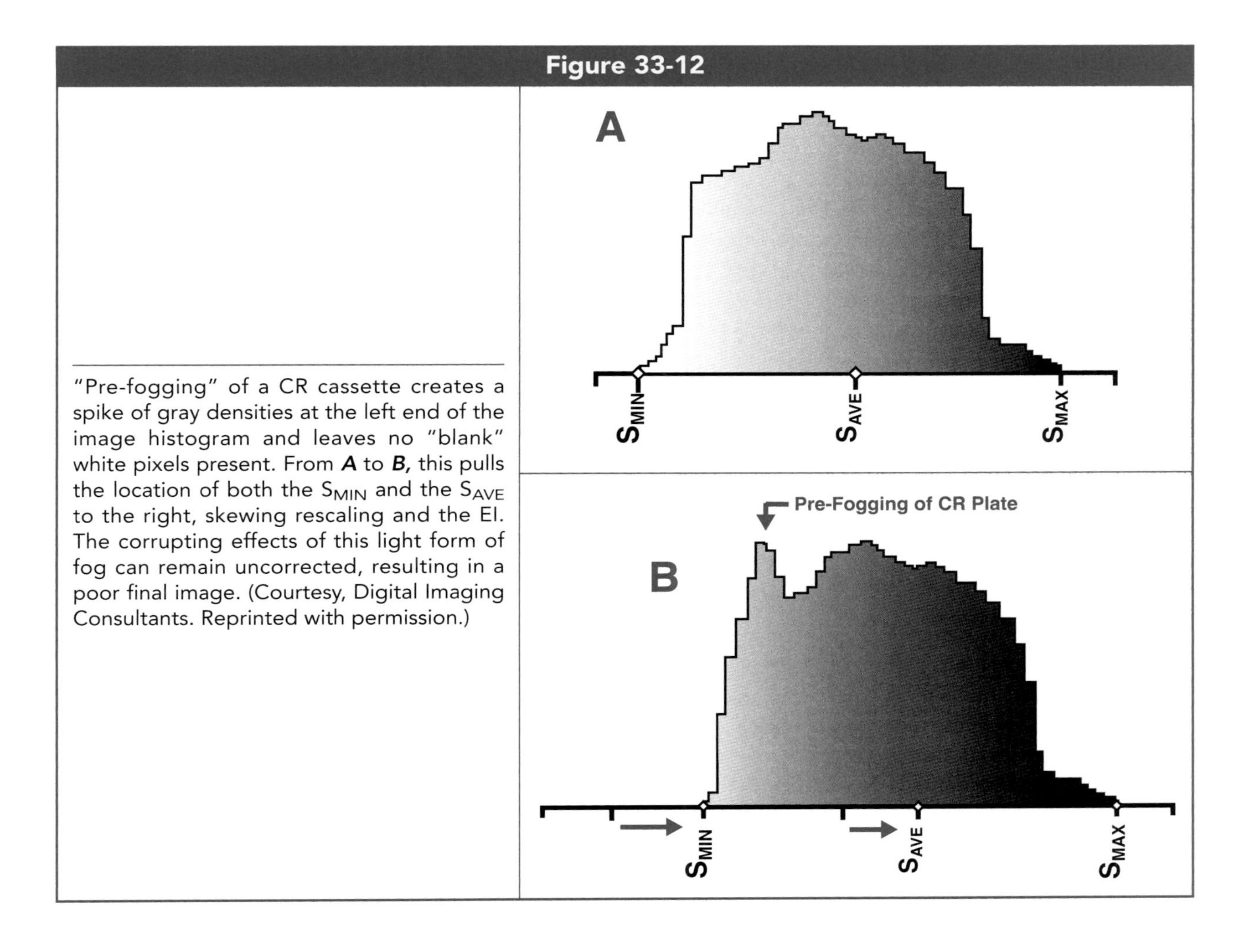

"Pre-fogging" of a CR cassette creates a spike of gray densities at the left end of the image histogram and leaves no "blank" white pixels present. From ***A*** to ***B***, this pulls the location of both the S_{MIN} and the S_{AVE} to the right, skewing rescaling and the EI. The corrupting effects of this light form of fog can remain uncorrected, resulting in a poor final image. (Courtesy, Digital Imaging Consultants. Reprinted with permission.)

produced from a completely independent source, follows proportional anatomy as presented in Chapter 26. Most discrepancies in the chart appear in the techniques listed for extremities. Thus, while refinement of some techniques will be needed, it seems that the proportional anatomy approach is still useful as a starting point in deriving techniques and in building technique charts.

It cannot be overemphasized that skills learned for "manual technique" are still not only useful, but essential. The "4-centimeter" rule for part thickness still applies. The 15 per-cent rule for kVp still applies. Distance rules of thumb are still valuable to memorize. All of these "manual technique" skills, though they may be declining in popularity, have the potential to save the radiographer much grief while setting techniques in the world of digitized imaging.

AUTOMATIC EXPOSURE CONTROLS (AECS)

When a CR system is installed, the automatic exposure control (AEC) for each radiographic unit must be re-calibrated for the CR system. This calibration must be based on the exposure index produced at the selected speed class, *not* on image density as was the case with film/screen systems. Once calibrated, technologists can set the optimum kVp for each procedure (see Table 16-2), and the AEC will terminate each exposure when the correct total mAs is reached.

With CR, there has been a trend among many radiographers to use only manual technique even for all bucky procedures. They report the frequent appearance of image mottle while using the AEC, and unpredictable results. If the AEC is properly calibrated for the CR system as described above, it will work with reasonable consistency. A likely cause of mottle is the obstinate use of low kVp's: High kVp must be used with the AEC to assure adequate penetration of the x-ray beam to the AEC detectors *as well as* adequate signal for the CR system to process. The importance of adjusting to the higher kVp recommendations presented in Table 16-2 cannot be overstated. Some radiographers insist on using kVp's that are not only below these recommended levels for CR, but even below those recommended for the older film/screen radiography (see Table 16-1 in Chapter 16). Historically, this has been fueled by an irrational preference (by both radiographers and some radiologists) for visually appealing high-contrast radiographs, images which actually lack the detail and information that a long-scale image can provide. We must break out of this paradigm which is now obsolete.

The selection of AEC detectors or "photocells" to activate for each anatomical projection is unchanged from the recommendations in Chapter 27. The "density" setting should generally not need adjustment. Using a +2 or +3 density setting with CR is tantamount to increasing the mAs unnecessarily, and only results in higher patient exposure (see Fig. 33-7). *If image mottle is a problem,* ***kVp*** *should be increased until sufficient signal is reaching the detector cells.*

Chapter 27 describes how essential it is, any time the AEC is used, to position the anatomy such that the tissue of interest completely covers the detector cell(s) that are selected. Some DR systems now have as many as five detector cells to allow more flexibility in positioning to ensure that the anatomy is over the cells. Since all DR projections are done "bucky," the automatic exposure control (AEC) is readily available tends to be used on nearly all procedures. Bilateral projections can be tricky and special attention must be given to which cells are activated. *Any time the AEC is used, whether on a CR or a DR system, off-centering of the anatomy can result in premature termination of the exposure. For digital systems, this is likely to result in image mottle,* as demonstrated in Figure 33-11. Since AEC tends to be the *modus operandi* for DR, image mottle due to underexposure is more likely with DR.

USE OF GRIDS

Grids are recommended with CR imaging for anatomical parts greater than 13 cm in thickness, just as they are used in conventional radiography. On DR systems, most procedures including small extremities are done with the grid in place.

In light of the previous discussion, note that the purpose of grids in digital radiography is to improve the signal-to-noise ratio (SNR) for the image data fed into the computer. With better quality data to begin with, the probability of mathematical processing errors is reduced.

The *Moire* artifact is a false presentation of diagonal lines as shown in Figure 34-21 in Chapter 34. Traditionally, this artifact could be caused by superimposing two linear grids. With CR imaging, the Moire artifact is actually much more likely to occur. This "electronic" version is called an *aliasing artifact.* It is due to the *Nyquist frequency*, the frequency at which the CR reader scans the plate line by line, interacting with the grid strips, which also have a "frequency." *Short dimension*, or SD grids have their grid lines running crosswise rather than lengthwise so that they may allow some vertical off-centering on horizontal beam projections such as cross-table hips and axial shoulders. If a short-dimension grid is used, such that the grid lines run parallel to the scan direction of the reader, and the grid frequency is close to the Nyquist frequency, the artifact can be created. The artifact is more common with stationary grids such as are used for mobile radiography.

When the Moire artifact is a recurring problem with a CR reader, either new grids must be purchased with a different frequency (grid lines per inch), or the scanning frequency of the CR reader would have to be changed. Some manufacturers recommend grids with a frequency higher than 140 lines per inch, but these are expensive and limit positioning latitude. Special *multi-hole* grids are available which solve this problem.

MARKERS AND ANNOTATION

Post-exam annotation (e.g., adding markers, "upright" or "decubitus" labels, post-injection times or tomographic cut levels *after* the initial image is processed) is simply more prone to human error, since it relies upon memory. Lead markers, such as timing markers during a urographic study, and "left" and "right" markers in particular, should be used in the radiographic room as the exam progresses, to minimize mistakes.

ALIGNMENT ISSUES

Centering of Anatomy

On a DR system, off-centering of the anatomy can result in unacceptable image mottle as shown in Figure 33-13. Although "manual technique" can be used with DR systems, they are most frequently operated with the automatic exposure control (AEC) engaged for all procedures, including extremities. When the anatomy is not centered properly over the AEC detector cell so that it is exposed to the direct x-ray beam, the AEC shuts off early and underexposure results. For a DR system, low exposure quickly results in the appearance of mottle.

CR systems are less likely to show mottle from off-centering of the anatomy within the field, but histogram errors can result. Figure 33-14 is a "Y" view of scapula in which there is an excess of the "raw beam" included within the collimated field and the anatomy is not centered. This resulted in histogram analysis error and a dark image.

Aligning Multiple Fields

CR is far more prone to segmentation errors than DR when it comes to field alignment. The inclusion of data from outside the collimated areas will cause a mis-calibration of the histogram and segmentation errors.

In practice, when multiple exposures are taken on one CR plate, it is often difficult for the computer to distinguish between the very light areas between exposure fields and similar light areas within the useful image, such as thick bones or barium. When a fog density is present between the collimated fields, such that the area takes on a light gray density instead of being "blank white," it is even harder for the computer to distinguish it from bone densities. Figure 33-15 illustrates histogram and segmentation errors that resulted when the computer read both fields and the space between them as one exposure.

In Figure 33-15, the single knee radiograph ***A*** was taken for comparison. Radiograph ***B*** employed identical radiographic technique but placed two collimated fields only one-half inch apart. Note the evident fog density between the two fields. Histogram analysis errors resulted in a lengthened gray scale for these images. In radiograph ***C***, to exaggerate any effects of scatter radiation, the kVp was increased to 110 with only 2 mAs for the two knee projections. The computer not only read both images as one, but now failed to recognize the upper edge of the collimated fields and treated that section of the plate as part of the image, not providing the usual masking at this upper border.

To assist the computer in making the distinction between bone images and blank spaces between fields, the body part must be centered well within each exposure field, all collimation must be parallel to and equidistant from the edges of the imaging plate, and all exposure fields must be well-separated from each other. Figure 33-16 illustrates this concept; in ***A***, two fields are properly aligned, parallel to the edges of the cassette and equidistant from the cassette edges and each other. In both ***B*** and ***C***, a single edge of a collimated field extends beyond the edge of the CR cassette; in ***D*** a single field is not centered on the cassette; and in ***E*** the collimated field is not parallel. All of these scenarios can cause segmentation failure resulting in improper processing of the image.

Exposing only two views on a single plate instead of three allows the fields to be separated by two inches. This reduces the chance of segmentation errors. The use of *careful* lead masking can also reduce errors by protecting the areas between the fields from scatter radiation. This is particularly recommended if the fields cannot be well separated.

Asymmetrical distribution of multiple fields, especially if any of these fields has no margin at one edge of the plate as described in the next section, contributes to segmentation errors. The persistence of processing errors in spite of the best efforts of manufacturers has led many radiographers and departments to abandon the practice of trying to get multiple exposures on one plate.

More to the point, there is no longer any compelling reason to attempt getting multiple fields on

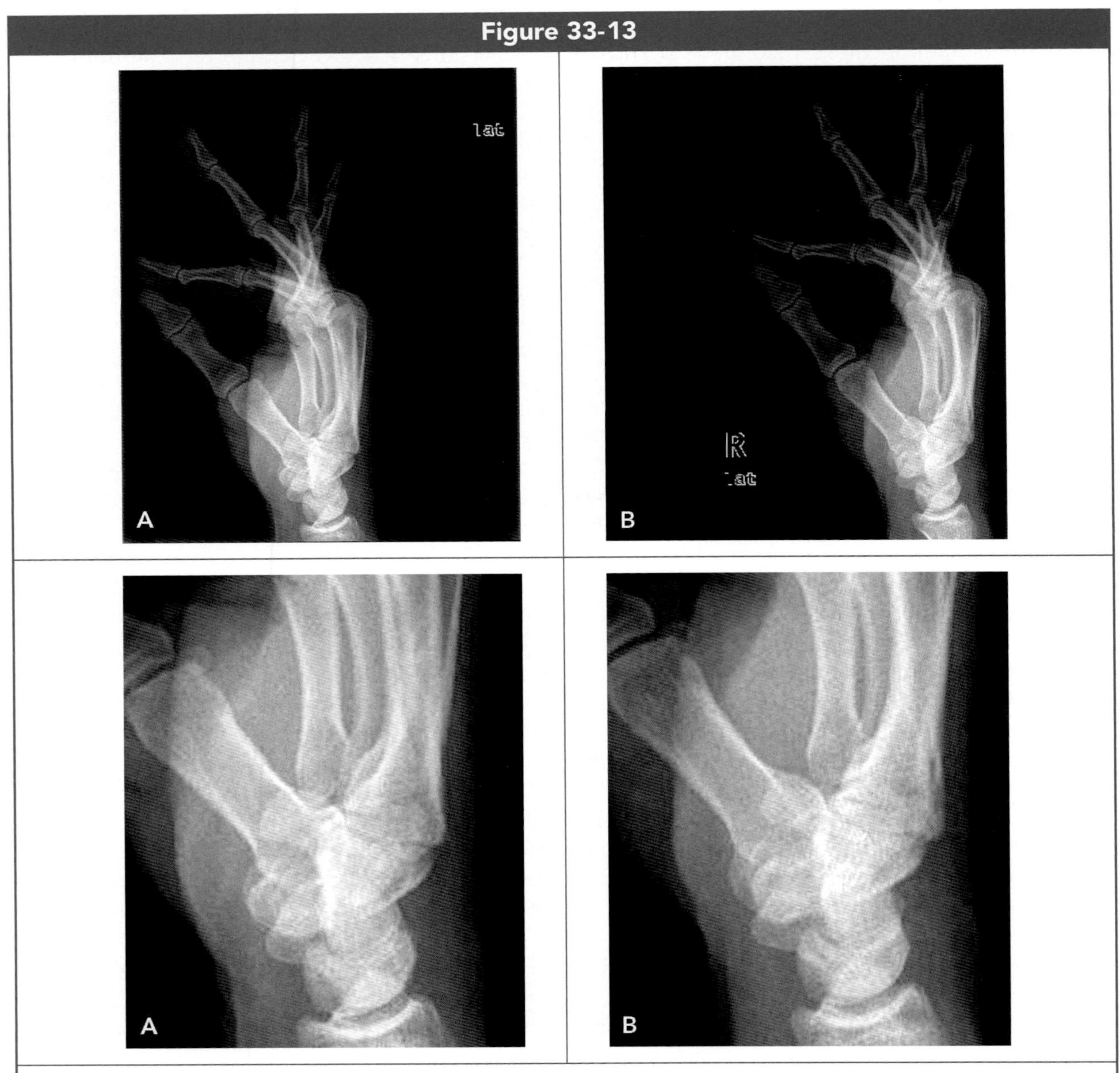

Figure 33-13

Effect of off-centering a fanned lateral hand projection on a DR system with the AEC engaged. Note that when the hand was off-centered in radiograph ***B***, exposure of the "raw" beam to the detector cell resulted in the AEC shutting off the exposure early. This underexposure resulted in mottle apparent in the magnified section below radiograph ***B***. (From Quinn B. Carroll, *Practical Radiographic Imaging*, 8th ed. Springfield, IL: Charles C Thomas, Publisher, Ltd., 2007. Reprinted by permission.)

one plate. With the old film-screen systems, squeezing two or three exposures onto a single film meant a savings in film costs. But, since CR plates are reusable, there is no economic disadvantage to using as many cassettes as needed, all of which will later be erased.

Exposing only one field per plate obviates segmentation errors and results in more consistent image quality. Most radiographers generally use a separate plate for each projection within a series. The one exception would be for projections of the individual digits of the hand, discussed next.

Overcollimation

A **30 per cent** rule has been demonstrated for CR: Exposure indicator errors are likely unless at least 30 per cent of the imaging plate is exposed. Although this situation is uncommon for single views taken on a 10" × 12" plate, it will be a problem for coned-down views of the fingers or thumb. For tightly-collimated views of these digits, it is recommended that two or three views be taken on one plate to ensure the minimum 30 percent plate coverage. The projections must be evenly spaced and well-separated and discussed in the preceding section. Large plates such as a 14" × 17" plate should not be used for single extremity projections such as an single elbow or ankle view.

Most CR systems scan the entire plate from the center outward during the reading process, whereas DR scans only the exposed area in a linear progression. Because of this difference, DR systems are *not* subject to this 30 per cent rule for plate coverage.

Figure 33-14

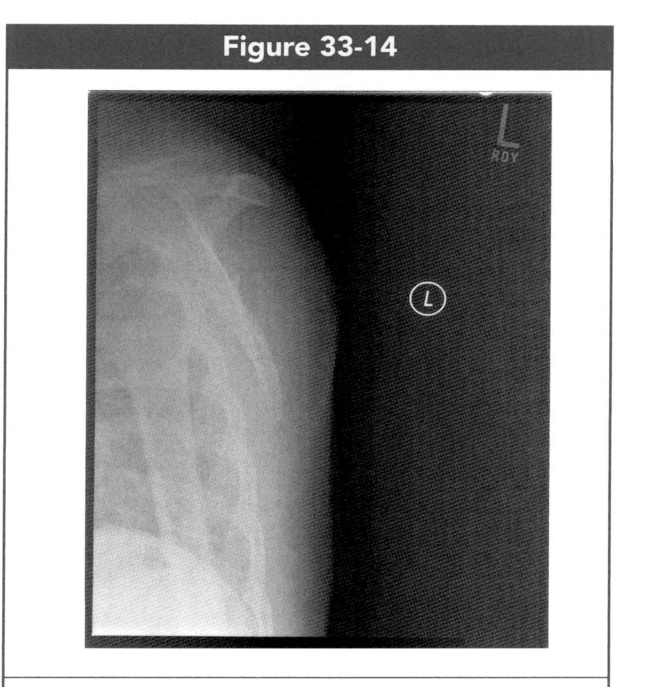

Histogram analysis error due to an excess of background radiation included within the collimated field on a scapular "Y" view. (From Quinn B. Carroll, *Practical Radiographic Imaging*, 8th ed. Springfield, IL: Charles C Thomas, Publisher, Ltd., 2007. Reprinted by permission.)

HORIZONTAL PROJECTIONS

Special consideration must be given to projections in which the image receptor is placed vertically in a cassette holder with a horizontal beam. These include the axial shoulder, the cross-table hip, and some cross-table projections of the spinal column. The collimated field often covers only the bottom two-thirds of the cassette, so that the bottom edge of the field is effectively "clipped" at the edge of the cassette while the upper edge is included (Fig. 33-17). As discussed above, this type of off-centering will lead to segmentation errors. There are two ways to

Figure 33-15

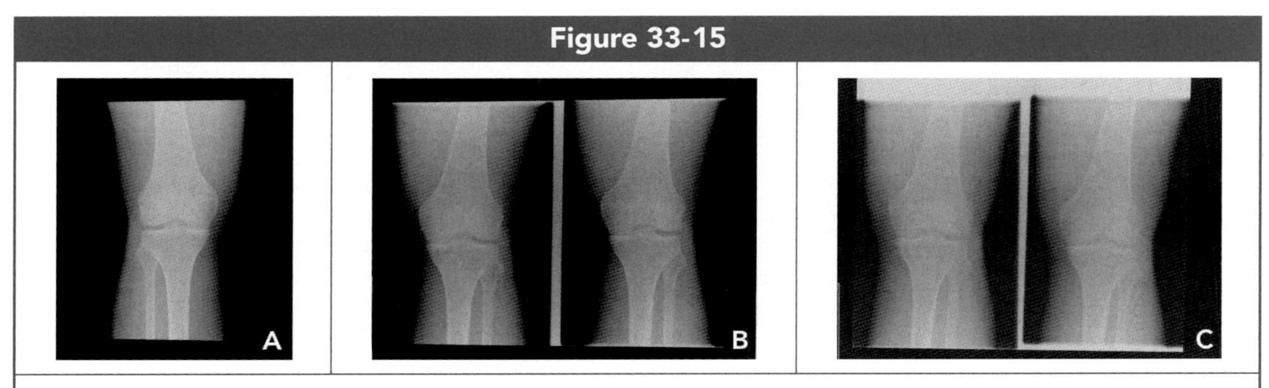

Radiograph ***B*** was taken with the same technique factors as ***A***, but with two fields positioned only ½ inch apart. The fog between the two fields resulted in histogram error which caused lengthened gray scale. For radiograph ***C***, 110 kVp was used to generate even more scatter, and resulted in *segmentation error* evident at the upper margin of the fields where there is a light strip. (From Quinn B. Carroll, *Practical Radiographic Imaging*, 8th ed. Springfield, IL: Charles C Thomas, Publisher, Ltd., 2007. Reprinted by permission.)

Figure 33-16

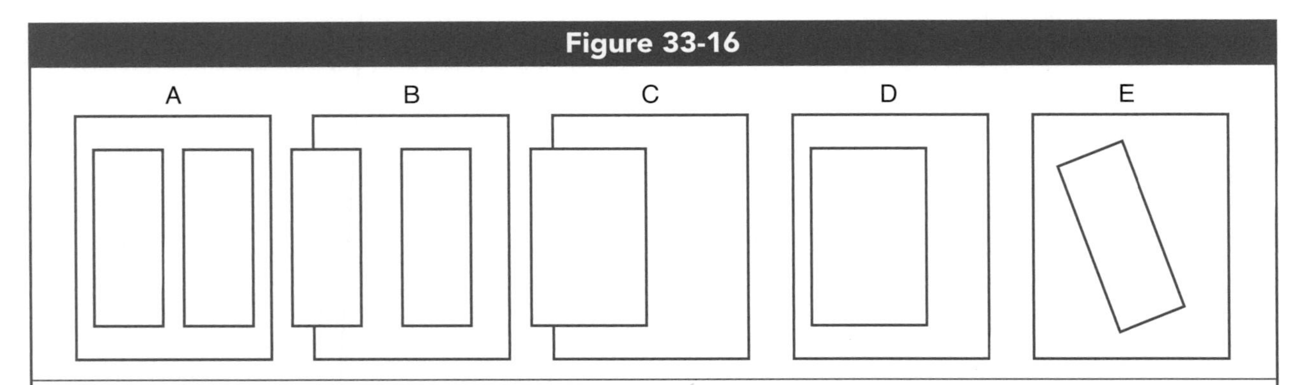

Collimated fields on a CR plate must have *symmetrical* collimated borders if there are any borders at all, parallel alignment with the plate, and ample spacing between multiple fields as shown in ***A***. If a field edge runs off the cassette, ***B*** and ***C***, fields are off-centered, ***D***, or not parallel to plate edges, ***E***, segmentation errors can occur. (From Quinn B. Carroll, *Practical Radiographic Imaging*, 8th ed. Springfield, IL: Charles C Thomas, Publisher, Ltd., 2007. Reprinted by permission.)

prevent this: A one-inch strip of lead can be taped to the bottom of the cassette as shown in Figure 33-18. The lead strip must completely cover the remaining lower cassette border. Alternatively, the patient must be built up to a point where the collimated field can be vertically centered to the cassette.

BILATERAL VIEWS

Figure 33-19 compares a single knee radiograph to a radiograph combining both knees using CR. The same size and type of cassette was used for both exposures, and no change was made in the algorithm applied. An identical "manual" technique was used for both exposures. Both were taken tabletop and non-grid so that the effects of any cross-scattered radiation would be taken into account. The image contrast, measured between the mid-patella and knee joint space, actually *increased* on the bilateral projection, from 3.2 to 4.2. The apparent brightness of the two projections is not visibly changed, and the exposure index increased only from 2260 to 2300, a completely insignificant change of 40 points. When "manual" technique is used, CR systems have no difficulty processing bilateral views such as this without histogram errors.

Figure 33-17

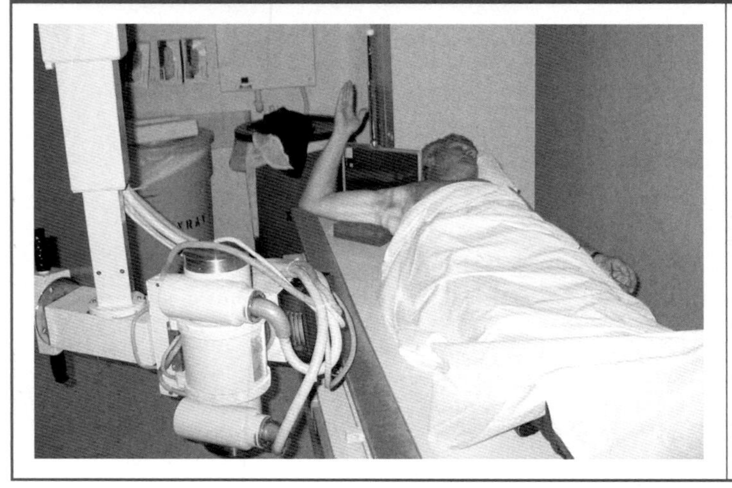

Horizontal beam projections such as the axial shoulder shown here, the cross-table hip, and some spine projections normally result in only the lower two-thirds of the image plate being exposed. This can result in segmentation error. (From Quinn B. Carroll, *Practical Radiographic Imaging*, 8th ed. Springfield, IL: Charles C Thomas, Publisher, Ltd., 2007. Reprinted by permission.)

On the other hand, when automatic exposure control (AEC) is used, special care must be taken on bilateral projections to activate two of the "side" detector cells and position the anatomy directly over them. Otherwise, an unacceptable amount of mottle can appear in the image. Since most DR procedures are done with the AEC engaged, this is a more frequent problem with DR than with CR. If a bilateral projection results in the detector cells being exposed to the "raw" x-ray beam, the AEC shuts off early, resulting in underexposure and subsequent mottle in the image.

Figure 33-18

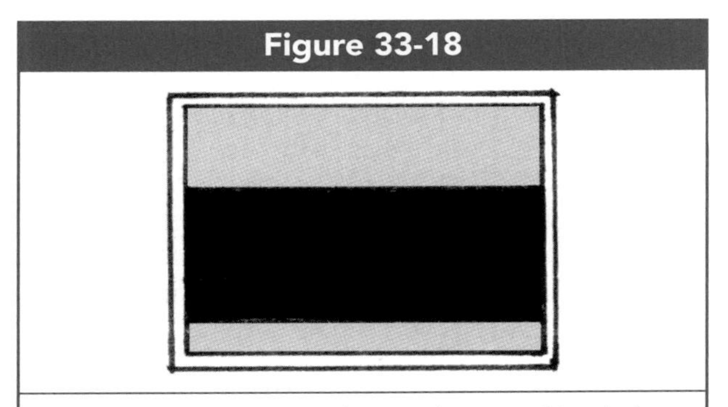

Diagram of placement of a 1-inch strip of lead along the bottom of the image plate to prevent segmentation errors for horizontal beam projections as illustrated in Figure 32-5. (From Quinn B. Carroll, *Practical Radiographic Imaging*, 8th ed. Springfield, IL: Charles C Thomas, Publisher, Ltd., 2007. Reprinted by permission.)

IMAGE RETENTION IN PHOSPHOR PLATES

Manufacturers caution that an exposed CR cassette will retain 75 percent of its exposure even after eight hours.

The experiment demonstrated in Figure 33-20 was conducted to determine the practical time limitations for processing an exposed CR plate. Index numbers were monitored to see how long it would take for them to drop below the recommended range, and the images were also visually evaluated. Thirteen tests were made, and only selected images are included in Figure 33-20. The results were impressive: Image ***A*** was fed through the reader immediately after exposure and developed an index number of 2260. This is somewhat above the recommended range and was used as a starting point to ensure conclusions that are not overstated.

Figure 33-19

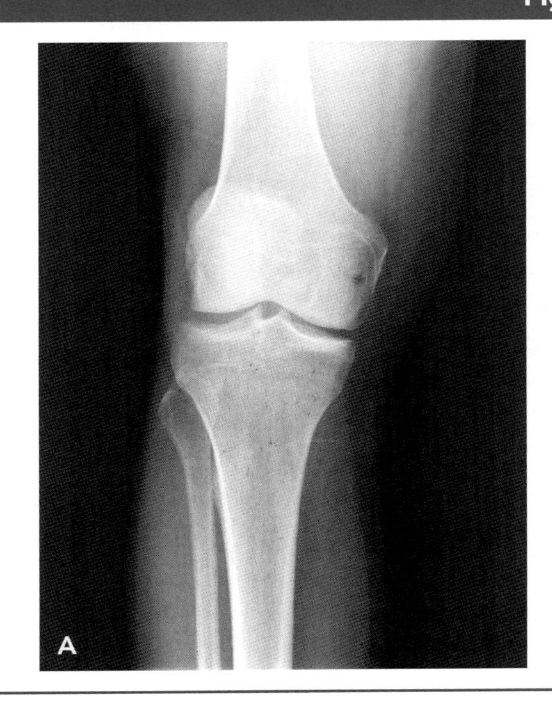

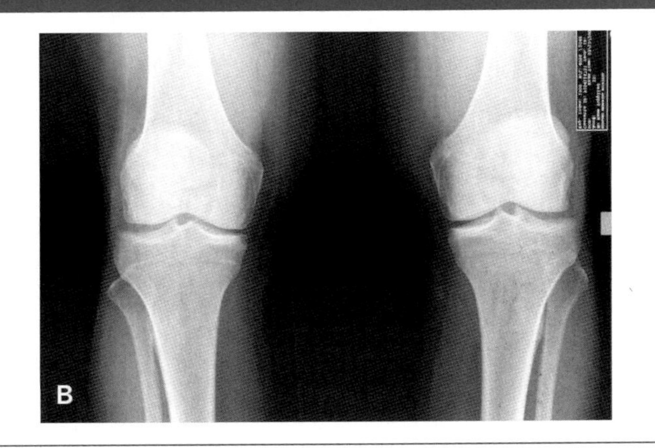

Comparison of a single knee projection, ***A***, with a bilateral knee projection, ***B*** using CR. Measured image contrast actually increased on this bilateral projection, and there was no significant alteration of brightness or exposure index, indicating no histogram errors. (From Quinn B. Carroll, *Practical Radiographic Imaging*, 8th ed. Springfield, IL: Charles C Thomas, Publisher, Ltd., 2007. Reprinted by permission.)

Image ***B*** was processed through the reader 56 hours, more than two days, after exposure, and still registered an index number of 1860, within the acceptable range. It did have visibly increased mottle, although this may not be obvious in the reprint presented here. At 64 hours, Image ***C***, the index number finally drops to 1740, below the recommended range and the mottle becomes obvious. Finally, image ***D***

Figure 33-20

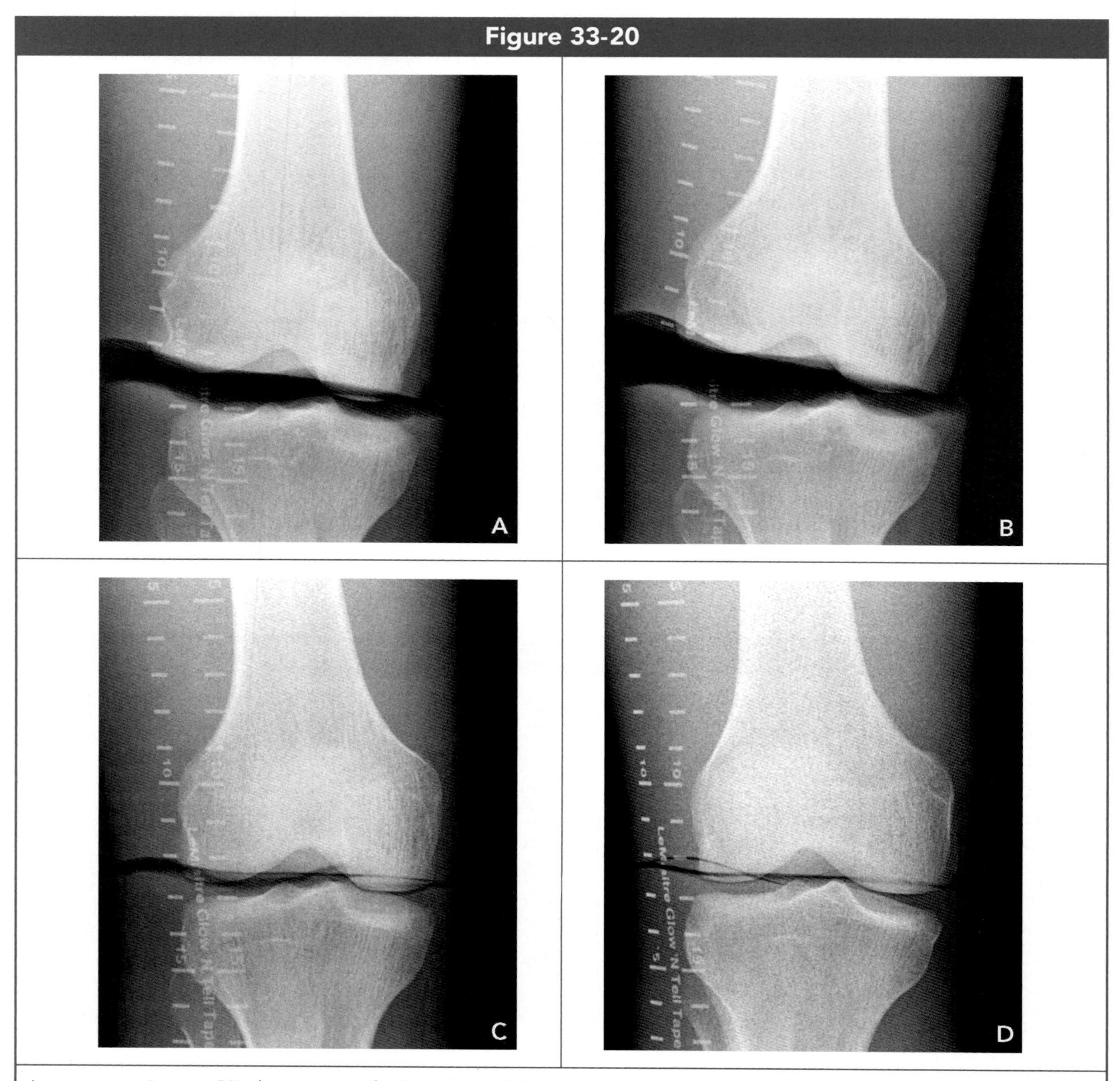

Image retention on CR plates. Magnified sections of these views are presented to better show mottle. Image ***A*** was read immediately after exposure at an exposure index of 2260. Image ***B***, processed 56 hours after exposure, registered an exposure index of 1860, still within the recommended range. Image ***C*** finally drops below the recommended range to 1740 after 64 hours. Remarkably, even after 10 days a mottled but substantial image still remains on the plate in ***D***. *There is no practical time limit over which it may be assumed that an image will have faded completely from a CR plate.* (From Quinn B. Carroll, *Practical Radiographic Imaging*, 8th ed. Springfield, IL: Charles C Thomas, Publisher, Ltd., 2007. Reprinted by permission.)

was processed 240 hours, or 10 days, after exposure. Mottle is severe and image quality has declined as expected. But, the important point to be made is that even after 10 days a very substantial latent image still remains on this plate. *There is no practical time over which it may be assumed that a latent image will have faded away completely on a CR plate.*

If there is any question at all as to the condition of a particular CR plate, it should be erased in the reader before use. Recall from the previous chapter that even unused plates should be erased prior to use if they have been in storage more than 24 hours, because of the accumulation of environmental fog.

SUMMARY

1. The digitization of radiography has *decoupled* the original technique factors used to acquire the image from the qualities of the final, displayed image. (With the exception of shape distortion, all of these qualities will have been adjusted by computer algorithms.)
2. In the digital age, radiographic technique plays one critical role: to ensure that *adequate signal* reaches the image detector system so that computer algorithms can successfully manipulate the image data.
3. Sufficient kVp and adequate *penetration* of the x-ray beam through to the digital receptor is more essential than ever. A CR or DR system cannot restore or adjust information which is simply absent from the remnant x-ray beam.
4. A survey of several manufacturers shows that a single 15 percent step increase in kVp can be generally applied with only very minor (often indiscernible) changes in contrast or mottle.
5. By increasing kVp at least 15 percent "across-the-board" for technique charts, and cutting mAs in half, actual patient skin exposure can be reduced by at least one-third. The new *optimum kVp's* in Table 16-2 are strongly recommended for digital imaging.
6. A substantially different approach to kVp may require re-setting of the target exposure indicator values.
7. The digitization of radiography has reduced the role of kVp from a major controlling factor (with screen/film radiography) to a mere *contributing factor* for displayed image contrast.
8. "Manual" *technique rules of thumb*, including proportional anatomy, the 4-centimeter rule, distance rules of thumb and the 15 percent rule are still not only applicable, but essential.
9. *Dose creep* has become an issue with computerized systems due to the legitimate fear of image mottle and the ability of CR and DR to correct the image for overexposure.
10. To minimize unnecessary patient exposure, radiographers must be discouraged from discarding views at the acquisition console and repeating exposures for minor positioning errors.
11. Although saturation can occur at *extremely high exposures*, there is no apparent decline in image quality with as much as three doublings of technique above normal when using CR or DR. On the other hand, relatively slight *reductions in technique* can result in visible image mottle.
12. With the use of *automatic exposure controls (AEC)*, centering of the anatomy is especially critical for digital systems, because off-centering quickly results in image mottle.
13. Although CR systems show sensitivity to scatter radiation "pre-fogging" a PSP plate prior to exposure, digital systems in general are impressively resilient to scatter radiation created *during* a radiographic exposure. Digital processing is able to *clean up* scatter "fog" patterns on projections notorious for them (such as the lateral L-spine) to a remarkable degree.
14. For digital imaging, the use of grids is still generally recommended to improve the signal-to-noise ratio for the input data reaching the image detector.
15. When multiple views are taken on one CR plate, good centering of the anatomy and symmetrical distribution of the exposed fields with ample separation between them will help avoid segmentation errors.
16. There is no compelling economical reason for placing *multiple* views on CR cassettes.
17. Unusual combinations of collimation, field orientation, or centering of the anatomy or field may cause segmentation errors for some CR

units. Errors can also be induced if a field is collimated so tightly that less than 30 percent of the CR plate is subjected to x-ray exposure.

18. *Bilateral projections* can generally be done with CR and DR systems. Care must be taken to use the correct detector cells if AEC is employed.

19. There is no practical time over which it may be assumed that a latent CR image will have completely *faded*. Over 24 hours only a slight loss occurs, and the image can still be processed. Plates must be erased prior to use if there is any question as to their condition.

REVIEW QUESTIONS

1. In the digital age, what is the major role of the radiographic technique factors set for an exposure?

2. What is the only image quality that will not have been "tampered with" by the computer prior to display?

3. Is it still true that "no amount of mAs can compensate for inadequate kVp"? Why?

4. Can the kVp be increased by one 15 percent step with compensating halving of the mAs, without compromising image contrast or increasing mottle to diagnostically unacceptable levels? Why is this a desirable goal?

5. For digital radiography, is kVp *the* controlling factor for displayed image contrast?

6. Can "manual" technique rules of thumb, including proportional anatomy, the 4-centimeter rule for part thickness, technique rules for changing distance and the 15 percent rule still be applied with digital imaging?

7. With CR systems, the tendency toward overexposure or "dose creep" results partly from radiographers' desire to avoid what characteristic in the image?

8. Is it true that for CR you can't collimate smaller than the plate size?

9. When the detector elements (dels) are overloaded with electric charge due to *extreme* overexposure (8 to 10 times normal), what effect may occur in the image?

(Continued)

REVIEW QUESTIONS *(Continued)*

10. Can a DR or CR system generally correct the image for too long a gray scale from over-penetration by the x-ray beam?

11. Can a DR or CR system correct the image for insufficient x-ray beam penetration?

12. List four factors which contribute to "dose creep" or the tendency to use too much exposure:

13. For CR, why is it important to keep multiple projections on one plate well-separated from each other?

14. With digital imaging, what economic advantage is there to exposing multiple views on one plate?

15. Generally, how much of the CR plate should be exposed as a minimum to guard against exposure indicator errors?

16. If an exposed CR plate has not been erased but has been stored for over a week, will the image have completely faded?

17. Can one generally perform bilateral views of the lower extremities using CR?

18. If two fields on the same plate cannot be well separated, what practice can help avoid histogram errors from any scatter radiation between the fields?

19. Manufacturers caution that after 8 hours ___________ percent of an exposure is still present on a CR plate.

20. What two types of radiation are CR cassettes especially vulnerable to during storage?

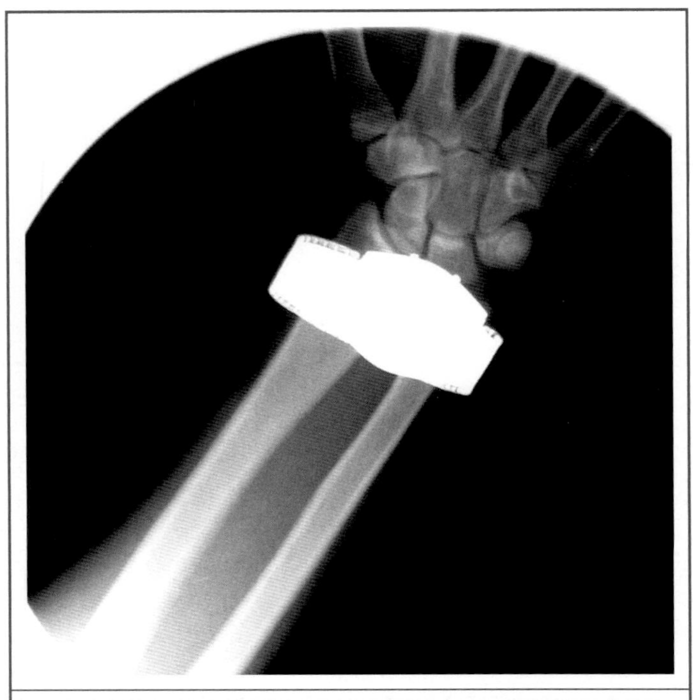

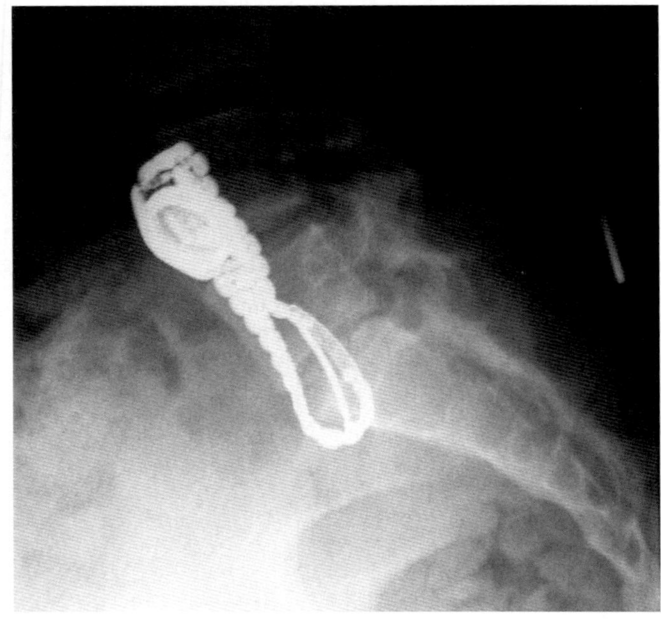

Top, an attempted wrist radiograph with the patient's watch still on. Bottom, this patient put her watch in her gown pocket, only to have it superimpose her lumbar spine views.

Chapter 34

CAPTURING THE DIGITAL IMAGE: DR AND CR

Objectives:

Upon completion of this chapter, you should be able to:

1. Compare the practical advantagtes and disadvantages of DR versus CR.
2. Describe the development of DR and miniaturized detector elements (*dels*).
3. Describe the components, function, detective quantum efficiency, and fill-factor of dels.
4. Describe the components of an *active matrix array (AMA)* and their functions.
5. Distinguish between the function, advantages, and disadvantages of direct-conversion versus indirect-conversion DR systems.
6. Explain the function of the CR cassette and phosphor plate in capturing the image.
7. Explain the functions of the CR reader/processor as it samples and processes the image.
8. Describe the limitations of CR and its sensitivity to "pre-fogging" of the phosphor plate.
9. State the formula relating pixel pitch to spatial resolution and define the Nyquist frequency.
10. Compare the absorption efficiency, conversion efficiency, and emission efficiency of CR, indirect-conversion DR, and direct-conversion DR systems.
11. Explain the impact of the *K-edge effect* on detector efficiency.
12. Define *detective quantum efficiency* and how it impacts the image.
13. Describe the causes and appearance of common digital image and printer artifacts.

COMPARING CR AND DR FOR CLINICAL USE

The fixed nature of the DR receptor system gives it both advantages and disadvantages when compared to CR. Of primary interest to the radiographer is that flexibility in positioning is limited on DR systems (Fig. 34-1): On some systems, the field size for each view in a series is preset by the quality control technologist in consultation with the manufacturer upon installation. This prevents any tighter collimation by the radiographer on smaller anatomy such as for pediatrics (and an attendant increase in scatter radiation). There is no ability to split fields for multiple views. There is little mobility with the receptor plate. For mobile DR units, the receptor plate is more bulky and may be attached by a cord to the machine, making it harder to maneuver around the patient as he/she lies. Modern cordless units use radio waves to transmit the image for immediate display.

Since it is awkward to remove the DR grid and risks damaging it, usually all DR procedures are done "bucky" or with the grid. Some DR units have as many as five AEC detector cells strategically placed across the board instead of the conventional three. The use of grids and AEC both limit positioning and require exact alignment of the x-ray beam.

Figure 34-1

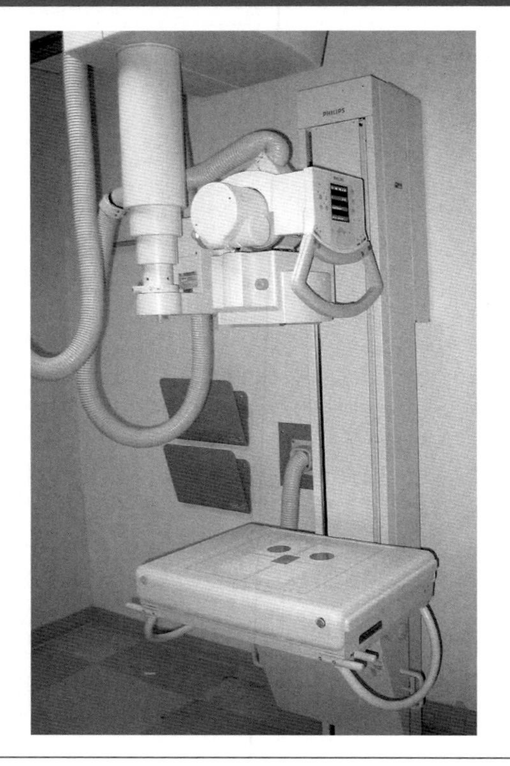

A vertical bucky for a DR unit receptor positioned horizontally to allow upper extremity positions. All DR projections must be done as single fields centered to the imaging plate detector cells. (From Quinn B. Carroll, *Practical Radiographic Imaging*, 8th ed. Springfield, IL: Charles C Thomas, Publisher, Ltd., 2007. Reprinted by permission.)

On the other hand, since DR operates at a consistent "speed class" for all procedures, radiographers are more confident when setting manual techniques and have less tendency to use higher technique than they think necessary.

Perhaps the two main advantages of DR are its high compatibility with PACS systems and increased departmental efficiency. Patients can generally be processed through the radiology department more quickly. The system is easier to use and less labor intensive. Radiographers do not spend time carrying image plates to and from processors. Images can be sent electronically direct from the imaging station into the integrated PACS system for storage. Provided that technologists do not waste time "tweaking" images at the workstation, general efficiency can be improved. DR systems also have higher detective quantum efficiency (DQE) and are less expensive to initially install, but are less serviceable and more expensive to maintain and repair. CR with its lightweight "cassettes" offers more familiarity for radiographers. While DR is more efficient and consistent, CR offers more adaptability for both positioning and technique.

DIRECT-CAPTURE DIGITAL RADIOGRAPHY (DR)

Direct-capture digital radiography (DR) refers to any digital imaging system which does not require the image receptor (cassette) to be physically carried and inserted into a processing unit that is separate from the exposure unit. For fixed DR units in the imaging department, the image receptor plate is built into the x-ray table or wall-mounted "chest" unit. Rather than removing it between exposures, the image receptor electronically sends information for each exposure directly to a computer processing system housed within the same x-ray machine, which then displays the image at the x-ray machine console.

Modern cordless mobile (portable) DR units use a detector plate that is physically separate from the machine, but the plate might be considered to be "directly connected" to the unit *electromagnetically*, because it uses radio waves to immediately send exposure information to the unit where it is displayed within a few seconds. The main distinction between DR and CR machines is this direct connection, either electronic or electromagnetic, between the exposure unit and the processing unit.

Two technological hurdles had to be overcome before direct-capture imagers could become feasible for mass production: First, since hundreds of individual pixel detectors would be required for a single image receptor plate, the detectors would have to be cheap to mass produce. Second, progress in the technology of miniaturization would have to reach a point where electronic *hardware* could be built at the pixel level. That is, individual x-ray and light detectors would have to be manufactured that were so close to the threshold of human vision that the generated image would appear as an *analog* image from a normal reading distance.

The first clinically useful DR system was developed in the 1980s as an offshoot from computerized tomography (CT). Referred to as *scanned projection radiography (SPR)*, it used a *bar* of rather large, CT-type detectors that was swept across the exposure field while a fan-shaped x-ray beam made a series of rapid x-ray exposures. Since it took some time for the x-ray tube and detector bar to complete their sweep across the image field, there was a high risk of patient movement or breathing. Because the detectors used were both expensive and large in size, not only was the resulting image resolution poor, but it was also not feasible to construct a complete array of detectors covering the entire exposure field. This would have to wait for the advancement of miniaturization technology.

Individual *detector elements (dels)* have now been developed that are truly microscopic, below the threshold of human vision. At this time, most DR systems use detector elements that are just at the threshold of human vision, but small enough to create an analog-appearing image at normal reading distance. The typical size of detector elements for a DR system is about 100 microns, or 1/10th of a millimeter (1/10th the size of a pinhead). Direct conversion and indirect conversion DR systems are both based upon the *active matrix array (AMA)*, a layer of microscopic detector elements each containing its own *thin film transistor (TFT)*.

The detectors used for scanned projection radiography (and CT) required a tubular photomultiplier to be attached to a scintillation crystal. This made the array of detectors too *deep* to form anything resembling an imaging *plate*. TFT detectors, on the other hand, are flat and extremely thin. This allows an entire array to be assembled into a panel thin enough for use as a "portable" plate for mobile radiography, called *flat panel* technology.

The Del

We have discussed in previous chapters how the *pixel* (picture element) is a two-dimensional representation of data acquired from a *voxel* (volume element) of tissue from within the patient's body. We have just raised the topic of the detector elements of DR systems, which are frequently referred to as hardware *pixels* which are square in shape and possess a surface area. Computer experts will insist that a "pixel" is *not* a square or any other shape or area, but that a true pixel is merely a numerical value assigned to a dimension-less *point* in an image.

To avoid confusing them with true pixels, the concise term for the hardware "pixel" element would be the *dexel*, a contraction of "detector element." In use, this term has been further contracted to the word *del*.

As an example, we might say that to form a digital radiograph, information from different *voxels* within the patient is collected by the *dels* of the imaging machine and computer-processed to become the *pixels* of the final image.

Direct Conversion Systems

Figure 34-2 illustrates the three main components of a single detector element or *del*. Most of the square area of the del is a thin semiconductor layer that is sensitive to x-rays or light. This is the capture area for detecting radiation, so the larger the area as a percentage of the whole square, the more efficient the del will be in absorbing x-rays or light. (This efficiency factor can be measured, and is referred to as the del's *detective quantum efficiency (DQE)*. The percentage of the square devoted to the semiconductor

Figure 34-2

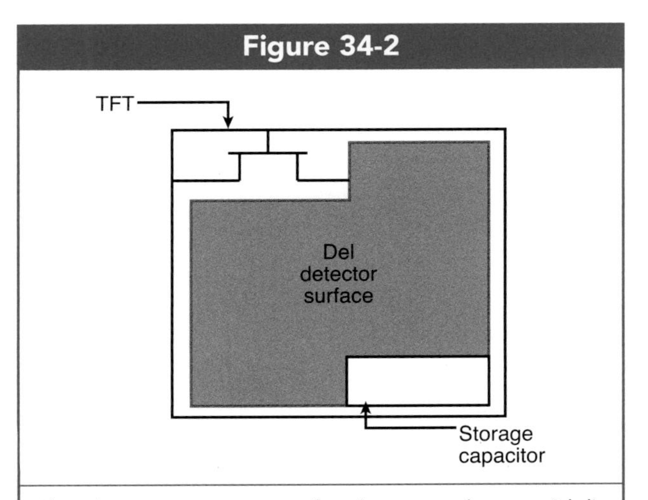

The three components of a detector element (del). The semiconductor detector surface area, the microscopic capacitor, and the thin film transistor (TFT) which acts as a switching gate. The *fill factor* for this dexel is about 80%. For direct conversion systems, this detector surface is made of amorphous selenium.

Figure 34-3

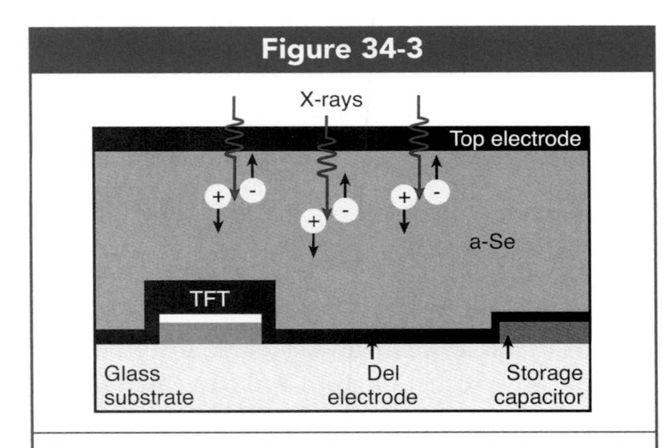

Cross-section of a del under exposure. Each x-ray absorbed within the semiconductor layer creates, by ionization, an *electron-hole pair*. Electrons drift upward to the top electrode, while positively-charged *holes* drift downward, building up an accumulation of positive charge on the capacitor.

detection layer is called the del's *fill factor*. A higher fill factor provides both higher contrast resolution (or signal-to-noise ratio) and better spatial resolution. One of the current limitations on del size is that the *TFT* and *capacitor* do not get smaller in size along with the overall dexel. Therefore, smaller dels have a lower *fill factor*, or less detector surface area, which then necessitates an increase in radiographic technique along with higher patient exposure.

In one corner we see a microscopic *capacitor*. This is the heart of the del, for it is the del's ability to *store electric charge* that makes direct-capture imaging possible. In the opposite corner we see the thin film transistor which acts as a switching *gate* to release the electrical charge when the del is read out.

Figure 34-3 is a cross-sectional diagram of the same del, showing how a thin (1 mm) layer of semiconductor material, in this case *amorphous selenium* (*a-Se*), is used to convert radiation energy into electrical charge. X-rays (or light rays) penetrating into this layer of selenium *ionize* its molecules, freeing up electrons. Each ionizing event creates an *electron-hole pair*, consisting of the freed electron and the positively-charged "hole" it leaves behind in the semiconductor molecule, that is, the *gap* in the molecule where an electron is now missing.

The top *electrode* in the diagram is an extremely thin conductor layer with a *positive charge* placed upon it. This charge attracts the freed electrons so they drift upward. At the same time, a negative charge placed on the *del electrode* below causes the positively-charged holes to drift downward. (This is just another way of saying that electrons from each successive layer below are pulled upward to fill holes, thus leaving holes in the lower layers—therefore the *holes themselves* appear to drift downward.) The net result is that *positive charge* builds up at the bottom of the semiconductor which is stored in the capacitor.

A nine-del section of the active matrix array is shown in Figure 34-4. A network of *data lines* and *gate lines* cris-crosses between the dels. *Gate lines* are controlled by the *address driver*, which controls the order in which the dels are read out. When the *bias voltage* along these lines is changed from -5 to +10 volts, it causes the TFT "gates" to open up sequentially

Figure 34-4

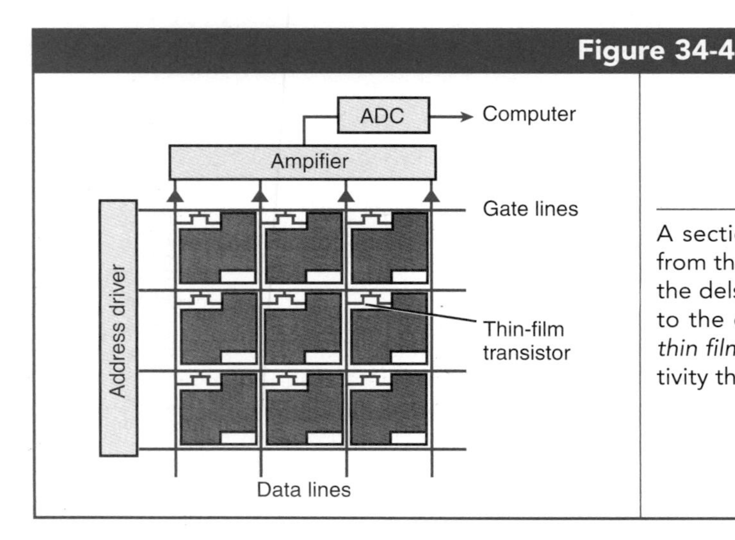

A section of the *active matrix array* (*AMA*). *Gate lines* from the address driver control the sequence with which the dels "dump" their charge into the *data lines* leading to the computer. The gate lines apply a charge to the *thin film transistors* (*TFTs*), creating a channel of conductivity through which stored charge can flow.

Figure 34-5

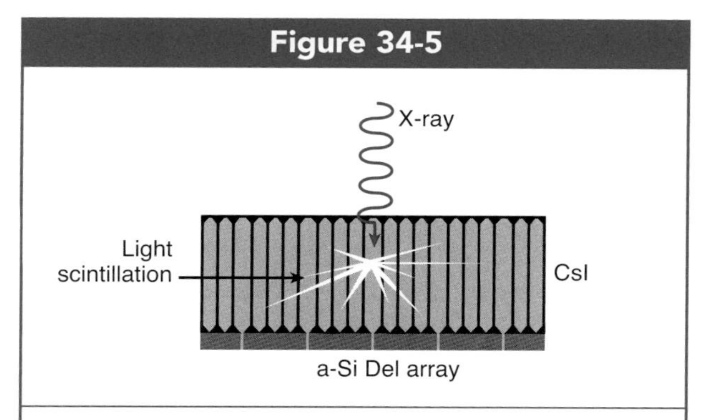

An indirect conversion system for a DR unit first uses a *scintillation* layer, usually tubular crystals of cesium iodide, that converts x-ray energy into fluorescent visible light. This light will be absorbed by an active matrix array below to generate an electronic signal.

Figure 34-6

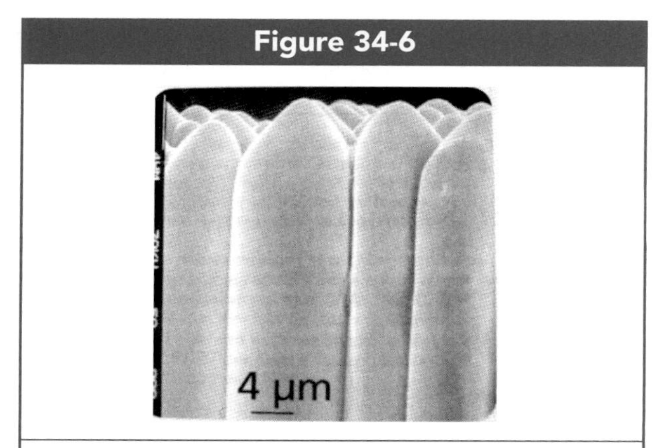

Crystalline cesium iodide forms channels which help control the dispersion of light. (Courtesy, Carestream Health.)

and dump the stored-up charge from each del in succession. For each particular TFT, the change in voltage causes a *channel* of conductivity to be opened up along the semiconductor material, which allows the stored electrical charge to flow out. The charge flows down a *data line* to an amplifier, which boosts the signal before sending it through an ADC into the computer.

Indirect Conversion Systems

Indirect conversion DR systems were developed before direct conversion systems. An indirect conversion system uses exactly the same active matrix array layout of TFT detectors or *dels* as a direct conversion system, only this array uses *amorphous silicon* rather than *amorphous selenium*, and the entire array is overlaid with a phosphor screen made of cesium iodide (or gadolinium oxysulfide) (Fig. 34-5). This phosphor screen *scintillates* or fluoresces when exposed to x-rays, emitting light that will strike the TFT detectors. Since light is normally emitted isotropically in all directions from a "turbid" or powder phosphor, which reduces image resolution, a method of forming vertical *crystalline* channels in the cesium iodide layer was developed such that the crystals lie *vertically*, forming light *channels* that confine the dispersion of the light somewhat (Fig. 34-6). (Nonetheless, the resolution achieved is still not as good as with the direct conversion system, in which electron holes drift directly downward to the TFT detectors.)

The light from the scintillation layer is then directed toward the detector elements in the AMA below (Fig. 34-7). By using amorphous silicon as a

Figure 34-7

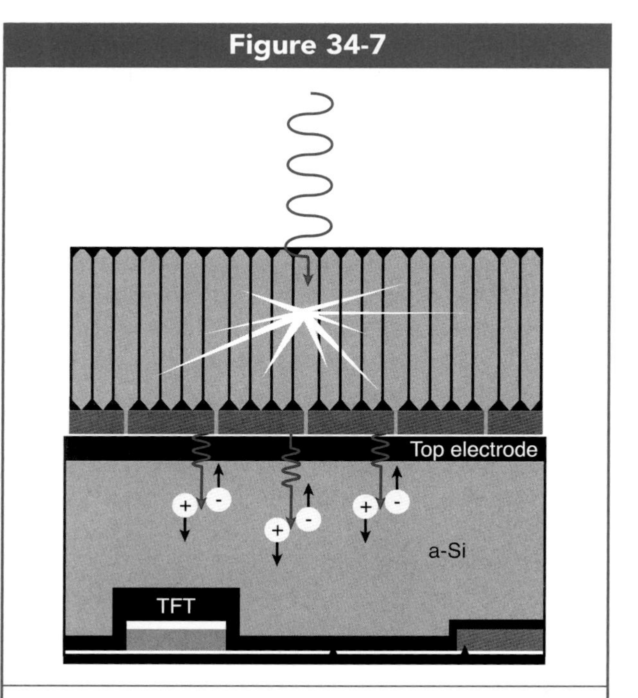

The active matrix array for an indirect conversion system uses amorphous *silicon* rather than amorphous *selenium*, because silicon is better at absorbing light rather than x-rays. Electron-hole pairs are created when light from the phosphor layer above ionizes the silicon atoms. Electric charge from the capacitors of the AMA is then used to generate an electronic signal.

detector element, these dels effectively become *photodiodes* which convert light into electrical charge. As with direct conversion systems, this charge is stored on a capacitor and released when the TFT gate opens for the del to be read out. *Amorphous* or non-crystalline forms of silicon and selenium are used for these TFT detectors because they can be coated onto the AMA in finely-controlled thicknesses. As shown in Figure 34-3, both types of detectors support the semiconductor layer, TFT and capacitor on a substrate that is usually made of glass.

In comparing the two approaches, direct conversion and indirect conversion, we find that the direct conversion system produces higher spatial resolution in the image, but the indirect system yields a higher detective quantum efficiency (DQE, to be discussed shortly) and therefore results in less patient dose. Since both characteristics are highly desirable, both types of systems have continued in use to date.

COMPUTED RADIOGRAPHY (CR)

The CR Cassette and Phosphor Plate

The CR cassette is designed in most respects to be used just as screen cassettes were used for film-based radiography. CR cassettes come in most of the same sizes (the 11" × 14" notably missing from this suite), and are very light for ease of use with mobile procedures. They can be used tabletop or placed in the bucky tray of a standard x-ray machine, and either manual or automatic exposure techniques can be employed.

Figure 34-8 illustrates the basic components of the CR image receptor plate which is inserted into the cassette. It is about one millimeter in thickness and somewhat flexible. The active phosphor layer is supported by a firm base, usually made of aluminum, and protected from scratches by a very thin coating of plastic. Light emitted by the phosphor crystals in a backward direction is reversed by the reflective layer, improving efficiency. However, the phosphor plate must be designed so as *not* to reflect the particular color of light used by the laser beam in scanning the plate for processing. An anti-halo layer prevents laser light from penetrating through to the reflective layer while allowing light emitted by the phosphor to pass through to it.

It is not necessary for the CR cassette which holds the phosphor plate to be light-tight as with film-based systems. The cassette is made of aluminum or plastic usually with a low-absorbing carbon fiber front. The back panel of the cassette may include a thin sheet of lead foil to reduce backscatter x-radiation from reaching the plate. The cassette also has a memory chip in one corner to download information on the examination and patient. Both the front and back panels are lined with felt material that minimizes the build-up of static electricity and cushions the plate from minor jolts.

Radiographers must be mindful that for most CR systems, the image plate housed in the cassette has only one single emulsion surface and must be placed facing forward in the cassette.

The CR receptor plate is a *photostimulable phosphor* (*PSP*) with the ability to store and release image data without an appreciable amount of lost information over time. It was found that a small number of barium-fluorohalide compounds, such as barium fluorobromide and barium fluorochloride, possess a unique property called stimulated phosphorescence.

Figure 34-8

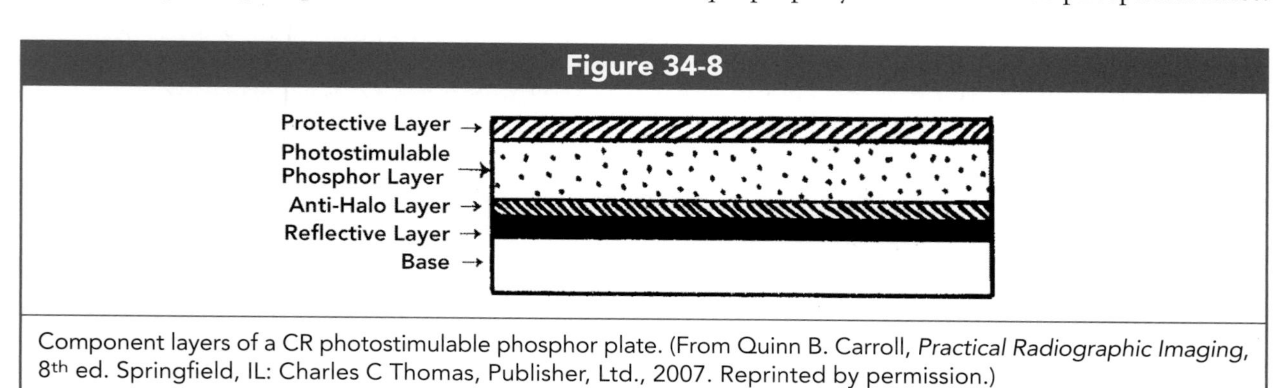

Component layers of a CR photostimulable phosphor plate. (From Quinn B. Carroll, *Practical Radiographic Imaging*, 8th ed. Springfield, IL: Charles C Thomas, Publisher, Ltd., 2007. Reprinted by permission.)

When a pure crystal of these compounds is "doped" or activated with small amounts of europium, the crystal develops a series of tiny defects called metastable sites or **F** centers (from the German farbzentren or "color centers") throughout its crystal lattice. These **F** centers act like small "electronic holes" in the crystal that can capture or trap electrons that are released from phosphor atoms when the exit radiation of the beam strikes the PSP plate (Fig. 34-10***A***, page 598). The imaging plate is thus able to store the energy of the remnant x-ray beam in the form of a *latent image* composed of electric charges stored within the **F** centers of the crystal lattice.

Because the energy retained by these latent electrons in the **F** centers is stable for relatively long periods of time, the image can be retrieved without any appreciable loss of information (fading) for many hours. It has been estimated that a typical PSP plate will retain up to 75 percent of the original latent image for 8 hours after the exposure. To release this trapped energy, the latent image of the photostimulable phosphor needs to be excited by an optical laser beam. Most CR processors (readers) use a red light helium-neon laser (Chapter 25), which scans the plate, adding energy to the trapped electrons in the **F** centers. This boost in energy enables the electrons to "jump" out of the trap and fall back into the shells of local atoms (Fig. 34-10***B***). As the electrons settle into atomic orbits, they lose energy which must be emitted in the form of radiation. This energy is manifested in the form of a blue-violet *glow*, or phosphorescence of the plate. The intensity of this blue-violet light is proportional to the amount of radiation originally received by the receptor plate under various tissues of the patient's body.

It may be noted that the barium fluorohalide compounds use for CR plates are the same types of chemicals that were formerly used for some types of radiographic intensifying screens whose function was to immediately convert x-ray energy into fluorescent light that exposed a film. One might ask whether the CR plates glow *during* the x-ray exposure. The answer is "yes," they do, and this can be demonstrated by removing a CR phosphor plate from its cassette and observing it under x-ray exposure in a darkened room.

This means that in the process of computed radiography, the photostimulable plate actually glows *twice*: Most of the energy absorbed by the x-ray beam is immediately emitted by fluorescence during the exposure, as shown in Figure 34-9. Exposure to an x-ray can *ionize* the atom, causing an electron to be ejected. In most cases, this electron immediately falls back into an atomic shell, with the accompanying loss of potential energy emitted as blue-violet light. For computed radiography, this originally emitted light is wasted.

However, a small percentage of the electrons freed by x-ray exposure become trapped in the **F** centers as shown in Figure 34-10. These remain trapped until they are exposed to laser light from the CR reader. At that time, the added energy of the laser light gives them the "boost" they need to escape the **F** centers and return to regular atomic orbits. As occurred during x-ray exposure, the potential energy lost from falling back into their orbits results in the emission of light. But, since this light emission was delayed, it must be classified as *phosphorescence* under stimulation of the laser beam.

A review of terminology might be useful here:

- *Luminescence* refers to any emission of light in general.
- *Fluorescence* refers to the *immediate* emission of light under stimulation.

Figure 34-9

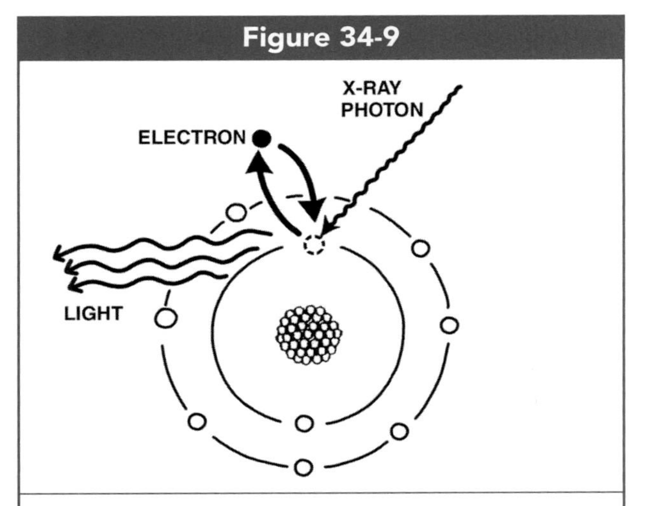

A CR plate *first* emits fluorescent light during x-ray exposure when most of the ejected electrons immediately fall back into their normal atomic shells. However, a portion of these electrons become trapped in *F centers* as shown in Figure 31-9. (From Quinn B. Carroll, *Practical Radiographic Imaging*, 8th ed. Springfield, IL: Charles C Thomas, Publisher, Ltd., 2007. Reprinted by permission.)

Figure 34-10

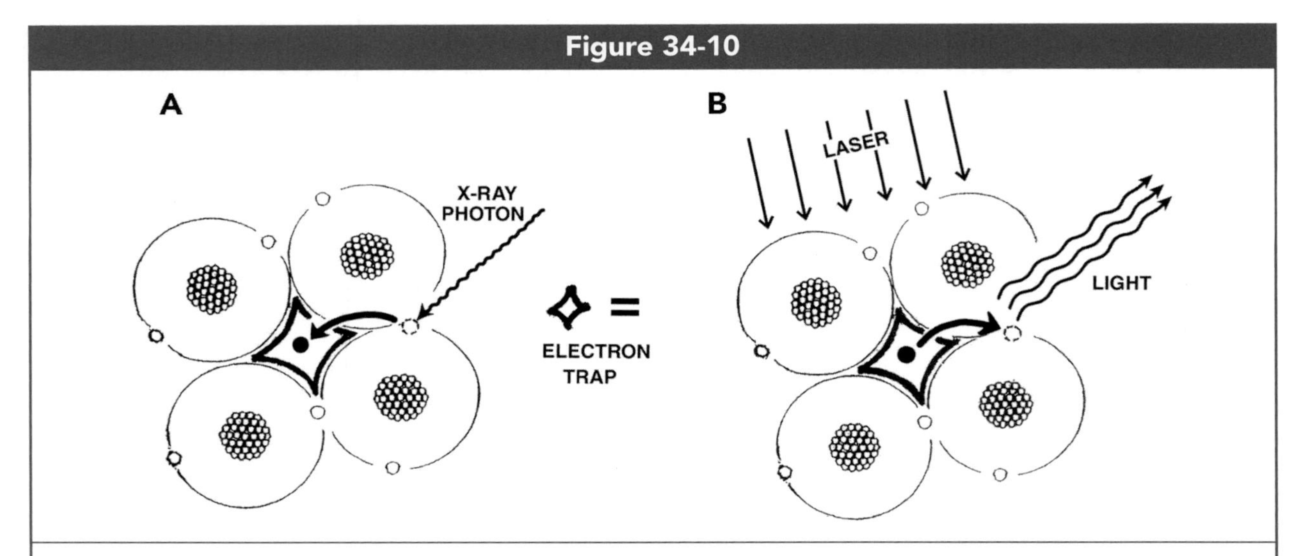

A portion of the electrons ejected during x-ray exposure become trapped in *F centers* within the lattice of the molecule, shown in ***A***. Later, when laser light from the CR reader adds energy to these trapped electrons, they escape the *F centers* and fall back into their normal atomic shells, emitting light *again* in the process. This dimmer light is amplified by the CR reader to form the image. (From Quinn B. Carroll, *Practical Radiographic Imaging*, 8th ed. Springfield, IL: Charles C Thomas, Publisher, Ltd., 2007. Reprinted by permission.)

- *Phosphorescence* is the *delayed* emission of light, some time after the original stimulus (exposure) has occurred. Phosphorescent materials "glow in the dark" without any added stimulation, but CR systems use *stimulated phosphorescence* which requires a laser beam to re-stimulate the phosphor.

Only a very small percentage of the electrons become trapped in the **F** centers, and the glow of the phosphor under laser stimulation is very dim indeed this second time around. But, the magic of computed tomography is that even this tiny amount of remnant light can be so amplified by the electronics of the reader that it can still produce a diagnostic image.

It should be noted that Figure 34-10 is a diagrammatic "trick" to help visualize the concept of an electron trap within a molecule, so it is not precisely accurate. A more accurate representation of how **F** centers work is presented in Figure 34-11, which shows the hierarchy of energy *bands* within an atom: As we work our way outward from the nucleus of an atom, we can label those inner atomic shells which have reached their full capacity of electrons, based on the $2N^2$ rule, as *filled bands* of energy. Outer shells hold their electrons so loosely that the electrons can undergo *valence bonding* or *ionic bonding* with other atoms in order to form molecules. The range of electron *energies* at which this bonding can occur is called the *valence band*. Above this level is the *conduction band*, that consists of electron energies at which they are freed from the atoms to flow as electrical current.

These energy bands are presented for a typical atom in Figure 34-11***A***. However, the structural organization of a whole molecule can affect the energy bands of the individual atoms within it. Some specific molecules are organized in such a way as to introduce an additional energy band *between the valence band and the conduction band*, shown in Figure 34-11***B***. While the conduction band is considered to be "outside" the atom, this new band is a *metastable energy state within the atom*. This is the **F** center or "electron trap," and it is created within molecules of barium fluorohalide when europium is added as a "doping agent" or impurity.

The CR Reader (Processor)

In the CR reader (or processor), an automated mechanical system uses suction cups to remove the PSP plate from its cassette, then moves it by a series of rollers into and through the area where an oscillating

Figure 34-11

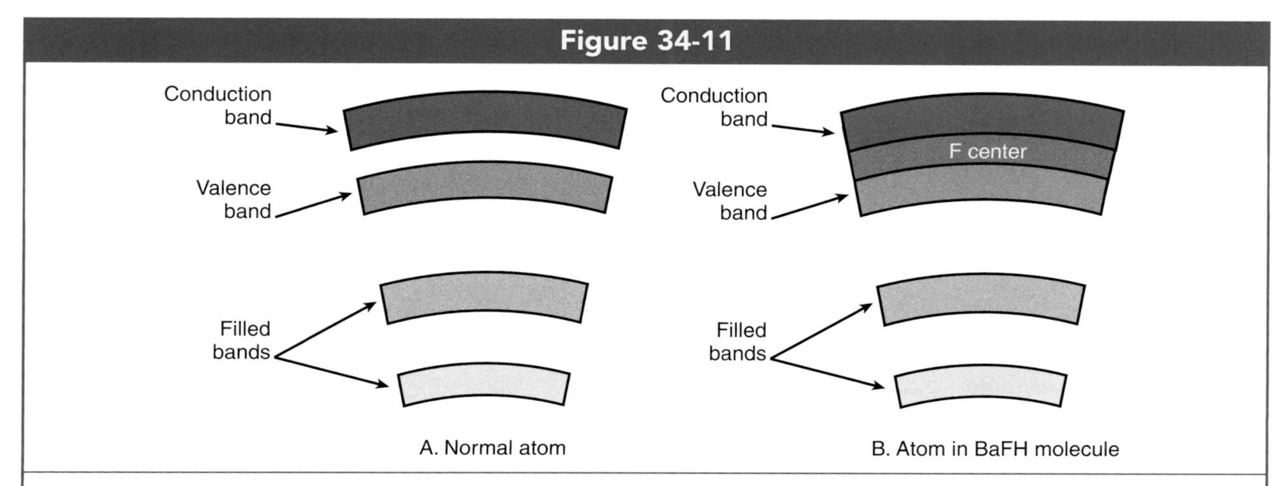

For a typical atom, ***A***, electron energy levels consist of a *filled* band, a *valence* band, and a *conduction* band which lies outside the atom's binding influence. When barium fluorohalide crystals are doped with europium, the molecular structure results in the creation of *F centers*, which actually consist of an additional energy band which is created between the valence and conduction bands. In the CR reader, exposure to the laser beam causes electrons in this band to fall back into their normal shells.

mirror deflects the laser beam to move rapidly across the plate, causing it to release its stored latent image.

The laser beam is circular in shape, with a typical diameter of 80 microns (micrometers). Because of this circular shape of the laser beam, the laser spot must overlap pixels that are being recorded (Fig. 34-12, ***B***). This is just the opposite of DR, in which the edges of the hardware dexels are well-defined, but each pixel is actually missing some of its detection area (Fig. 34-12, ***A***). Since the mirror steers the beam at a considerable angle to the peripheral edges of the plate, the circular beam elongates somewhat into a more oval shape at these extremes. This results in a slight loss of output intensity at the sides of the phosphor plate, which must be corrected electronically or with software.

The crosswise direction in which the laser beam scans across the plate is called the *fast scan direction*, whereas the direction the plate itself is moving through the reader is called the *slow scan* or *subscan direction* (Fig. 34-13). The visible light emitted from the PSP is directed through a light-channeling guide to a photomultiplier tube which records and amplifies the signal.

The conversion of the phosphorescent light from the PSP plate into digital data is a two-step process that is accomplished by changing light into an electronic signal and converting this into digital data that can be understood by the computer. The first step is performed by the *photomultiplier tube* (*PM tube*) (Fig. 34-14). The PM tube consists of a *photocathode* plate attached to an electronic amplifier. The photocathode is a layer of material which releases electrons when light strikes it, through the *photoelectric effect*.

The photoelectric effect was described in Chapter 11 as one of the interactions that occur when x-rays strike atoms. Specific chemical compounds can

Figure 34-12

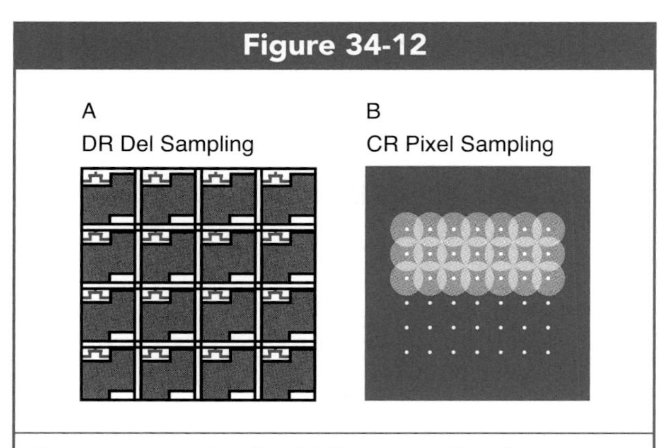

The hardware dexels of DR have well-defined edges but are missing some detector surface area, ***A***. By comparison, the software pixels sampled by a CR processor have poorly-defined edges, and since the laser beam is round, it must overlap the pixel areas detected in order to fill square pixels, ***B***.

Figure 34-13

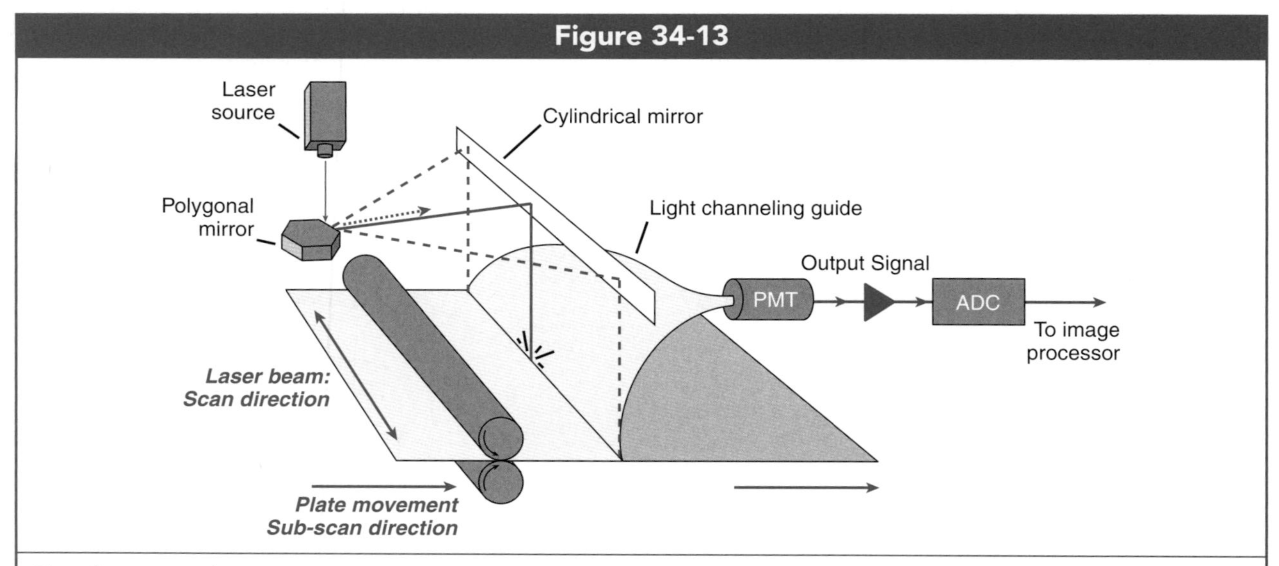

The "fast scan" direction as a PSP plate moves through the CR reader (processor) is the crosswise laser beam scan direction. The "slow scan or sub-scan direction" is the direction of the plate movement.

produce the photoelectric effect when *light* strikes them, even though light has much less energy than x-rays. A light photon is absorbed by a loosely-bound outer-shell electron, which is emitted from the atom, carrying this extra energy.

When the laser beam in the CR reader stimulates the PSP plate, the visible light emitted from it is directed through a light-channeling guide onto a photocathode layer on the input side of a PM tube. As shown in Figure 34-14, electrons are emitted from the photocathode. But, this electronic signal is too small to be detected by most other types of electronic devices, so it needs to be amplified.

This is accomplished through a series of *dynode* plates. *Dynodes* are electrodes which can be switched back and forth between positive and negative charges. With each strike, the dynode plate switches from positive to negative, repelling the stream of electrons toward the next plate, but with increasing effect such that the number of electrons in the stream is *multiplied*. Each collision of an incident electron with a dynode plate releases about 5 electrons from the plate. By having the electron beam pass through 10–12 of these dynodes in succession, the electronic signal can be amplified by more than a million times.

Figure 34-14

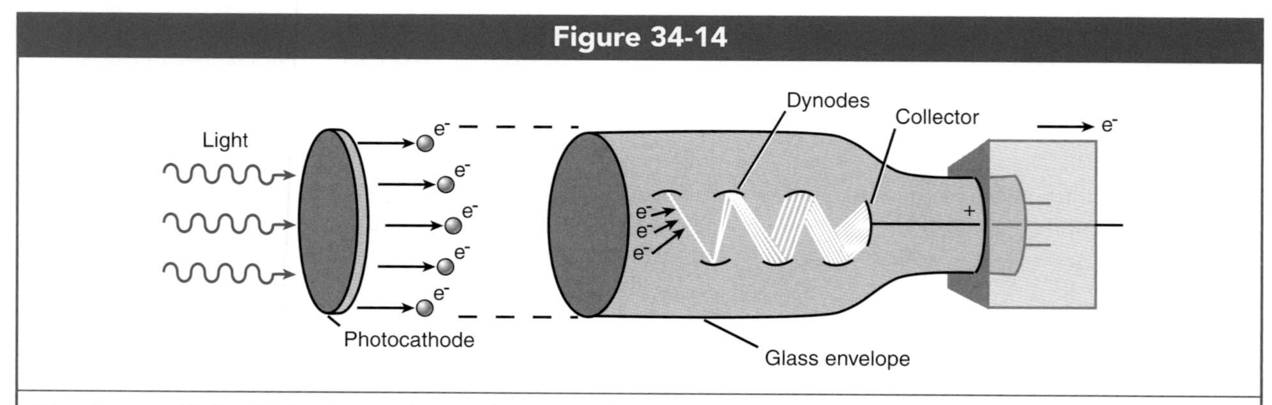

The photomultiplier (PM) tube employs a photocathode plate which emits electrons (by ionization) when struck by light. The electron stream is amplified by a series *dynodes* which sequentially boost the signal.

The PM tubes are specifically sensitive to the blue-violet end of the light color spectrum, whereas the *laser* light used to stimulate the plate is red. The graph in Figure 34-15 illustrates how these two wavelengths of light are far enough apart that there is effectively an "optical barrier" between them. Furthermore, while the laser is tightly focused, the blue-violet light emitted from the stimulated plate spreads out in all directions, so the PM tube can be positioned at an angle differing from that of any reflection of the laser beam.

Once the image plate has been scanned, it continues to move in the subscan direction toward the *eraser* section of the CR processor. In this section, the PSP plate is exposed to bright white light which removes any remaining information from the plate so that it can be reused. The "clean" image plate is now reloaded back into its original CR cassette. After the plate is reinserted into the CR cassette, the cassette is moved to the output tray of the processor for retrieval. The entire reading and erasure process normally takes about 90 seconds to complete. The properly erased PSP plate can be used thousands of times.

Figure 34-15

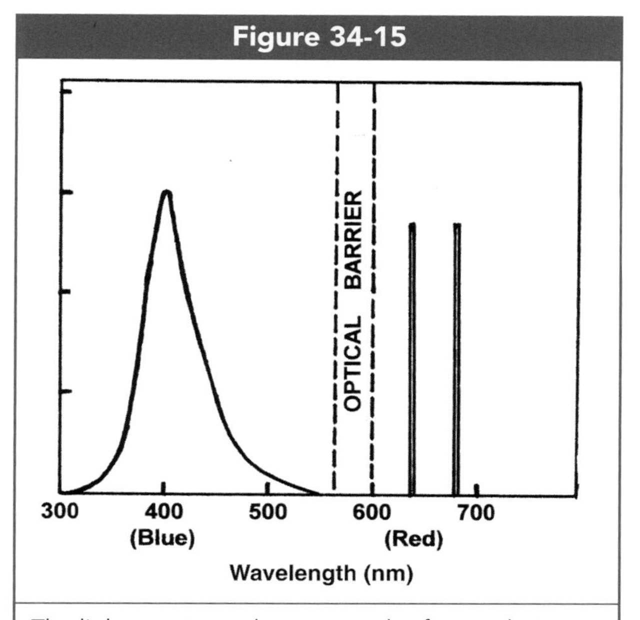

The light spectrum shows enough of a gap between the wavelengths of blue and red light to effectively constitute an "optical barrier" between these colors of light, allowing the CR reader to use one for laser stimulation and the other for read-out of the PSP plate.

Image Identification

In computed radiography systems, before or after the PSP imaging plate contained within the CR cassette has been exposed, it must be imprinted or marked with the appropriate patient information before the image is "read" in the processor. This includes the patient name, ID number, exam date, institutional ID, and position. The required information is typed from a keyboard or electronically transferred into the computer from a bar code scanner. Once the ID information is entered, it should be checked to ensure that each image is imprinted with the correct data. Each cassette must be identified prior to being placed into the CR reader. This can be most quickly done by scanning a bar code on the cassette.

Recent Developments in CR

A recent development for CR readers is *dual-sided reading*: The CR plates use a transparent base to support the phosphor layer, and there are two sets of detectors in the reader that capture light from both sides of the plate upon stimulation by the laser beam. Combined with a thicker phosphor layer on the plates, the signal-to-noise ratio is improved in these systems.

New *line-scan* readers have increased the speed of processing. They use a laser *line* source and a shaping lens to refine the beam into a fine line rather than a point. Stimulated light is then emitted from the phosphor line-by-line, captured by a lens array and fed to a CCD photodetector array.

To produce the laser beam, many newer CR readers now use solid-state semiconductor laser diodes rather than helium-neon gas lasers. These emit a slightly different wavelength of light, but are more reliable and consistent in the long run.

Background and Scatter Radiation

CR plates are approximately 10 times more sensitive than the older film/screen cassettes to accumulated background radiation during long periods of storage, and to scatter radiation during exposures. Typical background exposures can equal 70 to 80 μR per day, and it only takes 100 μR to produce a fog "density" on the plate. Note that over a weekend, more than 200 μR is likely to accumulate on the plate. Radiographers should be careful to erase any cassette prior to use if there is *any chance* it has been in storage for two days or more.

The experiment conducted in Figure 34-16 gives dramatic evidence to this sensitivity. The CR plate used for image ***A*** was erased just prior to processing for comparison. Image ***B*** is from a CR plate which had accumulated two days of background radiation, and processed without erasing it. Fog exposure is apparent. For image ***C***, the CR plate was erased just prior to use, then exposed to *only* the scatter radiation generated from an abdomen phantom during an exposure of 80 kVp and 30 mAs. It was placed in vertical position at 1 meter away from the x-ray table during exposure, then immediately processed. Fog densities are readily apparent.

The experiment in Figure 34-16 shows that CR *plates*, as hardware, are very sensitive to background and scatter radiation. When "fog" exposure is accumulated on these plates *before* using them for a regular radiographic exposure, computer software is often unable to correct for it, and the final digital image manifests a "fog" density. Ironically, computer software generally *is* capable of compensating for even high levels of scatter exposure generated *during* the exposure, as is shown in Figure 33-9, Chapter 33. The difference may be that in the case of "pre-fogging" the plate, the final histogram is effectively composed of two separately acquired histograms which are overlaid.

When exposing multiple images on one plate, fog densities between fields can lead to histogram errors due to segmentation failure. A fog density can accumulate across the plate not only from background radiation, but also from repeated exposure to scatter radiation within the radiographic room. It is more hazardous to leave CR cassettes out, leaning against walls in the x-ray room for example, than it was for film/screen cassettes. Many departments now prefer to store their CR cassettes in adjacent rooms or a centralized area well away from the x-ray machines.

When it is common practice to leave cassettes in the bin of a mobile x-ray machine, care must be taken to *rotate* the cassettes, placing fresh cassettes at the bottom of the stack and using cassettes from the top, so that no cassette is left in storage there for more than a day. For departments with a slower turnover rate, it may be advisable to assign a clerk or a radiographer to erase all CR cassettes each morning.

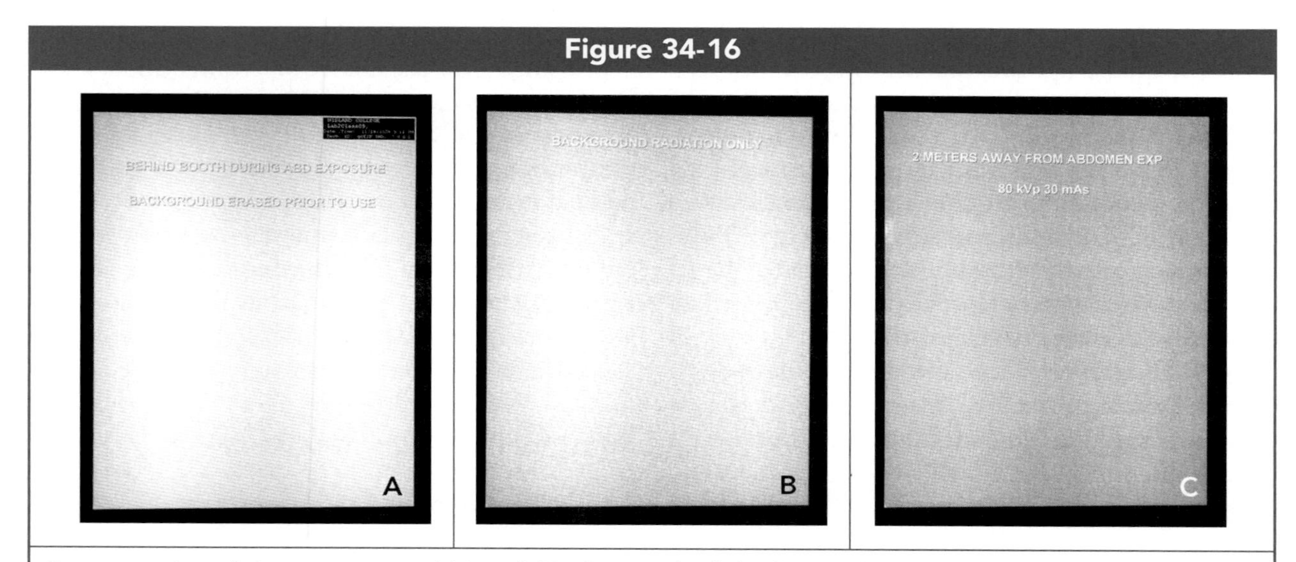

Figure 34-16

Demonstration of the extreme sensitivity of CR plates to both background and scatter radiation. Image ***A*** was processed from a CR plate that was erased and then stored behind the control booth during an abdomen exposure. By comparison, fog density from natural background radiation can be made out on image ***B*** from a CR plate which had not been erased after two days of protected storage. Image ***C*** is from a plate set 2 meters away from an abdomen phantom that was exposed to 80 kVp and 30 mAs. Scatter exposure is readily apparent.

SPATIAL RESOLUTION OF DIGITAL SYSTEMS

For all digital systems, the maximum spatial resolution is equal to the *Nyquist frequency* which is the sampling frequency expressed in line-pairs per millimeter (LP/mm). For DR systems, this image sampling frequency depends only upon the *del pitch*, defined as the distance from the center of one del to the center of the next, Figure 34-17***A***. The maximum spatial resolution is inversely proportional to a doubling of the del pitch.

$$SR = \frac{1}{2P}$$

where *SR* is the spatial resolution and *P* is the pitch of the hardware dels (for DR) or scanned pixels (for CR). Note that the dexel pitch is approximately the same distance as the *width* of each del, as shown in Figure 34-17***A***. Therefore, the maximum spatial frequency is also inversely proportional to a doubling of the width of the dels in millimeters.

(Technically for DR detectors, the del pitch includes any spaces *between* dels, shown in Figure 34-12***A***, whereas the del width does not include these spaces and is slightly smaller. However, for our purposes here, either one can be used to approximate the spatial resolution.)

For example, a del pitch of 0.1 mm results in a Nyquist frequency of 5 LP/mm as shown in Figure 34-17***B***, where the resulting image consists of five pairs of lines, each pair consisting of a white pixel and a black pixel, for a total of ten pixels within a space of 1 mm. A del pitch of 0.05 mm would yield 10 LP/mm, ($1 / 2 \times 0.05$ or 1/0.1).

For DR systems, spatial resolution is determined by the size of the detector elements, and is consistent regardless of plate or field size. These dels range in size from 100 to 200 microns. A 100 μm detector element provides a spatial resolution of about 5 LP/mm (less than a traditional 200-speed screen/film combination); A 200 μm del size yields about 2.5 LP/mm.

For CR systems, the Nyquist frequency is now the sampling frequency of the reader as its laser beam scans across the image plate. This sets the upper limit of the spatial resolution that can be produced. The same formula (above) is used to relate spatial resolution to the pixel size or pitch, but the "P" in the formula would now stand for the *pixel pitch* as sampled from the CR plate, rather than the *del pitch* of the detector elements in a DR receptor plate. However, due to light spread between the plate and the light guide, the net spatial frequency will be slightly less than the Nyquist frequency for a CR system, (about 4 LP/mm for 100 μm, 8 LP/mm for 50 μm).

Figure 34-17

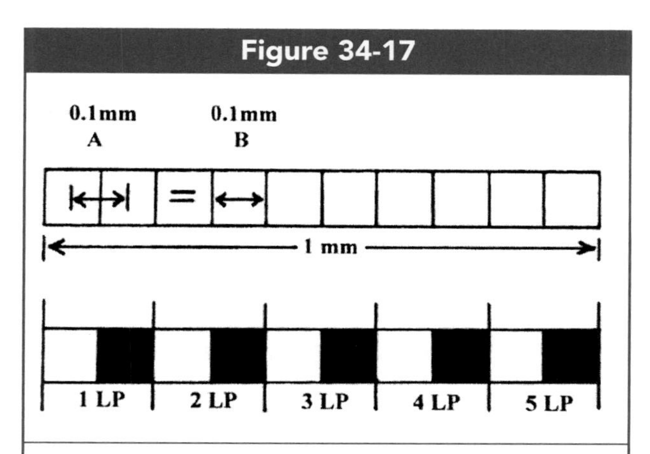

Pixel pitch, or *del* pitch, ***A*** is equal to pixel or del width, ***B***. Both are 0.1 mm in this example. The spatial resolution resulting here is 5 pairs of white and black lines, or 5 LP/mm.

(Good x-ray beam projection geometry, such as the use of small focal spots and long SID's, is still important to produce a sharp latent image on the imaging plate. But, if the scanning speed and sampling frequency of the reader results in larger measured pixels, this will reduce the original sharpness of the image.)

For *some* brands of CR equipment, smaller plates are scanned at a higher subscan indexing speed than larger plates, resulting in higher spatial resolution when the smaller plates are used. It becomes important to use smallest plate consistent with the part to be imaged. There is also less risk of histogram errors with higher Nyquist frequencies. However, most manufacturers have standardized their sampling frequency at 10 pixels per millimeter for all cassette sizes, so plate size is not an issue. This results in consistent image sharpness levels of about 4 line pairs per millimeter.

Some manufacturers encode the sampling frequency based on the anatomical study entered into the system rather than the plate size. Others offer special "extremity cassettes." These can produce *slightly* higher resolution in the final image, but as explained above, the sampling frequency overrides the original "cassette speed." Many departments have elected not to invest in variable-speed plates.

EFFICIENCY OF IMAGE RECEPTORS

Materials used for the *front panels* of image receptor plates, cassettes and x-ray tables must be as radiolucent as possible while still providing structural protection. Added layers in front of the actual *active matrix array* of detectors absorb more radiation and necessitate higher techniques.

CR Phosphor Plates

For *stimulable phosphors* (screens) such as are employed in computed radiography (CR), three characteristics determine the amount of radiographic technique needed. They are defined as follows:

1. *Absorption efficiency* is the ratio of x-ray photons absorbed by the phosphor crystals to the x-ray photons incident upon the phosphor layer. Elements with high atomic numbers, and compounds that take advantage of the *K-edge effect*, as explained in the following section, are better at absorbing x-rays. Also, the thicker the phosphor layer, the more x-rays are absorbed. The more x-rays are absorbed by the phosphor layer, the more light is emitted when the plate is stimulated by a laser beam in the digital reader (processor).
2. *Conversion efficiency* is the percentage of energy from absorbed x-ray photons that is converted into *light* rather than into infrared or heat energy which is wasted. It is a characteristic *inherent only to the particular chemical compound* used as a light-emitting phosphor.
3. *Emission efficiency* is the ability of the light produced by the phosphor crystals to *escape* the phosphor layer and reach the light guides in the CR reader that direct it to the light detector, a photomultiplier tube. Light emitted from a specific crystal must penetrate out past other crystals and through the chemical binder (plastic) layer in which the phosphors are suspended. When light is emitted *isotropically* (in all directions) from a turbid (powder) phosphor, the light emitted *laterally* within the phosphor layer is lost. Light emitted directly backward can be captured by using a *reflective layer* behind the phosphor layer, which reflects the light back to the light guides. Needle-shaped crystals have been developed for the phosphor, which act as tubular guides directing most of the light downward and improving the emission efficiency.

Any receptor system using a phosphor or scintillation layer to convert x-rays into light must be good at absorbing x-rays, converting their energy into light and emitting that light efficiently.

K-Edge Effect

Within the kV range characteristic of diagnostic x-rays, there is an irregularity in the photoelectric absorption of x-rays by most phosphors known as the *k-edge*. A graph of absorption by the photoelectric effect is shown in Figure 34-18, and shows this defect. The value of the k-edge is the *binding energy of the K-shell* for atoms of a particular element. The *k-edge effect* refers to a loss of absorption efficiency due to a *mismatching* of the k-edge for a particular phosphor element with the *average* energy range of the x-ray beam at which most x-rays are produced.

For example, the binding energy of the k-shell for *lanthanium* is 39 kV. This is the element plotted against the kV of x-rays in Figure 34-18. We generally expect that as kVp set by the radiographer increases, there will be more penetration and therefore, less absorption. We expect the curve for absorption by the photoelectric effect to drop steadily. The problem is that the photoelectric effect *cannot take place* unless the energy of the incident x-ray photon is slightly *higher* than the binding energy of the electron shell. (See Chapter 11.) As we examine the graph in Figure 34-18, we realize that all photoelectric interactions occurring *to the left of the k-edge* must be taking place only in the L-shell and M-shell of the lanthanium atom, since these energies are

Figure 34-18

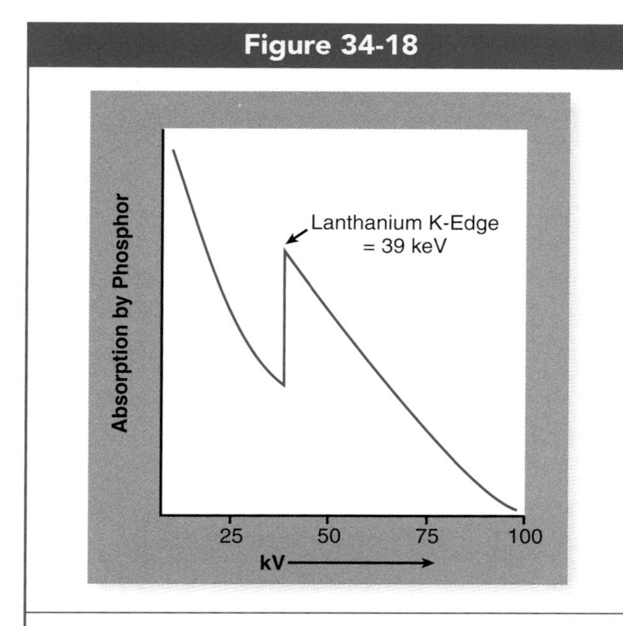

Radiation absorption curve for lanthanium shows a *k-edge* at 39 kV. The number of photoelectric interactions spike when the kV surpasses the k-shell binding energy. Interactions to the left of this spike are only occurring in the L and M shells.

Figure 34-19

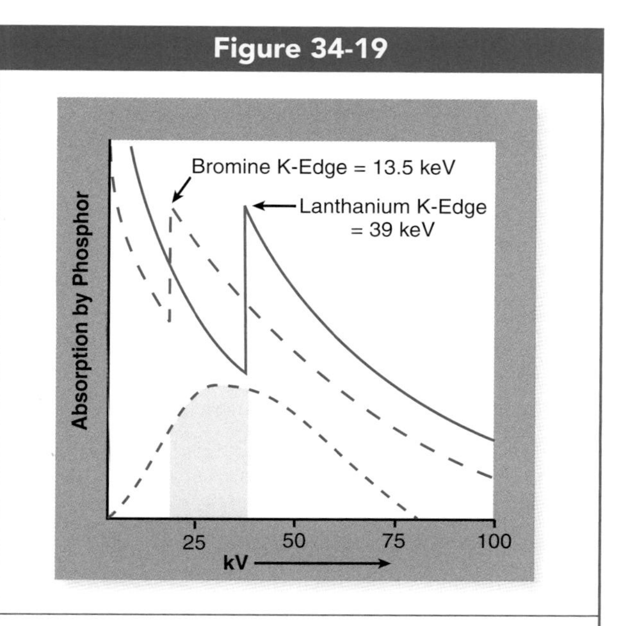

Superimposition of absorption curves for lanthanium and bromine over the energy spectrum curve for an 80-kVp x-ray beam shows that bromine better matches the *average* range of kV levels in the x-ray beam, taking advantage of the k-edge effect.

insufficient to dislodge an electron from the K-shell. As kV is increased, when the energy of the x-rays surpasses 39 kV, we see a sudden *spike* in photoelectric interactions. After this, *all* shells in the atom can undergo photoelectric absorption, so we see no more spikes in the graph, but again, the expected steady decline in photoelectic interactions as the kV continues upward.

Now, Figure 34-19 superimposes the k-edge graph for lanthanium and bromine with the *x-ray spectrum* curve for an 80-kVp x-ray beam. Recall that the *average energy* of an x-ray beam is approximately one-third of the peak energy or kVp, so the highest point in this curve is at about 27 kV. We could say that a large bulk of all the x-rays produced in this beam lie in the energy range from 17 to 37 kV. Yet, the k-edge for lanthanium is at 39 kV, so *none of these x-rays can be absorbed photoelectrically by the k-shell.*

Ironically, we find that if we use phosphor elements with *lower* atomic numbers, the k-edge can be better matched to typical diagnostic x-ray energies, improving absorption efficiency. *Barium* has a k-edge of 37 kV matching it a little better than lanthanium. Most stimulable phosphors are compounds of barium and fluorine with chlorine, or bromine. Chlorine has a k-shell binding energy of only 2.8 kV and absorbs few x-rays photoelectrically *at all.* The k-shell binding energy for *bromine* is 13.5 kV, and we see in Figure 34-19 that it lines up nicely with the lower energies in the x-ray beam, contributing to the absorption efficiency of the molecule.

The process of finding ideal phosphors for the photostimulable plates of CR is a trade-off between this *absorption efficiency* and the *conversion efficiency* for different compounds. This is also an issue for *indirect-conversion* DR systems, but not for *direct-conversion* DR, which does not employ light emission. It *is* important in DR, though, for the detection surfaces of the detector elements to be highly absorptive of x-rays, therefore, the trade-off between high atomic numbers and k-edge matching applies to these detectors as well.

DR Detector Panels

Phosphors are used for *indirect conversion* DR systems, and have precisely the same three characteristics for efficiency as the phosphor plates used in CR.

Both *indirect conversion* and *direct conversion* DR systems use an active matrix array (AMA) of detector elements or dexels. As we have described, each dexel has a detection surface which must absorb either x-ray photons directly or must absorb light photons from the phosphor layer above. In both cases, electrical charge must be released through ionizing events, and then stored in a capacitor. The detection surfaces in these AMA's share one characteristic in particular with CR plates, and that is *absorption efficiency*.

In the case of an active matrix array, *absorption efficiency* is the ratio of photons absorbed by the selenium or silicon detector surfaces to the photons incident upon these layers of photoelectric material. These layers must be kept very thin, so for a direct conversion DR system they are very dependent upon high atomic numbers and the *K-edge effect* to do a good job of absorbing x-rays. Even so, only a very small percentage of the incident x-ray beam is absorbed by these ultra-thin detector elements.

For indirect conversion systems, a much higher percentage of light photons from the phosphor layer above can be absorbed by the active matrix array. It is precisely for this reason that the indirect conversion system can save patient dose. Indirect conversion systems have higher overall absorption efficiency than direct conversion systems, first, because the phosphor layer can be thicker than the TFT detection surfaces, and second, because the TFTs absorb lower-energy light photons more readily than high-energy x-ray photons.

Once x-ray or light photons are absorbed in a del, the "conversion" of this energy into electric charge is nearly 100 percent, and the "emission" of this signal, which would translate to the percentage of electrons reaching the capacitors, is also nearly 100 percent, so there is no point in considering these other two types of efficiency as variable factors for active matrix arrays.

Detective Quantum Efficiency (DQE)

Detective quantum efficiency (DQE) is a measurement of the overall efficiency with which a detector can convert input exposure into a useful output image. Mathematically, it is the squared output signal-to-noise ratio divided by the squared input signal-to-noise ratio (SNR^2_{OUT}/SNR^2_{IN}). No imaging system can achieve a perfect DQE of 1.0 or 100 percent. At 70 kV, the DQE for CR plates is less than 30 percent, for direct-conversion DR it is about 67 percent, and for indirect-conversion DR systems it is about 77 percent.

High DQE is important, but does not always translate directly into a superior imaging system; other considerations, such as x-ray beam uniformity and effective energy, the latitude response of the system, sampling methods, display quality and viewing conditions also affect patient dose and final image quality.

DIGITAL ARTIFACTS

In direct-capture (DR) systems, the detector elements constituting individual pixels can suffer from various electronic faults which are not found in the reading process for computed radiography (CR) plates. These flaws introduce noise into the image or cause a loss of pixels. Additional software is configured to compensate for these electronic problems. Therefore, DR systems typically undergo more predisplay processes than those required in CR.

For CR, the most common source of artifacts is the image receptor plate. These artifacts are usually temporary white spots from dust or dirt on the plate, or *ghosting* which refers to residual images that were not fully erased from a previous exposure. The accumulation of permanent scratches on the plate, or nonuniform performance from aging phosphors can require replacement of the plate.

Artifacts which appear consistently on most or all images are likely the result of hardware or software problems in the CR reader (processor) rather than the image plate. Artifacts that run across the entire length of the image are most likely caused by objects on the mirrors or light guides within the CR processor; artifacts that do not are most likely from objects on the PSP plate, inside or on the cassette. Line or column drop-out can result from malfunction of the reader transport and scanning systems. Plates can become jammed in the reader, leading to artifacts. Dust particles sticking to the oscillating flat mirror, the rotating polygonal mirror, or the light guides in the reader can cause pixel drop-out (Fig. 34-20). Over time, the laser itself will need replacing due to loss of power.

Failure in segmentation or histogram analysis can be caused by accumulated background and scatter

Figure 34-20

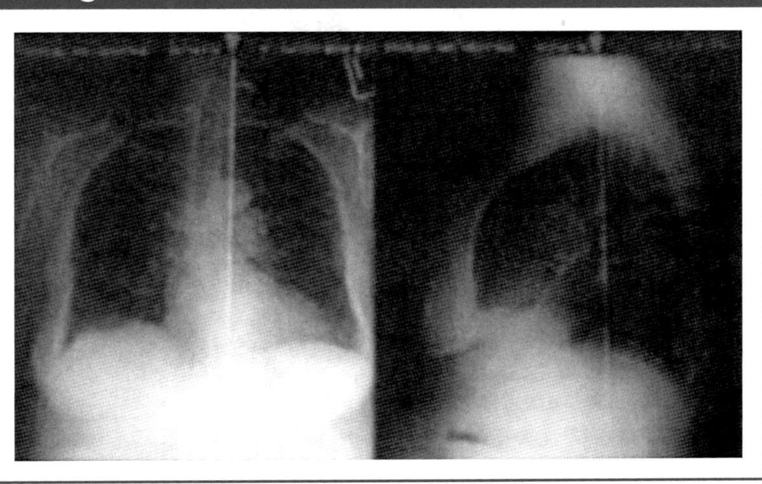

Dust on the optics (lenses and mirrors) of a CR processor can cause pixel drop-out artifacts, such as this one in which an entire line of pixels is deleted.

radiation on CR plates that have not been erased over a long period of time. Proper collimation is essential, and the use of lead strips is recommended to shield one side of the cassette when multiple exposures will be taken on the same plate.

Software artifacts can result from selecting the wrong processing algorithm, poor positioning, the presence of metal prosthetic devices in the body or other unusual anatomical variations, or segmentation failure in high-scatter situations. Edge enhancement algorithms can produce substantial *halo* artifacts; these appear as dark bands at the interfaces of adjacent high-contrast structures such as a metal prosthesis or a "solid-column" bolus of barium.

Artifacts involving missing lines or pixels, as demonstrated in Figure 30-3 in Chapter 30, can indicate memory problems, digitization problems, or communication errors between computer and imaging components.

Aliasing patterns like that shown in Figure 34-21 emerge in the image when the sampling rate (in LP/mm) approximates the line resolution (LP/mm)

Figure 34-21

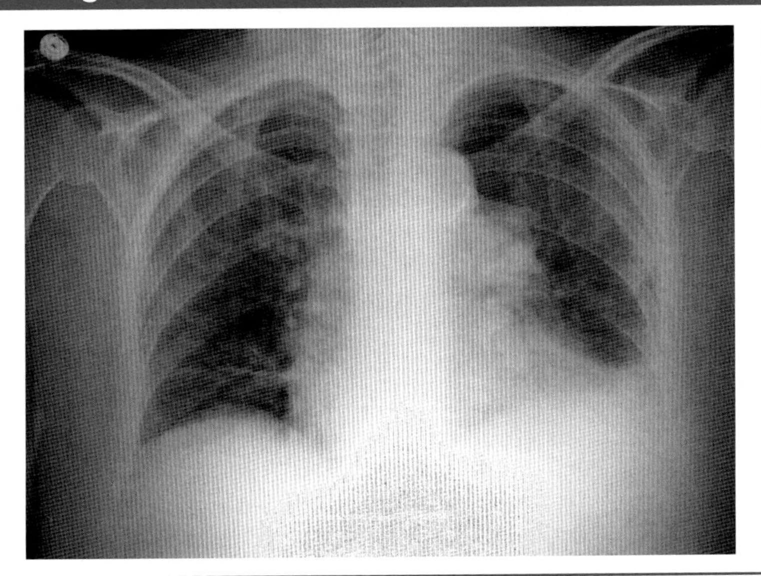

Aliasing, or *Moire* artifacts, are geometrical patterns like the one shown here, are caused when the scanning frequency of a CR reader or a TV monitor approximates the frequency of grid lines, resolution template lines, or the resolution of a photostimulable phosphor plate. (From Quinn B. Carroll, *Practical Radiographic Imaging*, 8th ed. Springfield, IL: Charles C Thomas Publisher, Ltd., 2007. Reprinted by permission.)

of the image plate, such that the two overlap at certain frequencies. (This is analogous to the way that spokes in a wheel spinning at a certain rate create the optical illusion of spinning slowly backward—it is due to *overlapping* of the spoke images at specific frequencies.) It also creates a problem when the line resolution is a multiple of the sampling frequency; if the plate resolution is 8 LP/mm and the sampling frequency is 4 LP/mm, inadequate pixel sampling will occur as every other line is missed. To prevent moire artifacts (aliasing), the *Nyquist Theorem* states that the sampling frequency must be at least 2 times the desired spatial resolution. Aliasing or *Moire* type artifacts can also be caused by the use of grids, and on display screens—these variations are discussed in later chapters.

Quantum mottle noise has been previously discussed, and may certainly be classified as a type of artifact caused by insufficient exposure. *Electronic* mottle can also be caused by fluctuations of electrical current passing through the imaging system.

Hardcopy Printer Artifacts

Hardcopy laser printers have many of the same issues that old-fashioned radiographic chemical processors suffered from. Artifacts which can show up on the hardcopy film include fog, pressure marks and static electricity marks. If the film conveyor system malfunctions, image distortion, abnormal shading or uneven scan line distribution can result. Artifacts can result from placing the single-sided emulsion of the film upside down in the printer.

SUMMARY

1. In comparing DR and CR imaging, the main advantages of DR are *efficiency and consistency*, while CR offers more *adaptability* for both positioning and technique.
2. All DR systems use an active matrix array of dels (detector elements) with their own TFTs. Currently, typical del size is about 100 microns. The size and shape of these detector elements has allowed flat detector plates to be developed.
3. For dexels (detector elements), since the capacitor and TFT cannot be reduced in size, the smaller the dexel, the lower the fill factor and the lower the detection efficiency.
4. Direct conversion DR systems use amorphous selenium to generate electric charge directly from x-ray ionization events. Indirect systems first convert x-ray energy into light using a phosphor, then use amorphous silicon to convert the light energy into electric charge.
5. While direct conversion DR systems produce higher image resolution, indirect systems save patient dose.
6. The *PSP* plate used in CR consists of a layer of photostimulable phosphor crystals protected by a thin plastic coat, with anti-halo and reflective layers behind, all supported by an aluminum base. The cassette which holds it usually has a low-absorption carbon-fiber front and a thin sheet of lead foil added to the back to minimize backscatter radiation.
7. During exposure, some of the ionized electrons within phosphor atoms are trapped in metastable energy bands created by the molecular structure. These "electron traps" hold their charges, forming a latent image, until they are stimulated by a laser beam in the CR reader. Added energy from the laser frees them to fall back into their normal atomic shells, emitting light in the process. This light is detected by a photomultiplier tube which converts it into an electronic signal representing the image.
8. In the CR reader, a round laser beam scans the PSP plate by indexing across it one line at a time. The number of these scanned lines per millimeter is the *Nyquist frequency*, which determines the spatial resolution of the images produced. Some manufacturers have higher Nyquist frequencies for smaller imaging plates. The higher the frequency, the sharper the spatial resolution.
9. In the CR reader, light emitted from the PSP plate is converted into electrical current and amplified by a photomultiplier tube. The electrical signal must then pass through an ADC for digitization.
10. CR plates are extremely sensitive to both background radiation and scatter radiation, so they must be protected and rotated, and must be

erased prior to use any time they have been stored for more than a day.

11. In all digital imaging systems the *maximum* spatial resolution is determined by the dexel pitch or pixel pitch, which is related to the size of the pixel. Maximum spatial resolution is always inversely proportional to a doubling of the dexel pitch or pixel pitch.
12. The phosphor layers used both in CR plates and in indirect-conversion DR systems have a characteristic absorption efficiency, conversion efficiency, and emission efficiency. The active matrix arrays of DR systems have a characteristic absorption efficiency. Proper matching of the *K-edge* of these materials to the average energies of the x-ray beam greatly improves absorption. Detective quantum efficiency (DQE) is a measure of a system's overall detection efficiency.
13. While DR systems suffer from more electronic and hardware artifacts, CR systems suffer from added artifacts inherent to the PSP plate. Both are subject to various *software* artifacts.

REVIEW QUESTIONS

1. In comparing CR to DR, which system offers more:
   a. flexibility in positioning?
   b. consistency of image quality?
   c. departmental efficiency?
   d. portability for mobile procedures?

2. For DR, both the direct conversion and indirect conversion types of detector plates have a layer of microsopic dels (detector elements), each of which is able to actively store and release its own electrical charge. This whole layer of elements is called an __________.

3. At present, typical DR del (detector element) size is about __________ mm.

4. As the size of detector elements gets smaller, what change, if any, is necessitated for radiographic techniques?

5. Through the __________ lines in the AMA, an address driver controls the order in which dexels are read out.

6. Compared to direct conversion DR systems, indirect systems produce lower resolution images because light from the phosphor layer tends to __________.

(Continued)

REVIEW QUESTIONS *(Continued)*

7. The *delayed* emission of light from a stimulated material is properly called __________.

8. In the CR reader, what are the *two* ways in which the photomultiplier light sensor can distinguish between light emitted by the laser and light emitted by the PSP?

9. Does it matter which way the PSP plate is facing when manually loaded into a CR cassette? If so, why; if not, why not?

10. In the entire radiograph production cycle, how many times does the PSP plate emit light, and when?

11. The metastable energy bands formed as "electron traps" by the molecular structure of the PSP are also called __________ -centers.

12. Eight hours after exposure, the PSP can retain up to __________ percent of the original latent image.

13. During the reading of a CR plate, the direction in which the plate moves during the scanning phase is called the:

14. What is the effect of a higher Nyquist frequency upon the sharpness of the image?

15. What is the measured spatial frequency in LP/mm for a DR detector panel with a dexel pitch of 0.08 mm?

16. In a photomultiplier tube, electrodes which can switch their charge back and forth are called:

17. How much more sensitive are CR plates to background and scatter radiation than conventional x-ray films?

(Continued)

REVIEW QUESTIONS *(Continued)*

18. For CR, what is the most common source of artifacts?

19. When the sampling rate of the CR reader approximates the line resolution of the PSP plate, what peculiar artifact can be caused?

20. List three aspects of a phosphor layer which affect its absorption efficiency for x-rays:

21. In terms of absorption efficiency, why do indirect-conversion DR systems save patient dose over direct-conversion DR systems?

22. The K-edge effect describes a __________ in the graph of absorption when kV is raised to the point where it just exceeds the K-shell binding energy of the atoms through which the x-rays are passing.

23. If CR artifacts occur consistently on most or all images, they are *not* likely to be caused by the:

24. The measure of an imaging detector's overall efficiency, expressed as a ratio for the SNR out over the SNR in, is called its:

Radiograph of a large snake with a broken neck.

Chapter 35

DISPLAY SYSTEMS AND ELECTRONIC IMAGES

Objectives:

Upon completion of this chapter, you should be able to:

1. Define light *polarization* and how it is used in liquid-crystal diodes (LCDs) to produce an image.
2. Explain how the pixels of an LCD are constructed and controlled.
3. Describe the light sources for passive-matrix and active-matrix LCDs.
4. Describe the resolution, advantages, and disadvantages of LCDs.
5. Describe the nature of pixels and subpixels in display monitors, their dynamic range, and options for resolution.

Nearly all radiographic images are now displayed electronically in their initial form. Electronic image display systems are used not only for viewing the images, but at the control consoles of most x-ray and other imaging machines, and also with every computer terminal. Therefore, radiographers should have at least some familiarity with electronic image display systems and how they work. The dominant type of display monitor is now the *liquid crystal diode* (*LCD*) screen.

LIQUID CRYSTAL DIODES (LCDS)

Liquid crystal diode (LCD) monitors are lightweight, portable, inexpensive, and available in a wide range of sizes. Their flat shape allows them to take up less space, they generate less heat, and they have a longer life with less maintenance than the older cathode-ray tube (CRT) monitors, if they are cared for properly. They have much lower reflection of ambient lighting off of their surface. On the other hand, a major setback for LCD monitors is the restricted angle from which the display screen can be viewed. This makes it difficult to point out image details to a bystander, who must move directly in front of the screen to appreciate the full-detail image.

To understand how LCDs work, it is first necessary to review the concept of *polarized* light. Polarizing sun glasses work to reduce reflective light glare because the plastic used in the lenses contains long, slender, aligned chains of iodine molecules which act somewhat like a radiographic grid. Only those light waves whose electrical component *vibrates* parallel to these long chains of molecules can penetrate through; light waves vibrating in all other directions are blocked by the molecular grid (Fig. 35-1). By placing two polarizing lenses perpendicular to each other, *all* light will be blocked, because those waves that make it through the first polarizing lens will be perpendicular to the second lens (Fig. 35-2). This is the starting point in constructing an LCD monitor screen. It consists of two thin sheets of glass, each containing a light polarizing layer which is perpendicular to the other sheet, Figure 35-3). The LCD process involves "tricking" these layers into allowing light to pass through when we wish, by using a special material in between them.

Inserted between the polarizing glass sheets of an LCD is a layer of *nematic liquid crystal* material, such as hydrogenated amorphous silicon. Liquid crystal compounds have a crystalline arrangement of molecules, yet they flow like a liquid. The term *nematic* means that the molecules have a thread-like shape and tend to keep their long axes aligned.

Figure 35-1

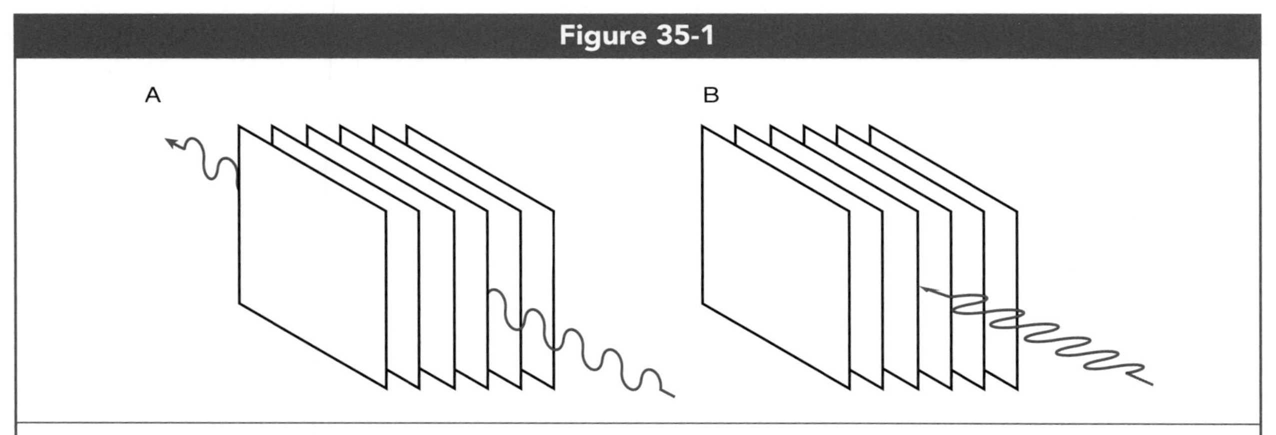

Only light waves vibrating parallel to the molecular chains in a polarizing lens can pass through it, ***A***. Light waves vibrating perpendicular to the molecular chains are blocked, ***B***.

Transparent conductors (made of indium tin oxide) act like flat-shaped "wires" to transmit electricity through the system. Rows of these transparent conductors are built into one glass plate, and columns of vertical conductors are put into the other plate. Each intersection of these two conductors constitutes a pixel. The surfaces of the electrodes that are in contact with the liquid crystal have been treated with a thin polymer layer. This layer has been rubbed in a single direction with an abrasive cloth material to create a finely *scratched* surface.

Normally, the "threads" of liquid crystal tend to line up with the direction of the scratches on the surfaces of the *uncharged* electrodes. But, the two sets of electrodes, front and back, are oriented with their scratched surfaces perpendicular to each other. As shown in Figure 35-4, ***A***, the result is that the liquid crystals line up in a *helical* or spiral pattern that *twists* 90 degrees between the two glass plates. Now, light passing into the liquid crystal layer will follow the orientation of the crystals, so that the light also twists 90 degrees and is able to pass through the second polarized sheet of glass (Fig. 35-4, ***A***).

Figure 35-2

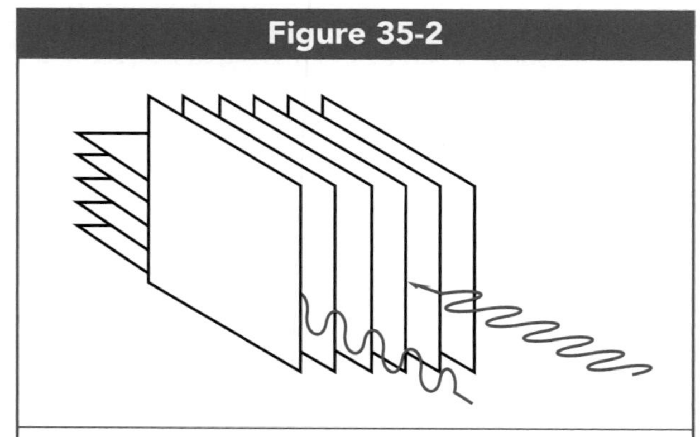

Two polarizing lenses placed perpendicular to each other will block all light from passing through.

The pixel is considered to be in the "on state" and light is transmitted through, as long as *no electrical current is applied* to the pixel electrodes. When an electrical charge *is* applied to the pixel, it is considered to be shut off, and the nematic crystals return to their normal state, all aligned with each other (Fig. 35-4, ***B***). The twisting effect on the light is lost, and light is unable to pass through the two perpendicular polarizing filters. This creates a dark spot in the LCD image. This is the basis for forming the light and dark areas required to produce an image. Figure 35-5 shows the electronics along one side of the plates which control current to the pixels.

The one remaining need is to supply a source of light. For a *passive-matrix LCD*, such as a typical LCD wristwatch or calculator, a reflective surface is placed behind the entire system of transparent sheets. This mirror simply reflects room lighting or sunlight, which then passes through the polarizing filters and liquid crystal layer on its way back to the viewer. A source of *side-lighting* can also be provided, as is done with a button on many watches. Passive-matrix LCDs are useful for smaller screens as employed in calculators and watches, but have very slow response times and poor contrast. This is because each row or column in the display is treated as

Figure 35-3

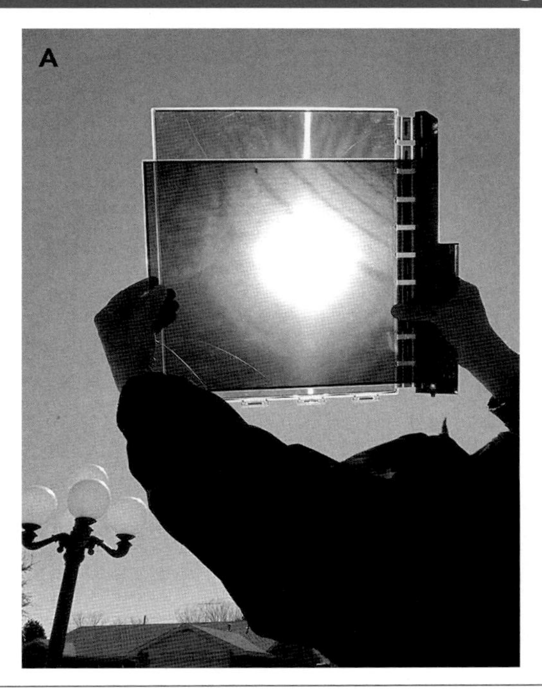

The two sheets of glass from an LCD monitor screen contain light polarizing layers that are perpendicular to each other. When the two sheets are held crosswise to each other as shown in ***A***, some light from the sun is able to pass through, but when they are held in their normal position as in ***B***, so that the polarizing layers are perpendicular to each other, all light is completely blocked. (Courtesy of Digital Imaging Consultants, Midland, TX. Reprinted by permission.)

a single circuit with an amplifier, addressing pixels one at a time by row and column addresses.

The displays needed for computer monitors and for imaging applications including radiography must be much brighter, with much sharper resolution, and must have high-speed response times. For this purpose, the *active matrix liquid crystal display* (*AMLCD*) was developed. In an active matrix system, *each pixel has its own dedicated thin-film transistor* (*TFT*). This allows each column line to access a pixel simultaneously to form a row, speeding up response and *refresh* times. (The *response* time is the time necessary for a pixel to change its brightness; the *refresh* time is the time it takes to reconstruct the next frame.) No matter how low the refresh rate, LCD monitors exhibit no flicker, because the pixels do not flash on and off between frames as those of a CRT do. The capacitor of each TFT also prevents charge from leaking out of the liquid-crystal cells, maintaining brightness and continuity.

A bright source of back-lighting is always supplied to the LCD layer in these systems. This lighting can come from either fluorescent bulbs or from light-emitting diodes (LEDs). The most common arrangement is to build in a pair of thin fluorescent light bulbs at the side of the monitor (Figure 35-6) and then use a sophisticated series of plastic light-diffusing filters to scatter the light evenly across the monitor panel. These filters (Figure 35-7) are so effective that the light intensity at the center of the display screen is only a few percent less than at the edges near the bulbs.

Digital processing can be executed on LCD images before display, including image scaling, noise reduction and edge enhancement, but too much processing can cause considerable *input lag* between the time the monitor receives the image input and the time it is actually displayed. LCD monitors can have refresh rates as high as 240 Hz, allowing interpolated frames to be inserted to smooth out motion.

Figure 35-4

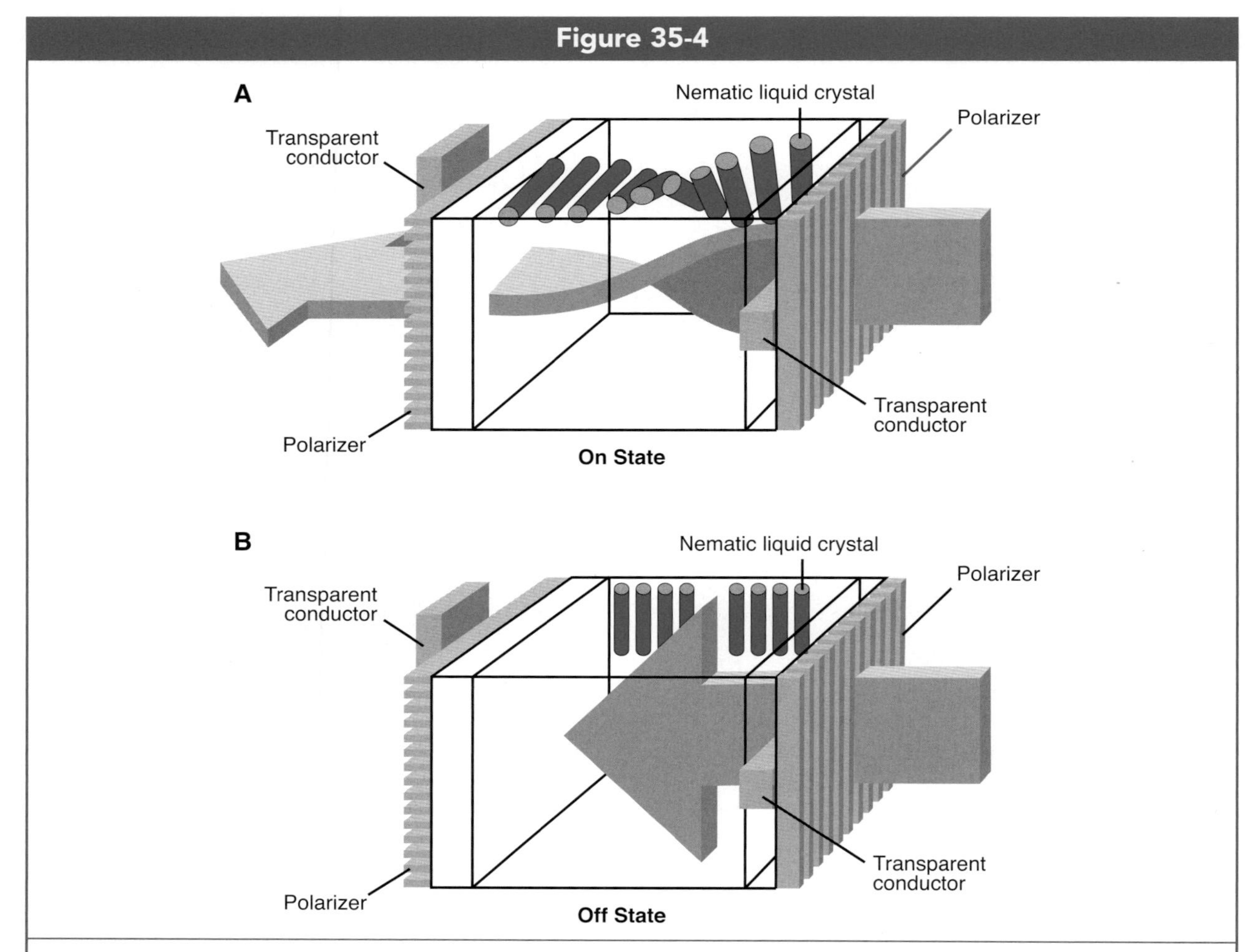

In an LCD, liquid crystals line up in a twisted, spiral pattern to match the scratched surfaces of uncharged electrodes at either end. In this "on" state, ***A***, light will be twisted along with the crystals such that it passes through both polarizing layers to the viewer. When an electrical charge is applied to the electrodes, ***B***, the crystals all align and the light is straightened out, such that it is unable to pass through the second polarizing filter.

By controlling the voltage applied across the liquid crystal layer in each pixel, light can be allowed to pass through in varying amounts, thus constituting different levels of gray. Increasing voltages *untwist* the column of liquid crystals more, until they are nearly parallel, and letting less and less light through. On very large monitor screens, the display must be *multiplexed*, grouping the electrodes in different segments of the screen together with a separate voltage source for each group, to ensure sufficient voltage to all pixels.

The resolution of an active display matrix is often stated in terms of the total number of pixels on the entire screen, or in each screen dimension, width and length (480 × 640, for example). LCD monitors may have resolutions as high as 3–5 megapixels. *Dot pitch* or *pixel pitch* is the distance between the centers of two adjacent pixels. The smaller the dot pitch, the less granularity (graininess) is present, and the sharper the image resolution.

An important item in caring for LCD monitors is to remember that the electrodes constituting the pixels are *built right into the screen*, and the flexible glass covering is extremely thin. *Poking or otherwise applying pressure to an LCD screen can damage and eventually destroy pixels*. Most "dead" pixels occur from this type of abuse, so manufacturers do not warranty against such damage. Truly "dead" pixels actually remain white, pixels that are permanently

Figure 35-5

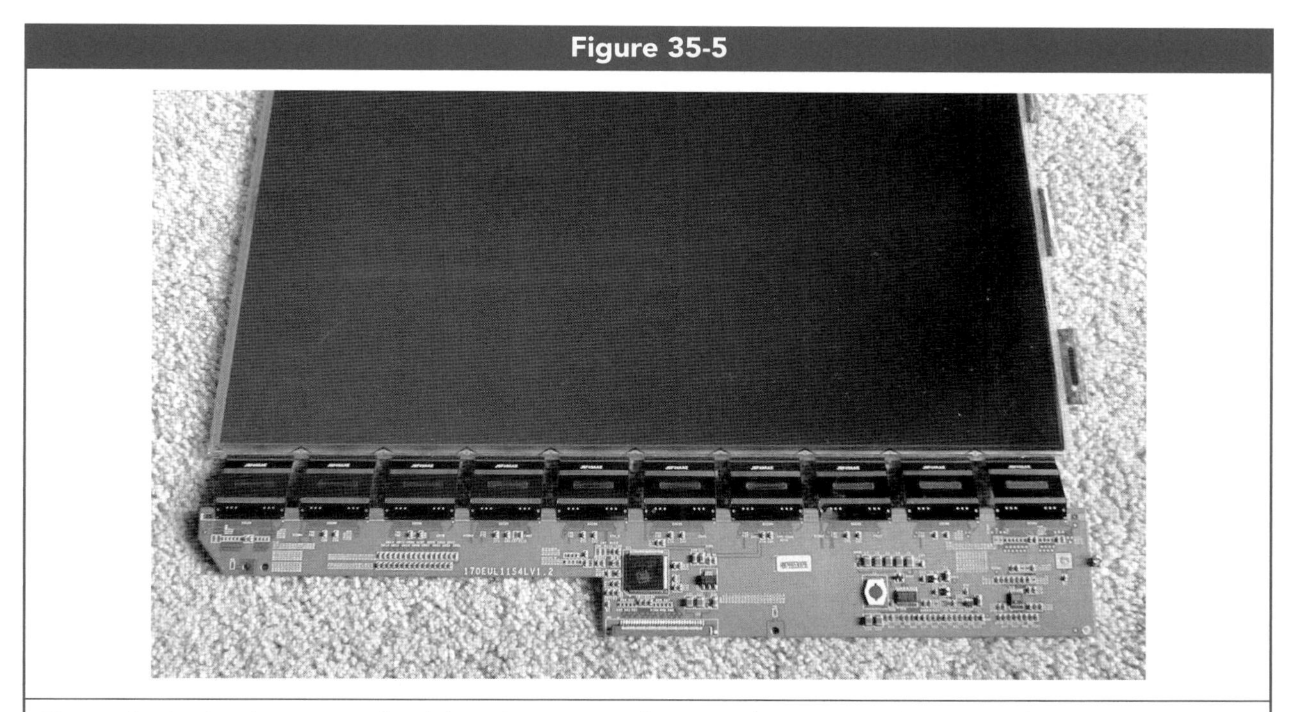

Electronic circuits along one side of the LCD monitor plates supply a controlled amount of electrical voltage to each pixel in each row. The greater the voltage to a particular pixel, the more the nematic crystals are *untwisted* and the darker the pixel appears. (Courtesy of Digital Imaging Consultants, Midland, TX. Reprinted by permission.)

Figure 35-6

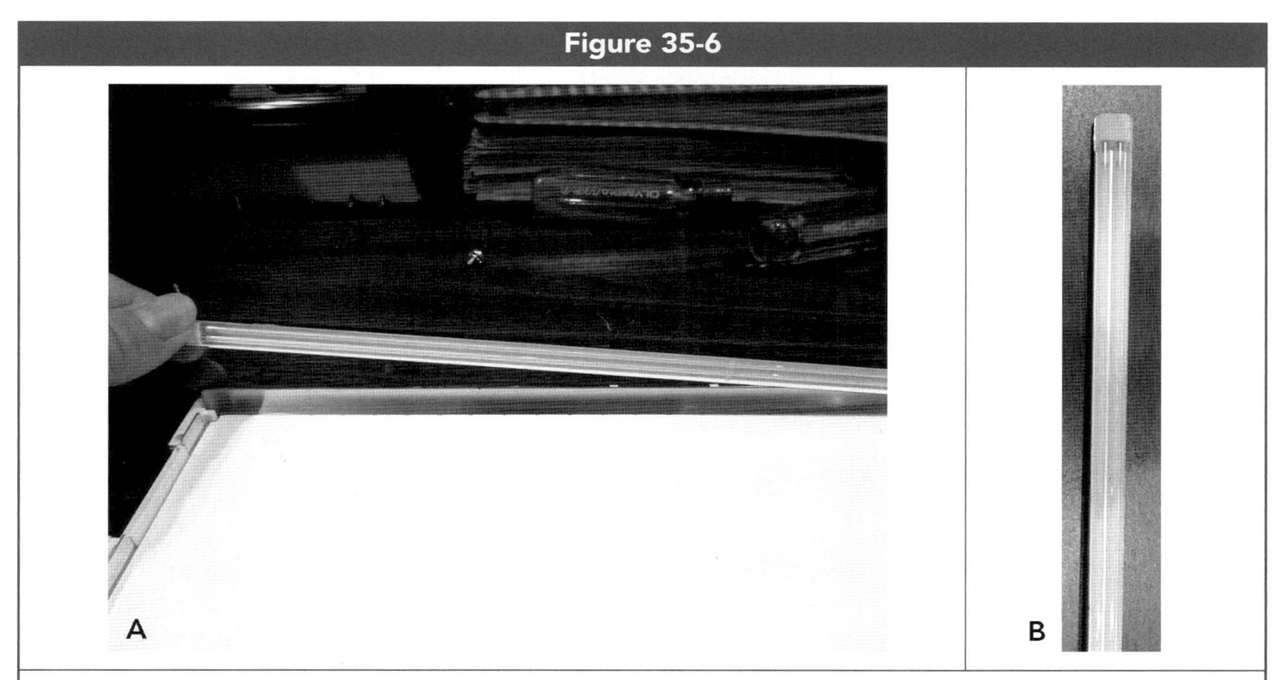

A pair of thin fluorescent light bulbs supply this LCD monitor with all the light it needs, from a *single side* of the screen, ***A.*** Both bulbs can be better seen in a close-up photo, ***B.*** (Courtesy of Digital Imaging Consultants, Midland, TX. Reprinted by permission.)

Figure 35-7

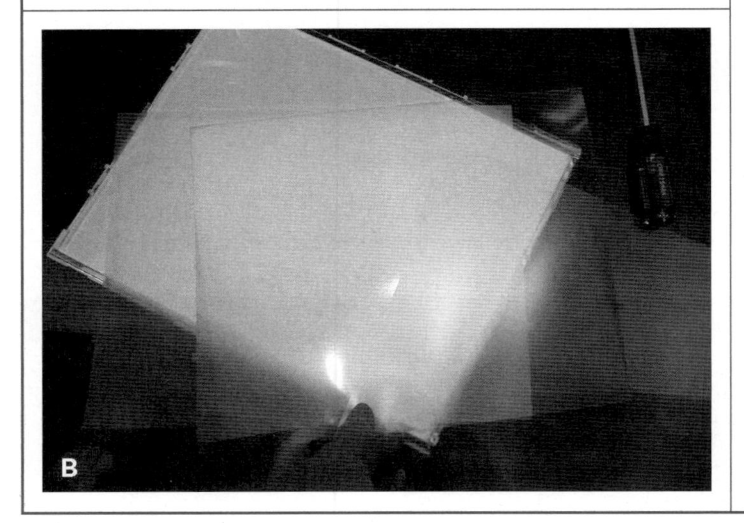

Several plastic light-diffusing filters scatter the light evenly across the entire surface of the LCD monitor screen. The thickest of these, ***A***, is about 1 cm and accounts for much of the thickness of the monitor. Each of the additional filters shown in ***B*** has a special texture to further diffuse the light. (Courtesy of Digital Imaging Consultants, Midland, TX. Reprinted by permission.)

dark are considered to be "stuck," since they are constantly receiving electrical charge.

OTHER FLAT MONITOR SYSTEMS

LCDs are an example of *nonemissive* display monitors in which optical effects are used to convert light from another source into graphics patterns. *Emissive displays* produce their own light, and include plasma panels, thin-film electroluminescent displays, and light-emitting diodes (LEDs) which consist of a matrix of very small diode "light bulbs." These are popular options for broadcast television, but have not found much application for medical imaging to date.

ADVANTAGES AND DISADVANTAGES OF LCDS

Compared to the cathode ray tube (CRT) TV monitor, which has become obsolete, LCD monitors have both very distinct advantages, and very distinct disadvantages in terms of the quality of the displayed image. Advantages include a smaller pixel size for better resolution, and a more even luminance across the screen. The degree of light reflection off the screen surface, both diffuse and specular, is much lower. The LCD has perfect geometry, whereas CRTs suffer from three types of geometrical distortion related to the curved shape of the phosphor surface and anomalies with controlling a charged electron beam. The sharpness of an LCD is uniform across the screen, and there is no flicker. Table 35-1 provides a summary.

There are at least three very substantial disadvantages to LCD display. The first is the limitation of off-angle viewing. When a person viewing the image is positioned at a moderate angle from the screen, the brightness is less, and it plummets as this angle is increased. *Luminous intensity* is the term used to describe the concentration of light at these different angles, and is mathematically defined as the *lumens per steradian*. The *steradian* is the angle, off of perpendicular, from which the screen is being viewed (see Figure 39-9 in Chapter 39, p. 683). Sharing such an electronic image requires the visitor to stand close to or behind the operator.

In the case of an LCD, the luminance affects not only the brightness of the image, but the contrast and the apparent spatial resolution as well. (Even from a perpendicular vantage point, the LCD has less than one-fifth the light intensity of a conventional viewbox. This requires the radiologist to use more windowing in order to optimize each image.)

The second disadvantage is that the LCD is not capable of transmitting a "true black" density. This can be seen by observing the blank image in a completely darkened room. It is due to the backlighting that is required for the LCD to operate, a small percentage of which always escapes the screen. The CRT was capable of presenting a much darker "background" density.

The third disadvantage of an LCD is that its brightness is both *time-dependent* and *temperature-dependent*; it takes 20 to 30 minutes to "warm up" and reach its nominal brightness output. Diagnoses should not be made until the LCD has been on for at least 20 minutes when the brightness output reaches about 95 percent. At 30 minutes it reaches 99 percent. LCDs operate well as long as the ambient temperature is a comfortable "room temperature" for humans. When the temperature varies more than about 10 degrees F (5° C), either below 60° F (16° C) or above 80° F (27° C), the brightness plummets.

Newer LCDs have self-calibrating software which keeps their brightness and contrast very consistent over time. Manufacturers claim that this software also compensates for warm-up time, but it is physically

Table 35-1

Advantages and Disadvantages of LCDs and CRTs

LCD	CRT
Advantages	Disadvantages
1. Perfect geometry	1. Geometric corrections needed
2. Uniform sharpness	2. Uneven sharpness
3. Low surface reflectance	3. High surface reflectance
4. No image flicker	4. Some image flicker
5. Uniform brightness	5. Uneven brightness
6. Slow brightness deterioration	6. Rapid brightness deterioration
7. No veiling glare	7. Veiling glare
Disadvantages	Advantages
1. Pixelation from black lines between pixels	1. Continuous image
2. 600:1 contrast ratio	2. Up to 3000:1 contrast ratio
3. No true black	3. Perfect black possible
4. Poor response speed	4. Instant response speed
5. Some image retention (ghosting)	5. No image retention
6. Limited off-angle viewing	6. Good off-angle viewing
7. Temperature dependence (brightness)	7. No sensitivity to ambient temperature
8. Long warm-up time	8. Short warm-up time

impossible for this to be a very substantial correction. Therefore, allowing at least 10 to 15 minutes warm-up time is still important for accurate diagnosis.

It is important to find the ideal brightness and contrast settings for an LCD monitor when it is installed, and then *never change them* to adjust images; rather, digital *windowing* should be used to adjust individual images. For an LCD, aging is primarily from deterioration of the backlighting source.

NATURE OF PIXELS IN DISPLAY SYSTEMS

To this point, our discussion of image pixels has related entirely to the *acquisition* of digital images. We think of CR pixels in terms of small round areas on the photostimulable plate which are sampled by the laser beam. We think of DR dels as hardware elements that are (roughly) square in shape. Technically, a "pixel" is any sampled point within a digital image. To the computer expert, a pixel has no particular shape or dimensions, but is a point *location* or address which has been sampled and assigned a numerical value.

Figure 35-8

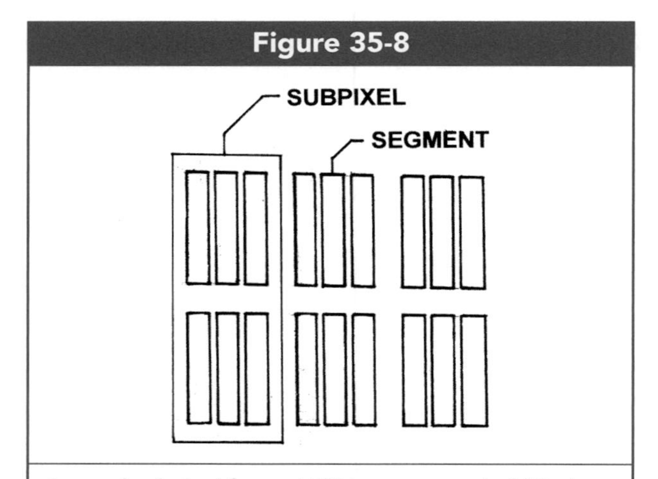

A standard pixel for an LCD is composed of 18 phosphor segments arranged in groups of 6 called *subpixels.* There are 3 individually addressable subpixels to each pixel. In a color monitor, all three subpixels are required to generate a "white" spot, but in a *monochrome* monitor such as can be used for radiographs, each subpixel can be used as a pixel, tripling the resolution.

However, we find that in order to minimize confusion in discussing medical images, we are compelled to refine the definition of a "pixel" in terms of its *display capabilities* in representing the full dynamic range of the imaging system (gray scale or color scale). In reconstructing a digital image for display, we define the pixel as *the smallest screen element which can represent all gray levels or colors within the dynamic range.*

For example, in a typical color CRT such as were used in older model home television sets, the phosphor screen is made up of *triads* of phosphor dots, red, yellow-green and blue. Light emitting diode (LED) systems use the same format. When all three dots are glowing, the effect at a distance to the human eye is to perceive a spot of white light, since white light is a mixture of all colors. Therefore, it requires the whole set of three phosphor dots to produce all possible colors as perceived by the viewer.

By our definition, then, each *triad* of phosphor dots on a TV screen would be considered as a single pixel, with the ability to depict any color within the dynamic range of the system. We are then forced to define each individual phosphor dot as a *subpixel*, a sub-element of a pixel. Some LCDs are constructed with a similar dot arrangement.

In most LCDs, a single pixel actually consists of 18 individual bar-like segments as illustrated in Figure 35-8. These are arranged in groups of three, called *domains*, with each *pair* of domains making up a *subpixel.* Thus, there are three subpixels in a pixel, each containing six segments. Note that the subpixels are in the shape of long rectangles. Now, each of the three subpixels can be treated as a separately addressed element by the computer. Thus, each subpixel can be programmed to glow red, yellow-green, or blue. When all three colors are emitted together, the effect is that the entire pixel appears as a white square area on the display screen.

Here is where we gain a tremendous advantage in standard radiographic imaging, because we can use *monochrome* LCDs instead of color LCDs. All we need for radiographs is a gray scale from black to white. The darkness of each subpixel in an LCD can be controlled by applying varying degrees of voltage through a resistor. Therefore, only a single subpixel is required to cover the dynamic range or gray scale. In effect, we acquire three functional pixels in the same space that a single pixel in a color monitor requires. Therefore, a monochrome display monitor, with the same level of technology and cost as a color monitor, has *three times* more pixels at one-third the size. A huge boost in resolution results from these smaller pixel units, and extremely sharp radiographs can be displayed.

Quality control of display monitors, including dead and stuck LCD pixels, is discussed in Chapter 39.

CONCLUSION: THE WEAKEST LINK

In the imaging chain, the weakest link sets the final limits for each image quality displayed. The display monitor is typically much more limited in both its resolution capabilities and in its dynamic range (bit depth) than the digital image processing system of a computer. This is why *class 1* monitors for the radiologist's diagnosis station are so expensive and why radiographers must be careful about judging images displayed on a *class 2* workstation monitor. A poor-quality monitor can effectively destroy the sharpness already achieved during image acquisition and processing.

(As alluded to in Chapter 14, one exception may apply when the image is displayed on a good-quality monitor, but was acquired with a large focal spot—perhaps made even larger by focal spot blooming. In this case, the focal spot can become the primary limiting factor for sharpness.) The dynamic range of the x-ray beam itself far exceeds that of the display monitor. When all image qualities are considered together, it is fair to state that, in general, the display monitor is the weakest link in the imaging chain.

SUMMARY

1. The display monitor is generally considered to be the weakest link in the imaging chain.
2. Liquid crystal diode (LCD) monitors have replaced CRTs, mostly due to their compactness and efficiency. Active matrix LCDs allow high response and refresh times which provide constant brightness and improved continuity of the image.
3. LCDs use electrical charge to untwist spiral columns of optical crystals between two polarizing filters. This blocks backlighting from passing through, thus creating a pattern of darker areas in the image according to the voltage applied.
4. Important advantages of LCDs over CRTs include uniform and long-lasting brightness, contrast and sharpness, low reflectance and no glare. Disadvantages include lower contrast, poor response speed, and limited off-angle viewing.
5. For display systems, we define a pixel as the smallest screen element capable of representing all gray levels (or colors) within the dynamic range. On monochrome LCDs, each of three rectangular subpixels can be used as a full pixel at one-third the size of a color monitor pixel, thus boosting resolution capability.

REVIEW QUESTIONS

1. The weakest link in the fluoroscopic imaging chain is the:

2. In terms of viewing quality, what is the greatest *disadvantage* of an LCD monitor?

3. What does the term *nematic* mean in reference to liquid crystals of amorphous silicon?

4. As long as *no* electrical current is applied to a pixel within an LCD, the pixel is considered to be in an ___________ state, and ___________ (does or does not) allow light to pass through.

5. In an LCD, each intersection of cris-crossing transparent flat "wires" forms a ___________.

6. The distance between the center-points of two adjacent pixels is called the ___________.

7. Even with a self-calibrating LCD, diagnosis should not be undertaken until it has warmed up for at least how long?

8. To adjust the brightness or contrast of an individual image, digital windowing controls should always be used and the ___________ controls should never be used.

(Continued)

REVIEW QUESTIONS *(Continued)*

9. In an LCD matrix, how many segments make up a *domain*?

10. In a monochrome LCD used for radiographic images, by using each subpixel as a "whole" pixel, the total number of pixels in the image matrix is multiplied by how much?

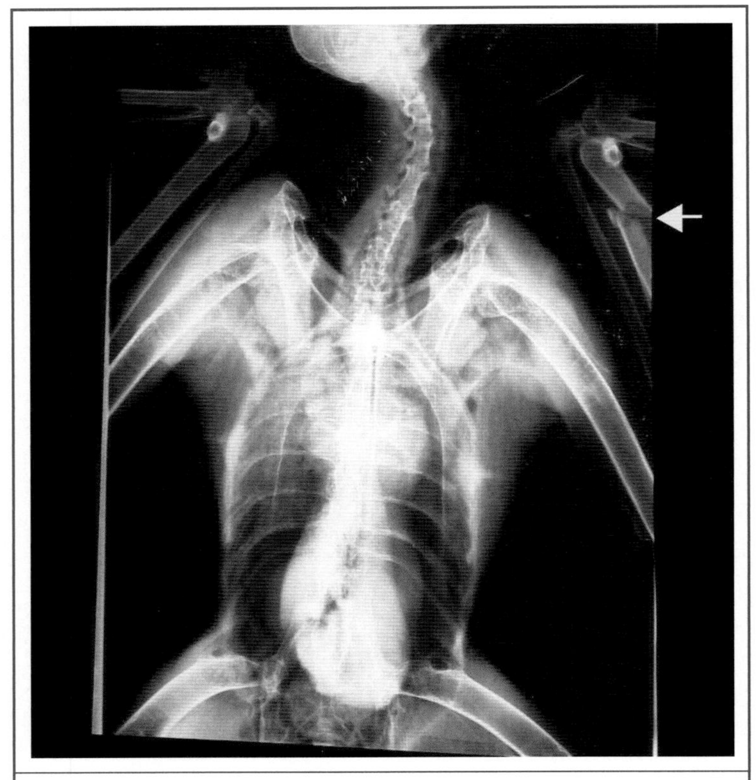

Radiograph of an eagle with a broken wing bone (arrow).

Chapter 36

PICTURE ARCHIVING AND COMMUNICATION SYSTEMS (PACS)

Objectives:

Upon completion of this chapter, you should be able to:

1. List the three broad components of every imaging system, and explain the positioning of the ADC and the DAC between them.
2. Describe the *PAC system*, its purpose, main functions, and typical features.
3. Define *service class users (SCUs)* and *service class providers (SCPs)*.
4. Describe the capabilities of *jukebox* storage and the ways the PAC system allows a user to access and manipulate the image.
5. Define the *control computer* and why it must smoothly interface with the HIS, the RIS, and computerized medical equipment at other hospitals and clinics.
6. Define the DICOM and HL-7 standards, and their purpose.
7. Define the DICOM header, and give several examples of *metadata*.
8. Define *lossless* and *lossy compression*, and how these apply to medical image storage and transmission.
9. Describe the purpose of DICOM viewers and why they should be included on CD or DVDs along with medical images, and why special websites with password-protected *URLs* are recommended for internet access to medical images.
10. Define the purpose and architecture of *RAID* and *SAN systems*.
11. Explain how a patient's *accession number* is used by the RIS for radiographic exams.

Every imaging system consists of at least three components: A machine for *image acquisition*, equipment for *image processing*, and devices for *image display and storage*. Figure 36-1 shows the basic data flow through these three components of the system. Note the conversion of images from analog to digital format between acquisition and processing, and the conversion back into analog format just prior to display.

Computed radiography (CR) is an interesting example of how newly-developed image receptor technology can be combined with conventional x-ray machines to create a new form of image acquisition that is compatible with computer processing. Direct-capture digital radiography (DR) requires total replacement of the image acquisition system. For all digital systems, image processing is performed by computers. Film-based systems required chemical processing. And, whereas conventional film radiographs could only be viewed on illuminators (viewboxes), digital images can be displayed as *soft copies* in the form of electronic images on monitor screens, while still offering the option of *hard-copy* printouts that can be viewed on illuminators.

Through digitization, images produced by all of the different modalities within a medical imaging department can be stored on magnetic tapes and

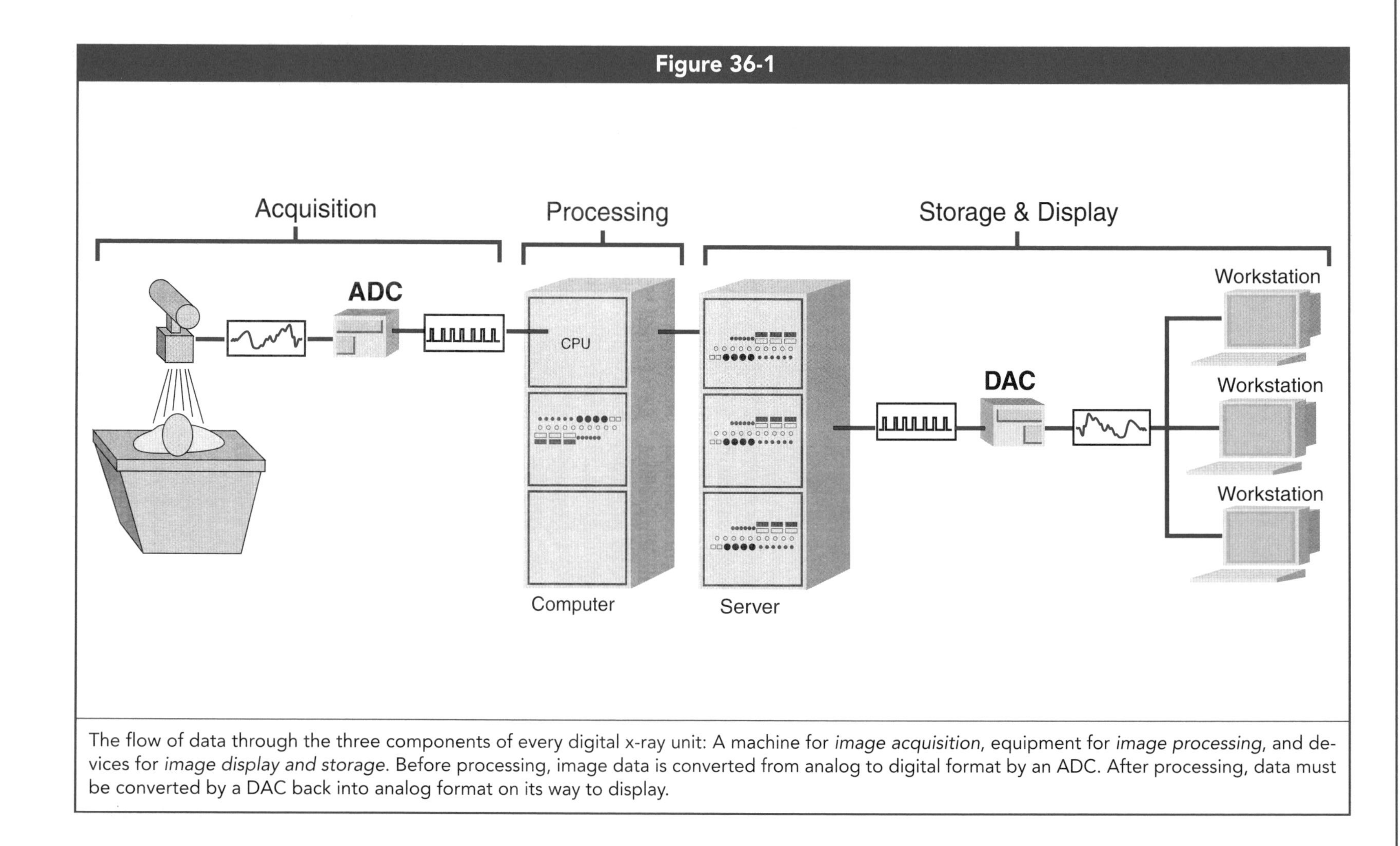

The flow of data through the three components of every digital x-ray unit: A machine for *image acquisition*, equipment for *image processing*, and devices for *image display and storage*. Before processing, image data is converted from analog to digital format by an ADC. After processing, data must be converted by a DAC back into analog format on its way to display.

discs or on optical discs, enabling a system to be developed that can retrieve them, display them for viewing on television-type monitors, transmit them to remote locations and provide archival storage. Such a system is called a *picture archiving and communications system* or *PACS*. Just as you might be the "user" of an internet service provider, image acquisition units with their workstations are considered as *service class users (SCUs)*, while the centralized storage and distribution services of the PAC system, including PACs workstations for accessing images, are considered as *service class providers (SCPs)*.

HARDWARE AND SOFTWARE

The typical components of a PACS are shown in Figure 36-2. The heart of the system is the *control computer*, or PACS *server* (Figure 36-3), which directs all digital traffic. All of the various image acquisition systems, including CR, DR, DF, CT, MRI, ultrasound and nuclear medicine, send their images to the control computer. There, they can be stored *en masse* using magnetic or optical *jukeboxes*, stacks of one hundred or more magnetic or optical discs. Modern PAC systems can store over one million medical images.

From storage, the images can be quickly accessed for printing or to be sent through a *local area network (LAN)* to various workstations within the hospital, to doctors' offices or to a radiologist's home. Newer technologies employ fiber optic LAN's which can transmit data at much higher speed than electronic lines. If the data is to be transferred to remote locations, then a *wide area network (WAN)* may be used.

It is essential that the control computer interface with both the hospital information system (HIS) and the radiology information system (RIS). For the HIS, the RIS, all of the image acquisition systems,

Figure 36-2

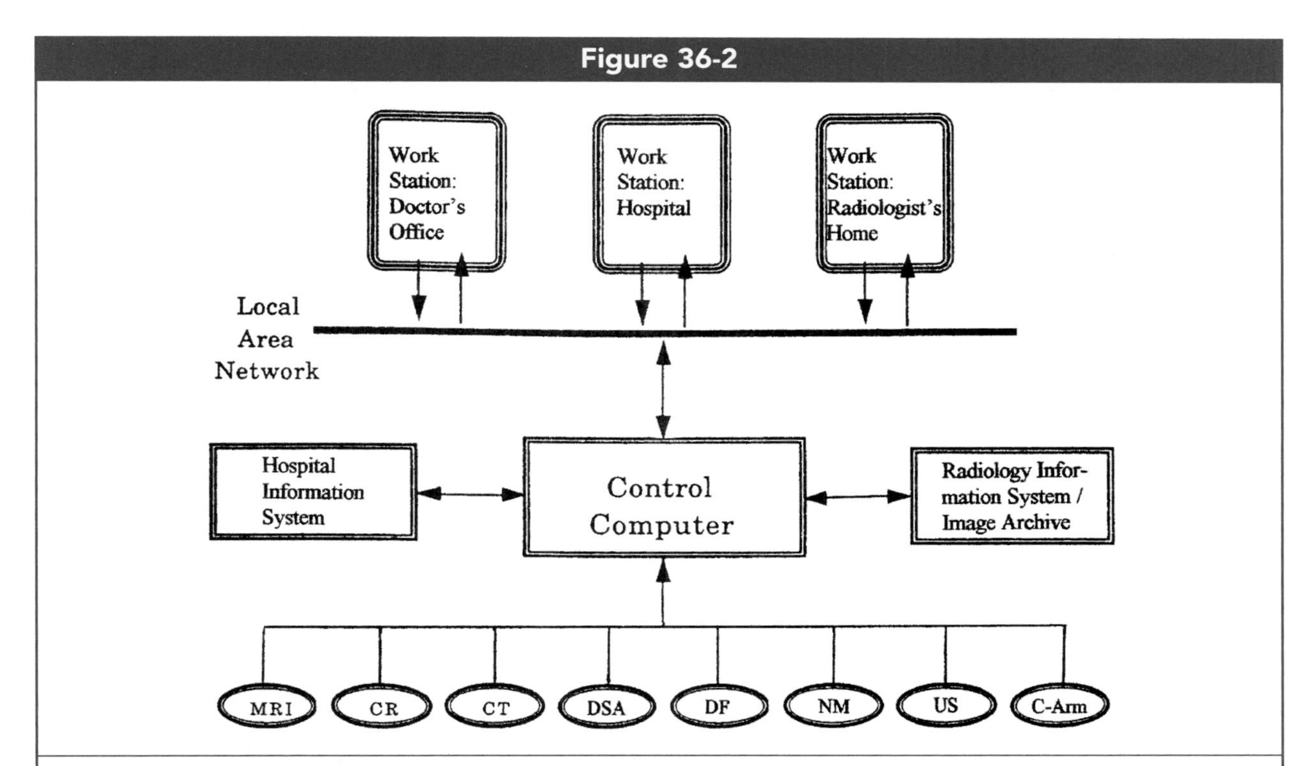

Diagram of a PACS. All digitized equipment within the department feeds acquired images and patient information into the control computer, which then allows universal access through the HIS, the RIS, workstations and display stations throughout the hospital, affiliated centers and offices, and even radiologist's homes. All input and output points of the system are called *nodes* of the *network* (a LAN). (From Quinn B. Carroll, *Practical Radiographic Imaging*, 8th ed. Springfield, IL: Charles C Thomas, Publisher, Ltd., 2007. Reprinted by permission.)

Figure 36-3

A PACS server.

control computer, and workstations to successfully send image data and patient information to one another, they must share a common high-level computer language. Further, it is essential that different brands of computerized medical equipment be compatible so that images and data can be sent between different hospitals and clinics. Therefore, in the 1980s a joint committee was formed between the American College of Radiology and the National Electronic Manufacturers Association to develop the standard called DICOM (*Digital Imaging and Communication in Medicine*). The DICOM standard does not specify the architecture or even the terminology used between different types of imaging equipment, but it does standardize the transmission *behavior* of all the devices within a PAC system.

FUNCTIONS

A main function of the PAC system is to act as a *database*. Lists of patient names or particular studies can be generated, and can be sorted according to their status (such as exam pending, under way, or completed). Various queries can be made, and users can initiate searches for specific images or exams. Examples of DICOM commands include *DICOM query/retrieve* which searches for a specific image, *DICOM get worklist* which imports patient information and study requisitions from the hospital or radiology information systems (HIS or RIS), *DICOM send* which conveys data to the general network, and *DICOM print*.

The DICOM *header* is a series of lines of information attached to each and every image, including identifying information, study information such as the number of images taken, date, time and procedure, characteristics of the image itself such as the image format, cassette size, position and body part, the various parameters used to digitally process the image, and display parameters. The header also provides information about data sources and where to locate the sources associated with the image.

Figure 36-4, ***A*** through ***E***, highlights several key types of data of interest to the radiographer that are part of the metadata stored in the DICOM header file. Figure ***A*** shows the first 25 lines of an actual header to show the typical format. This section includes the all-important accession number, and identifying information for the study, patient, referring and reading physicians, institution and department, manufacturer and even specific unit used to make the exposures. Figure ***B*** illustrates where very specific information on the patient and exam are listed, even including the patient position. Figure ***C*** lists the specifics of the exposure including the technique used, the digital processing code used and the derived x-ray exposure to the patient, which are bolded here for emphasis. Figure ***D*** shows how detailed the information can be on the detector and field orientation. Finally, Figure ***E*** demonstrates the digital processing parameters used including matrix size, window settings, and the type of rescaling curve and LUTs that have been applied.

All of this information is stored as *metadata*, the data "behind the image," on screens that can be accessed by a special menu. For many systems, key identification, study and image data from the header is displayed with each image as a *true header*, that is, along the top of the display screen in a bar above the image.

The PACS system allows manipulation of the image, including windowing of brightness and contrast, magnification or minification, and various types of annotation, at any workstation in the system. A web-based PACS allows access to images over the internet. Encryption software helps protect patient confidentiality for these types of services.

Compared to photographs on a standard digital camera, the digital storage capacity of PAC systems is astronomical. Each DR or CR image contains 4–5 megabytes. At only 200 images per day, this generates up to a gigabyte of memory consumption per day, 30 GB per month, for diagnostic radiology alone. Now consider that MRI scans consume up to 15 MB each, and CT scans up to 25 MB each. These are 3 to 5 *times* the size of CR and DR images. The PACS administrator must consider using image compression for long-term storage. Image compression is familiar to consumers who often receive or transmit images over the internet. "Lossy" compression ratios above 10:1 result in an irreversible loss of detail (sharpness) in the image and are not used in medical applications. *Lossless* compression ratios are those at less than 8:1, and have been defined as "visually acceptable" by the medical imaging community, assuming that the original image is a typical high-resolution radiograph.

Image storage is distributed among several nodes of the PAC system. The *acquisition workstation* at a particular CR, DR, CT MRI, NM or US unit has limited storage capability. The system is programmed to erase images on a regular time-frame, such as 24 hours after acquisition, in order to free up storage space for new studies. The *quality assurance workstation* acts as a hub for several acquisition units within a department. It must have much higher storage capacity and is typically programmed to erase images automatically after 5 days. Similar capacity and timing is required for each *diagnostic (or level I) workstation* used by the radiologist, because recently acquired images are often combined with previous studies which must be immediately available for the diagnostic analysis.

Within the PAC server itself, storage locations are divided into *on-line* or short-term storage and *archive* or long-term storage. Optical discs or tape media are used in for short-term storage, and *jukeboxes* stack many dozens of these devices in a separate location for long-term storage. As long as a particular file has been recently accessed, it is considered "active" and remains in on-line storage.

The PAC system can be programmed to automatically anticipate the need to compare previously acquired images and make those images available at particular workstations prior to a scheduled exam. This feature is called *prefetching*. The system searches the HIS and RIS for related studies and records for patients scheduled for the following day. The protocol for a pre-surgery patient might be to retrieve the previous two chest exams, that for a spine series to retrieve only the most recent exam. Retrievals can be executed during the night while demand on the system is low. Because it can take 5 minutes or more to retrieve a study from long-term jukebox storage, this feature is important for the radiologist's workflow. It also makes previous images immediately accessible to the radiographer as he/she prepares for a particular study.

IMAGE ACCESS

There are three different ways in which other clinics or departments within the hospital can access the PAC system: Perhaps the most common method is to install PACS software on any PC (personal computer) at a clinical site. This places some limitations on the viewer functions but is a simple and user-friendly approach. Or, actual PACS workstations can be located throughout a facility which provide more powerful options and quality for viewing. For either of these approaches, a password is needed to access the PACS, and the user must learn the basic commands to navigate the system. A third option is to integrate PACS images into the patient's electronic health record in the HIS. In effect, the images become part of the patient's *chart*. No special training is need for the user in this scenario, but manipulation of the image is limited to the actual display adjustments at the computer being used.

PACS images may include non-radiographic images such as photographs from a colonoscopy, and can store these in non-DICOM formats such as JPEGs. Viewing tools allow interactive consultation between multiple physicians who can be viewing the

Figure 36-4

	Tag ID	Description	Value
A	(0008,0021)	Series Date	20130709
	(0008,0022)	Acquisition Date	20130709
	(0008,0030)	Study Time	171352.000000
	(0008,0031)	Series Time	171411.000000
	(0008,0032)	Acquisition Time	171519.000000
	(0008,0050)	**Accession Number**	**884279001**
	(0008,0060)	Modality	DX
	(0008,0068)	Presentation Intent Type	FOR PRESENTATION
	(0008,0070)	Manufacturer	Canon, Inc.
	(0008,0080)	Institution Name	
	(0008,0090)	Referring Physician's Name	
	(0008,1010)	Station Name	CXDI-5821597MA
	(0008,1030)	Study Description	
	(0008,103E)	Series Description	CXR PA
	(0008,1040)	Institutional Department Name	
	(0008,1060)	Name of Physicians Reading Study	
	(0008,1070)	Operator's Name	
	(0008,1090)	Manufacturer's Model Name	CXDI
	(FFFE,E000)	Item	
	(0008,0000)	Unknown	98
	(0008,1150)	Referenced SOP Class UID	1.2.840.10008.5.1.4.1.1.1.1
	(0008,1155)	Referenced SOP Instance UID	1.2.392.200046.100.2.1.97032212825 .13081613020.1.1.1.1
	(FFFE,E00D)	Item Delimitation Item	
	(FFFE,E0DD)	Sequence Delimitation Item	
	(0008,2218)	Anatomic Region Sequence	
B	**Tag ID**	**Description**	**Value**
	(0008,2218)	Anatomic Region Sequence	
	(0010,0010)	**Patient's Name**	**John Doe**
	(0010,0020)	**Patient ID**	**139967010**
	(0010,0030)	Patient's Birth Date	19580105
	(0010,0040)	Patient's Sex	M
	(0010,0010)	Patient's Age	57
	(0018,0000)	Unknown	338
	(0018,0015)	**Body Part Examined**	**CHEST**
	(0018,1508)	Positioner Type	
	(0018,5101)	**View Position**	**PA**
	(0020,0010)	**Study ID**	**51080.1**
	(0020,0020)	**Patient Orientation**	**L\F**
	(0020,0062)	Image Laterality	U
	(0020,4000)	Image Comments	

Key types of metadata stored in the DICOM header file are bolded in these five figures taken from an actual header.

A is the first 25 lines showing the universal DICOM tag identifiers at the left, followed by a description of the type of information being given, then, in the third column the actual setting, status or value being used. Note the accession number.

B: Patient demographic information and exam information.

(Continued)

Figure 36-4 ***(Continued)***

	Tag ID	Description	Value
C	**(0018,0060)**	**KVP**	**+80.0**
	(0018,0000)	Device Serial Number	10000932
	(0018,1020)	Software Version(s)	V6.00.15
	(0018,1030)	Protocol Name	
	(0018,1050)	Spatial Resolution	0.160
	(0018,1052)	Exposure	+32
	(0018,1060)	Filter Type	
	(0018,1064)	Imager Pixel Spacing	0.160\0.160
	(0018,1180)	Collimator/grid Name	119cm/8:1/40/Al
	(0018,1200)	Date of Last Calibration	20131107
	(0018,1201)	Time of Last Calibration	110113.000000
	(0018,1401)	**Acquisition Device Processing Code**	**REX454Q2W3LP3GSS17,20Z1M2,1,3P1**
	(0018,1405)	**Relative X-ray Exposure**	**408**

	Tag ID	Description	Value
D	(0018,7004)	Detector Type	SCINTILLATOR
	(0018,7005)	Detector Configuration	AREA
	(0018,700A)	Detector ID	7e00182
	(0018,7030)	Field of View Origin	352\0
	(0018,7032)	Field of View Rotation	0
	(0018,7034)	Field of View Horizontal Flip	YES
	(0019,0016)	Unknown	Canon, Inc
	(0020,0011)	Series Number	1
	(0020,0013)	Instance Number	1

	Tag ID	Description	Value
E	(0028,0002)	Samples Per Pixel	1
	(0028,0004)	Photometric Interpretation	MONOCHROME2
	(0028,0010)	**Rows**	**2688**
	(0028,0011)	**Columns**	**2128**
	(0028,0100)	**Bits Allocated**	**16**
	(0028,0101)	Bits Stored	12
	(0028,0102)	High Bit	11
	(0028,0103)	Pixel Representation	0
	(0028,0301)	Burned on Annotation	NO
	(0028,1040)	Pixel Intensity Relationship	LOG
	(0028,1041)	Pixel Intensity Relationship Sign	1
	(0028,1050)	**Window Center**	**2048**
	(0028,1051)	**Window Width**	**4096**
	(0028,1052)	**Rescale Intercept**	**0**
	(0028,1053)	**Rescale Slope**	**1**
	(0028,1054)	**Rescale Type**	**US**
	(0028,2110)	Lossy Image Compression	00
	(0040,0555)	Acquisition Context Sequence	
	(2050,0020)	**Presentation LUT Shape**	**IDENTITY**
	(7ZFE,0010)	Pixel Data	00 00 00 00 00 ...

C: Specifics of the exposure including technique, digital processing code used and exposure to the patient.

D: Detailed information on the detector and field orientation.

E: The digital processing parameters used including matrix size, window settings, and the type of rescaling curve and LUTs that have been applied.

Figure 36-5

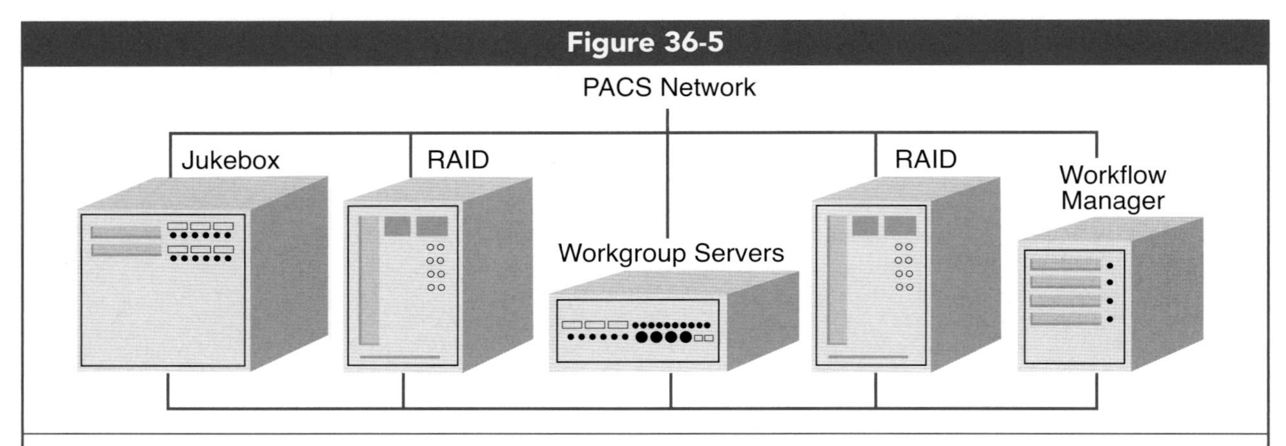

A *storage area network (SAN)*, a sub-network connecting several storage devices. Note that the diagram shows all of these components connected *in parallel*, so that if any one component fails, the connections between the other components are not broken.

same images simultaneously at different workstations. The efficiency of workflow is increased across the spectrum of diagnosis and treatment.

With film technology, physical copies of the original image always suffered from a progressive loss of sharpness as copies of copies were made. With digital images, the original quality is preserved when copies are made onto discs or other storage media. The main problem in preserving image quality occurs at the *receiving* end, because the receiving physician may not have the software to faithfully reproduce the image brought up on his/her own viewing monitor. To fix this problem, a DICOM viewer can simply be included on the CD or DVD along with the images sent. DICOM viewer software has become widely available and can be downloaded free of charge in many instances.

Because of their high resolution, medical images require much more storage space than normal e-mail services provide for attachments. Compression can seriously downgrade the quality of images and affect diagnosis. The PAC system can sometimes generate a unique *URL (uniform resource locator)* accessible only with a password. An e-mail can be sent to a physician with a link to this website address and the associated password, allowing access to the server that holds the files along with the viewer software needed.

Before PAC systems were developed, the rate of lost or misfiled medical documents was estimated at between 5 and 20 percent. Multimillion-dollar judgments and out-of-court settlements against doctors and medical institutions have resulted from misfiled or lost documents. Medical images and information must not be vulnerable to a computer drive "crashing," to electrical failure or file corruption on a single system. PAC systems distribute copies of the same data files across several computer hard drives *which are independent of each other* so this cannot happen. This distribution is referred to as a *Redundant Array of Independent Disks* or a *RAID* system. To protect against loss from natural disasters such as fires or floods, some institutions have backup storage facilities at remote geographic locations or other clinical sites. Figure 36-5 shows a *storage area network (SAN)*, a sub-network connecting several storage devices. Note that the diagram shows all of these components connected *in parallel*, so that if any one component fails, the connections between the other components are not broken.

INTEGRATION

A radiology information system or *RIS* is a data system for patient-related functions in the department, such as scheduling of appointments, tracking of patients, storage and distribution of reports, and billing. A hospital information system or *HIS* performs the same functions for the entire institution. *Health Level 7 (HL7)* is an important software standard for exchanging electronic text information in compatible formats between departments and institutions. It enables patient records and reports to be stored, accessed and viewed on different types of equipment, just as a PAC system standardizes the communication of images.

The HIS assigns a unique identifying number for each patient. For a radiology patient, this number is sent to the image acquisition unit, the PAC system, and the worklist server. As images and records are then moved about, the PAC system verifies a match between the images and the records prior to storing images. An *accession number* is an institution's unique number for identifying a particular procedure or exam. The accession number assigned by the RIS for each radiographic examination should be displayed in the DICOM header.

SUMMARY

1. Every imaging system consists of a machine for *image acquisition*, equipment for *image processing*, and devices for *image display and storage.*
2. A *PAC system* stores images from all of the different modalities in digitized format, enabling them to be retrieved, displayed or transmitted to remote locations. The main function of a PACS is to act as a *database*, allowing images, worklists and information to be retrieved by queries. A *prefetching* feature brings up previous studies related to a current requisition.
3. Image acquisition units and their associated workstations are considered *as service class users (SCUs)* of the PAC system, while the centralized storage, distribution devices and workstations of the PACs system itself are considered a *service class provider (SCP).*
4. Modern PAC systems store over one million medical images on magnetic or optical *jukeboxes*, stacks of one hundred or more magnetic or optical discs. The system allows retrieved images to be manipulated in a number of ways, including windowing.
5. The *control computer* is the heart of the PAC system; it must smoothly interface with the HIS, the RIS, and computerized medical equipment at other hospitals and clinics. For this purpose, the DICOM (*Digital Imaging and Communication in Medicine*) standard was created to standardize the transmission *behavior* of all these medical devices.
6. For each image, the DICOM header can be retrieved to access the *metadata* behind the image.
7. *Lossless compression*, at less than an 8:1 compression ratio, is generally acceptable to save computer storage space for medical images. Typical e-mail services do not meet this standard.
8. There are several ways to access PACS images, which may also include non-radiographic images. To preserve high resolution on any display system, DICOM viewers should be included on CD or DVDs along with the images, and on special websites with password-protected *URLs* for electronic access.
9. *RAID (Redundant Array of Independent Disks) systems* and *SANs (storage area networks)* protect image files from being lost due to human error, equipment failure and natural disasters by providing parallel storage of multiple copies at different locations.
10. The *accession number* assigned by the RIS for each radiographic examination and stored in its DICOM header, is used to smoothly integrate access and transmission of the image between the HIS, the PAC system, and image acquisition units.

REVIEW QUESTIONS

1. List the three universal components for any digital imaging system.

2. The workstation for a particular radiographic unit is considered as a *service class* ________ of the PAC system.

3. The heart of the PAC system is the ________.

4. A stack of one hundred or more magnetic or optical discs is called a ________.

5. So that different brands of computerized medical equipment might be compatible so that images and data can be sent between different hospitals and clinics, the ACR and the NEMA jointly developed the standard called ________.

6. In a word, the main function of a PAC system is to act as a ________.

7. Specific information on every aspect of the equipment, exposure, personnel, and digital processing parameters used for any image, along with demographic information on the patient and institution, is found in the DICOM header. All this information is collectively referred to as ________.

8. For an image, define *lossy compression.*

9. What is the typical megabyte size range for a CR or DR image?

10. Within a PACS server, storage locations are divided into ________ and ________ types.

(Continued)

REVIEW QUESTIONS *(Continued)*

11. The most common method for a doctor's office to access a PAC system is to ________ PACS ________ on PCs at the clinical site.

12. Without using DICOM viewer software, "copies of copies" made of any medical image will suffer from a progressive loss of ________.

13. What is the purpose of *RAID* and *SAN* systems?

14. The *HIS* for the whole hospital must conform to the software standard called ________.

Part IV

SPECIAL IMAGING METHODS

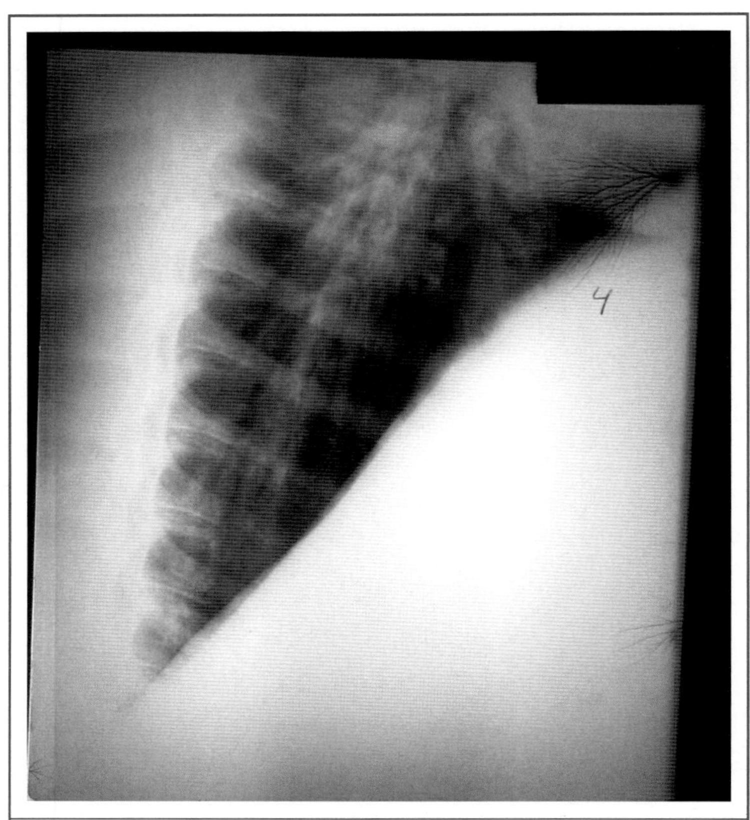

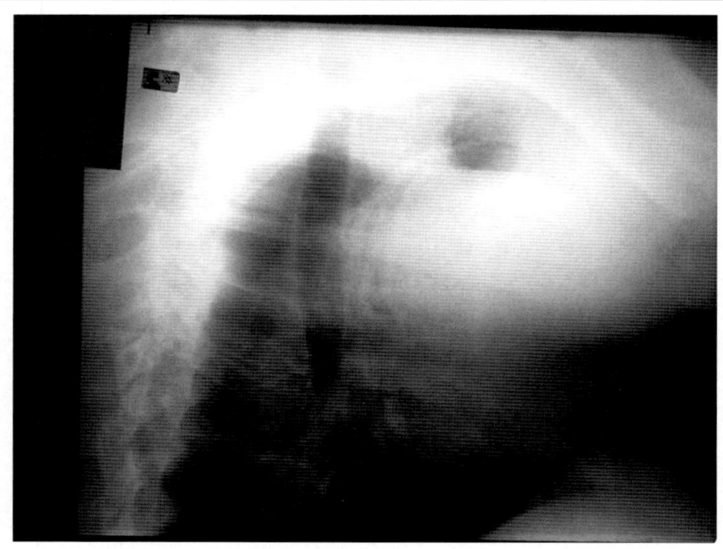

Chest radiographs of a horse, requiring two exposures on 14″ x 17″ (35 cm x 42 cm) plates to cover the costophrenic angles and upper lungs.

Chapter 37

MOBILE RADIOGRAPHY AND TOMOGRAPHY

Objectives:

Upon completion of this chapter, you should be able to:

1. Describe technique considerations, geometrical factors, and positioning and alignment considerations unique to mobile x-ray units.
2. Explain the principle of *parallactic shift* and how it is used to create tomograms.
3. Describe the relationships between tomographic amplitude, blurring effectiveness and focal plane thickness.
4. Describe the importance of movement patterns and orientation in preventing false images.
5. Apply the correct selection of focal depth and focal intervals for tomographic procedures.

MOBILE RADIOGRAPHY

Because of the many additional variables involved in mobile radiography, such as other medical equipment and furniture being in the way, limited space, and immobility of the patient, mobile or "portable" radiography frequently presents the greatest challenges faced by radiographers in securing good quality images. The adaptability and creativity required to maintain all radiographic variables at the optimum level possible for mobile and trauma procedures is what sets the most skillful radiographers apart. Following are special technique considerations that will help enhance image quality.

Mobile Generators

Although other types of generators exist, nearly all mobile x-ray units now use *constant potential generators (CPGs)* which are battery-powered. During a radiographic exposure, these units produce very efficient current with even less "ripple effect" than a three-phase, 12-pulse generator (see Chapter 8). Comparing the mobile techniques required with the three-phase, six-pulse stationary unit, *a modern battery-powered CPG x-ray unit typically requires a* reduction *of at least 8 kVp* to produce the same exposure level to the image receptor.

Most mobile units have a single "mAs" control, rather than mA and exposure time being selected separately (Fig. 37-1). The units are usually designed to operate at a high, fixed mA value, so that exposure time becomes the main variable by default. The mobile unit must be plugged into an electrical outlet between uses in order to recharge its bank of batteries. It is essential that these batteries be kept at optimal charge or the technique needed for a particular exposure may not be available.

Geometrical Factors

Radiographers must often perform "cross-table" projections during mobile procedures, and in working around other equipment and the patient with all attached devices it seems that every conceivable angle of tube and cassette plate can be encountered. In this process, all distances and alignment must be preserved to the extent possible to maximize image sharpness, and to minimize magnification and distortion.

Figure 37-1

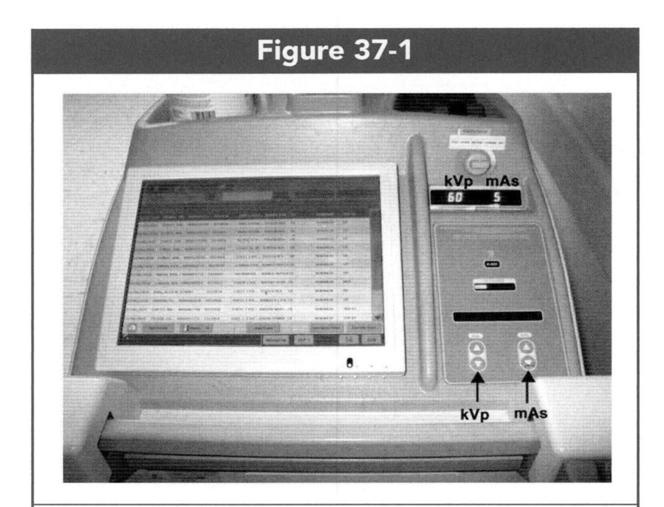

Console of a modern mobile x-ray unit, showing a selector for mAs rather than separate mA and exposure time settings.

Distance Considerations

The SID must often be estimated during mobile procedures. If a measuring tape is attached to the tube or collimator, it should be used, or the radiographer may wish to carry a compact tape measure in a pocket. If a tape measure is not available, a 72" (180 cm) SID may be estimated (easily within a 10% range of accuracy) as the distance from fingertip to fingertip of a 6-foot tall individual. This means that for the average male, both arms may be extended from the receptor plate to the x-ray tube *housing*, and for the average female, both arms may be extended from the receptor plate to the *collimator*, to estimate a 72" (180 cm) SID. To estimate 40" (100 cm), extend one arm and measure to the opposite axilla.

The radiographer must make every effort to keep the image receptor plate in contact with the part during mobile exposures. If this is not possible, the minimum OID possible should be achieved, and an increase in SID (a proportionate increase, if possible) should be considered to compensate.

Remember that any change in SID greater than 10 percent requires compensation in the set technique. (Remember, too, that any angle greater than 15 degrees requires a compensation in SID.) Rules of thumb for this very purpose were fully discussed in Chapter 22. As summarized in Table 22-2, from 40", the overall technique may be increased by ½ for 50", by 2 times for 60", and by 3 times for 72". It should be reduced to ½ for 30". Ratios can be made from these factors; for example, when changing from 60" to 72", using the associated factors to make a ratio, the new technique would be 3/2. From 72" to 60", it would be 2/3. Each factor can also be inverted when reducing distance; for example, when changing from 72" to 40", use 1/3 the original overall technique (1/3 the mAs).

Alignment and Positioning Considerations

The radiographer must become adept at visually estimating the alignment of the x-ray beam with the patient and cassette plate. This skill is necessary to minimize shape distortion and avoid superimposition of unwanted anatomy over the anatomy of interest. While the light field provides an easy approach to centering, it does *not* give any obvious indication of off-angling unless it is very severe. Therefore, the radiographer must always carefully observe the angle of the x-ray tube in relation to the patient, from *two perspectives.*

The first is to check the side-to-side angle as observed from behind the x-ray tube, siting right along the axis of the central ray. Mobile chest radiographs often turn out looking slightly rotated because of neglecting this fundamental point. It means that, when one stands directly behind the x-ray tube, it appears centered not only to the cassette and patient's chest, but *also to the foot of the bed, or, more accurately, to the patient's feet.* This will assure that the central ray is not only centered to the anatomy, but also perpendicular to it.

The second perspective is to check the cephalic/caudal angulation as observed from the side of the patient and the x-ray beam. It is always helpful in visualizing these angles to *stand back as far as possible* when observing them. These relationships are easier to see from a distance. The caudal angulation for sitting mobile chest radiographs provides a classic example of the degree of awareness needed for proper alignment; as illustrated in Figure 37-2, in spite of all positioning efforts, the lower back of the sitting patient is frequently not in contact with the cassette plate. This means that the long axis of the patient's body and the plate are at slightly different angles. Many a "lordotic" chest view has resulted from subconsciously placing

Figure 37-2

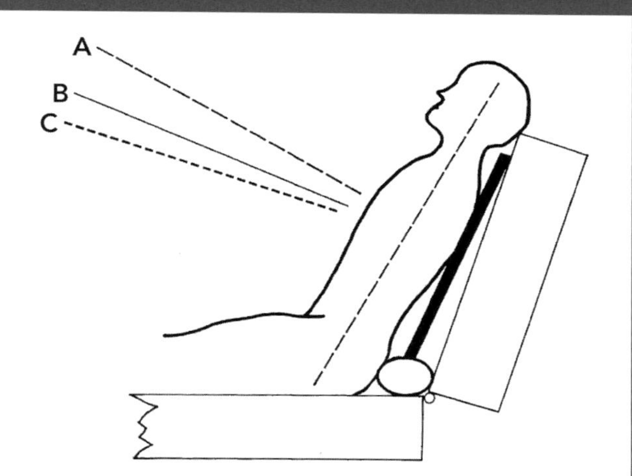

Typical alignment problems encountered during mobile AP chest radiography, when the axis of the body (dashed line) is not parallel to the receptor plate or bed. The proper CR angle is ***A***, perpendicular to the coronal plane of the body, rather than the receptor plate, ***B*** or the bed, ***C***. (For fluid levels, the patient must be fully upright.) (From Quinn B. Carroll, *Practical Radiographic Imaging*, 8th ed. Springfield, IL: Charles C Thomas, Publisher, Ltd., 2007. Reprinted by permission.)

the CR perpendicular to the plate (or even to the bed), rather than angling caudally until it is perpendicular to the actual coronal plane of the patient's body. (Note that, ideally, to demonstrate fluid levels, the patient should be seated fully erect when possible—in this case, the CR can be angled 5 degrees caudal and still show fluid levels well. If it is *known* that the primary indication for ordering the study was to rule out fluid levels, the CR should be kept horizontal.)

When long bones or extremities cannot be positioned parallel to the cassette plate, *Ceiszynski's law of isometry* should be used, as explained in Chapter 24, splitting the difference between the two angles. This minimizes shape distortion of the bones. However, many projections are designed specifically to demonstrate joint spaces rather than bones. For these, *it is always more important to keep the central ray perpendicular to the anatomy, rather than the cassette, if one must choose.*

Other Considerations

Low ratio grids (5:1 or 6:1) can be especially helpful for mobile radiography of larger body parts, because their low selectivity allows a much wider margin for error in angulation, centering, and especially distance variations. Any grid must be as carefully aligned as possible with the x-ray beam to avoid grid cut-off (see Chapter 20). For mobile procedures, watch for off-centering or off angling of the beam, tilt of the grid in either plane, keeping the SID within the grid radius, and upside forward placement for focused grids. It is not uncommon in surgery for the radiographer to place a gridded-cassette into a sterile cover held by a nurse, and then for the nurse to place the covered assembly upside-down under the patient.

Some manufacturers have experimented with AEC "paddles" placed behind the patient for mobile procedures to implement a form of automatic exposure control. These were generally found to be unworkable and caused many repeats. They have been widely discontinued, and are not recommended. Every radiographer should master the skills presented in earlier chapters enough to feel comfortable setting "manual" techniques. *Concise technique charts should also be developed and attached to every mobile x-ray unit.* These can be laminated or covered with transparent film, and taped or chained to the unit (Fig. 37-3). Although mobile radiography often challenges the radiographer's skills and knowledge, the same standards for image quality and minimizing patient exposure applied "in the department" should be sought after.

CONVENTIONAL TOMOGRAPHY

Parallactic Shift

Parallactic shift, or *parallax*, is the apparent change in position of an object against a background reference

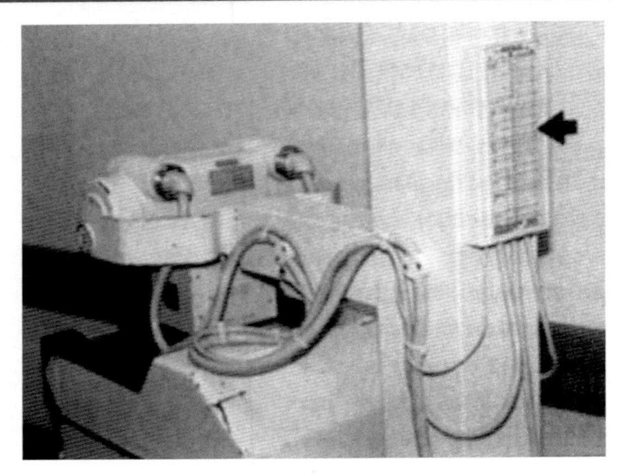

Modern mobile x-ray unit showing an attached manual technique chart on the back of the tube stand (arrow). (From Quinn B. Carroll, *Practical Radiographic Imaging*, 8th ed. Springfield, IL: Charles C Thomas Publisher, Ltd., 2007. Reprinted by permission.)

point when seen from two different points of view. You can demonstrate the parallax shift between your two eyes by holding a pencil close to your nose and alternating closing one eye and then the other. Although you are not aware of this shift consciously as you go about your daily business, your brain automatically and continually compares the parallax of objects seen by each eye to provide a sense of depth or distance.

Before CT and MRI were developed, *stereoscopic* radiographs were produced by exposing the anatomy from two different x-ray tube positions, and then viewing the two radiographs through a stereoscopic viewer that used mirrors and a partition to separate the view of each eye. Localization of the depth of foreign bodies could also be achieved using *triangulation* methods which measured the parallactic shift of an object between two projections, and then mathematically calculated the depth.

Conventional tomography continues to be used in spite of the advent of CT and MRI, particularly for urographic studies, for the compelling reason that it provides a perfectly adequate system which is much more economical. Tomography is broadly defined as the use of *movement* of the x-ray tube, the patient, and/or the image receptor plate in order to blur out contrasty anatomical structures which overlie the specific anatomy or pathology of interest, such as a kidney stone. In context, superimposed anatomy which obstructs the view of the anatomy of interest may be considered as a form of *noise* in the image that we wish to remove.

Unlike stereoradiography, visual depth perception, as such, is not created in these images. But, as we shall see, the tomographic *process* of blurring anatomical structures above and below a plane of interest is still based upon the same phenomenon of *parallactic shift*. The depth of the object or anatomy within the body is found by taking a series of tomographic images which are *focused* at different sequential depths, blurring out anatomy above and below these levels. The radiologist effectively sorts through the images like a deck of cards to determine the exact depth of the foreign object, pathology, or anatomy of interest within the body.

Tomography has gone by many names indicative of its ability to effectively take *layers* or planes out of the body, as if it were being autopsied in sections. These titles include *body section radiography*, *laminography*, *planigraphy*, and *zonography* which indicates the use of thick body sections for localization purposes. Although most tomography involves moving the x-ray tube and the image receptor plate in concert during exposure, it is also possible to obtain tomographic images by having the patient move a body part during the exposure, while the x-ray tube remains stationary. Referred to as *autotomography* ("self-tomography"), examples of this approach include breathing techniques that are used for the thoracic spine, sternum and SC joints, and the "wagging jaw" technique (Otonello method) for the upper cervical vertebrae.

The Focal Plane

The way that tomography uses parallactic shift is illustrated in Figure 37-4. The central ray of the x-ray beam is centered through the anatomy of interest to the image receptor plate. As the x-ray tube moves in one direction, it also rotates on its axis so that the central ray remains centered through the anatomy of interest, forming a *fulcrum point*. The receptor plate moves in the opposite direction to maintain its centering to the CR. Note in Figure 37-4 that in doing so, the anatomy of interest (the circle) is always projected to the center of the image receptor plate and is therefore not blurred.

However, anatomy above and below the part of interest is projected by peripheral rays within the x-ray beam that are *changing* throughout the exposure. This effectively moves these images across the receptor plate so that they are blurred. Note in Figure 37-4 that the triangle and the square exchange places on the receptor plate from one extreme of the tube movement to the other. Both of them have been blurred clear across the length of the plate, in opposite directions. During the exposure, they are always moving across the plate. The result is that the circle will always be visibly in focus on the resulting radiograph, while the objects above and below it will be blurred out of visibility.

The further an object is above or below the fulcrum point, the greater distance it will be blurred across the receptor plate. Surrounding the fulcrum point, at its exact depth, is a *focal plane* in which all the anatomy at this level is recorded in focus. We can imagine this focal plane as a thin sheet or slice of tissue within the patient, and since it is *infinitely thin*, actually *everything* above and below this hypothetical plane is blurred to a greater or lesser degree.

However, there is a limit to the capacity of the human eye to *see* microscopic amounts of blurring, so that an object must be blurred by a certain minimum amount before the blurring becomes visible to the eye. The practical effect of all this is that *the focal plane becomes a layer of anatomy with a distinct thickness, which will be recorded at the image receptor with apparently well-resolved sharpness* (Fig. 37-5). Within this focal plane thickness, no anatomy has been blurred enough for the human eye to detect the blur, so it all appears reasonably focused. Anatomy which has been visibly blurred is considered to lie outside of the focal plane.

In Figure 37-5, an imaginary minimum distance to cause visible blurring is diagrammed, (upper right). This distance represents the limit of the human eye to detect blur. On the *left* side of the diagram, note that the black box has been blurred across a distance greater than the minimum indicated by the "adequate blur" arrow. It will be blurred on the image and not considered to lie within the focal plane. The gray box, however, is close enough to the plane of the fulcrum point that it has not been blurred across the minimum distance. It will appear focused on the image and is considered to be within the focal plane. The resulting thickness of the focal plane is indicated by the hash-marked area which includes the gray box but not the black box. To blur out the gray box, the distance of the x-ray tube movement must be increased.

Figure 37-4

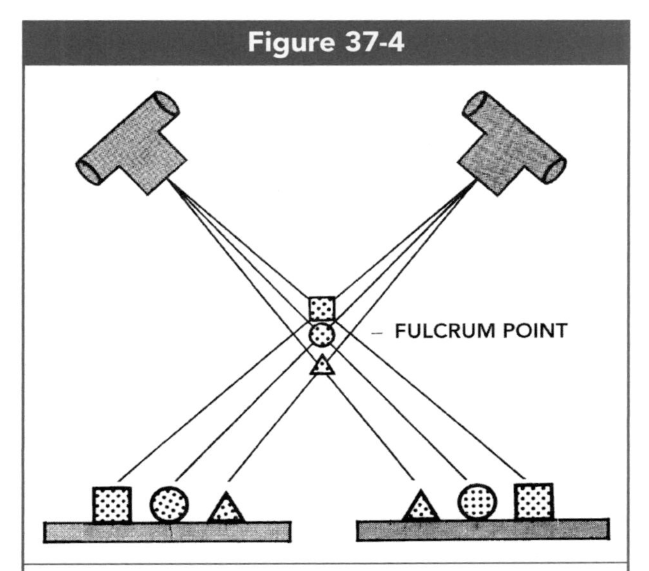

Diagram illustrating the displacement and blurring of images both above and below the fulcrum point of a tomographic tube movement. Note that during the course of the tube movement, the images of the triangle and square have moved across the surface of the receptor plate. (From Quinn B. Carroll, *Practical Radiographic Imaging*, 8th ed. Springfield, IL: Charles C Thomas Publisher, Ltd., 2007. Reprinted by permission.)

As shown on the *right* side of Figure 37-5, where the tube movement is increased, the gray box will no longer be considered to lie within the focal plane since it is now blurred from visibility (along with the black box). In effect, the focal plane, indicated by the hash-marked area, has been made much *thinner*. That is, centered around the depth of the fulcrum point, there is a thinner layer of tissue which is shown in apparent focus on the resulting radiograph.

The greater the blurring effectiveness obtained through the tube movement, the thinner the resulting focal plane. Two aspects of the tube movement bear upon its blurring effectiveness—the exposure arc and the amplitude (Fig. 37-6). The *exposure arc* refers to the angle between the extremes of the tube

Figure 37-5

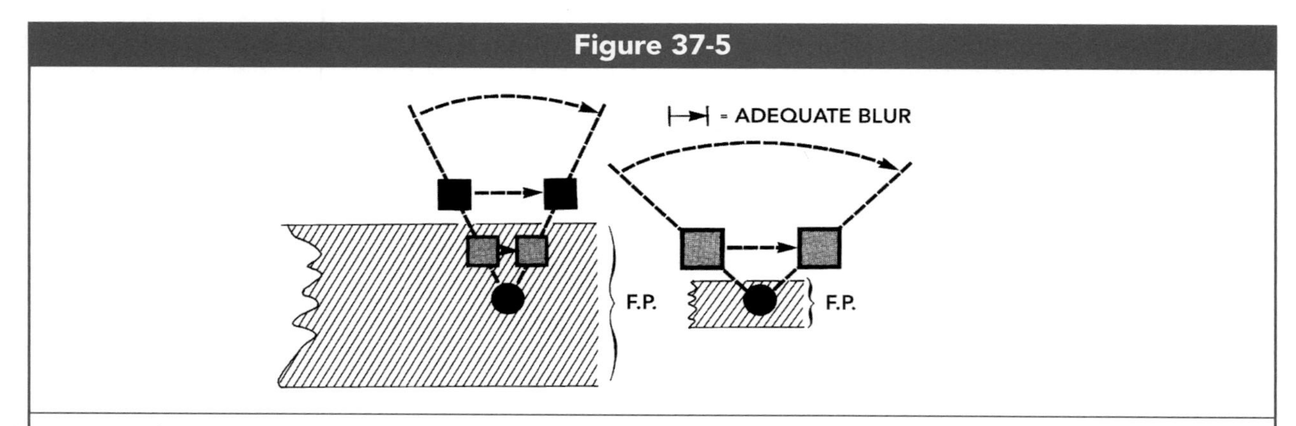

Diagram illustrating the definition of the tomographic focal plane as that thickness of tissue within which objects are not displaced enough to be visibly blurred. The distance of shift required for visible blurring is indicated by the arrow, upper right. *Left*, the black box is visibly blurred, but the gray box is not visibly blurred and is considered to lie within the focal plane. *Right*, increased tube movement now visibly blurs the gray box, and it is considered to lie outside the much thinner focal plane. (From Quinn B. Carroll, *Practical Radiographic Imaging*, 8th ed. Springfield, IL: Charles C Thomas Publisher, Ltd., 2007. Reprinted by permission.)

movement formed by the central ray, measured in degrees. The greater the exposure arc, the thinner the focal plane and the more effective the blurring. *Amplitude* refers to the actual linear distance traversed by the x-ray tube during the tomographic movement. The amplitude may be increased without changing the exposure arc by changing the pattern which the tube makes in moving. For example, the tube may trace out a 30-degree circle (with a diameter 15 degrees to each side of vertical) rather

Figure 37-6

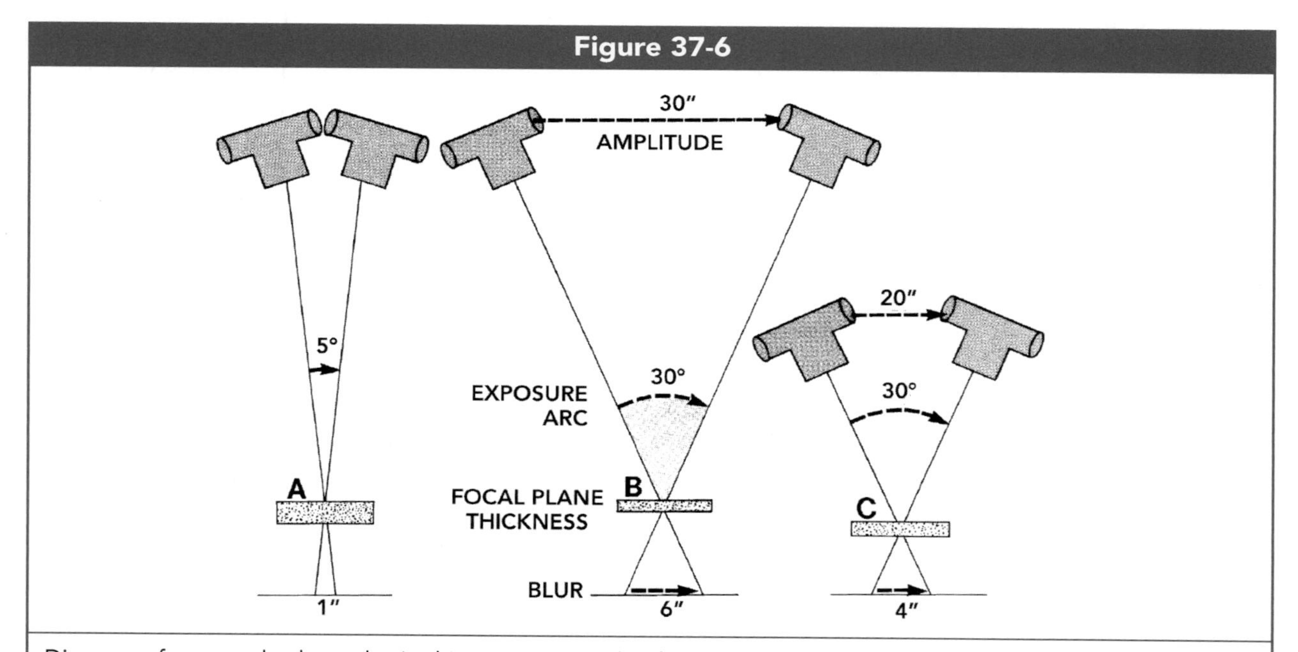

Diagram of two methods to obtain thinner tomographic focal planes. Plane ***B*** is thinner than plane ***A*** because of increased exposure arc (30 degrees versus 5 degrees). Plane ***B*** is thinner than plane ***C*** because, even though the exposure arcs are equal, amplitude (tube travel) is greater for ***B*** due to increased SID. (From Quinn B. Carroll, *Practical Radiographic Imaging*, 8th ed. Springfield, IL: Charles C Thomas Publisher, Ltd., 2007. Reprinted by permission.)

than a 30-degree line. In doing so, it travels more than three times the total amplitude distance. The blurring of unwanted structures in the resulting image is slightly more effective for circular movements than for linear movements of the same exposure arc for this reason.

Movement Patterns and False Images

Over the years, tomography units were invented with several different ingenious movement patterns which the x-ray tube would sweep out, including spirals, cloverleaf patterns, and circle eights. Increased complexity of the tube movement yielded increased amplitude, better blurring effectiveness, and ever thinner focal planes, the thinnest approaching 1 mm. These *pluridirectional* movements also had the advantage of greatly reducing *false images* such as streaks.

False images are categorized as a distinct form of *noise*. They should not be considered as images of real anatomy which has undergone shape distortion, and therefore would contain at least some useful information. Rather, a false image is a newly-created image produced by the relationship between the movement of the x-ray tube and the particular anatomical structure, which contains absolutely *no* useful information, but only serves to obscure information just as other artifacts would.

We have discussed how obscuring images are blurred across the area of the receptor plate during the tomographic exposure. A false image is created when an area is left on the receptor plate over which some portion of an object was always projected during the exposure. To prevent this effect, anatomy which we wish to blur out should be positioned such that it lies *perpendicular* to the linear x-ray tube movement whenever possible. This way, the blurring distance required to completely blur out the obstructing image is reduced to its own *width*, as illustrated in Figure 37-7.

Figure 37-7

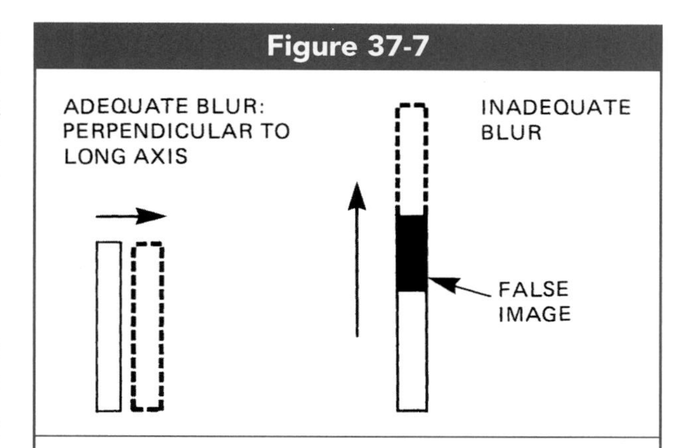

Placing obstructing anatomy perpendicular to the tube movement, *left*, is effective in eliminating streak artifacts. Streaks may be produced, *right*, when the tomographic movement is insufficient to completely displace the anatomical part lengthwise. (From Quinn B. Carroll, *Practical Radiographic Imaging*, 8th ed. Springfield, IL: Charles C Thomas Publisher, Ltd., 2007. Reprinted by permission.)

For example, during a linear tomogram in the torso such as might be employed during a urographic study, the *ribs* lie roughly perpendicular to the tube movement, which is lengthwise in relation to the x-ray table. The ribs will be well-blurred, because they only need to be moved one-half inch or so (their width) across the receptor plate. On the other hand, as can be seen in Figure 37-8, ***A***, cortical bone of the tibia and femur which runs vertically, parallel to a linear tomographic tube movement, will create linear *streaking* artifacts. Streaks are also commonly visible on urograms (Fig. 37-8, ***B***).

Since the development of CT and MRI, the use of tomography has been generally limited to linear tomography for urologic studies. These images typically manifest numerous streaking artifacts.

To avoid double-images and uneven exposure, it is best that the x-ray tube be already moving at a steady speed when the actual exposure begins and terminates. Therefore, when they machine is "prepped" for a tomogram, the x-ray tube is actually positioned farther than one-half of the selected exposure arc (in degrees) from vertical. Upon depressing the exposure switch, the radiographer will not hear the exposure begin until the tube has accelerated to a steady speed. Most tomograph machines are designed such that the x-ray tube and receptor plate move in straight horizontal lines rather than in an arc. This results in a changing SID throughout the exposure, which averages greater than 40" (100 cm), and, therefore requires an increase in overall technique. Generally, a tomogram requires about twice the technique of a routine overhead radiograph of the same anatomy.

Figure 37-8

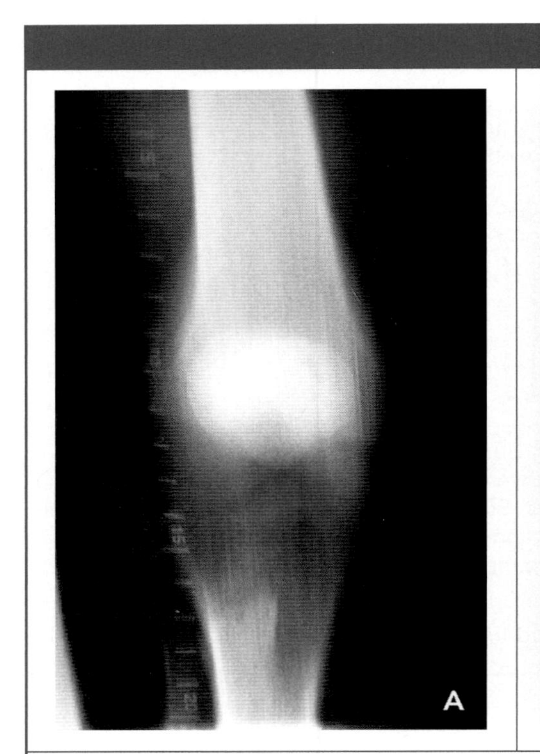

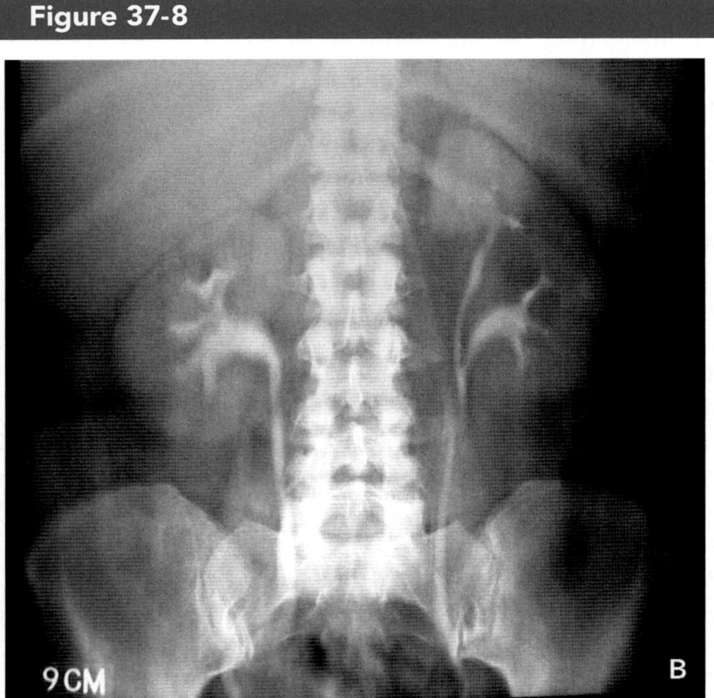

Vertical linear streaking is apparent in these typical tomographs of the knee and the iodinated urinary system.

Focal Depth and Focal Intervals

The *focal depth* refers to the location of the focal plane, in centimeters, from the *back* or *downside* of the patient's body upward. The focal depth is the location of the center of the focal plane, or the fulcrum point of the tube movement, for each cut. Usually, several tomograms are taken at different levels within the patient, commonly called *cuts* or *slices*. It is important to remember that, if a sponge or pad is placed under the patient, its thickness must be subtracted from the focal depth indicated on the machine to obtain the actual focal depth from the patient's downside—this is the focal depth that should be marked on the image. For example, if the machine is set to take a "slice" at a 7 cm level, but there is a 2 cm pad under the patient, this slice should be marked at *5 cm*.

Table 37-1

Tomographic Focal Plane Thicknesses by Linear Exposure Arc

Degrees of Linear Arc (Tube Movement)	Resulting Focal Plane Thickness
10°	1.2 cm
15°	1.06 cm
20°	9 mm
30°	6 mm
40°	3 mm

An very important practice that is often forgotten is that the *intervals* at which slices are taken must be correlated to the thickness of the slices. Table 37-1 lists the typical slice thicknesses for linear tomograms according to the exposure arc used. If the intervals between the slices are too small, unnecessary patient dose results as the slices overlap each other too much. If the intervals are too great, important anatomy or pathology may actually be missed between the slices taken. As a rule, the interval used should be slightly less than the thickness of the focal plane obtained. Consulting Table 37-1, for example, if a 40-degree exposure arc is used, slice thickness is about 3 mm. If cuts are taken every 1 cm, there will be gaps between these slices where some anatomy has been missed.

Slices should be taken every 2½–3 mm. *For the common practice of taking slices every 1 cm, a reduced exposure arc of 15 degrees is recommended rather than 30 degrees.*

For urography, in the absence of specific instructions from the radiologist, two rules of thumb are helpful in determining a starting point for focal depth; from the measured thickness of the patient's abdomen, you can take ⅓ of this thickness plus 2 cm, or ½ this thickness minus 2 cm. For example, if the patient measures 21 cm, begin taking cuts around 7 + 2 = 9 cm. Note that 11 – 2 cm yields about the same answer.

SUMMARY

1. Battery-powered mobile x-ray units require about 8 kVp less than 3ϕ fixed units. Manual technique charts should be provided on all mobile units.
2. During mobile procedures, the radiographer must be conscientious of variable SIDs and make appropriate compensations in technique.
3. Alignment of the x-ray beam is critical for mobile procedures. Alignment can always be better seen from a viewpoint at an increased distance.
4. When anatomy cannot be placed parallel to the receptor plate, it is generally more important to keep the CR perpendicular to the anatomical part rather than the plate. Ceiszynski's law of isometry should be used for long bones that are not parallel to the plate.
5. Tomography desuperimposes obstructing structures from the anatomy of interest by blurring objects above and below the focal plane. The greater the exposure arc or amplitude, the thinner the focal plane.
6. When possible, obstructing anatomy to be blurred should be positioned perpendicular to the tomographic tube movement.
7. Tomographic slice intervals should be slightly less than the slice thickness.

REVIEW QUESTIONS

1. If 80 kVp is normally used for a "wheelchair" AP chest at 72" (180 cm) SID in a fixed 3ϕ 6-pulse radiographic room, what kVp would be used on the same chest using a battery-powered CPG mobile unit?

2. If a 72" technique were used with a 60" actual SID, is this enough of a change to require a technique adjustment?

3. If the best you can do with a patient in traction is to place the receptor plate at a 15-degree lengthwise angle to the femur, what is the best x-ray tube angle?

4. For a projection of a joint space, when you have to choose, is it better to place the CR perpendicular to the anatomy or to the receptor plate?

5. Tomography works on the basis of __________ shift.

6. What is defined as moving the patient's body part in order to blur out obstructing anatomy?

7. The longer the exposure arc, the __________ the focal plane produced.

8. Pluridirectional tomographic tube movements do a better job of eliminating __________ than unidirectional movements.

(Continued)

REVIEW QUESTIONS *(Continued)*

9. What is the typical focal plane thickness obtained with a 30-degree linear tube movement?

10. For a patient that measures 25 cm in torso thickness, begin taking cuts at what level?

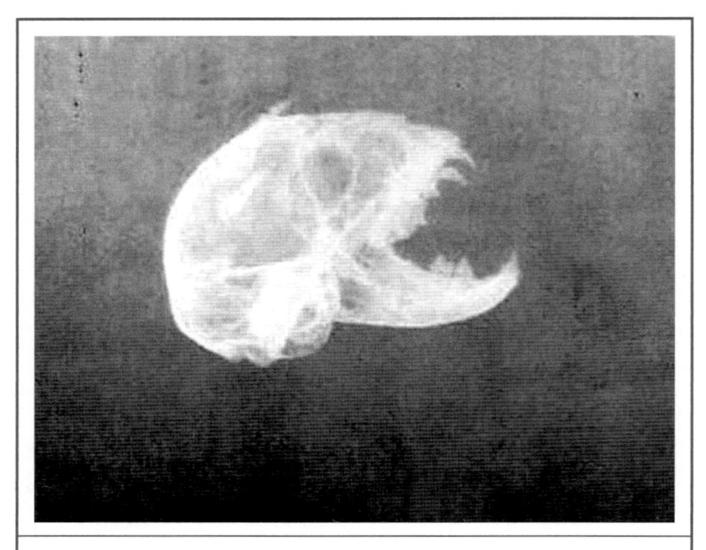

Mystery skull with "horns" found in the dessert: It is probably a cat skull with calcified ears.

Chapter 38

FLUOROSCOPY AND DIGITAL FLUOROSCOPY

Objectives:

Upon completion of this chapter, you should be able to:

1. Overview the history and development of fluoroscopy, especially in regard to patient exposure.
2. Describe the components of an image intensifier and how they work together to amplify the image.
3. Define the brightness gain and conversion factor for an image intensifier.
4. Explain how multiple field sizes are achieved, their effect on magnification and patient exposure.
5. Describe the two types of signal sensing and four approaches to stabilizing the brightness of the fluoroscopic image.
6. Explain the adjustment of proper fluoroscopic techniques.
7. Describe the qualities of the fluoroscopic image and the various forms of distortion that occur.
8. Describe the proper manipulation and positioning of a C-arm fluoroscopy unit.
9. Explain how patient exposure is minimized during fluoroscopy.
10. Describe the advantages and disadvantages of pulsed-progressive mode fluoroscopy, especially in regard to patient exposure.
11. Describe dynamic flat-panel detectors for fluoroscopy, their advantages and disadvantages.
12. Explain the application of temporal and energy subtraction methods to improve the image.
13. Describe the function and advantages of charge-coupled devices (CCDs) and complementary metal-oxide semiconductors (CMOSs).

DEVELOPMENT OF FLUOROSCOPY

Fluoroscopy is the production of *dynamic* radiographic images, in effect, moving pictures. Fluoroscopic images are obtained in *realtime* or immediately as they occur. Because fluoroscopy usually involves active diagnosis during the examination, it is usually performed by the radiologist or radiologist assistant, while the radiographer acts as an assistant and follows up with regular overhead views.

The great inventor, Thomas Edison, is credited with introducing the first fluoroscope in 1896, the year after Roentgen discovered x-rays. His device was a light-tight, hand-held metal cone (shown in Figure 40-1 in Chapter 40) with a fluorescent screen in the bottom and a viewing window in the top. With the x-ray tube operating from the opposite side of the patient, this cone would be held over the patient so that the remnant x-ray beam struck the fluorescent screen, making it glow and producing an image that could be viewed through the window.

Figure 38-1, a historical photograph, shows "open fluoroscopy" using a large intensifying screen suspended behind the patient. The fluoroscope further evolved by attaching the x-ray tube mechanically to

Figure 38-1

Historical "open fluoroscopy" involved watching a large intensifying screen placed behind the patient, without an image intensifier. This placed the operator directly in the x-ray beam.

the intensifying screen in a single, movable carriage system. The operator would then lean over the screen while the x-ray tube was energized to view the fluoroscopic image, placing the operator's head directly in the x-ray beam. This system had two major disadvantages. First, the image on the screen was extremely dim and required all of the lights to be turned off in the room. Second, it resulted in very high radiation exposure to both the patient and the operator.

In the late 1940s, the first electronic image intensification tube was introduced by inventor John Coltman. The first commercially available unit was marketed by Westinghouse Company in 1952 as the "Fluorex," and was only 3 inches in diameter. The image intensifier improved image visibility, drastically lowered patient and operator dose, and brought with it the ability to add multiple devices for recording permanent images. Fluoroscopic image intensification provides dynamic realtime imaging in which the physiological function of an organ can be observed. Various gastrointestinal organs can be observed with the use of contrast media. Static images generally referred to as *spot views* can also be obtained digitally.

Fluoroscopic examinations can require several hours with a total beam-on time up to an hour. During a fluoroscopic procedure, to reduce the radiation dose to the patient, the exposure rate in fluoroscopic image intensification is several orders of magnitude lower than in radiography. For example, an overhead abdominal technique for a large adult of 80 kVp and 600 mA at 0.1 sec. would result in a skin entrance exposure of about 1.0 R (roentgen) to the patient. If the same 600 mA of tube current were used for 10 minutes of fluorsocopic beam-on time, the patient skin entrance exposure would be about 5,900 R and would result in serious radiation injury to the patient.

An actual fluoroscopic image intensification examination of this adult would require only about 3 mA of tube current. Therefore, 10 minutes of fluoroscopic beam-on time would result in about 30 R to the patient. The exposure *rate* in fluoroscopy is much less than in radiography. However, the *total* x-ray exposure is usually much higher because of the extended amount of time the fluoroscopic beam is on.

Relatively few x-ray photons are used in forming a single fluoroscopic image; therefore, fluoroscopic images are statistically inferior to radiographic images. The radiographic image would be formed with 600 times more photons per second, but over a shorter period of time, than the fluoroscopic image. The fluoroscopic system needs to produce an image bright enough for the operator to see with less x-ray photons penetrating the patient. Therefore, fluoroscopic image intensification units must have a very high brightness.

The modern digital fluoroscopic imaging system (Fig. 38-2), consists of any x-ray system capable of continuous low mA output, an image intensification tube, and a closed-circuit camera with CCD or CMOS monitor. The collimator that limits the size of the x-ray beam is automatically adjusted to the proper field of view. When the SID is changed, as it often is when panning over anatomical regions of variable patient thickness, the collimator opens and closes to accommodate the differing height of the image intensifier.

The x-ray generator requires additional circuitry to operate in the fluoroscopic mode. In many systems, a three-phase radiographic system may have single-phase fluoroscopic circuitry. An additional electronic circuit, known as the *automatic brightness control (ABC)*, will alter the kVp or the mA, or both, with changes in anatomic part thickness and changes in the atomic number or density of tissues. The radiologist may select the brightness level desired, and the ABC will maintain it throughout a procedure.

Figure 38-2

Modern digital fluoroscopy tower, showing a digital photospot TV camera (arrow) attached to the image intensifier. From Quinn B. Carroll, *Practical Radiographic Imaging*, 8th ed. Springfield, IL: Charles C Thomas, Publisher, Ltd., 2007. Reprinted by permission.

THE IMAGE INTENSIFIER TUBE

Modern image intensifiers, though they are complicated devices, operate in a simple way. An x-ray tube, under the table, exposes the patient. The x-ray beam passes through the patient and is intercepted by the image intensification tube above. The objective of the image intensification tube is to convert the remnant radiation coming out of the patient into an amplified light image.

The image intensifier tube (Fig. 38-3), is an evacuated glass envelope, a vacuum tube that contains five basic parts. These are (1) the input phosphor, (2) the photocathode, (3) electrostatic lenses, (4) the accelerating anode, and (5) the output phosphor, diagrammed in Figure 38-4. The round tube can vary in size, with typical input diameters of 6, 9, 12 or 16 inches. The diameter of the output phosphor is always one inch (Fig. 38-3). The difference in size of the input phosphor to the output phosphor causes the light image produced at the input phosphor to be

Figure 38-3

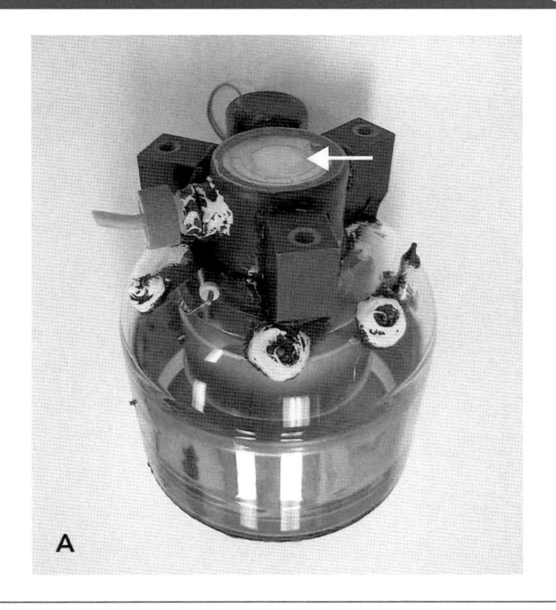

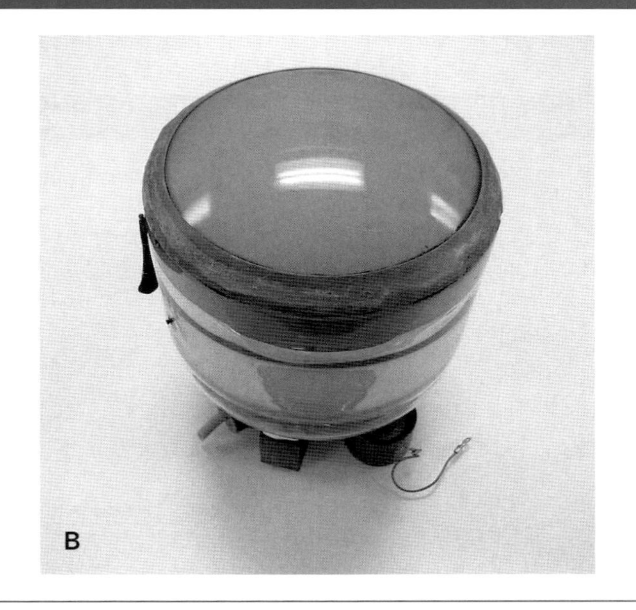

Photographs of an image intensifier tube, showing the output phosphor, ***A*** (arrow), and the input phosphor, ***B***.

thousands of times brighter at the output phosphor. *Increasing the physical diameter of the input phosphor results in increased image resolution*, because in effect a larger number of pixels are focused onto the same output size. As was discussed in the chapters on digital imaging, a larger number of pixels in an image matrix can only be accommodated by using smaller pixels, which enhances sharpness.

Input Phosphor and Photocathode

The input phosphor of a modern image intensifier is a layer of microscopic needle-shaped crystals of cesium iodide which are packed tightly together. These are fluorescent phosphors which transform the remnant radiation coming out of the patient into light. The light emitted is in the yellow-green wavelength. The columnar shape of the crystals helps prevent dispersion of the emitted light to preserve resolution.

The photocathode is a *photoemissive* metal, which is a combination of antimony and cesium compounds, that will receive the light emitted by the input screen phosphors and use the light energy to free up electrons. This is similar to the photoelectric effect, except that many light photons are required to cause a single electron to be emitted from the photocathode. However, since a single x-ray photon causes many light photons to be emitted from the cesium iodide phosphor, the net result is that a single 60 kV x-ray photon incident upon the input phosphor results in about 200 photoelectrons being emitted from the photocathode. Multialkalli photocathodes with compounds of potassium, sodium and cesium can triple this amount. Thus, a beam of millions of electrons is produced (Fig. 38-4).

Electrostatic Focusing Lens

The electrostatic focusing lens is a series of bands or rings of metal, which have varying positive voltage, through which the circular electron beam must pass. They have the capacity of pulling the electrons at the input side toward the output phosphor. The focusing rings have increasing degrees of positive electrical charge as they become narrower toward the anode, and are so arranged that they cause the electrons from the photocathode to be focused onto the much smaller output screen (Fig. 38-4). This process of focusing the electrons onto the output phosphor is called *minification*. Since the light image at the output screen is reduced in size, the electrons are *concentrated* onto the much smaller screen, so the screen glows more brightly.

Accelerating Anode

Located at the neck of the image intensifier tube (Fig. 38-4), the function of the accelerating anode is to attract the electrons from the photocathode and accelerate them toward the output screen. The

Figure 38-4

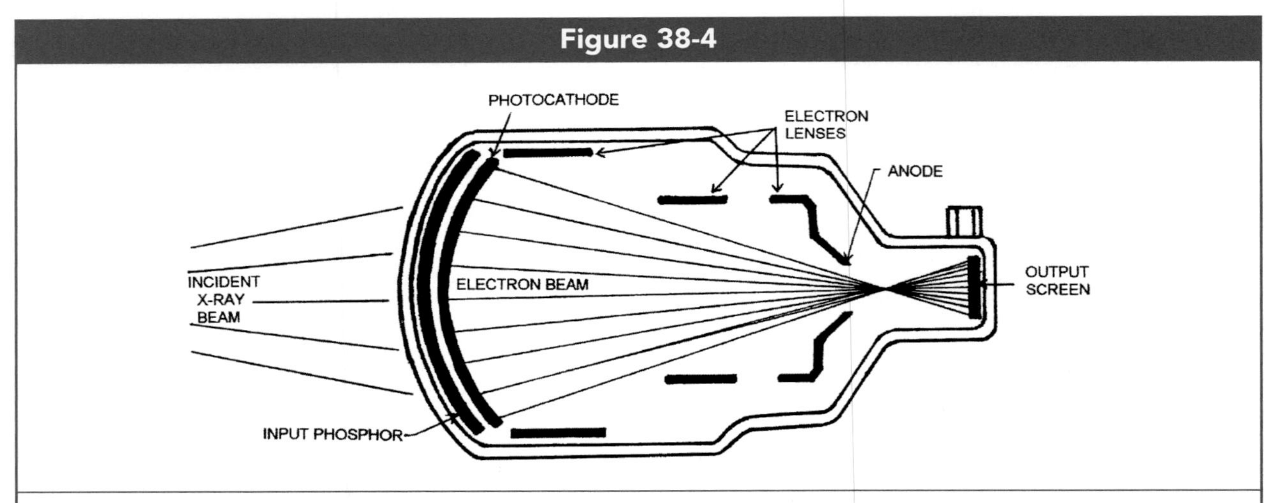

Diagram of components inside an image intensifier tube. From Quinn B. Carroll, *Practical Radiographic Imaging*, 8th ed. Springfield, IL: Charles C Thomas, Publisher, Ltd., 2007. Reprinted by permission.

accelerating anode is a small ring of metal, which has a positive charge of 25,000-30,000 volts. With this tremendous positive attraction to the anode, the electrons accelerate and strike the output phosphor with 50 to 75 times more kinetic energy than they left the photocathode with, all of which will be transformed into light. The ratio of the number of light photons emitted by the output phosphor to the number of *x-rays* striking the input phosphor is called the *flux gain*, (*flux* – "flow").

The metal housing around the image intensifier tube protects these internal electronics from outside magnetic fields, and also includes lead to reduce the leakage of scattered x-rays from inside the tube.

Output Phosphor

The output phosphor of the image intensifier tube (Figs. 38-3, ***A***, & 38-4), is made of cesium iodide. When the electrons interact with the cesium iodide phosphor their kinetic energy is transformed into light. Since the output phosphor is typically only 1 inch in diameter, the light emitted from it is very concentrated and bright. This light then passes to the CCD or CMOS camera tube.

Brightness Gain

Note from the above discussion that there are actually two distinct processes which both contribute to intensifying the image: flux gain and minification. It is the combination of these two processes that results in the final brightness achieved in the light image emitted from the image intensifier tube toward the television camera and other recording devices.

Minification gain can be easily estimated by forming a simple ratio from the areas of the input phosphor and the output phosphor. For example, the area of a 9-inch circular input phosphor, using the formula πr^2 is $3.14(4.5)^2 = 63.6$ square inches, that of the 1-inch circular output phosphor is 0.8 inches. These numbers round up to 64 and 1, respectively. If the entire 9-inch input phosphor is used, the minification gain will be approximately $64/1 = 64$. That is, the brightness of the light emitted from the image intensifier will be 64 times greater than the light originally emitted from the input phosphor, *from the minification process alone.*

If we now combine this effect with a *flux gain* of 70, for example, the total increase in the brightness of the image will be $64 \times 70 = 4{,}480$ times the original brightness of the input phosphor. Typical brightness gains can range from 5000 to as much as 20,000. This dramatic increase in the brightness of the image is the prime factor in reducing patient exposure as well as improving the visibility of the image.

Conversion Factor

The International Commission of Radiologic Units and Measurements (ICRU) recommends evaluating the brightness gain of image intensifiers based upon the *conversion factor*. The conversion factor is defined as the ratio of the luminance of the output phosphor to the input *x-ray* exposure rate:

$$\text{Conversion Factor} = \frac{\text{Candela / Meter}^2}{\text{Milliroentgen / Second}}$$

The *candela* is a unit for the direct measurement of light intensity or luminance. Since radiation quantity and output luminance are explicitly defined, the conversion factor method is more accurate and reproducible than the older "total brightness gain," which was based only upon multiplying the estimated flux gain by the minification factor.

Since the conversion factor is approximately 0.01 times the brightness gain, image intensifiers have conversion factors that typically range from 50 to 200.

Once the image is intensified it passes through a bundle of optic fibers to be received by a CCD or CMOS camera.

Multifield Image Intensifiers and Magnification Modes

Image intensifier tubes are identified by their *field of view* (*FOV*), a specification of the active *diameter* of the circular input phosphor. Large FOV image intensifier tubes are very useful for examinations in which a large area of the patient's anatomy must be viewed, for instance in lower gastrointestinal studies. A large FOV image intensifier allows more of the anatomy to be imaged simultaneously, resulting in less panning and potentially shorter studies.

However, it may be necessary to focus in and magnify a selected region of the patient's anatomy. This can be accomplished on most image intensifiers by changing the electric charge on the electrostatic lenses to bring the *focal point* of the electron beam closer to the input phosphor (Fig. 38-5). In this way, the beam is narrowed toward the input phosphor such that a smaller region from the input phosphor is projected onto the output phosphor. Since a smaller region of the patient's anatomy will be displayed on the same size of display monitor, the result is *magnification* of the image.

The magnified mode can be selected at the push of a button for image intensifier tubes with dual mode or trimode fields. For example, a trimode 9-inch (23 cm) input diameter tube usually has 7-inch (18 cm) and 5-inch (12 cm) magnification modes. Larger input phosphors have become popular, the most common being a dual field 25 cm/17 cm (10 inch/7 inch) combination.

When the size of the fluoroscopic image is unimportant, the magnified mode can be used for better spatial resolution: The number of pixels (picture elements) available in the final output image (the TV monitor) is the same, yet the area of anatomy being inputted to the image intensifier is smaller, so there are more pixels per area of anatomy. Just as with digital imaging, more pixels means smaller pixels per anatomical area. Thus, smaller, finer details can be made out.

However, for the same reason, the magnified image will be much dimmer. That is, due to the smaller field of view being inputted, there is less input light per output pixel. *Minification gain* is reduced. This loss of light intensity means that the *signal-to-noise ratio* (*SNR*) is lessened and noise will become more visible in the image unless the level of light intensity is restored. To compensate for the dim image, the automatic brightness control (ABC) will increase the mA (tube current).

In turn, the increased mA results in more patient skin dose. Exposure to the patient is increased by the ratio of the *change in area* of the input phosphor used. For example, when switching from a 25 cm field of view to a 17 cm FOV, the increase in patient dose will be

$$25^2 / 17^2 = 625 / 289 = 2.16 \text{ times higher}$$

You can see that these increases in patient dose will be significant—they can be as high as 3 or 4 times the normal exposure.

Figure 38-5

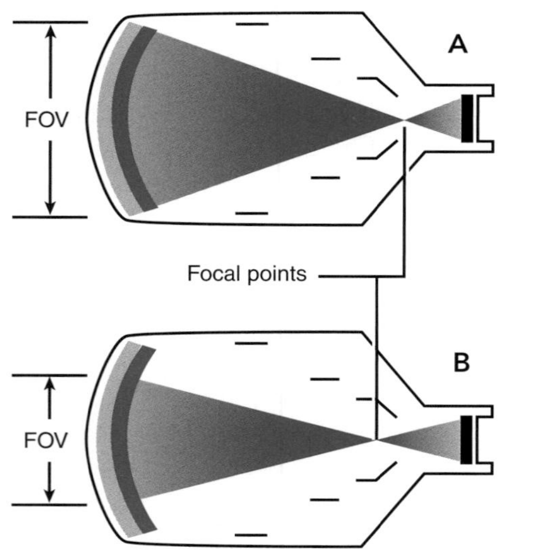

In magnification mode, ***B***, the active field of view at the input phosphor is reduced in size by pulling the focal point of the electron beam closer to the input phosphor.

AUTOMATIC STABILIZATION OF BRIGHTNESS

In image intensification, the brightness of the television monitor must be maintained at an acceptable level throughout the entire procedure. There are two general approaches to compensating for part thickness changes as the patient moves onto his or her side, or as the physical density of the part changes during the procedure.

The first is *automatic gain control* (*AGC*), in which amplification of the *electronic* signal is increased or decreased as needed after the image is acquired. The radiographic exposure factors remain unchanged. The system simply amplifies or reduces the amount of electric current ultimately flowing to the display monitor. Although patient exposure is not increased by this system, it may not provide the best possible fluoroscopic image. *If an insufficient number of x-ray photons are reaching the image intensifier, no amount of increased electronic amplification can overcome the mottle or noise that is produced.*

The second, more common approach is called *automatic brightness stabilization (ABS)*, in which actual *radiographic technique factors are adjusted* to compensate for fluctuations in signal. These fluctuations can be measured in the incident x-ray beam, the electronic signal produced, or the actual brightness of light at different stages in the process of image intensification.

Signal Sensing

The two most common methods of sensing and measuring the adequacy of signal flowing through the fluoroscopic system are to sample it at the photocathode of the I.I. or within the TV camera tube:

1. *Image Intensifier Photocathode Current.* The photocathode of the image intensifier is normally connected to the ground, while the accelerating anode is connected to a source of 25 to 35 kilovolts. The ground connection from the photocathode can be fed to a current amplifier so that the amplifier output is proportional to the incident x-ray input.
2. *Television Camera Signal Sensing.* Most television cameras have automatic gain control (AGC) circuits for controlling the camera tube target voltage or the gain in video amplifiers in order to provide a constant output signal over variations of image brightness. This AGC can be used to control the x-ray generator as well.

Types of ABS Circuits

Brightness stabilizers can be classified in terms of the variable controlled by the brightness sensor. There are four types of brightness stabilization circuits.

1. *Variable mA, preset kVp.* This system allows the operator to set the kVp value and the brightness sensor will control the x-ray tube current over a range up to 20 times multiplication. The operator can set this system to the required kVp for a particular examination and the brightness sensor will automatically adjust the mA to yield an image of sufficient brightness.
2. *Variable mA with kVp following.* This system will vary the mA as a function of the brightness sensor, but it has an additional circuit that senses if an upper or lower boundary of mA has been exceeded. If it has, the circuit then controls the adjustment of kVp through a motor-driven variable transformer. Therefore, if the mA rises above a certain preset value, then the motor will drive the kVp value higher.
3. *Variable kVp with selected mA.* With this system, the brightness sensor controls the kVp. The operator will have previously selected the value of mA required. If a motor-driven variable transformer is used to select kVp, the system has the additional advantage of remembering the last operating point as the operator energizes the system with the exposure switch. Therefore, restabilization of the system between scenes is very rapid. The operator can select a low mA that will force the brightness stabilizer to operate at a higher kVp for gastrointestinal examinations, or a high mA which will force the kVP of the system downward for best contrast when viewing iodine-based contrast media.
4. *Variable kVp, variable mA.* With this system, the output of the brightness sensor controls both the kVp and the mA in order to maintain either constant image contrast or constant suppression of image noise. Unfortunately, such systems make it difficult for the operator to select the mode of operation best suited for a particular examination.

A properly designed ABS system must accomplish the following objectives:

1. It must hold the image brightness constant for variations of patient thickness and attenuation.
2. It must ignore information at the image margins.
3. It must operate to preserve image contrast and minimize image noise.
4. It must keep the operation within the ratings of the x-ray tube.
5. It must effect a reasonable compromise between patient exposure and image quality.
6. It must keep the patient exposure within the radiation control regulations of 10 R/min, except when the operator selects an override mode of operation.

7. It must respond fast enough to track during an examination but slowly enough to avoid hunting between bright and dark portions of the image.
8. It must compensate for system variables, such as the magnification of the image, the intensifier, and the use of disc recorders.
9. It must be capable of being disabled or "held" at a particular equilibrium value prior to injection of contrast media.
10. It must be capable of being shut off to permit the manual control of factors.
11. It must display the operating factors and modes of operation to the operator.

FLUOROSCOPIC TECHNIQUE

Operation of the fluoroscopic image intensifier at higher kVp values will result in increased transmission of the x-ray beam through the patient so that less radiation will be required. *The brightness of the fluoroscopic image varies directly to the mA and roughly to the fifth power of the kVp.* For example, if the kVp were increased form 80 to 88 kVp, a 10 percent change, there would be more than a 50 percent increase in image brightness. Generally, high kVp and low mA are preferred except when the anatomical structures have very low inherent subject contrast. Table 38-1 lists the recommended kVp ranges for digital fluoroscopic imaging and spot-filming of various procedures.

There are some manufactured ABC units that need to have the kVp values set for the specific fluoroscopic procedure. Some will only automatically adjust the fluoroscopic kVp up or down plus or minus 10 kVp. It is possible for the fluoroscopic kVp to be too high or too low such that the image on the monitor may be too bright or too dim for the anatomical part. *Do not adjust the television brightness or contrast control on the monitor upward.* Increasing the brightness or contrast electronically also increases image noise. It may be possible to disengage the ABC system, manually adjust the kVp up or down 10 percent, and then turn the ABC back on, such that it makes automatic adjustments up to plus or minus 10 percent *around* this reset value.

Table 38-1

Recommended kVp Ranges for Digital Fluoroscopy and "Spot-Filming"

Examination	kVp Range
Barium Enemas (solid-column)	110–120
Small Bowel	110–120
Upper GI	110–120
Air Contrast Barium Enema	90–100
Abdomen and Pelvis	80–90
Iodine Procedures in Abdomen	75–85
Myelogram	75–85
Chest and Mediastinum	75–85
Joint Girdles	70–80
Extremities	65–75

From Quinn B. Carroll, *Practical Radiographic Imaging*, 8th Ed. Springfield, IL: Charles C Thomas, Publisher, Ltd., 2007. Reprinted by permission.

FLUOROSCOPIC IMAGE QUALITY

The number of x-ray photons absorbed by the image intensifier determines the statistical quality of a fluoroscopic image. No form of intensification can improve the image above the statistical level of the absorbed photons. The image quality of an image intensification system is defined by its resolution, scintillation, contrast and distortion.

Scintillation

Although the term *scintillation* is often used to generally indicate the emission of light, its more precise meaning is to "twinkle" or give off *varying amounts* of light in a fitful, inconsistent fashion. We have described *quantum noise* as the variation of intensity distribution within an x-ray beam. For a static image such as an "overhead" or "spot" radiograph, the result of quantum noise is quantum *mottle* in the image. For a continuous, dynamic image, however, the result

Figure 38-6

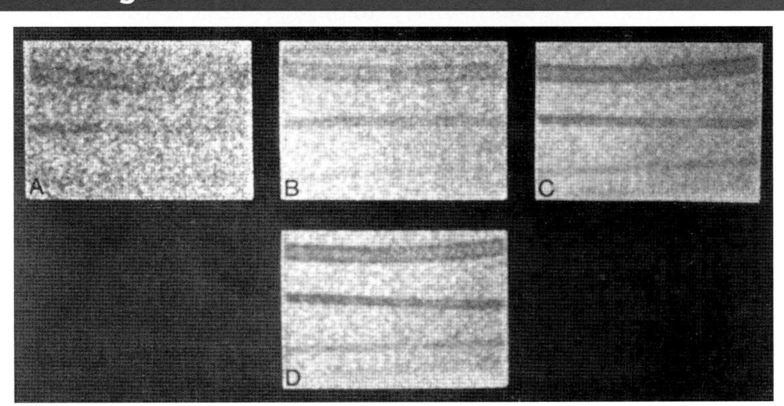

Four CRT monitor images of catheters demonstrate the effects of image noise or scintillation. As the signal intensity of the original x-ray beam was increased from ***A*** to ***D***, signal-to-noise ratio is steadily improved and the image becomes more visible. (Courtesy, Lea & Febiger, *Christensen's Physics of Diagnostic Radiology.*)

is scintillation, or small flaws in the distribution of brightness in the image.

Viewed in the final fluoroscopic display, which is the television monitor, the effect will be that of a random noise pattern superimposed on the fluoroscopic image, Figure 38-6. The noise pattern, since it is random in nature, has the tendency to appear to be moving, giving rise to the colloquial expression of "crawling ants" or "snow." This effect occurs when an insufficient number of x-rays per unit of time are absorbed at the input screen of the I.I. Quantum noise can be improved by high x-ray-to-light conversion efficiency. However, image quality can never be raised above that of the absorbed photons. *Therefore, the ususal method employed to eliminate quantum noise is to raise the x-ray tube current (mA) to generate more x-ray quanta in a given period of time.* Once this threshold is reached, the noise may disappear, the fluoroscopic display takes on a much more pleasing appearance and is generally easier to interpret.

Contrast

One of the areas in which the image intensifier tube does not perform well is in the preservation of image contrast. The contrast in the final image at the output phosphor screen is lowered or degraded by various effects. Any x-ray photon incident on the input screen of the image intensifier, which is *not* absorbed by the input phosphor, may pass through the intensifier tube and, if close to the intensifier tube axis, will strike the output phosphor screen. Since the x-ray photon has the property of exciting the output phosphor elements if it is absorbed, it will cause the output phosphor to emit light, producing an overall masking effect or a kind of inverse fog (brightness) on the output screen itself.

Another effect that would degrade the intensified image would be any light emitted at the output phosphor screen moving backward through the tube axis. This light can strike the *photocathode* at the input side, causing the photocathode to emit additional electrons. These electrons are in a random pattern and not part of the original image, yet they are focused and accelerated back to the output phosphor screen just as those which originated form the primary x-ray beam. This random contribution to the brightness of the output image diminishes image contrast.

Contrast tends to deteriorate as an image intensifier ages. The deterioration rate can be as high as 10 percent per year. Therefore, a periodic check of the image intensifier brightness is critical.

Distortion

Various types of distortion can occur in the fluoroscopically intensified image, including *pincushion distortion*, *veiling glare*, and *vignetting*.

Pincushion Distortion

This is a form of spatial distortion that warps the appearance of the image formed on a curved surface input phosphor to a flat output phosphor screen. Pincushion distortion results in slightly higher magnification of the input image toward the edge of the image. The amount of pincushion distortion is usually determined by an imaging grid or wire-mesh screen with regular spacing. Pincushion distortion is reduced when magnification modes are utilized, since only the central, less curved portion of the input phosphor is energized.

Because of pincushion distortion, the central region of the input phosphor produces better spatial resolution than the periphery. This means that operation in *magnification mode* also yields a sharper overall image. Between the two extremes in available input field of view diameters, 25 cm (10 inches) and 10 cm (4 inches), resolution can be increased from 4 LP/mm to 6 LP/mm.

Another kind of spatial distortion that can occur with the image from an intensifier tube is *s-distortion*, which can be caused by strong external magnetic or electrical fields in close proximity to the image intensifier tube. Both pincushion and the self-descriptive "s" distortion can be checked observationally by simply fluoroscoping a wire mesh.

Veiling Glare

Veiling glare is mainly the consequence of light scatter from the output screen window of the image intensifier. The scattered light, just like scattered radiation, adds to the background signal and degrades the contrast in the fluoroscopic image. The scattering of x-rays up through the image intensifier tube, and of electrons from the beam within the tube, also both contribute to veiling glare.

There is not much that can be done to eliminate the presence of veiling glare, although some improvement has been achieved through improvements in output phosphor design. The degree of veiling glare can be seen and compared between systems by centering a large lead disc in front of the input phosphor and energizing the fluoro unit. Any brightness in the middle area of the disc image, which should be dark, is indicative of veiling glare from scattered light, x-rays and electrons.

Vignetting

The brightness measured at the output phosphor will vary from the center to the periphery of the image, even if a homogeneous x-ray field was incident upon the image intensifier. The brightness will be greatest toward the center of the image and will fall off at the edges.

One source of vignetting is a consequence of pincushion distortion. With pincushion distortion the image is magnified to a greater extent toward the periphery. This means the minifying, electron-concentrating effect of the electronic optics is reduced at the periphery, causing less brightness there.

Vignetting also occurs in the optical coupling between the image intensifier tube and any recording device, because of scattered light effects. Vignetting can be checked for by simple observation of a "flat-field" image without any object or with a homogeneous object in the beam.

Processing the Image from the Intensifier Tube

Digital photospot cameras (Fig. 38-2) are high-resolution, slow-scan television cameras in which the television signal is digitized and stored in computer memory. The contents of the computer memory can be instantly displayed on a television monitor. Digital photospot cameras are usually 1024 scan-line television cameras. The television target is exposed to light emitted from the image intensifier, produced from a short pulse of radiographic x-rays. The electron beam in the camera tube scans the television target screen slowly, often using four conventional frame times for a total of 0.133 seconds (at 33 milliseconds each).

An analog-to-digital converter turns the voltage signal into a series of digital numbers stored in computer memory. Hard-copy images can be produced using multiformat cameras or laser imagers, which are commonly used to produce hard copies from CT and MRI units. As with all other digitized images, the spatial resolution is less than for older film-based systems, but is compensated for with enhanced contrast resolution.

MOBILE IMAGE INTENSIFICATION (C-ARM)

Mobile fluoroscopic image intensification units can be used in the emergency room, intensive care unit, coronary care unit, operating room, or fracture rooms. A "C-arm" unit with a closed circuit television attachment enables a physician to view the fluoroscopic image in "real time." An x-ray tube is mounted at one end of the C frame in alignment with an image intensifier, and the entire assembly can be rotated around a fixed axis in many directions to avoid moving the patient. Fracture reduction, needle biopsy, catheter placement, or hip pinning studies are often performed using the mobile I.I. unit.

Once the initial set-up is complete, it is important for the radiographer to test the unit to be sure of its operation and to orient the image on the screen so it is anatomically correct (upright, and left-to-left). An easy way to orient the image is to use right or left lead markers during the test image. A coin or paper clip taped to the intensifier tube will also work. The radiographer can use the screen rotation buttons and the screen reverse button to properly display the image.

During the fluoroscopic procedure the radiographer may be called upon to manipulate the intensifier to follow the position of a pacemaker wire or catheter. It is important to know how to manipulate the arm and overhead tube locks. There are four basic locks on a C-arm: Transverse, longitudinal, 360-degree circular rotation, and 180-degree AP to lateral rotation. These are illustrated in Figures 38-7 and 38-8. The transverse lock allows the C-arm to be extended and retracted across the table or patient. For example, it could be used to follow a guide wire from the should to the midline of the patient. The longitudinal lock allows the entire C-arm to *swivel* to the right or left. In situations such as a pacemaker or catheter placement, it may be easier for the radiographer to unlock both the transverse and longitudinal locks, giving the intensification tube nearly free movement in any direction whenever it is needed.

Figure 38-7

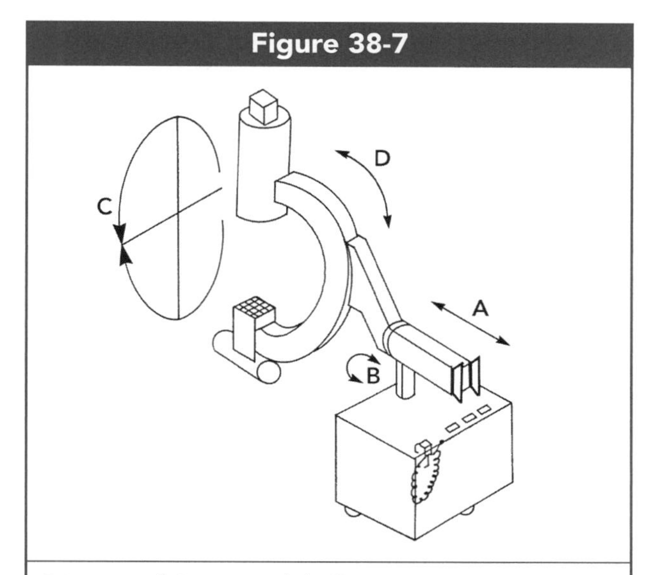

Diagram of C-Arm mobile fluoroscopy unit, showing, ***A***, transverse movement of the entire arm; ***B***, longitudinal swivel; ***C***, 360-degree rotation, and ***D***, 180-degree AP to lateral sliding movement. From Quinn B. Carroll, *Practical Radiographic Imaging*, 8th ed. Springfield, IL: Charles C Thomas, Publisher, Ltd., 2007. Reprinted by permission.

Figure 38-8

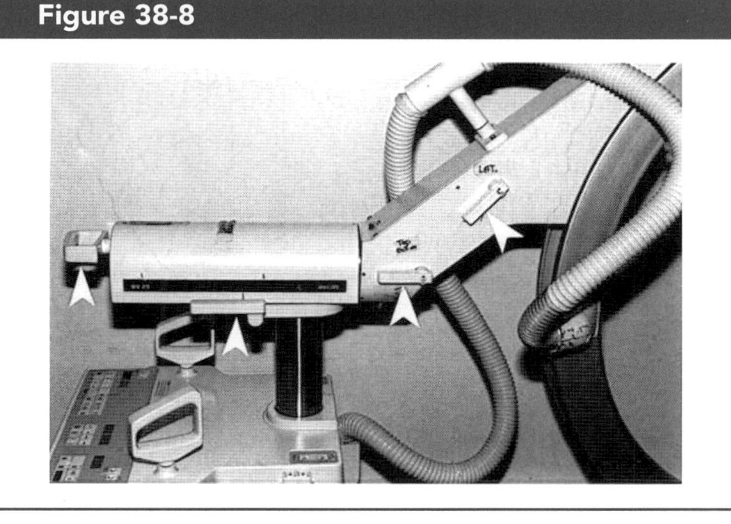

Close-up of typical locks on a C-arm unit. Arrowheads left to right: Transverse lock, longitudinal swivel lock, 360-degree rotation lock, and 180-degree AP to lateral lock. Compare with Figure 36-7. From Quinn B. Carroll, *Practical Radiographic Imaging*, 8th ed. Springfield, IL: Charles C Thomas, Publisher, Ltd., 2007. Reprinted by permission.

The 360-degree lock rotates the C-arm so that the intensifier tube is above or below the patient. *Less scatter radiation exposure is delivered to the head and neck of personnel by placing the image intensifier* above *the patient and with the x-ray tube* under *the table.* (This is more fully discussed in Chapter 44.) The 180-degree lock allows the x-ray tube and I.I. to move between frontal and lateral projections. The C-arm will move either over or under the patient, depending on where the image intensifier is positioned. If the intensifier is over the patient, the lateral movement will be below or under the patient, and vice versa. Care must be used when changing from frontal to lateral positions—the C-arm will have to be raised or lowered while rotating the arm over or under the patient.

It is important to note that in many surgical, intensive care, and emergency room procedures, a special table extension or dedicated intensifier table must be used. The metal frames of surgical tables, beds, and gurneys are not designed for use with the C-arm. A carbon-fiber type table, having a low absorption rate, will significantly reduce the required radiographic technique patient exposure.

MINIMIZING PATIENT AND OPERATOR EXPOSURE

Fluoroscopic image intensification represents a very large portion of the radiation dose delivered in diagnostic imaging because of continuous x-ray production and real time image output. While the exposure techniques are quite modest, such as 1 to 3 mA of tube current for many fluoroscopic studies, an exam usually takes several minutes and, in some difficult cases, can exceed hours of beam-on time. For example, a single-phase generator system delivering 1 mA at 80 kVp for 10 minutes of fluoroscopy beam-on time may produce a cumulative skin entrance exposure of 22 R. This is a substantial dose of radiation. (Exposure levels for comparison are discussed in Chapter 44.)

Fluoroscopic Exposure Time

Operators of fluoroscopic equipment must restrict the x-ray beam-on time to a minimum. The x-ray beam need not be operated continuously. A series of short bursts of exposure can be used for visual checks, called *intermittent fluoroscopy.* Five visual checks, assuming 12 seconds each, approximates one minute of accumulated exposure time. Assuming an exposure rate of 5 R per minute, this translates to approximately 400 mR skin entrance exposure to the patient per visual check. Regulations require that the exposure switch activate a cumulative manual reset timer, which typically emits an audible signal and interrupts the x-ray beam when 5 minutes of exposure time has accumulated. This is designed to make the operator aware of the beam-on time being accumulated. Related issues are further discussed in Chapter 44 on radiation protection practices. Intermittent fluoroscopy is the most important way to limit patient radiation exposure.

Further reduction of such exposure is achieved in several ways. "Last-image hold" devices using digital memory provide a continuous output image that can be examined by the fluoroscopist after a short burst exposure of x-rays. This device can reduce the fluoroscopy beam-on time by 50 to 80 percent in many situations.

Pulsed fluoroscopy can reduce patient dose. Lower frame rates than real time, such as 15 or 7.5 frames per second can be used in conjunction with digital image memory to provide a continuous output video signal with reduced x-ray pulsing and updating of the rates of the displayed information (Fig. 38-9). Synchronized to the frequency of the electrical power supply, which is 60 Hz in the U.S., the fluoroscopic image can be viewed at much lower framing rates than the common television frame rate of 30 frames/second.

The use of an optimal image intensification system with appropriate conversion gain, high contrast ratio, and acceptable spatial resolution and contrast sensitivity also reduce the patient dose. Restriction of the field size with collimation will help keep the dose-area product (DAP) to the patient as low as reasonably achievable (ALARA). Image quality will improve as the size of the x-ray beam is reduced because there is a reduction in the amount of scattered radiation reaching the I.I., limiting the amount of quantum noise. Proper filtration in the x-ray tube and collimator must also be maintained in accordance with regulations.

Figure 38-9

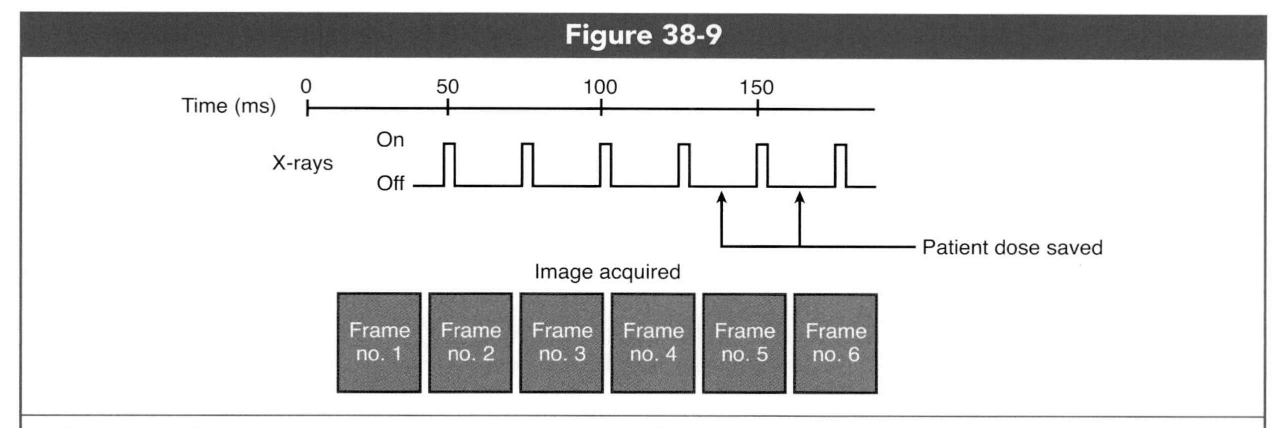

Pulsing of the fluoroscopic x-ray beam spares the patient from continuous radiation between frames, but requires increased mA. Using *shorter* pulse times further reduces patient exposure.

DIGITAL FLUOROSOPY (DF)

Although a digital fluoroscopy (DF) unit looks much like a conventional fluoroscopy unit, and procedures are conducted in much the same manner, a computer has been added to the equipment, and often there are two viewing monitors provided (Fig. 38-10). The right monitor is used to display subtracted images or to hold images other than the most recent one taken.

Technological advances have introduced *pulsed-progressive mode* fluoroscopy units which operate the under-table x-ray tube at an mA measured in the *hundreds* (like the overhead tube), but use a series of extremely short bursts of radiation with cooling periods in between them, rather than continuous exposure production. In continuous mode fluoroscopy, the radiation *between* frames is essentially wasted, exposing the patient while making no real contribution to the moving image. In the pulsed mode, a fluoroscope actually takes 30 "frame" exposures per

Figure 38-10

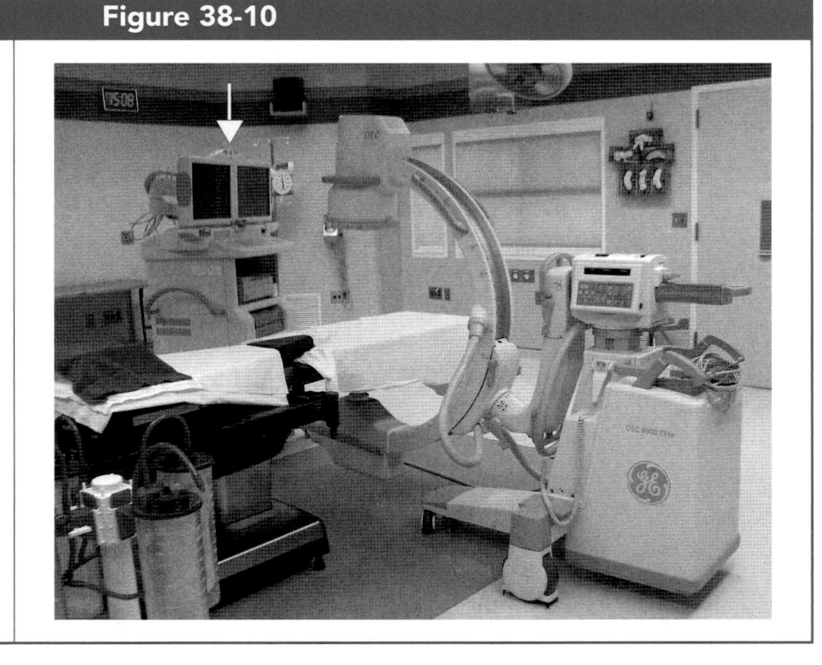

C-Arm unit set-up in surgery with the image intensifier properly positioned above the table, and two monitors provided (arrow) so that one image can be kept on "hold." (Courtesy, Trevor Morris, R.T.)

second and sends their images to the TV monitor screen to create the appearance of fluid motion in the dynamic image. The mA must be increased from *2 to 20 or more mA* in order to obtain adequate signal for each individual frame. But, the continuous radiation that would have been exposing the patient between frames is saved (and so is the x-ray tube, which would quickly overheat from continuous x-ray production at these mA levels).

The principle advantage of digital fluoroscopy over conventional fluoroscopy is that the acquisition of individual image *frames*, made possible by the pulsed mode, allows *image subtraction techniques* to be used which can be particularly advantageous for radiographic studies utilizing the intravenous injection of *contrast agents*.

The exposure time for each pulse, called the *pulse width*, can be adjusted on these units from 3 to 10 milliseconds. By selecting a 3 msec pulse width, rather than the common 6 msec pulse width, and keeping the mA at 20, *patient exposure can* be *reduced to one-half*. Newer fluoroscopy units combine the pulsed mode with high mA and spectral filters made of copper or a copper/aluminum alloy to achieve increased image quality while imposing some limitation on patient exposure. In spite of the potential to reduce patient dose by pulsing the x-ray beam, these units can be operated at mA stations as high as 200 or 300, and some features offered by manufacturers require these high levels of mA. The net result is that it is possible for patient exposure to become very excessive, which is further discussed in Chapter 44.

Dynamic Flat-Panel Detectors

Image intensifier tubes are being replaced by flat-panel digital detectors that are capable of recording dynamic (motion) images. Two types of *dynamic* FPDs are available—an indirect detector using a cesium iodide phosphor coupled to an active matrix array of amorphous silicon TFTs, and a direct detector using an AMA of amorphous selenium TFTs. The mechanics of how these systems work was fully described in Chapter 34 (see page 594), but *dynamic* FPDs have several differences from those used for static digital imaging.

The dynamic FPD typically has larger dimensions than an FPD used exclusively for static imaging, and larger matrix sizes up to 2048 × 2048 pixels. Pixel sizes of 200–300 microns are used digital fluoroscopy detectors, larger than those for digital radiography detectors (100–150 microns). *Dual-use* digital systems which allow both radiography and fluoroscopy to be performed, are capable of *binning* groups of four small pixels together to form an *effective pixel* with dimensions of 200 × 200 or 300 × 300 microns for the fluoroscopic mode of operation.

Dynamic flat-panel detectors can operate in either continuous or pulsed x-ray modes. Readout electronics must be able to handle high frame rates and fast data transfer rates.

Most DFPDs are also capable of a *zoom* feature, analogous to the "magnification mode" for an image intensifier.

In a dynamic FPD, a light-emitting diode (LED) array is located below the detector which has a function similar to the "erase" cycle of a CR processor—it produces a bright microsecond flash of light to erase images after each frame is taken in order to eliminate any ghost images. These ghost images are of a different nature than what we normally think of as "screen lag." They comprise a change in the *sensitivity* of the detector after exposure, which would *not* be observed as continuous glowing after a pulse exposure, but rather show up in the image only upon further exposure. The LED array "refreshes" the detector panel between each frame to prevent this phenomenon.

Radiographic grids for flat-panel detectors often are constructed with their lead lines oriented diagonally; this precludes aliasing artifacts from being caused. Most grids can be removed for imaging children or small body parts.

Digital flat-panel detectors have a much higher signal-to-noise ratio than conventional image intensifiers. One postprocessing technique called *temporal frame averaging* is able to reduce noise by as much as 44 percent. This is done by averaging, for each pixel value, one or two frames prior to, and one or two frames after, a particular frame with its own data. As with static imaging, the capability of applying postprocessing algorithms to dynamic images presents great advantages, including edge enhancement and last image hold.

Other advantages of dynamic flat-panel detectors over conventional fluoroscopy include their contrast enhancement of low subject-contrast anatomical structures, high DQE and dynamic range across all levels of exposure, freedom from various forms of distortion, and their rectangular shape which corresponds with a rectangular display screen.

Digital Subtraction Techniques

For any cardiovascular procedure involving the injection of an iodinated contrast agent, a drastic improvement in the visibility of the anatomy of interest is made possible through digital *subtraction* techniques. There are two very different approaches to subtraction, called *temporal subtraction* and *energy subtraction*. Temporal subtraction tends to be the more common of the two, because it is both more economical and technologically easier to achieve for the equipment.

Temporal Subtraction

Temporal subtraction refers to using the difference in *time* to distinguish between the exposures that will be subtracted from each other. In the most common form of temporal subtraction, an exposure is made before the injected contrast agent reaches the anatomy of interest. Automatic injectors can be programmed to function in tandem with the x-ray generator, such that the injection begins at a precise number of seconds before or after exposures are made.

From this image with no contrast agent present, a *mask* image is produced by the computer which is a reversed, *positive* image not unlike the "black bone" feature illustrated in Chapter 32 on digital processing. It is a simple matter for the computer to take all of the pixel values measured and invert them proportionally to their magnitude, such that blacks become whites and dark grays become light grays. When this mask image is superimposed over an original image, the result would be that all pixel values cancel each other out, leaving a featureless medium gray across the field—except that in this case, a contrast agent has been introduced into *only one of the two images*. Therefore, there is no canceling out of the contrast-agent portion of the image, and this anatomy appears without obstructing images of bony structures overlying it.

In a series of cardiovascular images, the mask image is subtracted from *each* "contrast" image. Modern digital equipment can complete this process so quickly that the subtracted images come up on the second viewing monitor within seconds after the exposures are made. Further, video noise may be reduced by having the computer combine four or five video frames to form each image in a process called *image integration*.

Although some authors construe the subtraction process as a type of image contrast enhancement, it is really a process of *desuperimposition* in which overlying anatomy that obstructs the visibility of the anatomy of interest is actually *removed*, which better fits the category of *noise reduction*.

Misregistration artifacts occur when the patient moves slightly between the mask image exposure and subsequent contrast exposures (Fig. 38-11). The

Figure 38-11

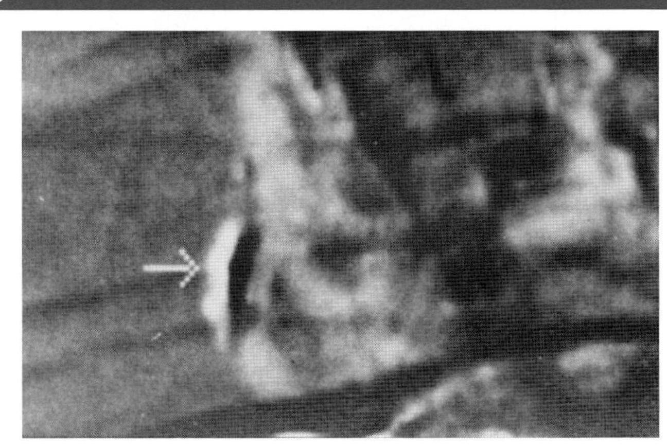

Misregistration artifact on a digital subtraction image of a blood clot (arrow) caused by lateral motion. This artifact can be corrected digitally by *pixel-shifting* the two superimposed images. From Quinn B. Carroll, *Practical Radiographic Imaging*, 8th ed. Springfield, IL: Charles C Thomas, Publisher, Ltd., 2007. Reprinted by permission.

computer can be directed to *reregister* these images by *pixel-shifting* the mask to the right or left, up or down, until the artifacts are minimized. *Remasking* is another option in which a different image in the series is selected to produce the positive mask.

For cardiac monitoring studies, motion becomes a persistent problem. A *time-interval difference (TID) mode* can be used which always subtracts the image preceding the current one by only three or four exposures, and repeats the process throughout the series. Image contrast is sacrificed in the process, but motion artifacts are reduced.

Energy Subtraction

Dual energy subtraction was also described in Chapter 31 as a digital processing feature. If two images taken with very different contrast levels can be provided to the computer, the computer is able to compare how much the pixel values (image density) *changed* between the two images for two different tissues. In this way, the computer can *identify* the tissues, and consequently subtract one from the image while leaving the other intact.

The use of iodine as a contrast agent facilitates this process greatly because of the *k-edge* in the x-ray absorption for iodine (Fig. 38-12). If an exposure is taken with the average kV of the x-ray beam just below the k-edge of iodine, and another with the average kV just above it, we see from the graph in Figure 36-12 that the *loss of photoelectric interactions* will be only slight for the iodine, yet very pronounced for body tissues. This means that the change in pixel value (image density) for the iodine will only be slight between the two images, but for the body tissues it will be large. This provides the computer a way to identify the iodine and tissue. Those pixels that recorded body tissues are all assigned a medium "background density" value by the computer, leaving the iodine image in relief against this background.

Energy subtraction requires rapid switching by the equipment of beam energy between exposures. This can be done by changing the kVp or by spinning different *filters* into place on a flywheel. The contrast quality of these images is not as good as for temporal subtraction, but high-quality images can be obtained by combining energy and temporal subtraction into *hybrid subtraction*.

IMAGE RECORDING DEVICES: CCDs AND CMOSs

In the modern fluoroscopic imaging chain, before the light image from the image intensifier tube is

Figure 38-12

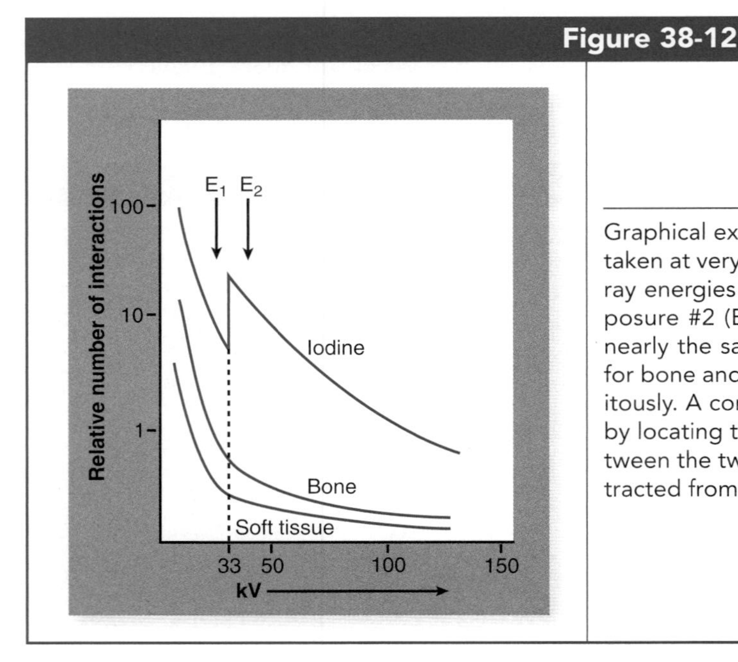

Graphical example of energy subtraction: Two exposures are taken at very different kVp levels, resulting here in *average* x-ray energies of 25 kV for exposure #1 (E_1) and 40 kV for exposure #2 (E_2). Occurrence of photoelectric interactions is nearly the same for iodine because of its *k-edge effect*, but for bone and soft tissue photoelectrics have dropped precipitously. A computer can thus identify the iodine in the image by locating those pixels whose value has not plummeted between the two exposures. The remaining pixels are then subtracted from the image processed in any way (sent to the viewing monitor as "spot views," recorded by a digital video recorder, saved onto a DVD), it is always transmitted through a bundle of optic fibers directly to a CCD or CMOS camera (Figure 38-13).

In the 1980s, miniaturization of all kinds of imaging devices, including the home *camcorder*, commercial TV cameras, telescopes and surveillance instruments was made possible by the development of the *charge-coupled device (CCD)*, which replaced the bulkier and more delicate TV camera tube. The CCD is a small, flat plate only about ½-inch in length and breadth for a home camcorder (Figure 38-14). The type used in fluoroscopy are about 1 inch in size to place atop an image intensifier tube, coupled to the 1-inch output phosphor by a short bundle of optic fibers. Its sensitive surface is made of crystalline silicon, a semiconductor.

As shown in Figure 38-15, when light photons enter the silicon layer, ionization of the molecules separates electrons from them, leaving positively-charged *electron holes* behind. A layer of microscopic electrodes beneath the silicon is given a negative charge, and a dielectric layer above acts as a *ground* for freed electrons.

The electrons drift toward the dielectric layer, and the positively-charged *holes* drift toward the electrodes. (This drift actually consists of a series of electrons moving up to fill each vacancy in the molecule above, in sequence, such that the vacancy itself, the "hole," appears to drift the opposite direction.) When a hole reaches an electrode, it pulls an electron from it, creating a positive charge on the electrode which can be measured by the circuit.

In Figure 38-15, we see that each electrode is connected to a storage capacitor and a thin-film transistor (TFT) gate (see *Direct Conversion Systems* in Chapter 34). Rows and columns of these TFTs form an active matrix array (AMA). The electronic signal, as it drains from each row, is boosted by an amplifier.

The CCD has high sensitivity or *detective quantum efficiency (DQE)*, so that less radiographic technique is needed and patient exposure can be kept relatively low. The dynamic range of a CCD is about 3000:1, ideal for fluoroscopy. The CCD produces high contrast and signal-to-noise ratio (SNR). It has almost no image lag, low blooming, and high resolution. Its resolution is determined by its pixel count and physical dimensions. For example, the CCD typically coupled to an image intensifier is about 1 inch in size and has a matrix of 2048 × 2048 pixels, making each pixel only 14 microns in size.

Using *progressive scanning*, in which each line of the raster pattern is read sequentially, the CCD can acquire images at 60 frames per second (compared to 30 fps for the older TV camera tubes using *interlaced scanning*). This makes the CCD highly suited to digital fluoroscopy.

Figure 38-13

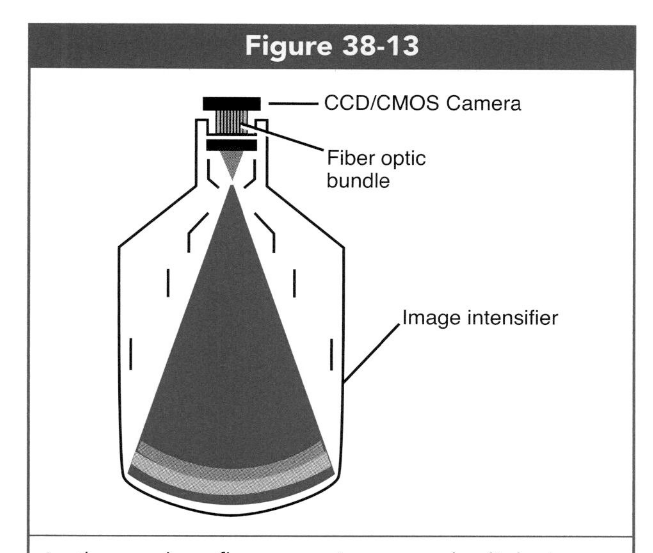

In the modern fluoroscopic tower, the light image from the image intensifier is transmitted directly to a CCD or CMOS camera atop the I.I. through a bundle of optic fibers.

Figure 38-14

A charge-coupled device (CCD). (Courtesy, Apogee Instruments, Inc.)

Figure 38-15

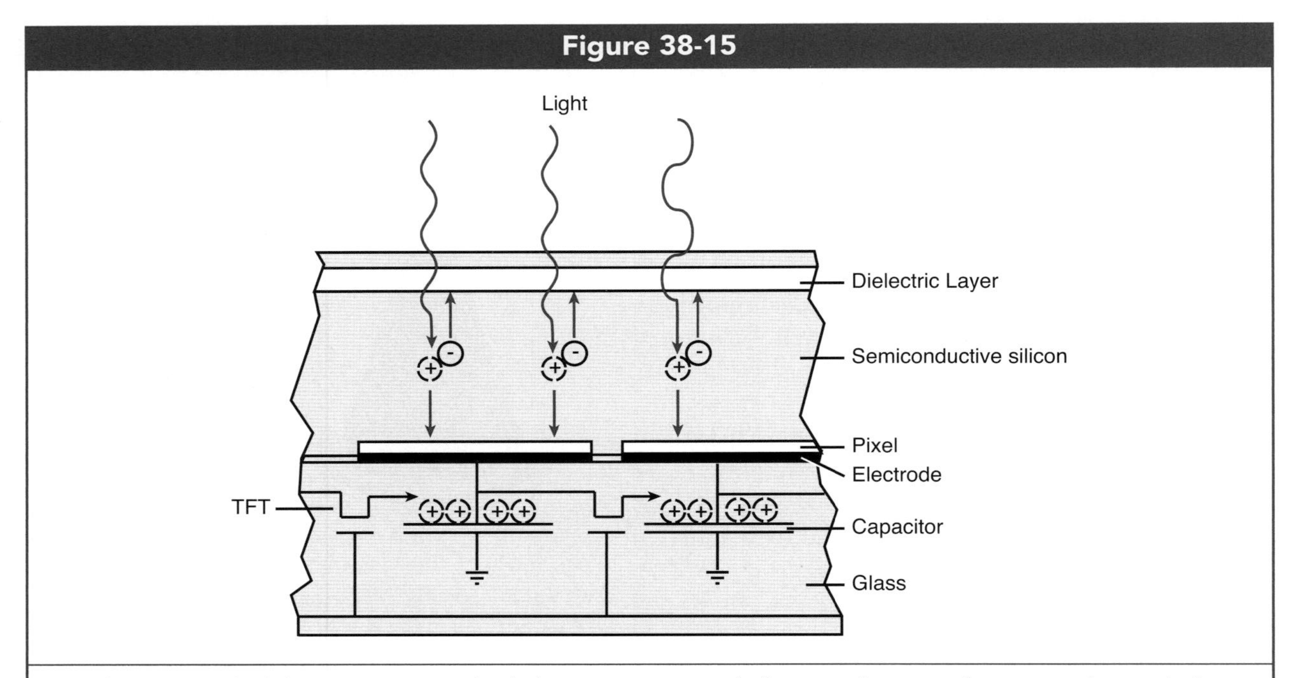

In a charge-coupled device, or CCD, the light image is recorded as a collection of positive electrical charge formed by "holes" drifting down to the electrodes when ionizations occur in silicon atoms. These positive charges are stored on capacitors. (Electrons freed from these ionizations drift upward to the dielectric layer.)

Complimentary metal-oxide semiconductors or *CMOS's* were pioneered in the early 1970s at about the same time CCDs were being developed. The name refers to CMOS's use of two metal oxide semiconductor field effect transistors, or *MOSFETs*, one p-type (positive) transistor and one n-type (negative) transistor, stacked together to form an electronic logic gate. This quickly became the dominant method for manufacturing integrated circuits for computers. However, for cameras, CMOS circuits at first produced far inferior images to the CCD. In the 1990s lithography (printing microscopic integrated circuits) developed to the point where CMOS imaging devices made a comeback, and they are currently in stiff competition with CCDs for imaging applications. The CMOS image sensor looks very much like the CCD in Figure 38-14.

The initial light image capture for the CCD and CMOS is identical—both use the photoelectric effect, where light or x-ray photons liberate electrons in a silicon layer as shown in Figure 38-15. Both then collect these electrons as an electrical charge in a layer of transistors below. The essential difference lies in the way that these charges are "read" off of the chip.

In a CCD, every del's charge is sent across the chip as an analog signal that is read at one corner of the array, where the electrical fluctuations are separated into individual readings that will represent pixels, and then digitized. In a CMOS sensor, each individual detector element (del) has its own transistor, amplifier, noise-correction circuit and digitization circuit so that it outputs digital data that moves down traditional wires to exit the sensor device.

In the CCD, because nearly all of the del can be devoted to light capture, signal uniformity is very high. In the CMOS, all of the added electronics reduce the sensitive surface area available for light capture. Some light sensitivity and signal uniformity is lost, and a higher percentage of noise results, but improvements made during the 1990s brought these qualities up to almost equal those of the CCD. The CMOS has much higher speed than the CCD, it consumes about *one-hundredth* the power, and it can be manufactured with traditional technology that makes it extremely inexpensive compared to the CCD. The current trade-off then is between somewhat higher image quality for the CCD and several economic advantages to the CMOS.

SUMMARY

1. Early fluoroscopy posed a serious radiation hazard to both operator and patient. The invention of the electronic image intensifier in the late 1940s drastically reduced this hazard. Nonetheless, by the accumulation of excessive beam-on times, modern fluoroscopy can still result in excessive radiation to the patient, and must be carried out with great discretion.
2. The modern image intensifier uses the processes of *minification* and *flux gain* in order to amplify the brightness of the fluoroscopic image from 5000 to 20,000 times. This results in *conversion factors*, defined as the ratio of luminance to x-ray exposure, from 50 to 200.
3. The magnification modes available on multifield intensifiers provide improved image resolution, but the loss in minification gain requires an increase in radiographic technique (to enhance the flux gain), which increases radiation exposure to the patient.
4. Unlike automatic gain controls, automatic brightness stabilization (ABS) systems maintain the brightness of the fluoroscopic image without amplifying image noise. After measuring the signal strength either at the photocathode of the I.I. or within the TV camera, the ABS can adjust kVp, mA or both in various formats.
5. The quality of the fluoroscopic image is a function of its resolution and contrast, and is adversely affected by scintillation and distortion. Distortion effects include pincushion distortion, vignetting and veiling glare.
6. The modern C-arm fluoroscopy unit allows flexible movement of the I.I. and x-ray tube around the surgery patient.
7. Intermittent energizing of the fluoroscope, the last-image hold feature, and pulsed mode can be used to keep patient exposure at a minimum.
8. Digital fluoroscopy units enable the use of image subtraction techniques, but can pose a substantial radiation risk to the patient and must be used with discretion. Subtraction can be accomplished by the *temporal* method which uses a precontrast-agent image as a mask, or by the *dual energy* method which takes advantage of the k-edge of iodine.
9. Dynamic flat-panel detector systems present many advantages over conventional image intensifiers, including ease of use, better image qualities, and postprocessing capability.
10. Charge-coupled devices (CCDs) provide a flat-shaped and compact light-sensing device that can replace the television camera tube atop an image intensifier. They have better DQE, SNR, contrast and resolution. Complementary metal-oxide semiconductors (CMOSs) consume one-hundredth the power of CCDs and are much cheaper, but they often produce slightly lower image quality.

REVIEW QUESTIONS

1. List the four basic components of the image intensifier tube:

2. What is the effect of increasing the physical diameter of the input phosphor on image resolution?

3. Minification of the electron beam is accomplished by what inner component of the I.I.?

4. What is the effect of using the magnification mode upon (a) image resolution, and (b) patient exposure?

5. What is the approximate conversion factor for an I.I. with a total brightness gain of 15,000?

6. What is the unit of luminance used in calculating the conversion factor?

7. The brightness of a fluroscopic image varies according to the __________ power of the kVp.

8. Contrast deterioration for an I.I. can proceed at __________ percent per year.

9. Which format for automatic brightness stabilization (ABS) has the advantage of quick restabilization between scenes?

10. The best way to reduce noise in the fluoroscopic image is to increase the _______.

11. Quantum noise which appears to be moving across the fluoroscopic screen is called __________.

(Continued)

REVIEW QUESTIONS *(Continued)*

12. Scattered light from the I.I. output screen due to random x-rays and electrons reaching the phosphor crystals is called __________.

13. Pincushion distortion is worst in what portion of the image?

14. In surgery, a C-arm fluoroscope should always be positioned with the I.I. __________ the patient.

15. The digital fluoroscope can reduce video noise by combining four or five sequential video frames. This is called:

16. Whereas conventional fluoroscopes operate at 1-3 mA, pulsed progressive mode digital fluoroscopes require __________ or more mA.

17. What is placed under a dynamic flat-panel digital detector in order to "erase" ghost images from each frame?

18. What is the term that describes the combining of four or more pixels to form a single larger "effective pixel"?

19. Like a direct conversion DR detector plate, a CCD is made up of an __________ of hardware pixels, each with its own capacitor and TFT.

20. In a CCD, what drifts toward the bottom layer of electrodes as ionizations occur in the crystalline silicon?

21. The main advantages of CMOS cameras over CCD cameras is in their (*economics* or *image quality*)?

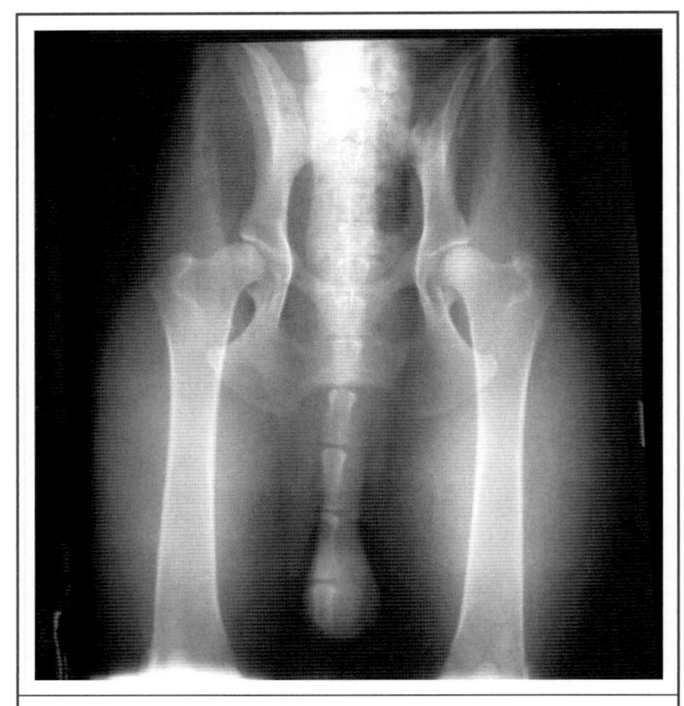

Radiograph of a goat's pelvis.

Chapter 39

QUALITY CONTROL

Objectives:

Upon completion of this chapter, you should be able to:

1. Define quality assurance and quality control as they relate to radiography, and list the components of a comprehensive quality control program.
2. List the methods and parameters for testing the exposure timer, mA linearity, exposure reproducibility, half-value layer, kVp calibration, collimator and distance controls, focal spot size and condition for radiographic units.
3. List the methods and parameters for testing the AEC, tomography units, and fluoroscopy units.
4. List the methods and parameters unique to testing digital image acquisition systems.
5. Describe the methods and parameters for testing electronic image display systems, including luminance, illuminance, contrast, ambient lighting and reflectance, noise and resolution.
6. Explain the function of photometers and densitometers.
7. Explain the nature, prevention and correction of dead and stuck pixels for LCD monitors.
8. Describe viewing angle dependence for LCDs.
9. Describe tests unique to LCDs and for a viewbox illuminator.
10. Describe the relevance and issues of repeat analysis in the age of digital imaging.

Quality assurance (QA) is a managerial philosophy that encompasses all aspects of patient care, image production and image interpretation. Elements of a QA program include a comprehensive written radiation protection program which addresses protection for both patients and personnel, radiation exposure monitoring for personnel, the provision of accurate and timely technique charts, surveys of satisfaction for patients and physicians, equipment calibration and monitoring, analysis of repeated exposures, and the encouragement and provision of continuing education.

Many of these elements are discussed in other chapters; this chapter will focus upon *quality control* (QC) which is generally understood to refer specifically to the calibration and monitoring of equipment, along with some discussion of the continuing value of repeat analysis.

Equipment calibration and QC have become increasingly the domain of specialized radiation physicists, more so since the advent of digital imaging, because many conventional tests required radiographic film which is no longer easy to access, and more sophisticated tests require specialized electronic instruments. Quality control technologists can be trained in these procedures and instrumentation. Although the average staff radiographer may never perform these tests, it is important that all radiographers have a basic understanding and appreciation for the required types of tests and why some acceptance parameters can be more stringent than others.

The components of a quality control program for radiographic equipment include (1) acceptance testing of the equipment at the time of purchase, (2) the establishment of baseline performance parameters for a particular unit during the first couple of months of use, (3) diagnosis of correctable deviations in performance [from the baseline parameters], and (4) documentation of actual corrections made.

RADIOGRAPHIC EQUIPMENT TESTING

Guidelines for testing and parameters for acceptable performance have been developed for all types of medical imaging equipment, digital processing equipment, and image display equipment by the American Association of Physicists in Medicine (AAPM) and related organizations. Key tests and ranges of acceptable accuracy for radiographic units are summarized in Table 39-1.

Radiographic Units

Exposure Timer

There are several ways to check the accuracy of the exposure timer. Most commonly, the radiation physicist uses an electronic ion chamber or photodiode device. The timer should be checked whenever a major repair or part replacement has been made to the generator or console. The ease and sensitivity with which the exposure timer can be adjusted allows a tight range of passing acceptance of ± 5 percent. (Exposure times less than 10 milliseconds are more difficult to set with precision, and are allowed a wider range.)

mA Linearity

Linearity is defined as the alignment of a particular mA station *relative to other stations* in its output of radiation. All other factors equal, changes in mA should result in directly proportional changes in exposure.

Milliamperage measures the rate of electrical current flowing across the x-ray tube. Various factors, including statistical variations in the generator, fluctuating demands on electricity throughout the hospital, and wear and aging of x-ray equipment can cause inconsistencies in the actual mA output over time. Collectively, these factors are more difficult to control than those affecting the exposure timer, so the acceptable range of accuracy for mA stations is set at 10 percent relative to each adjacent mA station, (the one below, and the one above the mA station being checked).

Note that if each of a series of increasing mA stations was 9 percent "hotter" than it should be relative

Table 39-1

Quality Control Tests and Parameters for Radiographic Units

Measurement	Frequency of Performance	Acceptable Range
Exposure Timer Accuracy	Annually	± 5%
mA Linearity	Annually	± 10% adjacent stations
Exposure Reproducibility	Annually	± 5%
Kilovoltage Peak Calibration	Annually	± 5 kVp
Half-Value Layer / Filtration	Annually	Exceed required minimum
Collimation	Semi-annually	± 2% of the SID
Vertical Beam Alignment	Semi-annually	± 2 degrees of vertical
Focal Spot Size / Condition	Annually	± 50% of nominal size
Image Receptor System	Semi-annually	Zero visible defects
Automatic Exposure Control	Annually	± 10%

to the previous station, the third station up could be as much as 27 percent off from the starting point and still fall within the limitations of accuracy. This is an indication of how difficult it is to precisely control this factor—*mA is by far the least reliable of the three electronic technique variables* used in radiography.

The most accurate way to check mA linearity is to use an ion chamber and calculate the mR/mAs obtained at each mA station, while holding the exposure time constant. Each time the mA is doubled, the mR/mAs should double within ± 10 percent. (Linearity may also be checked by setting the same mAs in a series of increasing mA stations combined with proportionately reduced exposure times, but this method is dependent upon the accuracy of the timer, which must be tested first and taken into account.)

To find the percentage by which any measurement is out of calibration from another, or from the set standard, subtract the difference between the two, divide this by the lesser of the two, and multiply by 100. For example, let us assume that for a fixed exposure time of 1 second, at the 100 mA station an ion chamber gives a reading of 640 mR, and at the 200 mA station it reads 1120 mR. Dividing the mAs values into the mR measurements, we obtain 6.4 mR/mAs for the 100 mA station, and 5.6 mR/mAs for the 200 mA station. The math for converting this difference into a percentage is as follows:

$$\frac{6.4 \text{ mR/mAs} - 5.6 \text{ mR/mAs}}{5.6 \text{ mR/mAs}} = \frac{0.8}{5.6} = 0.143$$

$$0.143 \times 100 = 14.3\%$$

The 200 mA station is 14% off calibration, outside the 10% limit.

A serviceman should be called to recalibrate this x-ray machine.

Exposure Reproducibility

Reproducibility is defined as the ability to repeat the same overall technique settings and obtain the same results in exposure. To test for reproducibility, we make ten exposures with identical settings for exposure time, mA and kVp. The average exposure is found by simply summing all the exposures and dividing by ten. Then, each individual reading is compared to the average, using the same formula just presented for mA linearity to find the percentage of deviation. Such exposures should be highly reproducible, within a margin of error of ± 5 percent.

Half-Value Layer

The most important quality of the x-ray beam is its ability to penetrate through the human body to expose the image receptor plate with sufficient signal to produce an adequate image. Protective filtration is used to remove low-energy x-rays from the beam because they cannot penetrate through the patient. Optimum kVp levels are necessary to assure adequate penetration through the anatomy. Beam penetration may be raised in two distinct ways: (1) by increasing the *minimum* energies present in the beam through the use of filtration, and (2) by increasing the *maximum* energies present in the beam through the use of higher peak kilovoltage (kVp). Either of these changes results in an increase in the *average* energy of the beam and in higher penetration capability. These concepts are fully discussed in Chapters 16 and 17.

The only true measure of actual x-ray beam penetration is the *half-value layer*, abbreviated *HVL*. Half-value layer is defined as that thickness of absorber material (usually aluminum) needed to reduce the intensity of the x-ray beam to one-half the original.

The most accurate approach is to expose an ion chamber to a set mA, exposure time, and kVp combination that will produce a substantial reading. (A practice lab is available in the instructor's manual disc.) Set to the mode labeled *dose* or *exposure* (*not* dose rate *or* exposure rate), an ion chamber will measure actual x-ray intensity in units of total mR or R that accumulated during the exposure. The ion chamber will hold this reading on the meter for a short time after the exposure has terminated—it should be read as soon as possible after exposure for accuracy, as the stored electrical charge will begin to "bleed off" after several seconds.

To produce a substantial and accurate reading, the technique should include a relatively long exposure time (e.g., 1 second), and the kVp should be set at a multiple of 10. The initial exposure is taken with no absorber placed over the ion chamber; this reading is logged and designated as a beginning level of 100 percent. One-half this reading is noted and designated as the 50 percent target level. Thin (1 mm, 0.5 mm, and 0.1 mm) sheets of aluminum are then added in increasing amounts (Fig. 39-1), the exposures repeated after resetting the chamber, and the measurements logged each time, until at least one measurement is obtained *below* the 50 percent level.

For accuracy, all of the measurements should be carefully plotted on graph paper and a *best-fit* line drawn in according to the instructions on graphs in Chapter 3. The accuracy of the measurement is dependent upon how carefully measurements are plotted and how well this best-fit curve is drawn in. Several tries are often needed, so a pencil is recommended.

When the graph is satisfactorily completed, the HVL may be read from it by extending an exactly horizontal line from the 50 percent mark on the left, over to intersect the plotted curve, then extending a perfectly vertical line from this point down to the axis of the graph. This amount is the HVL in thickness of aluminum.

Several other metals can be used to measure HVLs, including copper, tin, or even lead, depending on the voltage ranges of the radiation being measured, but aluminum is most commonly used in diagnostic radiology. When stating the HVL of a radiographic unit, the material should always be included. Also, each HVL is specific to the kVp used and the power of the generator. For example, one might state that, "the HVL for a single-phase unit operated at 80 kVp was 2.46 mm of aluminum."

The HVL is a direct indication of the penetration power of the x-ray beam—the higher the HVL, the more penetrating the beam. Minimum required HVLs are provided in tables published by governmental regulatory agencies and by scientific advisory groups such as the National Committee on Radiation Protection and Measurement (NCRP) (Fig. 39-2). For example, for a three-phase x-ray unit with the minimum 2.5 mm of aluminum filtration, the minimum HVL for operation at 80 kVp is 2.34 mm of aluminum. For a single-phase machine, it is less. Here are a few more examples of the requirements, given in millimeters of aluminum:

kVp	1 Φ HVL	3 Φ HVL
70	1.6 mm	2.0 mm
90	2.6 mm	3.1 mm
110	3.1 mm	3.6 mm

When the measured HVL falls below the minimum required, it can indicate *either* that the calibration of the kVp for this unit is off, or that there is insufficient filtration in the beam. If it is believed that the filtration is adequate, more sophisticated methods can be used to double-check the accuracy of the kVp control, such as voltage diode instruments or oscilloscopes. Once the kVp has been calibrated within guidelines, an insufficient HVL *can only indicate insufficient filtration*, regardless of whether the general minimum of 2.5 mm aluminum is present or not. That is, *sufficient HVL overrides the general minimum filtration rule.* Filters must be added until the minimum HVL is achieved.

We emphasize here the proper understanding of HVL and how it is measured because of its importance in protecting the patient from unnecessary radiation exposure, a professional issue every radiographer should be conscientious of.

Figure 39-1

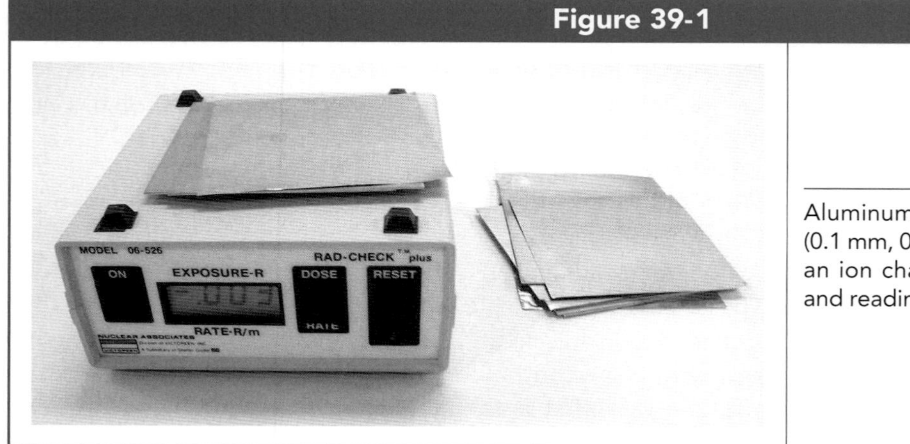

Aluminum slabs of variable thicknesses (0.1 mm, 0.5 mm. 1.0 mm) are placed over an ion chamber in increasing amounts, and readings taken, to ascertain the HVL.

kVp Calibration

To check the calibration of the kVp control, most medical physicists use an electronic device based on ion chambers or photodiodes with various amounts of filtration placed over them. This will immediately be recognized as a form of the *HVL* measurements just described. Indeed, the kVp level is effectively extrapolated from HVL tables in these devices for an *assumed* level of filtration in the x-ray tube and collimator. When there is a significant possibility that the filtration in the unit may be the cause of an inadequate HVL, it can be isolated as such by using more sophisticated (and time-consuming) measures of the kVp to confirm its accuracy. These include the use of voltage diode instruments or oscilloscopes by a medical physicist.

There are different guidelines for an acceptable range of accuracy for kVp; the one adopted here is a range of ± 5 kVp (not percent) of the kVp set at the console control.

Figure 39-2

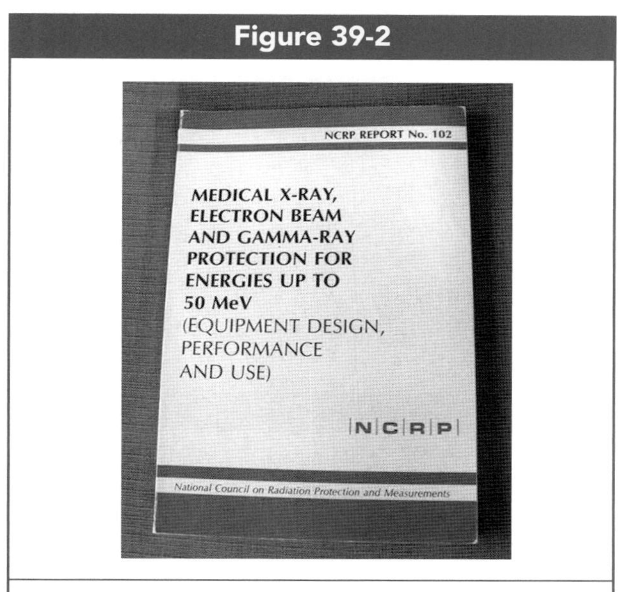

Report #102 of the NCRP, which includes tables of minimum HVL requirements for x-ray machines.

Collimator and Distance

When the actual size of the x-ray field is greater than that indicated on the collimator control knobs, unnecessary exposure to the patient occurs. If the actual field is less than indicated, anatomy of interest may be clipped off from the field of view, necessitating a repeated exposure. It is essential, therefore, that both the field size control knobs on the collimator and the projected visual light field accurately indicate that size and location of the actual x-ray beam.

As described in Chapter 18, it is not uncommon for the edges of the actual x-ray field to be as much as one-half inch off from the edges of the light field. Inaccuracies may be caused by slippage of the collimator shutters, which are often turned by gears or belts that may slip as the collimator ages, or by a crooked collimator mirror (Fig. 39-3).

Figure 39-4 shows a device which can be exposed to measure field size accuracy, field alignment, and vertical beam alignment all simultaneously. The x-ray tube must be carefully locked into its zero-degree position and precisely centered to the plate, then the light field is opened to match the edges as closely as possible to the rectangular outline on the plate, *without* re-centering the beam. On the resulting radiograph

Figure 39-3

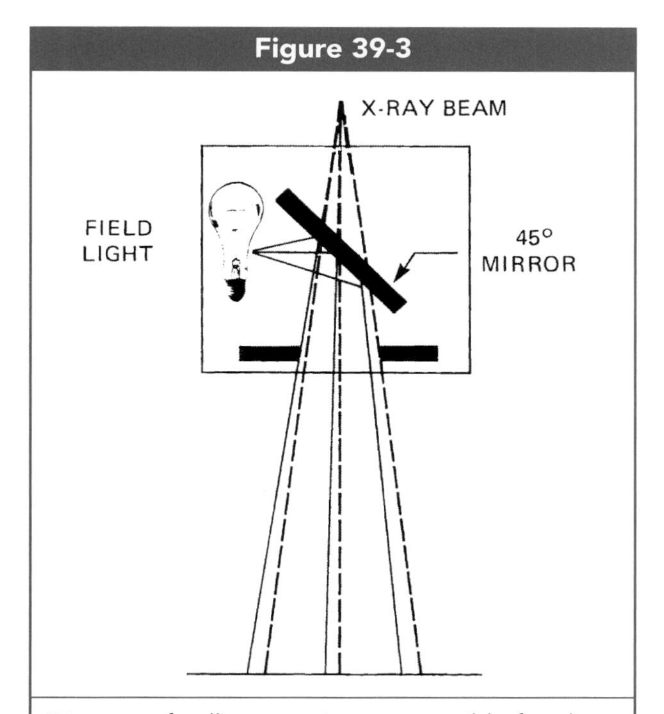

Diagram of collimator mirror responsible for alignment of the light field with the actual x-ray beam. It is not uncommon for the light field to be off more than one-half inch or 1 cm. From Quinn B. Carroll, *Practical Radiographic Imaging*, 8th ed. Springfield, IL: Charles C Thomas, Publisher, Ltd., 2007. Reprinted by permission.

Figure 39-4

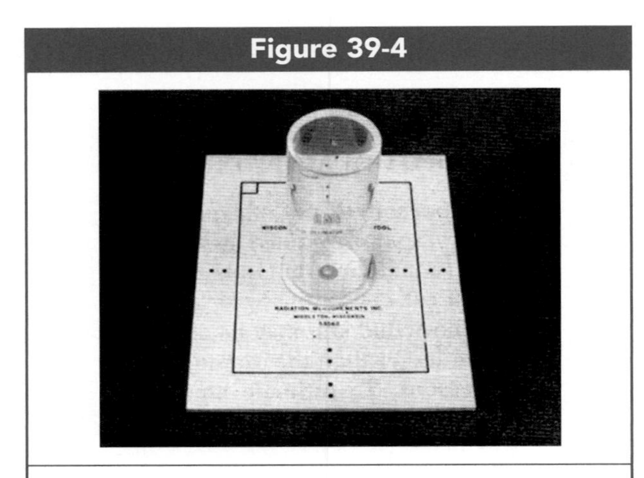

Photograph of an x-ray beam alignment test tool which includes checks for field size accuracy, location of field margins and the CR, and vertical alignment of the beam using the plexiglass cylinder with a screw embedded in the center of the top disc. From Quinn B. Carroll, *Practical Radiographic Imaging*, 8th ed. Springfield, IL: Charles C Thomas, Publisher, Ltd., 2007. Reprinted by permission.

(Fig. 39-5), remember that the demarcated edges represent the *light* field, while the black exposed area is the actual x-ray field, and that we are checking which way the *light field* is off from the actual x-ray beam, not the other way around.

Each dimension of the dark exposed area, length and width, should fall within 2 percent of the SID from the length and width of the demarcated rectangle on the device (± 0.8 inches for a 40" SID, ± 2 cm for 100 cm SID). The formula discussed under *mA linearity* above may be used to calculate the percentage of deviation. *Each* edge of the field should also fall within the 2 percent range from the indicated edge.

The verticality of the central ray is indicated by a metal screw, BB or other small marker embedded in the center of the plexiglass cylinder shown in Figure 39-5, which should fall with concentric rings indicated on the plate. With the central ray within 2 degrees of perfect vertical, the actual CR of the x-ray beam should fall within 1 percent of the SID from the indicated light field CR. The actual x-ray field CR is found by drawing an "X" across the diagonal corners of the exposed field area in the image.

Distance indicators on the x-ray machine may be checked by simply using a tape ruler. The measurement must be taken from the exact location of the focal spot, demarcated at the anode end of every x-ray tube casing by a red "+" sign or other mark. The SID indicator should fall within ± 2 percent of the SID from the actual SID.

Focal Spot Size and Condition

When a new x-ray machine is installed, the actual focal spot size for each of the two settings, "large FS" and "small FS," should be determined. It is important to understand that the *nominal* ("named") focal spot quoted by x-ray tube manufacturers is a measurement obtained under ideal conditions with a low mA setting. The fabrication of an x-ray tube is an extremely complicated process and it is difficult to precisely control the resulting focal spot size, which depends not only on the tube geometry, but also on the correct focusing of the electron beam. Therefore, manufacturers are

Figure 39-5

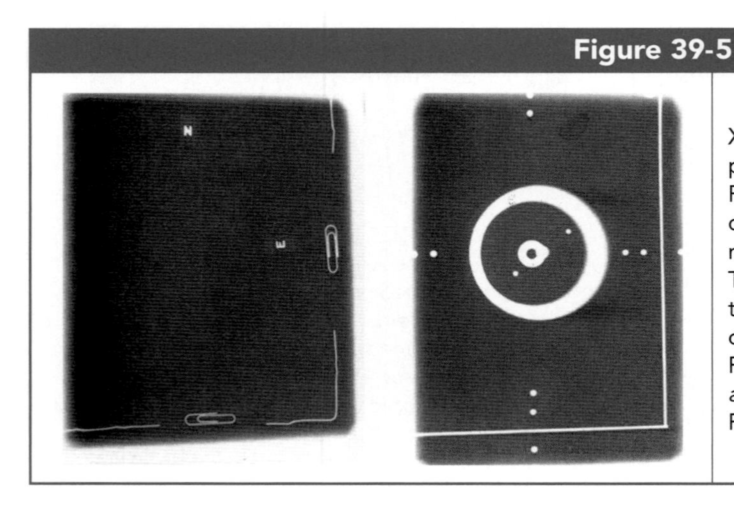

X-ray beam alignment test radiographs using paper clips and markers, *left*, and the device in Figure 37-4, *right*. Both show the light field (indicated by the white lines) to be misaligned to the north (up) and west (left) of the actual x-ray beam. The CR is also more than 2 degrees off from vertical to the east, indicated by the metal screw outside the diameter of the washer in the center. From Quinn B. Carroll, *Practical Radiographic Imaging*, 8th ed. Springfield, IL: Charles C Thomas, Publisher, Ltd., 2007. Reprinted by permission.

allowed considerable lenience in varying from their advertised focal spot sizes. Regulations allow focal spots under 0.8 mm to be as much as 50 percent larger than the nominal size, and those larger than 0.8 mm are allowed to be up to 40 percent larger than advertised.

As we described in Chapter 21, it is possible during an angiogram for the image of a small thrombus or embolism to completely disappear if the focal spot is somewhat larger than the pathology. So, it is important to know how small a lesion the x-ray unit is capable of resolving, and therefore to know the actual sizes of the two focal spots.

Focal spot *blooming* is a phenomenon in which at higher mA stations, the mutual repulsion of the electrons in the space charge that forms around the filament during the rotoring phase causes the cloud of electrons to swell. When they are propelled across the x-ray tube to strike the anode disc, the electron stream has a larger diameter and the resulting actual focal spot is larger. It may be desirable, then, to determine the focal spot sizes at high mA stations that are commonly used for a particular unit. It should be no surprise that the nominal focal spot is determined by the manufacturer at a low mA station.

A more important application for focal spot measurements is the *monitoring of the condition* of the focal spot over time. Once a baseline measurement is kept on file, any sudden deviation in its size is indicative of warping or damage to the anode which requires immediate attention. The focal spot should be routinely checked on an annual basis, but any sudden change in the sharpness of images should be followed up with a focal spot test.

The preferred tool for measuring focal spot size is the "slit camera," a metal template with a series of finely cut slits in groups of three which run crosswise and lengthwise (Fig. 39-6). This device must be carefully placed with the lines running perpendicular and parallel to the axis of the x-ray tube, at the precise SID indicated by the manufacturer. An exposure results in the pattern shown in Figure 39-7. Scanning from larger to smaller line patterns, the last resolved set of three before they become blurred is determined. From a table, this set will indicate the dimension of the focal spot. The dimension being measured, length or width, is *perpendicular* to the line pattern observed. Note from Figure 39-7 that the length and width of the focal spot are typically different.

Figure 39-6

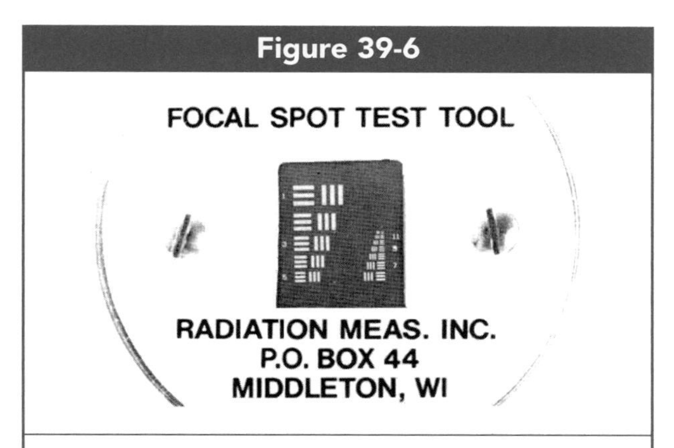

A "slit camera," the preferred method for measuring focal spot size and condition, has groupings of slits of decreasing size, which must be placed perpendicular to the dimension of the focal spot being measured. From Quinn B. Carroll, *Practical Radiographic Imaging*, 8th ed. Springfield, IL: Charles C Thomas, Publisher, Ltd., 2007. Reprinted by permission.

Automatic Exposure Control (AEC)

Several electronic and radiographic tests can be made for automatic exposure controls. Checks for reproducibility, kVp, and other parameters common

Figure 39-7

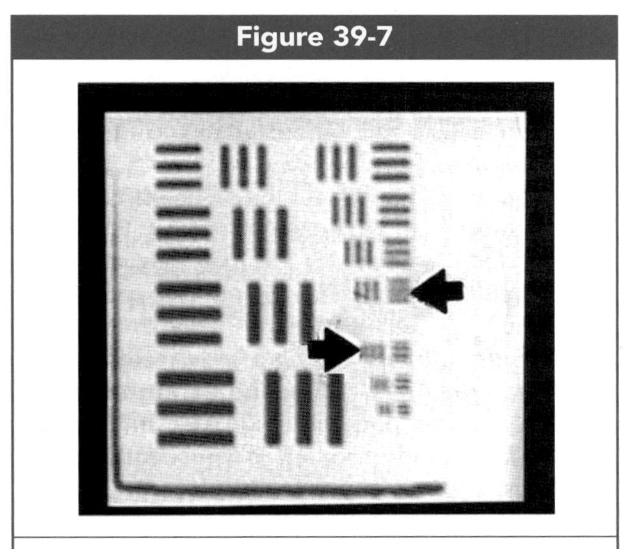

Example of a slit camera radiograph. Note that the blur points (arrows) indicating focal spot size are different for length and width. From Quinn B. Carroll, *Practical Radiographic Imaging*, 8th ed. Springfield, IL: Charles C Thomas, Publisher, Ltd., 2007. Reprinted by permission.

to regular radiographic units are essentially identical to those described in previous sections. AEC exposures must be repeatable within an accuracy range of ± 10 percent. To check repeatability, five exposures should be taken using an ion chamber to measure the output. AECs should also be linear within 20 percent between different rooms.

As fully described in Chapter 27, the "density control" can be easily checked for accuracy by a radiographer. The back-up timer can be tested by placing three sheets of leaded rubber over the activated detector cell(s). The exposure should automatically terminate at the set back-up time or at 600 mAs. Linearity and other tests require electronic equipment and are best performed by a medical physicist.

Tomographic Quality Control

Tomographic units must comply with the same general parameters as other radiographic units, in addition to tests that are specific to the tomographic function. Tests on exposure angle, section thickness, and resolution can be performed by a medical physicist. A simple test can be done by radiographers to determine if the level of the section is as indicated on the unit.

A typical section level test tool consists of wires or other objects placed at various levels of depth. (An aluminum or plexiglass step-wedge with lead numbers is adequate for a "homemade" test.) When a linear movement is to be evaluated, linear test objects should be placed at 45 degrees to the direction of the x-ray tube movement. The section level of the tomographic unit (e.g., 9 cm) is set to match one of the steps on the test device, and an exposure is made.

The resulting image should demonstrate the sharpest details at the level indicated (Fig. 39-8). The section level should be within ± 5 mm of the setting on the machine. When incrementing from one tomographic section level to the next, the section level should be accurate to within ± 2 mm.

Fluoroscopic Units

Fluoroscopic examinations can result in high radiation exposure to the patient, so QC checks are very

Figure 39-8

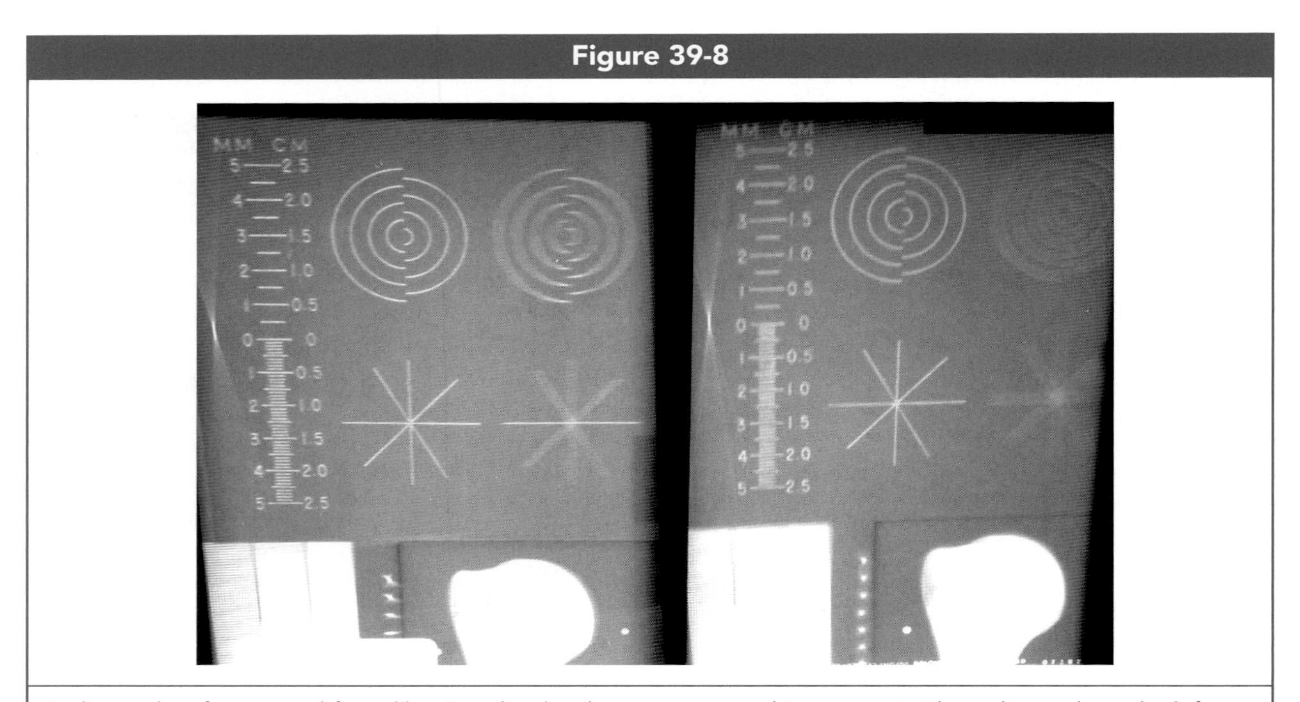

Radiographs of a test tool for calibrating slice levels on a tomographic x-ray unit. The radiograph on the left was taken with a 30-degree linear tube movement, the one on the right with a 30-degree hypocycloidal (clover-leaf) movement. From Quinn B. Carroll, *Practical Radiographic Imaging*, 8th ed. Springfield, IL: Charles C Thomas Publisher, Ltd., 2007. Reprinted by permission.

important and must be done accurately by a medical physicist using a calibrated ion chamber. Checks should be repeated after any major change is made in the x-ray tube, generator, or console. Regulations governing the allowable output of fluoroscopes are presented in Chapter 44.

Simple observational tests which a radiographer can perform for fluoroscopic image distortion (pincushion, s-distortion, veiling glare and vignetting) were described in Chapter 38.

Spot-film devices must also be regularly evaluated for exposure and for proper collimation. As described in Chapter 38, the active area of the input phosphor affects the exposure to the patient in inverse proportion to its diameter.

The automatic brightness control systems used in conjunction with image intensifiers are prone to deterioration over time. The radiation exposure to the input phosphor should be constant for different thicknesses of tissue-simulating material inserted in the beam.

Minimum source-to-table distance (15" or 37 cm for stationary units and 12" or 30 cm for mobile units), a primary barrier of 2 mm lead equivalent in the image intensifier housing, proper filtration, sufficient collimation to allow for an unexposed border on the display monitor, the presence of the bucky slot cover and fluoro curtain, a cumulative timer check, and x-ray output intensity should all be monitored yearly. In some states, the radiographer is required to record the amount of fluoro-on time for each procedure. In some cases, the number of overhead and spot radiograph exposures must also be recorded.

MONITORING OF DIGITAL ACQUISITION SYSTEMS

Although sophisticated testing of digital imaging systems (such as detective quantum efficiency or DQE) is within the purview of the medical physicist, there are several equipment checks that can be performed by radiographers or QC technologists. Once a baseline image or other data is established for a particular unit, it can be monitored for any sudden or dramatic deviations by simple, regular visual checks as follows.

Field Uniformity

Unlike the older screen/film systems, digital detectors are *all* inherently nonuniform. Uniformity corrections must be repeated on a regular basis ranging from daily to semi-annually, depending on the particular equipment. If a particular radiopaque (light) artifact is ever noted on more than one image, a nonuniform detector field may be the cause. The plate should be thoroughly erased prior to a test exposure. A "flat field" exposure may be taken using moderate technique settings with no object in the beam and opening the collimation to fully cover the detector plate area. A long SID (72 inches or 180 cm) should be used in order to minimize the anode heel effect.

The resulting image may simply be visually scanned for defects which would be reported to the medical physicist. For observation of the digital image, the contrast should not be adjusted to an unusually high level, since at extreme contrast settings any image will appear nonuniform.

Erasure Thoroughness and "Ghosting"

Inadequate erasure on a CR system can arise from burned-out bulbs in the erasing chamber of the processor, loss of lamp intensity, or too short duration of erasure. (Even with proper erasure, extreme overexposure from the previous exam can result in a residual signal.)

Expose the imaging plate with an aluminum step wedge or other homogeneous object in the beam, and process. Immediately reexpose the same plate without the object in the beam and with an inch or so of collimation from each edge. Visually examine the second image for any ghost image of the object.

"Ghosting," image lag, or "memory effect" in a DR system may be visually checked the same way. It is caused when electrical charge has been trapped in metastable states in the amorphous silicon or selenium during an exposure and is released only slowly over time. The silicon-based indirect conversion systems generally have shorter image lag than direct systems.

Intrinsic (Dark) Noise

A single plate can be erased and processed without exposing it to an x-ray beam. Visually scan the resulting image for unusual amounts of mottle or noise when compared to a baseline image.

Spatial Resolution

A wire mesh can be exposed and the image visually examined for any distortions or unsharpness. It is best to use a baseline image taken when the equipment was purchased for comparison. A standard "bar" test pattern, made of lead foil (Figure 14-22 in Chapter 14) can be used to obtain the spatial resolution in line-pairs per millimeter (LP/mm). A sharp-edged lead object, or a fine slit cut into lead foil may also be used for comparisons—physicists use these types of objects to obtain an "edge-spread function" or a "line-spread function" respectively.

MONITORING OF ELECTRONIC IMAGE DISPLAY SYSTEMS

As nearly all radiographic images are now viewed electronically, quality assurance for the display monitor has become extremely important. As with equipment QC for radiographic units, the staff radiographer may not be responsible for conducting these tests, but should be familiar with them and develop an appreciation for how important they are. We have stated that the display monitor is the weakest link in the modern medical imaging chain, yet misdiagnosis can result from poor display quality. It is critical that display systems be regularly and carefully evaluated.

Most diagnostic workstations consist of two or more monitors. Consistency between all display monitors within a particular workstation is essential; all monitors should have the same level of luminance and be set at the same contrast. Workstation monitors should be cleaned monthly. The frequency with which quality control tests are conducted depends on whether the monitor is a CRT or an LCD, whether it is self-calibrating, and its application (diagnostic workstation, technologist workstation, or display station), and are discussed in each following section.

Several scientific groups have published guidelines for monitoring the quality of electronic image display devices. Standards for medical images are found in DICOM Part 14. Tests and guidelines are available from the American Association of Physicists in Medicine (AAPM), the Society of Motion Picture and Television Engineers (SMPTE), the American College of Radiology (ACR), and others. These include tests for maximum luminance, luminance response, luminance uniformity, contrast, resolution, noise, reflectance, and others. We shall overview the major qualities of the electronic image using a conceptual, nonmathematical approach.

Task Group 18 of the AAPM and other scientific groups have developed all sorts of test patterns that can be downloaded from their websites for display on a particular monitor to check each specific quality. A generic test pattern was developed many years ago by the Society of Motion Picture and Television Engineers (SMPTE), which includes elements for most of the image qualities all in one comprehensive pattern (Fig. 39-12, p. 687).

Class 1 display monitors are defined as those at workstations used for diagnosis, and more stringent guidelines for quality control are applied than for *class 2* monitors, which may still be classified as "workstations" as we have defined them, but are not used for diagnosis.

While a QC technologist may not perform sophisticated measurements, it is a simple matter to monitor image deterioration over time. When a new monitor is installed, baselines can be established and logged for brightness, contrast and resolution, using just the SMPTE test pattern. Periodic checks can then be compared with the original.

Luminance

Luminance refers to the rate of light emitted from a source such as a CRT or LCD screen. The unit of luminous flux, the intensity of light as perceived by the human eye, is the *lumen* (*Lm*). The light emitted directly from a display monitor fans out in all directions within a dome-shaped hemisphere (Fig. 39-9). One way we can measure the volumes of space within this hemisphere is in units called *steradians.*

A steradian may be thought of as a three-dimensional angle, or a *solid* angle, which sweeps out a cone shape. The base of this cone has the area r^2, the square of its radius r from the center of the sphere, as shown in Figure 39-9. Light emitted from a point source isotropically (evenly in all directions) must spread out across this area in accordance with the inverse square law. The steradian is based on a specific angle, so that 4π steradians, or 12.57 of these cones, always fit within a sphere. Thus, approximately six steradians fit within the hemisphere of projected light depicted in Figure 39-9.

The unit *candela*, abbreviated *Cd*, is based upon the typical amount of light radiating from a candle, and is named after it. In terms of energy, the power of this amount of light comes to about 0.0015 watts per steradian. (Remember that the *watt* represents an energy rate *per second*, so time is already taken into account in this formula.) Strictly speaking, however, the candela unit describes the total rate of light emitted by the candle *in all directions*, whereas the lumen refers to the light flux within just one steradian.

We might say that the candela describes the power of the light *source*, while the lumen describes the power of the light flow traveling through *space*. The brightness of the source determines the intensity of the light beam. The relationship between these two units is that *one candela generates one lumen per steradian*:

$$1 \text{ Cd} = 1 \text{ Lm} / \text{sr}$$

The maximum luminance that an LCD is capable of is around 800 Lm or 800 Cd/m². The brightest setting for a typical CRT emits about 600 Lm. By comparison, conventional viewboxes used to view hard-copy printouts of radiographic images produce 1000–3000 Lm. The typical brightness settings preferred by radiologists for electronic display monitors fall in the range of 500 to 600 Lm. A conventional viewbox is five times this bright. The ACR requires a minimum brightness of 250 lumens for display systems.

The luminance of CRTs has been known to decay as much as 20 percent within 8 months, and one study showed CRTs failing luminance tests at an average age of only 1.7 years. Therefore, the output luminance of CRTs should be checked and logged monthly. LCDs maintain their luminance far longer, and may only need to be checked once a year.

The Photometer

The *photometer*, Figure 39-10, is a device used to measure light intensity. It can be built with photoresistors, photodiodes, or photomultiplier tubes which measure the light flux by the amount of electric charge or current it generates through the photoelectric effect. Most

Figure 39-9

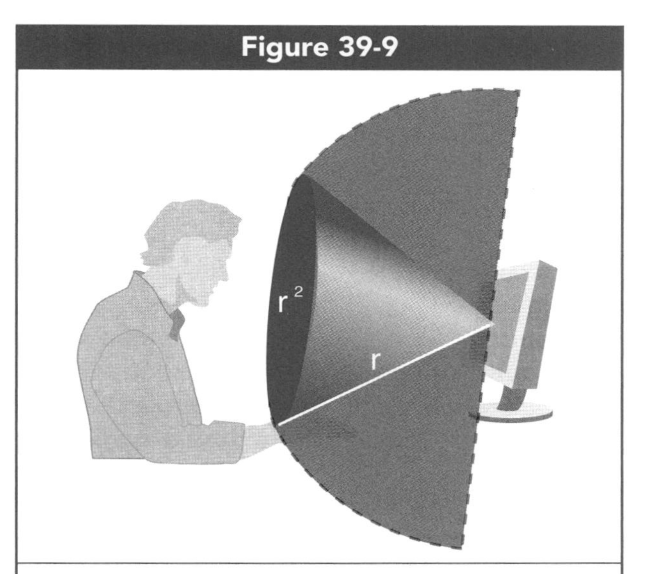

A *steradian* is the cone swept out by a 3D angle such that the area of its base is equal to the square of its radius from the source of light (r^2). There are roughly 6 steradians in a hemisphere. The brightness of light (luminance) from a display monitor is measured in candela per square meter (Cd/m^2) or in lumens (Lm). A 1-candela source of light emits one lumen per steradian, (or 0.0015 watts of energy per steradian).

Figure 39-10

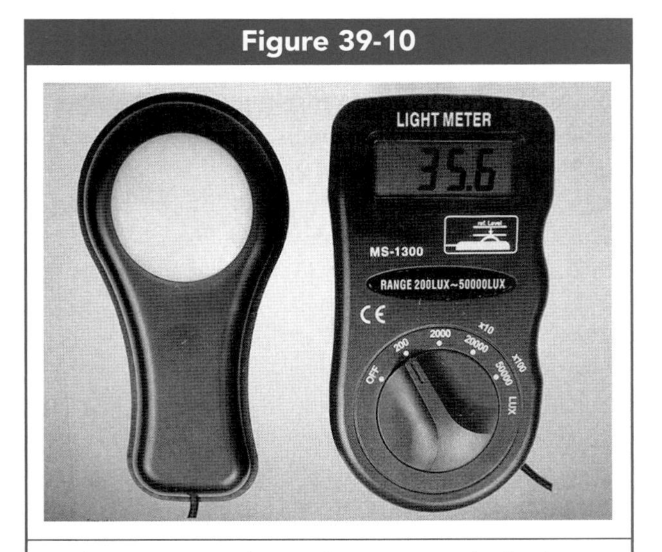

A photometer can be used to measure the luminance (light output) of a display monitor. Most photometers, like this one, read out in units of *lux*.

photometers are read out in units of *lux*, *lumens*, or *candela/m*^{2}, which are all equivalent units.

Many manufacturers of LCDs provide photometric instruments specifically designed for their monitors, which a QC technologist or medical physicist may use in accordance with the manufacturer's instructions.

The Densitometer

A densitometer measures the darkness of densities on a hard copy radiograph. It may be thought of as a sort of "inverse photometer." However, densitometers typically use a different scale numbering from zero to 4.0. This is an inverse logarithmic scale based on the fraction of light from a viewbox behind the radiograph which penetrates through it to the human eye or the densitometer. The readout is the *exponent* to which the base 10 is raised to obtain the *denominator* in the ratio of light being transmitted through the film. For example:

A density of 1.0 indicates that 1/10th of the light is transmitted. ($1/10^1 = 1/10$)

For a density of 2.0, 1/100th of the light is transmitted. ($1/10^2 = 1/100$)

For a density of 3.0, 1/1000th of the light is transmitted. ($1/10^3 = 1/1000$)

For a density of 4.0, 1/10,000th of the light is transmitted. ($1/10^4 = 1/10{,}000$)

A density of 4.0 appears pitch black to the human eye, a density of 2.5 appears as a medium gray. In this logarithmic scale, *each change of 0.3 in the density value corresponds to a factor of 2 in actual intensity*. For example, a density of 2.3 stops twice as much light as a density of 2.0. Similar scales may be used on some photometers.

Illuminance

Illuminance refers to the rate of light *striking* a surface. It describes how well objects in our field of view are *illuminated*. As you observe a computer screen, you would describe its brightness in terms of the luminance it is emitting, but when you read a paper on the desk in front of it, you would describe the apparent brightness of the paper in terms of the *illuminance* from the computer screen, and from the room lighting, which is making the paper visible. The concept of illuminance is applicable to the backlighting which strikes the polarizing layers of an LCD, or to room lighting which strikes a display monitor and can be reflected off of it.

The unit for illuminance is the *lux*, defined as 1 lumen per square meter of surface area.

$$1 \text{ lux} = 1 \text{ Lm} / \text{m}^2$$

The typical lux levels below will develop for the student some frame of reference for comparison:

Direct sunlight	=	105 lux
Overcast sunlight	=	103 lux
Full Moonlight	=	10 lux
Typical Office Lighting	=	75–100 lux
Conventional Reading Room Lighting	=	2–25 lux

We can see that *the maximum ambient lighting in a radiologic reading room should be less than one-fourth of normal office lighting.*

(Other units are used for the brightness of light, including the *nit*, defined as 1 cd/m^2, the *lambert*, the British unit *foot-candela*, and others, all of which will be avoided for this discussion.)

Luminance and Contrast Tests

The *maximum luminance* is simply checked at the same monitor setting over a period of time, using a photometer. New LCDs have self-correcting capability and may only need to be checked once a year. CRTs must be checked monthly because of their rapid deterioration.

Luminance response refers to the monitor's ability to accurately display different shades of brightness from a test pattern. As these must be compared to adjacent shades of brightness, this function is essentially identical to a *contrast* test, although different units of measurement might be incorporated. To test the sensitivity of the monitor, test patterns usually consist of adjacent dark gray squares at threshold contrasts that are just visible. On a monitor with good contrast, these squares will be resolved with a barely visible difference in density. Poor contrast results in two adjacent squares appearing as a rectangle with a single density.

On the SMPTE test pattern in Figure 39-12, there is a ring of density squares near the center. The two "50 percent" squares at the top of this ring should match. At the bottom of the ring there are two sets of squares marked "0/5 percent" and "95/100 percent." The observer should be able to distinguish the 95 percent white square embedded within the 100 percent white square, as well as the 5 percent brightness square embedded within the 0 percent square (black).

DICOM standards include a *grayscale standard display function* (*GSDF*). The GSDF was developed to bring the digital values assigned to the dynamic range into linear alignment with human perception, with increments called *JNDs* for "just noticeable differences." The AAPM recommends that luminance response should fall within 10 percent of the GSDF standard. The ring of density squares in the SMPTE pattern includes a "ramp" of squares in 10 percent increments from white to black. These should all be within 10 percent accuracy based on photometric measurements of the 0 percent and 100 percent areas.

Another contrast-related test is called the *luminance ratio* (*LR*). The LR compares the maximum luminance (L_{MAX}) to the minimum luminance (L_{MIN}). The value of this measurement becomes apparent when we realize that neither an LCD nor a CRT are capable of producing a "true black level" while they are energized. When no image is apparent on a CRT monitor, the screen is emitting about 0.5 candela per square meter. With a maximum luminance of 600 Cd/m^2, this translates to a luminance ratio of 1200. By comparison, even though a typical LCD may have a higher L_{MAX} , it has a very poor L_{MIN} (as can be seen in a completely darkened room). This results in an LR from 300 to 600, much lower than the CRT which can range from 1000 to 3000. On the universal SMPTE pattern, a white on black bar and a black on white bar are added specifically for checking the LR of CRTs.

Finally, *luminance uniformity* refers to the consistency of a single brightness level displayed across the area of the screen. Like the flat field uniformity test described in Chapter 30, five luminance uniformity measurements are usually taken, at each of the four corners of the screen and the center. By AAPM guidelines, these should not deviate more than 30 percent from their average.

Ambient Lighting (Illuminance) and Reflectance Tests

There are two kinds of reflection, or reflectance, off the surface of any monitor screen. These are *diffuse* reflectance (R_D) and *specular* reflectance (R_S). Diffuse reflectance is the ratio of the brightness of the general ambient light in the room that is being reflected off the surface of the screen to the output brightness of the display monitor itself. It is measured by physicists with an experimental setup that includes fluorescent lights behind the monitor which flood the room with diffuse light. Specular reflectance refers to the reflection of a specific, localized light source within the room, such as light bulb, placed behind the viewer of the display monitor. It is defined as the ratio of the reflected brightness to the brightness of the light source.

The importance of both is that *the ambient lighting within the room must be dimmed to a point where both types of reflectance are below any noticeable level, that would impede full visibility of the image.* Task Group 18 of the AAPM has published a table of recommended maximum illuminance values within a reading room based upon the measured specular and diffuse reflectance of the display monitor screens. Since CRTs have much higher reflectance than LCDs, the reading room must be dimmed more for CRTs. Of course, the illuminance of the reading room is most often adjusted subjectively, but it is important to point out that both the guidelines and the scientific methods are available to determine the ideal illuminance for reading rooms in an objective fashion.

Noise

No quantified guidelines for acceptable levels of noise in the electronic radiographic image have yet been published. However, the AAPM has made a test pattern (TG18-AFC) available for monitoring noise levels in the electronic image over time. These patterns consist of a series of circular areas within a black mask, which gradually increase brightness in JNDs (just noticeable differences) as they progress diagonally across a small square section of the screen (Fig. 39-11). There are 64 square sections within each quadrant of the screen. The observer

Figure 39-11

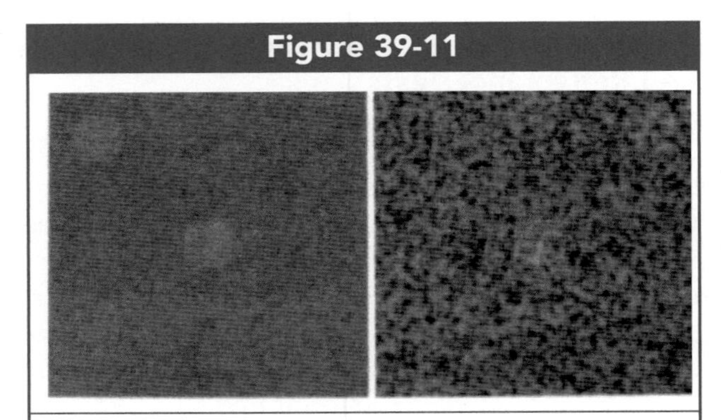

Sample radiographs using a QC test pattern for evaluation of image noise on electronic display monitors. Increasing noise reduces the visibility of very low-contrast circles embedded in the pattern.

counts the number of square sections in which the dimmest circle can be seen. The suggested "passing" point is that in at least three quadrants of the screen, the majority of square sections demonstrate the dimmest dot.

Resolution

Both the ACR and the AAPM recommend a minimum resolution for electronically displayed images of 2.5 LP/mm. The generic SMPTE test pattern has a series of high contrast bars and a series of low contrast bars of diminishing widths to check resolution. As shown in Figure 39-12, the entire pattern is repeated in the center and in each of the four corners of the screen. Similar to the "slit camera" used to check the focal spot of an x-ray machine, these are arranged in sets perpendicular to each other; horizontal bars are observed to check *vertical* resolution, and vertical bars to check *horizontal* resolution.

In all resolution tests, the observer inspects the bar images for blurring or "bleeding" of each white or black bar into the adjacent bars at the edges. The TG18-QC test pattern from the AAPM provides an excellent check for resolution across the area of the display screen. It consists of very small "C-X patterns," each consisting of a squared ⊏ with an x inside it, across the screen. By visually matching these with a set of larger C-X patterns that have been progressively blurred, the resolution can be read off from a correlated table. The screen is divided into 12 areas, and a "passing" score is attained when 4 or fewer of these do not meet the minimum resolution stardard. (This pattern also includes tests for contrast, noise and geometrical distortion.)

The resolution of an LCD is consistent and always uniform, and need not be regularly checked, but should be carefully determined upon purchasing display monitors. CRTs must be frequently checked for deterioration of their resolution. Electron beam width can "bloom" with aging, is exaggerated at higher luminance, reducing resolution for CRTs.

Dead and Stuck Pixels

We have described a "pixel" in a color CRT screen as a triad of phosphor dots, red, green and blue. In an LCD, a single pixel actually consists of 18 individual bar-like segments as was illustrated in Figure 35-8 in Chapter 35 (p. 620). There are three subpixels in a pixel, each containing six segments.

The LCD screen can be examined for bad pixels visually, and is aided by using a standard optical magnifying glass. With typical 12-point font, a period or the dot of an "i" can be seen to occupy one pixel. A bad pixel appears on the screen as a readily apparent dot. Failure of any one segment within a subpixel constitutes a "bad" subpixel, which is apparent on close inspection as a smaller defect.

Remember that in an LCD, electric charge is applied to turn a pixel "off" so that it is dark against a light background, so truly *dead* pixels are identified by bringing up a solid *black* background and examining it for white spots. On the other hand, it is also possible for pixels to become "stuck" in an "on" state. In this case, we look for black spots on a solid white field. For a complete check, both tests would be performed. (For color monitors, solid red, green and blue fields can also be used to examine the LCD for defective subpixels assigned to these colors.)

Pixels that are either dead or stuck can sometimes be fixed by very gently massaging them with the fingertip. However, remember that pixels can also be damaged by finger pressure and fingernails, so this procedure should not be attempted with expensive workstation monitors.

A standard has been recommended by the AAPM for a class 1 LCD, used for diagnosis, to "pass" inspection for defective pixels as follows:

Figure 39-12

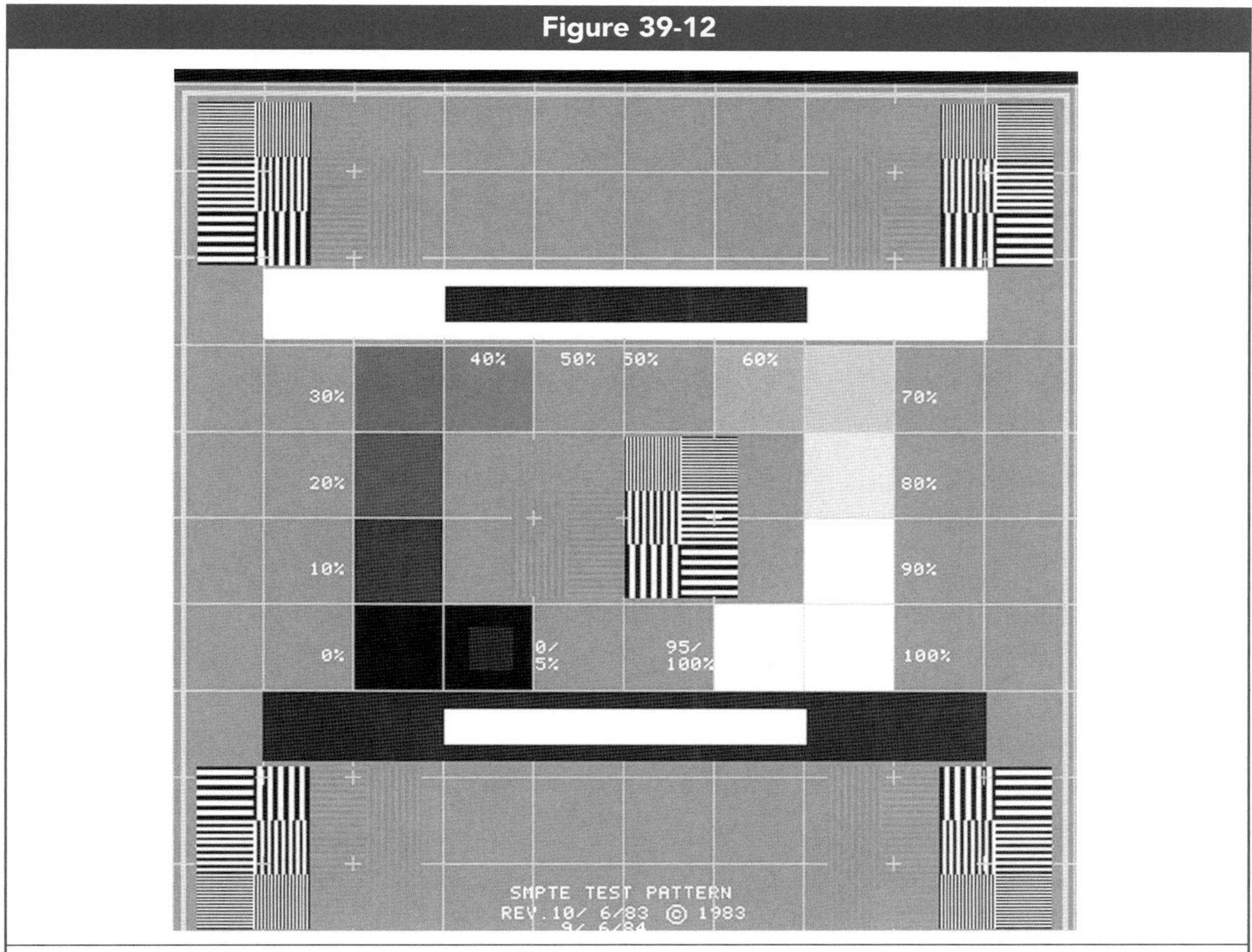

The universal SMPTE test pattern can be used to monitor the brightness, contrast, and resolution of an electronic image display. (Courtesy, Society of Motion Picture and Television Engineers.)

1. Fifteen or fewer bad pixels across the entire screen
2. Three or fewer bad pixels within any circle of 1 cm diameter
3. Three or fewer bad adjacent pixels

Manufacturers recommend that class 1 LCDs be "nearly perfect," but that class 2 LCDs, not used for diagnosis, be allowed 2 defective pixels, or 5 defective subpixels, per million.

Viewing Angle Dependence

We have described the limited viewing angle of LCDs as a significant disadvantage. Very expensive LCDs may include pixel elements that are slightly angled in order to increase the viewing angle. LCDs from different manufacturers may be compared subjectively by standing at set angles from the perpendicular, but would need to have their brightness equalized to begin.

An objective comparison can be made by carefully positioning a photometer at increasing angles from the perpendicular, maintaining a precise distance (1 m) from the center of display screen. The percentage of luminance lost can then be calculated by dividing the original perpendicular measurement into each of the subsequent measurements and multiplying by 100. Without the complications created by having to equalize the initial brightness between LCDs, these percentage figures can be validly compared to those obtained from the other manufacturers or models.

Stability of Self-Calibrating LCDs

The stability of a self-calibrating LCD should be checked by measuring the stabilized brightness every day for the first week, once a week for a month, and then each month for three months. This can be done with a photometer, ensuring that the settings are not changed. Luminance tests should be conducted on all LCDs annually.

The Viewbox Illuminator

The viewbox illuminator is used to view hard copy prints of radiographic images on transparent film. It consists of two 15-watt light bulbs behind a diffusing panel. For consistent and uniform backlighting, these bulbs should be regularly replaced. The illuminator must be regularly cleaned and examined for any stains, cracks or other artifacts on the surface which would interfere with diagnosis.

REPEAT ANALYSIS

Before the advent of digital radiographic imaging, most repeated x-ray exposures (about 60%) were necessitated due to images that were too dark or too light (from either improper technique or improper chemical processing). Digital imaging has dramatically reduced this figure to less than 10 percent. We can categorically state, therefore, that for digital radiography most retakes are due to positioning errors on the part of the radiographer.

With a digital imaging system, when a particular view is repeated by the radiographer, the new image is normally saved as a *replacement* file for the poor image. This means that the faulty image is erased and permanently lost. When the "reject image" option is selected, most PAC systems require the user to enter the reason for rejecting the image, from a menu. The menu typically has only a few very broad categories, such as "PACS problem, x-ray equipment problem, or positioning error."

A very unfortunate aspect of digital systems is that since these rejected images are lost, there is no way to follow up with more targeted repeat analysis. For example, with conventional film radiographs, it was possible to go back and subcategorize all "positioning" errors into specific procedures such as "skull" and "chest" positioning, which facilitated planning for continuing education. Inservice programs for staff radiographers could then be customized to specific areas for improvement. From the limited selections on the digital reject menu, many radiographers have simply formed a habit of entering "other" rather than "positioning" to cover their mistakes.

Some manufacturers are looking for ways to address this issue by incorporating more specific software for reject analysis. But, as long as images can be discarded at the acquisition workstation, any meaningful and accurate repeat analysis will be thwarted. For the present time, then, the primary value of the logging of rejected images by the PAC system has become purely budgetary in nature, and limited to the overall retake rates for staff and equipment. A realistic target for diagnostic x-ray departments is to maintain an overall repeat rate of 3–5 percent due to technologist errors.

SUMMARY

1. Quality control for radiographic equipment is an essential part of a quality assurance program. Guidelines for accuracy are published for each equipment performance parameter. To find the percentage by which any variable is out of calibration from its standard, divide the lower of the two numbers into the difference between them, and multiply by 100.
2. Some equipment checks should be strictly done by a medical physicist, but there are several that can be performed by radiographers, including beam alignment and collimation, AEC density control linearity, focal spot condition, and others. The least reliable of the electronic technique variables is the mA station.
3. The most important quality of the x-ray beam to monitor is the HVL, a direct measure of penetration.
4. The maximum luminance of CRTs used for diagnosis should be checked monthly, that of LCDs every year. Brightness and contrast settings between monitors at a particular workstation must

be consistent. Photometers are used to measure brightness of an electronic image, densitometers to measure darkness of a printed hard copy.

5. Ambient lighting in a reading room must be low enough to negate specular and diffuse reflection off of monitor screens.
6. Important luminance tests for a monitor include maximum luminance, luminance response (contrast), luminance ratio and luminance uniformity.
7. CRTs should be checked monthly for deterioration of resolution. The resolution of an LCD is consistent and need not be checked after installation.
8. Pixels in an LCD can become either "dead" or "stuck." An LCD should be repaired or replaced if there are more than 15 bad pixels present, or more than 3 bad pixels within 1 cm.
9. The stability of self-calibrating LCDs should be checked every day for a week, every week for a month, then every month for one quarter.
10. Since rejected digital images are permanently lost, the specific causes of repeats can no longer be followed up.
11. Several visual quality control checks for digital imaging equipment can be performed by the radiographer.

REVIEW QUESTIONS

1. List six of the elements of a comprehensive quality assurance program:
2. For a particular x-ray unit, when the collimator reads a 24 cm field width, the actual measured field width is found to be 26.5 cm. Is this within the acceptable range for an SID of 100 cm?
3. When an exposure timer is set to 0.05 seconds, the actual time is found to be 0.037 seconds. By what percentage is this timer setting out of calibration, *and* is it within acceptable guidelines?
4. What type of problem is indicated when there is a sudden change in the measured focal spot size?
5. At 80 kVp, the HVL for a particular 3Φ x-ray machine is found to be 1.9 mm of aluminum. If the kVp control is accurate, what specific change is needed as indicated by this HVL?
6. What part of a collimator is responsible for alignment of the light field to the actual x-ray beam?
7. Quality control standards are more stringent for *class 1* monitors, defined as those used for:

(Continued)

REVIEW QUESTIONS *(Continued)*

8. A 1 candela light source emits 1 lumen per __________.

9. On a densitometer, a readout of 3.0 means that what fraction of light from a viewbox is being transmitted through the image?

10. Correctly, the amount of light striking the polarizing layers of an LCD screen from its back-lighting source is defined not as luminance, but as __________.

11. The maximum ambient lighting in a reading room should be less than _________ that of normal office lighting.

12. Luminance response or contrast for a display monitor should be such that the _________ percent and __________ percent areas on the universal SMPTE test pattern can be resolved.

13. The DICOM grayscale standard display function (GSDF) is incremented in steps called *JNDs*, which stands for:

14. On the universal SMPTE test pattern, resolution bars are provided in what five positions?

15. To check an LCD for an electronically *dead* pixel, what color of solid background should be used?

16. According to the American Association of Physicists in Medicine, there should not be more than how many defective pixels across a class 1 LCD used for diagnosis?

17. What is a realistic target range for a departmental repeat rate, due to technologist error, in radiography?

Part V

RADIATION BIOLOGY AND PROTECTION

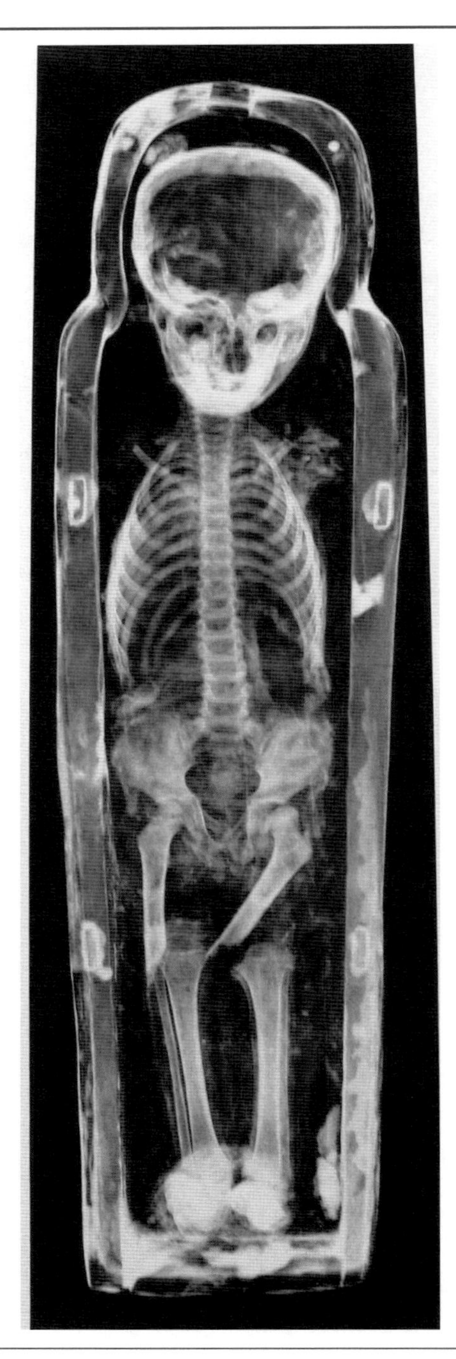

Radiograph of an Egyptian mummy, whose legs were broken in order to fit the body into the casket. (From R. Eisenberg, *Radiology: An Illustrated History*, 1992. Courtesy, Elsevier Health.)

Chapter 40

RADIATION PERSPECTIVES

Objectives:

Upon completion of this chapter, you should be able to:

1. Provide a balanced perspective on the radiation risk levels and standards of practice for the working radiographer.
2. Provide a balanced perspective on environmental radiation and its associated levels of risk.
3. Develop a scientifically-based frame of reference to compare the risks of radiography for patients to other common health risks.
4. Describe the sources of natural background radiation, man-made radiation, and their associated levels of risk.
5. State the proportional contribution of each source of radiation to our total annual exposure.
6. List the types of radiation and, for each, their associated levels of penetration, biological harm, mass and charge.
7. Explain why each of these types of radiation causes greater or lesser biological harm.
8. Define *half-life*, and calculate the levels of radioactivity associated with different half-lives and elapsed times.
9. Explain the relationship between the level of radioactivity and the half-life in determining risk.

PERCEPTIONS

On the Radiographer's Job

Every radiographer must be constantly mindful of two people for whom he/she is directly and primarily responsible for protecting from excessive radiation exposure: The patient and himself or herself. Radiation protection is part and parcel of the radiographer's job description, and legally it is considered part of the *standard of practice* for the profession. The *Code of Ethics* of the American Society of Radiologic Technologists (ASRT) places this responsibility squarely on the shoulders of the radiographer. Principles #5 and #7 state as follows:

#5. The radiologic technologist assesses situations, exercises care, discretion and judgment, assumes responsibility for professional decisions, and acts in the best interest of the patient.

#7. The radiologic technologist uses equipment and accessories, employs techniques and procedures, and performs services in accordance with an accepted standard of practice, and demonstrates expertise in limiting the radiation exposure to the patient, self, and other members of the health care team.

A healthy respect for the hazards of radiation came the hard way in the early days of radiography, and at a high cost. In 1904, the first historically recorded radiation fatality was Clarence Dally, an assistant to Thomas Edison. Figure 40-1 shows Edison trying out his newly developed hand-held x-ray fluoroscope to examine the bones in Dally's hand. After spending months and years experimenting with different x-ray apparatuses for Edison, Dally developed carcinomas on his hands (Fig. 40-2), and

Figure 40-1

Thomas Edison tries out his new hand-held x-ray fluoroscope to examine the hand of his assistant, Clarence Dalley, who later died from excessive accumulated radiation exposure. (From R. Eisenberg, *Radiology: An Illustrated History*, 1992. Courtesy, Elsevier Health.)

had to have both hands amputated. His early death was attributed to radiation exposure.

As other casualties to radiation research mounted, Dr. MK Kassabian developed 10 rules for radiation protection based on his experiments, including: *Never use the hands to test the intensity of x-ray.* He carefully documented the deterioration of his hands due to radiodermatitis, Figure 40-2***B***, and later died of metastatic cancer. For radiation workers, protective measures came to be embodied in three fundamental principles:

1. Minimize the amount of *time* you are exposed to radiation.
2. Whenever possible, increase your *distance* from the source of radiation.
3. Use lead *shielding* in protective structures (walls and windows), and apparel such as aprons and gloves.

In modern times, radiography is considered to be among the safest of occupations, as safe as working

Figure 40-2

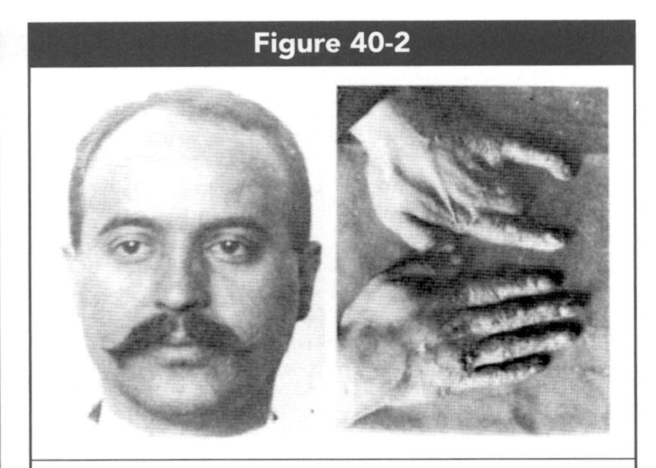

Clarence Dally and his hands showing amputations and radiation-induced radiodermatitis (cancer).

as a clerk or secretary. But, this is largely due to widespread application of the above guidelines. What we wish to accomplish in the next few paragraphs is to help the student develop a good professional *perspective* on the hazards of radiation. Armed with accurate information, radiographers can help the general public, patients, and other health care workers avoid irrational fears of radiation that deny society the vast benefits that can be realized from its proper use, while at the same time exercising the prudence that comes from an appreciation of its real risks.

On Environmental Radiation

The general public is largely misinformed about the relative hazards of radiation. Due to media hype, such issues as the Three-Mile Island atomic power plant radiation leak in 1979 (Fig. 40-3), the general risk of living near nuclear power plants, and the occupational risk of radiation workers both in medicine and in industry have been *extremely* overblown. On the other hand, the impact of such events as the Chernobyl nuclear plant meltdown in Russia in 1986 (Fig. 40-4), and the threat of nuclear weapons is often grossly underestimated by the public.

One survey asked people from three disparate groups representing a cross-section of American society to rank 30 sources of risk, from the greatest threat to life and health to the least; These groups were the League of Women Voters, college students,

Figure 40-3

Three-Mile Island nuclear power plant, Pennsylvania, where a partial core meltdown in 1975 released only 17 curies of radiation into the air, but media hype greatly exaggerated the event.

Figure 40-4

The Chernobyl nuclear power plant, Ukraine, where at least 5000 early deaths in Eastern Europe are attributed to a core meltdown in 1986 that released 50 million curies of radiation, a true environmental disaster.

and members of business and professional clubs. Actual risk for each category was ascertained by objective, statistical measurements of illnesses, accidents and deaths caused. Among the list of risks were medical x-rays and nuclear power plants:

> Medical X-Rays were ranked 22nd by the League of Women Voters, 17th by college students, and 24th by businessmen. The objective measurement of risk places medical x-rays in 9th place, after smoking, alcohol, automobiles, handguns, electric power, motorcycles, swimming, and surgery.
>
> Nuclear Power Plants were ranked as the highest source of risk among college students and the League of Women Voters as well. Businessmen ranked nuclear power 8th. The objective measures place nuclear power in 20th place, after all of those risks listed above for x-rays, and after railroads, flying, construction, bicycles, hunting, home appliances, fire fighting, police work, and contraceptives.

You can see that while the hazards of nuclear power were grossly distorted, those of medical x-rays were generally *underestimated* by the public.

Developing a Frame of Reference

For the purpose of introducing our topic, we will use the *mR* (*milliroentgen*) as the unit for radiation exposure. It is not necessary at this point to mathematically define the *mR* (this is done in the next chapter), but some general analogies can be made such that the unit can be used to get a feel for the range of risk levels involved from different radiographic procedures and environmental sources. We will select a handful of *landmark* comparisons which will provide a basis for further discussion.

We will make some broad comparisons to the natural background radiation levels, to the risks of driving and of smoking cigarettes. It must be acknowledged that such comparisons are inherently difficult to make with true statistical validity, and that studies vary widely in their estimates of equivalent risk. Nonetheless, by being conservative in making these conversions, we can make certain statements with a high degree of confidence.

Natural background radiation refers to radiation we all receive constantly from nature around us, from the earth, the sun and stars, and even from our own bodies which contain some radioactive minerals. Because the occurrence of radon gas around the world is so variable, natural background radiation can range from 100 to 500 mR per year. With radon gas exposure averaged worldwide, we obtain an average background exposure level of about *300 mR/year*. All living organisms on earth receive this amount of radiation, and so it might be considered as a comparatively *low* amount

For our first landmark, we can begin with the most common x-ray view taken, the single-view PA chest radiograph, which typically delivers about 20 mR to the skin surface of the patient. Dividing this amount into the natural background level, you can see that it would take perhaps 15 chest x-rays to equal your annual exposure from nature. We can say that getting a *suntan* is more hazardous than getting a single PA chest exposure. The overall risk can be compared to the risk of driving about 4 or 5 miles, or of smoking a few cigarettes and then quitting permanently.

A single AP abdomen radiograph makes a useful landmark for exposure levels. It falls in the range of 300-500 mR, which happens to be about the same as the *average annual occupational exposure for practicing radiographers*. It is also close to the natural background radiation exposure per year. Risks associated with this level of exposure are comparable to driving 100 miles or smoking 40 cigarettes and then quitting permanently. A single *sunburn* to the face causes more harm.

On the high end of diagnostic radiography procedures, we find that a typical barium enema series, assuming 4 minutes of fluoroscopy beam-on time and 11 overhead and spot-view projections combined, can deliver 12–15 R, that is, *12,000–15,000 mR* to the patient. (The upper GI procedure is somewhat less.) This exposure is ameliorated by the fact that it is *spread out* around the patient's abdomen. Nonetheless, it should be sobering to note that this dose is equivalent to *600 chest x-rays*, or to *40 years* of natural background radiation. As we shall see, this is an exposure level that should not be feared, but should be *respected* by taking all appropriate precautions.

Patients who receive multiple barium enemas over the years can *accumulate* enough radiation exposure to pose real health risks, so their doctors must be conscientious of the number of procedures ordered. Cardiac catheterizations, angiograms, other circulatory procedures, and C-arm fluoroscopy procedures in surgery typically deliver even more radiation to the patient. Poorly done, it is possible for a *single* procedure in these categories to burn the patient's skin.

By examining these three landmark procedures, the single PA chest, the single AP abdomen, and the multiview barium enema with fluoroscopy, we gain a perspective for the *substantial range of exposure levels the patient can be subjected to in routine diagnostic radiography*. Some procedures are truly almost negligible in their risk level, while others are of serious concern. Regardless of the risk level, the *ALARA* (*as low as reasonably achievable*) concept must always be followed by radiographers, because of the risks of *cumulative* radiation exposure over time.

In Chapter 1 an overview was given of the major historical developments of radiography. Those landmarks which had the most dramatic effect in reducing exposure to the patient include the invention of intensifying screens by Thomas Edison's group and the cassette/screen system by Michael Pupin in 1896, and the invention of the electronic image intensifier for fluoroscopy by John Coltman in 1948. The invention of the first automatic exposure control (AEC) by Russel H. Morgan in 1942, and the development of intensifying screens made from *rare earth phosphors* in 1972 also contributed to lowering radiation exposure.

Now, some *environmental* comparisons can be made relevant to our previous discussion on the general public's perceptions of risk. The Three-Mile Island accident was the worst release of radiation from a nuclear power plant in United States history. When a partial core meltdown occurred at this Pennsylvania plant in 1979, about 17 curies of radioactive gases were vented into the air. The average exposure to the local public was about 1.5 mR, an equivalent risk to driving less than a mile or smoking less than one cigarette.

It has been estimated that between 2 and 4 human deaths may have occurred prematurely due to exposure to radiation from Three-Mile Island. Media hype grossly exaggerated the danger, and fueled strong protests against nuclear power in general throughout the United States, and was responsible in large part for delaying the development of this important source of electrical power for several decades. (By comparison, the country of France produces more than 95% of its electricity from nuclear power.)

On the other hand, a real modern tragedy occurred in Chernobyl, Ukraine in 1986, the worst nuclear power plant disaster in world history, when a series of human errors combined with poor technical back-up resulted in a core meltdown that caused an explosion and fire. More than *50 million curies* of radiation were released into the air (compared to 17 for Three-Mile Island). Local villagers were receiving 25,000 mR per hour, and 116,000 people had to be evacuated from a 30-square mile area. Many firemen and victims accumulated more than 200 R, causing 31 acute deaths, 203 documented cases of acute radiation syndrome, and 500 hospitalizations.

A dense cloud of radioactive fallout drifted eastward across the Asian continent and Pacific Ocean as it dissipated. A spike in radiation levels could be measured on the California coastline. The average radiation exposure to *survivors* of Chernobyl was 45 R or 45,000 mR. The total number of deaths forecasted throughout Eastern Europe due to this event ranges from 5000 to 75,000. This was a tragic wake-up call to the level of oversight needed for nuclear power plants.

Japanese survivors of the atomic bombs which were dropped by the United States on Hiroshima and Nagasaki in World War II averaged about 200 R (200,000 mR), which caused many cancers and other health complications. It is sobering to note that the *hydrogen bomb* warheads that are placed on current nuclear missiles have over 1000 times the power of the *atomic bombs* that were dropped on Japan, and would spread radiation over a much greater radius when detonated.

Some human deaths will occur any time a single acute exposure of more than 100 R is delivered to a large enough population. At a point somewhere between 300 and 450 R, one-half of the humans exposed will die. Exposures above 700 R will be fatal to the entire human population exposed, regardless of medical intervention.

Table 40-1 is presented as a summary of these landmarks for comparative radiation risk.

Table 40-1

Landmark Radiation Levels and Comparative Risk

		Equivalent Risks		
Procedure / Source	*High-End Exposure*	**Cigarettes Smoked**	**Miles Driven**	**Equiv. Background Radiation**
Three-Mile Island	1.5 mR	0.2	0.33	2 days
Single PA Chest	**20 mR**	**2**	**4**	**3 weeks**
Single Abdomen	**400 mR**	**40**	**100**	**1.3 years**
Radiographer Annual Occupational Exp.	400 mR	40	100	1.3 years
Barium Enema Series	**12–15 R**	**1200**	**3000**	**40 years**
Chernobyl Survivors	45 R	3600	9000	150 years
Hiroshima Bomb Survivors	200 R	16,000	40,000	700 years

SOURCES OF RADIATION

We can divide the origins of radiation into *natural* and *man-made* sources. Radiation is all around us in nature, even in the water we drink. Cities must often report on the level of various contaminants in their treated water, which include some radioactive minerals. The sources of natural background radiation can be placed into three broad categories: Cosmic radiation, terrestrial radiation and internal radiation.

Natural Background Radiation

Cosmic radiation emanates from outer space, the largest source being our own sun. We are protected from much of the most hazardous types of cosmic radiation by the earth's magnetic field, and then the earth's atmosphere provides additional shielding before this radiation reaches us. Folks who live at higher altitudes receive less protection from the thinner layers of atmosphere above them. For example, people living in the "mile-high" city of Denver receive about 70 mR per year more than those living close to sea level. A transoceanic jet flight will bring you an additional 5 mR from the time spent at high altitude, and transoceanic flight crews can accumulate 170 mR per year, for which some airlines require radiation monitoring. The overall worldwide average exposure to cosmic radiation is about 30 mR per year.

Terrestrial radiation is the exposure we receive from radioactive minerals in the ground and water below us. Some parts of the earth's surface have very concentrated levels of radioactive minerals. For example, in parts of India the soil emits up to 4000 mR per year. In most cases, the higher levels of radiation are found in mountainous terrain where the earth's crust has been upthrust. Strike two for those in Denver. The worldwide average exposure to terrestrial radiation, *other than radon*, also averages about 30 mR per year.

Radon gas was found to be a big problem when a nuclear employee kept reporting to work in the mornings with higher exposure readings on his monitor than at the end of his previous work day. A subsequent investigation of his home found that high levels of radioactive radon gas had accumulated in his basement. Radon gas is a natural by-product of transmutation in the decay series for uranium. Unless the basement of a home is sealed (with a plastic liner, for example), radon gas seeps through cracks and flaws in the concrete foundation into the basement. Without good ventilation, the gas can accumulate to dangerous levels over a period of time.

Radon accumulation has since been found to be a worldwide problem, very acute in some areas, and almost nonexistent in others. The radiation levels can be so high, however, that when these figures are *averaged* for a continental population, they come to 100–200 mR per year. This must be added to our estimate of terrestrial radiation levels.

Finally, *internal radiation* refers to radioactive minerals that are permanently stored within the human body. Many minerals will be typically found with a small percentage being in the form of a radioactive isotope. The most prominent of these found in the human body is Potassium 40, which attaches to calcium in the bones. These minerals then radiate organs nearby which may be affected.

Strontium 90 is an example of a man-made source of radiation which became an internal source. It was released into the air along with radioactive fallout from nuclear bombs that were tested *above the ground* during the 1950s when the United States and Russia were locked in a race to develop nuclear weapons. The strontium-laced dust particles settled onto pasture grass which was eaten by cows. It then concentrated in their milk. Humans ingesting this cows' milk absorbed the strontium into their own bones and breast milk, where traces can be found to this day.

The total amount of radiation exposure from internal sources averages about 40 mR per year. When we sum up these sources (Table 40-2), we see that an overall average for natural background radiation would be 200–300 mR per year, but that those living in radon-free areas might receive only 100 mR per year.

Table 40-2

Annual Natural Background Radiation Levels

Cosmic Radiation	30 mR
Terrestrial Radiation (Non-radon)	30 mR
Internal Radiation	40 mR
Radon Gas	100–200 mR
TOTAL . . . . :	200–300 mR

Manmade Sources of Radiation

In addition to the medical use of radiation, we receive radiation daily from various consumer products, technology, waste, and building materials, which we might refer to as *manmade background radiation*. Compared to a wooden house, by dwelling inside a home made of brick we receive an additional 10 mR per year from potassium, uranium and thorium isotopes. People who work in buildings made of granite and marble, such as many government buildings and banks, receive still higher radiation exposures. Smoke detectors contain a radioactive source that adds a very small amount to our daily exposure.

The list of consumer products which irradiate us is truly fascinating; glossy magazines expose us to 4 mR per year because of uranium and potassium in the clay that is used. Enameled jewelry can deliver from 25 to 4000 mR per year, mostly from uranium. If you hold a Geiger counter up against the thorium-treated cloth *mantles* that are used in camping lanterns, they will set it wildly clicking. Our food also contains all kinds of levels of radioactive substances. For example, bananas contain enough potassium 40 to deliver 5 mR per year. Fertilizers, and foods containing them, can deliver over 100 mR per year.

We have previously discussed accidents at nuclear power plants. An interesting twist in this picture is found when we compare them under normal operation: By living within 50 miles of a *coal-fired* electrical plant, you will receive three times more radiation (0.03 mR per year) than from living the same distance from a nuclear power plant (0.01 mR per year). *Occupationally*, nuclear power plant and industrial radiation workers receive about 2 mR per year.

Cigarettes bear special mention: A person smoking one pack per day can receive from 2000 to 5000 mR per year from the elements polonium 210 and lead 210. About 22 percent of these isotopes is retained within the cigarette smoke, and about 15 percent of that is inhaled. The particulates from the smoke remain permanently within the lungs, exposing the surrounding tissues to deadly alpha radiation for the rest of the person's life. This is all in addition to the health effects of the tar and nicotine.

All of these manmade background sources combined only constitute about one-quarter of our manmade exposure to radiation. The other 75 percent is from *medical practice*, amounting to an average of about 55 mR per year. Nearly one-third of the entire population of the United States is subjected to a medical x-ray procedure each year. The pie chart in Figure 40-5 summarizes all of the types of radiation sources, natural and manmade, to which we are subjected.

Figure 40-5

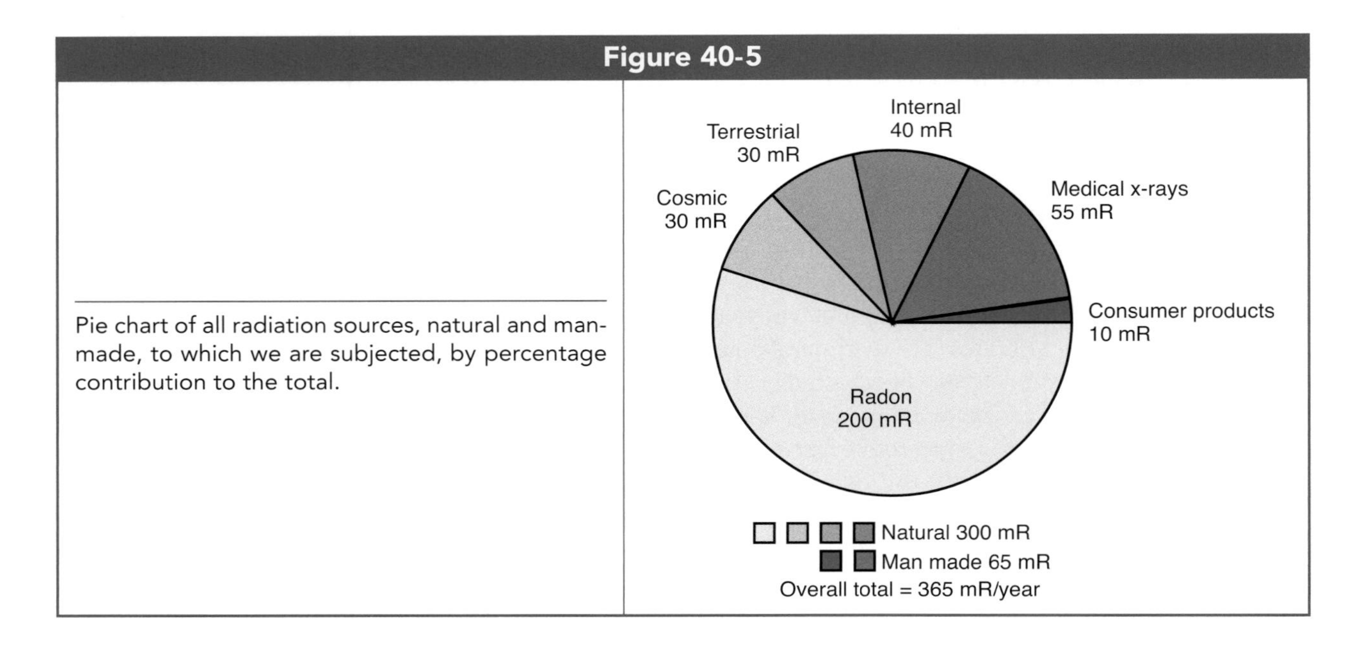

Pie chart of all radiation sources, natural and man-made, to which we are subjected, by percentage contribution to the total.

RADIOACTIVITY

Radiation in general can be classified into waves in media such as sound waves, particulate radiation such as alpha and beta particles, and electromagnetic waves such as gamma rays and x-rays. The spontaneous emission of alpha, beta and gamma radiation, and their effects on the atomic number and atomic mass of atoms, was discussed in Chapter 4. The nature of electromagnetic radiation and the electromagnetic spectrum were discussed in Chapter 5. It will be recalled that the alpha particle, consisting of two protons and two neutrons, is huge in mass compared to other particles, while the beta particle consists of a single, high-speed electron with diminishing mass and one-half the electrical charge. Both gamma rays and x-rays are electromagnetic waves possessing no mass or charge.

Table 40-3 summarizes the characteristics of all four types of radiation, in order of increasing penetration ability. Your attention is called particularly to the columns labeled *penetration* and *harm* for the current discussion. Note that those radiations which are more *harmful* to human tissue are also more *effective* in destroying tumors during radiation therapy.

Alpha particles have extremely low penetration, with almost none of them making it more than 1 *millimeter* into the body. Due to their extreme mass, roughly 8000 times larger than an electron, these behemoths *plow* through an atom reaping destruction in the form of multiple ionizations. Because they also have an electrical charge of +2, they need not

Table 40-3

Types and Characteristics of Radiation

Type	Symbol	Penetration	Harm	Mass	Charge	Source
Alpha	α	↓↓ (1 mm tis)	↑↑	~ 8000 e–	+2	Nucleus
Beta	β–	↓ (2 cm tis)	↑↑	1	–1	Nucleus
Gamma	γ	↑ (> x-rays)	↓	0	0	Nucleus
X-Ray	x	↑ (2% thru)	↓	0	0	OUTSIDE Nucleus

directly collide with any orbital electron to eject it from its orbital shell, but can attract and *pull* electrons out of their orbits as they pass by (Fig. 40-6).

The low penetration of alpha particles also means that the damage they do is more *concentrated* within that first millimeter of tissue, where all of their kinetic energy is deposited as they come to a complete halt in their movement. Alpha particles from external sources are therefore extremely harmful to the skin. The polonium 210 and lead 210 contained in tobacco smoke particles mentioned in the previous section are alpha-emitters. When these types of substances are aspirated into the lungs and remain there permanently, *all* of the alpha particles they emit, covering a 360-degree radius in all three dimensions, are absorbed by *local* tissue in concentrated form. This explains why this type of radiation exposure is particularly insidious in inducing lung cancer.

Beta particles also need not directly collide with the orbital electrons of atoms in order to ionize them. In this case, their negative charge *repels* orbital electrons out of their shells as they pass by (Fig. 40-7). However, since their charge only has a magnitude of 1 (versus 2 for alpha particles), they must pass by at a closer distance. Their extremely small size makes ionizing events more rare as they penetrate through the atom. Even so, as particles with mass, their penetration is limited to about 2 centimeters or less through human tissue. This suits beta radiation to the treatment of subcutaneous and testicular cancers, as the bulk of their energy is deposited around 1 centimeter into the tissue.

X-rays, as electromagnetic waves with no mass or charge, have so much higher penetration than beta particles that 1–2 percent of the x-ray beam passes all the way through the human body and exits the opposite side, making it possible to produce radiographic images with them.

Gamma radiation is used to treat cancers seated deeply within the body. Its penetration is higher than that of x-rays—as much as 10 percent of the beam can penetrate all the way through the body. Radiation dose is concentrated within a tumor mass by treating the patient from various *ports* or angles all passing through the same focal-point or center within the body. This also spreads out the distribution of radiation dose to the skin which spares it the harm done to the tumor. Since gamma rays have energies measured in hundreds of thousands to millions of volts, they can be very effective when they *are* absorbed within tissue.

Figure 40-6

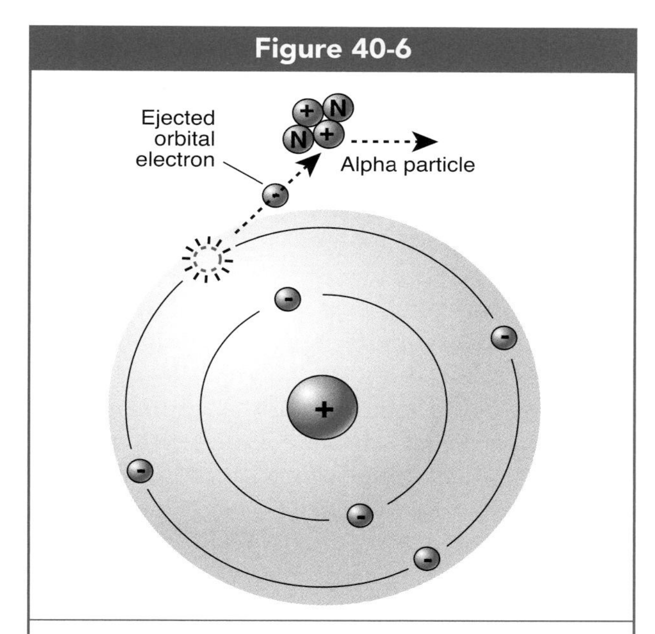

An alpha particle, with an electric charge of +2, pulls orbital electrons out of their atomic shells. Its massive size, nearly 2000 times that of an electron, also makes it much more likely to interact with several electrons as it passes through atoms.

Figure 40-7

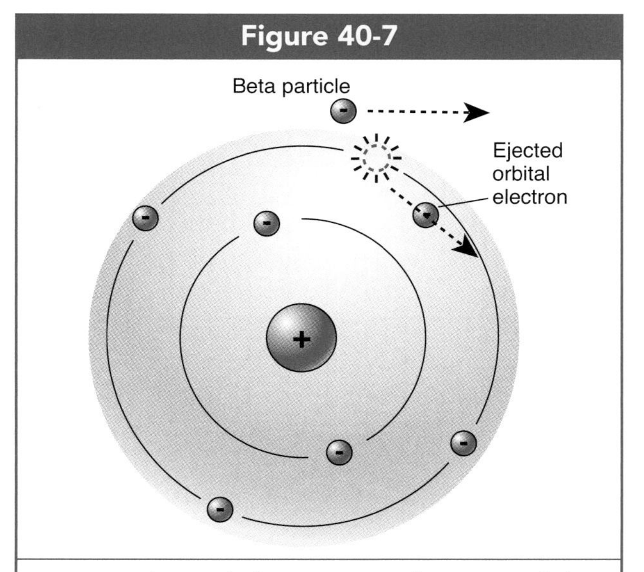

Beta particles, with their negative charge, *repel* electrons out of their orbital shells, but are less likely to encounter them than alpha particles.

As discussed in Chapter 4, *radioisotopes* are elements that have such an unusual number of neutrons that their atomic nucleus becomes unstable. The nucleus seeks a lower, more stable energy level by emitting different combinations of alpha, beta and gamma radiations, depending on the particular isotope. Some substances emit all three in the course of their radioactive decay.

There are actually two aspects of an unstable substance that make it more dangerous (or more effective); these are its *decay rate* or *radioactivity*, and its radioactive *half-life*. In its purest sense, the *radioactivity* of a particular substance is the *rate* at which it is decaying, usually measured as the number of decay events *per second*. On a Geiger counter, this is heard as the average frequency of "clicks" the device makes as each second passes.

This might be considered as a measure of the substance's *current* effectiveness in emitting radiation, whereas the radioactive *half-life* might be thought of as a measure of the substance's *long-term* effectiveness over time. We might say that a highly radioactive object is dangerous *right now*, but that a substance with a long half-life will *still be highly dangerous some time into the future.* A substance can be highly radioactive, but short-lived so that its radioactivity quickly dies out. Other substances can have lower rates of radioactivity, yet maintain that rate over long periods of time. Some can both be highly radioactive and have a long half-life. There is no direct relationship between the two concepts.

Figure 40-8

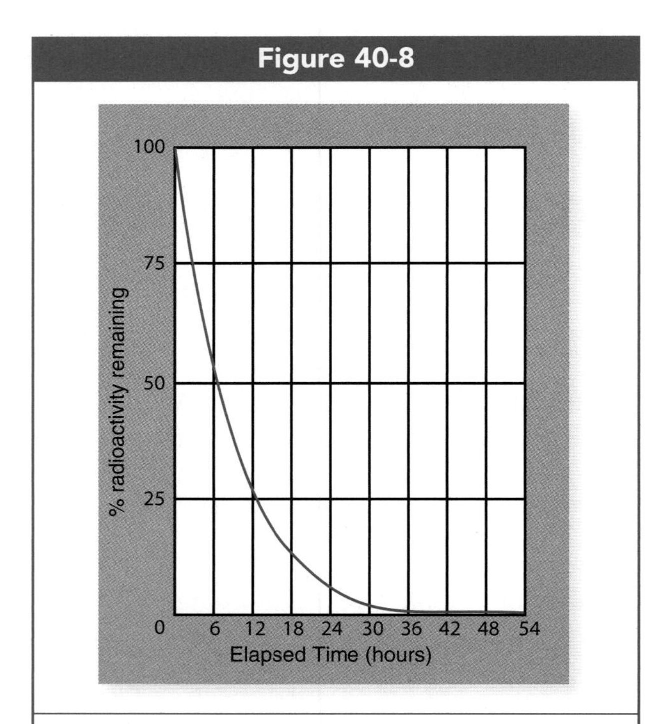

The radioactivity of *technetium 99m* and all other radioactive elements follows an *inverse exponential* curve, the same percentage of what remains decaying with each half-life.

HALF-LIFE

The *half-life* of a particular radionuclide (radioactive element), abbreviated $T^{1/2}$, is defined as the *time* required for the rate of its radioactivity to decrease to one-half of the original. The half-life is a characteristic of the particular radioisotope. That is, each radionuclide has a distinct half-life different from other radioactive elements. Unlike radioactivity, which decreases over time, the half-life is absolutely constant for a particular radionuclide.

For example, technetium 99m, abbreviated Tc^{99m}, is a radionuclide commonly used in nuclear medicine. It is an *isomer*, meaning that it emits only gamma rays to reach a lower-energy, more stable state in its nucleus. Tc99m has a half-life of approximately 6 hours. This makes it ideal for nuclear medicine imaging, because after the images are produced, the radioactivity fades away quickly so that the patient receives a minimal total dose of radiation.

Figure 40-8 shows that the actual *radioactivity* of technetium follows an inversely *exponential* curve over time. It is reduced to one-half the previous amount every 6 hours. Although in reality a time will come when the substance reaches *zero* radioactivity and is no longer radioactive, hypothetically this curve in Figure 40-8 never reaches zero because we *cannot predict* when the last radioactive atom in the sample will decay.

It is easier to read this exponential curve by plotting it on a semi-logarithmic graph as shown in Figure 40-9. Radioactivity is listed to the right in units of *microcuries* (μCi). Each microcurie represents about 37 thousand decay events, or gamma rays emitted, per second. For our example, we begin with a sample of Tc^{99m} whose radioactivity is measured

Figure 40-9

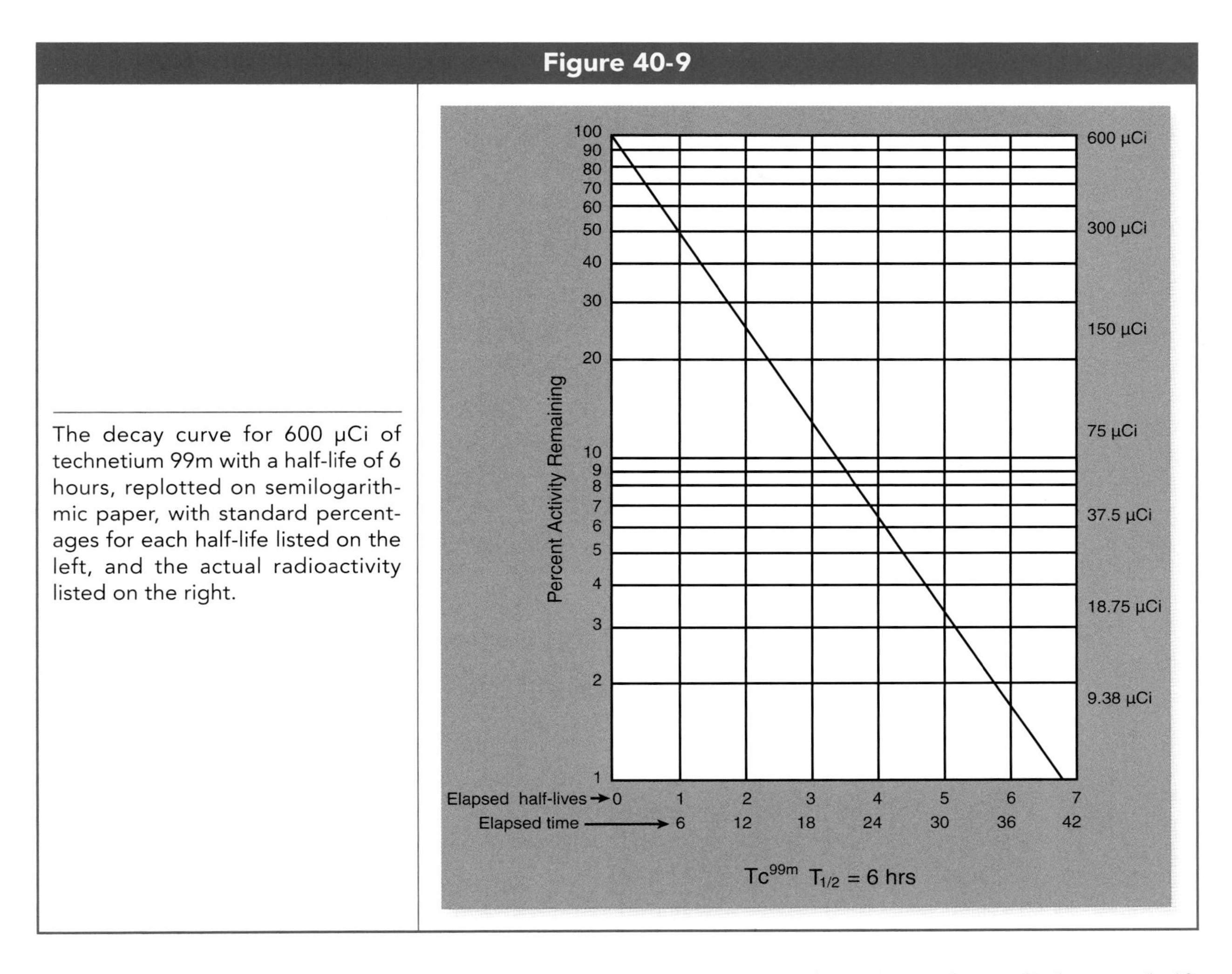

The decay curve for 600 μCi of technetium 99m with a half-life of 6 hours, replotted on semilogarithmic paper, with standard percentages for each half-life listed on the left, and the actual radioactivity listed on the right.

at 600 μCi. After 6 hours have passed, we see that, reading up to the curve and then to the *right*, there are 300 μCi of radioactivity remaining. After 12 hours, or *two half-lives*, 150 μCi remain. After 18 hours, or three half-lives have elapsed, 75 μCi remain, and so on. While the radioactivity is dropping, the half-life is always constant—6 hours.

On the left-hand side of the graph in Figure 40-9 we have listed the *percentage* of the original radioactivity remaining as time passes. Expressed in percentages, these numbers are always the same for a given number of elapsed half-lives, as shown in Table 40-4. You can see that after 10 half-lives have elapsed, any substance will have *less than one ten-thousandth* (less than 0.01%) of its original radioactivity remaining. It is generally defined that *after passing through 10 half-lives, a substance is no longer considered "radioactive."*

An algebraic formula can be applied to any half-life problem, but it is simpler to just apply the percentages from Figure 40-9 and Table 40-4. A "table method" for laying out these problems is presented in Table 40-5. By filling in the given information, in column #2 or #3, the mathematical set-up to solve the problem should become clear. (Be sure to always start with the original radioactivity listed at *zero* half-lives, not at half-life #1.)

Note that fractions of half-lives can also be factored in. For example, what percentage of radioactivity remains after 3 and 1/3 half-lives have elapsed? From Table 40-5, we see that at *3 half-lives* 12.5 percent remains, and at *4 half-lives* 6.25 percent remains. We need to find the point *one-third of the way down from 12.5 to 6.25*. To do this, first find the difference by subtracting 12.5 – 6.25 = 6.25. Now find *one-third* of this difference: 6.25/3 =

Table 40-4

Percentage of Original Radioactivity Remaining, by Half-Life

Half-Lives Elapsed	Radioactivity Remaining
0	100%
1	50%
2	25%
3	12.5%
4	6.25%
5	3.125%
6	1.575%
7	0.788%
8	0.394%
9	0.197%
10	0.098%

Table 40-5

Table Method for Solving Half-Life Problems

Half-Life#	Time Elapsed	Level of Radioactivity	Percent Radioactivity Remaining
0			100%
1			50%
2			25%
3			12.5%
4			6.25%
5			3.125%
6			1.575%

2.08. Finally, subtract this amount from the 12.5 percent that remained after 3 half-lives: 12.5 – 2.08 = 10.42. The remaining radioactivity after 3 and 1/3 half-lives will be 10.42 percent of the original radioactivity. The following practice exercises illustrate how to solve various types of problems using these relationships.

Practice Exercise #1

We have a sample of 89.88 μCi of Iowa Hawkium as of 1:15 p.m. In how many half-lives will exactly 12.5 percent of this amount remain?

Solution: *Radioactivity and time are irrelevant. 12.5% represents three halvings of 100%.*

Answer: *This amount of radioactivity remains after 3 half-lives.*

Practice Exercise #2

We have a sample of 80 μCi of Ohio Buckium with a $T\frac{1}{2}$ of 3 days. How radioactive will it be in 9 days?

Solution: *9 days = 3 half-lives*
Using the table method (Table 40-5):

At 3 HL, 12.5% remains: 12.5% × 80 = 10

OR, cut 80 in half 3 times: 80 -> 40 -> 20 -> 10

Answer: *After 9 days, its radioactivity will be 10 μCi.*

Practice Exercise #3

We have a sample of 80 μCi of Texas Longhornium with a $T\frac{1}{2}$ of 3 days. How radioactive will it be in 11 days?

Solution: *11 days = $3\frac{2}{3}$ half-lives*
Using the table method (Table 38-5):

3 HL = 10 μCi
4 HL = 5 μCi
The difference is 5 μCi

⅔ of 5 is 3.3. This will be the reduction from the 9th to the the 11th day.
10 – 3.3 = 6.7 μCi

Answer: *After 11 days, its radioactivity will be 6.7 μCi.*

Practice Exercise #4

A sample of Texas Longhornium with an original radioactivity of 300 Ci is found 26.67 hours later to have diminished to 31.25 Ci. How many half-lives have passed AND what is the $T\frac{1}{2}$ for Longhornium?

Solution: *Using the table method, cut 300 in half until the amounts just above and just below 31.25 are found:*

300 = 0 HL
150 = 1 HL
75 = 2 HL
37.5 = 3 HL
18.75 = 4 HL

The difference between 3rd and 4th HL =
37.5 – 18.75 = 18.75

The difference from the 3rd HL to final activity =
37.5 – 31.25 = 6.25

The proportion between 18.75 / 6.25 =
3:1, i.e., 6.25 is ⅓ of 18.75

Therefore, the final activity (31.25) is ⅓ of the way from the 3rd HL to the 4th HL

Answer: The Longhornium has undergone 3⅓ half-lives

It's T½ is 26.67 hours divided by 3.33 half-lives = 8 hours

The half-lives of various radionuclides can range to extremes in time. Some radionuclides have half-lives lasting only a few seconds or even a fraction of a second, others are measured in days, months, years, centuries, or even millennia. The uranium isotope U^{235} used in nuclear weapons, for example, has a half-life of about 700 million years. It remains radioactive for a very long time, indeed.

CONCLUSION

Radiation is ever-present in our environment and comes from both natural and manmade causes. Occupationally, radiography is a very safe profession. On the other hand, the range of exposures patients may receive in the course of medical treatment range widely from negligible amounts to levels of real concern for some procedures, and can become especially hazardous when accumulated over time from repeated procedures. While the general public tends to overestimate the hazards of industrial radiation, people generally *underestimate* the hazards of medical radiation. It is the legal and ethical responsibility of every radiographer to keep all exposures to both patients and personnel *ALARA*, as low as reasonably achievable, and to also be a source of accurate information for the public.

SUMMARY

1. Although in its early history radiography was a hazardous profession, by following the fundamental principles of *time*, *distance* and *shielding* modern radiography is classified among the safest of professions.
2. The range of radiation exposures for various medical procedures is wide, from practically negligible amounts for chest radiography to potentially hazardous amounts for some circulatory and surgical procedures.
3. The average annual occupational exposure to radiographers, the annual natural background exposure, and the exposure to a patient from a typical single-view abdomen projection, are all about the same (300–500 mR), and thus serve as a landmark for comparisons.
4. It takes a single, acute exposure of at least 100 R, or 100,000 mR, to risk death to a human being.
5. Natural background radiation includes cosmic, terrestrial, and internal sources, while manmade sources of radiation include consumer products and industry as well as medical practice.
6. The lower the penetration capability of a particular type of radioactivity, the more hazardous it is, because tissue damage becomes more concentrated. The introduction of alpha emitters into the human body by aspiration, ingestion, or other means poses the greatest health risk.
7. The half-life of a radioactive substance is the time required for its level of radioactivity to decline to one-half the original rate. The health hazard presented by a particular substance increases with both high radioactivity and long half-life.
8. Substances having undergone 10 half-lives are considered to be no longer radioactive.
9. While radioactivity declines over time, the half-life of a particular radionuclide is always constant.

REVIEW QUESTIONS

1. Legally, the reasonable provision of radiation protection for patients and personnel is considered to be part of the __________ of practice for radiographers.

2. The general public tends to __________-estimate the hazards of nuclear power and to __________-estimate the hazards of medical radiation.

3. What routine contrast agent procedure in diagnostic radiography delivers the highest exposure to the patient considering all views taken?

4. An exposure of __________ would be required to be fatal to an entire human population.

5. What source of natural background radiation poses the greatest hazard in modern times?

6. What percentage of all manmade radiation exposure is due to medical practice?

7. What specific type of radioactivity is most hazardous, and why?

8. After three half-lives have expired for any radioactive substance, what percentage of radioactivity remains?

(Continued)

REVIEW QUESTIONS *(Continued)*

9. We have a sample of 400 μCi of Oklahoma Soonerium with a half-life of 4 days. How radioactive will it be after 18 days?

10. Starting with a sample of 220 μCi of Nebraska Cornhuskium, if its half-life is 8 hours, how many hours will have passed by when 10.3 μCi of this radioactivity remains?

11. What is the *only* thing that determines the half-life of a particular radioactive substance?

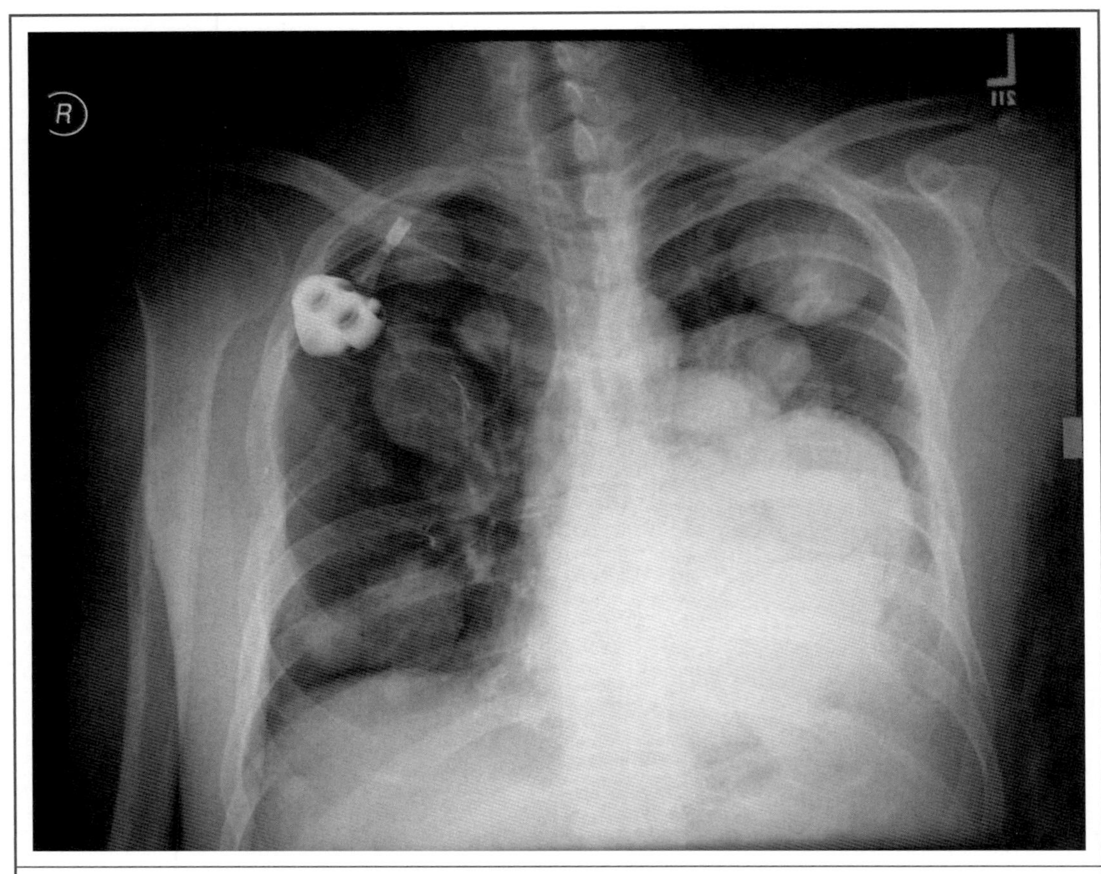

Massive cancer metastases within the lungs.

Chapter 41

RADIATION UNITS AND MEASUREMENT

Objectives:

Upon completion of this chapter, you should be able to:

1. Define the conventional units of measurement for radiation and how they are derived.
2. Explain *dose-area product (DAP)* and its importance for understanding biological harm.
3. List the *Systeme International* units for radiation and their conversion factors from the conventional units.
4. Calculate dose-area products and conversions between conventional and SI units.
5. List the dose-limiting organs of the body and how they are used to determine *total effective dose equivalents.*
6. Describe the historical evolution of dose equivalent limits (DELs) and its implications.
7. Define and calculate cumulative lifetime limits, prospective limits, and retrospective limits for different workplace scenarios, and state in each case which limit supersedes the others.
8. State by memory landmark occupational and public whole-body dose equivalent limits (DELs).
9. State by memory landmark partial-body DELs for different organs.
10. Define *genetically significant dose*; state the current estimate and percentage contributions by radiographic examination.
11. Describe the modes of operation and appropriate units for radiation detection instruments.
12. Define the sensitivity, accuracy, resolving time and range for radiation detection instruments, and how these apply to radiographic equipment.
13. Describe the four generic types of radiation detection instruments, how each works, and the advantages and disadvantages of each.
14. Describe the four types of gas-filled radiation detectors, how each works, and the advantages and disadvantages of each.
15. Describe the advantages and disadvantages of three different types of devices for personal radiation monitoring.
16. Explain the voltage-dependence of each of the four types of gas-filled detectors and how it affects calibration, accuracy, and validity of measurements taken.

RADIATION UNITS

As part of the worldwide conversion to the "metric system" or *systeme international, SI units* have been developed for radiation measurements which are more amenable to a base-10 number system than the conventional units, but the conventional units also continue to be widely used, particularly in the United States. The radiographer should be familiar with both systems. The conventional units will be used to introduce the *concepts* behind each type of measurement.

Conventional Units

The *Roentgen* (*R*) has been defined as an appropriate unit for measuring exposure to x-rays. *Exposure* is narrowly defined as *the intensity of radiation incident upon the surface* of an object, be it the skin of a human body or the wall of an ion chamber. How does one actually *count* the x-ray photons impinging across this area, *before* some of them are absorbed by the object? The best solution is to extrapolate their number from the ionizations they are causing within air molecules near the surface being considered. This process generates an *electric charge* within the air. Electrons that are freed by ionization can then be attracted to the positive anode of a detection device. There, they generate a small electrical current which can be measured.

In terms of actual ionizations in the air, 1R of x-rays passing through about one cubic meter of air liberates roughly *1½ quadrillion* electrons. This would be in one whole metric cube of air with about the same dimensions as the 40-inch standard SID, so the ionizations taking place within a 3-inch diameter detector chamber set on the tabletop would be only a fraction of this number, yet it still comes to around 10 billion ionizations, plenty to induce an electrical current.

The roentgen unit is only appropriate for measuring the *intensity of the "raw" x-ray beam* emitted from the x-ray tube at a particular distance. It is not a very good indicator for the *biological impact* of the radiation exposure. Asking how many roentgen one received from an x-ray exposure is liking asking what the temperature of a fire was that burned down your house—one is more vitally interested in *what actual damage was done.*

To determine this damage, we need a radiation unit that takes into account the amount of *energy* from the x-ray beam that was actually *absorbed* by body tissues, rather than how much radiation entered through the surface of the body. Some of that radiation will have penetrated clear through the body without undergoing any interactions, hence without causing any damage.

The unit that measures the amount of radiant energy deposited in body tissues is the *rad*, abbreviated with a small "r." *Rad* is an acronym for "Radiation Absorbed Dose." Although radiographers frequently use the terms *exposure* and *dose* interchangeably, the correct use of the term *dose* is always in connection with *internalized* or absorbed energy, just as a dose of medicine is internalized, whereas *exposure* connotes only what we were subjected to, or "exposed to" from the outside.

Specifically, one rad is defined as 100 *ergs* of energy absorbed per gram of tissue. Recall that the *erg* is a generic unit for energy, equivalent to one ten-millionth of a *Joule*. When 1 rad is absorbed by the body, *each* gram of exposed tissue absorbs 100 ergs of energy.

Since some x-rays penetrate all the way through the body tissues, we would expect the absorbed *dose* to be slightly less than the *exposure*. This is the case, as 1 R of exposure generates about 0.96 rads of dose. In other words, 1 R of exposure results in 96 ergs of energy being deposited into each gram of tissue through which the radiation passes. Since 0.96 rads is so close to 1, in practice we generally round it up and state that 1 R of x-ray exposure causes approximately 1 rad of absorbed dose.

$$1\text{R} \Rightarrow \sim 1\text{r}$$

The *rad* is a unit that applies to all types of radiation including particulate radiations such as alpha and beta particles, whereas the roentgen unit of exposure is designed only for use in reference to x-rays. The concept of absorbed dose still falls somewhat short, however, of getting to the point of *how much biological harm* has actually been inflicted upon an organism by an exposure to radiation. For this, we need a unit which takes into account two *weighting* factors, symbolized as W_r and W_t.

W_r refers to the relative harmfulness of the *type* of radiation, when compared to an x-ray beam having 250 kV. We know that, even though 1 rad of absorbed energy may be deposited within a gram of tissue, for extremely low-penetration radiation such as alpha particles, that energy will be more *concentrated* into small areas within that gram of tissue, which is more harmful. The W_r weighting factor is a ratio indicating how much more harmful these particulate radiations are, as follows:

W_r for x-rays and gamma rays = 1
W_r for electrons = 1
W_r for beta particles = 1.7
Average W_r for neutrons = 10 (range is 5 to 20)
W_r for alpha particles = 20

The second weighting factor, W_t, refers to the relative sensitivity of the type of *tissue* or organ being exposed, compared to other tissues. An overview of the W_t values is as follows:

Cortical bone, skin = 0.01
Organs in general = 0.05
Bone marrow, colon, lung, and stomach = 0.12
Gonads = 0.20

When the absorbed dose in rads is multiplied by these two weighting factors, we obtain the *dose equivalent*, a true measure of *biological harm* or *biological effectiveness*. The unit for dose equivalent is the *rem*. Like the rad, the term *rem* is also an acronym. It stands for the *R*adiation *E*quivalent in *M*an (or mammals). The rem is a true *biological unit*, whereas the roentgen and the rad are physics units. The rem is the most appropriate unit to use in conveying the effects of medical radiation exposure upon patients and personnel.

For a particular tissue, we can base comparisons solely on the W_r factor; we can then state that for *x-rays*, one roentgen of exposure deposits about 1 rad of absorbed dose into the body, causing one rem of harm or dose equivalent.

$$1 \text{ R} \Rightarrow 1 \text{ r} \Rightarrow 1 \text{ rem for x-rays}$$

For neutrons, a dose of 1 rad will generate, on average, 10 rems of dose equivalent. For alpha particles, 1 rad will cause 20 rems of damage.

Table 41-1 summarizes these *conventional units* for radiation, along with the unit for radioactivity, the *curie*. The curie is a simple count of the number of *decay events*, or emissions of any wave or particle from a radioactive object, each second. Thus, it represents the *rate* of decay, or radioactivity of the object. Specifically, one curie is equal to about 37 billion events per second.

Dose versus Dose-Area Product (DAP)

The unit of dose is the *rad*, defined as 1 rad = 100 ergs of energy deposited in *each* gram of tissue (1 rad = 100 ergs/gm). Since this definition includes "per gram," it is *independent of collimated field size*. For example, assume that a technique of 100 kVp and 30 mAs delivers a dose of 3 millirad (mrad). The same technique is used on 2 different patients, one using an 8 × 10-inch field and the other using a 10 × 12-inch field, as shown in Figure 41-1. We expect that both patients will receive the same dose, because

Table 41-1
Conventional Radiation Units

Unit	Measures	Defined as:	Abbreviation
Roentgen*	Exposure (surface)	Ionizations in Air (e-charge)	R
Rad	Dose	Energy absorbed (1r = 100 ergs/gm)	r
Rem	Dose Equivalent	Biological harm	rem
Curie	Radioactivity	Decays per second	Ci

*The Roentgen (R) is *not* used for particulate radiations.

Figure 41-1

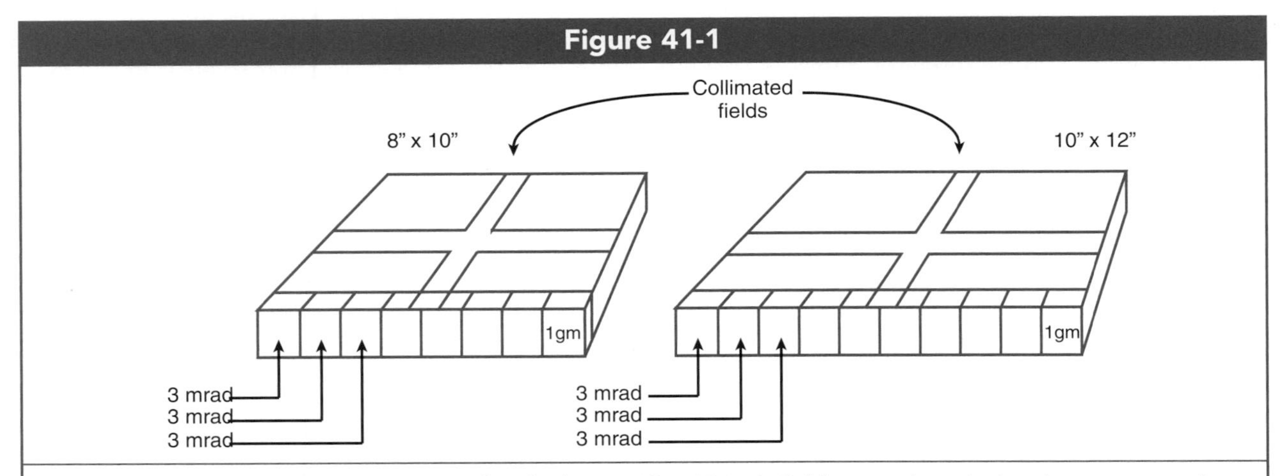

The *dose-area product* (*DAP*) is greater for the larger 10 × 12-inch field, even though the *dose*, measured per gram of tissue, is equal. DAP is a fuller indication of biological effect.

the same technique was used. This is true, since each gram of tissue absorbs 3 millirads.

However, we might ask, which patient is harmed more by these exposures? We know that the patient who had more of their body exposed (the 10 × 12" field) will be affected more *as a whole organism*. The concept of *Dose Area Product* takes this into account, by multiplying the dose times the field size.

For Patient A,

$$\text{the DAP} = 3 \text{ mrad} \times 8" \times 10" = 3 \times 80 \text{ in}^2 = 240 \text{ mrad-in}^2$$

For Patient B,

$$\text{the DAP} = 3 \text{ mrad} \times 10" \times 12" = 3 \times 120 \text{ in}^2 = 360 \text{ mrad-in}^2$$

Thus, the DAP is an indication of the total effect on or harm to a patient, taking into account the field size as well as the technique set. Insofar as the whole organism is concerned, DAP is a better indicator of overall harm (or effectiveness) than the dose in rads alone. In this example the resulting *units* were "milllirad-square inches." Any other distance measure may be used. If the field size were measured in centimeters, the resulting units from a DAP calculation would be "millirad-square centimeters."

Conversions for Systeme International (SI) Units

In the SI (metric) system, different units have been defined for exposure, dose, dose equivalent and radioactivity. These are presented in Table 39-2. The unit for exposure is the *coulomb per kilogram* (*C/kg*). As with the roentgen, the C/kg is a direct measurement of electrical charge generated by the ionization of air molecules. One coulomb of charge represents about 6 billion billion ionizations. A kilogram of air at typical atmospheric pressures is about one cubic meter of air, or a 40-inch cube of air space. The C/kg is measured with ion chambers just as the roentgen would be. Note that the C/kg *unit* is about 4000 times larger than 1 R.

It is always useful to remember which units are the larger when making conversions; it is easy to mistakenly invert a calculation, multiplying when you should have divided, or vice versa. By remembering which unit is larger, you can double-check your answer to see if you set the problem up correctly. When converting from small units to larger units, the answer must come out *less* than the original figure. Conversely, the answer should come out higher when converting from larger to smaller units.

We see in Table 41-2 that 1 R is equal to 2.58×10^{-4} C/kg. With the negative exponent, this number is a small fraction equal to the decimal number 0.000258. That is, 1 R is a very small fraction of a C/kg. When converting from R to C/kg, the answer should be much *less* than the input.

The SI unit for dose is the *gray*, abbreviated *Gy*. The gray is specifically defined as 1 joule of energy deposited per kilogram of tissue. The Gy_a connotes a joule of energy absorbed by one kilogram of *air*, whereas the Gy_t is defined for energy absorbed in *tissue*; however, the general use of the unit Gy is implied to refer to tissue unless otherwise specified. Since 100 ergs is one hundred-thousandth of a joule, and a gram is one-thousandth of a kilogram, we find the ratio of the gray over the rad to be 100,000/1000 or 100. *One gray is equal to 100 rads.* The gray is the larger unit.

The SI unit for dose equivalent is the *sievert (Sv)*. The conversion ratio is the same as for the gray: *One sievert is equal to 100 rems.* The sievert is the larger unit.

The SI unit for radioactivity is named after the discoverer of natural radioactivity, the *becquerel*, abbreviated *Bq*. A becquerel is a single decay event, one emission of an alpha or beta particle or a gamma ray from a radioactive object. One curie, then, is equal to 3.7×10^{10} becquerels, (37 billion Bq). This is such a large number that doses injected in nuclear medicine are generally measured in *microcuries*, equal to 37 thousand becquerels each.

In solving unit conversion problems, the *dimensional analysis* approach is recommended, and it is much easier if all numbers are converted into scientific notation format (see Chapter 3). Let's do two conversions for practice, one from large to small units, one from small to large:

Practice Exercise #1

How many roentgens are there 3 C/kg?

Solution: From Table 41-2:

$$3\ C/kg \times \frac{1\ R}{2.58 \times 10^{-4}\ C/kg}$$

Converting to scientific notation:

$$3 \times 10^{0}\ C/kg \times \frac{1\ R}{2.58 \times 10^{-4}\ C/kg} =$$

$$\frac{3 \times 10^{0}\ R}{2.58 \times 10^{-4}} = 1.16 \times 10^{4}\ R$$

Answer: This is 11,600 R.

Practice Exercise #2

How many Ci are in 500 becquerels?

Solution: From Table 41-2:

$$500\ Bq \times \frac{1\ Ci}{3.7 \times 10^{10}\ Bq}$$

Table 41-2

Systeme International Units

Unit	Abbreviation	Conversions
Coulomb per Kilogram	C/kg	$1R = 2.58 \times 10^{-4}$ C/kg
Gray	Gy	1Gy = 100 r
Sievert	Sv	1 Sv = 100 rem
Becquerel	Bq	$1\ Ci = 3.7 \times 10^{10}$ Bq

Converting to scientific notation:

$$500 \times 10^{0}\ Bq = \frac{1\ Ci}{3.7 \times 10^{10} Bq} =$$

$$\frac{500 \times 10^{0}\ Ci}{3.7 \times 10^{10}} = 135 \times 10^{-10}\ Ci$$

Answer: Reducing, this is 1.35×10^{-8} curies (0.0000000135 curies)

Try the following exercise, and check your answers from Appendix #1.

EXERCISE #41-1

1. 40 rad is equivalent to ________ gray:
2. 500 rem is equivalent to ________ sievert:
3. 50 Sv is equivalent to ________ Rem:
4. 8 R is equivalent to ________ C / kg:
5. 4 Bq is equivalent to ________ Ci:

DOSE EQUIVALENT LIMITS (DELS)

The first known recommendation for limiting occupational x-ray exposure was made in 1902 by a Boston dentist, William Rollins. Rollins also invented the first beam area restriction device, the aperture *diaphragm*, a lead plate with a hole cut in it that was connected in front of the x-ray tube. Rollins used the fogging of a photographic plate as an indication that a "reasonable" radiation limit had been reached. This equated to about 10 rem per working day or 50 rem (50,000 mR) per week.

In 1925, Sievert recommended a limit of 1 rem per week or 50 rem per year, a figure also adopted by the newly-formed U.S. Advisory Committee on X-Ray and Radium Protection in 1931. Five years later, the Committee cut this limit in half. In 1959, the National Council on Radiation Protection and Measurements (NCRP) revised the limit downward to 5 rem per year. Clearly, experience has resulted in more caution over the decades, with this limit being one-tenth what it was in 1925.

Although the International Commission on Radiation Protection recommended a 2 rem per year limit in 1991, and the NCRP made revisions in formulas in 1987 and modifications to partial body limits (to specific organs) in 1993, the NCRP has continued its recommendation of 5 rem per year as a prospective annual occupational limit to the whole body. These 1993 DELs have been adopted by state and federal regulatory agencies and have been made into law by the U.S. federal government. This is an annual total body limit for radiographers of 5000 mrem, which equates (for 50 work weeks per year) to a weekly limit of 100 mrem, or 20 mrem per day.

By comparison, practicing radiographers currently average 300–500 mrad per year of actual occupational dose, or less than one-tenth of the occupational whole body limit. This is a very safe margin. The limit is rarely exceeded, but in unusual circumstances it is certainly possible for diagnostic radiographers to exceed this amount.

Current guidelines are known as *dose equivalent limits (DELs)*, consistent with the unit *rem*. (DELs have replaced the older *maximum permissible doses (MPDs)* with which older radiographers are familiar.) The DEL is based on the concept of effective dose equivalents (EDEs), which are in turn based on summing and weighting estimated effects on essential organs to arrive at a total body equivalent, the *total effective dose equivalent or TEDE*. The *whole body dose limiting organs* upon which the TEDE is based are:

1. The gonads
2. The red bone marrow
3. The lens of the eye

As will be discussed later, these three organs are the most radiation-sensitive organs in the human body, each for a uniquely different reason.

There are three types of whole body occupational DELs: The *cumulative lifetime limit*, the *prospective limit*, and the *retrospective limit*. These all interrelate with each other in such a way as to ensure that both short-term and long-term exposure levels are controlled for the young and the old alike.

The Cumulative Lifetime Limit

The CLDEL for occupational radiation workers is 1 rem multiplied by the worker's age. However, it is illegal for any minor to be treated as an occupational radiation worker, so the limit does not become applicable until one is at least 18 years old. This immediately provides an example of why other types of limits are required—otherwise, a radiation worker who started employment at age 18 could receive 18 rems, an excessive amount, in his or her *first year* of work alone. The CLDEL does not protect very young workers well, but is designed for long-term control.

The CLDEL is an *absolute* limit, meaning that if it is exceeded, action is *required* which limits any further dose to 1 rem (10 mSv) per year until the CLEDE (cumulative lifetime effective dose equivalent) becomes less than the cumulative limit. The actions taken and the resulting dose levels must be documented.

The Prospective Limit

As the name suggests, a *prospective* limit "looks ahead" in the form of a future objective or goal. The prospective limit is designed to restrict doses at earlier ages and allow flexibility at older ages. This limit is set at 5 rem per year for occupational radiation workers. It is a *relative* limit which can be subjected to averaging over time. When it is exceeded, remedial action is strongly recommended but not absolutely required.

The Retrospective Limit

The retrospective limit "looks back" over the course of a year to monitor whether the radiation received might have been particularly concentrated over a short period of time, which is more harmful to the worker. It is used in some fields, such as nuclear power production, to provide further guidance on prospective limits. A typical retrospective limit is 3 rem per quarter. This ensures that all 5 rem of the annual limit is not received in a single quarter. When the retrospective limit is exceeded, corrective action is recommended in order to meet the prospective annual limit.

Retrospective limits are not generally used in medical radiography. But, they provide an example of the important concept that occupational radiation exposure should be more or less evenly distributed over time.

The following scenarios provide examples of how these different types of limits work together to meet this goal and philosophy. In all cases, remedial actions are addressed to the *more stringent guideline*, that is, the lowest applicable limit.

Scenario #1: A 20-year-old, after 1 year in the field, has accumulated 6 rem:

-Her CLL, equal to her age, is 20 rem
-Her PDL for the year is 5 rem

► The 5 rem limit applies. It was exceeded this year. Action is strongly recommended to bring the following years' dose levels below 4 rem so that the *average* is brought back down below the prospective limit.

Scenario #2: A 40-year-old has received 3 rem every year for 15 career years:

-His CLL is 40 rem
-His PDL is 5 rem

► The 40 rem limit applies. This man has accumulated 45 rems. Action is mandatory and must be documented, to bring his annual dose levels below 1 rem per year until the CLL is met.

Scenario #3: A nuclear plant technologist receives 4 rem in first quarter:

-Her PDL is 5 rem/year
-Her RDL is 3 rem/quarter

► The 3 rem limit applies, action is recommended to prevent the accumulation of more than 1 rem over the remainder of the year, so that the PDL for the year is not exceeded.

Current Limits

Tables 41-3 and 41-4 list current DELs in condensed form from the recommendations of the NCRP. Table 41-3 lists landmark limits for the whole body, both occupationally and for the general public. The original documents present all of these limits in *millisieverts*, the preferred unit internationally. The conversion from rem to mSv is simple to make: Multiply the rem by 10, (100 rem to a sievert; 0.1 rem to a *milli*sievert, a ten-fold difference). It is recommended that the student memorize all nine limits listed in these tables, in both unit systems.

From Table 41-3, the CCL and PDL have already been discussed. The *embryo/ fetus* DEL refers to the *occupational* dose received by the embryo or fetus of a pregnant radiation worker. *It has no application for patients or for the general public.* It bears mention that prior to 1993, this limit was set at a level 10 times higher but distributed over the entire 9 months of a typical gestation. To prevent the fetus from receiving the bulk of that amount in any one month, the limit is now listed *only as a monthly limit* at one-tenth the amount. Pregnant radiographers should wear a second radiation monitor at their waist level, from which fetal dose can be estimated. Policies for pregnant workers will be further discussed in following chapters.

The *student under 18* limit is targeted at educational situations, such as chemistry or physics laboratories,

Table 41-3

Whole Body Dose Equivalent Limits (DELs)

Occupational:	Cumulative Lifetime Limit: Implied Yearly Limit:	= Age 1 rem (10mSv)
	Prospective Yearly Limit: Implied Weekly Limit:	5 rem (50 mSv) 100 mrem (1 mSv)
	Embryo/Fetus/**Month**:	0.05 rem (0.5 mSv)
	Students under 18 Yearly Limit:	0.1 rem (1 mSv)
	Emergency—1 Event per Lifetime:	50 rem (0.5 Sv)
Public:	General Public/Year:	0.5 rem (5 mSv)
	Negligible Individual Dose (NID):	1 mrem (0.01 mSv)

and is not related to any paid-work limit. The *emergency* limit refers to situations which are *life-threatening* to a victim, in which a rescue worker may be exposed to large amounts of radiation while extracting or treating the victim. In such a situation, such as a fire in the nuclear power plant on a submarine, or a meltdown like the one in Chernobyl in 1986, firemen or rescuers should be sent in, if possible, in *shifts* designed to be of short enough duration to keep their dose below 50 rem. Such an exposure should only be allowed to occur once in each rescuer's lifetime.

It is important to distinguish the "general public" from medical patients. The DEL presented in Table 41-3 is intended for people who live near nuclear power plants or other radiation industries, or near radioactive waste sites.

There is no DEL for medical patients. The amount of radiation a patient receives in the course of medical treatment is entirely at the discretion of the physicians ordering the procedures. Patients always have the right to refuse a radiation procedure. But, it is impractical to attempt to impose limits where medical benefit must be weighed against medical risk. Through negligence and malpractice laws, patients do have legal recourse for excessive and unnecessary exposure to radiation.

The *negligible individual dose* (*NID*) at the bottom of Table 41-3 is not an upper limit, but a lower limit which effectively states that any amount of radiation claimed to have been measured at a level less than 1 mrem is *statistically insignificant* under all circumstances. Such readings are unreliable and can occur purely from electronic error in the detectors being used, so they cannot be directly connected with any biological effect, scientifically or legally.

Finally, Table 41-4 presents the DELs for specific organs of the body. You will note that these are both much higher than any of the whole body dose limits—this is because a radiation exposure limited to a small portion of the body is not as harmful to the whole organism as the same exposure delivered to the entire body. The lens of the eye, which has been associated with the formation of cataracts, stands out in its special sensitivity to radiation when compared to other organs.

Table 41-4

Occupational Partial Body Dose Equivalent Limits (DELs)

Lens of Eye:	15 rem
Red Bone Marrow, Breast, Lungs, Gonads, Skin, Hand, & Forearms:	50 rem

Table 41-5

Estimated Annual GSD Contribution by Radiographic Examination, 1970

Type of Examination	Percent Contribution
Lumbar Spine . . . . :	20%
Intravenous Urogram/Retrograde Pyelogram . . . . :	16%
Pelvis . . . . :	12%
Abdomen/KUB . . . . :	10%
Barium Enema . . . . :	10%
Hip . . . . :	5%
Other Abdominal Examinations . . :	20%
Examinations not listed . . . . :	9%

Genetically Significant Dose (GSD)

The *genetically significant dose* (*GSD*) is the gonadal dose that, if given to every individual, would cause the same genetic effects in the population as the existing distribution of radiation. In other words, it is an *averaged* quantity that gives us an indication of how much genetic harm is being caused to the entire human population due to the use of medical radiation. The GSD takes into account that some of the population are infertile, and not all exposed individuals receive measurable *gonadal* dose. It is an indication of the overall genetic harm to the population.

The current GSD is estimated to be approximately 20 mrem per year. That is, the rate of genetic mutations and other genetic effects throughout the entire population, beyond those caused by natural background radiation, is what would be expected if every person received 20 mrem of medical radiation every year. The GSD has been steadily rising over the decades and should continue to be monitored closely.

Table 41-5 gives the relative percentage contribution of different abdominal radiographic procedures to the GSD. You can see that the routine diagnostic exam making the greatest impact upon gonadal dose is the *lumbar spine series*, closely followed by intravenous urography (or the IVP).

Dose equivalent limits provide guidelines for corrective actions, but they *do not constitute a statement of "acceptable" levels of radiation*. The ALARA (As Low As Reasonably Achievable) concept ethically supercedes DELs. That is, regardless of what the DEL is for any particular situation, our *goal* is to always keep radiation levels at the minimum possible level.

RADIATION DETECTION INSTRUMENTS

All instruments designed to detect radiation operate on the basis of the ionization of atoms, which frees up electrons that can be measured as a charge or as a current in a circuit. There are three basic *modes* in which detection devices operate: The *detection only* mode is typical of some Geiger counters, where one hears "clicks" or perhaps beeps that indicate the presence of radiation. Since these clicks can overlap each other or occur in extremely rapid sequence, they can be impossible for a human to count. Without a meter or digital counter, such a device cannot be used to *measure* radioactivity, only to indicate its presence and general intensity.

In the *rate* mode, a meter can give a fairly accurate indication of the radiation rate, such as roentgen per hour (R/hr) *provided that rate is reasonably constant.* This is a useful application for x-ray machines or other manmade devices which generate a steady flow of radiation. But, most *natural* sources of radioactivity have wildly varying rates, which makes it impossible for a gauge needle or a digital counter meter to stabilize at an accurate reading. For such applications, the rate mode is only useful as an indicator of the presence of radiation.

Devices capable of the *integrate* mode are able to *accumulate* the count of radiation events over a set period of time. One can set the device to count radiation strikes for 10 seconds or for a full minute. Electronic devices are able to count these events at extremely high speed and with great accuracy. The read-out is the *total exposure* or the *integrated exposure* over that period of time, measured in units such as R or mSv. Rates can also be accurately derived from these read-outs, if they are taken over long enough periods of time, by simple mathematical division. In fact, this is the most accurate way to obtain exposure rates, to use integrated measurements taken over long periods of time and then divide by the seconds or minutes expired.

Characteristics of Radiation Detection Devices

Detection instruments all have four basic characteristics in common which describe their effectiveness and efficiency. These are their *sensitivity*, their *accuracy*, their *resolution (or interrogation) time*, and their *range*.

Sensitivity

The *sensitivity* of a radiation-detecting device is defined as its ability to detect *small amounts* of radiation. There are two main modifications that would make a detection instrument more sensitive. The first is to provide a larger *detection chamber* for initial absorption of the radiation (Fig. 41-2). The concept is quite simple: For low intensities of radiation the individual particles or photons within the beam are more *spread out*. In order to "catch" any of these, a larger "bucket" is needed. As shown in Figure 41-2, a small detection chamber may be completely missed by the few particles or x-ray photons that may be present.

Figure 41-2

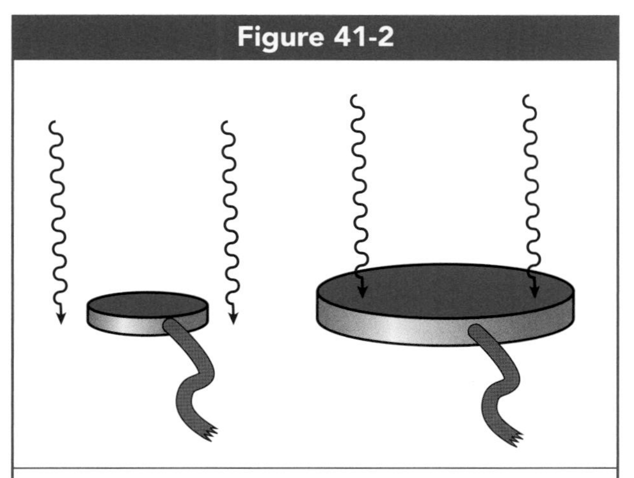

A larger detection chamber is more sensitive and accurate for measuring radiation, because it is more likely to capture photons from small exposures, that may completely miss a smaller chamber.

Figure 41-3

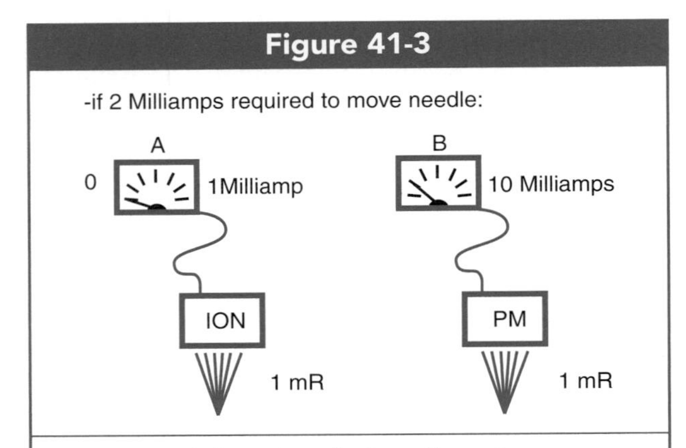

Sensitivity of detection devices: For a low exposure of only 1 mR, the simple ion chamber **A** is unable to generate enough electricity to move the gauge needle and read out an exposure. Because of electronic amplification, device **B** with a photomultiplier tube is *sensitive* enough to read out the amount.

The second way to enhance the detection sensitivity of a device is to increase its electronic *amplification* of the incoming signal. Figure 41-3 illustrates the importance of signal amplification: Any electronic detection instrument requires a certain *threshold current* in order to generate a read-out. A good example is provided by instruments that use a needle-type meter like the speedometer in many cars. To get a read-out, the needle must be induced to mechanically move, and since it has weight (albeit very small weight), it takes a certain *minimum* amount of electrical current supplied to an electromagnetic device to get the needle to move at all.

Figure 41-4

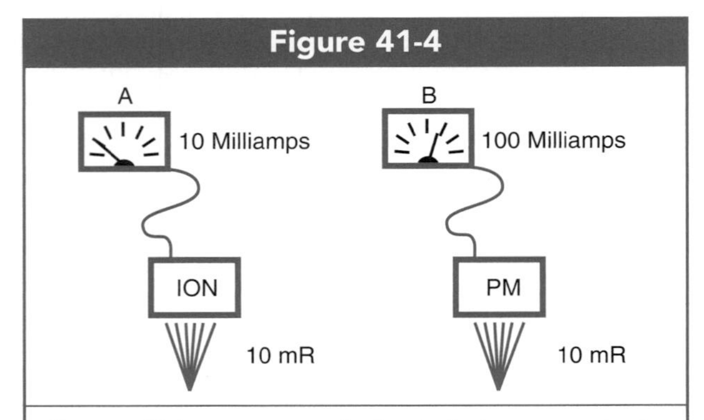

Given sufficient exposure for both devices to work, the more *sensitive* device, **B**, generates more electricity and moves the needle farther. This allows the increments on the meter to be adjusted for higher *accuracy*.

In Figure 41-3, we assume that for a particular meter, the threshold current required to move the needle is 2 milliamps of electricity. We then compare the sensitivity of two types of circuits using the same meter: Circuit ***A*** uses a simple ion chamber as a detector, which generates a current proportional to the exposure received, 1 milliamp for every milliroentgen. Circuit ***B*** employs a photomultiplier tube as a detector, which in this case *amplifies* the same signal by ten times. If both devices receive an exposure of 1 mR, note that device ***A*** will not have generated sufficient electrical current to move the needle of the meter. The read-out for this device will be *zero* radiation, misrepresenting the reality.

Device ***B***, on the other hand, will generate from the same exposure 10 milliamps of current and will move the needle to a certain point. We simply *adjust the scale* printed behind the needle so that it reads "1 mR" (rather than "10 mR") at this point, and arrange the rest of the scale proportionately. Device ***B*** has higher sensitivity than device ***A***.

In Figure 41-4 we take the same two devices and expose them to a greatly increased amount of x-rays, 10 mR each. This is more than enough to move the needle for device ***A***, and both instruments give us a read-out. However, we see that the needle for device ***B*** has moved physically much farther up the scale. Device ***A*** produced 10 mA of electrical current and moved the needle according to this amount of force, whereas device ***B***, with its amplifying circuitry, produced 100 mA of current. We conclude that *sensitivity* may also be defined as the *amount of electrical current or charge produced by a given radiation exposure.*

Now, each written scale behind the needles can be adjusted so that they both read-out the correct 10 mR of exposure. But, since the needle for device ***B*** has moved much farther, its *scale* can be subdivided into smaller units—this gives us a more *accurate* reading.

Accuracy

Accuracy is the *precision* with which measurements are obtained. As we have just seen, increasing *sensitivity* of a device is one way of improving its accuracy. But, it is not the only factor that determines how precise the read-outs from an instrument will be. For example, the actual printed scale behind the needle of a meter might be misaligned, so that the markings are off.

Electronic noise might be added to the signal, causing the read-out to be higher than the real exposure. There may be fluctuations in the amount of electrical *power* being supplied, such as batteries running down, that would affect the reading. The circuit must be designed to compensate for these types of electronic variables, and other aspects of the instrument must be carefully designed to maximize its overall accuracy.

Accuracy is the primary factor that determines the *reliability* of the information obtained.

Resolving (Interrogation) Time

There is a minimum time that must elapse *between* ionizations that can be detected. In order to "count" each ionizing event, the device must be able to *separate* it in time from other events. All types of devices require a fractional amount of time to "reset the detector" between sequential ionizations. Otherwise, when a particle or x-ray photon enters the detection chamber immediately behind another one, the second one will not be detected, and both will be counted as one. Resolving time, then, is yet another factor affecting the *accuracy* of the instrument.

Interrogation time is generally defined as the time required for a particular electronic circuit to *respond* to a particular stimulus, that is, the time for all of the switches to operate and for the electronic signal to pass from the original detection stage to the read-out stage. Although this may be distinguished technically from *resolving time*, insofar as radiation detection instruments are concerned the two terms are practically synonymous.

By using two sources of radiation with different but constant rates of activity, we can determine the resolving time of any detection instrument. To do this, the exposure rate from each source must be measured with the instrument and recorded separately, and then the rate must be measured from both sources *combined*. The results can be entered into the following formula to determine the resolution time for the device:

$$T = \frac{R_A + R_B - R_{(A+B)}}{2R_A \times R_B}$$

where T is the resolving time, R_A is the measured rate of radioactivity for source ***A*** and R_B is the rate for source ***B***.

For example, a sample of Cobalt 60 (C^{60}) and a sample of Cesium 137 (Ce^{137}) can be each placed in or near the detection device set to *integrate mode* (for accuracy) to be counted for a period of 10 seconds. For accuracy, each count should be repeated at least three times (recommended 10 times), and then averaged. The procedure is then repeated with both sources stacked together, and the averaged measurement recorded. To apply the formula, use the following practice exercise:

Practice Exercise #3

What is the resolving time for detection instrument yielding the following measurements taken from a source of C^{60} and a source of Ce^{137}:

C^{60} . . . . : 259 average counts in 10 seconds
Ce^{137} . . . . : 386 average counts in 10 seconds
C^{60} + Ce^{137} combined:
588 average counts in 10 seconds

Solution: Set up formula as:

$$\frac{259 + 386 - 588}{2(259)(386)}$$

$$= \frac{57}{199{,}948}$$

$$= 0.\ 000285$$

Rounding, $= 0.0003 = 0.3 \times 10^{-3}$ *seconds*

Answer: The resolving time for this device is 0.3 milliseconds.

Range

Also, to obtain a valid and accurate reading, *the sensitivity of the detection instrument must be matched to the expected intensity levels of radiation*, and the design of the detection chamber must accommodate the *type of radiation* to be detected and the *expected energy levels* of that radiation.

Remember that the *penetration* capability of beta particles is limited, and that of alpha particles is so low that even a sheet of paper can absorb a statistically significant number. Thus, for measuring naturally radioactive substances, it is often better to place them *inside* a detection chamber whose walls constitute an electrical anode for collecting charge released from the gas within. When *external* sources of particulate radiation or very low-energy x-rays are measured, particular attention must be given to the material used for the walls of the detection chamber to minimize absorption of the radiation before it can be measured. Geiger counters, for example, typically have a very thin, fragile sheet of mica film on the face of the detection chamber, which allows beta and alpha particles through. Detectors designed for high-energy x-ray or gamma ray sources can use a thin wall of aluminum at the chamber port.

If a typical Geiger counter is placed on an x-ray table and exposed using diagnostic radiography technique factors, you will find that it overloads the circuits of the counter, pinning the needle on the meter and usually sending the device into a "tilt" mode where it cannot be re-used without shutting it off first. The high *sensitivity* of Geiger counters is designed to measure intensity rates that are typical of *naturally occurring radioactivity*, not the much higher rates of a diagnostic x-ray beam, and the meters on these devices have scales that are calibrated according to these much lower *expected* amounts of radiation. The *range* of the Geiger counter is much too low for use on diagnostic x-ray machines.

Likewise, the range of a detection instrument may be set too *high* for the intensities of radiation expected, in which case low intensities will not be picked up and adequately displayed on the read-out. This is not just a function of the instrument's sensitivity, but also of its physical design.

These four aspects of detection devices may certainly affect each other, but are each defined as distinct functions. We can make a general statement on how *accuracy* is affected by the other three factors:

Generally, *higher accuracy* results from:
1. Increased sensitivity
2. Increased range
3. Faster resolving time

Unfortunately, sensitivity and range tend to work against each other because higher sensitivity requires the resolution of smaller units on the read-out meter, which limits the range that the read-out can cover.

Finally, note that all scientific measurements can be characterized by two critical aspects: *Validity* and *Reliability*. The *validity* of a measurement is its appropriateness to the concept that is being conveyed. We speak of "comparing apples and oranges" as an example of poor validity. Examples of invalid applications of radiation detection and measurements might be using the *roentgen* unit to try to measure beta radiation, or trying to use a Geiger counter to measure radiation from an x-ray machine, for which it was not designed.

Reliability is essentially the accuracy of measurement for the information provided. Radiation measurements must be both reliable and valid, based on the correct application of these four characteristics of the equipment used.

Types of Radiation Detection Instruments

Scintillation Detectors

Scintillation refers to the immediate emission of light (fluorescence) by a substance struck by x-rays or other radiation. Scintillation only occurs in special *crystalline* materials, when ionization causes orbital electrons to be elevated into higher energy levels called molecular *electron traps*, just as described in Chapter 34 for the photostimulable phosphor plates used in CR, but in this case the electrons *immediately* fall back out of the traps into their shells, releasing a burst of light in the process.

Most scintillation counters use a hermetically sealed scintillation crystal made of sodium iodide or cesium iodide. Note that both of these compounds contain iodine, with a high atomic number of 53 for effective absorption of x-rays. This scintillation crystal is coupled to a PM (photomultiplier) tube to amplify the electronic signal generated (Fig. 41-5). Light from the scintillation crystal strikes the *photocathode layer* of the tube, which emits electrons by the photoelectric effect. Electrons are then accelerated through a series of dynodes to magnify the pulse of electricity.

Scintillation-type detectors are more sensitive to x-rays and gamma rays than Geiger counters. They have both a very high sensitivity and a high range, making them useful as components of imaging machines such as CT and nuclear medicine units, but are sometimes used in portable radiation survey instruments.

Thermoluminescent Dosimeter (TLDs)

In other crystalline materials, when ionization from x-rays occurs in their molecules, the orbital electrons are elevated into electron traps and *remain* there for

Figure 41-5

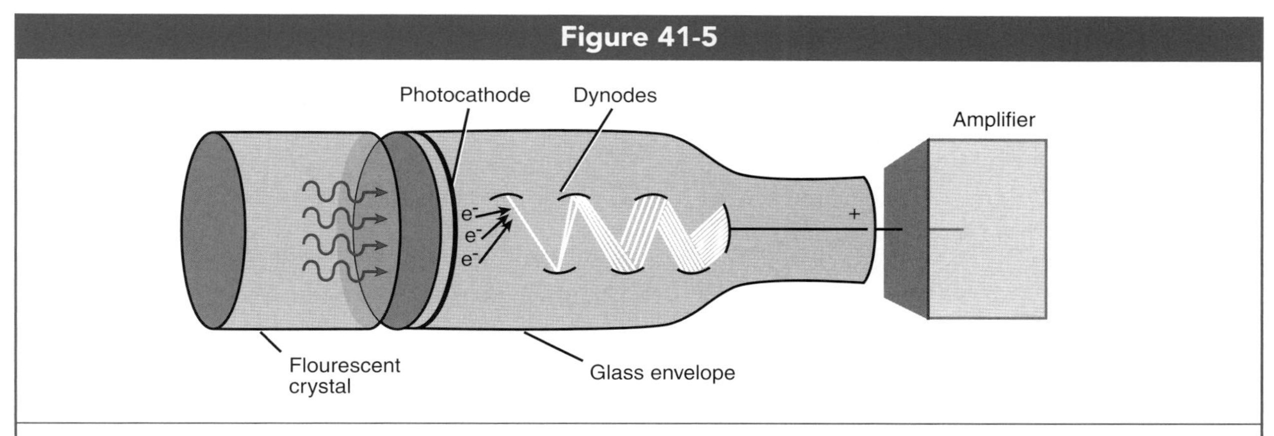

In scintillation detectors, a *photomultiplier tube* is attached to the fluorescent crystal, which converts the light into electrons and amplifies the electrical current through a series of dynodes.

an extended period of time. This is just as described in Chapter 34 for the photostimulable phosphor plates used in CR, but in this case, rather than use a laser beam to shake these electrons out of their traps so they can fall back into their orbital shells, *heat* is used to impart to them the extra energy needed to escape the traps. As always, a burst of light is released in the process (Fig. 41-6). This process of heating a crystalline substance to induce it to glow is called *annealing*.

The delayed emission of light (phosphorescence) which is induced when the crystal is annealed is referred to as *thermoluminescence*, meaning "from heat, light." A common application for these types of crystals is found in personal occupational radiation monitoring, in the form of *thermoluminescent dosimters* or *TLDs* (Fig. 41-7).

In the annealing oven, light emitted from the heated crystal is picked up and measured by a photomultiplier (PM) tube. A "glow curve" is plotted for the amount of light intensity as the temperature of the oven is increased (Fig. 41-8). The total area under this curve is proportional to the x-ray exposure accumulated by the crystal during the period of time the monitor was worn.

For personal monitoring purposes, it is more accurate to use (as an absorber) a chemical compound with x-ray absorption characteristics similar to those of the soft tissues of the body. The most common crystal used is composed of lithium fluoride, which

Figure 41-6

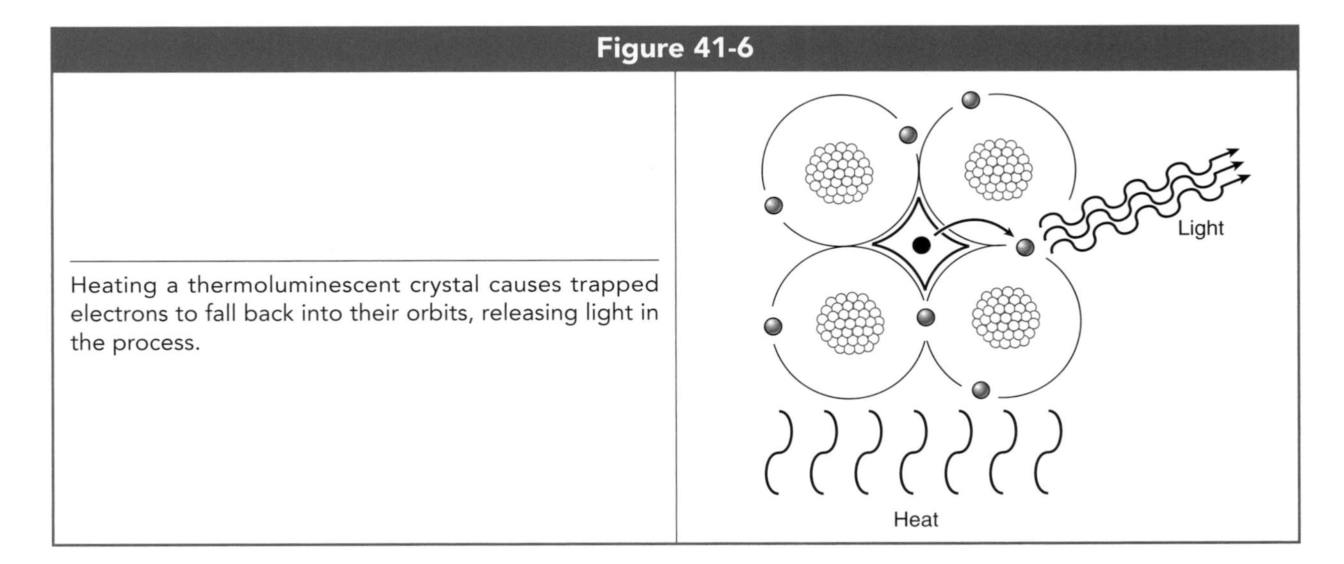

Heating a thermoluminescent crystal causes trapped electrons to fall back into their orbits, releasing light in the process.

Figure 41-7

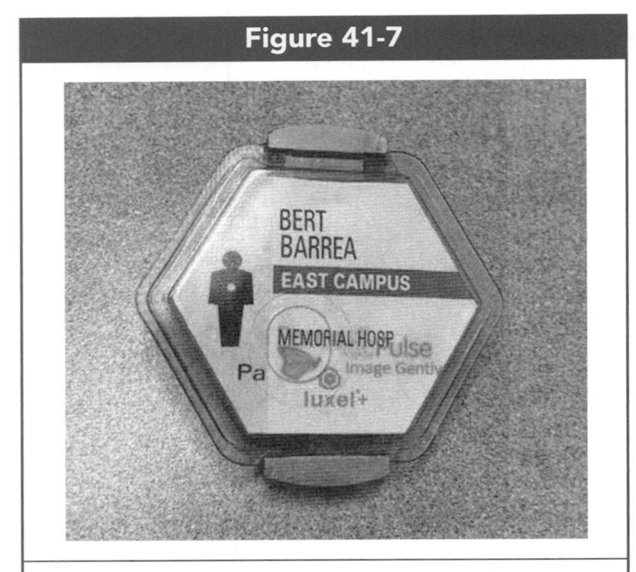

A personal thermoluminescent dosimeter or TLD.

also provides an accuracy of measurement considerably higher than the accuracy of film-based personal monitors.

In the TLD, a layer of crystalline detection material is situated under a series of filters, usually aluminum, copper, and tin (in increasing absorption effectiveness) (Fig. 41-9). An area with no filtration beyond the outermost light-tight covering is also designated. By making comparative ratios out of the readings recorded beneath each of these areas, an estimation of the *half-value layer* (*HVL*) can be made (see Chapter 17). From the HVL, the average energy level, in kV, of the original radiation can be derived (from a look-up table). The energy level, in turn, can be indicative of particular *types* of radiation. For example, energies in the millions of volts would have to be from *gamma rays* rather than *x-rays*, and might have been acquired in a radiation therapy department rather than in the diagnostic imaging department. Because of their extremely low penetration, exposure to particulate radiations can also be surmised by how quickly their intensity drops off through each subsequent filter.

Figure 41-8

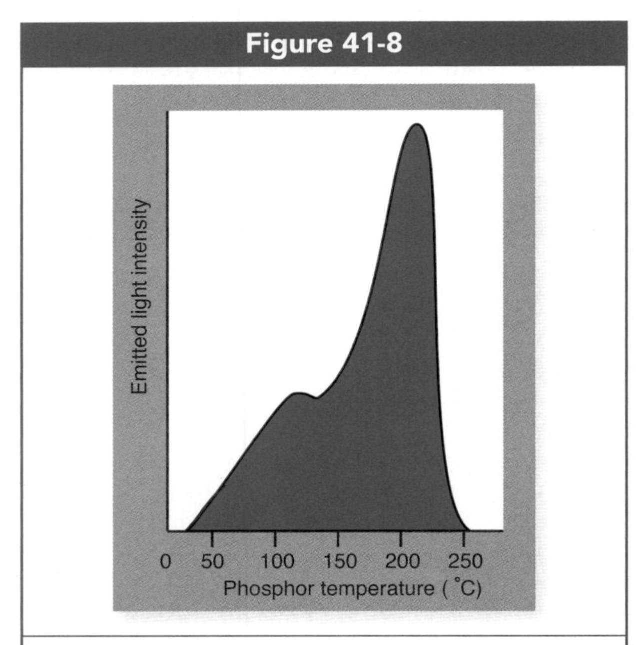

The *glow curve* for a TLD, showing the intensity of light emitted as the oven temperature increases.

Film Badges

The *film badge* is another general type of personal monitor. It uses a small packet of the same type of plastic film coated with a silver bromide emulsion that was used in traditional radiography. This film is hermetically sealed in a metal foil to protect it from moisture, chemical fumes and light (Fig. 41-9). However, it remains quite vulnerable to accidental exposure to heat, which can fog the film. The TLD is less sensitive to all these variables, but *extreme* exposure to heat or chemicals may affect it.

At the end of one month of occupational exposure, the film is chemically processed using strictly-controlled temperatures, chemical concentrations and timing. Processing results in a build-up of black metallic silver on the exposed areas, which is proportional in its darkness to the amount of radiation received.

A *densitometer* is an instrument (essentially a *photometer*) that measures the amount of light transmission penetrating through such a darkened film from an illuminator behind it. The intensity of this light must also be carefully controlled for purposes of obtaining exposure measurements. To convert the light measurement into a unit for *density* or darkness, the percentage of light transmitted through the film is inverted and the logarithm taken. This results in a density scale ranging from 0 for a blank white area (100% light transmission) to a "pitch black" area with less than 1 ten-thousandth of the incident backlighting penetrating through.

Figure 41-9

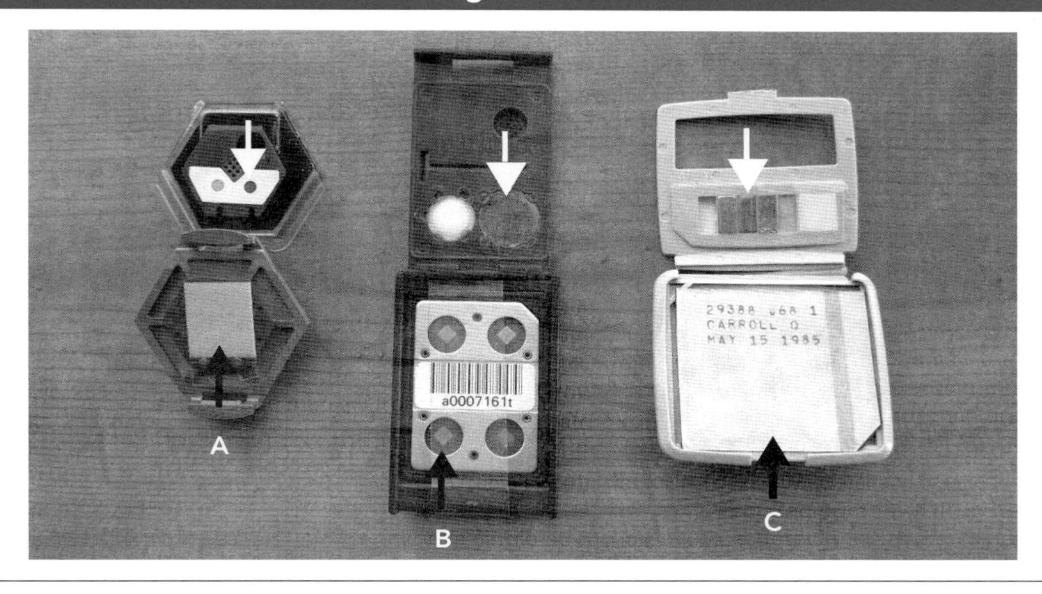

Black arrows (bottom) show a sheet of lithium fluoride in a radiographer's TLD, ***A***; crystals of lithium fluoride in a radionuclide waste technologist's TLD, ***B***; and a hermetically-sealed film packet in a film badge, ***C***. All three use various filters of copper, tin, aluminum, and simulated bone, and a space with no filter, to determine the HVL of the radiation (white arrows, top).

This was fully explained in Chapter 39 on "Quality Control."

As with a TLD, a series of filters are placed in front of the sheet of film to determine the HVL, and consequently the average kV level and type of radiation it was exposed to. The radiographer must take care to wear both devices *facing forward*, to ensure that the filters are in front of the film or detection crystals, for proper interpretation.

Film badges continue to be used in some areas because they are economical. However, their accuracy and reliability are greatly limited compared to those of a TLD.

Gas-Filled Detectors

In all gas-filled detectors, electrons freed from gas (usually room air) by x-ray ionization are then attracted to and strike a positively-charged anode plate or pin within the chamber (Fig. 41-10), generating electrical current or charge. This classification includes ion chamber devices, pocket dosimeters, proportional counters, and Geiger-Mueller tubes (more commonly known as "Geiger counters").

Pocket Dosimeters

Pocket dosimeters are based on the concept of the electroscope as described in Chapter 6 and generate a charge of static electricity rather than electrical current. The physics of the *ionization* which occurs

Figure 41-10

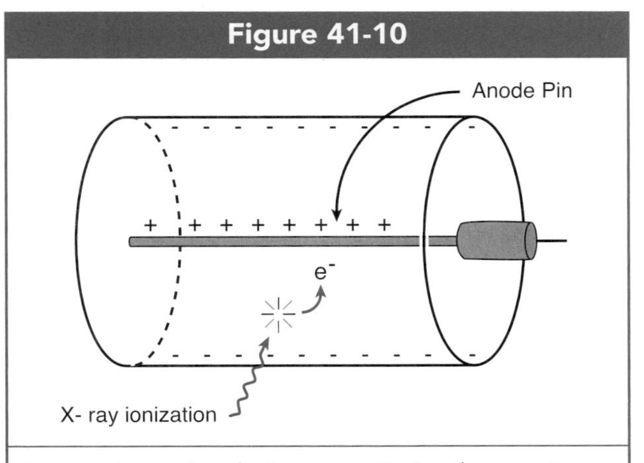

In a gas-ionization device, x-rays ionize the gas atoms, liberating electrons. These electrons are then attracted to a central anode pin or plate, where they accumulate to form an electrical current.

Figure 41-11

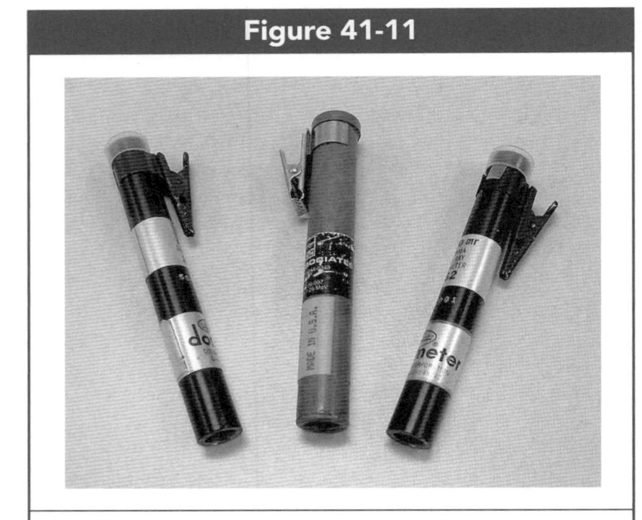

Pocket dosimeters, with windows on the end to see through.

within their chamber, however, is identical to that of any other gas-filled detector. This all occurs within a compact tube which, when held up to the light, can be seen through (Fig. 41-11).

Within the gas chamber, a fiber branches off an electrode on one side of the chamber which is given a positive charge (Fig. 41-12). This charge is produced when a contact at one end of the dosimeter is pushed onto the terminal of a small battery-powered circuit. Electrons are drawn from the dosimeter into the charger, leaving the dosimeter electrode with a positive charge that also extends out onto the loose fiber. Thus, the end of the fiber is *repelled* from the electrode and moves away from it across to the other side of the dosimeter chamber. This position corresponds to the *zero* on a scale imprinted on a window at one end of the dosimeter, as shown in Figure 41-13. When one is observing this scale through the window (held up to a light source), one is looking *end-on* at the fiber in Figure 41-12.

X-rays entering into the chamber *ionize* the gas within, freeing up electrons from their atoms. The electrons are immediately attracted to the positively-charged fiber or the wall electrode to which it is attached. In either case, the amount of *positive charge is neutralized as* these electrons fill vacancies in the atoms of the fiber or electrode. There is a certain amount of spring tension to the fiber, which splits off from the electrode. Therefore, as the positive charge is lessened, and the repulsive force diminishes between the fiber and the electrode, the fiber falls back toward the electrode wall (Fig. 6-16 in Chapter 6), and is seen from the window end to move across the imprinted scale (Fig. 41-13). With calibration of the scale, a read-out of the radiation received will be proportional to the amount of electrical charge released from the electroscope.

Pocket dosimeters are ideal for *short-term* personal monitoring in a manner that no other device can match, because they are always *immediately readable.* They can be checked every day or even every

Figure 41-12

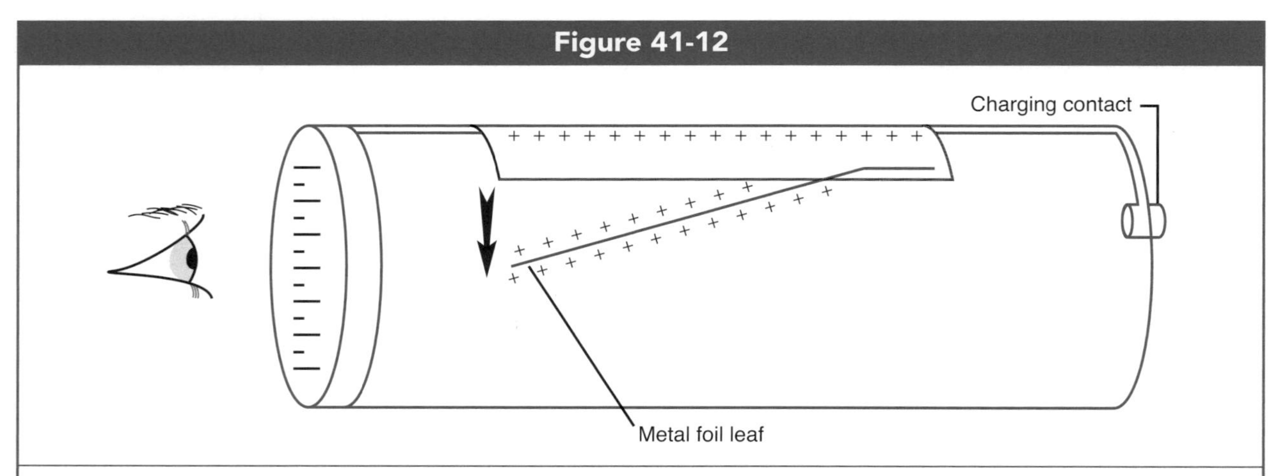

Diagram of the *electroscope* inside a pocket dosimeter. Both the wall of the chamber and a movable leaf of metal foil are given a positive charge, repelling the leaf across the chamber. As this charge is *drained* from radiation exposure, the leaf moves back toward the wall.

hour. They must be recharged whenever the fiber approaches the maximum mark, but if readings are recorded, sequential measurements can be taken by simply subtracting the previous reading from the present one. Unfortunately, what pocket dosimeters offer in convenience they lack in accuracy, which can be very unreliable. Slight blows to the device can knock the fiber over to an incorrect reading or jolt some of the electrical charge from the fiber and electrode.

Most pocket dosimeters are *self-reading*, as described, but *charger-read* dosimeters can only be read while in contact with the terminal on the charger box. For both types, one must look through the window while charging, and adjust the charging knob on the box until the fiber is seen to be at or close to the zero mark. If it is not exactly on this mark, the amount that it is located at will have to be subtracted from the first clinical reading taken.

Figure 41-13

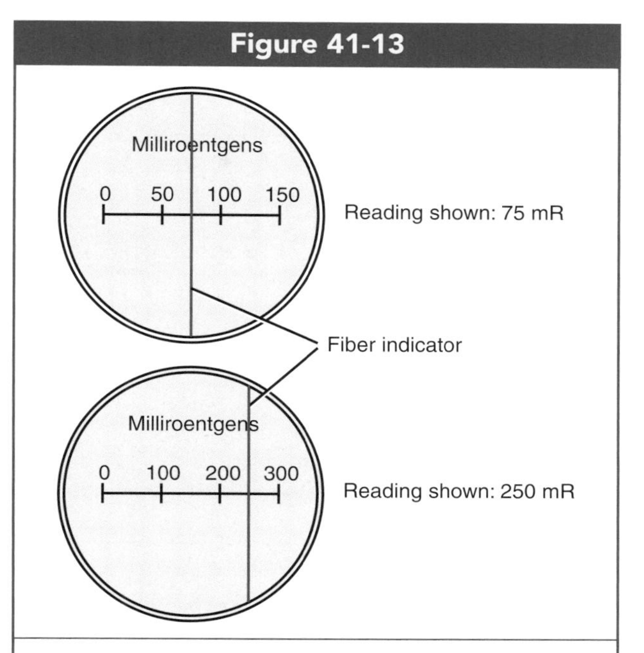

The scale seen through the window of a pocket dosimeter. The fiber is an end-on view of the movable leaf of an electroscope (see Figure 41-12).

Ionization Chambers

Ionization chambers include portable, hand-held devices like the "Cutie Pie" and the "R-Meter," tabletop models such as the one in Figure 41-14, and larger units which have the electronics contained in a briefcase and an extension cord with the ion chamber detector at the other end. These instruments typically have a wide range, from 1 mR per hour to 1000s of R/hr, and feature high accuracy. Each electron released from the gas by ionization is collected by the anode element or positive electrode.

As shown in Figure 41-15, in an ionization chamber, typically one electron released from the gas in the chamber reaches the anode pin for each x-ray that interacts within it. This proportionality gives ionization chambers high accuracy.

Ion chambers make excellent portable instruments for area radiation surveys around fluoroscopes, nuclear medicine generators and syringes, and brachytherapy patients. They are appropriate for checking the integrity of protective barriers, and the output and calibration of x-ray machines.

Proportional Counters

Proportional counters take advantage of something called the "cascade" effect," in which secondary electrons are produced after the initial ionizing event. The electron ejected from this first event has sufficient energy that it "knocks" *other* electrons out of their gas molecules, which in turn "knock" still others out (by repulsion). The result is a *cascade* in which several electrons eventually reach the

Figure 41-14

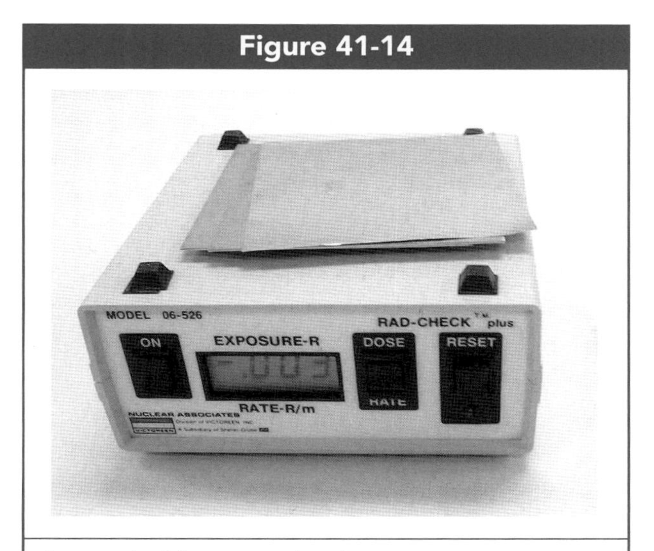

A typical tabletop ion-chamber survey meter, appropriate for use with x-ray machines.

Figure 41-15

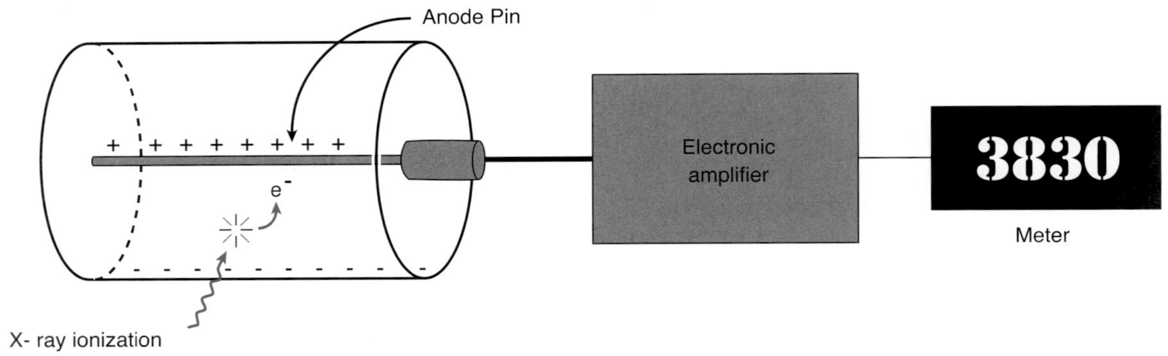

Anode Pin

Electronic amplifier

3830

Meter

X- ray ionization

For a simple ion chamber, one x-ray releases one electron from the gas, and only this one electron reaches the anode pin.

anode pin of the circuit (Fig. 41-16). This has the effect of magnifying the current generated from each x-ray.

Proportional counters manifest extremely high sensitivity, such that they have little application for clinical imaging. But they are able to distinguish between alpha and beta radiation, and are well-suited as a stationary laboratory instrument for assaying small amounts of radiation.

Geiger-Mueller Tubes

The Geiger-Mueller tube, or "Geiger counter" (Fig. 41-17), operates on the basis of *saturation* of the detection chamber. The "saturation effect" is identical in physical process to the "cascade effect," except that so much energy is imparted that *all* of the original gas molecules within the chamber are ionized from a single radiation exposure event. This is shown in

Figure 41-16

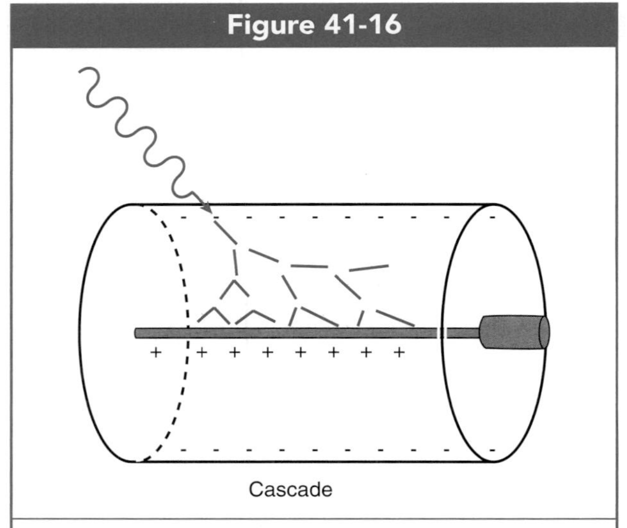

At higher voltages, the *cascade effect* occurs, in which the initial liberated electron causes multiple other ionizations of the gas on its way to the anode pin. This number is proportional to the supplied *voltage*, hence the device name *proportional counter*.

Figure 41-17

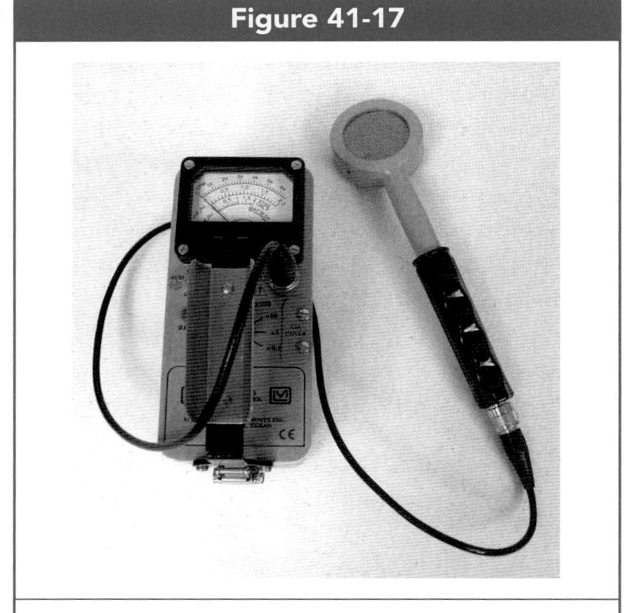

A typical Geiger-Mueller tube or *Geiger counter*. (Courtesy, Lynn Fyte.)

Figure 41-18, where the effect might more aptly be described as an "avalanche" rather than just a "cascade." A gas which is much more readily ionizable than air, such as *argon*, is injected into the detection chamber.

Of course, this greatly magnifies the electrical current generated from each impinging x-ray or gamma ray, such that a distinct "click" sound can be generated over a speaker in addition to obtaining a read-out from a meter. A problem arises in that, since the entire chamber is discharged with each event, the entire chamber must be *recharged*, which is to say, neutralized, before the next event can be detected. This requires more resolving time than it takes for ion chambers to reset for the next detection. By adding a "quenching agent" of ethyl alcohol vapor to the gas in the chamber, this process of restoring the argon gas to its original condition is speeded along.

Geiger-Mueller tubes have high sensitivity, but fairly low accuracy and a low and narrow range of less than 100 mR/hr for the intensity rate of the radiation. They work well as portable survey instruments for environmental radiation, and to detect radiation contamination on work surfaces in nuclear medicine and laboratories. An advantage for some applications is that they can detect single events when operated in the pulse mode, by generating a click or beep. On the other hand, Geiger counters are not generally designed to be capable of measuring an integrated (accumulated) dose. This is one reason for their limitation in accuracy.

Figure 41-18

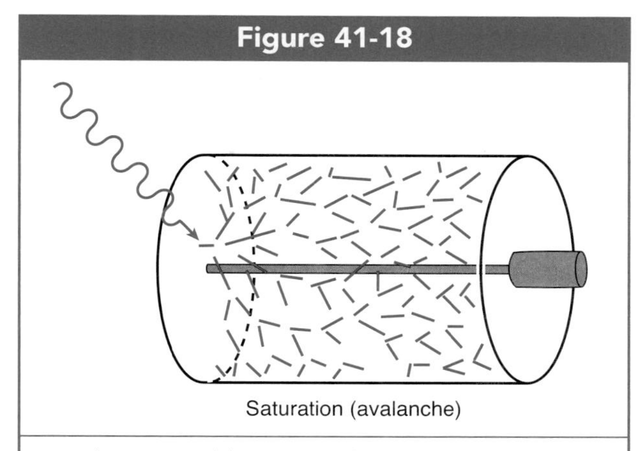

At voltages used by a GM tube, *saturation* causes all of the gas molecules in the chamber to be ionized with each incident x-ray. The entire chamber must be reset to detect the next event.

Personal Radiation Monitors

The types of radiation detectors that are commonly used for personnel monitoring have all been described in the preceding discussion, but will be listed here for some brief discussion relative to this purpose. They include:

1. TLDs, whose main advantage as personal monitoring devices is high accuracy, due to a sensitivity to less than 5 mR of exposure. These devices are so resilient to insult from heat, jolts, fumes, etc., that they need only be read once every quarter for their accumulated exposure. Their only disadvantage is slightly higher cost. The crystals in TLDs are reusable.
2. Film Badges, which are cheaper, but also have a lower accuracy, requiring more than 10 mR for accurate reading. Readings of less than 10 mR on a film badge can be due entirely to errors, and so are generally listed as *N* or "negligible" on radiation reports. Because of their susceptibility to insult from heat, fumes, moisture and light, film badges must be read and replaced each month.
3. Pocket Dosimeters, whose greatest advantage is their adaptability for making very short-term readings at will. This is offset by very poor accuracy, mostly due to their sensitivity to mechanical jolts. They must be worn on a pocket or other location where they are not likely to be disturbed.

Only TLDs or film badges should be used for regular monthly or quarterly monitoring. Be sure to wear them with front side facing forward, so that the filters used to determine energy levels and types of radiation are in front of the film or detection crystal.

All types of personal monitors are sensitive to background radiation, which skews their accuracy. In the case of film badges and TLDs, a *control monitor* must be used as an experimental control which is kept in a location well away from occupational radiation hazards, such as a manager's or secretary's office which is not adjacent to a radiation suite. At the end of a monitoring period, the background reading from the control monitor must be subtracted from all personally-worn monitors to obtain accurate readings. This is generally done by a professional service which supplies the monitors.

Voltage-Dependence of Electronic Detection Instruments

In most cases, we want the *response* of a radiation detection instrument to be proportional only to the actual amount of radiation received, and *not* dependent upon the voltage supplied by batteries or generators. For example, many devices operate on battery power; as the batteries begin to wear down, we do not want the read-out produced by the instrument to decline with the batteries—we would prefer that it just stop giving out readings and provide some indication that new batteries are needed, but keep giving accurate read-outs up to that point. Capacitors in the circuits of these instruments help keep the voltage supply constant until the batteries are just too low for further use.

Going in the opposite direction, neither would we want the read-outs to *increase* as the supplied voltage increases, but again, remain constant and accurate according to the radiation exposure received. In reality, as we operate detection instruments at higher and higher voltages, we find that in some ranges they are *voltage-independent* and therefore constant in their read-outs, but in other ranges they are *voltage-dependent*. For gas-filled detectors, what determines this relationship is the way in which the gas responds to radiation exposure with the particular voltage applied to the electrodes in the gas chamber; whether, for example, it responds by simple ionization, by the "cascade effect," or by the "saturation effect" which were described above.

Figure 41-19

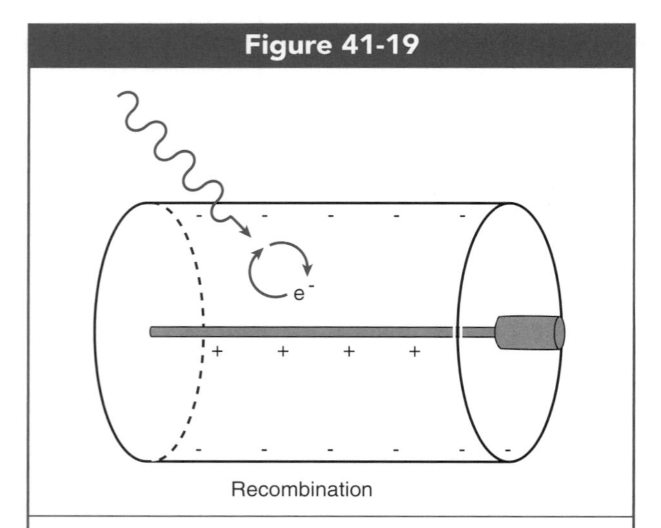

Recombination, in which the voltage supplied to the anode pin is too weak to attract electrons liberated from the gas in the chamber by x-rays. These electrons fall back into their orbital shells, and no current is generated at the anode.

To more fully examine this relationship, we begin with zero voltage and observe the response of a detector to a *constant source of radiation* as we gradually increase the voltage supplied to the gas chamber. We immediately discover that there is a *threshold voltage* for getting any read-out from the device at all. What is happening is illustrated in Figure 41-19. An x-ray may enter the chamber and knock an electron out of a gas molecule, but the positive charge applied to the central anode pin is just too weak to be "felt" by the electron. Instead of being pulled to the anode, the electron immediately falls back into its atom.

This effect is called *recombination*. Ionizations are taking place in the gas, but the freed electrons are recombining with their atoms rather than traveling toward the anode. Thus, no electrical current is generated within the circuitry in order to provide a read-out at the meter. The meter continues to read *zero*, inaccurately, until sufficient voltage is supplied to pull freed electrons to the anode pin.

As we increase the supplied voltage (for example, by using more powerful batteries), the operating threshold for the ionization chamber is reached and the instrument begins to count one electron flowing in its circuit for each x-ray absorbed within the gas chamber. As shown in Figure 41-15, each x-ray interaction liberates one electron which travels to the anode and becomes part of the electrical current generated there. A proportional read-out is produced. This effect is referred to as *simple ionization*.

The important thing to understand about simple ionization is that for a while, *as the voltage supplied to the chamber continues to be increased, only one electron continues to reach the anode pin for each ionization, so the read-out remains constant according to the radiation received*. This is generally desirable.

Figure 41-20

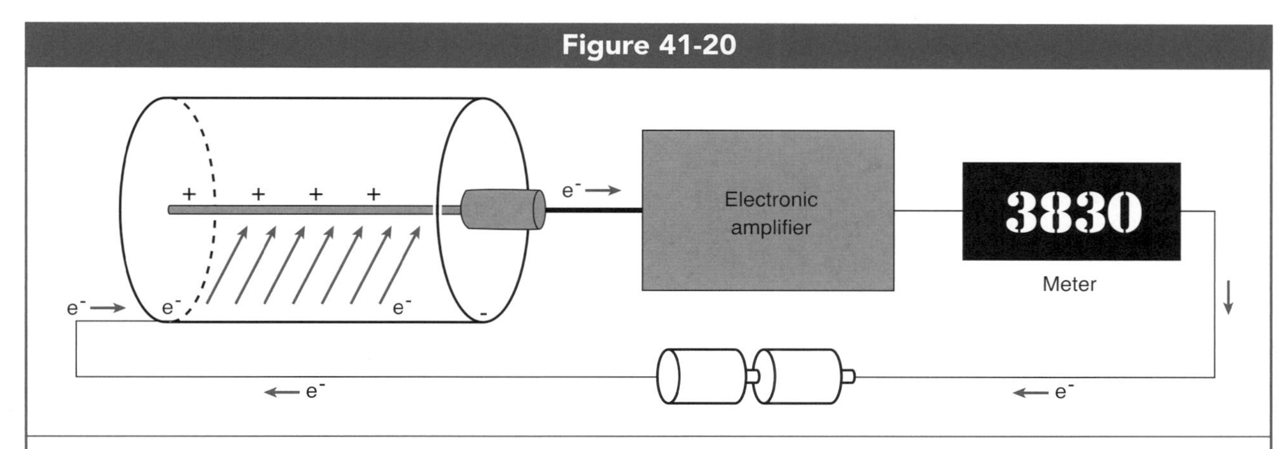

Continuous discharge, in which the supplied voltage is so high that a single event triggers the electrical current generated to complete the circuit and *spark* across the gas chamber to the anode pin continuously. This renders the meter unusable.

Continuing to increase the voltage, the amount of positive charge on the anode pin becomes very strong, and electrons freed from the gas molecules are attracted toward it with such force that they begin knocking other electrons out of their molecules along the way. The increased voltage has now passed the *threshold* level to bring about the cascade effect, Figure 41-16. Once the cascade effect begins, further increases in voltage cause an ever-increasing multiplying effect on the number of electrons reaching the anode pin. In other words, higher voltage causes greater cascades. This range of voltages is called the *proportional region*, because the amount of electrical current generated in the device is proportional to the *voltage supplied* rather than only to the radiation detected.

One can imagine that in the proportional region of voltages, the instrument would be rendered unusable for accurately measuring radiation exposure. However, the *degree* to which these higher voltages multiply the effects of radiation striking the chamber are known for each type of device, so compensating circuitry or software can be used to correct for these effects. For some applications in the laboratory, the detection *range* of proportional counters is suited to the types of radiation being measured. With the proper compensating circuitry or software, good accuracy can be maintained. These instruments derive their name, *proportional counters*, because their response is proportional to the *voltage* supplied as well as the radiation received. On some types, the voltage to the chamber can be increased or decreased at the touch of a knob.

Eventually, as we continue to increase the supplied voltage, we cross a threshold where the *saturation effect* begins to occur (Fig. 41-17). Above this level of voltage, since each single ionizing event discharges the entire chamber of gas, there are no more gas molecules available for further ionization. This means that the read-out will become constant once again, counting one complete discharge of the chamber for each x-ray detected, regardless of further increases in the voltage supplied. This range of voltages, where constant read-outs are again obtained, is called the Geiger-Mueller region.

There is, however, one last threshold of voltage to be considered: At a certain point, when the voltage supplied to the electrodes in the gas chamber reaches a high enough level, the electricity will simply *spark* across the gap between the negatively-charged walls of the chamber and the positively-charged anode pin. Upon the first ionizing event in the chamber, the electrical current simply begins to jump from the cathode to the anode disregarding the gas molecules in the space between (Fig. 41-20). Once this begins to occur, the electrical current will continue to flow constantly across the chamber and around the entire circuit of the detection device. The effect is called *continuous discharge*. It "pins the needle" at the meter and renders the device unusable, because at this point it is simply measuring its own electrical current rather than radiation.

Figure 41-21

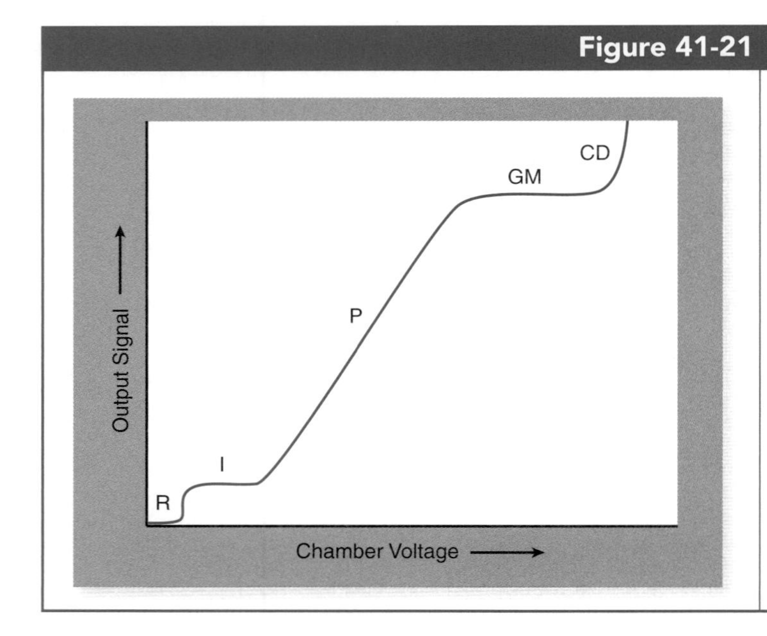

Gas ion chamber response curve for increasing supplied voltages, showing the *recombination* region, the *simple ionization* plateau, the *proportional* region, the *Geiger-Mueller plateau*, and *continuous discharge.*

All these types of responses to the supplied voltage in a radiation detection instrument can be plotted on a single graph, as illustrated in Figure 41-21. We see the response curve remain at zero until the operating threshold of a gas-filled chamber is reached, in the *recombination region* at the far left of the graph. When enough voltage is supplied that single electrons liberated from the gas reach the anode and a read-out is obtained, we see the curve spike vertically and then *plateau* for a while as the voltage continues to be increased. This is the *simple ionization region* of voltages, where read-out is constant even if the voltage goes up or down slightly. *Most instruments used for measuring diagnostic levels of radiation are ion chambers, which operate in the simple ionization region of voltages.*

If the batteries or other source of voltage are powerful enough, we see the cascade effect begin to occur. On the graph (Fig. 41-21), the *proportional region* of voltages produces a slope in the response curve, indicating that, for this region, the higher the voltage, the greater the current produced in the device. Once the supplied voltage reaches the *Geiger-Mueller region*, we find another *plateau* in the curve, indicating that the read-outs are constant (due to the saturation effect) even if the voltage goes up or down slightly. The gas chamber in a Geiger counter must be supplied with this higher voltage level so that the saturation effect will occur, which the circuitry is designed to measure.

Finally, when the supplied voltage is too high, we see that the response curve on the graph in Figure 41-21 turns sharply upward, to indicate that the meter on the device will read out at its maximum possible value, due to the *continuous discharge* of the current through the device.

Each type of radiation detection instrument is designed for operation within a specific range of supplied voltage or power.

SUMMARY

1. The conventional physics units for radiation exposure, dose, and radioactivity are the *Roentgen*, the *rad*, and the *Curie*, respectively. The equivalent SI units are the *Coulomb per kilogram*, the *gray*, and the *Becquerel*. To measure radiation harm to an organism, the conventional *rem* and the SI *Sievert* are used. These are biological, rather than physical units.
2. The dose-area product or *DAP* takes into account the field size and is a better indication of the overall effect of a radiation exposure to an organism than is the dose alone.
3. Dose equivalent limits (DELs) provide guidelines for corrective actions, but do *not* constitute acceptable levels of radiation exposure, which should

follow the *ALARA* philosophy. While prospective DELs protect younger workers, cumulative lifetime DELs protect older workers over the length of their career.

4. Dose equivalent limits are developed from the weighted effects of radiation on three dose limiting organs: The gonads, the red bone marrow, and the lens of the eye.
5. The genetically significant dose or *GSD* is at about 20 mrem per year and continues to climb.
6. Radiation detection instruments may operate in detection only mode, rate mode, or integrated mode. All may be evaluated by the four characteristics of sensitivity, accuracy, resolving time and range. These characteristics must be matched with the intensity, energy level and type of radiation being measured in order to obtain valid and reliable measurements.
7. Detection instruments include scintillation counters, thermoluminescent dosimeters, film badges and gas-filled devices. Gas-filled detectors are further subdivided into pocket dosimeters, ion chambers, proportional counters, and Geiger-Meuller tubes.
8. All electronic radiation detection instruments require a threshold voltage to operate, and fail above a certain maximum supplied voltage. Ion chambers and Geiger-Meuller tubes give consistent readings across a range of supplied voltages, but the readings from proportional counters depend on the supplied voltage as well as the radiation intensity.

REVIEW QUESTIONS

1. Which conventional radiation unit should only be used in reference to x-rays, and technically should not be used to indicate dose in tissue?

2. To obtain rems from rads (or sieverts from grays), the dose must be multiplied by __________ factors.

3. What is the dose-area product for a dose of 42 mGy delivered over a collimated field of 25 × 30 cm?

4. How many C/kg are in 5 R? (Use scientific notation.)

5. How many gray are in 200 rad?

6. How many rem are in 0.01 Sv?

(Continued)

REVIEW QUESTIONS *(Continued)*

7. How many curies are in 10 Bq? (Use scientific notation.)

8. When the current prospective annual dose equivalent limit is divided by 50 work weeks in a year, it comes out to about __________ mR per week.

9. What is the cumulative lifetime DEL, and what is the sum of the prospective annual DEL's for a radiographer who is 27 years old and has been working in the field for 6 years?

10. What is the monthly DEL for the fetus of a pregnant radiation worker?

11. What is the one-time emergency DEL for rescue workers?

12. What is the only body part for which the annual partial body DEL is only 15 rem rather than 50 rem?

13. The gonadal dose averaged among the reproductive population defines the __________.

14. What three characteristics of a radiation detection instrument affect its accuracy?

15. The amount of electrical current or charge generated by small amounts of radiation exposure is the definition for the __________ of a radiation detection instrument.

16. The ability of a radiation detection instrument to detect sequential ionizing events that occur one right after the other depends on the instrument's __________.

17. Which type of radiation detection instrument immediately fluoresces upon x-ray exposure?

(Continued)

REVIEW QUESTIONS *(Continued)*

18. To obtain a radiation measurement, the heating of a TLD crystal to induce it to phosphoresce is referred to as:

19. The "inverse photometer" used to measure the darkness of silver deposit on the film from a film badge is called a:

20. What is the greatest advantage, and what is the greatest disadvantage, of the pocket dosimeter?

21. In the voltage range of simple ionization, for each ionizing event within the gas in the chamber, how many electrons actually reach the anode pin?

22. If an ion chamber is not provided sufficient voltage to operate, what phenomenon occurs when ionizations of the gas within the chamber take place?

23. What is the term used to describe discharge of the entire chamber of gas from a single ionizing event?

24. For personnel monitoring, regardless of the type of monitor used, what must be kept in a location well away from radiation hazards in order to obtain reliable measurements?

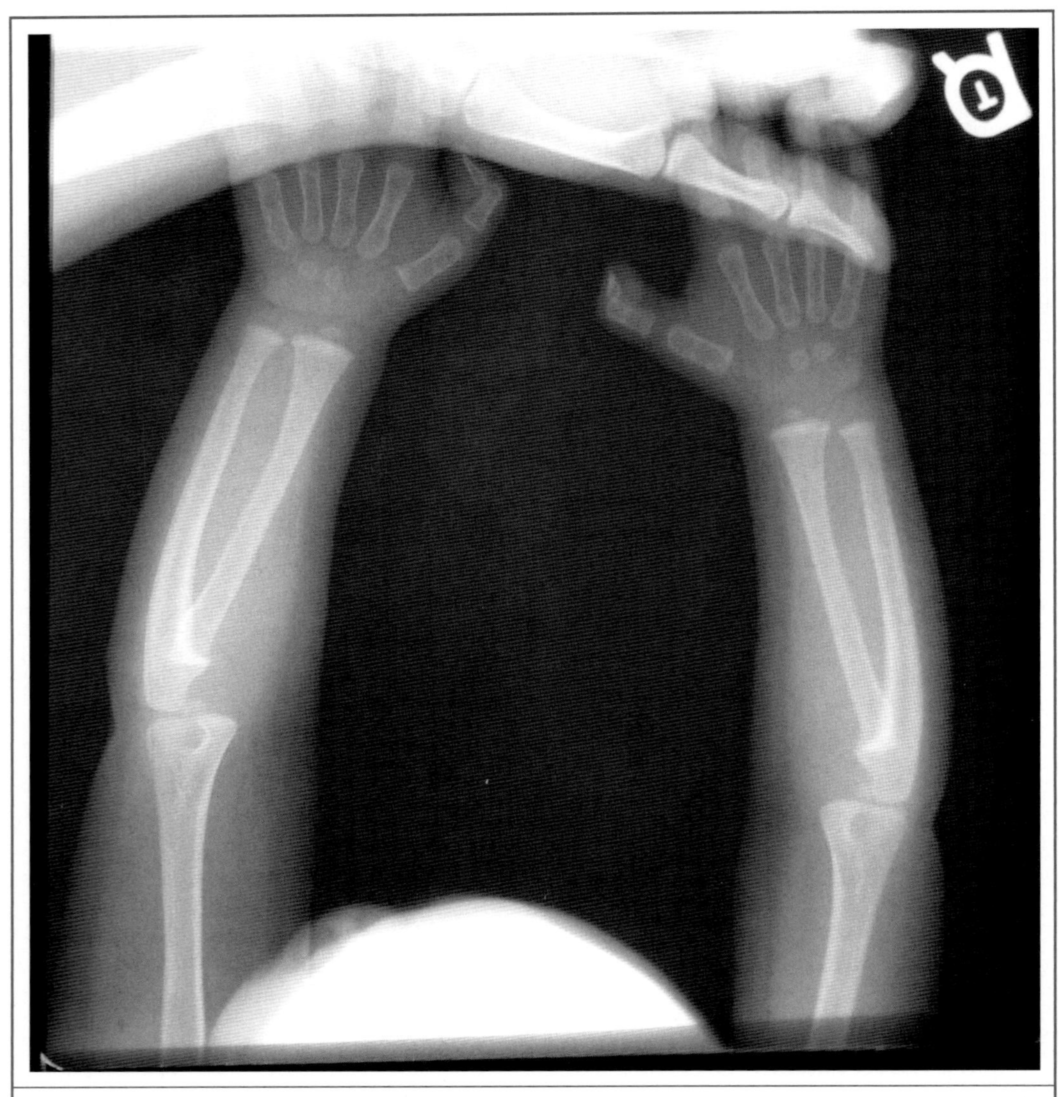

A misguided attempt to get both AP forearms on a single exposure with the wrists improperly turned into PA position not only resulted in a repeated exposure, but also pulled the small child's head into the direct x-ray beam, exposing the extra-sensitive lenses of the eyes to radiation at a very young age.

Chapter 42

RADIATION BIOLOGY: CELLULAR EFFECTS

Objectives:

Upon completion of this chapter, you should be able to:

1. Review the basic tissues of the human body, cell structure and metabolism.
2. Explain the roles of DNA and RNA in protein synthesis, and the three levels of genetic information transfer.
3. Describe the structure of DNA and chromosomes at different stages in cell's life cycle.
4. Describe in detail each stage in the life cycle of the cell, and at which stages the cell is most sensitive to radiation exposure.
5. Distinguish between the stages of mitosis and those of meiosis in cell reproduction.
6. Explain the two major propositions of the *Law of Bergonie and Tribondeau.*
7. Prioritize different types of cells according their sensitivity to radiation exposure.
8. Correctly interpret various radiation response curves, and survival curves, from graphs.
9. State the main propositions of the *target theory* for radiation damage to cells.
10. Describe the radiolysis of water and how it contributes to indirect damage mechanisms.
11. Explain how radiation can damage the cell membrane and its consequences.
12. Describe the specific types of structural changes that occur to the DNA molecule from radio-ionization, and their implications for mutations, repair, and disease.
13. Define and calculate linear energy transfer (LET) for different radiations.
14. Define and calculate relative biological effectiveness (RBE) for different radiations.
15. Describe the impact of dose rate, protraction, and fractionation on the effectiveness of a radiation exposure.
16. Describe the influence of oxygen enhancement ratio (OER), age and gender on the effectiveness of radiation exposure.
17. Relate LET, RBE, direct effect, indirect effect, single-strand breaks, double-strand breaks, point mutations, frameshift mutations, oxygen enhancement and reparability to the *penetration* of a particular type of radiation.

Figure 42-1

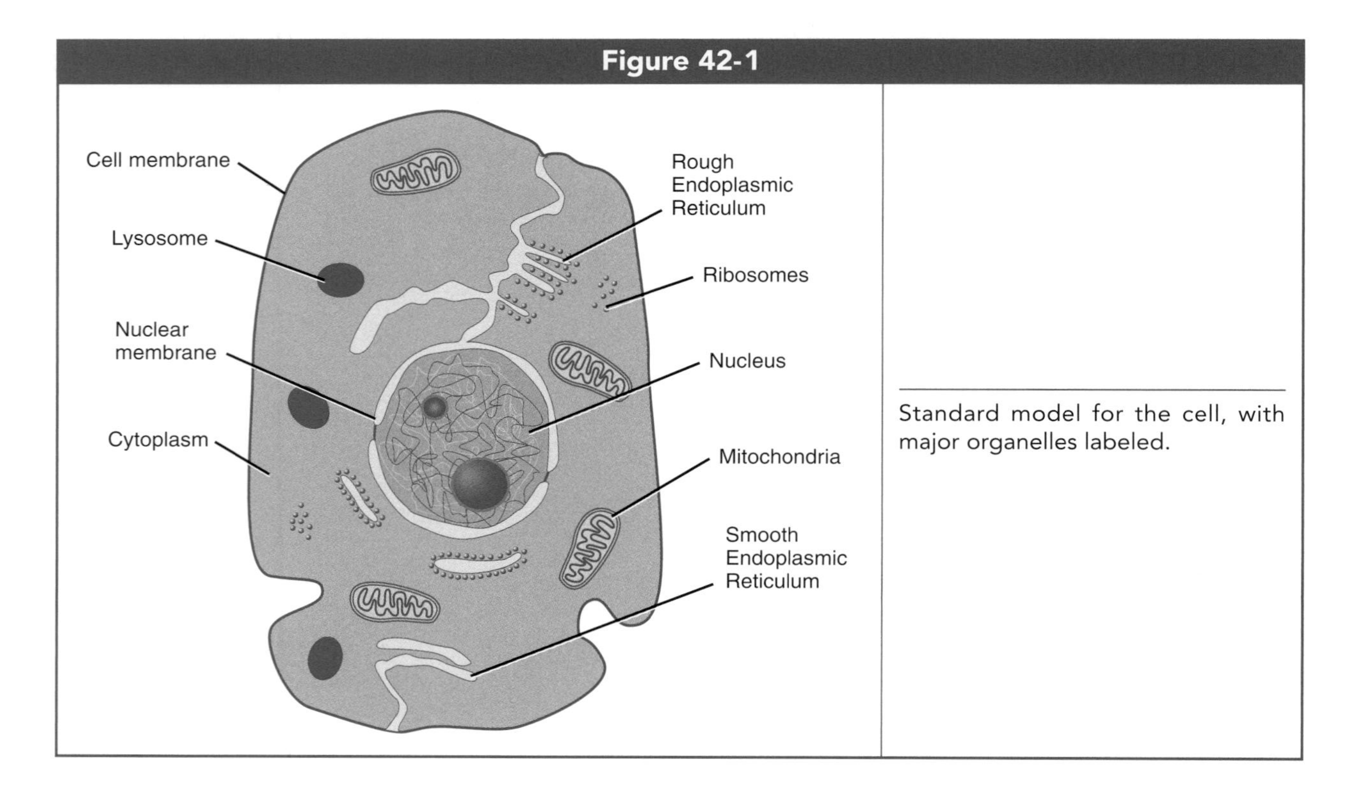

Standard model for the cell, with major organelles labeled.

BIOLOGICAL REVIEW

Tissues of the Human Body

The human body is composed of about 80 percent *water*. As will be explained, this is an important aspect for radiation effects. The remaining molecular composition of the body is about 15 percent proteins, 2 percent lipids (fats), 1 percent carbohydrates, and about 1 percent nucleic acids. These molecules are organized primarily within the living *cells* of the body, of which there are many types including epithelial (skin) cells, osteocytes (bone cells), nerve cells, and blood cells.

Within the blood, erythrocytes (*erythro* = "red," *cyte* = "cell") have a specialized "bowl-shape" designed to carry oxygen to the various tissues of the body, while leukocytes (*leuko* = "white") and lymphocytes from the lymph system fight off infectious organisms. Within each cubic millimeter (a cube about the size of a pinhead) are crowded 5 million erythrocytes and 7,500 leukocytes.

The suffixes *-blast* and *-clast*, when attached to a type of cell, refer to special cells whose function is to *generate* or *destroy* that particular tissue, respectively. For example, *erythroblasts* are erythrocyte-*forming cells* which are found concentrated in the bone marrow, whereas *erythroclasts* help disassemble worn-out and nonfunctional erythrocytes and tend to concentrate within the spleen. *Any cell type that ends with the suffix* -blast *is more sensitive to radiation than the mature cells it produces*, because of its rapid reproduction rate.

For example, *erythroblasts* are much more sensitive to radiation exposure than the mature *erythrocytes* produced by them. For this reason, the bone marrow, as a tissue, is of greater concern radiologically than the circulatory system. A *blastula* is a small bundle of such reproductive cells, also called *stem cells*, and describes the initial stages of a pregnancy before the organism is recognizable as a developing embryo. Blastulas are very sensitive to radiation.

Cells can also be broadly categorized as *somatic* (*soma* = "body") or *genetic* (*gen* = "beginning" or "originating"). While somatic cells specialize in a particular organ function for the survival and thriving only of the organism itself, genetic cells such as the sperm and ovum carry the task of preserving an entire species through reproduction.

Human Cell Structure and Metabolism

Highly specialized cells such as nerve cells can develop unique structures relating to their function, but all cells have certain basic components in common. Figure 42-1 (facing page) is the standard model for the basic human (eukaryotic) cell structure; most of the cell is composed of the cytoplasm, a watery medium in which the small *organelles* are suspended. These include the mitochondria which serve as an energy source for the cell, the ribosomes whose main function is the manufacture of proteins, and the lysosomes which execute the removal of waste products. The endoplasmic reticulum is a network of membrane-enclosed spaces that serve as a transport system. The r*ough endoplasmic reticulum* is bordered by numerous ribosomes.

Of particular interest for radiation effects are the membranes and the nucleus of the cell. There is a membrane around the entire cell, one around the nucleus, and one around every organelle. These hold the cell together and give it its structure. Most membranes consist of a double-layer of a fatty substance called a *phospholipid*. The outer cell membrane contains the moisture of the cytoplasm and, by its structure and chemical composition, only allows substances needed by the cell to pass through from the exterior.

The nucleus is the control center that directs the activities of the cell. It contains the *chromosomes*, long thread-like bodies that contain DNA (deoxyribonucleic acid) and certain proteins. Along the DNA the hereditary *genes* that determine all of the characteristics of the organism are lined up. *Nucleoli* are round bodies within the nucleus where RNA (ribonucleic acid) tends to concentrate. Nucleoli help in the formation of ribosomes, and RNA and the ribosomes both play critical roles in the production of *proteins*. It is the specific types of proteins produced by a cell that determine its function as part of an organ in the body. This assembly of large molecules, such as the synthesis of proteins, is known as *anabolism*. Protein synthesis requires the transfer of information from the nucleus (the blueprints) out to the cytoplasm (the factory). This occurs primarily along the endoplasmic reticulum.

In the course of its function, the cell must also break down large molecules into smaller units in a process called *catabolism*, which ultimately results in the waste products of water and carbon dioxide exuded by the cell. The term *metabolism* refers to the sum of all chemical transformations that occur within a cell.

Within the nucleus, there are 23 *pairs* of chromosomes, or a total of 46. Two of these are "sex" chromosomes which determine the gender of offspring, the other 44 are referred to as *autosomes* (*auto* = self) and determine the various traits and characteristics of the organism itself. Remarkably, the DNA within each cell contains all of the information necessary for the entire organism to develop, which is what makes cloning possible.

The DNA molecule looks somewhat like a twisted ladder (Fig. 42-2). The "rails" of this ladder structure consist of a double-helix made of chains of sugar phosphate molecules, while the "rungs" that cross between them are each made up of a pair of nitrogenous base molecules. These nitrogenous bases must be selected only from the following four—thymine,

Figure 42-2

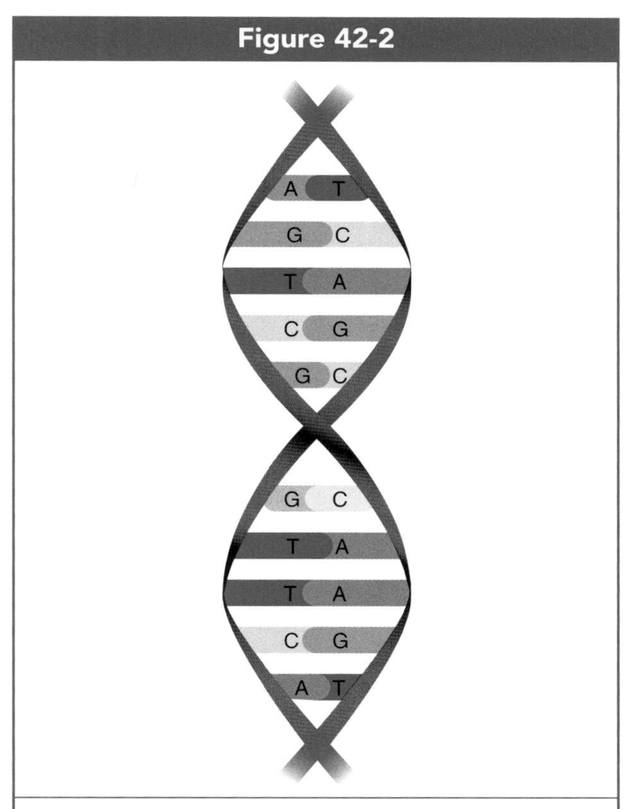

The "twisted ladder" structure of the DNA molecule, showing the paired bases of thymine, adenine, guanine, and cytosine that form the rungs.

adenine, guanine, or cytosine, abbreviated *T*, *A*, *G* and *C*, respectively. *T* can only chemically bond with *A*, and *G* can only chemically bond with *C*, such that the base pairs that make up each "rung" can only be of the four combinations *TA*, *AT*, *GC*, and *CG*.

There are many thousands of these base molecules along a strand of DNA, which constitute the genetic "code." The "letters" of the code are formed by triplets of the base molecules in sequence, such as *TAC*, *GCG*, or *AAG*. Each of these sets represents one "letter," called a *codon*.

We have stated that the physiological functions of every organ and the cells that make it up are controlled by the chemistry of *protein molecules*. A protein is a long chain-molecule, made up of a sequence of amino acids. The sequential order of these amino acids along the chain is what determines the specific chemical function of protein.

The synthesis (production) of protein molecules within the cell is a complex process, but may be simplified as follows: In the nucleus, a molecule of DNA splits or *unzips* down the middle of its "rungs," separating each pair of nitrogenous bases (*T*, *A*, *G* and *C*, Fig. 42-3). Alongside the "raw" exposed side of a single split DNA strand, other molecules that are capable of chemical bonds with the four bases will begin to attach to them in such a manner as to form a shorter chain molecule called *messenger RNA* (*mRNA*). Although the specific "rung" molecules for the mRNA are somewhat different from the original

Figure 42-3

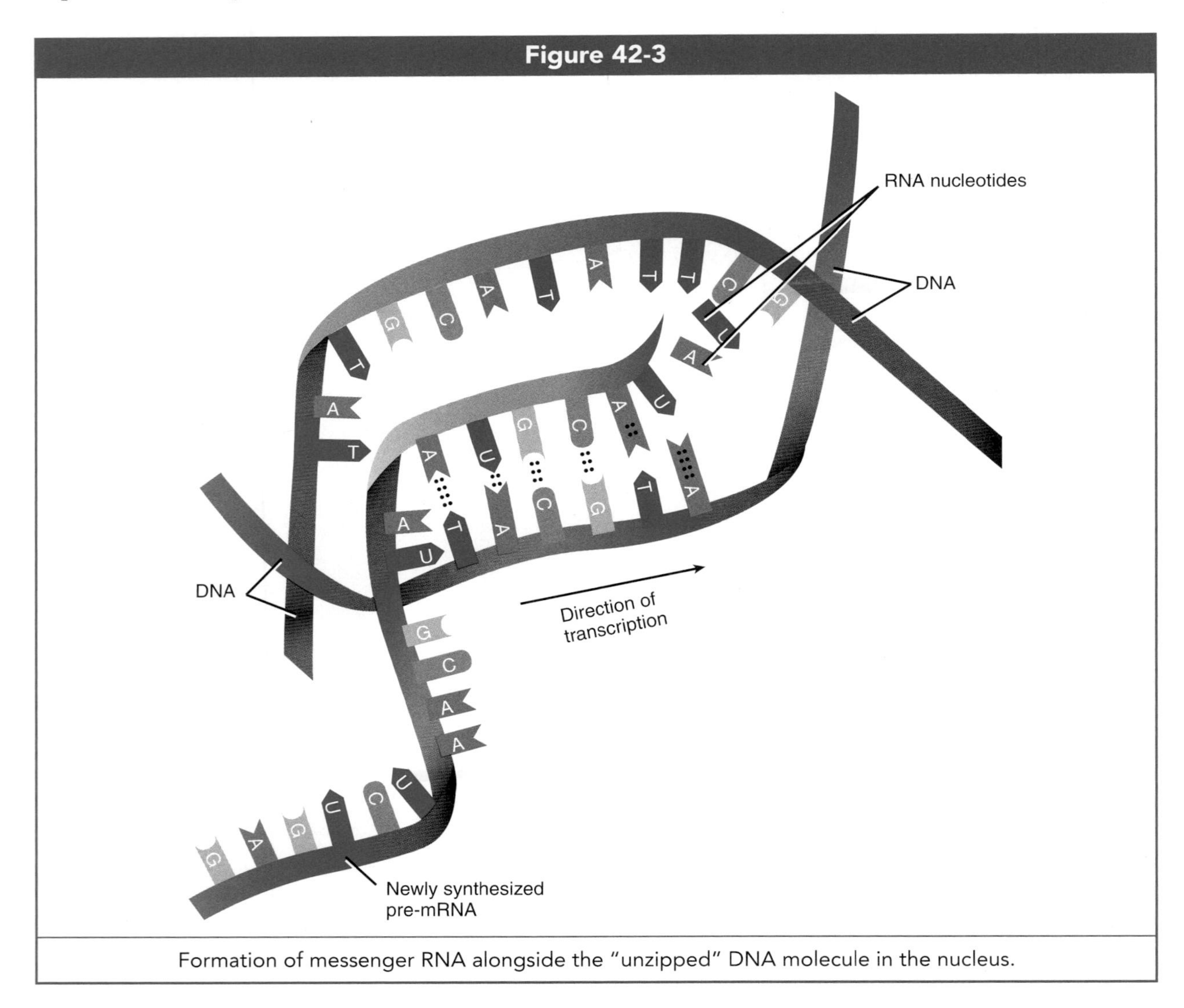

Formation of messenger RNA alongside the "unzipped" DNA molecule in the nucleus.

nitrogenous bases of the DNA, they nonetheless chemically *mirror* a section of code from the split DNA strand. These molecules are referred to as *RNA nucleotides*.

The mRNA then travels out of the nucleus into the cytoplasm, via the *rough endoplasmic reticulum* where ribosomes are waiting. In the vicinity of the ribosomes are free-floating molecules of *transfer RNA (tRNA)*. Each tRNA molecule has only a short segment of genetic code on it, and has a chemical structure that tends to bond with a *specific* amino acid. When a molecule of this particular amino acid comes into contact with the matching tRNA, the two connect. The mRNA may be considered as the *messenger* that physically brings the code from the nucleus to the ribosome, while the tRNA might be thought of as a *translator* in the process.

In the chemical reactions that follow, the ribosome acts as a protein-building machine. It moves along the mRNA strand, effectively "reading" the code and matching the short code segments of nearby tRNA molecules to the mRNA (Fig. 42-4). As the process continues, a *string* of tRNA molecules is brought into contact with the mRNA. This lines up the *amino acids* attached to the tRNA molecules in sequence. As these amino acids bond to each other in a chain, a complete protein molecule is built-up.

Note that the tRNA code segments mirror the code of the mRNA, which in turn is a mirror image of the original DNA code. The original code sequence of the DNA is thus replicated. In this way, a protein is formed whose sequence of amino acids has ultimately been dictated by the original DNA in the nucleus.

As the long protein molecule is formed, the amino acids separate sequentially from their tRNA molecules (Figure 42-5). In turn, the tRNA molecules detach from the mRNA strand. Each tRNA molecule is recycled—it separately attaches to another loose amino acid in the vicinity, and is reused by a ribosome in the fabrication of another protein. After a complete chain of amino acids has formed the correct protein, the protein moves off to become part of an organelle and perform its biochemical function.

Transfer of Genetic Information

There are three different ways in which the information contained within genes must be transmitted outward from the DNA molecule. First, this information must be passed from the nucleus of an individual cell out to its cytoplasm where it is used to create proteins vital to the survival of the cell itself and to its function within an organ. This is the process of *cellular metabolism* just described in the last section.

Figure 42-4

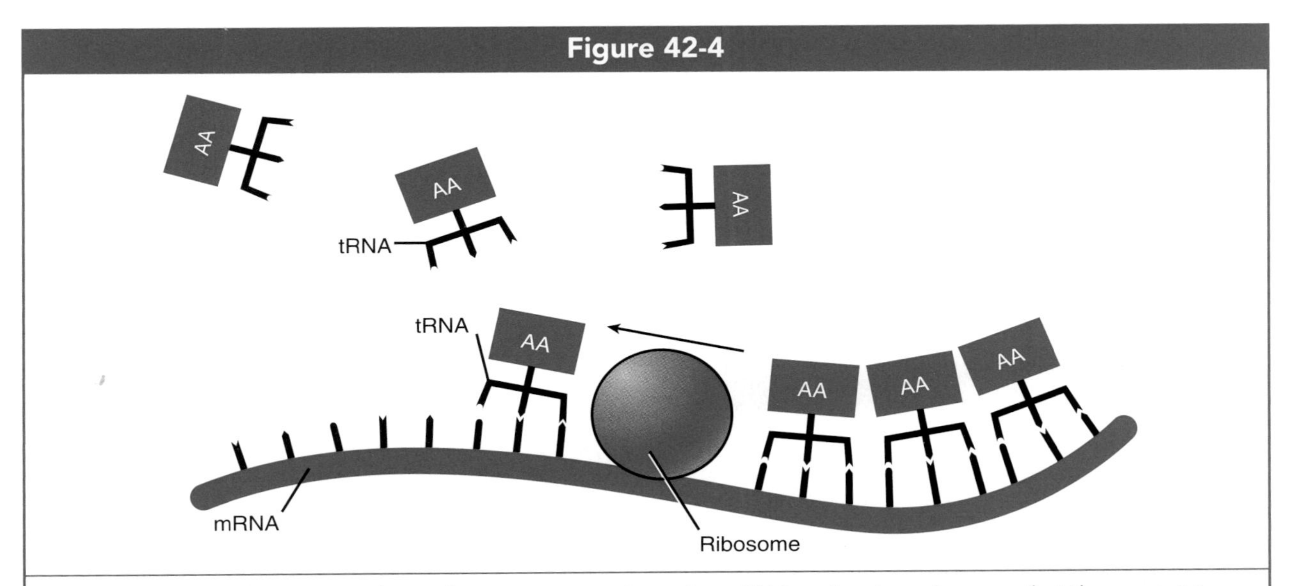

In the rough endoplasmic reticulum, *ribosomes* move along the mRNA molecule and ensure that the correct *transfer RNA* molecules attach to it in sequence. Each tRNA is attached to a specific amino acid, so the amino acids line up in sequence to form a protein molecule.

Figure 42-5

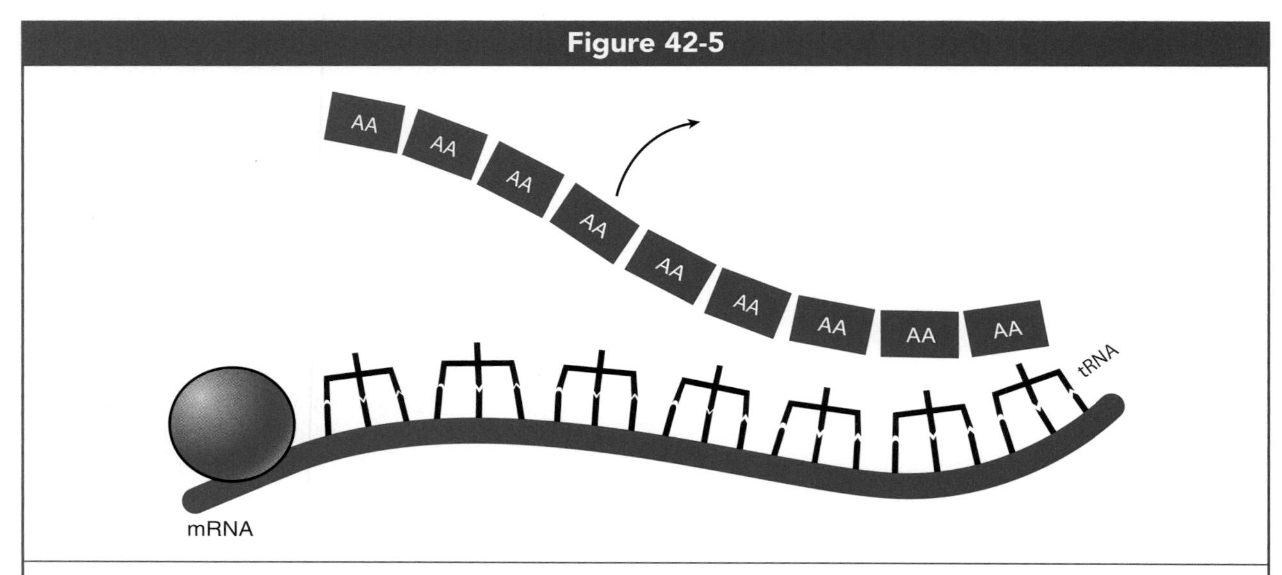

As a protein molecule is formed, its amino acids progressively break off from the tRNA molecules. In turn, the tRNA molecules will detach from the mRNA strand and be recycled in the transcription process.

Second, a whole organism must pass along genetic information to its offspring in the process of *inheritance*. By this mechanism, the next generation of the species acquires the information it needs for survival, along with traits that have evolved from previous generations. For humans, 23 chromosomes are inherited from each parent to form a complete complement of 46 chromosomes.

Third, genetic information must be passed along from a parent cell to progeny cells for the purposes of tissue growth and repair. *Stem cells* or *blast* cells must pass the information to daughter cells that will replace tissues, such as the skin which is sloughed off as the upper layers of cells die out. When an injury occurs which destroys local tissue, biochemical processes trigger the formation of blood clots and scar tissue, and kick into high gear the reproductive rate of surviving cells to generate replacement tissue.

Thus, the effective transmission of genetic information is essential for survival at the cellular level, the tissue level which preserves the whole organism, and at the species level (Table 42-1). Exposure of cells, tissues or whole organisms to x-rays can disrupt the transfer of genetic information because of the ability of x-rays to *ionize* atoms and molecules. When an atom loses one of its orbital electrons, *covalent* or *ionic* chemical bonds can be broken, causing a molecule to break up. When a protein chain is broken, that function within a cell is lost. When a molecule of DNA or RNA is broken or modified by ionizing events, *the genetic code can be altered*, harming the function of the cell or causing the next reproductive cycle of the cell to fail to produce viable offspring cells. Details of these processes will be described more fully.

Table 42-1

Summary: Transfer of Genetic Information
1. Within cell, from nucleus to cytoplasm = *metabolism*
2. From parent cell to progeny cells = *tissue growth and repair*
3. From parent organism to offspring = *inheritance*

Life Cycle of the Cell

Although the lifetimes of cells vary greatly, a typical body cell has a life cycle of only 24 hours. For the normal growth of the organism and repair of damaged tissues, somatic cells go through reproductive stages called *mitosis*. These periods of active reproduction are separated by interim periods called *interphase* (Fig. 42-6).

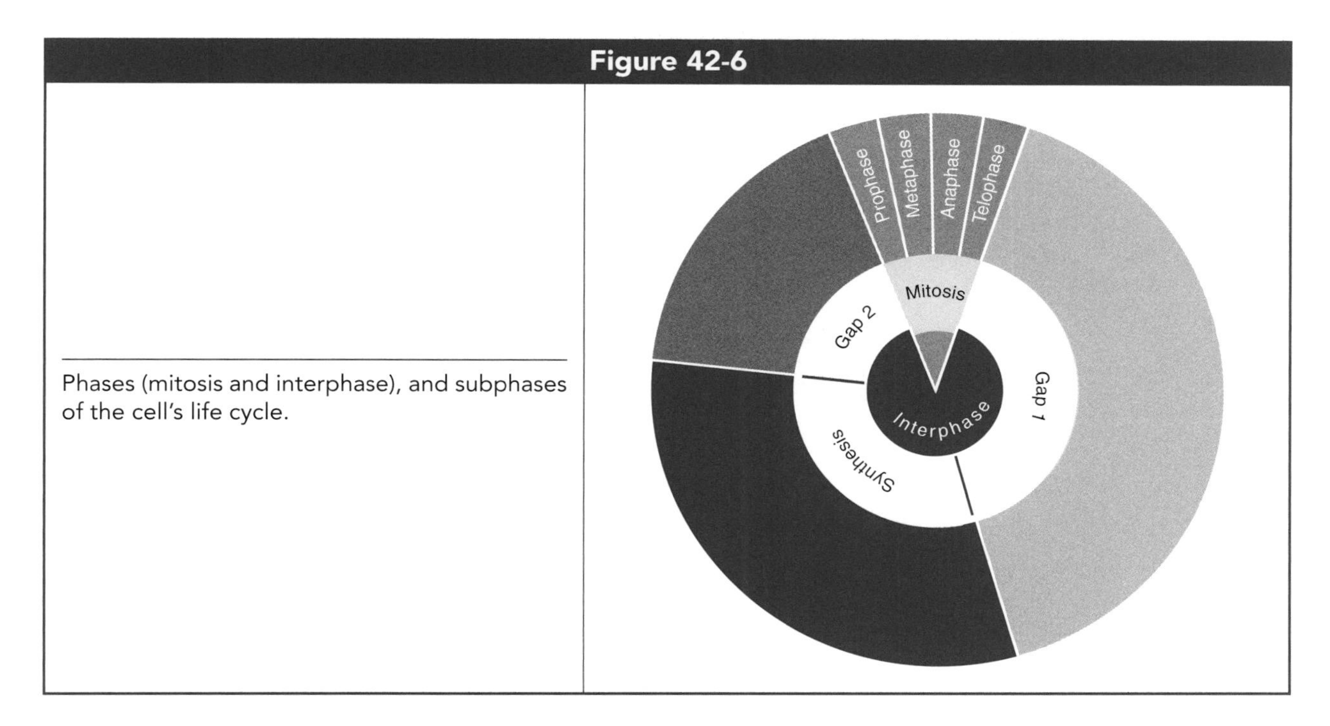

Phases (mitosis and interphase), and subphases of the cell's life cycle.

During interphase, when a cell is not dividing, the 46 chromosomes appear as a diffuse, granular mass within the nucleus called *chromatin*. This is because they consist of very loosely-coiled strands of *chromatin fiber*. The chromatin fiber is made like a rope, from twisting strands of chromatin together. The chromatin itself consists of a series of bead-like globules of proteins called *histones*, around which the DNA molecule wraps itself twice and then links to the next histone bead. As illustrated in Figure 42-7, there are *three levels of organization*, which might be summarized as follows:

1. DNA molecules wrap around histone beads and link them together to form chromatin strands.
2. Chromatin strands twist together to form chromatin fiber.
3. Chromatin fibers loop into the shape of two chromatids and the centromere.

Both mitosis and interphase can be further broken down into subphases or stages. Interphase consists of three stages called G_1 *(for Gap-1)*, the *DNA synthesis* phase (*S*), and G_2 *(for Gap-2)*. These subdivisions are illustrated in Figure 42-6.

During G_1 the cell is metabolically active, duplicating its organelles but not its DNA. At this time in the nucleus, the chromatin fibers holding the DNA are organized into a pair of rod-shaped structures called *chromatids*, held together by a constricted region called the *centromere* (Fig. 42-7). However, the loops of chromatin fiber making up this structure are so loose that it is difficult to make out the actual chromatid bars under a microscope. The G_1 stage typically lasts 8 to 10 hours, but can last from minutes to years in particular tissues. It is followed by the DNA synthesis phase, *S*.

The *S* phase also can be quite variable for special tissues, but typically lasts 6 to 8 hours. During the *S* phase, all genetic material doubles. Each DNA molecule is replicated into two identical daughter DNA molecules. These two continue to be held together by the centromere in such a way that the two-chromatid structure of the chromosome becomes a four-chromatid structure as illustrated in Figure 42-8.

The cell now enters into another "gap" phase, G_2, in which cell growth continues, and enzymes and proteins are formed in preparation for cell division. G_2 normally lasts from 4 to 6 hours. Even though the chromosomes have been doubled in their structure, the chromatin fibers that form them are still loosely looped together such that they remain difficult to observe under a microscope. The cell is now ready to divide through the process of *mitosis*.

Figure 42-7

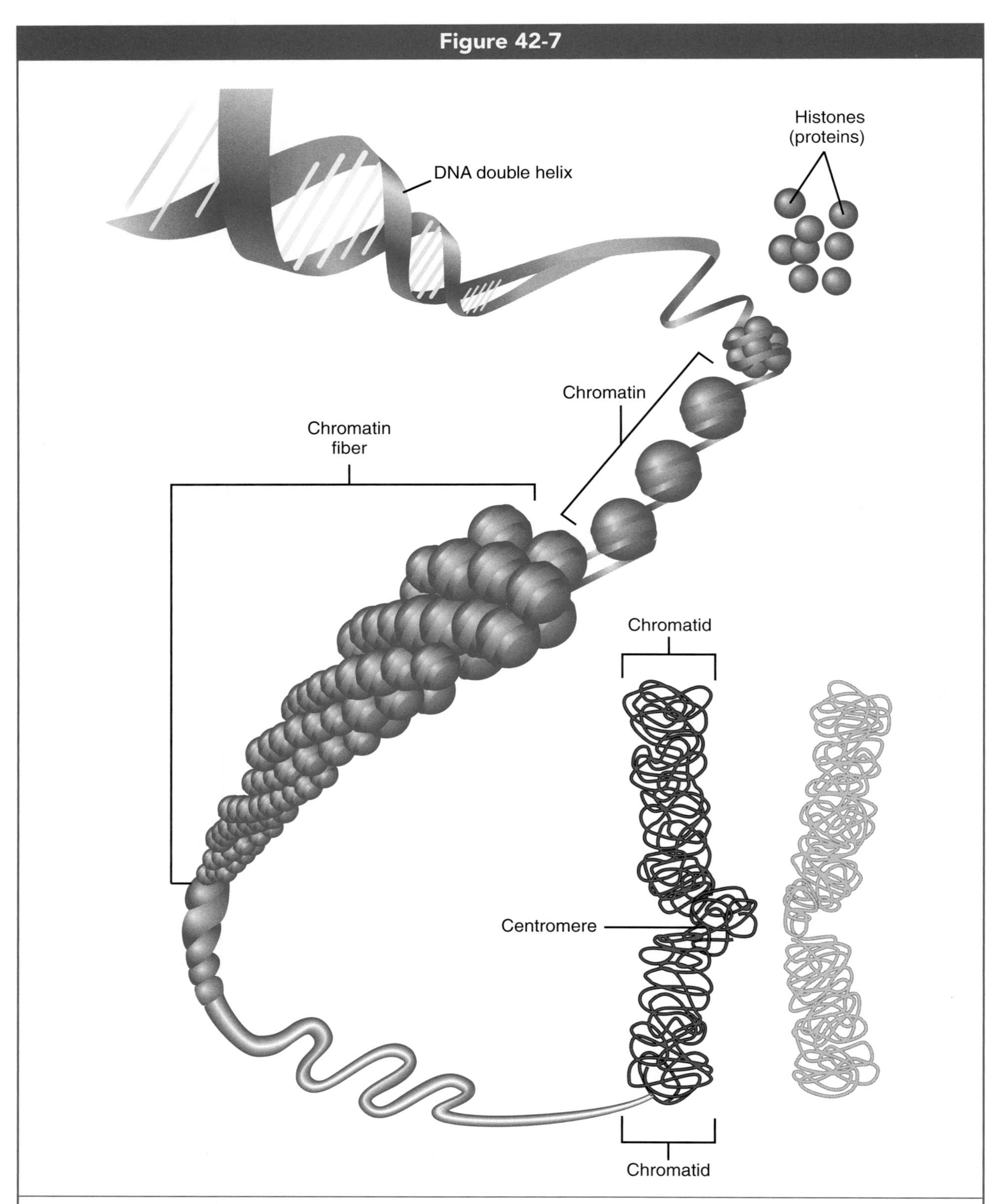

DNA molecules wrap around histone "beads," and link them to make strands of chromatin. These strands twist to form fibers, which then loop together in the shape of two chromatids connected by a centromere.

Figure 42-8

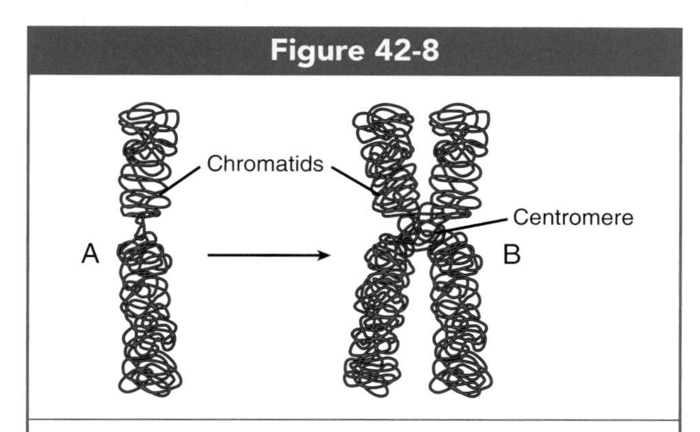

During the S phase, when all DNA is replicated to form a "second copy," the chromosome changes from a two-chromatid structure to a four-chromatid structure.

The sensitivity of the cell to radiation exposure varies with the stages of its life cycle. *The very beginning of the S phase is the most sensitive time during the cell's entire life cycle for exposure to radiation.* If radiation strikes the cell at this critical point in time and ionizes genetic molecules, various genetic mutations can be induced or biochemical changes can cause the next attempted cell division to result in nonviable daughter cells.

Out of a cell's entire life cycle, the very beginning of the S phase is when it is most vulnerable to ionizations from radiation exposure.

Mitosis

Mitosis is the cellular reproduction process used for the normal growth and repair of body tissues. Mitosis is also called *replication division*, because it results in a full complement of genetic material in each daughter cell that is an exact replica of the genetic material in the parent cell. There are four stages of mitosis listed in Figure 42-6. Figure 42-9 illustrates these four stages. They might be summarized as follows:

1. *Prophase.* characterized by a swelling in the size of the nucleus.
2. *Metaphase.* During metaphase, the loops of chromatin forming the chromosomes tighten and *condense* such that the chromosomes become readily visible under a microscope. The nucleus elongates, the chromosomes *line up* in the middle of the cell, and a network of *mitotic spindle fibers* (which has been developing all through the G_1 and G_2 phases) attaches to the centromeres of each chromosome (Fig. 42-9).
3. *Anaphase*, during which the double-chromosomes are split and each complete set of new chromosomes are pulled by the spindle fibers to each end of the nucleus, *polarizing* two identical copies of genetic material (Fig. 42-9).
4. *Telophase*, during which the nuclear membrane is temporarily dissolved while a new *cell membrane* forms through the middle of the cell mass, dividing it into two new cells. Following this, a new nuclear membrane is reconstructed around each nucleus, a process called *cytokinesis*.

Figure 42-10 is a stained photomicrograph in which some cells can be seen in *metaphase* in which the spindle fibers can be seen connecting to the

Figure 42-9

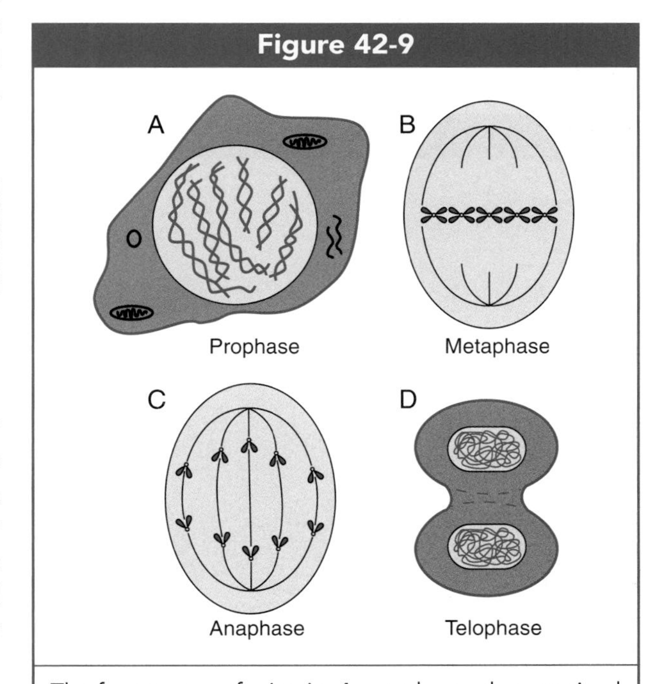

The four stages of mitosis: ***A***, prophase, characterized by swelling of the nucleus; ***B***, metaphase showing the chromosomes lined up in the middle; ***C***, anaphase showing the polarization of the split chromosomes toward each end of the cell; ***D***, telophase showing reformation of nuclear membranes and formation of dividing cell membrane.

Figure 42-10

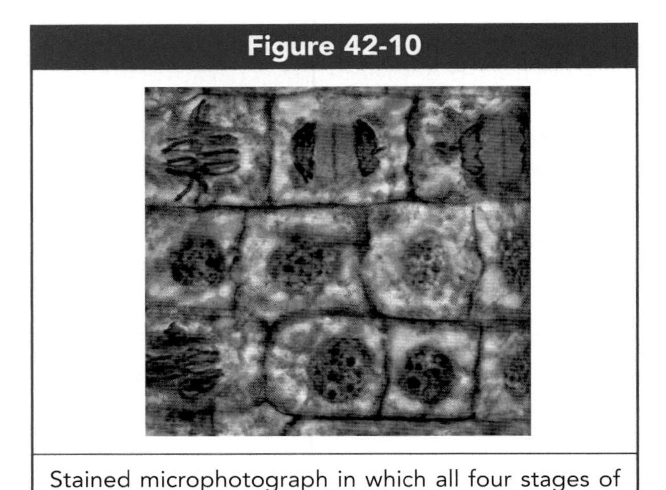

Stained microphotograph in which all four stages of mitosis can be made out.

centralized chromosomes, and others can be seen with their chromosomes separated at polar ends of the cell with a new membrane beginning to form between them.

Metaphase is the most radiosensitive phase during the process of mitosis (and the second most radiosensitive phase of the cell's entire life cycle). During metaphase, the densely-packed chromosomes are extremely vulnerable to damage from the ionizing interactions caused by x-rays or other radiations. Since the chromosomes can also be clearly seen during metaphase, structural changes representative of various mutations can actually be observed in the chromosomes themselves under a microscope.

> *During mitosis, metaphase is the stage when chromosomes are most vulnerable to damage from radiation exposure, which is visible.*

Meiosis

Meiosis is the term which describes the reproduction of *gametes* or sex cells. *Reduction division* is defined as a cell division in which only *one-half* of the chromosomes from the parent cell are preserved in each of the two daughter cells (Fig. 42-11). Meiosis may be described as a *replication division* followed by *reduction division*. The replication division is identical to that for mitosis, forming two daughter cells. But, during the interphase before in the next division

Figure 42-11

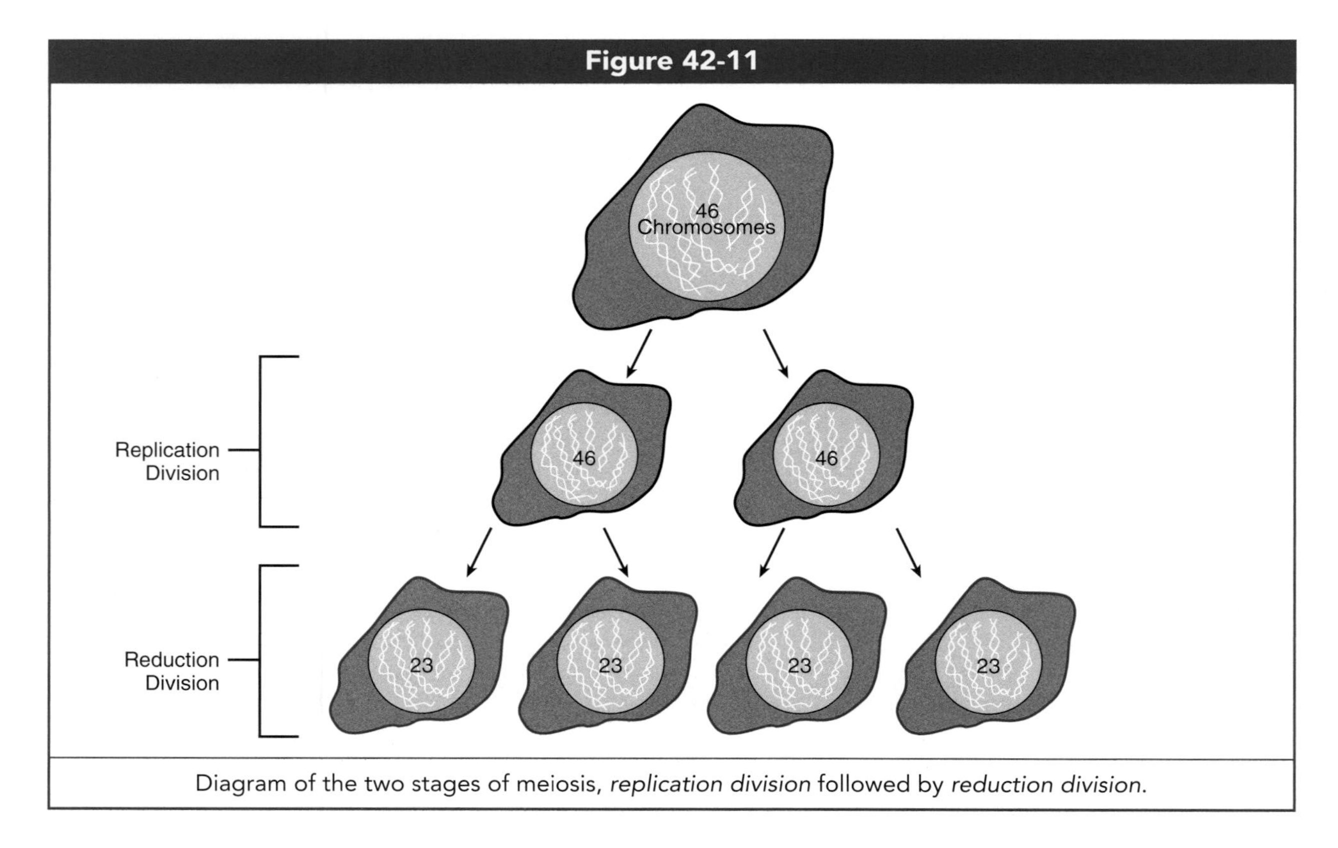

Diagram of the two stages of meiosis, *replication division* followed by *reduction division*.

Figure 42-12

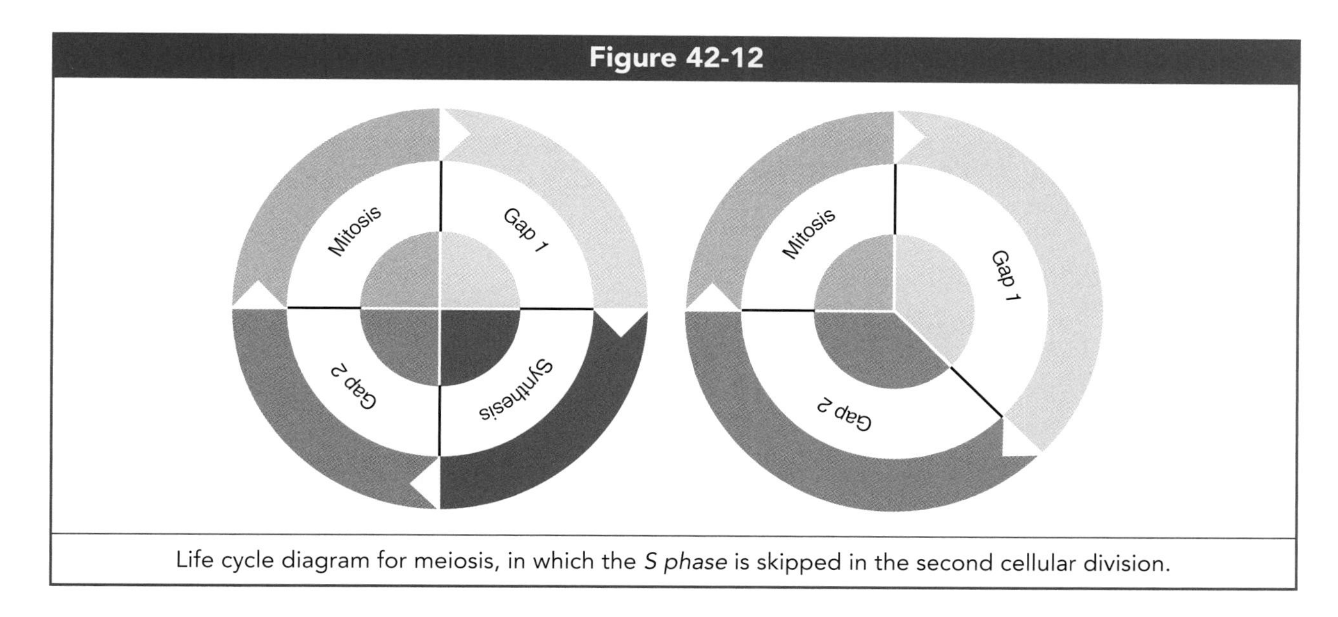

Life cycle diagram for meiosis, in which the *S phase* is skipped in the second cellular division.

takes place, the S phase is skipped (Fig. 42-12). This results in four daughter cells which each contain one-half of the chromosomes from the original parent cell when the second division takes place. These 23 chromosomes will be combined with those of an opposite gender gamete during sexual reproduction. In this way, offspring acquire traits from both parents.

CELLULAR RADIATION EFFECTS

Cell Sensitivity

Law of Bergonie and Tribondeau

While it is difficult to find "rules" that are *always* true in radiobiology, some useful generalizations can be made. Such is the case with the *law of Bergonie and Tribondeau,* two French scientists who made the observation in 1906 that characteristics of the cell itself affect its sensitivity to radiation. The law essentially states that cells which divide more, and cells that are less highly *differentiated*, tend to be more sensitive to radiation.

Cell *differentiation* refers to how specialized the structures and function of the cell are. An example of one of the most highly-differentiated cells in the body is the nerve cell, with its axon and dendrites specially designed for the conduction of electrochemical pulses. Another example is the *erythrocyte*, or red blood cell, which has a very unique shape designed to capture molecules of oxygen. An undifferentiated cell looks like the standard model in Figure 42-1 at the beginning of this chapter, with "typical" cell structures. A good example of an undifferentiated cell would be the epithelial cells of the skin and gastrointestinal linings. We might describe undifferentiated cells as having more *primitive* structure.

Cells we describe as more "reproductive" are not only those which divide more frequently, but those which have a longer "dividing future," such that the net result is that many daughter cells are produced over the lifetime of the parent cell. We might summarize Bergonie and Tribondeau's law as follows:

More Primitive → More Sensitive
More Prolific → More Sensitive

When one examines the list of tissues in Table 42-2, classified according to their radiosensitivity, one can immediately see that the law of Bergonie and Tribondeau holds considerable merit; generally, those tissues which top the list have more "basic" cells and reproduce more, whereas those which fall last on the list tend to be highly-differentiated and with low rates of proliferation. Note, for example, that *spermatogonia*, the basic stem cells from which sperm arise, fall in the highly-sensitive category, whereas *spermatazoa*, the mature sperm cells which are highly-differentiated in that they have a well-developed *flagellum* for locomotion, fall much farther down the list.

Table 42-2

Relative Sensitivity of Tissues and Cells to Radiation

Sensitivity	Subgroup	Tissue/Cell
High:		1. Lymphocytes
		2. Erythroblasts (in bone marrow)
		3. Lens of the eye
		4. Oocytes (egg cells)
		5. Spermatogonia
Intermediate:	Upper:	6. Myelocytes (bone marrow)
		7. Intestinal crypt cells
		8. Skin basal cells (internal linings)
	Mid:	9. Endothelial cells (internal linings)
		10. Glands in general
		11. Osteoblasts
		12. Spermatoblasts (spermatids, spermatocytes)
	Lower:	13. Spermatozoa
		14. Osteocytes
		15. Erythrocytes (mature red blood)
Low:		16. Muscle
		17. Connective Tissues (cartilage, ligaments, tendons)
		18. Nerve/Brain cells

The *eye lens* is one exception to this rule—the cells which make it up are extremely differentiated in order to provide a clear medium through which light will pass and focus, yet the eye lens is one of the most radiosensitive tissues in the body. Oocytes (egg cells) are another exception, in that they do not reproduce, yet they are very radiosensitive. The law of Bergonie and Tribondeau is but a general guide, which certainly has some exceptions.

We have described all "-blast" cells or "stem cells" as being highly sensitive to radiation. They meet both of Bergonie and Tribondeau's qualifications, as does a developing human embryo. The earliest stages of pregnancy, when the tissues are most primitive in structure and have an extreme reproduction rate, is the most hazardous period for the developing embryo to receive a radiation exposure. Roughly speaking, the embryo is about ten times more sensitive to radiation than an adult, while the fetus (after 3 months) is approximately two times more sensitive.

Cellular Response to Radiation

When different effects of radiation upon cells are studied, we find that they follow different patterns of proliferation with increasing radiation dose. For example, some effects require a certain threshold dose level before they begin to be manifested at all within a population, while others seem to occur in a steadily increasing, linear fashion with increasing amounts of radiation but appear to have no threshold dose—in other words, there is no amount of radiation *low enough* to be considered "safe" from causing these effects at least to some degree.

These relationships can be plotted on a graph called a *response curve*. In plotting the occurrence of various effects with increasing radiation dose, we find characteristic shapes to the response curves that result. Each plotted curve can be characterized as:

A. Either linear or non-linear
B. Either having a threshold or not (non-threshold)

Figure 42-13 demonstrates examples of these graphical characteristics. Curve ***A*** is a *linear, nonthreshold* curve. Its linearity suggests that the effect (the occurrence of leukemia, for example) increases *proportionately* to the amount of radiation received by a population. With every doubling of radiation dose, it may double, triple, or increase by some other proportion, but it *increases steadily* at the same rate for all exposure levels.

Figure 42-13

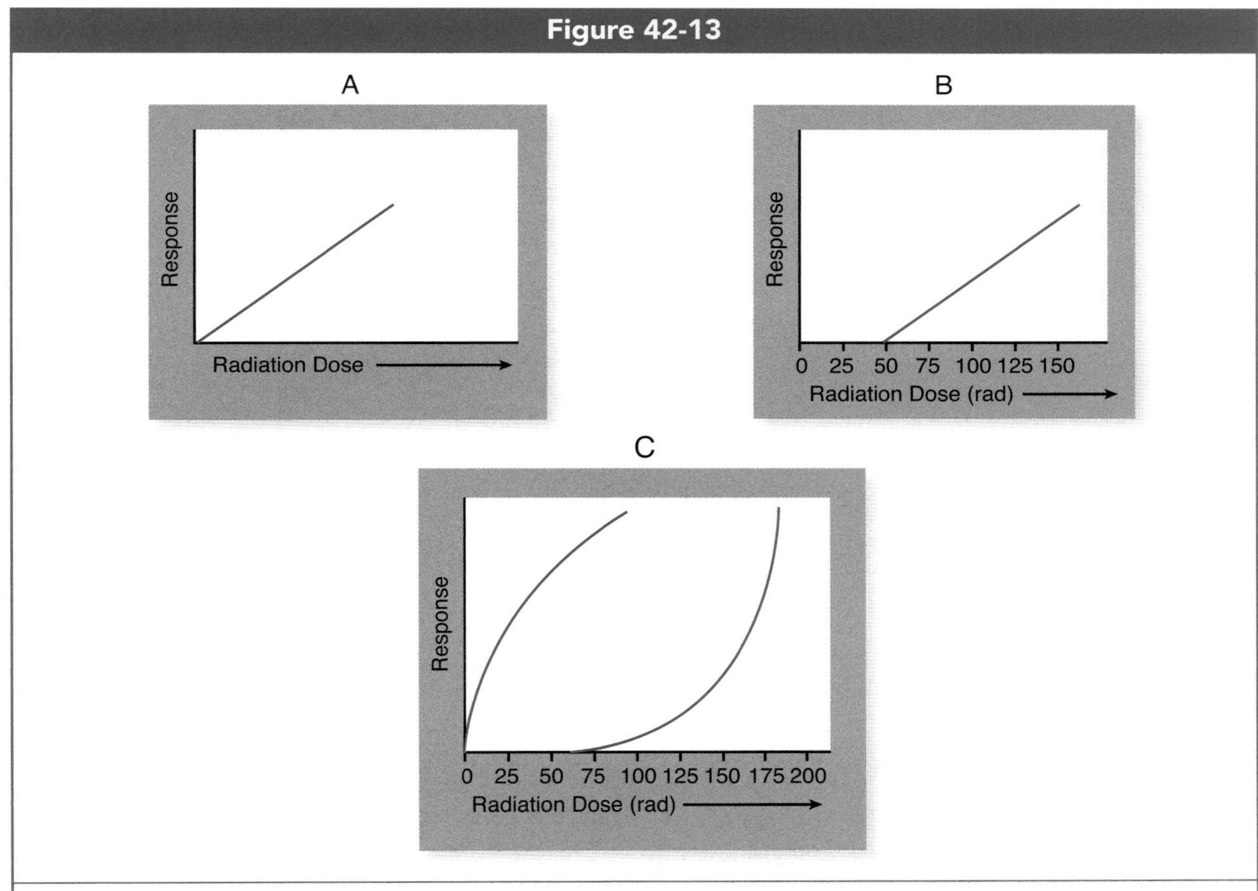

Response curves for increasing radiation dose. Graph ***A*** plots a *linear, nonthreshold* curve; Graph ***B*** is a *linear, threshold* curve; Graph ***C*** demonstrates two *nonlinear* curves with different rates of response.

Observing curve ***A***, is there a "safe" level of radiation exposure, low enough that we can be certain the effect will not occur at all? The answer is no, since going "backward" from right to left, we see that the response curve only reaches zero when the *radiation dose* reaches zero. A response curve may indicate the number of cases occurring within a population, or it may be an indication of the *severity* of a particular effect on a single organism if it is measurable. An example might be how dark a shade of *red* the skin has turned from a radiation burn (sunburn).

Curve ***B*** represents a *linear, threshold* curve. In this case, a *minimum* amount of radiation (50 R) must be delivered to the population before the effect (such as a disease) begins to be manifested. Once that amount of exposure is reached, however, further increases in radiation result in a proportional increase in the number of cases or in the severity of the effect.

In Figure 42-13, graph ***C*** plots two *nonlinear* curves. The *left* one is also a *nonthreshold* curve. The central feature about nonlinear curves is that the *rate* of the response changes at higher or lower exposure levels. Both curves represent greater response with higher exposures. But, for the logarithmic curve (*left*), the *rate* of the response slows down as high levels are reached. And, for the exponential curve (*right*), as exposure increases, not only is there more response, but the response increases more *quickly*.

Most actual biological responses follow one of the two curves illustrated in Figures 42-14 and 42-15. The graph in Figure 42-14 plots a *sigmoid* (or S-shaped) curve. It is a non-linear, threshold curve typical of many biological responses such as death within a selected population of organisms or cells,

Figure 42-14

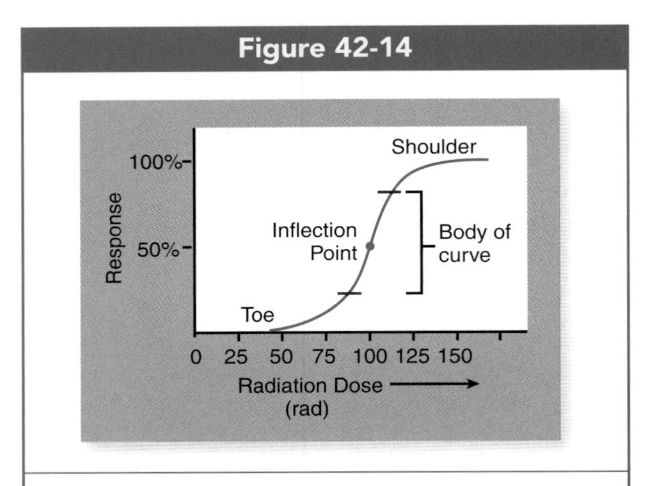

A *sigmoid* nonlinear curve with a threshold at 40 rad for death within a population of cells. Most somatic, deterministic radiation effects follow this type of curve.

which we shall use as an example. Below 40 rads in this case, we see no response for this effect. Less than 40 rads may be considered as "safe" dose *for this effect*, since all organisms survive it. When the threshold dose is exceeded, we begin to see deaths occur at a *slow rate*. These deaths represent the unusually weak or unhealthy organisms within the population that are particularly sensitive to further insult. This portion of the curve is referred to as the *toe* of the curve. The *body* of the curve is often close to a straight line whose angle of slope can be averaged. This slope represents the normal *rate* of increasing deaths for average, healthy organisms within the population as they are subjected to increasing amounts of radiation. The *shoulder* of the curve represents the deaths of those organisms which are unusually hardy or strong. The shoulder eventually levels off to a horizontal line, because *there are no more organisms left in the sample which have not submitted to the effect being studied*. In this case, all of the cells or organisms selected for our sample have died from radiation exposure.

By definition, a sigmoid-type response curve consists of a lower half in which incremental doses of radiation are becoming more and more effective at causing the response, and an upper half which begins to bend the other direction, indicating that incremental doses are now becoming less effective. The mid-point of the curve at which the *rate* of response reverses is called the *inflection point*, (Figure 42-14).

Figure 42-16 illustrates a linear, nonthreshold response curve, but a unique one in that there is already a response measured at zero radiation dose. The implication is that the effect being measured is already being caused within the population *by factors other than the radiation dose under study*. This is very common in epidemiological studies, since *many* diseases have several possible causes. For example, a *birth defect* rate of about 6 percent is already present for newborn babies who have not been subjected to any medical radiation procedures during their gestation. We find that when pregnant mothers have

Figure 42-15

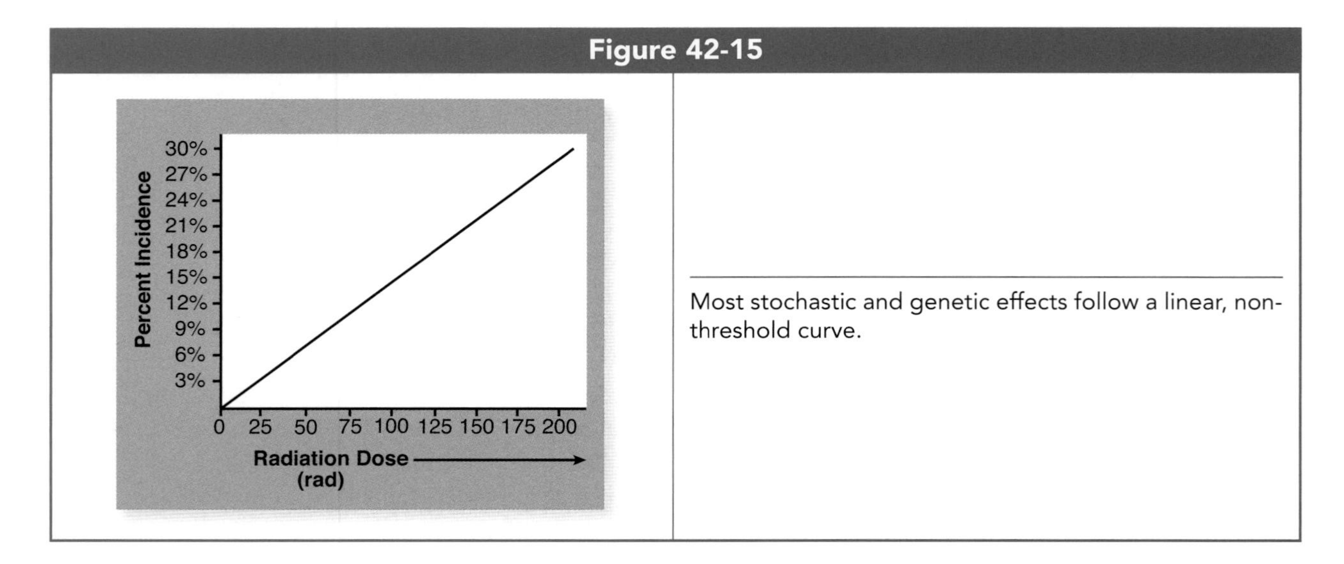

Most stochastic and genetic effects follow a linear, non-threshold curve.

Figure 42-16

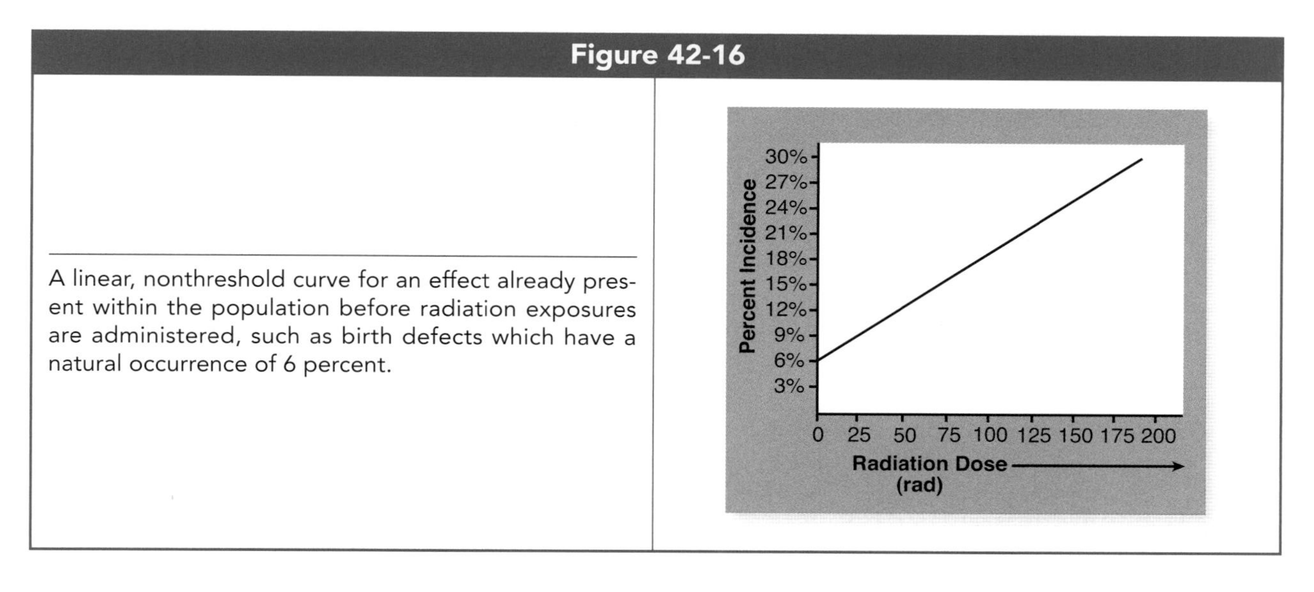

A linear, nonthreshold curve for an effect already present within the population before radiation exposures are administered, such as birth defects which have a natural occurrence of 6 percent.

undergone radiation procedures, the dose delivered to the developing fetus *contributes* to the rate of birth defects, elevating it in slight increments as the dose increases. To avoid misinterpretation of data, the incidence of defects from causes *other* than the medical radiation exposure must be reflected in constructing the graph.

These relationships will be used to help describe most of the biological effects that follow. Most somatic (visceral organ) effects follow a *nonlinear, threshold* curve. Genetic effects, life-span shortening and most cancers follow a *linear, nonthreshold* curve. Scientifically our main concern at dose levels typical of diagnostic radiology is for those *linear, nonthreshold* effects that occur to the population as a whole rather than for acute effects to the individual.

Theory of Cellular Damage

Much of our understanding with regard to the effects of radiation upon cells and tissues is based upon *target theory*. The premise of target theory is that there are certain molecules which are critical to the survival of the cell, and others that are not. Those which *are* critical are dubbed as "target molecules" insofar as radiation damage is concerned. When a nontarget molecule within the cell is ionized from radiation, any damage to the cell will be *sublethal*. When a *target* molecule, such as a particular gene on a chromosome, takes one or more "hits" from radiation, the cell may not survive.

A "hit" is *not* any single ionizing event, but rather the inflicting of *unrepaired, functional* damage to a chromosome, leading to *deactivation* of a portion of the genetic code. Experiments have demonstrated that for simple organisms such as bacteria, a single "hit" on a single target molecule can be sufficient to kill the organism. However, for more complex organisms including human beings, the mechanism of cellular damage appears to be more complicated. Human cells appear to be able to generally survive the deactivation of a single critical gene, but not *two or more.*

We come to this conclusion by examining the *survival curves* for cells from the different organisms, which plot the surviving fraction or percentage of the cells against increasing doses of radiation. Figure 42-17 presents the survival curves for, ***A***, bacteria, and, ***B***, human skin cells. We observe that for the bacteria, as soon as any dose of radiation is delivered, some of the organisms die. The entire curve is inversely proportional to the increasing dose. Clearly, if a single target molecule takes a *hit*, it can kill a bacterium.

But, the survival curve for human skin cells has a *shoulder* to it (Fig. 42-17***B***). Until a certain *threshold dose* is reached, no cells have died and there remains a 100 percent survival; yet, clearly some target molecules within these cells must have taken hits. We conclude that *more than one target molecule must be deactivated within these cells before any of them begin to die*. This is called the *multi-target/ single-hit model* of target theory.

Figure 42-17

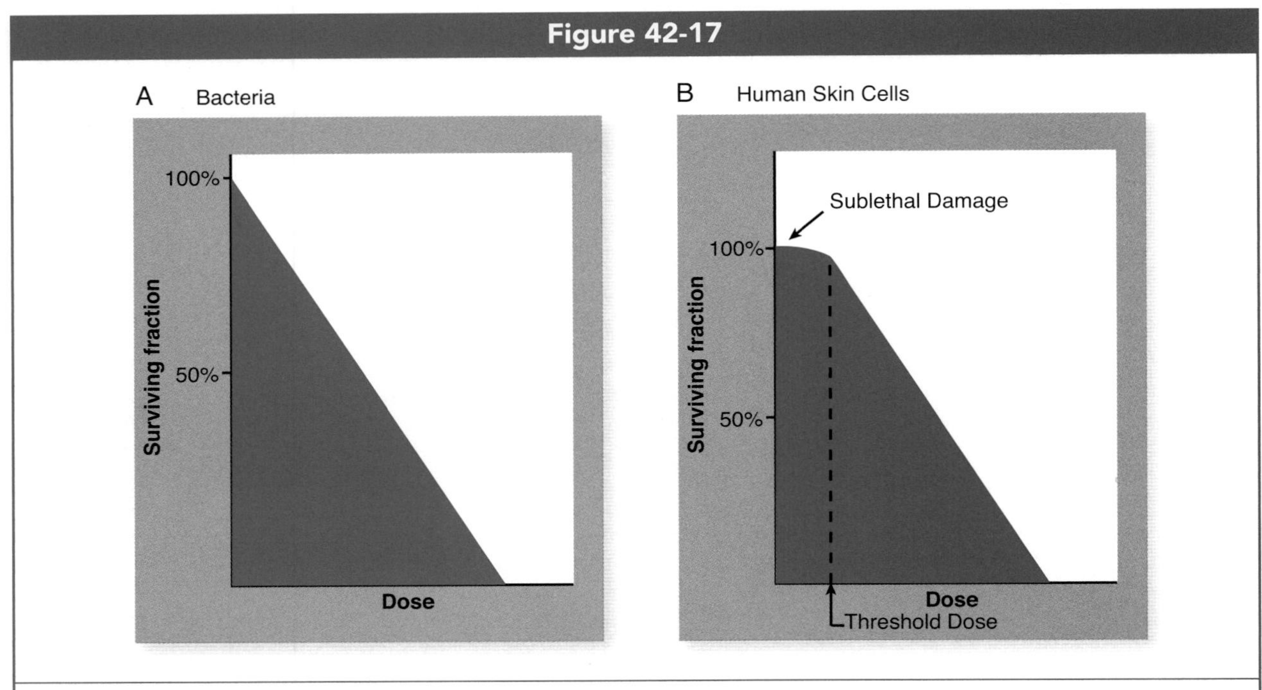

Survival curves for ***A***, bacteria showing a single-target relationship, and ***B***, human skin cells, showing a shoulder to the curve which indicates a multi-target response to radiation exposure.

It is easy to confuse the multi-target/singe-hit concept with what would amount to a "single-target/multi-hit" scenario; so, an analogy is provided in Figure 42-18 for clarification, using archers' targets and arrows. Remember that each target represents a *key* gene along the chromosome sequence. In example ***A***, we see the single-target/single-hit model which applies to bacteria. One deactivated target molecule is sufficient to kill the bacterium. Example ***B*** suggests that it takes two hits on the same target molecule in order to deactivate the molecule—this is *not* a model which has been found to describe any experimental results; rather, it is model ***C*** that we observe, in which it only takes a single hit to deactivate a molecule, but *two or more key molecules must be deactivated* to kill the cell.

Figure 42-18

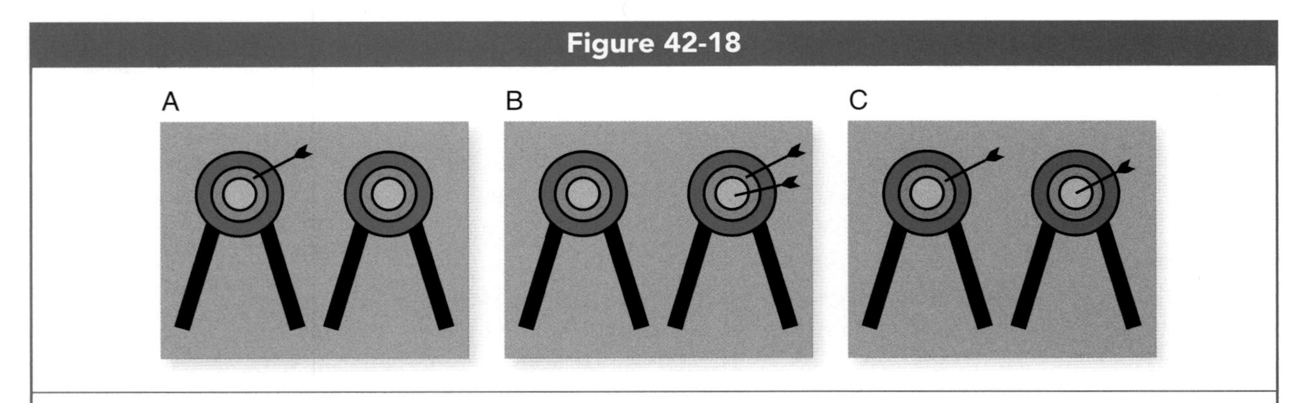

Archery analogy to clarify *hits* versus *targets* in target theory: ***A*** represents the single-hit, single-target model observed for death of simple organisms. ***B*** would represent a multi-hit, single target model which *we do not observe.* ***C*** is the single-hit, multi-target model we observe for complex cells and organisms.

The shoulder portion of the mammalian radiation response curve is further clarified in Figure 42-19. Each square in this illustration represents a cell. Note that the dots within the squares do not represent every ionizing event, but represent *deactivated target molecules* or *hits*. In ***A***, we see that at a low level of radiation exposure some cells have had one target molecule deactivated and others have not had any—all of these cells survive. In ***B***, increased radiation levels have resulted in *every cell* having sustained at least one deactivation of a key molecule, but the survival rate is still 100 percent. However, from this *threshold point* any further increase in radiation is bound to deactivate a *second key molecule* in at least some of these cells, as shown in ***C***, where five of the cells will not survive. It is at this point where the survival curve turns downward at the *shoulder* (Figures 42-17***B*** and 42-20).

Once the survival curve turns downward, it will follow a linear *slope*, the steepness of which indicates the relative *sensitivity* of the cells to radiation. As shown in Figure 42-20, endothelial cells lining the intestine, ***A***, are more sensitive to radiation than epithelial skin cells, ***B***. Curve ***A*** has a steeper slope and drops off more quickly as radiation increases, indicating a higher rate of decline for the survival of endothelial cells.

The deactivation of target molecules can occur from *direct hits* or *indirect hits*. A *direct hit* requires that a photon or particle of radiation directly ionizes a key gene on a chromosome. Various scenarios for this will be discussed in the next section. But, a key gene can also be deactivated *indirectly* when radiation ionizes water molecules nearby, creating *chemical substances which then attack the DNA or RNA* (Fig. 42-21). This process is referred to as radiation hydrolysis (*hydro* = "water," and *lysis* = "breakdown"), or the *radiolysis of water*.

Radiolysis of Water

We have mentioned the predominance of water as a component of living cells throughout the body. The ionization of water molecules from radiation exposure creates multiple *ions* and *free radicals*. A *free radical* is defined as any uncharged atom with a single, unpaired electron in its outermost shell. Such atoms are chemically highly reactive, since there is a very strong *valence* tendency to share this unpaired electron with another atom through covalent bonding, (see Chapter 4). The *ions* created also are highly reactive and seek ionic bonds with other chemicals.

Figure 42-19

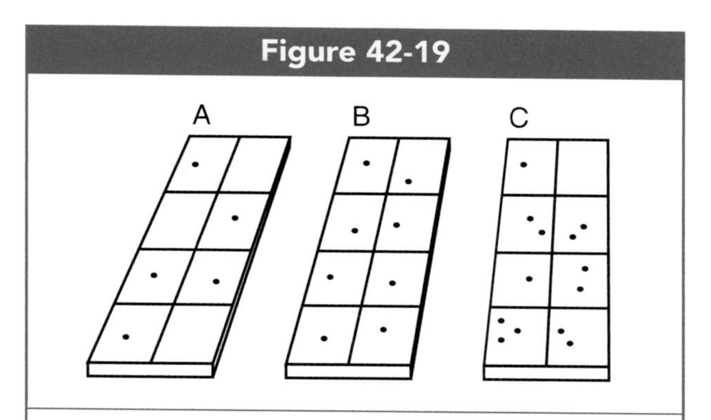

The multi-target model: Each square represents a cell, and each spot represents a *deactivated target*. As radiation exposure increases, in ***A*** there are no cell deaths, in ***B*** the survival rate is still 100 percent, and in ***C*** five cells expire.

Figure 42-20

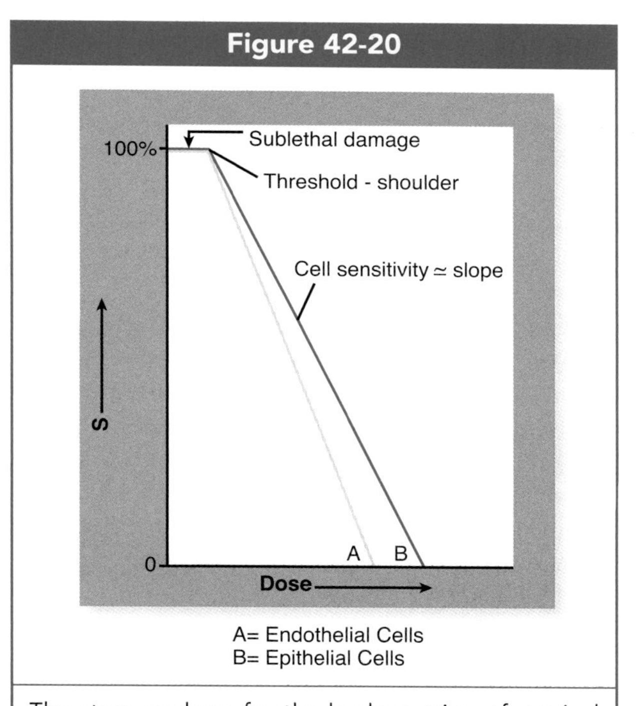

The steeper slope for the body portion of survival curve ***A*** indicates that endothelial cells are more radiosensitive than epithelial cells (***B***). The portion of each curve left of its *shoulder* represents sublethal damage to all cells from 0-1 targets molecules being deactivated. Along the slope of the curves, each additional *hit* deactivates a second critical molecule.

Figure 42-21

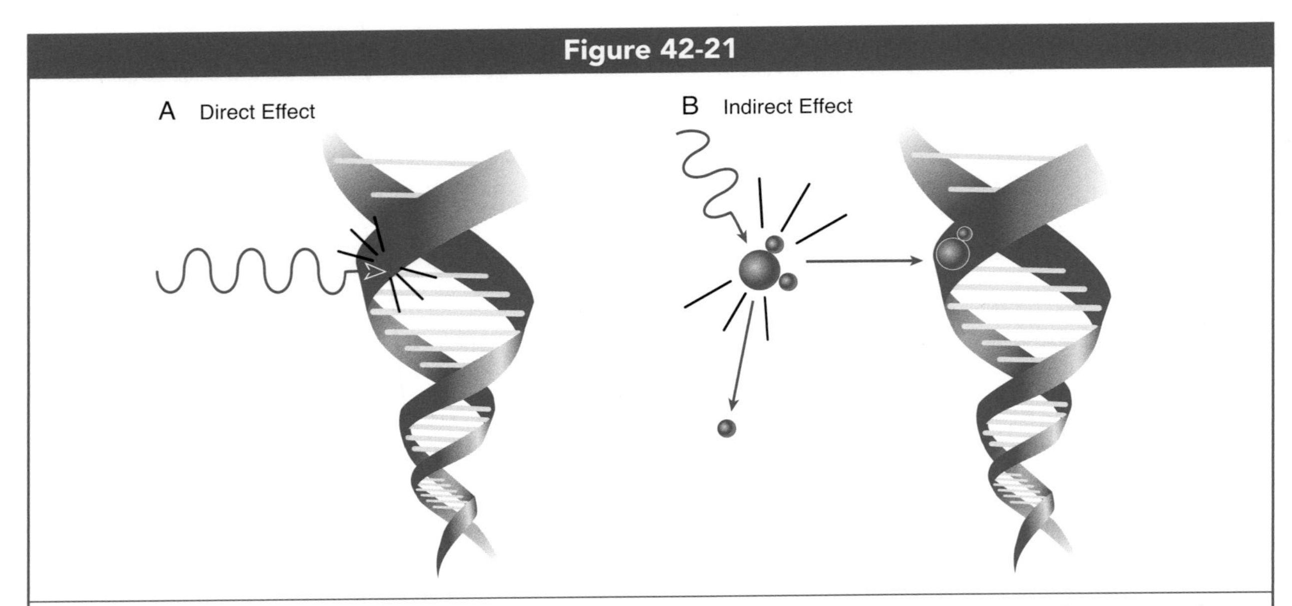

It is not necessary for an x-ray photon to directly strike the DNA molecule (***A***) to cause genetic damage. As shown in ***B***, it may ionize water molecules nearby, creating toxic compounds that then *chemically* attack the DNA.

Figure 42-22 illustrates how *all* of the immediate products of the radiolysis of water are either ions or free radicals. This is presented in pictorial form above the corresponding chemical formulas. Two molecules of water are depicted side-by-side. The ionization of one molecule leaves it with a positive-charge, and the freed electron "attaches" to the other forming a negatively-charged molecule. In the process, both molecules rearrange their physical structure. The negative molecule breaks up into a normal hydrogen atom (which is a free radical by definition, since it only has one electron), and a *hydroxyl* ion which carries the negative charge. The positively charged molecule dissociates into a hydrogen ion which is missing its electron, and a *hydroxyl radical.* This hydroxyl radical is single-handedly responsible for two-thirds of all cellular damage caused by the indirect effect.

Figure 42-23 illustrates further iterations of the chemical changes that can occur, in which hydrogen radicals attach to molecules of oxygen (O_2) to form hydroperoxyl (HO_2) radicals, and two of these combine to form H_2O_2 with the release of a molecule of oxygen. You may recognize H_2O_2 as *hydrogen peroxide*, which is extremely toxic to human cells. The ionic compounds produced are generally not unusual to normal body chemistry. However, the *free radicals*, even though they each exist for only a fraction of a second, can chemically attack tRNA in the endoplasmic reticulum or DNA and mRNA in the nucleus when they are nearby, causing changes in the genetic code and metabolism of the cell.

Direct hits of radiation to target molecules is a low probability event when compared to indirect hits. The cytoplasmic fluid around the DNA or RNA molecule provides a much larger "target area" to the incoming radiation. Most of the radiation damage caused to cells is inflicted through indirect hits.

Damage to the Cell Membrane

It is important to note that cell death can also occur from radiation damage breaching the cell *membrane.* Ionization of these critical molecules can cause them to break apart, literally leaving a rift in the membrane (Fig. 42-24). This allows external toxins to enter into the cell and leaves the cytoplasm exposed to effectively "dry out," killing the cell.

Types of Damage to Chromosomes

Studies of changes to DNA or RNA can be performed *in vitro*, ("in glass") or in a laboratory test tube in

Figure 42-22

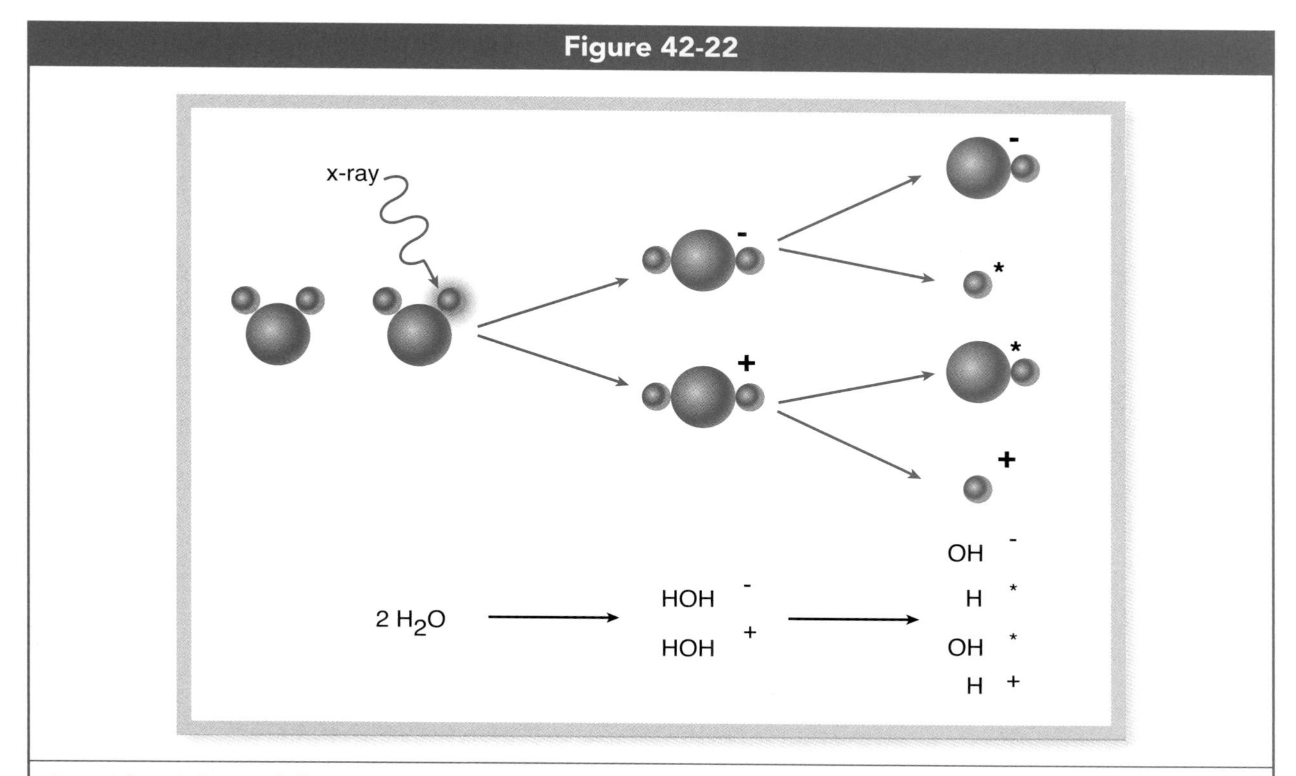

Pictorial and chemical illustrations of the radiolysis of water due to ionization from radiation. *All* of the products are either ions or radicals. The hydroxyl radical (OH*) is particularly destructive to DNA.

which the cells may be dead, or they can be done *in vivo* ("in life") by examining characteristics of the living organism or living cells in a petri dish or on a slide. We have noted that during *metaphase*, the chromosomes become visible under a microscope and can be observed for structural changes. This section will discuss the types of structural damage that can occur to the DNA (or RNA) molecule. We shall divide these broadly into *main chain scission* and *rung damage*.

The chemical changes of greatest concern are those called *frame shift mutations* which actually alter the sequence of the genetic code. This code is contained in the pairs of nitrogenous bases that make up the "rungs" of the ladder structure of the DNA. Changing it can be lethal to the cell.

This analogy comparing the molecular structure of the DNA molecule to a ladder (Fig. 42-25) helps understand why some types of DNA damage are more difficult to repair than others. In a chromosome, portions of the chromatids can be completely broken off. The ends of these chromatid fragments

Figure 42-23

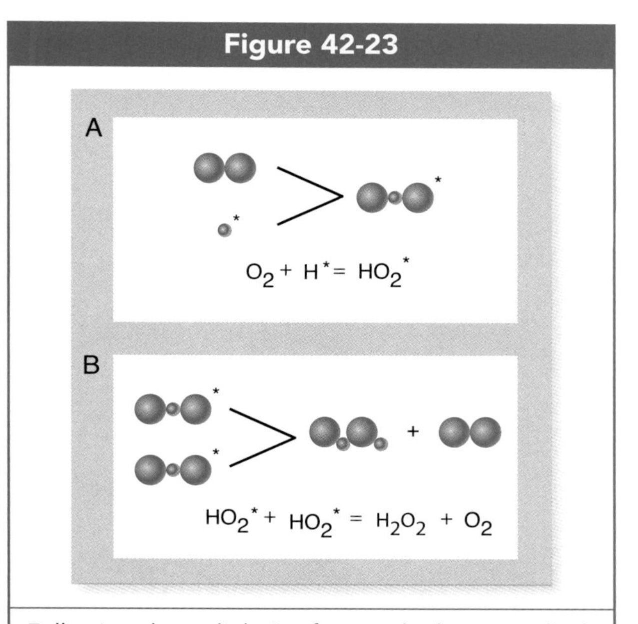

Following the radiolysis of water, hydrogen radicals combine with molecules of oxygen to form hydroperoxyl radicals, pairs of which form even more toxic hydrogen peroxide.

Figure 42-24

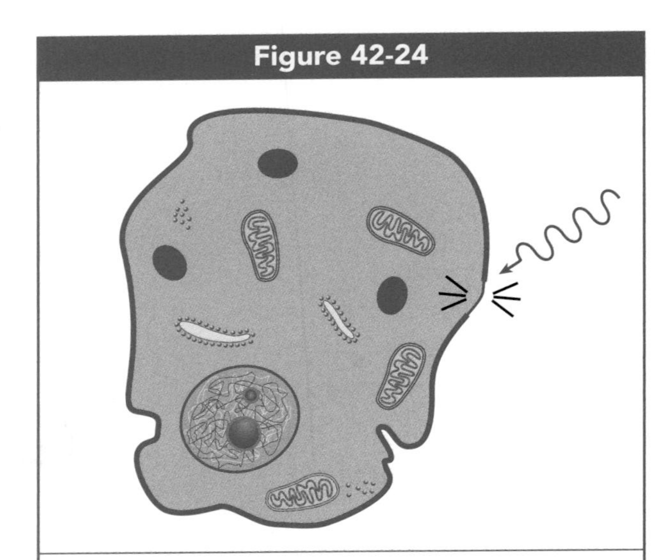

Ionizations from radiation can cause molecules in the cell membrane to dissociate, leaving a rift which allows toxins into the cell and leaks moisture out, killing the cell.

are "sticky," such that they can reattach to different chromatids, permanently altering the genetic code. In Figure 42-25, ***A***, we see a single break in one rail of the DNA "ladder," due to ionization. This has not changed the genetic code contained in the rungs. Nor can a fragment of another ladder be attached at the site of this single break. In ***B*** we see a scenario in which *both* rails of the ladder are broken by ionization. A fragment from another ladder (a foreign chromatid fragment) *can* attach itself to this site, connecting at both rails, such that the sequence of the genetic code is changed and there is a *frame shift mutation*. In ***C***, we have a case where ionization has disassociated a thymine molecule and broken it off from its rung. This simply leaves a *gap* in the genetic code, which constitutes a *frame shift mutation*.

Main Chain Scission

Main chain scission refers to any form of break in the "rail" of the ladder structure. *Single-strand breaks*, Figure 42-26***A***, and Figure 42-25***A***, can be quickly and easily repaired, and usually are, such that the rail at this point is chemically reattached with the correct molecule. However, it is possible for single-strand breaks to be misrepaired by inserting a similar but incorrect molecule. This results in a *point mutation*. Point mutations are usually undetectable, invisible under a microscope, and sublethal to the cell.

Double-strand breaks can occur when large alpha particles plow through the region, Figure 42-26***C***, or

Figure 42-25

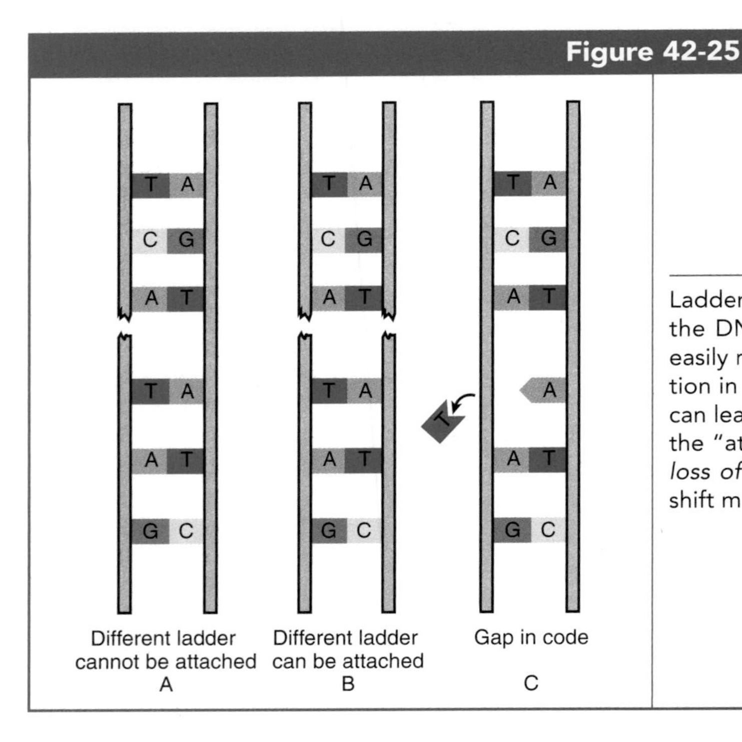

Ladder analogy for seriousness of radiation damage to the DNA molecule. The *single-strand break* in **A** is easily repaired and cannot result in a frame shift mutation in the genetic code. The *double-strand break* in **B** can lead to permanent frame shift mutations including the "attachment of another ladder" or chromatid. The *loss of a base* in **C** can also lead to permanent frame shift mutations in the genetic code.

Figure 42-26

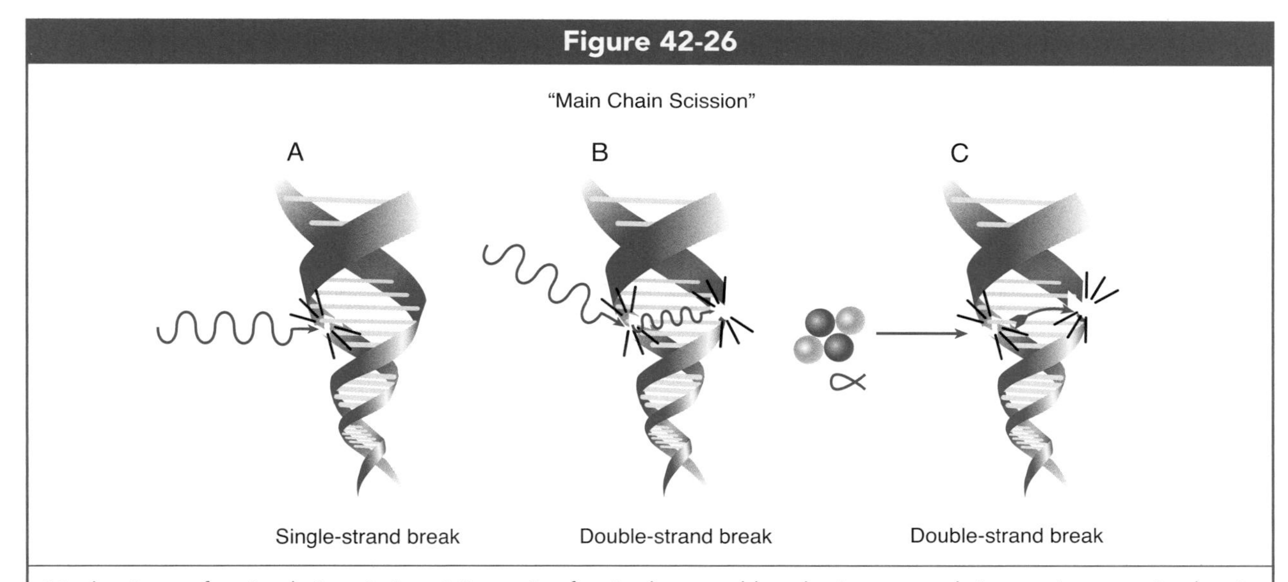

Mechanisms of main chain scission. Misrepair of a single-strand break, ***A***, can result in a *point mutation* but is sublethal to the cell. A more serious double-strand break is possible, but extremely unlikely, from scattered x-ray photons, ***B***. Most double-strand breaks are due to massive particles like the alpha particle, ***C***, which cause multiple ionizations along their path.

when by chance a scattered x-ray from the original break strikes the second rail of the DNA structure (Fig. 42-26***B***). The odds of perfect reattachment are slim, and most authors consider this to be an *irreparable* form of damage. It certainly can result in a *frame shift mutation*, causing permanent harm.

Double-strand breaks are more likely from the direct effect than from the indirect effect, since the indirect effect would require two different molecules to attack the DNA at points directly opposite each other in an unlikely coincidence.

Rung Damage

Unrepaired alterations of a rung can change the code sequence and thus have possibly lethal consequences to the cell. However, it is possible for simple *rung breakage* to occur (Fig. 42-27***A***), in which the two nitrogenous bases are separated by an ionizing event. Since these bases are matched to connect and remain near each other, held in place by the rails, simple rung breakage is reparable and usually recovered from.

The *loss of a base molecule*, however, in which it becomes detached from its rail, is irreparable (Fig. 42-27***B***). Sometimes a similar molecule will fit into this space, resulting in a change of base, but this still results in a *frame shift mutation* that is lethal to the cell.

Mutations and Chromosome Aberrations

Any unrepaired change to the genetic code may be considered as a genetic *mutation*. A few general observations about mutations are in order: First, *there are no mutations which are unique to radiation*. In other words, all of the effects of radiation exposure can also be caused by other factors in the environment (such as chemical toxins), or in the organism (such as hereditary predispositions to particular diseases). Second, although 99% of genetic mutations are harmful to the individual, most mutations are *recessive* from a hereditary standpoint. This means that even those which are irreparable on the cellular level are not likely to be manifest in the next generation—it is possible, but not probable for this to occur. Finally, evidence suggests that genetic mutations follow a *linear, nonthreshold* response curve. This means that *any* amount of radiation exposure, no matter how small, *may* cause a mutated gene, and that the incidence of genetic mutations within a large population increases proportionately to the amount of radiation exposure the whole population receives.

Figure 42-27

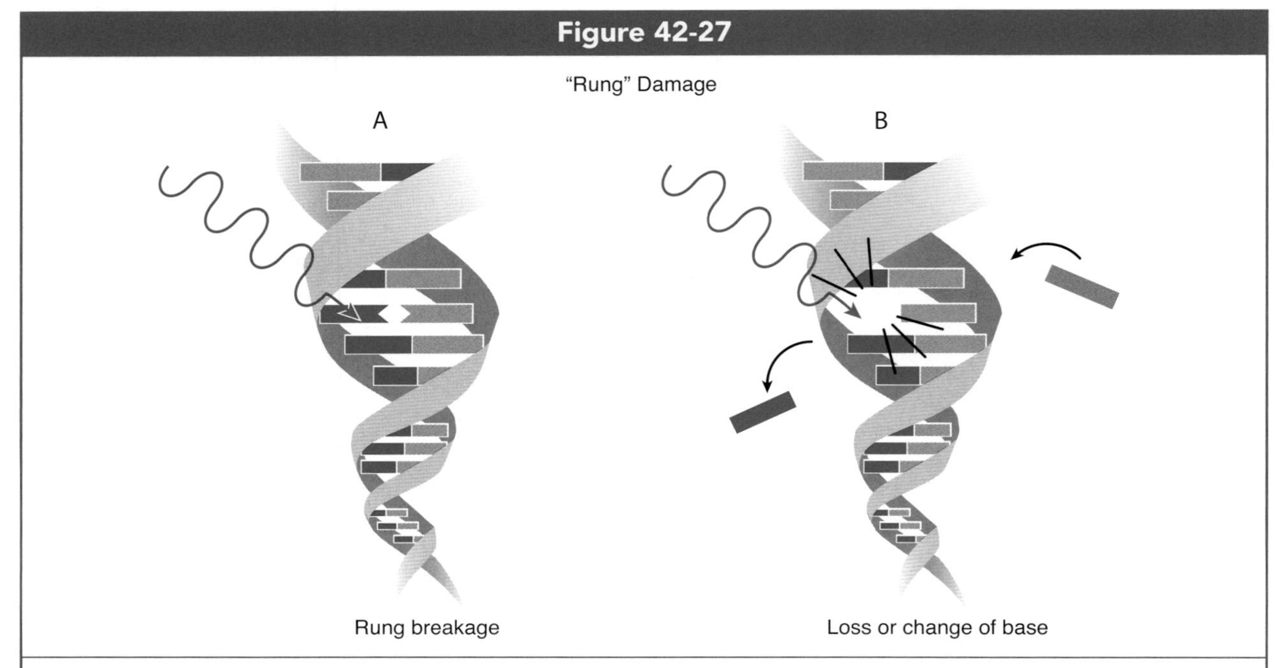

Rung breakage between the two nitrogenous bases, ***A***, is easily repairable by reattachment. However, the *loss* or *change* of a base molecule detached from its *rail*, ***B***, is irreparable and causes a frame shift mutation in the genetic code.

Visible Chromosome Aberrations

During metaphase, photomicrographs can be taken of the various chromosomes in the cell nucleus. By "cutting and pasting" such a photograph, the chromosome images can be aligned and organized from the longest chromosome pair to the shortest one. They are thus numbered for study and comparison with those from other individuals to check for abnormalities. Such a "map" of the chromosomes is called a *karyotype*, and is illustrated in Figure 42-28.

A chromosome "hit" is not any single ionizing event, but a *visible and functional* alteration in a chromosome that usually involves several changes on the molecular level.

It is fascinating to see the actual structural changes in chromosomes manifested under a microscope. Point mutations, responsible for most *late* effects of radiation, are not visible, but unrepaired single-strand breaks, if they affect more than one chromosome, can result in bizarre observable chromosome configurations that are rare but very harmful to the cell. So can double-strand breaks occurring along the same chromosome. These configurations all occur because the broken ends of chromosome fragments are biochemically "sticky," and will attach to the broken ends of other fragments, whether from the same chromosome or from other chromosomes.

Single-hit effects include *acentric* or *isochromatid* fragments which have no centromere, and *dicentric chromosomes* which have two centromeres, Figure 42-29. Note also in this figure that chromatid end fragments can be *translocated* by "swapping" chromosomes.

Multi-hit effects refer to those changed configurations arising from a single chromosome being broken in two locations. If the same chromatid is broken in two places, Figure 42-30, the middle fragment can be *deleted*, ***A***, or it may spin around and re-attach in reverse order, called a fragment *inversion*, ***B***. If two different chromatids are broken (Fig. 42-31), their endings can be *inverted*, ***A***, or the ends of the center section with the centromere can curl in on themselves and form a *ring*, leaving the two ends to form an *acentric fragment*, ***B***. Other variations of cross-linking can result in *multi-radius* configurations in which three or four pairs of chromatids stick together at their centromeres. The photomicrographs

Figure 42-28

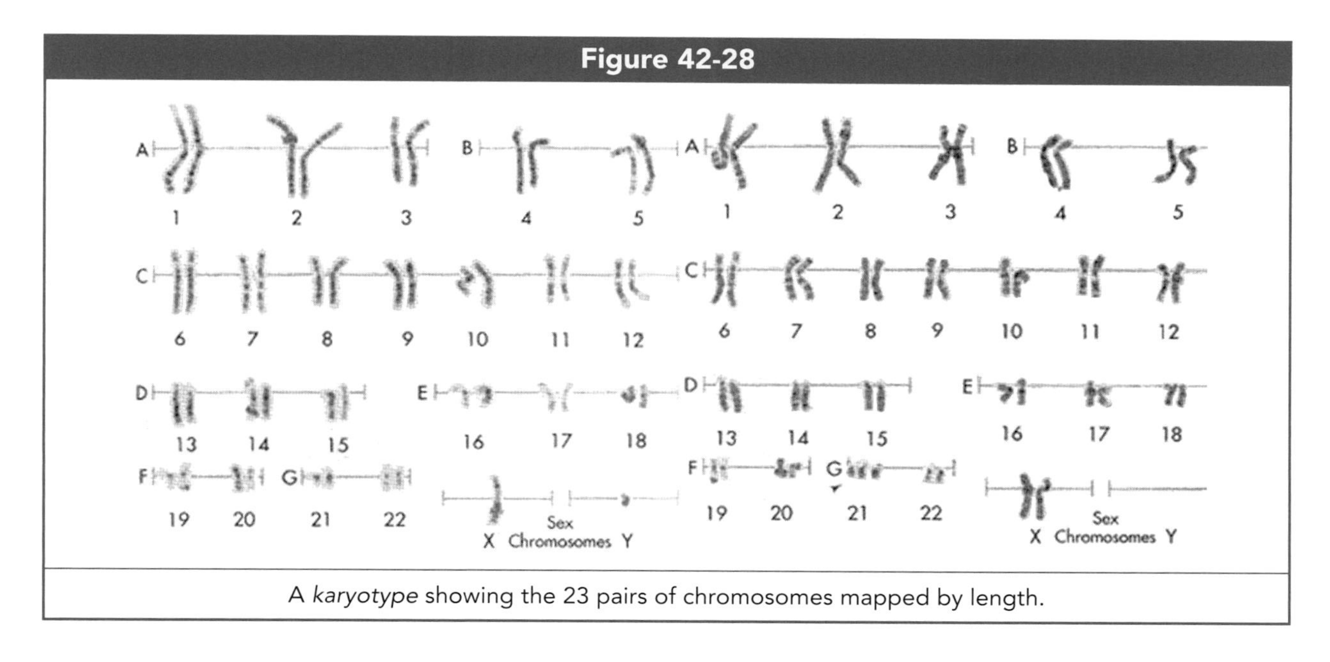

A *karyotype* showing the 23 pairs of chromosomes mapped by length.

in Figure 42-32 show actual chromosomes in these various states.

Linear Energy Transfer (LET)

Linear energy transfer (*LET*) is one indicator of *how harmful* (or how effective) a particular type of radiation is. LET is defined as the amount of *energy* deposited within tissue *per unit path length* of the radiation's travel. If one calculated the amount of gasoline burned *per mile* driven in a car, (rather than the number of miles per gallon), this would be a similar concept. It would be an expression of the *energy spent* per mile. We have defined the *rad* as an amount of energy deposited *per gram* of tissue. LET may be thought of as the amount of energy deposited *per millimeter* that a particle or x-ray photon *travels* as it penetrates into the body.

For diagnostic x-ray levels, the typical unit for LET is *kilovolts per micron*, or *kV/μm*. (The micron, or micrometer, is one-millionth of a meter.) In Figure 42-33 we give examples of two very different LETs for two types of radiation, even though both beams of radiation are set at 80 kilovolts of energy. In ***A***, an 80-kV x-ray photon penetrates 6 μm into tissue and then undergoes a compton scatter interaction. The scattered photon travels in a different direction for 8 μm distance and then is scattered again. The third photon in the series travels for 10 μm, whereupon it undergoes a *photoelectric* interaction which finishes absorbing all of the remaining photon energy. The total linear distance traveled by all of these photons is summed up as 24 μm. The total amount of energy

Figure 42-29

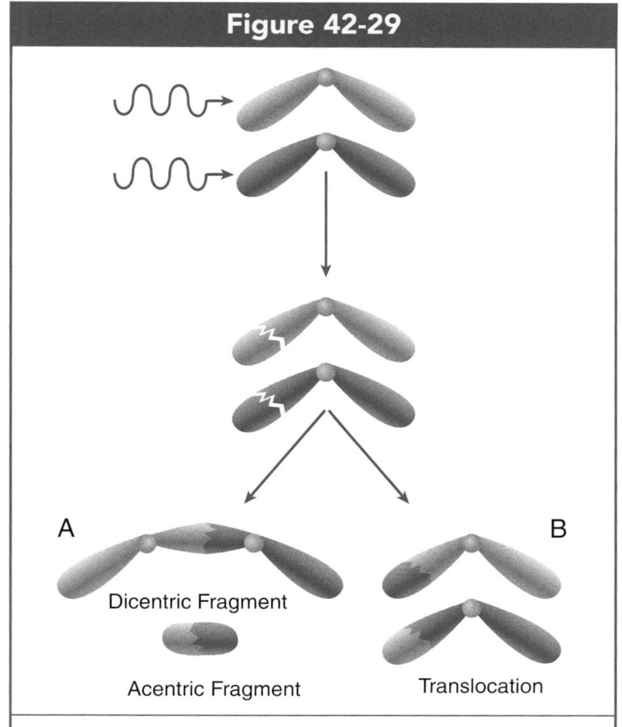

Diagram of three possible structural outcomes from a single hit to different chromosomes.

Figure 42-30

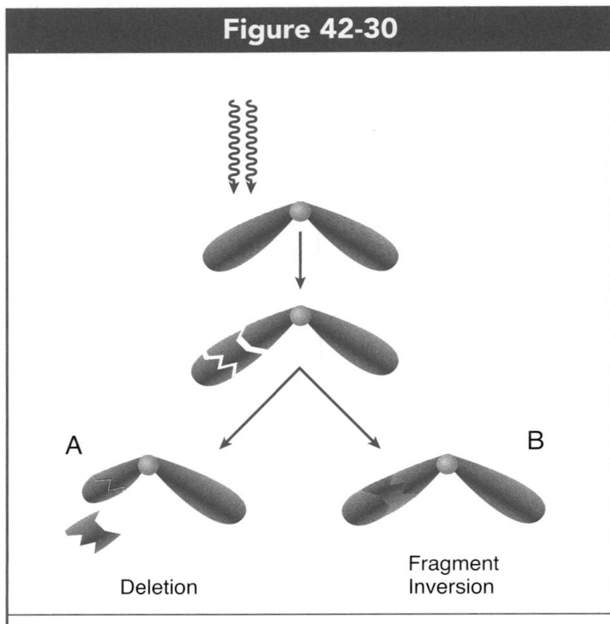

Diagram of two possible structural outcomes from double hits to one chromatid.

Figure 42-31

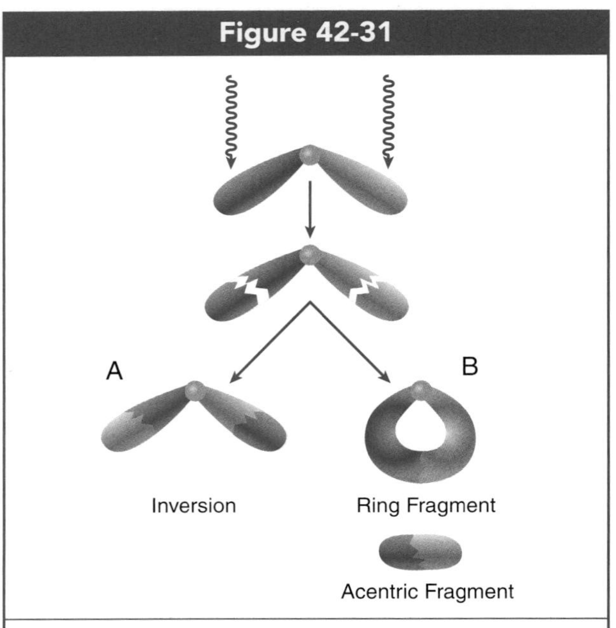

Diagram of three possible structural outcomes from single hits to *two* chromatids.

absorbed over that distance is 80 kV. The LET for this radiation is 80/24 = 3.3 kV/μm.

In ***B***, Figure 42-33, an 80-kV *alpha particle* penetrates into the same tissue. It only travels 3 μm before it "collides" with an atom, whereupon it is completely stopped in its motion. The LET for this particular alpha particle is 80/3 = 26.7 kV/μm. Compared to the 80-kV x-ray (LET = 3.3), the LET for the alpha particle is eight times greater. The implication is that the energy of the alpha particle was eight times more *concentrated* as it was deposited into the tissue, and therefore more harmful to the organism.

LET is always inverse to *penetration*—more penetrating radiations have *lower* LETs. Table 42-3 lists the typical LETs for four different types of radiation.

Table 42-3

LET by Type of Radiation

Type of Radiation	LET in kV/μm
Diagnostic X-ray	3
10 MV protons	4
5 MV alpha particles	100
Heavy Recoil Nuclei	1000

Relative Biological Effectiveness (RBE)

Relative biological effectiveness (*RBE*) is defined as the effectiveness of a certain type of radiation in causing a *specified* effect or disease. The formula for calculating RBE is

$$\text{RBE} = \frac{\text{Dose of 250 kVp x-rays required}}{\text{Dose of radiation z required}}$$

These dose measurements must be made in relation to causing a particular biological effect, which must be specified when discussing RBE. For example, suppose an experiment is conducted to determine how much radiation dose is required to kill 50 percent of the epithelial cells in a petri dish. Two different kinds of radiation are compared. For each, the dose is increased in increments until one-half of the cells are observed to be dead. The experiment shows that it takes only 2 grays of alpha radiation to kill half of the cells, but when the same number of cells are exposed to 250-kVp x-rays, it takes 6 grays to kill half of them. What is the RBE for alpha radiation in killing 50 percent of epithelial cells?

The formula is set up as: $\text{RBE} = \frac{6 \text{ grays}}{2 \text{ grays}} = 3$

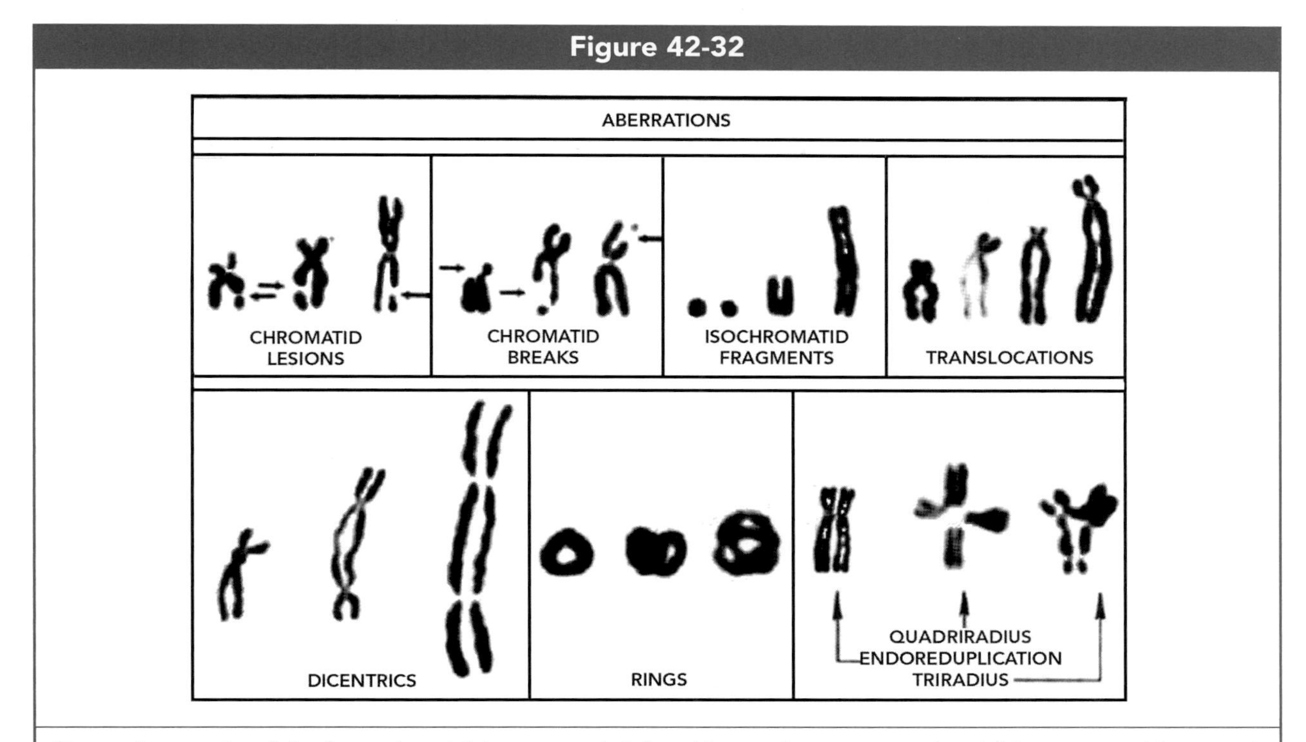

Photomicrographs of the foregoing visible structural deformities to chromosomes, in addition to more bizarre *multiradius configurations*. (From E. Travis, *Primer of Medical Radiobiology*. Year Book Medical Publishers, 1975.)

For killing 50 percent of the cells, the RBE of alpha radiation is 3.0. In plain language, this means that alpha radiation is 3 times more effective at epithelial "cell-killing" than x-rays are. If the same two types of radiation are used and even the same types of cells are used, but we are studying a different result such as the mutation rate in these cells, then the entire experiment must be repeated and the result specified *for the mutation rate*. The RBE for cell death cannot be applied to mutation rates.

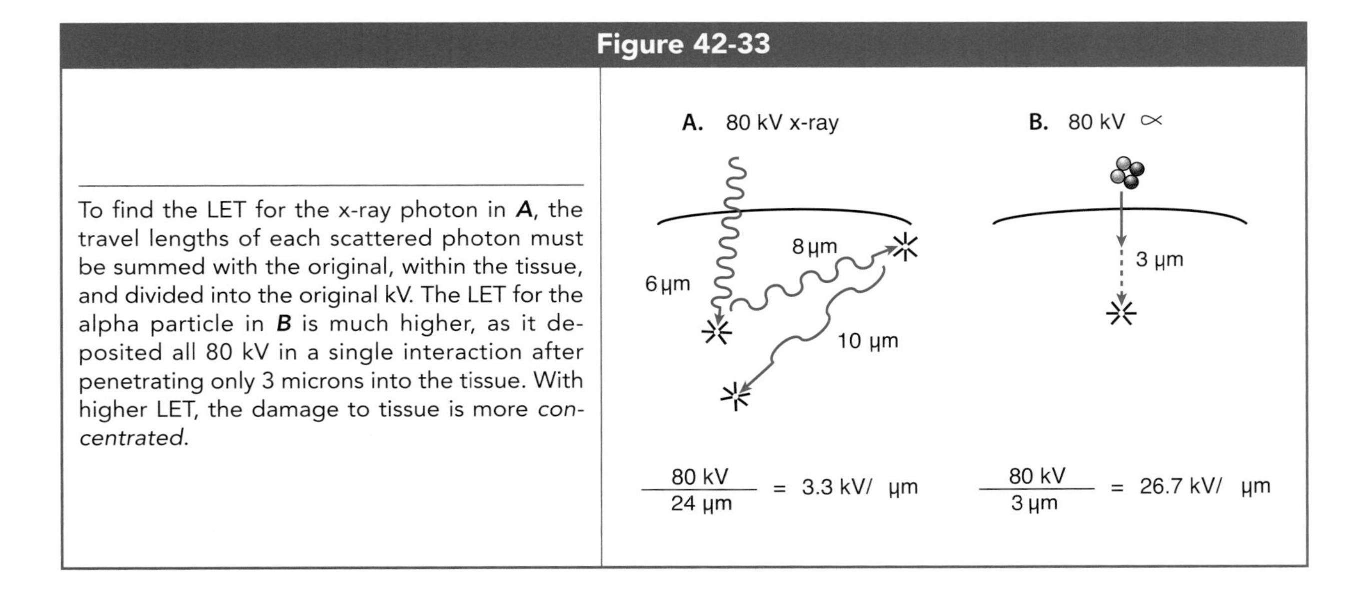

To find the LET for the x-ray photon in ***A***, the travel lengths of each scattered photon must be summed with the original, within the tissue, and divided into the original kV. The LET for the alpha particle in ***B*** is much higher, as it deposited all 80 kV in a single interaction after penetrating only 3 microns into the tissue. With higher LET, the damage to tissue is more *concentrated*.

Dose Rate

Dose rate, the rate at which radiation is delivered to cells or tissues, is directly related to the effect on the organism. Even though the same dose is delivered to two different individuals, if it is delivered *more quickly* to one of them, it will cause more harm. This can be directly compared to the effects of heat in causing a burn—if a certain amount of heat energy is delivered to the skin over an extended period of time, the result is comfortable warming, but if the *same* amount of heat is delivered quickly, a burn will result. *Acute* exposure to a deleterious substance or effect is always more harmful to the organism than *chronic* exposure.

For example, let us assume that two patients both receive a 100-rad treatment in radiation therapy, but the treatment time for patient A was 20 minutes, for patient B 10 minutes. Which patient will be more affected?

Dose rate for Patient A = 100/20
= 5 rads per minute

Dose rate for Patient B = 100/10
= 10 rads per minute

Answer: Even though the total dose was equal, patient B will be more affected because the *rate* of exposure was doubled.

Protraction of Dose

Protraction refers to *extending the time* over which a particular dose of radiation is delivered. It will have precisely opposite effects to *dose rate*, since the same amount delivered over more minutes must be delivered in smaller increments per minute. *Protraction reduces the dose rate*, therefore *protraction reduces the effectiveness of the dose.*

For example, let us assume that an AP L-Spine radiograph is taken on two patients of the same thickness, using the same total technique of 20 mAs at 80 kVp. But, on Patient A the 400 mA station is used, and for patient B the 200 mA station is used. Which patient will be more affected by this dose?

Dose rate for Patient A = 400 mA

Dose rate for Patient B = 200 mA, but *twice the exposure time* must have been used, thus *protracting* the dose.

Answer: Even though the total dose was equal, patient A will be more affected.

In radiation therapy, large amounts of radiation are delivered in order to destroy a cancerous tumor. To spare the individual, the dose must be *protracted* over a period of time. However, this protraction *also reduces the effectiveness* of the radiation in destroying the tumor. Therefore, a higher *total dose* must be calculated to compensate for this loss of effectiveness. The end result still spares the individual more than an acute exposure to radiation would.

Fractionation

Fractionation is defined as breaking the total delivered dose into several *discrete* portions, allowing a time period between each exposure. Fractionation is the very basis for radiation therapy treatments, because the time periods between doses allow the individual to *recover* somewhat from the effects of the radiation.

One might ask if the *cancer* doesn't also recover during these interim periods, and the answer is *yes, but as a rule, normal tissues recover faster than cancer tissue*. Therefore, with each treatment the cancer has a harder and harder time "bouncing back" when compared with the normal tissues of the body. These relationships are graphed as *survival curves* in Figure 42-34. Note that both the normal tissues and the cancer tissues become further weakened with each treatment. However, the number of cancer cells recovers to a lower percentage of the original than the normal cells do with each treatment. At a certain point, *all* of the surviving cancer cells are (hopefully) killed, while, (in this graph), perhaps 30 percent of normal cells still survive. The whole organism must then recover from this stage of decline.

Fractionation effectively *protracts* the total dose delivered over an extended time, reducing the effectiveness of the radiation exposure. In addition, the actual *separation* of the total dose into discrete treatments further reduces effectiveness. To compensate for these effects, the total dose must be recalculated to a higher amount to achieve the desired effect.

For example, we may find that 400 rads delivered in one dose will destroy a cancer mass, but will also be lethal to the patient. Fractionated into 10 separated doses, we find that 600 rads is now required to destroy

Figure 42-34

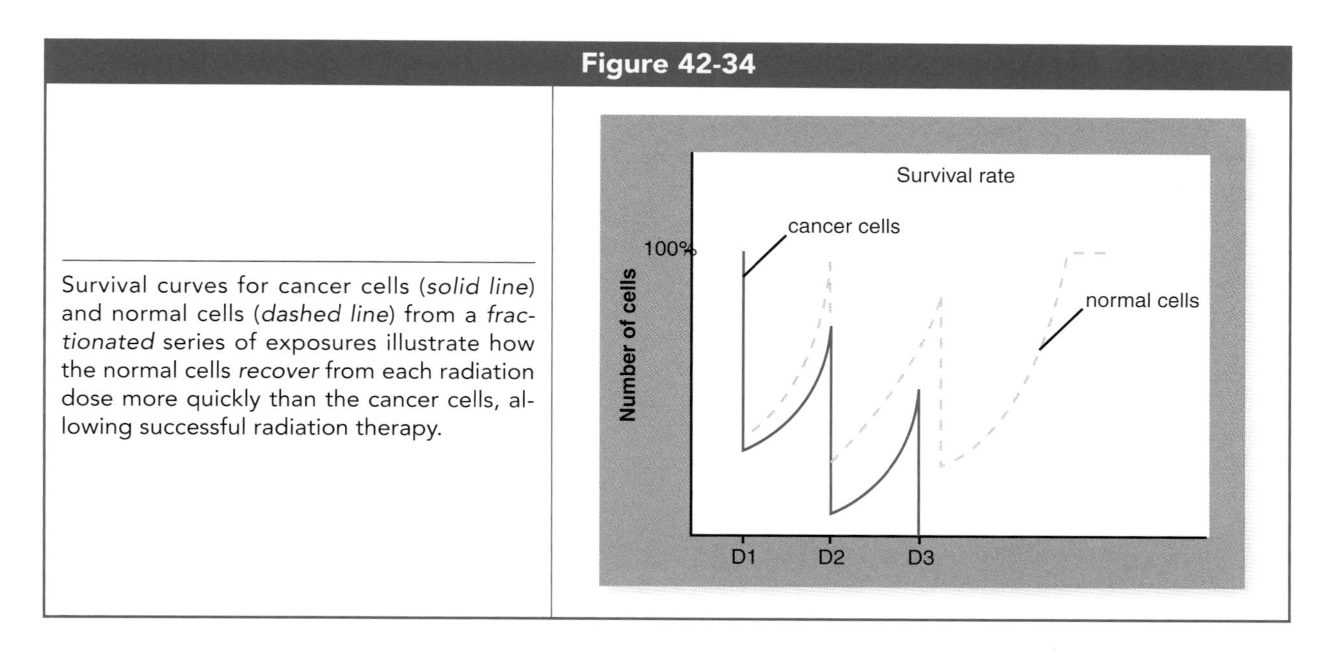

Survival curves for cancer cells (*solid line*) and normal cells (*dashed line*) from a *fractionated* series of exposures illustrate how the normal cells *recover* from each radiation dose more quickly than the cancer cells, allowing successful radiation therapy.

the cancer; however, the patient will survive this dose because of the recovery periods allowed.

Oxygen Enhancement Ratio (OER)

Recall that among the products of the radiolysis of water are hydroxyl free radicals and hydrogen peroxide, both of which are highly reactive and prone to chemically attack DNA. These molecules form from combinations of hydrogen and oxygen. We find that when the oxygen tension within tissues is increased, such as occurs in a hyperbaric oxygen chamber, the increased presence of oxygen within the cytoplasm of each cell encourages the production of these free radicals and toxins.

This amplifying effect of oxygen on the damage caused by radiation exposure can be quantified as the *oxygen enhancement ratio* (*OER*). The OER is specifically defined as the ratio of the radiation dose required to cause a particular biological effect *without* additional oxygen present to the dose required *with* added oxygen. The formula is:

$$\text{OER} = \frac{\text{Dose required under anoxic conditions to produce a given effect}}{\text{Dose required under aerobic conditions to produce the same effect}}$$

The OER applies only to *indirect damage*, since it is a chemical modifier, and it is much higher for low-LET radiations. (High-LET radiations cause more *direct* damage to target molecules, an effect which is independent of the presence of oxygen.)

Other Biological Factors Affecting Radiosensitivity

The *age* of an organism has an impact on its radiosensitivity. Either extreme, very young or very old, increases sensitivity, but for different reasons; very young organisms are more sensitive because of their rapid growth and development rate, in accordance with the law of Bergonie and Tribondeau. In the first 6 weeks of pregnancy, the human embryo is approximately ten times more sensitive to radiation than an adult. The late fetus and the newborn infant are about twice as sensitive as an adult. Very old organisms are more sensitive to radiation and all other risk factors because their degenerative state weakens the entire organism and its defenses.

In regards to *gender*, we find from epidemiological studies that women are somewhat more resilient to radiation exposure than are men, but for reasons not fully understood. Finally, certain *chemical* agents called *radiosensitizers* can enhance the effect of radiation on an individual. These include megadoses of vitamin K, pyrimidines, and methotrexate. Generally, these drugs make the individual about twice as sensitive to radiation than normal. A group of chemicals called sulfhydrils have a radioprotective effect, but are too toxic for human use.

Summary of Factors Affecting Radiosensitivity

Certain effects we have discussed tend to group together. For radiography, a high kVp set at the console generates a high-HVL, highly-penetrating x-ray beam. This gives it both low LET and low RBE, which tend to go together. We can state that generally:

Highly-penetrating radiations (such as x-rays) are predominantly associated with:

-low LET	-single-strand breaks
-low RBE	-sublethal and reparable cell damage
-indirect effect	
-oxygen enhancement	-point mutations

Low-penetrating radiations (such as alpha particles and neutrons) are predominantly associated with:

-high LET	-double-strand breaks
-high RBE	-lethal, irreparable cell damage
-direct damage	-frameshift mutations

"Radiation Hormesis" is the theory that *very small amounts* of radiation may actually produce health benefits, by stimulating vitamin D production and other chemical reactions that are beneficial to cell viability.

By way of summary, we can group all of the factors we have discussed into *physical factors* bearing upon the effectiveness of any exposure to radiation, and *biological factors* which render the organism more radiosensitive. These are listed in Table 42-4. Figure 42-35 is a pair of photomicrographs illustrating the actual effects of a high dose of radiation on bone marrow tissue.

SUMMARY

1. The cell is mostly made of water contained within a phospholipid membrane, within which various organelles function for survival. Information from the nucleus directs the metabolism of each cell, so that the entire organism may thrive.
2. The genetic code is made up of codons consisting of three nitrogenous base molecules each, in a sequence that directs the alignment of amino acids to form proteins. The code is transmitted from DNA in the cell nucleus to mRNA molecules, which travel out to the ribosomes in the cytoplasm, where the code is matched with tRNA molecules in order to align these amino acids in proper sequence.
3. Genetic information must be passed from nucleus to cytoplasm for cell survival, from parent cells to progeny cells for organism survival, and from generation to generation for species survival. The genetic code can be altered by exposure to radiation because of its ability to ionize atoms and thus break the covalent and ionic bonds that hold chemical compounds together.
4. The life cycle of a somatic cell includes the four phases of G_1 , S, and G_2 followed by mitotic replication division. The cycle for a genetic cell, called meiosis, consists of a replication division followed by a *reduction* division in which the S phase is skipped. The division process itself

Table 42-4

Factors Affecting Radiosensitivity of Organism

A. PHYSICAL FACTORS:	B. BIOLOGICAL FACTORS:
1. Weighting Factor (Quality Factor) of Radiation	1. Law of B & T: Type of Tissue
2. LET	2. RBE
3. Fractionation	3. Stage of cell in its life cycle
4. Protraction	4. OER
5. Dose Rate	5. Age of organism
6. Total Dose	6. Gender
7. Amount of Area Exposed (collimation)	7. Chemical Agents
8. Dose Are Product	

Figure 42-35

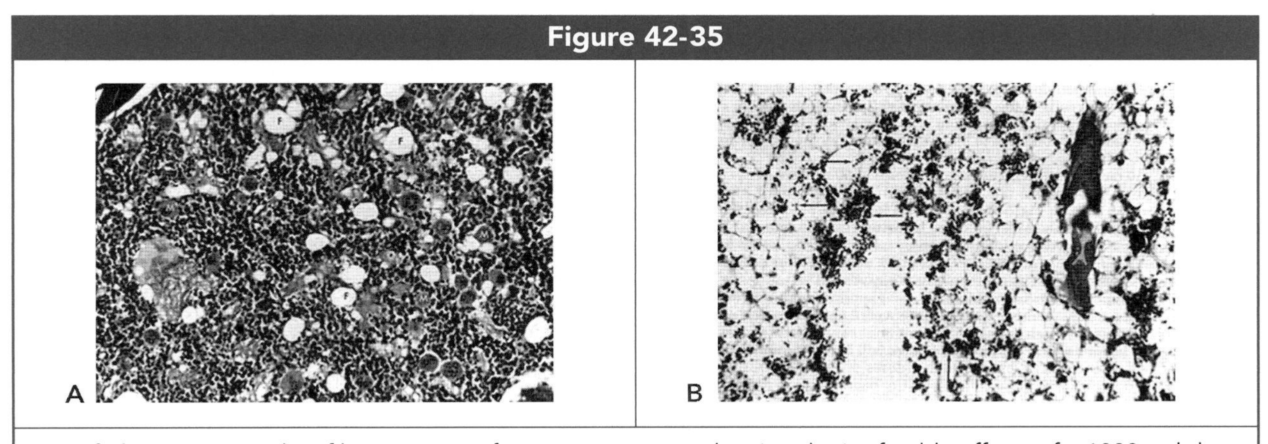

Pair of photomicrographs of bone marrow from a rat sternum, showing the irrefutable effects of a 1000 rad dose of radiation from ***A*** to ***B***. The functional megakaryocytes, pyknotic cells and red blood cells, which appear darker, are depleted and replaced by fat cells (white). (From E. Travis, *Primer of Medical Radiobiology*. Year Book Medical Publishers, 1975.)

consists of four stages: prophase, metaphase, anaphase and telophase.

5. The most radiosensitive time of the cell's entire life cycle is at the very beginning of the S phase. The most radiosensitive time during cell division is metaphase, when mutations in chromosomes can also be observed.
6. The law of Bergonie and Tribondeau indicates that in most cases, cells that are more primitive or more prolific in their reproduction tend to be more sensitive to radiation exposure.
7. Most somatic (visceral) effects of radiation follow a non-linear, threshold response curve as radiation dose increases. Genetic effects, life-span shortening, and most cancers follow a linear, non-threshold response curve. Since there are no biological effects that are unique to radiation, many effects are already present in a population before radiation exposure occurs, which must be taken into consideration.
8. Target theory indicates that lethal damage to cells only occurs when certain critical molecules undergo unrepaired damage from ionizing events. Molecular deactivation can occur either from direct "hits" by radiation or from indirect "hits" caused by local radiolysis of water.
9. Radiation hydrolysis produces free radicals whose atoms possess only a single, unpaired electron in their outermost shell, and ions, both of which are highly reactive chemically.
10. The cell membrane can be broken by ionizing radiations, causing cell death.
11. Radiation damage to chromosomes can be divided into main chain scission and rung damage. Main chain scission includes single-strand breaks, which result in point mutations if misrepaired, and double-strand breaks. Rung damage may consist of either a change of base or loss of base.
12. Highly-penetrating radiations tend to have a low LET and low RBE, are enhanced by the OER, and are associated more predominantly with indirect effect rather than direct, single strand breaks rather than double, sublethal cell damage and point mutations rather than irreparable damage.
13. The effectiveness (and harmfulness) of a radiation dose increases with dose rate, decreases with protraction, and decreases with fractionation of its delivery.

REVIEW QUESTIONS

1. Blast or stem cells have ___________ sensitivity to radiation exposure.

2. Cells whose purpose is the thriving of the organism itself through functioning as parts of organs within the body are broadly categorized as ___________ cells.

3. Specifically, the process of synthesizing proteins is called:

4. Using their letter designations, what are the four combinations of nitrogenous base molecules that are allowed in forming the "rungs" of the DNA molecule?

5. When, in the life cycle of a cell, does all of the genetic material double?

6. How long does the G_1 stage last?

7. Why does the chromatid and centromere structure of the chromosomes become visible under a microscope during metaphase?

8. What is the *second* most radiosensitive time during the entire life cycle of a cell?

9. After meiosis, how many chromosomes are contained in each daughter cell?

10. In which phase of cell division do the mitotic spindle fibers pull the aligned chromosomes into two different poles of the elongated cell?

11. What two types of cells mentioned are *exceptions* to the law of Bergonie and Tribondeau?

12. What type of radiation response curve implies that there is no "safe" level of radiation so low that it would not cause the effect?

13. For a sigmoid radiation response curve, the dividing point between increasing and decreasing rates of response is called the:

(Continued)

REVIEW QUESTIONS *(Continued)*

14. Target theory indicates that as radiation exposure increases, the cells of complex organisms follow a __________ -target, __________ -hit response for cell death.

15. Which type of chemical bonding would occur as a free radical molecule attacks a DNA or RNA molecule?

16. Which free radical molecule is responsible for more than two-thirds of all indirect-hit damage to a cell?

17. Unrepaired changes to the genetic code are called:

18. For main chain scission to result in a permanent change in the genetic code, what phenomenon must occur?

19. What is the term that describes the "swapping" of two chromatid end fragments between different chromosomes?

20. An 800-kV beta particle penetrates into the tissue 10 mm before its first interaction with an atom, "ricochets" and travels 6 mm to another interaction, and then "ricochets" again and is completely stopped after traveling another 4 mm. What is its LET?

21. Suppose it is experimentally found that for 250-kV x-rays, a threshold dose of 21 grays is required to cause epilation (loss of hair). For a beam of beta radiation to cause the same effect, it is found that a dose of only 16 grays is required. What is the RBE of this beta radiation for causing epilation?

22. Both fractionation and protraction extend the delivery of a radiation dose over time. What is the difference between them?

23. Why does the oxygen enhancement ratio have no relationship to direct-hit effects?

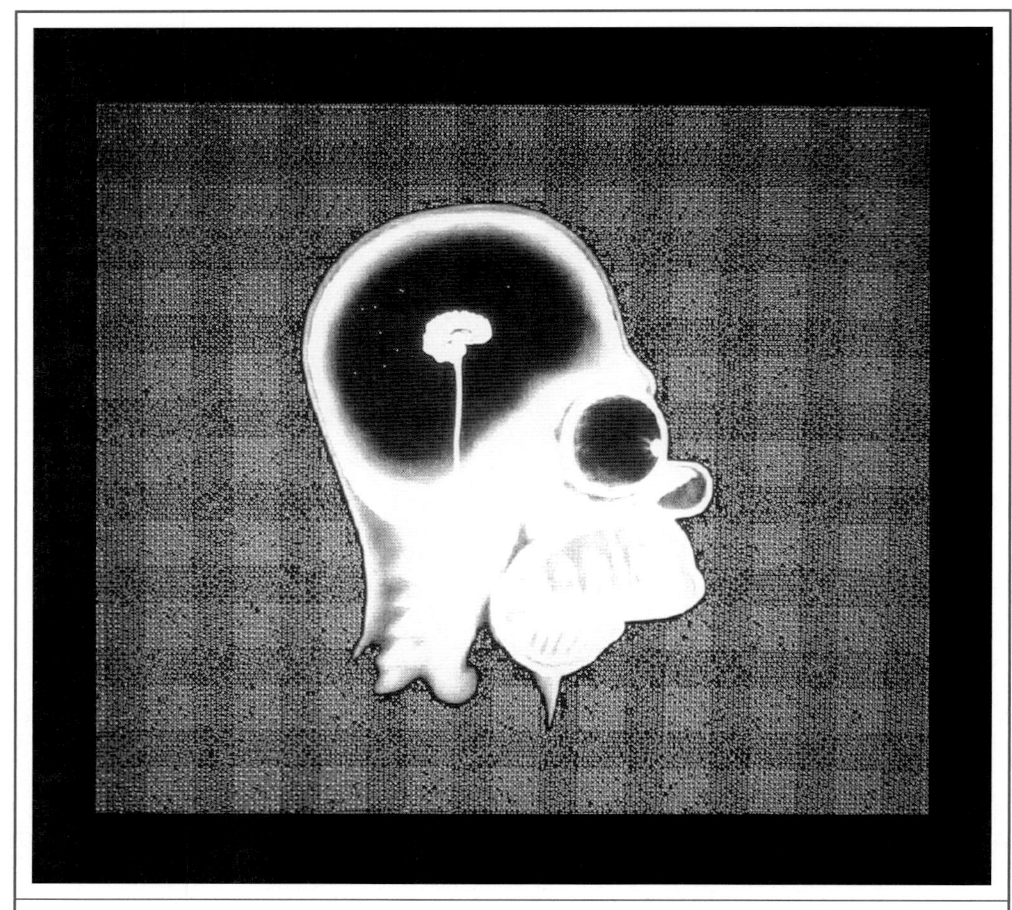

Lateral skull radiograph on Homer Simpson.

Chapter 43

RADIATION BIOLOGY: ORGANISM EFFECTS

Objectives:

Upon completion of this chapter, you should be able to:

1. Describe the roles, advantages and disadvantages of epidemiology, extrapolation and direct experimentation in measuring radiation risk.
2. Define *absolute risk*, *relative risk*, and *excess risk* for radiation exposure.
3. Distinguish between *stochastic* effects of radiation and *deterministic* effects.
4. List and describe various early effects of radiation.
5. Define common expressions of *lethal dose*, such as $LD_{50/30}$.
6. List and describe the stages of acute radiation syndrome and define the *N-V-D* syndrome.
7. For each type of acute radiation syndrome (ARS), state the mean survival time, the predominant cause of death, and the relationship between these two.
8. State by memory the *threshold dose* for each type of ARS.
9. Define erythema and epilation.
10. List and describe various late effects of radiation.
11. State for each period of human embryo and fetal development the expected *teratogenic* effects of radiation exposure.
12. Describe the *mutagenic* effects of radiation.
13. Quantify the life-span shortening effects of radiation.
14. Describe the risks of radiation for cataracts of the eye.
15. Elaborate on the carcinogenic effects of radiation, especially in regard to leukemia.
16. Describe the different carcinogenic risks of recreational and occupational exposure to radiation, including UV.
17. Describe the typical dose levels, risks and benefits of modern mammography, and the generally accepted guidelines for when mammograms should be performed.

MEASURING RISK

There are three methods by which radiation risks may be estimated; these are by epidemiology, by extrapolation, and by direct experimentation with laboratory animals. *Epidemiology* is the study of disease in populations. When we try to track the effects of radiation exposures to humans, we find that we have very little control over variables. Often, the exact doses received are not known. Statistical reliability requires large populations to be studied, but only on rare occasions do we have large populations of humans exposed to a high dose of radiation by accidents or by war. It can be difficult to track individuals over decades in order to follow up on long-term effects. Many other variables such as chemical toxins in the environment, lifestyle, or health habits (such as smoking), may cause the same effects as radiation, so it can be difficult to establish a reliable link between cause and effect.

The process of extrapolation takes known data from previous studies and *projects* how it would change at radiation levels that have not been studied. In effect, it takes a known response curve on a graph and *extends the curve* downward for lower levels of radiation that have not actually been observed, or upward for higher levels of radiation that have not been observed. While this may seem mathematically sound at first blush, the method has been shown to not always be *biologically* sound, as it is based on assumptions rather than scientific observation.

Direct experimentation with lab animals offers several scientific advantages; we are able to directly control all aspects of the experiment, and provide *control groups* for comparison, which are tightly monitored to ensure that variables other than radiation do not cause the effect under study. We can also use large populations which increases statistical reliability. These studies often lack *validity*, however, in that the response of mice, pigs or monkeys to radiation cannot necessarily be applied to humans. All three methods of studying radiation risk have disadvantages, so the best information is obtained when two, or all three methods have been applied to a particular phenomenon.

There are also three different approaches to *reporting* the results of radiation studies, or ways of expressing radiation risk. *Absolute risk* is simply the raw number of cases of a particular disease or condition. By convention, it is expressed as the ratio of cases *per million* population for a time period of one year. For example, the children of pregnant Japanese women surviving the atomic bombs had an absolute risk for birth defects of 100–200 per million per year. But, since variables other than radiation can cause *all* the effects that radiation does, absolute risk lacks specificity in linking the cases to radiation.

Relative risk is the ratio of the number of cases occurring in a population exposed to radiation to the number of cases occurring in a *control population of identical size*. We might state that the relative risk of breast cancer for 1000 women who have received mammograms is 2:1 when compared to 1000 women who have not had mammography procedures. *Excess risk* is defined as the number of cases *beyond* the normal occurrence expected within a particular group of people. For example, if 60 birth defects are generally expected to be identified within 1000 births, but 65 birth defects occur in a group of 1000 women subjected to a particular radiation exposure, we would state the excess risk as 5 for this group.

Stochastic Versus Deterministic Effects

The *stochastic effects* of radiation are defined as effects that increase in *probability* with increasing dose, while their *severity* is independent of dose level. Their occurrence is random, so it must be measured within populations. But, when a stochastic effect does occur, it is an "all or none" response for the individual. We might think of stochastic effects as *statistical* effects.

Generally, stochastic effects do not require a threshold dose to occur—even at very low dose levels, there is always a chance that the radiation dose received might cause the disease or effect. Yet, the odds of a *particular individual* experiencing the effect are small.

Most late effects of radiation exposure, which take months or years to manifest, are stochastic in nature. Examples of stochastic effects include most carcinogenic effects (cancers), and genetic (hereditary) effects. The incidence of leukemia provides a good specific example of what the term *stochastic* means. If *all of the individuals* in a large population receive a substantial dose of radiation, we will see a statistical increase in the number of cases of leukemia within that population. However, *it is impossible to predict which particular individuals will be affected*. When we single out an individual who does have leukemia, we find that the *severity* of the disease has no correlation to the total amount of radiation he or she has been exposed to as compared to other individuals. That is, each individual either gets leukemia or does not, but those who have accumulated higher dose levels do not experience *more severe* cases of leukemia. Higher doses of radiation to the population result in more *cases* of the disease, not in more *severe* disease.

It is the stochastic effects of radiation with which we are primarily concerned in the diagnostic radiology department. They are generally associated with low-level radiation. Stochastic effects usually follow a *linear, nonthreshold* response curve.

The *deterministic effects* of radiation, also called *nonstochastic effects*, are those which increase in *severity* with increasing dose above a certain threshold level.

The severity of the disease or effect is a function of an increasing number of cells which have been damaged. Deterministic effects occur only as a consequence of *large* doses of radiation, such as might be received in a radiation therapy department. However, *radiation levels from extended C-Arm fluoroscopy procedures or extended angiographic fluoroscopy can be high enough to cause deterministic effects.*

The term *deterministic* stems from the fact that *these effects are considered certain to occur to the exposed individual*, provided the typical threshold dose for the effect has been exceeded. Examples of deterministic effects include most early effects of radiation such as decreased blood cell counts, erythema (reddening of the skin), epilation (loss of hair), fibrosis, atrophy, sterility or reduced gamete count.

An unusual example of a *late* effect of radiation which is nonetheless deterministic is cataracts of the eye lens, which are certain to occur when their threshold dose is exceeded, but take many years to show up. While genetic effects, caused by mutations occurring *prior to conception*, are stochastic in nature, *en-utero exposure* effects where a developing embryo or fetus is exposed to radiation are deterministic. Deterministic effects typically follow a *nonlinear, threshold* response curve.

It is important to bear in mind that when we discuss conditions or diseases affecting the whole organism, as is done throughout much of this chapter, we assume that the radiation exposures received are *total body* or *whole body* doses. Scenarios in which we might envision whole body radiation exposures include such things as nuclear terrorism, nuclear warfare or nuclear power plant accidents, *not* medical applications to a patient.

Diagnostic x-ray procedures *and radiation therapy* are both generally limited to specific portions of the body, at which a collimated, restricted beam of radiation is directed. Thus, for medical applications, the *dose area product* (*DAP*) is generally limited, such that the total *dose* required to cause a particular disease would be much, much higher than those listed in the following sections. (Two exceptions, however, are the *erythema dose* and *epilation doses* received in radiation therapy, which can cause reddening of the skin and hair loss in a specific portion of the body at the listed doses.)

EARLY EFFECTS OF RADIATION

Early effects are those which become manifested within a period of time, after radiation exposure, that is measured in hours, days, or a few weeks. (*Late* effects appear in several months or years.) Most early effects are *somatic* (affecting the organism itself but not its progeny), and *deterministic* (described in the last section). They tend to follow a *nonlinear, threshold* response curve.

About 90 percent of somatic damage to an organism from radiation is biologically reparable. However, it is important to remember that the remaining 10 percent of harm to an organism becomes *cumulative* with repeated exposures to radiation.

Table 43-1 lists several examples of early radiation effects and the typical threshold doses required to cause them. The first practical biological measure of radiation harm to humans was a decrease in the white and red blood cell counts, which was used during World War I to assess whether a patient had received too much radiation. (The threshold dose to cause a

Table 43-1

Early Effects of Radiation

Whole Body	
25 rem	Decrease in blood cell count
50 rem	50% loss of lymphocytes
100 rem	Threshold for human death
350–400 rem	Human $LD_{50/30}$
600–700 rem	Lethal dose for human population
Partial Body	
5 rem	Chromosome aberrations (Can manifest as early *or* late effects)
200 rem	Erythema (reddening of the skin)
300 rem	Epilation (hair loss)
Gonadal	
10 rem to male	Decreased Sperm count after a few weeks*
10 rem to female	Suppression of menstruation
200 rem	Temporary infertility in both sexes
500 rem	Permanent sterility in both sexes

*Spermatozoa have a 3–5 week maturation period; already mature cells are radioresistant.

statistically significant and measurable drop in blood cell count is about 25 rem.) Since that time, the *lowest radiation dose* resulting in any measurable biological effect has been about 5 rem, which can cause chromosome aberrations that are visible under a microscope. To cause the effect, these doses must be received as an *acute*, one-time exposure and not protracted. (There is no correlation at all between the 5 rem/year occupational limit and a 5 rem acute exposure.)

It is worth noting that at 50 rem acute exposure (which is the DEL for a one-time, life-threatening emergency exposure), *one-half* of an individual's lymphocytes can be lost, leaving the person quite vulnerable to infections for a temporary period of time. Regarding radiation lethality to humans, it takes a threshold dose of 100 rem to be at risk of death, although only the weaker, unhealthy individuals within a population would be killed from this exposure.

We then see a steady increase in radiation deaths within the exposed population as exposure levels increase, until, at a level between 600 and 700 rem the entire exposed population will expire. What makes the difference within this range, beside the robustness of the individuals, is whether *medical care* is available to ensure adequate hydration of the body and protection from infections. If it is, a dose of 600 rem can be survived. Beyond 700 rem, however, even with good medical care it is likely that all individuals in the exposed population will die.

As low as 10 rem exposure can cause a measurable decrease in the male sperm count, or suppress menstruation in the female. At 200 rem, temporary infertility in both genders can be observed. It is significant to note that the radiation dose required to induce permanent sterility, 500 rem, is high enough that a majority of a human population would *die* from this exposure if it were delivered to the whole body. Under such circumstances, the risk of sterility would not be the very foremost, immediate concern. An exposure at this level, *localized to the gonadal area* of the body, such as might occur with radiation therapy treatments in the pelvis, is another matter.

Lethal Doses

As a practical matter, a statistically accurate method of estimating lethal dose levels for radiation is very difficult to establish experimentally on the basis of *total annihilation* of a population. The concept of *lethal dose 50/30*, abbreviated $LD_{50/30}$, was developed as a more feasible approach. It is defined as the amount of radiation that would cause 50 percent of an exposed population to die within 30 days. For example, the $LD_{50/30}$ for monkeys is 400 rad. This means that if a large group of monkeys is exposed to 400 rad, 50 percent of them will die from it within 30 days.

The $LD_{50/30}$ can be used in comparing the *resilience* of different species to radiation exposure. The $LD_{50/30}$ for rats and mice, (900 rad), is more than double that for humans, and the $LD_{50/30}$ for *goldfish* is 2000 rad, approximately five times the human $LD_{50/30}$. Insects also tend to have an extremely high $LD_{50/30}$.

The human $LD_{50/30}$ has been estimated at anywhere from 250 to 450 rad until the *Chernobyl* nuclear power plant disaster in 1986. This accident provided a rare opportunity to study human response to high levels of radiation (the only major one since the atom bombs were dropped on Hiroshima and Nagasaki in World War II). Scientists were surprised at the level of resilience shown by the exposed human populations around Chernobyl for all kinds of radiation effects, including death. Having had more than two decades to follow up on these populations, it can be stated with confidence that the human $LD_{50/30}$ falls between 350 and 400 rad.

Lethal doses are expressed as *LD x/y* where "x" is the percentage of the population that is expected to die and "y" is the number of days over which this is measured. Any value can be used for "x" and "y." Other commonly used lethal doses are the $LD_{100/60}$ (100% of the population expiring within 60 days, the $LD_{25/30}$, and the $LD_{75/30}$).

The highest radiation dose that any human has ever been known to survive is 850 rad. The human $LD_{100/60}$ is approximately 700 rem, even with medical treatment. When medical treatment is *not* available to a human population because of the breadth of a catastrophe, the $LD_{100/60}$ is thought to be about 600 rad. Death from acute radiation exposure follows a sigmoid, threshold response curve as shown in Figure 43-1.

Acute Radiation Syndrome

Humans subjected to high amounts of whole body radiation not exceeding 100 rad, (50–100 rad), may experience the *N-V-D syndrome*, referring to the symptoms of *nausea, vomiting, and diarrhea*, accompanied by a feeling of general *malaise* or weakness. These symptoms will last for a few days and then subside, with no further somatic implications for the individual.

Higher doses, above 100 rad, can induce the *acute radiation syndrome* (*ARS*), defined as a series of event stages that may lead to death. There are always *four* disease stages for ARS that occur in the following sequence:

1. Prodromal stage (prodrome)
2. Latent stage
3. Manifest illness stage
4. Death or Recovery

During the *prodromal* stage, the individual effectively suffers from *N-V-D syndrome*, which can last from a few hours to a few days. (The duration of all stages is inversely proportional to the amount of radiation exposure received—the higher the dose, the shorter the time periods for each stage.) The symptoms of nausea, vomiting and diarrhea may be accompanied by flu-like symptoms of fever and/or faintness, but they soon subside and the individual often feels a false sense of recovery from illness. This *pseudo-recovery* period is the *latent* stage.

The latent stage also lasts from a few hours to a few days, depending on the dose received. It is then followed by the *manifest illness stage*, in which prodromal symptoms return *in force*, accompanied by additional life-threatening symptoms that are characteristic of the *sub-syndrome* described in the next section. The manifest illness stage can last from several hours to about two months, at which time the individual either succumbs to the illness or recovers.

Based on *the predominant, most immediate cause of death*, ARS can be divided into three subsyndromes known as the *hematopoietic (or hematological) syndrome*, the *gastrointestinal (GI) syndrome*, and the *central nervous system (CNS) syndrome.*

The *hematopoietic syndrome* is defined by damage to the bone marrow being the primary cause of death; there is severe depression of all blood cell counts, resulting in anemia, hemorrhage, and serious infections. Ultimately, the individual is most likely to die from complications due to infections. The hematopoietic syndrome is also known as *bone marrow syndrome.*

Figure 43-1

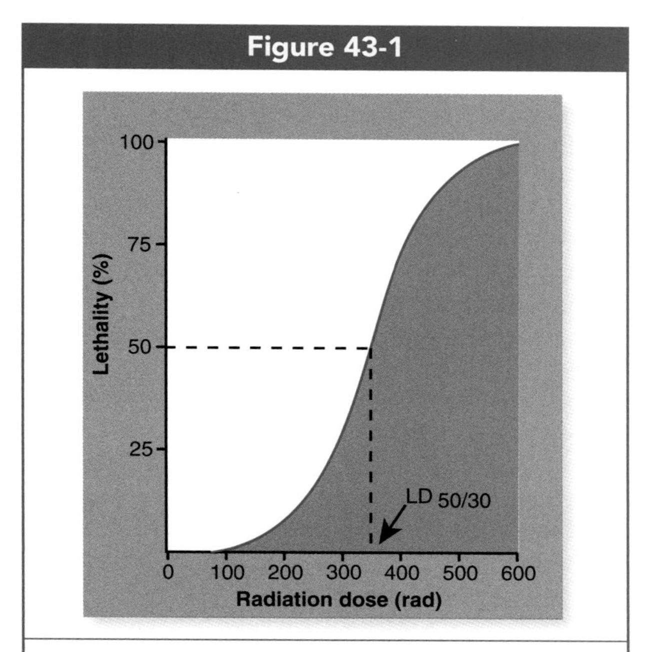

Sigmoid response curve for human death from acute radiation exposure, showing a threshold dose of 100 rad, an $LD_{50/30}$ of 350–400 rad, and an $LD_{100/60}$ of 600–700 rad.

The *gastrointestinal syndrome* is characterized primarily by damage to the small intestine. Irreparable desquamation of the GI lining results in very poor absorption of water and nutrients. Death results from dehydration, electrolyte imbalance, malnutrition and infections.

In the *central nervous system syndrome*, death occurs primarily from acute damage to the brain. It is believed that damage to blood vessels causes fatal cranial edema (pressure), along with some neural damage.

As the whole body dose increases, the average time between exposure to radiation and death generally decreases. The length of this period is known as the *mean survival time*. Each subsyndrome of ARS has a characteristic mean survival time. These are:

Hematopoietic (Hematologic) Syndrome:
2 to 8 weeks
Gastrointestinal Syndrome:
Between 4 and 10 days
Central Nervous System Syndrome: 2 to 3 days

The mechanism behind these mean survival times becomes clear when we consider the main causes of death for each syndrome. Most infections have to go unchecked for many weeks in order to result in death. The hematopoietic syndrome depletes the body's defense system, leaving it vulnerable to multiple infections from organisms that are ever-present in the environment and in the body.

In the gastrointestinal syndrome, the lining of the GI tract, particularly the small intestine, is severely damaged; the finger-like villi, designed for absorption of nutrients, normally slough dead cells every 24 hours. The mitotic activity of the cells in the intestinal *crypts*, designed to replace these cells, is drastically decreased by high exposure to radiation. The photomicrographs in Figure 43-2 show how with increasing radiation doses, the villi become increasingly blunted and denuded, such that there is a massive loss of absorptive surface. The tissue has effectively suffered a radiation *burn*, such that the functional absorption of nutrients into the blood system through those layers that remain is also decreased. Body fluids leak *out* into the lumen of the GI tract, resulting in dehydration.

This is all accompanied by overwhelming infections as bacteria that normally live within the GI tract break into the blood stream through the intestinal wall, further weakening the patient. However, it is essentially malnutrition and dehydration that will

Figure 43-2

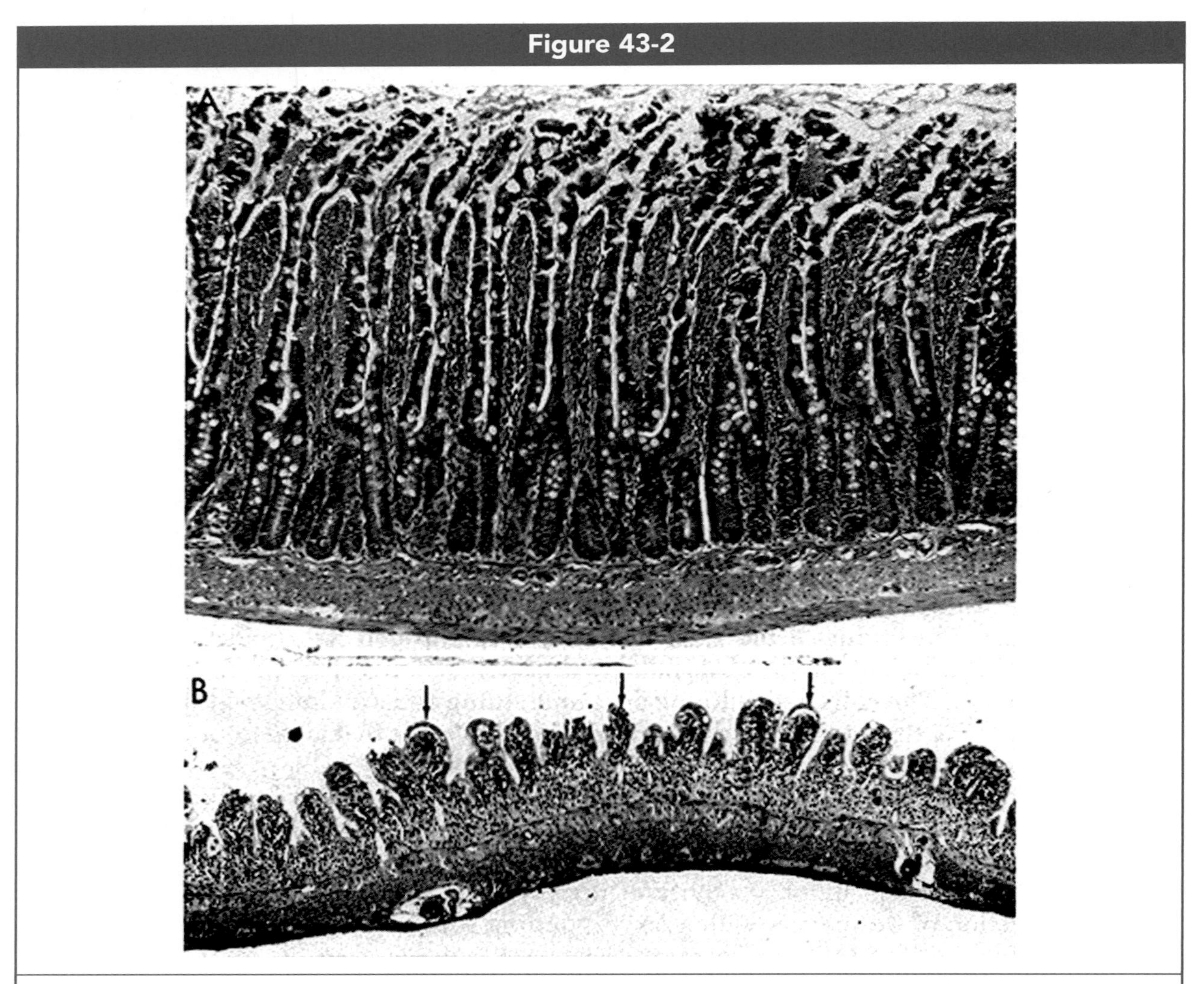

Small intestine of a rat exposed to 2000 rad shows ***A*** normal mucosa with long villi prior to exposure, and ***B***, severe blunting and edema of villi and denudation of mucosa after exposure. (From E. Travis, *Primer of Medical Radiobiology*. Year Book Medical Publishers, 1975.)

Figure 43-3

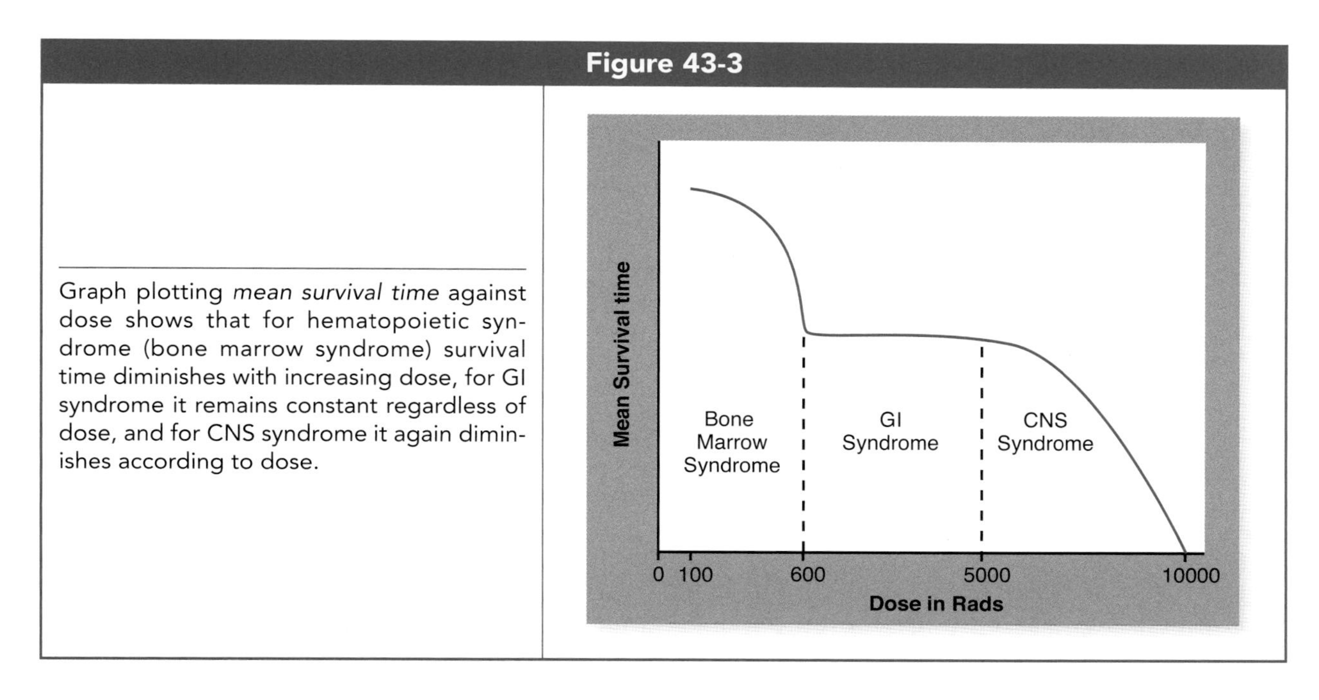

Graph plotting *mean survival time* against dose shows that for hematopoietic syndrome (bone marrow syndrome) survival time diminishes with increasing dose, for GI syndrome it remains constant regardless of dose, and for CNS syndrome it again diminishes according to dose.

kill the individual, before the infections have a chance to. One need only examine the effects of starvation to understand the survival time: A human can only live 7 to 10 days without nutrients, and only 4 to 7 days without water.

The radiation levels associated with the CNS syndrome are so high that the manifest illness stage of ARS begins within 5 to 6 hours of exposure, the first two stages correspondingly shortened to just a few hours each. At this time the victim suffers convulsions and eventually lapses into a coma prior to death. We have learned that the brain and nerve cells themselves are highly resistant to radiation. The ultimate cause of death is thought to be cranial pressure from massive accumulation of fluids, but it is not entirely understood.

Of the three subsyndromes of ARS, the *GI syndrome* stands out as unique for having a mean survival time that is relatively *constant* and *independent of dose* within its range. For both of the other syndromes, there is a steady decrease in mean survival time (MST) as radiation levels are elevated. This relationship is graphed in Figure 43-3. But, victims who die of GI syndrome all die in about the same period of time, regardless of which ones were exposed to higher or lower doses within the GI syndrome range. This is because *starvation* and *dehydration* take the same characteristic time, 4 to 10 days, to bring about death regardless of how severe the specific tissue changes or loss of blood cell counts might be. This constant MST shows up on the graph as a plateau for the range of exposure levels associated with the GI syndrome.

Some confusion surrounds the dose levels required to cause each of these syndromes, due to the misinterpretation of experimental data, in which dose levels required for the *majority of animals to die from the systemic cause of the subsyndrome* has been mixed up with the *threshold doses required* for the particular subsyndrome to begin to be manifested within the animal population. We wish to focus here upon the *threshold doses* for the three subsyndromes, which are presented in Table 43-2.

Table 43-2

Approximate Threshold Doses to Induce Acute Radiation Syndromes

Sub-Syndrome	Threshold Dose
Hematopoietic (Hematologic) Syndrome	100 rad
Gastrointestinal Syndrome	600 rad
Central Nervous System Syndrome	5000 rad

Properly interpreted, this table states that cases of hematopoietic syndrome *begin* to show up within an exposed human population at radiation levels of about 100 rad. Cases of GI syndrome *begin* to show up in some individuals at 600 rad, and cases of CNS syndrome are apparent in some individuals at dose levels above 5000 rad.

These threshold levels are *not* absolute cut-off points for the occurrence of one type of syndrome for all individuals. Above the 600 rad threshold level for GI syndrome, some of the population dies from GI syndrome and some still die from hematopoietic syndrome. The same relationship occurs in the transition from GI syndrome to CNS syndrome. What we can do in the way of further quantification is establish at what radiation level those dying from the higher-level syndrome exceed one-half of all deaths, or constitute the *majority* of deaths. (When we examine this question, we find that at levels above 1000 rad *more than half* of the deaths occurring are from GI syndrome with the remainder from hematopoietic effects. At levels of radiation above 10,000 rad the *majority* of deaths are from CNS syndrome rather than the other two types. It is these numbers that have sometimes been confused with *threshold* doses, which is a different concept.)

Figure 43-4

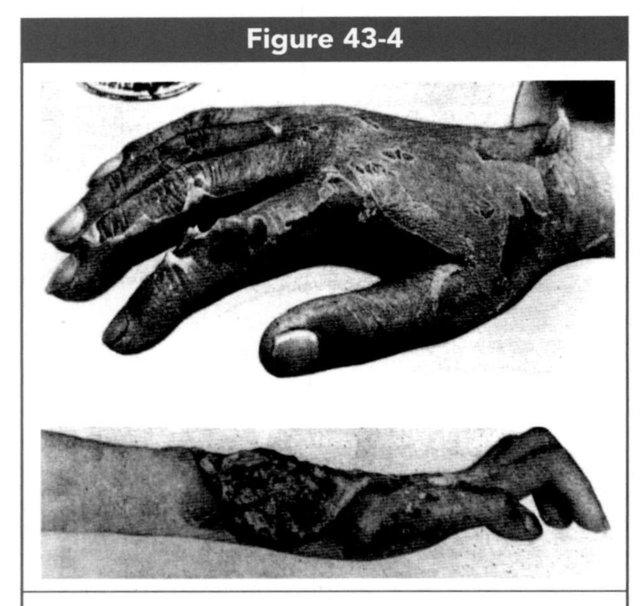

Radiodermatitis (top) is the early form of radiation-induced skin cancer (bottom). (From R. Eisenberg, *Radiology: An Illustrated History*, 1992. Courtesy, Elsevier Health.)

Other Early Effects

Chromosome aberrations were described in the last chapter. They have been shown to occur at dose levels as low as 5 rad. They are unique in that they can show up as *early or late* effects of radiation exposure.

There are several types of *local tissue damage* which can occur when only a part of the body is irradiated. Organs or tissues in that area can *atrophy* (shrink in size) and decline in their function, due to cellular death. With a high enough radiation exposure, an organ can be rendered completely nonfunctional, which can be life-threatening to the organism. (Kidney failure provides an example.)

Erythema refers to a sunburn-like reddening of the skin familiar to many radiation therapy patients. Roentgen himself, and many other early x-ray pioneers, suffered erythema "burns." Fractions of the *skin erythema dose* (*SED*) were used to measure radiation exposure until the unit *roentgen* was defined. Following a radiation therapy treatment, the patient may experience an initial wave of erythema which subsides within a couple of days, only to be followed by a second wave the following week. At higher doses of radiation, the second wave of erythema is followed by *desquamation* of the skin, in which the "raw" sublayers of the skin become exposed and open ulcers begin to form.

The threshold dose for skin erythema is about 200 rad (2 Gy). In accordance with the law of Bergonie and Tribondeau, the bottom layer of the epidermis, which consists of *basal (stem) cells* which rapidly reproduce to replenish the cells sloughed off from higher layers, is more sensitive to radiation than the uppermost layers of the skin; also, the cells of the intestinal lining which are replenished at a rate of 50 percent per day, are much more radiosensitive than those of the epidermis which is replenished at only about 2 percent per day.

Serious skin injuries to patients have been reported in modern times for cardiovascular and interventional procedures which often accumulate long fluoroscopy times. These have included massive ulceration from radiation burns (Fig. 43-4). High-dose fluoroscopy procedures continue to be of great concern, and ways of limiting the allowable dose rates are being scrutinized.

Epilation is the medical term for hair loss due to radiation exposure. The threshold dose for epilation is about 300 rad. Epilation is common in some radiation therapy procedures.

LATE EFFECTS OF RADIATION

The *late effects* of radiation exposure are those which take many months or years to become manifested. These include congenital (birth) defects, life-span shortening, cataracts and various cancers. Generally, late effects tend to follow a *linear, non-threshold* response; a few marked exceptions, however, include cataracts of the eye lens and radiodermatitis (skin cancer), which follow a nonlinear threshold curve.

Teratogenic Effects of Radiation

The *teratogenic* effects of radiation are those occurring from *en-utero* radiation exposure to a developing embryo or fetus. Teratogenic effects are not to be confused with mutagenic effects which occur from radiation exposure to gametes (sperm or egg) *prior* to conception—teratogenic effects are caused by exposure to radiation *after* conception. The dose-response relationship of most teratogenic effects is still unknown, but those that have been somewhat established appear to be *non-linear* and *threshold*.

A unique aspect of teratogenic effects is that the *type* of effect that will occur is dependent on the *stage of development* that the embryo or fetus is in at the time of exposure. We shall present these in four general periods of gestation. There is some overlap in effects between these periods, so the starting and ending dates should not be considered as fixed cutoff points.

Period #1: 0–2 Weeks Gestation

The *only* radiation risk to the developing *blastula* and very early embryo during this initial period is for *spontaneous abortion*, in which the embryo is resorbed. This is an "all or none" phenomenon, in which the embryo is either destroyed or survives. The dose is either completely lethal or completely inconsequential. If the embryo survives, there will be no identifiable effects from the radiation exposure. It has been estimated that a stout exposure to the early embryo of 10 rads increases its risk of spontaneous abortion from a normal rate of 25 percent to only 25.1 percent.

Period #2: 2–8 Weeks Gestation

This is the critical period of *organogenesis* for the embryo, in which tissues are differentiating and identifiable organs begin to form. The main radiation risk to the embryo during this time is for various congenital abnormalities to be induced. There is a natural incidence of congenital deformities of about 6 percent. Large doses of radiation appear to increase this ratio to about 7 percent.

The congenital deformities that occur during organogenesis can be further broken down into two types:

Earlier (2–5 weeks): During this period, skeletal defects are common, such as a stunted limb, a missing or extra finger, extended coccygeal segments, or other structural malformations.

Later (6-8 weeks): Neurological deformities are common during this stage. These can include such effects as *anophthalmia* where an eye fails to form properly (Fig. 43-5), *anencephaly* where the cerebrum fails to form in part or in toto, or *exencephaly* where the cranial vault fails to form over the brain (Fig. 43-6).

Figure 43-5

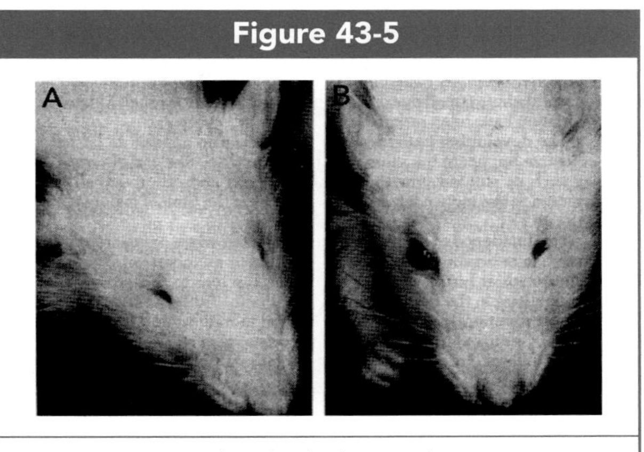

Two rats exposed to high doses of x-rays en utero demonstrate, ***A***, almost total anophthalmia of the eyes, and ***B***, anophthalmia in one eye. (From E. Travis, *Primer of Medical Radiobiology*. Year Book Medical Publishers, 1975.)

Figure 43-6

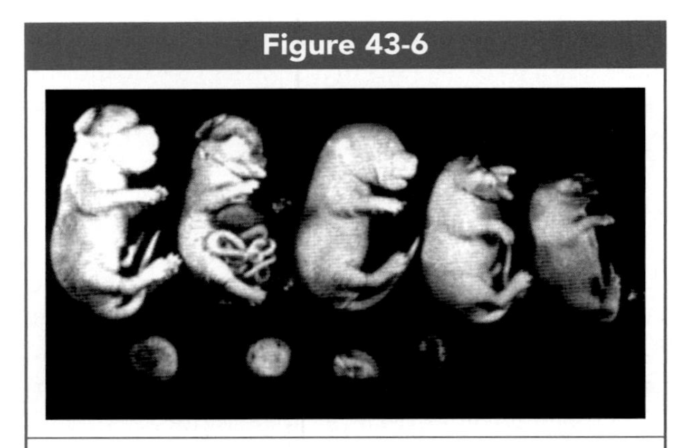

Litter of nine rats exposed to high doses of x-rays en utero demonstrate: Top, from left to right, exencephaly, exencephaly with evisceration, normal birth, and two anencephalics; and bottom, four resorptions. (From E. Travis, *Primer of Medical Radiobiology*. Year Book Medical Publishers, 1975.)

Period #3: 8–12 Weeks Gestation

During this period, the developing fetus becomes clear of risk for morphological (structural) deformities, but is susceptible to radiation-induced mental retardation. The natural frequency of mental retardation is approximately 6 percent, and large doses of radiation are believed to increase this figure to about 6.5 percent.

Combining the risks of the above stages in a pregnancy, it is apparent that the *first trimester is the most radiosensitive stage of a pregnancy* for the developing baby. During the first trimester, studies report that while there is "little evidence of injury" to the fetus from doses less than 1 rad, for a 5-rad dose there are "detectable abnormalities in the central nervous system, congenital malformations, and increased evidence of tumors," and at doses of 50 rads there is a "decrease in head size and stature, and a fivefold increase in mental retardation." During the first 8 weeks, the embryo is approximately *ten times* more sensitive to radiation than an adult.

Period #4: After 3 Months Gestation

After the first trimester, the developing fetus continues to be susceptible to one radiation risk in particular: Latent carcinogenic effects. Various types of cancer may occur later in life due to fetal radiation exposure. These malignancies can also be caused by a number of other factors, so it is difficult to establish *which* factor actually induced a particular neoplasm to proliferate. Of special concern is adolescent leukemia—leukemia which develops during the teen years. Because acquiring leukemia at this stage in life is less likely to be from several other potential carcinogens, it is *more* likely that it might be from fetal exposure to radiation.

Fetal dose must not be confused with entrance skin exposure (ESE) to the mother, because of the *filtering effect* of overlying tissues protecting the fetus. Fetal dose can be roughly estimated as one-third of the ESE at mid-gestation. As the fetus grows, this ratio is closer to two-fifths. These figures would not apply to a full-term fetus that resides only a couple of inches below the distended skin surface of the abdomen, when it would be more accurate to just use the measured or estimated ESE.

It must be stressed that *most diagnostic radiography procedures pose little or no measurable risk to the fetus*, since they do not place the fetus in the direct beam of x-rays. Even those that do, such as a single abdomen exposure, may be judged to be worth the limited risk to the fetus in the process of emergency care for the mother. Those diagnostic procedures which are of concern include c-arm fluoroscopy and general fluoroscopy over the abdomen, angiographic and cardiac cath lab procedures, barium enemas, upper GI studies, and multiexposure urograms. Generally, radiation therapy procedures are of high risk to a developing fetus.

Mutagenic Effects of Radiation

The *mutagenic* effects of radiation exposure are those effects upon the chromosomes of human gametes, spermatozoa and ova (egg cells), prior to fertilization, which may then contribute to congenital defects, disease present at birth, or latent disease. At least one study has shown an increase in congenital deformities for radiographers above that of the population (9% vs. 6%), but most show no statistically significant difference. This is a question that needs more study and for which current studies are under way.

What little is understood about mutagenic effects is based almost entirely on laboratory studies of flies and mice. It has been established that mutagenic effects follow a *linear, nonthreshold* response curve, and this was the impetus for lowering the occupational DELs in 1932. These are *stochastic* effects; as radiation exposure to the gametes increases, mutations climb in *frequency* but not in *severity*. Although this frequency is proportional to the dose, it is estimated to be low, perhaps one mutation per 1000 rad delivered to the *population* (not to an individual).

The *doubling dose* has been defined as that amount of radiation which, given to a population, doubles the spontaneous mutation rate. For humans, it is estimated to lie somewhere in the broad range between 50 rad and 250 rad. The term "congenital defects" refers to any defect present at birth, and may consist of either mutagenic effects or teratogenic effects.

Life-Span Shortening

There is an overall shortening effect on life expectancy for individuals who accumulate substantial radiation exposure. In the very early days of radiology, both radiographers and radiologists were clearly affected. Although no objective study of radiographers has been conducted until recently (one is currently under way), a study of medical doctors in the 1930s found that the life expectancy for radiologists was 5 years less than for other physicians.

Several studies of doctors have been done in modern times, which generally show no statistically significant difference between radiologists and other types of doctors in current practice. Nor was life-span shortening documented for survivors of the atomic bombs in World War II, but it is generally accepted that some small degree occurs with cumulative radiation exposure.

Current estimates are that lifetime radiation workers in general may lose about *12 days* of life on account of occupational exposure. This is to be compared with *74 days* of life lost on average for occupational accidents, and *435 days* for accidents in general. It is not at all likely that the ratio of life-span shortening for radiographers and radiologists exceeds one day per rad. The cumulative lifetime occupational exposure for a typical radiographer probably falls between 15 and 20 rad.

Cataracts of the Eye Lens

The risk of radiation-induced cataracts in the lens of the eye from occupational exposure was brought to the fore when it was found that many dozens of *high-energy nuclear research physicists* had acquired cataracts at a relatively early age. Further studies have indicated a threshold dose for cataracts of about 200 rads in a single, acute exposure, or approximately 1000 rads fractionated over time. Appearance of the cataracts follows a latent period of about 15 years on average. When *older* workers are exposed to high doses of radiation, the latent period becomes shorter and the cataracts themselves become more severe.

Typical levels of occupational radiation exposure for medical workers is much lower than for those in nuclear research. At the levels of scatter radiation experienced by most radiographers, it is probably not necessary to be concerned with eye protection. However, for those radiologists and radiographers who perform *daily cardiovascular or interventional procedures* which involve substantial amounts of fluoroscopic beam-on time, many now opt to keep some type of transparent shielding device in place or wear protective leaded glasses. While some would consider this an overreaction, it is in keeping with the ALARA philosophy and the better side of prudence.

Cancers

The general risk of acquiring some type of cancer in one's lifetime is very high, about 33 percent. It has been estimated that for every 10 rad of *total body* exposure, this risk is increased by another 1 percent. Current levels of occupational exposure for radiation workers is so low that *excess risk* for cancer in general is "undetectable." Even at the maximum DE limits for a working life, the increased risk of *dying* from cancer is not likely to be more than 1 percent, which is similar to other "safe industries."

> The risk of ever *acquiring* cancer in one's lifetime increases by about **1% for every 10 rad**. This is added to a natural occurrence already at 33%

Some occurrence of *lung cancer* was found in early radiologists, and in survivors of the atomic bombs. But, at diagnostic imaging levels of radiation exposure, there has never been any occurrence of *lung cancer* measured for radiographers, and there is none currently for radiologists.

At the other extreme is the case of native Indians in the American West who were hired by thousands to work in uranium mines during the mid-twentieth century. No protective measures were taken, and thousands of early deaths resulted. It is estimated that one-half of all uranium miner deaths were due to cancer of the lung. In Chapter 40 we discussed the *radiation hazard* for smoking cigarettes, which can deliver more than 5 rad per year to the lungs. Lung cancer can also occur in patients as a result of radiation therapy.

Skin cancer is at near-epidemic proportions in the United States. Once primarily restricted to those whose occupations kept them out in the sun for many hours each day, such as farmers, the current plague of skin cancer is resulting mostly from the *cosmetic use of radiation* and *recreational* activities. Tanning salons which employ ultraviolet radiation often advertise their safety, but continue to deliver enough *ultraviolet-A* and *ultraviolet-B* radiation to pose serious risk with repeated use.

UV-A radiation was once considered less harmful than UV-B, but has since been found to produce free radicals in skin cells, just as x-rays do, which can attack the DNA molecule. UV-B assists in producing healthy *vitamin D* in the skin, but excess doses can destroy *vitamin A*. Excessive exposure to *all types* of ultraviolet radiation is harmful and can contribute to skin cancer. Sun-screens with a high *SPF* (*sun-protection factor*) should be used liberally during all outdoor activities that can result in sunburn. A *single* sunburn can raise one's risk of skin cancer by a measurable degree. Repeated *suntanning* also produces a substantially higher risk for skin cancer.

Skin cancer has a latent period of 5–10 years, and in exception to most late effects of radiation exposure, shows a threshold response. The early stages of skin cancer occurring from exposure to radiation are known as *radiodermatitis* (Fig. 43-4).

The best opportunity we have had to investigate radiation-induced *bone cancer* was through follow-up studies of the "radium dial painters," a group of mostly women employed in factories during the two decades prior to World War II to paint watch and meter dials with glow-in-the-dark radium sulfate. These women would lick their paint brushes to obtain a fine point, constantly ingesting small amounts of radium that would eventually deposit in bone. Over the remaining decades of their lives, many accumulated doses as high as 50,000 rad from internal alpha radiation, resulting in many dozens of bone cancer deaths. Their overall relative risk for bone cancer was estimated at 122:1!

Our understanding of *thyroid cancer* comes primarily from the misguided treatment of children with enlarged thymus glands in the mid-twentieth century with irradiation, and a study of South Pacific island children exposed to nuclear fallout during atom bomb tests at about the same time. Thyroid cancer was found to have a latent period of 10–20 years, but the specific risk levels from radiation exposure remain uncertain. There appears to be some increased incidence of thyroid cancer from the Chernobyl nuclear power plant accident in 1986, but this is still under study.

The use of radioactive Iodine-131 (I^{131}) to treat hyperthyroidism, and for nuclear medicine imaging of the thyroid gland, has been discontinued because of its carcinogenic effects. One occupational study conducted in the 1980s indicated a possible increase in thyroid cancer among radiology *residents*. After this, *thyroid shields* consisting of a small leaded wraparound for the neck came into common use among radiologists and radiographers who perform daily cardiovascular and angiographic fluoroscopy.

Leukemia

Leukemia bears separate mention because it is the *earliest systemic disorder, requiring the least dose*, to be manifested from radiation exposure. We have learned that the bone marrow is among the most radiosensitive tissues in the body, so that the exposure of the marrow to high doses of radiation suppresses the formation of normal red blood-cell precursors as well as lymphocytes. Leukocytes, on the other hand, continue to accumulate relentlessly until they overwhelm the system in a life-threatening disorder whose name means "white blood," (*leuko* + *emia*).

Permanent damage to the bone marrow can cause leukemia to appear after a typical latent period of 4–7 years, although the individual remains at risk for 20 years after an acute exposure to high radiation levels.

Survivors of the atom bombs had a 3:1 relative risk for leukemia. Through the early 1940s, before radiation protection for personnel became the standard of practice, it was not uncommon for radiologists to accumulate radiation exposures 10–20 times those of modern radiologists, and their leukemia rate was well above that of other physicians. A conclusive link between occupational exposure and leukemia for *radiographers* has not been shown to date.

For patients undergoing numerous diagnostic x-ray examinations, there may be an increase in their leukemia risk as high as 12 percent. Many studies of radiation-induced leukemia have been conducted, and the data overwhelmingly support a *linear, non-threshold* response curve.

Mammograms and Breast Cancer

Radiation can *cause* breast cancer as well as detect it, and the risk/benefit analysis for mammography has been controversial for many years. During the 1960s, in the interest of obtaining high-resolution views for mammography, direct-exposure film and xeromammography methods were used which required high radiographic techniques. The relative risk for diagnostic mammograms under 40 years old was estimated variously from 3:1 up to 10:1. In the 1970s, the development of new high-resolution screen/film combinations allowed a reduction in dose to about one-fifteenth the previous levels. Even with this improvement, a 1976 study found the average dose for a mammogram to be 2 rad to each breast, leading to an *excess risk* of 370 cancers per million women, with 148 excess deaths per million each year. The study concluded that "the risk outweighed the benefit" of mammography at that time.

Entrance skin exposure (ESE) values can be misleading for mammography. Due to the low kVp's used, (26 kVp), and the associated low penetration of the x-ray beam, dose falls off very rapidly as the x-ray beam penetrates through the breast. The *glandular dose* toward the center of the breast, which is the primary concern for inducing breast cancer, is estimated to be 15 percent of the ESE. This is much higher than similar statistics for higher-kVp radiography in other portions of the body.

It is recommended that the *routine* for a mammogram series be restricted to the craniocaudad and mediolateral projections, and that the axillary view only be employed when a special follow-up view is deemed to be clinically essential.

The American College of Radiology has published the following recommendations for *dose limits* during mammography:

0.1 to 0.3 rad per view
0.4 to 1.2 rad per breast per exam

The American Cancer Society and the American College of Radiology jointly published the current guidelines for when *asymptomatic* women should have mammograms, which are as follows:

1. A *baseline* exam should be obtained between the ages of 35–40 years
2. A mammogram should be performed every 1–2 years from ages 40 to 50
3. Annual mammograms should be performed after 50 years of age

Some studies have concluded that there is "no or little clinical value" for mammograms in women under 50 years old, and radiologists in both Britain and Canada have been more conservative in performing mammograms than their counterparts in the United States. There is little question that for a *baseline* mammogram to be of adequate clinical value (to compare later mammograms to), it must be done by about age 40. There is also little question as to the risk/benefit ratio of mammograms for women over 50. With early mammographic detection in this age group, more than 90 percent of breast cancers are being cured.

The controversy is in regard to their value between the ages of 40 and 50. Certainly, if a woman has a *family history* of breast cancer, mammograms should be obtained during this period. *Women under 40 years of age should not be getting regular mammograms unless they have had clinical symptoms or signs of breast cancer.*

SUMMARY

1. Radiation risks may be estimated by epidemiological studies, extrapolation, and experimentation. All three methods have limitations. Risk

levels can be reported as absolute risk, relative risk, or excess risk.

2. Stochastic effects of radiation exposure increase in probability or occurrence with increasing dose, but their severity is independent of the dose. Deterministic effects increase in severity with increasing dose, and are certain to occur above threshold dose levels. With the low levels of radiation used in diagnostic radiology, we are primarily concerned with stochastic effects to the population.
3. Early effects of radiation are those occurring in less than several weeks. Most early effects are somatic, reparable, and follow a nonlinear, threshold response curve. Late effects occur after a delay measured in months or years from the time of radiation exposure. They include most genetic and carcinogenic effects, and usually follow a linear, nonthreshold response curve.
4. The threshold dose for human deaths is about 100 rad to the whole body. The human $LD_{50/30}$ falls between 350 and 400 rad. The $LD_{100/30}$ for the human population is considered to be at 700 rad total body dose.
5. Between 50 and 100 rad whole body dose, the N-V-D syndrome is expected. At a threshold dose of 100 rad, some exposed humans will begin to suffer the hematopoietic form of ARS, and may die primarily from infections. At 600 rad, cases of GI syndrome begin to occur, death resulting primarily from dehydration and malnutrition. At 5000 rad, cases of CNS syndrome begin to occur, death resulting from intracranial pressure. While the mean survival time for hematopoietic syndrome and CNS decreases with dose, it is independent of the dose level for the GI syndrome.
6. The threshold dose for skin erythema is about 200 rad, for epilation, 300 rad. A dose of 200 rad can cause temporary infertility in both genders.
7. Teratogenic effects of radiation to the developing human embryo or fetus are dependent upon the stage of gestation when exposure occurs. The predominant deformities are skeletal defects beginning at about 2 weeks gestation, neurological defects beginning at 6 weeks, and mental retardation beginning at 8 weeks. After the first trimester, the predominant risk is for latent cancer. In all these cases, the risk is increased slightly from an already existing substantial risk. Most teratogenic effects follow a nonlinear, threshold response curve.
8. Mutagenic effects from radiation exposure are little understood but known to be stochastic in nature, following a linear, nonthreshold response curve.
9. It is not likely that life-span shortening for radiographers exceeds one day per rad, and their cumulative lifetime occupational dose probably falls between 15 and 20 rads. This compares to an *average* life-span shortening of 74 days for occupational accidents in general.
10. Radiographers need not generally be concerned with cataracts of the eyes, but it is prudent for those working with high-exposure cardiovascular interventional procedures on a daily basis to use transparent shielding devices.
11. Some evidence of increased thyroid cancer among radiology residents has led to the common use of thyroid shields by radiographers and radiologists working daily with cardiovascular interventional procedures. General radiographers need not be concerned with this issue.
12. The risk of acquiring cancer in one's lifetime increases by about 1 percent for every 10 rads of radiation. This is added to a natural occurrence already at 33 percent. The overall cancer risk for radiographers and radiologists is not statistically different than that for the general public.
13. There is some concern for patients exposed to multiple high-level radiographic examinations, particularly in regard to leukemia risk for young patients and breast cancer risk for younger women having numerous mammograms. Physicians should order these types of procedures with careful deliberation, and radiographers performing them must consistentaly apply the *ALARA* philosophy.

REVIEW QUESTIONS

1. What is the main disadvantage for direct experimentation with animals in predicting radiation risks?

2. With increasing radiation dose to a population, deterministic or nonstochastic effects increase in their ___________.

3. Most deterministic effects of radiation follow a ___________ response curve.

4. What is the threshold dose to cause a significant drop in blood cell count?

5. What appears to be the threshold dose to see chromosome aberrations occur in a cell?

6. What would be the proper interpretation of the term, $LD_{75/60}$?

7. In what direction were our previous estimates of the human $LD_{50/30}$ revised, if at all, after data were compiled from the 1986 Chernobyl nuclear power plant accident?

8. What does *N-V-D* stand for?

9. The mean survival time for the GI syndrome of ARS is independent of the dose because death from ___________ always takes the same characteristic time.

(Continued)

REVIEW QUESTIONS *(Continued)*

10. List the four stages of acute radiation syndrome in order of occurrence:

11. What is the range of mean survival times for hematopoietic syndrome?

12. In radiobiology, what does *SED* stand for?

13. What two late effects of radiation exposure are unique in that they follow threshold response curves (which most late effects do not)?

14. When a developing embryo is exposed to radiation before 2 weeks of gestation, there is a small increase in the risk of ___________. This is an "all-or-nothing" effect.

15. At mid-gestation, fetal dose can be roughly estimated at ___________ of the mother's entrance skin exposure (ESE).

16. During the first trimester, the embryo is ___________ times more sensitive to radiation than an adult.

17. Effects caused by radiation exposure to the gametes prior to conception are called ___________ effects.

18. What type of cancer is considered by many to be at epidemic proportions in the United States?

(Continued)

REVIEW QUESTIONS *(Continued)*

19. What is the latent period of radiation-induced leukemia?

20. Generally, women under __________ years of age should not have regular mammograms unless they have had clinical symptoms or signs of breast cancer.

21. What is the dose limit recommended by the American College of Radiology for each view during a mammogram?

Before the hazards of radiation exposure were fully appreciated, we see two radiologists fluoroscoping their own hands to "test the hardness" or penetration of the x-ray beam. (From R. Eisenberg, *Radiology: An Illustrated History*, 1992. Courtesy, Elsevier Health.)

Chapter 44

RADIATION PROTECTION: PROCEDURES AND POLICIES

Objectives:

Upon completion of this chapter, you should be able to:

1. State typical *distributed skin exposures* for landmark radiographic procedures, and equivalent comparative risks.
2. Describe typical gonadal exposure levels for landmark radiographic procedures.
3. List methods of optimizing radiographic technique in order to minimize patient exposure.
4. Relate the roles of digital processing speed class and quality control programs to minimizing patient exposure.
5. Describe the types and methods of shielding appropriate for use on patients.
6. Describe those policies and practices which are appropriate for minimizing radiation exposure to the pregnant patient.
7. State guidelines for equipment monitoring related to patient exposure levels.
8. Explain the issues of fluoroscope technology and use, especially in relation to *high-level* or *pulse-controlled* equipment, that relate to minimizing patient exposure.
9. List those procedures identified by the *Safe Medical Devices Act of 1990* as high-risk procedures, and state the five recommendations made for addressing these risks.
10. Provide perspective on controversy over patient exposure levels for CT exams, especially for young patients.
11. Describe the elements of a good personnel monitoring program, and the minimum information that should be provided on occupational radiation reports.
12. List the three cardinal principles for occupational radiation protection, and the effectiveness of each on exposure level.
13. Correctly interpret *isoexposure* curves.
14. Describe the effectiveness of lead aprons, and guidelines for their use by personnel.
15. Define *tenth-value layer*.
16. State the requirements for radiographic equipment shielding.
17. Explain the philosophy and recommendations for holding of patients during an exposure.
18. Describe appropriate policies for limiting personnel exposure, especially in regard to pregnant working radiographers.
19. State general guidelines for equipment which minimize exposure to personnel.
20. Describe the factors and formula for the adequacy of structural barriers to radiation.

21. State the thickness requirements for *primary* and *secondary* barriers, and which applies to the control booth.
22. Define the types of radiation control areas, and the posted warnings required for each.
23. List the main advisory and regulatory agencies for radiation control and the primary focus of each.
24. Distinguish between legal requirements and the *standard of practice* in keeping all radiation exposures *ALARA*.
25. Define the role of the radiographer in being of professional service to other medical personnel and the public, as it relates to understanding and minimizing radiation exposure.

DIAGNOSTIC EXPOSURE LEVELS TO PATIENTS

Table 44-1 presents a compilation of averaged data that compares the *distributed* skin exposure for various diagnostic x-ray procedures to annual natural background levels of radiation and to the equivalent risks incurred by smoking cigarettes and by driving. These figures are presented to provide *perspective* in comparing the relative risk from different kinds of radiographic procedures. They assume that the patient is of average thickness and therefore an average radiographic technique is used.

Distributed skin exposure means that these are cumulative total exposures delivered by the entire radiographic procedure during the course of all fluoroscopy, spot films and overhead projections taken, exposures which may have been distributed or *spread out* by different projections to different specific portions of the skin. In other words, these figures should not be interpreted as scientifically ascertained *patient dose* delivered to a particular section of tissue—rather, they are summed totals representative of the *procedure.*

This data for the single-view PA chest, the single-view abdomen, and the barium enema series was discussed in Chapter 40 on *Radiation Perspectives.* The single-view PA chest, without question, causes the lowest radiation exposure to the typical patient in the diagnostic imaging department, about 20 mR. It is worth noting that adding the lateral view to a chest study more than quadruples this exposure, although the total is still extremely low. In Chapter 26 under the section on the *proportional anatomy* approach to radiographic technique, you will see that *the radiographic technique required for a lateral projection of the chest (and torso in general) is from 3 to 4 times that of the AP or PA projection.* If the mAs is used to make this adjustment, we expect the dose for the lateral projection to be 60–80 mR. Adding this to the 20 mR for the PA view, we obtain a range of 80–100 mR for the two-view chest series.

A 5-view cervical spine series generates 5 times this amount, the whole series being roughly equal to a single-view AP abdomen exposure (0.5 R). The single abdomen exposure is roughly in the same range as the annual occupational exposure for a radiographer, and also close to the annual natural background radiation exposure.

The radiographic "bone procedure" producing the highest exposure levels is clearly the lumbar spine series. Note that the listed procedure which can generate 5 R of distributed skin exposure is only the *3-view series.* When oblique projections are added to the lumbar spine series, this figure rises to approximately *7R (7000 mR).*

For a typical intravenous urogram (IVU or IVP), this abdominal skin exposure (0.5 R) may simply be multiplied by the number of views taken. This number varies widely and is critical to the actual exposure delivered to the urological patient. Seven views are assumed to estimate an exposure of 3.5 R. Note that, depending on the radiologist's instructions, as many as 12 views might be taken during a urogram which would result in a much higher patient exposure of as much as 6 R. It is sometimes possible to obtain sufficient diagnostic information from only 3 or 4 views, which would spare considerable patient dose.

Table 44-1

Radiographic Procedure Distributed Skin Exposure Totals* and Comparative Risks

Procedure/Source	Exposure	Equivalent Comparative Risks		
		Cigarettes Smoked	Miles Driven	Equivalent Background
Annual Natural Background	300 mR	30	60	–
3-Mile Island Accident (US)	1.5 mR	0.2	.33	2 days
Transoceanic Jet Flight	5 mR	0.5	1	6 days
Chernobyl Survivors (USSR)	45 R	3600	9000	150 years
Hrioshima Bomb Survivors	200 R	16,000	40,000	700 years
1-View PA Chest	20 mR	2	4	3 weeks
2-View Chest	100 mR	10	20	4 months
2-View Knee	150 mR	15	30	½ year
5-View C-Spine	500 mR	50	100	1.7 years
1-View Abdomen	500 mR	50	100	1.7 years
4-View Skull	1.2 R	100	250	4 years
7-View IVP	3.5 R	300	700	10 years
3-View L-Spine	5 R	400	1000	17 years
Upper G.I. Series 2/3 mins fluoroscopy and 8 OH's/spots	10 R	800	2000	33 years
Barium Enema Series* w/4 mins fluoro and 11 OH's/spots	15 R	1200	3000	50 years
Angiogram Series 2/5 mins fluoro and 16 overheads/spots	20 R	1600	4000	70 years
4 Dental Bitewings	1.2 R	100	250	4 years
2-View Mammogram	600 mR/breast	100	250	4 years

*Not to be interpreted as cumulative skin exposures, since these doses are *spread out* over the skin surface from different projections. Rather, these are general total exposures for *procedures*. Also, these are *local*, partial body exposures, not whole body or gonadal doses.

Averaged from BEIR Report (US Govt.), "BERT" (Background Equivalent Radiation Time) by Cameron, Hendee and Bushong, and an independent ion chamber survey for typical techniques.

The upper GI and barium enema series assume conservative exposure levels for both fluoroscopy (at 2.4 R per minute) and abdominal overheads (at 300 mR per exposure). Actual fluoroscopy exposure levels are often *much* higher than 2.4 R per minute, frequently double this amount, and sometimes triple.

In Table 44-1, the *cigarettes smoked* column should be interpreted to mean that the radiation risk is equivalent to a scenario in which this number of cigarettes was smoked by an individual, who *then stopped smoking for the rest of their life*, and had never smoked before. The *miles driven* column is meant to indicate an equivalent risk to either harm or death from the statistical possibility of having an automobile accident while driving this distance one time. The *equivalent background* column is the amount of time it would take to accumulate an equivalent exposure from natural background radiation—these figures are sobering for higher-exposure level procedures.

The intention of this presentation is to lend perspective and give the student an appreciation for those higher-exposure level diagnostic procedures, for which we should be particularly diligent. By communicating fully with the radiologist, the number of views for these high-dose procedures can sometimes be reduced. For the radiologist's or radiology assistant's part, *intermittent fluoroscopy* is perhaps the most effective measure that can be taken to spare patient exposure. This means the fluoroscopist avoids constantly watching the energized fluoroscope monitor between those times when peristalsis has placed barium or air in strategic locations for diagnostic information.

Gonadal Exposure

For most radiographic procedures, such as a knee or a chest radiograph, the gonads receive a small exposure from the *scatter* radiation produced. Scatter is emitted in all directions, only a small portion of which is directed toward the gonads. A general rule can be stated that, for those studies which do not include the gonads within the primary x-ray beam, the patient's gonadal exposure is of a magnitude about *one-thousandth* of the entrance skin exposure (ESE) within the x-ray beam.

This rule cannot be applied when the edges of the x-ray field comes within a couple of inches of the gonads, such as a lumbosacral spine on a male patient, in which case the intensity of scatter radiation will be considerably higher (1/100th of the in-beam ESE). Nor is it appropriate when the x-ray beam is directly over the gonads, such as a lumbar spine on a female patient. The ovarian exposure for a female patient is roughly estimated at *one-third* of the skin exposure (ESE) which is in the direct beam of x-rays.

Table 44-2 lists the typical gonadal doses to patients from common radiographic procedures.

Table 44-2

Typical Gonadal Dose to the Patient by Radiographic Exam*

	MRADS	
Type of Examination	**Male**	**Female**
Skull Series	0	0
Cervical Spine Series	0	0
Shoulder Series	0	0
Chest (2-view)	0	1
Upper GI Series	1	171
Barium Enema Series	175	903
Intravenous Urogram Series	207	588
Abdomen (1 view)	97	221
Lumbar Spine Series	218	721
Pelvis	364	210
Hip	600	124
Upper Extremities	0	0
Lower Extremities	15	0

*Taken from the X-Ray Exposure Study (XES) (Rockville, Maryland, Bureau of Radiological Health, 1975).

OPTIMIZING RADIOGRAPHIC TECHNIQUE

The entire focus of *radiographic technique* is to produce the best possible image quality *at minimum risk to the patient*. The implications of technique factors for patient exposure have been discussed in foregoing chapters, so this section will serve only as a *review* of each variable as it relates to protecting the patient from unnecessary radiation.

mAs and kVp

The *total* mAs used to produce a radiograph is *directly proportional* to patient exposure. Twice the mAs will deliver twice the skin entrance exposure (ESE) to the patient. Radiographically equivalent *increases* in kVp deliver less exposure to the patient, and although they used to alter image contrast with film-based systems, with *digital imaging* they no longer have a negative impact on image quality (except in great extremes). When it comes to *reducing* technique, it is more in the patient's interest to decrease the *mAs* rather than the kVp, since there will be a proportional reduction in patient dose. Furthermore, reductions in kVp can result in inadequate *penetration* through the anatomy and result in repeated exposures.

Because of these relationships, we can state that *in the interest of limiting patient dose while at the same time maintaining digital image quality, (1) Generally, high-kVp, low-mAs radiographic techniques are recommended, and, (2) Increases in technique should generally be made using kVp, while decreases in technique should be made by reducing the mAs.*

Generators and Filtration

A common misconception is that higher-power generators (3-phase or high-frequency), produce lower patient exposure because less mAs is needed. These designs are *electrically* more efficient in producing x-rays, generating more *mR per mAs*, so that the reduced techniques are merely compensating for a higher exposure rate. A *slight* savings in patient exposure is achieved because these generators also produce higher *energy levels* at a particular kVp setting. In practice, only a slight decrease in overall exposure to patients is normally attained by high-power equipment.

One promising opportunity which has not generally been taken advantage of relates to constant-potential generators used with battery-powered *mobile* units: We have mentioned that typical techniques for these machines, especially for chest radiographs, employ much lower kVp settings. In the previous section we advocated making *decreases* in technique with mAs rather than with kVp. *By using lower mAs values rather than lower kVp on CPG mobile units, patient dose could be significantly lowered.*

A minimum of 2.5 mm aluminum equivalency in total filtration is required on all x-ray equipment capable of operating above 70 kVp. Slab filters that have been placed in or above collimators should not be removed, as they are part of this required total. Proper x-ray beam filtration is critical to minimizing patient dose, because it removes from the x-ray beam low-kV x-rays that would otherwise only be absorbed within the patient, and have no diagnostic value since they never reach the image receptor anyway.

Field Size Limitation

Limitation of the field size by collimation or by the use of cones and cylinders contributes to image quality by reducing scatter radiation, but also is an essential component of good radiation protection practice. Highly radiosensitive organs may be spared exposure by collimation. This reduction is on the order of *100 times* less when the exposure just outside the edge of the primary x-ray beam is compared to the exposure within the field.

Radiographers should therefore be ever-mindful of proper collimation. *Over-collimation* can also be an issue when anatomy of interest is clipped off by collimating too tightly, necessitating a repeated exposure. Every retake *doubles* the radiation to the patient for acquisition of a particular view. Because field lights are not always accurate, *always allow at least one-half-inch of field light beyond each border of the anatomy of interest. Proper collimation is the most effective means whereby the individual radiographer can contribute to limiting radiation exposure to patients.*

Patient Status

Thicker body parts, casts, and additive diseases require more technique, and are outside the control of the radiographer. *Compression* using a paddle or other suitable device is one exception where intervention can reduce the technique required. Radiologists often use compression during GI studies to separate loops of bowel, and the ensuing automatic reduction in fluoroscopic technique is a side effect beneficial to the patient.

Grids and Image Receptors

High-ratio grids absorb more radiation, and require increased radiographic techniques in order to maintain sufficient signal to the image receptor. In the interest of minimizing patient exposure, *the minimum grid ratio which provides sufficient clean-up of scatter radiation should be used.* The level of scatter is difficult to determine for digital radiography where most of the fogging effects of scatter radiation has been compensated for through digital processing—it may need to be done with the help of a radiation physicist, so that the department can make appropriate purchasing decisions for grids for each radiographic unit.

Materials used for the *front panels* of image receptor plates or cassettes must be as radiolucent as possible while still providing structural protection. Added layers in front of the actual *active matrix array* of detectors absorb more radiation and necessitate higher techniques. For *stimulable phosphors* (screens) such as are employed in computed radiography (CR) *and* for indirect-conversion DR systems, three characteristics which determine the amount of radiographic technique needed are absorption efficiency, conversion efficiency, and emission efficiency as described in Chapter 34. Absorption efficiency is also critical for the surfaces of detector elements (dels) in all active matrix arrays used in DR systems. The greater the overall efficiency of the image receptor, the more patient exposure is saved.

Digital Processing Speed Class

The ability to *select* the speed class at which a digital processing system will develop radiographic images is unique in that it *further* empowers imaging departments to minimize patient exposure where there is the will to do so. Although for some radiographic procedures maximum signal-to-noise ratio (SNR) is critical, there are *many* cases where radiologists can agree to accept a minimal level of quantum mottle noise in the image in order to reduce patient dose, and some departments have established protocols on this basis. In the interest of patient protection, digital processing should generally be done *at the highest speed class which does not present an unacceptable level of mottle in the images.*

We have discussed the tendency for *dose creep* extensively in Chapter 33, which stems from the fact that for digital images, radiographic techniques far above those necessary may be employed without visible image degradation, while *too low* a technique can cause mottle to be manifested. This is purely an issue of *ethical and professional practice.* In order to prevent dose creep, individual radiographers *must* be cognizant of the exposure indicators provided by manufacturers of digital equipment. And, imaging departments should ensure that upon installation, manufacturers configure the equipment so that the exposure indicators are presented on image review screens, and then have a policy of keeping a record of these indices, either by hard copy or by having the software of the PACS configured to maintain permanent records.

Radiographic Positioning

The ability of digital units to display an image within just a few seconds after an exposure is taken has led to a disturbing new trend: The radiographer may leave a patient in position while quickly checking the image that comes up on the control console monitor for correct positioning, and then repeat the exposure to perfect the position. When this is done for relatively minor corrections to the position, the increased exposure to the patient is unwarranted, and this practice should be avoided.

Longer SIDs have been advocated for the best resolution of image *sharpness.* When the anatomy of interest can be placed very close to the image receptor, short SIDs can be intentionally used to *desuperimpose* overlying anatomy by their magnifying and blurring effects. These benefits should be the primary determining factors for the specific SID used. The technique is usually compensated for changes in distance, such that the end result is approximately the same exposure levels to the patient.

Most radiographic positions cannot be compromised to lower exposure to a particular portion of the body, but there are some important exceptions: *Scoliosis* series, which are often done repeatedly for a particular patient, can be performed in PA rather than AP position to take advantage of the natural filtering effect of overlying tissues to reduce exposure to the female breasts. Male gonadal dose can be similarly reduced for some lower abdominal projections. The skull can be done PA rather than AP to reduce exposure to the eye lens through tissue filtration.

Finally, the radiographer's own skills in both positioning and communicating instructions to the patient, especially for pediatric radiography, is essential to minimizing retakes and keeping public exposure *ALARA.*

Radiographic Technique and AEC

Systematic and scientific approaches to setting radiographic techniques will minimize errors that can result in additional exposure to the patient. The use of *technique charts* is strongly recommended to aid in this objective, and is required by law in most states.

When using automatic exposure control (AEC), all of the principles set forth in Chapter 27 should be carefully adhered to. The use of AEC in and of itself does not guarantee a reduction in patient exposure, and using the AEC improperly can cause repeated exposures to the patient. Optimum kVp, optimum mA, and an appropriate back-up time must be combined with *proper collimation of the x-ray beam* and proper positioning to minimize retakes. When an AEC seems to be malfunctioning, and in a number of positioning situations described in Chapter 27, it is often better to use *manual technique* than to risk repeated exposures to the patient. Radiographers should be comfortable with the principles of manual technique.

Quality Control and HVL

Quality control policies and procedures are covered in Chapter 39. By definition, quality control is the use of diagnostic tools to detect trends that can lead to either poor images or unacceptable levels of patient exposure *before* these undesirable results actually come about. Quality control programs bring increased *consistency* to radiographic practice, which, in turn, results in fewer retakes and fewer exposures to patients.

The *half-value layer (HVL)* refers to the amount of absorber material required to reduce the intensity of an x-ray beam to one-half the original output from the x-ray tube. It is a direct measurement of the *penetration quality* of the x-ray beam. Since poor penetration results in high absorbed dose to the patient, the HVL is an important indicator of the *safety* of x-ray equipment for the patient. When the HVL is insufficient, there is either an inadequate amount of protective *filtration* in the x-ray tube and collimator, or the kVp indicator is inaccurately reading out a higher kVp than the machine is actually producing. Either of these problems should be corrected immediately to prevent unnecessary patient exposure.

The step-by-step procedure for determining the HVL of an x-ray unit is described in Chapter 39, and a practice lab for measuring HVL is available in the instructor's manual (CD-ROM).

PROTECTING THE PATIENT

Patient Shielding

Various forms of lead protective devices are available for covering portions of the patient's body which may be exposed to scatter radiation, or to effectively function as a form of *collimation* by limiting the extent of the primary beam. Most of these are classified as *contact shields* which are in contact with the patient. "Lap" shields or half-shields come with a tie or clamping device at the top that fits around the waist, primarily for protecting the gonadal area. These are especially important to use on children, where the collimated field for a chest or other torso procedure can often extend down into the pelvis. *Shaped gonadal* shields are essentially like a baseball catcher's "cup" and form-fit around the genitals. Leaded rubber sheets can be laid directly across the patient, and can be customized into various shapes by cutting them with scissors. These precut circles, trapezoids, or "heart"-shapes are especially helpful for gonadal protection of pediatric patients, as they can be cut into a series of sizes corresponding to age.

Shadow shields are shaped lead tabs attached to an extension arm or "goose-neck" cord that connects to the collimator of the x-ray tube. By adjusting the extension, the lead form protrudes into the projected light field to block x-rays from a specific peripheral portion (Fig. 44-1). The primary advantage of shadow shields is that, by connecting them to the collimator rather than touching the patient with them, sterile fields in surgery can be preserved.

Radiation protection guidelines state that the *gonads* of the patient should be shielded any time they lie within 5 cm (2 inches) of the edge of a properly collimated x-ray beam (provided that diagnosis of anatomy of interest is not impeded thereby). Neither the male nor the female gonads can be protected from the substantial amount of scattered x-rays produced within the abdomen during torso procedures, but, gonadal shielding can reduce scatter

Figure 44-1

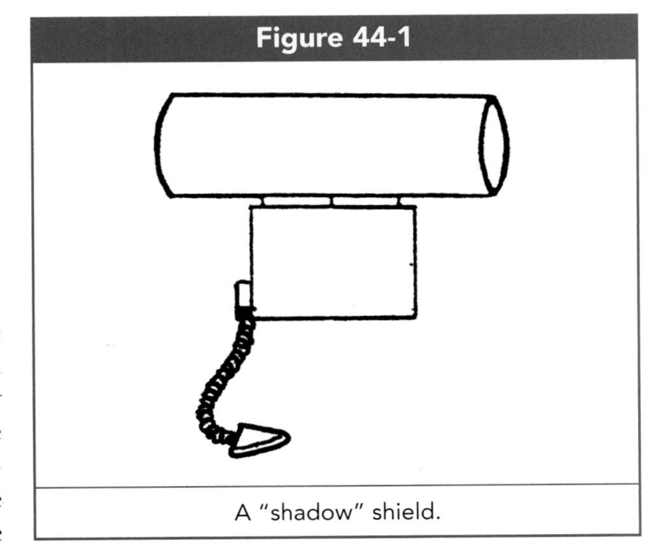

A "shadow" shield.

from other nearby objects and ensures that the edges of the *primary x-ray beam*, which frequently do not coincide exactly with the edges of the *light field*, do not include the gonads or scatter-producing tissue near them within the "raw" x-ray beam.

The lead aprons used by radiographers can also be used with patients. The effectiveness of lead shields depends upon the thickness of actual lead contained in them, as well as the *kVp* set at the console, and can vary from as little as 66 percent up to 99 percent. For a 75-kVp x-ray beam projected directly at an apron containing 0.5 mm of lead, about 88 percent of the radiation will be blocked. Overall, we can state that the types of lead aprons and gloves that are typically found in radiology departments have an *average effectiveness of about 85 percent* when used in the primary x-ray beam.

A philosophical word about shielding practice: Often patients may be concerned about shielding when the 5 cm rule does not apply. The radiographer may realize that neck, skull, and distal extremity procedures do not require lead shielding for the gonads from a scientific standpoint, and may explain this to the patient. Nonetheless, if the patient has expressed any concern at all, it seems like a wise practice to take a moment and provide the shielding anyway, if only for the sake of good public relations.

Policies for Patient Pregnancy

Patient education regarding the risks of diagnostic radiation to a developing embryo or fetus is strongly advocated, and can be provided in several ways including brochures or videotapes that are made available in patient waiting areas. Every female patient of child-bearing age should be asked if there is any chance she *could* be pregnant. If the answer is affirmative, a radiologist or supervising manager must be notified and must make the determination, in consultation with referring physicians, whether to proceed or not with radiography.

It should be emphasized that the question is *not* whether the patient is pregnant, but whether she could be. The age of puberty for young girls has been declining over the last several decades, and assumptions should not be made that a girl of ten years, for example, need not be screened before proceeding with x-ray procedures.

When the radiographic procedures ordered are not urgently needed, a policy of *elective scheduling* can be used in which the female patient schedules the procedure to be done during the 10 days following the *onset* of mensus (the menstrual cycle). This is the period of time when pregnancy is least likely to occur.

When a radiation exposure has occurred to a developing embryo due to unknown pregnancy at the time or to an emergency procedure, and the exposure occurs in the first trimester, the best estimation possible should be made of the total exposure received by the embryo. This will be approximately one-third of the mother's entrance skin exposure (ESE). If the estimated exposure to the embryo is less than 10R, *abortion* should *never* be counseled by medical personnel based on radiation risk to the embryo. There is a substantial risk for congenital defects if exposure to the embryo exceeds 25R, yet this would be an extraordinary occurrence for a *diagnostic* radiography procedure. The range from 10 to 25 R is a gray area insofar as any counseling is concerned in such a highly personal decision.

Guidelines for Equipment

Regulations governing radiation equipment are published by government health agencies and guidelines are published by scientific committees, in the hundreds of pages. Generally, a *should* statement is merely a recommendation, whereas a *shall* statement is typically required by government agencies with the power to assess fines or penalties. Following are highlights of equipment guidelines which help protect the patient. They have been paraphrased for clarification, and are not quotations.

1. All exposure switches *shall* be of the "dead-man" type. A "dead-man" switch is a switch that must be continuously held down to operate (Fig. 44-2). If the operator were to faint, for example, the switch would be released and the exposure to the patient would be terminated instantly.
2. The x-ray tube-to-tabletop distance *shall* not be less than 38 cm (15 inches) for a fixed radiographic unit. The TTD *shall* not be less than 30 cm (12 inches) for a *mobile* unit. Note that many collimators are built large enough that it is impossible to bring the x-ray tube too

close to the patient's surface because the collimator gets in the way. When the collimator is too small to prevent this, rails or plastic cylinders are typically connected to the front of the collimator to do so, as shown in Figure 44-3.

3. During fluoroscopy, exposure rates at the tabletop *shall* not exceed 10 R/minute, nor 2.1 R/minute/mA. They *should* not exceed 5 R/minute. Conventional fluoroscopes *should* not be operated above 10 mA. (No limit is applied to "recording" the image, to allow for unlimited "spot films" to be taken.)
4. During fluoroscopy, a cumulative timer *shall* emit an audible signal when 5 minutes of fluoroscopic beam-on time is reached. The radiologist *should* employ intermittent fluoroscopy and attempt to remain within this recommended exposure time limit for most procedures. However, this is at the radiologist's discretion. Radiographers should not impede the function of this timer. They can reset it immediately when it sounds, but should not interfere with it prior to serving its function.
5. A fluoroscope routinely operated above 90 kVp *should* have 3mm of aluminum equivalency filtration.
6. The fluoroscopic collimator shutters *shall* be visible on the TV monitor at all times the fluoroscopic beam is engaged. This ensures that the shutters are not excessively wider than the useful area of the input phosphor of the image intensifier, unnecessarily exposing portions of the patient's body which are not under examination.

Figure 44-2

All exposure switches on x-ray equipment must be of the *dead-man* type, which requires constant depression to continue the exposure.

Fluoroscope Technology

The amount of patient exposure delivered from a fluoroscopy unit depends a great deal upon its specific technology and use. A reasonable average for

Figure 44-3

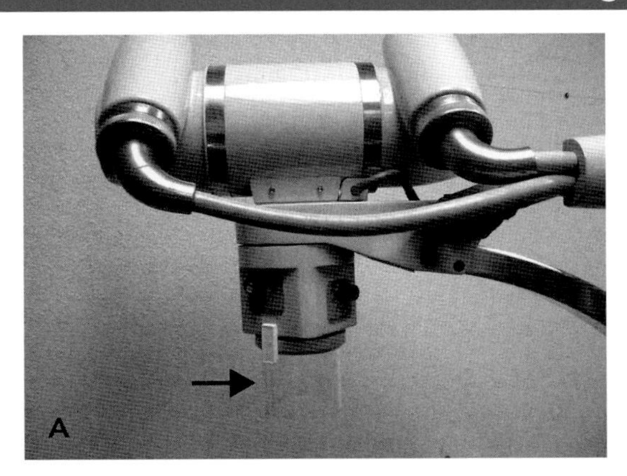

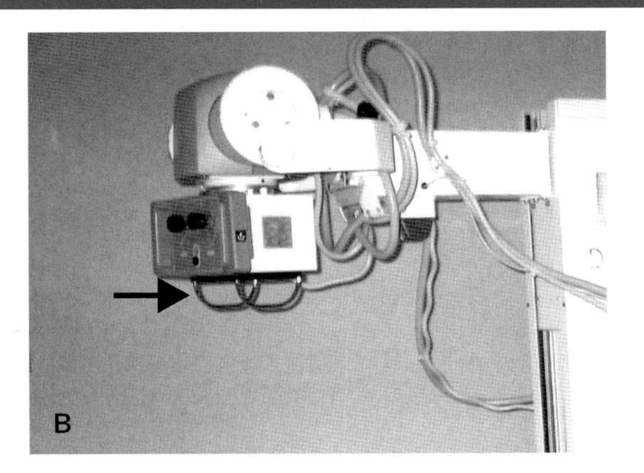

Plastic "cone" (***A***) or metal rails (***B***) used to prevent mobile x-ray tube from being brought closer to the patient than 12 inches (30 cm).

the actual skin exposure rate delivered by most conventional fluoroscopes is about 5 R per minute. Modern fluoroscopes make available to us, in addition to conventional fluoroscopy, *high-level control* fluoroscopy, *pulsed* fluoroscopy, and *digital* fluoroscopy.

"High-contrast" or "enhanced imaging" fluoroscopes are euphemistic terms for units that use higher dose rates to achieve somewhat improved images. The recommended operating limit of 10 mA does not apply to these machines, which are capable of operating from 20 mA to 300 mA. Nor does the 10 R/min exposure rate apply. No regulations were imposed on the first of these units to be marketed. Newer machines have a dose rate limit of "10 R/min for the first (AP) position, 20 R/min for the second (lateral) position." Some feature a louder warning signal that sounds after 10 minutes of fluoroscopic beam-on time. Political concessions to manufacturers should not veil the higher risk of these machines to patients.

The primary argument for allowing this type of technology onto the market is that the *physician* is ultimately responsible for the amount of radiation the patient receives, and would be counted on to properly weigh the risk versus the benefit of every radiation procedure. The fact is that all types of physicians can own and operate radiation equipment, and, as a rule, non-radiology physicians receive *no* training in radiation protection as a part of their medical education. In fact, a recent survey of several *radiologist* residency curricula found only one school that listed a formal course in radiation biology and protection.

On the other hand, technological advances have also introduced *pulsed mode* fluoroscopy units which can effectively reduce patient exposure rates, if used properly (see Chapter 38). When recording the fluoro image, 30 frames per second is the conventional frame rate to achieve a smoothly-flowing appearance to a dynamic (moving) image. (Less than 18 frames per second introduces *flicker*.) This concept can be applied not only to the playback of movies, but also to the *original viewing of the fluoroscopic image*. In continuous mode fluoroscopy, the radiation *between* frames is essentially wasted, exposing the patient while making no real contribution to the moving image. In the pulse mode, a fluoroscope actually takes 30 "frame" exposures per second and sends their images to the TV monitor screen to create the appearance of fluid motion in the dynamic image (Fig. 44-4). Unfortunately, the mA must be increased from *2 to 20 mA* in order to obtain adequate signal for each individual frame. But, the continuous radiation that would have been exposing the patient between frames is saved.

The exposure time for each pulse, called the *pulse width*, can be adjusted on these units by choosing between 6 or 3 milliseconds. By selecting a 3 msec pulse width and keeping the mA at 20, *patient exposure can be reduced to one-half.* Newer fluoroscopy units

Figure 44-4

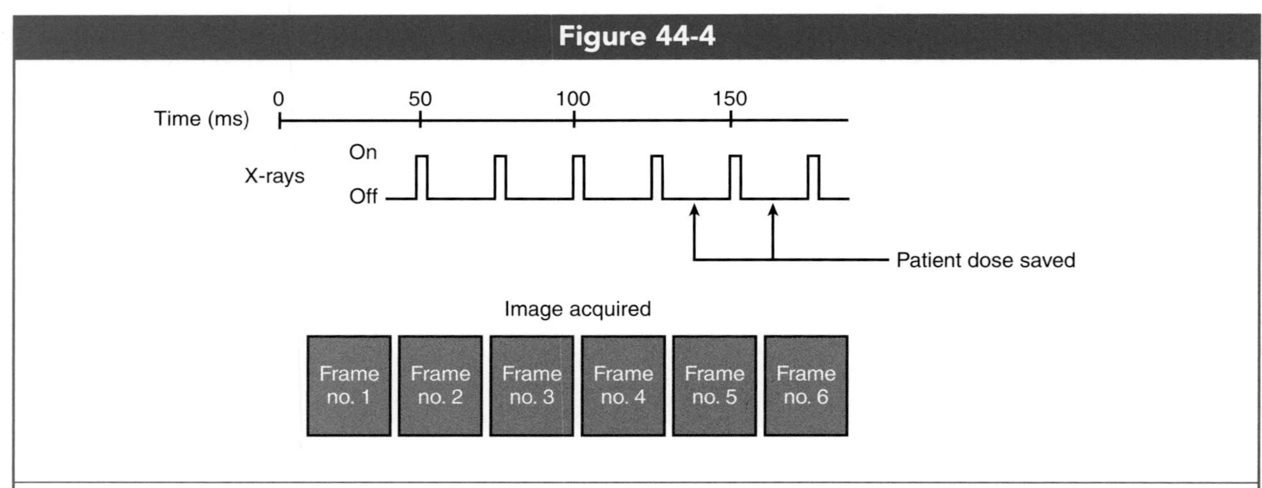

Pulsing of the fluoroscopic x-ray beam spares the patient from continuous radiation between frames, but requires increased mA. Using *shorter* pulse times further reduces patient exposure.

combine the pulsed mode with high mA and *spectral filters* made of copper or a copper/aluminum alloy to achieve increased image quality while imposing some limitation on patient exposure.

Radiology managers should work closely with medical physicists to monitor and control the actual radiation levels of fluoroscopic and C-arm equipment. Correlation to real practice is essential. For example, the physicist may measure the output of a C-arm fluoroscope in surgery as 3 or 4 R/min at a standardized 30 inch distance. But, to allow for a sterile field, the surgeon may move the image intensifier much higher, bringing the x-ray tube (on the opposite side) within 12 inches of the patient which would then result in an actual exposure rate to the patient as high as 18 R/min. Severe radiation burns *have* occurred with modern C-arm units, and it is critical that managers of medical imaging departments take ownership in this issue.

By way of illustration, let us take a typical exposure rate for a conventional fluoroscopic equipment and do a simple calculation. These units can emit from 2 to 10 R/min—C-Arm units generally deliver still higher doses. If we make a conservative estimate of an exposure rate of 4 R/min to the patient, and assume only 10 minutes of C-Arm beam-on time for a particular procedure, we see that *40 R* could be delivered to a localized portion of the patient's body. A whole body exposure of 25 R is sufficient to cause a measurable immediate drop in the blood cell count.

Doses as high as 80–100 rad are becoming commonplace among C-arm procedures. A radiation burn indicates that a dose of more than 300 rad has been received by the patient. Although no legal restrictions are imposed for patient exposure, it is strongly recommended that surgeons adopt a target maximum exposure to the patient between 20 and 40 R even for urgent situations. By practicing intermittent fluoroscopy, using state-of-the-art equipment, and working with the medical physicist, this should be an achievable goal.

We can summarize the factors that contribute to fluoroscopic exposure rates as:

1. The "mode" used at the image intensifier input phosphor, 5", 7", 9", etc. The smaller the diameter of the active input phosphor used, the more x-ray intensity is required to maintain the brightness of the output image. The automatic brightness control system automatically boosts the fluoroscopic technique factors, which increases patient exposure.
2. The mA selected.
3. Operation in continuous mA mode versus *pulsed mA* mode.
4. The actual source-to-skin distance (SSD) used. Remember that by simply rolling the patient into a lateral position, forcing the image intensifier to be placed higher, the SSD can be reduced such that exposure spikes.
5. The cumulative time of fluroscopy.
6. The number of digital (spot) exposures taken.

Current Issues

The *Safe Medical Devices Act of 1990* requires reporting to the U.S. Food and Drug Administration (FDA) of serious radiation injury or death. An FDA task force has identified the following procedures as *high-risk procedures for excessive radiation exposure*:

1. Percutaneous transhepatic cholangiography (PTCA)
2. Radio frequency ablation (RFAB)
3. Stent placement (STEN)

These procedures are described as presenting the "possibility of exceeding 100 minutes fluoro time at 2–5 R/min." One case study showed an average of 51 minutes fluoroscopic beam-on time for PTCA procedures, versus 7.1 minutes on average for angiograms conducted by radiologists. At 4 R/min, 51 minutes would deliver *204 R* to the patient, and 100 minutes would deliver *400 R*. The FDA task force concluded, "In cardiovascular interventional radiology, angiography and pain management procedures, new techniques requiring longer exposure times and higher dose rates to image smaller objects (emboli, etc.) are now resulting again in severe burns and lesions."

To help address this problem the FDA task force recommended the following procedures:

1. Recording in patient's charts the body part, field size and technique used, and an estimate of the dose received by the patient.

2. Obtaining written informed consent from the patient regarding possible radiation effects of each procedure.
3. A customized protocol delineating when the use of "high-level" fluoroscopy will be allowed.
4. Adoption of a maximum allowable cumulative fluoroscopy time.
5. Credentialing of *all* fluoroscopists (including MD's).

Another area of recent concern has been radiation overexposure to *adolescent* and *pediatric patients* from CT scans. Two primary causes have been identified by the FDA; the first is the *overuse* of CT scans, particularly by emergency room physicians. Physicians should strive to "eliminate inappropriate referrals for CT," since conventional radiography can often provide sufficient diagnostic information with less ionizing radiation delivered to the patient.

The second cause relates to the technique for CT scans employed by radiographers. It appears that *adult* technique levels, preprogrammed into the equipment, have been routinely used for children and other smaller patients. The FDA recommends "adjusting the CT scanner parameters appropriately for each individual's weight and size, and for the anatomic region being scanned," by reducing tube current (mA). Also, the development of *technique charts* based on patient weight or part diameter was advocated. It is estimated that a *33 percent reduction* in dose could be achieved *without loss of diagnostic information*, by increasing the table increment for axial scanning, or the *pitch* for helical scanning. (In effect, fewer "slices" are taken to reconstruct the image.) For contrast procedures, the number of multiple scans taken before, during, and after the injection of contrast material can often be reduced.

Radiographers must engage these types of issues and take a more active role in helping prevent unnecessary excessive exposure to the public. They may not always realize that they are effectively *the expert for radiation protection* present in a surgery suite or radiographic room. Radiographers should be more assertive in sharing this expertise, and realize that physicians are often *not* as knowledgeable in the area of radiation biology and protection as the radiographer might be.

PROTECTING PERSONNEL

Personnel Monitoring

The wearing of personal radiation monitoring devices is required if it is likely that any individual will receive more than one-quarter of the occupational DEL at any time during employment. (This limit would be 1.25 R in a year, about 100 mR in one month, or about 25 mR in any week.)

Federal regulations require that monitors be worn on the portion of the body likely to receive the most radiation exposure. For general radiographers, the conventional practice is to wear the monitor at the collar level and *outside* any lead apron or thyroid shield. The assumption is that lead aprons are being worn over the torso during fluoroscopic procedures, so that the highest dose area remaining will be the neck and head. Furthermore, this portion of the body includes the lenses of the eyes and the thyroid gland, both of which are of concern for radiation exposure. A reading from the collar area is taken as a good estimation of eye lens exposure. Because of the effectiveness of lead aprons, it is estimated that the head and neck receive 10–20 times the torso dose when aprons are used over the torso.

Doctors have been known to have their assistants wear radiation monitors *under* the lead apron, in order to minimize readings. This may be illegal. Rather, the philosophy of radiation physicists, generally adopted by regulatory agencies, is that lead aprons are so effective that there is little point in monitoring radiation exposure under an apron when sensitive organs in other parts of the body are being irradiated.

Thermoluminescent dosimeters (TLDs), or film badges provided by professional radiation monitoring service companies should be used for regular personal monitoring. Pocket dosimeters may also be used for short-term checks. These devices and their proper use were described in Chapter 41.

Personnel radiation *monitoring reports* must be made available by employers to all personnel at the end of each monitoring period. The maximum allowable reporting period is one quarter or 3 months. When a radiographer changes jobs, he or she should receive from the previous employer a report of cumulative total dose for the duration of service, and if known, the cumulative lifetime dose to that point in time. A copy of this report should be passed along to the new employer. However, *it is the ultimately the radiographer's responsibility to ensure this happens.* The *minimum* information provided on radiation monitoring reports must include the following:

1. Proper, full identification
2. The current period dose
3. The cumulative quarterly dose
4. The cumulative annual dose
5. The cumulative total exposure for the duration of service
6. The unused portion of the cumulative lifetime DEL

An example of a radiation report is given in Figure 44-5. It will be noted that in place of numerical figures, sometimes an "M" or "N" is reported. These respectively stand for *minimal dose* or *negligible dose*, and indicate that the exposure was less than the statistical error for the particular device (less than 10 mrad for a film badge, less than 5 mrad for a TLD). The categories of "penetrating or deep" dose versus "shallow" dose are provided primarily for nuclear industry workers to discriminate between alpha, beta and gamma exposure, and are not of much consequence for medical radiographers who work only with x-rays.

The Cardinal Principles: Time, Distance and Shielding

The greatest source of occupational radiation exposure to the radiographer is *scatter radiation from the patient.* As a rule-of-thumb, scatter exposure at 1 meter from

Figure 44-5

Radiation Dosimetry Report

RADIOLOGY DEPARTMENT

LANDAUER ®

Landauer, Inc. 2 Science Road Glenwood, IL 60425-1586
Telephone: (708)755-7000 Facsimile: (708)755-7016
Cust Serv: (800)323-8830 Cust Serv Tech: (800)438-3241
www.landauerinc.com

Electronic Representation

luxel

Account	Series	Analytical Work Order	Report Date	Dosimeter Recieved	Report Time In Work Days
054253	AN	1003210022	02/08/2010	02/01/2010	5

Part # / ID Number	Name / Birth Date	Sex	Dosimeter	Use	Radiation Quality	Dose Equivalent (MREM) For Periods Shown Below			Quarterly Accumulated Dose Equivalent (MREM)			Year to Date Dose Equivalent (MREM)			Lifetime Dose Equivalent (MREM)			Records For Year	Inception Date
						Deep DDE	Eye LDE	Shallow SDE	Deep DDE	Eye LDE	Shallow SDE	Deep DDE	Eye LDE	Shallow SDE	Deep DDE	Eye LDE	Shallow SDE		
For Monitoring Period:						10/15/2009 - 01/14/2010			Quarter 4			2009							
000AN			Pa	CNTRL		M	M	M										6	/
00227			Pa	WHBODY		M	M	M	M	M	M	M	M	M	1602	1611	1610	4	01/1979
00327			Pa	WHBODY		M	M	M	M	M	M	M	M	1	1664	1665	1674	4	03/1982
00418			Pa	WHBODY		M	M	M	M	M	M	M	M	M	91	91	90	4	03/1985
00420			Pa	WHBODY		M	M	M	M	M	M	M	M	M	699	701	728	4	03/1985
00438			Pa	WHBODY		M	M	M	M	M	M	29	31	32	1001	1064	1069	5	06/1985
00616			Pa	WHBODY	P	M	M	2	M	M	2	8	18	22	1635	1703	1786	4	04/1991
00621			Pa	WHBODY		M	M	M	M	M	M	M	M	M	514	514	523	1	06/1991
00722			Pa	CHEST								6	17	19	2547	2902	3015	5	04/1993
				NOTE		ABSENT													
00795			Pa	WHBODY		M	M	M	M	M	M	23	24	28	340	352	426	4	07/1995
01072			Pa	WHBODY		M	M	M	M	M	M	M	M	M	123	123	126	4	07/1999
01142			Pa	WHBODY		M	M	M	M	M	M	M	M	M	59	63	64	4	07/2000
01327			Pa	WHBODY		M	M	M	M	M	M	58	97	118	248	302	361	4	07/2003
01335			Pa	WHBODY	P	M	M	4	M	M	4	42	42	50	246	245	244	4	07/2003
01414			Pa	WHBODY		M	M	M	M	M	M	M	M	2	2	2	5	4	07/2005
01415			Pa	CHEST								M	M	M	33	33	33	4	07/2005
				NOTE		ABSENT													
01440			Pa	WHBODY	P	M	M	1	M	M	1	31	31	29	547	548	525	4	02/2006
01509			Pa	WHBODY		M	M	M	M	M	M	2	2	3	10	10	14	4	07/2007
01510			Pa	WHBODY		M	M	M	M	M	M	M	M	1	44	44	45	4	07/2007
01511			Pa	WHBODY		M	M	M	M	M	M	M	M	1	76	76	73	4	07/2007
01512			Pa	WHBODY	P	M	5	7	M	5	7	74	89	89	166	185	183	4	07/2007
01522			Pa	WHBODY		M	M	M	M	M	M	M	M	M	1	1	1	4	10/2007
01533			Pa	WHBODY		M	M	M	M	M	M	111	111	113	155	162	165	6	10/2007

M: MINIMAL REPORTING SERVICE OF 1 MREM

Quality Control Release: DRB 1 S PR 9438 RPT130 N1 C 03222

Accredited by the National Institute of Standards and Technology through NVLAP for the scope of accreditation under NVLAP Lab Code 100518-0**.

A typical radiation monitoring report (with names removed). (Courtesy, Mark Cranford, R. T.)

the patient, (perhaps an *average* location for the radiographer to be standing during a fluoroscopic procedure), is roughly estimated to be 1/1000th of the patient's in-beam skin exposure. The recommendations of the National Committee on Radiation Protection and Measurement (NCRP) from Report #116 state that whenever possible, "All personnel *should* stand at least 2 meters from the x-ray tube and patient."

The "cardinal principles" of radiation protection traditionally taught to radiography students are to minimize exposure *time*, maximize one's *distance* from the source of radiation, and use optimal *shielding* to protect oneself. Scatter radiation follows all the same laws as the primary x-ray beam in regard to all three of these relationships; scatter exposure to the radiographer is *proportional* to the time of exposure. It also follows the *inverse square law* just as the primary beam does, even though scatter is random in direction. Every doubling of the radiographer's distance from the patient and x-ray table reduces scatter exposure to one-quarter. Therefore, if the exposure one meter away from the table is 1/1000th of the patient's exposure, then by taking "one giant step back" to about 2 meters, the radiographer can further reduce this exposure to about 1/4000th of the patient's dose. (These may sound like extremely low ratios, but remember that the patient is not working around radiation 40 hours per week for 50 weeks a year.)

For perspective, let us take this reasoning one step further by adding shielding. Lead aprons average about 85–90 percent effectiveness for mid-range kVp levels, but are much more effective for scattered radiation since it has lower energies than the original beam. A radiographer standing one meter from the patient, with a lead apron on is not likely to receive more than 1/10,000th of the patient's exposure to the gonads under the apron. By standing behind the radiologist, the radiographer obtains the benefit of *two* lead aprons plus the filtering effect of the radiologist's body. By also backing up to 2 meters, the exposure for the radiographer (under the apron) is reduced to less than *1/500,000th* of the patient's in-beam exposure.

Isoexposure curves (Fig. 44-7), represent positions in the room of equal exposure from the radiation source, at about waist height. These work similar to the lines on a topographical map, which indicate locations of equal *altitude*. When the lines get closer together, it indicates steeper terrain or a quicker *rate* of change in the altitude (Fig. 44-6). Where isoexposure

Figure 44-6

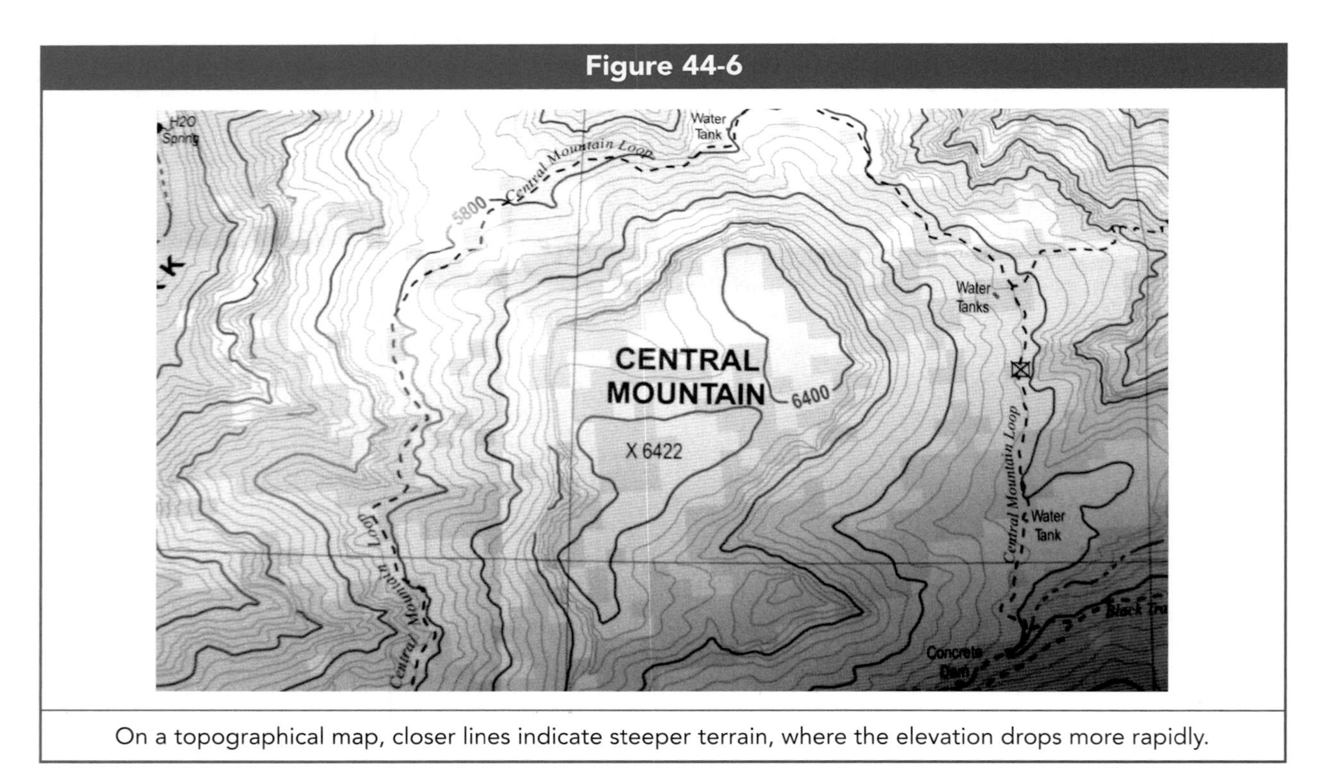

On a topographical map, closer lines indicate steeper terrain, where the elevation drops more rapidly.

curves are more concentrated in this way, the radiation exposure is dropping off more quickly from the source. The isoexposure curves for a typical x-ray room shown in Figure 44-7 indicate that the location of least exposure to personnel is to the side of the table; *generally avoid standing at the head or foot of the x-ray table unless it is necessary for the procedure or for patient care.*

Radiation exposure to personnel is always least at right angles to the CR of the x-ray beam. When the beam is vertical, as is the case in a radiographic room and with most C-arm projections, personnel are already at right angles to the CR. However, as described in Chapter 38, at diagnostic kV levels, backscatter radiation is more intense that forward scatter. By *always placing the image intensifier of a C-arm unit* above *the patient,* with the x-ray tube underneath the surgery table, less scatter radiation strikes the head and neck areas of personnel in surgery.

Figure 44-7

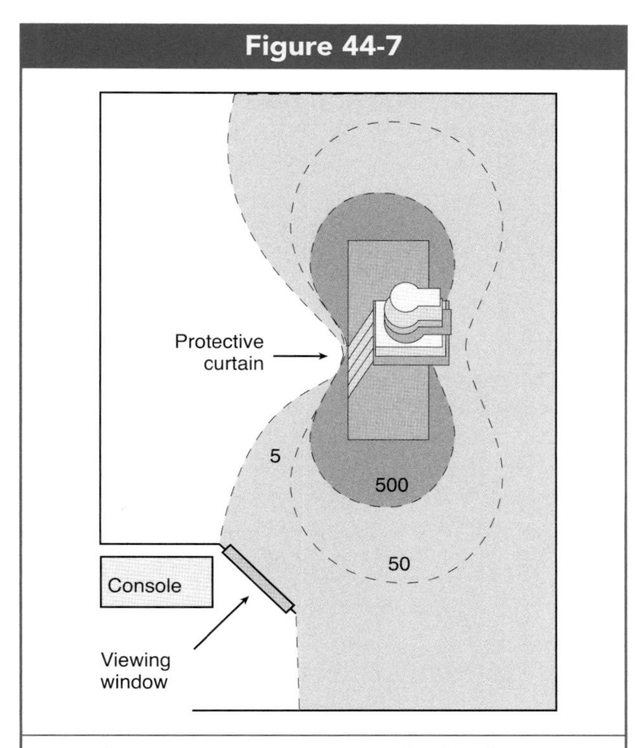

Typical isoexposure curves around a fixed x-ray unit. Where the lines are closer together, the radiation level drops off more quickly. The area of least exposure is to the side of the table.

Personnel Shielding Requirements

NCRP Report #102 states that,

> Only persons whose presence is necessary *shall* be in the diagnostic...x-ray room during exposure. All such persons *shall* be protected with aprons...Regulations also state that during lengthy fluoroscopic procedures...the leaded apron *shall* always be worn... Operators of mobile equipment should wear lead aprons. Apron and gloves should be worn when holding a patient or when closer than 2 meters (6 ft) from the beam.

Lead aprons must be worn by anyone whose occupational dose may exceed *5 mR/hr*. The thickness of lead must be 0.5 mm if the x-ray machine is routinely operated above 100 kVp, which generally applies to all "R&F" (radiographic and fluoroscopic) rooms, or "anywhere that sterile fields must be maintained." For nonfluoroscopic and low-kVp units, only 0.25 mm of lead is required, but 0.5 mm is still recommended. As we have mentioned, thyroid shields have become popular for personnel working in angiography or cardiac cath labs. Once each year, aprons should be placed under a fluoroscope and examined for any cracks that may have developed in the lead sheet.

For leaded gloves, only 0.25 mm thickness of lead is required. (The hands are among the least radiosensitive portions of the body, and flexibility is critical for dexterity.) Lead-impregnated gloves are approximately 20–30 percent effective during high-kVp fluoroscopy. One problem is that their use can lead to a false sense of security because on the fluoro screen the operator sees the effect of *two* thicknesses whereas the actual protection is only one thickness. Also, the presence of lead gloves in the fluoroscopic field can cause the automatic brightness control to increase the dose rate to compensate, adversely affecting the patient. Lead gloves must be used with careful deliberation.

The effectiveness of lead shields depends upon the thickness of actual lead contained in them, as well as the *kVp* set at the console, and can vary from as little as 66 percent up to 99 percent. The most common lead thicknesses used for aprons is 0.25 mm and 0.5 mm. For most radiographers who routinely participate in fluoroscopic procedures, 0.5 mm of lead is strongly recommended. This is sometimes obtained by a "wraparound" apron which actually has only

Figure 44-8

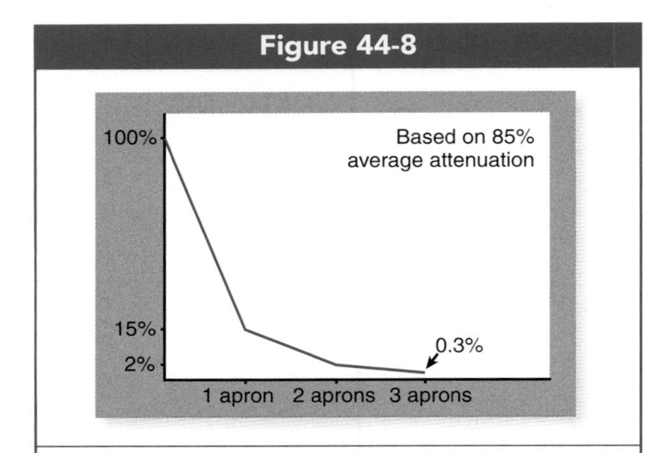

Graph plotting percentage of remaining x-ray exposure behind lead aprons that are sequentially stacked. This is an inverse exponential relationship, each apron reducing the exposure to about 15 percent of what penetrated through the previous apron.

0.25 mm of lead in it but overlaps in front of the body such that this portion is doubled. For a 75-kVp x-ray beam projected directly at an apron containing 0.5 mm of lead, about 88 percent of the radiation will be blocked. Overall, we can state that the types of lead aprons and gloves that are typically found in radiology departments have an *average effectiveness of about 85 percent.*

We have discussed half-value layer (HVL) as a measurement of x-ray beam penetration indicating the adequacy of filtration at a given kVp. It uses layers of absorbing material to do this. If the penetration of the x-ray beam is *fixed* by using a known amount of filtration and accurate kVp, then *HVL becomes a measure of the effectiveness of shielding* (x-ray absorbing material). We can describe filtering or shielding materials, such as aluminum, copper, tin, lead or concrete, as each having a characteristic HVL at a set kVp and initial filtration.

When describing shielding materials, *the lower the HVL of a material, the more effective it is at blocking x-rays*—it takes less material to "stop" one-half of the x-ray beam from penetrating through. For example, lead has a lower HVL than concrete; it takes *less* lead than concrete to block one-half of an x-ray beam. The lead is more effective.

Materials used for shielding, such as lead, copper or steel, are much more effective than materials used for filtering x-rays, and it is often less unwieldy to use the *tenth-value layer (TVL)*, rather than the HVL, as a unit. The TVL is defined as that amount of shielding material required to reduce the intensity of the x-ray beam to one-tenth of the original output from the x-ray tube. The lower the TVL, the more effective the shielding material.

Figure 44-8 graphs the *inversely exponential* relationship of shielding to exposure by way of *stacking* lead aprons one atop the other. We see that the first apron attenuates 85 percent of the radiation, leaving 15 percent which penetrated through it. If a *second* lead apron is placed over the first, the exposure does not go down to zero; rather, the second apron absorbs about 85 percent of the 15 percent remaining from the first apron. That is, it *leaves an exposure equal to 15 percent of 15 percent—this is 2.25 percent.* A third apron will leave 0.34 percent, and so on. Hypothetically, if the x-ray exposure continues indefinitely, the amount of radiation penetrating the aprons never quite reaches zero.

To be more precise, we must take into consideration the *hardening* effect of each lead apron on the x-ray beam. Each apron acts as a *filter*, removing the lower energies in the beam. In just the same way that filtration results in an increased *average kV* for the x-ray beam, each lead apron leaves a remnant beam that has higher average energy and therefore increased *penetration* power. Therefore, a higher percentage of this remnant beam will penetrate through the next apron, such that each apron is a bit less effective. For example, the first apron might remove 85 percent of the x-rays, the second apron 75 percent, and the third apron only 65 percent, and so on.

Lead aprons must be made available with every mobile unit and should be consistently employed by radiographers during mobile procedures. When performing a mobile procedure, standard material in walls do *not* provide adequate protection from radiation to *substitute* for a lead apron, nor do glass walls. However, it is a good idea to stand around such a barrier *with* a lead apron on, as it increases the probability of a second scattering event for the radiation, which reduces its energy.

The radiographer should always maximize both *distance* and *shielding* in combination for ALARA dose protection. The exposure cords on mobile machines are required to be designed to extend at least 6 feet from the machine for this very purpose.

Any person within 6 feet of the x-ray tube who cannot leave the area, such as a nurse in a newborn intensive care unit, *should be provided with a lead apron*. It is good practice to have at least two aprons attached to each mobile unit.

Equipment Shielding Requirements

Related guidelines for the fluoroscopic equipment itself include the following:

1. The *fluoro tower* is required to have the equivalent shielding of 2 mm of lead. (It is also considered a *primary* radiation barrier, discussed later.) This requirement refers to the entire bottom section of the tower which intercepts the raw x-ray beam, with the input phosphor of the image intensifier in the middle. Any radiation leakage of the primary beam beyond this barrier will be directed upward through the *ceiling* and may actually expose personnel working on the next floor of the building above.
2. Fluoroscopy must not be capable of operating in "park" position, where the image intensifier is placed behind the x-ray table. In this position, the *shielding* provided by the fluoro tower is completely removed, such that if the undertable x-ray tube were energized, the x-ray beam would be directed upward toward the ceiling, exposing medical personnel on the next floor.
3. The *bucky slot cover* (Fig. 44-9), must have at least 0.25 mm of lead, and must move into place whenever the bucky tray is moved to the foot of the table in preparation for a fluoroscopic procedure. This is critical for protection of the fluoroscopist's gonadal area, which is near the level of the table and patient.
4. The *fluoro curtain* (Fig. 44-9), must also have a minimum of 0.25 mm of lead in each slat. These slats overlap such that the benefit of 0.5

Figure 44-9

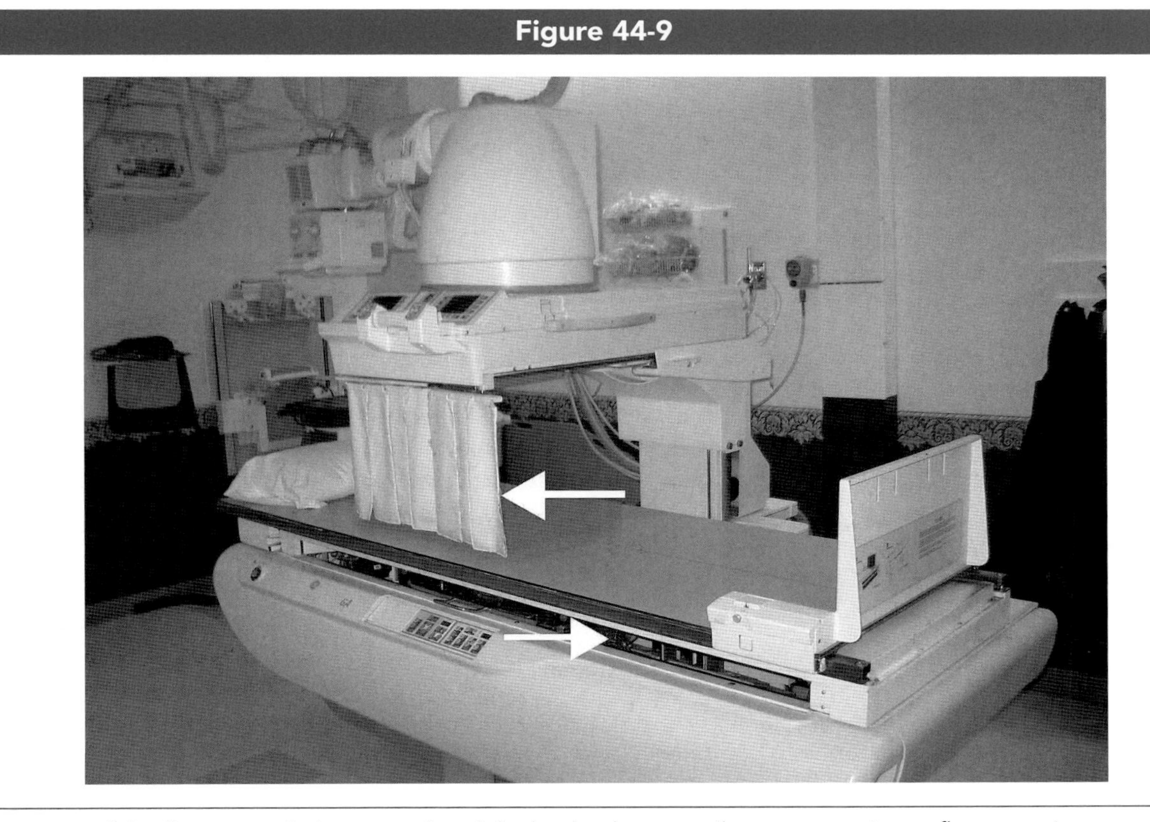

Location of the fluoro curtain (top arrow) and the bucky slot cover (bottom arrow) on a fluoroscopic x-ray unit.

Figure 44-10

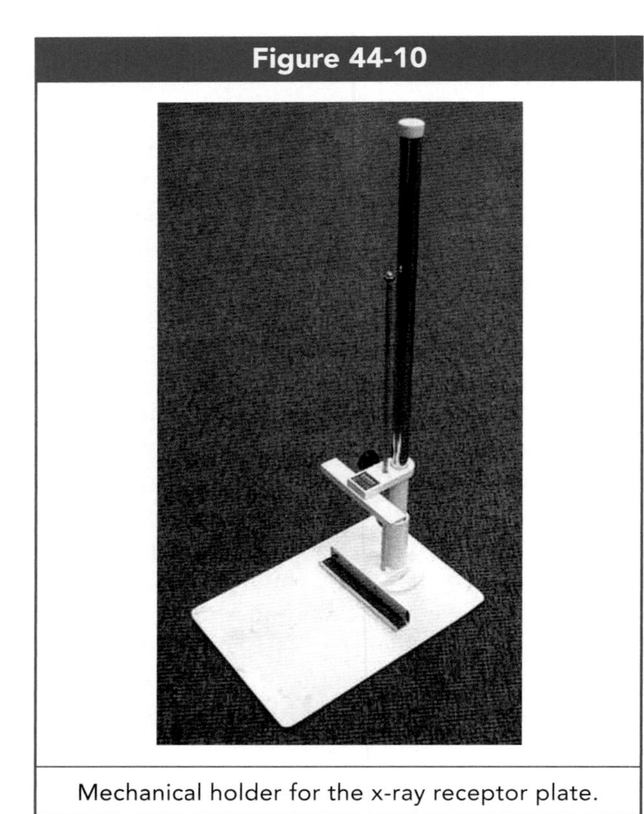

Mechanical holder for the x-ray receptor plate.

mm of protection is afforded to the fluoroscopist's mid-body.

Personnel Protection Policies

In the interest of the "as low as reasonably achievable" (ALARA) concept, the holding of patients or imaging plates during an exposure should always be minimized. When it is absolutely necessary to acquiring a successful radiographic view that the patient be held in place, two philosophical principles provide guidance in developing actual policies. Both concepts are based on the scientific fact that *for diagnostic levels of radiation exposure, we are primarily concerned about* stochastic *effects to the general population (such as mutation rates and cancer rates), rather than deterministic effects to the individual which require much higher doses.*

The first guiding principle, then, is that radiation exposure should be *distributed* among the population as much as possible. When possible, we look to those unlikely to have accumulated much radiation exposure to do the holding. The second principle is that less harm is done to the *population* if exposure is distributed to those people *known to be nonreproductive*. With these two concepts in mind, the following list of preferences for patient immobilization can be prioritized:

1st choice: Use a mechanical *device* to hold the plate or restrain the patient. Figure 44-10 shows an example of a cassette plate holder.
2nd choice: Have an *adult* relative or friend of the patient assist.
3rd choice: Have a *nonradiographer* health worker, such as a nurse, assist.
4th choice: Have a nonreproductive radiographer assist.

Using a radiographer of reproductive age and gender should always be the last resort, because this is a person who already has acquired occupational exposure and can contribute to the mutation rate among births in the population.

Whenever radiographers must be present in the radiographic room with the patient, and during all mobile procedures, remember that the least amount of scatter radiation is always at a right angle (90°) to the central ray.

Whenever possible, the radiographer should stand in a location, such as the control booth, where *radiation must scatter twice before reaching the radiographer* (Fig. 44-11). As a rule of thumb, *each scattering event reduces the intensity of radiation exposure to approximately 1/1000th the previous intensity*. The first scattering event is when the primary beam strikes the patient or tabletop. When the radiographer stands behind the protective wall of the control booth, scattered radiation passing the corner of the booth must strike the wall behind, and scatter passing over the top of the protective wall must strike the ceiling, scattering one more time before reaching the radiographer. This results in an exposure of ($1/1000 \times 1/1000 =$) one millionth the patient's in-beam exposure, which is further reduced over the *distance* to the control booth by the inverse square law. Mobile barrier shields on casters can be used in the same way.

Policies for Technologist Pregnancy

Each imaging department must develop its own policies for pregnant radiographers consistent with state and federal law. Involuntary leave for pregnant

Figure 44-11

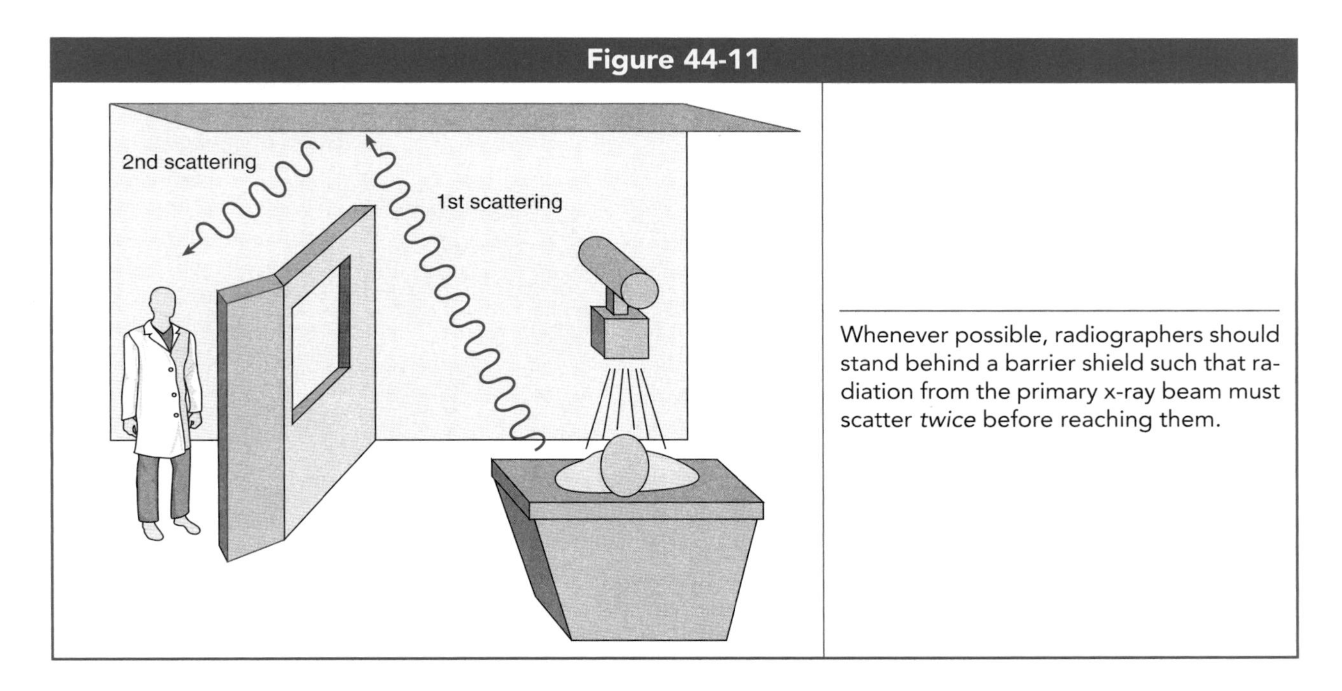

Whenever possible, radiographers should stand behind a barrier shield such that radiation from the primary x-ray beam must scatter *twice* before reaching them.

workers is discriminatory and unnecessary. It is also generally illegal to require the reporting of an early pregnancy to an employer. In all but the smallest departments, common sense measures can be taken to limit fetal dose to almost negligible amounts while at the same time allowing the pregnant worker to pull a full workload. The following policies are recommended:

1. Recommend, but not require the reporting of pregnancies at the earliest possible time.
2. For reported pregnancies, require a *second* personal exposure monitor to be worn at the waist level and under protective aprons, to obtain a more direct and therefore more accurate indication of fetal dose. (Until the last trimester, fetal dose generally may be estimated at 1/3 of this monitor reading.)
3. Limit the radiographer from performing *high-dose* procedures only. Each department must delineate what procedures are included in this definition, but among those that might be considered are surgical procedures, portable (mobile) procedures, and fluoroscopy (not overhead projections that are part of a fluoroscopic procedure).
4. Document that the radiographer has received instruction on DEL and ALARA guidelines.

Guidelines for Equipment

1. For a fixed radiographic unit, the exposure cord must be too short to allow the operator out from behind the protection of the leaded control booth walls.
2. The exposure cord for a mobile unit must be extendable to at least 6 feet. The operator should stand 6 feet (2 meters) away from the patient whenever possible.
3. *Leakage radiation* is defined as radiation emitted through the x-ray tube housing in any direction other than the *port*. Regulations require that leakage from the tube housing not exceed 100 mR/hr as measured at one meter from the x-ray tube with an ion chamber.
4. Generally, x-ray tube housings are required to have a minimum of 2 mm of lead throughout the casing.
5. All x-ray units are required to emit an audible or visible signal to indicate that exposure is taking place any time the beam is energized.
6. During C-arm fluoroscopy procedures, the fluoro x-ray tube should be positioned *under* the patient with the image intensifier over the patient (Fig. 44-12). It has been demonstrated that scatter radiation exposure to personnel is significantly reduced by placing the C-arm in

Figure 44-12

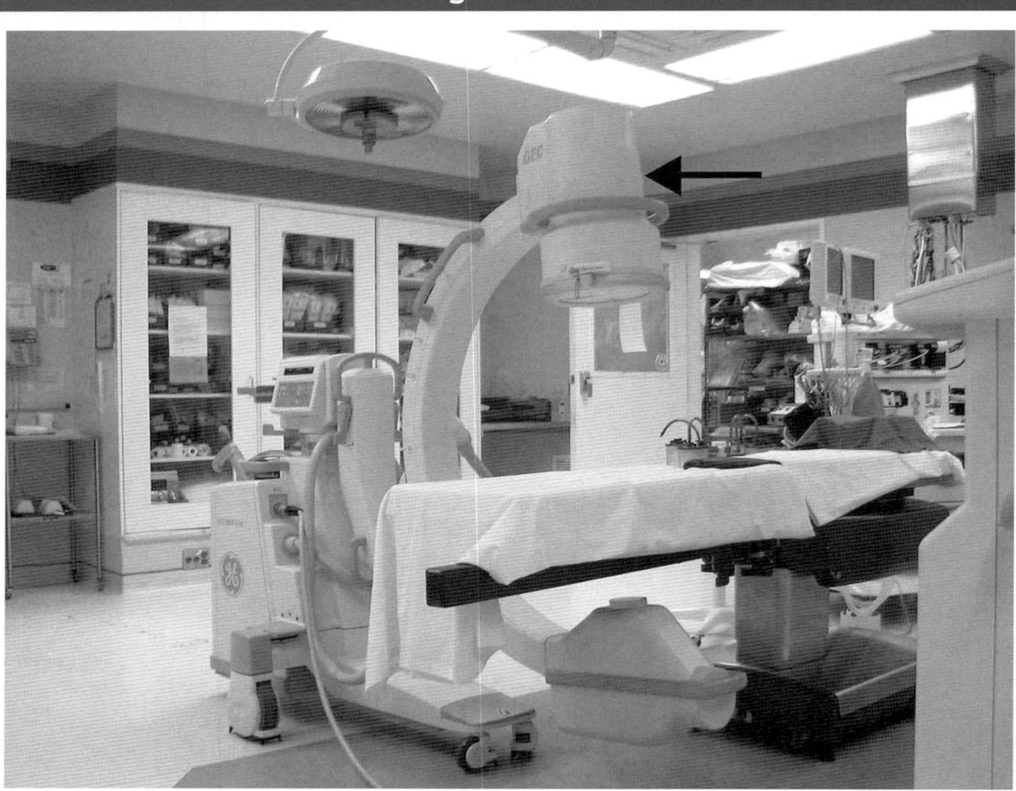

A modern C-arm fluoroscope, with the image intensifier positioned over the table and the x-ray tube properly placed underneath. (Courtesy, Trevor Morris, R.T.)

this position. One actual radiation survey was conducted using tissue-equivalent "phantoms" to produce realistic levels of scatter, concluded that exposure to the eyes and thyroid gland for personnel was reduced from 60 mR to about 12 mR by placing the x-ray tube *under* the table. This is because at *diagnostic levels of kV*, backscatter is more intense than forward scatter. In the same study, the *gonadal* exposure to personnel was about the same either way the tube was positioned because at this level *side scatter* is producing the exposure.

STRUCTURAL BARRIER SHIELDING

Structural barriers (floors, ceilings, walls, and windows) are divided into two types, primary and secondary. A *primary barrier* is defined as one which the *primary x-ray beam* strikes in the course of routine practice. A *secondary barrier* is normally only struck by *secondary (or scattered) radiation*. Regulations for practice forbid ever pointing the x-ray beam toward the control booth, so the control booth is always assumed to be a *secondary barrier* whose purpose is protection of personnel from scatter radiation.

A primary barrier must always have a shielding effectiveness equivalent to 1.5 mm (1/16 inch) of pure lead. A secondary barrier must have an equivalent of 0.8 mm (1/32 inch) of pure lead. There are various ways of achieving these requirements, but for most radiography rooms actual sheets of pure lead are laminated over typical panels of "sheet rock" (gypsum board) in the walls. Since the objective is only to ensure that radiation must scatter twice before reaching the operator, it is not necessary to have the lead extend all the way to the ceiling—rather, to reduce the weight of the panels, the lead is only required to extend *7 feet* (2.2 m) up the wall (Fig. 44-13), leaving a foot or more at the uppermost margin which is only sheet rock.

At the corners of the room and wherever there is a seam in the lead, additional strips of lead must be used such that *all joints overlap* at least 1 cm, or double the lead thickness, whichever is greater. During radiation surveys, these corners and seams are checked for leakage radiation penetrating outside the room.

Alternative shielding materials can be used for walls and protective windows. Specially leaded glass panels can have more than sufficient lead content to serve as secondary barriers and still provide visual transparency. They typically have a yellow tint to them. The radiation absorption efficiency of this glass is such that 1/4 inch (6.3 mm) is equivalent to about 2 mm of pure sheet lead. A good rule of thumb to use in determining appropriate thicknesses for windows is to use *4 times* the recommended sheet lead thickness. For example, the control booth is a secondary barrier requiring 1/32 inch (0.8 mm) lead equivalency. Four times this amount is 1/8 inch (3.2 mm) which should suffice for the control booth window. *Regular* glass typically has some lead and other minerals in it as well, and is roughly equivalent in its radiation absorption to concrete, which will be discussed next.

It takes a thickness of 2 inches of pure concrete to serve as a secondary barrier such as the control booth. Four inches of concrete is approximately equivalent in radiation absorption to 1/16 inch of pure lead. However, concrete can have flaws—air pockets within it compromising its consistency. For diagnostic radiography, the National Council on Radiation Protection and Measurements (NCRP) generally discourages the use of only concrete except where the "needed protection is minimal."

Masonry in general has about the same effectiveness as concrete. Two and a half inches of gypsum board (sheetrock) can serve as an adequate *secondary* barrier in many instances. To obtain this thickness, *two* layers of 5/8" board could be fixed on *each* side of a wall.

These foregoing lead thickness requirements should be considered only as *minimum* guidelines.

Figure 44-13

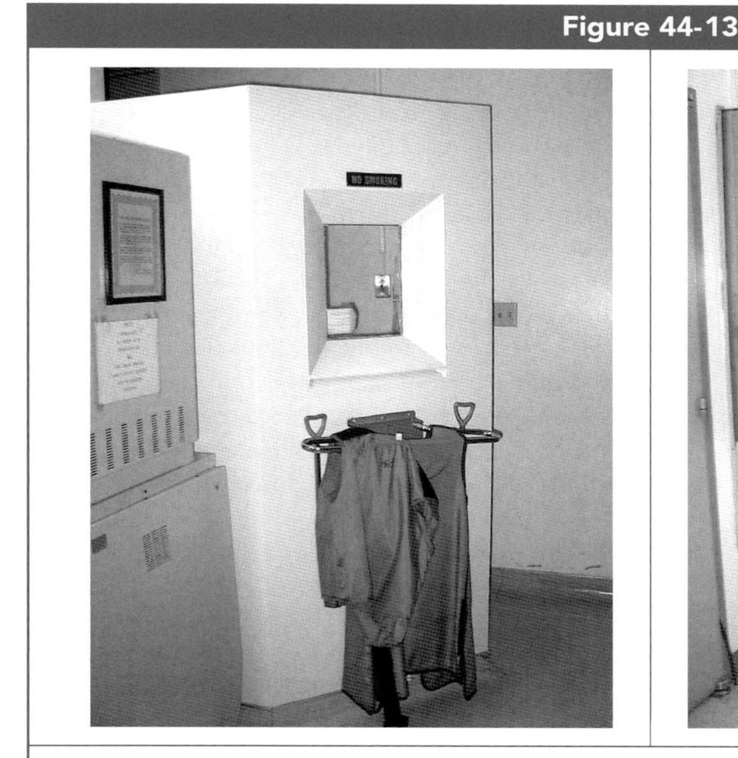

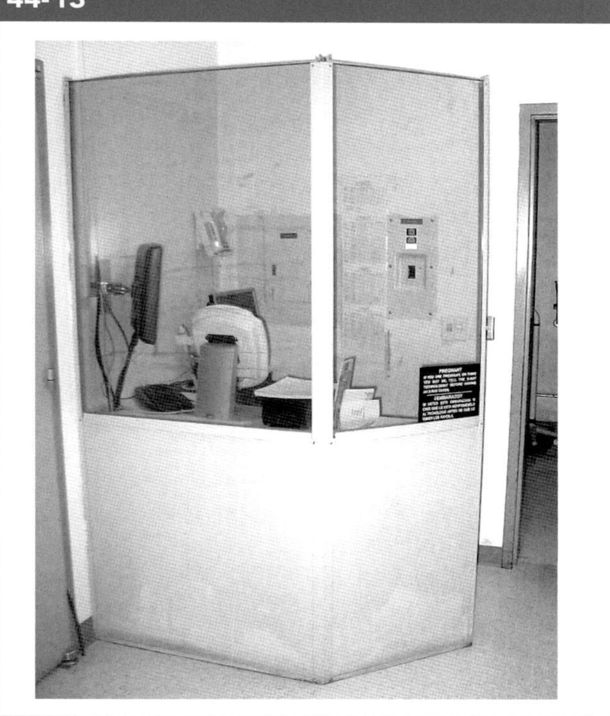

Lead for the protective barrier of a control booth, shown here, and all walls need only extend up to 7 feet.

There are several factors which determine the exact amount of lead required for a barrier. For example, more lead may be required if the *workload* of the room is very high or if the *occupancy* of adjoining rooms is high, or if the room is unusually small such that the *distance* from the x-ray unit to outer walls is shorter than normal. These and other factors are combined into the following *formula* used by radiation physicists to determine the exact amount of lead that should be included in any barrier:

$$\text{Lead required} = \text{proportional to } \frac{\text{W X U X O X E}}{\text{D}}$$

where *W* is the workload, *U* is the *use factor*, *O* is the occupancy, *E* is the x-ray energy, and *D* is the distance. Each of these factors will be defined and discussed.

Factors for Adequacy of Barriers

1. **W**: This is the *maximum workload* given in milliampere-minutes per week, mA-min./wk. It refers to the cumulative x-ray beam-on time for a 5-day week at an *averaged* mA setting for the machine, multiplied by the maximum number of exams performed and the typical beam-on time per exam.

   Simplified Example:

   Room A does 3 barium enemas and 6 C-Spine series per day.

   - For a 5 day week, this is 15 BE's and 30 C-Spines

   C-Spines = 5 overheads each × 30 = 150 overheads
   - Typical overhead exposure time = 0.1 seconds
   - Typical mA = 200
   - 150 × 200 × 0.1 = 3000 mA-seconds
   - 3000 mA-seconds/60 = 50 mA-min.

   BE's = 6 overheads, 6 spot-films, 4 minutes of fluoro each × 15 studies
   = 180 overhead/spots AND 60 minutes fluoro

   Overheads and Spot Films:
   - Typical exposure time = 0.1 seconds
   - Typical mA = 200
   - 180 × 200 X 0.1 = 3600 mA-seconds/60 = 60 mA-min.
   - Typical Fluoro = 5 mA × 60 minutes = 300 mA-min.

   Total for Room A = 50 + 60 + 300 = 410 mA-minutes per week.

2. **U**: This is the *use factor*, defined as the *ratio* of an 8-hour day shift that the primary beam is directed toward a particular wall or floor. *Standard use factors* are provided by regulatory agencies, so the calculation does not need to be made for every room. They are as follows:

   Standard Use Factors

Floor = 1	Other Walls = 1/4
Wall with a "chest board" = 1	Ceiling = <1/4

3. **O**: This is the *occupancy factor*, defined as the *ratio* of an 8-hour day shift that the public occupy adjoining rooms. Standard occupancy factors are also provided by regulatory agencies. They are:

   Standard Occupancy Factors

   Waiting Rooms, Offices = "full" = 1
   Halls, Restrooms = "partial" = 1/4
   Closets = "occasional" = 1/16

4. **D**: This is the *average* distance from the x-ray tube to each wall or barrier.

5. **E**: This is the *maximum* kVp (energy) that the unit is capable of. It is important as an indication of the penetration capability of the x-rays produced through walls and barriers.

As one can see from the formula, the amount of lead required in the barrier is *proportional* to *W*, *U*, *O* and *E*, but *inversely proportional* to *D*. We will not apply the actual formula here, which employs a look-up table once the math is done to convert the product of the factors into a thickness of lead. The objective is to simply understand the various factors that must be considered for radiation barrier protection.

Having said all this, the "bottom line" in barrier protection is that *dose equivalent limits* (*DELs*) for the public and for the radiographer must not be exceeded. These actual radiation levels for adjoining areas and rooms cannot be measured until after the radiology room is built, so the above formula and guidelines are necessary in *planning* construction. Once a room is built, actual surveys of the surrounding rooms can be performed by the radiation physicist, using TLDs for long-term cumulative exposures and ion chambers to check for any acute radiation leakage.

A practice exercise combining the aspects of both distance and shielding that apply to an adjacent room is provided in the instructor's manual.

Types of Radiation Areas

A *controlled area* is defined as being occupied primarily by radiation workers. The occupancy must be known, the working conditions must be supervised, and the radiation equipment must be inspected regularly. With these restrictions in place, somewhat looser limitations on the actual exposure accumulated may be imposed than would be used for the general public. Barriers must be sufficient to keep the occupational exposure rates in these areas below 100 mR/week.

> In controlled areas, structural barriers must be sufficient to keep occupational exposure rates below 100 mR per week.

An *uncontrolled area* is defined as a portion of the imaging department which non-radiation workers, such as secretaries and managers, waiting patients, or the general public can be expected to regularly occupy. These rooms and hallways are often immediately adjacent to x-ray rooms. The structural barriers between them must be sufficient to keep exposure rates below 10 mR/week in the uncontrolled areas. Surveys of these rooms and halls can be performed by a radiation physicist using TLDs for long-term cumulative exposures.

> In uncontrolled areas, structural barriers must be sufficient to keep exposure rates below 10 mR per week.

Posted Warnings

All radiation warning signs are required to be printed with the familiar "cloverleaf" radiation pattern (Fig. 44-14), printed against a background of yellow. The cloverleaf itself can be red, maroon, or black. If it is feasible that accumulated radiation exposure within a particular room might ever exceed *100 mR in 5 days, OR 5 mR in any one hour*, a standard warning sign reading "Caution—Radiation Area" must be posted near all entryways to the room.

If it is feasible that the accumulated exposure might ever exceed *100 mR in any one hour*, the posted sign must read, "Caution—High Radiation Area." Other signs are required in nuclear medicine, and tags for radioactive patients or corpses (Fig. 44-15). Transport vehicles carrying radioactive substances must have plainly visible signs on them indicating the hazardous nature of their content.

ADVISORY AND REGULATORY AGENCIES

Scientific agencies which regularly act in an advisory capacity for radiation protection include the National Council on Radiation Protection and Measurements (NCRP), the International Commission on Radiological Protection (ICRP), the National Academy of Sciences (NAS), and the United Nations Scientific Committee on the Effects of Atomic Radiation (UNSCEAR). In addition, the United States Food and Drug Administration (FDA), the Environmental Protection Agency (EPA), the Radiological Society of North America (RSNA), the American Medical Association (AMA), and the American Society of Radiologic Technologists (ASRT) have all provided guidance for radiation protection issues.

The NCRP and ICRP have published hundreds of reports which collectively provide an exhaustive coverage of all conceivable radiation exposure situations, both occupational and non-occupational. Medical imaging managers and quality control technologists should be familiar with at least the titles of

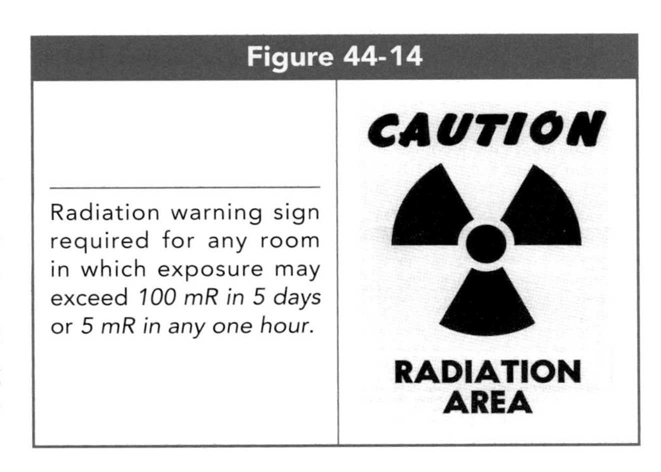

Figure 44-14

Radiation warning sign required for any room in which exposure may exceed *100 mR in 5 days* or *5 mR in any one hour*.

Figure 44-15

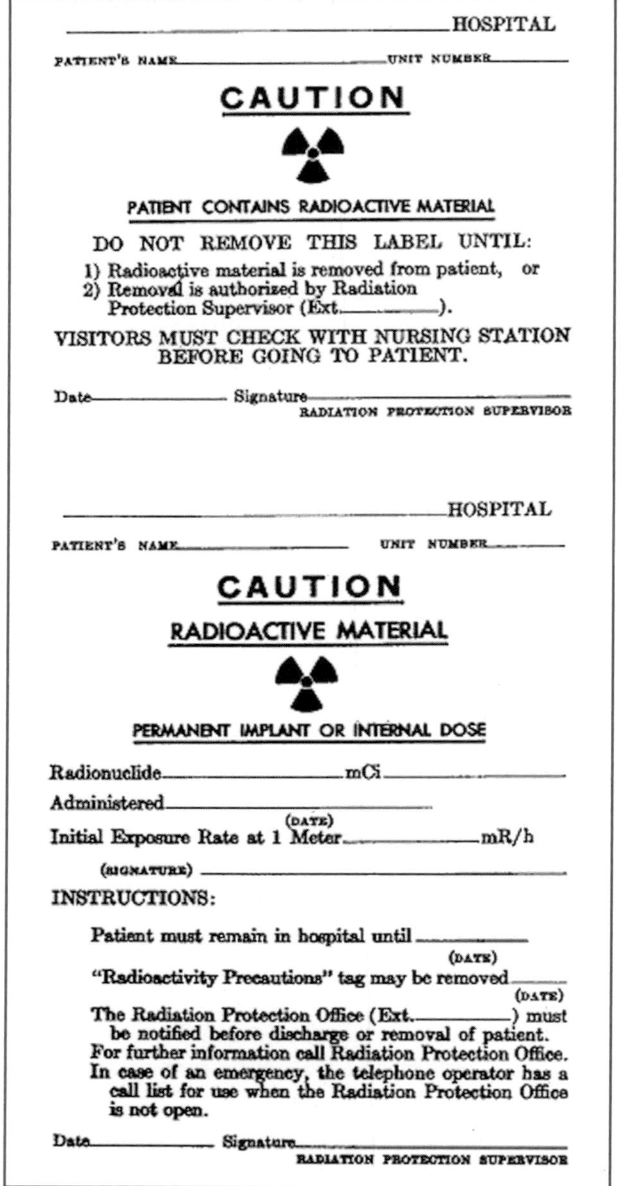

______________ HOSPITAL

PATIENT'S NAME ______________ UNIT NUMBER ______________

CAUTION

PATIENT CONTAINS RADIOACTIVE MATERIAL

DO NOT REMOVE THIS LABEL UNTIL:

1) Radioactive material is removed from patient, or
2) Removal is authorized by Radiation Protection Supervisor (Ext.________).

VISITORS MUST CHECK WITH NURSING STATION BEFORE GOING TO PATIENT.

Date ________ Signature ______________
RADIATION PROTECTION SUPERVISOR

______________ HOSPITAL

PATIENT'S NAME ______________ UNIT NUMBER ______________

CAUTION

RADIOACTIVE MATERIAL

PERMANENT IMPLANT OR INTERNAL DOSE

Radionuclide ______________ mCi ______________

Administered ______________
(DATE)

Initial Exposure Rate at 1 Meter ______________ mR/h

(SIGNATURE) ______________

INSTRUCTIONS:

Patient must remain in hospital until ________
(DATE)

"Radioactivity Precautions" tag may be removed ________
(DATE)

The Radiation Protection Office (Ext.________) must be notified before discharge or removal of patient.
For further information call Radiation Protection Office.
In case of an emergency, the telephone operator has a call list for use when the Radiation Protection Office is not open.

Date ________ Signature ______________
RADIATION PROTECTION SUPERVISOR

Radiation warning tags for patients.

these reports so they know where to find scientific guidelines for practice (Fig. 44-16).

Within the NAS, the National Research Council (NRC) has published several pertinent reports, the most prominent of which has been the *Biological Effects of Ionizing Radiation* (*BEIR*) report. Within the AMA, the American College of Radiology (ACR) has also made considerable contributions over the years.

The ASRT is currently conducting a comprehensive study of occupational exposure levels to radiographers, which will take a few decades to complete.

Regulatory agencies in the United States which have jurisdiction over radiation-related issues include the Nuclear Regulatory Commission (NRC), and the Bureau of Radiological Health (BRH). An important publication of the BRH was the *Nationwide Evaluation of X-Ray Trends* (*NEXT*). The NRC, while primarily focused on the industrial use and transportation of radioactive materials, also affects the medical use of radiation and publishes minimum guidelines for radiation protection. Federal guidelines define *agreement states* as those states that have agreed to establish their *own* radiation guidelines, which must be at least as stringent as federal guidelines. These states become exempt from oversight of the federal guidelines. *Nonagreement* states must meet the federal guidelines.

The EPA and the FDA function also as regulatory agencies, in that they can assess penalties or fines for certain infractions. Most state health departments include departments for radiological health, and most states have oversight agencies for the certification of medical radiation workers. Finally, the U.S. Occupational Safety and Health Administration (OSHA) publishes specifications relating to equipment used on the job, including radiation machines.

A FINAL WORD

The standard of practice for all personnel working around radiation is *ALARA*: "As Low As Reasonably Achievable." Legally, it is not sufficient to show that your exposure levels are within the maximum permissible limits. Rather, you must be able to show that all reasonable precautions were taken to keep your exposure as low as *achievable*. The reason for this is that the effects of radiation are cumulative, so our concern is for *effects of exposure accumulated over long periods of time* in your work.

Finally, every radiographer should have an active role in continuously *educating* not only himself or herself regarding radiation safety, but also other health care workers with whom the radiographer must collaborate. Regulations (NCRP Report #105) state that "Nurses *shall* be aware of the radiation safety policies

Figure 44-16

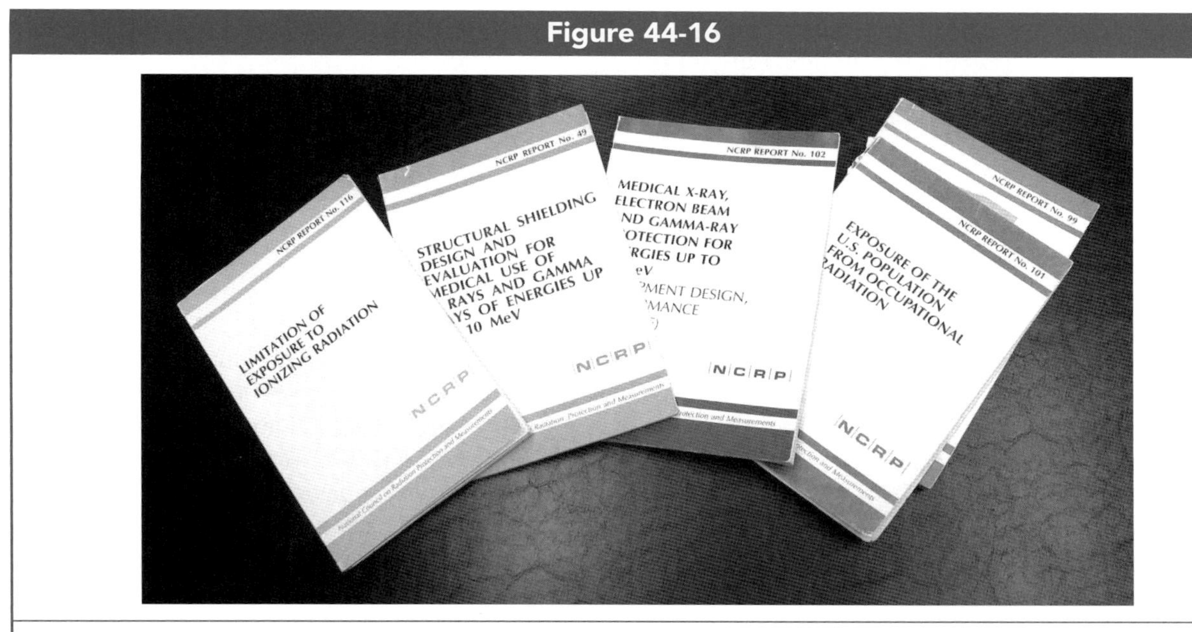

Various reports from the National Committee on Radiation Protection and Measurements (NCRP) regarding radiation biology, monitoring, instrumentation, equipment calibration, policies and safety.

regarding their specific work assignments." The report also states that "Technologists *shall* be aware of the approximate amount of radiation received by their patients," as well as "typical occupational doses for their assignment." In this modern day and age, few jobs require more constant updating of skills and knowledge than those in medical imaging.

SUMMARY

1. Radiographers are expected to be aware of the approximate amount of radiation received by their patients, presented in Tables 42-1 and 42-2, as well as typical occupational doses for their assignment.
2. The use of high-kVp techniques, careful collimation, low-ratio grids, efficient image receptors, high-speed-class processing, thoughtful positioning, knowledgeable use of the AEC, and a complete quality control program can help keep patient exposure levels *ALARA*.
3. Lead gonadal shielding should be used on patients whenever the collimated field size comes within 5 cm (2") of the gonads.
4. All female patients of reproductive age should be asked if there is a possibility they could be pregnant before proceeding with a radiographic exam. Pregnant patients in nonurgent situations may choose to follow the "10-day" rule for elective scheduling.
5. "High-contrast" or "enhanced imaging" fluoroscopes, the unbridled use of C-arm units in surgery, and excessive use of fluoroscopy by some cardiovascular centers all pose serious radiation hazards to patients. Pulsed mode and intermittent fluoroscopy help ameliorate these radiation levels. Percutaneous transhepatic cholangiography (PTCA), radio frequency ablation (RFA), and stent placements (STEN) have been identified as high radiation risk procedures for patients. More judicious use of CT scanning for small and young patients is needed.
6. Radiation monitoring devices for personnel are required if 1/4 of the occupational limit may be exceeded. Monitors must be worn outside lead aprons on the portion of the body likely to receive the highest exposure.
7. Lead aprons must be worn by anyone whose occupational dose may exceed 5 mR per hour. Lead aprons must be provided to any person

within 2 meters of the x-ray beam during exposure. Radiographers should also position themselves at right angles to the CR and to the x-ray table, and step back to a 2 meter distance, whenever possible. The proper application of radiation barriers forces radiation to be scattered two times before reaching personnel.

8. Guidelines for radiation equipment safety are available from the NCRP and other advisory groups, and should be carefully followed.
9. Mechanical devices should be used when possible for immobilizing patients. When holding of a patient is necessary, exposure should be distributed among non-radiation workers and nonreproductive adults.
10. Reporting or pregnancy by radiation workers must be voluntary and documented. A second monitor for fetal dose and restriction of the worker only from high-dose procedures are recommended.
11. To protect personnel from backscatter radiation which predominates at diagnostic kVp levels, C-arm fluoroscopes should be positioned with the image intensifier over the patient.
12. Fixed barrier shielding reaching 7 feet high must be provided according to minimum primary and secondary radiation parameters. Additional shielding thickness may be required according to the barrier shielding formula, or if any DELs for personnel or for the public are exceeded.
13. For *controlled* personnel areas, occupational exposure rates must not exceed 100 mR per week. For *uncontrolled* areas, rates must not exceed 10 mR per week. Radiation caution signs are required wherever exposure may exceed 100 mR in 5 days or 5 mR per hour.

REVIEW QUESTIONS

1. What is the most effective means of sparing radiation exposure to the patient during fluoroscopic examinations?

2. For studies which do not include the gonads within the primary x-ray beam, the patient's gonadal exposure averages roughly what fraction of the entrance skin exposure (ESE) within the x-ray beam?

3. Which type of lead shields are attached to the collimator to help preserve sterile fields?

4. To avoid accidental radiation exposure to the early developing embryo, female patients may wish to reschedule nonurgent procedures during what time period?

5. For a mobile unit, the tube-to-tabletop distance source-to-skin distance shall never be less than ___________.

6. For a conventional fluoroscope, the exposure rate at the tabletop should never exceed ___________.

7. What special mode for operating fluoroscopes and C-arm units can save considerable patient dose?

8. What types of physicians can own and operate radiographic equipment?

9. What three procedures were identified by an FDA task force for the possibility of exceeding 100 minutes of accumulated fluoroscopic time?

10. Personal radiation monitors are normally worn at the collar level and outside the lead apron in order to give an estimation of radiation exposure to which two organs in particular?

(Continued)

REVIEW QUESTIONS *(Continued)*

11. When the occupationally accumulated exposure over a reporting period is less than the sensitivity for the type of personal monitor being used, what initial is placed in the exposure column of the radiation report?

12. List the three practices which, combined, can reduce the radiographer's exposure to less than 1/500,000th of the patient's skin exposure?

13. Lead aprons should be provided to anyone positioned within __________ of the x-ray beam.

14. For any equipment that can be operated above 100 kVp, thus, in most situations, the recommended lead thickness for protective aprons is:

15. In a newborn intensive care unit, in the absence of a mechanical device or relative to hold the patient, who is the next best choice for helping with immobilization of the child, in terms of radiation biology?

16. It is appropriate to limit pregnant radiographers from:

17. Exposure cords on mobile units must allow the operator to be positioned at least _________ feet (or meters) from the x-ray beam?

18. For what type of radiation is a limit set at 100 mR per hour at a distance of 1 meter?

19. What equivalent thickness of pure lead must be met for a secondary barrier?

20. Leaded glass must normally be __________ times thicker than pure sheet lead to provide equivalent protection.

(Continued)

REVIEW QUESTIONS *(Continued)*

21. List the five factors that weigh into the formula for shielding requirements:

22. An *uncontrolled* area within a hospital or clinic which is not supervised must have structural barriers sufficient to keep exposure rates below __________ per week.

23. What must a posted warning sign read if more than 100 mR per hour of radiation exposure in the area could be exceeded?

24. What organization published the *Biological Effects of Ionizing Radiation* (*BEIR*) report?

Appendix #1

ANSWERS TO CHAPTER EXERCISES

Chapter 14

EXERCISE #14-1

1. 0.6 mm
2. 0.07 mm
3. 0.016 mm
4. 0.015 mm
5. 0.05 mm

EXERCISE #14-2

1. Relative sharpness of 3.3
2. Relative sharpness of 7.0
3. 1.5 times increase in relative sharpness
4. $\frac{2}{3}$ (0.67) of the original relative sharpness
5. 1.8 times increase in relative sharpness

EXERCISE #14-3

1. 64 cm
2. 120 in
3. 20 in
4. 66.7 cm
5. 15 cm
6. 100 cm
7. 36.7 cm
8. 12.5% magnification

Chapter 15

EXERCISE #15-1

1. 5 mAs
2. 25 mAs
3. 210 mAs
4. 7.5 mAs
5. 132 mAs
6. 15 mAs
7. 150 mAs
8. 4.2 mAs

EXERCISE #15-2

1. 2.5 mAs
2. 3.5 mAs
3. 25 mAs
4. 9 mAs
5. 1.8 mAs
6. 32 mAs
7. 2.5 mAs
8. 90 mAs

EXERCISE #15-3

1. 0.025 sec
2. 0.4 sec
3. 0.025 sec
4. 0.025 sec
5. 0.07 sec
6. 0.17 sec
7. 0.02 sec
8. 0.07 sec
9. 0.6 sec
10. 0.6 sec

Chapter 16

EXERCISE #16-1

1. 9
2. 102 kVp
3. 64.5 kVp
4. 58 kVp
5. appx. 118 kVp
6. appx. 37 kVp
7. appx. 87 kVp
8. appx. 65 kVp

Chapter 18

EXERCISE #18-1

1. 24 in
2. 15 in
3. 20 cm
4. 16 cm
5. 12 cm
6. 10 in
7. 30 in
8. 120 cm
9. 30 cm
10. 18 in

Chapter 19

EXERCISE #19-1

Part A:

1. 10 mAs
2. 5 mAs
3. 15 mAs
4. 7.5 mAs
5. 15 mAs

Part B:

1. 69 kVp
2. 63 kVp
3. 115 kVp
4. 59 kVp
5. 118–119 kVp

Chapter 22

EXERCISE #22-1

1. 0.31 relative exposure
2. 1.9 times relative exposure
3. 14 mR
4. 96 in

EXERCISE #22-2

1. 7.2 mAs
2. 4.4 mAs
3. 0.12 sec
4. 89 mA
5. 67.5 cm

EXERCISE #22-4

1. 75 mAs
2. 30 mAs
3. 15 mAs
4. 30 mAs

Chapter 23

EXERCISE #23-1

1. 40 cm
2. 7.5 cm
3. 18 cm
4. 40 in
5. 90 in
6. 12 in

EXERCISE #23-2

1. 4.0 relative sharpness
2. 20 relative sharpness
3. 5.33 relative sharpness
4. 18 relative sharpness

Chapter 26

EXERCISE #26-1

1. $\frac{1}{2}$
2. 3 ×
3. Equal
4. $1\frac{1}{2}$ ×
5. 4 ×
6. Equal
7. 2 ×
8. + 8 kVp
9. 2 ×
10. $\frac{1}{4}$
11. 2 ×
12. Equal

EXERCISE #26-2

KEY: Exercise 26-2: Constructing a Variable mAs Technique Chart from Scratch

EXERCISE: Assuming that the listed technique was good, complete the rest of the chart using proportional anatomy and the 4 cm rule of thumb.

Procedure	Projection	KVP	Ave CM	MAS						
Lumbar Spine	AP	80	22	18 cm	20 cm	**22 cm**	24 cm	26 cm	28 cm	30 cm
				12	18	**24**	36	48	72	96
	Oblique	80	26	22 cm	24 cm	**26 cm**	28 cm	30 cm	32 cm	34 cm
				24	36	48	72	96	144	240
	Lateral	80	30	26 cm	28 cm	**30 cm**	32 cm	34 cm	36 cm	39 cm
				48	72	96	144	192	240	384
Skull	PA	76	19	15 cm	17 cm	**19 cm**	21 cm			
				8	12	15	23			
	Lateral	76	15	11 cm	13 cm	**15 cm**	17 cm			
				4	6	8	12			
Shoulder	AP	76	14	10 cm	12 cm	**14 cm**	16 cm			
				3	4.5	6	12			
Knee	AP	70	12	8 cm	10 cm	**12 cm**	14 cm			
				2.3	3.4	4.5	6.8			

EXERCISE 26-3

KEY: Exercise 26-3: Constructing a Variable kVp Technique Chart from Scratch

EXERCISE: Assuming that the listed technique was good, complete the rest of the chart using proportional anatomy, the 4 cm rule, and the 15% rule in steps.

Procedure	Projection	mAs	Ave CM	KVP						
Lumbar Spine	AP	12	22	18 cm	20 cm	**22 cm**	24 cm	26 cm	28 cm	30 cm
				82	86	**96**	103	110	118	126
	Oblique	12	26	22 cm	24 cm	**26 cm**	28 cm	30 cm	32 cm	34 cm
				96	103	110	118	126	↑ mAs	↑ mAs
	Lateral	12	30	26 cm	28 cm	**30 cm**	32 cm	34 cm		
				110	118	126	↑ mAs	↑ mAs		
Skull	PA	10	19	15 cm	17 cm	**19 cm**	21 cm			
				72	78	85	98			
	Lateral	10	15	11 cm	13 cm	**15 cm**	17 cm			
				62	68	72	78			
Shoulder	AP	10	14	10 cm	12 cm	**14 cm**	16 cm			
				60	65	70	81			
Knee	AP	8	12	8 cm	10 cm	**12 cm**	14 cm			
				55	60	65	70			

EXERCISE 26-4

KEY: Exercise 26-4: Variable Time, Variable mA/Time, and Variable kVp Technique Chart Formats

Directions: Assuming that the listed technique was good, complete the rest of the chart using proportional anatomy, the 4 cm rule, and the 15% rule in steps.

Procedure	KVP	mAs for Each View		
WRIST	60	PA	OBL	LAT
		4	6	8

VARIABLE TIME CHART

Procedure	KVP	mA	Time for Each View		
WRIST	60	200	PA	OBL	LAT
			0.02 s	0.03 s	0.04 s

VARIABLE MA/TIME CHART

Procedure	KVP	mA/Time for Each View		
WRIST	60	PA	OBL	LAT
		200 / 0.02 s	200 / 0.03s	200 / 0.04 s

VARIABLE KVP CHART

Procedure	mAs	kVp for Each View		
WRIST	4	PA	OBL	LAT
		60	64*	69*

*Allow + or – 1 kVp

Chapter 28

EXERCISE 28-1

Part A:	Part B:
1. 13	1. 111
2. 50	2. 10011
3. 251	3. 111111

Appendix #2

ARRT STANDARD DEFINITIONS

Term	Film-Screen Radiography	Term	Digital Radiography includes both computed radiography and direct radiography *Computed Radiography (CR)* systems use storage phosphors to temporarily store energy representing the image signal. The phosphor then undergoes a process to extract the latent image. *Direct Radiography (DR)* systems have detectors that directly capture and readout an electronic image signal.
Recorded Detail	The sharpness of the structural lines as recorded in th radiographic image.	Spatial Resolution	The sharpness of the structural edges recorded in the image.
Receptor Exposure	The amount of radiation striking the image receptor.	Receptor Exposure	The amount of radiation striking the image receptor.
Density	Radiographic density is the degree of blackening or opacity of an area in a radiograph due to the accumulation of black metallic silver following exposure and processing of a film. $\text{Density} = \text{Log} \dfrac{\text{incident light intensity}}{\text{transmitted light intensity}}$	Brightness	Bightness is the measurement of the luminance of a monitor calibrated in units of candela (cd) per square meter on a monitor or soft copy. Density on a hard copy is the same as film.
Contrast	Radiographic contrast is defined as the visible differences between any two selected areas of density levels within the radiographic image. *Scale of Contrast* refers to the number of densities visible (or the number of shades of gray). *Long Scale* is the term used when slight differences between densities are present (low contrast) but the total number of densities is increased. Short Scale is the term used when considerable or major differences between densities are present (high contrast) but the total number of densities is reduced.	Contrast	Image contrast of display contrast is determined primarily by the processing algorithm (mathematical codes used by the software to provide the desired image appearance). The default algorithm determines the initial processing codes applied to the image data. *Scale of Contrast* is synomymous to "gray scale" and is linked to the bit depth of the system. "Gray scale" is used instead of "scale of contrast" when referring to digital images.

(Continued)

Film Latitude	The inherent ability of the film to record a long range of density levels on the radiograph Film latitude and film contrast depend upon the sensitometric properties of the film and the processing conditions, and are determined directly from the characteristic H and D curve.	Dynamic Range*	The range of exposures that may be captured by a detector. The dynamic range for digital imaging is much larger than film.
Film Contrast	The inherent ability of the film emulsion to react to radiation and record a range of densities.	Receptor Contrast	The fixed characteristic of the receptor. Most digital receptors have an essentially linear response to exposure. This is impacted by contrast resolution (the smallest exposure change or signal difference that can be detected). Ultimately, contrast resolution is limited by the dynamic range and the quantization (number of bits per pixel) of the detector.
Exposure Latitude	The range of exposure factors which will produce a diagnostic radiograph.	Exposure Latitude	The range of exposures which produces quality images at appropriate patient dose.
Subject Contrast	The difference in the quantity of radiation transmitted by a particular part as a result of the different absorption characteristics of the tissues and structures making up that part.	Subject Contrast	The magnitude of the signal difference in the remnant beam.

*For a full discussion of dynamic range, see Chapter 31.

GLOSSARY OF RADIOGRAPHIC TERMS

Absolute Risk: Number of cases within one year of a malignant disease, per million population per rem exposure.

Absorption Blur: Apparent blur at the edges of an image that is actually reduced x-ray absorption due to decreasing object thickness.

Absorption Efficiency: The ratio of photons absorbed by a particular material.

Accession Number: An institution's unique number for identifying a particular procedure or exam.

Active Matrix Array: A panel of electronic detector elements laid out in rows and columns, used to convert incoming light or x-ray photons into an electrical signal.

Actual Focal Spot Size: Area on the anode target surface that is struck by projectile electrons from the x-ray tube cathode.

Acute Radiation Syndrome: Collection of symptoms expected in humans for sudden, high exposures to radiation exceeding 100 rad of absorbed dose.

Added Filtration: Slabs of absorber, usually aluminum, placed in the x-ray beam after it exits the x-ray tube housing, to eliminate low-energy x-rays.

Air-Gap Technique: Use of a large OID to allow scatter radiation to spread out more before reaching the image receptor, thus improving subject contrast.

ALARA: *As Low As Reasonably Achievable:* Principle that radiation exposure should be kept at the practical achievable minimum in all situations, technical guidelines and limits notwithstanding.

Algorithm: A set of computer instructions to execute a specific task. In radiography, the "procedural algorithm" often refers to the default processing codes for each particular anatomical part.

Aliasing: See *Moire effect.*

Alpha Particle: Fragment of an atomic nucleus consisting of two protons and two neutrons, emitted as radiation.

Alternating Current: Oscillation of electricity in which electrons flow alternately in reversed directions.

Ammeter: Device that measures electrical current (amperage).

Amp, Ampere: Unit for the rate of electrical current equal to one Coulomb of charge per second.

Amplitude: Height or volume of a wave.

Anaphase: Stage in mitosis in which double-chromosomes separate.

Angstrom: Subatomic unit of distance equal to 10^{-10} meters, used to measure x-ray wavelengths.

Anode: Positively charged side of an x-ray tube, including the target.

Anode Heel Effect: The loss of x-ray output toward the anode end of the x-ray tube due to increased thickness of the anode itself, which acts as filtration.

Area Beam: X-ray beam projected in the shape of a circle, square or rectangle.

Artifacts: Any extraneous images that obscure the desired information in a radiograph.

Asthenic: Referring to a body habitus that is pathologically thin and frail.

Atomic Mass (A#): Total number of nucleons (protons plus neutrons) in an atomic nucleus.

Atomic Mass Unit (AMU): One-twelfth the mass of a carbon nucleus, roughly equivalent to that of a single proton.

Atomic Number (Z#): Number of protons in an atomic nucleus, identifying the element.

Attenuation: Reduction in the number or intensity of x-rays as a result of absorption and scattering.

Attenuation Coefficient: The ratio or percentage of the original x-ray beam intensity that is absorbed by a particular tissue area.

Automatic Brightness Control (ABC): Use of an electronic circuit during fluoroscopy to automatically maintain the brightness of the dynamic image by adjusting kVp, mA or both.

Automatic Exposure Control (AEC): The use of an electronic circuit and x-ray detectors to measure exposure rate near the image receptor and terminate the exposure when a pre-set level is reached.

Autotransformer: Electronic device that uses self-induction to control the kVp set at the console.

Background Radiation: Radiation from the surrounding environment which may expose image receptors or radiation monitors.

Backscatter Radiation: Scatter x-rays emitted at an angle greater than 90 degrees in a backward direction relative to the incident beam.

Band-pass filtering: Removal of selected frequency layers when a digital image is reconstructed by inverse Fourier transformation.

Becquerel (Bq): Standard international unit for radioactivity, equal to one decay event.

Beta Particle: A high-speed electron emitted as radiation from the nucleus of an atom when a neutron decays into a proton.

Bergonie & Tribondeau, Law of: Generally, tissues and cells that divide more frequently or are more primitive in structure tend to be more radiosensitive.

Binary: Having only two values, 0 and 1, yes or no, on or off.

Binding Energy, Electron: Amount of energy required to dislodge an orbital electron from its atomic shell; strength with which the electron is "held" in its orbital.

Binding Energy, Nuclear: Strong-force energy exchanged between protons and neutrons that holds them together in the nucleus of an atom.

Bit: *Binary digit*, a zero or a 1, the smallest unit of computer storage capacity.

Bit Depth: The maximum range of pixel values a computer, display monitor, or other digital device can store.

Bone Marrow Syndrome: See *Hematologic Syndrome.*

Bremsstrahlung: X-rays produced by the deceleration of projectile electrons as they pass near the nuclei of atoms in the x-ray tube anode.

Brightness gain: Degree to which an image intensifier is able to multiply the light intensity of an image.

Bucky: Tray in the x-ray table that holds a cassette or other portable image receptor.

Bucky Factor: Ratio of incident radiation to transmitted radiation through a grid.

Bucky Slot Cover: Protective lead slats under the tabletop that fold out when the Bucky is placed at the foot of the table, to shield the operator's gonadal area from scatter radiation during fluoroscopy.

Byte: A group of 8 bits used to represent a single alphanumeric character.

Calipers: Instrument with two legs used to measure the thickness of body parts.

Carcinogenic: Causing malignant cancer.

Cardinal Principles, Radiation Protection: Minimum exposure time, maximum distance from the source of radiation, and maximum radiation shielding.

Cascade Effect, Ion Chamber: Condition of a gas ionization detection device in which each ionization event results in a series of many additional ionizations through the gas toward the anode.

Cassette: Any rigid, flat rectangular container designed to hold a PSP plate or film.

Cathode: The negatively charged end of the x-ray tube, including the filament and focusing cup.

Cathode Rays: A stream of electrons passing through space or air.

Cathode Ray Tube: An image display monitor that uses a stream of electrons to stimulate phosphors.

Central Processing Unit (CPU): The computationally functional hardware of a computer system, including the control unit, arithmetic-logic unit, and primary storage.

Central Ray: The one ray in an x-ray beam that is precisely perpendicular to the x-ray tube axis, and delineates the center of the beam.

Characteristic Curve: A graph of the optical density produced in a displayed image plotted against the exposure level that produced the original latent image.

Characteristic X-rays: X-rays produced when electrons fall from higher to lower orbits in a "large" atom.

Charge-Coupled Device (CCD): A flat, compact light-sensing device that uses a single layer of silicon as its sensitive surface, used for recording images.

Charge, Electrical: State of being electrically negative or positive relative to another object.

Circuit, Electrical: A continuous circle of electrical conductors leading from a source of electromotive force (EMF) such as a battery or generator, through different resistive devices and back to the source.

Classical Scattering: See *Thompson Interaction.*

CNS Syndrome: Form of ARS in which deaths are generally attributable to damage to and shutdown of the nervous system, thought to be caused primarily by intracranial pressure.

Collimation: Restriction of the area of the x-ray beam using a lead device around its periphery.

Compensating Filtration: Material placed between the x-ray tube and the patient to absorb only portions of the field area such that the remnant x-ray beam is more uniform.

Complimentary Metal-Oxide Semiconductor (CMOS): A flat, compact light-sensing device that uses two semiconductor layers stacked together, one positive and one negative, used to record images.

Compton Interaction: Partial absorption of the energy of an incoming photon by an atom, resulting in a secondary x-ray and a recoil electron being emitted from the atom.

Compton X-ray: A secondary x-ray emitted from an atom that has absorbed only a portion of the energy of an incoming photon.

Computed Radiography (CR): Use of a photostimulable phosphor plate as an image receptor, which, after exposure, is scanned by a laser beam to release a latent light image. This light image is then converted into an electronic signal that can be fed into a computer for processing manipulation.

Computed Tomography (CT): Reconstruction of a cross-sectional image of the body using multiple projections from a fan-shaped x-ray beam.

Conductor: Material that allows the efficient transfer of heat or electricity.

Cone: Circular metal tube attached to the x-ray tube to limit the beam area.

Continuous Discharge: Condition of a gas ionization detection device in which an excessive supply of voltage, or high radiation exposure relative to the sensitivity of the device, results in a steady flow of electricity across the detection chamber and throughout the circuit, rendering it useless for readout.

Contrast: The ratio or percentage difference between two adjacent brightness (density) levels.

Contrast Agent: Chemical compound whose non-toxicity and x-ray absorption characteristics make it ideal for ingestion or injection in order to demonstrate body cavities on radiographic images.

Contrast Improvement Factor (CIF): Ratio of subject contrast in the remnant beam when a grid is used to the subject contrast without the grid.

Contrast Medium: See *Contrast Agent*.

Contrast Resolution: The ability of an imaging system to distinguish and display a range of attenuation coefficients from different tissues within the body; the ability to reproduce *subject contrast* with fidelity.

Controlled Area: Area occupied primarily by radiation workers, with supervision and known working conditions.

Convection: Transfer of heat by mixing of fluids.

Conversion Efficiency: The percentage of energy from absorbed x-ray photons that is converted to light by a particular phosphor material.

Conversion Factor: Ratio of output light intensity to input x-ray intensity for an image intensifier.

Coolidge X-ray Tube: A vacuumed tube that allows separate selection of voltage and current.

Cosmic Radiation: Radiation from extraterrestrial sources in space, having extreme energies greater than gamma rays.

Coulomb: Unit for electrical charge equal to 6.3×10^{18} free electrons.

Coulombs Per Kilogram: Standard international unit for radiation exposure, based on ionization of air just above the surface of the exposed object.

Covalent Bonding: Chemical union between two atoms formed by sharing a pair of electrons.

Crookes Tube: Forerunner of modern x-ray tubes, used in Roentgen's experiments when he discovered x-rays.

Cross-hatch Grid: Grid with two sets of lead strips running perpendicular to each other.

Cryogenic: Involving extremely cold temperatures.

Current, Electrical: Intensity of the flow of electrons.

Curie: Conventional unit for radioactivity, equal to 3.7×10^{10} decay events.

Cycle, Wave: Smallest component of a waveform in which no condition is repeated, consisting of a positive pulse and a negative pulse.

Cylinder: See *Cone*.

Cytoplasm: Protoplasm outside a cell's nucleus, composed mostly of water.

Dark Masking: Surrounding the image with a black border to enhance perceived contrast.

Decay, Radioactive: Emission of a photon of energy or a single particle.

Del: Acronym for "detector element," an individual hardware cell in a DR image receptor, capable of producing a single electronic readout from incoming photon (light or x-ray) energy.

Densitometer: Instrument that measures the degree of blackness or darkness on an exposed film.

Density, Electron: The concentration of electrons within a particular atom.

Density, Physical: The concentration of mass (atoms) within an object.

Density, Radiographic: The degree of darkness for an area in the image.

Detail: The smallest component of a visible image.

Deterministic Effects of Radiation: Effects that increase in severity for the individual with increasing dose, and whose occurrence within a population is predictable at a threshold dose.

Detective Quantum Efficiency (DQE): The ability of a detector element (del) to absorb x-rays or light; its sensitivity to photons.

Deviation Index: A standardized readout indicating how far an exposure falls outside the target or "ideal" exposure level expected for a particular projection.

Diamagnetic: Resistant to being magnetized.

Dielectric: Material that acts as an electrical insulator.

DICOM: Digital Imaging and Communications in Medicine guidelines which standardize the behavior of all the various digital devices used in PAC systems.

DICOM Header: A series of lines of information from the metadata of an image that can be displayed at the touch of a button at the monitor, or may be routinely displayed at the top (head) of each image.

Differential Absorption: Varying degrees of x-ray attenuation by different tissues that produces subject contrast in the remnant x-ray beam.

Differentiation, Cell: Degree to which a biological cell develops complex, specialized structures.

Diffusion (Dispersion): Scattering of photons or sound waves into random directions.

Digital Fluoroscopy (DF): Real-time (immediate), dynamic (motion) imaging with an area x-ray beam and image intensifier, in which the final electronic signal is digitized for computer processing.

Diode: Vacuum tube or solid-state electronic device with two electrodes, a negatively-charged cathode and a positively-charged anode, that conducts or blocks electrical flow depending on its condition.

Direct-Capture Digital Radiography (DR): Use of an active matrix array (AMA) as an image receptor, which converts x-ray exposure into an electronic signal that can be fed into a computer for processing manipulation. A DR system does not require the exposed image receptor (cassette) to be physically carried to a separate processing unit.

Direct-Conversion DR: A DR unit whose image receptor converts incoming x-rays directly into electrical charge with no intermediate steps.

Direct Current: Flow of electricity continuously in one direction through a conductor.

Direct Effect: Ionization of a target molecule (RNA or DNA) by an x-ray photon with no intermediate steps.

Display Station: A computer terminal restricted to accessing and displaying radiologic images.

Distortion, Shape: The difference between the length/width ratio of an image and the length/width ratio of the real object it represents, consisting of elongation or foreshortening.

Distortion, Size: See *Magnification.*

Dose: Amount of radiation energy absorbed by tissue or other material.

Dose Area Product: Radiation dose multiplied by the area of the exposed field.

Dose Equivalent: The biological harm caused by absorption of radiation energy, adjusted for the harmfulness of the type of radiation and for the sensitivity of the type of tissue exposed.

Dose Equivalent Limit (DEL): Guideline for the maximum allowable dose equivalent for cumulative, prospective or retrospective radiation dose.

Dose Rate: Energy absorbed from radiation per unit time.

Dosimeter: Any instrument for measuring radiation intensity.

Dot Pitch: See *Pixel Pitch.*

Doubling Dose: The amount of radiation expected to double the number of mutations occurring within a population.

Dual Energy Subtraction: The production of "bone-only" or "soft-tissue only" images by using two exposures at different average energies (average keV). This allows the computer to identify tissue types by the change in x-ray attenuation, then subtract bony or soft tissue from the displayed image.

Dynamic: Moving.

Dynamic Range: The range of pixel values made available for image formation by the software and hardware of an imaging system.

Dynamic Range Compression: Removal (truncation) of the darkest and lightest pixel values from the gray scale of a digital image.

Dynode: A diode that can switch electrical polarity, with the cathode and anode plates switching roles.

Early Effects of Radiation: Biological changes occurring within hours, days or a few weeks.

Edge Enhancement: Use of spatial or frequency methods to make small details, such as the edges of structures, more visible in the image.

Effective Dose Equivalent (EDE): A measure of the whole-body equivalent biological harm caused by radiation exposure, when sensitivity weighting factors for different exposed tissues are applied and the products summed.

Effective Focal Spot Size: The width of the projected focal spot for the central ray of the x-ray beam.

Elective Scheduling: Re-scheduling of radiation procedures for women during the 10 days following the onset of menstruation, to minimize the risk of radiation to a possibly developing embryo.

Electricity: Form of energy created by the activity of electrons or other charged particles.

Electrification: Acquiring an electric charge by the addition or loss of electrons.

Electrodynamics: Pertaining to moving electrical charges.

Electromagnetic: Pertaining to the electrical and magnetic fields surrounding any charged particle in motion.

Electromagnetic Radiation: Disturbance in both the electrical and magnetic fields surrounding any charged particle in motion, manifest as a double-wave with an electrical component and a perpendicular magnetic component.

Electromagnetic Spectrum: Range of electromagnetic radiations from low-energy radio waves through microwave, infrared, visible light, ultraviolet, x-rays and gamma rays to high-energy cosmic rays.

Electromotive Force (EMF): Electrical pressure caused by a potential difference, which causes electrons to move.

Electron: Extremely small elementary particle with a negative charge of 1.0 and mass of 1/1837 amu.

Electron-Volt (eV): See *Volt.*

Electrostatic: Pertaining to electrical charges at rest.

Element: Substance that cannot be further broken down by chemical means, composed of identical atoms.

Elongation: Projection of an image longer than the real object it represents.

Emission Efficiency: The percentage of light produced by phosphor crystals that escapes the phosphor layer and reaches a detector.

Energy: Ability to do work.

Energy Subtraction: See *Dual Energy Subtraction.*

Entrance Skin Exposure (ESE): Radiation exposure in Roentgens or C/kg incident upon the surface of the body.

Epidemiology: Study of the occurrence of disease within populations.

Epilation: Loss of hair due to radiation exposure.

Erg: Unit of energy used at molecular or atomic levels, equal to 1 ten-millionth of a Joule.

Erythema: Reddening of the skin due to radiation exposure, as in a sunburn.

Equalization: See *Dynamic Range Compression.*

Excess Risk: Difference between the observed and the expected number of cases of a disease.

Excitation: Temporary elevation of an orbital electron's energy by Thompson interaction.

Exposure Field Recognition: Ability of a digital imaging system to identify the "raw" background x-ray exposure field, so that data outside the anatomy may be excluded from histogram analysis and exposure indicator calculations.

Exposure Indicator: A readout estimating the exposure level received at the image detector as derived from initial pixel values in the acquired image histogram.

Exposure Latitude: The margin for error in setting radiographic technique that an imaging system will allow and be able to produce a diagnostic image.

Extrapolation: Estimation of a value beyond the range of known values.

Fan Beam: X-ray beam projected in a narrow slit shape such as in CT.

Farbzentren (F Centers): Energy levels within the atoms of a phosphor crystal that act as electron "traps" that can store an ionized electron for a period of time. Collectively, these captured electrons form a latent image.

Fast Scan Direction: The direction the laser moves across a PSP plate in a CR reader.

Ferromagnetic: Having very strong magnetic properties and high permeability, such that the material is easy to magnetize.

Field: Volume of space in which a force (such as electrical or magnetic attraction or repulsion) can affect objects.

Field, X-ray: The area exposed to a radiation beam.

Fifteen-Percent Rule: The exposure level at the image receptor can be approximately maintained while increasing x-ray beam penetration, by employing a 15 percent increase in kVp with a halving of the mAs, and vice versa.

Filament: Thin coiled wire in the cathode of the x-ray tube that emits electrons by thermionic emission.

Fill Factor: The percentage of a detector element's area dedicated to photon absorption.

Film Badge: Personal radiation monitoring device which is read out by processing a small sheet of photosensitive film. The density (darkness) measured on the developed film is proportional to the radiation dose absorbed by the film over the monitoring period.

Filtration: Removal of lower-energy x-rays from the beam by interposing aluminum or other attenuating materials, either for patient protection or to adjust the intensity of the x-ray beam across its area.

Filtration, Total: Inherent filtration plus added filtration.

Flat Field Uniformity: The consistency of pixel levels across the area of the image field when exposure is made with no object or anatomy in the x-ray beam.

Fluorescence: Immediate emission of light under stimulation.

Fluoroscopy: Real-time (immediate) viewing of dynamic (moving) radiographic images.

Flux gain: Ratio of the light output of an image intensifier to the x-ray input due to electronic amplification (using voltage or amperage) rather than geometric factors.

Focal Plane, Tomographic: Thickness of tissue centered on the fulcrum of a tomograph which appears with relative sharpness on the displayed image.

Focal Point, Grid: The point at which all lines extended from the lead strips of a focused grid converge.

Focal Spot: Area on the anode target surface where the electron beam is directed to produce x-rays.

Focused Grid: Grid in which the lead strips are canted (angled) to align with the diverging rays of the x-ray beam, reaching a focal point at a particular SID.

Focusing Cup: Metal shroud surrounding the filament in an x-ray tube; it is negatively charged to keep the electron beam emitted by the filament narrow and restricted.

Fog: An area of the image with an excessive loss of contrast and gray scale due to sources of noise.

Force: Anything that can change the motion of a different object; a push or a pull.

Foreshortening: Projection of an image shorter than the real object it represents.

Fourier Transformation: The mathematical process, in the frequency domain, that breaks the complex waves that form an image into their individual wave components.

Fractionation: Breaking a dose of radiation into separate exposures with time intervals between them.

Frame Shift Mutation: Change of a base molecule in the DNA structure that alters the genetic code.

Free Radical: An atom with a single, unpaired electron in its outermost shell that is highly reactive with other atoms and molecules.

Frequency: Rate of cycles per second for a moving waveform.

Frequency Domain Processing: Digital processing operations based upon the size of structures or objects in the image, which correlates to the wavelength or frequency of their wave function.

Fulcrum: Pivot point around which a tomographic tube moves.

Gamma Radiation: Naturally-occurring electromagnetic radiation with energies greater than x-rays, emanating from the nuclei of atoms.

Gastrointestinal Syndrome: Form of ARS in which deaths are generally attributable to damage to the intestinal mucosae with subsequent deficits in nutrition and hydration.

Gauss: Unit of magnetic field intensity for weak magnetic fields such as that of the earth or typical bar magnets.

Geiger-Mueller Counter: Radiation detection device that uses a gas ionization chamber to detect individual ionizations, subject to the saturation effect.

Geiger-Mueller Plateau: Range of supplied voltage levels to a gas radiation detection device that results in saturation (ionization of all gas molecules) for each initial x-ray ionization.

Genetic: Relating to traits that can be passed on to a subsequent generation of organisms.

Genetically Significant Dose (GSD): Gonadal dose that, if given to every individual in a population, would cause the same genetic effects as the existing distribution of radiation exposure.

Geometrical Integrity: See *Recognizability*.

Gradation (Gradient) Processing: Re-mapping of the gray scale range of a digital image by the use of look-up tables (LUTs). Intensity transformation formulas are applied to the range of pixel values to generate new LUTs.

Gray: Standard international unit for 1 Joule per kilogram (10,000 ergs per gram) of radiation energy absorbed by tissue.

Gray Scale: The range of pixel values, brightness levels, or densities present in a displayed digital image.

Grid, Radiographic: A plate consisting of alternating strips of lead and radiolucent material designed to absorb scatter x-rays.

Grid-Controlled X-Ray Tube: X-ray tube that uses a third electrode, usually in the form of a grid or mesh in front of the filament, to restrain the electron beam until exposure is needed, used for rapid sequence exposures such as in angiography.

Grid Cut-Off: Absence of density on portions of a radiographic image due to unintended x-ray absorption by a grid.

Grid Frequency: Number of lead strips per inch or per centimeter, in a grid.

Grid Lines: Linear radiopaque (light) artifacts on an image caused by absorption of x-rays by stationary lead strips in a grid when the grid is not moved during exposure.

Grid Ratio: Height of spaces between grid strips divided by the width of the spaces.

Habitus, Body: General size and shape of the patient.

Half-Life, Radioactive: The amount of time required for a radioactive source to decay to one-half its original rate of radioactivity.

Half-Value Layer (HVL): Thickness of a specified absorber required to reduce an x-ray beam to one-half its original intensity.

Half-Wave Rectification: Condition in which only every other pulse of AC electricity reaches the x-ray tube to generate x-rays.

Hardware: Physical components of a computer or related device.

Hematologic (Hematopoietic) Syndrome: Form of ARS in which deaths are generally attributable to damage to the bone marrow with subsequent deficits in the immune system.

Hertz: Unit of frequency; number of cycles or oscillations per second.

Heterogeneous: Having a range of different energies.

High-pass filtering: Removal of low-frequency layers when the digital image is reconstructed, such that higher frequencies (smaller details) become more visible.

HIS: Hospital Information System which stores and manages all images, lab results, reports, billing and demographic information on every patient.

Histogram: A graph plotting the pixel count for each brightness level (density) within an entire image.

Hit: Interaction of an x-ray photon with a target molecule that causes a harmful change.

Hormesis, Radiation: Theory that small amounts of radiation exposure can be biologically beneficial.

Housing, X-Ray Tube: Lead cylinder supporting the x-ray tube and providing thermal cooling and protection from leakage radiation.

Hybrid Subtraction: A combination of temporal and energy subtraction, where the two images are taken at different times and also at different energies, precluding the need of a contrast agent.

Hypersthenic: Referring to an obese body habitus.

Hyposthenic: Referring to a body habitus that is very thin but healthy.

Image Intensifier: Electronic vacuum tube that amplifies the brightness of the fluoroscopic image.

Image Receptor: Device that detects the remnant x-ray beam and records a latent image to be later processed into a visible image.

Indirect-Conversion DR: A DR unit whose image receptor first converts incoming x-rays directly into light using a phosphor layer, then converts the light into electrical charge.

Indirect Effect: Chemical changes to a target (RNA or DNA) molecule caused by free radicals or ions created when x-rays ionize nearby water molecules.

Induction, Electromagnetic: Creation of an electrical current or charge within, or magnetizing of, an object by electrical or magnetic fields emanating from other nearby objects.

Induction Motor: A motor that does not use any permanent magnet, but uses electrical current to magnetize both the stator and the rotor in order to induce the rotor to spin.

Infrared radiation: Electromagnetic radiation having energies less than visible light but greater than radio waves.

Inherent Filtration: Removal of lower-energy x-rays from the beam by materials in the x-ray tube itself and its associated housing.

Insulator: Material that inhibits the transfer of heat or electricity.

Integrate Mode: Operation of a radiation-detection instrument that keeps a running total of the radiation measured over a period of time.

Intensity Domain Processing: Digital processing operations based upon pixel values (brightness , or density levels), which form a histogram of the image.

Interface: Connection between computer systems or related devices.

Intermittent Fluoroscopy: Avoidance of continuous exposure to the patient during fluoroscopic procedures.

Internal Energy: The total of potential and kinetic energy contained by a molecule.

Internal Radiation: Radiation emitted from within the body itself.

Interpolation: Estimation of a value between two known values.

Interrogation Time: Time required for an electronic switch or detector to operate.

Inverse Fourier Transformation: The mathematical process, in the frequency domain, that re-assembles or adds together the individual wavelengths

Inverse Square Law: The intensity of radiation (and most forces) is inversely proportional to the square of the distance between the emitter and the receiver.

Inversion, Image: "Flipping" an image top for bottom and bottom for top.

Ion: Any electrically charged particle, including an ionized atom or a free electron.

Ionic Bonding: Chemical union between two charged atoms formed by electrostatic attraction.

Ionization: Removal of an orbital electron from an atom resulting in a charged state.

Ion Pair: Two particles of opposite electric charge created by the same event.

Isoexposure Curves: Map of locations in an x-ray room sharing equal exposure rates for a particular unit and technique.

Isomer: Atom with excess energy contained within its nucleus even though the neutron number is stable.

Isotope: Atom with an unusual number of neutrons in its nucleus.

Joule: Generic physics unit for energy, roughly enough energy to get a 1-pound object moving 10 miles per hour.

Jukebox: A stack of dozens of optical disc or tape storage devices.

Karyotype: A map of chromosomes based on their length.

K-Edge Effect: The surge of photoelectric light emission by a phosphor when the kV of incoming photons just exceeds the binding energy of the phosphor atoms K shell.

Kernel: A smaller sub-matrix that is passed over the image matrix executing mathematical operations on its pixels.

Kilovoltage (kV or keV): The standard unit measuring the amount of energy carried by photons or electrons.

Kilovolt-Peak (kVp): The maximum energy achieved by x-ray photons or electrons during an exposure.

Kinetic Energy: Energy of motion.

Laser: *Light amplification by stimulated emission of radiation:* Light emitted in phase (with all waves synchronized) when certain materials are electrically stimulated.

Late Effects of Radiation: Biological changes that occur in a period of months to years.

Latent Image: Image information that has not yet been processed into a visible form.

Latent Period: In acute radiation syndrome (ARS), the time interval between prodrome and manifest illness, during which symptoms improve temporatily.

Latitude: See *Exposure Latitude.*

Leakage Radiation: X-rays escaping through the x-ray tube housing in all directions other than through the intended window.

Lethal Dose (LD): Amount of radiation required to cause death within a specified number of days.

Light Guide: A fiber optic tube that can transmit light along a curved path.

Linear Energy Transfer (LET): Radiation energy deposited per unit path length within tissue.

Line-Focus Principle: Reduction of the size of the effective focal spot from the actual focal spot by inclining the surface of the anode target area.

Linearity, Exposure: Ability to produce consistent radiation output using various combinations of mA and exposure time.

Liquid Crystal Diode (LCD): A display monitor that uses a layer of liquid crystals between two polarized sheets of glass to control the blocking or emission of light from behind the screen.

Look-Up Table (LUT): A simple table of data that converts input gray level values to desired output values for the displayed image.

Lossless Compression: Digital image compression ratios less than 8:1 that preserve a diagnostic quality of image resolution.

Lossy Compression: Digital image compression ratios at 8:1 or greater that do not preserve a diagnostic quality of image resolution.

Low-Contrast Resolution: Ability to demonstrate adjacent objects with similar brightness (density) values).

Low-pass Filtering: Removal of high-frequency layers when the digital image is reconstructed, such that lower frequencies (larger structures and backgrounds) become more visible.

Luminescence: Any emission of light.

Magnetic Dipole: Orientation of a magnetic field around an entire atom, determined by the net spins of its electrons.

Magnetic Domain: A region of magnetically aligned atoms within a material.

Magnetic Moment: Orientation of the magnetic field around an electron or proton determined by its spin.

Magnetic Permeability: Ease with which an object may be magnetized by a surrounding field.

Magnetic Retentivity: Ability of an object to keep its magnetized state over time.

Magnification: The difference between the size of an image and the size of the real object it represents.

Main Chain Scission: Break in the "rail" of the DNA molecule.

Manifest Illness Stage: In acute radiation syndrome (ARS), the final stage of illness which leads to recovery or death.

mAs: See *Milliampere-Seconds.*

Mass, Physical: Amount of matter in an object or particle.

Matrix: The collective rows and columns of pixels or dels that make up the area of an image or image receptor.

Mean Survival Time (MST): Average time between exposure and death.

Meiosis: Germ cell division resulting in daughter cells with one-half the number of chromosomes in the parent cell.

Metabolism, Cell: Processes of synthesizing and breaking down molecules that make a cell thrive.

Metadata: An extensive database of information stored for each image in a PAC system, including information on the patient, the institution, the procedure, the exposure and the equipment used.

Metaphase: Stage of mitosis in which chromosomes visibly align in the middle of the cell.

Microfocus X-Ray Tube: X-ray tube with very small focal spot sizes at a fraction of a millimeter.

Microwaves: Radio waves emitted in phase (with all waves synchronized).

Milliamperage: The standard unit measuring the rate of flow of electricity (also frequently used by radiographers to describe the rate of flow of x-ray photons).

Milliampere-Seconds (mAs): The total amount of electricity used during an exposure (also frequently used by radiographers to describe the total amount of x-ray exposure used).

Minification: In the image intensifier, concentration of the electron beam onto an output phosphor that is smaller than the input phosphor, resulting in increased brightness for the output image.

Misregistration: Misalignment of two images for the purpose of subtraction.

Mitosis: Cell division resulting in daughter cells with the same number of chromosomes in the parent cell.

Modem: *Modulator/Demodulator:* A device that converts digital electronic signals into analog musical tones and vice versa.

Modulation: Changing of a video or audio signal.

Modulation Transfer Function (MTF): Measurement of the ability of an imaging system to convert alternating signals in the remnant x-ray beam into alternating densities or pixel values.

Moire Effect: False linear patterns produced in an image when the sampling rate of a processor/reader approximates the line resolution of the image receptor or the line frequency of a grid.

Monoenergetic: Having the same energy.

Mottle: A grainy or speckled appearance to the image caused by excessive noise, faulty image receptors or faulty display.

Multiscale Processing: Decomposition of the original image, by Fourier transformation, into eight or more frequency bands for individual digital processing treatments.

Mutagenic Effects: Biological effects caused by radiation exposure to the gametes prior to conception.

Mutation: Change or loss of a base molecule in the rung structure of a DNA or RNA molecule that alters the genetic code.

N-V-D Syndrome: Temporary *nausea, vomiting and diarrhea* caused by non-lethal but high doses of radiation exposure not exceeding 100 rad.

Nematic Crystals: Crystals with a long thread-like shape that tend to align together.

Neutron: Nuclear particle with no net electrical charge and a mass of just over 1 amu.

Noise: Any form of non-useful input to the image which interferes with the visibility of anatomy or pathology of interest.

Nucleon: Proton or neutron; massive particles making up the nucleus of an atom.

Object-Image Receptor Distance (OID): Distance from the structure of interest to the detector system for the latent image carried by the remnant x-ray beam.

Occupancy Factor (O): Ratio of an 8-hour dayshift that rooms adjoining radiation areas are occupied by the public.

Off-Focus Radiation: X-rays produced anywhere in the x-ray tube outside of the focal spot area.

Ohm: Unit of electrical resistance, typical of about 10 feet (3 meters) of ordinary wire.

Opaque: Not allowing the passage of light.

Operating System: Software that sets the general format for the use of a computer (e.g., home, business, science) including an appropriate interface ("desktop") for the type of input devices to be predominantly used.

Optical Disk: A removable plate that uses laser light to write and read data.

Overexposure: An area of the image excessively darkened due to extreme exposure or processing conditions.

Oxygen Enhancement Ratio (OER): Dose of radiation required to produce a specific effect under aerobic conditions (oxygen present) divided into the dose required under anoxic conditions (oxygen absent).

PACS: Picture Archiving and Communication System which provides digital storage, retrieval, manipulation and transmission of radiographic images.

Parallel Circuit: An electrical circuit with wires splitting into different branches to reach devices rather than placing the devices in a row along one conductor.

Parallel Grid: Grid with lead strips that are not canted (angled), but are equidistant at all points such that there is no focal point.

Paramagnetic: Having mild magnetic properties and moderate permeability, such that the material can be magnetized but not easily.

Particulate Radiation: Radiation consisting of particles with distinct, measurable mass such as electrons, protons, neutrons and alpha particles.

Partitioned Pattern Recognition: See *Segmentation.*

Penetration: The ratio or percentage of x-rays transmitted through the patient to the image receptor.

Penetrometer: An aluminum bar arranged in steps of increasing thickness, used as an absorbing object to measure x-ray beam penetrations characteristics.

Penumbra: Blur; the amount of spread for the edge of an image detail, such that it gradually transitions from the detail brightness to the background brightness.

Perceptual Tone Scaling: Gradation processing with LUTs based on the human perception of brightness levels in order to give the displayed image a more conventional appearance.

Phantom, Radiographic: An x-ray absorbing model designed to simulate human tissue or body parts.

Phase, X-Ray Machine: Degree of overlapping of electrical pulses from a generator; single-phase current has no overlap, three-phase 6-pulse current overlaps three cycles, and three-phase 12-pulse current overlaps 6 cycles before the first cycle is completed.

Phosphor: A crystalline material that emits light when exposed to x-rays.

Phosphorescence: Delayed emission of light.

Photoconductor: Material that conducts electricity when illuminated by light or x-ray photons.

Photodiode: Solid-state electronic device that converts light or x-ray energy into electrical current.

Photoelectric Interaction: Total absorption of the energy of an incoming photon by an atom resulting in the emission of a only a photoelectron.

Photoelectron: An electron emitted from an atom, having absorbed all of the energy of an incoming photon.

Photometer: Device that measures light intensity.

Photomultiplier Tube: A series of dynodes (reversible electrodes) that can magnify a pulse of electricity by alternating negative and positive charges in an accelerating sequence.

Photon: Electromagnetic energy in a discrete or quantized amount that acts as a "bundle" or "packet" of energy much like a particle.

Photospot Camera: Device that records only one frame at a time from a fluoroscopic image.

Photostimulable Phosphor Plate (PSP): The active image receptor used in computed radiograph (CR) whose phosphor layer can be induced to emit its stored light image, after initial exposure to x-rays, by re-stimulating it with a laser beam.

Phototimer: See *Automatic Exposure Control.*

Pixel: An individual two-dimensional picture element or cell in a displayed image, capable of producing the entire range of gray levels or colors (bit depth) for the system.

Pixel Pitch: The distance from the center of a pixel to the center of an adjacent pixel.

Planck's Constant: Fundamental physics quantity that relates the energy of radiation to its frequency.

Pocket Dosimeter: Personal radiation monitoring device based on a simple electroscope; radiation dose absorbed over the monitoring period releases electric charge from a fiber, which then moves across a window that can be read out at any time.

Point Lesion: Any change, temporary or permanent, to a single chemical bond in the rail structure of a DNA or RNA molecule from radiation exposure.

Polarization, Light: The filtering of light waves such that their electrical component is only allowed through in one orientation, vertical or horizontal, but not both.

Potential Difference: Difference between two states of electrical charge.

Preprocessing: All corrections made to the "raw" digital image due to physical flaws in image acquisition, designed to "normalize" the image or make it appear like a conventional radiograph.

Point Processing Operations: Digital processing operations that are executed pixel by individual pixel.

Polyenergetic: See *Heterogeneous.*

Positive Beam Limitation (PBL): Automatic collimation of the x-ray beam using an electronic system that senses the size of the image receptor.

Postprocessing: Refinements to the digital image made after preprocessing corrections are completed, targeted at specific anatomy, particular pathological conditions, or viewer preferences.

Potential Energy: Energy of relative position.

Power: The rate of work, or rate at which energy is spent per unit time.

Prefetching: An automatic search for and access of previous radiographic studies and records related to a scheduled exam many hours prior to the exam, so that they may be immediately available to the diagnostician at the time of the exam.

Preprogrammed Technique: Pre-set techniques for each anatomical part stored within the computer terminal of a radiographic unit, selected by anatomical diagrams at the control console.

Prereading Voltmeter: Meter placed for safety reasons in the primary circuit of the x-ray machine, which reads what the voltage will be after it is amplified by the step-up transformer, rather than indicating the actual voltage in its own circuit.

Primary Barrier: Structural shielding in any wall, floor or ceiling that the primary x-ray beam may strike in routine practice.

Primary Memory: Data storage that is necessary for a computer to function generally.

Primary X-Rays: X-rays originating from the x-ray tube that have not interacted with any subsequent object.

Projected Focal Spot Size: The width of the focal spot as seen from various viewpoints along the image receptor.

Prodrome: The initial period of ill symptoms from acute radiation exposure, which subside temporarily afterward.

Prophase: Stage in mitosis in which the cell's nucleus enlarges.

Proportional Anatomy Approach to Technique: The thesis that the required overall radiographic technique is proportional to the thickness of the anatomical part, such that comparisons between different body parts can be made to derive ratios for adjusting technique.

Proportional Counter: Radiation detection device whose readout is also affected by the voltage supplied to the device, used primarily in laboratory settings.

Proton: Nuclear particle with a positive charge of 1.0 and a mass of approximately 1 amu.

Protraction, Dose: Extension of the time over which a dose of radiation is delivered in a continuous exposure.

Pulse Mode Fluoroscopy: Breaking of the fluoroscopic exposure into a series of discrete, fractional exposures separated by time intervals.

Pyramidal Decomposition: Repeated splitting of the digital image into high-frequency and low-frequency components, with the lower-frequency bands requiring smaller and smaller file sizes for storage.

Quality Assurance: Overall program directed at good patient care, customer satisfaction, and maintenance of resources such as equipment.

Quality Control: Regular testing and monitoring of equipment to assure consistent quality images.

Quantization: Assigning digital values to each measurement from the pixels of an image matrix.

Quantum: See *Photon.*

Quantum Mottle: A grainy or specked appearance to the image caused by the randomness of the distribution of x-rays within the beam, which becomes more apparent with low exposure levels to the image receptor.

Rad: *Radiation Absorbed Dose:* Conventional unit for absorption of 100 ergs of radiation energy in each gram of tissue.

Radiation: Energy transmitted through space.

Radio Waves: Electromagnetic radiation of low energy and long wavelength.

Radioactivity: The rate of spontaneous emission of alpha, beta or gamma rays originating in atomic nuclei.

Radiographer: Medical imaging technologist who uses x-rays to produce single-projection images.

Radiography: Use of x-rays to produce single-projection images.

Radioisotope: Atom whose unusual number of neutrons results in an unstable state for its nucleus, resulting in radioactivity.

Radiologist: Medical physician specializing in the diagnosis of images.

Radiolucent: Allowing x-rays to easily pass through.

Radiolysis of Water: Breaking of water molecules into (possibly harmful) ions and free radicals by radiation.

Radionuclide: Any atomic nucleus emitting radiation.

Radiopaque: Absorbent to an x-ray beam, such that few x-rays pass through.

Radiosensitivity: Relative susceptibility of a tissue or organ to harm from radiation exposure.

Raleigh Interaction: Re-emisison of a photon whose energy has been temporarily captured by an atom as a whole.

Random Access Memory (RAM): Data that can be accessed from anywhere within storage in approximately equal amounts of time, without having to pass through other files or tracks.

Range, Radiation Detector: Levels of radiation energy and exposure that can be accurately read out by a detection device.

Read-Only Memory (ROM): Data storage that cannot be changed by the user.

Real-time: Immediately accessible.

Recognizability: The ability to identify what an image is, which depends upon its levels of sharpness, magnification and distortion.

Recoil Electron: An electron emitted from an atom, having absorbed only a portion of the energy of an incoming photon during a Compton interaction.

Recombination: Condition of a gas ionization detection device in which a low supply of voltage results in electrons released by radiation exposure falling back into their atomic orbits rather than reaching the anode, so that no readout is produced.

Rectification: Conversion of alternating electrical current into direct or one-way current.

Redundant Array of Independent Discs (RAID): A system that stores backup image files across several computer hard drives that are independent of each other, to protect against loss of files.

Reflection: General reversing of direction.

Refraction: Bending of the path of a light or x-ray in its generally forward motion.

Region of Interest (ROI): See *Volume (Values) of Interest.*

Relative Biological Effectiveness (RBE): The dose of a particular type of radiation required to cause a specific identified biological effect, divided into the dose of 250 kV x-rays required to cause the same effect.

Relative Risk: Ratio or percentage of the likelihood that a particular biological effect will occur due to radiation exposure, as compared to the occurrence of the effect already in the population from other causes.

Rem: *Radiation Equivalent in Man:* Conventional unit for biological harm caused by absorption of one rad of radiation energy, adjusted for the harmfulness of the type of radiation and for the sensitivity of the type of tissue exposed.

Remnant Radiation: Radiation having passed through the body but not having struck the image receptor or detector.

Reproducibility, Exposure: Ability to produce consistent radiation output from one exposure to the next, using identical exposure factors.

Rescaling: Re-mapping of the input data from the acquired image to a pre-set range of pixel values in order to produce consistent brightness and gray scale in the displayed output image.

Resolution: The ability to distinguish small adjacent details in the image as separate and distinct.

Resolving Time: Time required for a radiation detection device to re-set itself after an ionizing event such that a subsequent ionization can be detected.

Reversal, Image: Changing a (negative) radiographic image into a positive image by inverting pixel values.

Ripple, Voltage: Degree of variation in electrical pressure over time.

RIS: Radiology Information System which stores and manages all radiologic images and related reports, radiologic billing and demographic information on every patient.

Roentgen: Conventional unit for radiation exposure based on ionization of air just above the surface of the exposed object.

Rotor: The moving (rotating) part of a motor.

Sampling: Taking intensity measurements from each pixel or cell in an image matrix.

Saturation, Digital Radiography: Overwhelming a digital data-analysis system with electronic signal (from extreme exposure at the detector) such that the data cannot be properly analyzed, resulting in a diagnostically useless pitch black area within the displayed image.

Saturation Effect, Ion Chamber: Condition of a gas ionization detection device in which any single ionization results in an avalanche of ionizations throughout the entire volume of gas contained.

Scanning: Dividing of the field of the image into an array of cells or pixels, to format the matrix.

Scatter X-Rays: Secondary x-rays that are traveling in a different direction than the original x-ray beam.

Scintillation: Emission of light upon x-ray or electrical stimulation.

Series Circuit: An electrical circuit with devices paced in sequence or in a row along one conductor.

Secondary Barrier: Structural shielding in walls, floors or ceilings that the primary x-ray beam is never directed toward in routine practice.

Secondary Memory: Data storage that is not necessary for a computer to function generally.

Secondary X-Rays: X-rays produced by interactions within the patient or objects exposed by the x-ray beam.

Segmentation: Ability of a digital imaging system to identify exposure field borders and count multiple fields within the area of the image receptor.

Selectivity: (a) Ratio or percentage of scattered radiation transmitted through a grid; (b) effectiveness of tomography in producing a thin focal plane.

Semiconductor: Material that can act as an electrical conductor or resistor according to variable conditions such as temperature, the presence of light, or pre-existing charge.

Sensitivity, Radiation Detector: Ability to measure small amounts of radiation.

Service Class Provider: The centralized storage and distribution services of a PAC system.

Service Class User: The decentralized image acquisition units, workstations and display stations of a PAC system where the services of the system are accessed.

Shadow shield: Lead plate suspended above the patient so as not to be in contact with the patient, used to protect a particular body area from radiation.

Shape Distortion: See *Distortion, Shape.*

Sharpness: The abruptness with which the edges of an image detail stop as one scans across the area of an image.

Sievert: Standard international unit for biological harm caused by absorption of one Gray of radiation energy, adjusted for the harmfulness of the type of radiation and for the sensitivity of the type of tissue exposed.

Sigmoid: Having the shape of an "s".

Signal-to-Noise Ratio (SNR): The percentage (ratio) of useful signal or information to non-useful, destructive noise contained within the acquired image data.

Simple Ionization Plateau: Range of supplied voltage levels to a gas radiation detection device that results in one electron reaching the anode for each initial x-ray ionization.

Skin Erythema Dose (SED): Threshold radiation exposure required to cause reddening of the skin.

Slow Scan (Subscan) Direction: The direction a PSP plate moves through a CR reader.

Smoothing: Use of spatial or frequency methods to reduce contrast only at the level of small details (local contrast) to reduce noise or make the edges of structures appear less harsh in the displayed image.

Software: Mathematical and logic programming that directs a computer or related devices.

Solenoid: Electrical coil of current-carrying wire that produces a magnetic field.

Solid-State: Using crystalline materials rather than vacuum tubes to control electrical current.

Somatic: Relating to the internal body organs of an individual.

Source-to-Image Receptor Distance (SID): Distance from the x-ray tube target to the remnant (latent) image detector system.

Source-to-Skin Distance (SSD): Distance from the x-ray tube target to the uppermost (or first) surface of the body.

Space Charge: Electron cloud that hovers around the x-ray tube filament when thermionic emission is occurring.

Spatial Domain Processing: Digital processing operations based upon the locations of structures or pixels within the image matrix.

Spatial Frequency: Measure of the number of details per unit distance for an imaging system, usually expressed in line-pairs per millimeter.

Spatial Resolution: The ability of an imaging system to distinguish and display small pixels; the ability to produce *sharpness* in image details.

Spectrum: A continuous range of values, infinitely divisible.

Speed: The physical sensitivity of an image receptor system to x-ray exposure.

Speed Class: The programmed sensitivity of a digital processor or reader to the signal from the image receptor.

Spin: Angular momentum of a subatomic particle.

Spot View: A single-frame photograph taken from the fluoroscopic image.

Square Law: Principle that exposure at the IR can be maintained for different SIDs by changing the mAs by the square of the distance change.

Static: Still or stationary, not moving.

Stator: The stationary part of a motor; a coil of wire surrounding the rotor that causes the rotor to spin by electromagnetic induction.

Step-Down Transformer: Transformer that reduces electrical voltage proportional to its secondary windings.

Step-Up Transformer: Transformer that increases electrical voltage proportional to its secondary windings.

Step Wedge: See *Penetrometer.*

Sthenic: Body habitus that is average and healthy.

Stimulated Phosphorescence: Delayed emission of light upon stimulation.

Stitching, Image: Attaching several images together to display a large portion of the body.

Stochastic Effects of Radiation: Effects that increase in probability within a population, but not in severity, with increasing dose.

Storage Area Network (SAN): A small local network of storage devices connected in parallel so that if one device fails, all connections between the other devices are preserved.

Subject Contrast: Difference of intensity between portions of the remnant x-ray beam, determined by the various absorption characteristics of tissues within the patient's body.

Subpixel: A segment of a pixel that is not capable of producing the entire range of gray levels or colors (bit depth) for the system.

Subtraction: Removal of structures of a particular size from an image by making a *positive* (reversed image) mask of only those structures, then superimposing the positive over the original negative image to cancel those structures out. A contrast agent present in only one of the images will not be subtracted.

Target, X-Ray Tube: Surface on the x-ray tube anode struck by projectile electrons from the filament to produce x-rays.

Target Theory: Premise that certain molecules are critical to the survival of the biological cell, while others are not. Radiation "hits" taken by critical molecules may be lethal to the cell.

Technique, Radiographic: The electronic, time and distance factors employed for a specific projection.

Teleradiology: Transmission of medical images and reports to and from remote sites.

Telophase: Stage in mitosis in which a new cell membrane forms to separate the two daughter cells.

Temperature: Measurement of the kinetic energy of molecules.

Temporal Subtraction: Subtraction between two images taken at different times, the first prior to injection of a contrast agent and the second after the contrast agent is introduced.

Tenth-Value Layer (TVL): Thickness of a specified absorber required to reduce an x-ray beam to one-tenth its original intensity.

Teratogenic Effects: Biological effects occurring to the embryo/fetus from in-utero exposure to radiation.

Terrestrial Radiation: Radiation from the environment of the earth.

Tesla (T): Unit of magnetic field intensity for powerful magnets such as an MRI scanner; 1 T = 10,000 Gauss.

Thermionic Emission: Release of free electrons from the filament of the x-ray tube when it is heated.

Thermoluminescent Dosimeter (TLD): Personal radiation monitoring device which is read out by stimulating a crystal to emit light by heating it. The light emitted is proportional to the radiation dose absorbed by the crystal over the monitoring period.

Thin-Film Transistor (TFT): The electronic switching gate used in detector elements for direct-capture radiography.

Thompson Interaction: Re-emission of a photon whose energy has been temporarily captured by an orbital electron in an atom.

Threshold: Minimum intensity of a force or influence required to cause an effect.

Tomography: Use of a moving x-ray tube, image receptor, or body part in order to blur out layers of tissue above and below the plane of interest.

Total Effective Dose (TED): Accumulated lifetime occupational dose to a radiation worker, limited by guidelines to 1 rem (10 mSv) per year.

Transaxial, Transverse: Perpendicular to the long axis.

Transformer: Electrical device that uses induction between coils of wire to amplify or reduce voltage.

Translation, Image: "Flipping" an image geometrically left for right, and right for left.

Translucent: Allowing light to pass through.

Tube-Tabletop Distance (TTD): Distance from the x-ray tube target to the surface of the x-ray table.

Turns Ratio, Transformer: Quotient of the number of turns in the secondary coil to those in the primary coil of a transformer.

Ultraviolet Radiation: Electromagnetic radiation having energies greater than visible light but less than x-rays.

Umbra: The clearly defined portion of a projected image.

Uncontrolled Area: Area which non-radiation workers (secretaries, waiting patients, etc.) routinely occupy.

Unsharp Mask Filtering: Subtraction of the gross (larger) structures in an image to make small details more visible.

Unsharpness: The amount of blur or penumbra present at the edges of an image detail.

Use Factor (U): Ratio of an 8-hour day shift that the primary x-ray beam is directed toward a particular wall, floor or ceiling.

Valence: Balance of electrons versus open vacancies in the outermost orbital shell of an atom, which determines its electrical and chemical properties.

Video Display Terminal (VDT): Display monitor and connected input devices for a computer.

Vignetting: Loss of brightness at the periphery of an image.

Volatile Memory: Data storage that can be erased by the user.

Volume (Values) of Interest (VOI): Portion of an image histogram that contains data useful for digital processing and for calculating an accurate exposure indicator.

Visibility: The ability to see a structure in an image, which depends upon its brightness and contrast versus any noise present.

Volt: Unit for electrical energy or pressure required to move one ampere of current through one ohm of resistance (about 10 feet of typical wire).

Voltage: Electrical energy or pressure.

Voxel: A three-dimensional cube (or square tube) of body tissue that is sampled by an imaging machine to build up a displayed digital image.

Watt: Unit of electrical power, equal to voltage multiplied by amperage.

Wavelength: Distance between two similar points in a waveform.

Weight: Force on a mass caused by gravity.

Weighting Factors: Mathematical values by which a dose is multiplied to indicate the relative effectiveness of the specific type of radiation and the relative sensitivity of the type of tissue exposed.

Window Level (Level): The digital control for the overall or average brightness, density or pixel level of the displayed image.

Window Width (Window): The digital control for the range of brightness, density, or pixel levels in the displayed image.

Window, X-ray Tube: Port through which the intended x-ray beam is allowed to be emitted through a thinner section of glass.

Windowing: Adjusting the window level or window width.

Word, Digital: Two bytes of information.

Work: Force multiplied by the distance over which the force is applied.

Workload (W): Weekly sum of the product of mA and minutes of x-ray beam on-time for a particular x-ray unit.

Workstation: A computer terminal allowing access to and manipulation of radiologic images, and permanent saving of changes into the PAC system.

X-Rays: Electromagnetic radiation having energies greater than ultraviolet light but less than gamma radiation, able to penetrate through the human body for imaging purposes.

INDEX

C

D

E

F

J

K

L

M

N

O

P

Q

R

S

U

V

W

X

Z